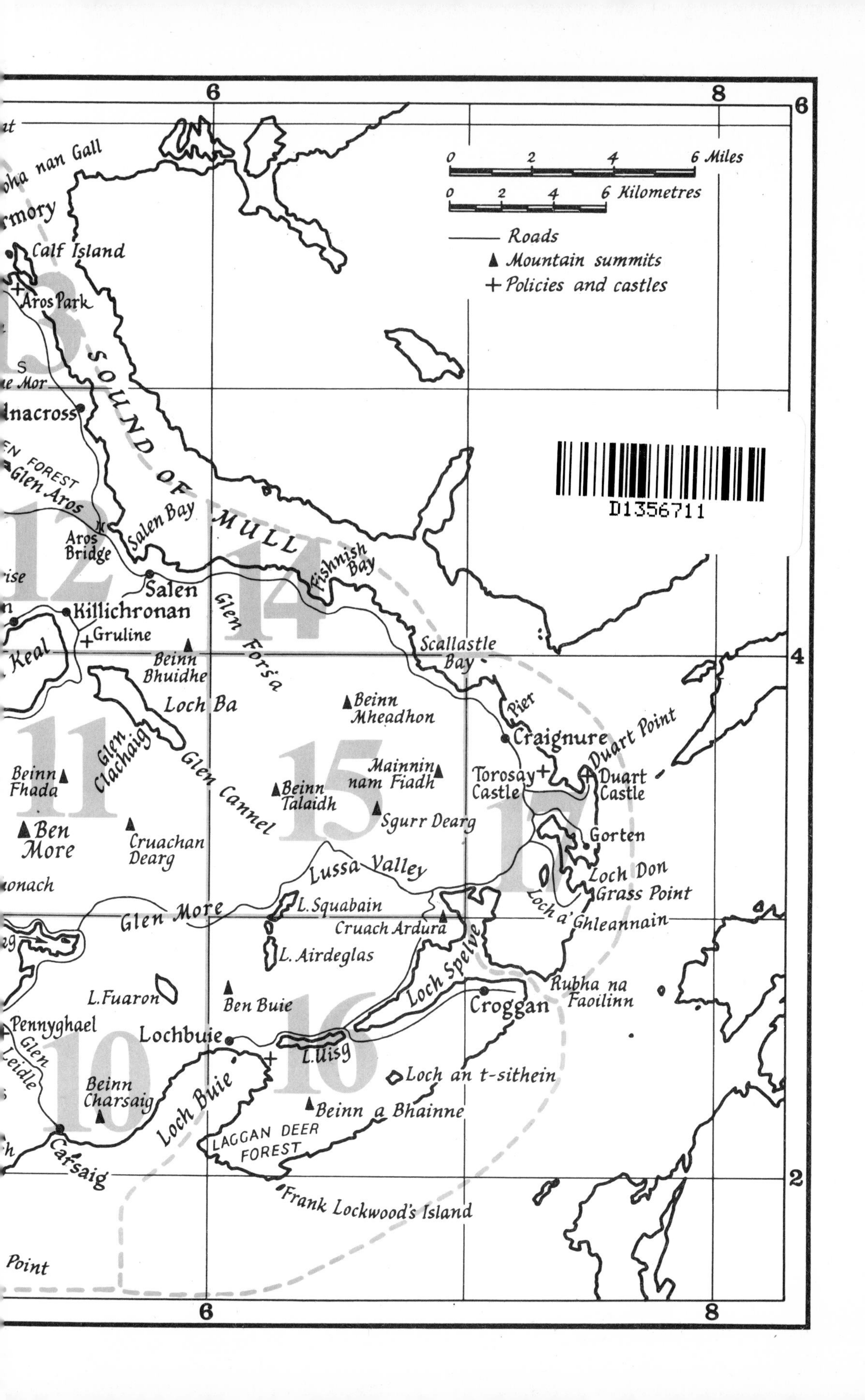

Calf Island
Aros Park
SOUND OF MULL
Salen Bay
Aros Bridge
Salen
Killichronan
Gruline
Beinn Bhuidhe
Glen Forsa
Loch Ba
Glen Clachaig
Glen Cannel
Beinn Fhada
Ben More
Cruachan Dearg
Fishnish Bay
Scallastle Bay
Pier
Craignure
Duart Point
Duart Castle
Torosay Castle
Gorten
Loch Don
Grass Point
Loch a' Ghleannain
Beinn Mheadhon
Mainnin nam Fiadh
Beinn Talaidh
Sgurr Dearg
Lussa Valley
Glen More
L. Squabain
Cruach Ardura
L. Airdeglas
Loch Spelve
Rubha na Faoilinn
Croggan
L. Fuaron
Ben Buie
Pennyghael
Glen Leidle
Lochbuie
L. Uisg
Loch an t-sithein
Beinn Charsaig
Loch Buie
Beinn a Bhainne
LAGGAN DEER FOREST
Carsaig
Frank Lockwood's Island
Point
0 2 4 6 Miles
0 2 4 6 Kilometres
Roads
Mountain summits
Policies and castles

[illegible]

The Island of Mull

The Island of Mull

a survey of its flora and environment

by the Department of Botany British Museum (Natural History)

Edited by A. C. Jermy and J. A. Crabbe

BRITISH MUSEUM (NATURAL HISTORY) LONDON 1978

ISBN 0 565 00791 2
Publication no. 791

Typeset in Times New Roman by
William Clowes (Great Yarmouth) Limited

Printed in Great Britain by
Fletcher & Son Ltd, Norwich
and bound by Richard Clay
(The Chaucer Press Ltd), Bungay, Suffolk

Preface

The proposal to launch a Departmental project to record, study and subsequently catalogue the plants in a delimited area in Great Britain was made with two points in mind. First, in the British Museum (Natural History) we had the resources, in terms of finance, staff, herbarium and library facilities, to cover all plant groups in a way which had not been attempted in a British Flora before. In doing this, the Department could make a contribution to the study of the British flora on a scale appropriate to the responsibilities of a national museum. Secondly, appreciating the need of the botanist to see plants alive and in the field, we have treated the fieldwork carried out in this survey as part of the training of junior staff, not only within the sphere of their particular duties, but also for an appreciation of the scope and the work of the Department as a whole.

The Island of Mull was chosen as the site for such a survey for several reasons. It appeared to be situated in an interesting position on the west coast of Scotland, such that its flora contained a number of southern elements, but at the same time had affinities with north-west highland Scotland; it had a reasonably wide range of habitats and varying land-use; it had not been subjected to intensive botanical research nor had a County Flora been prepared at any time.

In presenting this account of the Mull flora and its environmental background as a published work, we hope we shall promote further interest in Mull itself – as an island, an ecosystem and a community; and in the autecology and variation of the plants we have recorded. The basic data for the Flora are the results of an intensive programme of intermittent visits throughout the five years 1966–70, totalling some 150 man weeks. In 116 550 ha (450 sq. miles) of land and a similar area of sea-bed we found about 5280 species and varieties in over 1600 genera.

The following staff took part in fieldwork: E. B. Bangerter, R. Banks, Lorna F. Bowden (later Mrs I. K. Ferguson), J. F. M. Cannon, J. A. Crabbe, Gillian Dunford, A. Eddy, Dr I. K. Ferguson (Flora Europaea staff attached to BM), E. W. Groves, Margaret C. Guy (later Mrs B. Tebbs), Dr D. J. Hibberd, P. W. James, A. C. Jermy, Marilyn R. Jones, Lynn Kendrick, J. R. Laundon, Dr E. Launert (Flora Zambesiaca staff attached to BM), Sylvia A. Marshall, Dr A. Melderis, Jennifer A. Monk (later Mrs S. J. Moore), Elizabeth G. Moyes, J. M. Mullin, T. B. B. Paddock, Joyce H. Parker (later Mrs D. Holman), Theresa E. Parker (later Mrs M. Brendell), J. H. Price, Dr N. K. B. Robson, R. Ross, Mrs Shukla Sengupta, Patricia A. Sims, Christine J. Taylor, I. Tittley, A. R. Vickery and Susan White.

The Royal Botanic Garden, Edinburgh has co-operated with the project from its inception. In particular, the Regius Keeper, D. M. Henderson and Dr R. Watling, mycologist, spent several weeks on the island and are responsible for the account of the Fungi in Chapter 15. Dr R. W. G. Dennis, mycologist at the Royal Botanic Gardens, Kew, also substantially assisted in collecting data for this account.

It was fortunate for us that the Macaulay Institute for Soil Research, Craigiebuckler, Aberdeen, began a survey of Mull in 1966 as part of the Soil Survey of Scotland project. That memoir will appear in due course but we are indebted to the Institute's Director, Dr R. L. Mitchell, for allowing J. S. Bibby of his field staff to submit Chapter 5, and we are grateful to Mr Bibby for the work he has put into this concise account. Similarly, we are indebted to our colleague, Dr A. R. Woolley, of the Mineral Department, who has contributed substantially to the account of the geology of the Island (Chapter 4).

Throughout the period of the survey we have worked closely with H. T. Powell and other staff of the Scottish Marine Biological Association. It is my pleasure to record our thanks to Dr R. I. Currie who has enabled our phycologists to use the facilities of the Dunstaffnage Marine Research Laboratory and associated research vessels.

We include the Treshnish Islands in our survey as they are a natural part of the Mull assemblage. Logistically these could have proved difficult as they receive the full stretch of the Atlantic seas and even summer landings are unpredictable. With this in mind we approached the Ministry of Defence and we were grateful to the Flag Officer Naval Air Command, Vice-Admiral D.C.E.F. Gibson, for allowing a helicopter from his command to participate in the survey. We were pleased to have the full co-operation of the C-in-C at Lossiemouth RN Air Station, Capt. E. M. Brown, and his crews during the summers of 1968–69.

One of the encouraging aspects of this survey is the help we have received from amateur botanists either through the national societies like the Botanical Society of the British Isles, the British Lichen Society, the British Bryological Society and the British Pteridological Society, or working individually. We hope we have encouraged amateur botanists just as they have, in turn, stimulated us, thereby strengthening the link between the Museum and the public.

Lastly, but by no means least, it is my pleasure to record our gratitude to the people of Mull, to landowners who gave us permission to survey their property and their factors and employees who provided much helpful background information. We are especially grateful to Dr Flora MacDonald who for five years generously gave us facilities for preparing specimens and writing up field-notes in her house in Salen, and to Mr and Mrs Jim Carmichael of the Salen Hotel who co-operated so readily with the project and kindly tolerated our unconventional habits and muddy boots.

J.F.M. Cannon
Keeper of Botany

Editors' comments

This survey of the flora of Mull incorporates all known records with the results of this Department's attempt to collect and identify samples of all plant groups. It was inevitable that the coverage would be incomplete. Differing disciplines use various approaches, with consequent discrepances in comparative data; planktonic diatoms, for example, were collected from lakes but not from the inshore waters and the sea-lochs. Varying fieldwork time and logistic problems created gaps in coverage, for example terrestrial and freshwater algae. Gaps in naming occur, through lack of fertile material, for example in the freshwater algae, or of expert knowledge, as in the Cyanophyta. This latter group was the only major one which presented insuperable difficulties; there are few experts, and no-one in Britain had time to name our Mull material. We trust this presentation has used our data to the best advantage and that the gaps will be a stimulus to others.

Acknowledgment

All photographs, with the exception of that on the title-page to Part Two which is by J. S. Bibby, were taken by A. C. Jermy. The title page design (~~Pleistocene~~ landscape) Palaeocene is a copy of a drawing by Penny Laird made under the direction of C. J. Hill for display in the British Museum (Natural History). The map on the endpapers is based on the Quarter-inch Ordnance Survey with permission of the Controller of Her Majesty's Stationery Office, Crown copyright reserved.

We have great pleasure in recording our appreciation to all who have cooperated in the production of this book, especially: the authors who made every effort to comply with our editorial pleas, Dr G. S. Johnstone (IGS, Edinburgh) who checked the figures in Chapter Four, the artists Ann Davies and Valerie Jones, the BMNH Photographic Unit, Richard Cottingham (formerly of the BMNH Department of Public Services) who spent much time in designing the typography and lay-out, Bob Press for the indexing, and Keith Eady and his staff at William Clowes (Great Yarmouth) Limited for their great expertise and patience in typesetting.

A. C. Jermy and J. A. Crabbe

Synopsis of contents

Part one—prologue

Chapter 1 History of plant recording on Mull

J. F. M. Cannon

Chapter 2 Patterns of distribution within the flora of Mull

E. B. Bangerter, J. F. M. Cannon, A. Eddy, D. J. Hibberd and P. W. James

Part two—the environment

Chapter 3 General description and topography

A. C. Jermy

Chapter 4 Geology

A. R. Woolley and A. C. Jermy

Chapter 10 Terrestrial ecosystems

A. C. Jermy, P. W. James and A. Eddy

Part three—the flora

Chapter 15 Fungi

D. M. Henderson and R. Watling

Chapter 16 Freshwater algae (excluding diatoms)

D. J. Hibberd

Index to authors quoted in chapters 1-10

Index to geographical names in chapters 1-10

Part one
Prologue

1. History of plant recording on Mull

J. F. M. Cannon

Although the flora of Mull and the adjacent small islands was, by the standards of modern British floristics, poorly known previous to this survey, the number of botanists responsible for records from the area is much greater than might have been expected. As space prevents a full account of all these, only the most interesting and important records are considered here.

The earliest record attributed to the Mull area is one taken up by William J. Hooker (1821) from Sir Robert Sibbald's *Scotia illustrata . . . De plantis Scotiae* (1684) where *Cochlearia anglica* is recorded from the 'rocks of Inch Columb'. This was equated by Hooker with Iona, but in this he was almost certainly in error and the record properly refers to Inch Colm in the Firth of Forth. Thus Camden's remark about Mull in the footnotes to edition 2 of his *Britannia* (1695) that 'it abounds with wood and deer', becomes the first recorded comment on the vegetation of the area that is known to us. It is difficult to believe that Mull really 'abounded with wood' at that time, not least in the light of Dr Samuel Johnson's pungent comments made on a tour of Scotland in 1773 with James Boswell during which they visited Mull, Ulva, Inch Kenneth and Iona. Both, in due course, produced accounts of their journey in which the only references of a botanical nature are the comments by the great lexicographer who was clearly impressed by the paucity of woodland and desolate character of the Mull landscape. Boswell (1786) records the following typically peevish comments of his distinguished companion, but in evaluating them as a source of scientific information it must be remembered that Scot-baiting was a favourite pastime of the sage: 'Sir, I saw at Tobermorie what they called a wood, which I took to be a *heath*. If you show me what I shall take for *furze*, it will be something!' Further, after the loss of a walking-stick: 'No, no, my friend, it is not to be expected that any man in Mull, who has got it, will part with it. Consider, Sir, the value of such a *piece of timber here*'. The Welsh naturalist, Edward Lhwyd, writing to Richard Richardson in 1699, included a list of seven species from Mull; the presence of '*Alchimilla* [sic] *Alpina quinquefolia*' [*Alchemilla alpina*] and '*Cotyledon hirsut.*' [*Saxifraga stellaris*] among these suggests that Lhwyd was the first botanist to explore mountain habitats in Mull.

The importance of Iona as a pioneer centre of celtic Christianity, coupled with the unique geological interest of Fingal's Cave and its associated columnar basalt cliffs on Staffa, ensured that these two islands were frequently visited by men of learning and consequently soon became far better known botanically than Mull itself. The zoologist Thomas Pennant, in *A tour in Scotland and voyage to the Hebrides 1772* (1774), noted on the authority of his botanical companion the Rev. John Lightfoot, a few plants in his chapter on Iona. Lightfoot was indebted to Sir Joseph Banks, the outstanding scientific personality of his time, for a narrative of the latter's visit in the same year to Staffa. A record of *Radiola linoides* from Iona and a specimen of *Rhynchosinapis monensis* were until recently the only botanical commemorations that were known of Banks' visit to the area. The specimen is the earliest known gathering from the area, and forms a gratifying link between this survey and the Banksian Herbarium, which is one of the principal foundation collections of the British Museum Herbarium. Dr Averil Lysaght has recently drawn our attention to an entry in Banks' manuscript journal of his journey to Iceland, which is housed in the library of McGill University. Here, Banks writes that the churchyard on Iona was totally overgrown with the largest plants of *Petasites* that he had ever seen, 'which renders it impossible to search after inscriptions in the summer time'. Banks speaks facetiously of the guide who 'carried us under the ample shade of the *Petasites*, stopping here and there to inform us of the places where Kings and nobles had been interred . . .'. The well-kept grounds which surround the church today provide a strong contrast with that situation. Banks also records that '*Arundo arenaria* [*Ammophila arenaria*] which grows upon the sand hills near the shore is a favourite food of their cattle in winter and they reckon that the lands that produce it of great value for wintering their black cattle'. Lightfoot who published his *Flora Scotica* in 1777, based it substantially on the observations made during his travels with Pennant and included a number of additional records from the area.

Recording was still limited to Iona when a list of plants from there was published in the *Rural Essays* of the Rev. John Walker. This work (Walker, 1808) appeared after his death and a publisher's advertisement in the book suggests the essay on Iona was one of Walker's earliest and '. . . being probably written between the years 1764 and 1774' places it close to the date of Lightfoot's visit. Dr Thomas Garnett, Professor of Natural Philosophy and Chemistry at the Royal Institution, who visited the Mull area in 1798 in the course of a general tour of the Highlands and Islands, records many interesting botanical observations including several first records for the area. He gives a description of kelp-burning on Mull, specifically mentioning the use of *Fucus vesiculosus* and *F. serratus*. As is usual for the period, Iona and Staffa received a more extensive treatment than the remainder of the area, but some of the records he gives for Iona appear to be at least partially derived from Pennant's earlier publication.

The opening decades of the nineteenth century saw the visits of one of the foremost botanists of his time, namely William J. Hooker, accompanied by various naturalists. In 1807, Hooker went with Mr and Mrs Dawson Turner on a botanical tour in Scotland embracing Oban, Mull, Ulva, Staffa and Iona. Although it is possible that Hooker made other excursions in the area before taking up residence in Glasgow in 1820, this does not seem likely in view of an absence of information about any such trips in the life of William Hooker meticulously compiled by his son, Sir Joseph Hooker (1902). It is therefore a reasonable deduction that the number of original seaweed records that appeared in Turner (1809) and Hooker (1821) derived from the visits made in 1807. Two such records from Iona actually establish their derivation from the Hooker-Turner trip, since they are recorded in Hooker (1821) under the collectors' names of 'Mr Turner and Hook.' A third record appearing in volume two of Turner's *Fuci*, in a footnote to plate 90 *Fucus serratus*, is so close in time after the trip in 1807 that it can hardly have derived from other sources. Turner's footnote states: 'In the island of Mull, I was told that *F. serratus* goes by the name of black-wrack . . .'.

In the *Flora Scotia*, Hooker (1821) includes a few Mull localities for vascular plants, but none of them were new records. In July 1826, Hooker organised a party of students to visit the Western Isles including Mull, Iona and Staffa, on which Captain Dugald Carmichael and R. K. Greville accompanied him, the latter, and possibly Carmichael, collecting seaweeds at Iona and Staffa. Specimens of these algae are in the BM, in Glasgow and at the Royal Botanical Garden, Edinburgh, and some are mentioned in the *Scottish cryptogamic Flora* by Greville (1826–27) and *Algae Britannicae* (Greville, 1830). The Rev. M. J. Berkeley visited Carmichael in 1828 and we can assume, from the entry under *Codium fomentosum* in volume two of Hooker's *British Flora* (Harvey, 1833), collected on Iona. There are several further records from Staffa and Tobermory in that work and William Harvey himself visited Staffa in September and October 1832; there is a specimen of *Calithamnion arbuscula* in University College, Cork annotated by him to this effect.

Further knowledge of the flora of Mull is to be gleaned from J. MacCulloch *A description of the Western Isles of Scotland* (1819), a work primarily devoted to geology. L. Necker-de-Saussure, a Swiss scientist, published a delightful account of his visit to Mull in *Voyage en Écosse et aux Îles Hébrides* (1821). While his observations are confined to topography, vegetation and economic botany, lacking any precise records, they are nevertheless of considerable interest. A very infectious enthusiasm pervades the whole book and is particularly evident in his narrative of the visit to Staffa and the famous 'Grotte de Fingal'. The modern botanical traveller will understand and sympathise with his diary entry for 16 August: "Un vent impénetueux accompagné d'un déluge de pluie regna toute la journée". There is in the library at BM (NH) a manuscript list of plants dated 1829 by the Rev. G. Gordon, which was later expanded into his *Collectanea for a Flora of Moray* (1839) and which, despite its title, includes plants from as far away as Mull.

One of the most significant pioneer works on the British flora, *Cybele Britannica* by H. C. Watson (1847–59), will be familiar to all students of plant distribution in this country but it is not generally known that the manuscript slips accumulated by Watson as the basis of this work are preserved in the BM (NH) library. Unfortunately the records, although localised, are not always dated, but they must be earlier than the appropriate volume of the *Cybele* and thus provide for some species the first known reference to their occurrence in the area. A firm basis for some of the records can be traced to specimens in Watson's herbarium at Kew. In the herbarium of James T. I. Boswell-Syme (in BM) there are specimens collected by him during his visit to Mull in 1848. These substantiate most of the references to the Island in his third edition of Smith's *English Botany* (1863–86).

The particular interest in Iona and Staffa culminated in a list of the plants found there published in 1850 by W. Keddie, lecturer on Natural Sciences at the Free Church College, Glasgow, who was in the habit of visiting Iona regularly. This is the source of many first records and the first substantial catalogue for any part of the area; Keddie's algae from there are in the herbarium of Glasgow University. D. P. Maclagan, in an article in the *Scottish Gardener* (1855), contributed a few records of interest from south-east Mull. Marine algal records from Staffa and Iona were published by W. C. Dendy (1859, 1860) who himself may have visited the islands.

The systematic recording of the distribution of British plants, initiated by Watson, stimulated the production of numerous vice-county lists. As a contribution to the Flora of v.-c. 103, Mid Ebudes, George Ross published a list of plants from Mull in 1877 in the *Report of the Botanical Locality Record Club*, through which organisation numerous Ross voucher specimens were deposited at the British Museum (Natural History). Little is known about George Ross, who was undoubtedly the most important contributor to the early knowledge of the vascular flora. He was at one time a resident of Tobermory and later of Oban. In March 1878 he was elected an associate of the Botanical Society of Edinburgh, and in May of the same year a paper 'On the flora of Mull', was read at a meeting of that Society. This paper (Ross, 1879) substantially repeats his earlier paper with some minor additions and the provision of localities for the less common plants. He continued to contribute occasional records from Mull for the next few years, and his name appears for the last time, still as an Associate, on the Roll of the Botanical Society of Edinburgh in 1893.

In general terms, the geology and pedology of Mull is not of a kind to support a phanerogamic flora to inspire the particular interest of botanists, most of whom seem to have paused briefly while in transit to other Hebridean islands such as Skye or Jura. However, in August 1881, Mull was the chosen area of the Scottish Cryptogamic Society field meeting. The party, including Thomas King, George Ross, J. Stirton and F. Buchanan White, gave close attention to the bryophytes, fungi and lichens resulting in a list of 118 fungi and 102 mosses which were published by White in 1881 and 1882 respectively. No doubt a similar number of hepatics and lichens was also collected at the time, but the promised list never materialised. Stirton had earlier visited Mull and published (Stirton, 1877) on his collections of lichens in which he describes six new species of the genus *Lecidea*; of these taxa only one, *L. mullensis*, remains in the British check-list (James, 1965); he later (1899) published a further note on lichens from Carsaig.

Buchanan White and George Ross also collected around this time plankton samples from the Tobermory area and the material was named by John Roy (1883, 1893–4). Further algal samples were collected by John Murray whilst carrying out his *Bathymetric survey of fresh-water lochs of Scotland* (Murray & Pullar, 1908); the phytoplankton determinations were published by Borge (1897).

Other botanists at the end of the nineteenth century were concerned with bringing together miscellaneous records at the vice-county level for Scotland and published over the years in the *Annals of Scottish Natural History*. They included Arthur Bennett, J. W. H. Trail and Peter Ewing. The last-named published two editions of the *Glasgow Catalogue*, being a 'Comital Flora' for the west of Scotland (Ewing 1892, 1899). Ewing himself visited Mull (for example, he collected hepatics there in 1887) and indicates in his catalogues that voucher specimens were seen by him; some of these have been traced to the herbaria of Glasgow University and the British Museum. In most cases, however, it is not possible to say to which part of v.-c. 103 the record refers. The Rev. Symers M. Macvicar, a renowned bryologist and a resident of Salen, Ardnamurchan, collected vascular plants in the vice-county, but more particularly from Coll and Tiree. He collected together the hepatic records made by his daughter and by Ewing and Kennedy, and published them in 1910. H. N. Dixon (1899) made a brief visit in 1898.

In the herbarium of Miss Eleanor Vachell at the National Museum of Wales, Cardiff (NMW) are a few specimens gathered on Iona in 1899 by her father, C. T. Vachell. Both he and his daughter again visited the area in 1905. The plants observed are listed in her diary (also at NMW), with some voucher specimens in her herbarium. Eleanor Vachell was also a member of the Botanical Exchange Club (BEC) 1939 Excursion, and her herbarium includes some specimens from this visit.

Robert G. Davie, whose specimens are to be found at the BM and the Royal Botanic Garden, Edinburgh (E), was among the early twentieth-century collectors in Mull, visiting the area in 1906. A paper entitled 'The Flora of the Island of Iona, Argyllshire', left in manuscript at the British Museum (Natural History) by Alfred J. Wilmott who had recently joined the staff there, describes his visit to the island in 1912 and lists the plants he observed. It is accompanied by a map showing five recording areas and for the more interesting species precise localities are given. We have not been able to trace any specimens associated with this account. He appears to have been dissatisfied with this systematic list, which indeed contains some highly doubtful records, and therefore left it unpublished. In a general work on Mull, J. P. Maclean (1923, 25) published a floristic list, evidently based on earlier published sources and adding no new records. J. B. Duncan and William Young collected several bryophyte novelties in 1932; a few later records, mainly from Iona, were submitted to the British Bryological Society by L. B. C. Trotter in 1938.

The Botanical Exchange Club organised an excursion to Mull in 1939 under the leadership of John Chapple but, owing to the outbreak of war shortly after the visit, an account was not published until 1942. In the *Botanical Exchange Club Report* for 1939/40 (1942), Wilmott, who had not himself re-visited the area, attempted to collate the various notes made by members of the excursion into one systematic list. We have been able to locate only a few (Chapple (OXF) and Vaughan (NMW)) of the specimens which would have confirmed the several new vice-county records tentatively noted by Wilmott, who was himself unable to trace them under the prevailing wartime conditions. A member of the party, A. Templeman, stayed on after the meeting and was responsible for a short note (Templeman, 1942) in the same Report which includes for the first time records of plants from the Treshnish Isles.

In the marine algal field the coasts of Britain were being studied by ecologists who made small collections. J. A. Kitching in studying the coasts of Argyll visited various localities on Mull in 1934. Dr Lily M. Newton visited Mull during government surveys into the utilisation of seaweeds for agar between 1940–43. During the forties the only vascular plant collections of significance are of a few specimens (all at Glasgow) made on Iona by Lilias Small in 1941 and Professor K. W. Braid in 1948. Dr Heather Salzen (née Fairley) visited Ulva in 1948 and has kindly placed her notebook at our disposal.

The fifties were distinguished by the launching in 1954 of the Botanical Society of the British Isles Distribution Maps Scheme and a number of botanists visited Mull in connection with this. Through the courtesy of Dr F. H. Perring of the Biological Records Centre we have had access to the original record cards upon which the coverage of our area for the *Atlas of the British Flora* (Perring & Walters, 1962) was based. The following contributed to the records for the *Atlas* maps: Miss E. P. Beattie, Dr Ursula K. Duncan, R. E. C. Ferreira, B. Flannigan, Miss V. Gordon, Mrs S. Harris, J. McNeill, H. Milne-Redhead, J. Ounsted, J. E. Raven, Dr Heather Salzen, A. A. Slack, A. McG. Stirling and Miss B. E. Young. Other visitors to the area during this period included Miss M. B. Gerrans and Miss D. Hillcoat, both of the British Museum (Natural History). The former made a general collection in the area in 1958 and published an account of it in 1960, while the latter made a small collection while on holiday in Iona in 1959. Jean M. Tindale collected algae on the Treshnish Isles in 1953 and her specimens are in the Gatty Marine Laboratory, St Andrews.

In the sixties, previous to the beginning of the present survey in 1966, field excursions to Mull were organised in 1961 and 1965 by the Botanical Society of the British Isles and the Committee for the Study of the Scottish Flora. P. W. James and Dr Ursula K. Duncan made an extensive excursion to study the lichen flora prior to this survey in 1965.

The two papers of George Ross (1877 and 1879) must be regarded as the most important contribution to the knowledge of the flora of Mull up to this time. Subsequent fieldwork, including this survey, has however doubled the number of flowering-plants and ferns listed by him. Whilst the total includes micro-species (e.g. of *Hieracium* and *Euphrasia*) and recent introductions from gardens, it also includes a substantial proportion of undoubted native species such as *Corallorhiza trifida*, *Carum verticillatum* and *Najas flexilis*, all three of which have been discovered within the last 15 years. For some groups, e.g. fungi and lichens, we had virtually no records. After the completion of our own period of work in the field, Mrs Jean M. Millar published in 1972, a booklet entitled *Flowers of Iona* in which she lists in systematic order all the species of flowering plants, excluding grasses, sedges and rushes, known to her to occur on the island, together with a simple account of some of the habitats. Since the beginning of our Survey a number of groups have botanised on Mull and their records, when relevant, are quoted in Part 3. One party from the University of Bradford compiled a report on the flora of the Ardmeanach Peninsula (Crawley *et al.*, 1975).

I am indebted to a number of colleagues who have supplied information on collecting of their respective groups.

References

BORGE, O. (1897). Algologiska notiser 4. Süsswasser-Plankton aus der Insel Mull. *Bot. Notiser*, **1897**: 210–215.

BOSWELL, J. (1786). *The journal of a tour to the Hebrides with Samuel Johnson, LL.D.*, ed. 3. London.

CAMDEN, W. (1695). *Britannia*, ed. 2. London.

CRAWLEY, M. J., ROBINSON, H. D. SEAWARD, M. R. D. & SHAW. (1975). P. J. *Plant communities of the Ardmeanach Peninsula on the Isle of Mull.* Bradford University.

DENDY, W. C. (1859). *The wild Hebrides.* London.

———. (1860). *The beautiful islets of Britaine.* London.

DIXON, H. N. (1899). Bryological notes from the West Highlands. *J. Bot., London.*, **37**: 300–310.

EWING, P. (1892; 1899, ed. 2). *Glasgow catalogue of native and established plants, being a contribution to the topographical botany of the western and central counties of Scotland.* Glasgow.

GARNETT, T. (1800). *Observations on a tour through the Highlands and part of the Western Isles of Scotland, etc.* 2 vol. London.

GERRANS, M. B. (1960). Notes on the flora of the Isle of Mull. *Proc. bot. Soc. Br Isl.*, **3**: 369–374.

GORDON, G. (1839). *Collectanea for a Flora of Moray; or, a list of the phaenogemous plants and ferns hitherto found in the Province.* Elgin.

GREVILLE, R. K. (1826–27). *Scottish cryptogamic Flora*, vol. 5. Edinburgh.

———. (1830). *Algae Britannicae.* Edinburgh & London.

HOOKER, J. D. (1902). A sketch of the life and labours of Sir William Jackson Hooker (with portrait). *Ann. Bot.*, **16**: ix–ccxxi.

HOOKER, W. J. (1821). *Flora Scotica.* Edinburgh & London.

HARVEY, W. H. (1833). *Algae, Div. II Confervoideae, Div. III Gloiocladeae*, in Hooker, W. J., The English Flora of Sir James Edward Smith, vol. 5 (or vol. 2 of Dr Hooker's British Flora): 322–385, 385–401.

JAMES, P. W. (1965). A new check-list of British lichens. *Lichenologist*, **3**: 95–153.

KEDDIE, W. (1850). *Staffa and Iona described and illustrated* (Botany in Appendix pp. 139–148). Edinburgh, Glasgow & London.

LIGHTFOOT. J. (1777). *Flora Scotica.* London.

MACLAGAN, D. P. (1855). Notice of plants in the neighbourhood of Oban and in part of the Island of Mull. *Scottish Gardener*, **4**: 93–95.

MACLEAN, J. P. (1923, 1925). *History of the Island of Mull*, vols. 1 & 2. San Mateo, California.

MACCULLOCH, J. (1819). *A description of the Western Isles of Scotland.* London.

MACVICAR, S. M. (1910). The distribution of Hepaticae in Scotland. *Trans. Proc. bot. Soc. Edinb.*, **25**: 1–336.

MILLAR, J. M. (1972). *Flowers of Iona.* Baile Mòr.

MURRAY, J. & PULLAR, L. (1908). *Bathymetrical survey of the fresh-water lochs of Scotland:* 173–176. London.

NECKER-DE-SAUSSURE, L. (1821). *Voyage en Écosse et aux Îles Hébrides.* 3 vols. Geneva.

PENNANT, T. (1774). *A tour in Scotland and voyage to the Hebrides 1772.* Chester & London.

PERRING, F. H. & WALTERS, S. M. (1962). *Atlas of the British Flora.* London & Edinburgh.

ROSS, G. (1877). County catalogue of plants: Isle of Mull, Mid Ebudes. *Rep. bot. Locality Record Club*, **1876**: 188–192.

———. (1879). On the flora of Mull. *Trans. Proc. bot. Soc. Edinb.*, **13**: 234–242.

ROY, J. (1883). List of the desmids hitherto found on Mull. *Scott. Nat.*, **7**: 37–40.

———. (1893–94). On Scottish Desmidiae. *Ann. Scot. nat. Hist.*, **1893**: 106–111, 170–180, 237–245; **1894**: 40–46.

SIBBALD, R. (1684). *Scotia illustrata . . . De plantis Scotiae*, pars secunda **1**: 18. Edinburgh.

SMITH, J. E. (1863–86). *English Botany*, ed. 3 by J. T. I. Boswell-Syme. London.
STIRTON, J. (1877). Descriptions of new lichens. *Scott. Nat.*, **4**: 164–168.
———. (1899). The lichens & mosses from Carsaig, Argyll. *Ann. Scot. nat. Hist.*, No. 29: 41–45.
TEMPLEMAN, A. (1942). Iona and Lunga. *Rep. bot. Soc. Exch. Club Br. Isl.*, **1939–40**: 249–250.
TRAIL, J. W. H. (1909). Additional vice-county records from West of Scotland. *Ann. Scot. nat. Hist.*, **18**: 250.
TURNER. (1809). *Fuci*, vol. 2. London.
WALKER, J. (1808). *Essays on natural history and rural economy*. London & Edinburgh.
WATSON, H. C. (1847–59). *Cybele Britannica.* 4 vols. London.
WHITE, F. B. (1881–82). The cryptogamic flora of Mull. *Scott. Nat.*, **6**: 155–162, 210–212.
WILLIAMS, E. G. (1937). Notes on the plankton of some lochs in the Island of Mull. *N West. Nat.*, **11**: 23–29.
WILMOTT, A. J. (1942). Report on excursions, 1939: The Island of Mull. *Rep. bot. Soc. Exch. Club. Br. Isl.*, **1939–40**: 236–249.

2. Patterns of distribution within the flora of Mull

Introduction

The drawing of general conclusions from distributional patterns, both within and without Mull, from data such as has been obtained in the Mull Survey must necessarily be fraught with considerable difficulties, most of which stem from incompatibility between the various subsets of data on the major groups of plants recorded. This incompatibility may be summarised under three heads. First, our knowledge of the distributions of the species varies very much from group to group – from the much studied and relatively familiar flowering-plants, to some of the microscopic algae, for which our present knowledge provides no more than isolated chance records and presents no overall picture at all. The second factor is the relationship between the terminologies that have been adopted by students of different groups (and indeed within the same group) to describe plant distributions and phytogeographical relationships. While these are in broad agreement (cf. Ratcliffe, 1968), which is scarcely surprising since presumably they are all a reflection of the same spectrum of ecological factors, the precise definition of the terms used, coupled with the degree of subdivision into smaller units, creates difficulties when broad conclusions are sought. The resolution of phytogeographic data into one comprehensive system is one of the most important tasks that confronts British botanists, and must be solved if further substantial progress is to be made in the phytogeographic relationships of our flora. An attempt to do this for the flora of Skye by Birks (1973) is a positive move in this direction. Thirdly, the sampling techniques, which are dictated in part by the plants themselves, inevitably result in varying degrees of coverage for the different groups.

In the light of these problems the best approach appears to be an initial description of the distinctive distributional patterns seen on Mull, followed by a discussion of the phytogeographic relationships of the vascular plants of Mull, supported and supplemented by the increasing amount of information that is becoming available from the bryophytes and lichens (see p. 16). For the remaining cryptogamic groups, however, records are at best patchy, like freshwater algae (see p. 21), and at worst non-existent, and thus not capable of making any comprehensive contribution to the overall picture. No attempt has been made to incorporate information from the marine algae, since here the information is both limited and concerns organisms living in an entirely different ecological system; some notes on the broader distributional patterns of particular marine algae are included in the specific treatments in Chapter 19.

Selected distribution patterns shown by the vascular plants

E. B. Bangerter and J. F. M. Cannon

If distribution maps of all the vascular plants on Mull are reviewed, the majority show patterns of no general significance. There are, however, three groups of species which show distinctive patterns and these are discussed below.

Western distributions

A series of progressively-expanding patterns can be traced from the few species that are confined to Iona, through various intermediate stages to those which are widespread in the west, some of which also extend along the north and south coasts. In some cases the distribution is clearly linked to very restricted habitats, in others the habitats appear to be available in other localities, but some other factors, for example restricted availability of nutrients, may be preventing colonisation. Thus several species which are widespread in western Mull and on the coast of Morvern do not occur on the east coast of Mull.

In our area *Apium inundatum* is confined to Iona. In Britain as a whole it is not a coastal species, but does require a high base status and is widely distributed over the whole country. It is present on Coll and Tiree, but appears to be absent from Skye and its adjacent islands and the north-west mainland of Scotland. Another species limited to Iona is *Cakile maritima*, a coastal species whose limited distribution is probably associated with the shortage of suitable habitats in the area as a whole.

Until recently, it was thought that *Umbilicus rupestris* (Fig. 2.1) reached its northern limit in western Britain in Iona, but it has recently been found in several localities in the Ardnamurchan Peninsula to the north of Mull. We have also found it on Erraid and Eilean a' Chalmain (near Iona) and it has been recorded from Tavool House on the south side of the Ardmeanach. *Pimpinella saxifraga* (Fig. 2.2) is an example of a species limited by its requirements for a base-rich habitat, and occurs only on Iona, the dunes at Ardalanish Bay and on Inch Kenneth. *Hypericum elodes* (Fig. 2.3) is widespread on Iona, the Ross of Mull and western Brolass, with one widely-separated locality at Torosay in south-east Mull. In Britain as a whole, it is predominantly western in distribution and, like *Apium inundatum*, occurs in Coll and Tiree and the Outer Hebrides, being absent from north-west mainland Scotland. *Leontodon taraxacoides* (Fig. 2.4) and *Veronica anagallis-aquatica* (Fig. 2.5) have only been recorded from Iona and Calgary Bay. The former is probably associated with the machair conditions, but is generally very rare in western Scotland and apparently does not occur in the Outer Hebrides, while the latter is much more widely distributed. *Hieracium euprepes* shows a similar limited distribution. Calgary Bay probably enjoys a climate which, like that of Iona, is warmer, sunnier and drier than most other parts of Mull.

Many species of non-maritime plants show western coastal distributions in Mull. These include (in sequence of increasing range) *Catabrosa aquatica* (Fig. 6), *Scirpus cernuus* (Fig. 2.7), *Geranium sanguineum* (Fig. 2.8), *Carlina vulgaris* (Fig. 2.9), *Lemna minor* (Fig. 2.10), *Anagallis tenella* (Fig. 2.11), *Eupatorium cannabinum*, *Anthyllis vulneraria* (Fig. 2.12) and *Koeleria cristata* (Fig. 2.13). Some of these are well-known calcicoles, e.g. *Geranium*, *Carlina*, *Anthyllis* and *Koeleria*; it seems likely that their distribution in Mull is influenced by sea-spray carried by prevailing winds and bringing with it enough nutrients to permit the species to grow in otherwise hostile environments. This seems to apply particularly to *Geranium sanguineum* which in addition to its Iona habitats (sandy ledges on small cliffs associated with machair), also occurs in

Treshnish on highly unstable basalt-cliff screes directly exposed to the sea. The record for Tobermory, on the eastern side of the island, is thought to be of horticultural origin. *Hieracium caledonicum* (Fig. 2.14) occurs on Iona, the Treshnish Isles, Mornish and Inch Kenneth, and is also found on the mountains of the Ben More area. Many of the species extend along the south coast to Mull to Carsaig, Loch Buie and even to the Laggan Peninsula. *Eupatorium cannabinum*, which in Mull is a plant of sheltered cliff ledges is more frequent in western Scotland than in the east, suggesting a preference for the warmer microhabitats of less exposed situations.

Distribution of mountain species

The distributions of mountain and upland plants in Mull are perhaps best considered as a series of concentric ranges progressively extending out from the central massif. They are here considered in groups in order from the most restricted to the most widespread.

Ardmeanach. A few species are confined to the Ardmeanach (or only just extend to the western margins of the Ben More massif). Of these, *Koenigia islandica* (Fig. 2.15) is restricted to exposed habitats of an unusual kind (see Chapter 11). *Saxifraga oppositifolia* just crosses Gleann Seilisdeir to Creag Mhòr and Coirc Bheinn, while *S. aizoides* extends further to two localities in northern Mull – on Beinn na Drise and on the cliffs at Port Chill Bhraonain.

Ben More massif and Glen Forsa. *Cardaminopsis petraea* (Fig. 2.16) has the most limited distribution, being confined to Ben More and four localities in the mountains on the east and south-east sides of Glen Forsa; *Luzula spicata* (Fig. 2.17) and *Cryptogramma crispa* (Fig. 2.18) have similar patterns, but with the addition of several more stations in the central massif and round Glen Forsa. *Carex bigelowii* (Fig. 2.19) extends even further, reaching Ben Buie and Creach Beinn to the south of Glen More.

Ben More massif, Glen Forsa and Ardmeanach. This group shows similar patterns to the previous group, but with the addition of further localities on Ardmeanach (a high plateau to the west of the central massif). Here the species occur on steep seaward cliffs, as well as on the small summits of the plateau top. In more favoured situations, they may extend right down to sea level, as at the Gribun and on Creag Mhòr. *Saxifraga hypnoides* (Fig. 2.20) is a rare plant in widely-scattered localities throughout the area, while *Silene acaulis* shows a similar pattern but is more frequent. *Oxyria digyna* (Fig. 2.21) is similar again, with an extension to Creach Beinn. *Alchemilla alpina* (Fig. 2.22) is quite frequent throughout the mountains of central Mull, with several stations in the Ben Buie/Beinn Chreagach range, and an ultimate south-east extension to the north-facing slopes of the Laggan Peninsula.

Ben More massif, Glen Forsa, Ardmeanach and extensions to north and/or south. *Saxifraga stellaris* (Fig. 2.23) a widespread plant of central Mull, with a distribution like that of *Alchemilla alpina*, has one locality to the north of the central isthmus on Beinn na Drise, a station it shares with *Saussurea alpina* (Fig. 2.24) – a species with a similar overall pattern, but a much more limited number of localities. *Diphasiastrum alpinum* (Fig. 2.25), *Thalictrum alpinum* (Fig. 2.26) and *Salix herbacea* (Fig. 2.27) are similarly distributed, but with the addition of an isolated locality on Cruachan Min, one of the highest points in Brolass. In northern Mull, they occur also on Cnoc an da Chinn, instead of, or in addition to, Beinn na Drise. *Arctostaphyllos uva-ursi* (Fig. 2.28), which is widely scattered but rare in mountain habitats, also occurs frequently on low altitude cliffs on the Ross of Mull.

Maritime distributions

Some maritime plants occur in suitable localities all round the coasts of Mull and their distribution is an indication of the presence of the necessary habitat rather than a

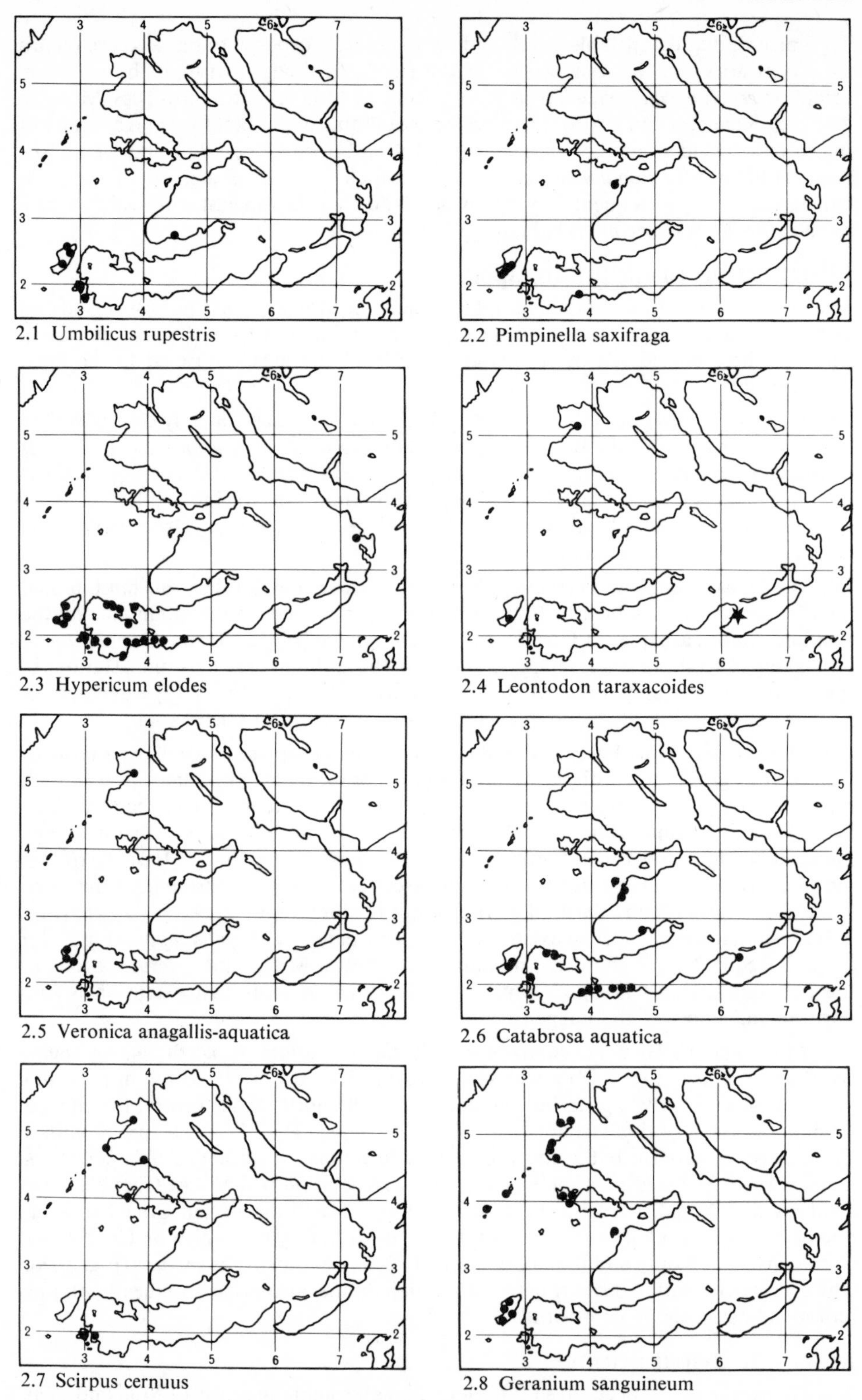

2.1 Umbilicus rupestris

2.2 Pimpinella saxifraga

2.3 Hypericum elodes

2.4 Leontodon taraxacoides

2.5 Veronica anagallis-aquatica

2.6 Catabrosa aquatica

2.7 Scirpus cernuus

2.8 Geranium sanguineum

Figs. 2.1–2.8 Distribution of selected species. ★ = pre-1930 record.

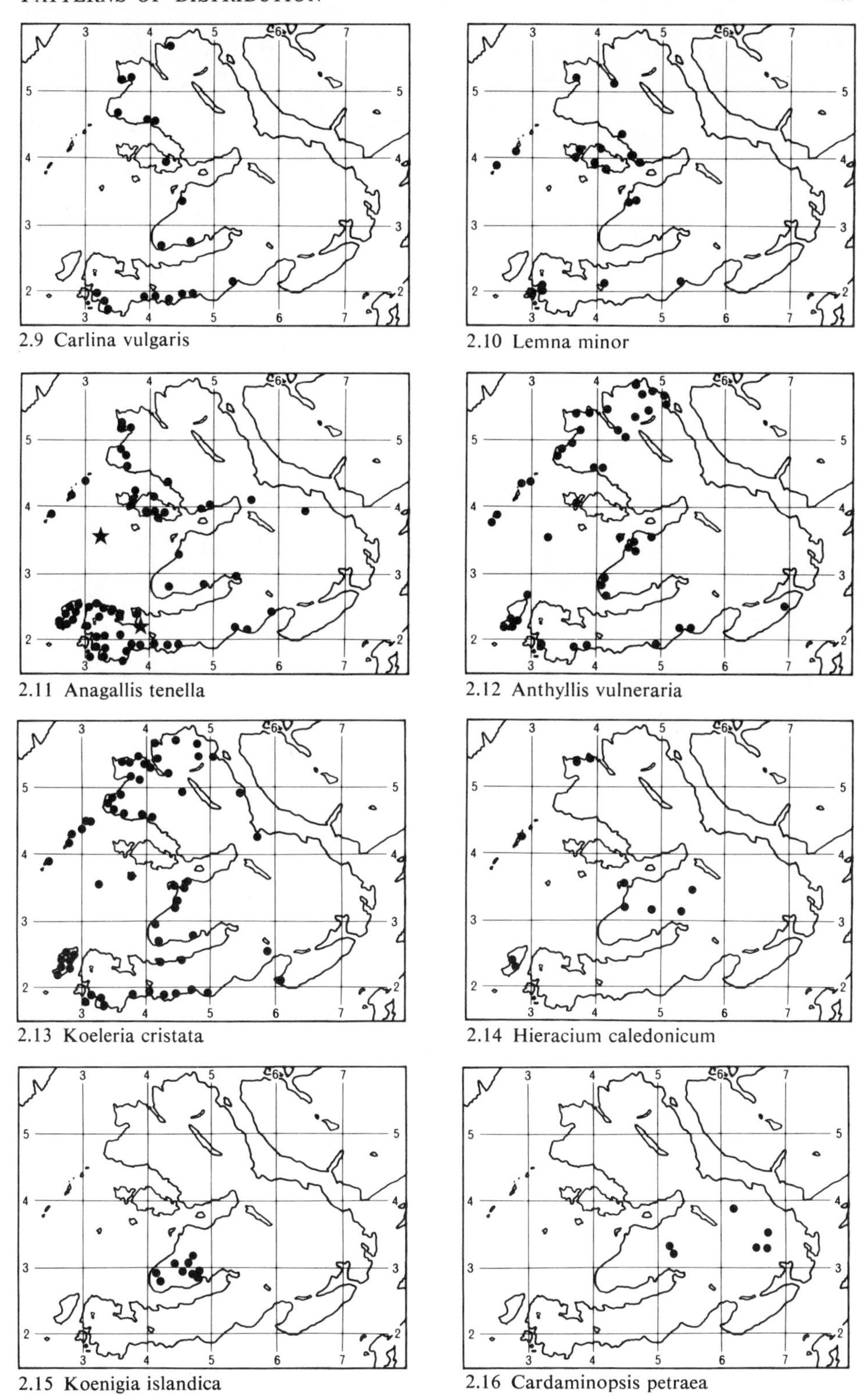

2.9 Carlina vulgaris

2.10 Lemna minor

2.11 Anagallis tenella

2.12 Anthyllis vulneraria

2.13 Koeleria cristata

2.14 Hieracium caledonicum

2.15 Koenigia islandica

2.16 Cardaminopsis petraea

Figs. 2.9–2.16 Distribution of selected species. ★ = pre-1930 record.

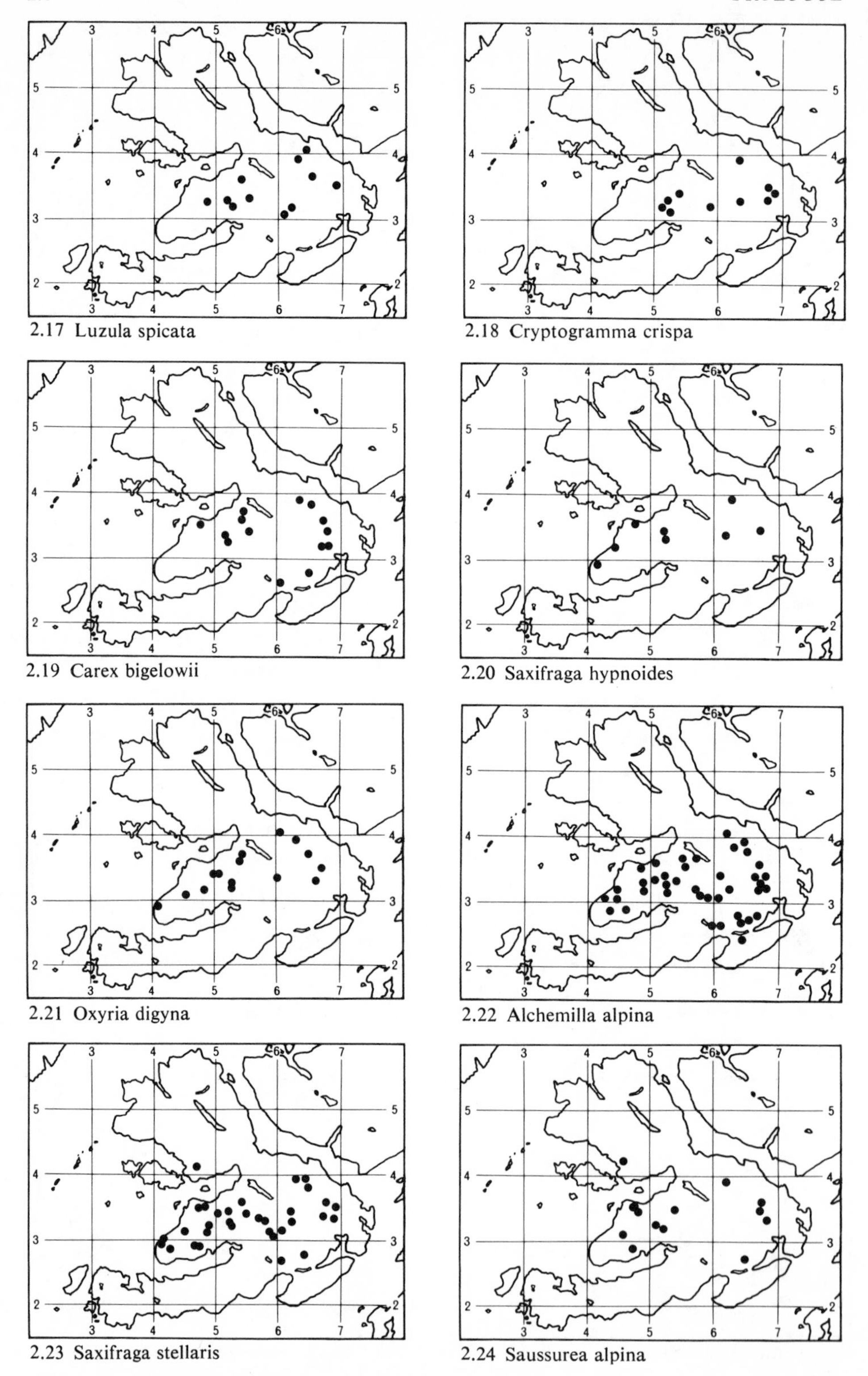

2.17 Luzula spicata

2.18 Cryptogramma crispa

2.19 Carex bigelowii

2.20 Saxifraga hypnoides

2.21 Oxyria digyna

2.22 Alchemilla alpina

2.23 Saxifraga stellaris

2.24 Saussurea alpina

Figs. 2.17–2.24 Distribution of selected species.

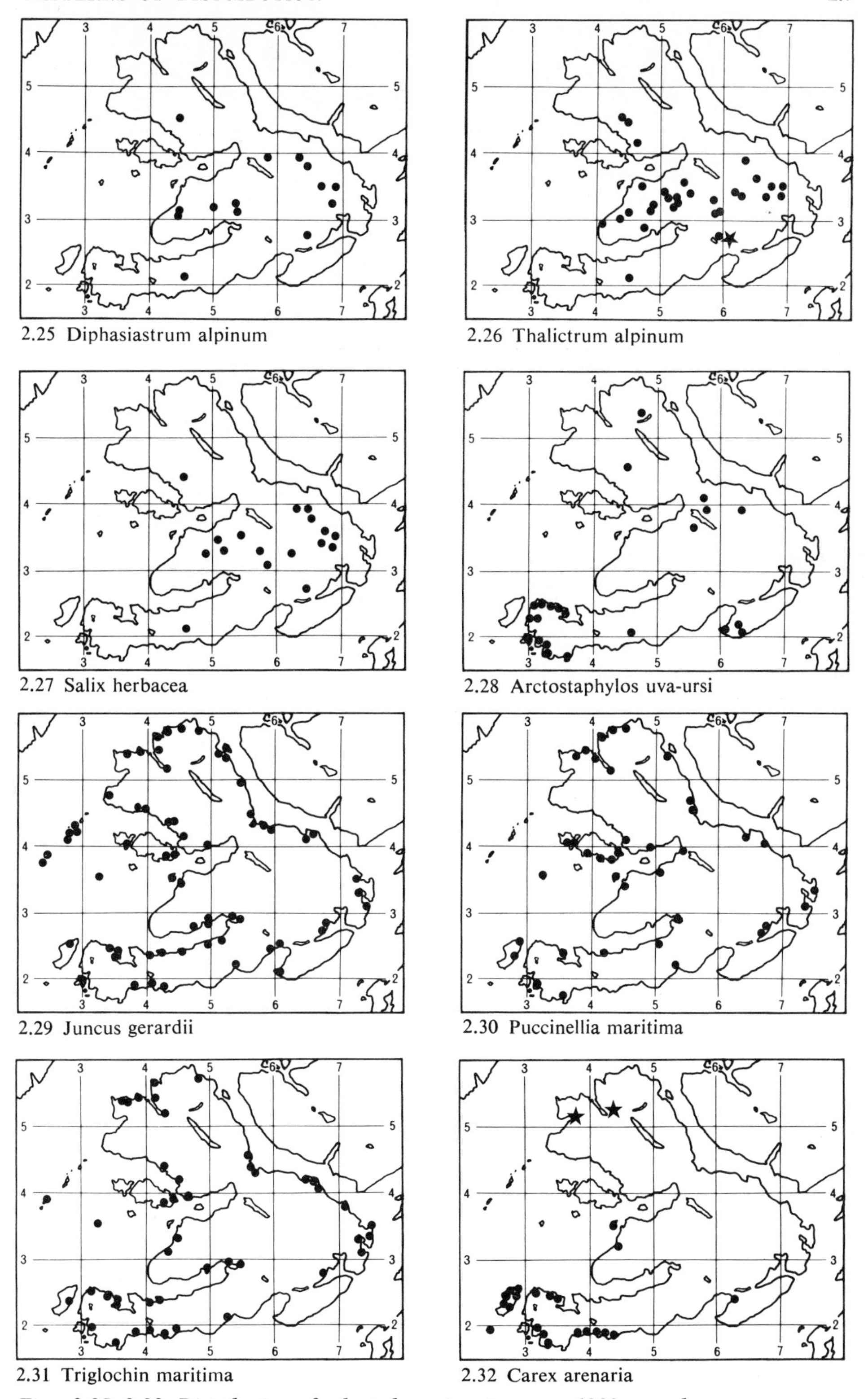

2.25 Diphasiastrum alpinum

2.26 Thalictrum alpinum

2.27 Salix herbacea

2.28 Arctostaphylos uva-ursi

2.29 Juncus gerardii

2.30 Puccinellia maritima

2.31 Triglochin maritima

2.32 Carex arenaria

Figs. 2.25–2.32 Distribution of selected species. ★ = *pre-1930 record.*

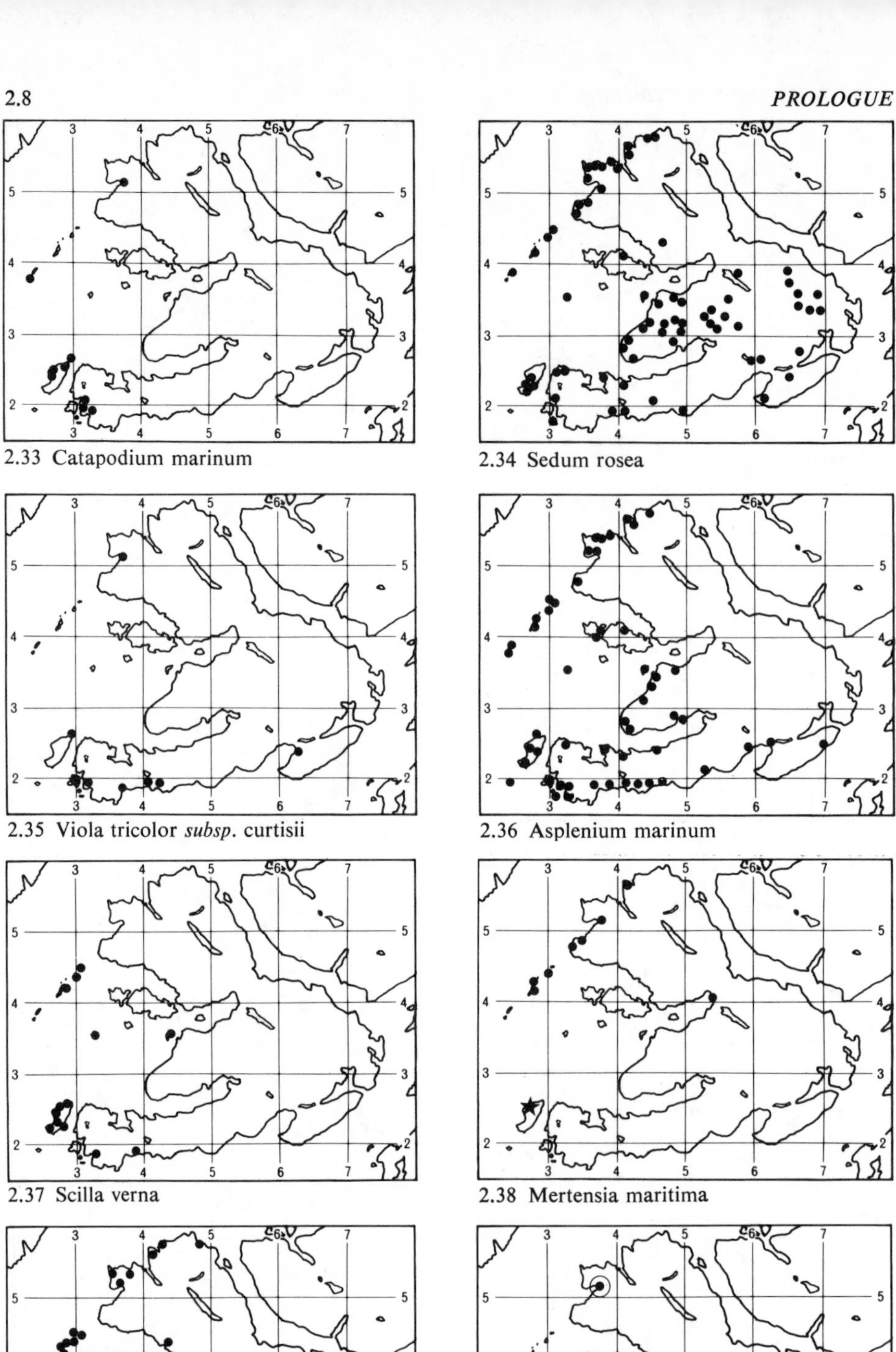

2.33 Catapodium marinum

2.34 Sedum rosea

2.35 Viola tricolor *subsp.* curtisii

2.36 Asplenium marinum

2.37 Scilla verna

2.38 Mertensia maritima

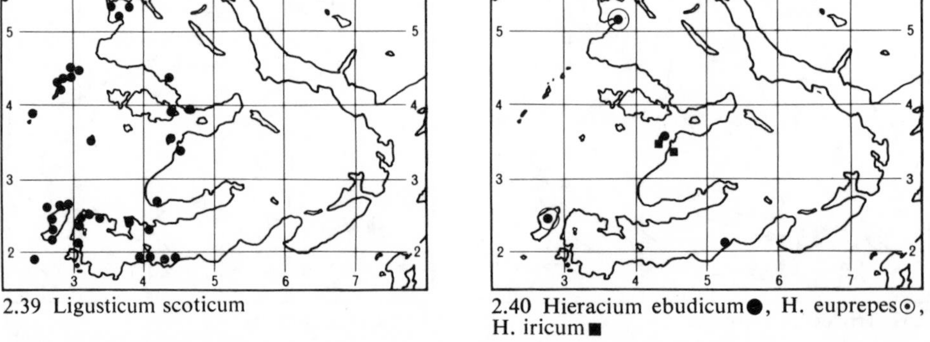

2.39 Ligusticum scoticum

2.40 Hieracium ebudicum●, H. euprepes⊙, H. iricum■

Figs. 2.33–2.40 Distribution of selected species. ★ = pre-1930 record.

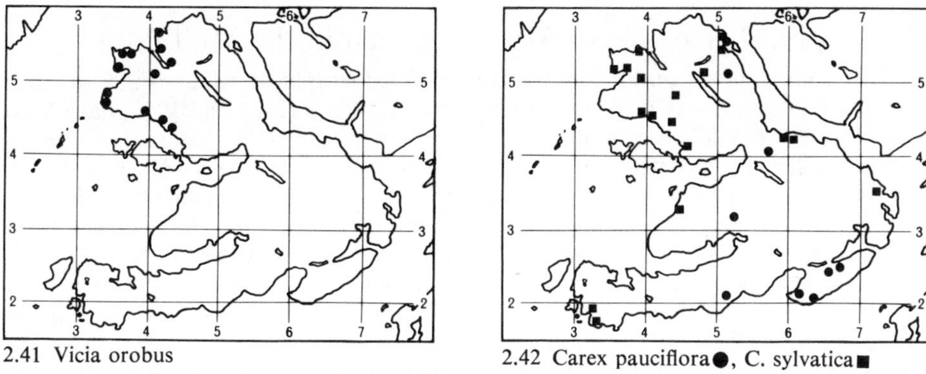

2.41 Vicia orobus

2.42 Carex pauciflora●, C. sylvatica■

Figs. 2.41–2.42 Distribution of selected species.

climatic phenomenon. These are chiefly salt marsh species like *Juncus gerardii* (Fig. 2.29), *Puccinellia maritima* (Fig. 2.30), *Triglochin maritima* (Fig. 2.31) and dune species like *Carex arenaria* (Fig. 2.32) and *Catapodium marinum* (Fig. 2.33). Some species in this group such as *Armeria maritima* and *Plantago maritima*, occur, as elsewhere in Britain, in open communities on mountains, often to 900 m. *Sedum rosea* (Fig. 2.34) behaves in a comparable manner occurring on the ledges and crevices of both maritime and montane cliffs. A second group of maritime species is confined to the western coasts, and related in distributional terms to the first group considered in this account. They are probably controlled by the same limiting factors such as the presence of base-rich shell sand e.g. *Viola tricolor* subsp. *curtisii* (Fig. 2.35) and the availability of warmer microclimates with fewer wet days. *Asplenium marinum* (Fig. 2.36) and *Scilla verna* (Fig. 2.37) reflect in Mull the pattern of their general British distribution. *Mertensia maritima* (Fig. 2.38) and *Ligusticum scoticum* (Fig. 2.39) although western on Mull have strongly northern affinities. The latter is clearly destroyed by grazing and always occurs on inaccessible cliff ledges, other than on the small ungrazed islands in the Treshnish group. The former appears to have been reduced since the times of the earliest observations on the Mull flora. It seems now to survive on the strand lines of some stony beaches, whereas Pennant (1774) noted that on Iona 'It makes the shores gay with its glaucous leaves and purple flowers'. Our own observations have disclosed only one depauperate, non-flowering individual there, the same being true of the station at the head of Loch na Keal and one near Treshnish Point, where it has been known for a considerable time.

Some other distribution patterns

A few strongly-calcicolous species have distributions that are directly linked to the very limited occurrence of strongly-base-rich habitats in Mull. These include *Dryas octopetala* (limited to outcrops along the cliffs to the south of the Gribun), *Erigeron acer* and *Equisetum telmateia* (confined to the cliffs west of Carsaig), *Hieracium iricum* (Fig. 2.40) (base-rich habitats on the Gribun and the adjacent Inch Kenneth) and *H. ebudicum* (Fig. 2.40) (Inch Kenneth and Carsaig). One of the most interesting patterns in the Mull flora is shown by *Vicia orobus* (Fig. 2.41) which, with the exception of one locality in the Ardmeanach (for which confirmation would be most welcome), is confined to north-west Mull. The reason for this distribution is obscure, but its clear-cut nature suggests that further work would be worthwhile. A species indicative of the deeper, basic woodland soils is *Carex sylvatica* (Fig. 2.42). *Carex pauciflora* (Fig. 2.42) is a bog plant of basalt terraces which is absent from the central volcanic massif.

Geographical elements of Matthews seen in the Mull flora

For the comparative consideration of the phytogeographical relationships of Mull plants, the geographical elements proposed by Matthews (1937 and 1955) have been used. In these publications he drew up lists of British flowering-plants that showed common distribution patterns over their whole global ranges (see Table 2.1) a departure from most earlier systems that had been mostly or wholly dependent on distributions within Britain. This approach has permitted the comparison of floras from different parts of Britain, so that Mull can be seen in its wider context. For the purposes of these comparisons, the lists provided in Matthews (1955) have been strictly used to insure complete compatibility, but in the floristic treatment in Chapter 11, where indications of geographic affinity not previously used by Matthews are given, they are based on an appraisal of their present-day distribution following, where appropriate, more recent works, e.g. Birks (1973). The general method adopted has been to compare the composition of the Mull flora, with those of other islands and similar comparable oceanic areas in the west of Britain. These are (in south to north sequence) Pembrokeshire, Anglesey, Isle of Man, Islay, Jura, Coll and Tiree, Rhum, Skye, Lewis and Harris, the Orkneys and the Shetlands. In addition to this sequence, two other less oceanic areas on the mainland of Scotland were included for contrast; these were the 100-km grid squares NN(27) and NO(37). The presence or absence of all the species listed by Matthews was checked for each of the above areas by reference to the *Atlas of the British Flora* (Perring & Walters, 1962), supplemented wherever possible by later published Floras or check-lists. These figures are presented as percentages of the total flora of their respective areas (Table 2.1). Histograms providing a more readily-appreciated picture of this data are given in Fig. 2.43.

The most important influence on the flora of Mull is its very strongly-oceanic climate (see Chapter 6), which it experiences as a result of its exposed position on the

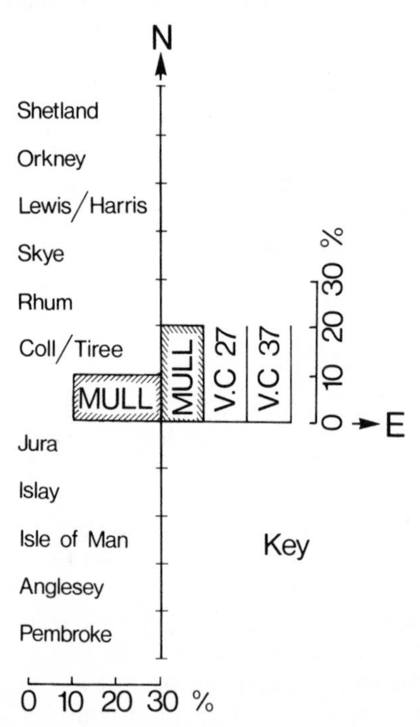

Fig. 2.43 Percentages of Matthewsian elements present in selected areas in western Britain (see Table 2.1).

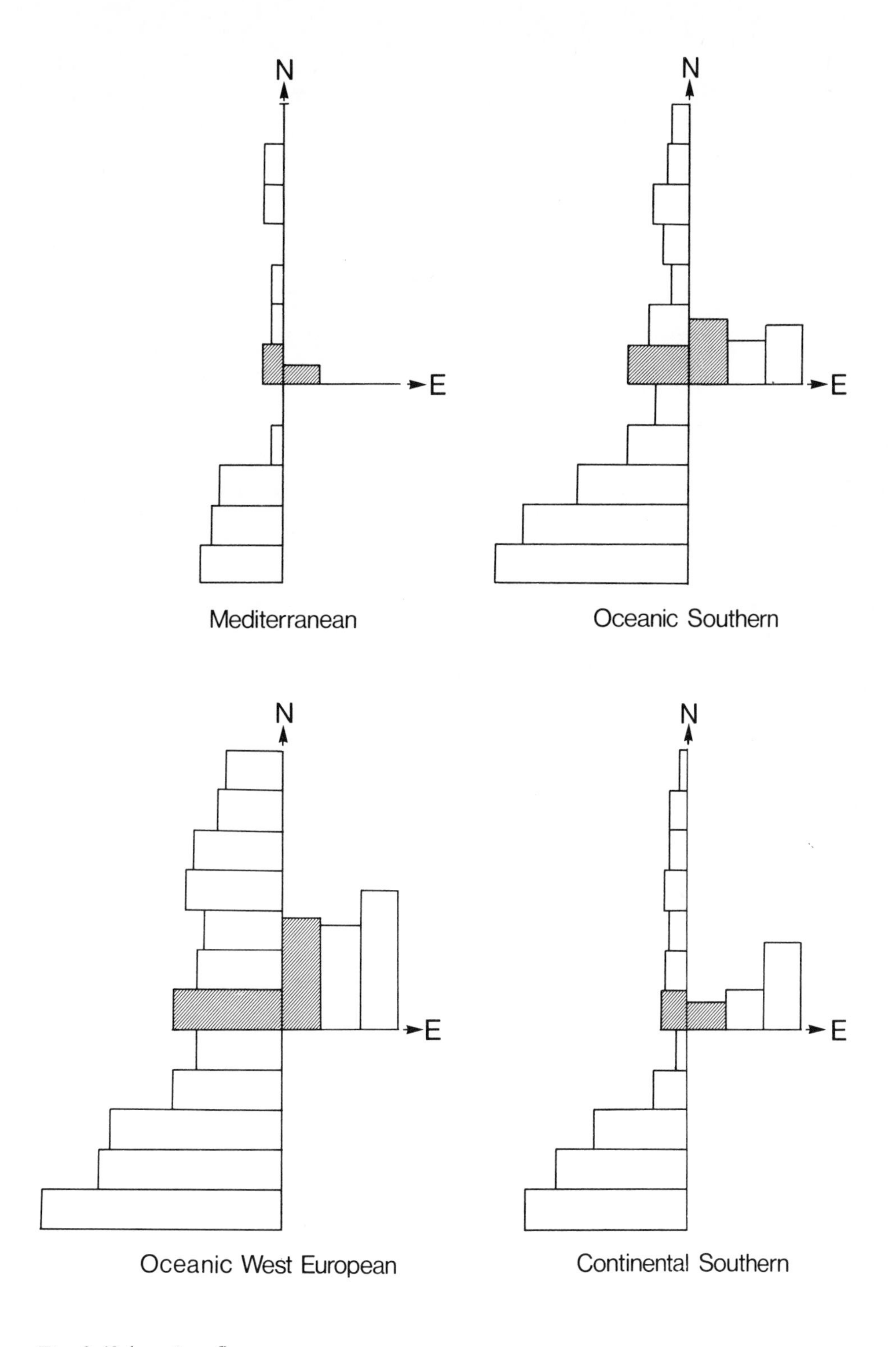

Fig. 2.43 (continued)

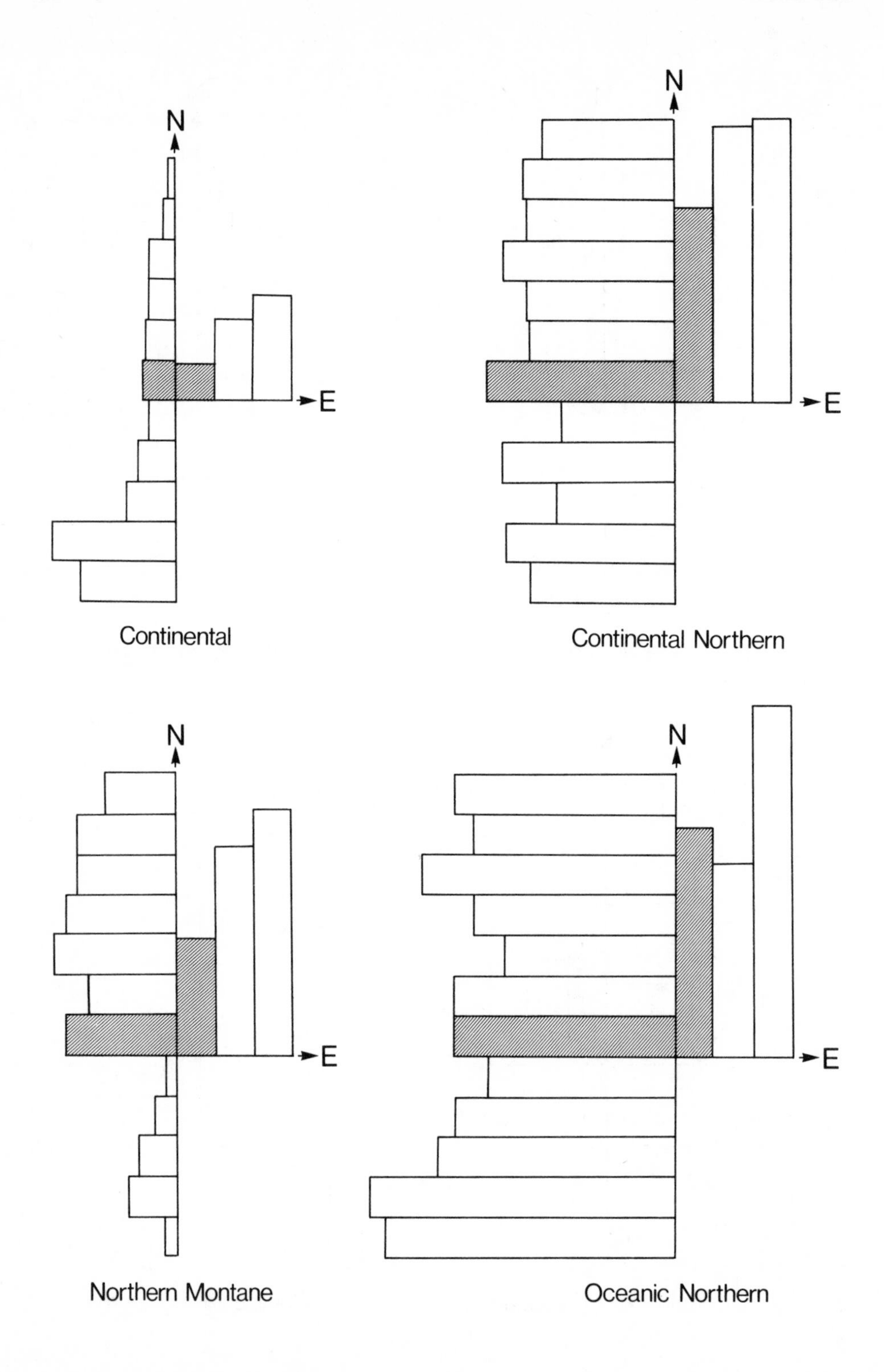

Fig. 2.43 (continued)

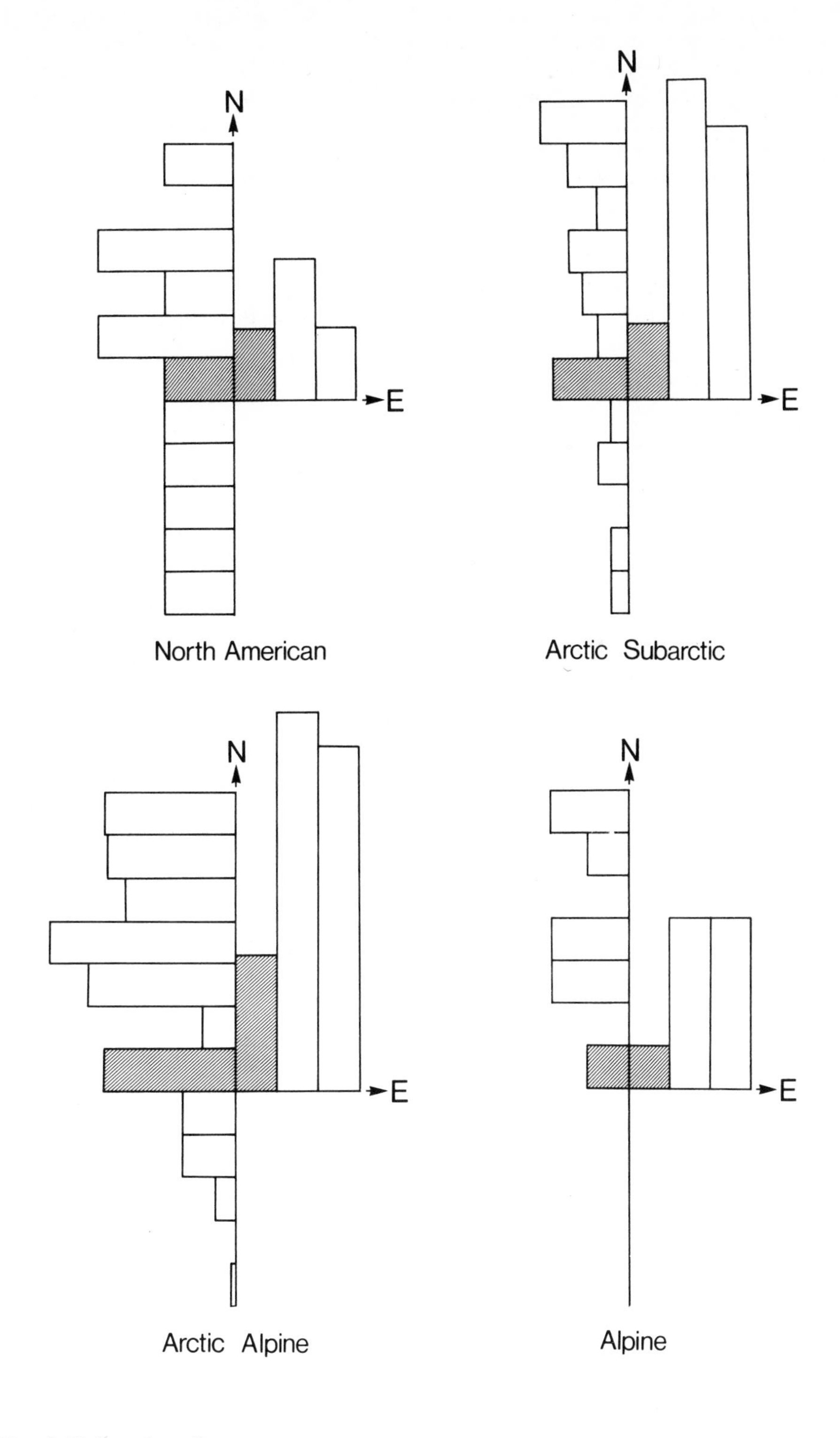

Fig. 2.43 (continued)

western seaboard of Britain, which itself has, by general European standards, a highly-oceanic environment. In very simple terms this implies a climate with a considerable rainfall, fairly evenly distributed throughout the year, accompanied by a relatively mild temperature regime and lacking violent extremes during the course of the annual cycle. From the histograms (Fig. 2.43), it can be seen that in Mull and areas to the north, the strongest oceanic element is the Oceanic Northern, whereas in areas to the south the Oceanic West European becomes progressively more dominant. As may be expected, the representation of the Arctic Subarctic, Arctic Alpine and Arctic elements is strongly influenced by the presence or absence of high altitude habitats, which in turn is to some extent dependent on the size of the areas concerned. Thus Mull has the second largest area of ground above 750 m, being exceeded only by Skye, which has nearly ten times as much ground above this height, although it is

Table 2.1 Percentages of Matthewsian elements present in selected areas† of western Britain.*

	P	A	IM	I	J	M	CT	R	Sk	LH	O	Sh	27	37
M	21	18	16	3	–	5	3	3	–	5	5	–	–	–
OS	50	43	29	16	5	16	11	5	7	10	6	5	11	15
OWE	62	47	44	28	22	28	22	20	25	23	17	15	26	35
CS	42	34	24	9	3	7	6	5	6	5	5	2	10	22
C	25	32	13	10	7	9	8	7	7	3	2	–	20	26
CN	37	43	30	44	29	48	37	38	44	38	39	34	68	70
NM	3	13	10	6	3	29	23	32	29	26	26	19	52	61
ON	74	78	61	57	48	57	57	44	52	65	52	57	48	87
NA	17	17	17	17	17	17	33	17	33	–	17	–	33	17
AS	4	4	–	7	4	14	7	11	14	7	14	21	75	64
AA	1	–	5	13	13	32	8	36	45	27	31	32	98	81
A	–	–	–	–	–	10	–	20	20	–	10	20	40	40

* Mediterranean (M); Oceanic Southern (OS); Oceanic West European (OWE); Continental Southern (CS); Continental (C); Continental Northern (CN); Northern Montane (NM); Oceanic Northern (ON); North American (NA); Arctic-Subarctic (AS); Arctic-Alpine (AA); Alpine (A).

† Pembroke (P); Anglesey (A); Isle of Man (IM); Islay (I); Jura (J); Mull (M); Coll and Tiree (CT); Rhum (R); Skye (Sk); Lewis and Harris (LH); Orkney (O); Shetland (Sh); Inverness/Perthshire, grid square NN (27); and Perth, Angus and Kincardine, grid square NO (37).

only twice as large in total area. Similarly the highest mountain on Mull, Ben More (951 m), is exceeded by Sgurr Alasdair (993 m) on Skye. Partly in consequence of this, Skye has a higher representation of Arctic Alpines (45 per cent) than Mull (32 per cent). These figures may be directly contrasted with the low-lying and very exposed islands of Coll and Tiree which have only 8 per cent of the Arctic Alpines. The Arctic Subarctic and Alpine elements provide similar results. The very much greater representation of these three elements in the 100-km grid squares 27 and 37 is very striking but understandable when one compares the extent of upland and the more continental climate of colder and more prolonged winters. Representatives of the Southern and Continental elements are naturally poorly represented in western and northern Scotland. This situation is in strong contrast to that of Pembrokeshire and Anglesey, which show a strong representation of southern elements, while grid squares 27 and 37 show a relatively large number of Continental species when compared to the western islands. The southern elements, with the Mediterranean, Oceanic West European and Continental become progressively reduced northwards, while there is a general corresponding trend in the reduction of the Northern Montane, Arctic Subarctic, Arctic Alpine and Alpine elements to the south, although, as previously noted, the occurrence of these is much influenced by the distribution of high ground. The Southern and Continental elements decline more or less regularly to the north, whereas the Northern Montane, Arctic Subarctic, Arctic Alpine and Alpine elements rise as sharply on Mull and then, except for relatively minor variations, remain as a fairly constant feature right up to the Shetlands. The group of species of

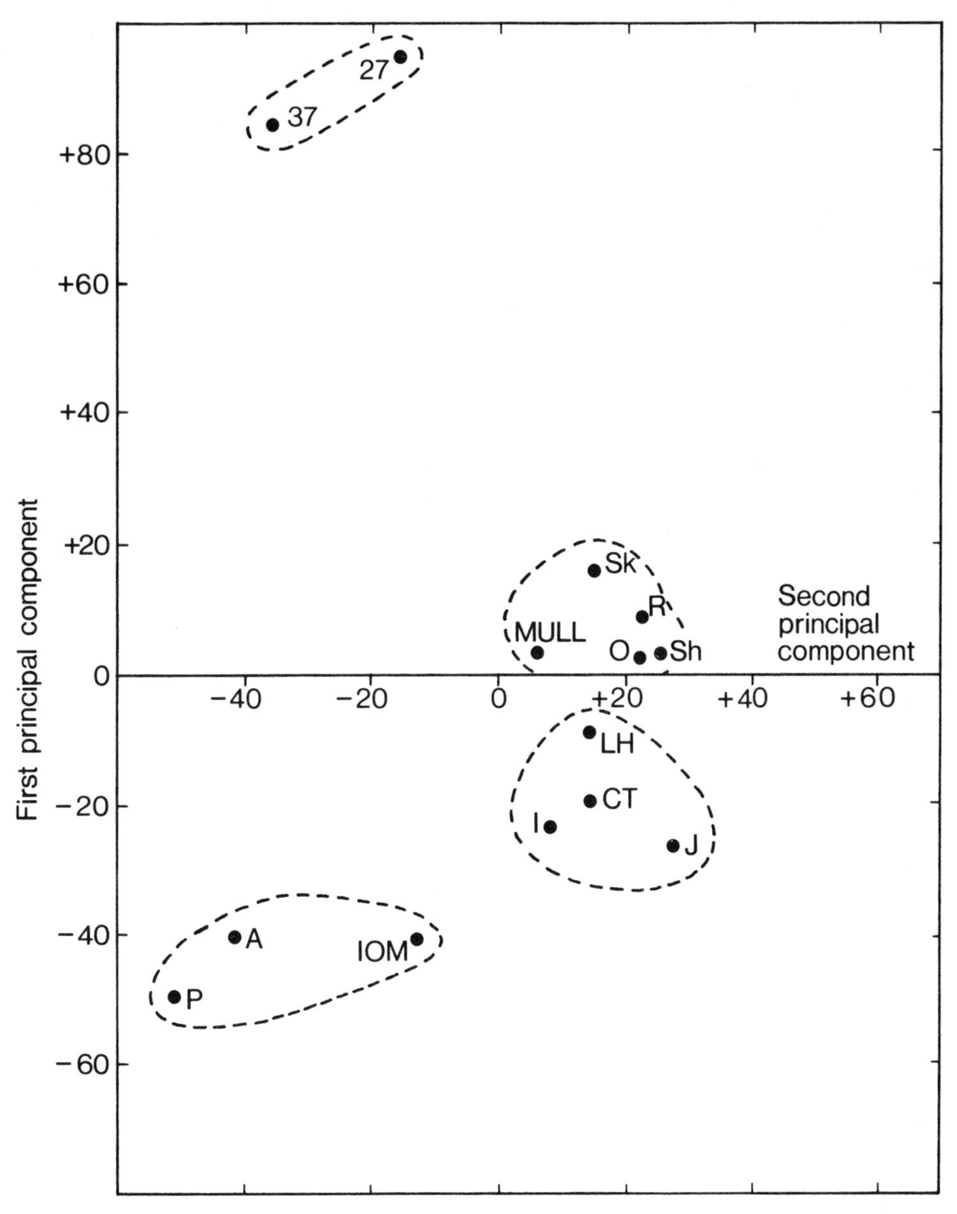

Fig. 2.44 Principal components analysis of the data in Table 2.1.

North American affinity is so small that its representation in percentage terms is rather misleading when compared to the other larger elements, and is of little significance in this context. The number of species of the Continental Northern element is strikingly constant from Pembroke right up the west coast to Shetland, while unexpectedly the Oceanic Northern has its greatest percentage in Pembroke and Anglesey. This seeming anomaly must be viewed in the wider context of Matthews' definition of his elements which were based on total distributions and such local minor divergences from anticipated patterns need not be regarded as invalidating factors.

We are specially indebted to Dr M. Hills of the Biometrics Section of the Museum's Department of Central Services, for general advice on the statistical consideration of the Mull floristic data, and in particular for the principal components analysis that he made from the data in Table 2.1, the results of which are plotted in Fig. 2.44. In the graphical representation, the distance of separation of the points can be taken as an approximate measure of the closeness of their phytogeographical relationship. It will be seen that, as might be expected, grid squares 27 and 37 stand right apart from the other areas considered; likewise Pembrokeshire, Anglesey and the Isle of Man, are clearly separated from the Scottish islands which, at this level of analysis, remain as a fairly tight cluster of dots. Within this grouping, there appears to be some affinity between Islay, Jura, Coll/Tiree and Lewis/Harris on the one hand, and Mull, Skye, Rhum, Orkney and Shetland on the other. It is suggested that the association of the outer islands (Coll/Tiree and Lewis/Harris) with the more southerly islands, may be due to the extreme oceanic conditions they experience, coupled with a complete lack of any high ground in Coll and Tiree. Under these conditions the more southerly elements may extend further to the north, and the elements that are more dependent on climatic conditions associated with high altitudes and high latitudes are supressed to some degree. Mull, Rhum and Skye all have substantial areas of high ground, which afford many habitats for plants with arctic or alpine affinities and at the same time may be marginally less oceanic in their climates. The Orkneys and Shetlands are of course very much further to the north and, although they lack substantial areas at high altitudes, have the climate required by arctic and montane species at lower levels. All the foregoing evidence suggests that in phytogeographic terms, Mull is near the pivotal point on the west coast of Britain, between dominance in the flora of Southern/Oceanic elements and Northern/Montane elements.

The distribution of bryophytes and lichens

A. Eddy and P. W. James

To European botanists the British Isles represent a major stronghold of those bryophytes and lichens which have, in relation to the continent, a markedly Oceanic (Atlantic) distribution. Furthermore, in these islands the east to west increase in numbers of such species is very marked, reaching their peak in Ireland and on the west coast of Britain. The Island of Mull, in its position closely adjacent to mainland Scotland, supports a bryophyte and lichen flora which can be regarded as a fairly typical of the Scottish Atlantic coast. It is characterised by a very high proportion of Oceanic or 'Atlantic' species (Ratcliffe, 1968). Together with the Ardnamurchan Peninsula, parts of Morvern and Kintyre, the lichen flora of Mull represents the best example of the predominance of the Atlantic element in the British flora, particularly in the abundance of many strictly Oceanic species. This abundance not only reflects the wide diversity of habitats favourable to lichens in the island, but also the relatively localised eutrophication and the absence of air pollution in any form. Unlike some other Scottish islands, Mull has retained enough old natural and planted woodland to ensure an adequate continuity of survival and succession for many of the more shade and moisture-loving species, many of which have become extinct further east and south in Britain.

Species with Oceanic or Atlantic distribution patterns

Ratcliffe (1968) provided a synthesis of the ecology and distribution of British bryophytes with Atlantic affinities, recognising several intergrading categories according to their northern or southern tendencies and degree of restriction within the broadly Atlantic overall pattern. Mull is situated in such a position on the west coast of Britain, that all of the categories recognised by Ratcliffe are represented to some degree, and lies between the latitudes in which the Northern Atlantic and Southern Atlantic groups merge in a pivotal situation like that shown by the flowering plants. The northern category is strongly represented, the southern less so. Though no equivalent to Ratcliffe's survey for Atlantic bryophytes has been produced for lichens, it is, nevertheless, possible to assign at least some taxa to his groupings. The intermingling of northern and southern lichens and bryophytes in the same localities is explained by the very equable climate with temperatures within the tolerances of both groups, with summers that are not too warm for northern species and winters that are never too cold for the southern. As a general rule the majority of lichens in this group are more characteristic of exposed more strongly illuminated habitats, often subject to greater fluctuations in humidity than is usual for many mosses. In the more sheltered damp habitats, lichens tend to come into direct competition with mosses, which often overgrow the crustose and smaller foliose species. Lichens are generally more intimately associated with substrates than mosses, corticolous species often showing preference for particular phorophytes, according to the texture, water-retaining capacity, and above all, pH of the bark (see Chapter 10). The predominance of shade-loving species on the eastern side of Mull is noticeable, but not so marked as with the bryophytes. Most species of this group of lichens are also frequent in old woodland (as at Torloisk House), in the *Salix* carr and on *Corylus* in the sheltered ravines on the western side of the island. The lack of suitable phorophytes, either in groves or in larger woodlands, is a notable feature of much of the lowland areas in the south of the island. Species with increasingly continental trends are rare or absent (e.g. *Caloplaca decipiens*), while widespread Atlantic and Sub-Atlantic groups make up the major part of the flora. Repetition in detail of the discussion in Ratcliffe (1968) would be inappropriate here, but a brief résumé of his distributional types, with notes on some of the species of lichens and mosses involved, is given below.

Widespread Atlantic. These are species for which the prime controlling factor is atmospheric humidity and have therefore a markedly western distribution in both Europe and Britain, but occur from the Iberian Peninsula (frequently also present in Macaronesia) to Scandinavia. Examples include the bryophytes *Glyphomitrium daviesii*, *Hedwigia integrifolia*, *Colura calyptrifolia*, *Mylia cuneifolia* and the lichens *Leptogium hibernicum*, *L. burgessii*, *Parmeliella atlantica*, *Pannaria pityrea*, *Parmelia crinita*, *P. laevigata*, *Usnea fragilescens*, *Pseudocyphellaria crocata*, *P. intricata*, *Sphaerophorus melanocarpus*, *Opegrapha sorediifera*, *Pyrenula laevigata*. The substrate varies but many of this group are saxicolous or corticolous, or both, and confined to habitats on Mull with a consistently humid microclimate. Most of the bryophytes are to be found in the woods and ravines on the east side of the island, where there is shelter from the frequently desiccating effects of the strong prevailing westerly winds, but many of the lichens, which require more light and are more resistent to desiccation, are more widely distributed. Other species such as the bryophytes *Fissidens* *curnowii* and *Harpalejeunea ovata* and the lichens *Leptogium burgessii* and *Sticta canariensis* (*S. dufourii* phase) are confined to wet rocks adjacent to streams or water trickles, while *Myurium hebridarum* is never more than a few metres from the sea. Many of the species are by no means rare in suitable sites, but *Saccogyna viticulosa* is the only abundant species in this group. A few mosses combine the need for high atmospheric humidity with a high light-demand, e.g. *Mylia cuneifolia* and *Plagiochila tridenticulata*, and are restricted to a few favoured sites in sheltered areas on the east coast.

Northern Atlantic. These are species with similar requirements to the previous group, but are restricted to regions with mean daily July isotherms below 19°C (66°F). They are strongly represented in Scandinavia and Western Scotland, decreasing southwards and absent from southern Portugal and Macaronesia. Examples include the bryophytes *Campylopus shawii* and *Trichostomum hibernicum* (Ratcliffe includes *Sphagnum strictum* but it is here excluded on the grounds of its North American range, where it extends southwards into Florida and Mexico). Comparable lichens include *Cladonia subcervicornis*, *Cavernularia hultenii*, *Parmelia omphalodes*, *Cladonia strepsilis*, *Cornicularia normoerica*, *Pilophorus strumaticus*, *Placopsis gelida*, *Stereocaulon delisei* and *Xylographa trunciseda*. Some probably represent extensions of a general boreal distribution and others often show tendencies to a circumboreal state. The bryophytes are generally speaking thermophobic (at least in some stage of their life cycles), and have a tendency to become increasingly montane in the southern parts of their range, but descend to sea level further north. They are to be found mainly on hill-sides with a shaded or northern aspect on Mull (e.g. *Herberta adunca*), or they are confined to subaquatic substrates where temperatures are depressed by evaporation (e.g. *Pleurozia purpurea*, *Campylopus shawii*). The lichens *Lecidea phaeops*, *L. taylorii* and *L. hydrophila* are often found near water and in seepages.

Southern Atlantic. Species in this category have an oceanic distribution but in contrast to the Northern Atlantic group are cryophobic and require average minimum temperatures above the January mean daily maximum isotherm of 0.6°C (33°F). This group is most abundantly represented in numbers of species in SW Ireland, Cornwall and SW Wales, extending southward to Macaronesia and being absent from Scandinavia. Numbers diminish northwards, and become increasingly confined to lowland, coastal districts and to sheltered or warm sites within them. Mull and the adjacent Scottish mainland lie at approximately the northern limit of many Southern Atlantic species. Ecologically, bryophytes of this distribution type favour warm coastal sites, with relatively high insolation combined with high humidity (e.g. *Marchesinia mackaii*) or, more usually, are found in moist, shaded lowland ravines and woodlands in areas protected from frosts and prevailing winds (e.g. *Adelanthus decipiens* and *Jubula hutchinsiae*). *Marchesinia mackaii* is the only frequent species on Mull, and it has some affinity with the Mediterranean group, which is expressed in its preference for more open situations and relatively high tolerance of desiccation. Lichens in this group are *Anaptychia obscurata*, *Parmelia sinuosa*, *P. reticulata*, *P. taylorensis*, *P. endochlora*, *Arthothelium ilicinum*, *Gomphillus calycioides*, *Porina coralloidea*, *Melaspilea ochrothalamia*, *Leptogium brebissonii*, *Graphis elegans*, *Ochrolechia inversa*, *Phaeographis dendritica*, *Ramalina portuensis*, *Graphina anguina*, *Lecidea tenebricosa* and *Lecanora jamesii*.

Mediterranean-Atlantic. These species have their centre of distribution in southern Europe, but with a slight Atlantic bias. Among them are some of rather wider tolerance, which reach their northern limits on Mull or adjacent areas. The group merges with continental types on the one hand, and with the Southern Atlantic group on the other, differing from the latter in the marked resistance of its species to desiccation. Mosses of this group are almost confined to open situations in the lowlands and are characteristically calcicolous or basiphilous, especially towards their geographical limits. Consequently on Mull, they are much more frequent around the exposed west coast than on the sheltered east coast. Temperature does not seem to be the main controlling factor for these species (in contrast to the Southern Atlantic group), and a rather high light regime and an effectively high base status in the substrate seem more important. Many species which seem indifferent to parent-rock chemistry in the Mediterranean, become strict basiphiles in the wetter parts of Britain where leaching is rapid. Basic rocks are rare on Mull, but this deficit is partly compensated for by wind-borne calcareous sand in some western areas and by sea-spray along the south, west and north coasts. Sites cleared of trees and in which competition by more luxuriant species is reduced (e.g. through agriculture or road making) sometimes provide suitable habitats for Mediterranean species (e.g. *Fossombronia caespitiformis*, *Anthroceros husnotii*), but such sites are usually too poor in base-status. Strictly Mediterranean and Continental mosses are few in number and abundance in Mull. A group of Mediterranean-Atlantic lichens can also be distinguished. These in Mull are coastal, sun loving and most frequent in the west, especially Iona, Ulva and Gometra. Though not generally calcicolous these are often confined to sheltered dry crevices and sunny slopes near the sea. They include *Buellia canescens*, *Diploschistes caesioplumbens*, *Sclerophyton circumscriptum*, *Bacidia scopulicola*, *Opegrapha cesareensis*, *Solenopsora holophaea*, *Arthonia lobata*, *Caloplaca arnoldii*, *C. littorea*, *Physcia tribacia*, *Lithographa dendrographa*, *Enterographa crassa* and *Rinodina roboris*, the last three species occurring in crevices in bark rather than on rocks.

Sub-Atlantic. The majority of species included in this group are widespread in Britain, although showing a slight western bias on the continent. They vary widely in their ecology and frequency, but include few epiphytes. On Mull, many species are abundant and may be locally dominant, including such conspicuous mosses as *Campylopus atrovirens* and *Breutelia chrysocoma*. Generally speaking the Sub-Atlantic assemblage is a heterogeneous respository for species which do not fit into the foregoing Atlantic types, and with a bias insufficiently pronounced to make further categorisation desirable or useful. Where a species has a marked preference for particular habitats, this is of course indicated in the text (e.g. *Grimmia maritima* and *Anaptychia fusca* which are confined to coastal rocks in the spray zone and *Oedipodium griffithianum* to the thin humus deposits under high-alpine crags). No attempt has been made to further divide the Sub-Atlantic group, although it is recognised that it would be possible to separate for instance Alpine and Arctic-Alpine elements (e.g. *Oedipodium griffithianum* and *Marsupella alpina*). Among the lichens, *Cetraria sepincola*, *Alectoria bicolor*, *A. pubescens* and *Cornicularia normoerica* could be grouped together as a northern element. Ratcliffe (1968) distinguishes a *Western British* group, which shows a marked western bias in Britain, but does not seem to have strong Atlantic affinities on the continent. In Europe, however, these are usually montane, occurring in regions where local topography has produced a climatic regime not unlike that in some lower altitude, more oceanic regions further west. As a distributional category, the Western British group of species seem to be most closely allied to the Northern Atlantic assemblage e.g. *Mylia taylorii* and *Bazzania tricrenata* are both found with *Herberta adunca* and *Mastigophora woodsii*, which are both clearly Northern Atlantic types. Among the Sub-atlantic lichens, the following may be noted: *Collema furfuraceum*, *C. flaccidum*, *C. fasciculare*, *Cladonia floerkeana*, *Arthonia tumidula*, *Parmeliella plumbea*, *Biatorella ochrophora*, *Nephroma laevigatum*, *Normandina pulchella*, *Peltigera collina*, *Ochrolechia tartarea*, *Parmelia perlata*, *Pannaria pityrea*, *Parmelia revoluta*, *Cladonia impexa*, *C. tenuis*, *Sticta limbata* and *S. fuliginosa*.

Endemic Atlantic lichens. There are a few lichens which so far as is known are endemic to the British Isles, and which have within this country an Atlantic type of distribution. The following species in this group are represented in Mull: *Lecanactis homalotropa* (SW Ireland and W Scotland), *Ocellularia subtilis* (also in SW Ireland and W Scotland), *Microglaena larbalestieri* (Connemara, Galway and Invernessshire), *Toninia pulvinata* (Lake District, Wales and Sutherland). Some of these species will probably be found to occur also in Scandinavia, France or Portugal.

Species in which the distribution is governed by altitude.

The lichens and bryophytes of Mull can also conveniently be considered in terms of their altitudinal zonation, and while the limitation of this approach is recognised (e.g. variation due to degree and direction of exposure), it does afford a useful framework upon which to hang further information. Three main zones are recognized here; it must be appreciated, however, that these terms are comparative and not necessarily the same as those applied by other authors.

Alpine zone. Above 900 m there are a number of species which apparently either require winter temperatures below freezing (e.g. *Oedipodium griffithianum* and *Conostomum tetragonum*) or are thermophobic and intolerant of temperatures in excess of a required minimum. These are not abundantly represented on Mull and are confined to damp rocks or humus-rich sites with a northern aspect. Prolonged snow-lie may encourage some of these species, though true snow-bed communities are not well developed here. Lichens in this group include *Parmelia alpicola*, *Thamnolia vermicularis*, *Umbilicaria cylindrica*, *U. torrefacta*, *Alectoria minuscula*, *Cetraria islandica* and *Ochrolechia frigida*.

Subalpine zone. The transition from lowland (here defined as below 200 m) to the alpine zone is a continuity of communities, the subdivision of which is highly subjective. In Chapter 13, for convenience, it is referred to as 'higher' and 'lower' subalpine. In the former are found species which have a wide distribution in mountainous areas and are often abundant above 500 m, but rare and scattered below that height. These include the mosses *Andraea alpina*, *Gymnomitrium obtusum* and *G. crenulatum* and the lichens *Arthrorhaphis citrinella*, *Cladonia bellidiflora*, *C. gonecha*, *Ochrolechia tatarea*, *Sphaerophorus fragilis* and *Lecidea demissa*. The lower subalpine group has many species which, although common in lowland areas of mountainous districts, are none the less almost confined to hilly areas where they demonstrate a wide altitudinal range, but do not usually extend up into the alpine zone. The bryophytes *Sphagnum fuscum*, *S. quinquefarium*, *Trichostomum tenuirostre*, *Marsupella emarginata* and *Nowellia curvifolia* are typical of this group. Over much of their range, intermittent drought may be the reason for the exclusion of subalpine species from the lowlands. In the north and west, however, the majority do descend locally to sea level. While this subdivision works effectively for the mosses, the distinction does not seem to apply with lichens, at least in Mull. The lichen species above occur equally in both subdivisions.

Lowland zone. Excluding those mosses which on Mull are exclusively lowland due to restrictive edaphic requirements, the species confined to lowland habitats belong to Southern and Mediterranean-Atlantic distribution types (*v.s.*). Among the mosses, the general British distribution pattern of these groups is inverted, i.e. Mediterranean species, which have a southerly or eastern trend in Britain as a whole tend to be restricted to the west coast of Mull; while the species of the Southern Atlantic group, which is exclusively western in Britain are to be found mainly in the sheltered gullies and woodlands along the east coast.

The distribution of freshwater algae

D. J. Hibberd*

Prior to the present survey there are only three publications dealing specifically with the freshwater algae of Mull. The most recent of these is that of Williams (1936) who examined the plankton of Lochs Bà, Poit na h-I, Lochan a' Ghurrabain (Aros Loch) and Carnain an Amais, Meadhoin and Peallach (the Mishnish lochs) and the Tobermory River and Reservoir. A total of about 120 species was recorded of which approximately two-thirds have been noted by us and will be discussed further below. Previous to this Roy (1883) had listed 103 desmids and Borge (1897) 13 freshwater algae for Mull.

In view of the dearth of previous records and the limited nature of the present survey, there is little value in a detailed comparison with these early records. More worthwhile is an assessment of the flora as a whole, with particular reference to the desmids and the euplankton, in the light of the early classic work of W. & G. S. West on the British freshwater plankton. On the basis of a detailed study of the lakes of western and north-western Britain, West & West (1903, 1905, 1909) established the following as being characteristic of the algal flora: (1) the lake phyto-plankton is unusual in the number and diversity of its constituent species; (2) it is dominated by desmids, these commonly accounting for about half of the total species recorded for any one area; (3) coccoid green algae are not very abundant and blue-green algae are very poorly represented, never normally forming blooms; (4) the phytoplankton is normally very sparse, containing a large number of species in small quantity. They point out that plankton of this type differs markedly from almost all other European freshwater plankton. Subsequent investigations have confirmed this point of view and in a survey of phytoplankton associations Hutchinson (1967) refers to plankton of this type as 'oligotrophic desmid plankton'. On account of its main location it is sometimes also referred to as 'Caledonian' (Williams, 1936).

While West and West did not attempt to determine experimentally the physical basis for this highly characteristic plankton flora, they stress that the lake basins in which it occurs are almost all on outcrops of older Palaeozoic or of Precambrian rocks often associated with intrusive igneous masses. They also point out that these basins lie almost entirely in areas of high rainfall and low summer temperature, but when compared with other parts of the British Isles and Europe they attribute this characteristic desmid plankton entirely to the geology. This premise still holds good and it now appears that these rocks produce very oligotrophic waters with low levels of nitrogen and phosphorus and very low levels of calcium, although the latter element may be increased by wind-blown sea-spray in coastal areas (see Chapter 9). Given this background, it would be expected from the geology and geographical position of Mull that its algal flora would be of the general type outlined above, and this indeed proves to be the case.

The plankton of the deepwater lochs lacking riparian vegetation.

Considering first the plankton of the deepwater lochs (see Chapter 9 and Tables 9.7 and 9.8) this is extremely sparse, the water appearing completely clear, desmids are dominant and blue-green algae and coccoid green algae very poorly represented. In addition, the species of Chrysophyta, Dinophyta and Bacillariophyta occurring in the plankton are the same as those listed by West & West (1909) as characteristic of the 'desmid plankton'. On Mull, this plankton association was found in Loch Assopol, Loch Bà, Loch Frisa, Loch a' Ghleannain and Loch Uisg, namely the five largest and

* Now at the Culture Centre of Algae and Protozoa (NERC), 36 Storey's Way, Cambridge.

deepest lochs on the Island, the waters of which are clearly oligotrophic (see Table 9.13). The total number of species noted in net samples from the open water of these five lochs is 56, with an average for each loch of 21. These figures combine records for the two occasions on which the plankton was sampled except for Loch Bà for which only one collection was examined. The plankton of Loch Carnain an Amais, Loch Fuaron, Loch an Tòrr and Loch Peallach was of a basically similar type though it was on average half as rich in species as the others. Only *Peridinium willei* was found in all nine of these lochs. The following species were found in four or more of the above lochs and were either rare or completely absent from other habitats; they are therefore considered as characteristic of the open water plankton community in Mull: *Asterionella formosa* var. *formosa*, *Botryococcus protuberans*, *Ceratium hirundinella*, *Closterium kuetzingii*, *Cosmarium depressum* var. *planctonicum*, *Dinobryon bavaricum*, *Eudorina elegans*, *Microcystis aeruginosa*, *Sphaerocystis schroeteri*, *Staurastrum anatinum*, *S. chaetoceros*, *S. lunatum* var. *planctonicum*, *Staurodesmus sellatus*, *S. spencerianus*, *S. subtriangularis* var. *inflatus*, *Xanthidium antilopaeum* var. *depauperatum*.

The total of 56 species recorded from the plankton of the deeper lochs is relatively low compared with the 161 species and varieties, exclusive of diatoms, listed for the Scottish plankton as a whole by West & West (1909). The present list would doubtless be extended considerably by the study of more material, and seasonal factors may be operative, though one of the features of this plankton type is the relative lack of seasonal variation apparently due to the low summer temperature of the water (West & West, 1909). It should also be pointed out, however, that every care was taken during the present Survey to collect only those species occurring in open water and which can therefore be considered as truly planktonic, and the absence of marginal species may somewhat have depressed the total.

A detailed consideration of the question of the existence of a truly euplanktonic phytoplankton community is not relevant to the present discussion and both Brook (1959) and to a lesser extent Duthie (1965) have reviewed the matter. It is sufficient to mention here that Brook has drawn up a list of desmids which he considered as true euplankton species, based on the assumption that any desmid found dividing in the open water was planktonic, after examination of the plankton of 98 large deep Scottish freshwater lochs. He also lists species which although normally found in other habitats, can occur in the plankton in significant amounts and are known to be capable of multiplying in open water (facultative plankters). The following 12 taxa, out of a total of 27 considered by Brook as being euplanktonic, have been recorded in the Mull deepwater lochs: *Cosmarium abbreviatum* var. *planctonicum*, *Staurodesmus iaculiferus*, *S. megacanthus* var. *scoticus*, *S. sellatus*, *S. subtriangularis* (as var. *inflatus*), *Spondylosium planum*, *Xanthidium antilopaeum* var. *antilopaeum* and var. *hebridarum-reductum*, *Staurastrum pingue*, *S. pseudopelagicum*, *S. lunatum* var. *planctonicum*, *S. anatinum*.

In addition, *Euastrum verrucosum* var. *alatum* was found only in the plankton of open water and this may be equivalent to the variety listed by Brook as *E. verrucosum* var. *reductum*. Of the species listed by Brook as facultative plankters *Micrasterias sol*, *Staurastrum denticulatum*, *S. arctiscon* and *S. brasiliense* var. *lundellii* have been found only in the open water of the deepwater lochs in the present Survey. Brook found *Staurastrum arctiscon* only in the plankton of the shallower lochs investigated and he therefore regards it as benthoplanktonic, although we found it truly planktonic in Loch Frisa; West, West & Carter (1923), however, regard it essentially as a planktonic species and record it from a number of deep Scottish lochs.

In contrast to the relatively rich euplankton of the deepwater lochs, a true plankton was found to be almost entirely absent from the great majority of the numerous small shallow lochs with vegetated margins. This confirms and extends the observations of Williams (1936) who described a rich euplankton flora (38 species) from an edge sample from Loch Bà, but noted only very few desmids in samples from the Mishnish Lochs, Lochan a' Ghurrabain and Loch Poit na h-I.

No detailed observations have been made on the euplankton flora of Lochan a' Ghurrabain but it appears generally to resemble the flora of the deepwater lochs listed above rather than that of Loch Poit na h-I as was inferred by Williams.

The plankton of Loch Poit na h-I

Loch Poit na h-I is a medium-sized, reasonably shallow loch (see Chapter 9) the plankton of which was recorded by Williams as relatively species-poor. Although not vastly rich in species (we recorded 31 species against 20 by Williams), the plankton of Loch Poit na h-I was found in the present Survey to be unique in character. Notwithstanding the shallowness of the loch and the existence of much rooted phanerogamic vegetation, there was a distinct euplankton flora on both occasions on which it was sampled (see Table 9.9) but in contrast to the deepwater lochs the plankton was concentrated and consisted largely of *Melosira italica* subsp. *subarctica* and *Synura spinosa*, and also contained relatively large quantities of coccoid green algae, particularly *Pediastrum boryanum*. However, a number of species of *Staurastrum* and *Staurodesmus* which were found to be characteristic of the deepwater lochs were also present, including *Staurastrum pingue*, one of the species regarded by Brook (1959) as euplanktonic, and *S. denticulatum*, one of Brook's facultative euplankters. Williams' list for Loch Poit na h-I does not include *Melosira* and the dominant plankton diatoms at the time of sampling appear to have been *Tabellaria fenestrata* and *T. flocculosa*, the latter species being only occasional there now. More significantly perhaps, Williams did not record any of the species of *Staurastrum* or *Arthrodesmus* (the latter mostly now included in *Staurodesmus*) which he found in Loch Bà. However, owing to its shallowness and therefore much greater temperature variation, a far greater floristic variation would be expected in Loch Poit na h-I than in the deepwater lochs. Williams designated this type of plankton as 'impure Caledonian', attributing its greater richness, particularly in blue-green algae possibly to eutrophication from sewage. Wind-blown sea-spray is regarded here as a more likely reason for this enrichment.

The algae of permanent bog-pools

Considering next the terraqueous habitats, West & West (1909) pointed out that the richness of the desmid floras of the western mountain regions was not confined to the plankton but was a general feature of most habitats. In common with Roy (1893) they noted that the richest areas of all were the small permanent bog-pools with clear water, one to several square metres in extent, often with floating *Sphagnum* around the margins. More recently Duthie (1965) also pointed out that the principal algal habitat, especially for desmids, in northern Wales was pools in the *Sphagnum* bogs, a few millilitres of the sediment often yielding over 50 species, the desmids numbering 1000–70 000 cells per cm^3. Our observations confirm the richness of the Mull desmid flora which may have been anticipated from this data, particularly in respect of the flora of the bog-pools. Species lists for two of these are given in Tables 9.10, 9.11 and 9.12, though these are comparatively rich, possibly as a result of the influence of wind-blown spray, the Salen pool being near the sea and that on Ulva being in an exposed westerly situation.

A total of 215 desmid species has been recorded here and the area may therefore be considered as desmid-rich using the criteria of West & West (1909), who defined a desmid-rich area as one in which 150–200, sometimes up to 300 species occur more or less abundantly. They also pointed out that this total included many species which occurred only rarely outside the western mountain regions. However, of the 43 desmid species considered to be virtually restricted in distribution to these regions, only 15 have been noted here, and while it is certain that the desmid list could be considerably extended, probably doubled by the examination of further material, it is also possible that the desmid flora may be only 'rich' rather than 'very rich' since West & West (1909) pointed out a falling off in the desmid flora of Skye when compared with much of the Scottish mainland.

Although the bog-pools are species-rich, their general uniformity is underlined by the greater diversity in pools which it is certain receive sea-spray. A series of such pools occurs at Carsaig in shallow rock platforms only a metre or so above high water mark in summer, but which receive continuous drainage from the high cliffs at the base of which they lie. A list of species found in these pools is given in Table 9.2. Desmids are present in greater variety and numbers than in the bog-pools, but it is the greater number of green and blue-green algae which give these pools such a distinctive character.

A total of 870 species of freshwater algae has been recorded for the Island, and in view of the preliminary nature of this survey pointed out in the introduction to Chapter 16, it is likely that this figure could be considerably increased by working through more desmid samples, by a critical examination of the blue-green algae and a serious attempt to deal with the flagellate groups. Even that figure is likely to fall short of the 1500 or so species recorded for Anglesey and Caernarvonshire by Woodhead & Tweed (1954–55) but this almost certainly reflects the uniformity of Mull which has comparatively few artificial habitats, few lowland areas and only one mountain over 900 m, and to some extent taxonomic concepts.

References

BIRKS, H. J. B. (1973). *Past and present vegetation of the Isle of Skye, a palaeoecological study*. Cambridge.

BORGE, O. (1897). Algologiska notiser 4. Süsswasser-Plankton aus der Insel Mull. *Bot. Notiser*, **1897**: 210–215.

BROOK, A. J. (1959). The status of desmids in the plankton and the determination of phytoplankton quotients. *J. Ecol.*, **47**: 429–445.

DUTHIE, H. C. (1965). Some observations on the algae of Llyn Ogwen, North Wales. *J. Ecol.*, **53**: 361–370.

HUTCHINSON, G. E. (1967). *A treatise on limnology*, vol. 2: *Introduction to lake biology and the limnoplankton*. New York, London, Sydney.

MATTHEWS, J. R. (1937). Geographical relationships of the British flora. *J. Ecol.*, **25**: 1–90.

———. (1955). *Origin and distribution of the British flora*. London.

PENNANT, T. (1774). *A tour in Scotland and voyage to the Hebrides 1772*. Chester & London.

PERRING, F. H. & WALTERS, S. M. (eds). (1962). *Atlas of the British flora*. London & Edinburgh.

RATCLIFFE, D. A. (1968). An ecological account of Atlantic bryophytes in the British Isles. *New Phytol.*, **67**: 365–439.

ROY, J. (1883). List of the desmids hitherto found in Mull. *Scott Nat.*, **7**: 37–40.

———. (1893). On Scottish Desmidieae. *Ann. Scot. nat. Hist.*, **1893**: 106–111.

WEST, W. & WEST, G. S. (1903). Scottish freshwater plankton I. *J. Linn. Soc. Bot.*, **35**: 519–556.

———. (1905). A further contribution to the freshwater plankton of the Scottish lochs. *Trans. R. Soc. Edinb.*, **41**: 477–518.

———. (1909). The British freshwater phyto-plankton with special reference to the desmid-plankton and the distribution of British desmids. *Proc. R. Soc.*, (B) **81**: 165–206.

——— & CARTER, N. (1923). *A monograph of the British Desmidiaceae*, vol. 5. London. Johnson reprint 1971, London & New York.

WILLIAMS, E. G. (1936). Notes on the plankton of some lochs in the Island of Mull. *NWest. Nat.*, **11**: 23–29.

WOODHEAD, N. & TWEED, R. D. (1954–55). The freshwater algae of Anglesey and Caernarvonshire. *NWest. Nat.*, **25**: 85–122, 255–296, 392–435, 564–601 (1954); **26**: 76–101, 210–228 (1955).

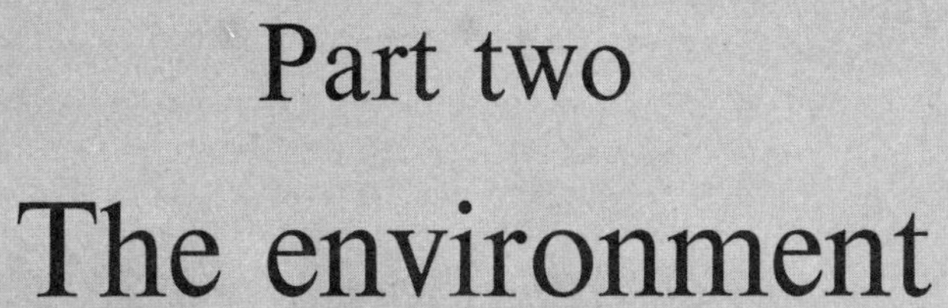

Part two
The environment

3. General description and topography

A. C. Jermy

Introduction

The Island of Mull lies between 5°40′ and 6°30′ longitude west and between 56°18′ and 56°40′ latitude north. Administratively it was, until 1974, part of the mainland county of Argyllshire but was placed by H. C. Watson (1832), together with Coll and Tiree, in his vice-county 103. It lies off the Argyll coast between Ardnamurchan and the Firth of Lorne, and its east coast hugs close to that of Morven, separated by the narrow Sound of Mull, just over 2 km wide at its narrowest point, Rubh' an t-Sean Chaisteil. It is 35 km (25 miles) from the south coast to the northernmost point and just over 42 km (30 miles) from east to west. The island is of an irregular shape being a series of four peninsulas, of which one is an archipelago of three sizeable islands (Ulva, Gometra and Little Colonsay), with the resulting sea-lochs in between. The coastline has been estimated at some 480 km (300 miles) and the total land area 116 550 ha (450 sq. miles).

The south-eastern corner of Mull is 6.5 km (3.5 miles) from the island of Kerrera and 4 km (2.5 miles) from Lismore and in projecting into the Firth of Lorne it protects Loch Linnhe and the mainland port of Oban from the onslaught of the westerly gales. The most westerly point – Stac an Aoneidh, on Iona – albeit protected to some extent from the north-westerlies by Tiree, 32 km (20 miles) away, is literally on the edge of the North Atlantic Ocean with nothing except the Skerryvore Light between it and Labrador. Caliach Point in the north-western corner is more protected by the island of Coll 11 km (7 miles) away but the full force of a north-westerly wind can blow in from the Little Minch through the gap between the Cairns of Coll and Ardnamurchan Point, hitting Mull between Quinish and Ardmore Points.

Geographically Mull is an extension of the Ardgour-Morven peninsula and owes its distinction to its variation of topography, its geology and soils, and its climate. On Mull, as elsewhere in Britain, man's use and management of the land for forestry, agriculture and sport has changed, and for the most part controls, the biological potential. For an interesting account of the history and peoples of the island the reader is directed to Mcnab (1970) and the earlier guides (Hannan, 1926; MacCormick, 1923) give a useful background to the island in the early part of this century.

Communications

A regular car-ferry service links Mull at Craignure with Oban and Lochaline (in Morvern). Latterly a car-ferry has been established between Lochaline and Fishnish Point. A small-boat passenger-only service runs irregularly from Oban to Grass Point on the southern shore. Day-trips for tourists run throughout the summer from Oban to Iona and then round Mull to Tobermory passing the island of Staffa. A small-boat tourist service also runs across from Ardnamurchan to Tobermory.

Tobermory was until recently the active commercial centre of the island community but its port facilities for landing both tourists and freight have been superseded by the new pier (opened in 1965) at Craignure from where both can continue their journey to Tobermory by road. Nevertheless the Sound of Mull and Tobermory Bay are welcome havens from westerly storms. Some general freight may still be delivered by boat and such ships will call intermittently at Fionnphort, Kintra, Bunessan, Carsaig, Croggan, Salen, Croig, Calgary and Ulva as well as Craignure and Tobermory. As on any island, coastal inhabitants at least possess boats and from time to time visit other islands. Apart from Iona and Staffa, traffic landing on the smaller offshore islands, e.g. Treshnish, Soa, or Eorsa, is from Mull itself, usually transporting sheep.

In the past, apart from more regular passenger services to some of the places mentioned above, the points of contact with the outside were the same as today. A principal eighteenth-century drove-road, along which the black cattle were driven to fairs and trysts in the south and east of Scotland, passed through Mull (Haldane, 1952). Cattle from Coll and Tiree were landed at Kintra on the Ross of Mull and driven through to Grass Point, by Loch Don, for shipment to Oban or diverted up to Salen for crossing over to Lochaline in Morvern. There were frequent sales at Salen and it is likely that traffic was in both directions across the Sound of Mull.

There is at the time of writing one grass airstrip on the coast at Glenforsa which has regular summer links with Glasgow and the Outer Hebrides.

Road communications are naturally restricted by the topography of the island. Because of the effective vehicular ferry calling four times each day in the summer months at Craignure, day-trips by coach from Oban to Iona are feasible. This has necessitated an improvement of the A849 (Craignure to Fionnphort; B8035 on pre-1962 maps) running down through Glen More and the north side of the Ross of Mull, and with it the resulting shoulders and verges of bared soil have been colonised by ruderal mosses and flowering plants. Minor roads for local traffic link the A849 with Uisken, Carsaig and Lochbuie on the south coast and in the south-east similar small roads go out to Croggan, Grass Point, Gorten and Duart Castle.

The B8035, formerly part of the A849, branches off the new road at the head of Loch Scridain and climbs up Gleann Seilisdeir to the Gribun and along the south shore of Loch na Keal to Knock and Salen. Here it links up with the A849 that has followed the coast along the Sound of Mull from Craignure. In other words the Ben More massif and the Torosay hills form an effective barrier between Salen and the SW peninsula. In the past there have been cart tracks up Glen Cannel to Gortenbuie and thence a foot track over to the Glen More, and up Glen Forsa to Tomsleibhe but it is doubtful that there was any regular travelling over the col between Beinn Talaidh and Sgùrr Dearg.

The northern part of the island is served by the B8073 which runs along the north shores of Lochs na Keal and Tuath via Ulva ferry junction. At Kilninian a minor road climbs over the moor to Dervaig on the north coast but the B8073 continues over the shoulder of Cruachan Odhar to Ensay Burn and Calgary and in doing so approaches 173 m (576 ft) – the highest road in Mull and one from which one can survey the whole western seaboard of the island and look down on the Treshnish

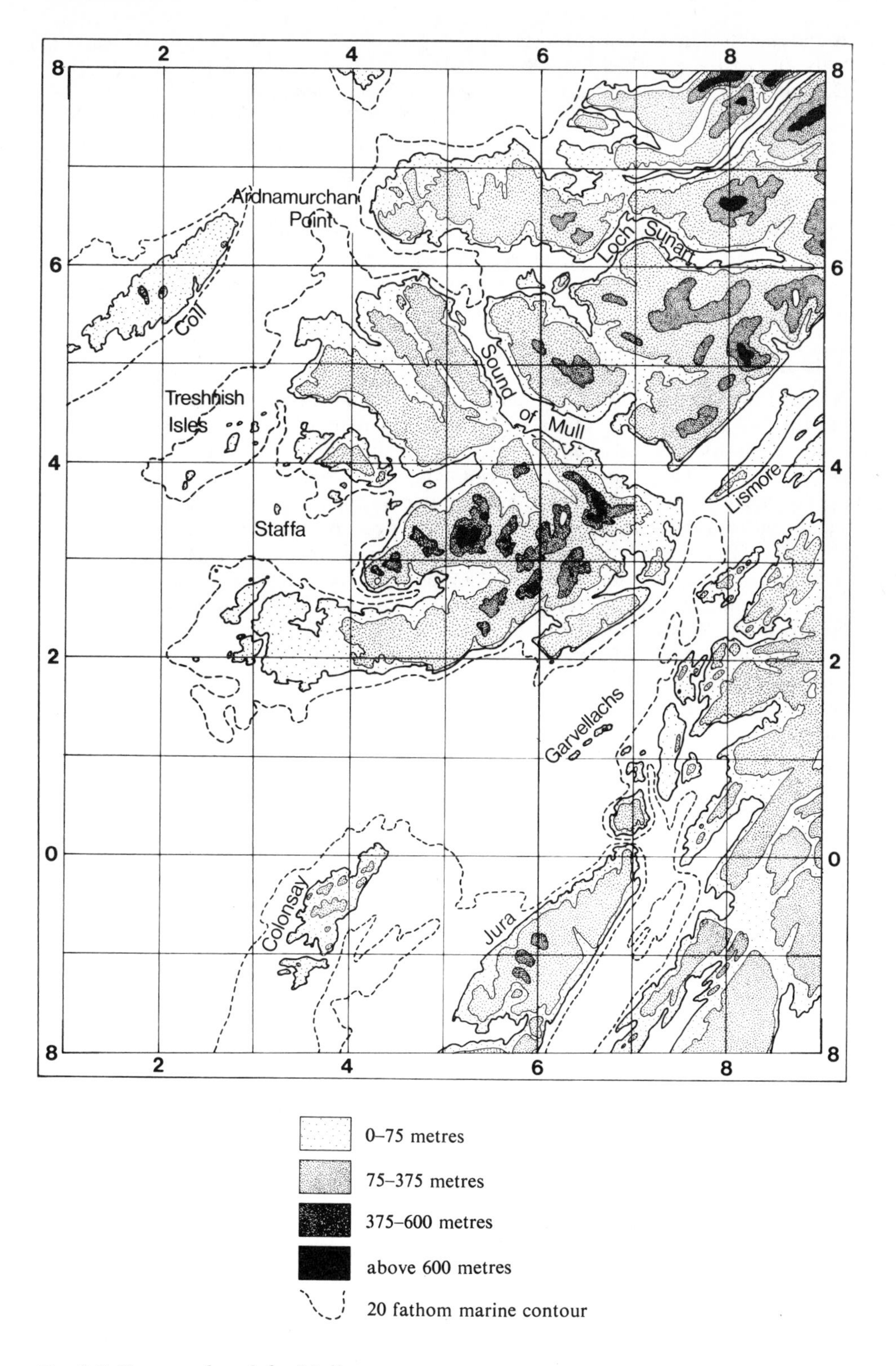

Fig. 3.1 Topography of the Mull area.

Islands and to Tiree. The road from Calgary goes overland to Dervaig and Tobermory to be linked there with the A848 along the Sound of Mull to Salen. A route from Dervaig to Salen runs up the Belart valley and down Glen Aros.

Topography

The general topographical features of the region covered by this survey are described in eleven units. It must be emphasised, however, that these are not necessarily phytogeographic units and do not correspond directly with those used in the floristic account later in this work.

The northern massif

North of the narrow neck of land that lies between the hamlet of Salen on the Sound of Mull and the head of Loch na Keal are approximately 28 000 ha (108 sq. miles) of peaty or basaltic hills, all of which lie within the civil parish of Kilninian and Kilmore or the Burgh of Tobermory. The plateau, lying at an average height of 225 m (750 ft) rises quite steeply from the narrow coastal strip (raised beaches) on the north of Loch Tuath and a similar low-lying area between Aros Bridge and Ardnacross on the Sound of Mull. Elsewhere, as along the northern shore of Loch na Keal, between Ardnacross and Tobermory and along the south side of Calgary Bay, the raised beach is lacking and the hills run steeply to the rocky shore. On the promontory heads of NE Mishnish, W Mornish and from Treshnish Point to Rubh'an t-Suibhein extensive and often impressive cliffs are formed although those at the latter site stand inland somewhat on the raised beach at Kilninian. Only in the north, between Caliach Point and Glengorm Castle is there a gentle decline into the sea although here, as throughout the basalt areas, the harder lava layers and the less easily eroded dolerite stand proud and form a series of terraces and the characteristic 'trap' topography (see Chapters 4 and 5).

The hamlet of Calgary lies at the head of the bay of the same name and is known for its sandy beach and small area of wind-blown sand – an incipient but species-poor *machair*. Other areas of shell and quartzite sands, smaller in extent, are found along the north coast between Port Langamull and Croig on Mornish. A sandy, but less calcareous, bar is seen at the head of Loch na Keal formed from the alluvium brought down by the River Bà.

The northern massif is cut deeply along its NW/SE axis by the Belart and Aros valleys. The River Belart has an extensive catchment and runs out at Dervaig on the north coast through the partially tidal freshwater basin, Loch na Cuilce and the sea-loch Loch a' Chumhainn. Complementing this valley on the SE of the watershed is the Aros River into which the Allt an Lon Biolaireich drains the uplands to the south and the Ledmore River runs out of Loch Frisa at the head of Glen Aros. The hills to the south-west of the Belart form a broad undulating ridge at 300 m (1000 ft) altitude, the highest being Beinn na Drise (471 m; 1391 ft). The rivulets running into the Belart and Aros fall over dolerite sills forming waterfalls that can be seen from some distance. The ridge top is broad with several small lochs and drops slightly as it approaches Treshnish Point and Beinn Duill is still 188 m (626 ft) within a kilometre of the sea. Most rivers are small but Ensay Burn running off Beinn Bhuidhe and Cruachan Odhar is the most notable exception. East of the Belart there is a narrow ridge nowhere above 270 m (900 ft) acting as the watershed between that river and Loch Frisa. Frisa is a long (7.4 km; 4.5 miles) narrow lake, the largest on the island and one of two on Mull included in the Bathymetric Survey of Scottish Lochs (Murray & Pullar, 1908). It is over 45 m (150 ft) deep. At the upper end of Loch Forsa the ground rises barely 30 metres before dropping into Loch an Torr which drains out to the north coast through Mingarry Burn in the valley known as Glen Gorm. This glen follows the line and presumably the strata of Loch Frisa. In Aros, on the east of Loch Frisa, the Altt na Torc and its tributaries drain Speinne Mòr, which at 437 m (1465 ft) is the highest hill in the northern massif, gathering enough water to form an impressive waterfall in Aros Park. In the north and along the road that joins Tobermory with Dervaig two smaller lochs flow through Loch Peallach into Tobermory River. North of the road in the area called Mishnish lies a castellated basaltic pinnacle, S'Airde Beinn (288 m; 959 ft), and from it more gentle slopes run to the sea. On this estate, Glengorm, considerable reclamation has been carried out and much of the rough grassland and moor has been scarified and re-seeded.

This region contains the oldest planted forest on the island – the Salen Forest, owned by the Forestry Commission and stretching up Glen Aros to the east of Loch Frisa around Lettermore House; it was planted during 1928–9 and is now reaching the stage when it can be cropped. The Forestry Commission own or lease other areas here in the north and many hectares have been planted in the last decade. A number of ancient 'policies' exist (Quinish, Calgary, Torloisk, Glenaros and Drumfin) and these are often sheltered by fine stands of both deciduous and coniferous trees.

Two islands may be associated with this area: Calve Island which lies off Drumfin, near Tobermory, in the Sound of Mull; and Eorsa, a sizable block of basalt in the middle of Loch na Keal. A farmer lives on Calve Island but Eorsa is uninhabited although regularly grazed.

The Ulva and Gometra archipelago

Onto the SW corner of the northern massif there abuts an archipelago, namely Ulva and Gometra and associated islets, which forms the southern shores of Loch Tuath. Ulva, at some 1800 ha, is the largest island of the group outside Mull. Its population is now reduced to a few employees who work on the estate but before the 'clearances' in 1875 there were over 400 crofters living on the island. The ruins of many crofts are seen on the south side and the vegetation there takes on an interesting pattern from the former 'lazy-beds' field pattern. Apart from a little-cultivated and afforested land around Ulva House at the east end, the island is moorland overlying basalt. Cliffs showing the typical hexagonal columns are impressive along the south coast but as on Mull itself the northerly declining dip of the lava flows makes for a gentler slope on the north side. There are twin peaks, Beinn Eolasary and Beinn Creagach, the latter reaching 308 m (1025 ft).

Gometra, a smaller island (400 ha), is separated from the west end of Ulva by Am Bru which becomes a tidal race on the flowing tide. In all physical aspects it is part of Ulva with an indented rocky coast and rising to just over 150 m (500 ft).

Off the south coast of Ulva lies an assemblage of rocks and islets with names like Eilean na Creiche (Devastation Island) and Garbh Eilean (Stormy Island) which remind one of the exposed nature of the coast here. A kilometre from these rocks lies Little Colonsay, just under a square kilometre and uninhabited but grazed throughout the year by sheep. It is a flat island with a single castellated buttress, Tòrr Mòr, rising to 60 m (201 ft.)

Further out, about five kilometres to the south-west, but arising from the same submarine basalt ridge as Little Colonsay, is Staffa, renowned for the impressive columnar basalt exposed around its coast. Sir Joseph Banks first saw it and brought news of its existence to the scientific world in 1772. The cliffs are penetrated by a number of sea-caves, the best known being Fingal's Cave which recedes into the cliff some 68 m (227 ft) and is 20 m (66 ft) high. The waves thundering into this cave inspired Mendelssohn to write his overture 'The Hebrides'. Also on the south of the island is another big cave (Boat Cave) and other caves like the Comorant's, Goat's and Clamshell Caves are seen further round to the east. The latter, as the name suggests is composed of curved, not straight vertical basaltic pillars. At the base of the cliffs between these caves wave-cut basalt columns form a level or stepped causeway (see Fig. 4.7). The cliffs on the south side reach 40 m (135 ft) but the ground slopes to within 15 metres of the sea at the north end. Like all other off-shore islands the acid grassland is grazed by sheep. Staffa is fully described by MacCulloch (1975).

The Torosay mountains

South of the Salen-Killichronan isthmus lies the greater and central massif of Mull stretching the whole width of the island. There is, however, variation not only in topography but also in geology and the area will be dealt with in three parts.

The Torosay mountains as here defined are those of Torosay parish lying to the east of Glen Forsa. They form a ridge, rising gently in the north with Maol Buidhe (360 m; 1201 ft), and comprising a number of peaks and saddles, which then runs south-east for some 6.5 kilometres. From Beinn Mheodhon (626 m; 2087 ft) through the twin peaks Dun da Ghaoithe (754 m; 2512 ft) and Mannir nam Fiadh (745 m; 2483 ft) the ridge drops steeply to the head of Loch Spelve. Dun da Ghaoithe is the second highest peak on the Island and this group of mountains has the climate (and plants) suggestive of similar high ground on the mainland. It is possible to walk for 6.5 kilometres on this rounded ridge without dropping below 555 m (1850 ft) and in the winter the whole area is exposed to the cold north-easterlies coming from the Morvern hills.

The slopes on the west are very steep, dry and mostly of loose scree; they become more gentle lower down Glen Forsa. The rocks here were formed closer to the centre of the extinct tertiary volcano and show a greater variation of petrographical structure than the more uniform basalt to the north and west, or the basalts that lie to the east along the Sound of Mull in Torosay. The result is a series of waterfalls in the upper reaches of the rivulets that drain to the east. Southwest of Mannir nam Fiadh across a 450 m (1500 ft) saddle is Sgùrr Dearg (729 m; 2429 ft). Lochs are few, but there are two notable ones at about 75 m (250 ft) in the south-east, namely Loch Bearnach and Lochan an Doire Dharaich. The slopes here contain a few remnants of oak woodland as the name of the latter loch suggests, and the Forestry Commission have established a sizable spruce forest between Fishnish and Scallastle Bays which is reaching maturity, developing a fauna and flora of its own.

The central massif

This area includes the central core of mountains from Beinn Talaidh by Glen Forsa west to Ben More, the Island's highest mountain, and out to the Gribun at the seaward end of Loch na Keal.

Beinn Talaidh (749 m; 2496 ft) rising abruptly on the western side of upper Glen Forsa is, when seen from any aspect, an impressive wedge-shaped mountain. Here and just to the west is the cauldron of the ancient volcano. To the north of Talaidh is a series of low, rocky hills: Beinn na Duatharach (448 m; 1493 ft) mainly composed of the course-grained volcanic rock, gabbro; as is also Na Binneinean, trough to lower hills of granophyres and agglomerates lying above the basalt at Salen. The western slopes of these hills fall to Loch Bà, a loch some 30 m (100 ft) deep also included in the Bathymetric Survey, and lying here in the lower part of Glen Cannel. It is fed by numerous streams from Beinn Talaidh and Cruachan Dearg (693 m; 2309 ft) and a complex of small mountains (*c.* 480–570 m) which lie on the southern edge of the volcano centre and of which the southern slopes fall into Glen More and the Coladoir River. See Chapter 4 for detailed maps of this area.

Running due west from Cruachan Dearg is a col, Creag Mhic Fhionnlaidh, linking at 326 m (1088 ft) with Ben More. It is the watershed between the River Clachaig (running into Loch Bà) and Allt Teanga Brideig, a tributary of the Coladoir River running into the head of Loch Scridain. North from the col is Beinn Fadha (691 m; 2304 ft) and Beinn a' Ghràig (582 m; 1939 ft) running out NE to Knock and the River Bà. These mountains, on the northern edge of the old cauldera are of granophyre and are impressive – as are those higher up Glen Cannel – for their steep barren scree slopes.

Ben More proper, at 949 m (3169 ft) the highest peak on the island, is a pyramidal basalt mountain with a flat windswept top and large-block scree falling away on all sides between 825 and 675 m (2750–2250 ft). The slopes to the north are more gentle and the easiest route to the summit is from the bothaidh, Dishig, up the shoulder An Gearna. Approaching from Glen Seilisdeir and the south, any route must weave around the vertical bluffs of the terrace escarpments. To descend to the south is even more tiring as one cannot see the vertical drops until right upon them.

The ridge that runs west from the shoulder of Ben More known as Maol nan Damh, does not drop below 300 metres until within a kilometre of the coast and there it broadens into a hammer-head of basalt that forms an impressive line of cliffs some 150 m (500 ft) high, from Creag Brimishgan in the north to those of Beinn Chreagach in the west. The ridge itself connects three rounded hills, the westernmost, A' Mhaol Mhòr being 425 m (1418 ft) high. Seen from the south across the valley these hills illustrate well the terraced benches and escarpments so typical of basalt topography. Both north and south slopes have many rivulets draining the almost bald tops, those of the south running into the 'Glen of the yellow Iris' (Glen Seilisdeir) which separates the central massif from the Ardmeanach.

At sea-level between the Rubh a' Ghearrain and Rubha Baile na h-Airde, at the base of the Beinn Creagach cliffs, lies an exposure of silicified chalk, worn smooth by wave action and appearing from a distance at low tide like white sand. It is the lower bed of Mesozoic sedimentary rocks which lie beneath the basalt on this coast but which are for the most part covered by landslip and glacial deposits. Across a narrow channel at this point is the island of Inch Kenneth, named after a contemporary of St Columba who established a monastery there (McNab, 1970). In 1773 when Dr Johnson and Boswell visited it, it was the home of Sir Alan MacLean, the Chief of the clan. It is now privately owned and farmed and, being composed of mainly calcareous sandstones and conglomerates it has a fertile soil. Its highest point is 49 m (162 ft) falling in vertical cliffs on the west coast which takes the full fury of the Atlantic gales.

Ardmeanach

On the southern side of Gleann Seilisdeir lies some 2500 ha of ground over 300 m (1000 ft) known as Ardmeanach. There is a Forestry Commission plantation in the Gleann itself and more low ground in the south-east may well be planted in the next decade. Along its south coast by Loch Scridain are a few pastures and sheltered hollows (e.g. Tiroran and Tavool House and some deep south-facing gorges but for the most part the slopes are steep. The exposed plateau is barren and with little soil and although much of it is owned by the Forestry Commission it is unplantable and poor sheep pasture. The gently undulating ground on Beinn na h-Iolaire and Beinn na Srèine (the highest point 321 m; 1704 ft) exhibits a feature seen nowhere else on Mull except on Beinn na Drise above Lagganulva in the northern massif, and then to a lesser degree. Here on Ardmeanach is an area of slightly convexed slopes of evenly-sized basaltic pebbles, some 2–4 cm diameter, set in a matrix of fine peaty silt, almost as if it had been graded by machine. The agent causing this is not clear but presumably frost action plays a part and so perhaps do the constant winds. It is the only habitat on Mull of the rare *Koenigia islandica*.

The western seaboard of this peninsula falls abruptly and in a series of terraces, to a rocky shore. In the south a remote and almost inaccessible area called The Wilderness exhibits a portion of the Staffa columnar basalt and some very fine cliff scenery. Embedded in the basalt flow at the time of eruption was a trunk of a deciduous tree almost 2 m across and 12 m high; all that remains is some mineralised charred wood or for the most part, where this was burnt away completely an infilling of bleached basalt. The resulting fossil tree was found by J. Maculloch, in 1819 and has been named after him (see Fig. 4.9). This part of the coast together with some 500 ha of hill ground is owned by the National Trust for Scotland. North from here the shore is almost impassable. A number of caves are formed at sea-level, that to which it is said that Abbott Mackinnon retreated as a recluse is the most famous. Further north one approaches that alluvial fan known as the Gribun and in doing so passes through a region where the Mesozoic sedimentary rocks underlie the basalt. For the most part they are covered by basalt rockfall but the sequence can be seen in the Allt na Teangaidh as it deflects from the line of the road to flow towards Balmeanach farm. Here, if the terraced cliffs of Creag a' Ghaill can be scaled, there is the possibility of interesting plants being found, for besides the calcareous Mesozoic rocks there are also dolerite sills and dykes, cut by the numerous small streams that fall over the cliffs, which are high in calcium- and magnesium-bearing minerals.

One very small island can be included here, that of Erisgeir, 3.5 km (2.25 miles) off the Ardmeanach coast and 4 km (2.5 miles) from Little Colonsay. It is low and flat and grazed from time to time by sheep.

The Loch Don area

At the eastern extremity of the island reaching out towards the island of Lismore is a stretch of low-lying ground in the parish of Torosay, at the base of the Torosay hills described above. To the north is the estate of Torosay Castle and south of this is the promontory of Duart Castle, built originally in the fourteenth century as the home of the clan and restored early this century by the father of the present Chief, Sir Charles MacLean. It has a commanding view of any shipping approaching the Sound of Mull from the south. Most of the land on the peninsula is arable except for a little forestry around Torosay Castle and Lochdonhead.

This eastern tip of the Island is cut into by Loch Don, a small sea-loch of irregular shape. The maritime grassland on its landward edge, in spite of being covered periodically by the tide, makes good pasture. At the western end is an estuarine basin called Leth-fhonn, into which run a number of rivers like Abhainn Lirein (the 'Stream of the green freshwater weed'), named possibly from the *Enteromorpha intestinalis* around its outflow from the Torosay mountains, and the River Rainich and Loch a' Ghleannain from the south. Loch a' Ghleannain is a loch surrounded by andesite – a rock not to be found elsewhere on Mull and through the bottom of it, but not seen on its shore, runs a band of the Dalradian limestone (of the same kind and age as that on Lismore). This promontory is cut off on the south by Loch Spelve and contains more moorland than that at Duart, rising to 270 m (800 ft) in Carn Ban. Loch Spelve is bigger than Loch Don but of similar shape. Its shores are steeper and more stoney; it has typical sea-loch shores and lacks true estuarine marshes as seen at Leth-fhonn. It is bounded on the south by the Croggan peninsula described below.

The Laggan–Croggan peninsula

Loch Spelve is a crescent-shaped loch running from the north to the south-west; it is linked with the Firth of Lorne by a narrow channel – Port na Saille. The southern side of this channel is formed by the Croggan peninsula. As this stretch of land curves to meet the Brolass coast it in turn is cut into by Loch Buie thus making a hammer-head pivotted to the south coast of Mull by a five-kilometre isthmus itself almost cut through by Loch Uisg which rises a kilometre from Loch Spelve and links westwards with Loch Buie. The area lying to the west by Loch Buie is the Laggan Deer Forest, a high and rugged plateau scattered with small lochs and stretches of interesting bog-land which drops steeply, and in places over 210 m (700 ft) cliffs, into the sea. The highest hill here is Druim Fada 399 m (1329 ft) and the plateau remains at around 300 m (1000 ft) only slightly dropping to the northeast as Croggan is approached. On the seaward side it is cut into only once in the form of Glen Libidil, due south of Kinlochspelve, but the effect of this narrow glen cannot be seen from the north. The northern slopes of the plateau that fall steeply to Loch Uisg are cut by streams and in places are densely wooded.

The Ben Buie massif

This area lies south of the River Lussa–Glen More divide and east of Brolass; it is bounded on the south by Lochs Buie, Uisg and Spelve. The two main mountains, Ben Buie itself (706 m; 2354 ft) and Creach Beinn (687 m; 2289 ft), are of mixed rock type but consist of considerable olivine gabbro suggesting they formed a vent of the Tertiary volcano. This coarse-grained rock gives rise to little soil and much of the massif is dry hill or scree. Even the north slopes are dry and relatively dull. The area is bounded on the Brolass side by the line of the basalt which runs through by Loch Fuaron, a fair-sized loch lying at 240 m (800 ft) on the shoulder of Ben Buie and draining to the north to the Coladoir River. Between Ben Buie and Creach Beinn is a col at 150 m (500 ft) on the north side of which is a series of Lochs (Airdeglais, an Ellen, an Eilein and Squabain) which form the headwaters of the river Lussa.

The eastern end of this massif that falls into Loch Spelve is dominated by Glas Bheinn (483 m; 1611 ft), a rounded hill of little note, but at the eastern extremity is a small wooded hill that juts out into Loch Spelve. This is Cruach Ardura (213 m; 709 ft); the peak is bare but there are interesting mossy woods on the east and west slopes. There are further natural woodlands around Loch Uisg and in the lower Lussa valley and some well-established plantations around Lochbuie House.

Brolass and the Ross of Mull

This south-western peninsula extends 12 km (7 miles) further west than the tip of the Ardmeanach, protecting to some extent Loch Scridain, which lies on its northern shore, from the south-westerly gales. The eastern end, the area known as Brolass, abuts onto the Ben Buie massif at Abhainn Loch Fuaron and Allt Mhic Slamhaich by Lochbuie. The underlying rock at this end is basalt but the contours look less stepped here because of the deeper peat in the shallow valleys and on the hills. The road to Iona passes down the north side of the peninsula and most habitation is confined to that area although there is a small settlement at Carsaig on an alluvium fan on the south coast. Just to the north-east of Carsaig is the highest area, Beinn na Crois rising to 495 m (1649 ft) and this hill is still close enough to the central mountains to have some elements of an alpine flora. West of this the hills get lower as the Sound of Iona is approached, although immediately west of Carsaig sizeable hills such as Beinn Chreagach (374 m; 1235 ft), Creachan Mòr (325 m; 1082 ft) and Cruachan Min (370 m; 1232 ft) still appear impressive and even more so from the south as they drop precipitously into the sea between Carsaig and Malcolm's Point and round to Cnoc nan Gabhar. Between Carsaig and Lochbuie to the east the coast is equally steep, chipped by numerous cascades and more deeply by Glenbyre Burn which cuts some 4 kilometres inland. The shoreline in the Carsaig area is interesting because here, as at the Gribun, are exposed the Mesozoic sedimentary rocks – calcareous sandstone, shales and cornstones (see Chapter 4), resting below the basalt. The differential erosion of these rocks forms caves, stacks and arches such as can best be seen at Malcolm's Point.

On the north side two major rivers drain the large peaty basins into Loch Scridain: the Leidle River coming out at Pennyghael and Beach River some 6.5 kilometres west. Into the head of Loch Scridain flows the River Coladoir in the lower reaches of which is the only remaining area of raised bog on Mull. There is a fine estuarine area with salt marsh at An Leth-onn at Scridain head.

At the point along the Ross when one draws level with the tip of the Ardmeanach across Loch Scridain, the landscape and topography change noticeably. Bunessan, the second largest settlement on the island, is situated at this point in a well-protected bay, Loch na Làthaich, on the north side of the peninsula. Running obliquely in a south-easterly direction from Bunessan is a fault line. It passes along the Abhainn Tir Chonnuill to Cnoc nan Gabhar. The rock that has been thrust up on the west of this fault line is a schist, a common rock type in the western Highlands of Scotland. Containing mica the rocky bluffs appear glistening in contrast to the matt surface of contrast to the matt surface of the basalt and the terraced topography is lacking. At this point the south coast becomes less formidable and has many small sandy bays, a notable one being at Uisken some three kilometres south of Bunessan.

On the east side of Loch na Làthaich is a low-lying promontory generally known as Ardtun. There is arable ground here and it is crofted as is the rest of the Ross from here westwards. Ardtun is known to palaeobotanists because of the fossil-bearing gravels and shales, interleaved between columnar basalt, which reveal something of the flora found late in the Tertiary times when the volcano on Mull had ceased pouring forth lava (see Chapter 4 and end papers). title page

The mica schist outcrop does not extend very far before its place is taken by an intrusion of hard granite which extends from Ardalanish Bay on the south and Loch Caol mouth on the north to the Sound of Iona. The topography is softer, with rounded low hills often strewn with pink and white mottled boulders of this attractive granite. There are broad basins of deep peat although crofters, who formed a stronghold in this area after the nineteenth-century clearances, have taken their toll and removed (and are still removing) considerable quantities. The peat diggings and benches are reflected in the vegetation patterns. Almost at the western end near Fionnphort is a large low-altitude lake, Loch Poit na h-I, interesting for its aquatic flora. Off the extreme south-west of the Ross but connected at low tide lies Erraid, a granite island no longer inhabited but, like most off-shore islands, regularly grazed.

Iona

The Island of Iona, lying off the western tip of the Ross of Mull across the 1.5 kilometre-wide Sound bearing the same name is probably the best known part of the whole of Mull. It was here in A.D. 586 that Columba arrived from Ireland and set up the monastery that was to influence the religious development of north and west Britain for the next 300 years. There was evidently much coming and going in the Iona settlement and, with the interchange of population, quite possibly an interchange of spores and seeds. Much remains of the past although mostly in ruins and thousands of tourists visit the island each year.

Iona is distinct for other reasons. It enjoys warmer and drier summers, in common with the tip of the Ross, than the greater part of Mull. Geologically it is more complex and best known for the Iona 'marble' – an attractive green felspatic rock of igneous origin and much used for ornaments and jewellery. The highest point on the island, Dùn, (100 m; 332 ft) at the north end and the high ground to the south (Druim an Aoineidh, 73 m; 243 ft) are composed of Lewisian gneiss as found on the Outer Hebrides. The large western bay, Camas Cuil an t-Saimh, is bordered by machair – the best seen in Mull but still not as good as that on the outer isles. It is now regularly grazed by cattle and there is a golf course on the northern part. A smaller bay and promontory on the north coast shows another good deposit of wind-blown shell-sand as do the islets of Eilean Chalbha and Eilean Annraidh.

Two offshore island groups might be mentioned here. Soa Island rising to 35 m (116 ft) a double granitic pillar, now grazed but uninhabited, lies about 4 km (2.5 miles) off the SSW. To the north-west of Iona is a basaltic island, Rèidh Eilean, 33 m (110 ft) high, a favourite haunt of numerous seabirds which produce an extensive and smelly guano deposit.

The Treshnish Islands

Off to the south-west of Treshnish Point and Rubh a' Chaoil on the west coast of the northern massif lie the Treshnish Isles. All are basalt and arise from a submarine basalt ridge which would have been above sea-level in Pleistocene times. The northernmost – Cairn na Burgh Beg – is just over 3.2 km (2 miles) from the mainland of Mull; close by it is Cairn na Burgh More which rises to 34 m (113 ft). On both were ancient forts and beehive cells. Less than a kilometre south-west lies the main group of the Treshnish – a series of rocky, often flat, islets of which some, like Fladda with an area of 25 ha barely rises above 15 m (50 ft) and still contains considerable peat. Others like the largest of the group, Lunga (approximately 2 km long) may rise higher, seen in this island in the form of a basaltic plug, Cruachan, 101 m (337 ft) high. Then 3 km (2 miles) further south-west lies Bac Mòr or Dutchman's Cap, shaped as the name suggests and rising to 75 m (284 ft) at the peak of the 'Cap'. The surrounding plateau, or 'brim' of the hat is composed of three layers of basalt with signs of a tropical soil formation after each eruption, and the high-rising 'crown' shows another six layers. Joined to this island by a causeway passable at low tide is Bac Beag, a smaller but equally rocky islet.

The entire group is part of the Treshnish Estate at Haun and most are grazed by a tenant farmer. Those few islands not grazed show a luxuriance of strand and cliff plants illustrating only too well how floristically rich the islands of this part of Britain might be without introduced animals.

References

HALDANE, A. R. B. (1952). *The drove-roads of Scotland.* London.

HANNAN, T. (1926). *The beautiful island of Mull.* Edinburgh.

MACCORMICK, J. (1923). *The Island of Mull, its history, scenes and legends.*

MACCULLOCH, D. B. (1975), *Staffa*, ed. 4. London.

MCNAB, P. A. (1970). *The Isle of Mull.* Newton Abbot.

MURRAY, J. & PULLAR, L. (1908). *Bathymetrical survey of the fresh-water lochs of Scotland.* London.

WATSON, H. C. (1832). *Outlines of the geographical distribution of British plants; belonging to the division of Vasculares or Cotyledones.* Edinburgh.

4. Geology

A. R. Woolley* and A. C. Jermy

Introduction

To the geologist interested in igneous rocks and volcanoes the island of Mull represents one of the classic areas of his science. The reason for this is that much of the south-eastern part of the island comprises the deeply eroded remains of a volcano, the exploration and description of which can be considered to have heralded the modern approach to igneous petrology and the understanding of the underlying constitution of volcanoes. This work is to be found in the Memoirs of the Geological Survey of Scotland (now part of the Institute of Geological Sciences) entitled 'Tertiary and Post-Tertiary Geology of Mull, Loch Aline, and Oban' (Bailey *et al.*, 1924). For complete coverage of the geology of the island it is necessary to consult a further four volumes (Cunningham Craig *et al.*, 1911; Lee *et al.*, 1925; Bailey *et al.*, 1925; and Richey *et al.*, 1930). The relevant Geological Survey of Scotland 1 inch to 1 mile geological maps are sheets 51, 52, 43, 44, and 35, of which sheet 44 is the famous sheet which covers the central volcanic complex. For a less-detailed but fine account of the geology of the island, which also places it in its geological context, reference should be made to the British Regional Geology handbook *Scotland: The Tertiary Volcanic Districts* (Richey *et al.*, 1961). There are also numerous research papers in the geological literature on particular aspects of the geology of Mull, but the wealth of data given in the memoirs and on the geological maps should be more than adequate for most readers.

A summary of the stratigraphical sequence of rocks found on Mull and Iona is given in Table 4.1, where the principal rock-types comprising the various units are also listed. The sequence can conveniently be considered in four groups (Fig. 4.1): a very old group, mainly of Precambrian age, of metamorphic and igneous rocks consisting of the Lewisian, Moine and Dalradian series; then a rather varied succession mainly of sedimentary rocks extending discontinuously from the Torridonian to the Upper Cretaceous; this is followed by igneous rocks of Tertiary age, which are themselves succeeded by unconsolidated sediments of glacial origin, and minor recent beach deposits.

Although this succession is a varied and interesting one, many of the rock-types are only found in small outcrops, particularly along the coast, and apart from Iona and the western part of the Ross of Mull, the geology is dominated by the Tertiary igneous rocks (Fig. 4.1). Although, as will be discussed later, there is considerable variation in the mineralogy, chemistry, and physical nature of the Tertiary igneous rocks, particularly in the south-east, there are large areas that show little variation in rock-type,

* Department of Mineralogy, British Museum (Natural History).

so that there will be little variation in any influence the rocks may have on soil formation.

The metamorphic and igneous rocks belonging to the Lewisian, Moine and Dalradian Series occupy two separate areas of Mull (Fig. 4.1). The Lewisian rocks are confined to the island of Iona while the Moine rocks, which are intruded by the Ross of Mull granite, occupy the westerly third of the Ross of Mull Peninsula. The Moines are sharply delimited from the Tertiary lavas which cover the rest of the peninsula by a fault which runs in a WNW–ESE direction. This fault is Tertiary in age, and it was probably movement on it which lifted the Lewisian and Moine rocks to their present relatively elevated position. Lesser outcrops of Moine are found on the Gribun coast and Inch Kenneth (Fig. 4.3) and in the eastern part of Mull (Fig. 4.4). Rocks belonging to the Dalradian metamorphic series, which is extensively developed on the Scottish mainland, outcrop in the vicinity of Loch Don on the eastern side of the island, close to the Tertiary volcanic centre. The developing volcano compressed, to some degree, the rocks around it throwing them into folds which have an arcuate outcrop concentric to the volcanic centre. In some places erosion of the anticlinal parts of these folds has revealed the underlying Dalradian and Moine rocks which were brought to higher structural levels by the folding. Within one of these outcrops a series of lavas occur which overlie the Dalradian rocks but are themselves overlain by rocks of Triassic age. These Old Red Sandstone lavas, like the Dalradian, are only exposed because of the folding around the Tertiary volcanic centre.

The Tertiary igneous rocks can be divided conveniently into two groups. The first group occupies the south-eastern part of the island and comprises a geologically very complex area of major and minor igneous intrusions and lava flows, which represent the remains of a very large, deeply eroded volcano (Fig. 4.1). This structure is readily apparent on the geological map (Sheet 44). The second group of rocks surrounds the volcano, and extends over the whole of the northern part of the island and across to Morven; the group comprises a thick pile of lava flows, which were originally poured out from the volcano. The lavas overlie the sedimentary rocks detailed in Table 4.1,

Table 4.1 The geological formations of Mull and principal rock-types occurring in them

Pleistocene and Recent	glacial and beach deposits
Tertiary	extrusive and intrusive basic to acid igneous rocks and minor sediments
UNCONFORMITY	
Upper Cretaceous	sandstone, shale, silicified chalk
UNCONFORMITY	
Jurassic	sandstone, shale, limestone
Triassic (including Rhaetic)	conglomerate, marl, sandstone, limestone
UNCONFORMITY	
Lower Old Red Sandstone	andesitic lavas, granite.
UNCONFORMITY	
Torridonian	shale, grit, conglomerate
UNCONFORMITY	
Moine – Dalradian	schist, quartzite, slate, limestone
Lewisian	granite, gneiss, minor marble

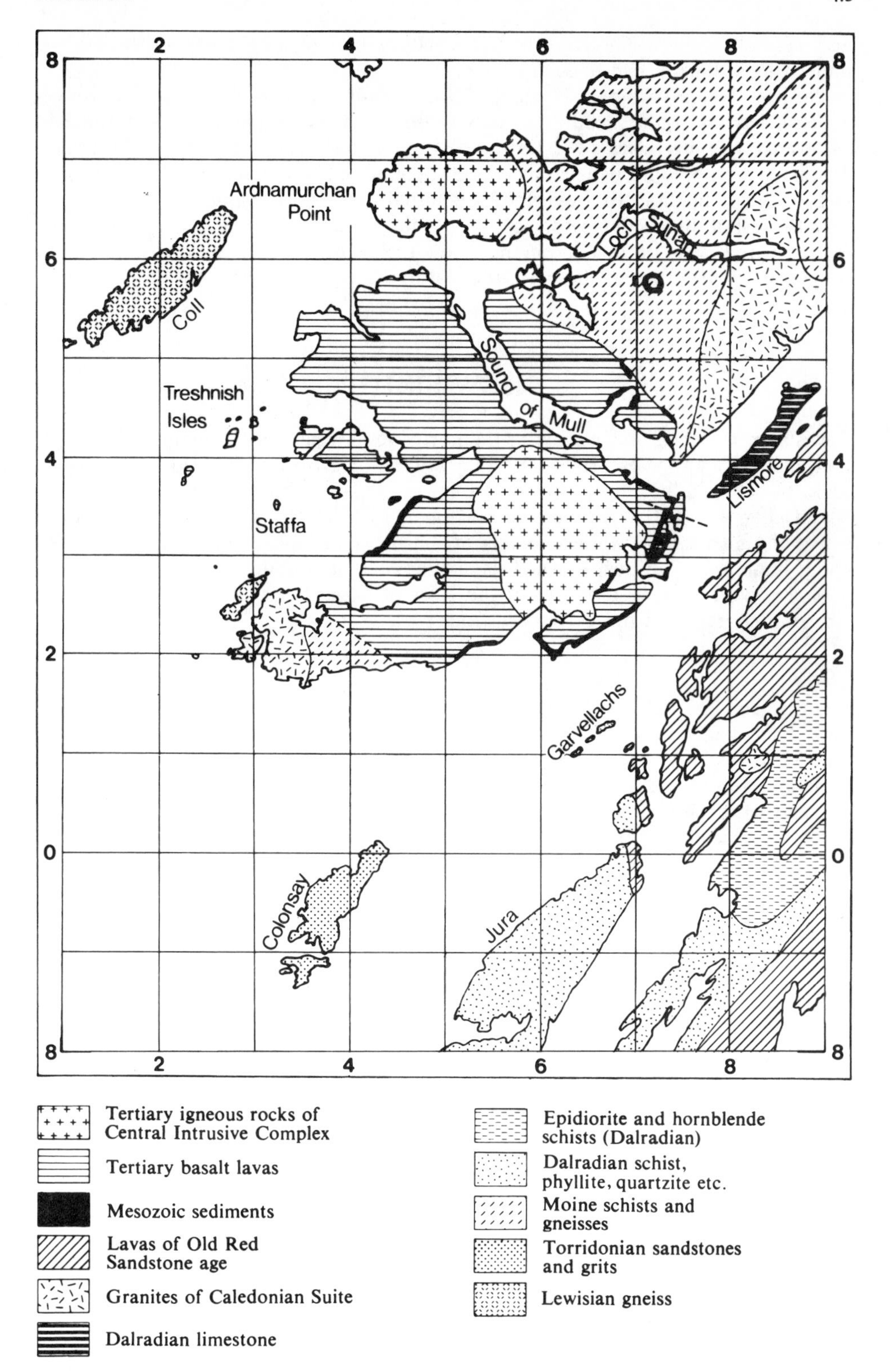

Fig. 4.1 The distribution of the major rock-types of the Mull area.

and it is only in a few places on the coast that the base of the lavas together with the underlying sediments can be seen.

The post-Tertiary geological history of Mull is mainly a story of the immense erosion which it has undergone, leading to the landscape which we know today. Intimately associated with this process was the distribution of glacial deposits and the formation of soils, but an account of this specialised topic, which is of considerable significance in considering the flora of the island, is deferred to the next chapter.

In the rest of this account of the geology of the island each of the major rock groups is considered with particular emphasis on their distribution, mineralogy, and to some extent chemistry, as these are the features likely to have the greatest influence on the flora.

Lewisian

Lewisian rocks occur extensively in the north-western part of the mainland of Scotland and through the Hebrides, and they are the oldest rocks exposed in the British Isles. In our area they are confined to Iona, comprising about three-quarters of that island (Fig. 4.2), and continue into the small islets to the south. These rocks are essentially gneisses with minor bands of marble, and have been eroded to a low-lying, hummocky landscape, which is similar to that produced by the granite and Moine rocks at the western end of the Ross of Mull, but quite different from the rugged topography of the Tertiary igneous rocks.

Gneiss

Granite gneiss constitutes about 90 per cent of the Lewisian of Iona, and is a grey or pink, finely-banded rock composed essentially of layers of quartz and feldspar, some 3 to 5 mm thick, separated by irregular thinner, darker layers. Interspersed amongst these granite gneisses are layers up to 65 m thick, though usually much thinner, of darker rocks rich in hornblende, though there are types gradational between these 'basic' hornblende gneisses and the 'acid' granite gneisses. The hornblendic rocks are of a coarser grain than the granite gneisses and locally contain very coarse patches with hornblende crystals up to 2 to 3 cm across. Veins of pegmatite up to 50 cm thick cut the gneisses and are coarse, granitic rocks in which quartz and feldspar are readily distinguished. There is also an extensive system of green epidote veins cutting the gneiss, which vary in thickness from mere films to veins some 50 cm across. The gneisses briefly described above were probably igneous rocks before they were metamorphosed, but there are a few small areas of Lewisian on Iona which were probably once sedimentary rocks. Marble is prominent in this group and is described later, but in addition it includes a garnet-biotite gneiss which can be traced along the shore on the north-west side of the island (Fig. 4.2), and a similar rock outcrops in the south-western corner of the island west and east of Druin and Aoineidh. These gneisses are relatively dark banded rocks in which a brown mica is readily apparent. Lying amongst the granite gneisses and defining the oval outcrop at the southern end of the island is a rock-type which was distinguished by the Survey Geologists as 'white rock' (Fig. 4.2). It is finer grained than the gneiss, white, and composed essentially of feldspar; to the inexpert it could be mistaken for marble.

Marble

Narrow bands of marble occur on the west side of Iona to the north of Dùn Cùl Bhuirg, and just west of Maol, and two outcrops occur in the south-east (Fig. 4.2). The main mass of marble in the west is called the 'Silverstone' and is about 2 m thick;

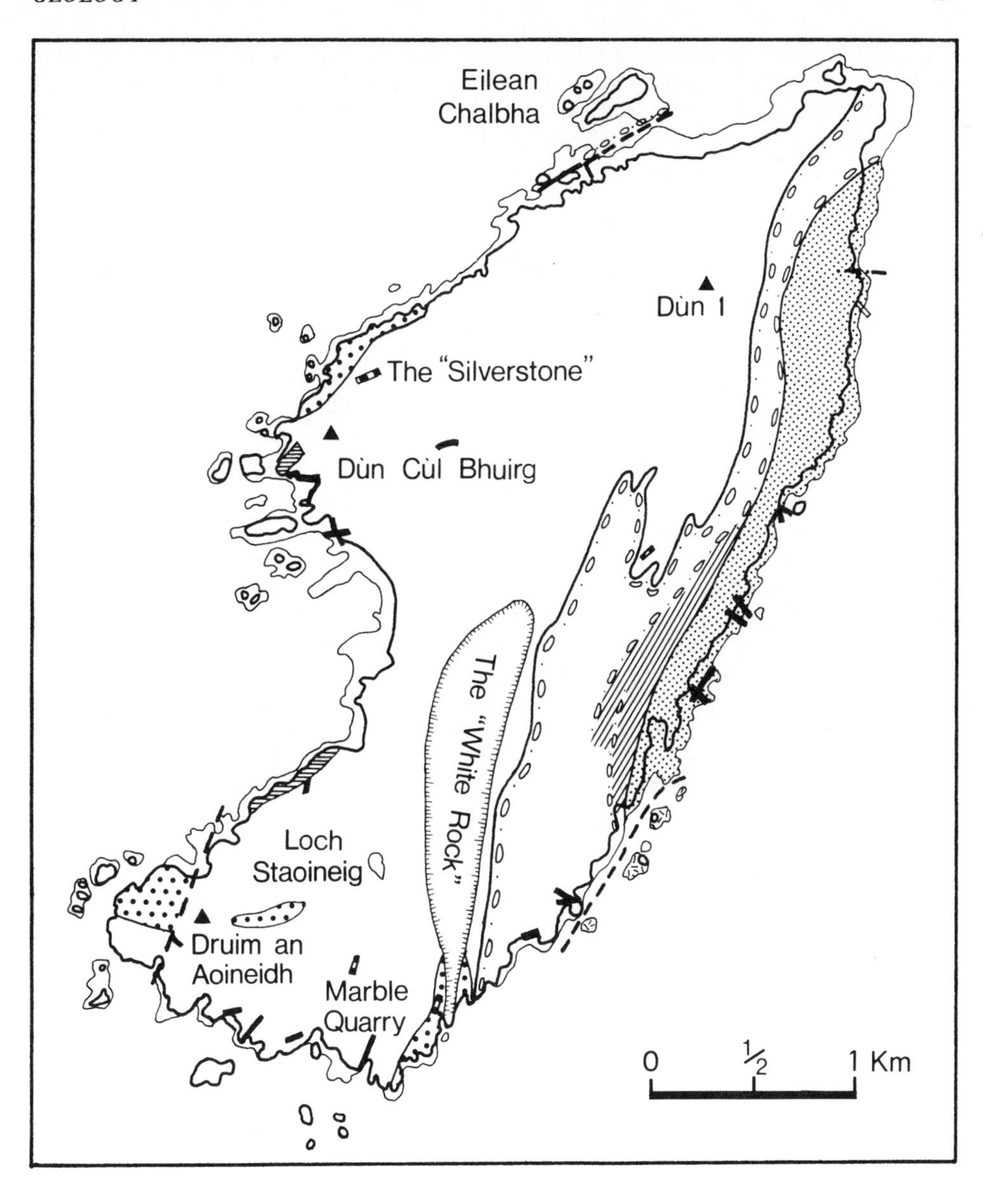

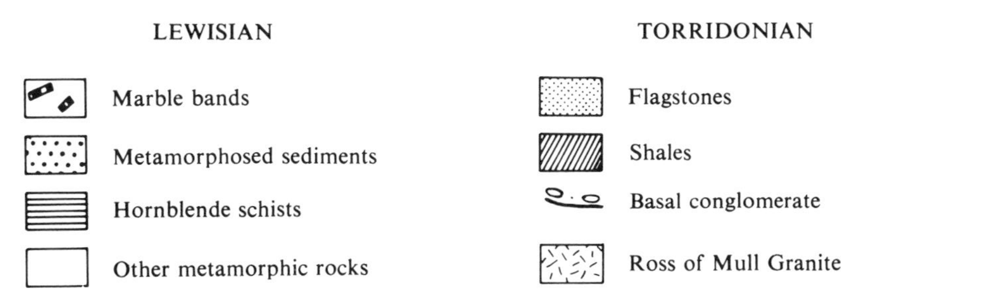

Fig. 4.2 Detailed distribution of the rock-types of Iona. Adapted from Bailey, Anderson et al. (1925).

it is pinkish-grey with green spots of serpentine and silvery flakes of talc in a carbonate matrix. The marble that outcrops close to the coast in the south-east has been quarried intermittently for some hundreds of years for 'Iona marble'. It comprises a vertical band from 6 to 10 m in thickness and is variably white, or white with a characteristic green mottling of serpentine.

Moine series

The metamorphic rocks of the Moine series are most extensively developed over a wedge-shaped outcrop south of Bunessan (Fig. 4.1). They comprise principally schists that readily cleave into slabs, and are of two types – psammites (metamorphosed sandstones) and pelites (metamorphosed shales). The psammites are relatively massive rocks with bedding indicated by thin micaceous layers and are mainly confined to the area south-east of Scoor and on the north-west coast of Loch na Làthaich. They are composed of grains of quartz, feldspar, micas and sometimes garnet; pebbles are often evident. The psammites sometimes grade into pure quartzites. The pelitic Moines are more abundant than the psammites and occupy all of the central part of the Bunessan Moine outcrop. These rocks are rich in mica, and hence cleave more readily than the psammites, and they contain abundant feldspar and commonly garnet. They often show small scale crumpling. In limited areas the minerals kyanite and tourmaline are found. With increase in quartz and feldspar the pelites grade into psammites. In a belt about 3 km wide running south-east from Loch Assapol are numerous narrow outcrops of hornblende schist, which contain abundant large garnets up to 4 cm across, which stand out prominently on weathered surfaces.

Throughout the Moines but more particularly in the area of Bunessan and Loch Assapol, calc-silicate bands between 3 and 15 cm thick occur. They represent original calcareous layers but now include, besides carbonate, garnet, hornblende and other minerals.

The Moine schists that outcrop in a strip along the coast south of Gribun (Fig. 4.3), on the western side of Inch Kenneth, and the islet of Erisgear some 7 km to the west, comprise principally schists of psammitic type similar to those described from the Bunessan vicinity. Moine rocks also outcrop at a number of localities in eastern Mull (Fig. 4.4) where they occur in the core of an anticline at Craignure, and form inliers in Tertiary igneous rocks to the north of the western end of Loch Spelve, and on the upper slopes of Sgùrr Dearg. The rocks of the inliers are very much shattered, and

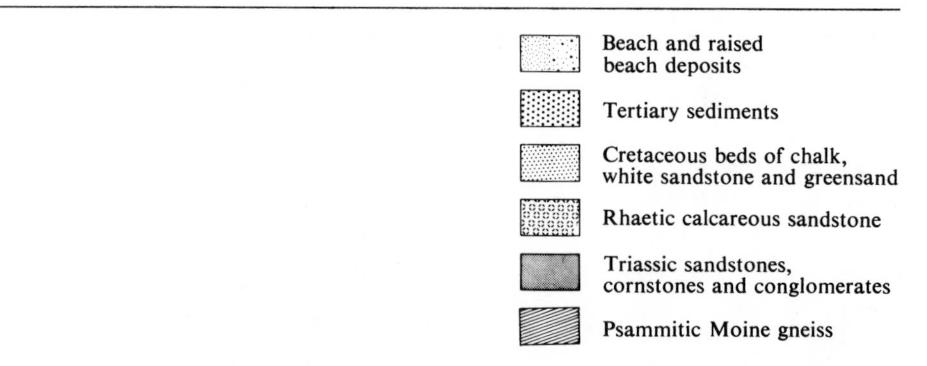

Fig. 4.3 Detailed distribution of the Mesozoic sediments of Inch Kenneth and Gribun. Drawn from original field maps with permission of the Director, Institute of Geological Sciences.

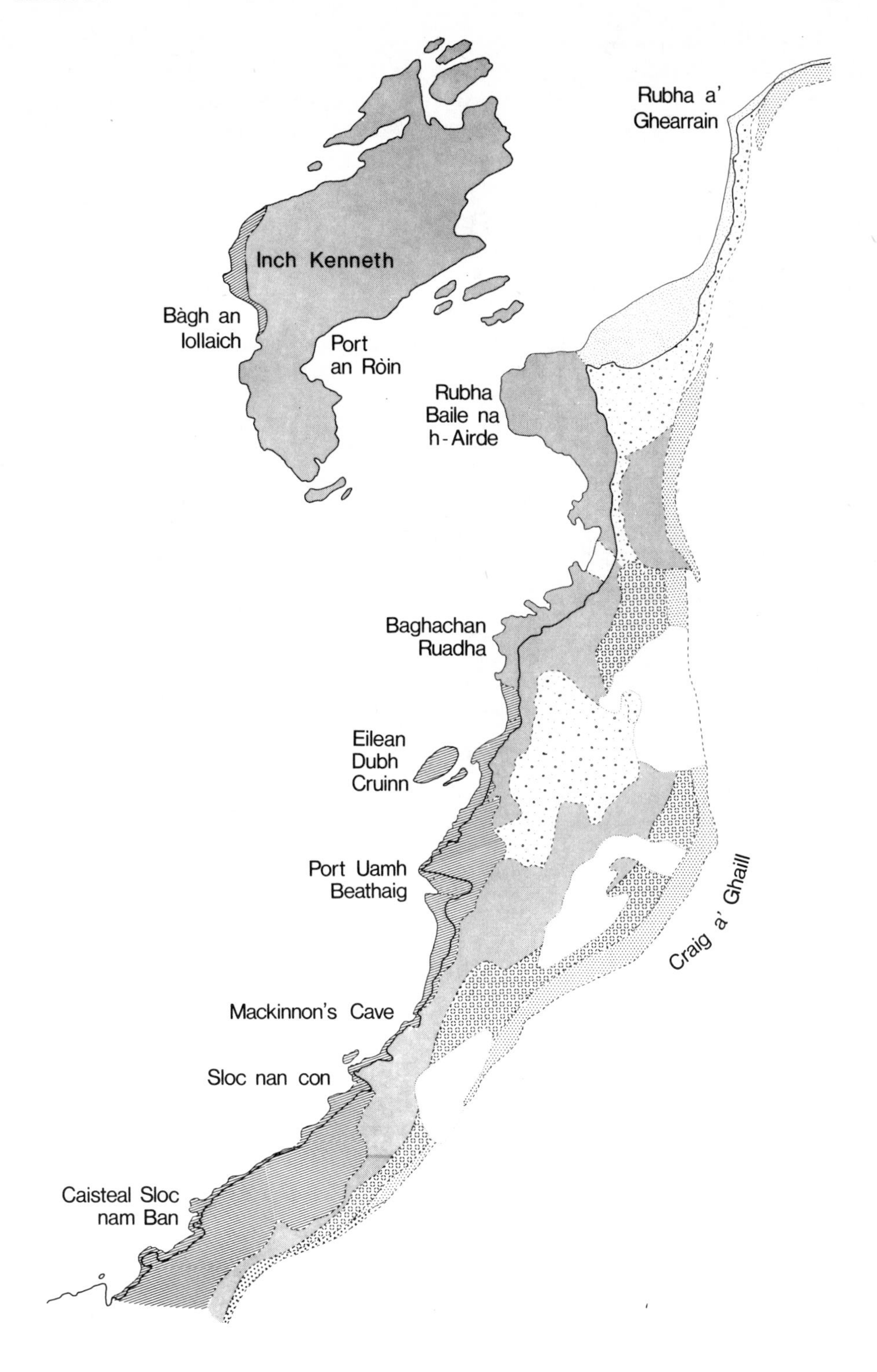

Rubha a'
Ghearrain
Inch Kenneth
Bàgh an
Iollaich
Port
an Ròin
Rubha
Baile na
h-Airde
Baghachan
Ruadha
Eilean
Dubh
Cruinn
Port Uamh
Beathaig
Craig a' Ghaill
Mackinnon's Cave
Sloc nan con
Caisteal Sloc
nam Ban

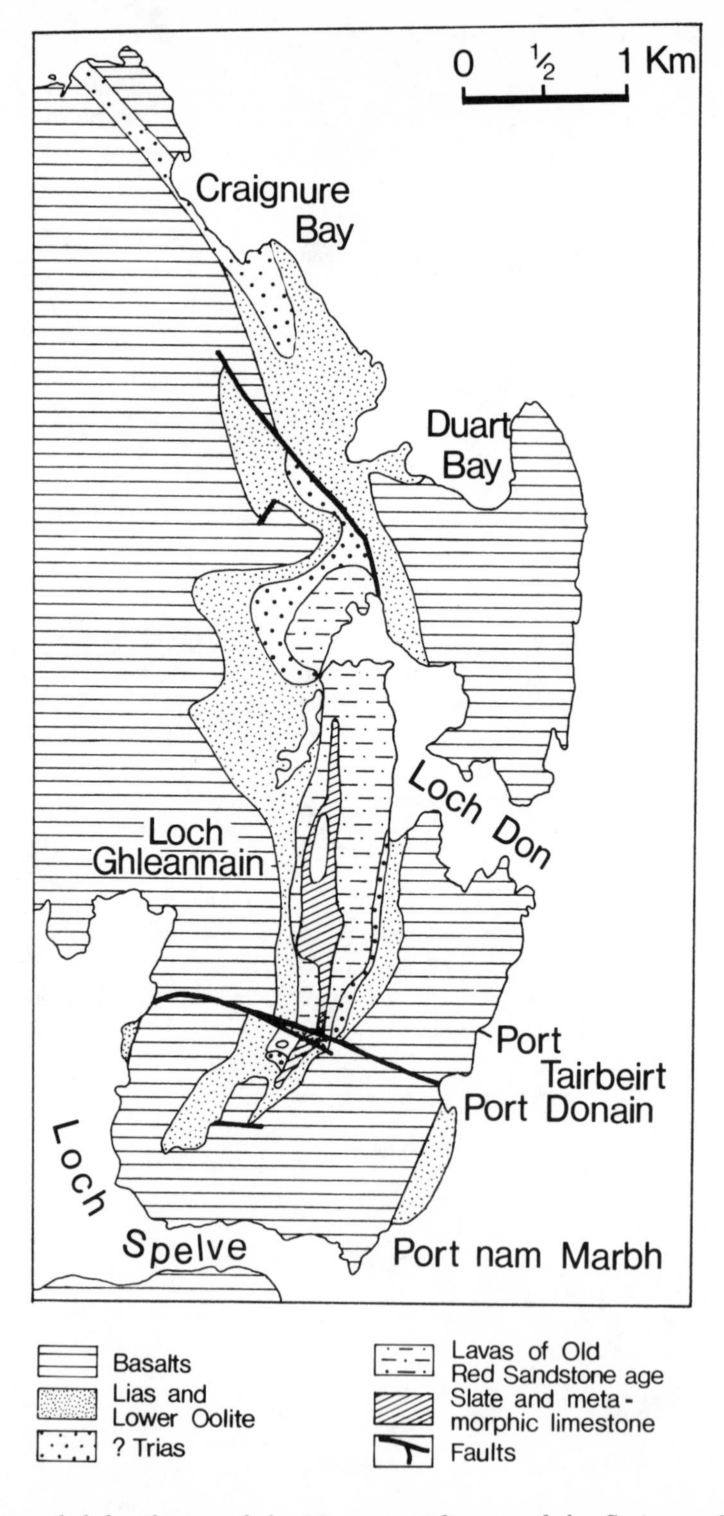

Fig. 4.4 Detailed distribution of the Mesozoic sediments of the Craignure–Loch Don area. Adapted from Richey (1961).

are cut by numerous igneous intrusions. Both pelitic and psammitic schists are found, similar to those described from the Bunessan area.

Dalradian

Dalradian rocks on Mull are confined to the core of the anticline which trends north-south in the vicinity of Loch Don (Fig. 4.4). The principal rock type is a grey to dark grey slate which has been tried for roofing material. There is also a dark, finely crystalline limestone which is the same rock that forms Lismore Island to the east. It is well exposed at three localities namely south of the road, 400 m south of Loch Don bridge; south of Loch a' Ghleannain; and on both sides of the fault crossing from Port Domain to Loch Spelve.

Torridonian

Sedimentary rocks which can be assigned to the Torridonian form a belt up to 800 m wide along the eastern side of Iona. They fall into two groups, a lower group of grits with a basal conglomerate and some thick beds of shale in the upper part, and an upper group of flags. The distribution of these groups is shown on Fig. 4.2. The basal conglomerate, which overlies the Lewisian, contains boulders of various metamorphic and igneous rocks set in a dark grey to dark green matrix which is essentially a grit. The conglomerate is interbedded with and overlain by pebbly grits, and is dark grey or dark green in colour. In the central part of the outcrop the grits pass up into shales one band of which is 90 m thick. At the northern end of the island, opposite Eilean Annraidh a layer of dark impure limestone is intercalated in the shale.

The flagstone group forms the shore from Sligneach almost to the northern point of the island. These rocks are characterised by rapid alternations of sandy and shaly layers. The rocks as a whole are dark in colour.

The Ross of Mull granite

The western extremity of the Ross of Mull is formed from a major intrusion of granite (Fig. 4.1). It is limited to the east by an intrusive contact with the Moine schists, and to the west it appears in small islets close to the east coast of Iona, so that this contact runs somewhere beneath the Sound of Iona. The Torran Rocks off the south coast are also of the same granite. The Ross of Mull granite is Devonian in age, the K-Ar dating method gives dates of 398–409 million years (Brown *et al.*, 1968), and it is part of an extensive suite of such granites found in the Scottish Highlands and in Ireland.

The granite country lies mostly below 140 m and forms smooth rocky hillocks isolated by areas of peat and raised beach deposits. This hummocky terrain is in strong contrast to the terraced slopes of the lava country to the east. The granite is pink or red and is a coarse, even-textured rock composed of quartz, alkali feldspar, some plagioclase feldspar, biotite and a little muscovite. Throughout the granite may be found inclusions of Moine schist and these may be two or three hundred metres across. In a few places the granite grades into a darker rock which should be classified as a quartz diorite. The most extensive area comprises Eilean a' Chalmain off the south-west coast, but other areas are distinguished on the geological maps sheets 43 and 35. These rocks contain less quartz than the granite, with plagioclase the dominant feldspar, biotite abundant and hornblende often present. A chemical analysis of the granite is given in Table 4.5.

Table 4.2 Description and location of Mesozoic sediments on Mull

Stratigraphical Divisions	Lithology	Where found in Mull (with thickness of bed in metres)
UPPER CRETACEOUS		
Upper Chalk (Senonian)	Chalk (silicified)	Gribun (3)
[Middle Chalk absent]		
Greensand (Cenomanian)	White sandstone with wind-blown rounded grains	Gribun (3)
	Greenish or greenish-grey sandstone with or without lime and glauconite	Loch Don (3)
	Occasional shell layers	Carsaig (7·5–14)
[LOWER CRETACEOUS & UPPER JURASSIC ABSENT]		
MIDDLE JURASSIC		
Great Estuarine Series	Blue shale	Loch Don (9)
Inferior Oolite	Sandstone grading into coarse crystalline limestone; micaceous and calcareous shales	Loch Don (30) Loch Spelve (Port nam Marbh, Androchet Glen, Port Donain) (30)
LOWER JURASSIC		
Upper Lias	Blue shale	Port nam Marbh (9) Croggan (2)
	Blue or black crumbling shale	Nuns' Pass, Carsaig (1)
Middle Lias: (Scalpa Beds)	Sandstone, rarely slightly calcareous with bands of micaceous shale	N of Loch Spelve (Port nam Marbh) (30) Loch Spelve to Lochbuie coast (6)
	White and brown sandstone	Carsaig Bay (60)
Lower Lias: (Pabba Beds)	Shales grading into calcareous sandstone and limestones	Torosay-Duart Bay, Abhainn Lìrein, Ardrochet Glen, Port Donain–Port nam Marbh, Croggan-Laggan peninsular (90) Tobermory River (9)
Broadford Beds	Shales, limestone and calcareous sandstone, red mudstone; with considerable shell beds.	Craignure Point, Loch Don anticline, E shore Loch Spelve (9) Croggan-Laggan peninsula, Loch buie (20) Sea coast of The Wilderness (1.8)
RHAETIC	Sandy limestone or calcareous sandstone	Gribun (12)
TRIASSIC	Conglomerates, sandstones, grits, breccias; occasionally calcareous	Craignure and Loch Don, Loch Spelve (extensive), Beinn Bheag (small) (2)
	Conglomerates, cornstones and sandstones	Gribun and Inch Kenneth (60)

Lavas of Old Red Sandstone age

In the anticlinal structure lying near Loch Don in the most easterly part of Mull are found Dalradian shales and limestones, as already described. Overlying these rocks is a group of lavas which itself is overlain by sediments of Triassic age (Fig. 4.4). The lavas are much shattered and difficult to identify, but they are correlated by the Geological Survey with much more extensively developed lavas of Old Red Sandstone age occurring near Oban on the mainland. They are fine grained and, if the correlation with the mainland succession is correct, are andesites. However, they are too altered to be readily identified with the microscope.

Mesozoic sedimentary rocks

The Mesozoic sedimentary rocks of Mull range in age from Triassic to Upper Cretaceous. Their mineral content is less variable than that of the volcanic rocks described below, but they do contain varying amounts of calcium and magnesium (usually as carbonates) which can become available to any plants growing upon them. For this reason and because these rocks are scarce on the island they are biologically important. The stratigraphical divisions and main locations are given in Table 4.2. In the following paragraphs notable outcrops of the different lithological types are described but it must be appreciated that these outcrops occur for the most part in cliffs or steep slopes and that the seepage or percolation of surface water through calcareous strata can affect lower non-calcareous beds. The texture (i.e. physical relationships and form of the mineral grains) of the various beds may be a critical factor in plant establishment.

Sandstones and shales

a) The Gribun and Inch Kenneth (Fig. 4.3) In the Gribun and on Inch Kenneth sandstones occur often interbedded with calcareous bands. In places the sandstones grade into sandy shale. A distinct brick-red sandstone is seen just south of Bagh an Iollaich on Inch Kenneth. Another well-defined band is seen between 20–30 m up the cliff in the gully below Tòn Dubh-sgairt on the Ardmeanach peninsula. A greenish grey shale some 1.5 m thick outcrops just south of the fault at Uamh nan Calman and in The Wilderness Cliff especially immediately south of Aird na h-Iolaire, dark shales interspersed with calcareous bands are common and red mudstone of the Broadford Beds series are found on the shore.

b) Carsaig area Glauconitic sandstones with layers of thin-shelled lamellibranchs and occasional patches of limestone occur above Carsaig House forming a series of waterfalls in a steep burn. Similar rocks interbedded with thin black shale occur above Aird Ghlas on the west side of Carsaig Bay. Middle Lias white sandstone some 60 m thick forms the main part of the cliffs either side of Carsaig Bay, and was at one time quarried for building purposes but its position prevented ease of transport. These cliffs show 'balls' up to a metre across of harder sandstone which becomes differentially eroded giving a pitted or mamillated surface (see Fig. 4.5). On the shore below the Nuns' Pass these harder 'balls' stand up like haystacks in a field.

c) Loch Don–Loch Spelve area (Fig. 4.4) Greenish, sometimes calcareous, sandstone outcrops for about 3.5 km between Leum na Muice Duibhe and Port a' Ghlinne. On the Croggan-Laggan peninsula the cliffs at Port Ohurnie are of a similar rock and make this stretch of coast difficult to negotiate from the land because of the easily eroded shale bands. On the peninsula lying between Lochs Spelve and Don the raised beach lying between Port nam Marbh and Port Donain is formed from a greenish

Fig. 4.5 Cretaceous greensands in the cliffs 1 km west of Carsaig. 1967.

indurated sandy shale sometimes lightly calcareous. Sandstones are also seen around Loch a' Ghleannain and in Gleann Rainich, and in the Abhainn Lìrein 400 m above Oakbank House similar strata grade into a sandy greenish mudstone. A fine-grained mostly rather hard, blue shale can be seen in the sea cliffs a kilometre to the east of Auchnacraig between Lochs Don and Spelve.

d) Tobermory area A red, current-bedded, compact sandstone of uncertain age, some 15 m thick is found between faults for 180 m in Bloody Bay, NE Mishnish. Sandstone is also found at sea-level north-east of Gualann Dubh on the Sound of Mull, 5 km south-east of Tobermory, and again 1 km west of Rubha nam Gall Lighthouse, north of Tobermory.

Limestones, calcareous sandstones and siliceous chalks

a) Inch Kenneth and Gribun (Fig. 4.3) The west cliffs of Inch Kenneth show good exposures of calcareous sandstone and limestone together with breccias and conglomerates which are seen again on the opposing Mull shore. Cornstones (limestones with concretionary nodules) are the dominant rock-type and are well exposed on the east of Inch Kenneth especially below the ruins of the old chapel and on the other side of the channel, near high-water mark on the Rubha Baile na h-Airde. There is much siliceous material in the cornstones here in the form of quartz grains and mamillated layers of chert. A silicified chalk of Cretaceous age is found on the Mull coast north of the Gribun. Much of it is covered by a scree of basalt blocks but there are signs of it, at low tide, to the south and landward side of Samalen Island. There is a narrow band in a landslip just above the road 500 m north of Clachandhu ruins. Further south there is another coherent landslip where the main road approaches Balmeanach Farm, where sandstone and concretionary limestone can be found in the stream-bed of Allt Teangaidh. The most distinctive series of calcareous rocks occurs in this stream in the deep gully below the main road at grid reference 17/455329. Similar calcareous sandstones and shelly beds (the fossils are mainly lamellibranchs) occur in exposed sections in various streams running off the cliffs in the Balmeanach Farm area. Some 20 m of pure concretionary cornstone is seen in The Wilderness between Tòn Dubh-sgairt and Caisteal Uamh an t-Saguirt.

b) Craignure–Loch Don area (Fig. 4.4) The Lower Lias outcrops on the coast to the south of Craignure Point as a white limestone. This limestone gets darker and more muddy upwards and the mollusc *Gryphaea* becomes an abundant fossil. The *Gryphaea* limestone is seen again on the other side of the Craignure anticline in the stream above Upper Achnacroish; here it is associated with a greenish mudstone. A similar hard dark-blue limestone is exposed where the path to Grass Point crosses the stream 1 km south-east of Ardnadrochet Farm and again on the east coast of Loch Spelve 1 km north-west of the old chapel.

On the eastern shore of the Croggan peninsula the *Gryphaea* limestone is seen again just on the north side of Altt Omhain 900 m south-west of Port a' Ghlinne where 15 m of impure limestone alternating with calcareous sandstone is exposed. Further west limestones are found at the base of the cliffs between Frank Lockwood's Island and Lord Lovat's Bay. Limestones of the oolite series are seen in Ardnadrochet Glen (see Fig. 4.4), in the stream-bed between the farms Ardnadrochet and Auchnacraig and again on the seaward side of that peninsula at Port nam Marbh and Port Donain. Smaller outcrops are found along the stream at Oakbank, and in Camas Mòr at Duart Bay.

c) Carsaig–Lochbuie area There is an exposure of Lower Lias consisting of a sandy limestone about 2 km west of Glenbyre Farm, Lochbuie. The Cretaceous greensands in the cliffs to the west of Carsaig (see Fig. 4.5) contain calcareous bands sufficient for there to be signs of a calcareous flora.

Conglomerates and breccias

Conglomerate is a rock consisting of rounded boulders or pebbles set in a finer grained sandy or limy matrix and thus because of its heterogeneous nature can be ecologically significant, especially for lithophytes such as lichens, because of the possible juxtaposition of calcareous and non-calcareous rocks within an essentially single rock unit. A breccia is a similar type of rock but the boulders and pebbles are angular.

a) SE Mull Several outcrops occur in this area. At Craignure Point there is a quartzite with calcareous pebbles embedded in it. There are numerous outcrops from Craignure Bay down to Maol an t'Searraich, south of Loch a' Ghleannain. At Rudha na Faing on the east side of Loch Spelve and for a considerable area on the west shore and up the Lussa valley, there are both Triassic conglomerates with quartzite boulders 30 cm or more in diameter, and also breccias with gneiss inclusions. The highest conglomerate on the island is on Beinn Bheag, east of Sqùrr Dearg, at 500 m altitude.

b) Inch Kenneth and Gribun Conglomerates are here interbedded with other Mesozoic rocks and especially at the base of the cliffs on Inch Kenneth and on the Gribun and The Wilderness.

c) Pebble beds and gravels of Ardtun and Carsaig Bay The famous Ardtun Leaf Beds (see p. 4.23) are included in a series of fluvio-lacustrine sediments outcropping for about 1.6 km close to the coast of the Ardtun peninsula between Lochs na Làthaich and Scridain. These deposits have been fully documented (Gardner, 1887; Bailey *et al.*, 1924; Seward & Holttum, 1924) and are sandwiched between basalt lavas of intermittent eruptions. Pebble layers within the Leaf Beds contain water-worn chalk flints and porphyritic lava pebbles different from the surrounding lava flows. These sediments re-appear at Malcolm's Point and again, as a very narrow seam, east of Carsaig at An Dùnan.

d) Lignites in the south-west A few seams of coal are interstratified with basalt lavas in south-west Mull, one of which was quarried for home consumption by the shepherd living at Shiaba, south-east of Loch Assopol (Bailey *et al.*, 1925). The other is associated with the gravel beds mentioned above at Gowanbrae and Ardtun and was worked in the 1870s but not for long. Neither outcrop has distinctive flora.

Tertiary igneous rocks and associated minor sediments

There are more than a dozen Tertiary volcanic centres along the western seaboard of Scotland and in Northern Ireland, and the Mull centre is arguably the most complex of these. As has already been pointed out, the Tertiary igneous rocks of Mull can be considered to fall into two groups, namely the extensive area of lavas covering about two-thirds of the island which are known as the 'Plateau Group', and the intrusive centre situated in the south-eastern part of the island which includes a number of major intrusions, as well as another group of lavas known as the 'Central Group' (Figs 4.1 & 4.6). Both the Plateau Group lavas and the rocks of the intrusive centre are cut by numerous minor intrusions. These major rock groups comprise a considerable number of rock-types which are summarised in Table 4.3. It is unfortunate that in this short general account it is necessary to introduce so many rock names, but in fact Table 4.3 represents a simplification from the even more extensive list of rocks described in the *Memoir*.

Associated with the Plateau lavas there is a small volume of sedimentary rocks which

Table 4.3 Summary of principal igneous rock-types of Tertiary age on Mull

Lavas	
	Olivine-poor (tholeiitic) basalts – Central Group Olivine-rich (alkaline) basalts – Plateau Group Mugearite
Rocks of major intrusions and vents	
	Olivine and quartz gabbro, diorite, felsite, granophyre. Vent agglomerate
Minor intrusions	
	Olivine and quartz basalt and dolerite Granophyre, felsite/rhyolite

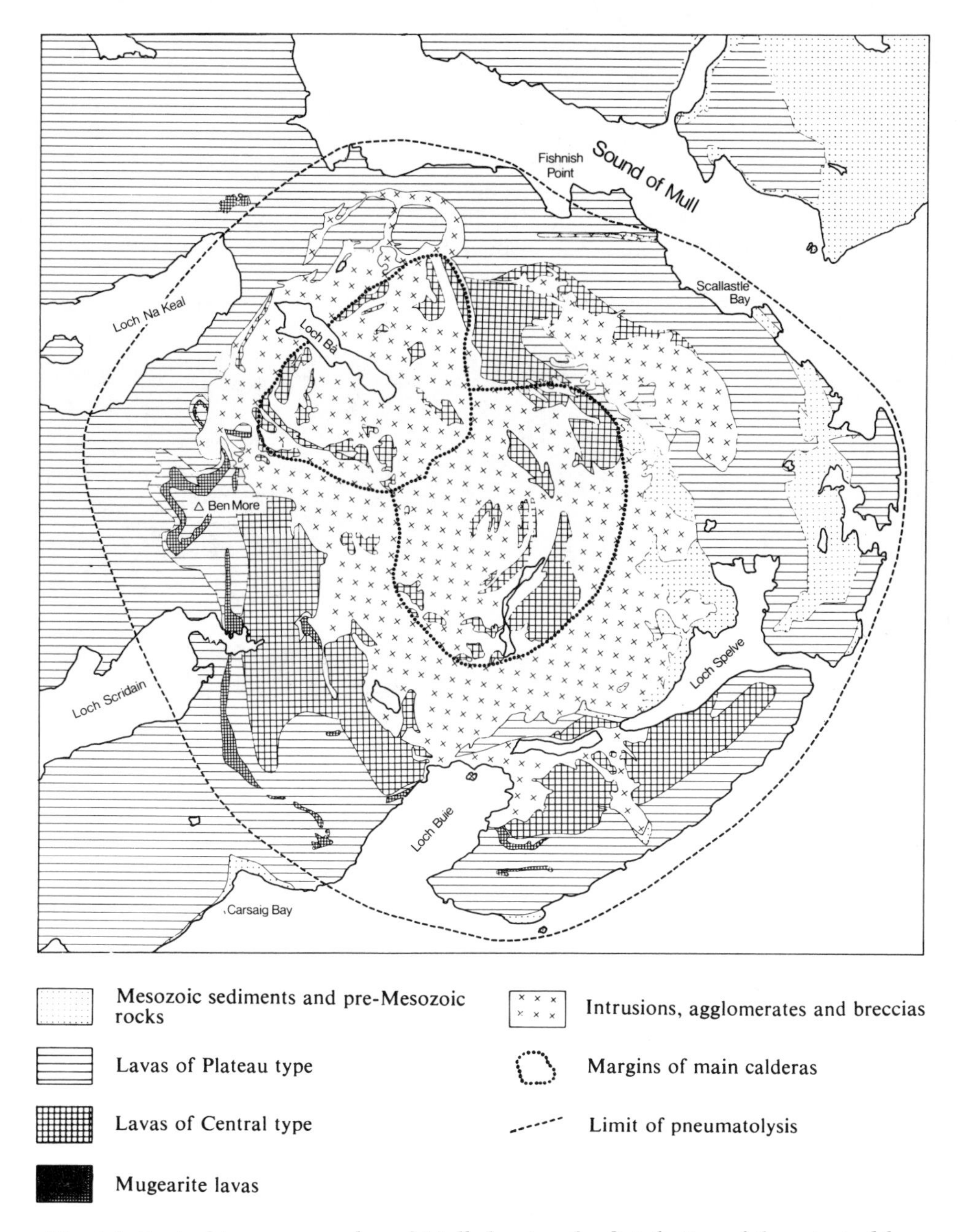

Fig. 4.6 Central igneous complex of Mull showing the distribution of the principal lava types, the igneous intrusions, and pre-Mesozoic rocks. Adapted from Bailey (1924).

must be of interest to the botanist, not only because of any effect they may have on the present flora, but also because they have proved a prolific source of palaeobotanical material, which has provided an unusually complete insight into the climatic conditions prevailing at the time the Mull volcano was active (see p. 4.22).

Lavas of Plateau Group

These lavas cover an area of some 73 000 ha on Mull (Fig. 4.1). Typically individual flows are about 15 m thick but may attain 30 m or more, while many flows are less than 15 m thick. Each flow comprises a lower massive portion and an upper thick layer of less massive, scoriaceous material. The massive parts of flows commonly display vertical joints which sometimes define distinct hexagonal columns (see Fig. 4.7 and Part 1 title page). The Plateau country develops a typical scenery of broad, flat-topped hills, with terraced hill-sides, produced by the more rapid weathering and erosion of the upper parts of individual flows. The tops of flows are often reddened due to weathering after extrusion and before burial by the succeeding flow.

The Plateau Group attains a thickness of over 900 m and consists mainly of alkaline

Fig. 4.7 Hexagonal columnar basalt showing vertical jointing, south end of Staffa. 1969.

olivine basalts. At the base of these lavas in western Mull, however, are a series of olivine-poor basalts which resemble some of the lavas of the Central Group, while near the top of the succession, lavas occur which are distinguished by the term mugearite. The distinction of the olivine-rich and olivine-poor members of the Plateau Group is readily achieved with the microscope but difficult in hand-specimens. However, the differences in the chemistry of the two groups are only slight and unlikely to be detectable in the flora.

The mugearite lavas are differentiated on the geological maps (Fig. 4.6 and Sheet 44), and there are extensive outcrops on Ben More and in an arc trending south from that mountain. They can be recognised by their pale weathering (weathered basalts are dark and often rusty-looking) and a marked flow texture, in which the constituent feldspar crystals are aligned; this feature is readily apparent under the microscope but more difficult to see in a hand-specimen.

The emplacement of the central intrusive complex caused some alteration of the surrounding lavas, the limits of which are shown on Fig. 4.6 as the 'limits of pneumatolysis'. Outside the limits the Plateau Group basalts weather with rusty-red surfaces, some break down spheroidally to soft red loam, while others may develop pustular surfaces left by a wearing away of intervening less resistant matrix (Bailey *et al.*, 1924: 95). Inside the limit, weathering yields sombre grey and brown surfaces, smooth except for veins and amygdales (gas cavities in lava filled by various minerals). These differences are reflected in the landscape, in that within the limit of pneumatolysis the lavas do not show such pronounced terrace-featuring as outside it.

Sedimentary rocks

Sediments occur both between individual flows of the Plateau Group, and at the base of the lava pile, but volumetrically they are insignificant by comparison with the lavas. The most widespread sediment is a mudstone which can be seen lying directly beneath the lavas at numerous localities. It is usually about 60 cm thick, but may attain 3 m. It is variably coloured purplish brown, or deep red, and sometimes black, pale green or buff; it is a fine-textured rock, sometimes unbedded. At many localities it is underlain by flint or silicified chalk conglomerates derived from underlying Upper Cretaceous sediments.

Sediments intercalated between lavas are principally red boles produced by weathering of the underlying basalt. At a few localities, however, the sediments include non-volcanic material brought in from elsewhere. The most noteworthy of these is a belt of sediments outcropping for about 1.6 km close to the coast of the Ardtun Peninsula between Loch na Làthaich and Loch Scridain which includes the famous Leaf Beds of Ardtun. The sediments vary between 4 and 15 m in thickness and include sandstones, gravels and shales.

Lavas of Central Group

The distribution of lavas assigned to the Central Group is shown in Fig. 4.6. The central part of the main intrusive complex subsided during the active volcanic episode along two main ring fractures which are shown on Fig. 4.6 and which produced large 'calderas' – a feature common to most volcanoes. The greatest thickness of Central Group lavas, some 900 m, are preserved within the limits of the caldera, but they also outcrop in the cores or arcuate folds around and concentric to the complex. That there were persistent crater lakes in the caldera is shown by the presence of 'pillow lavas' (rounded pillow-like masses of basalt formed by a rapid chilling of lava extruded into water) through much of the Central Group succession within the caldera limits. Apart from the pillow-lavas the Central Group is characterised by two basalt types – a highly-porphyritic type crowded with feldspar phenocrysts about 5 mm across, and a very compact non-porphyritic basalt.

Mineralogically the Central Group basalts are distinguished from those of the Plateau Group by the scarcity or absence of olivine, but often this difference is not easy to appreciate in hand-specimens.

Major intrusions of the Central Complex

The greatest complexity in the geology of Mull is encountered in the series of major and minor intrusions and lavas in the south-eastern part of the island, which marks the site of the Mull volcano. Most of the major intrusions are 'ring dykes' – arcuate-shaped intrusions concentric to the various intrusive centres. Although there is a relatively large number of these intrusions they are composed of only a few broad rock-types allowing the drafting of the relatively simple map given here as Fig. 4.6. The main rock-types are gabbros, diorite, granophyre, and felsite.

The gabbros are coarse-grained rocks in which the pale-coloured minerals, dominantly plagioclase, and the dark ones, mainly pyroxene, but sometimes olivine also, are distinguishable in hand-specimens. They grade into finer-grained variants which are referred to as dolerites. Some of the gabbros are characterised by the presence of olivine, whereas others contain quartz. These rocks are the intrusive, and therefore, coarser, equivalents of the Plateau and Central Groups of lavas, and are chemically the same. Grouped with the gabbros on Fig. 4.8 are rocks referred to by the Survey as augite diorite. These differ from the gabbros in containing a somewhat more sodium-rich plagioclase feldspar, and often the pyroxene is altered to other minerals such as hornblende.

Although chemically equivalent to the lavas, these rocks, particularly the gabbros, give rise to a rugged mountainous country quite different from that of the lava terrain. The reason for this probably lies in the tough physical properties of the gabbros, which, unlike the lavas, are not weakened by vesicles and numerous joints. Furthermore the gabbros, being intrusive rocks, were not exposed to prolonged weathering like the lavas. Any differences in the flora of the gabbro and lava country therefore are probably determined by the physical properties of the respective rocks.

The intrusive granophyres and felsites have only very minor equivalents amongst the lavas. They are grey to buff-coloured rocks, distinguishable from the gabbros by their paler colour. Little of their mineralogy can be discerned in hand-specimens, but they differ from the gabbros in containing an abundance of quartz, a more sodic plagioclase, and they commonly carry an alkali feldspar, biotite, and hornblende which replaces the original augite. However, almost all gradations occur between quartz gabbros and a rock composed solely of quartz and alkali feldspar.

It must be stressed that the south-eastern intrusive centre is of exceptional geological complexity and that although some of the major intrusions are of some considerable size, the cross-cutting of one intrusion by another, and the incorporation of numerous fragments of early intrusions by later ones means that the rocks are rarely the same over a significant area. The story is further complicated by the innumerable minor intrusions, yet to be described, which are so abundant as to be shown only formally on geological maps of all but the largest scales. Although the geological sheet 44 is complex the authors state that it is a considerable simplification of the outcrops themselves. From the botanical point of view, therefore, any interpretation of distribution related to geology in this part of the island must be made with great care.

Volcanic vents

There are many volcanic vents associated with the complex, and the larger ones are shown on Fig. 4.8. They are usually irregular in outline, and were probably drilled through the pre-existing rocks by violent gas-driven volcanic explosions. The vents are filled with breccias and agglomerates consisting of a wide size-range of blocks of

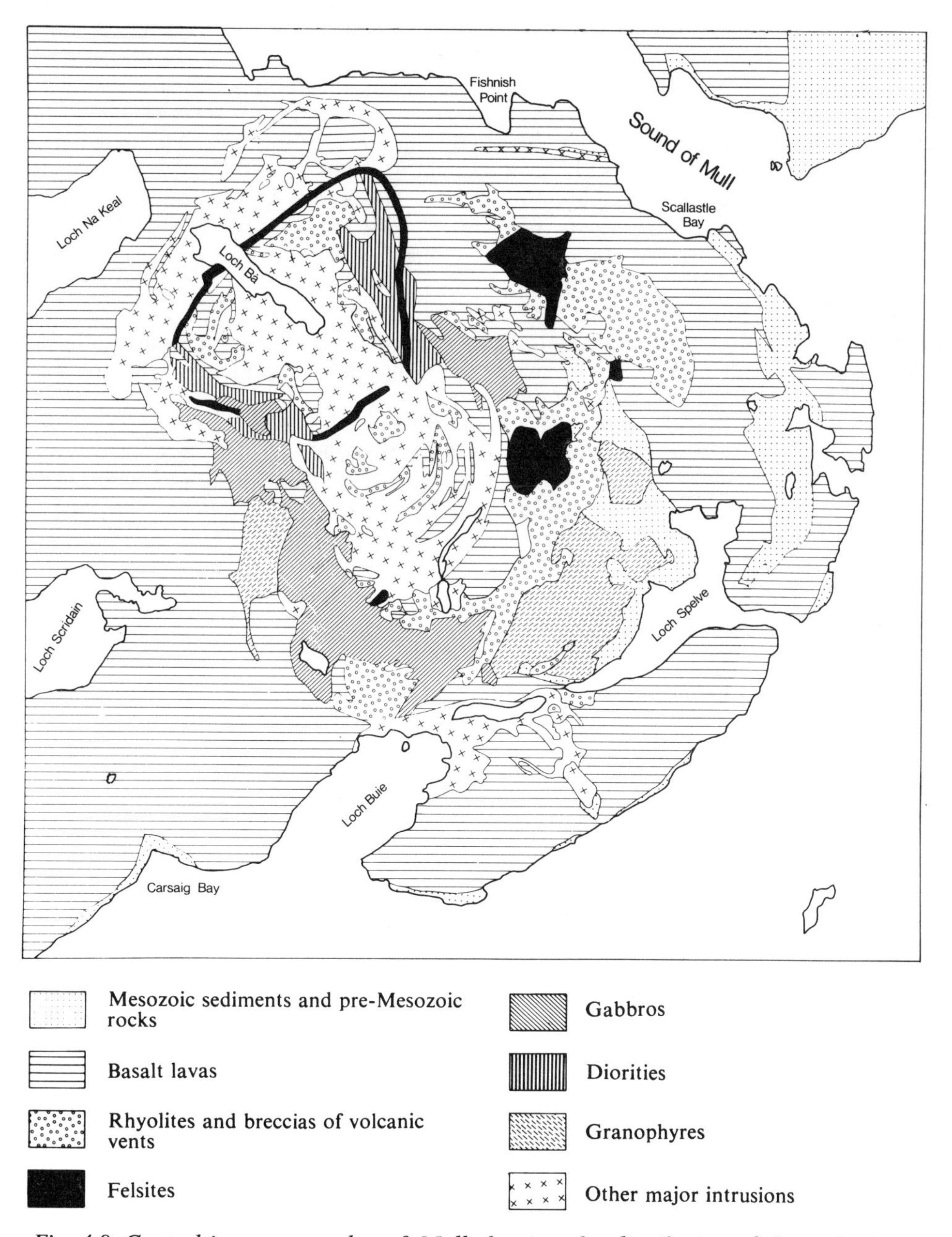

Fig. 4.8 Central igneous complex of Mull showing the distribution of the major intrusions. Adapted from Bailey (1924).

both Tertiary igneous rocks, such as granophyre and gabbro, and pre-Tertiary basement rocks from beneath the volcano.

Minor intrusions

The minor intrusions are one of the most spectacular manifestations of the Tertiary igneous activity of Mull. They take three principal forms – dykes, cone-sheets, and sills. The dykes trend generally north-west to south-east and cut most of the other rocks of the island. Because of their great abundance they are referred to as a 'swarm' and similar swarms are associated with the Tertiary igneous centres of Skye and Arran. Their abundance is shown by the fact that on a 20.2 km stretch of the south-eastern coast 375 dykes were counted averaging 167 cm in thickness. Members of this swarm can be traced across southern Scotland into the north of England. Most of the dykes are composed of olivine dolerite and quartz dolerite corresponding closely to the lavas of the Plateau and Central Groups, but on Mull itself there are also acid dykes, principally felsites and granophyres, though these are less abundant than the dolerites. Granophyre has already been briefly referred to; the felsites are pale-coloured buff or grey rocks, often with small phenocrysts of quartz or feldspar, and they are chemically similar to the granophyres.

Cone-sheets are dykes which have an arcuate outcrop arranged around an igneous centre, and dipping steeply inwards towards the centre. The cone-sheets of Mull are extremely abundant and comprise several sets. There are doleritic and granophyric sheets similar in composition to the dykes.

The majority of the sills do not occur within the main intrusive complex, but are concentrated to the west of it in the Ross of Mull, and between Loch Scridain and Loch na Keal. They dip more gently than the cone-sheets and sometimes are parallel to the lava flows into which they are intruded. The majority of the sills range between 4.5 and 9 m in thickness and are of dolerite, similar in composition to the non-porphyritic lavas of Central type. Some of them, however, range in composition towards andesite which is indicated particularly by the presence of more sodic plagioclase feldspar. Some of these sills are pitchstones comprising a non-crystalline rock (glass) with or without phenocrysts of feldspar and augite.

Chemistry

Although the geology of Mull is extremely complicated, not only because of the range of the rock-types present, but also because of the diversity of structural forms that these rocks take, investigation of any possible correlation between the chemistry of bed-rock and the flora is simplified by the fact that many of the rock-types, although different in their field relationships and perhaps grain size, are in fact chemically similar. In Table 4.4 the majority of the igneous rocks described in the previous pages

Table 4.4 Igneous rocks grouped according to similarity of composition

1	2	3	4
Granite (Ross of Mull)	Olivine gabbro	Quartz gabbro	Andesite
Granophyre	*Plateau Group:*	*Central Group:*	Diorite
Felsite	Olivine dolerite	Quartz dolerite	Pitchstone
Pitchstone	Olivine basalt	Basalt	

Table 4.5 Selected analyses of principal igneous rock-types

	1 Granite	2 Granophyre	3 Olivine gabbro	4 Plateau basalt	5 Quartz dolerite	6 Central-type basalt	7 Augite andesite	8 Mugearite
SiO_2	74·48	71·30	48·34	49·76	52·16	50·54	62·37	55·76
TiO_2	–	0·58	0·95	0·94	3·25	2·80	1·06	1·78
Al_2O_3	16·20	11·24	20·10	14·42	11·95	12·86	12·04	16·55
Fe_2O_3	0·20	1·80	1·97	3·95	4·86	4·13	1·87	3·10
FeO	–	2·84	6·62	7·77	9·92	8·75	5·81	6·02
MnO	–	0·31	0·32	0·20	0·18	0·32	0·24	0·22
(Co,Ni)O	–	not found	not found	not found	–	0·06	not found	not found
MgO	0·27	0·61	5·49	5·30	3·77	4·63	0·97	1·08
CaO	0·13	1·56	13·16	10·22	7·14	8·74	3·51	3·23
BaO	–	0·07	0·10	0·04	–	not found	0·07	0·07
Na_2O	3·78	3·44	1·66	2·49	2·36	2·89	3·47	6·28
K_2O	4·56	4·66	0·98	1·83	1·74	1·43	2·34	3·87
Li_2O	–	? trace	not found	trace	–	not found	not found	trace
H_2O†	0·60 (H₂O† + H₂O††)	1·04	0·44	1·03	1·95	2·25	5·54	0·95
H_2O††		0·39	0·02	2·04	0·56	0·17	0·44	0·80
P_2O_5	–	0·22	0·04	0·21	0·24	0·34	0·30	0·40
CO_2	–	–	0·11	0·06	0·18	0·33	–	0·03
FeS_2	–	not found	not found	0·04	0·18*	not found	not found	not found
	100·22	100·06	100·30	100·30	100·44	100·24	100·03	100·14

† driven off above 10°5C; †† driven off below 105°C; * sulphur

Locations and sources of analysis are as follows:

1 Ross of Mull Granite (Haughton, 1867: 30).

2 Granophyre, Craignure Bay, shore 45 m NNW of U.F.C. Manse (Bailey *et al.*, 1924: 20).

3 Olivine gabbro from Beinn na Duatharach 1 km NNW of summit (Bailey *et al.*, 1924: 24).

4 Plateau basalt which embeds Maculloch's Tree, Rudha na h-Uamha (Bailey *et al.*, 1924: 17).

5 Quartz dolerite from cone-sheet 63 m S of summit of Cruachan Dearg (Bailey *et al.*, 1924: 17).

6 Central-type basalt. Monadh Beag, stream junction 900 m N of Ishriff (Bailey *et al.*, 1924: 17).

7 Augite andesite (Inninmorite-pitchstone) from sheet; near head of stream from Tòm a' Choilich 800 m SW of Pennyghael (Anderson & Radley, 1916: 212).

8 Mugearite of the Ben More horizon; below road 260 m E of Kinloch Hotel (Bailey *et al.*, 1924: 27).

is grouped according to chemical composition, and within each of the four groups shown, although there is some range of chemical composition, this is strictly limited. In Table 4.5 are presented chemical analyses illustrating the compositions of some of the rocks of the four groups defined in Table 4.4. It is important to note that although the chemical differences between groups 2 and 3 are very slight, they are considered of profound significance by igneous petrologists. The greatest differences are between the granitic rocks of group 1 of Table 4.4 and the gabbroic and basaltic rocks of groups 2 and 3.

The flora of Mull as seen from geological sediments

The stratigraphy and petrology of the Mesozoic and Tertiary sediments are discussed above but the two other aspects, namely the climate and vegetation cover at the time, can be deduced from the physical characteristics and fossil contents of these rocks, and may be of interest to a student of the modern flora.

Fig. 4.9 "Maculloch's Tree" in the cliff-face of The Wilderness. 1967.

The late Cretaceous and early Tertiary sandstones contain a large proportion of rounded, wind polished grains suggestive of desert conditions prevailing at that time. The silicification of the chalk seen around the Gribun supports this idea; concentrated solutions result where rainfall is dissipated by evaporation rather than drainage (Bailey *et al*., 1925). These desert conditions passed away by the time basaltic ash was deposited from a subaerial volcano; these ashes then underwent lateritic decay that produced the red mudstone seen today. These mudstones are in part buried beneath later lava flows, the tops of which in turn commonly weathered to a red soil before being covered by further flows. Throughout this time moist warm climatic conditions must have prevailed which may have become cooler with the onset of vulcanism (Simpson, 1936). In places, sufficient plants must have existed to leave behind thin seams of lignite although Srivastava (1975) infers from the microspore content that these lignites are allochthonous and that the immediate area around the depositional site probably only had sparse vegetation. Phillips (1974) gives further evidence of the richness of Jurassic species in reworked deposits at Ardtun as compared with those in situ at Shiaba. Associated with these lavas are the leaf-beds and gravels of Ardtun, now famous for the well-preserved remnants of what was obviously a species-rich deciduous forest vegetation growing in a moist but warmer climate than that in Britain today. The base of a large angiospermous tree which was surrounded in situ by molten lava and subsequently mineralised is now exposed on the cliffs of The Wilderness (see Fig. 4.9).

The actual age of the Ardtun beds which are considered contemporaneous with the Mull lignites (Simpson, 1961) is debatable. The microspore assemblages of the lignites compare with those of North American deposits which span 69.5 to 65.2 million years (Srivastava, 1975). Thus, contrary to the prevalent view, Srivastava suggests that the igneous activity began in Mull within the Upper Cretaceous either at the end of the Campanian or the beginning of the Maastrichtian Stage.

Ardtun leaf-beds

The palaeobotany of the Ardtun leaf-beds is discussed by Seward & Holttum (1924). The deposits were discovered by Murdoch M'Quarrie of Bunessan and communicated to the Duke of Argyll who, realising their scientific importance, reported on them in 1851. Later and more extensive collections were made and studied by Gardner & Ettinghausen (1882) and Gardner (1885, 1886, 1887); some of this material and an unpublished catalogue (Seward, MS) is now in the Department of Palaeontology at the British Museum (Natural History), at the Royal Scottish Museum, Edinburgh and at the Hunterian Museum, Glasgow. Further material, including that from a hitherto unrecorded locality just east of Carsaig, was collected at the time of the Geological Survey (1907–1920) and deposited in the collections of what is now the Institute of Geological Sciences, at South Kensington, London.

The leaf-beds were formed by leaves falling into shallow pools or backwaters of a river system. The quantity of the leaves suggests that these pools remained undisturbed for some lengthy period. Associated with the beds are remains of an *Equisetum* (*E. campbellii* Forbes) which compares most closely with *E. telmateia* of the present day flora. This species and the fern *Onoclea hebridica* (Forbes) Seward & Holttum could well have been dominant species around the swampy margins of these pools. The sole extant species today of *Onoclea* (*O. sensibilis* L.) is native in eastern North America, Japan and eastern Asia. It has been introduced to Europe by horticulturists and has become established in several places; it has not been found on Mull. In its native environment in N. America at least, e.g. in Michigan, U.S.A., it grows on clay soil in wet hollows and around lakes with its rhizome very near or on the surface of the substrate and in spite of this it can withstand quite severe winters which may be colder than the climatic regime suggested for Eocene Mull by Seward & Holttum (1924). Fossil *Onoclea* has not been found elsewhere in Europe (Seward, unpublished MS).

The Gymnosperms are represented in the leaf-beds by a number of remains including a *Ginkgo*, *Sequoia*, *Cephalotaxus*, *Podocarpus*, *Abies* and an araucarioid species, referred to by Gardner (1887) as *Cryptomeria sternbergii* (Goepp.) Gardner but placed under a non-commital name, *Pagiophyllum sternbergii* by Seward & Holttum, who point out however, that the leaf arrangement is more akin to present day *Araucaria*. This gymnospermous flora suggests the warm temperate floras recorded elsewhere in Europe for that period and now confined to E. Asia and the southern hemisphere. An epiphyllous fungus of the predominantly tropical family Microthyriaceae was found on a *Podocarpus* leaf by Edwards (1922) which suggests high rainfall and warmer climate. One dwarf shoot of a *Pinites* sp. is also recorded; it is a five-needled species and similar in its foliage to *Pinus strobus* L. another eastern and central North American species, now widely cultivated in Europe as a timber tree.

Of the broad-leaved trees Seward and Holttum mention material from four families: Betulaceae (wood only), Corylaceae, Fagaceae and Platanaceae. Seward (unpublished MS) in a later study describes material from Caprifoliaceae, Cercidiphyllaceae, Cornaceae, Nyssaceae? Ulmaceae, Vitaceae. *Corylites hebridica* they describe as a new species from this leaf material mentioning that many of the leaves are like *Carpinus betulus* L. and some close to *Corylus rostratus* Michx of eastern N. America. *Quercus groenlandica* Heer. and *Platanus hebridica* (Forbes) Seward & Holttum, both with fossil representatives in N. America, were other tree species commonly found fossilised in the Ardtun leaf-beds. The ~~end papers~~ title page in this volume show an artist's reconstruction of that landscape drawn under the direction of Dr C. R. Hill. The vegetation appears to be of a kind widespread in the northern hemisphere at that time and, if anything, related to the present day species as seen in eastern U.S.A. and Canada rather than to the extant flora of Europe.

References

Anderson, E. M. & Radley, E. G. (1916). The pitchstones of Mull and their genesis. *Q. Jl geol. Soc. Lond.*, **71**: 205–217.

Bailey, E. B., Anderson, E. M. & contributors. (1925). The geology of Staffa, Iona & western Mull. *Mem. geol. Surv. G. B.* Sheet 43.

———, Clough, C. T., Wright, W. B., Richey, J. E., Wilson, G. V. & contributors. (1924). Tertiary and post-Tertiary geology of Mull, Loch Aline, and Oban. *Mem. geol. Surv. G.B.* Sheets 43, 44, 51, 52.

Argyll, The Duke of. (1851). On Tertiary leaf-beds on the Isle of Mull. *Q. Jl geol. Soc. Lond*, **7**: 89–103.

Brown, P. E., Miller, J. A. & Grasty, R. L. (1968). Isotopic ages of late Caledonian granitic intrusions in the British Isles. *Proc. Yorks. geol. Soc.*, **36**: 251–276.

Cunningham Craig, E. H., Wright, W. B. & Bailey, E. B. (1911). The geology of Colonsay and Oronsay, with part of the Ross of Mull. *Mem. geol. Surv. G.B.* 35, 27.

Edwards, W. N. (1922). An Eocene microthyriaceous fungus from Mull, Scotland. *Trans. Brit. mycol. Soc.*, **8**: 66–72.

Gardner, J. S. (1885). Eocene ferns from the basalts of Ireland and Scotland. *J. Linn. Soc. Bot.*, **21**: 653-664.

———. (1886). *Monograph of the British Eocene flora*, vol. 2, 159 pp. London.

———. (1887). On the leaf-beds and gravels of Ardtun, Carsaig, etc. in Mull. *Q. Jl geol. Soc. Lond.*, **43**: 270–300.

——— & Ettinghausen, C. (1882). *Monograph of the British Eocene flora*, vol. 1, 88 pp. London.

Haughton, S. (1867). Geological notes on some of the islands of the west of Scotland. *Jl R. geol. Soc. Ireland*, n.s. **1**: 28–31.

LEE, G. W., BAILEY, E. B. & contributors. (1925). The pre-Tertiary geology of Mull, Loch Aline, and Oban. *Mem. geol. Surv. G.B.* 35, 43, 44, 45, 62.

PHILLIPS, L. (1974). Reworked Mezozoic spores in Tertiary leaf-beds on Mull, Scotland. *Rev. Palaeobot. Palyn.*, **11**: 221–232.

RICHEY, J. E. (1961). *British Regional Geology: Scotland, the Tertiary volcanic districts*, 3rd ed. Edinburgh.

———, THOMAS, H. H. & contributors. (1930). The geology of Ardnamurchan, north-west Mull, and Coll. *Mem. geol. Surv. G.B.* Sheets 51, 52.

SEWARD, A. C. Unpublished MS *Catalogue of the fossil plants from the Ardtun leaf-beds*. Department of Palaeontology, British Museum (Natural History).

——— & HOLTTUM, R. E. (1924). Tertiary plants from Mull. Chapter 4 in E. B. BAILEY *et al.*, 1924, pp. 67–91.

SIMPSON, J. B. (1936). Fossil pollen in Scottish Tertiary coals. *Proc. R. Soc. Edinb.*, **56**: 90–108.

———. (1961). The Tertiary pollen-flora of Mull and Ardnamurchan. *Trans. R. Soc. Edinb.*, **64**: 421–468.

SRIVASTAVA, S. K. (1975). Maastrichtian microspore assemblages from the interbasaltic lignites of Mull, Scotland. *Palaeontographica*, Abt. B, **150**: 125–156.

5. Geomorphology and soils

J. S. Bibby*

Introduction

The Island of Mull owes its shape and relief to the *type* and *structure* of the rocks of which it is composed and to the way in which they have reacted to the influence of changing *climate* through *time*. Although rocks of Precambrian and Mesozoic ages are found in the island, the bulk of the rocks were formed during a period of vulcanicity in the early Tertiary and the evolution of the island must date from this time. Considerable speculation exists among geomorphologists over the sequence of events during the Tertiary period after the eruptions had ceased. There appears to be little doubt, however, that the major valley system of the island was established by subaerial erosion along lines of structural weakness at this time. Climatic conditions were deteriorating from warm and fairly moist to cold, culminating in the onset of the Pleistocene glaciation. Bailey *et al.* (1924) considered that Mull was an island long before glacial times and quoted the abrupt rise of the island from the continental shelf as evidence. He considered that late-glacial and post-glacial marine erosion had had little to do with shaping the coastline.

Whether Mull itself was separated from the mainland before the glacial overdeepening of the Sound of Mull is a moot point. There is evidence from several localities on the west coast of Mull for a pre-glacial wave-cut platform between 30 and 48 m however, and Synge & Stephens (1966) and Synge (1967) suggest that some raised beaches below 30 m may be the remains of pre-glacial beach levels. The available evidence suggests that there was considerable marine erosion prior to the major glaciation of the West Highlands and it would appear that the isles to the west of Mull (Iona, Staffa, Treshnish group) would almost certainly have become detached from Mull, even though Mull was still attached to the mainland, sometime during the Pliocene period. The area to the west of Ardmeanach peninsula was undoubtedly overdeepened as a result of the confluence of ice (Fig. 5.1) from the Mull 'sanctuary' (Bailey *et al.*, 1924) and the mainland and at the end of the major glacial period both Mull and the smaller western islands existed as separate entities.

During the major glaciation, Mull was completely inundated by ice. The higher parts of the island supported their own ice cap which resisted and diverted the flow of mainland ice (judged by the distribution of erratics of mainland origin) to north and south (Fig. 5.1). Glaciation had the effect of destroying the weathered mantle which would presumably have developed prior to glaciation, emphasising the differential erosion of the hard and soft rocks and the lines of weakness along which erosion had already commenced, and depositing a thin drift in the few parts of the landscape that were protected from the full force of the ice stream. In the vicinity of Bunessan a

* Macaulay Institute for Soil Research, Craigiebuckler, Aberdeen.

series of elongated drumlins were formed at this time, indicating that ice which had skirted the south flank of the Mull sanctuary had taken up a north-westerly direction of flow. The retreat of the major ice field was followed by a resurgence of valley glaciation confined to the central higher hills (Fig. 5.2), whilst surrounding areas were subject to a periglacial climate. Snowfield nivation, frost shatter of exposed rocks,

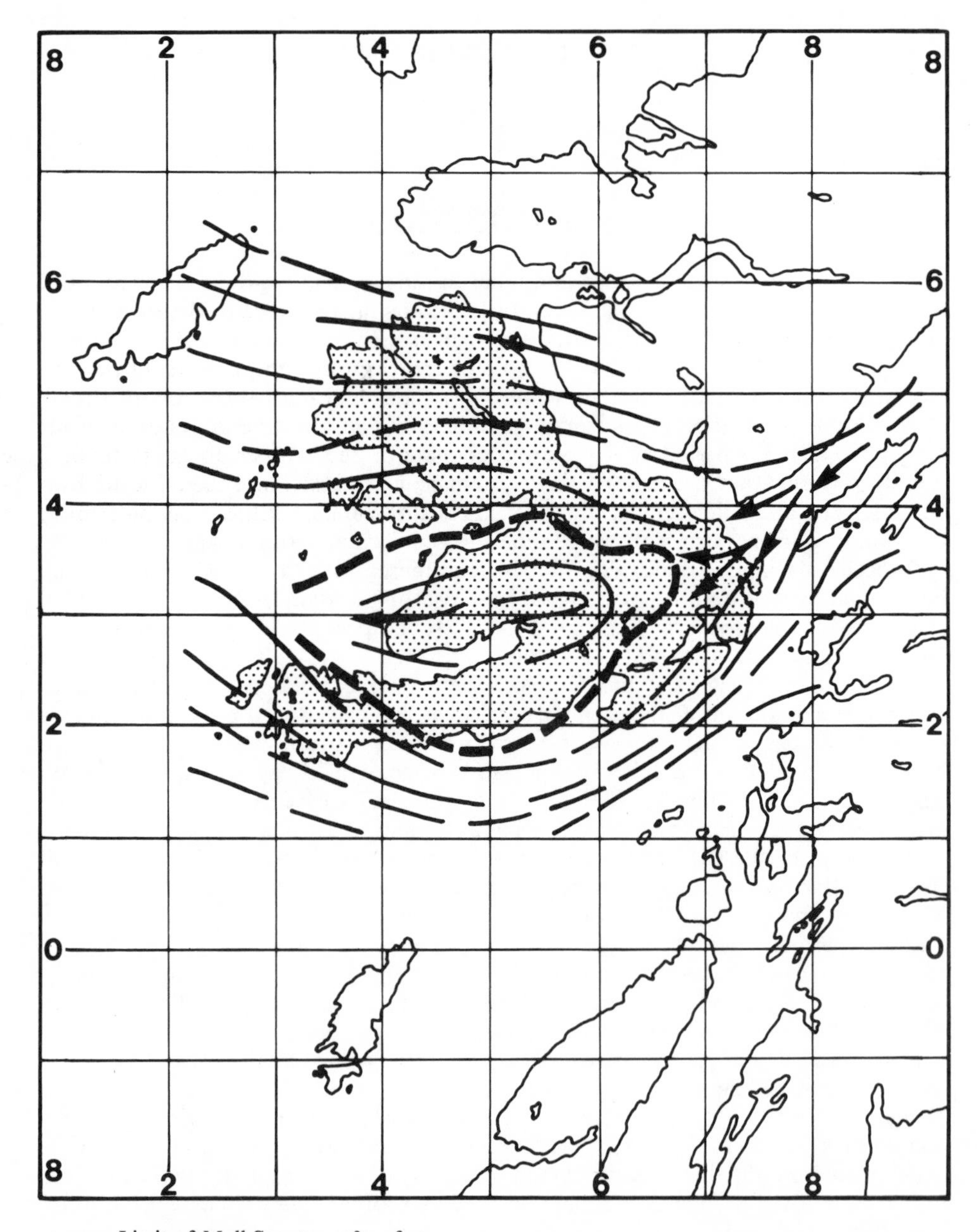

Fig. 5.1 Main glaciation of Mull. After Bailey (1924).

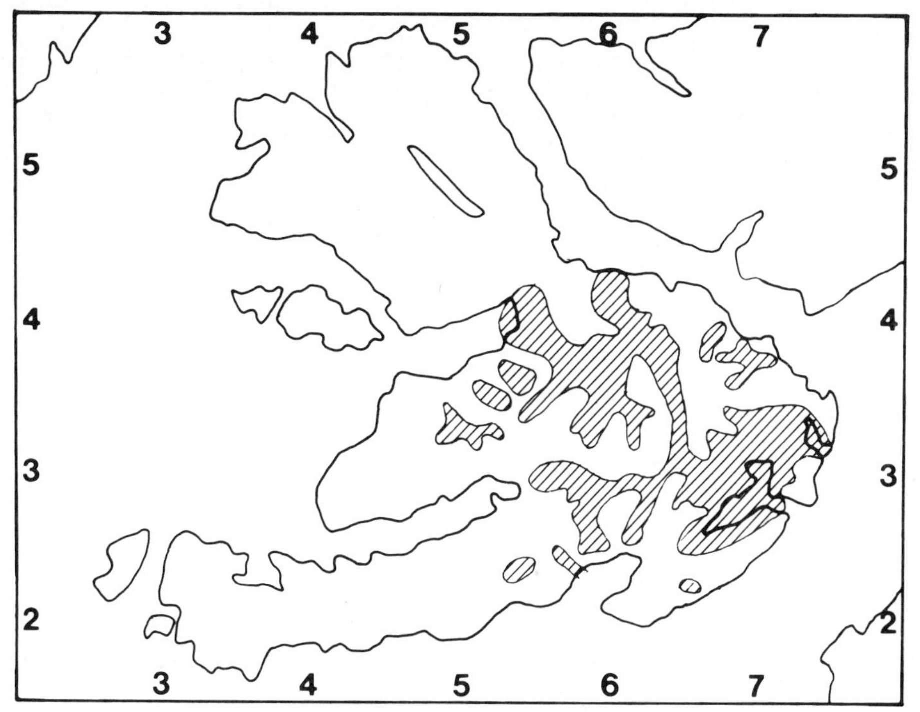

Fig. 5.2 Final glaciation of Mull. Areas cross-hatched show extent of ice; all other areas are affected by intense frost action and snowfields.

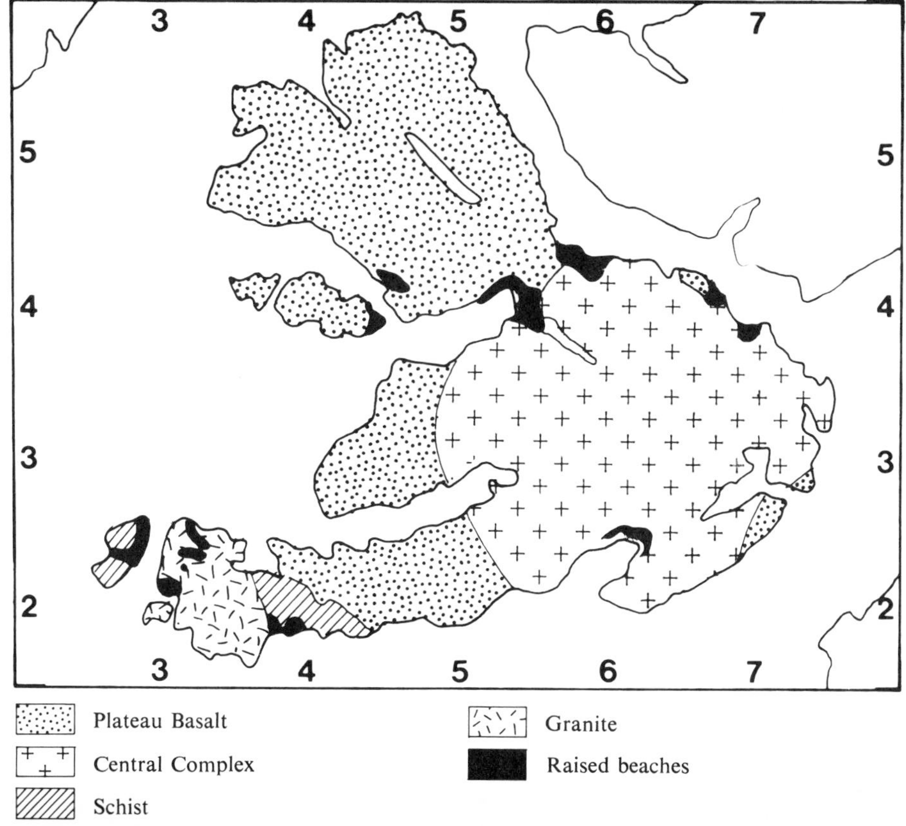

Fig. 5.3 Distribution of Land systems of Mull.

Table 5.1 Summary of the characteristics of the land systems of Mull and their relationships to soil associations and complexes as defined during the systematic soil survey of the island.

LAND SYSTEM	PHYSIOGRAPHY		DRIFT TYPE
PLATEAU BASALT Facet 1	Below 180 m; strong step topography divided into:	1. Riser sites	1. Loamy colluvial and morainic drift, occasionally indurated
		2. Tread sites	2. None, or shallow stony drift
Facet 2	Step topography, risers dominant; altitudinal range 180 m–360 m	1. Riser sites	1. Loamy colluvial and morainic drift, occasionally indurated
		2. Tread sites	2. None, or shallow stony drift
Facet 3	Step topography, treads dominant; altitudinal range 180 m–360 m	1. Tread sites	1. Organic deposits 1 m thick on rock or shallow stony drift
		2. Riser sites	2. Loamy, stony colluvium and moraine
Facet 4	Mountain summits and exposed ridges higher than 430 m		Loamy colluvium often capped by stone layers
CENTRAL COMPLEX Facet 1	Narrow flat often stony areas along major streams; various altitudes		Alluvial gravels, sands and sandy loams often with organic bands
Facet 2	Bottoms and lower slopes of glaciated valleys; various altitudes to 350 m		Fine sandy and sandy loam valley moraine, characterised by olive indurated horizons
Facet 3	Steeply sloping stabilised and active scree; altitude 0–360 m		Open angular scree; loamy or organic loam surface where stabilised
Facet 4	Mountain summits and exposed ridges higher than 430 m		Stony loamy colluvial and frost shattered rock debris
Facet 5	Weakly expressed step topography 0–360 m		Shallow stony loamy colluvium
ROSS OF MULL GRANITE Facet 1	Rocky hills with peat hollows; altitude 0–120 m		Clay loam till lodged in crevices; bouldery debris below outcrops
Facet 2	Dominantly peat flats with frequent rock outcrop 0–60 m; occasional drumlins		1. Frequently underlain by till or beach debris at depth
			2. Clay loam till drumlins
SCHIST Facet 1	Peat flats with drumlins and rock outcrops		1. Clay loam till drumlins
			2. Micaceous colluvium
Facet 2	Rocky hills with peat hollows		Micaceous colluvium on knolls
RAISED BEACH Facet 1	Level to gently sloping; below 30 m		Dominantly gravelly, occasionally sandy; upper levels have indurated horizons
Facet 2	Level to gently sloping; strongly undulating microrelief; below 10 m		Dominantly rocky with gravel pockets

SOIL GROUPS dominant	subsidiary	VEGETATION & LAND USE	SOIL ASSOCIATIONS & COMPLEXES
1. Brown forest soils; brown lithosols	Flush gleys	*Pteridium, Festuca, Agrostis, Calluna*; cultivation, improved grazing, forestry	DARLEITH Knockan
2. Organic soils; peaty gleys	Organic flush gleys; rock	*Molinia, Scirpus cespitosus, Erica tetralix*; rough grazing, forestry	Cruachan
1. Peaty podzols; iron podzols	Organic flush gleys; rock	*Calluna, E. cinerea*; rough grazing, forestry	Mishnish
2. Organic soils; peaty gleys	Organic flush gleys; rock	*Molinia, S. cespitosus, E. tetralix*; rough grazing, forestry	Cruachan
1. Organic soils; peaty gleys	Organic flush gleys; rock	*Molinia, S. cespitosus, E. tetralix*; rough grazing, forestry	Cruachan
2. Peaty podzols; iron podzols	Organic flush gleys; rock	*Calluna, Molinia, S. cespitosus*; rough grazing, forestry	Mishnish
Oroarctic soils	Rock	Montane heath; active frost heave and erosion; rough grazing (restricted)	Drise
Immature alluvial soils	Organic soils	Vegetation variable according to drainage status from *Juncus* to *Festuca-Agrostis* grassland; occasionally cultivated	ALLUVIUM
Peaty gleys with induration; organic soils	Peaty podzol Organic lithosol; rock	*Molinia* and *Agrostis-Festuca* grassland; rough grazing, occasionally improved grazing, forestry	KNOCKANTIVORE Gorten Knockantivore Talaidh
Brown forest soils; peaty podzols	Brown lithosols; peaty gleys	*Agrostis-Festuca* and *Molinia* grassland; rough grazing, occasional forestry	TOROSAY Cameron Odhar
Oroarctic soils; lithosols	Organic soils	Montane heath; rough grazing	Torosay Odhar
Peaty gleys; peat	Peaty podzols; alluvium; rock	*Molinia-S.cespitosus-E.tetralix* mire; rough grazing, forestry	Scarisdale
Organic lithosols	Rock	*Calluna* on dry rocky knolls, *Molinia* in hollows; rough grazing only	COUNTESSWELLS Erraid
Peat	Rock	1. *Molinia, S. cespitosus, Calluna* on knolls; rough grazing	Ghlinne Mhoir
Non-calcareous gleys	Peaty gleys	2. *Juncus, Agrostis, Festuca*; largely cultivated	TARVES Various soil series
Non-calcareous gleys	Peaty gleys	1. *Juncus, Agrostis, Festuca*; largely cultivated	
Peat	Rock	2. *Molinia, S. cespitosus*; rough grazing, forestry	STRICHEN Assapol
Rock	Peat; brown forest soil; podzols	*Molinia, S. cespitosus, Agrostis, Festuca, Calluna*; rough grazing	Gharbh
Brown forest soils; podzols	Peaty gleys	*Agrostis, Festuca*; dominantly cultivated	CORBY BOYNDIE FRASERBURGH GRULINE
Organic soils; peaty gleys	Rock	*Molinia, S. cespitosus*; rough grazing, forestry	Kilpatrick

landslips, mudflows and the sorting of the deposits by frost action were dominant throughout most of north and western Mull. Around the mouths of the glaciers outwash fans were formed of material brought down as moraine, and parts of these outwash fans were again resorted by changes in sea-level and the formation of raised beaches. The period was one in which the island finally assumed the form which we see today and during which the parent materials of its present-day soils were largely formed.

Land systems of Mull

The rocks of Mull reacted to the effects of glaciation in many ways according to their individual mineralogy and structure. Four major landform regions can be recognised, however, which have distinct, easily recognisable patterns (Fig. 5.3). To these a fifth may be added which is less compact in occurrence but no less easily recognised. Such landform regions have been designated 'land systems' (Christian 1957). The basic concept of a land system is that it is an easily recognisable unit of landscape which consists of a few component parts linked together in a recurring and consistent relationship with one another. The component parts have been called land facets. A general description of the physiography and geomorphic evolution of each of these land systems follows, with a comment on the major soil groups associated with it. This is summarised in Table 5.1.

Plateau Basalt land system

The distribution of this land system is indicated in Fig. 5.3; it occupies 45 percent of the island. The basalts are a sequence of flow lavas, often individually thin but cumulatively of great thickness. As each flow erupted it developed, due to gaseous escape by release of pressure, a slaggy top which, upon cooling, formed a vesicular basalt with porous, friable texture. The interior and base of the flow cooled and solidified relatively slowly to form a more massive less porous basalt. This initial differentiation in texture is of great importance in the subsequent weathering of the rock, because the vesicular basalts weather readily whilst the more massive types are resistant. The type of landscape to which this gives rise is often referred to as 'trap featuring' and is indicated diagrammatically in Fig. 5.4. The trap feature is the basic unit of the Plateau Basalt land system but the soils and vegetation found within it vary according to the climatic influences at work. This is most noticeable with increasing altitude and the sequence is represented in Fig. 5.5. Perhaps the land facets referred to in the diagram should be more strictly termed patterned facets, because none is uniform enough to meet the usual definition of the land facet. It should also be noted that the various facets are not always as strictly separate as they are shown in the diagram. Outliers of Facet 3 may occur within Facet 1, and of Facet 1 within Facet 2. Facet 2 may be absent altogether in places so that Facets 1 and 3 abut.

Soils of Facet 1 These are dominantly brown forest soils of low base status and loam or sandy loam texture. They are freely drained and may sometimes have compacted lower horizons. Other soils found are the shallower brown lithosol*, brown forest soils with imperfect drainage and the poorly drained gley, both the latter occurring in flush sites. Rock crops out frequently and peat and peaty gleys may occur where hard rock is close to the surface (see Fig. 5.4). The brown forest soils support an *Agrostis–Festuca* grassland with *Pteridium*; the shallow lithosols are frequently dominated by *Calluna* and *Erica cinerea*, whilst the peaty gleys and peats have *Molinia*, *Scirpus cespitosus* and *Eriophorum* spp. dominant in their vegetation.

* In revised Soil Survey terminology *lithosol* is replaced by *ranker*.

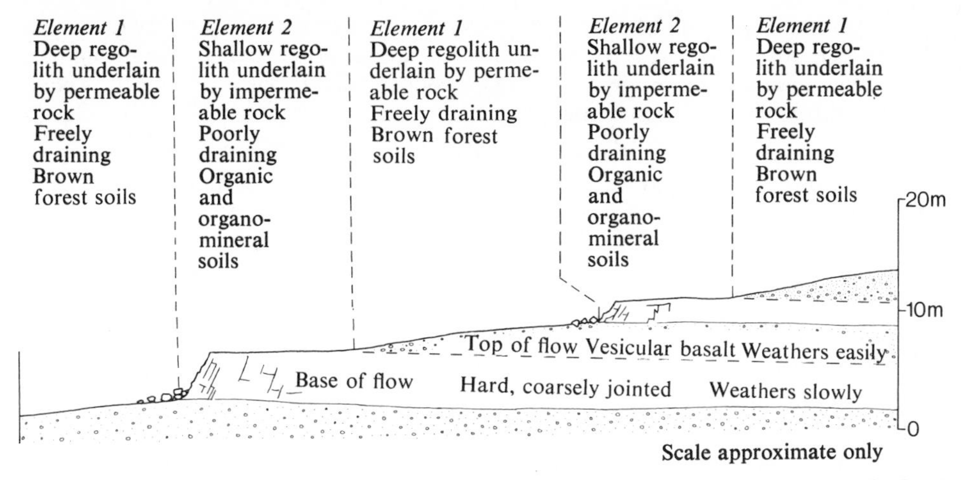

Fig. 5.4 Diagrammatic representation of two elements of the Plateau Basalt land system which indicate the differential weathering of lava flows.

Soils of Facet 2 Iron podzols, intergrading to brown forest soils as their surface horizon becomes less organic, and peaty podzols with thin iron pan are the major soil subgroups most frequently encountered. Despite the usually steep slopes, organic topsoils are common because of the higher rainfall and cooler conditions obtaining at altitudes higher than 220 m. There is reason to believe that but for the high basicity of the parent material podzolic soil types would be encountered at much lower altitudes than this. The mineral horizons are of loam or sandy loam texture and drainage is usually free below the surface humus layer of the iron podzol and below the thin iron pan of the peaty podzol. Other soils occurring are peaty gleys, usually in flushes or above hard rocks as in Facet 1 and thin layers of humus directly overlying rock. Rock outcrops are common.

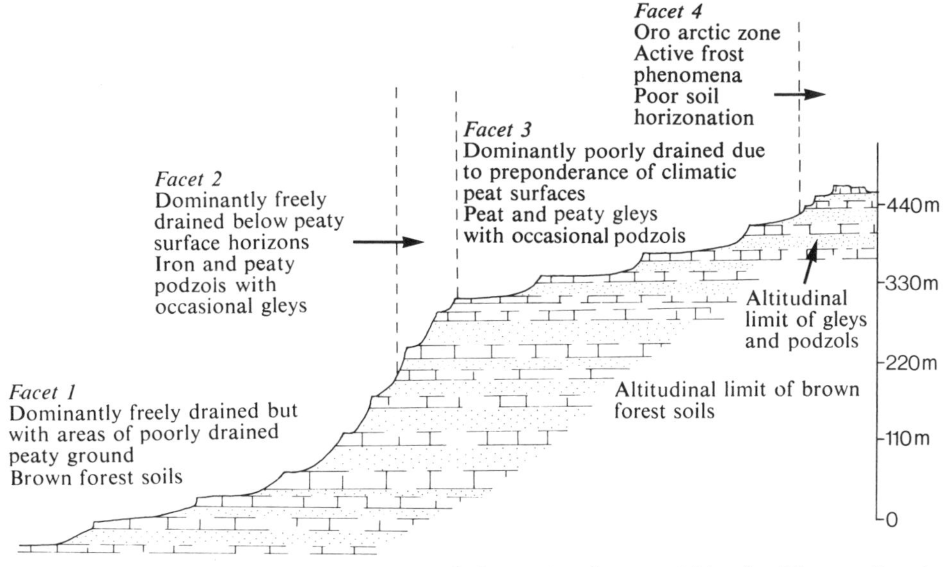

Fig. 5.5 Diagrammatic representation of the major facets within the Plateau Basalt land system.

The iron podzols are characterised by dry heath with *Calluna* and *Erica cinerea*, the peaty podzols support a slightly more moist vegetation with *Calluna* dominant and *Molinia* present. On some rocky ledges where shallow peat is developed *Molinia* and *Scirpus cespitosus* become dominant.

Soils of Facet 3 This facet has less regional slope than Facet 2 and consequently although podzols are found within it they are confined to the steeper scarp slopes and the dominant soils are peaty gleys and peats. A certain amount of flushing by spring water may locally enrich the surface but conditions are usually very acid and wet. Rock outcrop is frequent. Vegetation is *Molinia*, *Scirpus cespitosus* and *Calluna* mire associations.

Soils of Facet 4 Facet 4 occupies mountain summit sites above approximately 440 m although in exposed conditions it may be found at lower altitudes. The soil-types are very immature podzols and peaty gleys, with many lithosolic soils. The zone as a whole is characterised by the active frost heaving of the soils during the winter months, forming stone stripes and patterned ground and rupturing turf by creep and hydrostatic pressure. Loam and sandy loam textures are frequent, as are very coarse platy subsoil structures. Occasionally, however, the calcium-bearing secondary minerals in the vesicles enhance the calcium content of the basalt and slightly richer patches of soil occur. It is probably these that provide conditions suitable for *Koenigia islandica* on part of Ardmeanach peninsula. The major vegetation type of this facet is montane heath, with *Rhacomitrium lanuginosum*, *Dicranum* spp., etc.

Central Complex land system

The distribution of this land system, which occupies 35 per cent of the island, is shown in Fig. 5.3. The rock-types included within the central complex of Mull vary widely from acid granophyres through intermediate and basic cone sheets to basic gabbros (see Chapter 4). The main characteristic, however, is that all the rocks are very hard. Even the gabbros have extremely coarse structure and have not weathered extensively to provide the fertile residuum which characterises their occurrence in some other parts of Scotland. The resistance to erosion has ensured that central Mull remains a mountainous area. Colluvial creep of weathering products down the steep slopes results in mixing of material derived from both acid and basic sources although one or the other type may assume slight dominance locally.

The central mountain area with its higher precipitation provided a gathering ground for snow and supported a valley glaciation for a time at the end of the ice age. The ice has left classic evidence of its presence in the shape of morainic deposits in typical U-shaped valleys. Terminal and lateral moraines (Loch Don, Glen Forsa) medial moraines (upper Glen Forsa), push moraines (Kinlochspelve) and considerable areas of outwash fans and raised beaches (Gruline and Glen Forsa) provide detailed evidence of ice conditions in the area. The interfluves between glaciers suffered from intense frost action during the glaciation; crest sites suffered denudation whilst scree deposits formed against the edges of the ice and on lower slopes where these were not glaciated. Seasonal torrents from melting snow and ice built up alluvial cones on footslopes and in hill country. It is hardly surprising therefore that with such varied parent rocks and conditions of accumulation it is difficult to generalise about the regolith upon which the soils have been developed, for even in a small area several processes have usually been at work simultaneously. Some notes on the various facets (see Fig. 5.6) may however serve as a guide.

Soils of Facet 1 Areas of gravelly or sandy alluvium are frequently found along the major streams. In places where rather wider terraces have been cut the alluvium may now be overlain by peat more than 50 cm deep.

Soils of Facet 2 A single characteristic unites the soils of the valley moraines – the presence of an indurated horizon usually within 75 cm of the surface. The dominant soil is the peaty gley, the induration being responsible for impeding water percolation leading to anaerobic conditions and the development of a peaty surface horizon. Iron pans are frequently developed at the top of the induration further sealing the subsoil. Of the other soils peat less than 50 cm thick is frequently found directly overlying induration, peaty podzols are found often with imperfect drainage in the B horizons, due to induration impeding water percolation, and in the hollows between hummocks peat more than 1 m deep is almost universal. Textures in the mineral horizons are loams and fine sandy loams; the deposits are poorly sorted – angular stones and striated boulders occurring in loamy matrices. Structures are platy or massive in the indurated layers and platy or weak subangular blocky in the non-indurated layers. *Molinia* grassland is the dominant vegetation.

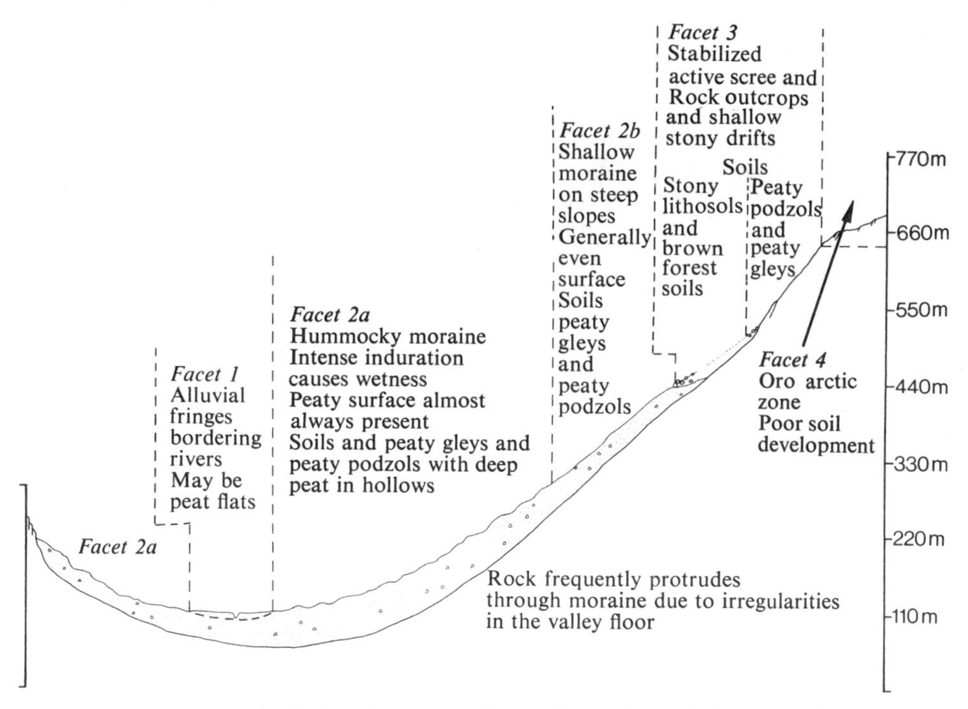

Fig. 5.6 Facets typical of the glaciated valleys of the Central Complex land system. The altitudinal limits shown in this diagram vary very widely.

Soils of Facet 3 Shallow soils over rock and excessively stony scree deposits are characteristic of this facet. The ground is almost always steeply sloping, but despite this, where rock intervenes at a short distance from the surface the soil surface is usually peaty and poorly drained. Where deeper coarse subsoils are present, as on the scree, peat may have failed to develop despite the extremely high rainfall and stony brown forest soils are found. These carry *Agrostis – Festuca* grassland which in spring and autumn contrasts sharply with the *Molinia* and *Nardus* grasslands of the peaty soils which frequently lie both above and below them. Rock outcrops and surface boulders are characteristic features and flushing is common.

Soil textures are gravelly loams and angular gravels, the humus content of the surface horizon is frequently high even when it is not peaty and soil structures are moderately to strongly developed subangular blocky. The stronger structures are due to grassland vegetation remaining undisturbed for a considerable time.

Soils of Facet 4 A stone pavement over thin loamy humose drift, becoming more stony again with depth is the typical sequence in these upland soils. Active frost heave occurs during the winter months. In hollows, where drainage is impeded by rock a fine black raw humus accumulates. Wind erosion features and stunting of the vegetation due to exposure are common.

Soils of Facet 5 This facet is not shown on Fig. 5.6. It occupies large tracts of ground on the periphery of the glaciated areas and ranges in altitude from sea-level to approximately 440 m. Its dominant parent rocks are flow basalts which are included in the Plateau Basalt land system. Late stage metasomatic influences caused a secondary hardening of the upper scoriaceous layers of the flows. In consequence these scoriaceous layers no longer weather readily, the soft brown loams characteristic of the Plateau Basalt land system are not produced and only thin drifts are found overlying hard impermeable rock. Peaty gley soils and peat directly overlying rock are found everywhere as a consequence of this impedance of drainage under the high rainfall. Occasional podzols are found and, rarely, brown forest soils at low levels where scree has accumulated to a sufficient depth to allow better subsoil drainage. Large areas of deep peat, frequently hagged in exposed positions, are also found. Small alluvial fans abound although the streams that formed them have frequently long disappeared – further evidence of the considerable amounts of melt water that flowed over these slopes at the end of the Pleistocene period.

The vegetation pattern varies according to depth of solum available for drainage, degree of run-off and flushing, but *Molinia* grassland is widely developed with *Calluna* becoming more noticeable on scarp slopes where drainage status is better.

Ross of Mull Granite land system

The soils overlying the outcrop of the Ross of Mull Granite form a very simple land system. Its occurrence is shown in Fig. 5.3. and it covers approximately 5 per cent of the surface of the island.

The parent rock is a hard, coarsely crystalline granite whose outcrop is dominated by widely spaced joint planes. This massive rock has produced little weathered material apart from a few quartz grains in the organic matter which covers the surface. However, lodged in crevices in the rock and found as drumlins is a till dominantly composed of debris carried by ice from the basalt and schist outcrops to the east. The surface of the granite is a gently undulating peneplain, the hollows in which are filled with peat, and the crests and slopes of the undulations are largely drift free and covered with a thin skin of peat. This gives rise to the two facets of the land system (see Fig. 5.7).

Soils of Facet 1 The rocky knolls which dominate this facet have shallow organic soils supporting a dense *Calluna – Erica cinerea* sward; in the wetter areas and channels *Molinia* is common. Lodgment till, usually severely restricted in extent, may sometimes be found in crevices or as a small low drumlin where a granite mass has proved an obstacle to ice flow. Such till areas are marked by *Agrostis – Festuca* grassland, *Calluna* moor and on occasion dense *Juncus effusus*.

Soils of Facet 2 Apart from the drumlins and lodgment till which occur more frequently than in Facet 1, the only soils found among the rocks are deep peats and thin peats usually less than 20 cm deep directly overlying rock. These organic soils are frequently drier than usual because of their coarse prismatic structure which allows water to escape to the rock interface and thence into the lower ground.

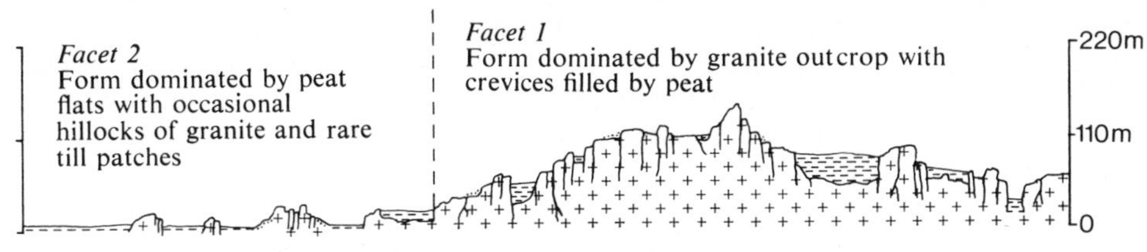

Fig. 5.7 The facets of the Ross of Mull Granite land system.

The till occupies slightly more than 1 per cent of the island's surface. In texture it is clay loam; below the topsoil, induration about 40 cm thick is frequently found overlying softer till with strong ochreous mottling. The soils are poorly and imperfectly drained gleys with a small amount of peaty gley.

Schist land system

The area within which the Schist land system will be found is shown in Fig. 5.3. It occupies only 3 per cent of the island. In essence it is very similar to the Ross of Mull Granite land system but differs in several important points (see Fig. 5.8). The pattern, although governed by rock structures, presents a sinuous ridged appearance very different from the blocky appearance of the granite. The proportion of detritus weathering directly from rock is much higher and results in hollows being filled, with a consequent smoother appearance and less rock outcrop. This is significant with regard to improvement potential. Many of the hollows have lodgment till derived from a mixture of schist and basalt and on the lower ground this may also be found as drumlins (e.g., at Scoor and Saorphin).

Soils of Facet 1 Facet 1 is dominated by peat flats with knolls of schist. The flanks of the knolls frequently have brown forest soils or imperfectly drained gleyed soils developed on them; the tops are either thin mineral soils on rock (brown lithosols) or thin peat on rock. The soils developed on the drumlins are similar in every way to those of the drumlins on the granite land system except that they do not have granite boulders incorporated.

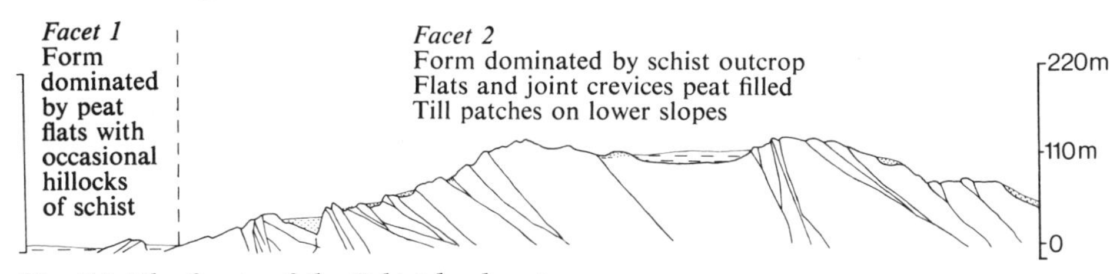

Fig. 5.8 The facets of the Schist land system.

Soils of Facet 2 The soils, although consisting principally of brown lithosols and thin peat on rock with some channels and basins of deep peat, have representatives of brown forest, peaty podzol and peaty gley soil subgroups. Mica occurs in all the mineral horizons but is more prevalent in those derived directly from the schists than in the tills.

Raised beach land system

The raised beaches of Mull are an obvious though minor part of the landscape, occupying about 4 per cent of its area. The more important occurrences are indicated on Fig. 5.3. Raised beaches are found on the east coast from just north of Salen right round the south coast and the Ross of Mull into Loch Scridain. There are few traces of beach levels in the upper part of Loch Scridain but they occur again at Burgh on Ardmeanach, on both shores of Loch na Keal from Gribun to Lagganulva and on the north side of Loch Tuath as far as Torloisk. North of this, beach deposits are found only in sheltered sites.

The content of the beach deposits varies widely; granite and schist with Mesozoic sandstones and occasional fossiliferous clay shales occur on the Ross of Mull, olivine-basalts predominate near Lagganulva and, where the beaches are formed from reworked moraine and outwash – as at Gruline, Glen Forsa and Scallacastle – granophyres, pneumatolysed basalts and dyke and cone-sheet materials predominate. Textures are gravelly except at Calgary, on Iona and on parts of the Ross of Mull where sands, sometimes shelly, are found.

On the east and south-east coasts from Lochbuie to north of Salen the raised beach is marked by a rocky platform at approximately 8 m. The platform has little beach sediment on it and is predominantly peat covered. Elsewhere on the island, however, two major beach levels separated by a low steep bank occur. The beaches appear to have a gradient to the north-west because the upper level is higher at the head of Loch na Keal and on the Ross of Mull than it is in the Lagganulva area. The soils of the upper beach are more mature than those of the lower and are frequently characterised by the presence of a compacted horizon. The soils developed upon the lower beach seldom show this feature.

The deposits of the beach, being dominantly coarse-textured, generally give rise to freely drained soils. Both gleys and peaty gleys are occasionally found in hollows on the upper beach level where the presence of induration prevents adequate drainage, and on the lower beach in sites where a ground water-table approaches the surface. In the vicinity of Fhionnphort poorly drained sands and gravels can be found where the cause of the wetness is an underlying clay loam till. Extensive peat deposits have formed upon raised beach gravels in this area probably due to the presence of the till at a depth of approximately 2 to 3 m.

Major soil groups of Mull

The system of soil classification referred to here is that currently used by the Soil Survey of Scotland. Soil profiles with similar types and assemblages of horizons are assigned to similar major soil groups and subgroups. A brief summary of their characteristics is given below and in Fig. 5.9.

Brown calcareous soils and brown earths

The **brown calcareous soils** of Mull are developed on shell sands. The content of free carbonate throughout the profile exceeds 5 per cent. A thin B horizon of brighter colour sometimes occurs beneath the dark-brown A horizon.

The **brown forest soils** have a moderately acid reaction and a mull or moder humus type. They are characterised by free drainage and uniformly brown colours. Those with **gleying** are found in flush sites or hollows, and B horizons exhibit weak mottling.

Gley soils

Gleys are mineral or peaty (O horizon less than 50 cm) soils which have developed under conditions of permanent or intermittent waterlogging. The **non-calcareous gleys** are infrequent on Mull, have no free calcium in the upper horizons and have an organic layer less than 3 cm thick. They are greyish brown in colour with greenish and bluish weathered stones and reddish yellow mottles in the profile. The **peaty gleys** are abundant but the strong mottle colours usual in such soils are muted by staining due to percolation of humus-laden waters. They are usually acid in reaction, shallow and stony.

Podzols

Podzols have a greyish brown bleached A_2 horizon, an O horizon of raw humus and a strongly acid reaction. The **iron podzol** on Mull is not typical of the subgroup of iron podzols throughout Scotland due to the influence of the basic parent materials of the soil. The A_2 horizon is usually heavily humus-stained and appears grey only upon drying. B_2 horizons are dark-brown rather than the browns or strong browns characteristic of the subgroup as a whole. The **peaty podzol** is characterised by the presence of a thin iron pan between the A and B horizons and impedance of drainage in the upper horizons. The A_2 may not be well defined but the B_2 is usually brighter than in the iron podzol. Under some circumstances it may be gleyed by water percolating through the thin iron pan or moving laterally downslope beneath it.

Lithosols

The **organic lithosol** consists of peat, sometimes above a thin A_1 horizon, on rock. It is strongly acid in reaction. The **brown lithosol** is similar to the upper horizons of a brown forest soil but with no B horizon development.

Organic soils

Organic soils contain more than 20 per cent organic matter and exceed 50 cm in depth. Most organic soils of Mull qualify as peat (>60 per cent organic matter) and result from conditions of waterlogging where, by the exclusion of oxygen, normal processes of decomposition have been restricted.

Soils of the Oroarctic zone

The soils of the Oroarctic zone (Birse 1971) show a wide range of profile characteristics from those verging on Arctic soils to intergrades with both the gley and podzolic groups. They are characterised by occurring on mountain summits and by being unstable, with only weak horizon development due to pedologic processes, though a layering, superficially resembling horizonation, may develop due to the action of frost sorting.

The soils of Mull have recently been the subject of a study by the Department of Soil Survey of the Macaulay Institute for Soil Research, as part of the systematic mapping of the soils of Scotland. A map at a scale of 1:63 360 has been published (Macaulay Institute, 1974) and a report is in course of preparation. Several of the land system units used as a basis for description in this chapter are further subdivided in terms of parent material to give soil associations. Apart from a few cases the soils of Mull occur in too intricate a pattern to resolve using the usual mapping unit – the soil series. It has been necessary to have recourse to the soil complex which is a group of defined units occurring in a defined pattern. For the purposes of the soil map the complexes have been based partly on the dominant major soil group and partly on physiography. An attempt has been made, by collecting information upon a basis

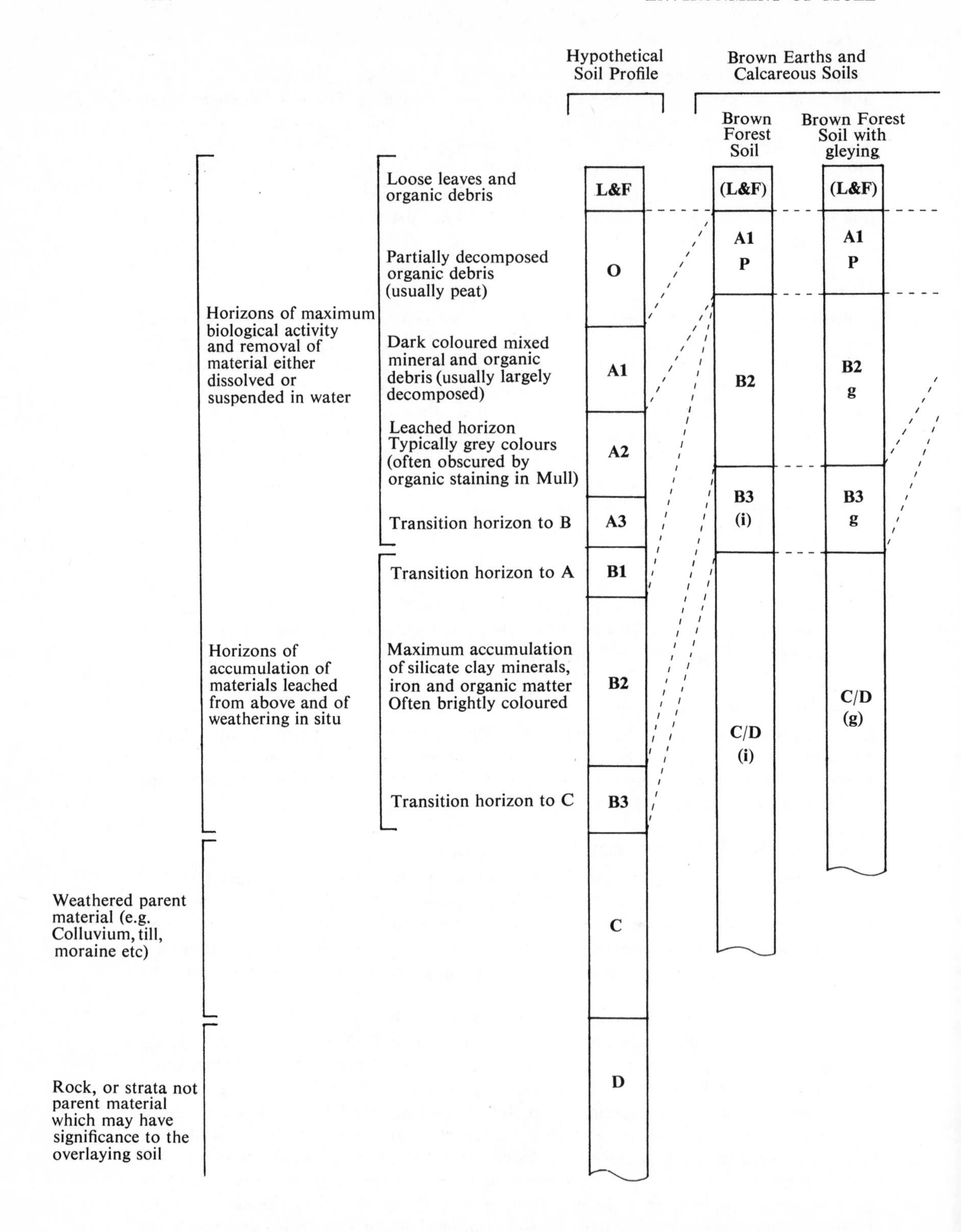
Hypothetical Soil Profile
Brown Earths and Calcareous Soils
Brown Forest Soil
Brown Forest Soil with gleying
Horizons of maximum biological activity and removal of material either dissolved or suspended in water
Loose leaves and organic debris
Partially decomposed organic debris (usually peat)
Dark coloured mixed mineral and organic debris (usually largely decomposed)
Leached horizon Typically grey colours (often obscured by organic staining in Mull)
Transition horizon to B
Horizons of accumulation of materials leached from above and of weathering in situ
Transition horizon to A
Maximum accumulation of silicate clay minerals, iron and organic matter Often brightly coloured
Transition horizon to C
Weathered parent material (e.g. Colluvium, till, moraine etc)
Rock, or strata not parent material which may have significance to the overlaying soil
L&F
O
A1
A2
A3
B1
B2
B3
C
D
(L&F)
A1 P
B2
B3 (i)
C/D (i)
(L&F)
A1 P
B2 g
B3 g
C/D (g)

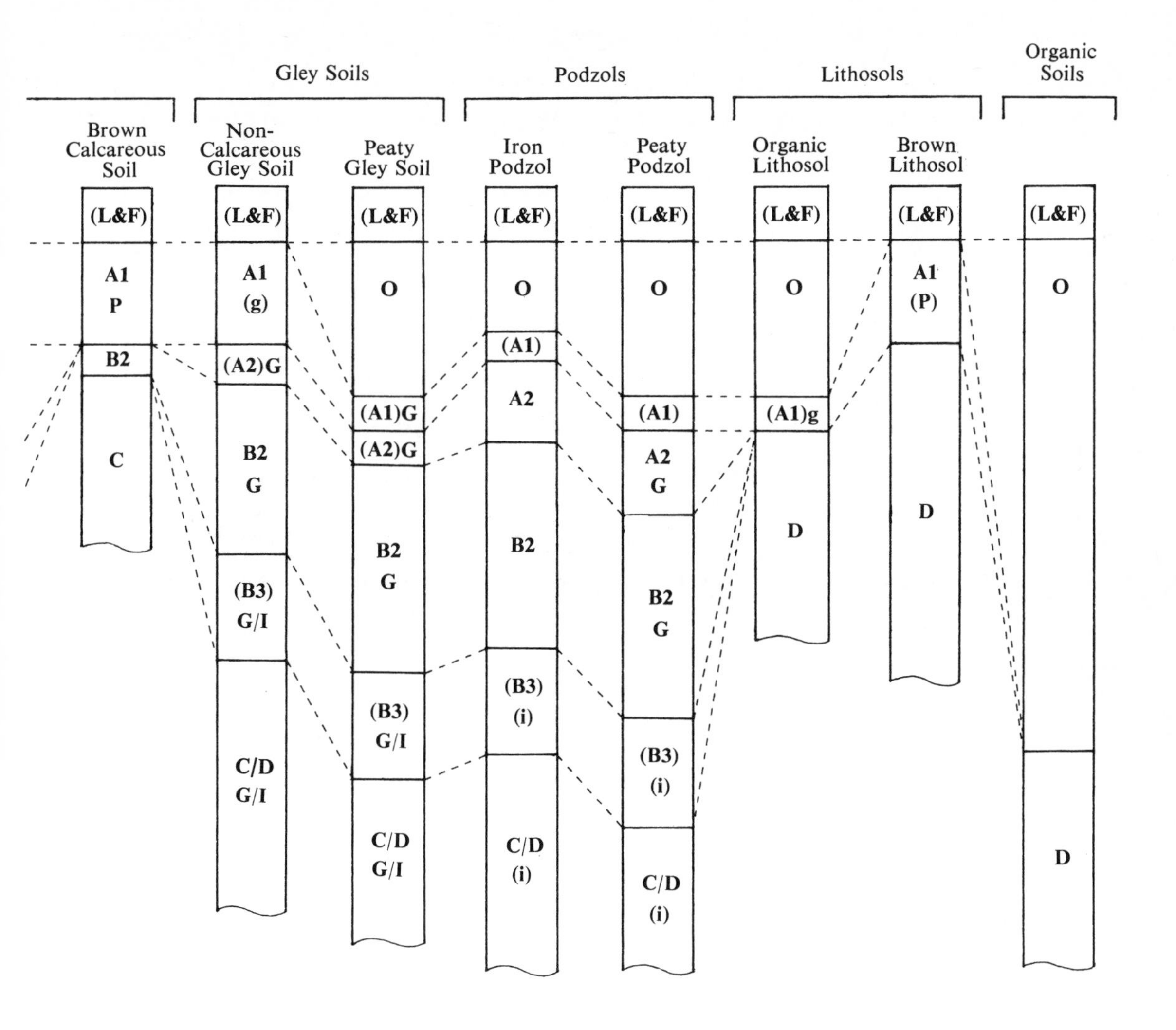

P	usually disturbed by ploughing
G	periodically waterlogged
I	compacted horizon

Lower case letters denote weakly developed characteristics. Symbol in brackets indicates that the characteristic is not uniformly developed throughout the group and may be absent.

A1	horizon symbol
(A1)	horizon may or may not be developed in the profile

Fig. 5.9 The salient characteristics of soil horizons and the horizon relationship within and between profiles of the major soil groups and sub-groups found on Mull.

suitable for statistical analysis, to indicate the relative proportions of soil of different types within each soil complex and to relate soil to site conditions. Studies of variation within the major soil types are being carried out and further details will be given in the *Memoir of the Soil Survey of Scotland* dealing with this area, which will appear in due course.

References

BAILEY, E. B., CLOUGH, C. T., WRIGHT, W. B., RICHEY, J. E., WILSON, G. V. *et al.* (1924). Tertiary and post-Tertiary geology of Mull, Loch Aline, and Oban. *Mem. geol. Surv. G.B.*

BIRSE, E. L. (1971). *Assessment of climatic conditions in Scotland 3. The bioclimatic sub-regions.* Macaulay Inst. Aberdeen.

CHRISTIAN, C. S. (1957). The concept of land units and land systems. *Proc. Pacif. Sci. Congr.* **20**: 74–81.

MACAULAY INSTITUTE OF SOIL RESEARCH. (1974). *Soil Survey of Scotland: Soil map of Island of Mull, 1:63 360.* Ordnance Survey. Southampton.

SYNGE, F. M. (1967). The relationship of the raised strandlines and main end-moraines on the Isle of Mull, and in the district of Lorne, Scotland. *Proc. Geol. Ass.*, **77**: 315–28.

——— & STEPHENS, N. (1966). Late- and post-glacial shorelines and ice limits in Argyll and North-east Ulster. *Trans. Inst. Br. Geogr.* no. **39**: 101–125.

6. Climate

A. C. Jermy

Introduction

Climate is one of the most important factors governing the distribution and frequency of plant species and its importance in Mull is discussed in Chapter 2. The climate of Mull and adjacent islands is typical for the near-shore Inner Hebridian islands off the coast of western Scotland and is affected in turn by the land-form of the island itself. The account given here is a generalised one, making use of the limited sources already familiar to the biogeographer (e.g. Darling, 1947; Green, 1962, 1964; Manley, 1952; Savidge, 1963), as the amount of statistical data collected on the island is minimal. There was, during our survey, only one co-operating climatological station sending in full monthly returns to the Meteorological Office, namely that at the Forestry Commission offices at Aros, beginning in November 1966. Additional to this, four rainfall gauging stations had been recording as from January 1961 being at the White House of Aros, at Gruline, at Ulva House and on Iona; all of these are virtually at sea-level. I am grateful to the Superintendent of the Edinburgh Meteorological Office for supplying copies of these records and other regional information on rainfall and the Director-General of the Meteorological Office for permission to publish them.

In general Mull enjoys a cool oceanic climate. It has a relatively small temperature range throughout the year and a high and evenly distributed rainfall allowing for the natural variations that topography will make. There is high cloud cover, on the hills at least, and the overall potential water deficit is insignificant and for most of the year more rain falls than water is lost through evaporation and transpiration from vegetation. There are frequent strong winds which may be the most significant climatological factor and most certainly that which controls the distribution and development of trees and woodland. The anomalous features for its geographical position stem from the warm North Atlantic currents and regional atmospheric circulation which leads to a cool but mild maritime airstream over the island. It is significant that in any parameter in which temperature is concerned the Calgary area (west of an approximate line from Rubha Àrd to Rubh' an t-Suibhein), the SW areas of Gometra and Ulva and the extreme west of the Ross peninsula, with Iona, are in a category (climatic type **ve** of Birse & Robertson (1967)) distinct from the rest of the island. This fact is reflected in the distribution of a number of plants (see maps in Chapter 2).

Rainfall and humidity

Rainfall

For all ecological purposes, as Green (1964) points out, rainfall must be considered in relation to evaporation which is in turn affected by the vegetation cover. The close relationship between rainfall and topography may be seen from a glance at the map in Fig. 6.1. Rainfall on Mull ranges from 1220 mm (48 in) at Iona to slightly over 3175 mm (125 in) on the higher peaks and where the topography falls gently to the coast; the western and northern seaboards have a rainfall of up to 1524 mm (60 in). The coast of the Ardmeanach peninsula rises abruptly to 300 m and over, and much of this area has a rainfall according to its altitude. Most of the central massif and the higher area of the Laggan–Croggan peninsular has a rainfall of 2540 mm (100 in) or more.

The duration of the rainfall is another aspect which has been discussed in relation to the distribution of bryophytes in Britain as a whole by Ratcliffe (1968). He points out that the distribution of rainfall which governs the length of drought periods is far more important than total amount. The number of 'wet days', i.e. a day with at least 1 mm (0.04 in) of rain, is used instead of total rainfall. The range on Mull varies from below 200 to over 220 'wet days'. It must be remembered however that these figures are based on meteorological data that refer to open sites, and that the microhabitat in which bryophytes and lichens especially are living will be much more humid and less likely to dry out irrespective of the distribution of the rainfall.

The seasonal distribution of rainfall is equally as important as the total amount. There is a maximum on Mull in winter (October–January, but with November appreciably lower than the other winter months) with a minimum in May. Fig. 6.2 and Table 6.1 show the monthly rainfall for the four Mull stations compared with those of Oban and Tiree. It will be seen that Tiree and Iona enjoy similar rainfall and this is most likely true of other climatic parameters.

Both in total amounts and in seasonal distribution the rainfall in Mull is similar to the western seaboard of Scotland to the north. Within altitudinal ranges the distribution on Mull is very similar to that of Skye, the percentage of rain in any one month being almost identical except for December when Mull is comparatively wetter.

Table 6.1 Average monthly rainfall (mm) for the four Mull stations (1962–71) compared with Oban and Tiree (1931–60). From Meteorological Office (1967–73, 1973).

	White House of Aros	**Ulva House**	**Cnoc Mòr Iona**	**Gruline**	**Oban**	**Tiree**
Jan	150	137	94	170	146	117
Feb	99	89	58·5	99	109	77
Mar	142	111·5	79	139·5	83	67
Apr	89	89	66	81	90	64
May	99	94	68·5	109	72	55
Jun	101·5	86·5	66	99	87	70
Jul	127	122	81·5	119	120	91
Aug	119	106·5	79	122	116	90
Sep	165	160	132	172·5	141	118
Oct	228·5	198	137	241	169	129
Nov	175	150	106·5	144·5	146	122
Dec	185	167·5	101·5	200·5	172	128

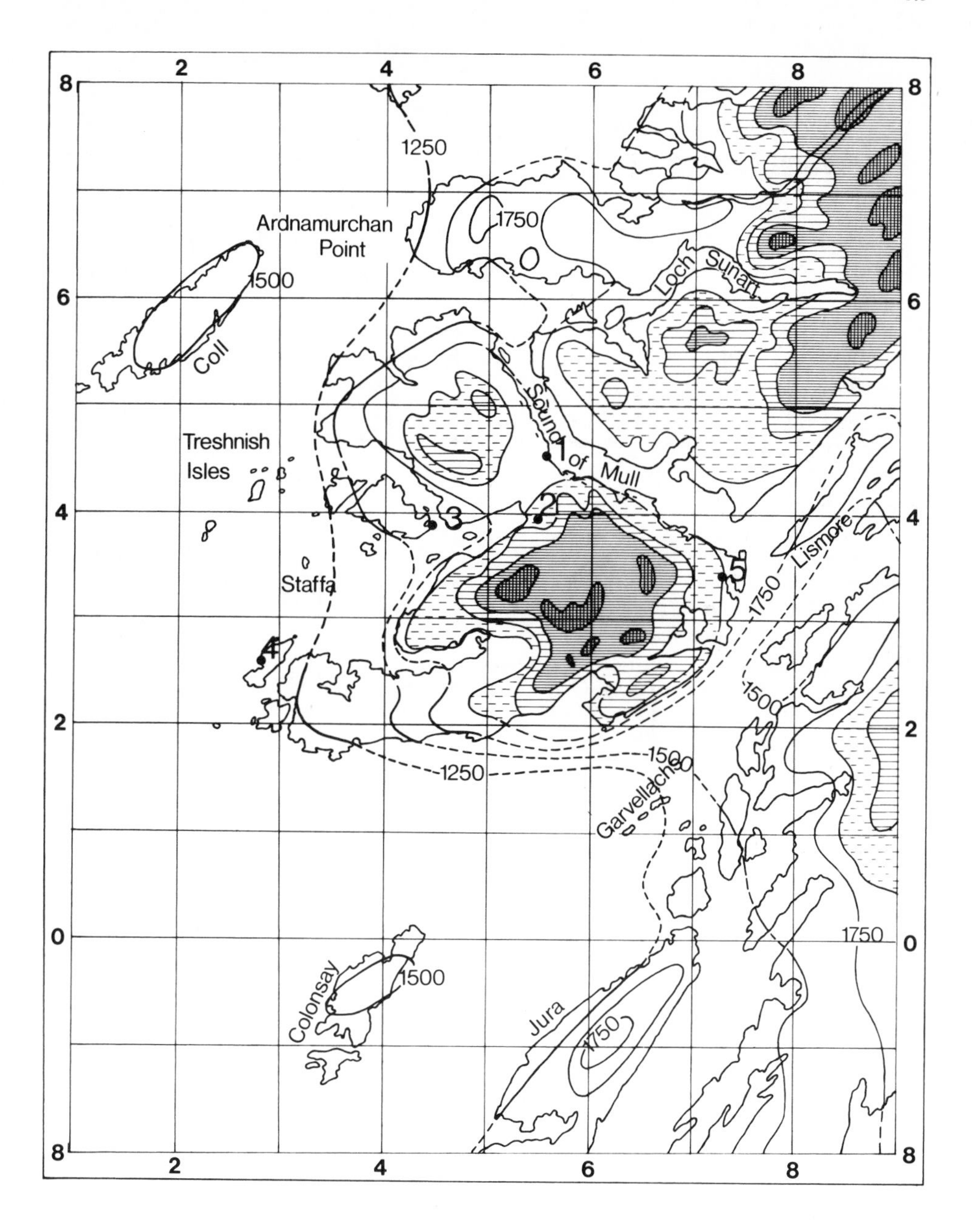

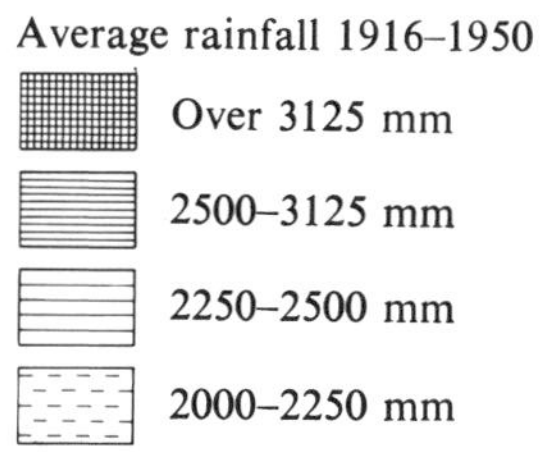

Fig. 6.1 The distribution of the average rainfall (1916–1950) in the Mull area.

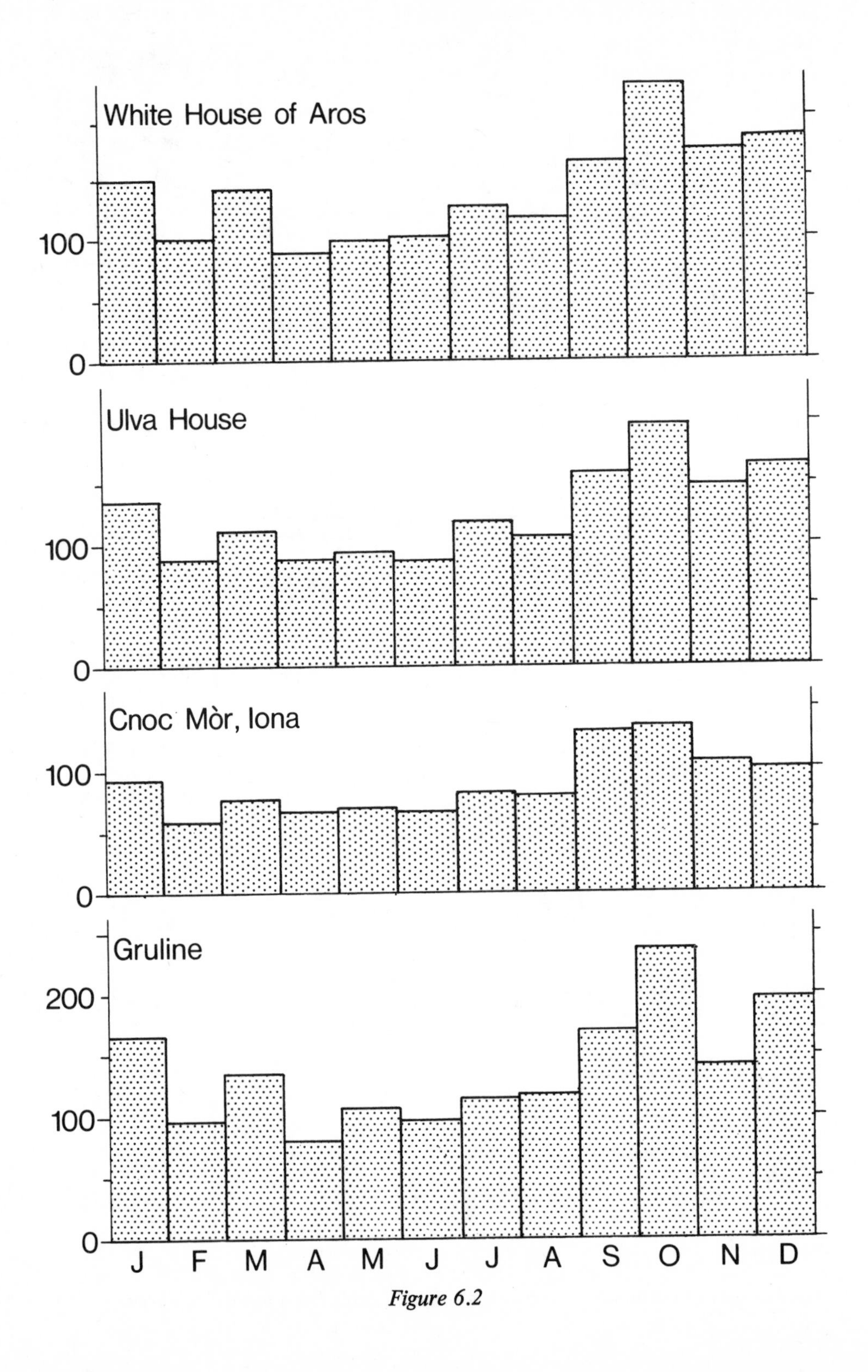
White House of Aros
100
0
Ulva House
100
0
Cnoc Mòr, Iona
100
0
Gruline
200
100
0
J F M A M J J A S O N D

Figure 6.2

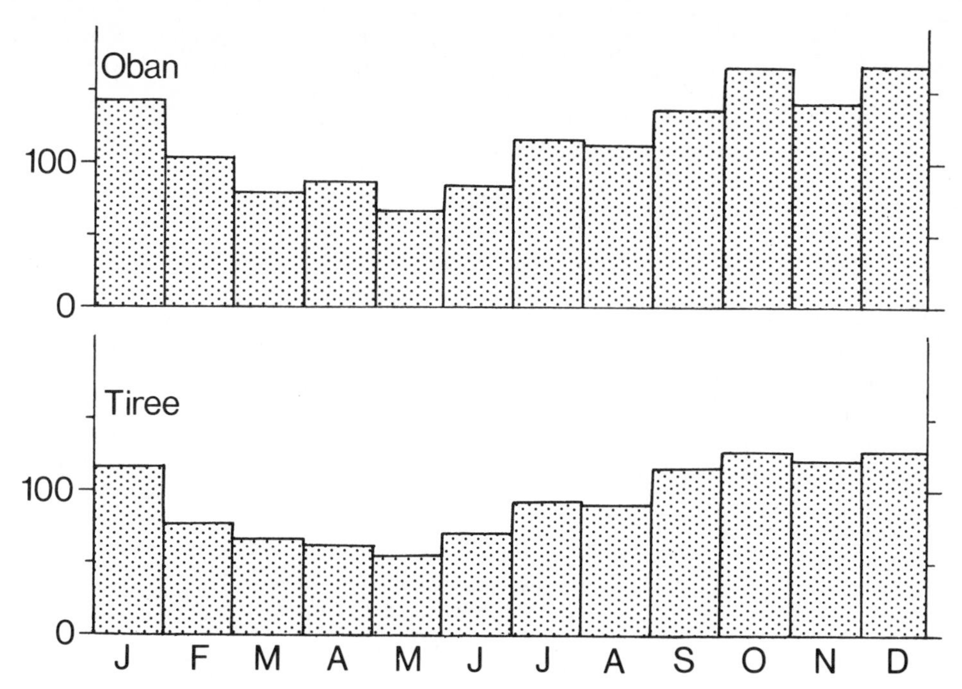

Fig. 6.2 Monthly rainfall (in mm) for the four Mull stations compared with those of Oban and Tiree.

Potential water deficit (P.W.D.) is another way of expressing the wetness of a climate. It is defined by Green (1964) as the total amount of water which would evaporate or transpire from a level, continuous surface of vegetation which covers the ground more or less completely, if available water were not limiting. Using this standard, Birse & Dry (1970) describe Mull as being wet to moist, i.e. having up to 50 mm P.W.D.

Fog and cloud

Fog and mist (cloud reaching the ground) are important during rainless periods because they prevent transpiration and therefore water loss from the plant. It is important to bryophytes and lichens which may be able to absorb water directly from the atmosphere. There are, however, side effects which affect the plant environment: the day-temperature is affected by fog as is the light intensity. Like the rest of the seaboard in western Scotland, Mull has annually only a few days of fog and these usually in late summer and autumn.

The diurnal variation in humidity can be considerable. Cloud on the mountain will often develop towards evening in August and in the latter half of the year. It will settle on the northern and eastern slopes away from prevailing winds and the distribution of certain alpine plant species may be controlled by this cloud, e.g. *Salix herbacea* reaching the lower altitude of 480 m (1600 ft) on the north-eastern shoulder of An Gearna (Ben More) may be due to this effect. The absence of cloud on the hills in spring and late summer can result in colder temperature and the formation of dew or hoar frost.

At low altitudes fog often results from smoke pollution but there is no industrial or even urban concentration on Mull sufficiently large to give this effect. Tobermory, the largest agglomeration of domestic housing, shows only slight effects of pollution by the presence of the lichen "indicator" *Lecanora conizaeoides* but the amount of fog there is minimal.

The density and type of cloud can also affect the insolation. The amount of light reflected from the grey stratus cloud is obviously less than that reflected from the white cumulus.

Sunshine

The amount of solar radiation, of which hours of sunshine is the commonest index, is a function of latitude, which determines day length and also the angle of elevation of the sun. The angle of elevation, in turn, determines the thickness of the atmosphere through which the radiation passes and the shadowing effect of hills (Green, 1964). It is the latter which is the more important factor on Mull. The high mountain masses, offset as they are to the south-east, shade the eastern slopes of the Torosay hills from the warm afternoon sun. The total hours of 'bright sunlight' recorded by the Meteorological Office for Tiree (see Table 6.2) is possibly very similar to that observed on Iona and the Ross of Mull, whereas that recorded at Onich, on the southern side of Loch Linnhe may reflect the pattern found on the eastern coast of Mull. For the most part Mull probably gets as much sunshine as south-western Scotland. On the other hand, in the winter, the long, narrow, north-facing valleys that run between the Torosay hills, Beinn Talaidh, Beinn a' Mheadhoin, Corra-Bheinn and Cruachan Dearg can get very little, if any, direct sunlight.

Table 6.2 The total hours of bright sunlight in Onich and Tiree 1931–60 (from Meteorological Office, 1973)

	Onich	**Tiree**
Jan	30	40
Feb	57	74
Mar	94	116
Apr	120	171
May	173	234
Jun	146	203
Jul	110	162
Aug	116	164
Sep	90	127
Oct	64	82
Nov	37	48
Dec	21	29

Wind

Wind speed (i.e. force) and its direction are major factors affecting island vegetation. As has been mentioned in Chapter 3 the westernmost point of Iona and the Treshnish and Ulva-Gometra archipelago are open to the full force of the Atlantic reach. Whilst again there are no records available it is likely that these areas on Mull have the strong winds that are associated with the Outer Hebrides where, at 10 m above the ground, speeds of 8 m/s (18 mph) are the average (Manley, 1952). Birse & Robertson (1970) indicate zones on Mull from "very exposed" (windspeed 6.2–8.0 m/s) to "moderately exposed" (2.6–4.4 m/s).

The prevailing wind is westerly along the whole of the western Scotland seaboard, and for most of the year this is in the south-western quarter. Autumn and winter gales tend to be from the north-west or occasionally round to the north-east. These winds have a threefold effect on the environment. In the direct sense the sheer physical force of the wind prevents establishment of or breaks struggling vegetation. Low-growing dwarf shrubs such as *Juniperus communis* subsp. *nana*, *Arctostaphyllos uva-ursi*, and *Salix repens* form close-knit communities on south-west-facing cliff-tops. Trees only thrive in gullies and sheltered ravines.

Another direct effect is discussed in Chapter 9 in relation to the chemistry of inland lochs. Winds coming off the sea are laden with droplets of salt water and both the terrestrial and aquatic environments near the sea are enriched with calcium, magnesium, sodium and potassium as a consequence. Fig. 9.6 and Table 9.12 (p. 9.18) show the amounts of total cations, calcium+magnesium and chloride in various lakes throughout the island. It is clear that the distance of the lake from the SW, W or NW coast determines the amount of cations deposited. Malloch (1971) found on the Lizard, Cornwall, salt deposition was correlated with wind speed, and there is every reason to suspect that this holds good for Mull. Another factor is also involved here, namely the altitude of the lake in relation to the coastal topography. Small lochs on Beinn na h-Iolaire at 450 m altitude on the Ardmeanach peninsula show a low cation content in spite of their proximity to a western coast (see Fig. 9.6); it appears that the salt water-laden winds deposit their spray on the cliffs of Creag a' Ghaill. In contrast the relatively low-lying Loch Staoineig on Iona which gets the full force of winds from south, west and north has the highest cation content of all. Here the higher ground rising 75 m between the sea and the loch is not high enough to act as a physical barrier. This may be due to wind speed which must be reduced considerably by a sudden rise of sea-cliff.

The toxic effect of salt has a serious and lasting effect when very strong gales lash the coastal vegetation. Stunted open birch-woods on the Ross of Mull show effects of salt-burn and some of the best ash-woods on the island, those in the southern valleys of on the Laggan Deer Forest, at Port Ohirnie and Uirigh na Salach, were killed completely by the winds in September 1962 when the tail-end of Hurricane Deborah hit the west coast of Scotland.

There are other indirect effects of the wind which may be more important especially to the cryptogamic plants, in particular, the effect on desiccation. In spite of the general high humidity, winds blowing through the woodlands in west-facing gullies tend to produce a drier microclimate than is found in the more sheltered woods on the east of the island, e.g. at Aros Park and this is reflected in the species composition of those sites.

Furthermore winds affect temperature and therefore frost and the action this has on soil formation. It has been suggested (Bibby, *pers. comm.*) that the unique soil phenomenon found on the Ardmeanach peninsula and which forms a habitat for the restricted *Koenigia islandica* may be the result of wind-action. A similar soil type is found on Beinn na Drise, near Lagganulva but so far *Koenigia* has not been found there. Near the coast the presence of an on-shore or off-shore wind can greatly affect the temperature and therefore the vegetation near the coast. Wind can also affect snow-lie and to a lesser extent the distribution of rainfall, most of the rain falling on the leeward side of the hills. Similarly, as said above, cloud will settle in the leeward hollows and corries.

Temperature

There are few records of temperature for Mull itself, the only daily recordings available are from Aros since 1966. They are standard screen temperatures but, as such, allow comparison with other localities. Obviously microclimatic conditions exist which affect both vegetation and individual plants to a much greater extent. Ecologists can look at two aspects of temperature: accumulated temperature and average monthly and diurnal ranges.

Accumulated temperature

Accumulated temperature is a measurement of the total temperature above or below certain thresholds over an extended period of time usually expressed in month-or day-degrees. Biologists usually consider 0°C (32°F) and 6°C (42.8°F) important thresholds. The importance of the first is obvious and plants must be in some state of dormancy either as seed or resting bud in order to survive. Maps showing average dates of first and last air-frost (*Climatological Atlas of the British Isles*, 1952) show the western tip of the Ross of Mull, with Iona and Treshnish Point/Calgary area where in the average year the first frosts arrive on low ground after 1 November and the last frosts by 15 April, as being more favourable to plant growth. On the eastern side of the island the first frosts are a fortnight earlier and the last about ten days later than on Iona. These statistics apply only to low ground 0–30 m; air-frost in the hills and upland valleys may well begin in September and go on until late May. Accumulated frost, i.e. number of day degrees below 0°C, is used by Birse & Robertson (1970) as an indication of severity of winter. Only the summits of the Ben More massif and the Torosay hills have over 470 day degrees C. The western lowland extremities on the other hand have less than 20 day degrees.

The 6°C threshold is not so much a matter of life and death to the plant but is generally accepted as the point below which plant growth is nil or insignificant. Values for the British Isles as a whole have been calculated by Gregory (1954) in terms of day-degrees using 42.8°F as the threshold. It is interesting to note that the seaboard of Mull, together with Coll, Tiree and the west and south coasts of Morven, have a range of 2000–2500 day-degrees per annum – similar to Fife, S Perth, Stirling, Lanark and N Ayr and coastal areas further south. Mull is virtually the northernmost area of this regime of warmth which may account for the southern and mediterranean element in its flora that is absent for instance from Skye and Rhum.

Average temperature ranges

Average monthly and diurnal ranges are similar on Mull as for the rest of western Scotland. Figures for the island are again only available for Aros and the monthly range variation is 12°C, the coldest months being January or February, the warmest July or August. Temperatures as low as −9.4°C have been recorded in February and as high as 26°C and slightly above in July and August. Altitude is the main factor affecting these temperatures. The fall-off in temperature with rise of elevation can be taken as a constant 1°F for every 300 ft rise (Anderson & Fairburn, 1955) and the range over the island due to altitude is within 6°C. The monthly temperatures at Aros (Table 6.3) are graphed in Fig. 6.3 and are compared with those of Oban and Tiree.

Snowfall

Snowfall itself is unlikely to be relevant to the distribution of plants on Mull because of its relative infrequency and short duration. Manley (1940) in a map of the British

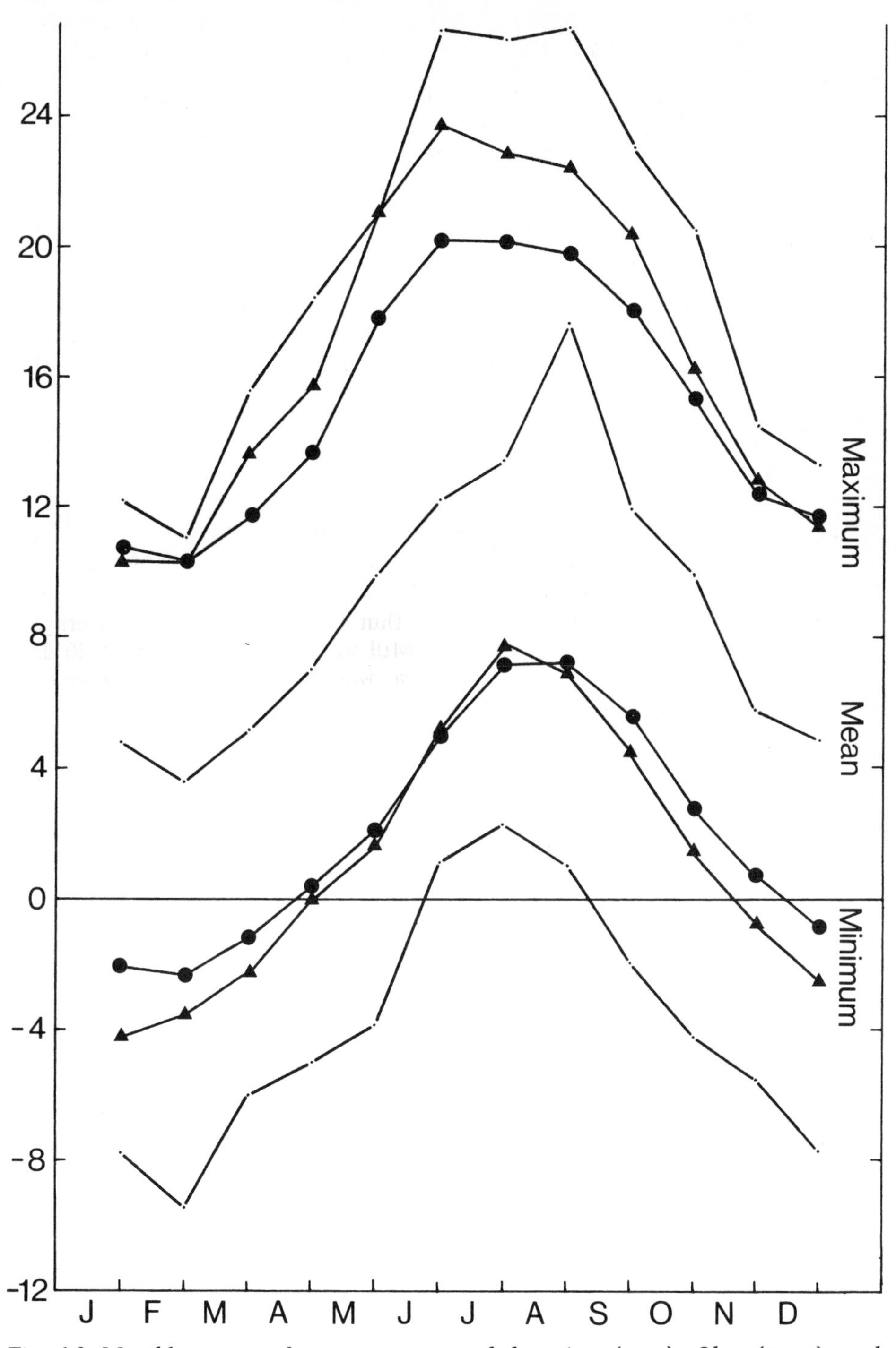

Fig. 6.3 Monthly ranges of temperature recorded at Aros (·—·), Oban (▲—▲), and Tiree (●—●).

Table 6.3 Average monthly temperatures (°C) at Aros Forestry Commission Office (1966–73) compared with those at Oban and Tiree (1931–60). From Meteorological Office (1967–73, 1973).

	Aros		Oban		Tiree	
	max.	min.	max.	min.	max.	min.
Jan	12·2	−7·8	10·4	−4·1	10·5	−2·1
Feb	11·0	−9·4	10·4	−3·6	10·2	−2·4
Mar	15·6	−6·0	13·8	−2·2	11·9	−1·2
Apr	18·4	−5·0	15·7	0·0	13·7	0·4
May	21·0	−3·8	21·1	1·6	17·8	2·1
Jun	26·6	1·0	23·8	5·2	20·1	5·2
Jul	26·4	2·2	22·8	7·5	20·0	7·4
Aug	26·7	1·0	22·4	7·1	19·6	7·3
Sep	23·0	−2·0	20·4	4·6	18·0	5·4
Oct	20·5	−4·2	16·3	1·4	15·4	2·7
Nov	14·5	−5·6	12·9	−0·8	12·6	0·7
Dec	13·3	−7·7	11·4	−2·4	11·3	−0·8

Isles showing the distribution of the mean annual number of days with snow or sleet observed to fall on ground beow 200 ft gives further evidence of the more clement climate of the Ross Peninsula and the Treshnish-Calgary seaboard, both of which have less than 15 days in which snow on average was recorded; east of this line lowland Mull was within the category of 15–20 days. In fact data from Aros (37 m alt.) in *The Monthly Weather Report* (*Meteorological Office*, 1967–73) shows the average annual figure for the years quoted as 23 in which snow or sleet fell and was still lying at 09 00 h on the average on 6 days each year. If for each 50 ft above 200 ft one adds a further day as Manley suggests, the area above 2000 ft (600 m) and 2500 ft (750 m) received snow or sleet on 36 and 40 days respectively.

Whatever amount of snow falls on Mull, it is only on the higher hills in the Ben More massif and the Torosay hills that snow persists for more than the single day or two. In the upper north-facing corries and similar slopes where snow can accumulate it may remain through April to early May. There are no areas of characteristic snow-bed vegetation as may be found on the Scottish mainland although on some north-facing slopes around Ben More a *Nardus*-bryophyte community suggests snow-lie. Hollows in the same area that pick up melt-water may give rise to a characteristic high-level *Scirpus cespitosus* mire on the strength of the melt-water. *Athyrium distentifolium* Tausch ex Opiz, the Alpine Lady-fern so characteristic of snow patches in western and central Scotland has not so far been verified for Mull although an unsubstantiated record has been submitted for the north corrie of Ben More. For the most part the relatively clement climate is against prolonged snow-lie and the few species requiring it which would otherwise find the right conditions are therefore absent.

References

ANDERSON, M. L. & FAIRBURN, W. A. (1955). Division of Scotland into climatic sub-regions as an aid to silviculture. *Bull. For. Dept. Univ. Edinb.*, **1**: 1–31.

BIRSE, E. L. (1971). *Assessment of climatic conditions in Scotland 3. The bioclimatic sub-regions.* Macaulay Institute: Aberdeen.

——— & DRY, F. T. (1970). Ibid. *1. Based on accumulated temperature and potential water deficit.* Ibid.

——— & ROBERTSON, L. (1970). Ibid. *2. Based on exposure and accumulated frost.* Ibid.

DARLING, F. FRASER. (1947). *Natural history in the Highlands and Islands.* London.
GREEN, F. H. W. (1962). *British rainfall 1958. Part III.* London.
———. (1964). A map of annual average potential water deficit in the British Isles. *J. Appl. Ecol.*, **1**: 151–158.
GREGORY, S. (1954). Accumulated temperature maps of the British Isles. *Trans. Inst. Br. Geogr.*, **20**: 59–73.
MALLOCH, A. J. C. (1971). Vegetation of the maritime clifftops of the Lizard and Land's End Peninsulas, West Cornwall. *New Phytol.*, **70**: 1155–97.
MANLEY, G. (1952). *Climate and the British Scene*, London.
METEOROLOGICAL OFFICE. (1952). *Climatological atlas of the British Isles.* London.
———. (1967–73). *The monthly weather report of the Meterological Office.* London.
———. (1973). *Tables of temperature, relative humidity, precipitation and sunshine for the world. Part III Europe and Azores.* London.
RATCLIFFE, D. A. (1968). An ecological account of Atlantic bryophytes in the British Isles. *New Phytol.*, **67**: 365–439.
SAVIDGE, J. P. (ed.) (1963). *Travis's Flora of South Lancashire.* Liverpool.

7. Marine physical environment

J. H. Price

Introduction

The length of the coastline of Mull and associated small islands is very considerable; we have no exact measure but it is certainly greater than 480 kilometres. This coastline, furthermore, is highly dissected and indented, providing thereby a vast range of habitat types that can be exploited by benthic marine algae. As with most of the outer Inner Hebrides, the local variations in habitat are superimposed on a major pattern change, especially affecting the intertidal, which results from the south and west being subject directly to the whole fetch of the North Atlantic whilst the east, and to a degree the north, coasts are generally more sheltered by their position and proximity to the mainland. This major patterning is similarly found on most of the larger islands off the Mull coast; Iona, Ulva, Gometra, Erraid, Eorsa, Staffa, and the other Treshnish Islands all show similar tendencies, with some local variations. Thus, within very restricted areas of coast profound differences occur in many aspects of the physical environment, their sum producing the extremes shown by the tremendously wave-beaten west-facing headlands and island coasts on the one hand, and the sheltered heads of lochs, virtually lacking water movement at times, on the other. Quantification of these differences, so far as is possible, is presented in detail in this chapter. Variations of this kind and degree can produce profound differences in floristics and community structure (see Chapter 8). Exposure to greater or lesser wave action affects primarily the intertidal and upper infralittoral, whilst the often associated differences in the superficial physical and mechanical nature, and the configuration and slope of the substrate affect all levels, except the deeper infralittoral, equally. Characteristics of freshwater run-off from the land are also of great importance in affecting the kind and distribution of marine flora, particularly in the intertidal areas of shore.

It is convenient to divide this treatment of the physical environment into two major sections: the open sea and the Sound of Mull; and the sea-lochs. The first of these owes much to Craig's treatment (1959) in the *Hydrography of Scottish coastal waters*, whilst the second is primarily based on the more recent *Hydrography of Scottish west coast sea-lochs* (Milne 1972).

The open sea and the Sound of Mull

Configuration and substrata

By far the major portion of the western seaboard of Mull, like the Sound of Mull shores, is composed of basaltic lavas, although there are certain area variations: (i) the Gribun and Inch Kenneth, which are of Mesozoic sedimentary rocks; (ii) Iona, the inner shore of Loch na Làthaich, and the south shore of the Ross of Mull from Rubh' Ardalanish to Rubha nam Bràithrean, which are of gneiss and schist; and (iii) the rest of the Ross of Mull shores west of the above, which, with Erraid and Eilean a' Chalmain, are granitic. (See Chapter 4 and Fig. 4.1.)

Although sheltered areas of the western coast, such as the depths of sea-lochs, or sheltered channels such as Ulva Sound, do reveal locations in which detrital conditions strongly affect the coastal (littoral) marine environment, the more open areas of the west include few intertidal areas of mobile substrata. There are a few primarily sandy bays flanked by variably steep rocky or rock platform shores (e.g. Calgary Bay, stns 77–82*) in the north-west of the island group, but the relatively low-lying Ross of Mull and protected parts of Iona are the main areas in which sandy bays occur. Between Fionnphort and Loch Buie entrance, the south coast of the Ross is variably cliffed, the land of the Ross rising from north to south. The western stretch of that southern coast, from about Rubha nam Bràithrean westwards, has many sandy beaches in locations that are somewhat protected from the strong western fetch. East of that, bays with sand and other detritus are fewer, although Carsaig has a large, principally sandy, beach. Sand is the predominant facet of the eastern and northern shores of Iona, and also occurs in the inner parts of the deep bay of Camas Cuil an t-Saimh, on the west. Even in such areas, long reefs and headlands of firm rock provide substrata for development of dense benthic algal growths.

Infralittorally, almost all firm rock areas around Mull show those substrata descending in variable depth into detritus of one form or another. It is usually these loose substrata, which range from sand, through sand/silt/small pebble mixtures to larger but still mobile small stones, that provide the effective factor in delimiting the lower fringe of both the *Laminaria hyperborea* growths and the general lower limit of colonisation by benthic marine algae.

The most impressive facet of the basaltic lava shores is provided by the fairly widespread occurrence of columnar basalt, which tends to provide precipitous to steep rocky shores, commonly in very wave-exposed positions. The classical examples on the island of Staffa are well known and have been described and photographed frequently (see Fig. 4.7). There are, however, many further examples on Mull, Gometra, Ulva, and various small islands seaward of Iona. On Mull itself, the cliffs of Carsaig (near stn 26) and Bloody Bay (stn 93) include examples of columnar basalt. Where such cliff examples abut immediately on the intertidal, the latter tends to be formed of fallen blocks of basalt, descending steeply to considerable depths (e.g. as at stn 65, on Ulva south shore).

The Sound of Mull, separating Mull from the mainland, is continuous in the south with Loch Linnhe and the Firth of Lorne. The presence of the adjacent mainland (the Sound is nowhere wider than 8 km) and the predominantly west to east thrust of the Atlantic fetch leads to the situation that most of the east coast of Mull from Bloody Bay (Mishnish) south along the Sound to beyond Grass Point, Loch Don, is with few exceptions much more sheltered from wave action and strong winds than the west coast. We did not encounter locations for which the exposure to wave action could be described as more than moderate. Generally, less precipitous shores, sheltered bays, and the outfall of considerable amounts of estuarine freshwater cause the deposition and retention of large quantities of mud and mixed silt in various locations, whilst shingle is a particular characteristic of others along this eastern coastline. The Tobermory–Calve Island area provides many examples of both these major characteristics. The strong estuarine influence in locations such as the Aros River outfall (stn 8) and Fishnish Bay (stn 11) has resulted in the consolidation there of

* These station numbers refer to collecting/recording sites listed on pp. 19.4 and localised in Fig. 8.1.

variable amounts of saltmarsh (e.g. stns 47 and 86; see Chapter 10 for further discussion on saltmarshes).

Water conditions

Although the whole mechanical and physico-chemical balance of the seawater medium and the environmental phenomena that impinge on it are of fundamental importance to organisms exploiting that medium, infralittoral benthic marine plants are primarily affected by three major facets of the environment. Assuming a seawater chemistry within 'normal' limits, the biogeographic distribution in terms of horizontal limits to the exploitable range is to be understood on a broad *temperature* basis. Of equally widespread importance to vertical limits but otherwise rather more local in their effects, the other two primary constraints are available levels of *salinity* and *light penetration*. Although the open sea responds, even in its surface layers, only sluggishly to temperature variation which comparatively rapidly affects the terrestrial and littoral environments, this is not true of the shallow coastal shelf areas that are the principal habitats of benthic marine forms. All shallow infralittoral organisms therefore are subject to wider and more rapid temperature and salinity changes than deeper growing species; this is even more emphatically true for littoral and supralittoral organisms, which also need the capacity to withstand regular and/or prolonged exposure to subaerial conditions. Characteristics of tidal fluctuation around the shores of Mull are considered later. It is unfortunate that physical environmental parameters are rarely recorded specifically for the shore or shallow infralittoral environment, and this is certainly the case for Mull. Meteorological data are most usually recorded for stations well inland, or at least remote from the shore itself. Hydrographic data are usually available for locations well out to sea, or at points well clear of the littoral in sea-loch systems. Since these are the best available data, they are presented and utilised here. For applicable terrestrial data such as insolation levels, wind strengths and directions, precipitation and air temperatures, see Chapter 6.

Temperature Craig (1959: figs 3, 4 and 6) has presented maps which demonstrate the distribution of sea surface temperatures for the whole Scottish coast. Typical values for the months of February, mid-July, and at about 1 October, are mapped. The areas of Mull and adjacent coasts are shown in Figs. 7.1A and B; based on Craig's fuller presentation, these show the typical February and mid-July values. The 1 October values have been omitted since mapping those for the Mull area would show little; the whole area then lies close to the 12°C isotherm, which describes a sinuous set of curves following the general north-south line of the western mainland coast but, when drawn as a continuous line, somewhat inland. No other isotherm approaches any part of the area. The relationship of these area surface temperatures to those for the rest of the Scottish coastline can be seen from Craig (*l.c.*), and to those for the north-western approaches to the British Isles from the presentation by Tulloch & Tait (1959). More general background data are available in Lewis (1964). Deep water bottom temperatures have very little relevance to the present work as they are outside the more important areas for benthic algal growth. Inshore bottom temperatures tend to vary little from those at the surface in the same area. Following Craig, Fig. 7.1A (surface temps. Febr.) shows a few conversion figures to add to the mapped surface temperatures to obtain the bottom temperatures at the same time in the same locations. The picture is more complex and the differences are wider for mid-July; for further data see Craig (*l.c.:* fig. 5).

Mapped temperatures demonstrate clearly that, towards the end of February, water shallowing towards a major land mass produces lowering average temperatures. Isotherms for 6°, 5°, 4° and 3°C all pass over, along, or very close to the shores of

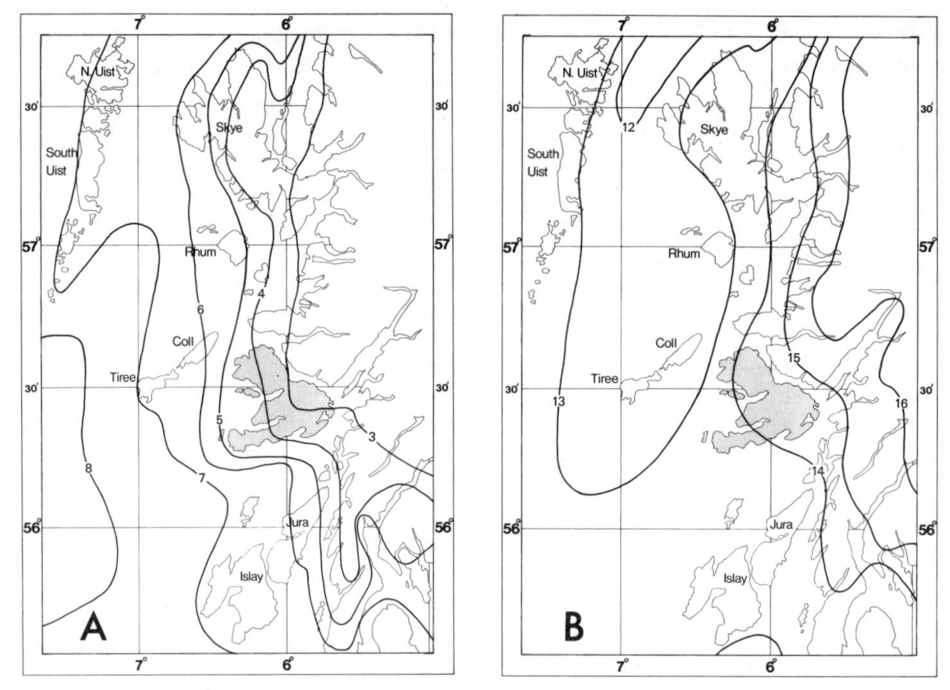

Fig. 7.1 Sea surface temperatures in February (A) and mid-July (B). After Craig, 1959, figs. 3 and 4.

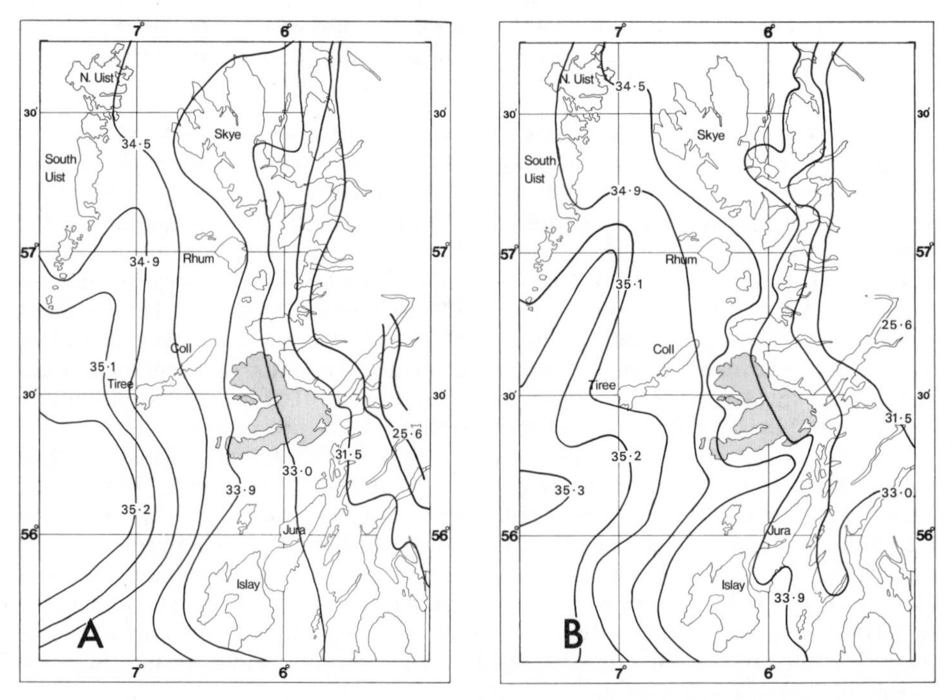

Fig. 7.2 Typical values of the salinity of surface water in winter (A) and of bottom water (B); seasonal changes are negligible in the latter. After Craig, 1959, figs. 1 and 2.

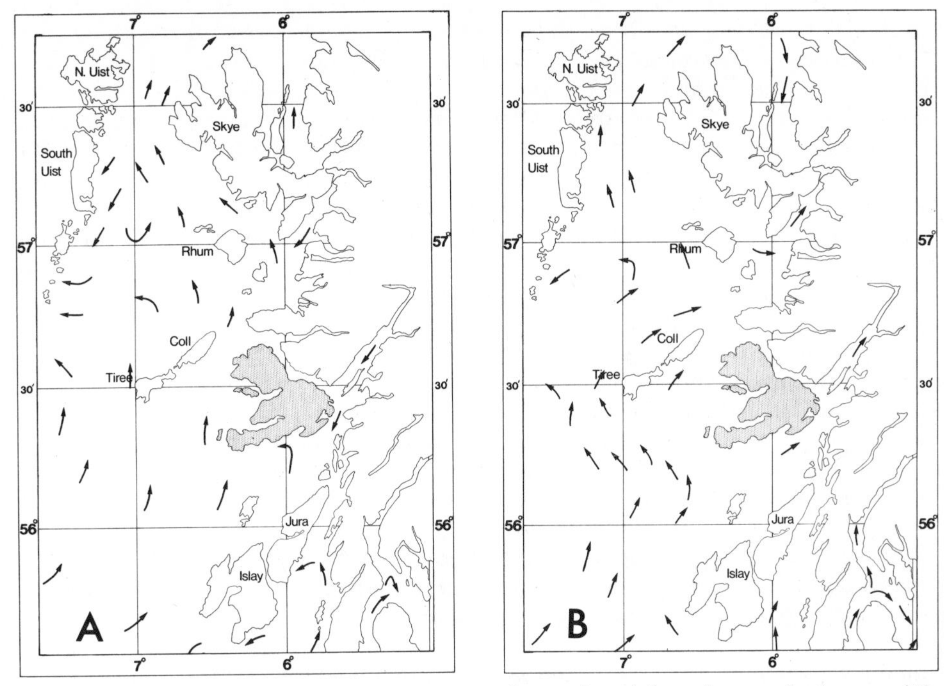

Fig. 7.3 Local currents in the Mull area at the surface (A) and near the bottom (B). After Craig, 1959, figs. 10 and 11.

Mull. At this time of year, water temperatures are at a minimum and there is vertical homogeneity throughout the water column, except in the presence of dilution of the surface water; then, the usually regular and smooth horizontal changes may be interrupted by patches of cold dilute waters close to river mouths. Around Mull, inshore waters are generally such in the western approaches that there is 0.8°C higher bottom temperature than surface temperature. Along the eastern Mull coastline, by contrast, in the Firth of Lorne and (presumably) in the Sound of Mull, greater effect of surface dilution due to increased run off from land mass drainage creates greater differential (1.5°C) between the warmer bottom and colder surface waters, apart from the generally lowering effect west to east at the surface itself. Temperature variations of this kind, whilst rarely critical at these levels, could be important in determining the presence, luxuriance and density of growth of benthic forms near the southern or northern limits of their temperature tolerance, in extreme years. They could also lead, in such years, to more widespread and vigorous growth along the western seaboard of species requiring as optima higher temperatures and salinities than were available along the eastern Mull coastlines. This differential west to east in February is generally more important to cool temperate benthic algal growth than differentials in summer; the growth and reproduction of a majority of the species involved are at their most vigorous earlier in the year. The greatest differences between surface and bottom temperatures, and indeed between in-shore surface and off-shore surface temperatures, tend to occur in estuarine regions or river mouths where tidal mixing is slight. On the west coast, this is principally in the sea-lochs (see later), whilst on the east coast it is manifest as further emphasis of the already lower temperatures and salinities in winter.

In July, the mapped temperatures demonstrate a reversal of the west-east drop of February. Then, the −2.5°C or so variation toward the east is replaced (Fig. 7.1B, surface temperature, mid-July) by an approximately +1.5°C variation in the same direction. The extreme west coast of Mull, Iona and the offshore islands therefore are subject to an average range of something like 7.5°C change in water temperature, mid-winter to mid-summer, whereas in the same period the eastern coastline experiences about an 11.5°C variation. Development of higher summer temperatures usually occurs over the period April to June, when surface temperatures in the offshore region rise steadily through about 7°C, or up to 9°C in extreme cases, as indicated for areas elsewhere than Mull in western Scotland. This is the west coast situation on Mull. Estuarine and in-shore temperatures – the east coast situation on Mull – show a greater rise everywhere. The rate of surface temperature rise depends on water depth and extent of vertical mixing due to wind and tide; there are therefore considerable variations, both from place to place and time to time.

Upper waters continue to gain heat from the sun until the end of August, but owing to patchy downward mixing there is a tendency for off-shore surface waters to show a decline in temperature during August. In lower depths, below about 40 metres, surface temperatures may not show a decrease until September. Craig (*l.c*: 12) indicates that in the Scottish coastal waters to a depth of 120 m there is a relatively simple temperature distribution by the end of September due to the breakdown of stratification. Vertical temperature change is negligible everywhere. Additionally, thenceforward, the large horizontal gradients of summer are absent, with temperatures steadily declining to the February minimum. From about late October, the estuarine and diluted surface water is again distinguishable by its relatively low temperature.

Salinity Distribution of surface and bottom salinities around Mull is mapped in Fig. 7.2A & B. These data are again modified from those presented by Craig (*l.c.*, figs. 1 & 2), although the concept of D.C.I. ('Defined coastal influence') is not used here. No doubt, the fact that D.C.I. is approximately proportional to the logarithm of the degree of dilution would, as explained by Craig, be advantageous in an area such as western Scotland, where salinity variations are principally due to dilution by freshwater run-off. However, this advantage would be largely offset for biologists by the need to convert back to parts per thousand, with which all are familiar, for real comprehension; that, in turn, would require republication of Craig's conversion table. Except in so far as concerns salinity patterns in lochs (see later), data required for a really detailed locally accurate presentation of Mull salinity patterns are simply not available to us, and in any case would be of limited use in the present context.

Every inflowing stream, of whatever size, will variably alter the prevalent salinity pattern of the marine locations in the vicinity of its mouth. Where the outflow is actually across an otherwise colonisable stable littoral area, the effects thus exerted in terms of presence, density, luxuriance and distribution of particular species are often profound. Where necessary, such effects are subject to comment in the marine ecological introduction, in the station lists, or in the species entries, as appropriate.

The marine region constituting the north-western approaches to the British Isles is one of relatively high salinities, reaching 35.45‰ in an area that bends eastwards sharply around Rosemary Knoll. The area of high salinities is sandwiched between coastal water of rapidly diminishing salinity on the east and open ocean water of slightly lower salinity than itself on the west (Tulloch & Tait 1959). The Hebridean area, of which Mull is part, forms a section of the diminished salinity coastal water to the east. Bottom salinities throughout the whole of the north-western approaches and the coastal waters tend unvaryingly to be rather similar; salinities vary in the more superficial water layers. Along the western coast of Iona, for example, bottom waters are at about 34.87‰; out where the surface salinities are high, deeper salinities still run at 34.90–34.95‰. At the surface in off-shore parts of the coastal area, salinities tend to decline in the summer period May to October as the extension of coastal water west of the Outer Hebrides increases by some 16 to 32 kilometres. This latter spread is then less inhibited than usual by vertical mixing, with the coastal water overlying the oceanic water and extending westwards beyond the edge of the shelf. In winter, vertical mixing is much greater, vertical stratification almost disappears, and the extension of coastal waters is therefore much less.

In in-shore regions, salinities are slightly higher in summer due to the general reduction in river flow, although there are some years in which the changes are negligible. In any case, the effect is small compared with day-to-day variations and with place-to-place variations already mentioned. Layering and local effect of freshwater runoff tend to be of much greater effect in sea-lochs (see later). The general effect of a closer approach to land on salinities is very similar to that on temperatures, in that both are lowered by the influence of freshwater run-off. From Fig. 7.2A (winter surface salinities), it is clear that passage west to east is accompanied by rapidly diminishing surface salinity. Deeper salinities do decline, relatively, as the water shallows along the inner half of Mull shores and the Firth of Lorne, but to a very much lesser extent than at the surface. From approximately equal salinities at bottom and surface (34.87‰ : 34.5‰) just off Iona, the situation towards the tip of the Ross of Mull is 34.5 : 33.9; near Loch Buie (south) and at Bloody Bay (north), 33.9 : 33.0; and along the mainland shore of the Firth of Lorne/Sound of Mull junction 33.0 : 31.5. By the inner end of Lismore, the situation has reached the much wider differential of 31.5‰ : 29.2‰.

Light Hellebust (1970: 149) has stated that light is probably the most important single factor determining the vertical distribution of marine attached plants. Although all the other factors described in this chapter are of considerable importance in influencing the distribution patterns of especially intertidal benthic algae, thus making it difficult to assess or comprehend the probably inconstant specific light effects in the supralittoral, littoral and shallow infralittoral, it is generally light intensity that determines the lower limits of infralittoral growth in the presence of adequate substratum. In other aspects of the vertical distribution, however, both quality and intensity of penetrating light are important and these change with water type, water depth and geographical region. Variations in the daylight spectrum caused by weather, angle of the sun, and other such modifying factors, only influence natural submarine illumination spectra to a small degree. The absorption characteristics of the water itself, and light attenuation caused by dissolved matter and suspended particles, are the over-riding determining factors in the submarine environment. Details of variation in daylight spectra, especially on sunny days, are effectively modified, smoothed out, and finally eliminated, relatively few metres below the sea-surface (Jerlov, 1951, 1968, 1970; Hellebust, 1970; Halldal, 1974). Submarine light conditions in open oceanic waters and those in coastal waters are customarily very different. Oceanic waters lack really significant amounts of dissolved 'gelbstoff' (yellow substance, the dissolved decomposition products of organic matter) and usually contain many fewer suspended particles than do coastal waters. 'Yellow substance' absorbs significantly in the yellow spectral bands (580–600 nm) and increasingly towards the shorter wavelengths (green, 500–560 nm; blue, 420–480 nm; violet, 400–420 nm; and ultraviolet, <400 nm); it is brought to the sea in large quantities by freshwater inflow such as rivers and is to some degree found in the sea itself. Suspended particles both absorb (more in the shortwave spectral bands) and scatter (all wavelengths). For the benthic marine algae, we are principally concerned with coastal waters, although open ocean conditions are sometimes important when algal growths on shallowly submerged outliers are under consideration. In general terms, for North Sea and north-east Atlantic coasts, it is rare to find benthic algal growths significantly below a depth of 40 m. It is worth mentioning that maximum depth for benthic algae tends to decrease into fiords and sea-lochs even when adequate depths and substrata are available for colonisation. In Trondheimfiord, for example, seaweeds were found in approximately 28 m maximum depth at the mouth, but only to 10 m in the innermost areas. Printz (1926) attributed this to the large amounts of 'yellow substance' brought by rivers to the surface layers of the inner fiord.

The main framework of classification of coastal and oceanic waters against which the above characteristics are usually graded for particular localities was provided by Jerlov (1951; 1968; 1970). The classification was considered by Jerlov (1951: 50–52) to be characteristic of coastal waters in the temperate zone, being based on experimental work in the Baltic, the Skagerrak, and various Norwegian fiords, *inter alia.* Of the nine types of coastal water recognised by Jerlov on the basis of penetration at a solar altitude of 45°, 1 is the clearest and 9 the least penetrated. All show optimum penetration (transmission) values percent/m in the 450–650 nm spectral bands, although the optimum percentages gradually shift from 450–550 nm (types 1–3) through 500–600 nm (types 4, 5, 6, 7, 8) to 550–650 nm (type 9). According to Halldal (1974), the water conditions at Kristineberg (western Sweden), with a total light quanta attenuation showing only 10 per cent left at 5 m depth, 1.27 at 10 m, 0.24 at 25 m, and 0.0037 at 50 m, and with the 570 nm band penetrating deepest (ultraviolet, violet, and most of the blue had disappeared by 30 m), have the typical quanta distribution of European coastal waters. These waters, acting as a spectral filter peaking at about 570 nm, belong to an optical type near to the type 9 of Jerlov.

In connection with British underwater conditions, Kain (1966: 327–331) has discussed many of the problems involved in measurement and comprehension of the wide variations in light penetration, although much of the variation remains difficult to explain. She indicates that, in general terms, horizontal visibility conditions underwater are worse in winter than in summer; this is common, apparently, to all temperate areas and especially to higher latitudes where seasonal variation in incident light is very large. Since reduction in visibility also normally indicates reduction in exploitable light penetration (the approximately 50 per cent of solar radiation in the photosynthetically usable range 350–700 nm corresponds roughly to the visible spectrum, except for the ultraviolet), poorer light penetration conditions are likely to prevail for much of the time underwater during winter.

The canopy of *Laminaria hyperborea* 'forest' itself cuts out some of the already drastically reduced incident irradiance; comparison of the latter with surface conditions off Port Erin, Isle of Man, for example, was presented by Kain (1966), who showed that even in the best conditions noted there, penetration of visible radiant energy was little more than 1 per cent of surface visible radiant energy at approximately 20 metres depth (near coastal water type 1 of Jerlov). Further similar data were presented by Kain (1971) and Kain *et al.* (1976). Although the extent to which the *L. hyperborea* canopy further reduces the available irradiance reaching the underflora is clearly variable (Kain recorded a value of 28 per cent amongst the 'forest' compared with conditions in open water at the same depth), irradiance could be rather low in the dense 'forests' of *L. hyperborea* off western Scotland. Kitching (1941), diving around the coast of Carsaig Island, Sound of Jura, to the south of Mull, assessed the illumination within the *Laminaria* 'forest' there at 1 per cent of that in open water at the same depth, although he noted that transmission could vary with depth and that the illumination at any one position in the 'forest' fluctuates continually with the movement of fronds overhead.

Clearly, under the usually-experienced ranges of conditions, this rather drastic reduction in available light in the infralittoral does not limit the growth of the colonising algae, at least so far as preventing colonisation is concerned. The presence in western Scotland of relatively hard rocky substrata and fairly deep waters means that turbidity under open sea conditions there is hardly a problem, although along the coast the situation is mostly much less suitable for benthic algal growth. However, except at the muddy, freshwater-offrun affected heads of sheltered sea-lochs, there is nothing here to compare with the extreme turbidity of certain areas of the east (Lincolnshire) and south-east (Kent; Sussex) of England. In fact, although the present survey did not assemble figures with which to support the argument as regards the Mull area itself (and there seems to be a similar lack of available published data for that particular area), it seems highly likely that at least the western coasts bear comparison with the nearby Arisaig, Inverness, slightly to the north, where Jupp & Drew (1974) have recently investigated biomass and productivity of *Laminaria hyperborea*. The site they chose, about 1100 m off shore, although sheltered somewhat from westerly gales by Rhum, Eigg and Muck, bears considerable comparison with some parts of the Mull coastline so far as general balance of the physical environment is concerned. Jupp & Drew noted that the clarity of water in their site area, a north-south shoal ridge with a westward slope of about 20°, approached that of Coastal Type 4 according to the Jerlov (1951) classification. It is probable that those similar areas around Mull have directly comparable light penetration conditions, with higher Jerlov grades applying in sea-lochs and on coasts with strong offrun and lesser water movement.

Although (see Chapter 8) it is clear that size and density of *Laminaria hyperborea* fall off with depth below a variable optimum at certain sites investigated on Mull, the impression is gained that the lower limits of these growths are in many (but not all) cases not light- but rather substrate-limited. This is directly analogous to the situation noted by Jupp & Drew (*l.c.*) at the Arisaig site. For a consideration of critical levels of irradiance in the ecology of *L. hyperborea*, and the depths at which they may be found, see Kain (1966 and 1971). According to Jerlov (1951: 52), even the clearest of coastal waters (type 1) has only 1.3 per cent of total incident light energy (as 100) in depths of 20 m, whilst types 4 and 9 are then down to between 0.29–0.20, and 0.0002 per cent, respectively.

Water movements

Water movements can be considered under the topic heads of (i) wave action and swell; (ii) current (including local eddies) and fetch; and (iii) tides and tide ranges.

Wave action The general topic of exposure to wave action and its effects on the flora have been considered elsewhere, as appropriate: in this chapter; in the ecological account (Chapter 8); in the station list and in the species entries (Chapter 19). The topic is not developed further here.

Currents That the western coastlines of Mull and the off-shore islands are directly subject to virtually unimpeded and strong, long Atlantic fetch of near-oceanic salinity has already been stated. The local currents and streams that result from this Atlantic fetch and the strong outflowing coastal stream of diluted water are indicated for both surface and bottom, as most usual net water movements in the Mull area, on Fig. 7.3, modified from Craig (1959: figs 10 & 11). This latter author emphasises that such water movements are variable in the Scottish coastal region, and that neither the mean motion nor the extremes are yet accurately known. In the region from the north of Ireland to Skye, the extra run-off from the land (Craig, 1959: 17, estimates this at 18 km^3/year) in general forms a characteristic wedge of lower salinity water outflowing at the surface and causing counter-currents and eddies. The resultant coastal stream still runs predominantly south to north. Restrictions on the escape path of the outflowing fresh water cause considerable dilution in the areas of great run-off; in the Firth of Lorne, across the southern coast of Mull, strong run-off leads to both diversion of some parts of the approaching northbound oceanic stream from the south-west, and the considerable dilution of the surface waters that do penetrate. The relationship of these water movements to those elsewhere along the western Scottish coasts is given in Craig (1959), who also emphasises that the known net water movements are not particularly consistent from day to day.

Tides The spring tide-range along the Scottish west coast increases south to north from the Sound of Jura (1.0 m) to Skye and other adjacent areas north of Ardnamurchan (4.5 m). The neaps range also increases over the same area, but much less rapidly on passing northwards (Sound of Jura, 0.5 m; north of Ardnamurchan, 2.0 m). Mull lies in an area for which the spring tide-range is about 3.5 m to 3.75 m and the neaps about 1 to 1.5 m. This range, and the actual tide heights between which it represents the difference, vary quite considerably around these median figures between locations along the Mull coastlines. The extent of uncovered littoral that the figures indicate also varies a great deal according to slope, configuration, aspect, swell and many other variables, from place to place. Tide heights and times are readily available for four representative locations along the western coast of Mull and for three locations along the eastern (Sound of Mull) coastline. These data, taken from the *Admiralty Tide Tables* for 1976, are presented in Tables 7.1. and 7.2. One of

Table 7.1 Tidal data for locations on Mull, based on the 1976 Admiral tidal predictions *(Hydrographic Department, 1975). W. coast (standard port Ullapool), E. coast (standard port Oban)*

Locations	**Mean heights (m)**				**Mean ranges (m)**	
	MHWS	MHWN	MLWN	MLWS	**Neaps**	**Springs**
West coast						
Carsaig Bay	4·08	3·14	1·83	0·58	1·31	3·51
Iona	3·99	2·96	1·52	0·58	1·43	3·41
Bunessan	4·27	2·99	1·77	0·64	1·22	3·63
Ulva Sound	4·39	3·17	1·83	0·64	1·34	3·75
East coast						
Tobermory	4·39	3·32	1·80	0·70	1·52	3·69
Salen	4·18	3·11	1·74	0·67	1·37	3·51
Craignure	4·02	2·96	1·68	0·64	1·28	3·38

Table 7.2 Times (GMT) of lower low water, based on the 1976* Admiralty tidal predictions *(Hydrographic Department, 1975)*

Locations	**Night**	**Day**
West coast		
Carsaig Bay	22·13–03·49	10·24–15·40
Iona	22·10–04·02	10·23–15·53
Bunessan	21·48–04·05	10·26–15·56
Ulva Sound	21·48–04·15	10·15–16·15
East coast		
Tobermory	22·47–04·31	10·33–16·40
Salen	22·53–04·36	10·38–16·45
Craignure	22·41–04·25	10·26–16·34

* 1 m or less above Chart Datum (i.e. lowest Astronomical Tide for the area), based on GMT for the periods 1 January to 20 March and 24 October to 31 December. BST is in force (1976) from 21 March to 23 October inclusive. Although the individual dates of beginning and end of the BST period vary a little year to year, the difference is never very great.

the more important characteristics of the tidal sequence in any area, so far as concerns benthic marine (intertidal) algae, is the time(s) of day when low waters usually occur. The tidal environment, for general data on which Lewis (1964: 18–35) should be consulted, imposes on the intertidal shore alternating periods of submersion and emersion. The balance of which of these alternating periods presents more favourable and which less favourable conditions for particular organisms is probably that which in large part determines the levels on the intertidal (littoral) or supralittoral shore normally occupied by those organisms, although many other physical and biotic factors are also involved.

Characteristics of the intertidal environment are generally more extreme when emersed, since the stabilising influence of the seawater medium is then temporarily absent. The effects of direct sunlight, higher and more fluctuating temperatures, and strong breezes can then cause bleaching or desiccation, which are clearly more extreme if the emersed period coincides with the hotter and brighter parts of the day. It is not practicable here to consider all the variables that affect this situation; Lewis (*l.c.*) gives a detailed general consideration in which he shows how the levels of high and low water of neap tides could be very important for many organisms.

Below low water of neap tides, conditions will clearly depend rather more on the presence of the sea for long periods, the times of emersion being both shorter and fewer. The restrictive or other effects of the degree of change imposed are likely therefore to be the greater in areas where one of the low waters of springs occurs during the day rather than in early morning and early evening. Mull, in common with much of the rest of the northern, western, and south-western coasts of the British Isles, is such an area, the exact times over which the lower waters of spring tides occur there being presented in Table 7.2. The generalised range of lower low-water times for the west coast is 21.45–04.15 h and 10.15–16.15 h, whilst for the east coast the equivalent ranges are 22.30–04.45 and 10.15–16.45 h.

As is clear, there is little difference between the two coasts in terms of times of lower springs, although the different characteristics of substrata, slope, aspect, degree of wave action, swell and so on, discussed elsewhere, indicate that rather lesser effects of desiccation are likely on the open parts of west- and south-west-facing Mull shores than on the eastern-facing shores or in sheltered conditions on the west. Local characteristics on either shore do reduce or reverse this trend, however, since configuration and similar factors often over-ride the more generalised circumstances. Specific cases receive comment elsewhere.

The sea-lochs

The sea-lochs of the western Scottish coast are in general sheltered and surrounded by high hills; those of Mull conform to this pattern. All such lochs tend to be deep and to have lower surface salinities than adjacent coastal waters due to freshwater run-off and differences in evaporation. Milne (1972) has presented a very useful account of the results of recent surveys by Strathclyde University that, apparently for the first time, included the sea-lochs of Mull and other areas south of Ardnamurchan. On Mull, such surveys were approximately contemporary with our own work there, being made over the period December 1964 to March 1968.

West coast sea-lochs are usually classified into estuarine and fiordic; of these, the latter are more numerous.

Estuarine sea-lochs

These possess deepwater entrances and lack a sill, so that full penetration of the salinity wedge from the nearby coastal sea occurs and gives a typical vertical estuarine circulation, with the bottom water in the loch resembling that of the coastal region. Tidal range and freshwater run-off affect the circulation–dilution relationship, but the outflowing upper water is always slightly diluted. Milne classifies these lochs as 'A'-type and indicates that the salinities and temperatures are similar to those of the local coastal waters. This is certainly true for Loch Tuath, the only 'A' type loch on Mull for which data are available at this time. Tuath, lying between Mull and Ulva, is continuous at its landward-end, through Ulva Sound, with Loch na Keal. The entrance to Loch Tuath is 3 km wide and 35 m deep; since it lies near the western extremity of coastal influence, the salinities recorded by the Strathclyde University team were high, ranging from 35·5‰ to 34·55‰ (surface to bottom). August temperatures at 15 m depth were 14·2°–13·0°C. Although the inner end is relatively sheltered from wave action, there is good circulation of water through Ulva Sound, so that oxygen content was little reduced with depth; values from 9·9 to 9·1 mg/l were recorded.

Fiordic sea-lochs

These lochs are partially separated from the main coastal oceanic waters by submarine barriers, or sills. Circulation of water in these sea-lochs depends on sill depth, tidal range, freshwater inflow, and the effects and direction of wind. The sill at the entrance prevents salinity wedge penetration because the bottom water is deflected upwards and thorough vertical mixing occurs. Since rainfall and run-off exceed evaporation, (i) the surface layer is of lower salinity, and (ii) there is a surface outflow of that relatively fresh water and a balancing inflow of deeper, more saline water. Inflow of saline water depends on sill depth and tidal range within the loch. With a shallow and narrow sill, sea water inflow is small, so that unless the coastal water is at higher salinity than the loch bottom water, bottom circulation will be small. This can cause stagnant conditions below a thermocline. If the tide range is considerable, spring tides may stimulate bottom circulation by deepening the water over the sill. Salinity can also be affected by wind; if the wind blows up the loch axis, less saline run-off can be blown back up to the loch-head, whereas in the reverse case the freshwater run-off may be blown out to sea before much mixing occurs. Annual extremes of salinity are emphasised by the rainfall in winter being heaviest, and the evaporation least, whilst the rainfall is least and the evaporation greatest in summer.

Temperature variation in fiordic sea-lochs depends on heat absorption by radiation; heat conduction; tidal range and currents; and prevailing winds, among lesser factors. Separation from direct coastal water effect at some levels by the sill increases the influence of air temperature on the loch. South of Ardnamurchan, including the Island of Mull, the tidal range is relatively small and there is little overall circulation; at neap tides, even water movement is small. Air temperatures have therefore considerable influence on water temperatures in lochs of the area, causing the water to be considerably colder than the nearby coastal waters in winter, warmer in summer. The heads of lochs show the greatest temperature range, although if there is direct run-off through a stream or river into the loch-head, the water there will be brackish and possibly colder in summer than the adjacent coastal waters. The warming-up or cooling-down process is also more rapid in the shallow upper reaches of a loch than it is in the main body of water. Deep water responds very slowly to changes in ambient air temperature, but slightly more rapidly to inflowing of higher temperature coastal water, or to mixing with more superficial layers through tidal rise and fall. The cold, increased melt-water run-off in winter becomes mixed with the then warmer bottom water through such tidal effects. Wind has a similar effect on the warmer (summer) or colder (winter) surface layers to that which it exerts on salinity, blowing them up the loch or out to sea, according to direction. Wind blowing down the loch may even cause circulation, creating vertical mixing.

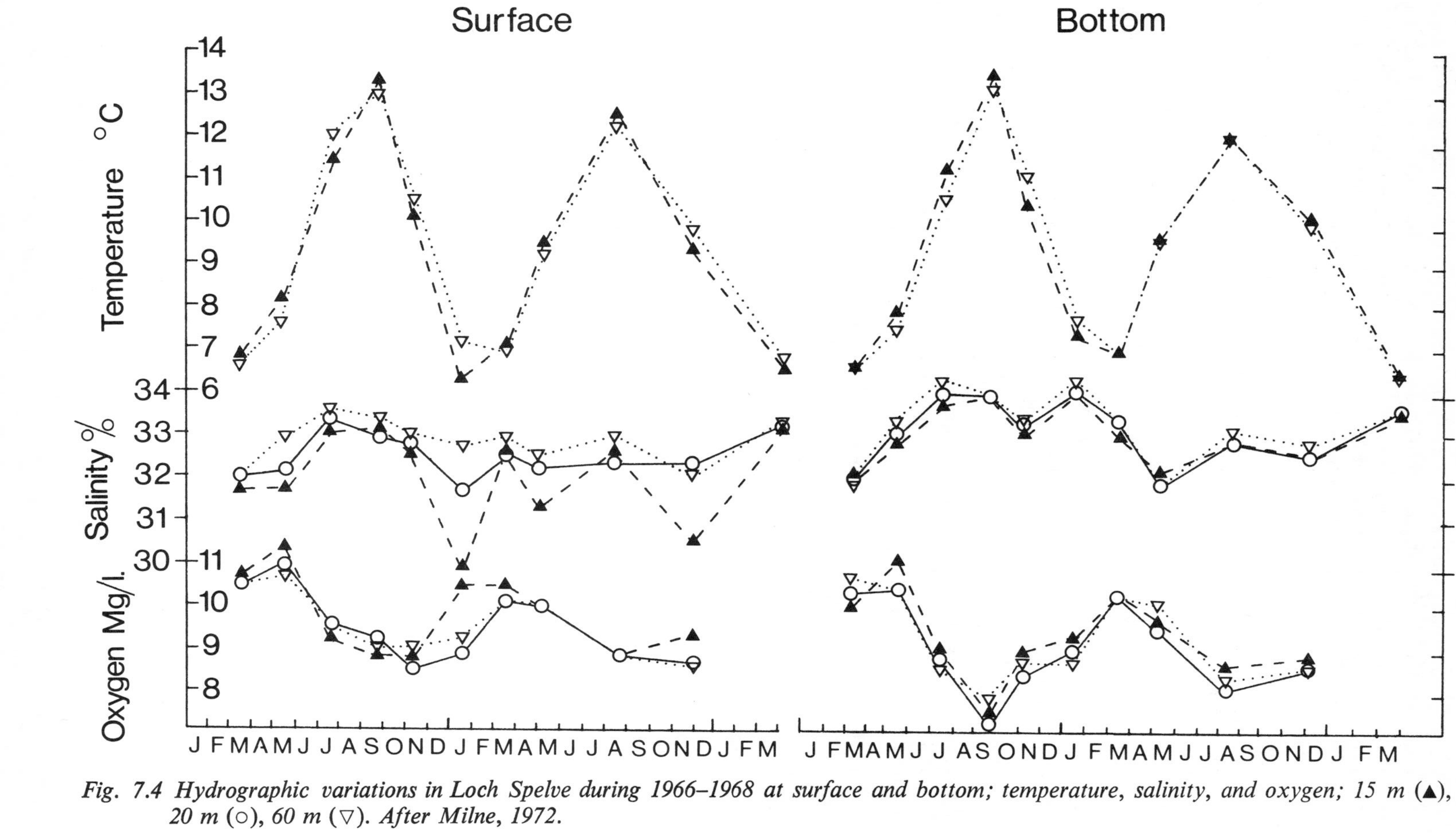

Fig. 7.4 Hydrographic variations in Loch Spelve during 1966–1968 at surface and bottom; temperature, salinity, and oxygen; 15 m (▲), *20 m* (○), *60 m* (▽). *After Milne, 1972.*

Milne (1972) has classified fiordic sea-lochs according to the complexity of their sills. Lochs with a single entrance sill are classified 'B'-type and those with two sills, one at the entrance and an internal sill dividing the inner (upper) and outer (lower) basins, as 'C'-type. On Mull, three fiordic sea-lochs have been investigated by Strathclyde University and all have emerged as 'B'-type (single sill) lochs; these are Lochs na Keal, Spelve and Don.

Loch Spelve This loch was the earliest to be examined, in the course of fish-farming surveys, observations being recorded at 10 stations in 1964–1968. The typical fiordic entrance is 6 m deep, and the main loch basin is 60 m deep. The facts that the loch has an entrance lying east and west, and that there are two arms of similar depth, one lying N–S, 2·6 km long by 0·6 km wide, and the other SW–NE, 4·1 km long by 1 km wide, both meeting in a deep basin inside the entrance channel 1·6 km long by 0.3 km wide, are stated by Milne not to be typical fiordic configurations. Fig. 7.4 (Milne, *l.c.*: fig. 15, modified) shows that there is a clear salinity wedge effect either side of the entrance sill. This general pattern is modified because the north arm (Strathclyde station 8) takes most of the catchment run-off and has lower salinities than the south arm (Strathclyde station 2). Station 5 was in the junction basin. Temperature stratification in Loch Spelve was typical of 'B' lochs, with variations of 3°C between surface and bottom (Fig. 7.3). Yearly temperatures varied between 6.5° and 13.25°C at the surface and 7·0–12·25°C at the bottom. Lower oxygen readings were recorded in autumn at 60 m, 6·7 mg/1, than at the surface, 9·5–10 mg/l.

Loch na Keal This loch was sampled from 14 stations by Strathclyde University in 1967 (Milne, 1972) and has an entrance 2·4 km wide and 48 m deep; it is 10 km long and the loch basin is 117 m deep. Salinities are therefore high, 33·0–34·85‰ surface to bottom. In August 1967, the top 15 m of the loch had temperatures of 13·7–13·0°C.

Loch Don This very small sea-loch, 0·4 km by 2·4 km, has a 2-metre deep entrance and basin depth of only 6 m; the survey was in 1966. Much of the loch surface is actually occupied by intertidal, so that salinities are low (30–32‰) and winter temperatures low, both when compared with Loch Spelve.

There are no data available in print for the remaining sea-lochs on Mull: Lochs Scridain, Buie, na Làthaich, and a'Chumhainn. From simple examination of maps and charts, it seems likely that Scridain is 'A' type; Buie possibly also 'A' type; and the other two rather similar in some ways to Loch Don ('B') although na Làthaich appears generally deeper.

The effect of growth in sea-lochs on the balance, distribution, density and luxuriance of the marine flora is variable. Apart from wave action, estuarine lochs provide habitat circumstances that, overall, vary little from the range provided by the open shores. Fiordic lochs, by contrast, provide conditions of light penetration, temperature and salinity that can clearly vary widely from those applying on adjacent coastal areas. In the extreme cases of lochs with as abnormally large a freshwater catchment area as Loch Etive (Milne 1972: 30, 33), salinities can be so low that the surface water at the head of the loch can become frozen over. Milne mentions stable salinities of 27–28‰ in the deep basin of the upper loch, with more brackish water near the surface, the 20‰ isohaline varying from 10–14 m deep in winter to being at the surface in summer. In February 1969, the head of the loch froze for 1 cm deep, had a brackish layer 5 m deep and the surface salinity was 10‰. Powell (1972: 273) has indicated that overall the surface 20 cm of Loch Etive varies from 26‰ after much dry weather to 0–10‰ after much rain, and that the effects on the ecology of the macroalgae there are profound. Various species of plants and animals present near the loch entrance gradually disappear with increasingly brackish conditions into the loch. Intertidal fucoids became much smaller in size, and *Fucus vesiculosus* and *F. serratus* are finally only found infralittorally in the inner loch, *F. serratus* penetrating to 4 m depth. This is similar to their situation in the Baltic Sea approaches. *Laminaria digitata* also becomes strictly infralittoral, and the shelter and low salinity causes it to become enfeebled anatomically, with brittle 'cape'-form (undivided) laminae.

Even in the fiordic lochs of Mull, there is no example of effects of anything like this extreme. The quantity of freshwater run-off is smaller than that into Loch Etive, and none of the lochs has so restricted an entrance with comparable run-off. Of the Mull fiordic sea-lochs examined by Strathclyde University, probably Loch Spelve was the nearest approach on physical environmental grounds to the Loch Etive situation, but even so the approach is a distant one; there is little evidence of any really deleterious effect on the marine flora of the intertidal areas of Loch Spelve, however, so far as our examinations have revealed. Salinity data for that loch have not been published in detailed form, although Milne (1972) indicates that the north arm has more run-off and lower salinities than the south arm.

References

CRAIG, R. E. (1959). Hydrography of Scottish coastal waters. *Mar. Res.*, **1959** (2): 1–30.

HALLDAL, P. (1974). Light and photosynthesis of different marine algal groups. Pp. 345–360, in Jerlov, N. G. & Nielsen, E. S. (eds), *Optical aspects of oceanography*. London & New York.

HELLEBUST, J. A. (1970). Light: plants. Pp. 125–158, in Kinne, O. (ed.), *Marine ecology*, vol. 1. London, New York, Sydney and Toronto.

HYDROGRAPHIC DEPARTMENT. (1975). *Admiralty tide tables vol. 1, 1976: European waters including Mediterranean Sea*. Taunton

JERLOV, N. G. (1951). Optical studies of ocean waters. *Rep. Swedish Deep-Sea Exped. 1947–1948*, **3** (1): 1–59.

———. (1968). *Optical oceanography*. Amsterdam, London & New York.

———. (1970). Light . . . general introduction. Pp. 95–102, in Kinne, O. (ed.), *Marine ecology*. vol. 1. London, New York, Sydney and Toronto.

JUPP, B. P. & DREW, E. A. (1974). Studies on the growth of *Laminaria hyperborea* (Gunn.) Fosl. 1. Biomass and productivity. *J. exp. mar. Biol. Ecol.*, **15**: 185–196.

KAIN, J. M. (Mrs N. S. JONES) (1966). The role of light in the ecology of *Laminaria hyperborea*. Pp. 319–334, in Bainbridge, R., Evans, G. C. & Rackham, O. (eds), *Light as an ecological factor*. Symposium no. 6 of the British Ecological Society. Oxford.

———. (1971). Continuous recording of underwater light in relation to *Laminaria* distribution. Pp. 335–346, in Crisp, D. J., *Fourth European Marine Biology Symposium*. Cambridge.

———, DREW, E. A. & JUPP, B. P. (1976). Light and the ecology of *Laminaria hyperborea* II. Pp. 63–92, in Bainbridge, R., Evans, G. C. & Rackham, O. (eds), *Light as an ecological factor: II*. Symposium no. 16 of the British Ecological Society. Oxford.

KITCHING, J. A. (1941). Studies in sublittoral ecology III. *Laminaria* forest on the west coast of Scotland: a study of zonation in relation to wave action and illumination. *Biol. Bull. mar. biol. Lab. Woods Hole*, **80**: 324–337.

LEWIS, J. R. (1964). *The ecology of rocky shores*. London.

MILNE, P. H. (1972). Hydrography of Scottish west coast sea-lochs. *Direct. Fish. Res. Rep. Scotl.*, **1972** (3): 1–50.

POWELL, H. T. (1972). The ecology of the macro-algae in sea-lochs in western Scotland. *Proc. int. Seaweed Symp.*, **7**: 273.

PRINTZ, H., (1926). Die Algenvegetation des Trondhjemsfjordes. *Skr. norske Vidensk-Akad. mat.-nat. Kl.*, **1926** (5): 1–273.

TULLOCH, D. S. & TAIT, J. B. (1959). Hydrography of the north-western approaches to the British Isles. *Mar. Res.*, **1959** (1): 1–32.

8. Marine ecosystems

J. H. Price and I. Tittley

Introduction

As Kitching (1935), Lewis (1957, 1964) and Lewis & Powell (1960, a & b) have variously shown, there can be recognised general patterns of events, in floral and faunal communities, that result from changes in exposure to wave action along the coasts of Argyll. The authors cited were concerned with the whole or major parts of the Argyll coast and were therefore considering both local and county biogeographical variation; we consider only the local changes occurring on the coasts of Mull and have therefore largely avoided subjects such as the relative amounts and distribution of the barnacles *Balanus* and *Chthamalus*.

A general system into which all rocky shores examined on Mull can be fitted with little difficulty is that given in Lewis (1964). Examples of individual shores that fit the various stages are presented later in this section. It should be stressed that each shore has individual characteristics that do not accord precisely with what might be expected of it in view of its form, its situation, and the extent of exposure. Where these are important variations, they have been indicated. Really exposed shores usually tend to be of steepish firm rock or rock slabs with large boulders; detritus and smaller debris are mostly removed elsewhere by the force of wave action. The debris often is deposited in areas of more sheltered conditions intertidally, or where water movement is less in the deeper infralittoral. Thus, although exceptions exist, there is a general correlation between substrate type and degree of exposure to water movement.

The intertidal

The real extremes of shelter from wave action on Mull shores occur at heads of lochs such as Spelve and Don (stns* 16, 17, 18); in these areas there is also in most cases the influence of strong inflow of fresh water. Such extremes of shelter from wave action are characterised by the presence of free-living populations of *Ascophyllum nodosum* ecad *mackaii* (the most extreme loss of usual *Ascophyllum* morphology) and ecad *scorpioides* (with some aspects still recalling attached sheltered-shore *Ascophyllum*). Gibb (1957) has considered at some length this situation in western Scotland and other areas of the British Isles; she noted large populations of several acres of *A. nodosum* ecad *mackaii* at Loch Spelve, north-west corner. These populations still exist (stn 18), to the west of Sàilean nan Each. They undoubtedly form the finest known beds of *A. nodosum* ecad *mackaii* in Mull. The substrate here is of mostly

* See Fig. 8.1 for location of collecting stations; details of localities are given on pp. 19.4–8.

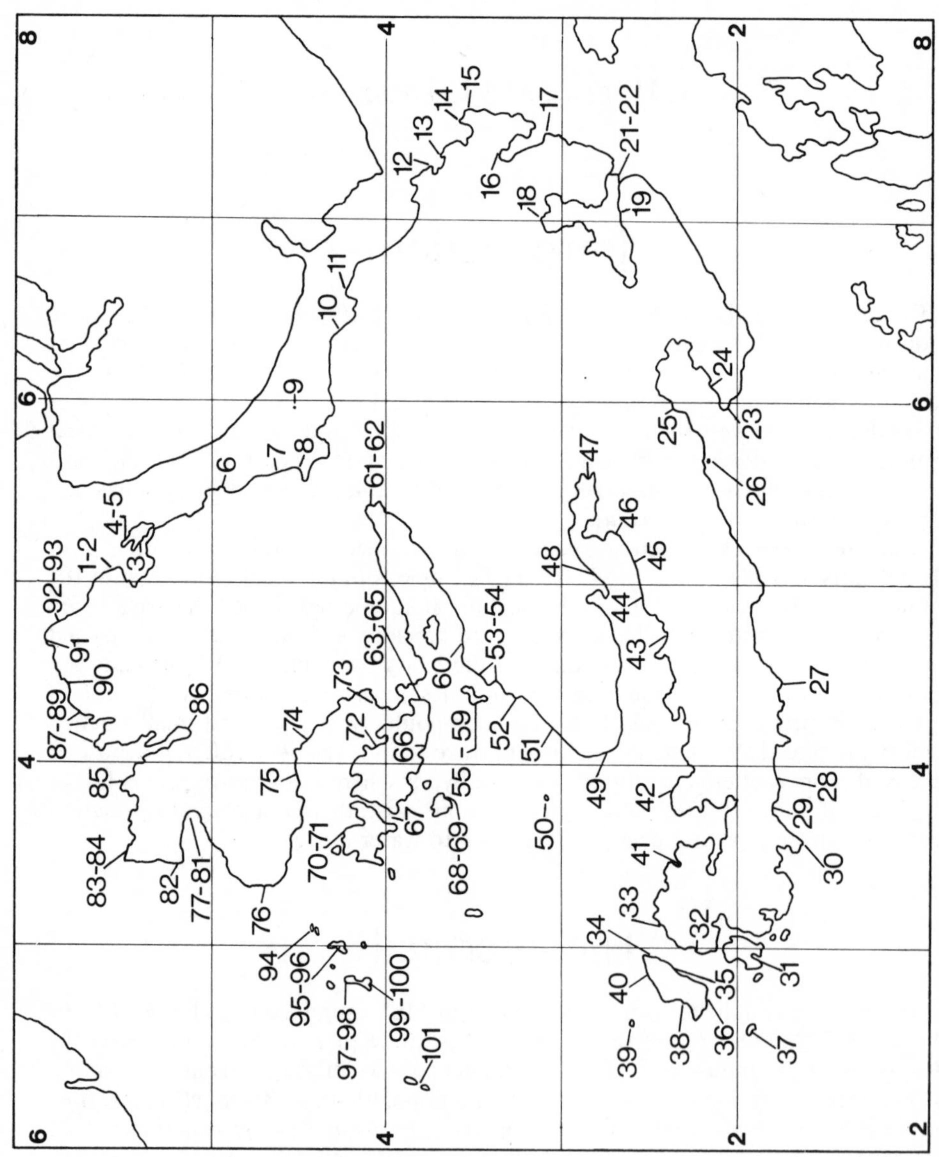

Fig. 8.1 Stations at which observations were made during the Survey. For full detailsof these localities see Table 19.1.

unconsolidated shingle on both sides of the outfall of the Lussa River into Loch Spelve; a few central ridges have undergone some degree of consolidation. On these firmer higher areas *Armeria* occurs in tufts, culminating above in an *Armeria* turf. Below this, a good growth of attached *Pelvetia* gives way to luxuriant *Fucus spiralis*, with equally luxuriant *F. vesiculosus* which intermixes with the *Ascophyllum nodosum* ecad *mackaii* in its main area of extensive lateral spread. *F. vesiculosus*, the sheltered form of the plant with many paired bladders, here carries, as an epiphyte, occasional very large plants of *Pilayella littoralis*. Also present at the same level as the concentration of *F. vesiculosus* are the typical attached form of *A. nodosum* grading through the ecad *scorpioides*, detached or attached but with still visible main axis, to the true beach form of ecad *mackaii*. In both forms, very occasional bladders are retained on a few exceptional plants. In May 1967, a few of the ecad *mackaii* plants farthest peripheral to the raised areas still bore evidence of fructiferous laterals.

Less impressive, but still large, areas of *A. nodosum* ecad *mackaii* on Mull are located in the very sheltered inner parts of Salen Bay, at An Leth-onn (a saltmarsh at Loch Scridain head), at Loch na Làthaich on the Ross of Mull, and at localities in Ulva Sound. Morphological trends towards ecad *mackaii*, generally categorised as ecad *scorpioides*, are also present in many of these listed stations, particularly where large areas of ecad *mackaii* occur. Other locations tending towards the same high degree of shelter from wave action support populations of plants recognisable as ecad *scorpioides* without, apparently, ecad *mackaii* currently being present; stations 16 (Loch Don head) and 70 (Acairseid Mhòr, Gometra) are cases in point, although the amount of ecad *scorpioides* present at the latter, a small stream head, is extremely small.

High degrees of shelter from wave action in which the whole of the 'classical' zonation sequence is manifest require firmer substrata than are commonly found in conditions carrying *Ascophyllum nodosum* ecad *mackaii* or ecad *scorpioides* populations. In such localities, the influence of freshwater inflow is commonly much less widespread (i.e. more restricted in area) than at heads of sea-lochs or bays. Many of the intertidal stations examined along the shores of the Sound of Mull – the generally less wave-exposed side of the island – carry such complete sequences and very brief details of a range of these can be obtained by consulting the station lists. As detailed examples we select two shores.

Calve Island, Tobermory is a location at which Kitching (1935) noted heavy growths of *Ascophyllum*, and which has a generally high degree of shelter from wave action; this is reflected in the dense growth of long fucoids. At station 5 (south part, near Bogha na Sruthlaig) examined in 1969, the density of *Ascophyllum* was at about the upper limit of potential density for shores. All the usual fucoid dominants were luxuriant, and this applied almost equally to most of the many stations we examined on Calve Island. *Alaria* was entirely absent from station 5, and *Himanthalia* was present as only a very few plants in standing water at the infralittoral fringe level. A similarly dense growth of *Ascophyllum* was present on a rocky shore (stn 7) of gentle slope halfway between Rubh Àrd Ealasaid and Kintallen, on the Sound of Mull. This shore showed vast areas of luxuriant growth of all the sheltered-shore fucoids; indeed Powell (*pers. comm.*) considered it to represent the best fucoid coverage and biomass of any shore he had seen anywhere near Oban. The shore was unusual in that percolation of fresh water in the area above the rocky outcrops produced marshy circumstances with brackish conditions just above the upper limit of supralittoral rock, thus carrying populations of *Armeria* and short, tufted, marsh forms of *Pelvetia* immediately above the normal *Pelvetia* of the zonation sequence. This normal *Pelvetia* was represented by a band of more than 10 metres horizontal width; *Enteromorpha intestinalis* was present where freshwater trickled down-shore, and the periwinkle *Littorina saxatilis* was frequent. *Pelvetia* graded abruptly or gently into a similarly

10-metre wide band of *Fucus spiralis*, with small and twisted *Fucus ceranoides* where the freshwater ran down. Down-shore of *F. spiralis* was a mixed band, of similar horizontal width, of *Ascophyllum nodosum* and *Fucus vesiculosus*, the *Ascophyllum* bearing much *Polysiphonia lanosa*. *Fucus serratus* was well represented below again and terminated in a richly developed infralittoral fringe of *Laminaria digitata*. Pools (shallow) under the normal *Pelvetia*, and the general damp under areas peripheral to the major fucoid vegetation, both carried a luxuriant subflora of chiefly red algae such as *Polysiphonia fruticulosa*, *Laurencia hybrida*, *Gigartina stellata*, *Dumontia incrassata*, *Lomentaria articulata*, *Chondrus crispus*, *Gelidium pusillum*, *Membranoptera alata* and *Phymatolithon lenormandii*. These red algae tended to come in more strongly the further down-shore one made observations; *Membranoptera alata* and *Phymatolithon lenormandii*, for example, were present only at the *Fucus serratus* level. Further up-shore, at *Pelvetia*, *F. spiralis* and *Ascophyllum* levels, small brown algae formed conspicuous elements in the underflora, or as epiphytes, or both. *Pilayella*, a case in point, was, after *Polysiphonia lanosa*, the major constituent of the epiphyte flora of *Ascophyllum*, and at that was only slightly more abundant than *Callithamnion hookeri*, then apparently restricted to *Ascophyllum*. Elements usually characteristic of other habitat circumstances were introduced down-shore at stn 7 by the presence of a shallow lagoon-like area in which occurred long yellowish fruiting fronds of *Himanthalia* (also detected in similar circumstances at station 54) and the 'cape' (undivided) form of *Laminaria digitata*. The latter, also detected by workers of the Scottish Marine Biological Association's Loch Etive research project in sheltered conditions in that and other similar lochs (Powell, 1970; 1972), seems to be here a response to sheltered standing water. Conditions of this type, i.e. lagoons retained at low water by rock ridges in otherwise sheltered densely *Ascophyllum*-covered shores, or circumstances in which such a lagoon or large pool does not receive strong direct wave-beat, whatever the conditions on the rest of the retaining shore, have been productive also of important new finds of elements hitherto rarely or never detected in the flora of Mull, or indeed of the southern part of western Scotland. *Cystoseira tamariscifolia*, the most important of these, is described more fully elsewhere in the species list (stns 78, 79, 101 are the more important here). *C. nodicaulis* should also be noted in the same context. It is worth mentioning here that the 'cape' form, or close approach to it, was found elsewhere in the more usual strictly infralittoral circumstances in fronds of *Laminaria digitata* (stns 5, 7, 70) and *L. hyperborea* (stns 6, 15, 41).

The first step in the gradation toward shores of intermediate exposure to wave action was on Mull often characterised by a reduction in general size and luxuriance of the fucoid growths, without loss of any particular species. Station 13 for example (Rubha na Sròine, near Craignure), was subject to a long fetch from the north-west, down the Sound of Mull. This station, some distance to the south down the Sound from the protected and sheltered station 7 (see above), revealed a complete fucoid zonation pattern but formed of plants all very much smaller and less luxuriant than those noted at station 7; the shore at station 13 was additionally rather steeper. This kind of general size reduction seems not always to occur, although it is customary as an effect on the susceptible fucoids. Often the first easily detectable sign of increasing exposure is the loss of *Ascophyllum nodosum*, although careful observation of any series of positions along shores with gradual increases in wave-beat normally reveals that there has occurred size reduction in at least the *Ascophyllum* prior to its elimination. The shore of stn 54, eastern end (profile, Fig. 8.2) which represented a virtually identical intertidal to that of the diving station 60 – Creag Brimishgan, Loch na Keal – gave an ideal non-*Ascophyllum* intertidal zonation sequence, including *Pelvetia*, *Fucus spiralis*, *F. vesiculosus*, *F. serratus*, *Laminaria digitata*, and continuing down beyond the infralittoral fringe with *L. hyperborea*, *L. saccharina*, and *Saccorhiza*. Infralittoral characteristics there are subject to later comment. Loss of *Ascophyllum*

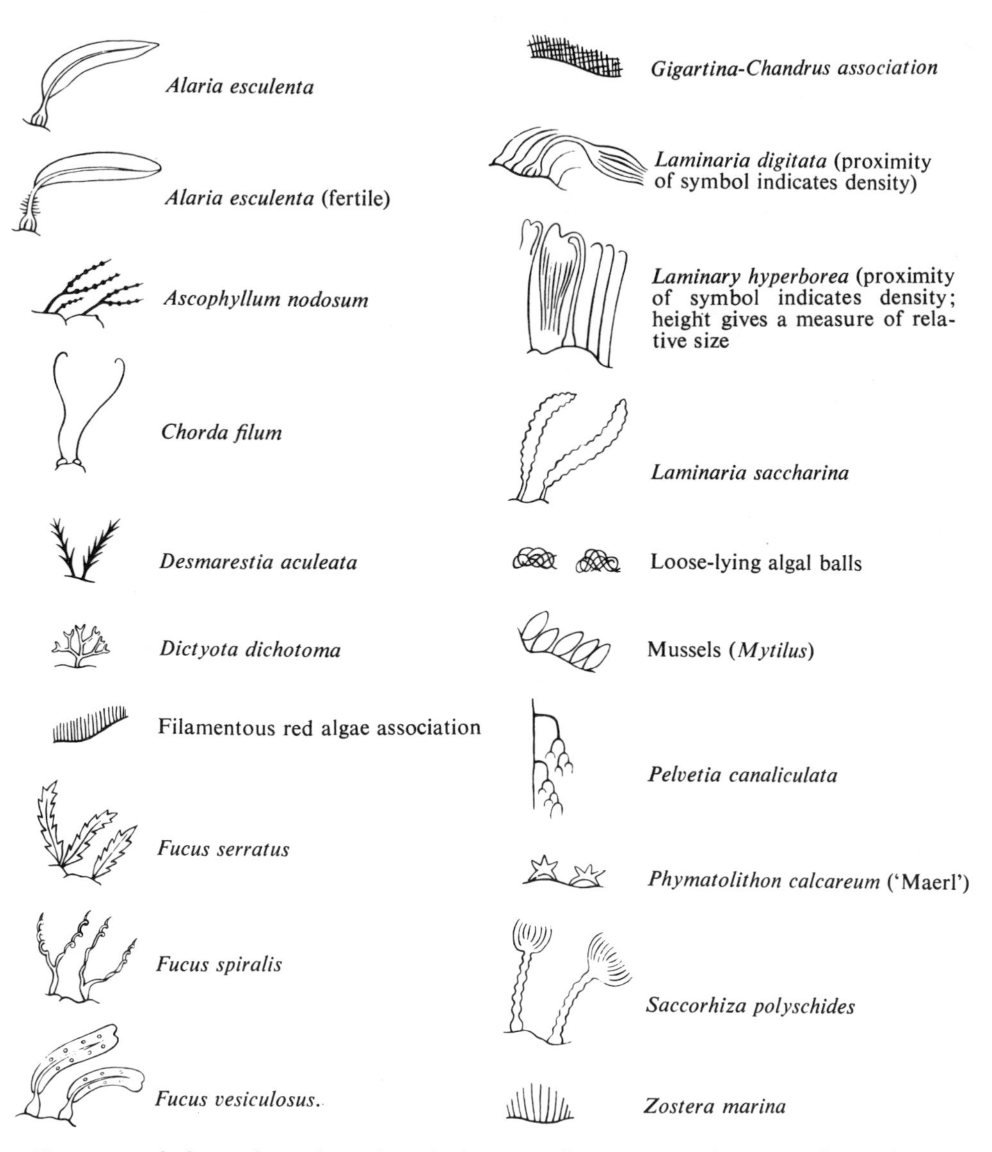

Key to symbols used in the selected shore profiles in Figs. 8.2–8.15. Symbols are exaggerated in relation to the vertical scale of the profile.

is customarily paralleled by loss of bladders from *F. vesiculosus*, at least to the extent that the classical form with paired vesicles along the laminae no longer occurs; occasionally, vesicles may occur on the fronds, but these latter are often not only evesiculate but also elaminate. When this stage is reached, commonly even *Pelvetia*, usually the normal form of *F. spiralis*, and often *F. serratus* are also missing, although the sequence can differ in that all these forms can occasionally persist alongside the

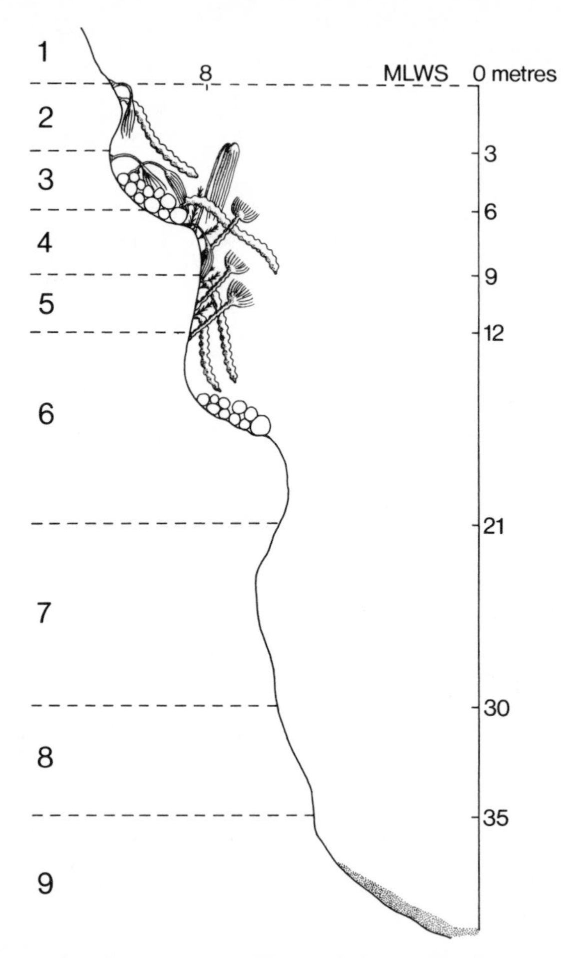

Fig. 8.2 Near-vertical cliff, stepped to 35 m, off Creag Brimishgan (data from stns 54 and 60).

Notes

1. Intertidal zonation classical but lacking *Ascophyllum*.
2. *Laminaria digitata – L. saccharina* codominant, with *Callophyllis, Cryptopleura, Dilsea, Lomentaria articulata, Odonthalia, Plocamium, Plumaria* and *Ptilota* as main subsidiaries.
3. *Laminaria digitata* dominant, with *Dictyota dichotoma, Nitophyllum punctatum* and *Phycodrys rubens* as main subsidiaries. Some *Laminaria saccharina.*
4. *Laminaria hyperborea – Saccorhiza* 'forest', with *Delesseria sanguinea, Nitophyllum punctatum*, and *Phycodrys rubens* as main subsidiaries. Some *Desmarestia aculeata, Laminaria saccharina* and *L. digitata.*
5. *Desmarestia aculeata – Laminaria saccharina* codominant, with *Dictyota dichotoma* as main subsidiary. Some *Saccorhiza*, very large, continues to > 10 m deep.
6. No laminarians present; flora consists of red algae (e.g. *Brongniartella byssoides, Delesseria sanguinea, Phycodrys rubens*) and *Dictyota dichotoma.*
7. *Lithothamnia* only.
8. No algae recorded below 30 m.
9. As slope eases, silt deposited; brittle star (*Antedon bifida*) abundant.

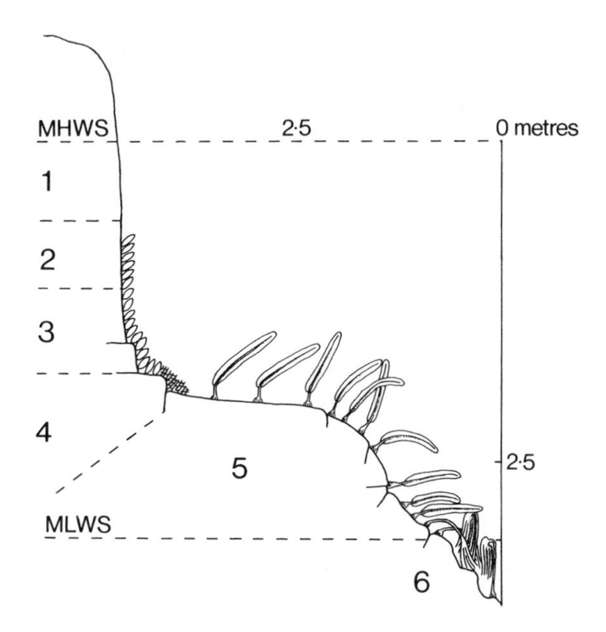

Fig. 8.3 Very exposed faces, S to SW facing, on the SE tip of Rubha na Faoilinn, Loch Buie (stn 23).

Notes

1. The lichen, *Lichina pygmaea*, with *Porphyra umbilicalis* immediately above; well developed barnacle zone thence down into *Mytilus* zone.
2. Clumps of *Ceramium shuttleworthianum* in upper *Mytilus* zone.
3. Young *Alaria* on *Mytilus*
4. Stunted and damaged *Chondrus crispus* and *Gigartina stellata*, with *Callithamnion arbuscula* on *Mytilus*.
5. *Porphyra* beneath *Alaria*: *Himanthalia* above.
6. Lower profile (not shown) consisting of *Laminaria* 'forest'.

linearis form of *F. vesiculosus*. *F. serratus* not infrequently is present at a slightly lower level at the same time, in the upper infralittoral fringe.

This stage in the effects of increasing exposure to wave-action is well illustrated by part of station 26, on the island of Gamhnach Mhòr, Carsaig Bay. The landward-facing shore of the island is sheltered, to a large extent, from the long fetch of the Atlantic which affects the seaward-facing shore. On the outer-facing shore, the maritime lichens show excellent development, the sequence reaching the top of this low-lying island. *Pelvetia* is absent from the open faces, but *Porphyra umbilicalis* occurs, associated with occasional *Littorina neritoides* and scattered *Chthamalus*. Overlying the latter is an excellently developed belt of the lichen, *Lichina pygmaea*, which also penetrates the upper *Balanus balanoides*. *Fucus spiralis* f. *nanus* occurs very sparsely at this level. Below, the middle shore is dominated by *Fucus vesiculosus* forma *linearis;* in June, we detected plants of 15–25 cms length in full fruiting condition. *B. balanoides*, small *Mytilus* (mussel) and *Patella vulgata* (limpet) were common. Both *Callithamnion arbuscula* and *Ceramium shuttleworthianum*, largely epizoic, were also common; these, in this area, are both excellent indicators of a high degree of wave-beat and generally occur on shores on which animals appear as the predominant organisms. In the lower mid-littoral, *Balanus* and *Fucus vesiculosus* f. *linearis* thinned out and patches of *Himanthalia elongata* and *Laminaria digitata*, representing the infralittoral fringe, began to appear. *F. vesiculosus* f. *linearis* at this level bore much epiphytic *Palmaria palmata*. Below this fringe, a mixture of *Laminaria digitata* and *Alaria esculenta*, the former still predominant, occurred, showing that this shore was by no means the ultimate in exposure to wave-beat. Pools in the boundary area characterised by *F. vesiculosus* f. *linearis*, *H. elongata* and *L. digitata* showed the usual upcarry of elements from below, and revealed other facets of the effects of stronger wave action, for example the presence of *Polysiphonia brodiaei*, small *Alaria esculenta*, and increasing amounts of *Gigartina stellata*.

In passing over the island crest and onto the sheltered side, the first signs of the effects of reduction in wave action occur in those inlets dissecting the crest from both sides. *Pelvetia*, in the form of abundant large plants, soon comes in and *Fucus spiralis* shows immediate increase, although plants of the normal form remain sparse on the sheltered side. *Fucus vesiculosus* f. *linearis* (up to 45 cm long) still predominates in the mid-shore levels, but plants with occasional bladders occur amongst the rest. *Balanus* is more sparse, and a great deal of *Patella vulgata* and *Nucella* (dog whelk) are present. Mid-littoral pools of adequate depth carry well-developed *Halidrys*, and gullies penetrating deeply into the mid-littoral have *Gigartina stellata*, *Chondrus crispus* and *Fucus serratus*. The latter, absent from the exposed side, is present in almost band-forming amounts below *Fucus vesiculosus* f. *linearis* and the infralittoral fringe again reveals *Himanthalia*, with *Laminaria digitata* and still some large amounts of *Alaria* below. Additional changes in the underflora of red algae on the two faces of the island include considerable increase in both amount and species of *Laurencia* on the sheltered side; only *L. pinnatifida* seems to penetrate the exposed face. By contrast, the red algal underflora is on the whole richer in damp places and in standing water in the lower reaches of the exposed face. Since both faces carry predominantly the fucoid *F. vesiculosus* f. *linearis*, both must be considered to fall within the intermediate exposure region of the scheme outlined in Lewis (1964). Whereas the exposed side of Gamhnach Mhòr shows strong tendencies towards the more exposed facets of the

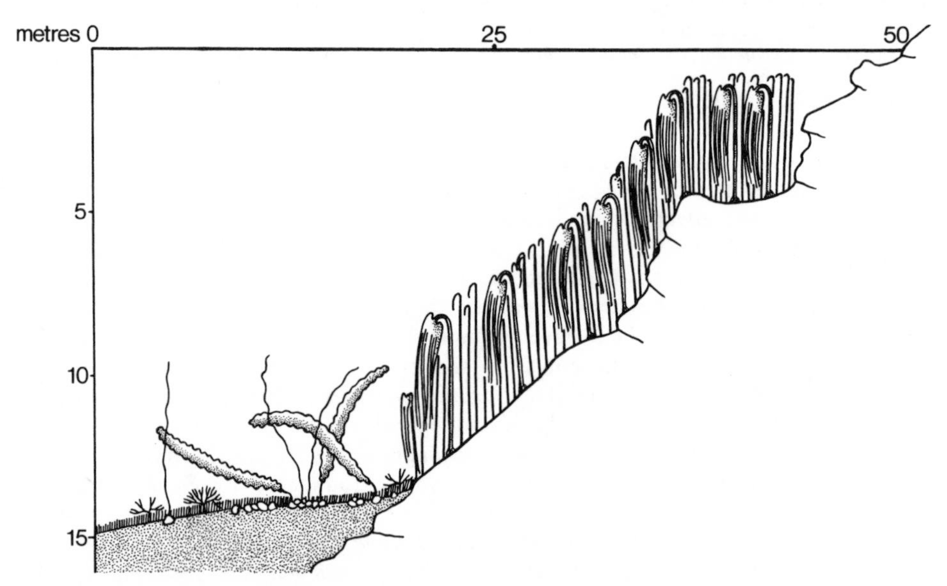

Fig. 8.4 Rocky shore, outside the harbour in Sound of Ulva (stn 67).

Note
The flora on the sand floor included *Ceramium rubrum*, *Cladostephus verticillatus*, *Cryptopleura*, *Furcellaria*, *Gracilaria verrucosa*, *Heterosiphonia* and *Phyllophora crispa*, with the following species both as epiphytes and on the sand: *Acrosorium uncinatum*, *Antithamnion plumula*, *A. spirographidis*, *Asparagopsis armata* [*Falkenbergia* phase]. *Bonnemaisonia hamifera* [*Trailliella* phase], *Brongniartella byssoides*, *Callithamnium corymbosum*, *Champia parvula*, *Chylocladia verticillata*, *Mesogloia vermiculata*, *Nitophyllum punctatum*, *Palmaria palmata*, *Plocamium cartilagineum*, *Polysiphonia nigrescens*, *P. urceolata*, *Porphyra*, *Pterosiphonia parasitica*, and *Spermothamnion repens*.

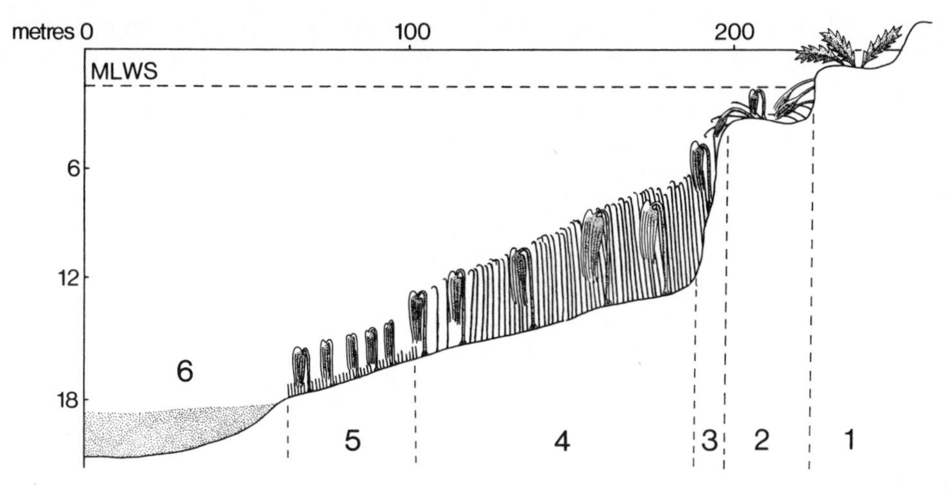

Fig. 8.5 Rocky shore on SE Lunga (stn 99).

Notes
1. Lower intertidal areas of *Fucus serratus, Himanthalia, Palmaria* and *Gigartina.*
2. *Palmaria* with epiphytic *Ptilota plumosa.*
3. Less *Laminaria hyperborea* on steeper area, with stipes to 1.4 m.
4. *L. hyperborea* 'forest', stipes up to 1.8 m.
5. Occasional *L. hyperborea, Halidrys, Phyllophora crispa* and small, mainly red, algae mostly epiphytic (e.g. *Acrosorium uncinatum, Antithamnion plumula, A. spirographidis, Brongniartella byssoides, Callophyllis, Ceramium rubrum, Cladophora* spp., *Cryptopleura ramosa, Dictyota, Heterosiphonia plumosa, Lomentaria clavellosa, Plocamium, Polysiphonia urceolata, Pterosiphonia parasitica* and *Spermothamnion repens*) with some species attached to stones (e.g. *Bonnemaisonia asparagoides, B. hamifera* [*Trailliella* phase] and *Callophyllis*).
6. Coarse white sand.

intermediate classification, the sheltered face reveals more affinity with the mozaic gradation from intermediate to sheltered shores. The presence of *Alaria* on both sides of the island is only anomalous if no account is taken of the steeply descending nature of the rock surface at that level, creating there more of the effects of strong water movement. A similar site, revealing the same sort of comparison between exposed and sheltered sides, was examined later at Coireachan Gorma (Ardmeanach Peninsula, stn 51) where a rock reef shows a very similar seaward face, in this case bearing also *Nemalion helminthoides*, and an exactly comparable landward (sheltered) face.

Shores in which ultimate degrees of exposure are attained in the intertidal on Mull are few and, by their nature, hard to examine in detail. In fact, we did not detect any shore for which it could really be said that over large consistent areas even *Fucus vesiculosus* f. *linearis* was entirely absent. Many stations examined had large stretches which closely approached this state, and smaller patches within which *F. vesiculosus* f. *linearis* was actually absent, but not sufficiently consistently to warrant application of anything approaching the ultimate in exposure classification.

Lewis (1964: 288–289) defines very exposed shores in fucoid terms as being characterised by the presence of *Fucus distichus* subsp. *anceps* and *Fucus spiralis* f. *nanus*, so far as the north and west is concerned. The limpet dominance of *Patella aspera* in mid and lower shore, the presence of dominant barnacles, limpets or *Mytilus*–Rhodophyceae communities, the latter occasionally replaced by a *Lithothamnion–Corallina*–Rhodophyceae belt just above *Alaria*, are also noted as important by Lewis. We did not detect with certainty *Fucus distichus* subsp. *anceps* anywhere on Mull or the adjacent westward-lying islands. *Fucus spiralis* f. *nanus* was certainly patchily noted at some stations, but in circumstances in which its significance as an exposure indicator was at best equivocal. For example station 38 – the exposed outer end of an un-named rock promontory in the bay of Camas Cuil an t-Saimh, W Iona – presented a rather mixed shore in which indicators of strong exposure to wave action were not developed to the exclusion of moderate exposure indicators. The presence of *Alaria esculenta*, *Callithamnion arbuscula*, *Ceramium shuttleworthianum*, *Polysiphonia brodiaei*, *Fucus vesiculosus* f. *linearis*, and *Himanthalia*, in varying amounts (although the first two were much scarcer than might have been expected) provided clear signs of exposure to fairly strong wave action. These were balanced by the opposing signs of well-developed lower-shore *Laminaria digitata* and high-upper-shore *Pelvetia canaliculata;* amongst the lower *Pelvetia* growths were occasional patches of *Fucus spiralis* f. *nanus*. On this shore, therefore, the presence of the latter could not be taken to indicate anything like the high degree of exposure that a more clear-cut set of characteristics would imply. A similar instance of the presence of *Fucus spiralis* f. *nanus* on a shore otherwise carrying characterising amounts of middle-level *F. vesiculosus* f. *linearis* and showing additional facets of only a moderately high degree of exposure was presented earlier above (Gamhnach Mhòr, exposed face, stn 26).

Patella aspera in large amounts was not frequently observed during the survey; station 51, Coireachan Gorma, was one of the few instances. This exposed rocky shore was backed by high cliffs and provided contrasts in exposure to wave action of the same type as shown by station 26. The rock reef constituting the shore was so orientated that the landward face was very much more sheltered than the seaward; the present remarks concern the latter. The seaward face carried a narrow but luxuriant band of *Alaria esculenta* at and just below infralittoral fringe level, above the *Laminaria digitata* communities. At this level and up to the middle-shore, *Patella aspera* was the predominant limpet and *Balanus balanoides* was well represented.

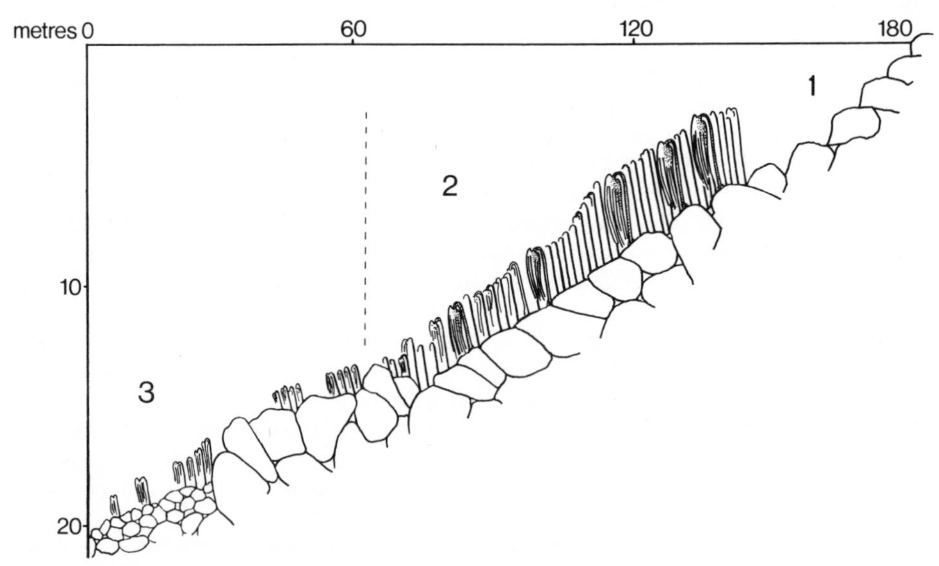

Fig. 8.6 Slope of stable boulders with smaller unstable stones towards the base; on east side of Bac Mòr, Treshnish Islands (stn 101).

Notes

1. Upper section of profile not recorded.
2. *Laminaria hyperborea* diminishing in density and plant size towards base of slope.
3. Mainly small red algae on small boulders and stones. Species include: *Acrosorium uncinatum*, *Bonnemaisonia asparagoides* (♀), *B. hamifera* [Trailliella phase], *Callophyllis laciniata*, *Chylocladia verticillata*, *Griffithsia flosculosa*, *Lithothamnia*, *Lomentaria clavellosa*, *Phycodrys rubens*, *Plocamium cartilagineum*, with *Cladophora* spp., *Dictyota dichotoma* and *Ralfsia* sp.

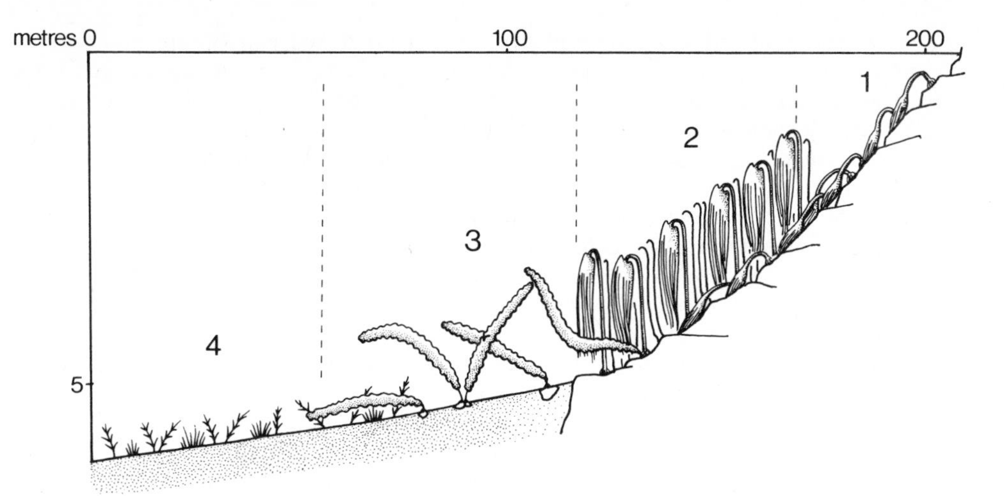

Fig. 8.7 Laminaria *associations on a rocky shore arising from a coarse sand bed. NW shore of Loch Buie near Glenbyre Farm (stn 25).*

Notes

1. *L. digitata* 'forest' rich in epiphytes; under and associated flora consisting of *Chorda*, *Dilsea carnosa*, *Halidrys*, *Himanthalia* and *Scytosiphon*.
2. Dense *L. hyperborea* 'forest' with luxuriant stipe-epiphytes including: *Callophyllis*, *Membranoptera*, *Phycodrys* and *Ptilota*.
3. *L. saccharina*, locally abundant, fronds up to 5 m; underflora including *Ceramium rubrum*, large *Cystoclonium*, *Dictyosiphon* and large *Ulva*.
4. *Desmarestia aculeata* locally abundant; *Zostera* occasional.

Immediately above the level of *Alaria*, well-developed communities of Lithothamnia–*Corallina*–Rhodophyceae, as indicated by Lewis in characterisation of the level, were indeed present here. *Laurencia pinnatifida*, *Lomentaria articulata*, *L. clavellosa*, *Hildenbrandia rubra*, *Nemalion helminthoides*, and *Polysiphonia brodiaei* were the main constituents of the Rhodophyceae at that level, and *Mytilus*, *Ralfsia* and *Bryopsis plumosa* also appeared. *Porphyra umbilicalis* and *P. leucosticta* grew among and on the other species. Towards the middle-shore levels, *Mytilus* and *Balanus balanoides* increased considerably, so that on the open steep rock surfaces there, *Balanus* and the *Mytilus*–Rhodophyceae communities were patchily and variously the predominant facets of the biota. Elsewhere and very locally, *F. vesiculosus* f. *linearis* was well developed and formed the predominant component. *Nemalion*, on open rock with *Balanus* and *Patella*, similarly increased in amount from the lower shore levels, and other red algae of the *Mytilus*–Rhodophyceae communities included *Callithamnion arbuscula* (on *Mytilus*), *Ceramium shuttleworthianum* (on *Mytilus*, abundant), *Laurencia pinnatifida* (occasional), *Palmaria palmata* (occasional), and *Corallina officinalis*. The presence throughout these mid-shore levels of large numbers of *Nucella lapillus*, taken together with the generalities described above and balanced against the presence of *Patella aspera* in (mainly) the lower shore, indicates a shore still falling within Lewis's (1964) 'exposed' category, although showing facets trending towards the 'very exposed' category.

A similar but perhaps slightly more extreme grading has to be allocated to station 23 – the headland of Rubha na Faoilinn – at the south-eastern tip of Loch Buie. Stations 24 and 23 reveal a nice gradation of exposure from south of Rubha Liath, below Aoineadh Fada, and along the south-eastern shore of Loch Buie to the tip of Rubha na Faoilinn. The semi-exposed shores of station 24, inner end, carried good fucoid growths, including *Fucus vesiculosus* with sparse bladders, and a well-developed

Pelvetia belt. The mid-shore showed at this point a mixed community of *Balanus* (of average density), *Patella vulgata*, and *F. vesiculosus*. *Alaria* already appeared in lower shore pools. Steeper rock faces appear as one moves towards the south-eastern tip of the loch shore and the density of *Balanus* began to approach levels suggested by Lewis for categorisation of 'very exposed' shores (100/5 cm^2). Both *Littorina neritoides* and *L. saxatilis* were common, a characteristic of Lewis's 'exposed' category; the presence of *Mytilis* bearing both *Callithamnion arbuscula* and *Ceramium shuttleworthianum* pointed in the same direction, although the latter two species were rather scarce. *Polysiphonia brodiaei* appeared in pools at this point in the transition to the really exposed south-eastern tip, and *Fucus vesiculosus* f. *linearis* was by then the predominant fucoid. Most of the rock faces of stn 23 (profile, Fig. 8.3) are south or south-west facing, thus receiving virtually the unbroken force of the long Atlantic fetch. The upper shore, at the higher fringe of the well-developed barnacle 'zone', carried a good density of *Porphyra umbilicalis* above and around abundant *Lichina pygmaea*. The middle-shore was dominated by *Mytilis edulis*, with occasional stunted *Fucus vesiculosus* f. *linearis* towards the upper reaches and with tufts of *Ceramium shuttleworthianium* and *Callithamnion arbuscula* fairly scattered throughout. The latter, with *Porphyra purpurea*, young *Alaria*, battered *Chondrus*, and *Gigartina stellata*, was more common towards the lower mid-littoral, still on *Mytilus*. *Corallina officinalis* appeared in pools in the lower half of the *Mytilus* levels. The shore eased off to a slightly more gentle angle just above the infralittoral fringe, at which level *Alaria* formed the predominant facet of the flora, with *Himanthalia* just above it and large amounts of *Porphyra umbilicalis* patchily amongst it. *Alaria*, the upper limit of which cannot have been very far below mean low water of neap tides, descended well into the infralittoral and there was thus clear depression of the Laminarian dominated levels (with *Laminaria digitata* and *L. hyperborea*) to below those levels usual for open shores without exposure to strong wave action. (This point is fully discussed below.) In their lower half, *Alaria* levels carried a strongly developed community in which

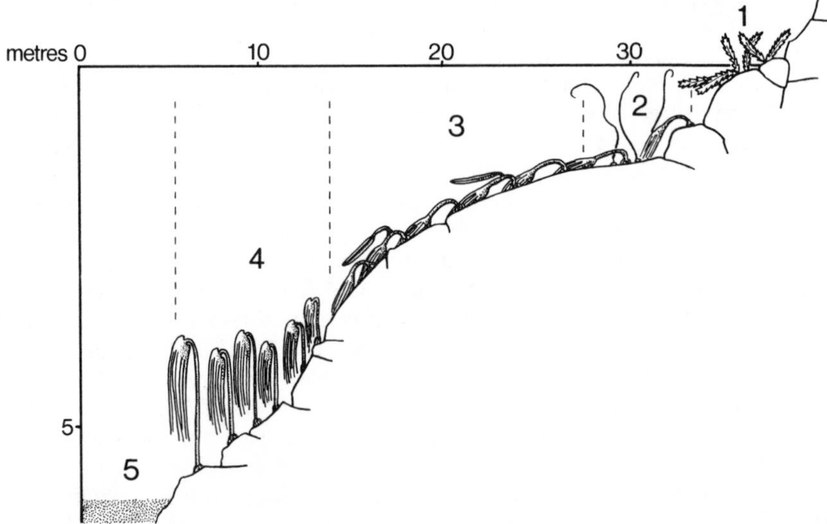

Fig. 8.8 SE shore of Inch Kenneth (stn 57); very similar to the NW shore but lacks infralittoral platform.

Notes

1. *Fucus serratus* with *Chondrus crispus* below.
2. *Himanthalia elongata* and *Chorda filum* commonly present around 1.5 m.
3. *Alaria* and many other species (mainly red algae) epiphytic on *Laminaria digitata*.
4. A few plants of *Dilsea carnosa* under *Laminaria hyperborea*.
5. Coarse sand.

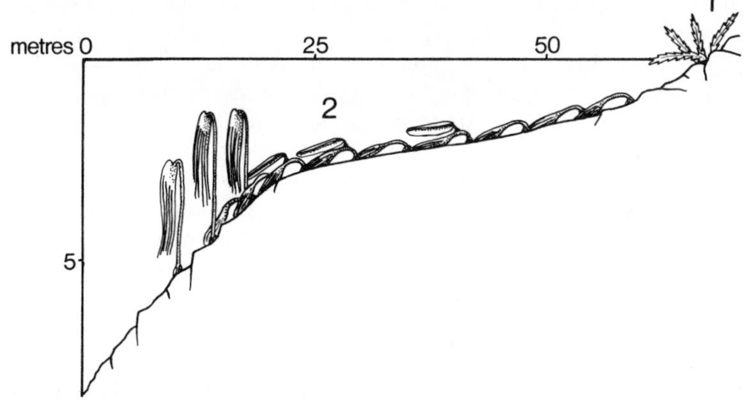

Fig. 8.9 A shallow infralittoral rock platform on the NW shore of Inch Kenneth (stn 59); wave-action enough to eliminate Laminaria saccharina *and encourage* Alaria.

Notes

1. Some shelter is indicated by *Fucus serratus*; latter gives way below to *Chondrus crispus*.
2. *Alaria* mostly epiphytic on *Laminaria digitata*; also abundant on *Laminaria* are *Chondrus* (on stipe), *Ectocarpus* (on lamina), *Melobesia* (on stipe), *Membranoptera* (on holdfast and stipe), *Ptilota* (on stipe), and *Palmaria* (on lamina and stipe).

Corallina officinalis and the Lithothamnia were major constituents. Therefore, ignoring local micro-differences in the environment that produce mosaic effects on the distribution of flora, the whole shore of station 23 was probably the nearest approach to a 'very exposed' shore, *sensu* Lewis (1964), examined by us on Mull. Areas of stations 39, 50, 51 and 69, as well as station 100 (see later), would probably merit the same classification.

The infralittoral

As indicated elsewhere, it is only within the last 35 years that really detailed descriptive study of the infralittoral has been practicable. Argyll was one of the first areas to be examined in this way (Kitching, 1941), before SCUBA techniques, which permit the optimum in study conditions generally possible in this restricting environment, became available. As a consequence of a spate of infralittoral studies made in the last 15 years, Argyll is currently one of the best known areas of the British Isles as regards its submerged flora. Most of the detailed data have been obtained by Norton and his colleagues at the University of Glasgow, who have published a summary (Norton & Milburn, 1972) based on selected sites, chiefly along the mainland coast of Argyll. In what follows, our findings are compared with theirs from this adjacent area.

On Mull, at different locations with apparently very similar environmental conditions, the same species were not always present, nor were their local distributions always the same. Norton and Milburn also found this to be the case in their studies. Chance was probably the greatest factor in producing this situation, not only in its effect on floristic distribution but also because it determined the times when we actually examined sites. The degree of similarity between comparable sites is probably greater in some seasons and years, less in others. This statement could, of course, apply equally well to intertidal as to infralittoral locations. It is the best possible argument for the examination of as many similar sites as practicable at as many possible seasons over the maximum possible period of time before conclusions are drawn. This does not imply that we doubt the effects of slope, aspect, water movement, substratum, turbidity, biotic factors, and other possible variables; all these have their undoubted influence, but it is their combination at different times and in different locations which we still have currently to regard as a matter of chance.

Substratum type certainly appears to be the major factor behind the appearance of recognisable communities characterised by the predominant larger plants (chiefly brown or red algae, or *Zostera*). Within the constraints of the geological nature of the area concerned, together with the local characteristics of detritus downwash off adjacent coasts, it is primarily the combined effects of current systems and wave-beat on shores that determine (a) the physical nature of the substratum temporarily or constantly presented to settling spores, gametes, zygotes or fragments, and (b) the turbidity of the water. This combination of characteristics does not only apply to the upper reaches of the infralittoral down to a depth dependent on the local extent of water movement, since the whole of the intertidal, within a range of variation already described earlier in this ecological section, is even more directly and strongly affected. Examples of the major infralittoral shore types on Mull that result from the locally varying combinations of these factors are discussed later and are depicted in the profiles in Figs 8.2 to 8.15.

The general picture that emerges is of infralittoral populations characterised on stable rock or on larger semi-stable to occasionally mobile substrata by larger brown algae above, grading down, at a depth variable according to the area, to various larger red algae. The more or less ubiquitous and continuous underlayer of Lithothamnia ends to be revealed at depths below the penetration levels of the larger red algae, the Lithothamnia being usually by some vertical distance the last forms to succumb to the adverse effects of increasing depth. Both Kain (1962) and Norton & Milburn (1972) have indicated that sites on stable rock surfaces in at least moderately strong wave action and of at least a minimum degree of steepness, examined by them in Argyll, were characterised by *Laminaria hyperborea* 'forest' (classically detected and described in 1941 by Kitching from work in 1932–36 down to 12 m depth off Carsaig

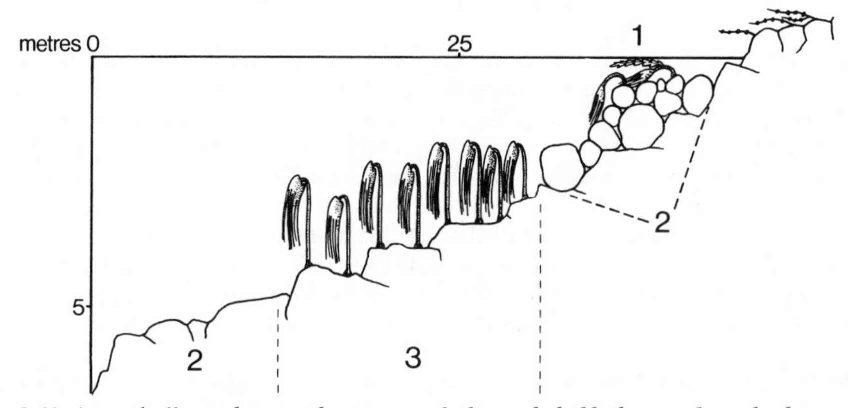

Fig. 8.10 Ascophyllum *shore with narrow infralittoral shelf, thence abruptly descending at least to 36 m, Bloody Bay (stn 93).*

Notes

1. The shelf area with *Fucus serratus*, *Melobesia* and *Palmaria*, epiphytic on stipes of *Laminaria digitata*.
2. Zone of greatest density of flora, with an abundance of small red algae.
3. *Laminaria hyperborea* zone, with epilithic *Delesseria*, *Dilsea*, *Phycodrys* and *Polysiphonia urceolata*. The following were recorded as epiphytes on the stipes of the *Laminaria*: *Chondrus*, *Cladophora rupestris*, *Cryptopleura*, *Delesseria* (large clumps), *Laminaria digitata*, *Lomentaria articulata*, *L. clavellosa*, *Melobesia*, *Membranoptera*, *Palmaria*, *Phycodrys* (in large clumps, with epiphytic *Alaria*) and *Ulva*.

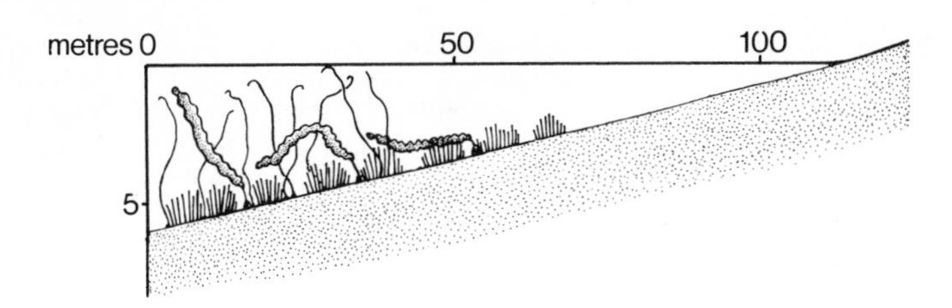

Fig. 8.11 A gently sloping shore on white coarse sand at the north end of the Sound of Iona (stn 34); Zostera – Chorda *community with some* Laminaria saccharina.

Note
Few attached algae were noted along the inner 40 m of the transect.

Island, Sound of Jura). We found this also to be true for Mull shores, as our observations at stations 1, 21, 52, 60, 65, 67, 88, 99, 101, and others show clearly; from these, examples (stns 60, 67, 99, 101) are illustrated as profiles in Figs 8.2; 8.4–8.6. We often noted a narrow zone or band of *Laminaria digitata* in (1–)3–6 m depth above the *L. hyperborea* 'forest', and to a variable degree overlapping with it (Figs 8.2, 8.3, 8.5, 8.7–8.10). Not infrequently, the *L. hyperborea* showed a relatively large admixture of plants of *Saccorhiza polyschides* (Fig. 8.2) and graded out in 9 or more metres to a band of *Laminaria saccharina* within the 3 metres, or little more, vertically below the *L. hyperborea* 'forest (Figs 8.4, 8.7). *Saccorhiza* plants occasionally penetrated this lower *L. saccharina* band on stable substrata (e.g. stn 65). A large admixture of *Desmarestia aculeata* (e.g., stns 60 [Fig. 8.2] and 25 [Fig. 8.7]) was quite common in the *L. saccharina* band. In areas of less stable substrata, with concomitantly lower wave action and general water movement, *L. saccharina* tended to widen very considerably the vertical levels that it colonised, moving up and taking over additionally much of the area vacated by the *Laminaria hyperborea*. The latter, as emphasised by many previous authors for western Scotland, shows little ability to tolerate instability in the substrate. By contrast, certain areas of considerable wave action lacked *L. saccharina*, presumably due to a high exposure factor (e.g. stn 59 [Fig. 8.9]). *L. saccharina* is, of course, not normally entirely absent from other levels than that in which it is band-forming when growing on stable substrata. Variable amounts of the species occur normally at all levels colonised by Laminarians, from the infralittoral fringe downwards. *Laminaria digitata* also shows occasional tendencies to a rather wider vertical ecological amplitude. In common with the previous workers, we found a general tendency for the *L. hyperborea* 'forest' to thin out, and for individual plants in the main to become smaller and more fragile, with increasing depth beyond 9–18 m, according to the conditions present. These tendencies stand out clearly in Table 8.1. The most extreme depths at which plants of *L. hyperborea* detected on Mull grew were rather greater than in the areas investigated by Norton & Milburn (*l.c.*), 26 m as against 24 m, but it is doubtful if significance can be attached to this probably chance difference. Although there is often, on stable substrata, a larger aggregation of plants of *L. saccharina* below *L. hyperborea*, the former therefore then characterising the lower levels, the lower limit for individuals of *L. saccharina* was in our experience less than that for *L. hyperborea*, 21 m as against 26 m. Despite penetration of *L. digitata* in appropriate circumstances into either the *L. hyperborea* 'forest' or the *L. saccharina* band, no plants of *L. digitata* were detected deeper than 9 m below Chart Datum.

The extent of substrata mobility tolerated by both *L. digitata* and *L. saccharina* is considerable, and generally greater in the latter. In conditions where *L. hyperborea*

Table 8.1 Size pattern in Laminaria hyperborea.

station no.	substrate	depth (m)	size (m) stipe	frond	density per m²	notes
6	boulders, sand	6	1·25	1·25	?	
	coarse sand	8	1·25	1·25	?	
	coarse sand	9	1·25	1·85	few	
	rock in sand	12	2·15	1·85	few	2
	rocks in sand	16	2·15	1·85	few	
15	rocks	3	1·85	1·25		
	rocks	6	1·85	1·25	5	3
	rocks	9	2·15	1·4	5	4
	sand, boulders	12	2·15	1·4	1	5
	sand, boulders	16	1·54	1·25	<1	5
	sand, pebbles	18	0·46– 0·62	0·31– 0·46	?	6
	sand, pebbles	19	0·46– 0·62	0·31– 0·46	<1	5
27	rocks	3	1·85	0·92	5	
	rocks	6	1·85– 2·15	0·92	5	
	rocks	9	1·25	0·62	5	
	rocks	12	0·92 0·62 0·92	0·31 0·31 0·62	5	
	sand, few rocks	14	–	–	few	
30	boulders	3	0·92	0·62	5	7
	boulders	5	1·25	0·92	5	
	rock gullies	6	1·25– 1·54	0·92	4–5	8
	sand, rock gullies	9	0·92	0·31	6–7	
	sand, rock gullies	14	(0·31–0·62 high)		?	7
	sand, boulders, pebbles	18	(0·31 high)		?	7
41	steep rock	3	1·85	0·62	5	3
	steep rock	6	1·85	0·62	5	3
	steep rock	9	1·85	0·62	5	3
	steep rock	14	0·92	0·62	4	6
	rock, sand interface	17	0·92	0·62	4	9
50	rock, boulders	6	1·54	0·62	6	
	rock, boulders	9	1·4	0·77	5–6	
	boulders, rock	12–14	1·08	0·62	4	
	boulders, stones	17–20	0·92	0·46	3	
51	rock	3	1·85	0·62	5	
	rock	6	1·85	0·62	5	
	rock	9	1·54	0·62	5	
	rock	12	1·08	0·77	6	
	gravel	16	0·62	0·77	?	5
	stones	19	0·46	0·46	1	
	sand, stones	21	0·31	0·31	4	

Table 8.1 (continued)

station no.	**substrate**	**depth (m)**	**size (m) stipe**	**frond**	**density per m²**	**notes**
71	sand, rocky outcrops	3	1·25	0·77	?	
	sand	6	–	–	few	
95	very steep boulders	3	0·31	–	?	
	very steep boulders	5	1·4	0·77	?	
	very steep boulders	6	1·54	1·25	8	
	very steep, large boulders	6–19	larger and denser			
	stones, shell-sand	16–19	0·31	0·46	?	9
100	rock	8	1·85	1·25	6	
	rock	9	0·92	0·92	4	
	rock	12	1·85	1·25	3	
	rock	16	1·85	1·85	3	
	rock, sand at base	19	1·25	1·25	3	9
101	boulders	6–16	1·85	1·25	5	
	boulders	16	1·25	0·62	4	6
	stones	22	<0·31	few cm	<1	9

Notes:

1 Size measurements represent largest *L. hyperborea* seen at level concerned along the line surveyed. Figures in parentheses give total plant height.
2 Age of sample plants estimated at 7–8 and 3–4 yrs.
3 Pseudocaped fronds.
4 Optimum depth for *L. hyperborea*.
5 Density variable.
6 Sudden diminution in size.
7 Plants very battered, stunted or with stipes lacking fronds.
8 With interspersed smaller *L. hyperborea*.
9 Lower limit of *L. hyperborea*.
? Density not estimated due either to mixture with *L. saccharina* or to difficulties with environment.

is locally absent, *L. saccharina* may dominate the whole of the *Laminaria*-colonised levels, even displacing, but rarely eliminating, the upper fringe of *L. digitata* (e.g. stn 72). *L. saccharina* colonises quite efficiently even small stones embedded in or resting on the very mobile sand/shell substrata present in shallow depths at certain locations around the island group. Station 67 (see profile, Fig. 8.4), where *L. saccharina* grows with *Chorda* and certain red algae on the more substantial debris and on *Pecten* over sand, in 14 m, is a case in point. Stations 56 and 34 (profile, Fig. 8.11) are also good examples; at both these stations the substratum is coarse sand and shell detritus, almost devoid of vegetation from as shallow a depth as about 6–7 metres. Above that, the communities are characterised by *Zostera marina*, with *Laminaria saccharina* intermixed. Present in considerable amounts also are large brown algae, for example *Halidrys siliquosa* (stn 56) and *Chorda filum* (stn 34). *Halidrys* colonises a wide range of substrata in deep standing water intertidally, or in the shallow infralittoral, but *Chorda* tends to be restricted to the substrata with a high degree of mobility, such as shell detritus, sand, and mud. In all these, it grows attached to the more substantial fragments present and may well cause the fragments to be transported locally backwards and forwards with the restricted water movements and gentle tidal undulation (as, e.g., at stns 44 [Fig. 8.12]; 35 [Fig. 8.13]; 54 [Fig. 8.14]). *Chorda* tolerates, but does not seem to require, more than a gentle water movement; for example, it is present as large plants in locations such as Ulva Sound, where the areas adjacent to the ferry landing (Mull side) carry tremendous plants of both *Chorda* and *Halidrys*,

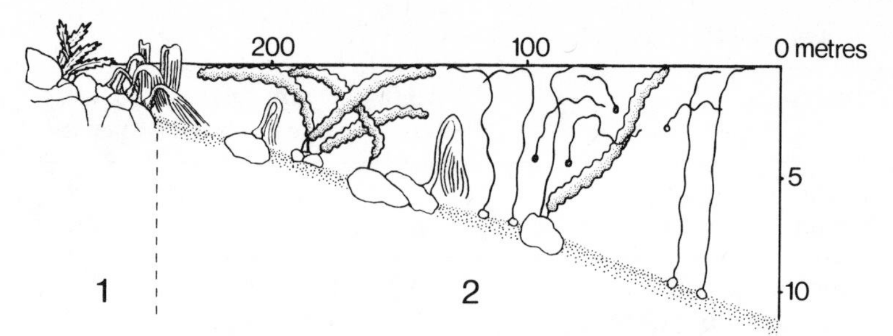

Fig. 8.12 Gently sloping shore of silt and black basaltic sand with scattered boulders and rock outcrops at Port na Gloidheig, south shore of Loch Scridain (stn 44).

Notes

1. *Laminaria digitata* abundant with young *L. saccharina* and *Desmarestia viridis; L. hyperborea* with epiphytes at lower rock fringe.
2. *Chorda* on small pebbles or shells often floating substrate.
3. With this slope, at 18 m depth, only *Lithothamnia*, filamentous diatoms and *Aglaozonia* (*Cutleria*) persist, attached on *Ascidiella* and slime tubes. Large amounts of loose-lying drift were also present.

the latter heavily epiphytised. These plants are attached to large fragments sitting on the detrital surfaces in shallow depths in areas just out of the main through-flow of water which is substantial and constant; wave-beat, however, is very restricted and the adjacent intertidal shows all the characteristics of a very sheltered shore. *Halidrys* in station 56 is also very heavily epiphytised as, at that location, are both *Zostera* and *L. saccharina*; the main epiphytes are *Punctaria*, Ectocarpoids, *Dictyota*, and a variety of smaller red algae, including *Polysiphonia fruticulosa*, *Plocamium* and *Jania*.

Zostera marina is undoubtedly the most obvious constituent of the shallow infralittoral community at station 34 (Fig. 8.11; Iona). *Chorda filum* is equally abundant there and *L. saccharina* very nearly so. Neither of the latter was much epiphytised, even the older plants of *L. saccharina* bearing only Ectocarpoids at the tips of the blade digitations. *Zostera*, by contrast, was (like that at stn 56) again loaded with epiphytes; particularly obvious were *Scytosiphon* and *Punctaria*. Circumstances in which *Zostera–Chorda–L. saccharina* populations occur are shown diagrammatically in the profiles in Figs 8.7 and 8.11. Important populations of *Zostera* were also noted at stations 25, 35, 52, 54, 70 and 75, always in shallow depths (usually less than 5 m).

Some of our observations indicate that, especially in circumstances where the growth vigour of the algae present is already adversely affected by diminution of light, but also occasionally elsewhere, grazing pressure of echinoderms may be an important factor in control of lower limit of colonisation; this phenomenon was also noted by Norton & Milburn (1972). Station 60 (Fig. 8.2) is a case in point, for on the silty rock from depths of 35 m up to 30 m very large numbers of *Antedon bifida* characterised the biota present. Fewer than usual upright algae penetrated beyond 21 m at this point, and even the Lithothamnia died out at 30 m, at which level higher density of *Antedon* commenced. Similarly on the steep rocky face of station 59 (Fig. 8.9) badly *Echinus*-browsed *Laminaria digitata* and *L. hyperborea* terminated shallowly with the latter species at about 7 m depth.

Whilst our data do not present any significant differences from those of Norton & Milburn (1972) with respect to the points considered above, they do differ considerably in other ways. Table 8.2 gives the numbers of species of each of the divisions of algae that we found confined either to the littoral or the infralittoral and those that occurred in both, and it also gives figures from Norton & Milburn (1972: Table 1) for comparison. It will be seen that we report 239 species in all against their 161, with a major difference, especially in the Rhodophyta, in the proportion of species confined

Table 8.2 Number of species of marine algae with depth.

	a) littoral	both a) and b)	b) infralittoral	total
Rhodophyta	28 (17)	69 (18)	30 (62)	127 (97)
Phaeophyta	25 (15)	37 (14)	23 (19)	85 (48)
Chlorophyta	19 (8)	8 (4)	0 (4)	27 (16)
Total	72 (40)	114 (36)	53 (85)	239 (161)

Species down to and including the infralittoral fringe have been counted as littoral; species rising to the infralittoral fringe but no higher have been considered infralittoral. Figures in brackets are from Norton & Milburn (1972) for comparison.

to the infralittoral: 53 to their 85 (Rhodophyta 30: 62). In those species found in both the littoral and infralittoral, however, we found 114 against their 36 (Rhodophyta 69 : 18). The definition of the boundary between the littoral and the infralittoral may differ between the two sets of data. The criteria we have adopted are indicated in the footnote to the table, but Norton & Milburn do not make clear precisely how they allocated species to these categories. The main reason for the difference, however, is almost certainly that our investigation of the littoral was more thorough than that of Norton and his collaborators. In an earlier taxonomic account of some of the algae collected during the same surveys (McAllister, Norton & Conway, 1967) the authors say that there were included only: '. . . the main intertidal species from the sites, thereby suggesting the degrees of exposure and wave action which were operative.' Differences of taxonomic opinion in determination may also contribute minor variations to the overall picture. These points suggest that, since our survey has been conducted over a longer period and with somewhat less of an infralittoral bias, perhaps the fuller picture emerging from our figures is nearer the general balance on the shores of Argyll.

In all adequately detailed studies, it has emerged that shallower levels in areas of only moderate wave action are the most species-rich of the infralittoral in western Scotland. These most species-rich bands for the individual location may well be depressed to lower levels in areas of greater wave action (cf. Milburn & McAllister,

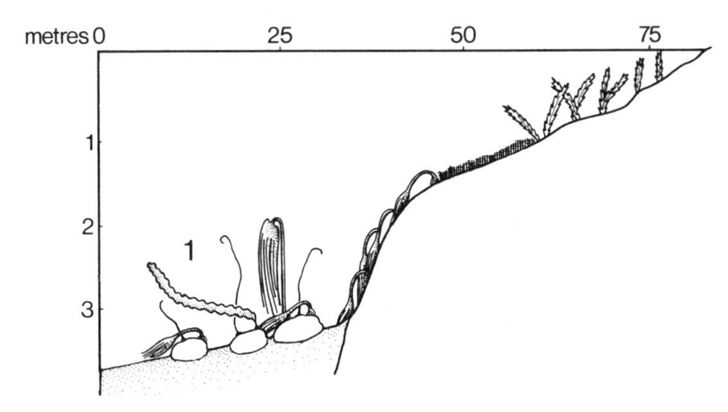

Fig. 8.13 A rocky face arising from white coarse sand on Iona at Baile Mòr jetty (stn 35).

Note
Algal cover in area 1, including sparse *Desmarestia viridis*, attached only to boulders.

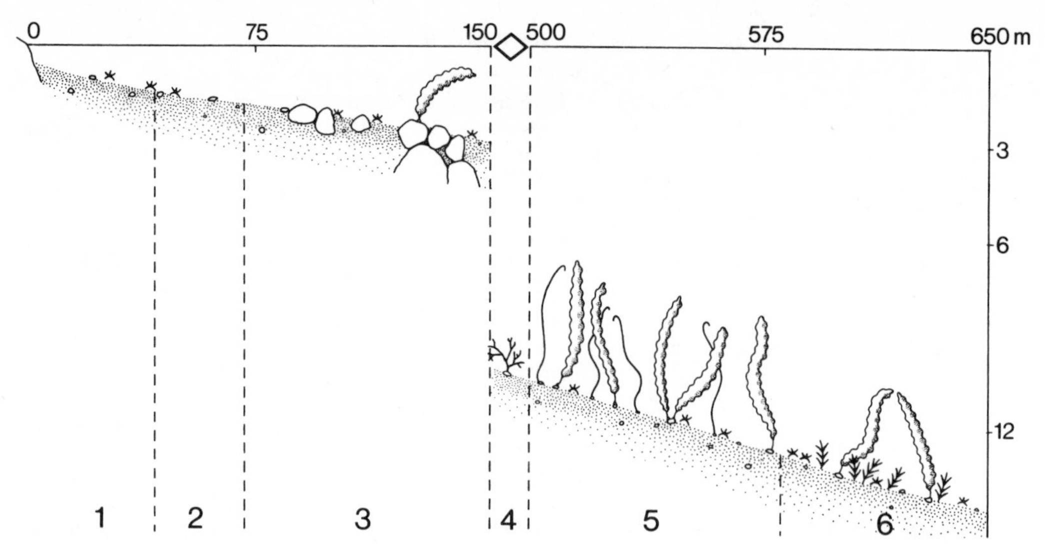

Fig. 8.14 Sheltered site with little water movement between Rubh' a' Ghearrain and Samalan Island (stn 54).

Notes

1. *Delesseria – Desmarestia viridis – Plocamium* association.
2. *Cystoseira nodicaulis* dominant at about 2 m.
3. *Halidrys* dominant on rocks; few laminarians.
4. Algae on shells: *Chorda filum*, *Corallina* spp., *Dictyota*, *Gracilaria* (♀) and *Laminaria saccharina*. Also recorded, but in silt were: *Chylocladia verticillata*, *Desmarestia viridis*, *Dumontia* and *Ulva*.
5. At 9 m depth *Asperococcus* and *Cystoclonium* appear with the *Laminaria*, *Chorda*, *Dictyota*, *Gracilaria* and *Ulva*. Throughout this zone were loose-lying algal balls of *Corallina*, *Desmarestia viridis*, *Ectocarpus*, *Halidrys*, *Odonthalia*, *Plocamium* and *Polysiphonia elongata*, with *Zostera marina*.
6. At 14 m *Chorda* no longer persists, *Desmarestia aculeata* appears with *Asperococcus*, *Chylocladia verticillata*, *Corallina* spp., *Gracilaria verrucosa* and *Plocamium*.

1967), as is true at our stations 4, 23, and 100 (profiles, Figs 8.3, 8.15, and 8.16) but in all such wave-exposed cases there are many fewer species than at the richest levels in conditions of lesser wave action but still clear water. Thus, increasing wave action reduces the number of species present and increasing depth generally tends to show progressive reduction in numbers. Increasing shelter beyond the moderate generally (as in lochs) leads to higher water turbidity and deposition of detritus over the rocky substrata; both these factors reduce the depth to which algae are present attached, and lessen the numbers of species even in those levels still carrying a relatively dense flora. Even under those optimum possible conditions for rich development of infralittoral flora in western Scotland, there are many species in all groups that do not appear in greater than certain depths. Norton & Milburn (1972) characterise this level generally as 12 metres in respect of species they refer to as dominant in shallow water. 'Dominance' is a difficult concept that we prefer to avoid by referring to species with more or less of an impact on the observer of the communites *in situ*; this may be through individual size of plants, density of growth, growth form of the algae concerned, position in which attached (especially as to epiphytes) or any combination of these. The idea of the complete dominance of one species over others, when interaction is undoubtedly the truer concept, is thus avoided. We accept in general terms the suggested level of *about* 12 metres as important in this matter; very similar indications emerge from some of our own data. However, the list of species presented by Norton & Milburn requires certain modifications before it can be applied to the infralittoral of areas examined around Mull. *Halidrys siliquosa*, for example, was detected, although atypically and unusually, down to 20 m; *Dumontia incrassata* was noted at

15 m; *Furcellaria lumbricalis* at 24 m; *Phyllophora pseudoceranoides* [*P. membranifolia*] at 18 m; and *Palmaria* [=*Rhodymenia*] *palmata* down to 13 m. Aside from these points, as will be obvious from the synoptic numbers below, we would need to add many more species to any such list for Mull. Some of these differences are in themselves insignificant, but they indicate in sum that details of the die-out level cannot be applied rigidly to the same species in similar, even closely adjacent, locations. On Mull and adjacent islands, of the commonly distributed infralittoral species (recorded from at least 10 stations) the situation is as follows: 27 species died out by or at 12 m depth; 34 species by or at 15 m depth; and 44 species by or at 18 m depth. Thus, since all the species concerned appeared at most infralittoral stations, any of the above three levels could legitimately have been chosen as the important limiting depth for shallower growing infralittoral species. In fact, they are all important levels, at one station or other, the balance changing somewhat from station to station within the overall summary figure presented. The die-out level is thus characterisable in exact terms only for the individual location or small group of similar locations, either as to species or as to most important level.

The lower limits that we found for some of the deeper growing, chiefly larger, red algae also differed from those found by Norton & Milburn (1972). This again illustrates local variation in the absolute levels at which various more or less standard events, such as species disappearance, or increase or decrease in population density, take place. Perhaps, since in most cases the observed limiting factor seems pre-eminently to be the disappearance of firm substrata below detritus, the island rocky coasts tend to shelve in slightly shallower depths and lesser horizontal distance than the rocky infralittoral of the Argyll mainland shores; most of the depths concerned were slightly less for species disappearance in Mull and area. The limiting depth, below extreme LWS, for many species was cited by Norton & Milburn as 27–30 m; by contrast, for the species noted in their list, we detected lower limits of the same forms on Mull as is shown in Table 8.3.

Despite their citation of 62 species of Rhodophyta that were restricted to the

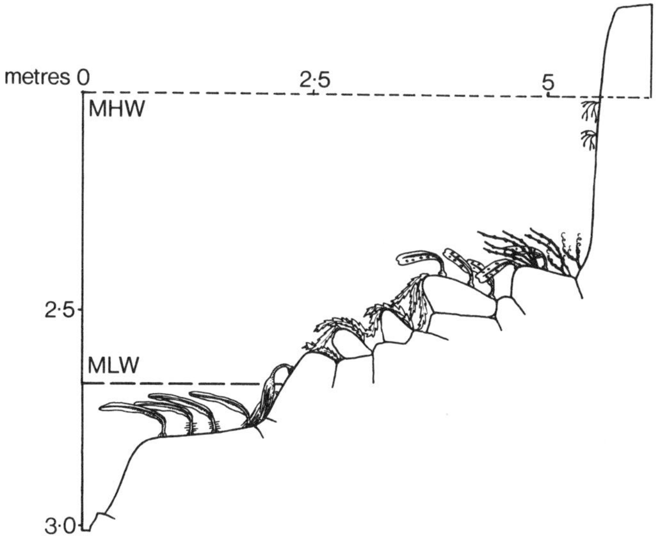

Fig. 8.15 Shore at Rubh an Righ, Calve Island (stn 4).

Note
The indications of shelter from wave-action (*Ascophyllum*; *Fucus vesiculosus* with bladders) are balanced against a strongly developed *Alaria* band in the shallow infralittoral, depressing the other laminarians present.

Table 8.3 The lower limits (in m) for selected algae on Mull.

Desmarestia viridis	18	Odonthalia dentata	18
Dictyota dichotoma	23	Phycodrys rubens	27
Ectocarpus siliculosus	21	Phyllophora crispa	21
Bonnemaisonia hamifera*	23	Pleonosporium borreri	22
Brongniartella byssoides	22	Plocamium cartilagineum	23
Callophyllis laciniata	22	Polysiphonia elongata	18
Cryptopleura ramosa	23	P. urceolata	21
Lithothamnia	30†		

* *Trailliella* phase. † Also detected by Norton & Milburn to beyond this depth. To this list should be added other deep-growing forms (see Table 8.4); only species noted from at least 8 stations are included, since the inclusion of more rarely recorded species would add a further 20 to the list and would not truly aid characterisation.

infralittoral, Norton & Milburn (1972) clearly indicate that they were unable to characterise the deeper infralittoral levels in terms of the presence of species that were not also present elsewhere in lesser depths or in the intertidal; with this, we concur (Table 8.2). To this extent, there is certainly no distinct deep-water algal flora. However, it *is* possible to characterise the deeper water flora very adequately by utilising the "... increasing preponderance of Rhodophycean species ..." that Norton & Milburn mention and that was clearly also present on Mull (see especially profiles in Figs. 8.2, 8.4, 8.5, 8.6). Since there is no disagreement regarding the general thinning-out of the *Laminaria hyperborea* 'forest' below about 10 m, and of the *Laminaria saccharina* band below about 15 m, it is clear that the persistent upright red algae, of which the majority are often non-foliose, form the characterising agents of the immediately lower levels, constituting thereby 'a distinct deep-water algal flora' in the sense of communities, although not of species confined to deep water. This is certainly most often the case on Mull. Later, we present also characterisation of the underflora of the Laminarian levels and of the epiphyte flora found on the different parts of the *Laminaria* plant. The red encrusting flora, present at most levels, comes into its own below those levels colonised by upright forms. As would be expected from the distribution figures we quote earlier (Table 8.2), the characteristic elements of underflora, epiphyte flora, and flora below the Laminarian levels do have species in common, although the balance changes.

Reference to Table 8.5 and profiles illustrated in Figs. 8.2–8.15, will illustrate the general situation for Mull; more detailed geographical descriptions of the stations

Table 8.4 The lower limits (in m) for selected algae on Mull: deep growing forms.

Antithamnion plumula	22	Heterosiphonia plumosa	22
Bonnemaisonia asparagoides	23	Hypoglossum woodwardii	21
Callithamnion hookeri	27	Laminaria hyperborea	26
Ceramium rubrum	21	L. saccharina	21
Chylocladia verticillata	23	Lomentaria clavellosa	22
Desmarestia aculeata	21	Membranoptera alata	27
Delesseria sanguinea	24	Nitophyllum punctatum	24
Enteromorpha intestinalis	21	Polysiphonia nigrescens	21
E. prolifera	21	Pterosiphonia parasitica	21
Erythrotrichia carnea	22	Ptilota plumosa	27
Griffithsia flosculosa	23	Saccorhiza polyschides	22
Halidrys siliquosa*	20	Ulva lactuca	21

* Rarely so deep.

Table 8.5 Characterising flora, excluding the Lithothamnia, *occurring epilithically at the lower fringes of, and below, the* Laminaria saccharina/L. hyperborea *bands.*

Rhodophyta:	
Antithamnion plumula	*Griffithsia corallinoides*
A. spirographidis	G. flosculosa
Asparagopsis armata	*Heterosiphonia plumosa*
(*Falkenbergia rufolanosa stage*)	Hypoglossum woodwardii
Audouinella floridula	Lomentaria clavellosa
Bonnemaisonia asparagoides	L. orcadensis
B. hamifera (*Trailliella* stage)	Odonthalia dentata
Brongniartella byssoides	*Phyllophora crispa*
Calliblepharis ciliata	*Pleonosporium borreri*
Callithamnion hookeri	Polysiphonia elongata
Ceramium rubrum/diaphanum	P. nigra
Chylocladia verticillata	P. nigrescens
Cystoclonium purpureum	*Pterosiphonia parasitica*
Delesseria sanguinea	Spermothamnion repens
Gracilaria verrucosa	
Phaeophyta:	
Cladostephus spongiosus f. verticillatus	*Desmarestia aculeata*
Cutleria multifida (Aglaozonia stage)	*Dictyota dichotoma*
	Giffordia sandriana/G. secunda

Species that occurred in the levels concerned at only one of the stations examined have been ignored for the purposes of general characterisation. Those in *italics* are constants. Table based on data from stations: 1, 6, 13, 15, 21, 26, 27, 30, 41, 43, 44, 50, 51, 54, 60, 65, 67, 71, 72, 84, 95, 99, 101.

are given in Chapter 19 (p. 19.4). It is enough here to emphasise the differences below the lower fringe of the *L. hyperborea* 'forest'. Lunga (stn 100) is a SW-facing station of very considerable exposure; the presence of strongly developed *Alaria* and the depression of the *L. digitata* and *L. hyperborea* areas show this clearly. In the lower levels where *L. hyperborea* and *L. saccharina* are only occasional, the epiphytic flora shows diminution in species and amounts, but to nothing like the same degree as the epilithic flora is restricted at those levels. Only six species, listed at stn 100, were commonly detected attached on rock at 12–15 m, for example. Comparison of this situation with the contrasting circumstances of open but relatively less exposed infralittorals, as those profiles in Figs 8.5 and 8.6, immediately reveals a tremendous increase in the numbers of species (especially of filamentous red algae) occurring epilithically around the lower fringes of *L. hyperborea* and *L. saccharina*, and in the levels immediately below, where these larger forms have virtually disappeared. Even considering only the more frequently detected species, and excluding those found at that level only epiphytically on *Phyllophora crispa*, station 99 (Fig. 8.5) showed ten species of red algae and one green occurring commonly and densely. If the *P. crispa* epiphytes are also included, the numbers increase to sixteen reds, one green and one brown. The distribution of all these filamentous forms is mosaic at station 99, and the reduction of wave action is reflected in the absence of *Alaria*. A different but analogous situation at slightly greater depth, in station 101 (Fig. 8.6), showed more than 10 species of smaller red algae, at least two brown algae and one green alga commonly present. At station 67 (Fig. 8.4), where the substratum changes at about 14 m to sand in large part so eliminating the *L. hyperborea* 'forest', levels immediately below that showed only a very open canopy of mixed *Laminaria saccharina* and *Chorda filum* that had but little effect on the underflora. The latter included at least 23 red algae, with two common brown species; some of the 23 red algae (see Fig. 8.4) were predominantly epiphytic on the other forms, of which the major hosts are indicated in the figure, but all occurred epilithically on the more substantial fragments at intervals in the levels concerned.

The stations 67, 99 and 101, illustrated by profile, serve as examples of the general situation in which, under conditions of adequate water movement and therefore water

Table 8.6 Major stipe epiphytes on Laminaria hyperborea. (Continued on next page.)

	number of occasions noted at:								total sites* at which observed
depths (m)	**0–3**	**4–6**	**7–9**	**10–12**	**13–15**	**16–18**	**19–21**	**22–24**	
Species of frequent occurrence									
Callophyllis laciniata	6	12	15	9	6	5	2		18
Cryptopleura ramosa	14	25	10	5	8	5	2	1	26
Ectocarpoids†	3	8	4	4	4	3	1		11
Lomentaria articulata	9	8	1	1					12
Lithothamnia	5	4	1	1	1	2	2	2 }	20
Melobesioids	7	10	4	0	0	1	0	0 }	
Membranoptera alata	14	36	10	4	5	5	3		31
Palmaria palmata	15	19	6	1	1				19
Phycodrys rubens	15	31	19	9	9	6	3		29
Plocamium cartilagineum	4	7	6	3	3	1	1		12
Polysiphonia urceolata	11	20	5	2	2	3	1		21
Ptilota plumosa	11	32	4	2	2	1	1		20
Species of occasional occurrence									
Alaria esculenta	3	2							3
Antithamnion plumula	0	1	1	1	0	1	2		4
Audouinella purpurea	1	3	1	0	1	1			5
Callithamnion hookeri	3	2	2	1					4
Ceramium rubrum	3	3							5
Chondrus crispus	5	3							6
Cladophora rupestris	3	5							7
Delesseria sanguinea	4	2	1						5
Desmarestia viridis	3	5	3	1					6
Dictyota dichotoma	4	3	1	0	0	1	1		6
Griffithsia flosculosa	0	1	0	0	3	4	1		5
Laminaria digitata	4	7							9
Laminaria hyperborea	3	6	1	1	1	1	1		9
Laminaria saccharina	3	5	1						7
Litosiphon filiformis	3								3
Lomentaria clavellosa	5	3	3	1	1	1			8
Nitophyllum punctatum	1	1	3	1	1	1			5
Odonthalia dentata	1	2	1	1					5
Polysiphonia elongata	1	2							3
Polysiphonia nigrescens	4	3	1						4
Pterosiphonia parasitica	0	5	1	1					3
Punctaria tenuissima	2	4	3						4
Rhodomela confervoides	4	8	2	1					8
Sphacelaria cirrosa	5	2	3						5
Sphacelaria fusca	3	2	1	1					5
Ulva lactuca	6	7							10

* These are sites observed, not stations which may contain several sites; each site may include several depth observations.

† Most of these records relate to *Ectocarpus fasciculatus*, frequently fruiting. Ectocarpoids were mostly on the upper stipe or near the lamina tips (laminar records not included here).

Table 8.6 (continued) Major stipe epiphytes on Laminaria hyperborea. Species of rare occurrence recorded from one or two stations only, although many of these species are possibly of more frequent occurrence locally. In a few cases, collections were lost and misdeterminations in the field cannot be ruled out.

	depth (m)	**no. of stations**		**depth (m)**	**no. of stations**
Antithamnion cruciatum	2·5–5	1	Eudesme virescens	3–6	1
Apoglossum ruscifolium	6–9; 22	2	Fucus serratus	0–3	1
Brongniartella byssoides	5–8	1	Fucus sp. (very young plants)	3–5·5	1
Callithamnion roseum	3–6	1	Heterosiphonia plumosa	16–19	1
Callithamnion tetragonum	0–3	1	Kallymenia reniformis	5–8;	
Chaetomorpha melagonium	3–5·5	1		15–18	2
Chylocladia verticillata	0–3	1	Phyllophora pseudoceranoides	0–3	1
Cladophora sericea	3–5·5	1	Phymatolithon lenormandii	6–9	1
Cladophora sp.	2–5	1	Polyneura gmelinii	3–9	1
Corallina officinalis	0–3	1	Polysiphonia violacea	0–3	1
Cutleria multifida	0–8;		Porphyra leucosticta	0–3;	
(Aglaozonia)	9–12	2		3–6	2
Cystoclonium purpureum	0–3	1	Porphyra miniata	0–3;	
Desmarestia aculeata	0–3;			12–15	2
	5–8	2	Ptilothamnion pluma	6–9;	
Dilsea carnosa	0–3	1		9–13;	
Enteromorpha intestinalis	0–3	1		16–19	1
Enteromorpha linza	3–5·5	1	Spongomorpha arcta	2·5–5	1
Enteromorpha prolifera	0–3	2			

clarity below the shallowest levels of the infralittoral, the largely filamentous epilithic species overlapping with and spreading below the Laminarian 'forest' develop to such a degree that they characterise these lower 'forest' fringes and levels below. Table 8.5 lists the characterising, largely filamentous and red, forms. Those most frequently detected in these situations are in italics; many of the others occur almost as frequently. Because of variations in physical levels between stations, and because of mosaic variations in substrate types at different levels, no attempt has been made to quote depths for the individual species. Where needed, ranges of depths tolerated can be obtained from entries in the species lists. The presence of some characterising species depends to some extent on substrate nature where the Laminarian 'forest' thins and dies out. *Gracilaria verrucosa*, for example, is customarily present only where there is detritus. Many species listed have, however, a wide spectrum of substrate tolerance. The list omits species which are also major characterising facets of the epiphyte flora of *Laminaria hyperborea* stipes (see Table 8.6, summarising Laminarian epiphytes, for details of these). Those species that are primary characterisers amongst the epiphyte flora tend also to occur in large amounts at forest fringe and lower levels, on rock and other substrata. However, because of the presence at these latter levels, in even greater amounts, of more filamentous red or brown algae, the species that characterise the epiphyte communities do not to anything like the same degree characterise the communities below the level of the Laminarians. These latter lower level communities therefore possess a recognisably distinct community structure of sufficient consistency to be of use in ecological description. Species of the filamentous red and brown algae that do characterise the communities below the Laminarians also in some cases occur as epiphytes on *L. hyperborea* stipes, but generally much less frequently than they appear epilithically at levels below the *L. hyperborea* 'forest'; they are certainly less frequent or luxuriant as epiphytes than are the species noted as characterising the epiphytic flora.

Apart from the Lithothamnia, already indicated above as very widespread beneath the 'forest' of *L. hyperborea* and/or the cover of any other large brown algae that may be predominant in conditions of stable substrata, there is a non-crustose underflora again characterisable by relatively few species. The upper fringes of Laminarian cover tend to include as underflora some of those algae tolerant of the changing circumstances of the lower littoral and infralittoral fringe but not apparently able to colonise more deeply into the infralittoral. There is thus an unusual element at those levels (e.g. *Plumaria elegans*, *Cladophora rupestris*, *Gelidium* spp.), but this in no way confuses the otherwise general homogeneity of the underflora of *Laminaria* spp. Unstable

Table 8.7 The depth ranges of the underflora to Laminarians. (Continued on next page.) The lower limits of the characteristic underflora elements commonly lie below the level to which *L. hyperborea* and *L. saccharina* penetrate.

	number of occasions noted at:									total sites* at which observed
depths (m)	0–3	4–6	7–9	10–12	13–15	16–18	19–21	22–24	25–27	
Species of frequent occurrence										
Brongniartella byssoides^{+}	1	0	2	3	4	5				7
Callophyllis laciniata	4	4	2	3	3	4	2	1		13
Chondrus crispus (E)	7	4	1	1						8
Cryptopleura ramosa	6	6	4	3	4	5	1	1		15
Delesseria sanguinea^{+}	5	7	4	7	4	7	4	2		16
Desmarestia aculeata^{+}	3	3	3	4	4	1	2			9
Desmarestia viridis	3	4	4	4	1	2				8
Dictyota dichotoma^{+}	3	5	1	3	4	4	3	2		11
Heterosiphonia plumosa^{+}	0	0	1	1	3	3				8
Membranoptera alata^{++}	4	3	3	2	0	1	0	0	1	8
Nitophyllum punctatum^{+}	1	7	5	3	2	4	2	1		8
Odonthalia dentata^{+}	7	5	1	2	3	2	0	1		12
Phycodrys rubens	5	6	8	5	4	6	3	1		15
Phyllophora crispa^{+}	1	0	2	2	6	3	1			6
Plocamium cartilagineum	6	6	6	2	4	6	2			17
Pterosiphonia parasitica^{+}	2	3	3	3	3	3	1			6
Species of occasional occurrence										
Apoglossum ruscifolium^{+}	0	0	1	0	1	1	1			4
Bonnemaisonia asparagoides^{+}	0	0	1	0	2	3	1			4
Ceramium rubrum	2	0	1	2	2	2	2			7
[1]Chorda filum^{+}	1	3	1							4
Cladophora rupestris (E)	3	0	1							4
Cutleria multifida (Aglaozonia stage)	0	0	0	2	0	0	2			4
[2]Dilsea carnosa^{+} (E)	5	4	0	0	1					7
[3]Ectocarpoids	0	0	3	1	2	1	2			4
Furcellaria lumbricalis^{+}	3	2	0	1						5
Gigartina stellata^{+} (E)	7	1								7
Gracilaria verrucosa^{+}	3	2	0	1						3
[4]Halidrys siliquosa (E)	3	2	0	0	0	1				4
Hypoglossum woodwardii	0	0	2	0	1	2				4
Laurencia hybrida (E)	3	1	1							4
Lomentaria clavellosa	1	1	2	2	1	3	1			6
Phyllophora pseudoceranoides^{+}	2	1	0	0	1	1				4
Polysiphonia urceolata^{++}	2	3	2	0	0	3				6
Porphyra miniata/leucosticta (E)	1	0	2	1						3
Ptilota plumosa^{++}	4	2	1	0	1	0	0	0	1	8
Ulva lactuca	3	3	3	1	1	2	1			6

For notes, see next page.

Table 8.7 (continued) The depth ranges of the underflora to Laminarians.

	number of occasions noted									total sites* at which observed
depths (m)	0–3	4–6	7–9	10–12	13–15	16–18	19–21	22–24	25–27	
Species of rare occurrence										
Rhodophyta										
Ahnfeltia plicata⁺	×	×								3
Antithamnion plumula			×	×	×	×	×			3
Audouinella floridula⁺			×	×	×					2
[5]Bonnemaisonia hamifera⁺⁺ (Trailliella stage)						×	×	×	×	2
Calliblepharis ciliata⁺					×	×	×			1
Callithamnion hookeri			×	×		×				3
Ceramium diaphanum	×			×	×					2
Chondria dasyphylla	×					×				2
Chylocladia verticillata (E)	×		×	×						3
Corallina officinalis (E)	×	×	×	×						2
Cystoclonium purpureum	×	×		×	×	×				2
Dumontia incrassata (E)	×	×								3
Gelidium latifolium (E)	×									2
Griffithsia corallinoides		×		×		×				2
Griffithsia flosculosa						×		×		2
Halarachnion ligulatum							×			1
Jania rubens⁺⁺	×									1
Kallymenia reniformis⁺						×				1
Laurencia pinnatifida⁺ (E)	×									2
Lomentaria articulata⁺⁺ (E)	×									3
Lomentaria orcadensis					×					1
Palmaria palmata⁺⁺	×			×			×			2
Phyllophora truncata			×							1
Plumaria elegans (E)	×									2
Polyides rotundus (E)	×	×	×	×	×					3
Polysiphonia elongata		×	×	×		×				3
Polysiphonia nigra⁺						×				1
Polysiphonia nigrescens (E)	×	×				×				2
Rhodomela confervoides	×	×				×				3
Phaeophyta										
Cladostephus spongiosus⁺	×	×	×							3
Desmarestia ligulata			×	×						1
Eudesme virescens (E)	×		×	×						3
Pseudolithoderma extensum			×		×					2
Tilopteris mertensii							×			1
Chlorophyta										
Bryopsis plumosa							×			1
Codium fragile	×									1

Notes:
* These are sites observed not stations which may contain several sites; each site may include several depth observations.
1. Mostly on unstable substrata with *L. saccharina/Halidrys*.
2. Large plants to 60 cm occur occasionally on open shores where forest thins.
3. Many of these records relate to *E. fasciculatus*.
4. The deeper records are unusual: ? reattachment of fragments.
5. More widespread *below* Laminarian levels.

Symbols:
⁺ = more frequent and abundant as epilithic underflora than as epiphyte.
⁺⁺ = more frequent and abundant as epiphyte (see Table 8.6).
unmarked = of equal frequency at those levels as epiphyte or on rock.
(E) = incursion from infralittoral fringe/eulittoral.

circumstances in the infralittoral, supporting as major flora algae like *Chorda filum*, *Halidrys siliquosa* and *Laminaria saccharina*, do carry rather different underflora, including a few elements rarely found in the underfloras of stable surfaces (see the profiles in Figs 8.4, 8.7 and 8.14 for examples).

As indicated earlier, and in agreement with Norton & Milburn (1972), there is certainly considerable common floristic element between the epiphytic flora of the Laminarians (chiefly considering *L. hyperborea*) and their underflora. Nevertheless, there are differences in kind and degree amongst the other species that make up those most commonly present in these different circumstances. Tables 8.6 and 8.7 illustrate this clearly. Amongst the underflora, certain more or less constantly characteristic species occur at all depths with much the same frequency, so far as our samples are concerned. *Delesseria sanguinea*, *Dictyota dichotoma*, *Desmarestia aculeata*, *Nitophyllum punctatum* and *Pterosiphonia parasitica* provide immediately obvious examples; the first four of these are species large enough to provide visual impact when the populations are viewed in situ, but *Pterosiphonia parasitica* is of little effect in this way. *Delesseria* was detected sporadically, and never abundantly, as an epiphyte on *L. hyperborea*, although it seems to occur fairly regularly in small amounts such as single plants, widely scattered among the more abundant epiphytic species. *Dictyota dichotoma* was almost equally rare and similarly widely scattered when epiphytic; although frequent in the underflora, it was not as consistently present, nor so evenly present at all depths, as was *Delesseria*. *Nitophyllum punctatum*, sparse but of considerable depth range as an epiphyte, as an underflora constituent showed distribution with some slight emphasis around the 3–9 m depth levels. This was not sufficiently clear, nor was the sample of sufficient size, to do more than indicate a possibly interesting topic for further work. Its noted frequency on stipe bases and holdfasts, as an epiphyte, also needs confirmation. *Desmarestia aculeata* was little detected as an epiphyte; most of the records from the underflora were grouped into the first 15 m depth, although the species was noted down to 21 m; with a higher representative sample, some definitive tendency to relative absence from deeper waters may emerge. *Pterosiphonia parasitca*, more frequently noted as a stipe base and holdfast epiphyte in the 3–6 m level, occurred over a much wider depth range when epilithic. Certain other species characteristic of the underflora and rare as epiphytes revealed depth restrictions or variations to degrees worthy of comment. *Heterosiphonia plumosa* was noted only once as an epiphyte, from 16–19 m depth. Depths in the 12–21 m range accounted for 11 of 13 underflora observations or collections of the species, a rather high proportion for this to be merely a matter of chance. *Brongniartella byssoides*, like *Heterosiphonia* rare as an epiphyte, also showed a firm tendency (12 out of 15 observations) to more frequent occurrence in the underflora at depths between 9 and 18 m. Unfortunately, for neither *Heterosiphonia* nor *Brongniartella* were the available samples large enough to be sure that no aspect of collecting method or subjective selection had influenced the results. A rather larger sample is available for one of the very characteristic species of the underflora, *Odonthalia dentata*. Although this species was more or less consistently present at all levels examined, there was a strong tendency for more consistent occurrence in the first six metres of the infralittoral (12 of 21 observations). The few collections of the species as an epiphyte reveal little by way of distribution pattern; numbers are too small for a conclusion to be drawn.

Thus, of the twelve species most constantly encountered in the underflora of the Laminarians and other major browns, eight can reasonably be employed, either generally throughout all examined levels or tending to certain levels as specified above, in characterising the flora in this habitat. The remaining four show a frequency of occurrence of approximately the same order both as epiphytes and in the underflora. These species are *Callophyllis laciniata*, *Cryptopleura ramosa*, *Phycodrys rubens*, and

Plocamium cartilagineum. In this context, it should be borne in mind in examining Tables 8.6 and 8.7 that the general size of the sample in the underflora was less than that in the epiphyte flora; the sheer numbers involved are thus misleading. As an epiphyte, *Cryptopleura ramosa* was not only frequent, generally distributed around Mull, present at all infralittoral levels examined, and often represented all along stipe and holdfast; it was also often abundant, individually luxuriant, and fruiting. Tetrasporangia and carposporophytes were detected with some frequency, the former being more characteristically present. By contrast, *Callophyllis* was rarely detected in a reproductive state, although large plants were common on lower parts of stipes and on holdfasts. *Phycodrys rubens* was also more frequent basally on *L. hyperborea*, although detected at all levels along the stipe. Like *Cryptopleura*, luxuriant individuals were noted and, at least in the first 9 m of the infralittoral, *Phycodrys* was often prolifically tetrasporangial. Unlike *Cryptopleura*, however, which was rarely detected as showing many signs of grazing by animals, *Phycodrys* had been strongly grazed at some locations. *Callophyllis* and *Plocamium* were most frequently observed as epiphytes along the more open, western shores of the islands. *Plocamium* was never as abundant epiphytically as *Cryptopleura*, *Phycodrys* and *Callophyllis*, although small numbers of very large plants were widespread at levels down to about 15 m and still present occasionally below that; plants were usually carried median or below on the *L. hyperborea* stipes, and on the holdfasts. Curiously, *Plocamium* was never noted in the reproductive state as an epiphyte.

Characterisation of the epiphytic flora can be readily seen from Table 8.6. Leaving aside the four common species already named above, those involved are *Lomentaria articulata*, *Membranoptera alata*, *Polysiphonia urecolata*, *Ptilota plumosa* and *Palmaria palmata*. Some of these latter species can only be viewed as characterising at certain depths, often very restricted. *Lomentaria articulata*, for example, was not noted below 12 m and was rare, epilithic or epiphytic, below 6 m; the three epilithic observations were all in 3 m depth or less. It was not noted in the reproductive state. A general pattern of much greater frequency and luxuriance in shallower depth was shown by *Membranoptera alata*, whether as epilithic underflora or as an epiphyte; even though quite frequent as underflora, the species is so vastly more frequent as an epiphyte as to be useful in general characterisation at depths less than 9 metres. *Membranoptera* was often noted as small and depauperate, as well as being heavily grazed, as an epiphyte; it was only once detected in reproduction, then bearing tetrasporangia. *Polysiphonia urceolata* showed a very similar distribution of observations at different depths to that of *Membranoptera*, but differed radically in that on rather more than 50% of the occasions when it was noted as an epiphyte, *P. urceolata* was in the reproductive state (Table 8.8). *Ptilota plumosa*, also, is a good characterising epiphytic

Table 8.8 Number of observations of reproductive states in Polysiphonia urceolata *as an epiphyte at various depths.*

depth (m)	reproductive state				total
	♀ (filled)	♀	♂	⊕	
0–3	6	1	1	2	10
4–6	8	–	1	1	10
7–9	3	–	–	1	4
10–12	1	–	–	–	1
13–15	–	–	–	–	0
16–18	–	–	–	1	1
	18	1	2	5	26

species for the shallow sublittoral, at least to 6 m and probably to 15 m; very frequently recorded for many sites as an epiphyte, it was relatively little detected in the underflora. *Palmaria palmata* manifests similar tendencies to *Ptilota* and certain other red algal epiphytes (see Table 8.6) in its distribution with depths. On the *L. hyperborea* stipes, it was commonly noted only above the middle of the stipe, its distribution there hence paralleling its depths distribution. It was often the predominant form apically on the stipe. Rarely noted in fruit, it showed a marked decrease in size when growing elsewhere on the stipe than towards the apex.

Of the species less generally but still frequently noted as epiphytes, *Alaria* shows a very similar depths distribution to that applying when, as more usually, it is epilithic. By contrast, many species (e.g. *Ceramium rubrum*, *Cladophora rupestris*, *Callithamnion hookeri*, *Chondrus crispus*, *Delesseria sanguinea*, *Desmarestia viridis*, *Dictyota dichotoma*, *Nitophyllum punctatum*, *Odonthalia dentata* and *Pterosiphonia parasitica*) were found epilithically at depths pronouncedly greater than their apparent lower limits as epiphytes. *Audouinella purpurea* was largely restricted to holdfasts over its fairly wide (0–18 m) depths range, whilst species such as *Nitophyllum punctatum*, *Laminaria hyperborea* itself, *Odonthalia dentata*, and *Pterosiphonia parasitica*, were most frequently on the lower stipe and holdfast. The presence of large populations of *L. hyperborea* epiphytic on the same species in depths shallower than 6 m brings up interesting problems of vertical distribution, of density of breeding populations, and of population dynamics generally, which we were unable to investigate. The juveniles detected were also mostly on the lower parts of stipes, or on the holdfasts. Like *Alaria*, *Laminaria digitata* showed similar common depth distribution whether epiphytic or epilithic; the greatest density of *L. digitata* as an epiphyte occurred, not surprisingly, in the overlap area *L. digitata*↔*L. hyperborea*. There was no apparent restriction of *L. digitata* to any parts of the stipe or holdfast. Interestingly, the largest size (1.5 m or above) of epiphytic *L. digitata* was consistently noted in waters sheltered from strong wave-action. Since all the epiphytic open water *L. digitata* plants appeared to be both young and small, perhaps size above a certain level may lead to detachment of the epiphyte, or both epiphyte and 'host', in all but conditions of minimal water-movement. A similar phenomenon was noted for *Laminaria saccharina*, which was never detected epiphytically on its own species or on *L. digitata*.

In summary, eight species are sufficiently dissimilarly distributed partially or wholly to characterise the underflora; four species (excluding the Lithothamnia) are approximately equally likely to occur with high frequency as both underflora and Laminarian stipe epiphytes; five species are sufficiently dissimilarly distributed partially or wholly to characterise the epiphytic flora. Ectocarpoids, being recorded only as a group of species under field conditions, have been omitted from this consideration although treated in both Table 8.6 and Table 8.7.

References

GIBB, D. C. (1957). The free-living forms of *Ascophyllum nodosum* (L.) Le Jol. *J. Ecol.*, **45**: 49–83.

KAIN, J. M. (1962). Aspects of the biology of *Laminaria hyperborea* I. Vertical distribution. *J. Mar. biol. Ass. U.K.*, **42**: 377–385.

KITCHING, J. A. (1935). An introduction to ecology of intertidal rock surfaces on the coast of Argyll. *Trans. R. Soc. Edinb.*, **58**: 351–374.

———. (1941). Studies in sublittoral ecology III. *Laminaria* forest on the west coast of Scotland: a study of zonation in relation to wave action and illumination. *Biol. Bull. mar. biol. Lab. Woods Hole*, **80**: 324–337.

LEWIS, J. R. (1957). Intertidal communities of the northern and western coasts of Scotland. *Trans. R. Soc. Edinb.*, **63**: 185–220.
———. (1964). *The ecology of rocky shores.* London.
——— & POWELL, H. T. (1960a). Aspects of the intertidal ecology of rocky shores in Argyll, Scotland I. General description of the areas. *Trans. R. Soc. Edinb.*, **64**: 45–74.
——— ———. (1960b). Aspects of the intertidal ecology of rocky shores in Argyll, Scotland II. The distribution of *Chthamalus stellatus* and *Balanus balanoides* in Kintyre. *Trans. R. Soc. Edinb.*, **64**: 75–100.
MCALLISTER, H. A., NORTON, T. A. & CONWAY, E. (1967). A preliminary list of sublittoral marine algae from the west of Scotland. *Br. phycol. Bull.*, **3**: 175–184.
MILBURN, J. A. & MCALLISTER, H. A. (1967). Contributions to the sublittoral ecology of the west of Scotland. *Br. phycol. Bull.*, **3**: 407.
NORTON, T. A. & MILBURN, J. A. (1972). Direct observations on the sublittoral marine algae of Argyll, Scotland. *Hydrobiologia*, **40**: 55–63.
POWELL, H. T. (1970). The new S.M.B.A. laboratory at Dunstaffnage, near Oban; and a preliminary account of the ecology of the macro-algae in Loch Etive. *Br. phycol. J.*, **5**: 270.
———. (1972). The ecology of the macro-algae in sea lochs in western Scotland. *Proc. int. Seaweed Symp.*, **7**: 273.

9. Brackish and freshwater ecosystems

A. C. Jermy, D. J. Hibberd* and Patricia A. Sims

Brackish systems

Brackish water habitats, normally defined as those with a salinity between 00·3–16·5‰, are scarce on Mull for two reasons: the high rainfall and quick run-off ensure that rivers are for the most part fast-flowing, and thus have a flushing effect on the small estuarine regions; and there are no maritime lowland marshes that could be drained for agriculture so that ditches which are often brackish are absent. There are however, pools, either at the base of the cliff line or on the low-lying islands and promontories, which are influenced directly by spray and occasional high tides.

One such pool at sea level on Garmony Point south of Salen, on the Sound of Mull, within 300 m of the high tide mark had a typical brackish reed-swamp flora of *Scirpus tabernaemontani*, *Blysmus rufus* and *Iris pseudacorus*. Diatom species characteristic of brackish water were also present. In an April gathering from the pool, *Fragilaria pinnata* and *F. virescens* var. *subsalina* were dominant. Other important constituents indicating brackish conditions included *Caloneis subsalina*, *Brebissonia*

Table 9.1 Chemical analysis (in mg/l) of low altitude pools on three Treshnish Islands collected July, 1969.

island	**area sq m**	**depth cm**	**total dissolved solids**	**Ca**	**Mg**	**Na**
Bac Mòr	120	100	1820	90	460	625
Fladda			8810	300	1380	2800
Sgeir Errionaich	30	30	14900	440	2240	4300

	K	**Cl_2**	**PO_4**	**SO_4**	**SiO_2**	**Total N**	**pH**
Bac Mòr	2·5	1130	0·60	91	2·4	5·65	6·7
Fladda	110	5110	0·02	620	abst.	0·63	7·6
Sgeir Errionaich	180	8430	0·16	1050	1·5	4·40	6·4

* Now at the Culture Centre of Algae and Protozoa (NERC), 36 Storey's Way, Cambridge.

boeckii, *Mastogloia smithii*, *Navicula crucicula*, *N. elegans*, *Nitzschia littoralis*, *Rhoicosphenia curvata* and *Synedra pulchella*. The freshwater species *Achnanthes microcephala* and *Navicula jaernefeltii* were also present.

On the Treshnish Isles there are shallow pools often not more than ten metres across and three to six metres above high tide. An analysis of the ionic content of

Table 9.2 Combined species list for supralittoral pools at Carsaig (coll. nos. H39–43; 45; 46; 48).

CYANOPHYTA

Anabaena cylindrica
Aphanocapsa montana (*enormous growths of this gave bluish-green colour to the bottom of the pool*)
Oscillatoria sp.
O. irrigua
O. sancta
Tolypothrix tenuis

CHRYSOPHYTA

Chrysamoeba radians
Chrysopyxis stenostoma
Lagynion ampullaceum

BACILLARIOPHYTA

Achanthes hauckiana
A. saxonica
Caloneis subsalina
Melosira juergensii
Navicula avenacea
N. elegans
N. variostriata
Rhoicosphenia marina
Synedra pulchella

XANTHOPHYTA

Ophiocytium sp.

EUSTIGMATOPHYTA

Chlorobotrys regularis

CHLOROPHYTA

Ankistrodesmus falcatus
Binuclearia tectorum
Bulbochaete sp.
Crucigenia irregularis
Dictyosphaerium pulchellum
Gonium formosum *f.* suecicum
Mougeotia sp.
Oedogonium sp.
Oocystis spp.
Pediastrum boryanum *var.* cornutum

CHLOROPHYTA (cont'd)

P. boryanum *var.* longicorne
P. tetras
Scenedesmus acuminatus
S. denticulatus *var.* linearis
S. quadricauda
S. tibiscensis
Schizochlamys compacta
Spirogyra sp.
Tetraedron minimum
Zygnema sp.

Desmids

Closterium acutum
C. dianae
C. gracile
Cosmarium angulosum
C. blyttii *var.* novae-sylvae
C. difficile *var.* subimpressulum
C. formosulum
C. humile
C. impressulum
C. notabile
C. reniforme *var.* reniforme
C. sphagnicolum
C. sportella
C. subtumidum
C. tinctum
Euastrum ansatum
E. binale *var.* hians
E. elegans
E. oblongum
E. pectinatum
Gonatozygon brebissonii
Hyalotheca dissiliens
Netrium digitus
Penium exiguum
Pleurotaenium ehrenbergii
Staurastrum denticulatum
S. lunatum
S. punctulatum *var.* kjellmanii
Staurodesmus sp.
Tetmemorus granulatus

water from pools on three islands, Fladda, Bac Mòr, and Sgeir Errionaich is given in Table 9.1. That on Bac Mòr is the largest and attracts a number of seabirds, which accounts for the high phosphate content of the water. All are devoid of phanerogams except that on Bac Mòr where a depauperate plant of *Ruppia maritima* was found. Macro-algae consisted solely of *Enteromorpha* spp. The diatom flora of one similar pool, that on Sgeir a' Chaisteil, was investigated and contained the following species: *Achnanthes hauckiana*, *Amphipleura rutilans*, *Amphora holsatica*, *A. pseudohyalina*, *Melosira nummuloides*, *Navicula digitoradiata*, *N. gregaria*, *Nitzschia apiculata*, *Opephora martyi*, *Rhoicosphenia curvata*. As might be expected all the species recorded are known to be tolerant of seawater or are true marine species.

On the main island small pools form at the base of some of the larger cliffs especially where sedimentary rocks outcrop on the shore. These pools, some 50–100 m from the tide line, appear mainly to be the result of freshwater trickles or drips from the cliffs but which are enriched by sea water, probably from splash or spray. The algae recorded for such pools from the raised beach at Carsaig are listed in Table 9.2. Many of the species are typical of freshwater pools, but the flora is particularly rich in number of species and contains many coccoid green algae; the diatoms recorded are mainly brackish water species The diatoms listed in Table 9.3 were found in a detailed study of a wet rock-face in the same locality and appear to indicate influence of sea spray since most of the dominant species are known to prefer base-rich conditions.

Table 9.3 Diatoms found in gelatinous matrix of blue-green algae on wet coastal rocks at Rubha Dubh, Carsaig cliffs (coll. nos. F.6 and 7); * = dominant species

Achnanthes affinis	E. argus
A. clevei	E. reticulata*
A. flexella	Eunotia arcus
A. linearis	E. pectinalis
A. trinodis*	Fragilaria brevistriata
Amphora veneta	F. capucina
Anomoeoneis styriaca	F. leptostauron
A. zellensis	F. pinnata
Caloneis alpestris	F. vaucheriae
C. ventricosa	F. virescens
Cymbella affinis	Gomphonema gracile
C. angustata	G. intricatum (*incl. vars.*)
C. cistula*	G. parvulum
C. delicatula	G. tergestinum
C. helvetica	Hannaea arcus
C. hustedtii	Mastogloia smithii *var.* lacustris*
C. leptoceros*	Navicula fragilarioides
C. parva	N. vulpina
C. perpusilla	Nitzschia denticula*
C. prostrata*	N. linearis
C. sinuata*	N. sinuata
C. ventricosa	Rhopalodia gibba*
Didymosphenia geminata	R. gibberula
Denticula elegans*	Synedra amphicephala
D. tenuis	S. rumpens
Epithemia adnata*	S. ulna

Freshwater systems

Systems of flowing (lotic) water

Using the definitions of Elton & Miller (1954), the majority of rivers and streams on Mull are of 'medium' to 'small' size and usually have a 'fast' rate of flow. The larger rivers are the Beach, Lussa, Coladoir, Glencannel (Bà), Forsa, Aros and Bellart; their individual catchment averages 2,600 ha. Since their chemical composition relates to some extent to the rock and the soil type they drain, the amount of dissolved solids is likely to be proportional to the area of catchment. For example, the Forsa drains from the peaty gleys, oroarctic soils and lithosols of the Central Complex Land System, a fact reflected in the generally low content of calcium, magnesium and total dissolved solids (Table 9.4); the water sample analysed was taken halfway up the valley before the river picked up further nutrients from the arable ground lower down. The Aros and Bellart, on the other hand, collect from deeper peat and more basic soils, and the analysis of the latter shows some signs of high sodium and chlorine suggesting a back flow from Loch a' Chumhainn (Table 9.4). The chemical analyses of a small moorland stream running off Plateau Basalt and shallow peat near Carsaig by Cnoc a' Bhragad and of a peat runnel on the Ardmeanach on the west side of Maol Mheadhonach are also given in Table 9.4. These show that the water draining from this basaltic soil is richer than that collected by the Forsa.

The vegetation in the rivers is sparse and usually confined to small bays which have become cut off from the main flow of water. Here *Potamogeton polygonifolius* may flourish with other oligotrophic lake species such as *Littorella uniflora* and *Myriophyllum alterniflorum*. *Carex nigra* may be a common riparian species in a niche where, in the more eutrophic waters in parts of eastern Scotland, *Carex aquatilis* Wahlenb. or further south *C. acuta* L., would grow. The margins of these rivers are constantly being eroded bringing a certain amount of peat and alluvium on to the river bed. For much of their area, however, the floor of the river is stony, often with large boulders, 20 to 100 cm across. The following species of lichens are found on rocks which are

Table 9.4 Chemical analysis of lotic waters on Mull (in mg/l)

river	mid Forsa	lower Aros	lower Belart	streamlet Maol Mheadhonach
Catchment area (ha.)	4,100	4,400	3,300	5
Total dissolved solids	41·5	65	86	51
Ca	8	14	14	10
Mg	4	10	16	10
Na	7·8	10·5*	18·5*	8
K	0·4	0·5	0·7	0·2
Cl_2	13	18	24*	21
PO_4	0·12	0·12	0·16	0·04
SO_4	4·9	3·7	6·6	1
SO_2	3·9	6·6	12·2	0·7
Total N	0·07	0·35	0·79	0·38
pH	6·8	7·15	7·4	6·0

* High salinity here probably due to backwash from estuary at high spring tides.

usually submerged: *Polyblastia cruenta*, *Verrucaria aquatilis* and *V. kernstockii*. On boulders and pebbles which are more frequently exposed *Catillaria chalybeia*, *Verrucaria aethiobola*, *V. aquatilis* and *V. elaeomelaena* are often common. The undersides of all these rocks are often covered with a brown velvety film of diatoms.

A few smaller lowland rivers are peaty and more slowly flowing and in these *Chara globularis* var. *virgata* is common. Some, slightly more mesotrophic, contain aquatics such as *Rorippa nasturtium-aquaticum*. In the peat swamps of Loch na Cuilce, at one time part of the Bellart main stream but now bypassed, there are relatively large areas of *Phragmites*, *Scirpus tabernaemontani*, *Sparganium erectum* and the established alien *Crocosmia* × *crocosmiflora*, the latter more frequent in the marginal drainage ditches.

On the basalt, the rivers and their tributaries flow into the main alluvial plains over the benches and terraces of the lava flows. The smaller tributary streams begin on the upland plateau and there the flow through the peat is slower. *Potamogeton polygonifolius* is the dominant vascular plant and this and the *Carex* and *Molinia* roots that emerge in the open water are often covered with species of *Tabellaria*, *Frustulia*, *Eunotia*, *Bulbochaete*, *Microspora*, *Mougeotia*, *Spirogyra*, *Ulothrix* and *Zygnema*. The

Fig. 9.1 The Scallastle River descending dolerite sills in the Torosay hills at Coire nan Each, c. 480 m alt. 1967.

smaller tributaries on the open hillside flowing through shallow or skeletal inorganic soils or running on the bed-rock itself are shallow, often fast-flowing and well-aerated. Those, for example, on Creag a' Ghaill have a diatom flora usually dominated by species of *Achnanthes* and *Cymbella*, and where small waterfalls are present the rock surface is covered with *Didymosphenia geminata*. In these fast-flowing streams, lichens such as *Verrucaria aquatilis*, *V. kernstockii* and *Polyblastia cruenta* are often present on submerged pebbles and bed-rock. A few streams flow through richer pastures which are grazed by cattle; a sample from such a stream taken in north Iona, near Arduara, showed *Nitzschia communis*, a characteristic species requiring high nitrogen (Cholnoky, 1968) to be a co-dominant diatom with *Nitzschia hantzschiana* and *Achnanthes lanceolata*. Other associates were *Achnanthes affinis*, *Amphora ovalis* var. *pediculus*, *Cymbella ventricosa*, *Meridion circulare* and *Nitzschia linearis*. The vascular plant flora included *Apium inundatum*, *Veronica beccabunga* and *V. anagallis-aquatica*.

Where the streams flow through the steeper rocks, narrow gorges are cut which confine and increase the force of water. Waterfalls, often three times the width of the stream and from three to ten metres high, form over the harder sills. Lines of such waterfalls are seen on the east side of Mannir nam Fiadh (see Fig. 9.1) and on the many tributaries of the Bellart running off Beinn na Drise. An impressive waterfall, about twenty metres high, can be seen by the road from Lagenulva to Achleck at

Table 9.5 Commoner diatoms found on wet rock surfaces at Kilbrenan waterfall

Achnanthes linearis	Epithemia argus
A. microcephala	E. reticulata
A. minutissima	Gomphonema gracile
A. saxonica	G. parvulum *var.* exilis
Anomoeoneis styriaca	G. intricatum *var.* intricatum
A. zellensis	G. intricatum *var.* pumilum
Cocconeis placentula	Hannaea arcus
Cymbella affinis	Synedra radians
C. cesatii *var.* capitata	S. ulna
Diatoma tenue	Tabellaria flocculosa
Didymosphenia geminata	

Kilbrenan. Here the vertical rock slabs are for the most part devoid of vascular plant vegetation, although the ferns *Cystopteris fragilis* and *Polystichum aculeatum* var. *cambricum* may withstand the mainstream of water. A list of the more common diatoms found at Kilbrenan is given in Table 9.5.

On the rocks of waterfalls and on those in the splash zone of fast-flowing streams, filamentous green algae, e.g. *Spirogyra* spp., *Ulothrix zonata* and *Rhizoclonium hieroglyphicum*, diatoms (mainly attached species) and the moss *Hygrohypnum luridum* are common. Those rock surfaces which are only intermittently inundated are frequently covered with a thick black mat of *Stigonema* spp. (mainly *S. mamillosum*) and also *Gloeocapsa* spp. The lichens *Catillaria chalybaea*, *Gyalidea fritziae*, *Polyblastia cruenta*, *P. strontianensis* and *Verrucaria hydrela* also occur in this habitat. On ledges, mosses such as *Dichodontium pellucidum* var. *flavescens*, *Gymnostomum recurvirostrum*, *Trichostomum hibernicum*, *T. tenuirostre* and the liverworts *Pellia endiviifolia* and *Scapania undulata* are found. Small pools formed at the base of waterfalls may contain the moss *Fontinalis antipyretica*.

Where streams flow through open moorland, the environment of the rocks in the splash-zone combines high wetness (humidity) with exposure to bright sunlight; the lichens *Catillaria chalybeia*, *Lecanora lacustris*, *Lecidea hydrophila*, *Rhizocarpon laevatum*, *Verrucaria aethiobola* and *V. praetermissa* are the dominant plants in these open streams. The vascular plant flora on rocks in the gorges is similar to that of the surrounding moorland. Isolated peat pockets may have remnants of bog or mire flora with *Narthecium ossifragum*, *Scirpus cespitosus*, *Carex flacca*, *C. demissa* and *C. panicea*.

A few ferns, e.g. *Asplenium trichomanes* subsp. *trichomanes*, *Cystopteris fragilis*, *Hymenophyllum wilsonii* and *Phegopteris connectilis* can tolerate constant splashing. Mostly the flora consists of mosses and liverworts, e.g. *Scapania undulata*, *Diplophyllum albicans*, *Heterocladium heteropterum*, *Lejeunia patens*, *Rhacomitrium aquaticum* and *Trichostomum tenuirostre*. Open wet gullies are dominated by *Montia* spp., *Philonotis* spp. and other species which enjoy the constant wetness or the replenishment of scarce nutrients and here become luxurious, e.g. *Cochlearia* spp.

In the more shaded ravines *Vaccinium* spp. overhang the water and on ledges in the splash zone *Phegopteris connectilis*, *Gymnocarpium dryopteris* and *Polystichum aculeatum* are mixed with bryophytes such as *Isothecium myosuroides* and *Dicranum scoparium*. *Hymenophyllum wilsonii* and *Polypodium* spp. may cover rocks and tree boles whilst *Sphagnum subnitens* may festoon the banks of smaller streams. On the rocks in the splash zone, blue-green algae such as species of *Dichothrix*, *Lyngbya*, *Rivularia* and *Scytonema* form a continuous and often gelatinous film amongst the bryophytes, and the lichen *Bacidia inundata* may be common on bare rock. The lichens *Enterographa hutchinsiae*, *Opegrapha gyrocarpa*, *O. lithyrga* and *O. zonata* are also found in these moist, deeply shaded crevices.

Still-water (lentic) systems

These include all the open bodies of water from lakes to puddles. There are some, in all size categories, into which there is a significant in-flow, but in such cases the through flow is slow, and they may therefore be regarded as static waters in which movements will mostly be promoted by wind currents producing wave action.

Lentic bodies of water are, as in any west Scotland landscape, frequent on Mull. Those sampled are divided here first on size and then on nutrient status. Generally speaking, if a lake contains over 100 p.p.m. (i.e. 100 mg per litre) of dissolved substances (total cations and anions) and over 20 p.p.m. of calcium, it is, for Mull, relatively rich. There are no lakes on Mull that can be equated with the truly calcareous, eutrophic lochs of Sutherland (e.g. Croispol) and the majority of them are what Spence (1964; 1967) would class as 'nutrient poor' (oligotrophic) with only a few as 'moderately rich' (mesotrophic). Water chemistry and its dependence in most cases to proximity to the sea is discussed at the end of this chapter. All the lochs on Mull have originated as a result of glacial activity, either in glacial or tectonic rock basins or, more rarely, in drift basins. Those in rock basins formed by ice-scour are steep sided and do not have much soil at the shore-line for redistribution by wave action.

The soils of the loch basins vary with the basin and the surrounding soil-type. Bottom substrates may be more important to the nutrient uptake and productivity of submerged aquatics than hitherto thought (M. Denny, *pers. comm.*). Those lochs in skeletal soils have a graded shore-line from coarse gravel through to clay. Eroding by wave action of the windward shore will produce further aquatic soils, often reworked (eroded) peat, which in most cases intergrades with the inorganic deposits. Generally, soil texture becomes finer as the water deepens and the distance from the shore increases. In some lochs, e.g. those on the western end of the Ross peninsula, or some mountain lochs, e.g. Loch Bearnach, riparian species such as *Phragmites australis*, *Carex rostrata* and *Cladium mariscus* will form peats beneath the water level. The surface layer of this peat, especially in shallow waters between 30 and 50 cm in depth, becomes oxidised and somewhat gelatinous, thus preventing passage of nutrients from the peat to the overlying water. Most lochs will contain varying depths of mud produced from decaying plant material, much of which will have been washed into the loch; these lochs will tend to have brown-coloured water, characteristic of nutrient-poor lochs (Spence, 1967). A few lochans, e.g. those high-level shallow ones high above the coast on the Ardmeanach and Laggan Deer Forest will have clear water and black mud at the bottom. This mud is suggestive of the decay of planktonic algae probably desmids and is usually indicative of richer waters, in this case from wind-blown sea-spray.

It should be pointed out that the terms lake, tarn, pool etc., would be preferable when comparing with standard works (e.g. Elton, 1966), but we have used the Gaelic *loch* and its diminutive *lochan* because they are common parlance with all who carry out field-work in Scotland. Unfortunately their adoption by the Ordnance Survey follows local usage and not always a logical size difference.

Large lakes (lochs). These are over 40 ha (100 acres) in area and range from 20 to 50 m deep, of which three, Uisg, Bà and Frisa, were included in the Bathymetrical Survey of Scottish Lochs (Murray & Pullar, 1908). Loch Assopol was not surveyed at that time but lies along a fault line and is possibly within this depth range, and Loch an Tòrr is similarly placed in line with Loch Frisa. Lochs Bà, Frisa and Assopol have a distinct species-rich but dilute phyto-plankton flora composed mainly of desmids which are characteristic of oligotrophic lakes (see Table 9.6). The littoral

Table 9.6 Euplankton recorded in Loch Frisa (coll. nos. A1 and H55)

CYANOPHYTA

Chamaesiphon curvatus (*on floating debris*)
Microcystis aeruginosa
Oscillatoria irrigua
Rhabdoderma gorskii

DINOPHYTA

Ceratium hirundinella
Peridinium willei

CHRYSOPHYTA

Dinobryon bavaricum
Synura spinosa *f.* spinosa

BACILLARIOPHYTA

Asterionella formosa *var.* formosa
Cyclotella comensis
C. comta
C. glomerata
C. kuetzingiana
C. socialis
Fragilaria crotonensis
Melosira varians
Rhizosolenia eriensis
Stephanodiscus rotula *var.* minutula
Surirella robusta
Synedra fasciculata
S. ulna *var.* danica
Tabellaria fenestrata
T. flocculosa

CHLOROPHYTA

Botryococcus protruberans
Eudorina elegans
Pediastrum sp.
Sphaerocystis schroeteri

Desmids

Closterium acerosum *var.* angolense
C. kuetzingii
Cosmarium abbreviatum *var.* planctonicum
C. botrytis
C. depressum *var.* planctonicum
C. subarctoum
Cosmarium spp.
Staurastrum anatinum
S. arctiscon
S. cumbricum *var.* cambricum
S. gracile
Staurastrum lunatum *var.* planctonicum
S. pelagicum
S. pingue
S. pseudopelagicum
Staurodesmus sellatus
S. spencerianus
S. subtriangularis *var.* subtriangularis
S. subtriangularis *var.* inflatus
S. subtriangularis *var.* limneticus
Xanthidium antilopaeum *var.* depauperatum

algal communities are dominated by diatoms. These often form a rich brown film covering rocks, stones, mud and submerged aquatic plants and also occur as flocculent brown masses in the shallow marginal waters. A single gathering from such a habitat may contain 60–100 taxa characterised by small species of *Navicula* and *Achnanthes*, thus showing that, although the main body of the water is oligotrophic, these marginal waters show signs of local enrichment. This may be due to seepage from surrounding alluvium, from streams entering the loch or the possible replenishment of nutrients by wave action. The diatom flora is generally mesotrophic in character and the habitat is referred to thus in the systematic list (Chapter 17). In addition to the diatoms, blue-green algae, for example *Dichothrix orsiniana* which forms a felt, and *Scytonema* spp., are often common at loch margins, and the exposed boulders at the edge of the water are often covered in *Gloeocapsa* spp. and *Stigonema* spp. The latter community resembles that on rocks in the splash zone of waterfalls, and is probably the result of variable water level and splash from wave action which produces conditions of intermittent wetting. There are few species in the macroflora. *Isoetes lacustris*, *Potamogeton gramineus*, *Sparganium minimum* are frequent, and when stony margins occur an open turf of *Littorella uniflora* and *Lobelia dortmanna* associations may be seen, with *Myriophyllum alterniflorum* in small amounts. *Sparganium erectum* is characteristically seen near outflows, and swamp species like *Carex rostrata*, *Eleocharis palustris* and *Equisetum fluviatile* are rare in small bays.

Medium lakes (lochs) These are 1–40 ha (2·5–100 acres) in area; they can vary in nutrient status but can generally be divided into two depth categories: those clearly shallow lakes with much rooted vegetation and those of uncertain depth with only marginal vegetation. In general the deeper lakes have a higher nutrient level than the large lakes mentioned above. The shallower ones on the other hand are distinctly nutrient-poor.

The deeper lakes include Caol Lochan, Lochan a' Ghurrabain, Loch a' Ghleannain, Lochan na Guailne Duibhe, Lochnameal and Loch an Tòrr. The phytoplankton in all

Table 9.7 Euplankton recorded in Loch a' Ghleannain (coll. nos. A81–83 and H73)

CYANOPHYTA	CHLOROPHYTA
Chroococcus limneticus	Eudorina elegans
Microcystis aeruginosa	Sphaerocystis schroeteri
DINOPHYTA	*Desmids*
Ceratium carolinianum	Closterium kuetzingii
C. hirundinella	Cosmarium depressum
Peridinium willei	Staurastrum anatinum
CHRYSOPHYTA	S. lunatum *var.* planctonicum
Dinobryon bavaricum	S. pingue
Mallomonas sp.	S. pseudopelagicum
BACILLARIOPHYTA	S. subcruciatum
Asterionella formosa *var.* formosa	Staurodesmus megacanthus *var.* scoticus
Cyclotella comta	S. spencerianus
C. comensis	S. subtriangularis *var.* inflatus
C. glomerata	Xanthidium antilopaeum *var.* depauperatum
Fragilaria crotonensis	
Tabellaria flocculosa	

these lochs was generally similar, though somewhat poorer in yield and species numbers than that in the larger lochs (see Table 9.7). The littoral algal floras were also similar to those of the larger lochs and the diatoms at least showed indications of local enrichment. The vascular flora was characterised by *Littorella uniflora*, *Lobelia dortmanna* and *Isoetes lacustris* around the more stony shores and usually by the outflow, with *Eleocharis palustris* as a marginal reed-swamp species. *Scirpus lacustris* is frequently seen at the lower (i.e. outflow) end of these lochs in a water depth of one to two metres. *Phragmites*, with *Carex rostrata* and *Potamogeton natans* as common associates, is found around the mouths of inflowing streams.

The shallow nutrient-poor lakes include Loch Bearnach, Loch Airdeglais, Loch an Eilein, Loch an Ellen, Loch Fuaron, Loch Squabain and Loch an t'sithein. Most of these lochs have fine muddy bottoms, and often leaves of large phanerogamic populations cover areas of the water surface. Loch Bearnach is a typical example of this type with *Phragmites australis*, co-dominant with *Carex rostrata*, standing in about 40 cm water over much of its area. *Potamogeton natans*, *Nymphaea alba* and *Nuphar lutea* are common open-water associates. Where stony margins are present, *Littorella uniflora*, *Lobelia dortmanna* and *Subularia aquatica* establish. *Elatine hexandra* a rare species on Mull, was recorded here. The euplankton collections from these lochs were always extremely poor. Sometimes, however, large numbers of crustaceans were caught so that any phytoplankton present would have been consumed.

In direct contrast to these examples Loch Poit na h-I while of medium size and reasonably shallow is the most base-rich lake on the Island (138 p.p.m. T.D.S.; 25 p.p.m. Ca). Although situated on the granite, it is at the exposed western end of the Ross of Mull and receives nutrients from the salt-laden winds. It has a varied and interesting phanerogamic flora and a species-rich and dense phytoplankton (see Table 9.8). Around its shallow, stony margins *Littorella uniflora–Lobelia dortmanna–Chara vulgaris* f. *contraria–Isoetes echinospora* associations do not suggest the variety to be

Table 9.8 Euplankton recorded for Loch Poit na h-I (*coll. no. B51*)

CYANOPHYTA

Anabaena sp.
Chroococcus limneticus
Microcystis aeruginosa
M. incerta

DINOPHYTA

Ceratium hirundinella

CHRYSOPHYTA

Synura spinosa

BACILLARIOPHYTA

Asterionella formosa *var.* formosa
Cyclotella comta
C. glomerata
C. meneghiniana
Diatoma tenue *var.* elongatum
Fragilaria crotonensis
Melosira italica *subsp.* subarctica
Stephanodiscus hantzschii

BACILLARIOPHYTA (cont'd)

Surirella robusta
Synedra delicatissima
S. ulna *var.* danica
Tabellaria fenestrata
T. flocculosa

CHLOROPHYTA

Eudorina elegans
Pediastrum boryanum *var.* cornutum
Scenedesmus acuminatus
S. serratus *f.* minor

Desmids

Staurastrum cumbricum *var.* cambricum
S. denticulatum
S. pelagicum
S. pingue
Staurodesmus crassus
S. dejectus
S. patens
S. spencerianus

Fig. 9.2 Loch an Sgalain, 3 kms SW of Bunessan. Typical lake basin in granite area of the Ross of Mull. 1967.

found in the deeper areas where rare hydrophytes, e.g. *Najas flexilis*, *Potamogeton lucens*, *P. berchtoldii*, *P.* × *zizii*, *Nitella flexilis*, were recorded. *Cladium mariscus*, *Phragmites australis* and *Scirpus lacustris* are also seen on Mull at their best around this loch. The plankton, on both occasions the loch was sampled, was of an unusual type unique on the Island. It consisted mainly of *Melosira italica* subsp. *subarctica* and *Synura spinosa*, though green algae, particularly *Pediastrum*, were also important. In addition, however, several of the spinous desmids characteristic of the oligotrophic lochs were also present. The littoral diatom flora was the richest of all sampled.

Other smaller lakes on the granite (e.g. Loch an Sgalain, Fig 9.2) have a similar swamp flora to Loch Poit na h-I consisting of *Cladium mariscus*, *Scirpus lacustris*, *Sparganium erectum*, and the only Mull localities for *Cladium mariscus* and *Chara vulgaris* f. *contraria* are on this peninsula. Loch Staoineig, on the gneiss of Iona, has the largest amounts of electrolytes of all the lochs on Mull (237 p.p.m. T.D.S.; 19 p.p.m. Ca) but the vascular flora of this loch is numerically poor. No good sample of euplankton has been collected but the littoral diatoms are typically those of a base-rich lake, i.e. they compare with Loch Poit na h-I although numerically poorer.

Ponds or pools (lochans) These are less than 1 hectare (2·5 acres) and can again be divided on ionic content. Those situated in deep peat close to the sea are enriched by salt-laden winds carrying sea-spray and shell-sand. These are similar to the medium deeper lakes mentioned above in both vascular and diatom floras. One such lochan on the Ross of Mull, Loch an t'Suidhe, close to the road 1 km west of Bunessan, has a flora of *Nymphaea alba*, *Nuphar lutea* and *Potamogeton natans* and surrounding it is dense reed-swamp of *Carex rostrata*, *Menyanthes trifoliata*, *Phragmites* and *Potentilla palustris* with *Equisetum fluviatile* common in the outer fringes. The number of hydrophyte species is proportional to the area of water. Similar lochans are seen elsewhere on the Ross of Mull, e.g. Linne nan Ribheid (Fig. 9.3) and Loch Cholarich.

Away from the Ross of Mull such base-rich lochans are rare. The only example studied was one on the Killichronan Estate just north-east of the farm. This pond is surrounded by improved pasture and probably derives its high nutrient content from fertilizers and from the use of the pond by cattle. Reedswamp is reduced to a narrow belt on one side where *Phragmites – Carex rostrata* is the dominant community with frequent *Sparganium erectum*. *Nuphar lutea* and *Nymphaea alba*, with *Potamogeton natans*, cover half the open water.

The second category of lochan is represented by the oligotrophic moorland pools

Fig. 9.3 Linne nan Ribheid, 3 kms SE of Fionnphort. Floating mat of Phragmites-Rhynchospora-Sphagnum *community in peat basin between granite outcrops. 1967.*

so characteristic of the landscape of Mull. These are frequently found on the ridges and near the summits of the lower hills where they may form in rock hollows produced by frost shattering or by glacial scouring. The diatom flora of such pools is species-poor and vascular plants are sparse and characteristic (cf. Pearsall, 1918), only *Isoetes echinospora*, *Littorella* and *Menyanthes* and the common liverwort *Scapania undulata* being recorded.

Lochans of a slightly higher base content are found around 240–300 m on the Laggan Deer Forest and, again at about the same height, above Torloisk and along the north side of Loch Tuath. Where the shore is pebbly *Littorella uniflora* and, farther out, *Lobelia dortmanna* may form continuous lawns. In one typical example, Lochan na h-Earba, at a depth of about 70 cm a band of *Sparganium angustifolium* was

Fig. 9.4 Typical bog-pool on the Coladoir River raised bog system. 1967.

Table 9.9 Algae commonly found in bog-pools and amongst wet Sphagnum. *Species listed were noted ten times or more in the 70 collections of this type.*

CYANOPHYTA

Chroococcus turgidus
Various minute coccoid colonial species, often in quantity
Various filamentous species

DINOPHYTA

Peridinium spp. *especially* P. inconspicuum

BACILLARIOPHYTA

Anomoeoneis exilis
A. serians
A. serians *var.* brachysira
Cymbella perpusilla
C. rabenhorstii
Eunotia alpina
E. bidentula
E. denticulata
E. exgracilis
E. exigua
E. flexuosa *var.* eurycephala
E. incisa
E. lunaris
E. pectinalis *var.* minor
E. serra *var.* diadema
Fragilaria virescens *and varieties*
Frustulia rhomboides *var.* saxonica
Navicula bryophila
N. festiva
N. hoefleri
N. mediocris
N. subtilissima
Pinnularia abaujensis
P. appendiculata
P. biceps
P. hilseana
P. microstauron
P. parvula
P. rupestris

BACILLARIOPHYTA (cont'd)

P. subcapitata
P. sudetica
P. viridis
Stenopterobia intermedia
Tabellaria flocculosa
T. quadriseptata

CHLOROPHYTA

Mainly filamentous species: Binuclearia tectorum, Bulbochaete spp., Oedogonium spp., Microspora sp., Mougeotia spp., Spirogrya spp., Ulothrix spp., Zygnema spp., *with a few coccoid forms, e.g.* Oocystis spp.

Desmids

Bambusina brebissonii
Closterium dianae
C. gracile
C. intermedium
Cosmarium cucurbita
C. margaritiferum
C. pyramidatum
C. subtumidum
Cylindrocystis brebissonii
Euastrum ampullaceum
E. crassum
E. elegans
E. pectinatum
Hyalotheca dissiliens
Netrium digitus
Penium spp.
Pleurotaenium minutum
Staurastrum margaritaceum
S. tetracerum
Staurodesmus extensus
Tetmemorus brebissonii (*incl. var.* minor)
T. granulatus
T. laevis
Xanthidium spp.

dominant, and *Myriophyllum alterniflorum* appeared in isolated colonies. At this point the bottom consisted of boulders often very large and up to 1 m diameter, between which were pockets of deep mud colonised by *Isoetes lacustris*. In the centre of the lochan, the bottom was flat, muddy and devoid of rocks, and there *Isoetes lacustris* formed a lawn through which *Potamogeton natans* grew to cover the centre of the lochan with its floating leaves. In some of these pools, e.g., Loch an Daimh on the Laggan Deer Forest, the *Isoetes lacustris* attains a very large size and is suspected to be of considerable age.

Table 9.10 Algae recorded from a permanent bog-pool on Ormaig, Ulva (coll. no. H–123). Only the more common species of diatoms (Bacillariophyta) are listed.

CYANOPHYTA	CHLOROPHYTA
Chroococcus turgidus	Eremosphaera viridis
Synechococcus aeruginosus	Oocystis sp.
	Desmids
DINOPHYTA	Closterium dianae
Peridinium inconspicuum	C. lunula
	Cosmarium angulosum
RHAPHIDOPHYTA	C. brebissonii
Vacuolaria viriscens	C. cucurbita
	C. difficile *var.* difficile
CHRYSOPHYTA	C. difficile *var.* sublaeve
Dinobryon cylindricum	C. punctulatum *var.* subpunctulatum
Phalansterium consociatum	C. ralfsii *var.* montanum
	C. subtumidum *and other unidentified species*
BACILLARIOPHYTA	Cylindrocystis brebissonii
Anomoeoneis serians	Euastrum ampullaceum
A. serians *var.* brachysira	E. bidentatum
Cymbella perpusilla	E. binale
C. rabenhorstii	E. crassum
Frustulia rhomboides *var.* saxonica	E. dubium
Navicula bryophila	E. elegans
N. festiva	E. gayanum
N. mediocris	E. oblongum
N. subtilissima	E. pectinatum
Pinnularia biceps	E. ventricosum
P. microstauron	Hyalotheca dissiliens
P. parvula	Netrium digitus
P. rupestris	Staurastrum irregulare
P. viridis	S. simonyi
	Staurodesmus extensus
EUGLENOPHYTA	S. quadratus
Euglena sp.	Tetmemorus brebissonii
Trachelomonas abrupta	T. granulatus

Bog-pools The pools found on deep peat are of two kinds. One is formed by the gravitational movement of the bog surface resulting in the splitting of the upper peat layers along an irregular line which roughly follows the contours (Pearsall, 1956). Pools which may have formed in this way are seen on the raised bog in the Coladoir river valley (see Chapter 10 and Fig. 9.4); they are steep-sided except sometimes along their shorter sides where the bog surface may be regenerating. The macrophyte flora is sparse and is composed of *Eriophorum angustifolium*, *Menyanthes trifoliata*, *Utricularia minor*, *U. ochroleuca* and *Sphagnum cuspidatum*. Similar pools may be found where a peat slope slumps away from the thinner covering on the summit of a lowland peat-covered knoll. Some of the smaller pools on the hill summits where the peat is almost completely denuded may be little more than 20 cm deep. The bottom is usually covered in flocculent detritus, sometimes coloured green by the large number of desmids it contains. Filamentous algae, e.g. species of *Binuclearia*, *Bulbochaete*, *Hyalotheca*, *Microspora*, *Oedogonium* and *Tribonema* commonly occur either suspended in the water or tangled amongst the stems of rooted vegetation. A tychoplanktonic flora is found in the bottom detritus and amongst the filamentous algae mentioned above and the majority of desmids listed in Chapter 16 occur in this habitat, although rarely in great quantity. Diatoms, however, were often present in considerable quantity but represented by only a few species in any one pool (see Table 9.9). Pools in bogs close to the west and south-west coasts probably contain more nutrients (anions

Table 9.11 Algae recorded from a Sphagnum pool (coll. nos. H10; A4) on Garmony Point, close to the sea. Only the more common species of diatoms (Bacillariophyta) are listed.

CYANOPHYTA	CHLOROPHYTA
Aphanocapsa sp.	Ankistrodesmus falcatus
DINOPHYTA	*Desmids*
Cystodinium sp.	Bambusina brebissonii
CHRYSOPHYTA	Closterium baillyanum
Dinobryon sertularia	C. intermedium *var.* intermedium
BACILLARIOPHYTA	C. lunula
Anomoeoneis serians *var.* brachysira	Cosmarium brebissonii
Cymbella rabenhorstii	C. cucumis *var.* magnum
Eunotia alpina	C. difficile *var.* subimpressulum
E. denticula	C. punctulatum
E. exgracilis	C. pyramidatum
E. incisa	C. quadratum
E. lunaris	Desmidium coarctatum *var.* cambricum
E. pectinalis *var.* minor	Euastrum ansatum
E. pectinalis *var.* undulata	E. crassum
E. serra *var.* diadema	E. didelta
E. tenella	Euastrum elegans
E. valida	E. oblongum
Pinnularia appendiculata	E. pectinatum
P. rupestris	E. subalpinum
P. stomatophora	Gonatozygon brebissonii
P. subcapitata	Hyalotheca dissiliens
P. viridis	Micrasterias denticulata
Tabellaria flocculosa	M. thomasiana
EUSTIGMATOPHYTA	M. truncata
Chlorobotrys regularis	Netrium digitus
EUGLENOPHYTA	Penium spirostriolatum
Trachelomonas sp.	Tetmemorus granulatus
	Xanthidium armatum

and cations) than those inland for the same reasons discussed under water chemistry (see also Fig. 9.6) since the algal flora of these pools was found to be richer in numbers of individuals and species. An example from a permanent bog-pool on Ormaig, Ulva, is listed in Table 9.10. The flora from a *Sphagnum* pool at sea level on Garmony Point on the Sound of Mull is listed in Table 9.11.

The second type of bog-pool may be a later stage of that described above or a hollow formed between *Sphagnum* hummocks or where peat diggings have become inundated (Fig. 9.5). As in the first type these pools are permanent and may have a continuous cover of *Sphagnum cuspidatum* and/or *S. auriculatum* (aquatic forms) on the bottom. The eventual build-up of *Sphagnum* stems may result in a succession of

Fig. 9.5 Sphagnum auriculatum, Potentilla palustris *and* Menyanthes trifoliata *colonising flooded peat cutting on the Ross of Mull.* 1967.

Sphagnum species, of which *S. auriculatum* (emergent forms) and *S. recurvum* are the first to supplant the *S. cuspidatum* (cf. Osvald, 1949) but this aspect has not been studied on Mull. *Carex limosum*, *Utricularia* spp. and *Menyanthes trifoliata* are common, with *Eriophorum angustifolium* occasional. Desmids and diatoms are common, although poorer in species and quantity compared with the former type of pool (Table 9.9).

Table 9.12 Physical and chemical characteristics of selected lakes showing relationship with proximity to coast and catchment area. (Chemical analyses in mg/l.)

lake	alti-tude (m)	distance from coast (km)		substrate and rainfall (mm)	catchment area (ha)	loch dimensions		pH	TDS	Mg + Ca	Ca	Na	Cl	PO_4	hardness ratio
		SW	NW			area (h)	depth (m)								
Loch Staoineig*	53	1·2	0·8	Peat over gneiss; 990	75; controlled outlet*	1	1–2	5·6	237	47	19	40	104	0·43	0·78
Loch Poit n'a h-I	12	2·0	2·0	Drift and granite; 1270	2500; outlet small	40	3–4	7·0	138	40	25	27	37	0·02	1·45
Loch Assopol	24	2·4	2·4	Drift over schist; 1473	4000; outlet small	54	3–4	7·1	104	37	15	16	33	0·06	2·18
Loch Binnein Ghorrie	270	0·4	9·0	shallow peat over basalt; 1778	20; no outlet	2	1–2	5·3	82	12	8	19	41	0·04	0·60
Loch a' Ghleannain	45	9·6	>20	Andesite basin with limestone; 2032	1000; outlet medium	10	3–4	7·1	60	22	14	9	21	0·22	2·20
Loch Frisa	120	12	6·4	In basalt valley 2032	14000; outlet large	4050	>45	7·0	56	22	14	12	20	abs.	1·53
Loch Bearnach	83	11·2	>20	Deep peat shelf on igneous rock 2286	2000; outlet small	5	1–2	6·6	37	8	4	9	9	0·16	0·90
Loch Bà	12	>20	>20	Glacial lake in caldera igneous complex; 2540	43000; outlet large	2980	>30	6·6	30	8	4	6	12	abs.	1·33
Loch an t'sithein	300	2·4	>20	Deep peat over basalt 2540	1000; outlet small	13	1–2	5·7	21	14	10	9	19	0·80	1·60

* Used as a reservoir for domestic supplies

Water chemistry

Water from 58 sites has been analysed for pH, alkalinity (as $CaCO_3$), calcium, magnesium, sodium, potassium, chloride, sulphate, silicate, phosphate and nitrogen as nitrate, nitrite, albuminoid and ammonia. The metals lead, copper, zinc and iron were also determined. Samples were taken either in April/May, June/July or in September over the period of fieldwork (1966–1970) and in some cases in two seasons on two- or three-yearly occasions. Selected analyses are given in Tables 9.1, 9.4 and 9.12. The analyses were carried out by the Laboratory of the Government Chemist and we are grateful to the successive Superintendents of that Laboratory for their co-operation.

As was to be expected, water samples varied according to year, season and weather conditions prior to sampling. It is therefore appreciated that these analyses can do little except show the general nutrient-poor nature of loch water on Mull and its similarity with that of the rest of north-west Scotland other than on limestone hills. Nevertheless, there is a variation in the chemistry of loch water which correlates for the most part with distance from the coast in the direction of the prevailing south-west wind (see Fig 9.6.) as Holden (1961) pointed out. The correlation is more complex than Holden suggested, however, and not only is distance from the sea important, but also the area of lake and its catchment. Holden also mentions wind velocity as another important parameter, and whilst we have no figures to support this, our observations show that lakes close to the north coast of the Island, which are separated from the south-west coast by at least 16 km of moorland also have relatively high ionic concentrations. This suggests that the less frequent but higher velocity north-west or north-east gales are having a similar effect to the prevailing but calmer south-west winds. Sea-spray mostly affects concentrations of sodium, magnesium and chloride in the lakes. It affects also the conductivity in as much that this is the result of the total dissolved solids, i.e., cations and anions. For this reason, as stated by Spence (1967), conductivity cannot be directly correlated with richness and plant distribution. Loch Staoineig on Iona, with a T.D.S. of 237 mg/l of which Na^+ and Cl^- accounts for 164, has six vascular plants and a relatively species-poor diatom flora. Loch poit na h-I has a T.D.S. of 138 of which the Na^+ and Cl^- content was recorded as 74 mg/l; it contains 22 macrophytes (vascular plants, *Fontinalis* and charophytes) and a very characteristic and rich planktonic flora (see Table 9.8).

Alkalinity may be measured as the total quantity of dissolved bicarbonate and carbonate expressed as milli-equivalents per litre. In a hundred Scottish lochs sampled by Spence (1967) only two showed dissolved CO_3 ions and it is likely that alkalinity there is synonymous with dissolved HCO_3. In his earlier (1964) account, Spence used parts per million $CaCO_3$ as does Brook (1964) in his account of the phytoplankton of Scottish lochs. Those waters with 16–100 p.p.m. calcium carbonate Spence terms 'moderately rich', and 29 Mull loch samples fall into this category, 21 falling below. Only Lochnameal, Meall nan Gabhar and the Treshnish sites have readings of over 60 p.p.m. and are nutrient-rich in the sense of Spence. It is difficult to explain these high readings in those lochs of eastern Mull, which are certainly not affected by sea-spray. They are in line with each other in a N–S direction and it is just possible that an underlying fissure of calcium-rich mineral may be the cause. It is more probable that glacial deposits in this area, at least in the basin of Lochnameal, contained calcium-bearing rocks.

Water 'hardness', being the amount of calcium and magnesium precipitated as carbonate has been recorded for the Mull lochs. Pearsall (1922) draws attention to ionic ratios of calcium and magnesium with those of sodium and potassium. Seddon (1972) prefers the inverse ratio i.e. the weight of Ca and Mg precipitated as carbonate

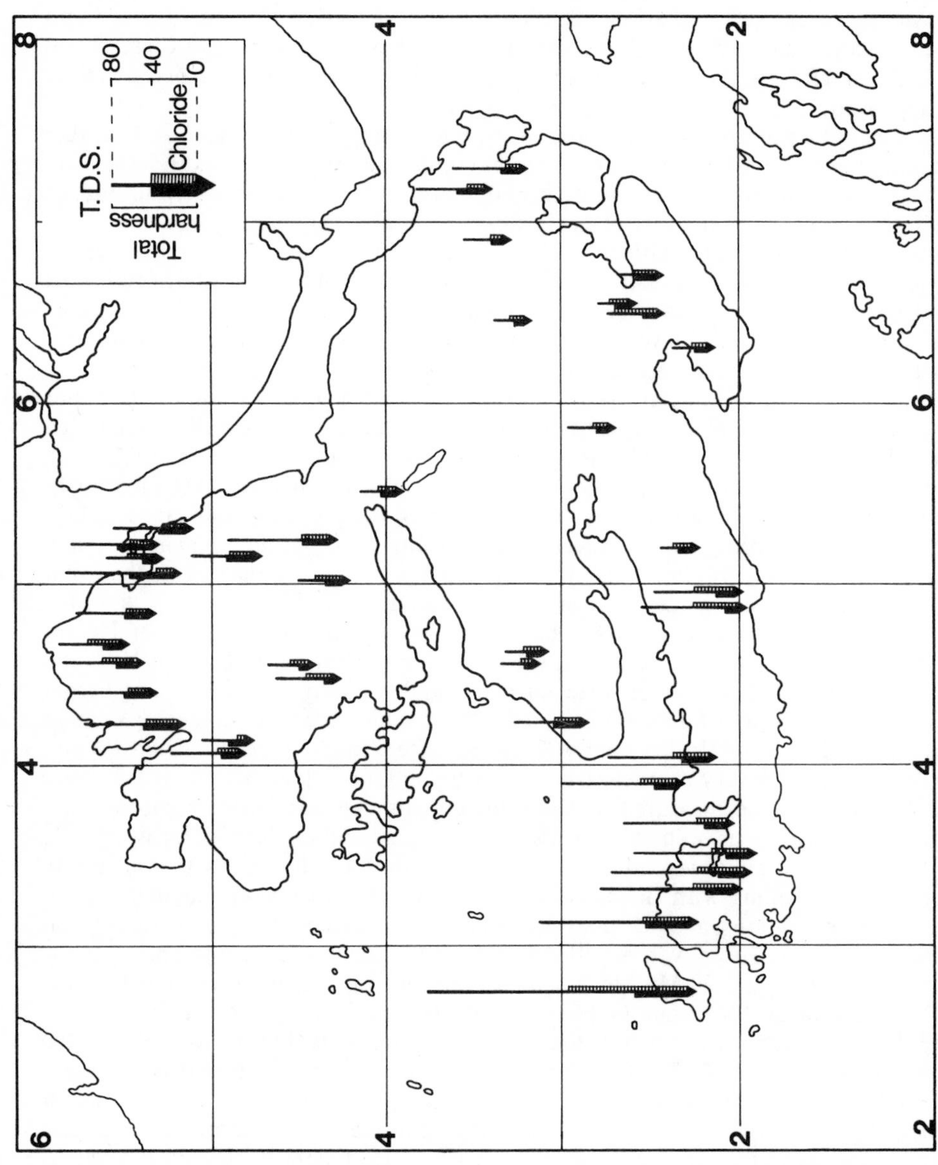

Fig. 9.6 Histograms showing the amount of total dissolved solids, total hardness, and salt (measured as chloride), in selected lakes.

divided by the weight of elemental Na and K. This is referred to as the hardness ratio in Table 9.12.

The pH of natural waters may vary considerably throughout the year and in consecutive years, but our readings are, on the whole, reasonably constant. They range from 4.4 in the peaty lochs, e.g. Lochan nan Daimh, to 7.0 and above on some of the deeper lakes in rocky substrates, e.g. 7·4 in Lochan a' Ghurrabain and 7·5 in Loch an Tòrr and Loch na Cuilce. Heavy metals were rarely present in sufficient quantity to be biologically significant. Copper, however, was present in measurable quantities (0·04 mg/l) in the Ben Buie – Beinn na Croise area although it is unlikely that this small concentration is sufficient to affect either plant or animal populations.

In conclusion, it can be stated that the chemical factors controlling the distribution of macrophytes in Mull lochs agrees with that suggested for Scotland as a whole by Spence, and to a similar extent to the situation in Wales reported by Seddon (1965; 1972) the following species being confined to nutrient-poor water: *Elatine hexandra* (only two sites on Mull); *Isoetes lacustris; Juncus kochii* 'f. *fluitans*'; *Lobelia dortmanna; Potamogeton polygonifolius; Scirpus fluitans; Sparganium angustifolium; Subularia aquatica.*

The presence of *Isoetes echinospora* in Loch Poit na h-I upholds the suggestion of Seddon (1965) that this species, and possible others, can withstand higher trophic waters if the competition from other macrophytes is not too intense. On the other hand, *Potamogeton lucens*, recorded from the same loch, is said by Spence to be confined to rich waters with the alkalinity not less than 1·2 m-eq HCO_3/l. (i.e. 60 p.p.m.); Loch Poit na h-I has 40 p.p.m. magnesium and calcium and a hardness ratio of 1·45.

References

BROOK, A. J. (1964). The phytoplankton of the Scottish freshwater lochs. Pp. 290–305 in Burnett J. H. (ed.), *The vegetation of Scotland.* Edinburgh & London.

CHOLNOKY, B. J. (1968). *Die Ökologie der Diatomeen in Binnengewässern.* Lehre.

ELTON, C. S. (1966). *The pattern of animal communities.* London & New York.

——— & MILLER, R. S. (1954). The ecological survey of animal communities: with a practical system of classifying habitats by structural characters. *J. Ecol.*, **42**: 460–496.

HOLDEN, A. V. (1961). Concentration of chloride in freshwaters and rainwater. *Nature, Lond.*, **192**: 961.

MURRAY, J. & PULLAR, L. (1908). *Bathymetrical survey of the fresh-water lochs of Scotland.* London.

OSVALD, H. (1949). Notes on the vegetation of British and Irish mosses. *Acta phytogeogr. suec.*, **26**: 5–62.

PEARSALL, W. H. (1918). On the classification of aquatic plant communities. *J. Ecol.*, **6**: 75–84.

———. (1922). A suggestion as to factors influencing the distribution of free-floating vegetation. *J. Ecol.*, **9**: 241–253.

———. (1956). Two blanket-bogs in Sutherland. *J. Ecol.*, **44**: 493–516.

SEDDON, B. (1965). Occurrence of *Isoetes echinospora* in eutrophic waters in Wales. *Ecology* **46**: 747–748.

———. (1972). Aquatic macrophytes as limnological indicators. *Freshw. Biol.*, **2**: 107–130.

SPENCE, D. H. N. (1964). The macrophytic vegetation of freshwater lochs, swamps and associated fens. Pp. 306–425 in Burnett, J. H. (ed.), *The vegetation of Scotland.* Edinburgh & London.

———. (1967). Factors controlling the distribution of freshwater macrophytes with particular reference to the lochs of Scotland. *J. Ecol.*, **55**: 147–170.

10. Terrestrial ecosystems

A. C. Jermy, P. W. James and A. Eddy

Introduction

Ecosystem classification

Terrestrial ecosystems may be classified in one of three general schemes. The earlier British ecologists, especially Tansley (1911; 1949) and Pearsall (1950), divided vegetation, a tangible unit of the ecosystem, according to main habitat types and altitudinal zonation, taking into account land-use factors and seral changes due to succession. Provided the habitat types are defined, this system is the most useful to, and most used by, zoologists (cf. Elton, 1966). It was such a descriptive method that was used by Robert Smith in one of the first accounts of Scottish vegetation, that of the Tay basin (Smith, 1898, 1899). Later, as part of a project launched as the *Botanical Survey of Scotland*, Smith classified and mapped the vegetation around Edinburgh (Smith, 1900a) and North Perthshire (Smith, 1900b) the latter giving him sufficient experience to write Chapter 13 "Arctic-alpine vegetation" in Tansley's *Types of British Vegetation* (1911).

Inspired by Smith, the geologist and ecologist C. B. Crampton surveyed the vegetation of Caithness 'in relation to the ecology' (Crampton, 1911). Patton (1924) wrote an account of the vegetation of Beinn Laoigh, but little further work in Scotland was carried out on these lines until the Cairngorm studies by Watt & Jones (1948), Metcalfe (1950), Burges (1951) and Ingram (1958). Poore & McVean (1957) discussed an ecological framework for Scottish mountain vegetation based on the five factors – altitude, oceanity, snow cover, base status and soil moisture.

A second scheme is that developed by Braun-Blanquet (1932) and used by many European ecologists over the past forty years. It is a sociological classification based on floristic composition and synecology, in which the fundamental unit is the *Association* arranged collectively in a hierarchy of *Alliances*, *Orders* and *Classes*. Poore (1955a, b) discusses the relationships of the Continental schools of phytosociology and its possible application to British ecological work. Other accounts using this classification in British vegetation can be found in Birse & Robertson (1967, 1977), Eddy *et al.* (1969), Shimwell (1971), and Birks (1973) where a full discussion on the usefulness of the method may be found.

The third approach is to some extent a reappraisal of the phytosociological methods in terms of life-form (forest, dwarf-shrub, grass-heath, moss-heath etc). This method is at least compatible with the Continental approach. It was first discussed by Poore (1955c) and brought to the fore in the Nature Conservancy Monograph *The plant communities of the Scottish Highlands* (McVean & Ratcliffe, 1962). This work, which does not include Mull or the Western Islands, describes nevertheless a number of moorland types found on the Island. On a more local basis, the techniques of Poore are used by Birse & Robertson (1967) in an accompanying account of the vegetation in the Soil Survey Monograph on the area around Haddington and Eyemouth. The latter authors have recently developed their own ideas in a major work on plant communities in lowland Scotland and the Southern Uplands (Birse & Robertson, 1977).

In 1964, the meeting of the International Botanical Congress in Edinburgh occasioned a comprehensive and scholarly work on Scottish vegetation as a whole, namely *The vegetation of Scotland* (Burnett, 1964). This work treats the marine littoral (maritime) areas as far as shingle-bank and saltmarsh but the algae of the rocky shore are excluded. The reader is advised to consult this work if interested in comparing vegetation types in Scotland and for a broad account of the climate and geology of the country.

The account of the vegetation types given in this chapter follows for the most part that given in Burnett (1964) believing that this gives a sufficient framework around which the floristic account in Chapters 11–15 may be presented. The Soil Survey of Scotland Memoir on Mull (Bibby, in prep.) will, we understand, contain a chapter on the vegetation of the Island by E. L. Birse and J. S. Robertson, with which the reader will be able to assess the plant communities from a phytosociological viewpoint.

The following terrestrial systems are discussed in this chapter: maritime communities (p. 10.5), forest communities (p. 10.21), dwarf shrub communities (p. 10.47), mire communities (p. 10.50), ombrogenous bog communities (p. 10.56), grassland and mountain summit heath communities (p. 10.64), anthropogenic communities (p. 10.72).

A note on ecological terminology used

Throughout the ecological literature concerned with the British Isles there is a degree of confusion and loose usage over the terms *marsh*, *swamp*, *fen* and *bog*. Over the past decade, and since more comparisons are being made with Continental vegetation types, a further term *mire* has come into use.

Marsh As Tansley (1911, 1949) defined the term, this is a vegetation type on predominantly inorganic soil in which the summer water-table is at, or near, the ground level. It is here regarded as superfluous and is used only in the context of coastal communities, i.e., saltmarsh and its variants. The term *marshy* is used in Part 3 to denote a habitat of which the substrate has a high percentage of inorganic matter with a high water-table throughout much of the year although this may be related to some extent to rainfall, e.g. an alluvial terrace on a stream-side.

Swamp The term swamp is restricted in this account to associations on inorganic or organic substrates, but most likely on the latter, which are found at the margins of permanent bodies of open water. The water level is mostly above the substrate surface at all times except in periods of abnormal drought. These riparian communities are frequently associated with the Common Reed (*Phragmites australis*), hence the term reedswamp, but large stands of sedge (*Carex* spp.) may also be dominant in these situations. On the whole the water movement is nil or imperceptibly moving in a lateral direction. Seasonal fluctuations, or in the case of estuarine swamps, diurnal (i.e. tidal) fluctuations may well be seen in water level in which case water movement is vertical as well as lateral.

Fen In a normal succession from open water to a closed tall herb or dwarf shrub community the swamp communities described above accrete sufficient silt or build up sufficient peat to raise the substratum level above the water-table. It is to these dynamic communities that Pearsall (1918) restricted the term *fen*. Tansley (1911; 1949) had a broader concept and defined fen as a 'decidedly alkaline, nearly neutral, or somewhat, but not extremely, acid peatland' in which the summer water level was close to the surface. Because of the nature of the source of the water, its chemical composition can vary and in accordance with Tansley's definition both 'rich' and 'poor' fen are found on Mull. The term is used here almost exclusively in the sense of Pearsall (1918) to describe those communities in the upper level of the hydrosere in which the water level in summer is at or around substratum level. The distinction between fen and swamp lies in the water-table alone but this in itself determines the nature of the secondary associates. In fen, mosses and liverworts and an increasing number of herbs are found as the substrate becomes higher. The semi-aquatic swamp is thus sufficiently different as a habitat to maintain that term. Fens around basin lakes may be fed with ground water that drains through them to the lake itself. In this case the term *mire* described below could be used as there would be an, albeit slow, lateral water movement; but where a stage in the hydrosere is implied the term *fen* is used.

Mire The term mire has been used in a very wide context, especially by Continental ecologists, to denote wet-lands with a lateral flow of ground-water. It is used here in the sense of Ratcliffe (1964a) for habitats with a water-table at or around the surface but in which the water maintains a lateral flow through the substratum. The factors controlling this kind of wet-land are mainly topographical, i.e. the slope and configuration of the land, and whilst it is the rainfall that controls the amount of water initially available, it is the size and configuration of the catchment area that delimits the mire. In the hill-country of Mull, as elsewhere in Scotland, the lateral seepage of (drainage) water can often maintain sites on gentle slopes, in hollows or valley bottoms, in which the water-table is always at the surface. The term *soligenous* is applied to this kind of association. The water having possibly flowed through various substrata can be again rich or poor in chemical content, hence the terms *base-rich* or *base-poor mires.*

Associated with soligenous mires are areas in which the lateral flow of water is faster and the water itself of higher base-status than the surrounding area. Such an area may be the result of a doleritic sill or some other base-rich rock. Depending on the nutrient status of the area as a whole these sites may have a distinctive plant community. Such areas are termed *flushes* and although only a form of mire the term is retained here to indicate an area (and usually a plant community) with a higher nutrient (usually base) status than the surrounding mire. As Pearsall (1950) points out, the rate of flow of the nutrients through a flush is often more important than the chemical content of the water at any one time.

Bog The term bog is used here in the sense of Ratcliffe (1964a & b). It is an area of vegetation that has all the characteristics of a mire except in the origin of its water supply. In the mire the water is run-off, i.e. drainage water, whereas in the bog it is rain-water which is held in the peat, hence the term *ombrogenous bog.* As rain is the sole source, the amount of nutrients is low and rainfall itself is a major factor (usually over 127 cm) but topography determines whether the land will drain or not. Because of the complex topography on Mull, mosaics of bogs and mires exist which can only be differentiated by the principal plant species they contain. As the main water source will be that flowing through these areas they are included under mire.

Major factors affecting the vegetation

The major factors affecting the vegetation and the distribution of individual species within it are summarised below.

Land use It is doubtful if much of Mull had ever been densely forested at least within the present climatic regime; such woods as did exist were situated in the sheltered coves and valleys and remnants of these are seen today. It is likely that the crofter of the sixteenth or seventeenth centuries found the mosaic of bog, mire and rough grassland as still existed at the beginning of the twentieth century. The major difference was the introduction of sheep which together with the 'clearances' between 1850 and 1870 heralded the decline of the small black cattle as grazers from the hills.

The population was considerably higher in the middle of the nineteenth century (Macnab, 1970). There were many more crofts throughout the glens and the outline of their lazy-beds may be seen around the ruined cottages. For the most part, however, it is the twentieth century land use practices which have the major effect on vegetation throughout the island and are as follows:

drainage of wetlands or semi-wetlands;
burning of heather and ling communities to increase grazing potential;
peat cutting, thereby raising or lowering the substrate level in relation to the water-table;
grazing for various periods and at various intensities preventing development to climax woodland where such is the climatic and edaphic climax;
afforestation and protecting from all grazing animals (including wild goats and deer) leading to a woodland climax with alien species as dominants.

Altitude The highest land-mass (Ben More) is 950 m (3169 ft) and it is likely that a considerable upland area was devoid of natural (i.e. native) forest. It is difficult to estimate the altitudinal extent of pine or oak forest at optimum conditions on Mull. Remains of pine as fossilized stumps in the peat have been found on the hills to the west of Carsaig at just over 300 m (1000 ft) and on the Laggan Deer Forest at 360 m (1200 ft). There is no native pine forest on Mull today but the highest oak/birch woodland is at Coillan Fhraoich Mhoir in Torosay reaching 285 m (930 ft).

Exposure Exposure cannot be divorced from altitude and is likely to be the deciding factor as to the altitude colonized by tree species in hilly districts. Exposure is a combination of two factors: wind and the sheer physical force it emits, and the lowering in temperature a constant cool air stream can effect. On Mull it is an important factor on the higher hills and on the coastal lowlands. In coastal areas winds can carry considerable sea spray which affects the nutrient status of the coastal habitats (see Chapter 9).

Oceanicity Mull is situated on the western seaboard not only of the British Isles but also of the European land mass. The factors which contribute towards an overall effect which has been termed the Index of Oceanicity (Poore & McVean, 1957) are complex and interacting and are as follows:

rainfall – a high rainfall considerably greater than evaporation with a large number of wet days (see Ratcliffe, 1968) and therefore sustained high humidity;
soil leaching – a secondary effect of the high rainfall is the leaching of soil;
insolation – this is reduced in seaboard areas because of high cloud cover;
temperature – the difference between summer and winter temperatures is less and gradation from one to the other less abrupt than in a continental area;
snow and frost – because of high late winter and spring rainfall and periodic warm winds any snow cover is quick to melt (although often to be followed by a hard frost).

Soil moisture The water content of the soil and the ability of the soil-type to maintain that content is a deciding factor in vegetation. It is directly related to soil type and indirectly to the amount of rainfall. Iron pans form an impervious layer which can hold water in the surface layers of soil. This gives rise to ideal conditions for peat formation which in itself forms a retaining sponge. Well-drained soils and those with high quantities of sand and pebble matrix can maintain more xerophytic communities.

Base status The chemical content of soils and ground waters is a fundamental factor in differentiating plant communities. In particular base status, i.e. the calcium, magnesium, sodium and potassium ion content, is important and in the basaltic soils of Mull this can be high (see Chapter 5). It is likely that trace elements or some of the toxic metals, e.g. zinc, copper, may be of importance also. The chemical content of lake waters is discussed in Chapter 9.

Snow cover In Mull this is generally a minor factor as even on the higher hills snow cover is short-lived. On the highest north-facing slopes of Ben More, Ben Buie and the Mannir nam Fiadh range snow may persist until May and affect the vegetation. The absence of snow-cover in the winter months, because of exposure to wind preventing the accumulation of snow can also be a deciding factor on the high mountain vegetation.

Maritime communities

Cliff and rock vegetation

Lichen associations of the littoral and supralittoral In the maritime ecosystem, lichens may be seen as a continuous series of overlapping bands of individual species; their position is not related to the sea, in the sense of submersion, however, but according to the wetness and shelter of the environment. In reality therefore, the zonation is sometimes obscured by topographical irregularities which cause specific environmental conditions of wetness and dryness to be duplicated in other zones of the shore. Exposed sunny situations show the greatest variation in this respect. Species associated with wet places penetrate upshore in crevices; those of dry places, downshore on rapidly draining ridges. The zonation pattern showing indicator species as seen on an average exposed shore on Mull is shown in Table 10.1, adapted from that proposed by Fletcher (1973a & b). The total cover of lichen thalli is much smaller on exposed shores, due to the physical action of the waves abrading developing thalli, arresting growth or denying any establishment at all. Whilst there is a diminution in thallus number there is no corresponding diminution of species numbers. All species, especially *Verrucaria striatula* and *V. maura*, take advantage of crevices in extending their upshore range. Similarly penetration downshore into the barnacle zone occurs on the ridges where barnacles do not become established. Littoral lichens appear to compete with barnacles for substrate but once established can resist colonisation by competing animals and algae, although *Arthopyrenia halodytes* often establishes itself on barnacles. Large algae create cover for lichens, increased shelter enabling them to survive in habitats which would otherwise be unsuitable.

The lowest of the littoral lichen zones is marked by *Verrucaria maura*, as a black band which contrasts with the red of the bare basalt around most of the Mull shores. On north-facing shores, where there is less chance of extreme drying during low tide, this species will advance higher up the shore to above the barnacle zone. Otherwise *V. maura* is submersed with each high spring tide if not at the neaps. The fructicose species, *Lichina confinis*, may become established within the *Verrucaria* zone and the two species may compete for space but the *Lichina* will rise above the *Verrucaria* and up to 5 m up a cliff-face on an exposed shore. However, neither species can withstand extreme exposure. A second species, *L. pygmaea*, has a narrower range usually associated with the barnacle or just above it. It prefers more sunny situations than *L. confinis* and is regularly submerged by the tide. It is absent from sheltered and north-facing sites and is almost entirely absent from the east side of the Island.

In the absence of limestone at sea level one calciphile lichen, *Arthopyrenia halodytes*, commonly establishes itself by immersing its thallus in the shell of barnacles, limpets, and several other molluscs (23 species were noted as substrates on Mull). *A. halodytes* can also occur on acid rock and is quite common but often over-looked on this substrate. This species becomes progressively rarer on sheltered shores.

The maritime cliffs, i.e. those directly affected by the sea, have been termed supralittoral by Lewis (1964). Fletcher (1973b) redefined the lower boundary of the supralittoral as being indicated by *Caloplaca marina* and *Lecanora helicopis* and divided the zone into three sub-zones (see Table 10.1) as follows:

(i) *mesic* – in which all the species have a high requirement for sea-water; on Mull this includes *Caloplaca marina*, *C. microthallina*, *C. thallincola*, *Catillaria chalybeia*, *Lecania erysibe*, *Lecanora actophila* and *L. helicopis*, constituting the 'orange belt' of Knowles (1913);
(ii) *submesic* – the highest position in which foliose species can survive and only rarely submerged by sea water, characterised by *Xanthoria parietina*;
(iii) *xeric* – in which the moisture derived from the sea or the soil is minimal; a zone which rapidly dries out and an area of high osmolarity. This zone includes *Anaptychia fusca*, *Lecidella subincongrua*, *Ramalina cuspidata*, *R. siliquosa*, *Rhizocarpon constrictum* and *Rinodina luridescens*.

The region above these is called by Fletcher (*l.c.*) the terrestrial region, inhabited by those species that are either tolerant towards sea-water (halophilic; e.g. *Lecanora polytropa*, *Lecidea sulphurea* and *Parmelia saxatilis*) or those that are intolerant (halophobic; e.g. *Lecidea macrocarpa* and *Parmelia omphalodes*). The former species being able to tolerate sea-water are found on the seaward-facing sides of rocks and promontories; the halophobes will come close to the shore but only in sheltered locations.

Microclimatic factors play a part in the regional distribution of certain species

Table 10.1 A summary of the zonation pattern of maritime lichens showing indicator species (after Fletcher, 1973a, b).

Terrestrial	Halophobic	*Parmelia omphalodes*
	Halophilic	*Parmelia saxatilis*
Supralittoral	Xeric	*Anaptychia fusca* *Parmelia pulla* *Rhizocarpon constrictum* *Ramalina siliquosa* agg.
	Submesic	*Xanthoria parietina*
	Mesic	*Lecanora helicopis* *Caloplaca marina* *Lichina confinis*
Littoral	'Fringe'	*Arthopyrenia halodytes* *Verrucaria maura*
	'Eu-littoral'	*Lichina pygmaea* *Verrucaria amphibia* *Verrucaria microspora* *Verrucaria mucosa* *Verrucaria striatula*

and the following are more frequent on sunny shores where their distribution may be determined by temperatures: *Buellia aethalea*, *B. subdisciformis*, *Caloplaca marina*, *Fuscidea cyathoides*, *Lecanora actophila*, *L. fugiens*, *Parmelia conspersa*, *P. delisei*, *P. loxodes*, *P. pulla*, *P. verruculifera*, *Ramalina cuspidata*, *Rinodina atrocinerea*, *R. luridescens*.

Crevices in the cliff-face often have distinct floras, e.g. *Bacidia scopulicola*, and *Solenopsora vulturiensis*, in damper situations and *Caloplaca arnoldii*, *C. littorea*, *Opegrapha* spp. and *Sclerophyton circumscriptum* in drier crevices.

The cliffs of Mull, suitable as they may seem, do not possess large colonies of breeding sea-birds and the effect of high nitrogen associated with bird guano is small. Nevertheless, gulls, gannets and cormorants use coastal rock ledges as perching areas and droppings accrete sufficiently to encourage the development of nitrogen-loving lichen communities. Provided salt spray is within reach *Lecanora leprosescens* is common; with it are *Buellia canescens*, *Caloplaca verruculifera*, *Lecanora caesiocinerea*, *L. dispersa*, *Rinodina subexigua* and *Xanthoria parietina*. The Treshnish Isles do have a larger bird population (kittywakes, fulmars, auks, storm petrels and cormorants) and these lichens are found there abundantly in keeping with a specialised angiosperm flora.

Fig. 10.1 Granite cliffs on the south shore of Erraid, Ross peninsula. 1967.

The vegetation of cliffs arising from exposed shores This habitat is sparsely vegetated and those plants that are present established themselves initially in the shelter afforded by cracks and crevices. *Plantago coronopus* and *P. maritima*, both of which can become exceedingly fleshy in exposed habitats, and *Asplenium marinum*, are species which can withstand salt-drenching. The latter species has not been found on shores where strong wave-lashing is absent. *Cochlearia officinalis* and *C. danica*, *Tripleurospermum maritimum*, *Silene maritima*, *Ligusticum scoticum* and *Festuca rubra* (the ecotype often referred to as var. *littoralis*) can also be within wave-lash. Cliffs of over 75 m may have a typical maritime angiosperm flora as follows, with the first three species forming a dominant sward: *Plantago coronopus*, *P. maritima*, *P. lanceolata*, *Armeria maritima*, *Silene maritimà*, *Sedum anglicum* and *Festuca rubra* and with *Koeleria cristata* out of the reach of constant salt-spray. On the south-west granite cliffs (see Fig. 10.1) *Umbilicus rupestris* may be found often with annual species of *Cerastium*. Similar associations are found on the basalt pavement cliffs on Staffa (Fig. 10.2).

In places along the coasts the effects of fresh-water run-off may be seen as water-

Fig. 10.2 Truncated hexagonal basalt columns showing cliff mosaic vegetation on east shore of Staffa. 1969.

drips over a cliff. A notable example is seen on the steep cliffs of the Ardmeanach near Mackinnon's Cave where complex algal communities of *Rivularia* spp. and green filamentous species festoon the cliff. Marine rock pools are at the cliff base and in the littoral zone but some may be above the highest spring tide level and maintain their fresh-water from drainage and rain. Such pools may have *Carex distans*, *C. demissa* and *C. otrubae*. The last species can withstand immersion in salt water and is a frequent cliff-base crevice plant in the west of Mull. It must be remembered however that some shallow pools on the uppermost terraces out of reach of normal neap tides may lose water by evaporation and have a very high concentration of salts. These are discussed in Chapter 9.

Terrestrial cliffs influenced by the maritime environment Much of the coast of Mull has, 6–7·5 m above sea-level, signs of a raised beach which may now only extend a few metres or may be as wide as 25 m. Where high cliffs arise on its landward side the effect of windblown spray may give rise to a lush community comparable to the base-rich, flushed cliffs inland. Ferns like *Athyrium filix-femina*, *Dryopteris filix-mas*,

Fig. 10.3 Basalt cliffs at Port a' Bhàta, Ulva, with Osmunda regalis *and dwarf shrub species. 1967.*

D. pseudomas, *Osmunda regalis* (rarely seen inland), *Polypodium interjectum* and *P. vulgare*, and tall herbs such as *Angelica sylvestris*, *Luzula sylvatica* and *Mercurialis perennis*, and the woody species *Hedera helix*, *Lonicera pericylmenum*, *Populus tremulans*, *Prunus spinosa* and *Solanum dulcamara* form a characteristic association (see Fig. 10.3). It is interesting to note that *Dryopteris oreades*, common on inland cliffs, appears to be intolerant of the maritime conditions. A number of species, e.g. *Vicia orobus* are found on Mull only on such maritime cliffs.

On the open ledges of such cliffs, the grasses *Koeleria gracilis*, *Festuca rubra* and *Dactylis glomerata* form a broken turf with *Brachythecium rutabulum*, *Tortella flavorirens*, *Frullania microphylla*, *F. fragilifolia*, *Cladonia rangiformis*, *C. subcervicornis*, *Peltigera canina* and *P. polydactyla*. Where the bases of such cliffs stand out of reach of tidal scouring, talus ranging from fine rock powder to large scree blocks will accumulate, stabilise and become colonised by vegetation. For the most part these cliffs are of basalt but in the Gribun area and around Carsaig Bay Mesozoic sediments may be included in the talus slopes. On the former cliffs, glacial and basaltic soils have slipped from the original cliff-top and now obscur the basic Triassic rocks underlying the basalt. It is a slope with a well-established hazel-blackthorn scrub (see p. 10.28) which takes the full force of the SW gales. Further south on the Ardmeanach peninsula at about 50 m above sea level, the better-drained slopes being enriched with calcium and magnesium from the salt-laden winds, contain *Brachypodium sylvaticum*, *Carex caryophyllea*, *Cynosurus cristatus*, *Euphrasia* spp. and *Koeleria cristata*. Where calcium-bearing rocks form part of the talus, a species-rich sward may result with *Anthoxanthum odoratum*, *Agrostis stolonifera* (especially where drainage water seeps out) and *Festuca ovina* with *Campanula rotundifolia*, *Prunella vulgaris*, *Thymus drucei* and *Trifolium repens*. In the basic flushes on such cliff slopes west of Carsaig *Cirsium palustre*, *Epilobium* spp., *Equisetum palustre* and *E. telmateia* are characteristic, the latter in its only locality on Mull.

The early colonisation of these talus slopes is often by weed species, e.g., *Cirsium arvense*, *Equisetum arvense*, *Polygonum* spp., *Senecio jacobaea*, *Silene cucubalus*, *S. maritima* and *Tripleurospermum maritimum* followed by grasses like *Agrostis* spp., *Arrhenatherum elatius*, *Bromus mollis*, *Cynosurus cristatus*, *Dactylis glomerata* and *Festuca* spp.

Boulder strewn pebble beaches

Much of the west coast of Mull is formed from the deep sea-lochs of Loch Scridain, Loch na Keal and Loch Tuath. They open to the west or south-west and receive a certain amount of storm-wash, especially on their northern shores. These coasts are low-lying and the shores, apart from where dolerite sills or similar volcanic intrusions outcrop across the shoreline, are boulder strewn and gently sloped and the marine algal communities may abutt directly onto the species-rich grassland of the raised beach (see Fig. 10.4).

Two most important factors on this type of shore are the mobility of the shingle and the resulting particle-size distribution. A number of species can withstand the salt water e.g. *Rumex crispus* var. *trigranulatus* and *Mertensia maritima*, but periods of calm, when the young seedling can become established, must be few although once a tap root is formed there is every chance of the plant remaining in situ. In both these species copious seeds are formed – Salisbury (1942) estimated 29,000 seeds per plant for the *Rumex* and Scott (1963) estimated 756 for *Mertensia* – which float and can be distributed by water currents. The colony of *Mertensia* at Ensay Burn is unlikely to colonise further however as it is sandwiched between rocky promontories and the narrow bay takes the full force of the Atlantic swell. It is therefore unlikely that seeds can float out even on a low spring tide.

Fig. 10.4 North side of Calgary Bay looking towards Rubha nan Oirean. Laminaria digitata *exposed at low tide with fucoid zones at base of cliff detritus. 1966.*

Particle-size becomes smaller higher up the beach and the *Mertensia* community is associated with a considerable amount of fine silt or quartz sand. It grows on a north-east facing shore on Sgeir a' Chaisteille in the Treshnish group, (see Fig. 10.5) in the 'littoral fringe', with *Fucus spiralis* and *F. vesiculosus*, where it is obviously submerged with each tide. The upper zones of this community lacked the fucoids but had instead *Rumex crispus*, *Atriplex glabriuscula*, *Tripleurospermum* and *Ligusticum*

Fig. 10.5 Mertensia maritima *on basalt boulder beach on Sgeir a' Chaisteille, Treshnish Isles. 1968.*

scoticum. On stabilised pebbles in such situations *Buellia stellulata*, *B. verruculifera*, *Lecanora polytropa* and *Lecidella scabra* are frequent.

On the sea-loch shores with a minimum of shingle movement in the upper tide-levels *Honkenya peploides* is one of the few species to be established below the drift line. The drift itself, mainly seaweed detritus, is an important source of nutrients and around this line and just above it a characteristic association can be seen along all the sea-lochs and in places along the south and north coasts where pebble-beaches are partly protected from direct scouring. This is dominated by *Potentilla anserina* with *Rumex crispus*, *Polygonum hydropiper*, *Stellaria media* and *Tripleurospermum maritimum*. Upshore from this there is frequently a band of *Iris pseudacorus* with other eutrophic mire species e.g. *Lythrum salicaria*, *Lycopus europaeus*, *Scutellaria galericulata*, *Filipendula ulmaria* and *Juncus effusus*. In places where silt is washed down from above, weed-species like *Polygonum aviculare*, *Trifolium repens* and *Stellaria media* will become established. At this point there is an ecotone with the natural community of the loch side, often bog or mire on thin peaty soil or acid grassland on a more siliceous soil derived from raised beach deposits. The effect of salt water on the pH of the soil will be to decrease the acidity and to increase the sodium and potassium cations but not the calcium. A detailed study of this aspect of coastal soils on Mull was made by Gillham (1957) who suggests the *Iris* could not only tolerate the salt water around its roots but also benefit from the mitigated acidity or increased base status of the soil. It may also be that fresh-water drains from the upland at this level, albeit floating on a salt water table. When drainage water is obviously seeping through, the moorland vegetation supports *Oenanthe crocata*, and the pebbles of the shore will be colonised by *Enteromorpha* spp.

On the medium-exposed shore, salt-spray will increase the base status of the marginal peat and the frequency of the salt-intolerant species will be drastically reduced; e.g., *Calluna vulgaris*, *Erica* spp., *Scirpus cespitosus*, *Juncus acutiflorus* and *Sphagnum* spp. In places, marine erosion combined with out-flow of drainage water from the upland breaks up these coastal mire communities, and boulder- or pebble-strewn channels result. It is possible at that stage for the mire to degenerate and the halophobes mentioned above to die out. *Myrica gale*, *Molinia* and *Eriophorum augustifolium* can withstand salt water and the resulting increase in pH, at least round their roots; on sheltered shores such as the Sound of Mull these species may form a mixed association with *Armeria maritima*, *Juncus articulatus*, *J. gerardii*, *Plantago* spp., *Potentilla erecta* and *Succisa pratensis*.

On pebble beaches exposed to the south west or west few if any plants can establish, but on those less exposed or where the main shingle movement is during the winter (e.g., on the north coast) some annual species may establish themselves on or just above the drift line; *Galium aparine*, *Polygonum aviculare* and *Stellaria media* are examples.

The Treshnish and other small islands

Some fifty rocks or islets exposed for several days at a time during spring tides may be considered separately; they contain only isolated patches of terrestrial vegetation. All of the islets in this category are of basalt and are flat-topped. If silt or wind-blown debris accumulates it is extremely thin and fills small crevices only. In spite of this *Armeria maritima*, *Glaux maritima*, *Juncus gerardii* and *Plantago maritima* frequently establish themselves and *Asplenium maritimum* is seen in rock crevices. Littoral and mesic zone lichens are present, but often compressed, and rarely do the rocks stand high enough to escape wave-splashing and permit the establishment of submesic and xeric species.

Only nine of the larger islands in the Treshnish group are grazed, leaving a number of smaller ones, rising to 8–12 m, to develop a rocky shore and cliff vegetation without the pressure of sheep or goat grazing. *Ligusticum scoticum*, *Angelica sylvestris*, *Anthriscus sylvestris*, averaged well over a metre in height and formed robust plants and dense stands from the upper tide-marks upwards. Associated was *Rumex crispus* var. *trigranulatus*, *Potentilla anserina*, *Tripleurospermum maritimum* with an under-layer of *Honkenya* and *Glaux maritima*.

The pebble-sand substrate with *Mertensia*–fucoid association seen on Sgeir a' Chaisteille is described above. The larger islands have vertical cliffs, those on the west side extremely exposed. As remarked above, the bird populations are considerable especially on Lunga, Bac Mòr and Fladda. On the smaller islands too the cliff-top grasslands and *Armeria–Plantago* turf are encircled by perimeter heaps of guano. The flora on and around these is characteristically of *Holcus lanatus*, *Poa annua*, *P. subcaerulea*, *Rumex acetosa* and *Stellaria media* (cf. Gillham, 1956). Older hummocks no longer used are colonised by *Armeria* and *Festuca ovina*. The more extensive grassland and flush communities are described elsewhere but the lower islets, where remnants of a deeper soil remain, are covered with a thick turf of *Agrostis tenuis*,

Anthoxanthum, *Carex flacca*, *Festuca ovina* and *F. rubra; Koeleria cristata* is a frequent associate. Such grassland relies on salt-spray for much of its nutrient.

Silty shores and saltmarshes

Expanses of alluvial silt built up in estuarine situations are few and relatively small around the coast of Mull. There are large rivers which drain extensive areas of glacial drift or upland skeletal soils. Of these the Colodoir River flows into the largest mire and saltmarsh system on the Island at the head of Loch Scridain, at An Leth-onn. In the south-east and flowing through similar soils on the south of Beinn Bheag is Abhainn Lirein ('the stream of the green weed'). This flows into a tidal basin known as Leth-fhonn, in the western arm of Loch Don. On the east coast a number of rivers run off the Torosay hills into the Sound of Mull and in places into sheltered bays, e.g., Duart and Fishnish, but in the former at least, the submarine coast of the Sound is steep and winter storms are strong enough periodically to scour the beaches and prevent silt accretion. At Salen Bay, between Rubha Àrd Ealasaid and Rubha Mòr, silt brought from Glen Aros is deposited in sufficient quantity to form a small saltmarsh. Again over the watershed in Glen Aros the Belart River drains an extensive area of moorland and alluvium to flow through the rocky narrows of Loch na Cuilce into Loch a' Chumhainn forming there a small but interesting saltmarsh. These four sites are described in more detail below. A fifth site not studied in detail was that on the eastern end of Ulva by Ulva House. Here incipient saltmarsh is forming along similar lines to that of Leth-fhonn; it is one of the few localities for *Salicornia* on the Island. There is an incipient saltmarsh in Fishnish Bay which is similar to but less extensive than that in Salen Bay and not described here.

Another area, where from the geomorphology and topography one might expect some saltmarsh formation, is at the northern tip of Loch Spelve where the Lussa River flows in, but the scouring is too efficient and only enclaves of incipient saltmarsh are formed. The soil at the loch margins here is formed from a narrow outcrop of Mesozoic sediments and is base-rich. It supports a *Festuca ovina–Armeria–Plantago* turf which soon becomes a typical base- and herb-rich community above the influence of the salt water. It is good pasture and regularly grazed.

Another geomorphological feature situated on a sea-loch head is at the eastern end of Loch na Keal where the River Bà has deposited extensive but nutrient-poor silt in a typical deltaic formation. The loch-shore line is composed of large to medium boulders and is typical of that described above; the silt is now well above and beyond the influence of salt water (except for salt-laden winds) and has been absorbed into the adjacent arable farm system.

Loch Scridain–An Leth-onn The River Coladoir, on entering the sea-loch by way of the small bay called Loch Beg flows over An Leth-onn, a low-lying area influenced by the daily tide-rhythm. Silt accretion is helped by the rocky bar running from the Aird of Kinloch across to Ardvergnish and approximately a square kilometre of saltmarsh is formed here. Up to the beginning of this century a track and man-made ford existed over the rock bar, thus accentuating the ponding action upstream. On the north side of the delta a brackish marsh forms at the ecotone between the saltmarsh and the fresh-water mire or meadow forming on the alluvium. It is likely that the area is stable and that further accretion is offset by wave-action erosion at high tide or flow from the River Coladoir.

Figure 10.6 summarises lines of possible succession on the saltmarsh. The fucoid sea weeds attached to the larger pebbles accrue plant debris and silt in which *Suaeda maritima* may germinate. Also on the larger pebbles and boulders are the lichens *Caloplaca marina*, *C. thallincola*, *Lecanora atra*, *L. helicopis*, *Lichina confinis* and *Verrucaria maura*, the two latter species often in abundance. Sooner or later *Puccinellia*

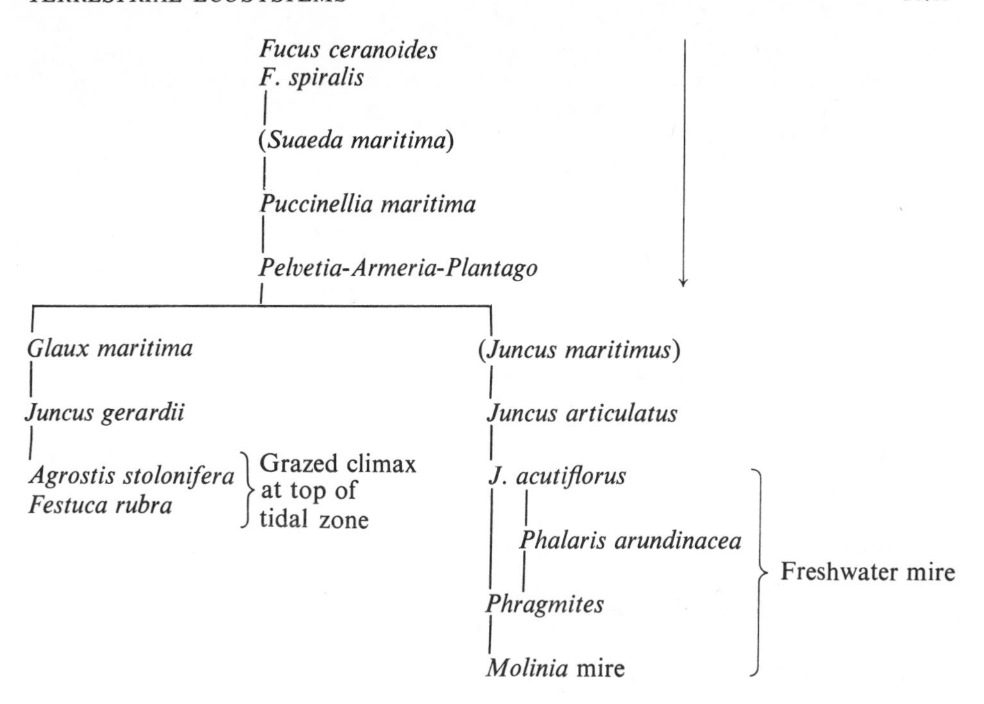

Fig. 10.6 Postulated plant succession at An Leth-onn.

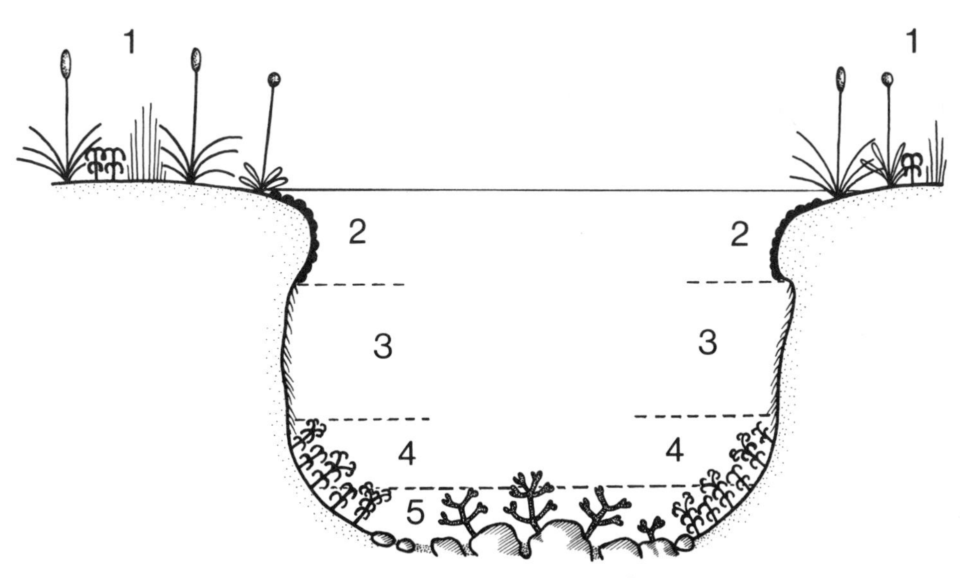

Fig. 10.7 Profile of typical saltmarsh channel at An Leth-onn.

Notes

1 *Plantago-Armeria* marsh with erect *Puccinellia* and free-living *Pelvetia*.

2 *Rivularia nitida* on lip of pool.

3 *Vaucheria* spp.

4 *Pelvetia* attached to substrate.

5 *Fucus spiralis* on rocks at bottom of pool.

maritima establishes itself and in the ensuing tillers further coarse silt accrues. The stages that follow are typical for most British salt-marshes (Chapman, 1934,; Yapp & Johns, 1917) and in the developing turf of *Armeria maritima-Plantago maritima* and *P. coronopus* the free-living *Pelvetia canaliculata* ecad *libra* is a co-dominant. It is interesting to note that here and elsewhere on Mull the other common saltmarsh algae, *Bostrychia scorpiodes* Kütz. or *Catenella repens* Batt., are absent. As in other salt-marshes numerous channels and occasional pans are formed which contain salt-water for varying periods. The deeper pans and channels contain *Fucus ceranoides* and *F. spiralis*; the shallower ones are often filled with filamentous green algae e.g. *Rhizoclonium* species. Figure 10.7 shows a profile of a typical channel. In many of the wider channels a *Pelvetia* turf may form at the base of the vertical edge which is often covered in *Vaucheria* spp. and *Rhizoclonium* spp. At the saltmarsh rim of the channel the blue green alga *Rivularia australis* commonly covers the fine silt. It is at this position that *Carex scandinavica* should be searched for. *Carex demissa* is found, and *C. serotina* could be expected to be more common, with *C. extensa* and *Blysmus rufus* on the upper more pebbly parts of the saltmarsh near points of freshwater seepage. The volume of freshwater flowing down the Coladoir combined with the fact that the main westerly flow of salt water is stopped by Lùb na Cloiche Duibhe probably means this marsh is not as saline as could be expected and this may account for the absence of *Aster tripolium* and *Salicornia.*

Glaux maritima rapidly fills-in extinct rivulets in the saltmarsh mosaic and at this stage *Juncus gerardii* invades. The silt deposited here is coarse with a number of pebbles and the resulting substrate is reasonably firm. When such upper reaches are grazed by cattle or sheep this solid-based pasture can withstand the trampling and few pools are made by hooves breaking the turf. *Agrostis stolonifera* is the dominant species with *Festuca ovina* agg. only rarely seen.

On the north side of the river-mouth the land is broken up by ancient river channels and the resulting islands are not so easily grazed; *Juncus maritimus* is seen especially along the wave-washed shore. The *J. gerardii* association is lush and contains *Glaux maritima* and *Juncus articulatus.* When flushes enter the basin *J. acutiflorus*, and then *Phalaris arundinacea*, become established and *Phragmites australis* soon becomes dominant. Mosses like *Archidium alternifolium*, *Campylium stellatum* and *Physcomitrium pyriforme* can be found here. When mire conditions are minimal but just apparent these species will disappear but *Phragmites* will remain, becoming less in stature as the nutrient supply is reduced. When lateral water movement in the soil is almost nil *Phragmites australis* gives way to *Deschampsia caespitosa–Holcus mollis–Molinia caerulea–Carex panicea* association with mire species like *Hydrocotyle vulgaris*, *Samolus valerandi*, *Filipendula ulmaria* and *Scutellaria galericulata.* The pH is around 6·0 and a certain amount of peat accumulates in what rapidly becomes freshwater 'mixed fen' or mire.

Loch Don head–Leth-fhonn Loch Don is a small sinuous sea-loch which receives regular tidal flushing and has for the most part a typical pebbly shore throughout. There is one sizeable river, the Abhainn Lirein, and a number of smaller ones including the outflow for Loch a' Ghleannain, that flows into a low tidal basin Leth-fhonn. The neck of this basin at the Loch Don side is constricted by a dolerite sill over which a bridge of 16th Century origin has been constructed narrowing the outlet even further. The result is a silty saltmarsh area drastically affected at its margins by the fresh water infiltrating. Figure 10.8 summarises the major associations in this area. There are small areas only of *Plantago-Armeria* marsh, the land being sufficiently high to be immersed only on the peak of the incoming tide, a zonation preferred by *Juncus gerardii.* The silt is over a pebbly shore and *Carex demissa* and *C. extensa* form open stands in these areas. Where silt is deeper especially in side bays which have no

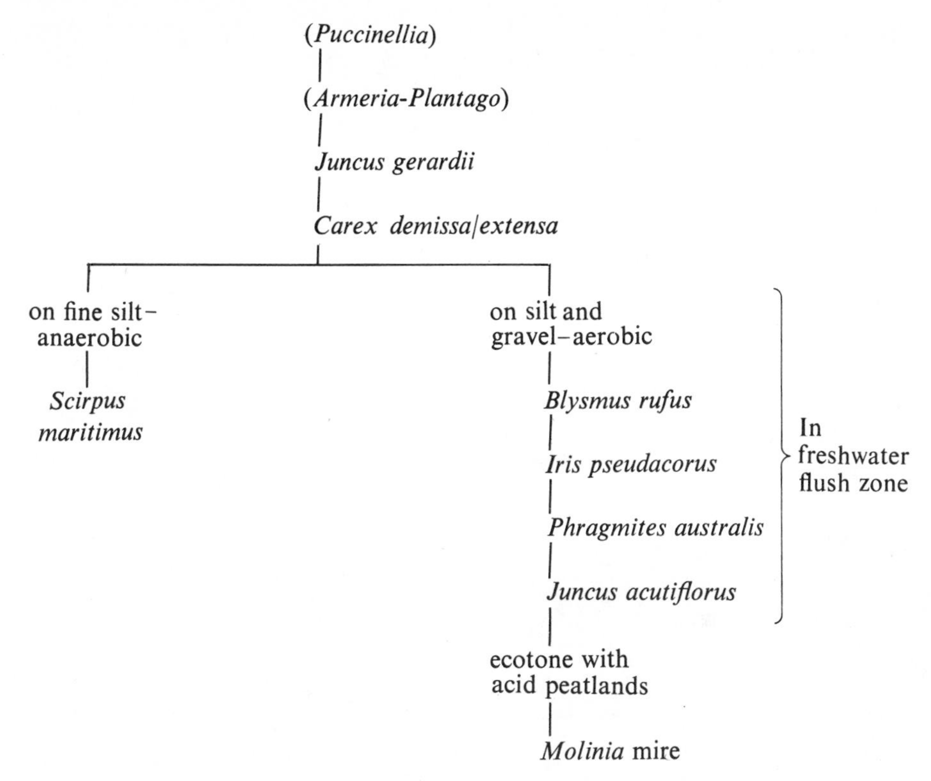

Fig. 10.8 Halophyte succession at Leth-fhonn. Species in brackets are infrequent.

through-flow of salt or fresh water *Scirpus maritimus* becomes the dominant and often sole plant. As one gets progressively closer to a freshwater inflow *Blysmus rufus*, *Iris pseudacorus* and *Phragmites australis* are found in that order. At the edge of the *Phragmites* and *Scirpus* and where freshwater seepage occurs *Juncus acutiflorus* may be abundant.

On the east of the bridge and at the head of the loch itself there is an area of silt colonised by a turf of *Puccinellia maritima – Festuca rubra* which regularly gets covered at the highest tides. Slightly higher land still has *Armeria maritima* and *Plantago* species which with the *Festuca* form a plagioclimax kept in this stage by sheep grazing and regular tidal washing; doubtfully further silt accretion will take place here. In the deeper water by the freshwater outflow *Fucus ceranoides* is locally abundant and *F. vesiculosus* occasional together with *Ascophyllum nodosum* forma *scorpioides*.

Salen Bay The saltmarsh at Salen Bay (see Fig. 10.9) is small but significantly more saline than either described so far. Two species, *Suaeda maritima* and *Aster tripolium*, occur here in quantities not found elsewhere on the island. The outwash from Glen Aros and Loch Frisa is met by southerly currents that flow down the Sound of Mull and which directs the finer silt into Salen Bay. Only one small stream, Altt na Searmoin, flows into this marsh and the volume of outflow is such that it does not move the silt further; it is likely that this saltmarsh may develop and increase in area. The area is bounded by the road and the *Juncus gerardii* marsh develops into a narrow *Agrostis stolonifera* zone which rapidly meets rubble and waste detritus tipped at the roadside edge. *Enteromorpha intestinalis* and *Rhizoclonium implexum* are seen amongst the grass stems. Near the Aros River outflow are large banks of *Fucus ceranoides* and on stones and pebbles in the silt *Rhodocorton purpureum* may occur. On its southern

Fig. 10.9 Suaeda maritima–Aster tripolium *saltmarsh on Salen Bay looking south down the Sound of Mull, with Rubha Mòr and Salen pier in the distance. 1970.*

rim the saltmarsh meets raised beach deposits on which an acid grassland or neutral more association form and meet the saline ecotone abruptly. Figure 10.10 shows the succession observed here.

Dervaig – Loch a' Chumhainn There is a restricted area of maritime meadows just north and west of the road-bridge over the River Belart. Much of the silt brought down by that river settles in Loch Cuilce but a certain amount passes through the bridge

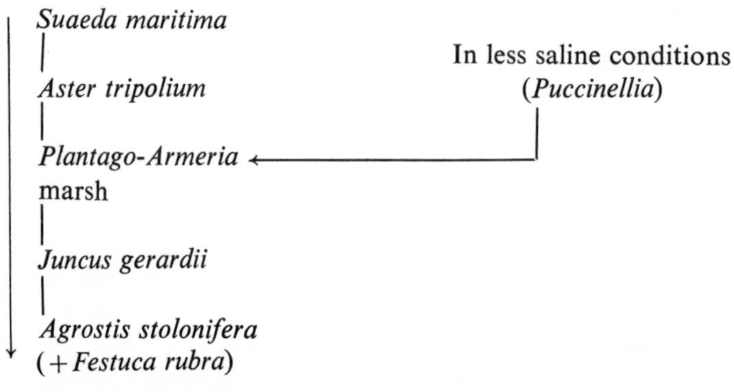

Fig. 10.10 Saltmarsh succession at Salen Bay. Species in brackets are infrequent.

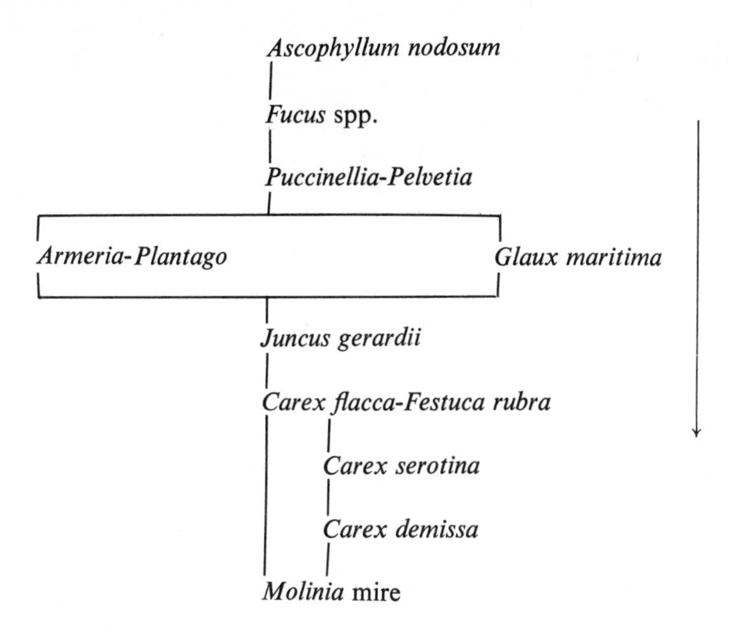

Fig. 10.11 Plant succession around the head of Loch a' Chumhainn.

and is washed back into small appendix-like bays at the southern end of the sea-loch. That nearest to the road by the bridge supports a small area of *Puccinellia maritima* marsh with *Fucus vesiculosus* and *F. serratus*. Silt accretes around the *Puccinellia*, and islets of *Plantago-Armeria-Juncus gerardii* association are developed. At the edge of the loch is a *Myrica-Molinia* mire which grades into a *Carex* association of *C. demissa*, *C. flacca* and *C. serotina*. The succession is tabulated in Figure 10.11.

By the inflow *Enteromorpha intestinalis* and *E. prolifera* are mixed with *Puccinellia* and the herb sward. A mosaic of pools of varying depths occurs here and an algal flora is well developed between the tussocks of grass. Figure 10.12 shows the flora associated with these high-level tidal pools in the upper *Puccinellia-Armeria-Plantago* marsh.

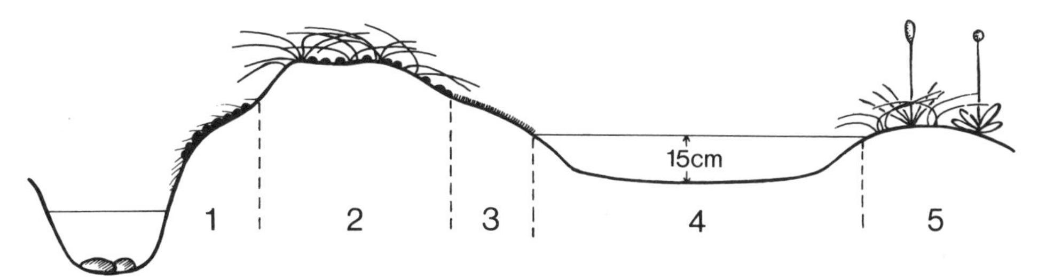

Fig. 10.12 Profile of high-level tidal pool in the upper Puccinellia–Armeria–Plantago *marsh near Dervaig.*

Notes

1 *Vaucheria* spp. and *Rivularia nitida.*

2 *Puccinellia maritima* with *Rivularia.*

3 *Stictyosiphon tortillis.*

4 Pool with *Microcoleus* sp., *Percursaria* sp., *Rhizoclonium viparium* and *Schizothrix* sp.

5 *Puccinellia-Armeria-Plantago community* with *Enteromorpha* spp.

Sandy shores and machair

Mornish and the Ross of Mull The distribution of sandy bays is confined to the northern coast of Mornish and the granite and schistose areas of the Ross of Mull. In some places the strand-line vegetation of annuals like *Galium aparine* and *Atriplex hastata* is present and behind it *Potentilla anserina*, *Polygonum aviculare* and *Stellaria media*.

Where wind-blown sand accumulates *Agropyron junceiforme*, followed in a few places by *Ammophila arenaria*, may be established. In bays in the north is a mature turf of *Carex arenaria*, *Agrostis canina*, *A. stolonifera*, *Festuca rubra/ovina* with *Koeleria cristata*, *A. tenuis* and *Anthoxanthum odoratum*. Base-rich grassland herb species like *Prunella vulgaris*, *Trifolium repens*, *Bellis perennis*, *Galium verum*, *Leontodon taraxacoides*, *L. autumnalis*, *Ranunculus acris*, *Galium verum*, *Thalictrum minus*, *Plantago coronopus* and *P. lanceolata* are frequent. In the more open areas *Lotus corniculatus* and *Thymus drucei* form large clumps, the latter becoming co-dominant in places with *Koeleria cristata* and *Festuca* spp. Lichens such as *Cladonia rangiformis*, *C. chlorophaea*, *C. pocillum*, together with *Toninia coeruleonigricans* and *Lecidea decipiens* help to bind the soil as do *Tortula ruraliformis*, *Ceratodon purpureus*, *Bryum pendulum* and *Barbula fallax*. *Pteridium aquilinum* is an early coloniser in these

Fig. 10.13 Port nan Ròn. Granite moorland with sandy foreshore. Ophioglossum vulgatum *was found under the colonising bracken stand. 1968.*

sandy pastures and if the grazing pressure is low this species soon becomes dominant. *Ophioglossum vulgatum* is found in the shade of bracken in such colonised sand in its only Mull localities (see Fig. 10.13).

The sandy areas described above are basically quartzite crystals with varying proportions of shell or foraminiferous sands. Ritchie (1974) reporting on the shell content of various Hebridean machair systems gives a figure of 81+% $CaCO_3$ for Ardalanish Bay and Port Langamull. The largest area of open pasture of this kind is seen at Calgary. Here cars are allowed to drive onto the sand often starting wind erosion channels. The area is not true machair, although a few species like *Rhodobryum roseum* and *Thuidium philibertii* can be found there. At Calgary there is an *Agrostis tenuis – Festuca ovina* turf with *Poa subcaerulea*, *Holcus lanatus*, *Trifolium repens*, *T. arvense*, *Veronica chamaedrys*, *Rhytidiadelphus squarrosus* and *Rumex acetosella*, running right to high tide level where a wave shelf is formed; below this is a *Honkenya peploides* zone with *Silene maritima*.

Sandy areas of Iona Two areas on Iona stand out for their sand-dune and machair associations. The largest one, which is about one kilometre long by half a kilometre wide, is on the shore of Camas Cuil an t'Saimh. Here considerable quantities of the substrate are foraminiferous deposits or composed or molluscan shell deposits. The resulting sandy soil is highly basic and with a pH of 7·5–8·2. Early colonisation of the mobile sand is mainly by *Agropyron junceiforme*, *Festuca ovina* and *F. rubra* which together with *Agrostis tenuis* and *Carex caryophyllea* very quickly form a closed turf. The mosses *Rhodobryum roseum*, *Thuidium philibertii*, *Pottia heimii*, *Amblystegium serpens* var. *salinum* and *Entodon concinnus* may often form dense turf with some of the *Cladonia* species mentioned above. On the seaward edge *Tortula ruraliformis*, *Ceratodon purpureus* and *Barbula fallax* will cover the more open patches, with *Peltigera rufescens* occasionally forming large mats.

Like machair elsewhere in the Hebrides this pasture is common-land and has been grazed since the mid-18th Century at least. The area to the north of the bay is now utilised as a golf course.

The second extensive sand area on Iona is that on the northern tip by Caolas Annraidh. Here the sand is mobile and forms appreciable dunes of *Agropyron* and *Ammophila*. *Carex arenaria* may form a heathy zone with *Thymus drucei*, and when the moorland itself is being covered by blown sand *Erica cinerea* and *Calluna vulgaris* will continue to grow through. There are patches of *Festuca*-mixed-herb grassland which are regularly grazed but the area is constantly having wind-blown sand deposited. The nearby island Annraidh is a dense stand of *Ammophila-Agropyron* with luxuriant *Angelica sylvestris*, *Lotus corniculatus*, *Atriplex hastata*, *Tripleurospermum maritimum*. Although close to Iona it is separated by a deep channel and grazing animals including rabbits or hares are absent. Part of Soa Island, 3 km to the SSW of Iona, has a little blown sand colonised by a base-rich type of *Festuca-Agrostis* grassland, but the amount is insignificant.

Forest communities

The deciduous woodland habitat on Mull is nowhere extensive and it is doubtful whether forest cover was anywhere continuous on the island during the present climatic regime. Soil-type, topography and exposure allow woodland to develop in comparatively small pockets and such woods have a mixture of shrubs and more mature trees. The ground flora is rarely distinctive and is often predominantly of the vegetation types (dwarf shrub, grassland or mire) surrounding the wood. In fact such dwarf shrubs like *Calluna* and *Vaccinium* may, because of the shelter afforded by the trees, grow to a greater stature. On hillsides that are otherwise favourable for woodland conditions, biotic pressures, mainly grazing, may prevent spread of the tree species.

A closed tree cover can have two main effects on the habitat: it will reduce light-intensity and more importantly it will increase the humidity within the wood. Strong winds can have an effect on this situation however: they can penetrate the wood and circulate the air and in the spring and summer such wind can have a drying effect. Such winds will come from the west and south and affect woodlands exposed in that direction; woods facing east are likely to be more sheltered and therefore have a more consistent high humidity in spite of the area having a lower rainfall (see Fig. 6.1). Furthermore the western part of the Ross of Mull has more sunshine and the humidity fluctuations are therefore larger because of this.

Wood and scrubland may best be discussed under eight headings:

(i) native birch and oak woods which are widespread;
(ii) ash woods which are poorly developed on the Island and restricted to the more basic soils on the south coast;
(iii) mixed deciduous woods on brown earth soils;
(iv) older woodlands, usually small and scattered in lowland situations;
(v) hazel-blackthorn scrub usually on steep escarpment slopes or in exposed situations;
(vi) *Salix* associations;
(vii) mixed deciduous plantations around the old established estates;
(viii) recent conifer plantations.

Birch-oak native woods

Birch and oak woods are discussed together as they often occupy similar habitats and are mixed in varying degrees. Three birches are present: *Betula pubescens* subsp. *pubescens* and subsp. *odorata* are the commonest taxa but their full distribution still requires to be investigated on Mull; *B. pendula* is seen as a component in birch woodland near habitation and on the larger estates and is likely to be introduced in the first instance. The oak species are similarly little known but *Quercus petraea* is the commonest species. It is possible that *Q. robur* may also have been introduced in historic times. Hybridization between the two species is certainly common.

Ecologically, little distinction between the taxa of both groups is seen on Mull. *Betula* as a genus on the other hand show signs of establishing itself more readily on the base-poor, often wet, moorland soils than do the oaks. They will often colonise uphill on a steep shoulder on very leached soils whereas oaks prefer the deeper soils with the accumulated nutrients at the base of the slope. This is well seen on the south side of Lochbuie. Another factor must be that the seeds of birches are easily carried uphill by wind, whereas the acorn can only be carried uphill by a vole or similar rodent. The absence of young seedlings generally, except on the more rocky bluffs, is due to the commonly open access to grazing animals like deer, sheep and goats. Although this pressure is low (more than 4 hectares per sheep; King & Nicholson, 1964) for Scotland as a whole, it is a major factor for the establishment or continuance of woodland on Mull. Many birch woods in exposed situations on the south coast are old and moribund with twisted trunks infected with *Fomes fomentarius* and *Piptoporus betulinus*, with very few oaks and some hazel associated with them; they show little signs of regeneration although hazel has remarkable power of regenerating from almost moribund stools.

Associations within the birch woods are not distinct but the following may be distinguished as noda, if not clearly delimited communities. The *Vaccinium*-rich birchwood of McVean (1964) is the commonest type with *V. myrtillus* dominant (occasionally *V. vitis-idaea*), *Luzula sylvatica*, *Blechnum spicant*, *Oxalis acetosella* (where soil is deep enough) and the mosses *Hylocomium splendens*, *Thuidium tamariscinum* and *Rhytidiadelphus loreus*. Shrubs are scarce, *Lonicera periclymenum*, *Corylus* and *Sorbus aucuparia* may become common especially near gorges or small ravines. This type of woodland reaches to 150 m (500 ft) in Scarisdale wood and to over 225 m

Fig. 10.14 Scarisdale Wood, Loch na Keal. Open oak wood with Molinia-*herb-rich field layer and moss-covered boulders. 1966.*

(750 ft) around Loch Bà. In the lower, more open and less steep situations at these localities oak (*Quercus petraea*) often becomes dominant, with the ground flora containing much *Molinia* amongst the herbs (Fig. 10.14).

There is considerable woodland which intergrades with the above which contains *Quercus petraea* as a common if not co-dominant species. In many places the soil is deeper, with a higher base content and with a ground flora of species like *Oreopteris limbosperma*, *Dryopteris pseudomas*, *Conopodium majus*, *Viola riviniana*, *Holcus lanatus*, *Anthoxanthum odoratus*, *Lysimachia nemorum*, *Primula vulgaris*, *Anemone nemorosa* *Endymion non-scriptus*, *Pteridium*, *Sanicula europaea*, *Hylocomium splendens*, *Mnium hornum*, *M. undulatum*, *Plagiochila asplenioides*, *Thuidium tamariscinum* and *Peltigera* sp. As before, *Corylus* and *Sorbus aucuparia* may be common trees reaching 8–12 m in height respectively. *Ilex aquifolium* and *Salix* spp. may also be present. *Salix cinerea* subsp. *oleifolia* forms a distinct facies with or without *S. aurita* around spring-heads in tributary valleys. This wetter situation is associated with *Carex echinata*, *C. remota*, *C. nigra*, occasionally *C. paniculata*, *Crepis paludosa*, *Equisetum sylvaticum*, *Filipendula ulmaria* and *Angelica sylvestris*.

Birch-oak woodlands described above are on steep hillsides facing, in particular, north and east, but also, on the Ross peninsula, south and east. On steep and more basic soils they will grade into *Corylus* scrub described below. Basic soils of sheltered gullies may have occasional ash (*Fraxinus excelsior*). These woods are often colonising talus slopes and will have larger boulders on the woodland floor which are colonised by a characteristic association of mosses and liverworts. Birks (1973) distinguished three associations on the boulders of oak-birch woodland two of which are seen in Mull woods. The *Hymenophyllum wilsonii–Isothecium myosuroides* association which is found on the sides of large boulders with *Dicranum scoparium*, *Plagiochila spinulosa* and *Scapania gracilis*, is an unstable mat that occasionally falls away leaving a bare rock surface. The other association, that usually on top of the large boulders, is a moss mat of *Hylocomium splendens*, *Pleurozium schreberi*, *Rhytidiadelphus loreus* and *Thuidium tamariscinum; Cladonia furcata*, *Agrostis* spp., *Deschampsia flexuosa* and *Oxalis acetosella* are also frequent. This type of woodland with a luxuriant fern flora (*Dryopteris pseudomas*, *D. austriaca*, *D. aemula*, *Phegopteris connectilis*, *Hymenophyllum wilsonii*, *H. tunbrigense* and *Oreopteris*) is seen in the area south of Loch Uisg, an area of steep hillsides dissected by many small streams.

There is a further type of mossy oak birch woodland which is on less steep, but rocky, situations and is best developed on the granite around Loch Caol. Here many of the mosses mentioned above, *Dicranum scoparium*, *Pleurozium*, *Rhytidiadelphus*, *Thuidium* etc. and the lichens *Cladonia furcata*, *C. gracilis* and *C. subcervicornis* cover

Fig. 10.15 Ash wood in Allt Ohirnie, Laggan peninsula, devastated by Hurricane Deborah on 22 September 1962. 1967.

narrow flat terraces on which *Quercus petraea* is dominant with *Sorbus aucuparia*, *Ilex aquifolium* and *Betula pubescens* subsp. *pubescens* frequent and *Corylus* as a subdominant understorey. This woodland is rarely above 8 m high and on exposed sites down to 3 m. There is a rich fern flora of *Dryopteris aemula*, *D. assimilis*, *D. austriaca*, *D. pseudomas*, *Hymenophyllum tunbrigense* and *Phegoteris connectilis*.

Ash woodland

Woodland in which the ash (*Fraxinus excelsior*) is dominant is now virtually absent in Mull. Previous to 1962, when Scotland had the effects of the tail-end of Hurricane Deborah, small isolated open woods of a few hectares only were established in which ash up to15 m high was the dominant tree with *Crataegus monogyna*, *Sorbus aucuparia* and *Ilex aquifolium* occasionally, and in which oak (*Q. petraea*) is rare. One such wood in Allt Ohirnie (Fig. 10.15) received the full force of the SW gale and most trees were killed. A few only were showing signs of life in 1967 and as a herd of 400 goats roam this area as well as a considerable flock of sheep, regeneration from seedlings has been for a long time out of the question except in inaccessible situations. Another similar wood exists further along that coast at Port na Muice Duibhe. Oaks are established on soils of high base status formed from and on Mesozoic limestone or cornstone sediments. The ground flora that still persists is *Circaea lutetiana*, *Mercurialis perennis*, *Brachypodium sylvaticum*, *Anthoxanthum*, *Viola riviniana*, *Agrostis tenuis*. This may well represent a depauperated state of the *Fraxinus-Brachypodium sylvaticum* association developed more luxuriantly on limestone pavement in Skye (Birks, 1973) or elsewhere in NW Scotland, e.g. Rassal Wood (McVean & Ratcliffe, 1962; McVean, 1964). On the other hand it may be an extreme form of the mixed deciduous woodland described below which contains ash as a characteristic tree, developing on basic soils and with a similar ground flora. The ash woodland remnants are however in south facing gullies and are more exposed to desiccating winds than the mixed woods.

Ash is an indicator of basic soils and is confined to the Mesozoic sediments or plateau lava of basalt and apart from that on the Laggan is a constituent in the mixed deciduous woodland described below.

Mixed deciduous woodlands

Scattered through the lowland basaltic areas of the northern half of Mull are pockets of a medium deep basic loam which supports a richer ground flora than that described above for the birth or birch-oak woods. These woodlands are never very extensive and are for the most part in the lower reaches of stream valleys, e.g. that to the east of Kilninian school and the similar valley of the Allt Hostarie just to the west. The woodland at Kilninian is in a steep-sided valley almost at sea-level and has *Ulmus glabra* and *Fraxinus excelsior* as dominant trees with oaks, mainly *Quercus petraea* becoming dominant on the upper parts of the slopes. The trees are erect and often 15–18 m high. Hazel is the dominant understorey shrub. The ground flora contains *Brachypodium sylvaticum* (often dominant), *Holcus mollis* (f) on flatter ground, *Endymion* (f), *Arrhenatherum* (o), *Sanicula europaea* (o), *Lysimachia nemorum* (o), *Viola riviniana*, *Fragaria vesca* (o), *Primula vulgaris* (r), *Allium ursinum* (r). Ferns like *Dryopteris pseudomas* and *Athyrium filix-femina* are abundant but on the steeper leached slopes.

In a deep ravine in Kellan wood (see Fig. 10.16) some 18 m (60 ft) above sea-level a similar wood may be found, again with *Corylus*, and here, ash saplings as an understorey, *Sorbus aucuparia* being rare. *Allium ursinum* is the dominant ground species with a very rich fern flora: *Athyrium filix-femina* (c-f), *Dryopteris austriaca* (f), *D. pseudomas* (f), *Blechnum spicant* (f-o), *Polystichum aculeatum* (o), *Dryopteris filix-mas* (o). Similar but somewhat less rich woods are seen on the south side of the Ardmeanach peninsula. Higher up in some of these gorge-like valleys isolated trees of ash, elm,

Fig. 10.16 Kellan Wood on north side of Loch na Keal, showing alders on alluvium (on left) grading into a mixed deciduous community on rising rocky hillside. 1966.

birdcherry and oak stand as remnants of a previously more extensive woodland whose ground flora still remains on the shaded ledges of the steepsided valleys. Mixed deciduous woodland is discussed by Tansley (1949) under the general heading 'Quercetum petraeae or sessiliflorae' and describes woods on the acid soil of western Wales and on the calcareous brown loams of the west of England. Woods on the basalt, which in their tree-species and the floristically rich field-layer are similar to those described by Tansley and related to, and only one stage removed from, the *Fraxinus-Brachypodium* association described above; it is listed by Birks (1973) for the deep brown earths or organic rendzinas of Skye. Pearsall (1950) describes deciduous woodlands of flushed soils and gives the example of Naddle Forest in Westmorland where 'ash-streaks', as he calls them, are confined to the flushed gullies or streambeds. These mixed woodlands are rarely found above 30 m (100 ft) but an old ash-oak wood (N of Kingharair) in a shallow tributary valley on the west side of the River Belart, i.e. facing NE, at about 75 m (250 ft) could be mentioned here. Some of the large and old ash-trees are dying and there is no regeneration. There is no obvious reason for this except perhaps that the surrounding birch wood is being gradually eroded by man and grazing animals and the microclimate is changing. The ground from under the ash is typical and as described above. The lower part of this wood is very wet due to seepage running out over harder lava flows or impervious sills and *Salix cinerea* subsp. *oleifolia* and *Salix aurita* are locally common, with *Prunus spinosa* frequent; this lower woodland as a whole is a hazel-birch association (*Corylus-Betula pubescens*) with occasional *Sorbus aucuparia.* The ground flora is that characteristic of wet woodland and includes *Filipendula ulmaria, Cirsium heterophyllum, Deschampsia cespitosa, Geum rivale, Valeriana officinalis, Galium odoratum, Primula vulgaris, Equisetum sylvaticum* and *Phegopteris connectilis* and the mosses *Hylocomium brevirostre, Mnium undulatum* and *Rhytidiadelphus triquetrus.*

Alder woodland

Alder (*Alnus glutinosa*) woods are rare on Mull in spite of the occurrence of suitable imperfectly drained brown forest soils or non-calcareous gleys. The few seen are in waterlogged but oxidising soils supplied throughout the year with water by a seepage from an uncomformity in the underlying rock, usually basaltic lava beds. Isolated trees or small stands may be seen along the courses of several rivers especially where they approach the coast, e.g. Altt Ardnacross, and along the alluvial banks of the rivers like Forsa and Glencannel and in east Ulva. At Ardnacross considerable regeneration is taking place but for the most part these stream-side occurrences are of rather stunted trees.

An example of such a wood where alder is dominant is seen on the NE tip of Loch na Keal at sea level on the edge of Kellan Wood (see Fig. 10.16). The area is small and the alder trees up to 7 m high. As the ground rises above the seepage line and onto a steep rocky slope the wood and its field layer changes character to a mixed deciduous (oak-ash) wood described above .The field layer in the alders is dominated in the wetter areas by *Juncus kochii* with *J. effusus, Ranunculus flammula, Cirsium palustre, Mentha aquatica, Lythrum salicaria, Carex demissa, C. echinata, C. remota, Athyrium filix-femina,* and in the ground layer *Pellia epiphylla, Acrocladium cuspidatum* and *Mnium pseudopunctatum.* In the drier areas *Brachypodium sylvaticum, Holcus lanatus, Dryopteris austriaca, Ranunculus acris, Oreopteris limbosperma* and *Ajuga reptans* are frequent.

Deschampsia cespitosa is absent from this wood, perhaps due to the relatively high base content of the soil. A similar wood near Duart castle is less basic and distinctly peaty, showing the characteristic species of *Juncus acutiflorus, Dryopteris carthusiana* (found only in this association on Mull), *Filipendula ulmaria, Iris pseudacorus, Betula pubescens* and *Salix cinerea* subsp. *oleifolia* but the stand is too small to draw comparisons. This may be similar to alder dominated woods described for Berwickshire by Birse & Robertson (1967) and less akin to those described for highland and western Scotland by McVean & Ratcliffe (1962) and Birks (1973). That described from Kellan Wood is similar in associates to that described along the Great Glen by McVean (1956).

Hazel scrub-woodland

Hazel (*Corylus avellana*) is the most abundant woody species on Mull and, as shown above, is a shrub or understorey tree in birch, oak and mixed deciduous tree communities. Its ability to sucker and regenerate from almost moribund rootstocks ensures its prolonged existence in many sites. Two types of hazel wood will be discussed, easily demarcated from the *Betula-Vaccinium* association already described.

Herb-rich hazel woodland This is a woodland where the hazel may reach 5 m in height although often much less. In it *Vaccinium myrtillus, Sphagnum* spp. and *Deschampsia flexuosa* are absent and in this it can be separated from the birch-dominated association mentioned above. *Betula pubescens* (especially "subsp. *odorata*") may be frequent in places, however it can be found on the flushed brown earths on steeper hillsides

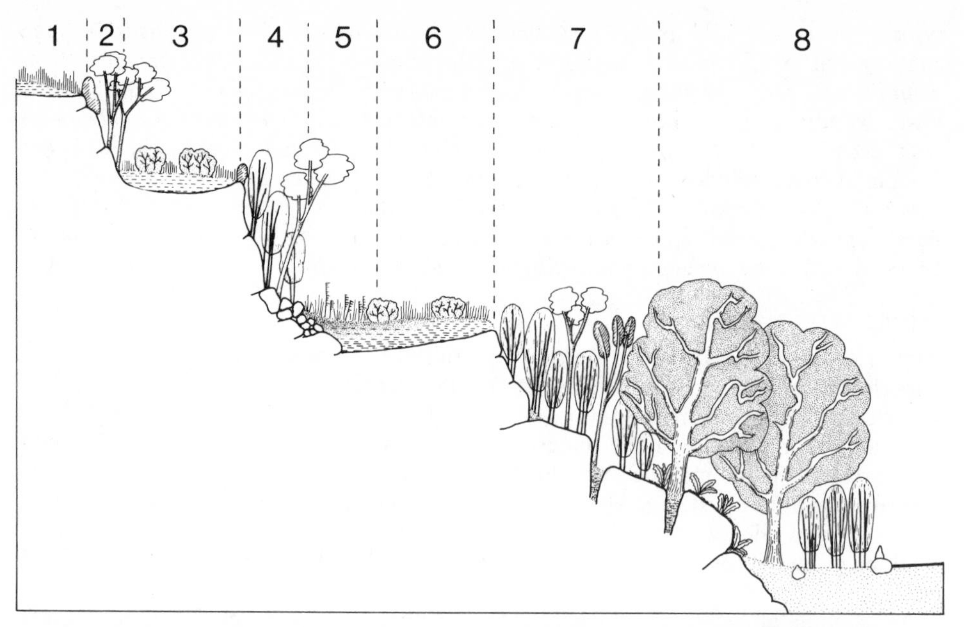

Fig. 10.17 Profile through Druimghigha Woods. Site of special scientific interest.

Notes

1, 3, 6 *Calluna-Erica* associations with *Scirpus cespitosus* on deep peat, overlying terraces;

2 Escarpment slope *Betula* woodland with *Vaccinium myrtillus* on moorland edge.

4 Similar woodland with *Corylus*.

5 *Deschampsia cespitosa–Molinia* on inorganic downwash at base of escarpment.

7 & 8 Mixed deciduous woodland on lowermost escarpment, with rich herb layer.

throughout the Island and a good example is seen on the second terrace at Druimghigha wood S.S.S.I. (see Fig. 10.17). Its field layer includes *Anthoxanthum odoratum*, *Circaea intermedia*, *Deschampsia cespitosa*, *Galium saxatile*, *Oxalis acetosella*, *Primula vulgaris*, *Viola riviniana*. This moss layer contains *Hylocomium brevirostrum*, *Thuidium tamariscinum* and *Hypnum callichroum*.

Hazel-blackthorn scrub is predominantly *Corylus*, often with *Prunus spinosa* co-dominant on steep cliffs especially facing the sea. Good examples are seen at the base of the cliffs of Beinn Creagach above Clachandhu (see Fig. 10.18) on the north end of the Gribun and again further south above Balmeanach farm. In the former case they are growing on an ancient landslip of basic soils; in the latter, brown forest soils formed on the Mesozoic sediments. The associated field layer may be depauperate due to the steepness and exposed nature of the site but has species characteristic of the basic woodlands described above, e.g. *Conopodium majus*, *Primula vulgaris*, *Scilla non-scripta*, *Sanicula europaea* and *Veronica chamaedrys*.

On open cliff ledges amongst open hazel scrub small, often stunted trees of *Populus tremula* are occasionally seen but they never grow to an appreciable size because of the exposure they have to endure.

Sallow-carr and other willow associations

Bog or mire communities that may become established at the, usually windward, edge of upland lakes and slowly colonise the open water rarely show on Mull the climax stages of that hydrosere. The mire stages are dealt with elsewhere in this chapter but one willow woodland which has formed in the mire around a small inflow stream at the

Fig. 10.18 Looking south to Clachandhu and the gulch running up to Beinn Creagach, the only site for Asplenium viride *on Mull.* Corylus avellana *scrub on landslip detritus is particularly well established here. 1966.*

north end of Lochnameal is worthy of note here. The inflow is relatively base-rich when compared with most other waters in the north of Mull (see Table 9.12). This sallow (*Salix cinerea* subsp. *oleifolia*) carr has the remnants of the herb-rich mire, but very soon the canopy becomes closed and the mire plants are shaded out. Dominant species in the field layer are *Filipendula ulmaria*, *Iris pseudacorus*, *Mentha aquatica* and *Myosotis scorpioides*. *Lythrum salicaria*, *Lycopus europaeus*, *Phragmites australis*, *Caltha palustris* and *Scutellaria galericulata* are frequent associates. The water level in the summer is about 50 cm above the substrate of shallow anerobic peat floating on lake muds. Many of the sallows have fallen over and as their mass increases with age the substrate is further depressed. This community may be cyclic with shoots arising from prostrate stems but no regeneration was observed and it may eventually develop into a *Molinia-Eriophorum* mire. Islands of peat are formed around the boles of larger sallows on which *Agrostis stolonifera*, *Poa trivialis*, *Cardamine hirsuta*, *Lychnis flos-cuculi* and *Brachythecium rutabulum* have become established.

Salix aurita may often form dense stands in mixed herb mires (see p. 10.51).

Planted woodland around large houses

Mixed woodland of considerable maturity is found around the old-established houses on Mull. Species introduced, in the first place for ornament and secondarily as windbreaks include *Abies* spp., *Acer pseudoplatanus*, *Aesculus hippocastanum*, *Cupressus* spp., *Fagus sylvestris*, *Tilia × europaea* and *Tsuga heterophylla*. Some trees may be 150 years old or more and are of enormous stature. There is little documentation as to planting dates although the history of the Aros Estate is briefly mentioned by Macnab (1970). The area south of Tobermory was originally owned by the MacLeans of Coll up to 1855 who built a house, originally called Drumfin House, in the early nineteenth century and who apparently planted the beeches by the loch (A. H. Allan, *pers. comm.*, who possesses a print of that age showing the trees). The Allan family have owned the estate since 1872 and were responsible for planting mixed hard woods and conifers, including European larch, in the 1880–90 period. Some of these trees, and especially the larches were felled during the 1914–18 war but when Mr A. H. Allan took over the estate in 1946 all the beeches were standing and 'a large number of very big silver firs' (Allan, *in litt.*, 1974). There was also at that time some of the original European Larch which had attained a very large size. Drumfin House, re-named Aros House was vacated by the Allan family in May 1952 and the whole woodland was badly hit by a northerly gale in the December of that year. In 1950 considerable cutting had taken place to make way for the main hydro-electric line for the Island and for minor lines to the estate cottages. This opening up presumably had made the wood vulnerable. The shrubs, including rhododendrons and the several exotics were introduced around 1900; their provenance was, whenever possible, around Aberdeen as Alexander Allan maintained that 'any plant that *lived* in Aberdeen would *thrive* in Mull' (Allan, A. H., *in litt.*, 1974). In 1954 the wooded area around Aros House was sold to the Forestry Commission and the estate is now managed partly as a commercial forest and partly by the Burgh of Tobermory as a public park.

In several places the ground flora is typical of mixed deciduous woods on rich brown earth soils at least where the *Gaultheria shallon* and *Rhododendron ponticum* have not shaded it out. There is a small valley in which is found a fern-rich association of *Blechnum spicant*, *Dryopteris aemula*, *D. austriaca*, *D. pseudomas*, *Gymnocarpium dryopteris*, and *Phegopteris connectilis* with *Circea lutetiana*, *Luzula sylvatica* and *Oxalis acetosella*. The mosses include *Dicranum scoparium*, *Hylocomium splendens*, *Pleurozium schreberi* and *Rhytidiadelphus squarrosus*; in the more basic areas, *Cirriphyllum piliferum*, *Mnium undulatum*, *Plagiochila asplenioides* and *Rhytidiadelphus triquetrus* are common.

Other good examples of woodland around large houses are found at Torloisk, Pennyghael and Torosay Castle. The ground flora is similar to the above where the same moisture-soil conditions are found. Elsewhere and especially where dense summer foliage give rise to deep shade, a sparse flora of *Deschampsia flexuosa*, *Luzula multiflora*, *Circaea lutetiana* etc. prevails.

In all of these woodlands, the large and old trees with varying bark textures, have a well developed lichen flora which is discussed under the phorophyte section below.

Planted conifer woodlands

Until 1924 and the formation of the Forestry Commission there was little monoculture of conifers. Where conifers had been planted for timber they were usually in small stands associated with the larger farms or houses. Often the European larch (*Larix decidua*) was the sole species planted for example in Aros and at Lettermore.

The Forestry Commission purchased ground in Mull shortly after 1925 in the Glen Aros area. Their first plantings, now in the early 1970's being cropped, were made in 1928–30 and were of Sitka spruce (*Picea sitchensis*), the lodgepole pine (*Pinus contorta*)

Fig. 10.19 Salen Forest in Glen Aros. Thirty-year old spruce plantation showing a carpet of Sphagnum palustre *and* S. fallax *in glade opened up by storm damage. 1968.*

and the scots pine (*P. sylvestris*). *Larix decidua* was also planted and where surrounded by other conifers this has formed a unique habitat with a characteristic lichen flora and shows the widest range of *Usnea* spp. on the Island, e.g. *U. extensa*, *U. filipendula* and *U. fragilescens* amongst others; bryophyte epiphytes are poorly represented. Other larches are now planted by the Commission, e.g. *L. kaempferi* and *L.* × *eurolepis*.

Characteristic bryophyte communities have formed under the spruce forest in Glen Aros: *Dicranum majus*, *Hypnum cupressiforme* var. *ericetorum*, *Pleurozium schreberi*, *Polytrichum commune*, *P. formosum* and in damper areas, *Sphagnum spp.* especially *S. palustre* and *S. fallax* (see Fig. 10.19).

The history of woodland on Mull

There is little recorded data on the status of forest on Mull in early historical times. It is likely that the general Norse invasion of the eighth to tenth centuries accounted for considerable destruction of woodland by burning as elsewhere in Scotland. As the population increased, reaching nearly 11,000 in the early nineteenth century (Macnab,

1970), woodlands would have been cut for house timber and to some extent for home fires. The crofters of this period would have had the short stocky black cattle roaming the uplands and preventing regrowth. It is unlikely, however, that Mull was ever covered with forest to any great extent because of considerable exposure in many western sites to winds. Bogs on Brolass and Ross of Mull contain numerous *Pinus* boles and roots suggesting the forest of the Boreal Zone died in situ, presumably due to climatic changes (see Chapter 4). In the account of his and Johnson's 'Tour of the Hebrides', Boswell (1786) remarks on the barrenness of the island (see Chapter 1). The 'clearances' of the crofters took place between 1825 and 1865 and the subsequent management of the land for sheep may have accounted for the loss of further open woodland through burning to attract new heather growth.

Apart from around the houses of the larger estates there was little planting on a large scale. On the setting up of the Forestry Commission in 1924 land in Glen Aros was taken over and in 1928–30 the eastern shores of Loch Frisa around Lettermore House and in Fishnish Bay near Craignure. These are now the oldest conifer plantations on the Island. Mull Forest now extends to over 4800 ha (12 000 acres) of planted woodland which will be increased by 1980 to 9400 ha (23 500 acres) on land already owned by the Commission. Most of this forest is young, 2400 ha (6000 acres) planted in the last fifteen years and only 46 ha (115 acres) being more than fifty years old. Of the total of 15 374 ha (38 000 acres) now owned by the Forestry Commission 6000 ha (14 500 acres) are not likely to be planted being either good agricultural ground (which is leased out for that purpose), or being upland and unplantable for either trees or crops. The future of forestry on Mull is discussed in a Report by the Highland and Islands Development Board (1970). Its expansion is likely to be controlled by economic rather than ecological factors, in spite of the close proximity of the Wiggins Teape pulp mill at Fort William.

Epiphytic communities

It is a feature of the vegetation of the western sea board of Scotland that, locally at least, epiphytic mosses and lichens are conspicuous and often abundant on various host trees and shrubs (phorophytes). Mull is no exception to this and a rich epiphytic flora is found in woodland and in sheltered, more rarely isolated trees throughout the Island. There are more epiphytic species of bryophytes on the eastern side of the Island possibly due to the more constant humidity and lack of salt-laden winds. Most lichens on the other hand, require the humidity fluctuation and the light conditions that are found in the more open western woods. That more species are found on the older (and therefore often larger) phorophytes, e.g. those in policy deciduous woodland, reflects the slowness in establishment of these distinct epiphytes, i.e. they are restricted to certain phorophytes for one reason or another; others are facultative and not only establish themselves on several different phorophytes but may also grow over rocks in the near vicinity. A list of the major phorophytes and the epiphytes recorded on them is given in Table 10.2.

J. J. Barkman has published a detailed account of the ecological factors that affect epiphytes, including the dynamics, structure and chorology of epiphytic associations in Europe (Barkman, 1958). He gives a classification of the phytosociological associations according to the system of Braun-Blanquet. We have made no attempt to equate the epiphyte associations found in Mull with those of Barkman; the ecological factors on the other hand are summarised here in relation to our observations on Mull and are given in the order we consider to be of importance. It should be stressed that nowhere on Mull is the atmosphere polluted by industrial discharge; or urban smoke sufficient to influence the distribution of those lichen species which are particularly sensitive.

Table 10.2 Major species of bryophytes and lichens recorded as epiphytes of the more widespread tree species in Mull. B = on boles and major branches; T = on secondary branches and twigs*; brackets( ) indicate presence as occasional only. The ecological conditions of each phorophyte are listed under the following: (1) bark texture; (2) pH; (3) moisture retention properties of bark; (4) light regime; (5) epiphyte density and variation.

* In *Corylus avellana*, *Salix* spp., *Sambucus nigra* and *Sorbus aucuparia* there is no ecological significance between B and T.

Acer pseudoplatanus: (1) smooth to medium rough, peeling; (2) 5·6–6·5; (3) not absorbent; (4) partially open branching; (5) medium-high, many crustose spp. medium to dense shade in summer.

Species		Species	
Hepaticopsida		L. confusa	BT
Frullania dilatata	BT	L. expallens	B(T)
Lejeunea ulicina	B	Lecidea symmicta	(B)T
Lophocolea heterophylla	B	Lecidella elaeochroma	BT
Metzgeria furcata	B	Leptogium burgessii	B
		L. lichenoides	B
Bryopsida		Lobaria amplissima	B
Hypnum cupressiforme	BT	L. laetevirens	B
Orthotrichum lyellii	BT	L. pulmonaria	BT
O. stramineum	B	L. scrobiculata	B
Tortula laevipila	B	Micarea cinerea	B
T. papillosa	B	Microglaena muscorum	B
		Nephroma laevigatum	BT
Lichenes		N. parile	B
Acrocordia alba	B	Normandina pulchella	BT
A. biformis	B	Ochrolechia androgyna	B
Alectoria fuscescens	B	O. turneri	B
Anaptychia fusca	B	O. yasudae	B
Arthonia punctiformis	BT	Opegrapha atra	BT
A. radiata	BT	O. herbarum	BT
A. tumidula	B	O. varia	B
Arthopyrenia fallax	BT	O. viridis	B(T)
A. punctiformis	BT	O. vulgata	BT
Bacidia naegelii	B	Pannaria pityrea	B
B. phacodes	B	P. rubiginosa	B
B. rubella	B	P. sampaiana	B
Buellia disciformis	B(T)	Parmelia exasperata	T
B. punctata	B(T)	P. crinita	B
Calicium viride	B	P. glabratula	B
Caloplaca citrina	B	P. perlata	B
Candelaria concolor	B	P. revoluta	BT
Cetrelia cetrarioides	B	P. saxatilis	BT
Collema furfuraceum	B	P. subaurifera	(B)T
C. subfurvum	B	P. sulcata	BT
Enterographa crassa	B	Parmeliella atlantica	B
Evernia prunastri	BT	P. corallinoides	B
Graphis elegans	B	P. plumbea	B
G. scripta	BT	Peltigera collina	B
Lecania cyrtella	BT	Pertusaria albescens	B
Lecanora chlarotera	BT	P. albescens var. corallina	B

Table 10.2 (*continued*)

Lichenes (*continued*)			
P. hymenea	BT	Pseudocyphellaria crocata	B
P. leioplaca	BT	Pyrenula nitida	BT
P. pertusa	BT	P. nitidella	BT
Phaeographis dendritica	BT	Ramalina calicaris	BT
Physcia adscendens	BT	R. farinacea	BT
P. aipolia	BT	R. fastigiata	BT
P. leptalea	BT	R. fraxinea var. calicariformis	B
P. orbicularis	BT	Rinodina sophodes	T
P. tenella	BT	Thelopsis rubella	B
P. wainioi	B	Usnea fragilescens	BT
Physciopsis adglutinata	B	U. inflata	BT
Physconia grisea	B	U. subfloridana	BT
P. pulverulenta	BT	Xanthoria candelaria	B
		X. polycarpa	T

Alnus glutinosa: (1) smooth or rough; (2) 2·6–4·5; (3) partially absorbent; (4) closed branches, medium shade; (5) medium to dense cover, species-poor. In boggy sites. All species recorded on boles and major branches only.

Hepaticopsida

Colura calyptrifolia
Frullania dilatata
F. tamarisci
Lejeunea patens
L. ulicina
Lepidozia reptans
Lophocolea cuspidata
Metzgeria furcata
Plagiochila punctata
P. spinulosa
P. tridenticulata
Radula complanata
Saccogyna viticulosa
Scapania gracilis

Bryopsida

Dicranoweisia cirrata
Dicranum scoparium
Hypnum cupressiforme
H. cupressiforme var. filiforme
Isothecium myosuroides
Orthotrichum affine
O. stramineum
Tetraphis pellucida
Ulota bruchii
U. crispa
U. phyllantha
Zygodon viridissimus

Lichenes

Alectoria fuscescens
Arthonia punctiformis
A. radiata
Arthopyrenia fallax
A. punctiformis
Bacidia endoleuca agg.
Buellia disciformis
Catillaria griffithii
C. lightfootii
Cetraria chlorophylla
Cetrelia cetrarioides
Dimerella lutea
Evernia prunastri
Hypogymnia physodes
H. tubulosa
Lecanora chlarona
L. chlarotera
Lecidea granulosa
L. symmicta
L. uliginosa
Lecidea elaeochroma
Normandina pulchella
Opegrapha sorediifera
Parmelia exasperatula
P. glabratula
P. laevigata
P. perlata
P. revoluta
P. saxatilis
P. sulcata
Pertusaria hymenea
P. multipuncta
P. pertusa

Table 10.2 (continued)

Phlyctis argena	Usnea fragilescens
Physicia aipolia	U. inflata
Ramalina farinacea	U. rubiginea
Stenocybe pullatula (obligate)	U. subfloridana
Sticta limbata	

Betula pendula, B. pubescens and 'intermediates': (1) young parts smooth with papyraceous peeling, older areas very rough, corrugate with large smooth plateaux which peel, very old parts corrugated; (2) 4·6–5·50; (3) not absorbent; (4) open crown, low to medium shade; (5) very low on smooth parts, medium on rough areas, species-poor. Contains betulin.

Hepaticopsida		Cladonia coniocraea	B
Aphanolejeunea microscopica	BT	C. digitata	B
Douinia ovata	BT	C. macilenta	B
Frullania dilatata	BT	C. polydactyla	B
F. germana	BT	C. squamosa	B
F. tamarisci	BT	C. squamosa subsp. allosquamosa	B
Lejeunea patens	BT	Evernia prunastri	B
L. ulicina	BT	Graphis elegans	B
Metzgeria furcata	BT	G. scripta	B(T)
Mylia cuneifolia	BT	Haematomma elatinum	B
Plagiochila punctata	BT	Hypogymnia physodes	BT
P. spinulosa	B	H. tubulosa	BT
P. tridenticulata	B	Lecanora chlarotera	(B)T
Scapania gracilis	B	Lecidea cinnabarina	B
		Leptorhaphis epidermidis (obligate)	B
Bryopsida		Mycoblastus affinis	B
Dicranum scoparium	B	M. sanguinaris	B
Hypnum cupressiforme var. filiforme	B	Ochrolechia androgyna	B
		O. tartarea	B
Isothecium myosuroides	B	Pannaria pezizoides	B
Ulota bruchii	BT	Parmelia endochlora	B
U. crispa	BT	P. laevigata	B
U. phyllantha	BT	P. saxatilis	BT
		P. sulcata	BT
Lichenes		P. taylorensis	B
Alectoria fuscescens	B	Pertusaria multipuncta	BT
Arthonia punctiformis	T	P. multipuncta var. ophthalmiza	B
Arthopyrenia cerasi	T	Platismatia glauca	BT
Bacidia umbrina	T	Pseudevernia furfuracea	BT
Buellia griseovirens	B(T)	Sphaerophorus globosus	B
Catillaria lightfootii	BT	S. melanocarpus	B
C. pulverea	B	Usnea extensa	BT
C. sphaeroides	B	U. filipendula	BT
Cavernularia hultenii	BT	U. fragilescens	BT
Cetraria chlorophylla	BT	U. subfloridana	BT
C. sepincola	(T)	U. rubiginea	BT

Table 10.2 (continued)

Corylus avellana: (1) smooth but with epicormic shoots or tumerous areas which are rougher; (2) 5·6–6·5; (3) not absorbent; (4) partially open branching, medium shade in summer; (5) low to medium, species-rich.

Hepaticopsida

Colura calyptrifolia
Frullania dilatata
F. germana
F. tamarisci
Lejeunea patens
L. ulicina
Metzgeria furcata
Plagiochila punctata
Radula aquilegia
R. complanata

Bryopsida

Orthotrichum affine
O. striatum
O. tenellum
Ulota bruchii
U. crispa
U. phyllantha
U. vittata

Lichenes

Arthonia aspersella
A. lurida
A. punctiformis
A. radiata
A. spadicea
A. stellaris
A. tumidula
Arthopyrenia antecellans
A. fallax
A. punctiformis
Arthothelium ilicinum
A. spectabile
Caloplaca ferruginea
Catillaria griffithii
C. pulverea
Cetrelia cetrarioides
Cladonia caespiticia
Collema fasciculare
Dermatina swinscowii
Evernia prunastri
Graphina anguina
Graphis elegans
G. scripta
Lecanactis homalotropum
Lecanora chlarotera
L. confusa
L. intumescens
L. jamesii
Lecidella elaeochroma
Leptogium brebissonii
L. burgessii
L. hibernicum
L. lichenoides
Lobaria laetevirens
L. pulmonaria
L. scrobiculata
Melaspilea lentiginosa
Microthelia micula
Nephroma laevigatum
N. parile
Normandina pulchella
Ocellularia subtilis
Ochrolechia turneri
O. yasudae
Opegrapha atra
O. herbarum
O. servdiifera
O. vulgata
Pachyphiale cornea
Pannaria pityrea
P. rubiginosa
Parmelia endochlora
P. laevigata
P. perlata
P. saxatilis
P. sinuosa
P. sulcata
Parmeliella atlantica
P. corallinoides
P. plumbea
Peltigera collina
Pertusaria hymenea
P. leioplaca
P. leucostoma
P. multipuncta
P. pertusa
Phaeographis dendritica
Porina chlorotica var. carpinea
P. leptalea
Pseudocyphellaria crocata
P. thouarsii
Pyrenula laevigata

Table 10.2 (continued)

Lichenes (continued)	S. limbata
P. nitida	S. sylvatica
P. nitidella	Thelotrema lepadinum
Pyrenula sp.	Tomasellia gelatinosa
Sticta fuliginosa	T. ischnobela

Fagus sylvatica: (1) persistently smooth but often superficially roughened by the fungus *Polymorphum rugosum*; (2) 4·6–5·5; (3) slightly absorbent; (4) open crown, deep shade in summer; (5) medium, many crustose spp., species-rich.

Hepaticopsida		C. coniocraea	B
Drepanolejeunea hamatifolia	B	Collema fasciculare	B
Frullania dilatata	B	C. flaccidum	B
Lejeunea ulicina	B	Dimerella diluta	B
Metzgeria furcata	B	D. lutea	B
		Enterographa crassa	B
Bryopsida		Evernia prunastri	B(T)
Dicranoweisia cirrata	B	Haematomma coccineum	B
Hypnum cupressiforme	B	H. coccineum var. ochroleucum	B
Orthotrichum affine	B	Hypogymia physodes	BT
O. lyellii	B	H. tubulosa	BT
Ulota crispa	B	Lecanactis abietina	B
Zygodon viridissimus	B	Lecanora carpinea	T
		L. chalarona	BT
Lichenes		L. chalarotera	BT
Acrocordia alba	B	L. confusa	BT
A. biformis	B	L. expallens	B(T)
Alectoria fuscescens	B	L. intumescens	(B)T
A. subcana	B	L. jamesii	B
Arthonia punctiformis	T	Lecidea cinnabarina	B
A. radiata	(B)T	L. symmicta	(B)T
Arthopyrenia antecellans	B(T)	Lecidella elaeochroma	BT
A. fallax	T	Leptogium brebissonii	B
A. punctiformis	(B)T	Lobaria laetevirens	B
Arthothelium ilicinum	(B)T	L. pulmonaria	B
Bacidia chlorococca	T	L. scrobiculata	B
B. endoleuca agg.	B	Menegazzia terebrata	B
B. sphaeroides	B	Micarea prasina	B
B. umbrina	T	Mycoblastus sanguinarius	B
Buellia disciformis	B(T)	Ocellularia subtilis	(B)T
B. erubescens	BT	Opegrapha atra	(B)T
B. punctata	B	O. herbarum	B
Calicium viride	B	O. niveoatra	B
Caloplaca ferruginea	B	O. vulgata	B(T)
Catillaria atropurpurea	B	Pannaria pityrea	B
C. griffithii	B	Parmelia exasperata	B
C. lightfootii	B(T)	P. glabratula	B
C. pulverea	B	P. reticulata	B
Cetraria chlorophylla	(B)T	P. revoluta	B
Cetrelia cetrarioides	B	P. saxatilis	BT
Cladonia chlorophaea	B	P. sinuosa	T

Table 10.2 (continued)

Lichenes (continued)		Pseudevernia furfuracea	B
P. subaurifera	(B)T	Pseudocyphellaria crocata	B
P. sulcata	BT	P. thouarsii	B
Pertusaria amara	B	Pyrenula nitida	B(T)
P. hemisphaerica	B	P. nitidella	B(T)
P. hymenea	B(T)	Ramalina calicaris	BT
P. leioplaca	BT	R. farinacea	BT
P. multipuncta	B(T)	Sphaerophorus globosus	B
P. pertusa	BT	Thelotrema lepadinum	BT
Phaeographis dendritica	(B)T	Sticta limbata	B
Phlyctis argena	B	Usnea inflata	B
Platismatia glauca	BT	U. subfloridana	B

Fraxinus excelsior: (1) young branches smooth, flat plateaux between deep furrows which are only on older trunks; (2) 5·6–6·5; (3) absorbent; (4) open crown, thin shade; (5) high, species-rich.

Hepaticopsida		B. canescens	B
Cololejeunea minutissima	B	Calicium viride	B
Drepanolejeunea hamatifolia	B	Caloplaca aurantiaca var. flavovirescens	B
Frullania tamarisci	BT		
Lejeunea ulicina	BT	C. cerina	B
Plagiochila asplenioides	B	Candelaria concolor	B
Porella laevigata	B	Candelariella reflexa	B
		Catillaria sphaeroides	B
Bryopsida		Catinaria grossa	B
Dicranoweisia cirrata	B	Collema flaccidum	B
Hypnum cupressiforme var. filiforme	B	C. furfuraceum	B
		C. subfurvum	B
Isothecium myurum	B	Evernia prunastri	BT
Leucodon sciuroides	B	Gomphillus calycioides	B
Neckera pumila	B	Graphis elegans	B
Orthotrichum lyellii	BT	G. scripta	BT
O. stramineum	BT	Gyalecta truncigena	B
Tortula virescens	B	Lecanactis homalotropum	B
Ulota phyllantha	BT	Lecania cyrtella	(B)T
Zygodon conoideus	B	Lecanora chlarotera	BT
Z. viridissimus	BT	L. confusa	BT
		L. expallens	B(T)
Lichenes		L. jamesii	B
Acrocordia alba	B	Lecidea symmicta	(B)T
A. biformis	B	Lecidella elaeochroma	BT
Arthonia lurida	BT	Leptogium azureum	B
A. punctiformis	BT	L. burgessii	B
A. radiata	BT	L. cyanescens	B
Arthopyrenia fallax	BT	L. teretiusculum	B
A. punctiformis	BT	Lobaria amplissima	B
Bacidia rubella	B	L. laetevirens	B
Biatorella monasteriensis	B	L. pulmonaria	B(T)
B. ochrophora	B	L. scrobiculata	B
Buellia alboatra	B	Microglaena muscorum	B

Table 10.2 (continued)

Nephroma laevigatum	B	P. pertusa	B
N. parile	B	Phaeographis dendritica	BT
Normandina pulchella	B(T)	Phlyctis agelaea	B
Opegrapha atra	(B)T	P. argena	B
O. sorediifera	B	Physcia adscendens	BT
O. varia	B	P. aipolia	T
O. vulgata	B(T)	P. tenella	BT
Pannaria mediterranea	B	Pseudocyphellaria crocata	B
P. microphylla	B	P. thouarsii	B
P. pityrea	B	Pyrenula nitida	B
P. rubiginosa	B	P. nitidella	B
P. sampaiana	B	Ramalina calicaris	(B)T
Parmelia caperata	B	R. fastigiata	BT
P. exasperata	T	Rinodina roboris	B
P. glabratula	BT	R. sophodes	B
P. perlata	B	Thelotrema lepadinum	B
P. reticulata	B	Usnea fragilescens	BT
P. subaurifera	T	U. inflata	B
P. subrudecta	BT	U. rubiginea	B
Pertusaria albescens	B	U. subfloridana	BT
P. amara	BT	Xanthoria candelaria	B(T)
P. hymenea	BT	X. parietina	BT
P. leioplaca	(B)T	X. polycarpa	(B)T
P. leucostoma	B		

Ilex aquifolium: (1) smooth; (2) 5·6–6·5; (3) repellent; (4) densely shaded, evergreen; (5) medium but scattered, species-poor. All species recorded on boles and major branches only.

Bryophyta

None recorded

Lichenes

Arthonia lurida
A. punctiformis
A. radiata
A. spadicea
A. stellaris
A. tumidula
Arthopyrenia antecellans
A. fallax
A. punctiformis
Arthothelium ilicinum
Bacidia sphaeroides
Graphina anguina
Graphis elegans
G. scripta
Haematomma elatinum
Hypogymnia physodes
Lecanora chlarona
L. chlarotera
L. intumescens
L. jamesii
Lecidea elaeochroma
L. symmicta
Lepraria incana
Ocellularia subtilis
Ochrolechia androgyna
Opegrapha atra
O. sorediifera
O. vulgata
Parmelia glabratula
P. saxatilis
P. sulcata
Pertusaria amara
P. hymenea
P. leioplaca
Porina chlorotica var. carpinea
Stenocybe septata (obligate)
Thelotrema lepadinum
Usnea inflata
U. subfloridana

Table 10.2 (continued)

Pinus sylvestris: (1) rough, resinous flaking; (2) 3·0–4·5; (3) not absorbent; (4) partially open branching to densely shaded in plantations; (5) poor, especially in crowns, species-poor. All species recorded on boles only.

Bryopsida
Hypnum cupressiforme

Lichenes
Alectoria fuscescens
A. subcana
Bacidia beckhausii
B. endoleuca agg.
B. nitschkeana
Buellia schaereri
Calicium abietinum
C. subtile
Catillaria griffithii
C. sphaeroides
Cetraria chlorophylla
Chaenotheca chrysocephala
C. ferruginea
Cladonia coniocraea
C. digitata
C. polydactyla
C. squamosa var. allosquamosa
Evernia prunastri
Haematomma elatinum
Hypogymnia physodes
H. tubulosa
Lecanactis abietina
Lecanora chlarona
L. confusa
L. expallens
L. piniperda
Lecidea cinnabarina
L. granulosa
L. scalaris
L. turgidula
L. uliginosa
Lepraria incana
Micarea cinerea
M. prasina
M. violacea
Mycoblastus sanguinarius
Ochrolechia androgyna
O. tartarea
Parmelia laevigata
P. reddenda
Parmelia saxatilis
P. sulcata
Platismatia glauca
Sphaerophorus globosus
Usnea distincta
U. extensa
U. fragilescens
U. hirta
U. inflata
U. rubiginea
U. subfloridana

Quercus petraea and **Q. robur:** (1) very rough throughout, rarely with flat plateaux, younger trees smooth to medium-rough; (2) 3·6–5·5; (3) partially absorbent; (4) closed branches but medium shade; (5) dense to medium, species-rich. Contains tannin.

Hepaticopsida	
Adelanthus decipiens	B
Barbilophozia attenuata	B
Cephaloziella starkei	B(T)
Cololejeunea minutissima	BT
Frullania dilatata	BT
F. germana	BT
F. tamarisci	BT
Lejeunea patens	BT
L. ulicina	BT
Lepidozia reptans	B
Mylia cuneifolia	T
Metzgeria furcata	BT
Plagiochila punctata	BT
P. spinulosa	BT
P. tridenticulata	BT
Scapania gracilis	B
S. nemorea	B
Bryopsida	
Dicranoweisia cirrata	B
Dicranum scoparium	B
Hypnum cupressiforme	BT
Isothecium myosuroides	B
I. myurum	B

Table 10.2 (continued)

Orthotrichum affine	BT	L. quernea	B
Tetraphis pellucida	B	L. symmicta	(B)T
Ulota crispa	(B)T	Lecidella elaeochroma	T
U. phyllantha	(B)T	Lobaria laetevirens	B
		L. pulmonaria	B
		L. scrobiculata	B
Lichenes		Lopadium pezizoideum	B
Acrocordia biformis	B(T)	Melaspilea ochrothalamia	B
Arthonia didyma	N	Nephroma laevigatum	B
A. dispersa	T	Ochrolechia androgyna	B
A. lurida	T	O. inversa	B
A. punctiformis	T	Opegrapha atra	BT
A. radiata	T	O. ochrocheila	B
A. tumidula	B	O. vulgata	B(T)
Arthopyrenia fallax	T	Parmelia caperata	B
A. punctiformis	T	P. crinita	B
Arthothelium ilicinum	T	P. laevigata	B
Bacidia endoleuca agg.	B	P. perlata	B
Calicium viride	B	P. revoluta	B
Caloplaca sarcopisioides	B	P. saxatilis	BT
Catillaria atropurpurea	B	P. sulcata	BT
C. lightfootii	T	P. subaurifera	(B)T
Cetrelia cetrarioides	B	P. taylorensis	B
Chaenotheca ferruginea	B	Pertusaria albescens	B
Cladonia chlorophaea	B	P. amara	BT
C. coniocraea	B	P. hemisphaerica	B
C. polydactyla	B	P. hymenea	BT
C. squamosa	B	P. multipuncta	BT
C. squamosa var. allosquamosa	B	P. pertusa	BT
Dermatina quercus (obligate)	T	Phaeographis dendritica	BT
Evernia prunastri	B(T)	Phlyctis argena	B
Graphis elegans	B(T)	Platismatia glauca	B(T)
G. scripta	BT	Pseudocyphellaria crocata	B
Haematomma elatinum	B	P. intricata	B
Hypogymia physodes	(B)T	Ramalina calicaris	(B)T
H. tubulosa	BT	R. fastigiata	BT
Lecanactis abietina	B	R. farinacea	BT
Lecanora chlarona	BT	Thelopsis rubella	B
L. chlarotera	BT	Thelotrema lepadinum	B
L. confusa	BT	Usnea filipendula	BT
L. expallens	B(T)	U. rubiginea	B
Lecidea granulosa	B	U. subfloridana	BT

Salix species: (1) smooth; (2) 5·6–6·5; (3) not absorbent; (4) partially open branching, shading slight to medium; (5) low to medium, many crustose spp., species-rich. Usually in boggy sites.

Hepaticopsida

Frullania dilatata	L. ulicina
Lejeunea patens	Plagiochila punctata

Table 10.2 (continued)

Bryopsida

Dicranoweisia cirrata
Hypnum cupressiforme var. filiforme
Orthotrichum affine
Ulota crispa
U. phyllantha

Lichenes

Anaptychia obscurata
Arthonia punctiformis
A. radiata
Arthopyrenia punctiformis
Arthothelium ilicinum
Bacidia endoleuca agg.
B. sphaeroides
Buellia griseovirens
Caloplaca ferruginea
Catillaria atropurpurea
C. lightfootii
C. pulverea
Cetraria chlorophylla
Cetrelia cetrarioides
Cladonia coniocraea
C. polydactyla
Collema nigrescens
C. subfurvum
Dimerella diluta
D. lutea
Evernia prunastri
Graphis scripta
Gyalideopsis anastomosans
Haematomma elatinum
Hypogymnia physodes
H. tubulosa
Lecanactis homalotropum
Lecanora chlarona
L. chlarotera
L. confusa
L. expallens
L. jamesii
Lecidea cinnabarina
L. granulosa
L. symmicta
L. tenebricosa
Lecidella elaeochroma
Lobaria laetevirens
L. pulmonaria
L. scrobiculata
Melaspilea ochrothalamia
Menegazzia terebrata
Nephroma laevigatum
N. parile
Normandina pulchella
Ochrolechia androgyna
O. turneri
Opegrapha atra
O. sorediifera
Pachyphiale cornea
Pannaria pityrea
P. rubiginosa
Parmelia exasperata
P. glabratula
P. laevigata
P. revoluta
P. saxatilis
P. sinuosa
P. subaurifera
P. subrudecta
P. sulcata
Parmeliella atlantica
Peltigera horizontalis
Pertusaria amara
P. hymenea
P. leioplaca
P. multipuncta
P. pertusa
Phlyctis agelaea
P. argena
Pseudevernia furfuracea
Pseudocyphellaria crocata
P. thouarsii
Ramalina calicaris
R. farinacea
R. fastigiata
Sticta dufourii
S. fuliginosa
S. limbata
S. sylvatica
Usnea extensa
U. filipendula
U. fragilescens
U. subfloridana

Table 10.2 (continued)

Sambucus nigra: (1) rough, spongy; (2) 6·6–8·0; (3) very absorbent; (4) partially open branching; (5) medium, crustose spp. predominating.

Hepaticopsida

Frullania tamarisci
Lejeunea patens
L. ulicina
Metzgeria furcata
M. fruticulosa
Radula complanata

Bryopsida

Cryphaea heteromalla
Dicranoweisia cirrata
Dicranum scoparium
H. cupressiforme var. filiforme
H. cupressiforme var. mamillatum
H. cupressiforme var. resupinatum
Isothecium myurum
Neckera pumila
Orthotrichum affine
O. diaphanum
O. pulchellum
O. stramineum
O. striatum
O. tenellum
Tetraphis pellucida
Tortula virescens
Ulota bruchii
U. crispa
U. phyllantha
Zygodon conoideus
Z. viridissimus agg.

Lichenes

Arthonia radiata
Bacidia arceutina
B. endoleuca agg.
B. friesiana
B. naegelii
Caloplaca cerina
C. cerinella
C. citrina
Candelariella reflexa
Catillaria atropurpurea
Collema furfuraceum
C. occultatum
Coniocybe sulphurea
Evernia prunastri
Gyalecta truncigena var. derivata
Graphis scripta
Lecania cyrtella
Lecanora chlarotera
L. dispersa
L. sambuci
Lecidella elaeochroma
Lepraria incana
Leptogium lichenoides
Micarea prasina
Opegrapha atra
Pachyphiale cornea
Peltigera collina
Physcia adscendens
P. aipolia
P. orbicularis
P. tenella
Ramalina farinacea
Rinodina sophodes
Usnea subfloridana
Xanthoria parietina
X. polycarpa

Sorbus aucuparia: (1) smooth, often with small areas of linear roughness; (2) 4·6–5·5; (3) repellant; (4) open crown; (5) low.

Hepaticopsida

Colura calyptrifolia
Frullania dilatata
F. tamarisci
Lejeunea patens
L. ulicina
Mylia cuneifolia
Metzgeria furcata
Plagiochila punctata
P. spinulosa

Bryopsida

Hypnum cupressiforme
Orthotrichum stramineum
Ulota bruchii
U. crispa

Table 10.2 (continued)

Bryopsida (continued)	L. symmicta
U. drummondii	Lecidella elaeochroma
U. phyllantha	Lobaria laetevirens
	L. pulmonaria
Lichenes	L. scrobiculata
Arthonia dispersa	Micarea cinerea
A. lurida	Normandina pulchella
A. punctiformis	Ocelluria subtilis
A. radiata	Ochrolechia pallescens
A. tumidula	Opegrapha atra
Arthopyrenia fallax	O. herbarum
A. punctiformis	O. vulgata
Arthothelium ilicinum	Pachyphiale cornea
Bacidia sphaeroides	Parmelia exasperata
Buellia disciformis	P. revoluta
B. griseovirens	P. saxatilis
Caloplaca ferruginea	P. sinuosa
Candelariella reflexa	P. sulcata
Catillaria pulverea	Peltigera collina
Cetrelia cetrarioides	Pertusaria leioplaca
Evernia prunastri	P. pertusa
Graphis elegans	P. multipuncta
G. scripta	Phaeographis dendritica
Hypogymnia physodes	Physcia aipolia
H. tubulosa	Pseudocyphellaria crocata
Lecanora carpinea	P. thouarsii
L. chlarona	Thelotrema lepadinum
L. chlarotera	Tomasellia gelatinosa
L. confusa	Usnea fragilescens
L. intumescens	U. inflata
Lecidea cinnabarina	U. subfloridana

Tilia ×europaea and **T. platyphyllos:** (1) smooth, often superficially roughened; (2) 5·6–6·5; (3) not absorbent; (4) closed branches but medium shade; (5) low, species-poor. All species recorded on boles only.

Bryophyta	Parmelia exasperatula
None recorded	P. glabratula
	P. laciniatula
	P. sulcata
Lichenes	P. subrudecta
Bacidia naegelii	Physcia adscendens
Caloplaca citrina	P. orbicularis
Candelariella aurella	P. tenella
C. reflexa	P. wainioi
Lecanora chlarotera	Platismatia glauca
L. dispersa	Ramalina farinacea
Lecidella elaeochroma	Xanthoria candelaria

Table 10.2 (continued)

Ulmus glabra and **U. procera:** (1) rough, but more finely furrowed than old oak or ash; (2) 6·6–8·0; (3) absorbent; (4) partially open branching; (5) high.

Species	
Hepaticopsida	
Frullania dilatata	B
F. tamarisci	B
Lejeunea ulicina	B
Metzgeria furcata	B
Bryopsida	
Dicranoweisia cirrata	B
Hypnum cupressiforme	B
Isothecium myosuroides	B
Ulota crispa	BT
Lichenes	
Acrocordia alba	B
A. biformis	B
Arthonia punctiformis	(B)T
A. radiata	BT
Arthopyrenia pyrenastrella	T
Bacidia friesiana	B
B. rubella	B
B. sabuletorum	B
Biatorella ochrophora	B
Buellia alboatra	B
B. canescens	B
B. punctata	BT
Calicium viride	B
Caloplaca citrina	B
C. holocarpa	B
Candelariella reflexa	B
C. xanthostigma	B
Catillaria sphaeroides	B
Catinaria grossa	B
Collema flaccidum	B
C. furfuraceum	B
C. sufurvum	B
Evernia prunastri	BT
Gomphillus calycioides	B
Gyalecta flotowii	B
G. truncigena	B
G. truncigena var. derivata	B
Lecania cyrtella	BT
Lecanora carpinea	(B)T
L. chlarotera	BT
L. confusa	BT
L. expallens	B(T)
L. sambuci	(B)T
Lecidea symmicta	(B)T
Lecidella elaeochroma	BT
Lepraria incana	B
L. membranacea	B
Leptogium azureum	B
L. burgessii	B
L. cyanescens	B
L. lichenoides	B
L. teretiusculum	B
Lithographa dendrographa	B
Lobaria amplissima	B
L. laetevirens	B
L. pulmonaria	B
L. scrobiculata	B
Nephroma laevigatum	B
N. parile	B
Normandina pulchella	B
Opegrapha atra	BT
O. vermicellifera	B
O. vulgata	B(T)
Pannaria pityrea	B
P. rubiginosa	B
Parmelia crinita	B
P. saxatilis	B
P. sulcata	B
Parmeliella atlantica	B
P. corallinoides	B
P. plumbea	B
Peltigera collina	B
Physcia adscendens	BT
P. aipolia	T
P. orbicularis	B
P. tenella	BT
Physconia farrea	B
Sticta dufourii	B
S. fuliginosa	B
S. limbata	B
S. sylvatica	B
Thelopsis rubella	B
Usnea subfloridana	BT
Xanthoria candelaria	B
X. parietina	BT
X. polycarpa	BT

Chemical composition of the bark Nutrients dissolved in the bark can be the major source for appressed lichens, creeping bryophytes and, occasionally the rhizomatous *Hymenophyllum wilsonii*. Fruticose lichens and bryophyte species which form an open weft or mat rely more on the dust or sea spray-laden winds for nutrients. Barkman (*l.c.*) divides trees into three groups; (a) those with eutrophic bark, i.e. an ash content of (5–) 8–12% of the dry bark which include *Acer pseudoplatanus*, *Sambucus nigra; Tilia* spp. and *Ulmus campestris*; (b) with mesotrophic bark, the ash content being (2–) 3–5% and including *Quercus robur*, *Q. petraea*, *Fagus sylvatica* and *Fraxinus excelsior*; (c) with oligotrophic bark, ash content 0·4–2·7% and including *Alnus glutinosa*, *Betula* spp., *Picea abies* and *Pinus sylvestris*.

There are no analyses for cation or anion content of tree bark on Mull. Those isolated analyses quoted by Barkman indicate *Sambucus nigra*, by far the richest phorophyte on Mull, to have the highest amounts of CaO, K_2O, MgO, P_2O_5 and SO_3 and Barkman (*l.c.*: 97) goes on to suggest that 'nearly all iron, manganese, silicate and phosphate and most of the calcium, magnesium and carbonate must be present in solid form'. This means that Barkman's analyses may be entirely meaningless when considered in terms of what is really available to the epiphyte itself. Available ions are obviously affected by environmental factors such as amount of rain and by acidification of the bark by the lichens growing there (excretion of lichen acids and through respiration). The amount of K and Na may also vary with species and the ratio of their oxide (K_2O/Na_2O) gives in the following decreasing order: *Fagus sylvatica* – *Sambucus nigra* – *Tilia* spp. – *Fraxinus excelsior* – *Ulmus campestris*. Some epiphytes or lithophytes that need high sodium and are generally confined to coastal sites, e.g. *Ulota phyllantha*, may be found on *Ulmus campestris* inland.

The presence of betulin, resins and other tree-specific substances would partially explain the presence of reduced and/or specialised epiphytic floras which occur on certain tree species, e.g. *Betula*, *Alnus* and conifers.

Nutrients are also supplied from wind-blown dust, sand and clay particles, salt-laden winds from the sea (see Chapter 9), from man (artificial manures) and from animals in the form of excreta and urine from dogs and birds in particular. River water, which may inundate the epiphyte on the bark of trees in narrow valleys and ravines, can bring (albeit in small amounts) constant replenishment of certain ions. Dust blown onto wayside trees from cart-tracks in dry weather is less now than it was fifty years ago before the roads were metalled. Fine base-rich shell-sand particles can be blown onto otherwise acid barks, e.g. *Betula*, and influence the epiphytes.

The acidity of the bark is a very important factor in epiphyte ecology and readings for the tree species encounted on Mull are given in Table 10.2. Differences in pH on the same trunk have been repeatedly observed; the bark in the upper part may be more acid than that at the base or vice-versa; it may also vary according to aspect, degree of inclination and age of the bark itself. Kershaw (1964) on discussing the distribution of lichens in relation to the pH on three deciduous trees showed that the bark of *Fraxinus* may have a higher pH under lichen thalli (as for example *Parmelia sulcata*) and suggested the lichens themselves have pronounced modifying effect on the pH. On the other hand bark with no epiphyte cover may be more easily leached of bases and can have an acidifying effect. Epiphytes are also able to regulate the pH themselves by the exchanging of hydrogen ions and vice-versa (Barkman, 1958).

Moisture relations The amount of moisture available to an epiphyte may vary according to a number of factors.

(a) The amount of rain falling onto the trunk or branches is governed in the first instance by the severity and duration of the rain shower and the periodicity or frequency that such showers have. This will directly effect the degree to which a habitat or epiphyte community will dry out.

(b) The amount of water reaching the epiphyte may be controlled by the microrelief or rugosity of the bark; those in furrows or rain channels obviously are at an advantage. The angle of inclination and general topography of the trunk surface may also cause rain shadows.

(c) The degree to which fog (cloud) is easily formed will affect the general humidity. There are few high altitude woodlands but a number of wooded ravines are high enough for more frequent enveloping by cloud. The fact that water is close by, e.g. in a stream, in a ravine or as a seepage flush in a wet *Alnus* or *Salix* carr, can increase the humidity.

(d) The actual porosity of the bark varies with the phorophyte and some species e.g. *Sambucus nigra*, are very spongy. Others, e.g. *Betula* species, may have water repellent ridges on the bark surface.

Bark texture. The roughness of tree bark will be an obvious factor deciding the ease with which lichen or moss propagules can lodge and become established. Most trees, however, will have the bark surface scarred with lenticels or epicormic shoots thus allowing even on the smoothest bark some means of attachment for moss and lichen spores. Establishment is more difficult and here the rugosity of the bark is all important in allowing attachment of rhizoids, hyphae or protonemata. Some trees e.g. *Acer pseudoplatanus*, *Betula* spp. and some conifers, lose their outer bark layers, often irregularly, and whilst preventing long term establishment of epiphyte communities, present a new surface for repeated colonisation.

Age of phorophyte individual and of the phorophyte community. The texture and possibly the chemical nature of the cortical layers change with age. Thus the younger branches (i.e., the twigs) may have a different assemblage of species, some of which die out as the branch ages and the texture of the bark changes. Gradients of nutrients, pH, light intensity and humidity may be found from the crown to the base and this will be affected by the age and composition of the phorophyte community (i.e. the wood) itself. An older wood will contain a greater diversification of micro-niches and therefore species of epiphytes.

Dwarf shrub communities

The dwarf-shrub communities (heaths) of Scotland have been fully discussed by Gimingham (1964). He defines heath as a distinct regional formation occurring naturally where exposure, soil immaturity and soil infertility limit the entry of trees; elsewhere, it has originated only after forest clearance and is maintained by management practices that prevent the re-establishment of trees, such as periodic burning and grazing by domestic animals. Heath communities can occur on the brown forest soils of low base content, gleys, podsols and drained peat. On Mull there are no instances where heath is forming on immature blown-sand areas although there may be marginal encroachment from the heathland at the edge of some sandy bays.

Where *Calluna* is growing on peat with impeded drainage the wetness coupled with a high rainfall can produce an actively growing peat substrate and bog conditions. These communities are dealt with under that heading (p. 10.62).

It is by no means certain that forest originally covered the whole of Mull; it is most likely that exposure to winds in the NW–SW sector prevented the growth of trees and there there was only scrub. The eastern slopes of the Torosay hills, and the southern slopes of Ben More through to Ben Talaidh are the most likely to have been forested, and the heath communities which are well established there can probably be regarded as semi-natural. Grazing pressure from sheep, feral goats and deer is heavy enough to prevent any re-establishment of the forest although birch (*Betula* spp.) and rowan (*Sorbus aucuparia*) is established on cliff ledges which are largely inaccessible.

The stratification seen in *Calluna*-dominated communities and the range of life-forms involved give the maximum number of habitats. The *Calluna* will usually form a dense canopy between 25 cm and 60 cm above the ground; when grazing is eliminated e.g. in new forest plantations, *Calluna* can attain a metre in height. The second stratum (10–30 cm above ground) is composed of subordinate shrubs e.g. *Arctostaphyllos uva-ursi*, *Empetrum nigrum*, *Vaccinium* spp., and grasses and sedges e.g. *Deschampsia flexuosa*, *Carex* spp., *Eriophorum* spp., *Scirpus cespitosus*. The third stratum is that of robust mosses like *Hylocomium splendens*, and mat-forming herbs like *Galium saxatile* between 5 and 10 cms. The fourth is the true 'ground' stratum with mat-forming (pleurocarpous) mosses e.g. *Hypnum cupressiforme*, the small erect (acrocarpous) mosses e.g. *Pohlia nutans* and lichens e.g. *Cladonia* spp., *Lecidea* spp.

Not all *Calluna* communities are so constructed, however, and periodic burning to produce young shoots for sheep to graze will naturally alter the structure of both the individual plant (i.e. its morphology) and that of the community. In communities where *Empetrum* or *Arctostaphyllos* are dominant *Calluna* may emerge as isolated bushes in an otherwise closed canopy which is in fact the second stratum described above. The floristic composition is further varied by the changes in topography and substrate, and mosaics of complex structuring will result.

The microclimates within a stratified heath have been described by Delany (1953) and Gimingham (1964). Light intensity is reduced by the *Calluna* canopy to less than 20 per cent of that in the open. Both temperature and humidity relate to the stratification. Little solar radiation penetrates the canopy so that temperature within remains lower and more constant and humidity is high (80 per cent often remaining at 90 per cent).

Calluna vulgaris associations

This is the widespread dry heather moor named 'Callunetum vulgaris' by McVean & Ratcliffe (1962). It occurs throughout the basalt regions of Mull but only in isolated stands on the steeper slopes of the mountains of the Central Complex ascending to 450 m (1500 ft). Soils are well drained, mostly peaty podsols, often with considerable raw humus in the surface layers. Whilst *Calluna vulgaris* is dominant, *Erica cinerea* and *Potentilla erecta* are constant, the former often co-dominant. *Hypnum cupressiforme* var. *ericetorum* and *Pleurozium schreberi* are common throughout. *Carex binervis* and *Agrostis tenuis* are frequent and *Carex nigra*, *Dactylorrhiza ericetorum*, *Hypericum pulchrum* occasional. On coastal sites where soils of a slightly more base-rich nature occur a 'herb-rich' facies develops with *Antennaria dioica*, *Euphrasia* spp., *Hypochoeris radicata*, *Sieglingia decumbens*, *Solidago virgaurea* and *Succisa pratensis*, as well as an increase of *Carex nigra*.

This association and its variants are discussed by Birks (1973) under *Calluno-Ulicetalia*.

Lichen associations in Callunetum. These are poorly developed on Mull and only of scattered occurrence. They are chiefly confined to drier, well-drained, south-facing areas, edges of peat hags and hummocks of decaying tussocks. The high water-table and high rainfall encourage rapid development of terrestrial algae which is followed quickly by a phanerogamic succession. The most successful lichens in wetter sites are *Cladonia strepsilis*, *C. impexa* and *C. uncialis*. In drier sites *Cladonia* spp. predominate, e.g. *C. arbuscula*, *C. chlorophaea*, *C. coccifera*, *C. crispata* var. *cetrariiformis*, *C. floerkeana*, *C. gracilis*, *C. impexa*, *C. squamosa* (incl. subsp. *allosquamosa*), *C. tenuis*, *C. uncialis* and *C. verticillata*. On higher ground above 100 m additional species occur, including *C. bellidiflora*, *C. capitata* (on one site only), *C. cornuta* and *C. gonecha*. Additional species occur where the Callunetum abuts rock outcrops, e.g. *C. furcata*, *C. rangiferina*, and particularly *C. subcervicornis* with *Cornicularia aculeata* and

C. muricata. Lecidea granulosa and *L. uliginosa* are abundant, especially on recently burnt areas and sometimes are accompanied by *L. oligotropha.* The edges of peat hags are the major sites for *Omphalina ericetorum* (*Botrydina vulgaris* agg.) and *Omphalina hudsoniana* (*Coriscium viride*); on the squamules of the latter, *Thelocarpon epibolum* may occur. This is also the site for the rare *Lecidea glaucolepidea.* Other species occurring on stony ledges between *Calluna* are *Baeomyces roseus, B. rufus* and, at higher altitudes *B. placophyllus* and *Lecidea demissa.*

***Calluna* associations of cliff ledges** The *Calluna–Vaccinium* associations described above become more luxurious on cliff ledges where they are protected from grazing and burning. Such ledges may be found from sea level to around 480 m (1600 ft). The principal species with *Calluna* are *Vaccinium myrtillus* and *Luzula sylvatica;* frequently *Calluna* is restricted to isolated bushes. Constant species are *Blechnum spicant, Deschampsia flexuosa, Oxalis acetosella* and *Thuidium tamariscinum* (Birks 1973). On Mull the more common species are the ferns *Dryopteris austriaca, Gymnocarpium dryopteris, Phegopteris connectilis* and more rarely *Hymenophyllum wilsonii,* and the bryophytes *Breutelia chrysocoma, Dicranum majus, D. scoparium, Diplophyllum albicans, Herberta adunca, Hylocomium splendens, Hypnum cupressiforme* var. *ericetorum, Lophozia ventricosa, Plagiochila spinulosa* and *Plagiothecium undulatum* with *Sphagnum quinquifarium* in quantity in the wetter flushed areas.

Calluna–Erica association

A characteristic association of open dry heath results from periodic burning on the schistose and granite rocks area of the Ross of Mull. The soils are shallow with little podsolization tending towards the formation of gleys in the hollows. The *Calluna* is co-dominant with *Erica cinerea,* with *E. tetralix* sparse on the knolls, becoming equally co-dominant in the hollows. *Potentilla erecta, Scirpus cespitosus* and *Succisa pratensis* are common with *Festuca rubra* with *Agrostis tenuis–Sieglingia* association on the drier mounts. *Molinia caerulea* is occasional but probably increasing due to the higher phosphate levels occurring on the burnt ground. This is a variant of the *Calluna–Erica tetralix* wet heaths which occur where drainage conditions maintain more or less permanently wet soils.

Calluna–Arctostaphylos association

The distribution of *Arctostaphylos uva-ursi* on Mull is shown in Fig. 2.28; it forms an often dense dwarf-shrub association with *Calluna vulgaris* on the southerly facing slopes of the basaltic areas on the Laggan and Brolass. *Salix repens* is often common in this community.

Calluna–Vaccinium myrtillus associations

This is a bryophyte-rich association uncommon on Mull. It is seen on slopes of old block scree, the interstices now often filled by morainic debris or raw humus, between sea-level and 270 m (900 ft) on the Mishnish moorland especially on those slopes which face north. Besides the two dominants, *Blechum spicant, Dryopteris austriaca* and *Potentilla erecta* are frequent. Mosses and liverworts include: *Campylopus atrovirens, Hypnum cupressiforme, Rhytidiadelphus loreus, Sphagnum capillaceum, Diplophyllum albicans, Herberta adunca, Mylia taylori* and *Plagiochila spinulosa. Hypogymnia physodes* is frequently epiphytic on *Calluna* stems. Ratcliffe (1968) suggests this association, rich in 'Atlantic' (oceanic) bryophytes, falls within the zone of 220+ 'wet days' a year. Birks (1973) confirms this for Skye and, from what meteorological data are available for Mull, the isolated examples from Mishnish are likely to receive such conditions. The association there is possibly less rich in bryophytes than those of north slopes in Skye.

Similar stands are found on shallow or skeletal soils on the north and east sides of the Ben More massif. At approximately 210 m (700 ft) this association locally contains considerable *Empetrum nigrum*. Fertile *Empetrum hermaphroditum* was found on a few occasions in this community but for the most part, *Empetrum* here was sterile and impossible to identify with certainty. *Deschampsia flexuosa* is a constant species but rare, with *Pleurozium schreberi*, *Hylocomium splendens*, *Cladonia arbuscula*, *C. impexa* and *C. uncialis*. Above 360 m (1200 ft) *Vaccinium vitis-idaea* becomes as abundant as *Empetrum* but again only locally. Here *Calluna* becomes less frequent and the association grades into the *Vaccinium* associations of the higher mountain tops.

Mires

Mires, as defined on p. 10.3, is the term used here for all wetland vegetation depending on a *lateral* flow of water. It is possible, through variation in geological and topographical formations, to find small mires (flushes according to some, e.g. Pearsall, 1950) in amongst large expanses of bog. Such mires are usually nutritionally poorer and intergrade with the bog communities; they are considered at the end of this section. Mires are however essentially determined by physiography and are fully discussed by Ratcliffe who stresses (1964a, p. 429) that they depend on a topography and geological structure which give localised concentration of gravitational seepage

Fig. 10.20 Southern slopes of the A' Mhaol Mhòr–Coirc Bheinn ridge of the Ben More massif seen from Ardmeanach. 1967.

water close to or at ground surface during at least part of the year. On Mull they are found on most peaty gentle slopes where there is a sufficient water catchment area above. Often on the basaltic terraces water is impeded by the less pervious and more slowly eroded basalt beds of lava (see Fig. 10.20). Where these strata are level, small bogs may result but for the most part they are sloping and water drains laterally through the peat. Several mire communities may thus occur in this situation. In the steeper valleys mires may be found where the slopes flatten into the valley floor. They are common in the small river-valleys and on the alluvial terraces where there is seasonal flooding which brings silt particles that become incorporated into the peat layers. Such flooding has a distinct effect on the floristic composition of the vegetation. The mineral (ionic) content of the water flowing through the mire may also affect the vegetation. For the most part, even over the more basic basalt areas, calcium and magnesium levels are low (less than 5 p.p.m.).

Springs, rock-trickles and flushes in the sense of Ratcliffe (1964a) (not of Pearsall, 1950), are a special kind of mire in which the flow of water is more localised and usually more rapid. Springs may be differentiated by the emergence at ground level of a more or less permanent flow of free drainage water (Ratcliffe 1964a, p. 538). Again the quality of water is the dominant factor. Those on Mull range from oligotrophic to a rich mesotrophic; the latter, arising from seepage lines at the base of calcareous Mesozoic sediments, are few but may be termed eutrophic and contain a number of plants otherwise rare on Mull, e.g. *Equisetum telmateia*. There are unfortunately no eutrophic montane flushes on Mull.

Juncus acutiflorus–Acrocladium cuspidatum association

This is a rare community confined to loch edges, in south Mull especially around the outflow channels, which are subjected to seasonal changes in water level; it is never extensive and according to Ratcliffe (1964a) best regarded as an anthropogenic type of vegetation. The soil is a highly humified peat containing some silt. These areas are usually grazed by cattle on Mull. The flora is suggestive of mildly base-rich conditions and the example seen at Loch a' Ghleannain contains *Achillea ptarmica, Carex nigra, Carex panicea, Crepis paludosa, Epilobium palustre, Holcus lanatus, Hydrocotyle vulgaris, Prunella vulgaris, Ranunculus acris* and *Scutellaria galericulata*.

McVean and Ratcliffe (1962, Map 27, p. 385) suggest that a *Juncus acutiflorus–Acrocladium cuspidatum* nodum is typical of the east and south of Scotland indicating a climatic correlation, thus it could be that Loch a' Ghleannain is an outlier with a cold basin micro-climate. Lochs on the basalt of N Mull occasionally have *Juncus acutiflorus* around their outflow; they are less herb-rich and *Molinia caerulea* is often co-dominant.

Carex paniculata–mixed herb mire

There are a number of base-rich flushes that arise from the basalt between sea-level and 225 m (750 ft) and these are best seen in the northern half of the island on the moorlands to the west of Glens Aros and Belart. The peat contains considerable inorganic silt and is saturated throughout the year. The flush is typified by *Carex paniculata* tussocks which may reach 75 cm, often with characteristic grassland species established upon them, e.g. *Galium saxatile*. Most of the sites have a dense or open scatter of *Salix aurita* and *S. cinerea* subsp. *oleifolia* and occasionally the ground flora may be severely reduced through the lack of light. The following species may be frequent: *Caltha palustris, Carex nigra, Crepis paludosus, Filipendula ulmaria, Geum rivale*; occasional species will include *Cirsium palustre, Equisetum sylvaticum, Scutellaria galericulata, Viola riviniana*, and the bryophytes: *Brachythecium rivulare, Ctenidium molluscum, Dicranum majus, Eurhynchium striatum, Frullania tamarisci, Hylocomium splendens, Mnium hornum, M. undulatum, Plagiochila spinulosa, Saccogyna viticulosa* and *Thuidium tamariscinum*.

Schoenus nigricans 'flushes'

Schoenus nigricans is for the most part local and rarely dominant in any community outside the Ross of Mull. Occasional flushes on the basalt on inorganic soil, and thus more base rich, contain *S. nigricans*, *Carex dioica*, *C. hostiana*, *Eleocharis quinqueflora*, *Pinguicula vulgaris*, *Campylium stellatum* and *Scorpidium scorpioides*. The distribution of *Schoenus* on peat has been correlated with proximity to the sea where the peat surface can derive additional ions from the salt-laden winds (Sparling, 1967). Distribution of *Schoenus* on Mull follows the described pattern and the species can be found as a common component on the deep-peat lands of the Ross of Mull. In all cases investigated, *Schoenus* was most abundant where lateral water movement was obvious. In the deeper peat cuttings between the benches it was the dominant species or co-dominant with *Myrica gale; Molinia caerulea* was frequent and with the species mentioned above were *Carex demissa*, *C. panicea*, *Drosera rotundifolia*, *Erica tetralix*, *Pinguicula lusitanica* and *Potamogeton polygonifolius*.

Carex rostrata–'brown moss' association

This association is infrequent, being confined to lowland sites in the south of the Island, usually on deep peat, and may be correlated with a topographical configuration which channels and moves water laterally in what is otherwise a bog situation. The moss flora which is so characteristic includes *Campylium stellatum*, *Drepanocladus revolvens* and *Scorpidium scorpidioides*. *Carex nigra*, *C. panicea* and *Eriophorum angustifolium* are also frequent associates. Sometimes, e.g. on the Ross of Mull granite, it is a community around portion of a loch where again it may be associated with water movement into the basin. In these situations it often grades into a *Carex rostrata* mire in which *Aulacomnium palustre* becomes the dominant moss species.

Scirpus cespitosus–Carex–Selaginella mire

There are areas of sloping shallow peat on the steeper sides of streams which are constantly wet and which contain a varied assemblage of plants, *Scirpus cespitosus*, *Carex echinata*, *C. panicea*, *Eleocharis multicaulis*, *Molinia caerulea*, *Narthecium ossifragum*, *Pinguicula vulgaris*, *Selaginella selaginoides*, *Succisa pratensis* and *Sphagnum palustre*, and *S. subnitens* are characteristic. It is difficult to place this association. It grades on inorganic better drained soils into *Agrostis–Festuca* 'Herb-rich' grassland. *Hammarbya paludosa* was found in this association in its sole station on Mull. *Schoenus nigricans* was absent from these sites, *Erica tetralix* was rare. This may be the same as Birks' (1973) *Trichophorum cespitosum–Carex panicea* association. It is infrequent on Mull and is probably medium-rich in its nutrient requirement.

Molinia dominated mires

Molinia–Myrica association This is a widespread and often extensive community especially on the basalt. It is to be found in broad valleys (see Fig. 10.21) where flushing or periodic flooding is frequent, also on slopes where an impervious lava bed results in lateral seepage. Besides *Molinia* and *Myrica* the following are common: *Potentilla erecta*, *Succisa pratensis*, *Narthecium ossifragum* and *Scirpus cespitosus*. The following *Sphagna* are often abundant: *S. fallax*, *S. papillosum*, *S. subnitens*, and *S. subsecundum* agg.

Molinia–Carex association Ratcliffe (1964a) defines this as an oligotrophic mire on shallow peats overlying gley mineral soils. He give *Erica cinerea* and *Calluna vulgaris* as common constituents with *Carex echinata*, *C. panicea*, *Drosera rotundifolia* and

Fig. 10.21 Molinia–Myrica *mire on schist of the Ross of Mull. 1967.*

Eriophorum angustifolium, *Narthecium ossifragum* and *Potentilla erecta* as constants. This association may be seen in the basaltic areas merging into the *Molinia–Myrica* community in several instances.

Sphagnum dominated mires

Juncus effusus–Sphagnum–Polytrichum association *Juncus effusus* is a species associated with disturbance (Ratcliffe 1964a). There is little evidence on Mull to support this; however, where willow scrub is cut on flushed hillsides *J. effusus* does tend to establish itself. The dominant *Sphagnum* is *S. fallax* with *S. palustre* and *Polytrichum commune* common, *S. girgensohnii*, *S. squarrosum* and *S. teres*, *Carex nigra*, *Galium saxatile* and *Potentilla erecta* frequent to occasional. This association is usually found in the smaller tributary valleys and along seepage lines on the moorland.

Carex-Sphagnum 'flushes' These mires are similar to and occur with the above; resembling them floristically except for the presence of *Juncus effusus*. The *Sphagnum* species may include *S. papillosum* and *S. auriculatum*. *Carex echinata* is usually associated with *C. nigra*. The high altitude equivalent of this, the alpine *Carex–Sphagnum* mire of Ratcliffe (1964b), is not seen on Mull.

Carex lasiocarpa–Menyanthes–Phragmites association

This is a community confined on Mull to the Coladoir river bog; small stands from 5–10 m are situated in hollows, the under-lying topography causing the very slight lateral flow of water, only slightly richer in nutrients, than the bog itself. *Menyanthes* is confined to the bottoms of the hollows; for the most part *C. lasiocarpa* is co-dominant with *Phragmites* with *C. rostrata* as a common component. The moss layer is almost entirely *Sphagnum* fallax with local patches of *S. palustre*.

Bryophyte-dominated mires

Campylium–Carex panicea association A frequent but seldom extensive mesotrophic or eutrophic association related to, but more base-impoverished than, the *Cratoneuron commutatum* communities. Dominance of herbs and sedges is prevented by grazing, i.e. the nodum is biotically maintained. The characteristic (constant) species are *Carex panicea, Campylium stellatum, Drepanocladus revolvens* with other hygrophytic mosses e.g. *Bryum pseudotriquetrum, Philonotis fontana, Cratoneuron filicinum*, and several other sedges.

Philonotis–Saxifraga stellaris association These are oligotrophic bryophyte 'flushes', related ecologically to the base-poor *Sphagneto-cariceta subalpina* on the one hand and richer *Cratoneuron* or *Campylium* flushes on the other. It is mainly a subalpine nodum. *Philonotis fontana* is the characteristic dominant associated with *Dicranella palustris, Scapania undulata* var. *dentata* and *Drepanocladus* spp. *Sphagnum auriculatum* is often present, and may become dominant on more base-impoverished sites.

Fig. 10.22 Typical Koenigia *'flush' with pebbles over fine frost-heaved peat on northern slope of Beinn na h-Iolaire. 1967.*

Anthelia–Deschampsia 'flush' On the NW shoulder of Ben More, at Coire nam Fuar-on, lies an extensive block-scree falling from the summit ridge at 750 m (2500 ft) to 675 m (2250 ft). At the lower level there is an unconformity between the mugearite cap and the basalt and this, together with the water from the gradual melting of snow which lies later in this coire, gives rise to a small but distinct spring which is dominated by *Anthelia julacea. Deschampsia cespitosa* is occasional with occasional plants of *Thalictrum alpinum, Carex bigelowii* and *Agrostis canina.* In the wetter parts is a characteristic form of *Carex nigra.*

Koenigia–Oligotrichum mire

This association is found only on the Ardmeanach between 420–510 m (1400–1700 ft) on the plateau area between Fionna Mhàm, Beinn na Streine and Beinn na h-Iolaire. The plateau is only slightly undulating and there is much bare soil and gravel (see Fig. 10.22) with occasional seasonal open water pans. There are signs of frost-heaving with incipient polygon formation, platelike rocks from 30 to 50 cm across. Beneath this surface layer of pebbles is a fine evenly-graded silt into which the minute plants

Fig. 10.23 Close-up of pebble 'flush' on Figure 10.22. Koenigia *(arrowed), with* Oligotrichum hercynicum, *is amongst the pebbles. 1967.*

of *Koenigia islandica* extend long fine roots. The pH ranges from 5.25 to 5.80. Associated with *Koenigia*, which may be only one centimetre high, is *Oligotrichum hercynicum* (see Fig. 10.23); *Carex demissa*, *Juncus bulbosus*, *J. triglumis*, *Poa annua*, *Saxifraga stellaris* and *Sedum villosum* are occasional; *Trapelia moorei* is a frequent associate. *Koenigia* is found only in this situation on Mull although on Skye, its other British location, it is found also on basalt screes (Birks 1973). It is difficult to see what the main physical factors are that produce this very characteristic terrain. Bibby (pers. comm.) has suggested wind may play a significant part. The only other area also similarly exposed which shows this type of soil structure is on Beinn na Drise, but here it occurs at a slightly lower altitude c. 390 m (1300 ft); so far *Koenigia* has not been discovered there.

Cliff ledge communities

A number of tall herb communities growing in gullies and on wet ledges may be included here, as a dominant ecological factor is moving ground water, and in some, flushing may be periodic. *Luzula sylvatica* with *Vaccinium* spp. is described on p. 10.49. Fern-rich associations, in which *Athyrium filix-femina*, *Blechnum spicant*, *Dryopteris austriaca*, *D. oreades*, *D. pseudomas*, *Gymnocarpium dryopteris* and *Phegopteris connectilis* form large stands, are common. In the base-rich gullies, *Angelica sylvestris*, *Cochlearia* spp., *Filipendula ulmaria*, *Geum rivale*, *Mercurialis perennis*, *Trollius europaeus* and *Tussilago farfara* are common. An ecotype of *Carex binervis*, very characteristic in size and habit, with culms often reaching 120 cm, occurs on ledges where drainage-water flows through shallow peat. On steep, often vertical cliffs with narrow ledges on the Ardmeanach are a number of facies which are difficult to separate: *Alchemilla glabra*, *Cystopteris fragilis*, *Deschampsia cespitosa*, *Festuca rubra*, *Oxyria digyna*, *Pinguicula vulgaris*, a small ecotype of *Ranunculus acris*, *Saxifraga aizoides*, *S. oppositifolia*, *Sedum rosea*, *Selaginella selaginoides* and *Thalictrum alpinum*, are common or frequent. A few Mull rarities e.g. *Asplenium viride*, *Dryas octopetala*, are on these or nearby Gribun cliffs. The calcareous cliffs exposed to the south coast are for most of their length relatively dry and do not have mire communities. Constantly irrigated cliffs approximately 7 m high and 60 m long can be seen at 450 m (1500 ft) on Beinn Bheag in the Torosay hills. Few vascular plants are able to established there and only *Cochlearia officinalis*, *Huperzia selago* and *Sedum rosea* were recorded.

Ombrogenous bogs

The definition of this complex of communities on Mull is the same as used by Ratcliffe (1964a). The height of the water-table, its source and nutrient content are the major factors controlling such bogs. A high rainfall, as is seen over most of the western sea-board of Scotland together with the low evaporation rate, contributes to a high water-table in the moorland terrain of impervious rock outcrops and basins where it is easily trapped. Such bog-land rapidly develops in the centre and gives rise to the characteristic convex surface appearance of the 'raised' bog. The other main type of bog, that called blanket bog in which the gently undulating topography is covered with the peat 'blanket-fashion', is not seen extensively on Mull; the rainfall is too low in the areas where it might occur, e.g. on the gently undulating granite and schist lands of the Ross peninsula, and the wetter parts of the islands are too steep. There are, on the other hand, a few examples of typical lowland raised bog and much upland bog on the Laggan, Brolass, the SE Torosay hills, and on the northern half of the island is intermediate in character.

On the Ross peninsula itself, basins within the granite at one time contained peat to a considerable depth. This area, given over to the crofters by the Duke of Argyll when he and other landlords were introducing sheep on a large scale, has been extensively dug for turf. Crofting by the smallholder still continues on the western Ross, and peat-digging for fuel is still carried out. Old diggings lower the water table differentially and both bog and mire associations may be seen adjacent to each other separated only by a peat bench. In other parts of Mull peat-cutting undoubtedly took place but it is burning, to produce sheep-grazing holdings, which has chiefly affected the bog surfaces and composition. Added to this was an early attempt to open up drainage channels, and over the past 30 years, deep-furrow ploughing to cut through the iron pan of the podsol has had a drastic effect in drying out the habitat. The following three types of bog are seen on Mull.

Lowland raised bogs

One lowland raised bog, that in the lowest reaches of the Coladoir River at 15–30 m (50–100 ft), at the head of Loch Scriddain, falls into this category and is worthy of special mention. The valley floor is only slightly uneven and the bog, a *Scirpus–Eriophorum* association sensu Birks (1973), is built up on alluvium. There are three main deep peat areas each bounded by a 'lagg' stream which takes the upland water through to the river. There are other areas of similar deep peat adjacent (see Fig. 10.24) which have since been furrowed and drained. These bog systems have a complex pattern of anastomosing pools and furrows with ridges between covered with *Rhacomitrium lanuginosum*, *Calluna vulgaris*, *Erica tetralix*, *Eriophorum vaginatum*, *E. angustifolium*, *Cladonia* species, and *Sphagnum auriculatum S. papillosum*, etc. Seen from the

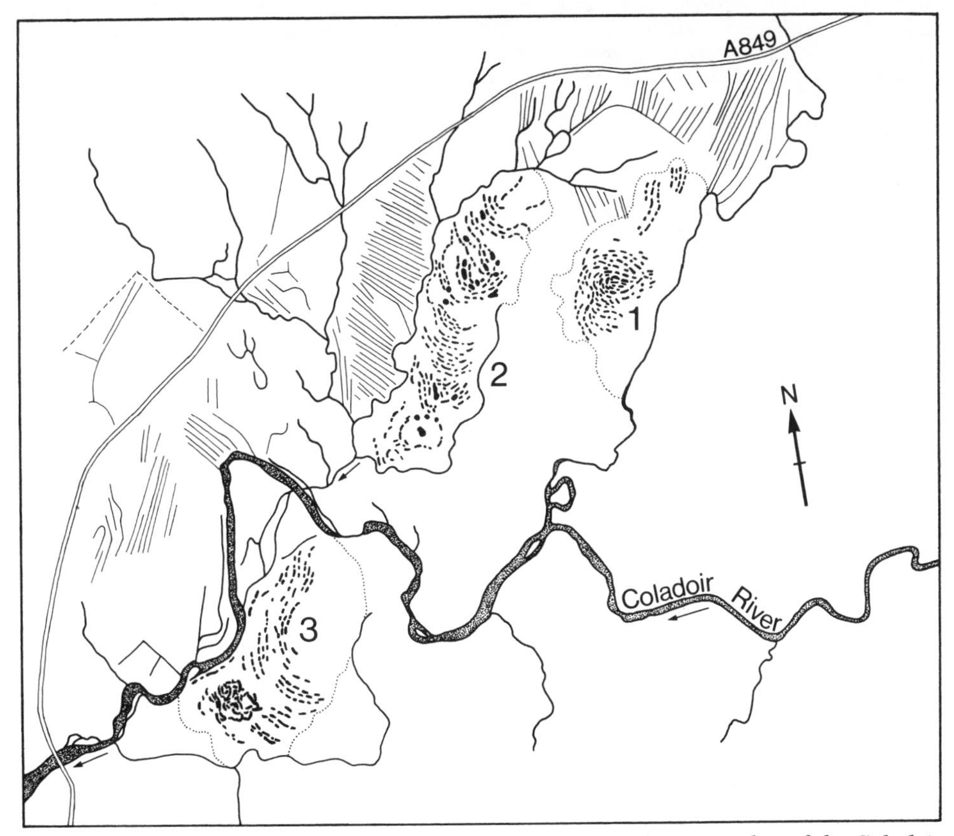

Fig. 10.24 Sketch-map showing extent of raised bog in the lower reaches of the Coladoir River (cf. Fig. 10.25).

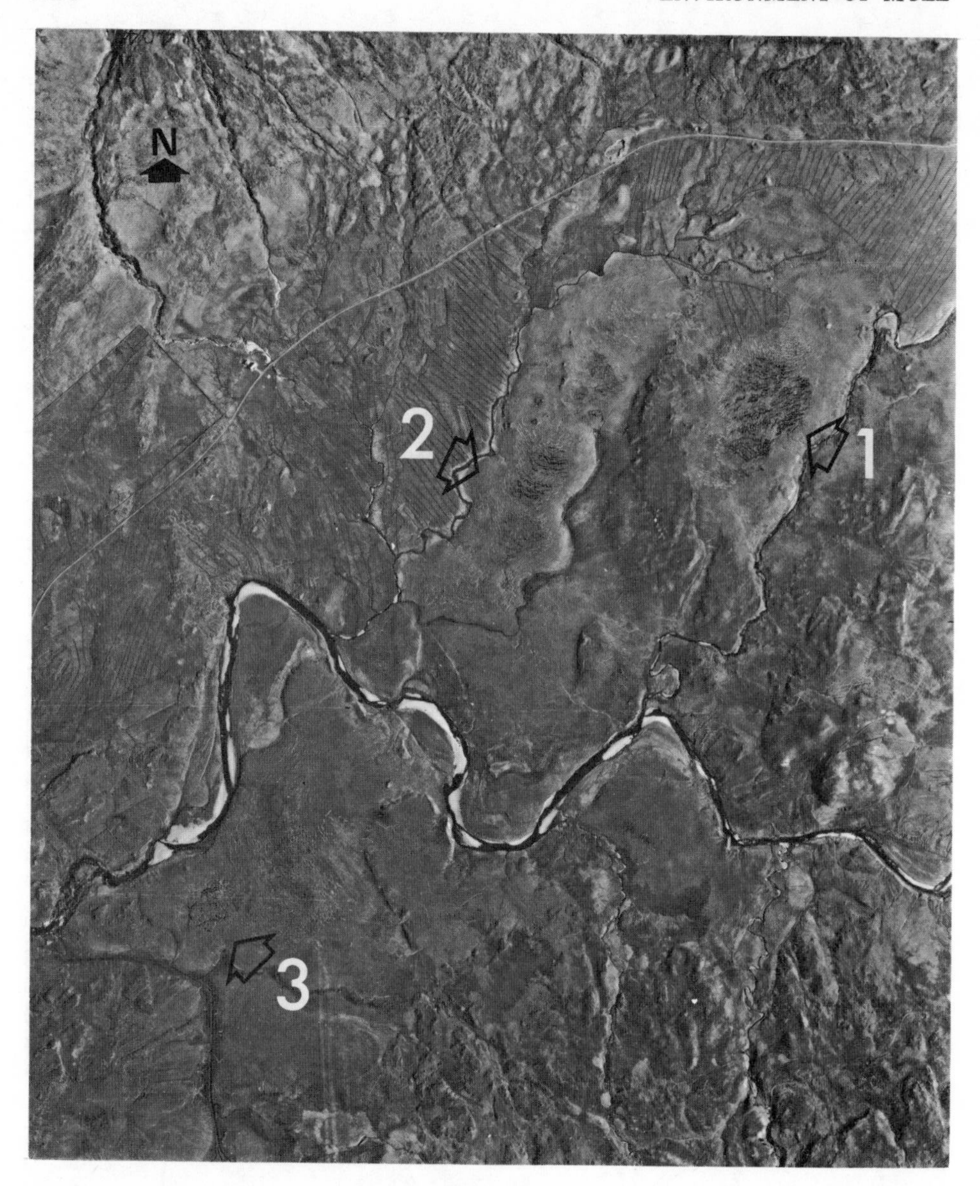

Fig. 10.25 Vertical aerial photograph of the lower reaches of the Coladoir river showing raised bogs (arrowed). See Figure 10.24 for interpretation. 1967. Crown Copyright Reserved.

air (Fig. 10.25) these small bogs may be compared with the more extensive Claish Moss in Sunart (Ratcliffe, 1964a) and the Strathy bogs of East Sutherland described by Pearsall (1956). Unlike the Strathy bogs, however, there is only a very slight slope and, whilst many of the pools are elongated with their long axes orientated parallel to the general surface-contours of the bog (see Fig. 10.26), it is doubtful that any 'pressure folding' of the surface, as postulated by Pearsall, has occurred here. The underlying peats, on a superficial examination, appeared highly humified and almost fluid. Nor is this bog likely to be a comparatively recent flooding of a more compact type as suggested by Ratcliffe & Walker (1958) for the Silver Flowe bogs in Galloway.

Fig. 10.26 Raised bog pools in the lower Coladoir River valley looking towards Beinn nam Feannag. 1967.

The Coladoir River bog has much in common topographically with the bog at Rhiconich, Sutherland described by Boatman & Armstrong (1968).

The pools vary between 3–8 m in their longest dimension and are on the whole shallow (20–50 cm) and well vegetated; the pH is 4·2–4·4 but in the *Sphagnum papillosum* hummock as low as 2·8. Those that apear devoid of vegetation are steep-sided and on closer examination often show a high desmid flora (see Chapter 9). In others, *Sphagnum cuspidatum* and *S. auriculatum* are dominant with *Carex limosa*, *C. panicea*, *C. rostrata* (rarely, and then when the pool is connected to a flow of mineralotrophic water), *Eriophorum angustifolium*, *Rhynchospora alba* and *Utricularia minor* occasional. It is in these pools with phanerogamic plants that the *Sphagnum* is densest. From time to time underground channels form (naturally, by shrinkage?) and a pool may be without water. Such bare peat is colonised by algae especially *Zygogonium ericetorum* and then by *Sphagnum* spp., *Drosera anglica*, *Narthecium ossifragum*, *Rhynchospora alba* and bryophytes like *Campylopus* spp., *Cephalozia* spp., *Calypogeia sphagnicola* and *Riccardia latifrons*. Within these bogs the underlying rock topography or the presence of old drainage ditches will produce areas of flowing water. Such conditions will give rise to mire communities in a 'mosaic' of bog types. Hence *Sphagnum fallax* may fill old ditches or flushed hollows (fen 'windows') and *Phragmites australis*, *Carex rostrata* and *C. lasiocarpa* may become locally dominant.

Fig. 10.27 Raised bog community, Coladoir River, showing Rhacomitrium lanuginosum *hummocks being colonised by* Molinia *and* Calluna. *1967.*

Fig. 10.28 Calluna–Erica tetralix–Scirpus cespitosus *bog showing old peat-cutting face, Allt Linne nan Ribheid. 1967.*

In some of the broader, flatter hollows between ridges and on a quaking substrate a community of *Sphagnum compactum*, *S. cuspidatum* and *S. auriculatum*, with *Drosera anglica*, *Erica tetralix*, *Eriophorum angustifolium*, *Rhynchospora alba*, *Cladonia uncialis* and *Pleurozia purpurea*, is found with considerable bare peat. Similar very wet associations, often with *Drosera intermedia* in abundance, may be seen at the margins of lochs where the water-table is virtually at the peat surface. *Sphagnum fallax* may also occur (with or without *S. palustre* and *S. subsecundum*) with *Carex echinata* in similar slightly flushed hollows in the bogs of the Ross Peninsula, Brolass and the Laggan-Croggan area. *Carex paupercula* is found in its only station in Mull in this association.

The ridges and hummocks between the pools and hollows are predominantly made up of *Rhacomitrium lanuginosum* over a foundation of *Sphagnum* (*S. papillosum* and *S. subsecundum* agg.). Older hummocks frequently erode (by wind action?) and *Cladonia arbuscula*, *C. impexa* and *C. rangiformis* become interwoven with the *Rhacomitrium* (see Fig. 10.27). *Calluna* and *Empetrum nigrum* may be associated on more stable hummocks with *Sphagnum capillaceum*, *S. magellanicum* and *S. papillosum*.

In other parts of Mull this kind of bog no longer exists to any large extent mainly due to the land management: burning for producing sheep and deer moor and, latterly, draining for forestry. An area of bog at c. 75 m (250 ft) on the eastern side of Strath Bearnach just to the west of the loch of the same name may have once been an active raised bog. The area is flat and the shallow pools mostly circular or anastomosing. It has been burnt periodically and the whole area is a carpet of *Sphagnum* (*S. cuspidatum*, *S. magellanicum*, *S. papillosum*, *S. subsecundum* agg.) with isolated hummocks only of *Rhacomitrium lanuginosum*. *Cladonia* spp., e.g. *C. arbuscula*, *C. impexa*, *C. rangiformis*, *C. tenuis* and *C. uncialis* are common. The area is still wet enough for *Carex limosa* although the community is becoming changed to a *Molinia–Calluna* bog with considerable *Eriophorum vaginatum*.

Lowland *Calluna–Erica tetralix–Scirpus cespitosus* bog

This is the common bog type on the deeper peats overlying hollows in the schist and granite of the Ross peninsula. Much of this area has been dug for fuel in the past by crofters giving rise to pool communities of *Sphagnum* in the lower levels (see Fig. 10.28). *Calluna vulgaris* and *Scirpus cespitosus* are co-dominant with *Erica tetralix*, *Eriophorum angustifolium* and *Narthecium ossifragum*. This rapidly grades on the shallower peat-ranker soils, lying over emerging rock outcrops, to a drier facies of *Calluna–Erica cinerea* heath described on p. 10.49. In similar sites on shallow inorganic soils especially where flushing may be prevalent and on the deeper peat-bog where burning is practised, *Molinia caerulea* becomes established and common, possibly through increased phosphate, and further drying out will produce *Molinia* grassland.

Fig. 10.29 Area around the Fontinalis Lochs, Laggan Deer Forest. The paler community is Scirpus cespitosus–Eriophorum–Carex pauciflora *association. 1968.*

Fig. 10.30 Beinn Chreagach, Brolass. Summit bog communities at 370 m containing Carex pauciflora, *with basalt topography of Beinn Chàrsaig in the background. 1968.*

This bog type is also frequent on the flat upland terraces of the typical basalt trap topography.

Scirpus cespitosus–Eriophorum–Carex pauciflora association

This is found around 150 m (500 ft) on very wet, but shallow peat, overlying the more impervious basalt lavas of the Laggan Deer Forest (see Fig. 10.29). It is included here as an ombrogenous bog type but as impeded drainage is implied there may well be some lateral flow of water where the lava bed inclines. It is further found in some quantity on the hills around Beinn Chreagach in Brolass (see Fig. 10.30). *Carex pauciflora* is often co-dominant with *Scirpus cespitosus*, *Calluna* is often abundant but stunted, and *Drosera anglica*, *D. rotundifolia*, *Eriophorum angustifolium*, *Molinia caerulea* (rare) and *Narthecium ossifragum* are associates with the following bryophytes: *Drepanocladus exannulatus*, *Scorpidium scorpioides*, *Sphagnum auriculatum*, *S. papillosum*. There is often considerable water lying on the surface. The pH is about

4·3 but the proximity to the sea leads one to suspect that the ionic content of the water is comparatively high (see Chapter 9).

Grassland and mountain summit heath

The grasslands on Mull are, for the most part, anthropogenic in origin, being derived from the felling and burning of the forest and shrub zones or by the effects of grazing by domestic animals, deer, feral goats and rabbits and hares. They are characteristically on brown forest soils or gley soils derived from colluvial or morainic drift, alluvium or on the more gravelly or sandy soils of raised beaches. On Brolass and the Ross of Mull they are to be found on the till and loam of the drumlins. In the south-east (on the Central Volcanic Complex) they are on the lower hill-slopes and on alluvium and gravels in the major stream-beds. On the steeper slopes they will be found only on the stable scree where this has the interstices filled with brown forest soils. Rarely are any true grasslands found above 600 m (2000 ft), above that height they grade into the grass-heath or summit moss heaths described below.

Over most of the plateau basalt area grasslands are confined to the lower step or terrace up to 180 m (600 ft) where brown forest soils or occasionally flushed gleys form on the loamy colluvial and morainic drift. Some of the coastal areas may never have supported forest or even scrub of any density and the natural climax there may have been a herb-rich sward on the cliff tops which on grazing (initially by wild herbivores and black cattle) has been reduced to grassland.

The soils that now produce grassland are often some of the more fertile. It was those areas that were cultivated by the eighteenth century crofters. They are located in the bottom of the large valleys like Glen Forsa, Glencannel, Glen More and the valley of the River Belart; on the raised beaches on the south of Ardmeanach and on Croggan; and on the lower step of the plateau basalt in Quinish, Mishnish, Mornish and Treshnish. Grasslands in upland Britain have been studied in detail in the Scottish Highlands (McVean & Ratcliffe, 1962; King & Nicholson, 1964); in Skye (Birks, 1973); in Southern Scotland (Birse & Roberton, 1967, 1977; King, 1962); in N. Britain (Eddy *et al.*, 1969; Williams & Varley, 1967) and N. Wales (Edgell, 1969; Ratcliffe, 1959). No detailed account is so far available for Mull but a fuller account will appear with the Soil Survey Monograph (Birse & Robertson, in prep). A broad outline of the grassland types encountered on Mull is given in this Chapter.

Animals affecting grasslands by grazing

Domestic animals During the 'clearances' in the middle of the nineteenth century the crofters were evicted and their small arable holdings reverted to grass, soon to be grazed and trodden by the newly imported Blackface sheep. Some of the areas were taken in by the new lairds as arable, e.g. Glengorm and Quinish. On the Ross of Mull the land was owned by the eighth Duke of Argyll who cared more for his crofting tenants than did many contemporary landlords. The richer areas of soil, mainly drumlins, between the schist or granite outcrops made good pasture and today are still grazed frequently by cattle as well as the ubiquitous sheep.

The present situation with regard to grazing of domestic animals is still a dominant factor in maintaining grasslands in the forest zone. Whilst full-time occupations in agriculture have declined by probably more than 50 per cent in the last 20 years (Highland & Islands Development Board, 1970) cattle and sheep stocks have increased. Stocking densities are often meaningless when quoted in general terms; regional variation of the three parishes are given in Table 10.3. Comparative statistics are also given for the three Highland counties Argyll, Ross & Cromarty and Inverness. The

Table 10.3 Stocking densities in each of the parishes of Mull (in bold typeface) compared with other Highland counties.

	Kilninian	**Torosay**	**Kilfinichan**	Argyll	Ross	Inverness
Total acres per ewe	4·4	6·6	3·8	4·8	9·7	8·2
Total acres per cow	60·0	168·0	59·0	46·0	96·0	86·0

figures reflect the lack of grassland and large estates in Torosay Parish which contains the northern half of the Ben More massif and the volcanic Central Complex In each of the other two there are considerably more cattle; in Kilninian this is due to the large estates like Glengorm, Killichronan and Quinish; in Kilfinichan (with Kilvickeon) due to the 58 small crofts on the Ross.

Deer Mull carries a considerable population of Red Deer (*Cervus elaphus*). A census carried out in March 1969 by the Red Deer Commission showed the total population to be 3,361, made up of 885 stags, 1796 hinds and 680 calves (Highland & Islands Development Board, 1970). Of this 300 are distributed around the northern sector of the Island, i.e. north of the Salen-Gruline road. The bulk of the population is in the mountainous centre and the Torosay hills, the Laggan peninsula, the Ardmeanach and Ross of Mull as far west as Beach. A few animals may forage west of this especially in the winter. A suggestion to extend the deer-proof fence of the Forestry Commission area at Assopol to both the north and south coasts, if it is acted upon, will prevent marauding of the crofts and cut down grazing density on natural grassland. The seasonal movements of the deer are known, many coming off the higher hills and to the south-west in winter. Deer are stalked on all the estates containing herds in the south and although deer forests are not managed in any way to give a future population of strong animals they are a major sporting amenity which may well be scientifically, and thereby economically, managed in the future. In 1968/69 211 stags and 136 hinds were shot.

There is a small herd of Fallow Deer (*Dama dama*) in the natural woodland at Gruline (Highland & Islands Development Board, 1970) but this is of little consequence ecologically.

Hares and rabbits The Brown Hare (*Lepus capensis*) is scarce on Mull. Flux (1962) found the Scottish Mountain Hare (*L. timidus scoticus*) associated with well-managed grouse-moor, a habitat which is almost non-existent in Mull. The Rabbit (*Oryctolagus cuniculus*) on the other hand is common wherever a light soil is extensive enough to allow warren formation. Where windblown sand has encouraged the growth of *Festuca-Agrostis* turf, e.g. on the south coast of the Ross peninsula, it has usually been the rabbit which has prevented the total spread of bracken, as much by their burrows as by grazing. Statistics on the rabbit population are not available for Mull but if five rabbits are considered to be the equivalent of one sheep as Hume (1939) suggests, rabbits must have a considerable effect on maintaining grassland.

Feral goats The population of feral goats is estimated at 1000, 400 of which are on the Laggan Deer Forest (J. Corbett, *pers. comm.*, 1969). Although they compete for good grazing, farmers are content to leave small herds to roam the hills as they can graze off luxurious and otherwise tempting plants on ledges too inaccessible for sheep; the temptation thus removed, less sheep fall to their death on the cliffs.

Agrostis–Festuca association

This is the *Agrostis–Festuca*-species poor grassland of McVean & Ratcliffe (1962) found on freely drained acid brown forest soils with a relatively low base status and is common, up to 350 m (1150 ft) in the wider glaciated valleys of the Central Complex, e.g. Glen Forsa and Glen More. The soils are usually sandy valley moraine, or those formed from weathering of the parent rock, usually a range of the nutrient-richer volcanics. The dominant species are *Festuca ovina* and *Agrostis tenuis* although other grasses like *Anthoxanthum odoratum*, *Agrostis canina*, *Sieglingia decumbens* and to a lesser extent *Nardus stricta* occur, and the herbs *Galium saxatile* and *Potentilla erecta* and the mosses *Hylocomium splendens* and *Rhytidiadelphus squarrosus* are constants.

Herb-rich *Agrostis–Festuca* grassland

When brown forest soils are derived from the basalt the base status is somewhat higher, there is usually present a considerable amount of humus and the soil has a good 'crumb' structure and is freely drained. The number of herbs in such *Agrostis-Festuca* grassland increases and becomes what has been termed by McVean & Ratcliffe (1962) 'species-rich *Agrosto–Festucetum*'. *Agrostis tenuis*, *Anthoxanthum odoratum* and *Festuca ovina* are common or co-dominant; *Carex pilulifera*, *C. caryophyllea* (especially on maritime cliffs), *Plantago lanceolata*, *Potentilla erecta*, *Prunella vulgaris*, *Thymus drucei*, *Viola riviniana* and the bryophytes *Hylocomium splendens*, *Hypnum cupressiforme* and *Pseudoscleropodium purum* are frequent. *Pteridium* often invades this community, seriously decreasing its agricultural value.

There is on Mull a maritime facies of this grassland type on the stabilized scree slope or basalt cliff ledges on the northern coast and on the Ardmeanach (see photograph on title page to Chapter 5). Here *Carex carophyllea*, *C. flacca*, *Koeleria cristata* are commonly associated with the other grasses, and *Galium verum*, *Linum catharticum*, *Lotus corniculatus*, *Plantago coronopus* and *P. media* are frequent.

Deschampsia–Festuca–Nardus association

These grasslands contain *Nardus stricta* as a dominant or co-dominant with *Festuca ovina* and *Deschampsia flexuosa* on bleached podsolic or peaty gley soils. *Agrostis canina*, *A. tenuis*, *Anthoxanthum odoratum* may also be present. *Carex binervis*, *C. pilulifera*, *Galium saxatile*, *Potentilla erecta* and the bryophytes *Hylocomium splendens*, *Hypnum cupressiforme*, *Pleurozium schreberi* and *Rhytidiadelphus squarrosus* are associates.

On the more gentle slopes and lower ridges (up to 500 m) residual blanket peat, mixed with the brown lithosols or forming very acid peaty gleys, will be indicated by a high frequency of *Juncus squarrosus*. At the bottom of the steep-sided hills of the Central Complex, drainage from the upper 300 m (1000 ft) will produce partial mire conditions in the *Nardus* grassland and *Molinia caerulea* will show itself often as a co-dominant; it differs from other *Molinia* grassland in having a higher proportion of *Festuca*, *Deschampsia* and, to a lesser extent, *Nardus*. A shallow peat layer is usually formed, the flushing becoming apparent in the gley formation seen in the A_2 layer.

Luzula sylvatica–grassland association

McVean and Ratcliffe (1962) describe what they term a *Luzula sylvaticum*-grassland nodum with *Agrostis tenuis*, *Deschampsia flexuosa*, *Luzula sylvatica*, *Rumex acetosa*, *Rhytidiadelphus loreus* and *R. squarrosus*, as constants. Such a community is found on steep slopes between 90–210 m (3–700 ft) on the north-west and west sides of the Laggan Deer Forest and in some of the narrow valleys running off Brolass to the south e.g. Glen Byre. In a number of other similarly exposed slopes *Vaccinium myrtillus* is a co-dominant with the *Luzula*. There is a reasonably high grazing and manuring influence on these slopes from sheep, goats and deer. Fragmentary stands are also seen at higher altitudes, e.g. 650–750 m on the north slopes of Ben More.

High-nitrogen grasslands

Grassland communities depending on heavy deposition of dung and urine are frequent in the sheep country of the Southern Uplands (King & Nicholson, 1964). The characteristic species are *Festuca ovina, F. rubra, Agrostis tenuis* and *Poa pratensis* with *Holcus lanatus, Anthoxanthum odoratum, Rumex acetosella, Rhytidiadelphus squarrosus.* This type of grassland may result on the windblown sand areas on the south coast of the Ross peninsula where a high density of sheep aggregate. Such grassland is soon colonised by *Pteridium* and it is in such a community that the only station of *Ophioglossum vulgatum* is found on Mull.

A grassland of similar composition with *Poa subcaerulea, Rumex acetosa, Cerastium diffusum* and *C. semidecandrum* is found on the cliff-tops of the Treshnish Isles especially in hollows on the lee-side where sheep congregate. There is also the added nitrogen from bird guano which gives rise to characteristic guano-communities. Apart from the herbs already mentioned and *Stellaria media,* the grasses appear to withstand and benefit from the high nitrogen, and *Poa annua* and *P. subcaerula* are locally dominant. The latter species (as *P. pratensis*) was reported as rare in the Pembrokeshire islands in similar situations (Gillham, 1956) where *Glechoma hederacea* was a common herb. This species was not recorded from this habitat in the Treshnish Isles or Mull. Gillham (1956) suggests *Festuca* species are most resistant to manuring and both *F. ovina* and *F. rubra* are common in such habitats on the Treshnish Isles. Similar cliff-top grassland was recorded for Hirta, St Kilda, by McVean (1961).

Another association which forms a dense community on islands no longer grazed is one in which *Arrhenatherum elatius* is dominant and *Dactylis glomerata, Poa trivialis* and *Holcus lanatus* are frequent associates. It is usually associated with shelter and derelict human dwellings or sheep pens. It is obviously on high nitrogenous soils and the nitrophilous herb *Urtica dioica,* and *Anthriscus sylvestris,* form dense patches. When such grassland is at sea level at the head of a bay it may well reach down to the strand from which other species will invade. Examples of these are *Angelica sylvestris, Ligusticum scoticum,* often robust and over 150 cm high and replacing *Smyrnium olusatrum* in similar niches on Puffin Island, Anglesey (Gillham, 1956), *Tripleurospermum maritimum* and *Rumex crispus.*

On Bac Mòr, now ungrazed, on a level plateau some 15 m (50 ft) above sea-level, there is an *Agrostis stolenifera–A. tenuis–Anthoxanthum odoratum–Festuca* spp. grassland in which *Carex flacca* is a common associate. This community was not seen elsewhere.

Bracken communities

Bracken (*Pteridium aquilinum*) is developed to its maximum on brown forest soils 25 cm or deeper. It is found colonising the deeper pockets of soil in the partial shade of the birch-oak woodland, and is therefore often a relict when such woodland is cut and grazing maintains a grassland community. The upper limit of *Pteridium,* here 300 m (1000 ft), often indicates the upper limit of primeval forest (Pearsall, 1950) or corresponds to the potential forest limits (Birks, 1973). Other edaphic influences besides depth and type of soil are also important; soil aeration resulting from free drainage being one recently discussed (Poel, 1961). From its original woodland habitat or relict sites *Pteridium* quickly colonises *Argostis-Festuca* grasslands. The main reason for increase of bracken in the last 100 years in Mull is due to change of land-use patterns. In the eighteenth and early nineteenth centuries the small black cattle and, to a lesser extent the horse, were the grazing animals of both lowland and, in summer, of hill pasture. Besides eating very young bracken, cows will damage emerging shoots by trampling, and horses will likewise break shoots with their hooves. Even as late as 1890 bracken was so scarce on the Lochbuie Estate that the one stand near the farm was protected for winter bedding for the cows (J. Corbett, *pers. comm.*).

Pteridium will invade the first two types of grassland discussed above. *Festuca*, then *Agrostis* species will disappear as frond density increases; *Anthoxanthum odoratum* will persist for some time together with *Holcus mollis*, which may even increase in such communities (King & Nicholson, 1964).

Molinia grassland

On acid peaty soils which are wet but not permanently waterlogged *Molinia caerulea* may increase almost to the exclusion of everything else. Such grassland is small in extent along burn sides and on flat alluvial terraces. Often however it is associated with a flushing and it becomes associated with *Myrica* described in the section on mires above. In a few extensive basins, e.g. in Gleann Airigh na Searsain in Brolass, there are extensive and uniform stands of tussocky *Molinia*. Tussocks, often 75 cm high, have a deep litter which prevents the establishment of other plant species and are

*Fig. 10.31 Looking towards Glen Forsa down Abhainn an t-Sratha Bhàin. Basic cone sheets of igneous complex forming steeper slopes with morainic drift in valley bottom, with grassland–*Calluna *mosaic. 1966.*

frequented by adders. Towards the edge of the basin it grades into a *Molinia-Callunetum.*

High-level *Nardus* grassland (grass heath)

McVean & Ratcliffe (1962) and McVean (1964) discuss a number of high-level *Nardus stricta* communities above 600 m (1800 ft) which they suggest are affected more by snow-cover than by pressure of grazing. Snow-cover in Mull (see Chapter 6) is neither extensive nor prolonged but where corries are formed, notably on the north face of Ben More and on the east of A' Chioch, on An Cruachan and Beinn Talaidh, a chionophilous *Nardus* community rich in bryophytes is found; *Hylocomium splendens, Pleurozium schreberi, Rhytidiadelphus loreus* and *R. squarrosus* are characteristic.

On the more gentle slopes of the ridges and the shoulders of the summits of the central and south-east mountains on a loamy colluvial, *Nardus* is present only in small

Fig. 10.32 Mainnir nam Fiadh from Beinn Bheag. Sparse vegetation on the drystone screes of breccias and other rocks formed in the vents of the old volcano. Harder rocks of dolerite and gabbro stand out with lichens relieving the similar sparse vegetation. 1966.

quantities, in places with *Vaccinium myrtillus* and *Empetrum nigrum/hermaphroditicum* but in an open community and often in a mosaic of heath dominated with mosses (see below and Fig. 10.31). *Carex bigelowii*, *Galium saxatile*, *Diphasiastrum alpinum*, *Huperzia selago*, *Potentilla erecta*, *Salix herbacea*, *Rhacomitrium lanuginosum*, *Rhytidiadelphus loreus*, *Cetraria islandica* and *Cladonia uncialis* are frequent associates as this community grades into the true summit heath.

Sometimes where gullies are seen running east-west *Nardus* may be found on the north-facing side with *Juncus squarrosus* on the opposite side suggesting either differential effect of insolation or, more likely, the effect of snow-lie. On the steeper slopes in the Torosay hills dry stone screes are devoid of vegetation with patches of *Nardus* and *Rhacomitrium* on the more stable areas (see Fig. 10.32).

Fig. 10.33 Ben More at 600 m. North-facing block scree of mugearite showing lichens (note Rhizocarpon geographicum *top right) and the moss* Andreaea *on rocks, with* Alchemilla alpina–Carex bigelowii *turf between larger boulders. 1967.*

Summit moss heath

The summit plateau associations are those dealt with here. When grass species are dominant they are described as grass heaths (see above) but for the most part grasses and herbs are scattered and cushion-forming mosses common (moss heaths); in other places, pleurocarpic mosses such as *Rhacomitrium lanuginosum* cover frost-shattered rock debris. Where large block screes form, *Andreaea* spp. and many lichens including *Rhizocarpon geographicum* are conspicuous with *Alchemilla alpina* co-dominant in the *Carex bigelowii* turf between the rocks (see Fig. 10.33).

The commonest association is one that corresponds to the *Cariceto-Rhacomitrietum* of McVean & Ratcliffe (1962). It is found on the flat or barely sloping rounded summit of Ben More, Cruachan Dearg and Beinn a' Ghraig and in the south-east, on Beinn Bhearnach (Fig. 10.34), Mainnir nam Fiadh and north to Beinn Mheadhan. The six constant species are *Carex bigelowii, Cladonia uncialis, Deschampsia flexuosa, Galium saxatile, Rhacomitrium lanuginosum* and *Vaccinium myrtillus. Carex bigelowii* rarely flowers in this situation unless in the shelter of a rock. Apart from on Ben More this community is never extensive and exposure to frost and wind often bares

Fig. 10.34 Beinn Bhearnach, north-western slopes at 600 m. Cliffs of gneiss arising out of summit plateau of cone sheets covered with Rhacomitrium *heath. 1968.*

considerable areas leaving islands of mosses with occasional *Carex bigelowii*, *Festuca vivipara*, *Huperzia selago*, *Luzula spicata* and *Saussurea alpina*.

Meadowland and reverted pasture

Characteristic meadowland grazed communities on damp, fertile, basic or slightly acid soils are situated throughout the Island in lowland sites. They are more common on the basalt areas in the north, from Mishnish to Mornish, on the north shore of Loch Tuath and Loch na Keal and to a lesser extent on the Sound of Mull coast. In the south-east some such meadows are to be found on the brown forest soils of the raised beaches and on the peninsula between Duart and Loch Don. Isolated meadows may be found on the Gribun associated with the basic flushed soils there. On the Ross of Mull similar meadowland is associated with croft holdings on the south shore of Loch Scridain, increasing in the Ardtun and Kintra areas. The association is dominantly of the grass species *Anthoxanthum odoratum*, *Cynosurus cristatus*, *Festuca rubra*, *Holcus lanatus* and *Poa pratensis* and the following herbs are common or frequent: *Centaurea nigra*, *Dactylorrhiza fuchsii*, *D. purpurella*, *Leucanthemum vulgare*, *Plantago lanceolata*, *Ranunculus repens*, *Rhinanthus minor*, *Senecio jacobaea* and *Trifolium repens*.

The land-use of this community is two-fold: many fields are mown for hay in the summer and lightly grazed by sheep and cattle in the winter. On some larger estates, e.g. Glengorm, some fertilising and reseeding with grasses such as *Alopecurus pratensis*, *Dactylis glomerata*, *Lolium perenne* and *Phleum pratense* has increased the yield and agricultural value. Other meadows are grazed and then only lightly; they are usually found on a mosaic of soils with pockets of peaty gleys or more clayey soils where there is impeded drainage. Here *Juncus effusus* or *J. conglomeratus* with *Ranunculus acris*, *Cirsium palustre* and *Lychnis flos-cuculi* may form extensive areas. Such communities are transitional to a semi-natural herb-rich meadowland dominated by *Juncus acutiflorus*. They are characteristically on more silty peats which are always moist and seasonally flooded but on the whole partially aerated, not to be confused with the *Juncus–Acrocladium* mires of more upland sites, and occur occasionally around the outlet stream of lowland lakes e.g. Loch a' Ghleannain. Frequent associates are *Achillea ptarmica*, *Equisetum palustre*, *Filipendula ulmaria*, *Hypericum tetrapterum*, *Lotus uliginosus*, *Lychnis flos-cuculi*, *Holcus lanatus*, *Ranunculus repens*, *Valeriana officinalis*, *Angelica sylvestris*, *Cirsium heterophyllum*, *C. palustre* and *Galium boreale* are frequent. In more marginal areas which become frequently flooded *Carex nigra*, *Crepis paludosa*, *Equisetum fluviatile* and *Senecio aquaticus* may indicate a transition to mire communities. This community is often derived from fen associations of loch margins by drainage.

Arable fields abandoned for some years may revert to a mixed herbaceous pasture with *Agrostis* species and considerable *Holcus lanatus* and *Anthoxanthum odoratum*; *Rhinanthus minor*, *Leucanthemum vulgare*, *Achillea ptarmica*, *Trifolium repens* and *Plantago lanceolata* are frequent. In the occasional fields where this has occurred and has subsequently been left ungrazed, *Calluna* has colonised to form a dwarf-shrub community or on more basic soils a *Salix-Betula* scrub may form.

Reversion of anthropogenic communities when biotic pressures are released was not studied in detail and could be an interesting project for future work.

Anthropogenic communities

Habitats which have been created by man are few in Mull although the history of land-use is such that most of the natural vegetation of the Island has been affected in one way or another. The habitats discussed here are those which have been formed as the direct result of man's technology, e.g. road and bridge building.

Road side embankments

During the past 20 years one route, the A849 between Fionnphort and Pennyghael and in particular that road originally classified as B8035 and now upgraded to be the continuance of the A849, has been engineered to make a good road through to the ferry at Craignure. In doing so cuttings and embankments have been made on the new section of the road best seen in the lower part of Glen More. The road down Glen More passes through uninhabited and uncultivated country and ruderals are not nearby to colonise open habitats. It would be an interesting study to document the development of the ruderal associations here. The species that colonise the bare evenly graded sloped shoulders vary with the substrate. Where peat is involved and especially where there is a run-off from peat-land adjacent to the road, moorland species such as *Calluna vulgaris*, *Deschampsia flexuosa*, *Erica* spp., *Juncus bufonius* and *Potentilla erecta* commonly colonise together with *Ceratodon purpureus*, *Funaria obtusa*, *F. hygrometrica*, *Nardia scalaris*, *Oligotrichum hercynicum*, *Pohlia annotina*, *Polytrichum aloides* and *Solenostoma crenulatum*. Where glacial soils or shattered igneous rock debris form a shoulder, annuals such as *Aira praecox*, *Cardamine* spp., *Sagina* spp., and perennials (or biennials) *Capsella bursa-pastoris*, *Ranunculus repens*, *Rumex acetosa*, *R. acetosella*, *Sonchus asper*, *Trifolium repens* and *Tussilago farfara* are quick to colonise.

Mortared walls and roofing materials

Walls of bridges, barns, outhouses, dwellings and those beside gardens, although made with local rock, contain mortar in which a number of lime-loving (calciphilous sensu Ferreira, 1963) species become established. The ferns *Asplenium adiantum-nigrum*, *A. ruta-muraria*, *A. trichomanes* subsp. *quadrivalens*, are frequent, with more rarely *A. scolopendrium*, *Cystopteris fragilis* and *Polystichum aculeatum*. *Asplenium ceterach*, a common wall plant in SW England, is recorded for this habitat although no longer extant probably through weeding. A number of phanerogams may also be characteristic of walls: *Aira praecox*, *Cymbalaria muralis*, *Dactylis glomerata*, *Festuca ovina* agg., *Geranium robertianum*, *Taraxacum unguilobum* and *Veronica* spp. and the mosses *Anomodon viticulosus*, *Barbula revoluta* and *Ctenidium molluscum*. The following lichens were recorded from mortar or cement walls: *Acrocordia salweyi*, *Caloplaca heppiana*, *Lecidella stigmatea*, *Lempholemma chalazanum* and *L. chalazanellum* (both on mortar exclusively), *Leproplaca chrysodeta*, *Polyblastia tristicula*, *Protoblastenia monticola*, *Toninia aromatica*, *Verrucaria glaucina*, *V. hochstetteri*, *V. muralis*, *V. nigrescens* and *V. viridula*. *Xanthoria elegans*, *Lecania erysibe* and *Lecanora muralis* were frequent on asbestos roofing on Iona. Slate roofs, a common material for domestic houses formed a habitat for *Acarospora fuscata*, *A. smaragdula* and *Lecidella scabra*. A few flowering plants, e.g. *Centranthus ruber* and *Erinus alpinus*, are confined to walls.

Churchyards and cemeteries

Like elsewhere in Britain gravestones and monumental slabs form possible surfaces for lichen colonisation and may introduce into an area stone of a different chemical or physical composition from that of the country rock. There are few introductions in the churchyards on Mull where granite and other igneous rocks are the general gravestone materials. The lichen flora is not distinct, but tombstones allow the spread of a species where otherwise lack of suitable habitat would prevent it.

Pavements, jetties and pathways

A number of ruderals and colonists are found in particular in cracks in tarmac between stone slabs in pavements or jetties; *Taraxacum cordatum* and *T. duplidentifrons* are two species which have been recorded only in this habitat. Near the coast, e.g. on old jetties, *Plantago coronopus* and *P. maritima* become a dominant component between stone pavement blocks. Regular pathways over grasslands or rough pasture, moorland or edges of bogs and mires will produce a number of species characteristic of tramped areas; *Plantago lanceolata* and *P. major* quickly get established with *Poa annua*, Festuca *ovina* and *Agrostis* spp. On base-rich clay paths *Barbula cylindrica*, *Calypogeia fissa* and *C. muellerana* are frequent.

Arable fields and gardens

The following species have been recorded only on waste ground and in arable fields or gardens as ruderals: *Aethusa cynapium*, *Agrostemma githago* (now extinct), *Anagallis arvensis*, *Brassica napus*, *B. rapa*, *Capsella bursa-pastoris*, *Cardamine hirsuta*, *Chenopodium album*, *Chrysanthemum segetum*, *Cichorium intybus*, *Conium maculatum*, *Coronopus didymus*, *Euphorbia helioscopia*, *E. peplus*, *Fumaria bastardii*, *F. muralis* subsp. *boraei*, *F. officinalis*, *Galeopsis speciosa*, *G. tetrahit*, *Lamium amplexicaule*, *L. molucellifolium*, *Meconopsis cambrica*, *Papaver dubium*, *P. rhoeas*, *Polygonum convolvulus*, *Raphanus raphanistrum*, *Senecio vulgaris*, *Sinapis arvensis*, *Sisymbrium officinale*, *Tanacetum vulgare*, *Trifolium hybridum*, *Veronica agrestis* and *V. filiformis*. When old hay meadows revert to an undisturbed closed community *Rhinanthus minor* often becomes an abundant component. *Leucanthemum vulgare* similarly becomes established in such situations.

Old gardens around derelict habitation

Old crofts and derelict cottages soon revert to the acid grassland or dwarf shrub communities characteristic of the site and soil. Certain species are found in these sites which would not have been there if man had not originally tilled a patch of soil. Such derelict crofts may be found on the Ross of Mull and on Ulva in particular. On the latter island the lazy-beds topography can be still seen and the whole settlement of Ormaig with over 50 houses gives a complex patterning to the semi-natural vegetation now present, giving rise to variation in drainage and subsequent water-tables. Around the crofts, ruderals like *Arctium minus*, *Bellis perennis*, *Capsella bursa-pastoris*, *Matricaria matricarioides*, *Potentilla anserina*, *Rumex obtusifolius*, *Sagina* spp., *Sambucus nigra* and *Tussilago farfara* are to be found. A number of garden decorative plants e.g. *Aconitum napellus*, *Aquilegia vulgaris* and *Cicerbita macrophylla*, and herbs e.g. *Mentha spicata*, linger in derelict gardens for a considerable time. The record for *Atropa bella-donna* is possibly a remnant of earlier cultivation.

Fence posts

Fence posts, especially when long-standing, give the same habitat as decorticated trees. They are often found in areas where trees are scarce, e.g. on saltings or pasture near the sea, and in bog communities. The flora associated is almost entirely that of lichens although *Dicranoweisia cirrata* is a characteristic moss. On oak posts *Cladonia* spp., *Lecanora expallens*, *Lecidea symmicta*, *Parmelia* spp., *Ramalina* spp. and *Usnea* spp. are frequently found. On conifers a characteristic list would contain *Biatorella moriformis*, *Calicium* spp., *Lecanora chlarona*, *L. piniperda*, *Lecidea granulosa*, *L. uliginosa*, *Micarea chrysophthalma*, *M. melaena* (the latter on rotting wood especially) and *Xylographa vitilago*.

References

Barkman, J. J. (1958). *Phytosociology and ecology of cryptogamic epiphytes*. Assen.

Birks, H. J. B. (1973). *Past and present vegetation of the Isle of Skye: a palaeoecological study*. Cambridge.

Birse, E. L. & Robertson, J. S. (1967). The vegetation. Chapter 5 and Appendix VI in Ragg, J. M. & Futty, D. W., *The Soils of the country round Haddington and Eyemouth*. Edinburgh.

———. (1977). *Plant communities and soils of the lowland and southern upalnd regions of Scotland*. (*Monogr. Soil Surv. Scot.*) Aberdeen.

———. (in prep.) The vegetation. In Bibby, J. S., *The soils of the Island of Mull*. (*Mem. Soil Surv. Scot.*)

Boatman, D. J. & Armstrong, D. (1968). A bog-type in north-west Sutherland. *J. Ecol.*, **56**: 129–141.

Boswell, J. (1786). *The journal of a tour to the Hebrides with Samuel Johnson, Ll.D*. ed. 3. London.

Braun-Blanquet, J. (1932). *Plant sociology* (English translation by Fuller & Conard). New York & London.

Burges, N. A. (1951). The ecology of the Cairngorms III. The *Empetrum-Vaccinium* zone. *J. Ecol.*, **39**: 271–284.

Burnett, J. H. (ed.) (1964). *The vegetation of Scotland*. Edinburgh.

Chapman, V. J. (1934). The ecology of Scolt Head Island. In J. A. Steers (ed.) *Scolt Head, Island*. Cambridge.

———. (1964). *Coastal vegetation*. Oxford & New York.

Crampton, C. B. (1911). *The vegetation of Caithness considered in relation to the geology*. Cambridge.

Delany, M. J. (1953). Studies on the micro-climate of *Calluna* heathland. *J. Anim. Ecol.*, **22**: 227–239.

Eddy, A., Welch, D. & Rawes, M. (1969). The vegetation of the Moor House National Nature Reserve in the Northern Pennines, England. *Vegetatio*, **16**: 239–284.

Edgell, M. C. R. (1969). Vegetation of an upland ecosystem: Cader Idris, Merionethshire. *J. Ecol.*, **57**: 335–359.

Elton, C. S. (1966). *The pattern of animal communities*. London & New York.

Ferreira, R. E. C. (1963). Some distinctions between calciphilous and basiphilous plants. *Trans. Proc. bot. Soc. Edinb.*, **89**: 399–413.

Fletcher, A. (1973a). The ecology of marine (littoral) lichens on some rocky shores of Anglesey. *Lichenologist*, **5**: 368–400.

———. (1973b). The ecology of maritime (supralittoral) lichens on some rocky shores of Anglesey. *Lichenologist*, **5**: 401–422.

Flux, J. F. C. (1962). *The ecology of the Scottish mountain hare*, Lepus timidus scoticus *Hilzheimer*. Ph.D. thesis. Univ. Aberdeen.

Gillham, M. E. (1956). Ecology of the Pembrokeshire islands V. Manuring by the colonial seabirds and mammals. *J. Ecol.*, **44**: 429–454.

———. (1957). Coastal vegetation of Mull and Iona in relation to salinity and soil reaction. *J. Ecol.*, **45**: 757–778.

Gimingham, C. H. (1964). Dwarf-shrub heaths. Pp. 232–287 in Burnett, J. H. (ed.), *The Vegetation of Scotland*. Edinburgh & London.

Highlands and Islands Development Board. (1970). *The future of forestry on Mull*. MS ined. Inverness.

Hume, C. W. (1939). The rabbit menace. *Emp. J. exp. Agric.*, **7**: 132–138.

Ingram, M. (1958). The ecology of the Cairngorms IV. The *Juncus* zone: *Juncus trifidus* communities. *J. Ecol.*, **46**: 707–737.

Kershaw, K. A. (1964). Preliminary observations on the distribution and ecology of epiphytic lichens in Wales. *Lichenologist*, **2**: 263–276.

King, J. (1962). The *Festuca-Agrostis* grassland complex in S. E. Scotland. *J. Ecol.*, **50**: 321–355.

——— & Nicholson, I. A. (1964). Grasslands of the forest and sub-alpine zones. Pp. 168–231 in Burnett, J. H. (ed.), *The vegetation of Scotland*. Edinburgh & London.

KNOWLES, M. C. (1913). The maritime and marine lichens of Howth. *Scient. Proc. R. Dublin Soc.*, n.s. 14: 79–143.

LEWIS, J. R. (1964). *The ecology of rocky shores.* London.

MACNAB, P. A. (1970). *The Isle of Mull.* Newton Abbott.

McVEAN, D. N. (1956). The ecology of *Alnus glutinosa* (L.) Gaertn. V. Notes on some British alder populations. *J. Ecol.*, **44**: 321–330.

———. (1961). Flora and vegetation of the islands of St Kilda and North Rona in 1958. *J. Ecol.*, **49**: 39–54.

———. (1964). Grass heaths. Pp. 499–513 in Burnett, J. H. (ed.), *The vegetation of Scotland.* Edinburgh & London.

——— & RATCLIFFE, D. A. (1962). *Plant communities of the Scottish Highlands. Monogr. Nat. Cons. 1.* London.

METCALFE, G. (1950). The ecology of the Cairngorms II. The mountain Callunetum. *J. Ecol.*, **38**: 46–74.

PATTON, D. (1924). The vegetation of Beinn Laoigh. *Rep. bot. Soc. Exch. Club Br. Isl.* 268–319.

PEARSALL, W. H. (1918). On the classification of aquatic plant communities. *J. Ecol.*, **6**: 75–84.

———. (1950). *Mountains and moorlands.* London.

———. (1956). Two blanket-bogs in Sutherland. *J. Ecol.*, **44**: 493–516.

POEL, L. W. (1961). Soil aeration as a limiting factor in the growth of *Pteridium aquilinum. J. Ecol.*, **49**: 107–111.

POORE, M. E. D. (1955a). The use of phytosociological methods in ecological investigations I. The Braun-Blanquet system. *J. Ecol.*, **43**: 226–244.

———. (1955b). Ibid. II. Practical issues involved in an attempt to apply the Braun-Blanquet system. *Ibid.*, **43**: 245–269.

———. (1955c). Ibid. III. Practical application. *Ibid.*, **43**: 606–651.

———. (1956). Ibid. IV. General discussion of phytosociological problems. *Ibid.*, **44**: 28–50.

——— & McVEAN, D. N. (1957). A new approach to Scottish mountain vegetation. *J. Ecol.*, **45**: 401–439.

RATCLIFFE, D. A. (1959). The vegetation of the Carneddau, North Wales I. Grasslands, heaths and bogs. *J. Ecol.*, **47**: 371–414.

———. (1964a). Mires and bogs. Pp. 426–478 in Burnett, J. H. (ed.), *The vegetation of Scotland.* Edinburgh & London.

———. (1964b). Montaine mires and bogs. Pp. 536–558 in Burnett, J. H. (ed.), *The vegetation of Scotland.* Edinburgh & London.

———. (1968). An ecological account of the Atlantic bryophytes in the British Isles. *New Phytol.*, **67**: 365–439.

——— & WALKER, D. (1958). The Silver Flowe, Galloway, Scotland. *J. Ecol.*, **46**: 407–445.

RITCHIE, W. (1974). Spatial variation of shell content between and within "machair" systems Pp. 9–12 in Ranwell, D. S. (ed.), *Sand dune machair.* (*Coastal Ecology Research Station Symposium.*) Norwich.

SALISBURY, E. J. (1942). *The reproductive capacity of plants.* London.

SCOTT, G. A. M. (1963). *Mertensia maritima* (L.) S. F. Gray (Biol. Fl. Brit. Isles). *J. Ecol.*, **51**: 733–742.

SHIMWELL, D. W. (1971). *Description and classification of vegetation.* London.

SMITH, R. (1898). Plant associations of the Tay basin. *Proc. Perthsh. Soc. Nat. Sci.*, **2**: 200–217.

———. (1899). On the study of plant associations. *Nat. Sci. Lond.*, **14**: 109–120.

———. (1900a). Botanical Survey of Scotland I. Edinburgh District. *Scot. geogr. Mag.*, **16**: 385–416.

———. (1900b). Ibid. II. North Perthshire. *Ibid.*, **16**: 441–467.

———. (1911). Arctic-alpine vegetation. Chapter XIII in Tansley, A. G., *Types of British vegetation.* Cambridge.

SPARLING, J. H. (1967). The occurrence of *Schoenus nigricans* L. in blanket bogs. *J. Ecol.*, **55**: 1–13, 15–31.

TANSLEY, A. G. (1911). *Types of British vegetation.* Cambridge.

———. (1949). *The British Islands and their vegetation.* Cambridge.

WATT, A. S. & JONES, E. W. (1948). The ecology of the Cairngorms I. The environment and the altitudinal zonation of the vegetation. *J. Ecol.*, **36**: 283–304.

WILLIAMS, J. T. & VARLEY, Y. W. (1967). Phytosociological studies of some British grasslands I. Upland pastures in Northern England. *Vegetatio*, **15**: 169–189.

YAPP, R. H. & JOHNS, D. (1917). The salt marshes. Part II in Yapp, R. H., Johns, D. & Jones, O. T., The salt marshes of the Dovey estuary. *J. Ecol.*, **5**: 65–103.

——— & JONES, O. T. (1916). The salt marshes of the Dovey Estuary Part I. Introduction. *J. Ecol.*, **4**: 27–42.

Part three
The flora

11. Flowering plants and conifers

E. B. Bangerter and J. F. M. Cannon

Introduction

Collection of data

Field-recording was planned and records were amassed on a 2-km square basis. As all except critical material could be identified in the field the standard Biological Records Centre field-scoring card was used. The accumulated data, amounting in the end to over 50 000 individual species entries were listed on sheets in a series of looseleaf binders and plotted on base maps. The Survey aimed to provide an even cover of the area under review but no attempt was made to sample equally every tetrad as has been so successfully accomplished in some other, especially lowland, counties. To have done so would have prolonged the work and produced results that would have not justified the expenditure of time and resources. Areas studied were chosen subjectively for their likely habitat potential, less attention being paid to some of the wide areas with comparatively uniform conditions.

For publication purposes, it was decided that a unit system of modified 10-km squares would allow the best deployment of the data at our disposal (see end-papers) and each entry contains a list of those unit numbers where the species has been recorded. To have used the National Grid 10-km square, whilst directly compatible with the records held at the Biological Records Centre, would have necessitated the use of four squares for one natural area in some instances, e.g., the island of Ulva, whilst on the other hand one square might include two separate areas as in square 17/42 which spans Loch Scriddain and includes parts of Brolass in the south and the Ardmeanach in the north.

In addition to the Museum's own fieldwork and the contributions from outside the Museum, the following sources of records have been used. All published records in literature known to us have been extracted, and all recording cards relevant to our area have been made available to us through the courtesy of Dr F. H. Perring of the Biological Records Centre. The following herbaria have been fully searched for Mull specimens: British Museum (Natural History) (BM); Department of Botany, University of Glasgow (GL); Royal Botanic Garden, Edinburgh (E); the National Museum of Wales, Cardiff (NMW); and the H. C. Watson Herbarium at the Royal Botanic Gardens, Kew (K-Wa). In association with our study of the last herbarium, a complete search was made of Watson's manuscript slips which are preserved in the library of the Department of Botany, British Museum (Natural History) and which were the basis of the records in his *Cybele Britannica* (1847–1859). In some cases the slips provide fully localised data for the records published in this pioneer work on the distribution of British plants.

Arrangement and nomenclature

The arrangement follows Dandy (1958) with the interpolation of introduced aliens and critical species as necessary and appropriate. The nomenclature also follows Dandy (1958), with the corrections provided by Dandy (1969), although in a few cases the systematic treatments of *Flora Europaea* have been preferred. Where any doubts might have arisen, a basic synonymy has been given to explain the choice of name. The English vernacular names used throughout are those recommended by Dony, Perring and Rob (1974); Gaelic plant names are not given. The first known record is given below the specific name. For most Mull species these are of no great interest, dating only from the published lists of Keddie and George Ross. There are, however, a number of considerable historical interest (see Chapter 1 which covers the history of botanising in Mull). The dates shown for recent introductions may be of interest to students of alien plants. The status of the species in Mull is indicated by the use of a standard series of terms detailed below and following the usage of Allen (1969). Out of a total of 816 flowering plants, 116 are deliberate introductions (i.e., hortal or denizens).

Native Occurring at least in part in non-artificial habitats and consistently associated with certain other species in these.

Colonist An invader allowed in unintentionally as a result of human activities (usually as a weed of cultivation) and now only occurring in artificial or semi-artificial habitats.

Denizen Growing wild but known or suspected to have been introduced for cultivation as crops or herbs.

Hortal An ornamental species escaped into the wild from obvious cultivation.

Following the status term, an indication, when applicable, is given of the phytogeographical element to which the species belongs, following Matthews (1955) and later amendments by Ratcliffe (1968). This topic is discussed in Chapter 2 which summarises our findings on the phytogeography of Mull. There follow general statements on the habitats occupied by the species in Mull, and a general comment on its frequency and distribution within the area. For species with eight or fewer localities, details are given with the records arranged in sequence by recording area numbers. The following conventions have been adopted: dates without brackets refer to the actual year records were made or specimens collected; dates within brackets refer to the publication dates of records in the literature. Where no collector or recorder is cited, the record or specimen results from the work of our own Survey during the five-year period 1966–1970. With the exception of first records, dates are only given for specimens and records from before 1930, which, as it happens, provides a convenient division between the older workers and more recent times; and also fits in with the convention established by Perring & Walters (1962) in their distribution maps of British plants. The location of voucher specimens is indicated by the use of the standard abbreviations of *Index herbariorum* (Lanjouw & Stafleu, 1964) and *British herbaria* (Kent, 1958). In some cases a short note on any point of special interest concludes the entry. The much-appreciated specialist help we have had with certain critical genera is acknowledged in the text.

Abbreviations

Herbaria and recorders

BEC Botanical Society and Exchange Club of the British Isles excursion 1939 (Wilmott, 1942)
BM British Museum (Natural History)
BMMS British Museum Mull Survey
BSBI Botanical Society of the British Isles excursion 1961 (Ratcliffe, 1963)
CGE Botany School, University of Cambridge
CSSF Committee for the Study of the Scottish Flora excursion 1968
DMS BSBI Distribution Maps Scheme records
E Royal Botanic Garden, Edinburgh

GL Botany Department, University of Glasgow
HAMU Hancock Museum, Newcastle-upon-Tyne
K Royal Botanic Gardens, Kew
K-Wa H. C. Watson herbarium at the Royal Botanic Gardens, Kew
NMW National Museum of Wales, Cardiff
OXF Department of Botany, University of Oxford
SSS Soil Survey of Scotland (Macaulay Institute, Aberdeen)
TTN Somerset Archaeological and Natural History Society, Taunton

Phytogeographical elements

M Mediterranean
OS Oceanic Southern
OWE Oceanic West European
ON Oceanic Northern
WSA Widespread Sub-Atlantic
SSA Southern Sub-Atlantic
NSA Northern Sub-Atlantic
CS Continental Southern
C Continental
CN Continental Northern
NM Northern Montane
NA North American
AS Arctic-Subarctic
AA Arctic-Alpine
A Alpine
W Widespread
E Endemic to the British Isles

Acknowledgements

We are indebted to the following for help with critical groups: Jean K. Bowden (*Dactylorhiza*); R. K. Brummitt (*Calystegia*); C. D. K. Cook (*Sparganium*); J. Cullen (*Anthyllis*); J. E. Dandy (*Potamogeton, Ruppia* and *Najas*); E. S. Edees (*Rubus fruticosus* agg.); R. M. Harley (*Mentha*); P. F. Hunt (*Dactylorhiza*); A. G. Kenneth (*Hieracium*); H. A. McAllister (*Campanula rotundifolia, Deschampsia*); R. Melville (*Rosa*); A. Neumann (*Salix*); F. H. Perring (*Arctium, Symphytum*); P. H. Raven (*Onagraceae*); A. J. Richards (*Potentilla erecta, Taraxacum*); R. H. Roberts (*Mimulus*); J. P. Savidge (*Callitriche*); P. D. Sell (*Fumaria, Hieracium, Pilosella*); A. McG. Stirling (*Hieracium*); B. T. Styles (*Polygonum aviculare* agg.); P. G. Taylor (*Utricularia*); T. G. Tutin (*Zostera*); C. West (*Hieracium, Pilosella*); P. F. Yeo (*Euphrasia*).

We have received many records from individual botanists and in particular would like to thank Dr Ursula Duncan, A. G. Kenneth and A. McG. Stirling.

Index to genera

CONIFEROPHYTA

CONIFEROPSIDA

CONIFERALES

PINACEAE

Larix Mill.

L. decidua Mill.
Larch. Denizen
Macculloch (1819).
Occasional in scattered localities around policies; also planted for forestry.
3–7, 10–13, 17

Pinus L.

P. sylvestris L.
Scots Pine. Denizen
Macculloch (1819).
Doubtless originally introduced, but regenerates easily and may spread naturally.
Occasional; in scattered localities, especially near houses and farms.
4, 6, 8, 11–13, 17
There are some recently planted *Pinus* seedlings on **1** Sgeir 'a Chaisteil but it is most unlikely that these will survive to maturity in the climatic conditions that prevail there.

CUPRESSACEAE

Juniperus L.

J. communis L. subsp. **nana** Syme
Juniper. Native; NM
2 Iona, Garnett (1800).
Cliff tops, well-drained acid grassland and dry heather moorland up to 500 m.
Occasional; widespread, especially in the south-west.
1–11, 15, 16

TAXACEAE

Taxus L.

T. baccata L.
Yew. Hortal
G. Ross (1877).
Planted trees in park woodland and in cemeteries.
Recorded from three localities in SE Mull.
16 Loch Buie; Loch Spelve Burial Ground.
17 Torosay Castle.

MAGNOLIOPHYTA

MAGNOLIOPSIDA

RANALES

RANUNCULACEAE

Caltha L.

C. palustris L.
Marsh-marigold. Native; W
Keddie (1850).
Lowland mires, flushes and streamside to 500 m.
Frequent; widespread.
1–17
Our observations in Mull support the views of Woodell & Kootin-Sanwu (1971) that subsp. *minor* cannot satisfactorily be separated from subsp. *palustris*. Most plants at higher altitudes and in more exposed situations may be referred to subsp. *minor*, but in the absence of meaningful morphological discontinuities we prefer to follow the opinions of the above authors.

Trollius L.

T. europaeus L.
Globeflower. Native; NM
'Glen Grant', 1876, *G. Ross* (BM).
Streamsides, wet pastures and flushed ledges to 500 m.
Occasional; in scattered localities but not recorded from the SW, Ulva or the Treshnish Islands.
5–9, 11–15
G. Ross (1879) stated 'Common all over the north end of the island and apparently so in the south'. This would not be an appropriate comment today. It has been suggested that this species may have decreased following the change from cattle- to sheep-grazing.

Aconitum L.

A. napellus L.
Monkshood
Hortal. Not recorded other than by Ewing (1890) but frequently cultivated in gardens.

Anemone L.

A. nemorosa L.
Wood Anemone. Native; W
G. Ross (1877).

Deciduous woods and shady ravines, sometimes surviving in relict woodland situations in more open habitats.
Frequent to abundant; widely distributed.
2–17

A. × hybrida Paxton (*A. japonica* Hort.)
Japanese Anemone
Hortal. Recorded as an escape or outcast in **4** Ulva.

Clematis L.

C. vitalba L.
Traveller's-joy
Hortal. Naturalised or planted in hedgerow at **6** Achnadrish, 1967, *Jermy* (BM). Far north of the natural range of the species in Britain.

Ranunculus L.

R. acris L.
Meadow Buttercup. Native; W
Keddie (1850).
In many habitats from grass heath, waste ground and roadsides to marsh runnels and *Eleocharis/Carex* mire.
Frequent to abundant; widespread.
1–17
This species shows a wide range of variation in our area. Plants with rather broad leaf lobes and long hairs on the upper leaf surface (var. **villosus** (Drabble) S. M. Coles) have been collected from the more exposed outer small islands – Soa, Staffa, Bac Mòr and Sgeir a' Chaisteil, while the forms noted on the mainland of Mull all seem to be attributable to the more widespread var. **acris**. In this way the strongly oceanic pattern shown by var. *villosus* in its European and British distributions seems to be reflected on a small scale in Mull.

R. repens L.
Creeping Buttercup. Native; W
G. Ross (1877).
Roadsides, waste ground and damp pastures, wetter parts of dune areas and submaritime grassland.
Frequent to abundant; widespread.
1–17

R. bulbosus L.
Bulbous Buttercup. Native; W
Keddie (1850).
In machair and other sandy submaritime habitats.
Rare; an infrequent plant almost confined to the west. The localities in E Mull probably represent ruderal occurrences.
2 Iona, near Calva (BM); Arduara, *SSS*. **3** Eilean a' Chalmain (BM); Uisken Bay (BM). **4** Gometra House (BM). **5** Port Langamull (BM). **6** Near Tobermory, *Young*. **8** S Ardmeanach, *Flannigan*. **15** Scallastle.

R. auricomus L.
Goldilocks Buttercup
Recorded by *BSBI*, 1961, in flushed stabilised scree at 150 m in **11** Glen Cannel, but is without confirmation. The species may occur elsewhere in our area in similar habitats but possibly never flowering and thus be overlooked. It is however some distance from its presently known distribution.

R. flammula L.
Lesser Spearwort. Native; W
Keddie (1850).
Flushes, lowland mesotrophic mires, shallow water in lochs, river banks, streams and ditches, ascending at least to 650 m.
Abundant; widespread.
1–17
Plants referable to both subsp. **flammula** and subsp. **scoticus** (E. S. Marshall) Clapham occur in our area together with intermediates which cast doubt on the validity of these taxa – at least as far as Mull is concerned. Forms of subsp. *scoticus* appear to predominate.

R. sceleratus L.
Celery-leaved Buttercup. Native; C
10 Kinloch–Scridain, 1877, *G. Ross* (BM).
Margins of ponds with a high nitrogen content, normally near farms or other settlements.
Rare; in four widely scattered localities in the west.
8 Gribun, *Ribbons*; Inch Kenneth (BM). **10** As above. **11** Kellan, *Temperley*. There is also a doubtful record for **4** Ulva.

R. hederaceus L.
Ivy-leaved Crowfoot. Native; OWE
Keddie (1850).
Ditches, muddy tracks and springs. Rarely in coastal streams at high tide mark where occasionally subjected to salt water.
Occasional; of scattered distribution, most frequent on Iona.
2, 3, 5, 7, 12, 17

R. trichophyllus Chaix
Thread-leaved Water-crowfoot
The *Ewing* (1899) record for v.-c. 103, an unlocalised specimen labelled *R. drouetii* (BM), is almost certainly from Tiree.

R. aquatilis L.
Common Water-crowfoot
Recorded by *Keddie* (1850) but not subsequently other than in Ritchie & Ritchie (1945), which appears to be based on Keddie's list.

R. baudotii Godr.
Brackish Water-crowfoot
The Ewing (1899) record for v.-c. 103 is almost certainly based on a specimen (BM) from Tiree.

R. ficaria L.
Lesser Celandine. Native; W
G. Ross (1877).

Deciduous woodland, shady ravines and maritime cliff ledges.
Occasional to frequent; widespread.
1–17
The Mull records are all subsp. **ficaria** despite careful searching for subsp. *bulbifer* (Marsden-Jones) Lawalrée.

Aquilegia L.

A. vulgaris L.
Columbine. Hortal
13 Aros Park, 1966, *Cannon & Groves.*
Roadsides and old walls.
Occasionally naturalised near houses.
12 Kellan Mill; Salen (BM). **13** As above; woods N of Tobermory.

Thalictrum L.

T. alpinum L. [Map, Fig. 2.26]
Alpine Meadow-rue. Native; AA
Ben Buie, 1877, *G. Ross* (BM).
Rock ledges in ravines and in flushed montane grass heath, descending to 300 m.
Occasional; widely distributed in the central massif, Ardmeanach and the mountains round Glen Forsa, with a few outliers elsewhere.
7–11, 15, 16

T. minus L.
Lesser Meadow-rue. Native; WSA
Y-Columb-Kill [Iona], *Lightfoot* (1777).
On sand dunes, coastal cliff ledges and lochside shingle.
Occasional; widely distributed in W Mull: Iona, Soa and Gometra and with an outlying inland population on the NE shore of Loch Bà. There is an old record Maclagan (1855) from **16** Loch Buie.
2–5, 9, 11
Both subsp. **minus** and subsp. **arenarium** (Butcher) Clapham are represented in Mull. The coastal populations e.g., **2** Iona dunes, **3** Uisken Bay, **5** Calgary Bay are usually subsp. *arenarium*, but subsp. *minus* also occurs coastally on cliff ledges e.g., at **3** Knockvologan and on **2** Iona. The population on **2** Eilean Annraidh is intermediate, while the inland population at **11** Loch Bà is subsp. *minus*.

BERBERIDACEAE

Berberis L.

B. vulgaris L.
Barberry
Hortal. Listed by *Ewing* (1890) but not subsequently recorded as an escape.

NYMPHAEACEAE

Nymphaea L.

N. alba L.
White Water-lily. Native; WSA
G. Ross (1877).
In lochs and deeper lochans, especially those with a deep muddy bottom.
Occasional; in the north, the south-east and the Ross, but unrecorded from the central areas.
2, 3, 6, 7, 13, 15–17
Plants referable to subsp. **occidentalis** Ostenf. occur at: **7** Lochan SE of Loch Tràth (BM), **13** Glac Uamharr (BM), **16** Lochan an Gleann Beag. It is probably more common than the records indicate but is not clearly differentiated from subsp. **alba** and in some places e.g., **13** Lochan a' Ghurrabain both subspecies seem to be present.

Nuphar Sm.

N. lutea (L.) Sm.
Yellow Water-lily. Native; W
G. Ross (1877).
In lochs and larger lochans with muddy bottoms.
Rare; in a few localities in N Mull and one locality in the south-east.
6 Glengorm Castle, *V. Gordon.* **7** Drimnacroish. **12** Meall Damh Àrd. **13** Lochan a' Ghurrabain, *Gerrans* (BM); Glac Uamhar (BM); Caol Lochan. **17** Lochan Doire Dharaich.
N. pumila (Timm) DC. has not been recorded from Mull. The specimens in BM labelled 'Ben More' are almost certainly from Perthshire (v.-c. 88).

RHOEADALES

PAPAVERACEAE

Papaver L.

P. rhoeas L.
Common Poppy. Colonist
2 Iona, 1912, *Wilmott.*
An arable weed.
Rare; known only from **2** Iona and refound W of Baile Mòr during the survey.

P. dubium L.
Long-headed Poppy. Colonist
Keddie (1850).
An arable weed and in waste places.
Rare; known only from two areas.
2 Iona, near the village, *James.* **13** Tobermory, near the Golf Course.

P. argemone L.
Prickly Poppy. Colonist
Keddie (1850).
Not subsequently recorded. The original record is probably from Iona.

P. somniferum L.
Opium Poppy. Hortal
2 Iona, near village, 1912, *Wilmott.*
On waste ground and rubbish dumps.
Rare; in scattered localities.
2 As above. **12** Salen between village and pier, *U. Duncan.* **14** N of Fishnish, *Kenneth* (BM).

Meconopsis Vig.

M. cambrica (L.) Vig.
Welsh Poppy. Hortal
16 Lochbuie, 1939, *BEC.*
In hedgerows and on old walls.
Occasional; naturalised in scattered localities,

particularly in the N half of Mull.
5–8, 12, 13, 16

Glaucium Mill.

G. flavum Crantz
Yellow Horned-poppy
Has been recorded from the old pier at **5** Calgary Bay but we have been unable to substantiate this record. Since *Meconopsis* is well established in this locality, it seems likely that the record is an error.

FUMARIACEAE

Dicentra Bernh.

D. formosa (Haw.) Walp.
Naturalised in a wood at **8** Kilfinichen, 1968, *U. Duncan & James* (BM); the sole record.

Corydalis Medic.

C. claviculata (L.) DC.
Climbing Corydalis. Native; OWE
16 Loch Buie, 1939, *BEC.*
In deciduous woods and on shady cliffs.
Rare; known from two localities in SE Mull.
16 Loch Buie, cliffs to W of village (BM); Cruach Ardura (BM).

Fumaria L.

All determinations by P. D. Sell
F. bastardii Bor.
Tall Ramping-fumitory. Colonist; M
2 Iona, 1887, *Somerville* (BM).
An arable weed.
Rare; known only from Iona and Uisken.
2 Iona, fields S of Baile Mòr (BM). **3** Uisken sands, *Kenneth & A. Stirling* (BM). The Iona specimens have been determined as var. **hibernica** Pugsl.

F. muralis Sond. ex Koch subsp. **boraei** (Jord.) Pugsl.
Common Ramping-fumitory. Colonist; OS
2 Iona, fields S of Baile Mòr, 1968, *James* (BM).
An arable weed.
Rare; in Iona and Mornish.
2 As above. **5** Sunipol (BM); Port Langamull (BM).

F. officinalis L.
Common Fumitory. Colonist; W
Keddie (1850).
An arable weed.
Rare; in scattered localities.
2 N central Iona; fields S of village (BM). **11** Loch na Keal, 1889, *J. Stirling & Kidston* – as *F. bastardii* (GL). **13** Tobermory, *G. Ross* (1879).
A specimen from Iona among Rothery's plants (1839) at Edinburgh and labelled *F. capreolata* is probably this species.

CRUCIFERAE

Brassica L.

B. oleracea L.
Cabbage
Denizen. A relict of cultivation and not the wild plant. Known only from **2** Iona, near Baile Mòr, 1912, *Wilmott* MS.

B. napus L.
Rape. Denizen
2 Iona, 1912, *Wilmott.*
On disturbed ground and waste places near farms and houses.
Rare; in scattered localities as a relict of cultivation.
2 Baile Mòr. **5** Sunipol (BM). **9** Cnoc Mòr; An Crosan; Loch Assopol.

B. rapa L.
Turnip. Denizen
12 Salen, 1966, *U. Duncan.*
On disturbed ground, rubbish dumps and near farms and houses.
Rare; in scattered localities as a relict of cultivation.
5 Haun (BM). **10** Carsaig (BM). **12** Salen, *U. Duncan* (BM); shingle spit at head of Loch na Keal (BM).

Rhynchosinapis Hayek

R. monensis (L.) Dandy
Isle of Man Cabbage. Native; E
2 Iona, 1772, *Banks* (BM).
Known only from the above specimen which has the following data: 'Iona Scotiae insula in littoribus arenosis prope mare.'

Sinapis L.

S. arvensis L.
Charlock. ? Colonist
G. Ross (1877).
An arable weed of waste places and disturbed ground round farms.
Rare; in scattered localities.
2 Near Cnoc-nam-bràdhan, *James*; near Baile Mòr, 1912, *Wilmott*; N Iona. **4** SE Ulva, *Salzen.* **6** Track to Glengorm Castle. **8** Gribun, *A. Stirling.* **12** Ledmore; Salen, *U. Duncan.* **13** Tobermory. **16** Lochbuie. **17** Lochdonhead, *U. Duncan.*

S. alba L.
White Mustard. Denizen or colonist
8 Inch Kenneth, 1969, *Jermy* (BM).
On grassy cliff with seabird droppings. Known only from the above locality.

Raphanus L.

R. raphanistrum L.
Wild Radish. Native or colonist; W
G. Ross (1877).
In waste places and along roadsides; an arable weed.
Rare; in scattered localities.
2 S of Baile Mòr (BM). **3** An t-Àrd. **5** Haun (BM). **13** Tobermory.
Both specimens at BM are yellow-flowered.

R. sativus L.
Garden Radish. Denizen
10 Carsaig, 1967, *Bowden* (BM).
On a rubbish dump; the only record.

Cakile Mill.

C. maritima Scop. subsp. **maritima**
Sea Rocket. Native; WSA
2 Iona, 'White Sands', 1906, *Davie* (E).
In loose sand above the strand line.
Rare; probably confined to Iona and the adjacent Eileann Annraidh.
2 SE of Baile Mòr, *Hillcoat* (BM); N Iona; Eileann Annraidh (BM).
There is also a doubtful record from **5** 'shore at Croig'. (Bowerman, pers. comm.)

Lepidium L.

L. heterophyllum Benth.
Smith's Pepperwort. Native; OWE
G. Ross (1877).
On roadsides, disturbed ground and waste places.
Rare; in scattered localities.
6 Laorin, *G. Ross* (1879). **7** Kellan Mill (BM). **10** Carsaig, *Kenneth & A. Stirling*. **12** Between Salen and Gruline, *Gerrans* (BM); Glen Forsa House, *Gerrans* (BM); spit at head of Loch na Keal (BM); Aros Castle, *J. Duncan* (Hb. J. Duncan). **16** Loch Buie, *H. Milne-Redhead.*
The record for *L. campestre* (L.) R.Br. in Maclagan (1855) for Loch Buie is assumed to be this species.

Coronopus Zinn

C. squamatus (Forsk.) Aschers.
Swine-cress. ? Colonist
2 Iona, 1880, *Ewing* (GL).
No ecological data given. Only one (?) record and presumably restated by *DMS*.

C. didymus (L.) Sm.
Lesser Swine-cress. Colonist
2 Iona, 1880, *Ewing* (GL).
Rare; only one recent record: **12** Glen Forsa House, *U. Duncan* (BM), where it occurs as a weed in a gravel path.

Thlaspi L.

T. arvense L.
Field Penny-cress. Colonist
13 Tobermory, 1967, *Melderis*.
An arable weed and in waste places.
Known only from the above record. Confirmation is desirable.

Teesdalia R.Br.

T. nudicaulis (L.) R.Br.
Shepherd's-cress. Native; C
11 Ben More, Coire nan Each, 1939, *Templeman* (BM).
Rare; known only from three localities: **11** As above. **12** Quarry track, 1·5 km SW of Salen, *Clark*. **14** Lower Glen Forsa, shingle bank in river.

Capsella Medic.

C. bursa-pastoris (L.) Medic.
Shepherd's-purse. Native; CS
G. Ross (1877).
In waste places, arable fields, by waysides.
Occasional; in scattered localities.
2–13, 16, 17
G. Ross (1879) records this as 'common'.

Cochlearia L.

In Mull, as elsewhere, this genus presents very considerable difficulties and the account presented here must be regarded as tentative. In order that our records can be evaluated in relation to those of other workers (e.g., Pobedimova 1968, 1969), we are including a review of the characters that we believe to be of value in the recognition of the taxa represented.

C. officinalis subsp. *officinalis*
Plants usually large of 10 cm or more often straggling in habit; basal leaves tending towards cordate, distinctly fleshy; cauline leaves prominent, sessile, stem-clasping; fruiting pedicels at right angles to the stem, fruit spherical rather than elliptical; in maritime habitats.

C. officinalis subsp. *alpina* (including *C. micacea*)
Very variable in size; leaves not fleshy; fruit ovoid elliptical; in alpine habitats.

C. danica
Plants of up to 15 (–20) cm, usually of a compact habit; basal leaves tending towards cordate, fleshy; upper cauline leaves petiolate and more or less ivy shaped; fruiting pedicels at an acute angle to the stem, fruit elliptical; flowers more likely to be pigmented than in the other species; in maritime habitats.

C. scotica
Plants of up to 10 cm, with a compact habit; basal leaves truncate or reniform (very rarely tending to cordate), fleshy; cauline leaves sessile but not stem clasping (very rarely shortly stalked); fruiting pedicels at right angles to the stem, fruit elliptical; in maritime habitats.

C. atlantica
Plants of up to 20 cm, often forming rather dense clumps with many wiry stems; basal leaves ± truncate, sometimes tending to reniform, somewhat fleshy; cauline leaves sessile to stem clasping; fruiting pedicels at an acute angle to the stem; many flowers failing to set fruit, fruit nearly orbicular, but occasionally slightly elliptical; seeds with rather scattered tubercles (especially round the keel) and with a ± retained exterior whitish membrane (*C. officinalis*, the species that this taxon most closely resembles, has seeds with a dense covering of tubercles which lack the membrane).

C. officinalis L. subsp. **officinalis**
Common Scurvy-grass. Native; ON
2 Iona, 1839, *Rothery* (E).

On cliff ledges, boulder beaches, in grass above shore lines and in upper levels of saltmarsh. Not recorded on Mull as an alpine.
Locally frequent; widely distributed around the coast, especially on the W side.
1–10, 12–14, 16, 17
Very large, thin-leaved, forms occur in deep shade in cave entrances and luxuriant fleshy forms are found in nitrogen-rich habitats associated with bird ledges.

C. officinalis subsp. **alpina** (Bab.) Hook.
C. alpina (Bab.) H. C. Wats; *C. micacea* E. S. Marshall.
Alpine Scurvy-grass. Native; NM
11 Ben More, 1939, *Templeman* (BM).
On damp rock faces in ravines and corries at 400–950 m.
Rare; on the Ben More massif and some of the higher mountains of Glen Forsa.
11 Abhainn Dhìseig and Coire nam Fuaran, *Templeman* (BM); Ben More, NW corrie (BM); ibid., NE corrie (BM); ibid., N corrie, *U. Duncan* (BM). **15** Sgùrr Dearg, *Stace*; Beinn Talaidh, Coire Ghaibhre (BM); Dùn da Ghaoithe (BM); Beinn Chreagach Mhor.

C. scotica Druce
Scottish Scurvy-grass. Native; E
2 Iona, rocks below abbey, 1958, *Gerrans* (BM).
In coastal turf, on shore rocks, cliff ledges and sides of saltmarsh runnels.
Occasional; widely distributed round the coast, possibly commoner in the west.
1–7, 9, 12, 17

C. atlantica Pobed.
Native; E
12 Between Salen and Aros bridge, 1958, *Gerrans* (BM).
In upper level saltmarsh pools, surrounding turf, adjacent sandy patches and crevices in shore rocks.
Rare.
7 Ballygown (BM). **12** As above; Salen Pier, *U. Duncan* (BM). **17** Loch Don, Gorton (BM).
This species based on a holotype from Lewis (BM), and recorded only from the Outer Hebrides and Arran, appears to be a recognizable entity. However, the high level of fruit abortion suggests that hybridity may be involved. Further observations on this taxon are most desirable.

C. danica L.
Danish Scurvy-grass. Native; ON
2 Iona, 1939, *Templeman.*
On seaward margins of machair and cliff ledges.
Rare; in the extreme SW but probably also in Ulva and N Mull.
1 Bac Beag (BM). **2** Iona, Calva (BM); by Eilean Maol Mhartuin, Arduara, *Kenneth & A. Stirling* (BM); Culbuirg (BM); S of Baile Mòr (BM). Also recorded from **4** Ulva, *Salzen.* **5** Croggan, *Bowerman.* **12** Ardnacross, *U. Duncan & James*, but these three records are unsupported by specimens and need confirming.

C. anglica L.
English Scurvy-grass
Recorded for **13** Aros Castle, 1939, by *BEC.*
Lack of confirmation leads us to suspect that this record may have been based on a misidentification, but an occurrence of this species in Mull is possible, although unlikely. The record in Sibbald (1684) for Inch Columb was taken up by Hooker (1821) who assumed the locality to be Iona. We are indebted to A. McG. Stirling for pointing out with strong evidence that the island concerned is almost certainly Inchcolm in the Firth of Forth.

Subularia L.

S. aquatica L.
Awlwort. Native; NM
6 Loch Peallach, 1876, *G. Ross* (BM).
Towards margins of lochs and lochans with stony bottoms or on fine nekron muds between pebbles.
Rare; in scattered localities, probably under-recorded.
6 As above; Tobermory Reservoir, *Temperley* (BM); Lochan Dearg (BM). **7** Loch a' Ghael (BM); Loch Frisa. **9** Creachan Mòr lochans (BM). **11** Loch Bà, *CSSF.* **12** Coire Mòr (BM). **15** Loch Squabain (BM). **17** Loch a' Ghleannain.

Draba L.

D. incana L.
Hoary Whitlow-grass. Native; AA
2 Iona, Port Ban, 1951, *Ferreira*
On coastal screes and cliffs and (?) machair, associated with localised base-rich conditions.
Rare; in three widely separated localities and possibly spasmodic in occurrence; the Iona specimen could not be refound in 1955.
2 As above. **8** Tòn Dubh-sgairt (BM). **10** Carsaig Bay, *U. Duncan.*

Erophila DC.

E. verna (L.) Chevall.
Common Whitlow-grass. Native; W
5 Calgary, 1965, *Kenneth.*
On machair and old walls.
Rare; in scattered localities but absent from the south-east.
2 Arduara (BM); Cùl Bhuirg (BM). **3** Knockvologen, *SSS.* **5** Calgary Bay (BM). **6** Mingary, *U. Duncan* (BM). **12** Meall Damh Àrd. **13** Aros Park; Tobermory.
The specimens are all subsp. **spathulata** (Lang) Walters, and it seems likely that all records are of this subspecies.

Armoracia Gilib.

A. rusticana Gaertn., Mey. & Scherb.
Horse-radish
Denizen. Naturalised in one locality: **12** Glen Forsa House, 1958, *Gerrans* (BM).

Cardamine L.

C. pratensis L.
Cuckoo-flower. Native; W
Keddie (1850).
By streams and ditches and in damp pastures.
Frequent; widespread.
1–17
This species does not appear to be very variable in Mull.

C. flexuosa With.
Wavy Bitter-cress. Native; WSA
13 Drumfin, 1876, *G. Ross* (BM).
By damp tracksides and streamsides, usually under shade.
Frequent; widespread.
1–17

C. hirsuta L.
Hairy Bitter-cress. Native; W
Keddie (1850).
By road and tracksides, on rocks, stone walls; rarely as an arable weed.
Occasional; widespread but much less common than *C. flexuosa.*
2–9, 11–17

Barbarea R.Br.

B. vulgaris R.Br.
Winter-cress
It is likely that the record **13** Tobermory, *G. Ross* (1879) is an error for *B. intermedia.*

B. intermedia Bor.
Medium-flowered Winter-cress. Colonist
13 Tobermory, 1968, *U. Duncan & James* (BM).
On waste places, roadsides and rubbish dumps.
Rare; around centres of habitation in the NE.
6 SW of Tobermory, *Kingston* (BM). **12** Salen (BM). **13** Tobermory rubbish dump (BM).
See also *B. vulgaris.*

Cardaminopsis (C. A. Mey.) Hayek

C. petraea (L.) Hiit. [Map, Fig. 2.16]
Northern Rock-cress. Native; AA
11 Ben More, 1825, *Trevelyan* (E).
On wet flushed rock ledges and large scree at higher altitudes.
Rare; scattered localities from Ben More to the mountains on the E side of Glen Forsa.
11 Ben More, N side; Coire nam Fuaran, *BEC*; Ben More, S screes and NE rocks, *BEC.* **15** Beinn Bheag (BM); Dùn da Ghaoithe; Sgurr Dearg, *Stace*; Beinn Chreagach Mhor.

Arabis L.

A. hirsuta (L.) Scop.
Hairy Rock-cress. Native; CS
16 Loch Buie, *Maclagan* (1855).
In dune grassland and on coastal rocks.
Rare; a few localities in S Mull and in Iona.
2 Iona, *Braid* (GL). **3** Port nan Ròn (BM); Knockvologan, *SSS.* **8** Gribun, *Ribbons.* **9** Cnoc Mòr. **16** As above.

Rorippa Scop.

R. nasturtium-aquaticum (L.) Hayek
Nasturtium officinale R.Br.
Water-cress. Native; W
2 Iona, below abbey, 1958, *Gerrans* (BM).
In shallow streams and slow-flowing ditches.
Rare; in scattered localities.
2 Opposite Eileann Mòr (BM); near Ruanaich (BM); N of Dùn (BM). **3** Ardtun (BM). **4** SE Gometra, *U. Duncan* (BM). **5** Tostarie (BM). **8** Inch Kenneth (BM); Balmeanach; Tòn Dubh-sgairt, *Ribbons.* **10** Carsaig, *Kenneth & A. Stirling* (BM).

R. microphylla (Boenn.) Hyland.
Nasturtium microphyllum (Boenn.) Reichenb.
Native; W
10 Carsaig, 1965, *U. Duncan.*
In shallow streams and runnels.
Rare; in scattered localities.
2 Near Calva (BM). **4** Ulva, near A' Chrannag (BM). **5** Calgary Bay (BM); Sunipol. **6** Lower river Bellart. **7** Drimnacroish. **10** Carsaig (BM). **17** Duart.

R. × sterilis Airy-Shaw
R. microphylla × nasturtium-aquaticum
Native; W
5 Calgary Bay, 1966, *Cannon & Groves* (BM).
In shallow streams and runnels.
Rare; in scattered localities.
3 Port Uisken Bay, *Kenneth & A. Stirling* (BM). **5** As above; Tostarie; Haun (BM). **6** Between Rubha nan Gall and Ardmore Point, *Stace.*

In addition to the above there are three records in the aggregate sense (first noted by *G. Ross*, 1877) from **16** Loch Buie area.

Hesperis L.

H. matronalis L.
Dame's-violet. Hortal
Ewing (1890).
In hedgerows near houses and rubbish dumps.
Rare; a garden outcast naturalised in a few localities.
6 Near Glengorm, *Kenneth.* **12** Salen (BM); Aros Castle. **13** Aros Park; Tobermory.

Erysimum L.

E. cheiranthoides L.
Treacle Mustard. ? Colonist
12 Head of Loch na Gaul [? Loch na Keal], *Hooker* (1821).
Otherwise known only from: **13** Tobermory, *Matthews* (1923).

Sisymbrium L.

S. officinale L.
Hedge Mustard. Native; W
G. Ross (1877).
On waste ground and roadsides.
Rare; an infrequent weed in scattered localities.
2 Baile Mòr (BM). **4** Ulva, *Salzen.* **12** Salen, *U. Duncan*; Aros Castle. **13** Tobermory (BM); Tobermory Golf Course.

Arabidopsis (DC.) Heynh.

A. thaliana (L.) Heynh.
Thale-cress. Native; CS
13 Tobermory, rocks below battery, 1877, *G. Ross* (BM).
A garden weed and on cliff ledges.
Rare; in scattered localities.
4 Ulva, Cragaig (BM); near Gometra House (BM). **7** Torloisk, *CSSF.* **10** Carsaig, *Kenneth & A. Stirling*; Kilfinichen (BM). **12** Glen Forsa House, *Gerrans* (BM); Aros Castle (BM). **13** Tobermory. **16** Loch Buie, *H. Milne-Redhead.*

RESEDACEAE

Reseda L.

R. luteola L.
Weld
Recorded by Ewing (1892) for v.-c. 103 but not otherwise.

VIOLALES

VIOLACEAE

Viola L.

V. odorata L.
Sweet Violet
Hortal. Recorded only from **13** Tobermory, 1970, *Robson*; established on roadside bank near houses.

V. riviniana Reichenb.
Common Dog-violet. Native; W
13 Runa Leap [Rubha na Leip], 1876, *G. Ross* (BM).
In open deciduous woodland and scrub, shady ravines and cliffs, and drier areas of Callunetum.
Abundant; widespread.
1–17

V. canina L.
Heath Dog-violet
Recorded from **8** S Ardmeanach Peninsula, 1956, by *Flannigan* but in the absence of a voucher specimen its occurrence must remain in doubt.

V. palustris L.
Marsh Violet. Native; CN
13 Allt na Meal, 1876, *G. Ross* (BM).
In *Molinia* flushes, streamsides and wet gullies.
Abundant; widespread.
1–17

V. lutea Huds.
Mountain Pansy
A specimen from **12** Glen Forsa House 1958, *Gerrans* (BM) has been redetermined as *V. tricolor*. In view of this and other evidence we believe that this species is unlikely to occur in Mull and that the *G. Ross* record from **5** Calgary and the original *Keddie* record (1850) from **2** Iona were probably *V. tricolor* also.

V. tricolor L.
Wild Pansy
Keddie (1850).

V. tricolor subsp. **tricolor**
Native or colonist; W
13 Erray, 1876, *G. Ross* (BM).
On arable land and in derelict gardens.
Rare; in scattered localities.
9 Loch Assopol. **12** Glen Forsa House, *Gerrans* (BM). **13** As above; Tobermory; Rubha na Leip; grounds of Aros House.

V. tricolor subsp. **curtisii** (E. Forst.) Syme
Native; W [Map, Fig. 2.35]
Ewing (1890).
On sand dunes, especially in machair.
Rare; scattered coastal localities in the west and south.
2 Eilean Annraidh (BM); Iona, 'Temple Glen', *Davie* (E). **3** Erraid; Knockvologan (BM); Ardalanish dunes, *Kenneth & A. Stirling* (BM). **5** Calgary Bay, *Chapple* (OXF); *U. Duncan*; *Richards.* **9** Cnoc Mòr; An Crosan. **17** Sands by Laggan Lodge (BM).

V. arvensis Murr.
Field Pansy. Native or colonist; W
G. Ross (1879)
An arable weed and on roadsides and other disturbed ground.
Rare; in scattered localities.
2 Near Shian, 1912, *Wilmott.* **3** Uisken, *Kenneth & A. Stirling.* **5** Sunipol (BM). **7** Lagganulva, *J. Duncan* (Hb. J. Duncan). **9** Tràigh nam Beach (BM). **10** Carsaig, *U. Duncan & Beattie.* **12** Spit at mouth of River Bà (BM). **13** Tobermory. **16** Loch Buie, *H. Milne-Redhead.* **17** Torosay Castle.
Described by G. Ross (1879) as 'common'.

POLYGALALES

POLYGALACEAE

Polygala L.

P. vulgaris L.
Common Milkwort. Native; WSA
Keddie (1850).
On grassy cliffs, roadsides and cliff ledges, in more base-rich situations.
Occasional; widespread, but more frequent in the west.
1–10, 12, 13, 15–17
Undoubtedly much less common than *P. serpyllifolia*, for which suitable habitats are more widely available in Mull. Flowers usually blue, but pink and white flowered forms have been recorded.

P. serpyllifolia Hose
Heath Milkwort. Native; WSA
13 ? 'Tom-a-Voulin' [? near Tobermory], 1876, *G. Ross* (BM).
In grass heath, rough upland grazing, in *Calluna*/*Molinia* associations and on ledges in gorges.
Abundant; widespread.
1–17

Flowers usually blue, but more rarely pinkish-mauve or white.

CISTIFLORAE

GUTTIFERAE

Hypericum L.

H. androsaemum L.
Tutsan. Native; OS
16 Loch Spelve, 1848, *Syme* (BM).
In open communities banks and cliffs and along woodland margins. Especially characteristic of sheltered ledges on sea cliffs.
Occasional to frequent; widespread especially round the coast.
2–17

H. calycinum L.
Rose-of-Sharon
Hortal. Recorded in: **2** Iona, 'Sandeels Bay', 1906, *Davie* (E). **7** Torloisk House. **13** Calve and Lovage Islands, *Harris.*

H. perforatum L.
Perforate St John's-wort. Native or colonist; W
G. Ross (1877).
Beside roads and tracks.
Rare; in three localities in N Mull.
5 Calgary Bay, *G. Ross* (1879). **6** Penmore Mill, *G. Ross* (1879); Achnadrish, *Kingston* (BM). **13** Rubha na Leip.

H. maculatum Crantz subsp. **maculatum**
(*H. dubium* Leers)
Imperforate St John's-wort. Native; C
Known only from **12** Aros Castle, 1889, *J. Stirling & Kidston* (GL).

H. tetrapterum Fr.
Square-stalked St John's-wort. Native; WSA
Penmollach [? Penmore], 1876, *G. Ross* (BM).
In damp coastal meadows and marshy pastures.
Occasional; locally frequent in the north-west.
2–10, 13, 15–17

H. humifusum L.
Trailing St John's-wort. Native; WSA
Aros, Erray and Druimfin, *G. Ross* (1879).
In open communities on pathsides, banks and in clearings.
Occasional; widely scattered in more sheltered localities.
3, 4, 6–8, 12–17

H. pulchrum L.
Slender St John's-wort. Native; WSA
G. Ross (1877).
On grassy slopes, dry banks and roadsides.
Frequent; widespread.
1–17

H. elodes L. [Map, Fig. 2.3]
Marsh St John's-wort. Native; OWE
2 Iona, 1887, *Somerville* (BM, GL).
In marshy areas, pool margins and runnels, occasionally in quite fast-flowing streamlets.
Occasional to locally frequent; confined to Iona, the Ross and W Brolass with an outlier at Torosay.
2, 3, 17

CENTROSPERMAE

ELATINACEAE

Elatine L.

E. hexandra (Lapierre) DC.
Six-stamened Waterwort. Native; WSA
15 Loch Bearnach, 1967, *Eddy* (BM).
On muddy bottom in shallow water.
Rare; only from two localities but possibly under-recorded.
11 Loch Bà, *CSSF.* **15** As above.

CARYOPHYLLACEAE

Silene L.

S. vulgaris (Moench) Garcke
Bladder Campion. ? Native; W
2 Iona, 'Shelly Bay', 1906, *Davie* (E).
The only record apart from an unlocalised record by the *BEC* in 1939.

S. maritima With.
S. vulgaris subsp. *maritima* (With.) A. & D. Löve
Sea Campion. Native; ON
1/2 Iona and Staffa, *Keddie* (1850).
On coastal shingle and cliffs. Inland on montane screes and cliffs.
Frequent around the W and S coast and on the islands; but very rare on the E coast. Also at high altitudes in the mountains of the central massif and those forming the E side of Glen Forsa.
1–16
A possible earlier first record exists in the Watson, *Cyb. Brit.* ms. slips, for **1** Staffa, *Churchill Babington.*

S. acaulis (L.) Jacq.
Moss Campion. Native; AA
Hope in Hudson (1778).
On rock ledges or forming cushions on steep flushed slopes, reaching down nearly to sea level at **8** Creag Mhòr.
Locally frequent in Ardmeanach, the central massif and the mountains on the E side of Glen Forsa.
8, 11, 15

S. dioica (L.) Clairv.
Red Campion. Native; W
Keddie (1850).
Among grass at side of tracks and under shade of trees in gullies and wood margins.
Occasional; widely scattered.
1–10, 12, 16, 17

S. alba (Mill.) E. H. L. Krause
White Campion
Reported by *Keddie* (1850) but not otherwise recorded apart from Ritchie & Ritchie (1945) which may be based on the above.

Lychnis L.

L. flos-cuculi L.
Ragged-Robin. Native; W
Keddie (1850).
Beside stream runnels, in open lowland mire associations and on damp cliff ledges.
Occasional; widely distributed at low altitudes, but more frequent in the west.
1–17

Agrostemma L.

A. githago L.
Corn-cockle. Colonist
G. Ross (1877).
Otherwise known only from one old specimen, but perhaps formerly more widespread before the introduction of improved farming methods.
2 NW of Baile Mòr on the coast, 1906, *Davie* (E)

Cerastium L.

C. tomentosum L.
Snow-in-summer. Hortal
Naturalised near houses on **2** Iona, 1955, *McNiell*; it has also been noted from **13** Tobermory (BM).

C. arcticum Lange
Arctic Mouse-ear. Native; AS
11 NE corrie of Ben More, 1967, *Eddy* (BM).
On wet rocks in spray from small cascade.
Known only from the above record.

C. holosteoides Fr.
C. fontanum Baumg. subsp. *triviale* (Link) Jalas
Common Mouse-ear. Native; W
Keddie (1850).
On roadsides, shingle, shell sand, wate ground and weedy pasture. Also in species-rich upland grassland.
Frequent to abundant; widespread.
1–17

C. glomeratum Thuill.
Sticky Mouse-ear. Native; W
13 Strongarbh, Tobermory, 1876, *G. Ross* (BM).
On roadsides, in farmyards and arable fields.
Occasional; widespread but much less common than *C. holosteoides*.
1–17

C. diffusum Pers.
C. atrovirens Bab.
Sea Mouse-ear. Native; OWE
Keddie (1850).
On shell sand, machair and in shallow soil in crevices of shore rocks and cliff ledges.
Frequent in the west, particularly in Iona and the small islands; a few scattered localities in the east.
1–5, 8, 9, 12, 13, 16, 17

C. semidecandrum L.
Little Mouse-ear. Native; W
G. Ross (1877).
In sandy places and waste ground.
Rare; in scattered localities.
1 Sgeir a' Chaisteil. **2** Iona, *Ferreira*. **3** Coast opposite Erraid; Meall nan Càrn; Beinn nan Gabhar. **13** Runa na Leip, *G. Ross* (1879).
These records are not satisfactory and further confirmation is desirable.

Stellaria L.

S. media (L.) Vill.
Common Chickweed. Native; W
G. Ross (1877).
On roadsides, waste ground, farmyards and gardens, cliffs and other habitats near the sea.
Frequent; widespread.
1–17
On the Treshnish Isles very large succulent forms occur on cliffs with sea birds' nests and around gull roosts. Similar forms are also found in other nitrogen-rich habitats, e.g. farmyards.

S. holostea L.
Greater Stitchwort. Native; W
G. Ross (1877).
Along roads and tracks and among rocks by beach.
Rare to occasional; in scattered localities particularly in the north.
4–8, 10–13, 15, 17

S. palustris Retz.
Marsh Stitchwort
Recorded for v.-c. 103 by *Ewing* (1892) but almost certainly an error for any part of that vice-county.

S. graminea L.
Lesser Stitchwort. Native; W
G. Ross (1877).
By road and tracksides, in grassy vegetation.
Rare to occasional; widely scattered, especially in the north.
5–8, 11–13, 17

S. alsine Grimm
Bog Stitchwort. Native; W
G. Ross (1877).
In marshy runnels, river gravels and among rocks; often in high altitude flushes which it sometimes dominates.
Occasional to frequent; widespread, especially in the west.
2–17

Sagina L.

S. apetala Ard.
Annual Pearlwort. Native; WSA
2 Iona, 1912, *Wilmott*.
On walls, and in short turf.
Rare; in scattered localities.
2 As above. **4** Ulva, *Salzen*. **5** Calgary, *U. Duncan and Beattie*. **13** Tobermory.
Possibly under-recorded.

S. maritima Don
Sea Pearlwort. Native; OS
1 Staffa, pre-1847, *Churchill Babington* in Watson, *Cyb. Brit.* ms. slips.

Short turf on cliff tops and among stones above shore.
Rare; in scattered localities in the west and north, especially on the small islands.
1 Bac Beag (BM); Bac Mòr (BM); Fladda (BM); Staffa (BM). **2** Iona, N central. **5** Sunipol. **6** Mingary. **13** Tobermory to Rubha nan Gall, *G. Ross* (1879).

S. procumbens L.
Procumbent Pearlwort. Native; W
Keddie (1850).
Along roads and tracks, in short turf on cliff tops and on rock ledges.
1–17

S. subulata (Sw.) C. Presl.
Heath Pearlwort. Native; WSA
'Mull', 1848, *Syme* (BM).
In short turf and on track margins; at higher altitudes in compacted scree and rock crevices.
Frequent; widespread.
1–17
On Mull this species is sometimes glabrous and has been mistaken for *S. saginoides*.

S. nodosa (L.) Fenzl
Knotted Pearlwort. Native; CN
Keddie (1850).
On coastal rocks and mountain stony flushes, possibly in the more base-rich situations.
Rare to occasional; widespread but recorded most frequently in the west.
1–3, 5, 6, 8, 9, 11, 12, 16

Minuartia L.

M. verna (L.) Hiern
Spring Sandwort. Native; AA
11 Ben More, 1956, *Sutherland* (E).
Known only from a rock ledge at 500 m on the above. Determination confirmed by G. Halliday.

Cherleria L.

C. sedoides L.
Minuartia sedoides (L.) Hiern
Cyphel. Native; A
8 Beinn na Srèine, 1961, *BSBI*.
Above 500 m in stony *Agrostis* heath.
Rare; in two widely separated localities.
8 As above; Coirc Bheinn. **15** Mainnir nam Fiadh, *Stace*; Dùn da Ghaoithe, *Stace* and *CSSF*.

Honkenya Ehrh.

H. peploides (L.) Ehrh.
Sea Sandwort. Native; ON
Keddie (1850).
In loose sandy shingle and above strand line on beaches.
Rare to occasional; scattered localities in W and S Mull.
2, 3, 5, 8, 10, 12, 16

Moehringia L.

M. trinervia (L.) Clairv.
Three-nerved Sandwort. Native; W
13 Druimfin, 1876, *G. Ross* (BM).
On roadsides, wall tops and under scrub.
Rare; in scattered localities.
3 Port nan Ròn (BM). **7** Torloisk House (BM); Lagganulva, *J. Duncan*. **10** Carsaig Bay, *U. Duncan*. **13** As above. **17** Between Torosay Castle and Craignure (BM).
There is also a doubtful record for **4** Ulva.

Arenaria L.

A. serpyllifolia L.
Thyme-leaved Sandwort. Native; W
5 Calgary Bay, 1876, *G. Ross* (BM).
In loose sand and unfixed dunes, and on coastal rocks and old walls.
Rare; in five scattered localities.
2 Iona, Cùl Bhuirg (BM). **3** Coast opposite Erraid Knockvologan, *SSS*. **5** Calgary Bay, *U. Duncan*; Treshnish Point. **8** Inch Kenneth (BM).

A. leptoclados (Reichb.) Guss.
Slender Sandwort
The occurrence of this species in Mull seems unlikely (see Perring & Sell 1968). In the absence of voucher specimens we cannot confirm the few records that have been made.

Spergula L.

S. arvensis L.
Corn Spurrey. Colonist
G. Ross (1877).
An arable weed; in waste places and roadsides.
Occasional; widespread and locally abundant.
1–17
All specimens are var. **sativa** (Boenn.) Mert. & Koch, except one from **8** Inch Kenneth which is var. **arvensis.**

Spergularia (Pers.) J. & C. Presl

S. rubra (L.) J. & C. Presl
Common Sandspurrey
Listed by *G. Ross* (1877) and stated by him (1879) to be 'common'. Recorded in 1939 by *Evetts* (BEC) but no subsequent records have been made and it seems likely therefore that these records are errors.

S. media (L.) C. Presl
Greater Sandspurrey. Native; WSA
13 Dóirlinn, 1876, *G. Ross* (BM).
In upper levels of saltmarsh.
Rare to occasional; in scattered localities.
3, 4, 6, 8–10, 12–15

S. marina (L.) Griseb.
Lesser Sandspurrey. Native; W
12 Salen, 1877, *Ewing* (BM).
In upper levels of saltmarsh and salt-sprayed turf on rock ledges.
Rare; in scattered localities.
5 Sunipol (BM). **6** Dervaig, old pier (BM); W

of Glengorm Castle. **7** Lagganulva, *J. Duncan*. **8** Aird Kilfinichen, *Flannigan*. **11** Tòrr a' Bhlàir. **12** As above; Eilean Bana.

PORTULACACEAE

Montia L.

M. fontana L. subsp. **fontana**
Blinks. Native; NC
G. Ross (1877).
In flushes, slow-flowing streams, damp tracks and *Salix* carr, ascending to at least 850 m.
Frequent; widespread.
1–3, 5–9, 11–13, 15–17
The above distribution is based on critically examined material; this subspecies may be assumed to occur in all areas.

M. fontana subsp. **variabilis** Walters
Native; WSA
G. Ross (1877) [as *M. rivularis*].
In flushes, damp patches on cliffs and ditches.
Rare; distributed in W Mull, but apparently much less frequent than subsp. *fontana*.
1 Cairn na Burgh More (BM). **3** Coast opposite Erraid (BM); Dearg Phort (BM). **6** NW Loch Frisa. **8** Gribun; Coire nan Griogag (BM). **10** Carsaig (BM). **11** Tòrr a' Bhlàir. **13** Rareick, *G. Ross* (1879); Tobermory, 1883, *Bailey* (E).

M. sibirica (L.) Howell
Pink Purslane. Hortal
8 S Ardmeanach Peninsula, 1956, *Flannigan*.
On roadsides and wood margins near houses.
Rare; naturalised in scattered localities.
3 Ardchiavaig; Bunessan, *Kenneth & A. Stirling* (BM). **8** As above. **12** Salen (BM); Forsa bridge (BM). **13** Aros Park, *U. Duncan*; Tobermory.

CHENOPODIACEAE

Chenopodium L.

C. bonus-henricus L.
Good-King-Henry. Denizen or colonist
Keddie (1850).
On roadsides and rubbish dumps.
Rare; established in two localities, one of which may be the same as Keddies record.
2 Near Maclean's Cross (BM). **10** Carsaig, rubbish dump (BM).

C. album L.
Fat-hen. ? Native; W
Keddie (1850).
On waste ground, gardens and arable land.
Rare to occasional; scattered, but not recorded from SE Mull.
2–6, 8, 12, 13

C. rubrum L.
Red Goose-foot
Only known from **16** Loch Buie, *Maclagan* (1855) and possibly an error.

Beta L.

B. vulgaris L. subsp. **maritima** (L.) Archangeli
Sea Beet. Native; OS
2 Iona, 1912, *Wilmott*.
On banks above saltmarsh, raised beaches and amongst coastal rocks.
Rare; in scattered localities; not recorded from the southeast.
2 Traigh Mòr, *Braid* (GL); near Baile Mòr, 1912, *Wilmott*. **3** Ardalanish, *Kenneth & A. Stirling*. **5** Calgary Bay. **12** Salen. **13** Aros Park.

Atriplex L.

A. patula L.
Common Orache. ? Native
Keddie (1850).
On waste ground, roadsides, rubbish dumps and coastal shingle.
Rare; scattered localities in the north and west.
2 Baile Mòr. **3** Coast opposite Erraid (BM); Cnoc an t-Suidhe. **4** A' Chrannag; SE of Ulva House (BM); Gometra, *U. Duncan*. **6** Laorin Bay; Loch Peallach. **13** Tobermory (BM); Erray House; Rubha nan Gall (BM).

A. hastata L.
Spear-leaved Orache
A field record from **13** Calve and Lovage Islands, 1955, *Harris*, has not been confirmed. It seems likely that confusion has arisen with *A. glabriuscula* in an immature state.

A. glabriuscula Edmondst.
Babington's Orache. Native; ON
G. Ross (1877).
On stony beaches and sand above drift line, river shingle, upper levels of saltmarsh.
Frequent; widely distributed around the coast.
1–10, 12–14, 16, 17
A record for *A. rosea* L. from **16** Loch Buie, *Maclagan* (1855) may represent an earlier first record for this species.

Suaeda Forsk. ex Scop.

S. maritima (L.) Dumort.
Annual Seablite. Native; SC
13 Druimfin, 1876, *G. Ross* (BM).
In saltmarsh, with *Armeria*, *Plantago maritima* and *Glaux*.
Rare to occasional; in scattered localities around the coast.
5, 6, 8, 10, 12, 13, 17

Salsola L.

S. kali L.
Prickly Saltwort. Native; WSA
16 Loch Buie, *Maclagan* (1855).
In loose sand above drift line on beaches.
Rare; scattered localities on the S and W coasts.
2 SE of Baile Mòr, *Hillcoat* (BM); Sandeels Bay, *Braid* (GL). **3** Coast opposite Erraid (BM); Knockvologan; Ardalanish Bay (BM); Uisken. **5** Calgary Bay, 1876, *G. Ross* (BM); *BMMS* (BM). **16** Laggan Sands, *Kenneth*; shore W of Laggan Lodge (BM).

Salicornia L.

S. europaea L.
Glasswort. Native; SC
6 Quinish, *Bird* in *G. Ross* (1879).
In saltmarsh in *Plantago-Festuca* turf.
Rare to occasional; in scattered habitats round the coast.
3–6, 8, 10, 12, 17

MALVALES

TILIACEAE

Tilia L.

T. platyphyllos Scop.
Large-leaved Lime. Hortal
Recorded only from park woodland in **17** Torosay Castle, 1968, *James* (BM).

T. × europaea L.
T. cordata × platyphyllos
Lime. Hortal
G. Ross (1877).
Occasional; in scattered localities, usually associated with large houses.
4–7, 10–13, 16, 17

MALVACEAE

Malva L.

M. sylvestris L.
Common Mallow
Recorded by the *BEC* in 1939 and not subsequently confirmed.

GERANIALES

LINACEAE

Linum L.

L. usitatissimum L.
Flax. Denizen
13 Druimfin, *G. Ross* (1879).
In a cornfield.
Known only from the above record. Perhaps a survivor from earlier cultivation.

L. catharticum L.
Fairy Flax. Native; W
Keddie (1850).
In rough pasture, grass heath and on tracksides, especially on drier slopes.
Frequent to abundant; widespread.
1–17

Radiola Hill

R. linoides Roth
Allseed. Native; WSA
2 'I Columb Kill' [Iona] 1772, *Banks* in Watson, *Cyb. Brit.* ms. slips.
In bare sandy ground.
Rare; in a few isolated localities. Only recorded once during the period of the survey but possibly overlooked.
2 Iona, 1887, *Somerville* (BM, E); *Templeman*; *A. Stirling*; Dùn I, *G. Ross* (1879). **3** Fionnphort, *BEC*. **12** Near Kellan, *BEC*; Salen.

GERANIACEAE

Geranium L.

G. pratense L.
Meadow Cranesbill. Hortal
3 Bunessan, 1877, *G. Ross* (BM).
By roadsides.
Rare; in scattered localities, probably always as an escape.
2 Baile Mòr (BM). **3** As above; Ardalanish Bay (BM); Uisken (BM). **9** Cnoc Mòr. **10** Carsaig to Nun's Pass, *Kenneth & A. Stirling*. **12** Aros Castle, *Gerrans* (BM). **13** Tobermory. **16** Lochbuie, near old castle, *BEC*.

G. sylvaticum L.
Wood Cranesbill. Native or Hortal; NC?
8 S Ardmeanach, 1956, *Flannigan*.
Along roadsides and tracks.
Rare; in scattered localities, possibly as an escape.
3 Near Bunessan, *U. Duncan & Beattie*. **8** Creag a' Ghaill, *U. Duncan*; as above. **13** Tobermory.
The record from **2** Iona (Millar, 1972) is in error for *G. pratense*.

G. endressii Gay
French Cranesbill. Hortal
Naturalised in grounds of large house. **12** Glen Forsa House, 1958, *Gerrans* (BM).

G. sanguineum L. [Map, Fig. 2.8]
Bloody Cranesbill. Native and hortal; W
2 Iona, *Lightfoot* (1772).
Basiphilous; on grassy cliff ledges and loose cliff scree.
Occasional; in the west including the small islands. As an escape only at Tobermory.
1, 2, 4, 5, 8, 13

G. dissectum L.
Cut-leaved Cranesbill. Native; W
Keddie (1850).
In machair; also on roadsides, walls and dry banks.
Rare; in scattered localities.
2 Central Iona (BM). **3** Ardchiavaig. **4** E side of Gometra. **6** Dervaig. **12** Salen. **13** Tobermory (BM); Aros Park; cliffs N of Tobermory. **14** Faoileann Ghlas. **16** Loch Buie; SW arm of Loch Spelve.

G. molle L.
Dovesfoot Cranesbill. Native; W
Keddie (1850).
In machair turf, loose shell sand and by roadsides.
Occasional; widespread in the west but very rarely recorded in the east.
2–13, 16

G. pusillum L.
Small-flowered Cranesbill
The extreme rareness of this species in W Scotland leads us to suspect that the record in G. Ross (1877) may be an error.

G. lucidum L.
Shining Cranesbill. Native; WSA
6 Dervaig, 1958, *U. Duncan & Beattie.*
By roadside on sandy bank and on dry *Calluna* hummock.
Rare; recorded only from one area.
6 As above; Dervaig, Kenneth (BM); Loch na Cuilce (BM).

G. robertianum L.
Herb Robert. Native; W
G. Ross (1877).
In deciduous woodland, shady gullies and by roadsides; also on coastal cliffs and stony shores.
Frequent; widespread.
1–17
We have followed *Flora Europaea* in not giving formal recognition to subsp. *maritimum* (Bab.) H. G. Baker.

Erodium L'Hérit.

E. cicutarium (L.) L'Hérit.
Common Storksbill. Native; WC
2 Iona, *Walker* (1808).
In fixed sand and machair, also in loose sand above the strand line; rarely in disturbed ground and tracksides.
Occasional; in scattered coastal localities in the west and south, but recorded only from Tobermory on the Sound of Mull.
1–5, 8, 9, 13, 16
We have followed *Flora Europaea* in not giving formal recognition to subsp. *dunense* Andreas. Forms approximating to that subspecies predominate, especially in the machair habitats, but satisfactory separation of these from subsp. *cicutarium* seems impracticable.

OXALIDACEAE

Oxalis L.

O. acetosella L.
Wood-sorrel. Native; W
G. Ross (1877).
In deciduous woodland, by shady stream banks, in dwarf-shrub communities and amongst block scree.
Frequent to abundant; widespread, but absent from the small islands, possibly due to the lack of sheltered habitats.
2–17

O. europaea Jord.
Hortal
17 Lochdonhead, 1966, *U. Duncan* (BM).
Naturalised near houses at the above and at **12** Salen (BM).

BALSAMINACEAE

Impatiens L.

I. noli-tangere L.
Touch-me-not Balsam. Hortal
7 Torloisk, 1967, *Kenneth.*
By a roadside.
Rare; the above sole population recorded was still persisting in 1970 (BM).

SAPINDALES

ACERACEAE

Acer L.

A. pseudoplatanus L.
Sycamore. Hortal
G. Ross (1877).
In deciduous woodland, by roadsides and around houses.
Occasional; widespread particularly in association with habitations, being frequently planted as shelter. Seedlings are seen, but it is probably not appreciably extending its range by natural means.
2–17
The Scottish name for Sycamore is 'Plane' (cf. Murray, 1966) and should not be confused with *Platanus* sp. which is not recorded for Mull.

A. platanoides L.
Norway Maple. Hortal
6 Glengorm Castle, 1960, *V. Gordon.*
Known only from the above locality and **17** the grounds of Torosay Castle, *U. Duncan* (BM).

A. campestre L.
Field Maple. Hortal
2 Iona, 1955, *Ferreira.*
Rare; introduced in a few places near houses.
2 As above. **10** Pennycross woods, Carsaig (BM); E of Carsaig. **11** NE Loch Bà. **12** Salen. **16** Lochbuie.

HIPPOCASTANACEAE

Aesculus L.

A. hippocastanum L.
Horse Chestnut. Hortal
13 Cliffs N of Tobermory, 1966, *Cannon & Groves.*
Rare; introduced in a few places near houses or on estates.
4 Ulva House. **6** Dervaig, old pier. **7** Torloisk House. **8** Gribun. **10** Pennyghael (BM); E of Carsaig. **11** Knock. **12** Ardnacross, *U. Duncan & James.* **13** As above. **16** Lochbuie, Marian and Moie Lodge. **17** Between Craignure and Torosay Castle.

CELASTRALES

AQUIFOLIACEAE

Ilex L.

I. aquifolium L.
Holly. Native; OS
16 Loch Spelve, 1848, *Syme* (BM).
In ravines, on roadsides and cliff gullies; in *Quercus-Betula-Sorbus aucuparia* woodland in SE Mull.
Occasional; widespread but more frequent in SE Mull.
2, 4, 6–8, 10–13, 15–17

RHAMNALES

RHAMNACEAE

Rhamnus L.

R. catharticus L.
Buckthorn. Hortal
17 Rubh' a' Ghuirmein, 1968, *Groves* (BM).
Presumed introduction; known only from the above locality where one plant is established in an area which also has *Hippophae*.

ROSALES

LEGUMINOSAE

Ulex L.

U. europaeus L.
Gorse. Native; OWE
G. Ross (1877).
By roadsides, round farm buildings and on river banks.
Occasional; widely distributed, particularly in the coastal areas.
3, 4, 6–17

Sarothamnus Wimm.

S. scoparius (L.) Wimm. ex Koch
Cytisus scoparius (L.) Link
Broom. Native and ? Hortal; W
G. Ross (1877).
On roadsides and riverbanks.
Rare; scattered localities in the north and east.
6 Dervaig (BM). **10** Carsaig, *U. Duncan & Beattie*. **12** Kellan; between Salen and Gruline (BM). **13** Cliffs N of Tobermory; Aros Park. **14** Lower Glen Forsa. **15** Glen Forsa, Gaodhail (BM). **16** Loch Buie.

Medicago L.

M. lupulina L.
Black Medick. Native or colonist; W
4 Ulva, 1948, *Salzen*.
On waste ground and cultivated places.
Rare; in scattered localities. Possibly under-recorded, but perhaps confused with *Trifolium dubium* which is certainly a much more frequent species in our area.
4 As above. **6** Dervaig. **8** S Ardmeanach Peninsula, *Flannigan*. **11** Loch Bà, *H. Milne-Redhead*. **12** Cnoc an Teine. **13** Tobermory Golf Course and cliffs. **16** Loch Buie.

Trifolium L.

T. pratense L.
Red Clover. Native and denizen; W
G. Ross (1877).
On roadsides, in meadows and grassy places, probably frequently as a result of introduction.
Occasional to frequent; widely distributed apart from the central massif.
1–17

T. medium L.
Zigzag Clover. ? Native; W
G. Ross (1877).
On roadsides and ledges on clay cliffs.
Rare; scattered localities in the west.
2 Iona, *Ferreira*. **3** Coast opposite Erraid. **5** Ensay Burn, *U. Duncan*; near Sunipol (BM).

T. hybridum L.
Alsike Clover. Colonist
4 Ulva, field near Church, 1965, *U. Duncan*.
In dry pasture and waste ground.
4 As above. **6** Glac Mhòr (BM). **7** Torloisk *CSSF*. **13** Tobermory (BM).
All plants examined are subsp. *hybridum*.

T. repens L.
White Clover. Native and denizen; W
2 Iona, *Pennant* (1774) [as 'natural clover'].
In relatively base-rich grassy places to at least 800 m; also roadsides and maritime cliff tops.
Abundant; widespread.
1–17
The populations include introduced agricultural strains as well as native forms.

T. campestre Schreb.
Hop Trefoil. Native and colonist; W
Keddie (1850).
Along tracksides and in waste places.
Rare; in a few scattered localities.
2 Near Cnoc-nam-Bràdhan; Traigh Mòr, *Braid* (GL); Sandeels Bay, *Small* (GL). Coast opposite Erraid; Knockvologan; Loch Poit na h-I; Uisken, *Kenneth & A. Stirling*. **10** Carsaig, *U. Duncan*. **13** Tobermory.

T. dubium Sibth.
Lesser Trefoil. Native; WSA
G. Ross (1877).
On disturbed ground and roadsides.
Occasional; widely distributed but nowhere frequent.
2–13, 16, 17

Anthyllis L.

A. vulneraria L. [Map, Fig. 2.12]
Kidney-vetch. Native; WSA
2 Iona, 1839, *Rothery* (E).
On cliff ledges and tops, sub-maritime *Festuca* grassland.
Occasional; widely distributed around the coasts on the west and north, with a few inland localities in Mishnish.
1–10, 13, 16
Our material has been examined by Dr J. Cullen. The majority of the specimens are referable to subsp. **lapponica** (Hyl.) Jalas, varieties **lapponica** and **langei** Jalas. Subspecies **vulneraria** is also represented, together with forms intermediate between the subspecies. An unusual form of subsp. *lapponica* occurs that is reminiscent of subsp. **corbieri** (Salm. & Trav.) Cullen. From the data available no useful conclusions can be drawn about the ecological preferences and factors determining the distribution of this very complex and confusing group.

Lotus L.

L. corniculatus L.
Common Birdsfoot-trefoil. Native; W
G. Ross (1877).
In herb-rich grassland, often sprawling over boulders and forming mats.
Abundant; widespread.
1–17
The form of this species with thick, fleshy leaves, found on the Treshnish Isles, reverted to the normal thin-leaved form under cultivation in SE England.

L. uliginosus Schkuhr
L. pedunculatus auct. non Cav.
Greater Birdsfoot-trefoil. Native; CS
G. Ross (1877).
In marshy grassland and streamsides, base-rich mires and peaty alluvial terraces with *Molinia* and *Holcus*.
Occasional; scattered localities, never common but possibly more frequent in the north-east and round Craignure and Torosay.
3, 4, 6–9, 11, 13–17
The majority of the Mull plants are glabrous.

Coronilla L.

C. varia L.
Crown-vetch. Hortal
Included by Turner and Finlay (1967) in their list, but seen by us only as a garden plant.

Vicia L.

V. hirsuta (L.) Gray
Hairy Tare. Native or colonist; W
G. Ross (1877).
On waste ground by seashore.
Rare; known today only from Iona.
2 Baile Mòr (BM). **6** Sorne, *G. Ross* (1879).

V. tetrasperma (L.) Schreb.
Smooth Tare.
Noted by *G. Ross* (1877, and in 1879 from **3** Tom-a Voulin [Tom-a Mhuillin]). We agree with Perring & Walters (1962) in regarding this as an error, though its occurrence as a casual cannot be ruled out.

V. cracca L.
Tufted Vetch. Native; W
G. Ross (1877).
In roadsides, hedgerows, field margins and on sheltered cliff ledges.
Occasional; widely distributed, especially to the west, but never common.
1–13, 15–17

V. orobus DC. [Map, Fig. 2.41]
Wood Bitter-vetch. Native; OWE
6 Loch a' Chumhainn, Dervaig, 1947, *J. Muirhead* (E).
On grassy slopes and banks near the sea, rarely on cliff ledges.
Occasional in NW Mull especially Mishnish, Mornish and Quinish. A record in the *Atlas*, unlocalised from **8** SW Ardmeanach needs confirming.
5–8
This species has one of the most interesting distribution patterns revealed by our survey. Further detailed work on its autecology should be very rewarding.

V. sylvatica L.
Wood Vetch. Native; CN
8 Tòn Dubh-Sgairt, 1939, *Templeman* (BM).
Shady cliffs and streamsides.
Rare; in scattered localities.
4 Ulva, *Salzen*. **6** Quinish, *BSBI*. **8** As above; Creag a' Ghaill, *Ribbons*; Allt na Teangaidh, *McAllister*. **10** Carsaig, *U. Duncan*; between Cameron and Glenbyre, *Ribbons*.

V. sepium L.
Bush Vetch. Native; W
G. Ross (1877).
On roadsides and grassy banks.
Occasional to frequent; widely distributed especially in the west.
1–13, 15–17

V. sativa L.
Common Vetch. Denizen
G. Ross (1877).
On roadsides and near farms; in herb-rich turf on banks.
Occasional; a few scattered localities.
2, 3, 5, 7, 8, 10, 12, 13
The Mull plants seen are all subsp. **sativa.**

V. lathyroides L.
Spring Vetch. Native; W
3 Knockvologan, 1972, *Murray* (BM).
Known only from the above in sandy grassland by the shore.

Lathyrus L.

L. aphaca L.
Yellow Vetchling
Known only from the record **2** Iona, Shian Farm, 1912, *Wilmott* for which confirmation is most desirable.

L. pratensis L.
Meadow Vetchling. Native; W
G. Ross (1877).
On roadsides and other herb-rich grassy communities.
Occasional; widely distributed.
2–17

L. sylvestris L.
Narrow-leaved Everlasting-pea. Native; W
16 Glen Liebetal [Glen Libidil], 1848, *Syme* (BM).
Known only from the above specimen. Stated by Syme to be 'certainly wild in this area'.

L. montanus Bernh.
Bitter Vetchling. Native; WSA

Keddie (1850).
In a wide variety of habitats from roadside banks to grassland, heathland and stream banks up to 900 m.
Frequent to abundant; widespread.
1–17
A complete range of leaf widths is found from the very narrow var. **tenuifolius** (Roth) Garcke (recorded as long ago as 1879 by G. Ross) to very broad-leaved forms.

Pisum L.

P. sativum L.
Pea. Denizen
16 Laggan Bay, 1967, *J. Duncan* (BM).
When found in the above locality it was stated to be an arable weed; otherwise seen by us only in gardens.

ROSACEAE

Filipendula Mill.

F. ulmaria (L.) Maxim.
Meadowsweet. Native; NC
G. Ross (1877).
In deep peat or shallower loamy peat at riversides and streamsides, lowland flushes and ravine ledges in the mountains.
Abundant; widespread.
1–17

Rubus L.

R. chamaemorus L.
Cloudberry
Has been recorded from **4** Ulva, Beinn Chreagach, but not confirmed and is very unlikely.

R. saxatilis L.
Stone Bramble. Native; NM
13 Tobermory, 1829, *Christy* (E).
On rock ledges, stream banks and among boulders; also in open *Quercus-Corylus* scrub.
Occasional; widely distributed.
2–17

R. arcticus L.
Druce (1920) reports Walker as noting [in 1782] 'In rocky mountainous parts of the Isle of Mull'. Syme (1864), who had visited Mull himself, records 'Ben More in the Isle of Mull'. There are specimens (BM & E) labelled 'Isle of Mull' in James Dickson's *Hort. Sicc. Brit. Fasc.* 19: 6 (1802). R. M. Harley has confirmed the identity of the BM specimen. In spite of these early records, we must continue to regard the locality of this species as dubious, at least until its status in the British flora is clarified.

R. idaeus L.
Raspberry. Native & ? denizen; WS
Keddie (1850).
On waste places and roadsides. Margins of damp woods.
Occasional to locally frequent; widespread often as an escape from gardens.
2–17

R. spectabilis Pursh
Hortal
17 Kilpatrick crossroads, 1965, *Kenneth.*
On roadsides and wood margins.
Rare; naturalised in two areas of SE Mull.
16 Lochbuie (BM). **17** Between Craignure and Torosay Castle (BM); Duart.

R. fruticosus L. agg.
Bramble. Native; WS
G. Ross (1879).
On roadsides, hill slopes, wood margins.
Locally frequent; widespread but never very common and probably limited by grazing.
1–17
All our specimens of *R. fruticosus* agg. have been examined by E. S. Edees and the account of the segregate species which follows is based on his determinations. It cannot be considered as a complete evaluation of the group in Mull and further investigation in late July and August, a period not well covered by our parties, will doubtless reveal additional species. For this reason, we have omitted general distribution and ecology.

R. scissus W. C. R. Wats.
3 Dearg Phort (BM).

R. plicatus Weihe & Nees
12 Between Salen & Glen Forsa, *U. Duncan* (BM). **17** Between Craignure & Torosay Castle, *Cannon & Groves* (BM).

R. sublustris Lees
Glen 'Aroin' [? Aros], 1887, *Ewing* (GL). Determined as *R. corylifolius* var. *sublustris* but not subsequently confirmed.

R. selmeri Lindeb.
R. nemoralis P. J. Muell. sec Wats.
4 Ulva, Cragaig (BM). **7** Fanmore, *U. Duncan* (BM). **10** Carsaig, *U. Duncan*; W of Carsaig (BM). **12** Tòrr nan Clach (BM). **16** Lochbuie, *C. Muirhead & Shaw* (E). **17** End of Torosay peninsula (BM).

R. laciniatus Willd.
6 Penmore Mill (BM).

R. lindleianus Lees
8 Ardmeanach, Burgh (BM).

R. insularis Aresch. sensu Wats.
3 Erraid (BM). **4** Ulva, near the pier, *U. Duncan* (BM). **8** Ardmeanach, Burgh (BM).

R. polyanthemus Lindeb.
6 Glengorm Castle (BM). **12** Allt Ardnacross (BM). A specimen 'Mull no 7' 1894, *Macvicar* (GL) was determined by Moyle Rogers as *R. pulcherrimus* Neum. (*non* Hook.) a synonym of this species.

R. dumnoniensis Bab.
R. rotundatus P. J. Muell. ex Genev.
16 N side of Loch Uisg (BM). **17** Lochdonhead, *U. Duncan* (BM).

R. cf. **mucronulatus** Bor.
13 Aros House woods, *U. Duncan* (BM).

R. radula Weihe ex Boenn.
13 Near Tobermory, 1894, *Macvicar* (E) [det Moyle Rogers].

Rubus sp. A specimen from Kilninian (BM) has been attributed to section *Triviales* by Edees, with the following comment 'Between *R. conjungens* and *R. latifolius* but answers to no named species'.

Potentilla L.

P. palustris (L.) Scop.
Marsh Cinquefoil. Native; CN
Keddie (1850).
In ditches, on boggy shores of lochs and lochans and in moorland flushes.
Occasional; most frequent on Iona and the Ross.
1–10, 12–15, 17

P. sterilis (L.) Garcke
Barren Strawberry. Native; W
G. Ross (1877).
In short turf at tracksides and along wood margins.
Occasional; widespread but never very common. Possibly under-recorded through confusion with *Fragaria.* Not recorded from the granite of the Ross or from Iona.
3–12, 15–17

P. anserina L.
Silverweed. Native; W
G. Ross (1877).
On sandy foreshores and machair, tracks and roadsides on well-drained soils.
Frequent; widespread at low altitudes.
1–17
Forms on foreshores tend to be large and fleshy.

P. norvegica L.
Ternate-leaved Cinquefoil. Hortal
Known only from **13** Aros House woods near old stone pier, 1939, *BEC.*

P. erecta (L.) Räusch
Tormentil. Native; W
Keddie (1850).
In rough grazing, along roads and tracksides, wood margins and cliff tops; in acid grassland and dwarf shrub communities.
Abundant; widespread. The most frequently recorded vascular plant species in Mull.
1–17
An unusual specimen from the rubbish dump E of **4** Ulva House (BM) was tentatively re-referred to *P. anglica* Laichard., but A. J. Richards suggests (pers. comm.) that it is more probably *P. anglica* × *erecta* (*P.* × *suberecta* Zimmet.).

Fragaria L.

F. vesca L.
Wild Strawberry. Native; W
G. Ross (1877).
On dry grassy slopes, roadsides, tracksides and cliff tops.
Occasional to frequent; not recorded from the north of the Ross and only one record from Iona.
1–17

Geum L.

G. urbanum L.
Wood Avens. Native; W
G. Ross (1877).
In deciduous woodland and along shady roadsides.
Occasional; widely distributed but apparently absent from the Ross peninsula, W of Carsaig.
2, 4–8, 10–13, 15, 16

G. rivale L.
Water Avens. Native; W
'Isle of Mull', 1834, *Shiels* (BM).
Along shady stream and river banks, in ditches and wet woodland.
Frequent; widespread.
2–17

G. × **intermedium** Ehrh.
G. rivale × *urbanum*
Native; W
S Ardmeanach Peninsula, 1956, *Flannigan.*
In addition to the above record it has been recorded from **13** Aros Park and Rubha na Leip.

Dryas L.

D. octopetala L.
Mountain Avens. Native; AA
8 N of Ton Dubh Sgairt, 1939, *Templeman* (BM).
Unstable ledges in gullies and cliff tops on local calcareous strata.
Rare; confined to a small area from the Gribun to Ton Dubh Sgairt.
8 Caisteal Sloc nam Ban, *Ribbons*; Mackinnon's Cave (BM); Creag a' Ghaill; Allt na Teangaidh.

Agrimonia L.

A. eupatoria L.
Agrimony.
G. Ross (1877).
Has been recorded 15 times (areas **3–5, 7, 8, 10, 12, 13**) but we are of the opinion that the Mull *Agrimonia* populations are probably all *A. procera.* However in our area the characters used to distinguish between these species do not seem well correlated.

A. procera Wallr.
A. odorata auct. non Mill.
Fragrant Agrimony. Native; W
5 Calgary, 1939, *Chapple* (OXF).
On grassy tracksides, banks and cliff ledges, margins of deciduous woodland.
Occasional; widespread in scattered localities, especially in the west.
3–5, 7–10, 12, 13

Alchemilla L.

A. alpina L. [Map, Fig. 2.22]
Alpine Lady's-mantle. Native; AA
Lhwyd (1699).
On more stable scree margins, sides of boulders in upland grassland and in rock crevices above 450 m, descending to lower altitudes on stream banks.
Occasional; widespread and locally frequent on Ardmeanach, the Central Massif, and the mountains of Glen Forsa and Loch Spelve area.
8, 10, 11, 14–16
There are also records from **2** Iona and **4** Ulva but these are most likely in error.

A. vulgaris L. agg.
Lady's-mantles.
G. Ross (1877).

A. filicaulis Buser subsp. **filicaulis**
Native; AA
8 Allt na Teangaidh, 1965, *Kenneth.*
In well-drained, base-rich habitats.
Rare; in widely scattered localities.
7 Beinn na Drise. **8** As above. **15** Beinn Talaidh, *Birse & J. Robertson* (SSS). **16** Creach Beinn, *Kenneth.*

A. filicaulis subsp. **vestita** (Buser) M. E. Bradshaw.
Native; CN
17 Near Craignure, 1958, *U. Duncan* (BM).
On streamsides and rock ledges, under drier conditions than those favoured by *A. glabra.*
Rare to occasional; widespread.
3, 6–10, 12, 14–17

A. xanthochlora Rothm.
Native; W
7 Kilninian, 1958, *U. Duncan* (BM).
Along streamsides, roadsides and on wet flushed ledges to 450 m.
Occasional; widespread but not recorded from Ardmeanach peninsula or Iona.
3–7, 9–17

A. glabra Neygenf.
Native; CN
13 Near Tobermory, 1883, *Bailey* (HAMU).
Along stream and river banks, roads and track-sides, damp gullies.
Frequent; by far the most frequent *Alchemilla* in our area, and the only species to have been recorded from a small island (Lunga).
1–17
Dwarf plants have been noted in dry *Festuca* grassland.

Aphanes L.

A. microcarpa (Boiss. & Reut.) Rothm.
Parsley-piert. Native or colonist; CN
13 Strongarbh, 1876, *G. Ross* (BM).
An arable weed and along tracksides.
Occasional; widespread.
2–13, 16, 17
All the specimens collected have been determined as *A. microcarpa* and we believe that field records of *A. arvensis* L. should be referred to this species, confirming the situation shown in Perring and Sell (1968).

Acaena Mutis ex L.

A. inermis Hook. f.
A. microphylla var. *inermis* (Hook. f.) Kirk
Hortal. Well established at the side of the south drive to **6** Glengorm Castle, 1959, *C. Muirhead* (E) and *BMMS* (BM) where it has persisted for at least ten years.

Rosa L.

All our specimens have been examined by R. Melville and the following account, based mainly on the treatment of Clapham, Tutin and Warburg (1962), has some modifications necessary to accommodate his views. The present state of knowledge of this very complex genus prevents a satisfactory evaluation of the populations on Mull, and the following account must be regarded as provisional.

R. pimpinellifolia L.
R. spinosissima auct.
Burnet Rose. Native; C
1 Staffa, *Macculloch* (1819).
On grassy slopes and among boulders, especially on open cliff tops and near the sea.
Frequent; widespread, but apparently less common in the SE.
1–13, 15, 16
Both var. **pimpinellifolia** and var. **hispidula** Rouy have been recorded.

R. ×gracilis Woods
R. pimpinellifolia × sherardii
Native
12 Aros Castle, 1968, *Robson* (BM).
On roadsides and rocky slopes.
Known only from the above locality and at **10** Pennyghael (BM).

R. rugosa Thunb.
Japanese Rose. Hortal
12 Rubha Àrd Ealasaid, near White House of Aros, 1958, *Gerrans* (BM).
Along roadside, and on open ground.
Naturalised in four localities in NE Mull.
12 As above; track to Aros Castle (BM); between Salen and Forsa river, *Gerrans* (BM). **13** Aros Park, *U. Duncan.*

R. canina L.
(including *R. dumalis* Bechst.)
Dog Rose. Native; W
Keddie (1850).
Along roadsides and on cliff edges and in crevices.
Occasional; widespread.
1–17
Var. **dumalis** (Bechst.) Dum [*R. dumalis* Bechst.] and var. **fraxinoides** H.Br. are widespread. Forma **viridicata** (Pugsl.) Rouy has been collected near **13** Tobermory (BM). Introgression

with *R. sherardii* has been detected in a specimen from **14** Faoileann Ghlas (BM). Other varieties and forms were noted by Heslop-Harrison & Bolton (1938) without precise localities, but it has not been possible to correlate this data with our own.

R. canina × sherardii
Native
11 Creag Mhòr, 1966, *Cannon & Groves* (BM).
Along roadsides, mostly near sea-level.
Five scattered records: **10** Carsaig, *U. Duncan.* **11** Torr a' Bhlair (BM). **13** Tobermory (BM). **16** Lian Mòr near Croggan (BM).

R. afzeliana sensu Wolley-Dod
Native; W
12 Between Salen and Forsa river, 1958, *Gerrans* (BM).
On roadsides, pathsides and in rocky stream gorges.
Rare; widely scattered.
12 Allt na Searmoin, *U. Duncan.* **17** Lochdonhead, *U. Duncan.*
Forma **adenophoria** (Gren.) Wolley-Dod: **6** Poll Athach (BM).
Var. **glaucophylla** (Winch) Wolley-Dod: **10** Carsaig, *U. Duncan* (K).
Var. **oonensis** (R. Kell.) Wolley-Dod: **8** Allt na Teangaidh (BM).
Forma **transiens** (Gren.) Wolley-Dod: **12** between Salen and the Forsa river, *Gerrans* (BM).
Heslop-Harrison & Bolton (1938) noted this species in Mull as *R. glaucophylla* Winch, recognising two varieties, *reuteri* (Godr.) Heslop-Harrison and *subcristata* (Bak.) Heslop-Harrison. No precise localities were given.

R. afzeliana × sherardii
Native
Known only from **10** Carsaig, 1965, *U. Duncan* (K).

R. sherardii Davies
Downy Rose. Native; W
Heslop-Harrison & Bolton (1938).
On roadsides, wood margins, rocky banks and streamsides.
Occasional; widespread.
3–17
Specimens among our collections have been determined as: var. **omissa** (Deségl.) Wolley-Dod; var. **suberecta** (Ley) Wolley-Dod and forma **resinoides** (Crép.) Wolley-Dod.
Field records for *R. villosa* agg. have been referred to this species.

R. rubiginosa L.
Sweet Briar. Native or hortal; W
13 Aros Park, 1966, *Cannon & Groves* (BM).
Along roadsides and deciduous wood margins near houses.
In addition to the above locality, known only from **16** Cnoc na Faoilinn (BM).
Recorded without locality in 1939 by the *BEC* as 'in need of confirmation'.

R. rubiginosa × sherardii
Native
11 Rubha Ard nan Eisirein, 1966, *Cannon & Groves* (BM).
Along roadsides.
In addition to the above locality, known only from **8** Creag Mhòr (BM).

Prunus L.

P. spinosa L.
Blackthorn. Native; W
G. Ross (1877).
Along roadsides, wood margins and colonised scree with *Corylus*.
Occasional; more frequent in N Mull.
1–9, 11–13, 15, 16

P. domestica L. subsp. **insititia** (L.) C. K. Schneid.
Wild Plum. Denizen
6 Mingary Burn, 1970, *L. Ferguson* (BM).
Under cover of low *Corylus* at side of burn, remote from houses. The only locality.

P. avium (L.) L.
Wild Cherry. ? Native; W
G. Ross (1877).
Along roadsides and wood margins.
Occasional; in widely scattered localities.
3 Bunessan area, *U. Duncan & Beattie.* **11** Knock. **13** Strongarbh, *G. Ross* (1877). **17** Lochdonhead, *U. Duncan*; 1 mile NW of Craignure, *U. Duncan* (BM).
The record from Iona (Millar, 1972) is an error for *Malus sylvestris.*

P. cerasus L.
Dwarf Cherry. Hortal
Persisting in derelict garden at **9** Knockan, 1970, *James* (BM).

P. padus L.
Bird Cherry. Native; W
13 Druimfin, 1876, *G. Ross* (BM).
Roadsides, stream and river banks.
Rare; but almost certainly under-recorded. Present records suggest an eastward distribution in Mull.
3 Bunessan area, *U. Duncan & Beattie.* **6** Bellart estuary; Allt Tòrr a' Bhacain. **12** Salen (BM); Kintallen. **13** Lochan na Guailne Duibhe; Aros Park *BEC*; Tobermory. **15** Strathcoil (BM). **16** Sròn nam Boc. **17** Ardchoirk.

P. laurocerasus L.
Cherry Laurel. Hortal
Known only from **13** Aros Park, 1966, *Cannon & Groves.*

Cotoneaster Medic.

C. simonsii Bak.
Himalayan Cotoneaster. Hortal

16 Lochbuie, 1966, *Cannon & Groves* (BM).
On roadsides and tracksides.
In addition to the above locality, known only from **6** Dervaig (BM).
Recorded without locality in 1939 by the *BEC* as 'awaits confirmation'.

C. microphyllus Wall. ex Lindl.
Small-leaved Cotoneaster. Hortal
13 Near Aros Cottage, 1939, *BEC*.
On old walls and ruined buildings.
Rare; naturalised in a few localities.
5 Tostarie. **6** Dervaig. **7** Torloisk. **12** Glen Forsa House, *Gerrans* (BM); Aros Castle, *Gerrans* (BM). **13** Tobermory; Aros House (BM).

C. frigidus Wall. ex Lindl.
Hortal
Known only from **17** woods by Kilpatrick House, 1967, *James* (BM).

C. bullatus Bois.
Hortal
Known only from **12** track to Aros Castle, 1968, *Robson* (BM).

Crataegus L.

C. monogyna Jacq.
Hawthorn. Native; W
2 Iona, *Keddie* (1850).
On river and stream banks, roadsides and wood margins.
Occasional; widespread and locally frequent.
2–17
Keddie noted 'only one specimen known' – presumably referring to Iona. In Mull as a whole it has possibly become more frequent with the decline of crofting.

Sorbus L.

S. aucuparia L.
Rowan. Native; W
G. Ross (1877).
In open deciduous woodland along roadsides and on sides of mountain gorges up to 500 m.
Frequent; widespread.
2–17

S. intermedia (Ehrh.) Pers.
Swedish White-beam. Hortal
2 Iona, 1955, *Ferreira*.
Besides tracks and wood margins.
Apart from the above locality known only from **8** Tavool House (BM); **10** Pennyghael (BM); and doubtfully from **13** Tobermory.

S. aria (L.) Crantz
Common White-beam. Hortal
11 Knock, 1968, *James* (BM).
Trackside near houses; more isolated specimens possibly bird-sown.
In addition to the above known only from **16** Lian Mòr; Loch Uisg, *Kenneth & A. Stirling* (BM).
In Watson's *Topographical Botany* MS at Kew, there is an undated entry under v.-c. 103 for this species. (*sine loc.*) attributed to Miss Harvey, which may refer to Mull.

S. latifolia (Lam.) Pers.
Broad-leaved White-beam. Hortal
Known only from **10** Pennyghael, 1971, *Kenneth & A. Stirling* (BM).

S. rupicola (Syme) Hedl.
Native; ON
8 Mackinnon's Cave, 1961, *A. Stirling* (GL).
In addition to the above record known only from **17** Port nam Marbh, *Slack & A. Stirling*.

Malus Mill.

M. sylvestris Mill.
Crab Apple. ? Hortal
13 Aros Park, 1967, *L. Ferguson*.
Woodland near houses.
In addition to the above locality, known only from **1** Iona, *Millar* (1972) and **4** Ulva House.
Recorded without locality in 1939 by the *BEC* as 'needs confirmation'.

Kerria DC.

K. japonica (L.) DC.
Hortal.
Naturalised in deciduous woodland near **10** Pennyghael House, 1968, *I. & L. Ferguson* (BM).

CRASSULACEAE

Sedum L.

S. rosea (L.) Scop. [Map, Fig. 2.34]
Rhodiola rosea L.
Roseroot. Native; AA
Lhwyd (*in litt.*, 1699).
On maritime and montane cliff ledges and gullies, to at least 750 m.
Occasional to locally frequent; especially on the more exposed coasts in the west.
1–11, 15–17

S. telephium L.
Orpine. Hortal
2 Iona, 1955, *Ferreira*.
Rarely naturalised on buildings and old walls.
In addition to the above, known only from: **6** Dervaig, *Bowerman* and **13** Tobermory.

S. anglicum Huds.
English Stonecrop. Native; OWE
1 Staffa, pre-1847, *Churchill Babington* in Wats., *Cyb. Brit.* ms. slips.
Rocks and cliff ledges especially near the sea, less frequently in short turf on cliff tops.
Frequent to locally abundant; widespread.
1–17

S. acre L.
Biting Stonecrop. Native and ? hortal; W
2 Iona, *Keddie* (1850)
On unconsolidated shell-sand, blowout margins, machair, more rarely on rocks or walls.

Occasional; widespread in the west, most frequent on Iona. Absent from the E except at Tobermory, where it may be an introduction.
2–10, 12, 13, 16

S. villosum L.
Hairy Stonecrop. Native; CN
8 Beinn na Srèine, 1961, *A. Stirling.*
Gravel flushes with *Koenigia, Juncus triglumis* and *Sagina subulata* at above 500 m.
Rare; confined to the mountains of Ardmeanach.
8 Beinn na h'Iolaire and Beinn na Srèine, *A. Stirling*; Creach Beinn (BM). E summit of Bearraich (BM).

Umbilicus DC.

U. rupestris (Salisb.) Dandy [Map, Fig. 2.1]
Navelwort. Native; OS
2 Y-Columb-Kill [Iona], *Lightfoot* (1777).
Rock crevices in cliff gullies.
Apparently confined to Iona, the adjacent islands and the SE corner of the Ross, with the exception of a single record from **8** Tavool which may be an introduction.
2 Iona, Dun I (BM); W and S of Baile Mòr (BM); SE coast; N central; near the Ruins, *Harvey* (BM, K-Wa), *Stables* (K-Wa); NW coast, *U. Duncan & Beattie.* **3** Erraid; Ross coast opposite Erraid; Eilean a' Chalmain (BM). **8** Near Tavool House, *Flannigan.*
On the northern edge of its range; since found on the Ardnamurchan peninsula.

SAXIFRAGACEAE

Saxifraga L.

S. stellaris L.
Alpine Saxifrage
Only known from *D. Robertson* (in Hudson, 1778), without exact locality, which needs confirmation.

S. stellaris L. [Map, Fig. 2.23]
Starry Saxifrage. Native; AA
Lhwyd (*in litt.*, 1699).
On damp scree margins, gully ledges, sides of high altitude runnels. Normally above 300 m, but sometimes carried to lower altitudes by streams.
Occasional to locally frequent; widespread through the mountains of Ardmeanach, the central massif, Glen Forsa and Loch Spelve to Loch Buie, with an outlier in N Mull at Beinn na Drise.
7, 8, 11, 15, 16

S. × urbium D. A. Webb
S. spathularis × umbrosa
London-pride. Hortal
8 Tavool House, 1968, *James & U. Duncan* (BM).
By derelict croft and streamside in gully.
Known only from the above record and in **4** Ulva, *Salzen.*

S. hirsuta L.
Kidney Saxifrage. Hortal
Naturalised in deciduous woodland at **7** Torloisk House, 1970, *James* (BM).

S. tridactylites L.
Rue-leaved Saxifrage. Native; W
3 Knockvologan, 1971, *Birse & J. Robertson* (SSS).
Known only from fixed dune at the above locality. In similar habitats in Coll and Tiree and in the Outer Hebrides.

S. hypnoides L. [Map, Fig. 2.20]
Mossy Saxifrage. Native; NM
Lhwyd (*in litt.*, 1699).
On wet rock faces, rocks by streams in gullies, wet scree margins to 850 m on Ben More.
Rare; in scattered localities from the Ardmeanach coast to Glen Forsa.
8 Tòn Dubh-Sgairt, *Templeman*; Creag a' Ghaill (BM); Aoineadh Thapuill (BM); Creag Brimishgan, *Templeman* (BM). **11** Ben More, summit ridge, *Gerrans* (BM); NW Corrie (BM). **15** Beinn Talaidh, Coire Ghaibhre (BM); Dùn da Ghaoithe, *CSSF*; Beinn Chreagach Mhòr and Bheag.

S. aizoides L.
Yellow Saxifrage. Native; AA
8 Gleann Seilisdeir, 1877, *G. Ross* (BM).
On flushed rock ledges and wet rock faces.
Rare; apparently confined to the mountains and cliffs of Ardmeanach and adjoining areas, with two outlying stations in the north.
6 Port Chill Bhraonain. **7** Beinn na Drise. **8** Creag a' Ghaill, *U. Duncan*; Tòn Dubh-Sgairt (BM); Allt na Teangaidh, *Kenneth*; S of Mackinnon's Cave; Clachandu, *CSSF*; Creag Brimishgan, *BEC*; Coirc Beinn, *Kenneth.*

S. oppositifolia L.
Purple Saxifrage. Native; AA
8 Creag Brimishgan, 1939, *BEC.*
On rock and unstable skeletal soil patches.
Rare; apparently confined to Ardmeanach and adjoining areas. The *Atlas* dot for 17/32 is based on an error.
8 Creag a' Ghaill, *Duncan*; Tòn Dubh-Sgairt (BM); Beinn na h'Iolaire, *Kenneth*; S of Mackinnon's Cave; above Tavool and Coire Buidhe (BM); Coirc Bheinn, *Kenneth*; Creag Brimishgan, *BEC.*

Tolmiea Torr. & Gray

T. menziesii (Pursh) Torr. & Gray
Hortal.
Known only from a ditch near cottage in **13** Aros Park 1966, *Cannon & Groves* (BM).

Chrysosplenium L.

C. oppositifolium L.
Opposite-leaved Golden-saxifrage. Native; WSA
G. Ross (1877).

In ditches, streamsides and riversides, waterfalls, damp shady gorges, sometimes with *Montia* at heads of springs.
Occasional to frequent; widely distributed.
2–17
There is a record in Maclagan (1855) for *C. alternifolium* L. from Loch Buie; this seems most unlikely and it is probable that *C. oppositifolium* was the species involved.

PARNASSIACEAE

Parnassia L.

P. palustris L.
Grass-of-Parnassus. Native; CN
G. Ross (1877).
Marshy streamsides and lowland flushed areas.
Occasional; widely distributed.
2–13, 15–17

GROSSULARIACEAE

Ribes L.

R. rubrum L.
R. sylvestre (Lam.) Mert. & Koch
Red Currant. Denizen
12 Salen, 1968, *Mullin.*
In field margins near houses.
In addition to the above record known only from **5** Sunipol (BM).

R. spicatum Robson
Downy Currant. Denizen
12 Glen Forsa House, 1967, *U. Duncan* (BM).
In damp woodland and waste ground near houses.
In addition to the above record known only from **3** Kintra.

R. nigrum L.
Black Currant. Denizen and ? native; W
7 Lagganulva, 1967, *J. Duncan.*
Usually near derelict buildings, houses or farms.
Rare; in scattered localities.
3 Kintra (BM). **4** Near Gometra House. **7** As above. **12** Below Meall nan Caorach; Kellan Wood. **13** Tobermory; Aros Park. **17** Ardchoirk.
This species may be native in *Alnus* carr on the N side of Loch na Keal at Kellan wood.

R. uva-crispa L.
Gooseberry. Denizen
13 Cliffs N of Tobermory, 1966, *Cannon & Groves.*
On roadsides and near houses.
Rare; naturalised in scattered localities.
6 Loch a' Chumhainn. **11** Knock (BM). **12** Salen, *U. Duncan*; Kellan Mill; Aros Castle. **13** Tobermory; cliffs N of Tobermory.

SARRACENIALES

DROSERACEAE

Drosera L.

D. rotundifolia L.
Round-leaved Sundew. Native; CN
Keddie (1850).
In bogs, peaty pools, especially with *Sphagnum* spp. ascending to 600 m and possibly higher.
Frequent to abundant; widespread.
1–17

D. anglica Huds.
Great Sundew. Native; CN
'Mull', 1830, without collector (K-Wa).
In boggy ground, margins of peaty pools, and especially *Carex/Sphagnum* oligotrophic flushes.
Occasional to frequent; widely distributed, but not recorded from the west except from Iona and the Ross of Mull.
2, 3, 6, 7, 9–14, 15–17
Stated by G. Ross (1879) to be 'equally common, in some places it seems to have starved out *D. rotundifolia*'. Today this statement would certainly be quite untrue as this species is not nearly so common as *D. rotundifolia.*

D. × obovata Mert. & Koch
D. anglica × rotundifolia
Native; CN
13 Near Tobermory, 1887, *Ewing* (BM).
In bogs and peaty areas with parents.
Rare; in scattered localities, possibly under-recorded.
3 Meall nan Càrn; Ardtun (BM). **9** Cnoc Mòr; An Crosan; Loch Assopol. **11** E side of Loch Bà, *U. Duncan.* **12** Loch Dairidh, *U. Duncan.* **13** As above. **14** Lower Glen Forsa.
The *Atlas* dot for 17/45 is here treated as the record for **13.**

D. intermedia Hayne
Oblong-leaved Sundew. Native; CN
Keddie (1850).
Often almost floating in peaty pools of bog complex, or in stagnant peaty channels with *Potamogeton polygonifolius.*
Occasional; in scattered localities but apparently absent from the west except for Iona and the Ross.
2, 3, 7–11, 13, 15–17

MYRTALES

LYTHRACEAE

Lythrum L.

L. salicaria L.
Purple Loosestrife. Native; W
Keddie (1850).
By streamsides and ditches.
Occasional; widespread but never very frequent; apparently somewhat more common in the west (Iona, the Ross and Ulva).
2–12, 14, 16, 17

Peplis L.

P. portula L.
Lythrum portula (L.) D. A. Webb
Water Purslane. Native; CN
G. Ross (1877).
Local ecology not known.
Rare; confined to Iona.
2 Iona, *Templeman*; between Dùn and Boineach, *Kenneth & A. Stirling.*

Stated by G. Ross (1879) to be 'common'. It seems unlikely that this species has decreased dramatically and probable that Ross was in error, confusing this species with another, perhaps *Montia* which is very common.

ELAEAGNACEAE

Hippophae L.

H. rhamnoides L.
Sea Buckthorn. Hortal
17 Càmas Mòr near Torosay Castle, 1967, *James* (BM).
About 20 old trees of both sexes, along 30 yards of foreshore by adjoining woodland. Known only from the above locality. See Groves (1971).

ONAGRACEAE

Critical advice given by P. H. Raven.

Epilobium L.

E. parviflorum Schreb.
Hoary Willowherb. Native; C
G. Ross (1877).
In ditches and seepage flushes.
Rare to occasional; widely scattered but absent from SE Mull.
2–8, 12

E. montanum L.
Broad-leaved Willowherb. Native; W
G. Ross (1877).
By roadsides, on gorge ledges and waste ground.
Frequent; widespread.
1–17

E. roseum Schreb.
Pale Willowherb
Known only from **2** Iona, Baile Mòr, 1912, *Wilmott*, which we must regard as doubtful.

E. tetragonum L. subsp. **tetragonum**
E. adnatum Griseb.
Square-stalked Willowherb
Has been recorded by *G. Ross* (1877) and from **4** Ulva and **6** Dervaig but without reliable confirmation.
Specimens under this name have proved on re-examination to be *E. obscurum.*

E. obscurum Schreb.
Short-fruited Willowherb. Native; WSA
Ewing (1890).
By roadsides, in ditches and on walls, often associated with houses or farms.
Occasional; scattered throughout the area at lower altitudes.
1–4, 6–10, 12–17

E. palustre L.
Marsh Willowherb. Native; CN
G. Ross (1877).
In shallow water of streamlets, ditches and loch shores and in lowland fen communities. Ascending to about 350 m.
Frequent; widespread.
1–17

E. anagallidifolium Lam.
Alpine Willowherb
Has been recorded from **15** Dùn da Ghaoithe, *CSSF*, but confirmation is necessary to insure that confusion with the next species, which is known with certainty in Mull, has not been involved.

E. alsinifolium Vill.
Chickweed Willowherb. Native; AA
11 Abhain Dhiseig, 1939, *Templeman* (BM).
In spring-line flushes associated with calcareous bands in the basalt.
In several localities in NE Ardmeanach, otherwise known only from two widely separated localities.
8 N of Tòn Dubh-Sgairt, *Templeman* (BM); The Wilderness (BM); Coireachan Gorma (BM); Bearraich to Creach Beinn (BM). **15** Beinn Chreagach Mhor.

E. brunnescens (Cockayne) Raven & Engelhorn
E. nerterioides A. Cunn.
New Zealand Willowherb. Hortal
13 Lakeside, Aros Park, 1958, *Gerrans* (BM).
In well-drained sites with periodic surface water flushing e.g., streamsides, moist tracksides, and banks.
Occasional; widely distributed but absent from the west.
6–17
Probably still spreading in Mull. It has been found in some very natural habitats, e.g., at c. 300 m in stream gullies on **10** Sròn Bhreac.

E. angustifolium L.
Chamaenerion angustifolium (L.) Scop.
Rosebay Willowherb. Colonist and ? native; W
13 Tobermory, 1952, *Macleay.*
Waste ground, roadsides and wood clearings.
Occasional; in scattered localities and never very common.
2–8, 10, 12–14, 17
Normally in disturbed ground, though one record, **8** E side of Maol Mheadonach, *U. Duncan*, may represent a native population. Otherwise the species appears to be a recent addition to the Mull flora, and was not noted by any of the earlier recorders.

Fuchsia L.

F. magellanica Lam.
Fuchsia. Hortal
2 Iona, W of Baile Mòr, 1958, *Gerrans* (BM).
By roadsides and in derelict gardens.
Occasional; in scattered localities.
2, 3, 5, 6, 9, 10, 12, 13, 16, 17
A large well-established population at Carsaig, by the river mouth, is probably the result of natural seed dispersal.

Circaea L.

C. lutetiana L.
Enchanter's-nightshade. Native; C
G. Ross (1877).

In shady places in mature woodland, on cliffs and on stream banks.
Occasional; in scattered localities.
3–8, 10, 13, 16, 17

C. × intermedia Ehrh.
C. alpina × lutetiana
Upland Enchanter's-nightshade. Native; CN
8 N of Tòn Dubh-Sgairt, 1939, *Templeman* (BM).
In shady places in gullies, scrub woodland and at streamsides.
Occasional; widely distributed and much more frequent than *C. lutetiana.*
3–13, 15–17

C. alpina L.
Alpine Enchanter's nightshade
Recorded by *G. Ross* (1877) and from **13** Druimfin (1879); also by *Templeman* in 1939 from **8** Tòn Dubh-Sgairt (BM). The record in Turner & Finlay (1967) is based on a specimen collected by Bowerman. Both specimens have been re-identified as *C. × intermedia.* In Raven's opinion, *C. alpina* is unlikely to occur on Mull.

HALORAGACEAE

Myriophyllum L.

M. verticillatum L.
Whorled Water-milfoil
Keddie (1850).
Known only from *Keddie* (1850), which appears very unlikely, and cannot be accepted without confirmation.

M. spicatum L.
Spiked Water-milfoil. Native; W
G. Ross (1877).
In shallow pools, demanding a higher cation level than the next species.
Rare; known certainly only from **1** Fladda (BM).
Recorded by G. Ross (1879) from **6** Mishnish Lochs and **13** Lochnameal. Ross, however, does not record *M. alterniflorum,* which is well known from the Mishnish Lochs and generally common in Mull. It seems likely that he was mistaken in his identification.

M. alterniflorum DC.
Alternate Water-milfoil. Native; WSA
6 Loch Peallach, 1965, *Kenneth* (BM).
Lochs, lochans and peaty pools, fast and slow-flowing streams or sometimes roadside ditches. Ascending to at least 400 m.
Occasional to frequent; widely distributed.
2, 3, 6, 7, 9–17
Very variable in size, both overall and in length of leaves and leaf segments, probably in association with rate of water flow and availability of nutrients.

CALLITRICHACEAE

Critical determinations by J. P. Savidge.

Callitriche L.

C. stagnalis Scop.
Common Water-starwort. Native; W
12 Near Salen, 1887, *Ewing* (GL).
In ditches, pools and streams, both stagnant and fast flowing; on organic mud and wet peat, to at least 300 m.
Frequent; widespread.
1–13, 15–17
An early record, Keddie (1850) as *C. verna,* probably represents this species.

C. intermedia Hoffm.
Intermediate Water-starwort. Native; WSA
Ewing (1899) for v.-c. 103; almost certainly Mull.
In ditches, shallow streams and loch margins, from fast-flowing streams to drying pools.
Rare; widely distributed but mainly in the north of Mull.
5 Calgary Bay (BM); E of Ensay (BM). **6** Tobermory Reservoir, *Chapple* (OXF). **7** Loch a' Ghael (BM); Loch Trath (BM); Cnoc an da Chinn. **8** Inch Kenneth (BM). **12** Lochan on Meall nan Gabhar (BM). **16** Lochbuie House lodge (BM).
All the material examined is subsp. **hamulata** (Kütz.) Clapham.

UMBELLIALES

CORNACEAE

Swida Opiz

S. sanguinea (L.) Opiz
Thelycrania sanguinea (L.) Fourr.
Dogwood. Hortal
Known only from **11** Knock, 1966, *Cannon & Groves.*

ARALIACEAE

Hedera L.

H. helix L.
Ivy. Native; W
8 Mackinnon's Cave, *Macculloch* (1819).
In woods, on sea cliffs and sheltered gullies, less commonly on open rock faces, to at least 700 m.
Frequent; widespread.
1–17

UMBELLIFERAE

Hydrocotyle L.

H. vulgaris L.
Marsh Pennywort. Native; WSA
Keddie (1850).
In marshy areas and ditches, showing a marked preference for nutrient-rich coastal freshwater mires.
Frequent; widespread at lower altitudes.
1–17

Sanicula L.

S. europaea L.
Sanicle. Native; WSA

G. Ross (1879).
In relatively base-rich woodland, especially in relict shady woodland conditions in gorges.
Occasional to frequent; widespread.
3–8, 10–13, 15–17

Astrantia L.

A. major L.
Astrantia. Hortal
Ewing (1890).
On roadsides near gardens and graveyards.
Rare; naturalised in two localities.
7 Kilninian church. **13** Tobermory outskirts, *Stace*.

Eryngium L.

E. maritimum L.
Sea-holly. Native; OS
Pennant (1774).
Loose sand above strandline on beaches.
Rare; recorded with certainty only from **2** Iona, where it was described by Pennant (1774) and Walker (1808) as frequent and by Keddie (1850) as abundant. Not found by Templeman in 1939, but noted as rare by Ferreira in 1955. It was not found during our Survey although there is no obvious reason why this species should not still occur on Iona where there are many suitable habitats.
Turner and Finlay (1967) include this species in their list, on the basis of a record **5** Treshnish Point which subsequent search has failed to confirm; the area appears to lack suitable habitats.

Anthriscus Pers.

A. caucalis Bieb.
A. neglecta Boiss. & Reut.
Bur Chervil
Recorded for v.-c. 103 by *Ewing* (1892); no direct evidence but probably not based on a Mull record.

A. sylvestris (L.) Hoffm.
Cow-parsley. Native; W
Keddie (1850).
Hedgerows, banks, usually associated with farms or houses.
Occasional; never very abundant except on some of the Treshnish Isles. This species is dominant on half the small, low-lying islet Sgeir an Fheòir. Its frequency in these habitats is probably influenced by the availability of phosphates.
1, 2, 5, 6, 8–10, 12–16

Myrrhis Mill.

M. odorata (L.) Scop.
Sweet Cicely. Denizen; CS
Ewing (1890).
Roadsides and tracksides associated with houses.
Rare; in scattered localities.
8 Tiroran, *Flannigan*. **12** Aros Castle; Salen crossroads (BM).

Torilis Adans.

T. japonica (Houtt.) DC.
Upright Hedge-parsley. Native or colonist; CS
G. Ross (1877).
Waste places and in grass at sides of roads and tracks.
Rare to occasional; stated by G. Ross (1879) to be 'common', but certainly not so today and apparently absent from SE Mull.
4–9, 12, 13

Conium L.

C. maculatum L.
Hemlock. Colonist
2 Iona, grounds of nunnery, *Keddie* (1850).
Cultivated areas and waste ground.
Rare; in scattered localities.
2 Iona, Culbuirg (BM); Tonighurt, *Braid* (GL). **6** Croig, *G. Ross* (1879); Dervaig, *Rae* (BM). **8** Inch Kenneth (BM); S Ardmeanach peninsula, *Flannigan*.
This species may have been cultivated in the past for its medicinal properties.

Apium L.

A. graveolens L.
Wild Celery
Recorded in 1939 by the BEC and noted by Wilmott (1942) as 'needs confirmation'. An improbable record judging from its general British distribution, but possible as a garden outcast.

A. inundatum (L.) Reichb. f.
Lesser Marshwort. Native; C
2 Iona, 1948, *Braid* (GL).
Shallow marsh runnels or immersed in the water of deeper lochs.
Rare; confined to Iona.
2 Iona, marshy area N of Golf Course (BM); Loch Staoineig (BM).

Carum L.

C. verticillatum (L.) Koch
Whorled Caraway. Native; OWE
10 Lower reaches of Coladoir river, 1956, *Corner* (BM).
Marshy slopes, stream banks and in runnels.
Rare; in scattered localities in SE Mull.
10 As above; Glenbyre burn (BM). **14** Corrynachenchy, *Kenneth* (BM). **16** Loch Spelve (BM); Loch Uisg, *J. Duncan*; road to Loch Buie, *Bowerman* (Herb. Bowerman).

Conopodium Koch

C. majus (Gouan) Loret
Pignut. Native; OWE
G. Ross (1877).
In deciduous woodland and in deep soils on open hillsides with *Pteridium* – presumably as a relic of past woodland.
Frequent; widespread.
1–17

Pimpinella L.

P. saxifraga L. [Map, Fig. 2.2]
Lesser Burnet-saxifrage. Native; W
2 Iona, 1955, *Ounsted et al.*
Grassy slopes on base-rich soils and in partially fixed dunes with shell fragments.
Rare; limited to Iona, the Ross and Inch Kenneth.
2 Iona, in several localities SE of Baile Mòr (BM); Loch Staoineig, *Stace*; SE coast. **3** Ardalanish Bay, *Kenneth & A. Stirling* (BM). **8** Inch Kenneth (BM).
Presumably limited by calcicole requirements, but other potential habitats exist in Mull, e.g., the shell-sand areas on Iona, at Calgary and at Port Langamull.

Aegopodium L.

A. podagraria L.
Ground-elder. Colonist
G. Ross (1877).
Always in vicinity of houses or farms.
Rare to occasional; in scattered localities.
2–17

Berula Koch

B. erecta (Huds.) Coville
Lesser Water-parsnip
Recorded by Ewing (1899) for v.-c. 103 but more likely from Coll or Tiree rather than Mull.

Oenanthe L.

O. lachenalii C. C. Gmel.
Parsley Water-dropwort. Native; WSA
8 Balnahard, 1939, *BEC.*
In wet meadows and ditches near the sea.
Rare; in four widely scattered localities.
8 As above; Port Uamh Beathaig, *Flannigan.* **12** Cnoc an Teine. **17** Lochdonhead, *U. Duncan.*

O. crocata L.
Hemlock Water-dropwort. Native; OWE
12 Salen, *Evans* in Maclagan (1855).
In ditches, streams and marshy pastures at low altitudes and especially near the sea.
Frequent; widely distributed.
2–10, 12–14, 16, 17

O. aquatica (L.) Poir.
Fine-leaved Water-dropwort
Recorded in Turner & Finlay (1967) almost certainly in error for *O. crocata.* Mull is well beyond the range in Britain for *O. aquatica.*

Aethusa L.

A. cynapium L.
Fool's-parsley. Colonist
12 Salen, 1968, *Mullin.*
Known only from the above record where it was growing in the new housing estate on the way to the pier. Very rare in W Scotland.

Ligusticum L.

L. scoticum L. [Map, Fig. 2.39]
Scots' Lovage. Native; AS
2 Iona, *Lightfoot* (1777).
Usually in damp, often shady crevices of coastal rocks, but in the absence of grazing, as on some of the smaller Treshnish Isles, this species occurs commonly in the open, often in luxurious forms.
Occasional; widespread in the west but never common except on some of the small islands.
1–9
Although unrecorded from the E side of Mull, it does occur on the adjacent mainland.

Angelica L.

A. sylvestris L.
Wild Angelica. Native; CN
Keddie (1850).
In ditches, wet places, probably associated with moving ground water.
Frequent to locally abundant; widespread.
1–17
Occurs as a dominant on the flat tops of the Treshnish Isles in the absence of heavy grazing. Here the plants are of short stature and with strikingly developed leaf bases. These features are presumably correlated with the very exposed habitat.

Heracleum L.

H. sphondylium L.
Hogweed. Native; W
G. Ross (1877).
Roadsides and tracksides. Absent from higher areas and places remote from human activity.
Occasional to locally frequent; widespread but apparently rarer in the south-east.
1–13, 15–17

H. mantegazzianum Somm. & Levier
Giant Hogweed. Hortal
Only known from a colony of 8 plants at **17** Torosay Castle, 1968, *James* (BM).

Daucus L.

D. carota L. subsp. **carota**
Wild Carrot. Native; W
Keddie (1850).
On coastal rocks, grassland and waysides.
Occasional to locally frequent; widespread but not recorded from SE Mull.
1–10, 12, 13, 16

EUPHORBIALES

EUPHORBIACEAE

Mercurialis L.

M. perennis L.
Dog's Mercury. Native; W
G. Ross (1877).
In deciduous woodland and in relict gully woods up to 450 m.
Frequent; widespread but most common in NW Mull.
2–17

Euphorbia L.

E. helioscopia L.
Sun Spurge. Colonist

G. Ross (1877).
Waste ground, arable fields and rubbish dumps.
Rare to occasional; widely scattered in the north and west, apparently most frequent in Iona.
2–7, 8, 12, 13
Stated by G. Ross (1879) to be 'common in cultivated ground'. This is not true today and may reflect a general reduction in the area of cultivated land.

E. peplus L.
Petty Spurge. Native or colonist; W
G. Ross (1877).
Known only from roadsides on Iona.
2 Iona, *McNiell*; Baile Mòr (BM).
Also stated by G. Ross (1879) to be 'common in cultivated ground' – see comments under previous species.

POLYGONALES

POLYGONACEAE

Polygonum L.

Critical determinations by B. T. Styles.

P. aviculare L. sensu lato
Knotgrass. Native; W
G. Ross (1877).
Occasional; widespread.
1–10, 12, 13, 16, 17

P. aviculare L. sensu stricto
P. heterophyllum Lindm.
Native; W
17 Lochdonhead, 1966, *U. Duncan.*
On muddy tracks and waste ground.
Widespread and doubtless more common than the records suggest, though possibly less frequent than *P. arenastrum.*
1 Staffa (BM). **3** W end of Loch Assopol, *U. Duncan.* **4** Ulva by pier, *U. Duncan.* **13** Tobermory, *Stace.* **17** As above.

P. arenastrum Bor.
P. aequale Lindm.
Native; W
16 Loch Buie, 1959, *C. Muirhead* (E).
On muddy tracks, farmyards, roadsides.
Widespread and possibly more frequent than *P. aviculare* s.s.
6 Loch Peallach (BM). **7** Ledmore (BM); Camas an Lagain, *Richards* (OXF). **8** Inch Kenneth (BM); Burgh, SW Ardmeanach (BM). **16** As above.

P. oxyspermum C. A. Meyer ex Ledeb. subsp. **raii** (Bab.) D. A. Webb & Chater
P. raii Bab.
Ray's Knotgrass. Native; OS
5 Calgary Bay, 1939, *BEC.*
Sandy shores and coastal shingle above drift line.
Rare; in scattered localities in the west.
2 Iona, *Ferreira.* **3** Uisken (BM); *Kenneth & A. Stirling*; near Loch Mòr (BM); Cnoc an t'Suidhe (BM). **5** Calgary Bay, *U. Duncan.* **9** Cnoc Mòr.

P. viviparum L.
Alpine Bistort. Native; AA
11 Coire nam Fuaran, 1939, *Templeman* (BM).
Mountain pastures, especially along spring lines, wet rock ledges and screes.
Rare; scattered localities from W Ardmeanach, through the central massif to the mountains of Glen Forsa, with an outlier in the Ross.
8 Caisteal Sloc nam Ban, *Ribbons*; Fionna Mhan. **9** Cnoc Mòr. **11** Ben More, summit, N side. **15** Mainnir nam Fiadh; Beinn Talaidh, *J. Robertson & Birse.*

P. bistorta L.
Common Bistort. Hortal
Known only from Druimfin, where G. Ross (1879) noted it as a garden escape.

P. amphibium L.
Amphibious Bistort. Native; W
2 Iona, 1891, *Kidston* (GL).
Margins of lochans and streams, sometimes spreading on to land in terrestrial forms.
Rare to occasional; in scattered localities, especially in Iona and the Ross.
2, 3, 8, 9, 11, 12, 16

P. persicaria L.
Redshanks. Native or colonist; W
G. Ross (1877).
In waste ground, arable fields and rubbish dumps.
Occasional; widespread in scattered localities.
2–17

P. lapathifolium L.
Pale Persicaria. Native or colonist; W
4 Ulva, 1948, *Salzen.*
In waste ground, arable fields and rubbish dumps.
Rare; in a few scattered western localities. Probably under-recorded and possibly confused with *P. persicaria.*
2 Iona, *Ferreira.* **4** As above; near A' Chrannag. **5** Langamull (BM).

P. hydropiper L.
Common Water-pepper. Native; W
G. Ross (1877).
On stony loch margins and in marshy pasture and ditches, especially near farms.
Occasional; widespread, but not recorded from Ardmeanach.
1–7, 9–17

P. convolvulus L.
Black Bindweed. Native or colonist; W
G. Ross (1877).
On arable land and waste ground.
Rare; in scattered localities.
2 Iona, *Ferreira*; field in centre of island. **3**

Uisken. **12** Glen Forsa House (BM). **13** Tobermory, *Stace*.
Stated by G. Ross (1879) to be 'common'.

P. aubertii L. Henry
P. baldschuanicum auct., non Regel
Russian Vine. Hortal
Known only from **6** Penmore Mill, 1968, *Watling* (BM), where it occurs with road-side shrubs.

P. cuspidatum Sieb. & Zucc.
Japanese Knotweed. Hortal
12 Between Salen and Forsa river, 1958, *Gerrans* (BM).
Roadsides and waste ground.
Occasional; well established in scattered localities.
3, 5–9, 11–13, 16, 17

P. sachalinense F. Schmidt
Giant Knotweed. Hortal
6 Dervaig, 1967, *Kenneth* (BM).
In addition to the above record also naturalised at **13** Tobermory.

P. polystachyum Wall. ex Meisn.
Himalayan Knotweed. Hortal
13 Tobermory, 1967, *Kenneth* (BM).
In addition to the above locality also naturalised by the lake in **13** Aros Park (BM).

P. campanulatum Hook. f.
Lesser Knotweed. Hortal
5 Near Calgary, 1965, *Kenneth*.
In addition to the above record, also naturalised near **6** Dùn Auladh on the Dervaig to Penmore road (BM).

Koenigia L.

K. islandica L. [Map, Fig. 2.15]
Iceland Purslane. Native; AS
8 Maol Mheadonach and Beinn na Srèine, 1956, *Corner* (E).
In open flushed basaltic gravel in eroded patches on mostly N and NE facing slope 300 m; continuing frost action and wind erosion results in the formation of soil polygons, and the habitat remains open and suitable for annual growth.
Rare; confined to the mountains of the Ardmeanach, but there locally frequent.
8 Maol Mheadonach, as above and *Gerrans* (BM); Beinn na h'Iolaire (BM); Beinn na Srèine, as above; Coire Bharr Reamhair (BM); Creach Bheinn (BM); E summit of Bearaich (BM); Fionna Mhàm (BM); above Coire nan Dearcag, *Ratcliffe*
One of the most interesting species in the Mull flora, occurring elsewhere only in Skye in the British Isles. It may be that the peculiar conditions necessary to produce this habitat, among which the climatic factors of rapid freezing and thawing coupled with salt laden winds are probably important, are confined in Mull to the area now occupied by *Koenigia*. Its local frequency suggests that here its special needs are being well met by the environment.

Oxyria Hill

O. digyna (L.) Hill [Map, Fig. 2.21]
Mountain Sorrel. Native; AA
11 Ben More, 1939, *BEC*.
On flushed rock ledges and unstable slopes above 450 m, rarely on sea cliffs where it may be adventive.
Rare to occasional; widespread in central Mull, from W Ardmeanach, through the central massif to the mountains of Glen Forsa and Creach Beinn.
8, 11, 14–16
Has also been recorded from **4** Ulva, but is unlikely in this area.

Rumex L.

R. acetosella L.
Sheep's Sorrel. Native; W
G. Ross (1877).
In base-poor well-drained soils at roadsides and tracksides, in open woodland, grassland, shingle beaches and in Callunetum.
Frequent to abundant; widespread.
1–17

R. acetosa L.
Common Sorrel. Native; CN
G. Ross (1877).
In deep or peaty loams or in shallow soils with a high nitrogen content. Roadsides, gullies and rough grassland.
Abundant; widespread.
1–17
On the small islands it is often associated with bird-nesting sites, where it grows luxuriantly.

R. crispus L.
Curled Dock. Native; W
G. Ross (1877).
In maritime shingle, among rocks, roadsides and waste places.
Frequent to abundant; widespread at low altitudes.
1–17

R. obtusifolius L.
Broad-leaved Dock. Native; WSA
Ewing (1890).
On waste ground, by roadsides and in wet meadows.
Occasional; widely distributed.
1–13, 15–17

R. sanguineus L. var. **viridis** Sibth.
Wood Dock. Native; W
Ewing (1890).
On stony and sandy shores above the strand line, waste ground and roadsides.
Rare to occasional; mainly in the southwest.
2–6, 9–11, 17

R. conglomeratus Murr.
Clustered Dock. Native; WSA

G. Ross (1877).
In damp woodlands and waste places.
Rare to occasional; in scattered localities.
2–4, 6–9, 12, 16, 17
The available information on this species in Mull is unsatisfactory and further data is most desirable.

Rheum L.

R. rhaponticum L.
Rhubarb. Denizen
12 Cnoc an Teine, 1970, *Robson & Taylor*.
Known from the above locality where it is naturalised by a derelict farmyard and on rocks near sea at **8** Allt na Teangaidh, *J. Raven*.

URTICALES

URTICACEAE

Urtica L.

U. urens L.
Annual Nettle. Native; W
2 Iona, *Keddie* (1850).
On arable land and waste ground; rarely above the strand line on sandy beaches.
Rare; in widely scattered localities.
2 As above. **3** Erraid; coast opposite Erraid; Cnoc an t'Suidhe (BM); Ardtun. **6** Glengorm Castle, *V. Gordon*. **8** SW Ardmeanach Peninsula. **12** Kellan Mill. **13** Tobermory. **17** Torosay Castle.

U. dioica L.
Common Nettle. Native; W
Macculloch (1819).
In waste places and by roadsides and derelict buildings, in nitrogen-rich habitats.
Frequent; widespread.
1–17

ULMACEAE

Ulmus L.

U. glabra Huds.
Wych Elm. Native; W
G. Ross (1877).
In deciduous woodland and especially in relict woodland at lower altitudes in gorges, ascending to c. 250 m.
Occasional; widely distributed.
2–8, 10–13, 15–17

U. procera Salisb.
English Elm. Hortal
10 Carsaig, 1958, *U. Duncan & Beattie*.
In addition to the above locality also recorded from **8** SW Ardmeanach Peninsula and **17** Torosay Castle.

MYRICALES

MYRICACEAE

Myrica L.

M. gale L.
Bog Myrtle. Native; ON
16 Loch Spelve and Loch Buy [Buie], 1848, *Syme* (BM).
In boggy places with moving ground water, forming a characteristic association with *Molinia*, often on sloping ground near lochs and streams. Ascending to about 250 m.
Abundant; widespread, but not recorded from the more exposed small islands.
2–17

FAGALES

BETULACEAE

Betula L.

B. pubescens is represented by a wide range of forms, the extremes of which in the direction of *B. pendula* we have recorded as 'intermediates'. These are probably variants of the tetraploid *B. pubescens*, rather than the simple hybrids with *B. pendula* that they would appear to be on morphological grounds. Cytological investigations would be essential to reach a definite conclusion. To quote Walters (1968) '. . . if the tetraploid has a polytopic origin, we may be pursuing a will-o-the-wisp in looking for pure *pubescens* at all.' We believe that the majority of the scrubby birch on Mull is *B. pubescens* plus intermediates, and that *B. pendula* is much more limited in distribution and possibly associated with the less natural habitats.

B. pendula Roth. (*B. verrucosa* Ehrh.)
Silver Birch. Native
G. Ross (1877).
By roadsides and streamsides, in deciduous woodland and in gullies at lower altitudes.
Frequent; widely distributed, but probably over-recorded although certainly more widespread than suggested by Perring & Walters (1962).
3–17

B. pubescens Ehrh.
Downy Birch. Native; CN
Ewing (1890).
In scrubby woods on mountain slopes, gullies and wet streamsides; often forming large trees in sheltered sites.
Frequent; widely distributed, but records include intermediate forms.
2–17
Macculloch (1819) gives the first known record for 'birch'.

B. nana L.
Dwarf Birch
The locality on a specimen in the Taunton Museum, 1879, *Slater* (TTN) is 'Ben Mohr Mull', written presumably in the collector's own hand. However, in the absence of any other record from the area, and because there are Ben Mores in other vice-counties, we cannot exclude some error of labelling although Slater had a reputation for careful work. Efforts by ourselves and others have failed to confirm its occurrence, and suitable habitats are not conspicuous in our area.

Alnus Mill.

A. glutinosa (L.) Gaertn.
Alder. Native; W
Macculloch (1819).
Along riversides and streamsides at lower altitudes, sometimes forming small areas of woodland along seepage lines.
Occasional to frequent; widespread but apparently more frequent in N Mull. Noted as an introduction in Iona.
2–17

CORYLACEAE

Carpinus L.

C. betulus L.
Hornbeam. Hortal
Known only from *G. Ross* (1877).

Corylus L.

C. avellana L.
Hazel. Native; W
Macculloch (1819).
With birch and oak in woodland and forming scrub in gullies and on cliff screes. Ascending to 400 m.
Frequent to abundant; widespread.
2–17

FAGACEAE

Fagus L.

F. sylvatica L.
Beech. Hortal
G. Ross (1877).
In park woodland and around larger houses.
Rare to occasional; in scattered localities.
4–7, 10–14, 16, 17

Castanea Mill.

C. sativa Mill.
Sweet Chestnut. Hortal
G. Ross (1877).
Noted also at **10** Carsig and at **17** Torosay Castle.

Quercus L.

We believe that *Q. robur* is more frequent in our area than *Q. petraea*. Intermediates have been recorded both by ourselves and others, but there seems to be no agreement among the specialists as to the likely status of apparently hybrid populations. Thus Cousens (1965) states 'Introgression in the pedunculate oak (*Q. robur*) and sessile oak (*Q. petraea*) in Scotland was known to be so extensive that neither could be defined satisfactorily'.

Q. robur L.
Pedunculate Oak. Native; C
16 N of Loch Spelve, 1848, *Syme* (BM).
In deciduous woods with birch and in coastal gullies, roadsides and around houses, where often planted.
Occasional to frequent; widespread.
2–17

Q. petraea (Mattuschka) Liebl.
Sessile Oak. Native; C
7 Torloisk, 1939, *BEC.*
In scrubby woodland on slopes with birch, in gorges and at roadsides.
Occasional; widespread but probably less frequent than *Q. robur.*
3–7, 10–17
The above distribution includes records of plants of intermediate character, and the Mull oaks which have been recorded as this species as a whole do not conform to the normally accepted concept of this taxon.

SALICALES

SALICACEAE

Populus L.

P. alba L.
White Poplar. Hortal
G. Ross (1877).
In addition to the above record, it has recently been reported from **12** Kellan Mill.

P. tremula L.
Aspen. Native; W
Lightfoot (1789).
In gullies with relict woodland on mountains at lower altitudes, also on coastal cliffs colonising new habitats with *Corylus.*
Occasional; widely distributed in the west and along the S coast.
2–10, 16

P. nigra agg. (including *P. serotina* Hartig)
Black Poplar. Hortal
Recorded only in **11** on the Gruline policy at Loch Bà, 1956, *H. Milne-Redhead.*

P. gileadensis Rouleau
Balsam Poplar. Hortal
Known only from **12** Ardnacross, 1968, *James & U. Duncan.*

P. trichocarpa Torr. & Gray
Hortal.
Known only from **7** Torloisk House grounds, 1970, *James* (BM).

Salix L.

Critical determinations by A. Neumann.

S. pentandra L.
Bay Willow. ? Native; CN
13 Aros Park, 1969, *Jermy* (BM).
Known only from the above where it occurs near the lake in *Phalaris* swamp.

S. alba L.
White Willow. Hortal
G. Ross (1877).
On roadsides and near houses.
Rare; in scattered localities.
5 Calgary, *U. Duncan & Beattie.* **10** E of Killiemore House, *U. Duncan*; Aird of Kinloch (BM).
13 Tobermory; Calve and Lovage islands, *Harris.*

Sometimes used as fencing stakes and possibly established in this way.

S. fragilis L.
Crack Willow. Hortal
12 Salen, 1966, *Cannon & Groves.*
By roadsides and near houses.
Rare; in scattered localities.
6 Dervaig, *Beattie.* **12** As above. **17** Lochdonhead, *U. Duncan.*
Also possibly used as fence posts.

S. purpurea L.
Purple Willow. Hortal
Ewing (1890).
By roadside near houses.
Rare; in scattered localities, possibly as a result of planting for osiers.
3 Uisken, *Mullin.* **6** Dervaig, *Beattie.* **7** near Ballygowan (BM); near Fanmore (BM); N of Kilfinichen Bay (BM). **12** SW of Salen, *U. Duncan* (BM); Aros Castle. **17** Lochdonhead, *U. Duncan.*

S. viminalis L.
Osier. Hortal
6 Dervaig, 1956, *Beattie.*
On roadsides near houses.
Rare to occasional; Probably planted as an osier with subsequent natural spread.
3, 4, 6–10, 12, 13, 16, 17

S. caprea L.
Goat Willow. Native; CN
G. Ross (1877).
In deciduous woodland, beside rivers, streams and roads.
Occasional; in widely scattered localities.
2–9, 11–14
Subsp. **sericea** (Anderss.) Flod. occurs in similar habitats to subsp. *capraea* and is not recorded separately.

S. × laurina Sm.
S. capraea × viminalis
Native; CN
9 Tràigh nam Beach, 1965, *U. Duncan & James* (BM).
By roadsides and rivers.
In addition to the above record only known from **14** Dail Bhàite (BM).

S. cinerea L. subsp. **oleifolia** Macveight
S. cinerea subsp. *atrocinerea* (Brot.) Silva & Sobrinho
Grey Willow. Native; OWE
16 N of Loch Spelve, 1848, *Syme* (BM).
In deciduous woodland, roadside ditches and at loch sides. Forming a carr with birch and alder, colonising *Carex rostrata* mire.
Occasional; in scattered localities.
3–8, 10–17
Subsp. *cinerea* has not been recorded. A specimen in Herb. H. C. Watson (K-Wa) labelled 'Ardjura, 1848, *J. T. Syme*' is presumably a duplicate of the above. The record from Iona (Millar 1972) as *S. atrocinerea* is an error for *S. aurita.*

S. aurita L.
Eared Willow. Native; CN
'Mull', 1848, *Syme* (BM).
By riversides, streamsides and lochsides, in flushes on hillsides, among rocks and on cliff ledges.
Abundant; widely distributed and by far the most frequent willow on Mull.
1–17
Can occur as dwarf forms at high altitudes. Hybrids with *S. cinerea* are well known and probably occur in our area, although they have not been detected. **S. ×ambigua** Ehrh. (*S. aurita × repens*) is recorded in Hooker (1830) as 'Isle of Staffa, Mr Borrer', but has not been otherwise noted.

S. repens L.
Creeping Willow. Native; CN
2 Iona, *Walker* (1808).
In Callunetum, in dry grassy slopes and wet grass heath, especially on slopes near the sea.
Frequent; widespread.
1–17
Subsp. **argentea** (Sm.) G. & A. Camus has been recorded from **2** Iona, **3** Ardalanish Bay and **6** the Mishnish Lochs. It is probably characteristic of damp areas in fixed shell-sand dunes.

S. lapponum L.
Downy Willow
Recorded only on **2** Iona (*Garnett*, 1800) which must almost certainly be an error, and by Wilmott (1912), who himself clearly felt doubtful as he noted 'no flowers, named on leaf characters'.

S. herbacea L. [Map, Fig. 2.27]
Dwarf Willow. Native; AA
16 Creach Bhinn [Creach Beinn], 1848, *Syme* (BM; K-Wa).
Stony mountain tops, nutrient poor summit grassland and established scree, rarely descending to as low as 480 m, under localised special conditions below the N corrie of Ben More, where wind funnelling may effect the microclimate.
Occasional; mountains of the central massif and Glen Forsa, with outliers on Creach Beinn, at Cnoc an dà Chinn and at Cruachan Min in Brolass.
7–9, 11, 15, 16

ERICALES

ERICACEAE

Rhododendron L.

R. ponticum L.
Rhododendron. Hortal
8 S Ardmeanach Peninsula, 1956, *Flannigan.*
By roadsides, near houses and in deciduous woodland.
Occasional; widespread in scattered localities, especially round the large estates. Certainly

spreading naturally, particularly around Loch Spelve and Loch Uisg.
3, 5–13, 16, 17

Gaultheria L.

G. shallon Pursh
Shallon. Hortal
13 Aros Park, 1958, *U. Duncan & Beattie*.
In margins of deciduous woodland and on low clay cliffs.
Rare; in a few localities. Presumably originally introduced for game cover.
12 Rubha Àrd Ealasaid, *Gerrans* (BM). **13** As above. **17** Rubha na Sròine (BM).

Pernettya Gaudich.

P. mucronata (L.f.) Gaudich. ex Spreng.
Prickly Heath. Hortal
13 Near Tobermory, 1934, *Storr* (E).
In roadside and deciduous woodland.
13 Tobermory crossroads, *U. Duncan & Beattie*; Aros Park. **16** Lochbuie.

Arctostaphyllos Adans.

A. uva-ursi (L.) Spreng. [Map, Fig. 2.28]
Bearberry. Native; AA
Lightfoot (1777), but see below.
In cliff grassland, Callunetum, rocky outcrops and gorges.
Occasional; widespread, but most frequent on the granite of the Ross.
2, 3, 6, 7, 9, 11, 12, 15, 16
A record for *Arctous alpinus* by Lhwyd in Ray (1724) from 'That end of Mull next to Y-Columb-Kill [Iona]' is almost certainly this species and is the earliest record from Mull known to us.

Arctous (A. Gray) Nied.

A. alpinus (L.) Nied.
Alpine Bearberry
Lhwyd in *Ray* (1724) but see notes under previous species.

Calluna Salisb.

C. vulgaris (L.) Hull
Heather. Native; W
G. Ross (1877).
Forming large areas of moorland, especially on the better drained areas and ascending to at least 650 m.
Abundant; widespread.
1–17
White-flowered forms have been recorded.

Erica L.

E. tetralix L.
Cross-leaved Heath. Native; OWE
G. Ross (1877).
The wetter moorland areas, e.g., boggy depressions with *Eriophorum vaginatum* in heather moorland. Possibly more frequent on the granite of the Ross, though this may be due to its colonising old peat cuttings.
Frequent to abundant; widespread.
1–17

E. cinerea L.
Bell Heather. Native; OWE
G. Ross (1877).
In drier moorland areas forming dwarf shrub communities with *Calluna*, also on rock ledges in gullies.
Abundant; widespread.
1–17

E. vagans L.
Hortal
Recorded only from **6** SE of Glengorm Castle, 1969, *James* (BM).

Vaccinium L.

V. vitis-idaea L.
Cowberry. Native; AA
Lhwyd (*in litt.*, 1699).
In Callunetum and grassheath at higher altitudes, but probably descending to c. 150 m.
Occasional; widely distributed.
6–11, 15, 16

V. myrtillus L.
Bilberry. Native; CN
G. Ross (1877).
In better drained habitats, from deciduous woodland, through heathland to ledges on cliffs and in gorges.
Frequent to abundant; widespread.
2–17

V. uliginosum L.
Bog Blaeberry
'Low boggy grounds in the island of Mull', *Lightfoot* (1777) but in the absence of other records and in view of the habitat noted, it seems likely to be an error.

V. oxycoccus L.
Common Cranberry. Native; CN
G. Ross (1877).
In *Sphagnum* hummock in Eriophoretum near small lochan.
Rare; known from one locality in N Mull, where it occurs as a very small population.
6 Loch na Criadach Moire (BM).
Also noted by Wilmott in 1912 on Iona, but this record must be regarded as doubtful.

Enkianthus Lour.

E. deflexus (Griff.) Schneid.
Hortal
Only known record **13** Aros Park, 1967, *James & L. Ferguson* (BM).

PYROLACEAE

Pyrola L.

P. minor L.
Common Wintergreen. Native; CN
13 Tobermory, 1829, *Christy* (E).
In well drained habitats in deciduous woodland and in Callunetum.

Rare; scattered localities in the north with outliers in Ardmeanach.
6 Loch Mingary (BM). **7** Above waterfall near Tòrr an Damh. **8** S Ardmeanach Peninsula, *Flannigan*. **13** Aros Park (BM); Tobermory as above and *G. Gordon* (K-Wa); Rubha na Leip; opposite Calve and Lovage islands.

P. media Sw.
Intermediate Wintergreen. Native; CN
13 Tobermory, 1829, *Christy* (E, K-Wa).
In deciduous woodland.
Rare; existing records, of which only one is recent (1970), are confined to the Tobermory area.
13 Tobermory, as above, and 1851, *Wilson* (BM); Heatherfield, 1876, *G. Ross* (BM); Druimfin and Baliscate, *G. Ross* (1879); Aros Park, *Clark*. A specimen, 'Loch Frisa', 1905, *C. & E. Vachell* (NMW), cannot be precisely localised to a recording square.

P. rotundifolia L. subsp. **maritima** (Kenyon) E. F. Warb.
Round-leaved Wintergreen
Listed in Turner & Finlay (1967) but its occurrence has not been confirmed and is unlikely (see Perring & Sell, 1968). There is also an old record for Mull of *P. rotundifolia* in the MS list of G. Gordon (1829).

Orthilia Raf.

O. secunda (L.) House
Serrated Wintergreen. Native; CN
14 Allt Mòr Coire nan Eunachair, 1968, *Stace* (BM).
On vertical rocks by stream. The discovery of this on Mull constitutes an extension of range from the adjacent parts of Morvern. Known only from the above locality.

EMPETRACEAE

Empetrum L.

E. nigrum L.
Crowberry. Native; AA
16 Loch Spelve, 1848, *Syme* (BM).
On exposed cliff tops, in bogs and dwarf shrub communities and at higher altitudes in montane heathland and among screes.
Occasional to locally frequent; especially on Iona, the Ross and adjacent Brolass, otherwise widely distributed.
1–17

E. hermaphroditum Hagerup
Mountain Crowberry. Native; AA
11 Beinn nan Gobhar, 1968, *McAllister* (BM).
Confirmed only for the N-facing cliffs on the above mountain. Two field records for which confirmation is desirable are from **11** the saddle between Beinn Fhada and Ben More and **15** Beinn Creagach Mhor.

PLUMBAGINALES

PLUMBAGINACEAE

Armeria Willd.

A. maritima (Mill.) Willd.
Thrift. Native; ON
Keddie (1850).
On coastal rocks and cliff ledges, on silt at middle saltmarsh levels, and in montane habitats in crevices, on ledges and in screes.
Frequent all round the coasts, at higher altitudes on the mountains of the central massif, Ardmeanach and Glen Forsa.
1–17

PRIMULALES

PRIMULACEAE

Primula L.

P. veris L.
Cowslip
Stated to be naturalised at Druimfin, *G. Ross* (1879), but not otherwise recorded.

P. vulgaris Huds.
Primrose. Native; W
Keddie (1850).
In deciduous woodland, on roadside banks and grassy cliffs. Ascending shady mountain gullies to 300 m.
Frequent; widespread throughout the area apart from the higher habitats.
1–17.

Lysimachia L.

L. nemorum L.
Yellow Pimpernel. Native; WSA
13 Tobermory, 1829, *Christy* (E).
By sides of streams, wet tracksides and in damp deciduous woodland.
Frequent; widespread.
2–17

L. vulgaris L.
Yellow Loosestrife. Native; C
Lightfoot (1777).
On deep peat in mixed fen with *Juncus* spp. and *Phragmites*.
Rare; known at present with certainty from only one locality.
2 Iona, *Ritchie & Ritchie* (1934). **13** Lochnameal (BM).
Said by Lightfoot to be 'by the side of lochs in the island of Mull', noted also by Syme (1864) 'reaching north to Mull in Argyllshire' and by G. Ross (1877). There is some evidence that the Ross record may be an error for *L. nemorum* and it does not seem likely that there has been a great reduction in the distribution of this species in Mull since the time of Lightfoot.

L. punctata L.
Dotted Loosestrife. Hortal
12 1 mile south of Salen, 1967, *Kenneth*.
In roadside ditches near houses.
In addition to the above record also known

from **8** Kilfinichen (BM) and **17** between Torosay Castle and the sea (BM).

Anagallis L.

A. tenella (L.) L. [Map, Fig. 2.11]
Bog Pimpernel. Native; OS
2 Iona, *Garnett* (1800).
In lowland mires, especially at the sides of runnels and in other relatively well drained habitats.
Occasional to locally frequent; widespread in the west, especially on the granite of the Ross and on Iona.
1–5, 7–10, 12, 15

A. arvensis L.
Scarlet Pimpernel. Native or colonist; W
Keddie (1850).
On arable land, waste ground and tracks.
Rare; in scattered localities in the west and north.
2 Iona: near Baile Mòr (BM); north central; Sandeels Bay, *Braid* (GL). **3** Uisken. **5** Calgary, 1876, *G. Ross* (BM). **6** Sorne, *G. Ross* (1878). **7** Lagganulva, *J. Duncan*. **8** Tiroran, *Flannigan*; SW Ardmeanach (BM). **13** Tobermory; Erray House.
Decreasing in the area owing to changes in land-use.

A. minima (L.) E. H. L. Krause
Chaffweed. ? Native; C
7 Kellan, 1939, *Temperley* (BM).
Moist sandy places and muddy tracks.
Rare to occasional; in scattered localities but apparently absent from SE Mull.
Probably under-recorded.
2–4, 6, 8, 10–12

Glaux L.

G. maritima L.
Sea-milkwort. Native; CN
Keddie (1850).
In upper level of saltmarsh, in splash-zone sandy turf, and at side of streamlets near the sea.
Frequent; widespread round the coasts.
1–17

Samolus L.

S. valerandi L.
Brookweed. Native; C
1 Staffa, 1829, *G. Gordon*.
At top of saltmarshes, among rocks at upper levels of maritime communities and by the sides of streams and runnels near the sea.
Occasional; in scattered localities.
1–6, 8, 10, 12, 17

CONTORTAE

BUDDLEJACEAE

Buddleja L.

B. davidii Franch.
Butterfly-bush. Hortal
2 Iona, 1955, *Harris*.
Naturalised in scattered localities near gardens. In addition to the above locality known also from **10** Pennyghael woods. **11** Loch Bà, *H. Milne-Redhead*. **13** Aros Park.

OLEACEAE

Fraxinus L.

F. excelsior L.
Ash. Native; WSA
8 Mackinnon's Cave and **11** Scarisdale, *Macculloch* (1819).
Associated with *Quercus petraea* in sheltered valleys and on the lower slopes of basalt terraces, usually on the more base-rich soils.
Occasional to frequent; widely distributed.
2–17
The ash is a favourite tree for planting round houses and farms. An extensive natural ashwood, by Mull standards, existed on the Laggan coast at Allt Ohirnie until the catastrophic gales of September 1962, when it was almost totally destroyed.

Syringa L.

S. vulgaris L.
Lilac. Hortal
Known only as an introduction from **12** Salen, 1966, *U. Duncan*.

Ligustrum L.

L. vulgare L.
Wild Privet. Hortal
Strongarbh, *G. Ross* (1879).
Usually near farms or houses.
Rare; in scattered localities and probably always as a result of original deliberate introduction.
5 S of Caliach Point. **8** S Ardmeanach, *Flannigan*. **9** near Lee (BM). **10** Glenbyre. **12** Salen. **13** Tobermory and cliffs to N of town.

L. ovalifolium Hassk.
Garden Privet. Hortal
Recorded only from **12** Salen, 1966, *U. Duncan*, where it persists as a relic of an old hedge planting.

APOCYNACEAE

Vinca L.

V. minor L.
Lesser Periwinkle. Hortal
13 Druimfin, *G. Ross* (1879).
In addition to the above record known only from one recent locality: **12** Kellan Mill.

V. major L.
Greater Periwinkle. Hortal
Known only from **12** Salen, 1958, *Gerrans* (BM), where it was persisting in 1968.

GENTIANACEAE

Centaurium Hill

C. erythraea Rafn
Common Centaury. Native; WSA
1 Staffa and **4** Ulva, 1805, without collector's name (E).

In machair grassland and short turf on coastal cliffs.
Occasional; widely distributed in the west.
1–10
The record for *C. latifolium* auct. (Staffa, 1829, *G. Gordon*) quoted by Hooker (1830) and Watson (1837) must be assumed to be this species as *C. latifolium* (Sm.) Druce has only occurred in Lancashire where it is now extinct.

C. littorale (D. Turner) Gilmour
Seaside Centaury
Recorded at **13** Sorne by G. Ross (1879) but not substantiated. A specimen of *C. erythraea* from **1** Staffa, 1843, Stables (K-Wa) labelled '*? littoralis*', is probably the basis for the record in the *Cyb. Brit. MS* slips. The species could, however, occur in our area.

Gentianella Moench

G. campestris (L.) Börner
Field Gentian. Native; CN
2 Iona, *Walker* (1808).
In machair and in short turf by the sea, on cliffs and at roadsides.
Occasional; widespread but absent from the higher areas of central Mull and from Brolass.
1–8, 11–17

G. amarella (L.) Börner
Autumn Gentian
Recorded by *Garnett* (1800); presumably an error for the preceding species.

MENYANTHACEAE

Menyanthes L.

M. trifoliata L.
Bogbean. Native; CN
2 Iona, *Garnett* (1800).
In pools and the shallower parts of lochs, more rarely in streams and seepages.
Frequent; widespread, but known from **8** the Ardmeanach only by one record.
2–17

TUBIFLORAE

BORAGINACEAE

Symphytum L.

S. officinale L.
Common Comfrey. Hortal ?
Ewing (1890).
On roadsides and similar disturbed localities.
Rare.
Has been recorded from **8** Tavool House, *BEC*; S Ardmeanach Peninsula, *Flannigan* (probably same locality as the preceding). **12** Salen, *U. Duncan & Beattie.*
It is very probable that these records should be applied to the following species. A specimen from a rubbish dump at **10** Carsaig (BM) has been tentatively identified by Perring as *S. officinale*, but in that situation was very probably of recent garden origin.

S. × uplandicum Nyman
S. asperum × officinale
Russian Comfrey. Hortal
12 Track to Aros Castle, 1958, *Gerrans* (BM).
Along roadsides and wood margins.
Rare; in scattered localities near houses.
2 Iona, *Ferreira*. **8** Kilfinichen (BM). **12** As above and Glen Forsa House, *Gerrans* (BM).

S. tuberosum L.
Tuberous Comfrey. Hortal?
Recorded only from **16** Lochbuie, 1956, *H. Milne-Redhead.*

Pentaglottis Tausch

P. sempervirens (L.) Tausch
Green Alkanet. Hortal
13 Aros Park, 1966, *Cannon & Groves.*
On roadside banks.
Rare; established in a few places near gardens. In addition to the above locality also recorded from **2** Iona and **13** Tobermory (BM).

Lycopsis L.

L. arvensis L.
Bugloss. Colonist
2 Iona, 1841, *Harvey* (K-Wa).
An arable weed and in disturbed machair.
Rare; known only from Iona and the Ross.
2 Iona, Culbuirg (BM); field S of Baile Mòr (BM); N central Iona. **3** Ardalanish; Uisken, *Bowerman.*

Pulmonaria L.

P. officinalis L.
Narrow-leaved Lungwort
Listed by Turner & Finlay (1967) but otherwise noted only in cultivation.

Myosotis L.

M. scorpioides L.
Water Forget-me-not. Native; W
Keddie (1850).
In ditches, runnels and wet marshy pastures.
Occasional to frequent; widespread in Mull, but possibly over recorded at the expense of *M. caespitosa.*
2–7, 9–17

M. secunda A. Murr.
Creeping Forget-me-not. Native; CN
13 Tobermory, 1883, *G. Ross.*
In ditches, flushes, wet pastures, by stream margins and under *Salix* scrub.
Occasional to frequent; widespread at lower altitudes.
2–17

M. caespitosa K. F. Schultz
Tufted Forget-me-not. Native; CN
G. Ross (1877).
In ditches, runnels, flushes, seepages and on rocky shores.
Occasional; widespread.
1–13, 15–17

M. sylvatica Hoffm.
Wood Forget-me-not. Hortal
8 Port Uamh Beathaigh, 1956, *Flannigan.*
In addition to the above record also recorded from **8** Kilfinichen (BM) and doubtfully from **4** Ulva.

M. arvensis (L.) Hill
Field Forget-me-not. Native; W
G. Ross (1877).
On waste ground and along wood margins, and on machair pasture and tracksides near the sea.
Occasional; in scattered localities, especially round the coast.
2–13, 15, 17

M. discolor Pers.
Changing Forget-me-not. Native; WSA
Keddie (1850).
By sides of streamlets, on waste ground, sandy shores and marshy pastures.
Occasional to frequent; widespread.
1–17

Mertensia Roth

M. maritima (L.) Gray [Map, Fig. 2.38]
Oysterplant. Native; AS
2 Iona, *Pennant* (1774).
On shingle beaches above the strand line, less commonly on pebble beaches overlying sand. In one locality it occurs on shingle banks at the side of a river outlet at the head of a sea-loch. Usually within reach of sea-spray.
Rare; in scattered localities in the west and especially on the Treshnish Isles and from Treshnish Point to Calgary. Some populations are quite large, e.g., Sgeir a' Chaisteil and Ensay Burn, but many others are very small, in some cases only single depauperate plants being found. Populations may fluctuate due to water-moved substrates.
1 Fladda; Sgeir a' Chaisteil (BM); Lunga, *Richardson* (BM). **2** Iona, *A. Stirling*; *BMMS* and numerous old records. **5** Port Haun, *I. Vaughan* (NMW); Ensay Burn (BM); Calgary Bay, *Raven*; Tostarie, *Backus*. **6** Quinish, *Bird* in G. Ross (1879). **11** Shingle spit at head of Loch na Keal, 1889, *Kidston* (GL) and *BMMS*, 1967. **16** Loch Buie, *Maclagan* (1855).
Almost certainly substantially reduced during the last two hundred years since 1772, when Pennant, referring to Iona, stated 'The beautiful Sea-Bugloss makes the shores gay with its glaucous leaves and purple flowers'; we have located only one stunted plant on Iona. Observations by J. Raven suggest that the colony at Calgary Bay has increased from a single plant in 1965 to a population of at least 125 shoots in 1973. Most of these were very small and only one was mature enough to flower and it is probable that a few, or even only one clone was involved.

CONVOLVULACEAE

Convolvulus L.

C. arvensis L.
Field Bindweed. Native or colonist; W
16 Loch Buie, *Maclagan* (1855).
In waste places, gardens and arable fields.
Rare; in scattered localities.
2 Iona, 1912, *Wilmott*. **5** Calgary, *G. Ross* (1879) **7** Ballygown. **8** Tiroran, *Flannigan*. **10** Carsaig, *U. Duncan & Beattie*. **13** Tobermory harbour (BM). **16** As above.

Calystegia R.Br.

Critical material determined by R. K. Brummitt.

C. sepium (L.) R.Br.
C. sepium (L.) R.Br. subsp. *sepium*
Hedge Bindweed. ? Native; W
G. Ross (1877).
Along roadside hedges, on waste ground and sprawling on shingle.
Rare; in scattered localities.
7 Torloisk, *Bowerman*. **10** Carsaig, Innamore Lodge, *Kenneth & A. Stirling* (BM). **12** Salen, *U. Duncan* (BM); Lower Glen Forsa; Ardnacross (BM). **13** Tobermory. **16** Loch Buie.
No pink-flowered plants (subsp. *roseata* Brummitt) have been observed.

C. pulchra Brummitt & Heywood
C. sepium subsp. *pulchra* (Brummitt & Heywood) Tutin; *C. dahurica* (Herbert) G. Don
Hairy Bindweed. ? Hortal
13 Tobermory, 1959, *C. Muirhead* (E).
On roadsides and in derelict gardens.
Rare; in scattered localities.
2 Iona, near nunnery (BM). **6** Dervaig. **12** Salen (BM). **13** Tobermory (BM).

C. silvatica (Kit.) Griseb.
C. sepium subsp. *silvatica* (Kit.) Maire
Large Bindweed. ? Hortal
13 Cliffs N of Tobermory, 1966, *Cannon & Groves*.
Along roadsides and in derelict gardens.
Rare.
2 Iona, *Harris*. **9** Balevulin (BM). **10** Kilfinichen (BM). **12** Salen, *V. Gordon*. **13** As above.

C. soldanella (L.) R.Br.
Sea Bindweed
Included by Turner & Finlay (1967), based on a record from **5** Treshnish. We have searched for it without success and the available habitats seem very unlikely for this species. It is present on Coll and may occur on Mull.

SOLANACEAE

Lycium L.

L. barbarum L.
L. halimifolium Mill.
Duke of Argyll's Tea-plant. Hortal
Known only with certainty from **2** Iona, Baile Mòr, 1968, *James* (BM). An earlier record from

Iona, 1912, *Wilmott* as *L. chinense* Mill., presumably refers to this species.

Atropa L.

A. bella-donna L.
Deadly Nightshade. ? Denizen
2 Iona, *Pennant* (1774).
In association with ruined buildings.
Known only from old records from Iona. Recorded by several of the early authors (probably following the original references by Pennant and Lightfoot). Recorded by Keddie (1850) and by Ritchie & Ritchie (1947), but these two are also probably merely repetitions. It was probably originally the remains of cultivation as a drug plant at the monastery.

Solanum L.

S. dulcamara L.
Bittersweet. ? Colonist
10 Carsaig to Nun's Pass, 1967, *Kenneth & A. Stirling* (BM).
Known only from the above record where it occurs in a crevice in sandstone on a rocky shore.

Lycopersicon Mill.

L. esculentum Mill.
Tomato
Known only from **2** Iona, 1968, *James* (BM) where it occurs on a beach near the sewage outlet from Baile Mòr, presumably only as a transient casual.

SCROPHULARIACEAE

Verbascum L.

V. thapsus L.
Great Mullein. Hortal
Known only from **8** Tavool House, 1939, *BEC*.

Linaria Mill.

L. repens (L.) Mill.
Pale Toadflax. Hortal or colonist
12 Salen, 1966, *U. Duncan* (BM).
On roadside near houses and on old walls.
Rare; apart from the above locality recorded only from **13** Tobermory.

Cymbalaria Hill.

C. muralis Gaertn., Mey. & Scherb.
Ivy-leaved Toadflax. Hortal or colonist
13 Tobermory, 1952, *Macleay*.
On old walls.
Rare; with a scattered distribution, but may be a quite large population where it occurs.
2 Iona, nunnery walls, *Gerrans* (BM); Baile Mòr. **6** Poll Athack, *Ribbons*. **7** Torloisk House. **8** Tiroran (BM). **13** Aros Park; Tobermory (BM); cliffs N of Tobermory.

Scrophularia L.

S. nodosa L.
Common Figwort. Native; W
G. Ross (1877).
On streamsides, hedgerows, lochsides and among rocks at top of beaches.
Occasional to frequent; widely distributed at low altitudes.
2–17

Mimulus L.

M. guttatus DC.
Monkeyflower. Hortal and colonist
13 Druimfin, 1876, *G. Ross* (BM).
In ditches and streamsides.
Occasional; in scattered localities often near houses.
3, 6, 8, 9, 11–13, 17
The Atlas record for **6** Dervaig is somewhat suspect, as the hybrid, *q.v.*, may have originated in Dervaig through planting, and not spontaneously.

M. guttatus × cupreus
Critical determination by R. H. Roberts.
Hortal
8 Gribun, S of Clachandu, 1967, *Kenneth* (BM).
In roadside ditches.
Rare; apart from the above record, recorded only from **6** Dervaig (BM).

M. moschatus Dougl. ex Lindl.
Musk. Hortal
12 Near Kellan, 1939, *BEC*.
Streamsides.
Rare; apart from the above locality recorded only from **11** Garbh Choire, N of Gruline House (BM).
The first record must be treated with reservation, as *M. guttatus* is well known from the locality concerned.

Erinus L.

E. alpinus L.
Fairy Foxglove. Hortal
13 Aros Park, 1966, *Cannon & Groves* (BM).
On old walls.
Rare; in addition to the above locality known from: **7** Torloisk House; **13** entrance to Aros House (BM); and Tobermory.

Digitalis L.

D. purpurea L.
Foxglove. Native; WSA
G. Ross (1877).
On roadsides, among rocks, in wood and plantation margins and on grassy slopes with *Pteridium*.
Frequent to abundant; widespread.
1–17

Veronica L.

V. beccabunga L.
Brooklime. Native; W
Keddie (1850).
In small streams and runnels and surrounding mire associations.
Occasional; scattered but more frequent on the Ross of Mull and on Iona.
2, 3, 5, 8, 10, 15

V. anagallis-aquatica L. [Map, Fig. 2.5]
Blue Water-speedwell. Native; W
Keddie (1850).
In ditches and runnels.
Rare; apparently confined to Iona and Calgary Bay.
2 Iona, behind Golf Course (BM); coast S of Baile Mòr, *Hillcoat* (BM); W of the Dùn (BM). **5** Calgary Bay (BM).

V. catenata Pennell
Pink Water-speedwell
Included in the list in Turner & Finlay (1967), but regarded as a probable error, although this species has been recorded from Coll and Tiree.

V. scutellata L.
Marsh Speedwell. Native; W
G. Ross (1877).
In small streams, runnels and ditches.
Occasional; widespread in scattered localities at low altitudes.
2, 4–7, 9, 11, 17

V. officinalis L.
Heath Speedwell. Native; W
G. Ross (1877).
In margins of deciduous woodland, on gorge ledges, rocky outcrops and roadsides.
Frequent; widespread.
1–17

V. montana L.
Wood Speedwell. Native; C
G. Ross (1877).
In shade, by tracksides and margins of deciduous woodland.
Rare; in scattered localities.
2 Iona, *Harris*. **12** Salen. **13** Tobermory (BM); *J. Robertson & Birse*; Rubha na Leip; Calve and Lovage Isles, *Harris*; Erray and Druimfin, *G. Ross* (1879).
Further observations on the distribution of this species on Mull are desirable.

V. chamaedrys L.
Germander Speedwell. Native; W
G. Ross (1877).
In deciduous woodland and on roadsides and tracksides.
Frequent; widespread.
1–17

V. serpyllifolia L.
Thyme-leaved Speedwell. Native; W
G. Ross (1879).
In deciduous wood margins, roadsides and tracksides.
Frequent; widespread.
1–17
Subsp. *humifusa* (Dicks.) Syme, a plant of mountainous areas in Wales, N England and Scotland has not been recorded. It could occur in Mull and should be looked for in suitable habitats.

V. arvensis L.
Wall Speedwell. Native or colonist
G. Ross (1877).
In arable fields, farmyards and gardens, tops of sandy beaches and machair grassland.
Occasional; widespread at low altitudes.
1–17
Minute forms occur in the machair on **2** Iona and at **5** Port Langamull.

V. hederifolia L.
Ivy-leaved Speedwell
Known only from **8** S Ardmeanach, 1956, *Flannigan*, and confirmation is desirable.

V. persica Poir.
Common Field-speedwell. Colonist
2 Iona, 1955, *Ferreira*.
In arable fields and gardens.
Rare; in scattered localities.
2 Iona, Baile Mòr (BM); near Clachanach. **6** Glengorm Castle. **10** An Leth-on. **12** Achad nan Each. **13** Tobermory.

V. polita Fr.
Grey Field-speedwell
Known only from **2** Iona, 1955, *Ferreira*; confirmation is desirable.

V. agrestis L.
Green Field-speedwell. ? Colonist
Keddie (1850).
An arable and garden weed.
Rare; in scattered localities.
2 Iona, nunnery garden, *U. Duncan* (BM). **7** Opposite Eorsa. **8** Inch Kenneth (BM); S Ardmeanach, *Flannigan*. **12** Salen.

V. filiformis Sm.
Slender Speedwell. Colonist
2 Iona, 1955, *Ferreira*.
A garden weed.
Rare; in scattered localities.
2 Iona, *Ferreira*. **7** Torloisk House (BM). **10** Pennyghael House (BM). **13** Tobermory. **16** Loch Buie near the castle, *U. Duncan & Beattie*.

Pedicularis L.

P. palustris L.
Marsh Lousewort. Native; C
Keddie (1850).
On loch shores and in lowland flushed habitats. In more eutrophic conditions than *P. sylvatica*.
Occasional to frequent; widely distributed.
2–17

P. sylvatica L.
Lousewort. Native; WSA
Keddie (1850).
Along streamsides, in wet grassland, Callunetum, runnels and open Molinietum.
Frequent to abundant; widespread and ascending to higher altitudes than *P. palustris*.
1–17

Plants referable to subsp. *hibernica* D. A. Webb occur in mixed populations with subsp. *sylvatica*. Intermediates are also found, and, for Mull at least, subsp. *hibernica* seems unlikely to merit recognition at this taxonomic level.

Rhinanthus L.

R. minor L.
Yellow-rattle. Native; W
G. Ross (1877).
By roadsides, and tracksides and in damp pastures near the sea.
Occasional to frequent; widespread at low altitudes.
2–17
A form comparable to subsp. **borealis** (Sterneck) P. D. Sell has been recorded from **8** Creag a' Ghaill, *U. Duncan* (BM) but the majority of the Mull populations are probably that ecotypic variant referred to as subsp. *stenophyllus* (Schur.) O. Swartz. The situation is clearly very complex and needs further study.

Melampyrum L.

M. pratense L.
Common Cow-wheat. Native; W
13 Druimfin, 1876, *G. Ross* (BM).
In deciduous woodland, under *Pteridium* and on grassy cliff-tops.
Occasional; widespread but not recorded from Iona or the Ardmeanach peninsula.
3–7, 9–17

Euphrasia L.

The following account is based on critically determined specimens and specialist advice from P. F. Yeo.

E. micrantha Reichb.
Native; CN
2 Iona, 1912, *Wilmott* (as *E. gracilis*).
Short turf on cliff tops, grazed heather moorland and rough grazing.
Occasional; in scattered localities, but not recorded from NW Mull.
1–4, 7–12, 14–17

E. × electa Townsend
E. micrantha × scottica
Native; ON
4 Ulva, E of Sailean Ardalum, 1965, *U. Duncan* (BM).
Grass heath near the sea.
Rare; in scattered localities.
3 Uisken Bay (BM). **4** Ulva, E of Sailean Ardalum, *U. Duncan* (BM). **14** An Carnais (BM).

E. scottica Wettst.
Native; ON
11 Glen Cannel, 1961, *A. Stirling* (CGE).
In grass heath and flushes.
Occasional; in scattered localities, but apparently absent from Brolass, the Ross and Iona.
1, 5, 7, 8, 10–17

E. tetraquetra (Bréb.) Arrond.
E. occidentalis Wettst.
Native; OWE
4 Ulva, E of Sailean Ardalum, 1965, *U. Duncan* (BM).
In grazed dune pasture, cliff tops and other maritime short turf communities.
Rare; in scattered localities.
1 Cairn na Burgh More (BM). **2** Iona, N end beyond road (BM); machair on W coast, *U. Duncan*. **4** Ulva, E of Sailean Ardalum, *U. Duncan* (BM). **17** Duart peninsula (BM).

E. nemorosa (Pers.) Wallr.
(including *E. curta* (Fr.) Wettst.)
Native; CN
16 Creach Beinn, 1965, *Kenneth*.
In short turf of base-rich grassland and by the sea.
4 Gometra, E side near bridge, *U. Duncan* (BM). **5** Calgary Bay (BM). **6** Rubha nan Gall to Bloody Bay, *Stace*. **7** Camas an Lagain, *Richards*. **8** Creag Mhor, *Richards*; Rubha Baile na h-Airde, *Richards*; Tavool, *Richards*. **11** Loch Bà. **15** Beinn Bheag (BM); Beinn Talaidh, *Birse & J. Robertson* (SSS). **16** As above. **17** Craignure, golf course, *CSSF*.

E. confusa Pugsl.
Native; OWE
13 Near Tobermory, 1960, *V. Gordon*.
In machair and grass heath near the sea, also in inland communities in *Calluna* and short turf.
Occasional; widely distributed.
1–3, 5, 6, 8–14, 16, 17

E. arctica Lange ex Rostrup subsp. **borealis** (Townsend) Yeo
E. borealis (Towns.) Wettst.; *E. brevipila* auct.
Native; OWE
2 Iona, 1936, *Callen* (det. Pugsley).
In short turf near the sea and at tracksides, in open *Betula* scrub and grass heath.
Occasional to frequent; the most frequent and widespread *Euphrasia* in our area.
2–17
Can apparently survive in longer grass with more competition than the other species.

E. × difformis Townsend
E. arctica subsp. *borealis × micrantha*
Native; OWE
4 Gometra, E side near bridge, 1966, *U. Duncan* (BM).
Tracksides and on grassy cliff top.
Rare; known only from the above and **16** Port à Ghlinne (BM).

E. × venusta Townsend
E. arctica subsp. *borealis × scottica*
Native; OWE
5 Calgary Bay, 1965, *U. Duncan* (BM).
In machair and *Festuca*/*Agrostis* grassland.

In addition to the above known only from **15** bridge over the Abhainn Bhearnach (BM).

E. rostkoviana Hayne
Native; W
Known only from a grass bank at roadside at **8** Kilfinichen Bay, 1968, *David* (CGE).

Odontites Ludw.

O. verna (Bellardi) Dumort.
Red Bartsia. Native; W
G. Ross (1877).
In machair grassland, short turf and wet meadows.
Occasional; widespread, but apparently absent from the central massif and most of Glen Forsa.
2–9, 12, 13, 15–17
All plants observed are subsp. **verna**, the more frequent subspecies in N Scotland.

OROBANCHACEAE

Orobanche L.

O. alba Steph. ex Willd.
Thyme Broomrape. Native; CS
2 Staffa, 1829, ? *Christy* (in *Ed. Forster's Herb.*; BM).
On *Thymus*, on cliff ledges and open cliff tops.
Rare; scattered distribution in the west.
1 Staffa, several records from the 19th century. **3** Ard Mòr (BM). **4** Ulva, *Salzen.* **5** Calgary Bay, *J. Duncan.* **6** Bloody Bay, 1877, *G. Ross* (E). **8** Inch Kenneth (BM); Creag a' Ghaill, *U. Duncan.* **9** Malcolm's Point (BM). **10** Carsaig, *U. Duncan*; 1 mile SW Carsaig, *Clark.*

LENTIBULARIACEAE

Pinguicula L.

P. lusitanica L.
Pale Butterwort. Native; OWE
Without locality, *Borrer* (1810).
On wet peat in heathland and margins of bog-pools; also rocky streamsides, flushes and seepage flows.
Occasional to frequent; widespread.
2–4, 6–13, 15–17
Stated by G. Ross (1877) to be 'pretty frequent all over the island'. Our records would scarcely support this, but it may be under-recorded as it is inconspicuous when not in flower.

P. vulgaris L.
Common Butterwort. Native; CN
13 Balliscate, 1876, *G. Ross* (BM).
By streamsides, runnels in peat, bog pool margins and in flushes to at least 900 m.
Frequent to abundant; widespread.
1–17

Utricularia L.

Critical advice by P. Taylor.

U. vulgaris L.
Greater Bladderwort. Native; W
Known only from **6** Loch na Cuilce, 1967, *Kenneth* (Hb. Kenneth).

U. ochroleuca Hartm.
U. intermedia Hayne
Intermediate Bladderwort. Native; CN
3 Linne nan Ribheid, 1939, *Templeman* (BM).
In bog-pools, acid streams, loch margins, shallow peaty lochans and stagnant ditches.
Occasional; apparently confined to the south.
2, 3, 8–12, 15–17

U. minor L.
Lesser Bladderwort. Native; CN
2 Iona, 1939, *Templeman.*
In bog-pools, peaty summit pools and lochans.
Occasional; widespread in scattered localities.
2, 3, 5–10, 12, 15, 16

LABIATEAE

Mentha L.

Critical determinations by R. M. Harley.

M. arvensis L.
Corn Mint. Native; W
G. Ross (1877).
On damp roadsides, rocky shores and in derelict gardens.
Occasional; in scattered localities.
2–6, 10, 12, 13, 17

M. aquatica L.
Water Mint. Native; W
G. Ross (1877).
By stream and ditches and in marshy areas round lochs and lower altitude lochans.
Frequent; widely distributed at lower altitudes.
2–17

M. × verticillata L.
M. aquatica × arvensis
Native; W
12 Allt na Searmoin, 1966, *U. Duncan.*
On river banks.
Rare; in addition to the above record known only from **11** Between Gruline bridge and Loch Bà, *U. Duncan* and **12** Allt Ardnacross (BM).

M. spicata L.
Spear Mint. Denizen
Ewing (1890).
Waste ground and roadsides.
Rare; naturalised in three localities near houses.
8 Kilfinichen Bay (BM). **12** Salen, *U. Duncan.* **13** Tobermory.

M. × villosa Huds.
M. longifolia (L.) Huds. × *rotundifolia* (L.) Huds.; *M. × niliaca* Juss. ex Jacq.
? Denizen
10 Carsaig near Innamore Lodge, 1967, *Kenneth & A. Stirling* (BM).
By streams and near houses.
Rare; apart from the above record known only from **12** Ardnacross (BM) and Tenga House (BM).

Lycopus L.

L. europaeus L.
Gipsywort. Native; W
12 Kellan, 1939, *BEC*.
Streamsides and ditches, marsh and carr round lochs and tops of stony shores.
Occasional; widespread in scattered localities around the coast.
3–12, 16, 17
An old undated sheet in Hb. Glasgow – 'Mull' *Stark* is certainly earlier than the first record cited above.

Thymus L.

T. pulegioides L.
Large Thyme
Recorded by *G. Ross* (1877) [as *T. chamaedrys*] but very unlikely. The G. Ross specimen (BM), so named, is *T. drucei*.

T. serpyllum L.
Breckland Thyme
Listed by *G. Ross* (1879) but doubtless used in the old aggregate sense; most unlikely in Mull.

T. drucei Ronn.
Wild Thyme. Native; ON
Keddie (1850) [as *T. serpyllum*].
Around rocks in dry grassland and on cliffs. A characteristic species of machair.
Abundant; very widespread with a wide altitudinal range.
1–17
White-flowered forms have been noted in Mull.

Prunella L.

P. vulgaris L.
Selfheal. Native; W
G. Ross (1877).
In wood margins, grasslands and roadsides; a prominent species of machair.
Abundant; very widespread, with a wide altitudinal range.
1–17
White-flowered forms have been noted at **17** Torosay.

Stachys L.

S. arvensis (L.) L.
Field Woundwort. Native or colonist; W
13 Erie [Erray], 1877, *G. Ross* (BM).
Arable and trackside weed.
Rare; scattered distribution in N and W Mull.
2 Iona, Baile Mòr, *U. Duncan*. **3** Bunessan, *Kenneth & A. Stirling*; near Uisken, *Kenneth & A. Stirling*. **4** Ulva, Calinish. **5** Sunipol. **6** Glengorm Castle. **7** Lagganulva, *J. Duncan*; near Kellan Mill (BM). **8** SW Ardmeanach Peninsula (BM). **13** As above; Balliscate, *G. Ross* (1879).

S. palustris L.
Marsh Woundwort. Native; W
G. Ross (1877).
In ditches, beside streams and lochs.
Occasional; widespread and locally abundant at lower altitudes.
2–10, 12–17
Has occurred at **12** Salen in a relatively dry arable field as an aggressive and abundant weed.

S. × ambigua Sm.
S. palustris × *sylvatica*
Native; W
12 Salen, 1966, *U. Duncan* (BM).
Waste ground, roadsides and streamsides.
Rare; in addition to the above also recorded from **3** Uisken; **9** Aoineadh Beag; and **10** Carsaig Bay, *J. Duncan* (BM); Beinn nan Gobhar.

S. sylvatica L.
Hedge Woundwort. Native; W
G. Ross (1877).
Along margins of deciduous woodland, shady places in stream gullies and cliff crevices.
Frequent; widespread at lower altitudes.
1–17

Lamium L.

L. amplexicaule L.
Henbit Deadnettle. Native or colonist; W
G. Ross (1877).
An arable and garden weed.
Rare; confined to Iona and the adjacent Ross (with an old record from Tobermory).
2 Iona, *G. Ross* (1879); Ferry Beach, *Stace*; Bay on W side of island (BM). **3** Near Bunessan, *U. Duncan & Beattie*. **13** Tobermory, *G. Ross* (1878).

L. molucellifolium Fr.
Northern Deadnettle. Native or colonist; W
2 Iona, 1829, *Christy* (E).
Arable and garden weed.
Rare; apparently confined to Iona and the adjacent Ross.
2 Iona, fields S of Baile Mòr (BM); farm below the Dùn. **3** Uisken.

L. hybridum Vill.
Cut-leaved Deadnettle
Recorded only on **2** Iona, 1955, *Ferreira*; confirmation is desirable as confusion with the previous species cannot be excluded.

L. purpureum L.
Red Deadnettle. Native; W
Keddie (1850).
Arable and garden weed and in waste ground.
Rare to occasional; in scattered localities at low altitudes.
2–8, 10, 12, 13, 16, 17

L. album L.
White Deadnettle. Native or colonist; W
8 S Ardmeanach Peninsula, 1956, *Flannigan*.
Along wood margins and other shady places.
Rare; apart from the above record also recorded from: **4** Ulva, *Salzen*; **13** Tobermory; and **16** Lochbuie.

A rare plant in NW Scotland, possibly spread in Mull from gardens.

Galeopsis L.

G. tetrahit L.
Common Hemp-nettle. Native; W
G. Ross (1877).
In arable fields and farmyards.
Occasional; widespread at low altitudes.
2–17
Specimens very much larger than normal have been noted from silage pits and manure heaps.

G. speciosa Mill.
Large-flowered Hemp-nettle. Native; C
G. Ross (1877).
In arable fields and farmyards.
Rare; in four scattered localities.
2 Iona, *Ferreira*. **5** Haun. **12** Kellan, *BEC*. **13** Tobermory, *G. Ross* (1879) and *BMMS* (BM).

Glechoma L.

G. hederacea L.
Ground-ivy. Native or ? colonist; W
G. Ross (1877).
As garden weed and under deciduous woodland.
Rare; in scattered localities, possibly always in association with habitations.
2 Iona, 1912, *Wilmott*. **4** Ulva, E of Cille Mhic Eoghainn (BM). **6** Near Loch Mingary. **13** Erray, *G. Ross* (1879); Tobermory; cliffs to N of Tobermory. **17** Torosay Castle.

Scutellaria L.

S. galericulata L.
Skullcap. Native; CN
6 Croig, 1876, *G. Ross* (BM).
Usually among rocks at top of seashores, but rarely also on stony loch sides.
Frequent; widespread around the coasts.
1–14, 16, 17

S. minor Huds.
Lesser Skullcap. Native; OWE
11 Base of Ben Graig [Beinn a' Ghraig], Loch na Keal, 1876, *G. Ross* (BM).
In ditches and marshy streamsides.
Frequent; widespread at lower altitudes.
2–17
More frequent than the preceding species in inland habitats.

Teucrium L.

T. scorodonia L.
Wood-sage. Native; W
Keddie (1850).
In deciduous woodland, and margins of coniferous woodland, mountain ledges and rocky areas at low and middle altitudes, roadsides.
Frequent to abundant; widespread.
1–17
An unusual form with laciniate leaves has been collected from **13** Aros Park (BM).

Ajuga L.

A. reptans L.
Bugle. Native; WSA
G. Ross (1877).
In deciduous woodland, shady gullies and streamsides.
Frequent; widespread.
1–17
White-flowered forms were noted by G. Ross (1879).

PLANTAGINALES

PLANTAGINACEAE

Plantago L.

P. major L.
Greater Plantain. Native; W
G. Ross (1877).
On tracksides and roadsides, waste ground and in open communities above shores.
Frequent; widespread.
1–17

P. media L.
Hoary Plantain
Keddie (1850), a very unlikely record. Also noted by Ritchie & Ritchie (1945) presumably on the basis of the Keddie record and by Maclagan (1855) from **17** Duart Castle.

P. lanceolata L.
Ribwort Plantain. Native; W
Keddie (1850).
On roadsides and tracksides, in arable fields and disturbed grassland.
Abundant; widespread.
1–17

P. maritima L.
Sea Plantain. Native; W
Keddie (1850).
On coastal rocks but also in open habitats inland, including tracksides, rock ledges and screes.
Abundant; widespread.
1–17

P. coronopus L.
Buckshorn Plantain. Native; CS
1 Staffa, *Hooker* (1830).
On coastal rocks and machair, roadsides and tracksides inland.
Frequent around the coast and rarely in inland localities.
1–17

Littorella Berg.

L. uniflora (L.) Ascher.
Shoreweed. Native; CN
16 Loch Uisg, *Maclagan* (1855).
The shallower parts of oligotrophic lochs and lochans usually in depths of under 1 m. Rarely

forming beds in fast flowing rivers (e.g., the lower Forsa).
Frequent; widespread.
2, 3, 5–17

CAMPANULALES

CAMPANULACEAE

Campanula L.

C. latifolia L.
Giant Bellflower. ? Hortal
10 Carsaig, 1939, *BEC*.
In deciduous woodland and on shady banks.
Rare; in scattered localities all near large houses.
7 Torloisk, *BEC*. **10** As above; Pennyghael (BM). **17** Torosay Castle.

C. trachelium L.
Nettle-leaved Bellflower. Hortal
A rare introduction, known only from **12** Aros Cottage, 1939, *BEC*, where it still occurs having persisted for at least 27 years.

C. rapunculoides L.
Creeping Bellflower. Hortal
Only known from ? **13** Hedge at F.C. manse, *G. Ross* (1879).

C. rotundifolia L.
Harebell. Native; W
G. Ross (1877).
In grassland, especially on the drier slopes, in machair, ledges on cliffs and in gullies. Ascending to the highest altitudes.
Frequent to abundant; widespread.
1–17
We are indebted to H. A. McAllister for the following note: *Campanula rotundifolia* in Mull is hexaploid ($2n = 102$). This is the cytotype which occurs in western and northern oceanic areas of the British Isles (Cornwall, Isle of Man, Ireland, Hebrides and west and north mainland of Scotland) and on isolated mountain tops (Teesdale, Merrick, Ben Alder and Ben Nevis). In its few, large flowers and often hemispherical ovaries it resembles the arctic *C. gieseckiana* ($2n = 34, 68$).

LOBELIACEAE

Lobelia L.

L. dortmanna L.
Water Lobelia. Native; ON
12 Salen, '*Mr Evans*' in Maclagan (1855).
In shallow water of oligotrophic lochs and lochans avoiding purely organic substrates.
Occasional to frequent.
3, 5–14, 16, 17

RUBIALES

RUBIACEAE

Sherardia L.

S. arvensis L.
Field Madder. Native; W
5 Calgary Bay, 1876, *G. Ross* (BM).
In species-rich shell-sand grassland and as a rare arable weed.
Rare; in widely scattered localities in the north and west.
2 Iona. **5** Calgary Bay (BM). **6** W of Glengorm Castle (BM). **7** Near Kellan Mill; near Fanmore, *U. Duncan & Beattie*. **8** Inch Kenneth. **13** Tobermory, *G. Ross* (1878).

Cruciata Mill.

C. laevipes Opiz
Galium cruciata (L.) Scop.
Crosswort
Known only from *G. Ross* (1877). Its occurrence on Mull is unlikely.

Galium L.

G. odoratum (L.) Scop.
Woodruff. Native; W
13 Druimfin, 1876, *G. Ross* (BM).
In deciduous woodland, particularly in the relict woodland of shady mountain gullies.
Frequent in N Mull, occasional in the south.
3–14, 16, 17

G. boreale L.
Northern Bedstraw. Native; CN
13 Tobermory, 1838, *Churchill Babington* (K-Wa).
On rock ledges on inland cliffs, especially by streams and waterfalls.
Frequent; widespread but absent from some of the more exposed western areas.
4, 6–17

G. mollugo L.
Hedge Bedstraw. ? Native; W
G. Ross (1877).
Rare; in scattered localities.
2 Iona, 1912, *Wilmott*. **5** Calgary, *U. Duncan & Beattie*. **8** Tiroran, *Flannigan*.
No voucher specimens have been examined and confirmation is desirable.

G. verum L.
Lady's Bedstraw. Native; W
G. Ross (1877).
On dry grassy slopes and rock ledges, especially on cliff tops.
Frequent; widespread but absent from the central massif and the mountains around Glen Forsa and Loch Spelve.
1–17

G. saxatile L.
G. harcynicum Weigel
Heath Bedstraw. Native; WSA
G. Ross (1877).
In rough grassland, in Callunetum and open Pteridetum, ascending to the tops of the mountains.
Abundant; widespread.
1–17

G. pumilum Murr.
Slender Bedstraw
Recorded by Ferreira, 1958, for **1** Lunga but no specimens are available and in the absence of confirmation must be regarded as an error.

G. palustre L.
Common Marsh-bedstraw. Native; W
Keddie (1850).
On river and stream banks and in ditches and lowland mires but lacking in the more oligotrophic habitats.
Frequent; widespread.
1–17

G. uliginosum L.
Slender Marsh-bedstraw. Native; CN
G. Ross (1877).
In very wet mire with *Juncus* and *Potentilla palustris.*
Rare; known only from a few localities in NE Mull and at Carsaig.
6 Creag nan Croman; Sgùlan Breac; Loch Peallach; NW Loch Frisa. **10** Carsaig, *J. Duncan.* **13** Lochnameal (BM).

G. aparine L.
Cleavers. Native; W
G. Ross (1877).
On shingle and boulder beaches, roadsides and waste ground.
Frequent; widespread around the coasts.
1–17

CAPRIFOLIACEAE

Sambucus L.

S. nigra L.
Elder. Native; W
G. Ross (1877).
On roadsides and in open deciduous woodland; also in some of the wooded gorges. Often associated with buildings.
Occasional; widespread at lower altitudes, especially in N Mull.
2–8, 10–13, 16, 17

S. racemosa L.
Red-berried Elder. Hortal
10 Carsaig, 1967, *U. Duncan* (BM).
Established on a rubbish dump and on roadside.
Apart from the above record recorded also from **12** roadside S of Salen.

Viburnum L.

V. lantana L.
Wayfaring-tree. Hortal
Known only from **2** Iona, 1955, *McNiell.*

V. opulus L.
Guelder-rose. Hortal and native; W
G. Ross (1877).
In deciduous woodland, especially in the richer gorges.
Rare to occasional; widespread.
3–6, 10–13, 15, 16

Symphoricarpos Duham.

S. rivularis Suksd.
Snowberry. Hortal
6 Dervaig, 1956, *Beattie.*
Along roadsides.
Rare; naturalised in scattered localities in N Mull.
In addition to the above record also recorded from **12** Salen (BM); Ardnacross, *James & U. Duncan*; Kellan Mill. **13** Tobermory.

Lonicera L.

L. periclymenum L.
Honeysuckle. Native; W
Keddie (1850).
In deciduous woodland and on cliff ledges and in gorges.
Frequent; widespread.
1–17

L. nitida E. H. Wilson
Hortal
12 E Glen Forsa, 1969, *I. & L. Ferguson.*
Persisting in old derelict hedgerows.
Also recorded from **13** Aros Park.

Leycesteria Wall.

L. formosa Wall.
Himalayan Honeysuckle. Hortal
Known only from roadside at **13** Tobermory, 1967, *Kenneth* (BM).

VALERIANACEAE

Valerianella Mill.

V. locusta (L.) Betcke
Common Cornsalad. Native; W
Keddie (1850).
In fixed or loose shell sand and in open communities on cliffs.
Rare; in scattered localities in the west.
2 Iona, beach near Golf Course (BM); Culbuirg (BM); near Baile Mòr, 1912, *Wilmott.* **3** Tòrr Fada; Àrd Mòr; Knockvologan *SSS.* **5** Calgary Bay (BM); Frackadale [Frachdil], *G. Ross* (1879); Port Langamull (BM). **6** Loch Mingary; Penmollach [? Penmore Mill], 1876, *G. Ross* (BM). **10** Carsaig (BM).

Valeriana L.

V. officinalis L.
Common Valerian. Native; W
G. Ross (1877).
On stream banks, in ditches, wet pastures and lowland flushes.
Frequent; widespread.
3–17

Centranthus DC.

C. ruber (L.) DC.
Red Valerian. Hortal
2 Iona, nunnery, *Campbell* in Molony (1951).
On old walls and ruins.
In addition to the above locality known only from **13** Tobermory.

DIPSACEAE

Knautia L.

K. arvensis (L.) Coult.
Field Scabious. Native; W
13 Cliffs N of Tobermory, 1966, *Cannon & Groves*.
On grassy bank by track.
Rare; in addition to the above record also noted from **2** Iona, near Baile Mòr.
Included by Wilmott in his account of the 1939 BEC excursion as 'no locality, needs confirmation, would be NCR'.

Succisa Hall.

S. pratensis Moench
Devilsbit Scabious. Native; W
G. Ross (1877).
In roadsides and tracksides, in pastures, rough grazing and on cliff tops, ascending to at least 800 m.
Abundant; widespread.
1–17

ASTERALES

COMPOSITAE

Helianthus L.

H. annuus L.
Sunflower. Hortal
Known only from drift zone on beach near abandoned village: **9** An Crosan, near Shiaba, 1968, *Mullin*.

Senecio L.

S. jacobaea L.
Common Ragwort. Native; W
G. Ross (1877).
In rough grassland, roadsides and waste places, cliff tops and beaches above the strandline.
Frequent; widespread at lower altitudes.
1–17
Dwarf forms of only 15–20 cm occur in exposed habitats on the small islands.

S. aquaticus Hill
Marsh Ragwort. Native; W
G. Ross (1877).
On streamsides and lochsides, in *Juncus acutiflorus* and herb-rich mires.
Frequent; widespread at lower altitudes, probably more frequent in the west.
1–17

S. × ostenfeldii Druce
S. aquaticus × jacobaea
Native; W
17 Grass Point, 1968, *I. Ferguson* (BM).
In addition to the above locality known only from **4** Ulva, Ardalum, but probably more widespread than these records suggest.

S. sylvaticus L.
Heath Groundsel. Native; W
G. Ross (1877).
Above the strand line on beaches of basaltic sand or pebbles, but some old records from inland localities.
Rare; recorded by G. Ross from the Tobermory area, but recent records are all from the south.
10 Carsaig (BM); Rubha Buidhe (BM). **13** Druimfin and Strongarbh, *G. Ross* (1879). **16** Loch Buie, *H. Milne-Redhead.*

S. viscosus L.
Sticky Groundsel. ? Colonist
4 Ulva, near pier, 1966, *U. Duncan.*
On waste ground and roadsides.
Rare; in scattered localities near houses.
4 Ulva, near pier, *U. Duncan.* **10** Glen Leidle, at top of pass. **13** Tobermory (BM). **17** Craignure, telephone kiosk, *Kenneth & A. Stirling* (BM).

S. vulgaris L.
Groundsel. Colonist
G. Ross (1877).
An arable and garden weed on roadsides and waste places.
Occasional; in scattered localities.
2–13, 16, 17

Doronicum L.

D. pardalianches L.
Leopardsbane. Hortal
Established at the side of stream in ravine at **8** Tavool House (BM), presumably the same locality as that recorded by *Flannigan*, 1956.

Tussilago L.

T. farfara L.
Coltsfoot. Native; W
G. Ross (1877).
Roadsides, waste places, stream banks in disturbed habitats.
Occasional to locally frequent; widespread at lower altitudes.
1–10, 12, 13, 15–17

Petasites Mill.

P. hybridus (L.) Gaertn., Mey. & Scherb.
Butterbur. Native; W
2 Iona, burying place of Oran, *Pennant* (1774).
Streamsides and ditches.
Rare to occasional; in scattered localities but apparently absent from the SE.
2, 4–10, 12, 13

P. albus (L.) Gaertn.
White Butterbur. Hortal
Naturalised by a streamside, at the bridge near **17** Torosay Castle, 1966, *U. Duncan.*

P. fragrans (Vill.) C. Presl
Winter Heliotrope. Hortal
Naturalised in woods near **7** Torloisk, 1968, *U. Duncan & James* (BM).

Inula L.

I. helenium L.
Elecampane. Denizen

2 Iona, monastery, *Walker* (1808).
On roadsides and waste ground near houses.
Rare; in scattered localities in the west.
2 Iona, near cathedral. **3** Scoor near Bunessan, 1913, *Pember* (E). **4** Ulva, near church, *U. Duncan*. **6** Dervaig, near church (BM). **8** Gribun, *Clegg*. **10** Pennyghael (BM); Carsaig, *U. Duncan*. **13** Tobermory, upper village (BM). There is also a record 'near Aros' in Hooker (1821). Presumably originally introduced as a medicinal or culinary herb.

Filago L.

F. vulgaris Lam.
F. germanica (L.) L.
Common Cudweed
No confirmation of Keddie's (1850) record has been discovered (the mention in Ritchie & Ritchie (1945) is presumed to be based on the Keddie record). From the general distribution it seems unlikely and may be an error for ***Gnaphalium uliginosum*** which Keddie does not list.

Gnaphalium L.

G. sylvaticum L.
Heath Cudweed. Native; W
G. Ross (1877).
On roadsides, sand dunes and in open woodland.
Rare; in scattered localities mostly in the E of Mull.
8 The Wilderness. **12** Allt na Searmoin, *U. Duncan*; Aros Castle, *J. Duncan*. **13** Aros Park (BM); Golagoa and Erie [Erray], *G. Ross* (1879). **14** Lower Glen Forsa. **15** Loch Buie, *Kenneth & A. Stirling* (BM); Moie Lodge (BM). **17** Duart.

G. uliginosum L.
Marsh Cudweed. Native; W
12 Salen, 1876, *G. Ross* (BM).
Weed of wet parts of arable fields and tracksides.
Rare to occasional; in scattered localities.
1–8, 11–17

Anaphalis DC.

A. margaritacea (L.) Benth.
Pearly Everlasting. Hortal
7 Torloisk to Ulva road, 1963, *Bowerman* (Hb. Bowerman).
On roadsides and disturbed ground near houses.
In addition to the above locality also known from **13** Tobermory, near the Western Isles Hotel (BM).

Antennaria Gaertn.

A. dioica (L.) Gaertn.
Mountain Everlasting. Native; NM
Lhwyd (*in litt.*, 1699).
In short turf on well-drained soils and on rocks, from sea-level to at least 600 m.
Frequent; widespread, especially in the west but not recorded from the granite of the Ross.
1–13, 15–17
The Mull specimens are all var. **dioica**, but some show a tendency towards the broad leaves and tomentose upper surface of var. **hyperborea** (D. Don) DC.

Solidago L.

S. virgaurea L.
Golden-rod. Native; CN
G. Ross (1877).
On well-drained grassy banks, ledges in gullies and wood margins.
Frequent; widely distributed.
2–17
Most Mull plants are referable to var. **cambrica** (Huds.) Sm., but var. **virgaurea** also occurs in more sheltered conditions, e.g., in woods near **13** Tobermory, at **16** An Coire and in a coastal gully at **3** Slugan Dubh.

S. gigantea Ait.
Hortal
Included by Turner & Finlay (1967) but not otherwise recorded.

Aster L.

A. tripolium L.
Sea Aster. Native; W
G. Ross (1877).
On silt in upper saltmarsh levels, in both very sheltered E coast estuaries and exposed habitats on the W coast and Treshnish Isles.
Rare to occasional; in scattered localities but apparently absent from Brolass and the Ross.
1, 4, 5, 7, 8, 10, 12, 13, 17
Often very small dwarf plants and always the typical rayed var. **tripolium**.

Erigeron L.

E. acer L.
Blue Fleabane. Native; W
10 Carsaig, 1961, *A. Stirling* (BSBI).
On calcareous cliff slopes. Known only from this locality where it was also noted in 1965 by *U. Duncan*. A very rare species in Scotland.

Bellis L.

B. perennis L.
Daisy. Native; W
2 Iona, *Pennant* (1774) [as 'Daisies'].
Under short turf conditions on roadsides and tracksides, species-rich grassland and montane grass heaths.
Frequent; widespread.
1–17

Eupatorium L.

E. cannabinum L.
Hemp-agrimony. Native; W
5 Calgary Bay, 1939, *BEC*.
On damp cliffs and sides of steep coastal gullies.
Occasional; coasts of Mull, but apparently absent from the granite of the Ross and from Iona.
4–10, 12, 16

Anthemis L.

A. arvensis L.
Corn Chamomile. Colonist
12 Between Salen and Gruline, 1958, *U. Duncan.*
An arable weed.
Rare; known only from two localities.
12 As above; *Gerrans* (BM). **13** Tobermory, Erray House and Golf Course.

Achillea L.

A. millefolium L.
Yarrow. Native; W
G. Ross (1877).
On roadsides and tracksides, and in species-rich grassland.
Abundant; widespread at lower altitudes.
1–17

A. ptarmica L.
Sneezewort. Native; W
G. Ross (1877).
In *Juncus effusus* and *J. acutiflorus* associations; in damp meadows and at lochsides.
Frequent; widespread at lower altitudes.
1–17

Tripleurospermum Schultz Bip.

T. maritimum (L.) Koch
Scentless Mayweed. Native; W
1 Staffa, *Macculloch* (1819).
On rock ledges and gullies in sea cliffs, coastal shingle, upper margins of saltmarsh. Subsp. **inodorum** (L.) Hyland. ex Vaarama occurs in disturbed inland habitats.
Frequent around the coasts (subsp. *maritimum*) especially in the west and on the small islands.
1–10, 12, 13, 16, 17
Subsp. **inodorum** has been collected in arable land at **16** Loch Buie (BM) and recorded from **17** Lochdonhead. Plants somewhat intermediate in character have been collected at **2** Iona, Port Ban, *Hillcoat* (BM). **3** Uisken (BM). **6** Loch Peallach (BM).

Matricaria L.

M. matricarioides (Less.) Porter
Pineapple-weed. Colonist
2 Iona, 1942, *Small* (GL).
On tracks and in gateways, especially around farms.
Occasional to frequent; widespread in the inhabited areas of Mull. Occurs on Staffa but not recorded from the Treshnish Isles.
1–14, 16, 17

Chrysanthemum L.

C. segetum L.
Corn Marigold. Colonist
Keddie (1850).
In arable fields and roadsides.
Occasional; in scattered localities.
2–7, 9, 12, 14, 16

Leucanthemum Mill.

L. vulgare Lam.
Chrysanthemum leucanthemum L.
Ox-eye Daisy. Native; W
G. Ross (1877).
In grassland at lower altitudes.
Occasional; widely distributed, but less frequent in the south-east.
2–17

Tanacetum L.

T. parthenium (L.) Schultz
Chrysanthemum parthenium (L.) Bernh.
Feverfew. Hortal or denizen
8 Tavool House, 1939, *BEC.*
On roadsides and in derelict gardens.
Rare; in scattered localities associated with houses.
2 Iona, *Ritchie.* **3** Ardtun. **6** Dervaig, *DMS.* **8** As above. **12** Kellan; Salen (BM). **13** Tobermory, Baydoun (BM) and Golf Course.
The *Atlas* map dot (Perring & Walters, 1962) 17/43 should be on 17/42. Possibly originally introduced into Mull as a medicinal plant.

T. vulgare L.
Chrysanthemum vulgare (L.) Bernh.
Tansy. Hortal or denizen
G. Ross (1877).
On waysides and waste ground.
Rare; in scattered localities, always associated with houses or farms.
3 Kintra; Ardtun (BM); Uisken, *Kenneth & A. Stirling.* **5** Calgary Bay. **8** S of Gribun. **12** Salen, road to Pier; near Kintallen. **13** Tobermory, *G. Ross* (1878).

Artemisia L.

A. vulgaris L.
Mugwort. Colonist or native; C
Keddie (1850).
On waste ground and roadsides.
Rare; in several localities in Iona, but otherwise very scattered.
2 Iona, in Baile Mòr (BM); Cnoc-nam-Bràdhan, *U. Duncan*; U.F. Church, *Braid* (GL); Martyr's Bay, *Small*; Druim Dhughaill. **3** Kintra (BM); Ardchiavaig. **7** Lagganulva, *J. Duncan.* **13** Tobermory (BM).

Carlina L.

C. vulgaris L. [Map, Fig. 2.9]
Carline Thistle. Native; C
5 Calgary Bay, 1939, *BEC.*
In base-rich areas and in habitats where sea-spray or blown sand cause localised slight increases in base status.
Occasional; widespread in the west, most frequent along the S coast of the Ross and Brolass but notably absent from Iona.
3–10

Arctium L.

Specialist advice from F. H. Perring.

A. lappa L.
Greater Burdock
Has been recorded (*DMS*) but in view of the distribution of this species in Britain it seems almost certain that errors for *A. minus* were involved.

A. minus Bernh.
Lesser Burdock. Native or colonist; W
? **13** Bedarach Wood, c. 1878, *G. Ross* (BM).
Along roadsides and on waste ground.
Occasional; widespread but apparently more frequent in the N and W and extremely scarce in the SE.
1–13, 16, 17
All specimens collected are subsp. **nemorosum** (Lej.) Syme, which is probably the only subspecies present in Mull, as might be expected from general British distribution patterns.

Carduus L.

C. tenuiflorus Curt.
Slender Thistle
Only known from **8** Aird Kilfinichen and Tiroran; both records made 1956 by *Flannigan*. The general distribution pattern of this species suggests that these are likely to be errors.

C. acanthoides L.
C. crispus auct.
Welted Thistle. Colonist or native; C
2 Iona, 1948, *Braid* (GL).
Roadside.
Rare; known only from Iona and the Tobermory district.
2 Iona, S of Baile Mòr (BM). **13** Tobermory, Erray House and Golf Course.

Cirsium Mill.

C. eriophorum (L.) Scop.
Woolly Thistle
The record from **12** Salen, *Gerrans* (1960) is based on a misidentified specimen (BM) of *C. vulgare*.

C. vulgare (Savi) Ten.
Spear Thistle. Native; W
G. Ross (1877).
On roadsides and in pastures, waste ground and open grassy communities on cliffs.
Frequent; widespread.
1–17

C. palustre (L.) Scop.
Marsh Thistle. Native; W
G. Ross (1877).
In marshy pastures, rich mires and montane flushes, ascending to at least 800 m in sheltered gullies.
Frequent to abundant; widespread.
2–17

C. arvense (L.) Scop.
Creeping Thistle. Native; W
G. Ross (1877).
In pastures, waste places, roadsides and gardens. Rarely in semi-natural habitats, e.g., montane rough grazing at c. 300 m near Carsaig.
Frequent; widespread but absent from the higher areas.
1–17

C. heterophyllum (L.) Hill
Melancholy Thistle. Native; CN
13 Tobermory, 1829, *Christy* (E).
On well-drained grassy slopes and streamsides, soil pockets on cliff ledges and stabilised scree.
Occasional to locally frequent; widely distributed but apparently less common in the SE.
2–16

Saussurea DC.

S. alpina (L.) DC. [Map, Fig. 2.24]
Alpine Saw-wort. Native; AA
11 Ben More, 1887, *Ewing* (BM).
In subalpine heath, rock ledges and fine scree, rarely below 400 m.
Rare to occasional; the mountains of Ardmeanach, Ben More massif, the E side of Glen Forsa, Creach Beinn and Beinn na Drise.
7, 8, 11, 14, 15

Centaurea L.

C. montana L.
Perennial Cornflower. Hortal
3 Kintra, 1967, *Cannon*.
Roadside banks near houses.
In addition to the above locality known also from **13** Tobermory (BM).

C. nigra L.
Common Knapweed. Native; W
G. Ross (1877).
On roadsides and tracksides, waste places, in pastures, margins of Pteridetum and in cliff grassland.
Frequent; widespread.
1–17

Cichorium L.

C. intybus L.
Chicory. ? Colonist
2 Iona, 1963, *Kenneth*.
On roadsides and waste ground.
Rare; in addition to the above locality also recorded from **5** Calgary, *U. Duncan & Beattie* and **13** Tobermory, Erray House and the Golf Course.

Lapsana L.

L. communis L.
Nipplewort. Native; W
Keddie (1850).
On roadside banks, wood margins.
Occasional; scattered distribution at low altitudes.
2–7, 9, 10, 12, 13, 16, 17

Hypochoeris L.

H. radicata L.
Catsear. Native; W
2 Iona, 1836, (K-Wa).
On roadsides and waste ground, and semi-natural grassy slopes.
Frequent to abundant; widespread.
1–17
On the smaller islands it can flourish in high-nitrogen habitats around bird colonies.

Leontodon L.

L. autumnalis L.
Autumn Hawkbit. Native; W
G. Ross (1877).
On roadsides, grassy banks, cliff tops and dunes.
Frequent; widespread.
1–17

L. hispidus L.
Rough Hawkbit
Recorded by Keddie (1850), Ross (1877) and listed by Ritchie & Ritchie (1945), but we have been unable to confirm its occurrence. It is likely to be an error for *L. taraxacoides*.

L. taraxacoides (Vill.) Mérat [Map, Fig. 2.4]
Lesser Hawkbit. Native; WSA
16 Loch Buie, *Maclagan* (1855).
On fixed shell sand and rocky outcrops near the sea.
Rare; in addition to the above unconfirmed old record known from the following recent records: **2** Iona, Loch Staoineig, *Stace*; **5** Calgary Bay (BM).

Sonchus L.

S. arvensis L.
Perennial Sow-thistle. Native or colonist; W
G. Ross (1877).
On coastal rocks and seashore.
Rare to occasional; in scattered localities.
1–4, 7, 9, 12, 13, 17
Said by G. Ross (1879) to be 'common'. Its present distribution would scarcely support such a statement, and it seems likely that it has become less frequent as a result of reduction in arable farming since the time of Ross.

S. oleraceus L.
Smooth Sow-thistle. Native or colonist; W
Keddie (1850).
In waste places and on roadsides and tracksides.
Occasional; in scattered localities.
2–10, 12, 13, 15–17

S. asper (L.) Hill
Prickly Sow-thistle. Native or colonist; W
G. Ross (1877).
A field weed, on roadsides and tracksides and grassy cliff ledges.
Occasional; in scattered localities.
1–10, 12, 13, 16, 17

Cicerbita Wallr.

C. macrophylla (Willd.) Wallr.
Blue Sow-thistle. Hortal
13 Tobermory, 1968, *Stace*.
Derelict gardens and waste ground.
Apart from the above known only from **8** Kilfinichen (BM).

Hieracium L.

Critical determinations and advice by P. D. Sell and C. West. Additional special help from A. G. Kenneth and A. McG. Stirling.
The following account is based entirely on collected specimens with the exception of the relatively frequent *H. vulgatum* for which some field records have been accepted. Because of lack of critical distribution records phytogeographical elements have been omitted.

H. anglicum Fr.
The record in G. Ross (1877) is almost certainly based on a specimen collected by him at **13** Loch na Meal in 1876 (BM). This has been redetermined as *H. ampliatum.*

H. ebudicum Pugsl. [Map, Fig. 2.40]
Native
8 Inch Kenneth, 1969, *Jermy* (BM).
On grassy ledges on base-rich cliffs.
Rare; in addition to the above locality known only from **10** Sròn nam Boc, Carsaig (BM).

H. hebridense Pugsl.
Native
8 Coire nan Griogag, 1970, *Jermy* (BM).
Known only from cliffs in wet gully facing sea, at the above locality.

H. ampliatum (W. R. Linton) A. Ley
Native
13 Loch na Meal, 1876, *G. Ross* (BM).
On cliff ledges.
Rare; apart from the first record known only from **8** Ardmeanach, E side of the peninsula, *U. Duncan* (BM); Allt na Teangaidh, *Kenneth* (CGE); Beinn na h'Iolaire, *Kenneth* (Herb. *Kenneth*).

H. langwellense F. J. Hanb.
Native
15 Beinn Bheag, 1966, *Jermy* (BM).
On ledges on base-rich outcrops.
Rare; in addition to the above record known also from **9** Creachan Mòr (BM). **10** Nun's Pass (BM) and Coire nan Each (BM).

H. flocculosum Backh.
Native
8 Creag a' Ghaill, 1965, *U. Duncan* (BM).
On rock ledges.
Rare; in addition to the above record known only from **15** Coire nan Each (BM).

H. shoolbredii E. S. Marshall
Native
8 Gribun, 1956, *H. Milne-Redhead* (CGE).

On ledges of maritime and inland cliffs, stream gullies and in fine talus.
Occasional; widely distributed but apparently absent from the Ross and Iona.
4–13, 15–17

H. iricum Fr. [Map, Fig. 2.40]
Native
8 Gribun shore, 1965, *Kenneth* (CGE).
Confined to the base-rich conditions of the Gribun and Inch Kenneth.
Rare; only known from **8** with the following additional records: Creag a' Ghaill, *Kenneth* (BM); Inch Kenneth, NW cliffs (BM).

H. sommerfeltii Lindeb.
Native
6 S'Airde Beinn, 1965, *Kenneth* (CGE).
Known only from rock ledges on the above.

H. argenteum Fr.
Native
11 Near Dishig, 1965, *U. Duncan* (BM).
On cliff ledges.
Rare; in addition to the above record known also from **4** Little Colonsay (BM). **5** Rubha nan Oirean (BM). **6** S'Airde Beinn (BM). **16** Lord Lovat's Cave (BM).

H. chloranthum Pugsl.
Native
6 S'Airde Beinn, 1965, *Kenneth* (Herb. Kenneth).
On ledges in gullies and on cliffs.
Rare to occasional; in widely scattered localities. One of the commoner and more easily recognisable Mull hawkweeds.
6–8, 10–12, 16

H. orimeles F. J. Hanb. ex W. R. Linton
Native
8 Coirc Bheinn, 1967, *Kenneth* (CGE).
Known only from a mildly basic gully on SW-facing exposure on the above.

H. sarcophylloides Dahlst.
Native
17 Near mouth of Loch Spelve, 1964, *A. Stirling & Slack* (CGE).
Known only from ledges on sea cliffs at the above.

H. uistense (Pugsl.) Sell. & West
Native
10 Sròn Bhreac, 1969, *Cannon & Launert* (BM).
Known only from damp clay ledges in stream gorge on the above.

H. subtenue (W. R. Linton) Roffey
Native
11 W side of Beinn nan Gobhar, 1966, *U. Duncan* (BM).
Known only from a cliff ledge on the above.

H. cymbifolium Purchas
Native
10 Sròn nam Boc, Carsaig, 1970, *James* (BM).
Known only on a fallen calcareous sandstone boulder on the above.

H. duriceps F. J. Hand.
Native
13 Rubha nan Gall, 1967, *Melderis* (BM).
Cliff ledges.
In addition to the above known only from **8** Creag Brimishgan (BM).

H. exotericum Jord. ex Bor.
Colonist
Known only from stonework of bridge on **12** Forsa river, 1958, *Gerrans* (BM). In the opinion of Sell & West this species is certainly an introduction on Mull.

H. pictorum E. F. Linton
Native
8 Glen Seilisdeir, 1971, *Kenneth & A. Stirling* (BM).
Known only from a ravine of south-flowing river in the above glen.

H. subhirtum (F. J. Hand.) Pugsl.
Native
13 Erie [Erray], 1877, *G. Ross* (BM).
On rock ledges.
In addition to the above record known also from **5** S'Airde Beinn, *Kenneth* (BM). **8** Allt na Teangaidh, *Kenneth* (Herb. Kenneth); Creag a' Ghaill (BM). **11** Abhainn na h'Uamha, *Kenneth* (Herb. Kenneth).

H. uisticola Pugsl.
Native
16 Allt na Cuinneige, 1968, *Jermy* (BM).
Known only from damp ledges on the above.

H. euprepes F. J. Hand. [Map, Fig. 2.40]
Native
2 Iona, 1948, *Braid* (GL).
Rare; known only from Iona, above and 1 mile SW of Baile Mòr (BM) and **5** Calgary Bay (BM).

H. anfractiforme E. S. Marshall
Native
11 Beinn Fhada, 1967, *Kenneth* (BM).
Known only from gully on the above.

H. cravoniense (F. J. Hanb.) Roffey
Native
10 Carsaig, 1965, *U. Duncan* (BM).
On cliff ledges.
Rare; known only from the above locality and **5** Port Langamull (BM).

H. rubiginosum F. J. Hand.
Native
6 S'Airde Beinn, 1965, *Kenneth* (CGE).
Cliff ledges.
Rare; in addition to the first record, also known from **4** SW of Gometra House (BM); **8** Creag Mhor, *Richards* (Herb. Richards); **14** stream valley near Loch Bhearnach (BM).

H. caledonicum F. J. Hand. [Map, Fig. 2.14]
Native
11 Ben More, 1958, *U. Duncan* (BM).
On cliff ledges.
Rare; in widely separated localities in the west. **1** Lunga, *Richardson* (BM). **2** Iona, behind Baile Mòr (BM); wall of nunnery (BM); Cnoc nan Bradhan, *U. Duncan* (BM). **5** near Sunipol (BM); Rubha an Àrd (BM). **8** Inch Kenneth (BM); Creag a' Ghaill (BM); Coirc Beinn, *Kenneth* (BM). **11** Ben More, *U. Duncan* (BM); Beinn Fhada, *Kenneth* (BM).

H. vulgatum Fr.
Native
G. Ross (1877).
On rock ledges on cliffs and in gorges, on old walls. In both open and shady situations.
Occasional; the most frequent Mull hawkweed. Widespread but apparently absent from the granite of the Ross and Iona.
3–8, 10–17

H. sparsifolium Lindeb.
Native
5 Rubha an Àrd, 1968, *Jermy* (BM).
Known only from cliff ledges near the sea at the above.

H. latobrigorum (Zahn) Roffey
Native
11 Bridge over the river Bà near Gruline, 1966, *U. Duncan* (BM).
On stonework of old bridge and gorge ledges.
Rare; in addition to the above known also from **5** Ensay Burn, *U. Duncan* (BM). **12** Allt na Searmoin, *U. Duncan* (BM); Aros bridge, *Kenneth* (BM).
In his 1877 list G. Ross recorded *H. rigidum*, a synonym of this species.

H. subcrocatum (E. F. Linton) Roffey
Native
13 Erie [Erray], 1877, *G. Ross* (BM).
On open hillside with *Pteridium* and *Rubus*.
Rare; in addition to the above known only from **6** Laorin Bay (BM).

H. strictiforme (Zahn) Roffey
Native
2 Iona, S of landing stage, 1963, *Kenneth* (Herb. Kenneth).
On open rocky base-rich habitats.
Rare; known only from the above and **8** Allt na Teangaidh (BM).

H. maritimum (F. J. Hand.) F. J. Hand.
Native
8 Coirc Beinn, 1967, *Kenneth* (BM).
On mildly basic rocks on SW facing stream sides.
Rare; known only from the above **8**.

H. umbellatum L.
Native
13 Heatherfield, 1877, *G. Ross* (BM).
Rare; known only from the above and **12** Beinn nan Càrn (BM).

H. 'boreale'
A specimen at BM from Erie [Erray] collected by *Ross* (G. Ross 1879) and named *H. boreale* has been redetermined as *H. subhirtum*.

Pilosella Hill

Critical determinations and advice by P. D. Sell and C. West.

P. officinarum C. H. & F. W. Schultz
Mouse-ear Hawkweed. Native
Keddie (1850).
In short turf in well-drained habitats.
Frequent; widespread.
1–17

The following account of the subspecies is based entirely on critically determined specimens.

P. officinarum subsp. **officinarum**
Native
5 Rubha an Àrd, 1968, *Jermy* (BM).
On grassy cliff tops and machair.
Rare; recorded only from the west.
2 Iona, Culbuirg (BM). **3** Dearg Phort (BM). **5** As above. **6** Croig (BM). **7** Fanmore, *U. Duncan* (BM). **8** Inch Kenneth (BM); above Tiroran (BM).

P. officinarum subsp. **micradenia** (Naegeli & Peter) Sell & West
Native
8 Creag a' Ghaill, 1967, *Kenneth* (BM).
Along roadsides, on wall tops and machair.
Rare; in addition to the above known only from **2** Cnoc nam Bràdhan, *U. Duncan* (BM); **8** Inch Kenneth (BM); **12** Aros bridge, *Kenneth* (BM); **16** Creach Beinn, *U. Duncan* (BM).

P. officinarum subsp. **trichosoma** (Peter) Sell & West
Native
8 Inch Kenneth, 1969, *Cannon* (BM).
On grassy cliff ledges and rocks near the shore.
Rare; in addition to the above recorded only from **7** Cruachan Ceann a' Ghairbh (BM) and **9** Knockan (BM).

P. officinarum subsp. **tricholepia** (Naegeli & Peter) Sell & West
Native
12 Aros bridge, 1958, *Gerrans* (BM).
On roadsides, machair and grassy cliff tops.
Rare; in scattered localities in the north and west.
1 Staffa (BM). **2** Iona, Carraig an Daimh (BM). **5** Treshnish Point (BM). **6** Port Langamull (BM). **7** Torloisk House (BM). **8** Inch Kenneth (BM). **10** Pennyghael crossroads (BM). **12** As above; Kintallen (BM). **13** Entrance to Aros Park, *U. Duncan* (BM).

P. officinarum subsp. **trichoscapa** (Naegeli & Peter) Sell & West
Native

8 Allt na Teangaidh, 1965, *Kenneth* (Herb. Kenneth).
Roadsides and rock ledges.
Rare; in addition to the above locality **8**, recorded only from **5** Beinn na Sgiathaig (BM), Langamull (BM); **6** S drive to Glengorm Castle (BM); and **10** Nun's Pass (BM).

P. officinarum subsp. **melanops** (Peter) Sell & West
Hieracium pilosella subsp. *melanops* Peter
Native
3 Uisken, 1971, *Kenneth & A. Stirling* (BM).
In stabilised sand and on roadside.
Known also from **3** Ardalanish, *Kenneth & A. Stirling* (BM) and **7** Kilbrenan, *Kenneth & A. Stirling* (BM).

P. officinarum subsp. **euronota** (Naeg. & Peter) Sell & West
Hieracium pilosella subsp. *euronotum* Naegeli & Peter; '*Un-named subspecies*' of Sell & West in Kenneth & A. Stirling (1970)
Only known specimen is from **5** Treshnish Point, 1969, *Cannon & Launert* (BM).

Crepis L.

C. capillaris (L.) Wallr.
Smooth Hawksbeard. Native; W
13 Druimfin, 1876, *G. Ross* (BM).
On road and tracksides, as arable weeds, on machair and open grassy habitats near the sea.
Occasional; widely distributed.
1–10, 12, 13, 15–17
Both erect forms and small prostrate forms in short turf occur.

C. paludosa (L.) Moench
Marsh Hawskbeard. Native; W
13 Druimfin, 1876, *G. Ross* (BM).
On shady ledges and banks in gullies and stream valleys.
Occasional to frequent; widespread but more frequent in the east.
3–17

Taraxacum Weber

Critical determinations and advice by A. J. Richards.

T. officinale sens. lat. Dandelion. Native
G. Ross (1877).
Frequent; widespread.
1–17

The following account is based entirely on specimens seen by Richards and on his own field observations. We are indebted to him for the following general statement. 'The *Vulgaria* [V] are local and confined to man-made habitats, a most unusual situation, except in the extreme north and west of the British Isles, while the Atlantic Section *Spectabilia* [S], which is local over the country as a whole, is dominant. Also the generally subxerophilous and thermophilous Section *Erythrosperma* [E] is restricted to sandy spots, as in the off-shore islands.'

T. cordatum Palmg. [V]
? Colonist
10 Pennyghael House, 1969, *Jermy* (BM).
In tarmac of courtyard.

T. raunkiaerii Wiinst. [V]
? Native
12 Salen, Dr MacDonald's driveway, 1966, *Jermy* (BM).
In crevice in tarmac.
In addition to the above locality, known also from **13** Tobermory jetty (BM).

T. subcyanolepis M.P.Ch. [V]
? Native
6 W of Tobermory, 1967, *U. Duncan* (BM).
At side of a ditch.

T. euryphyllum (Dahlst.) M.P.Ch. [S]
Native
4 Ulva, Cragaig, 1968, *U. Duncan & James* (BM).
On sea cliff crevices, loch shores and in stream beds.
In addition to the above locality known from **5** Calgary Bay, *Kenneth & A. Stirling*. **10** Pennyghael, *Kenneth & A. Stirling*. **11** bridge at Knock, *Richards*.

T. eximium Dahlst. [S]
Native
8 S of Mackinnon's Cave, 1967, *R. Ross & Sims* (BM).
In crevices in wet rocks, stream gorges and on tracksides.
In addition to the above locality also known from **8** N of Coirc Beinn (BM). **9** Knockan (BM). **11** Creag Mhic Fhionnlaidh (BM). **13** Lochan na Guailne Duibh (BM).

T. faroense (Dahlst.) Dahlst. [S]
Native
6 Loch Carmain an Amais, 1967, *Bangerter & Moyes* (BM).
In wet flushes, grass heath, Callunetum and on rock ledges.
Widespread but not recorded from Iona, the Ross and Ulva.
1, 6–9, 12, 14, 15, 17

T. fulvicarpum Dahlst. [S]
Native
2 Iona, S of Baile Mòr, 1968, *Jermy, Cannon & T. Parker* (BM).
In arable fields and grass heath near the sea.
Known with certainty only from the above record and doubtfully from the small island of Soa (BM) in the same area.

T. laetifrons Dahlst. [S]
Native
4 Ulva, S coast track, 1967, *Bangerter* (BM).
Between rocks in harbour wall and in short cliff-top turf.

Apart from the above record known also from **1** the small island of Cairn na Burgh More (BM). **2** Iona, Dùn, *Kenneth & A. Stirling*. **13** Tobermory (BM).

T. landmarkii Dahlst. [S]
Native
12 Aros, 1970, *James* (BM).
In boggy patch in pasture near the shore.
10 Pennyghael, *Kenneth & A. Stirling*. **11** Allt Tenga Brideig, *Kenneth & A. Stirling*. **12** Aros (BM); Kintallen (BM).

T. maculigerum H. Lindberg f. [S]
Native
2 Iona near Culbuirg, 1968, *U. Duncan & James* (BM).
On grassy slopes near sea, cliff screes, machair and trackside.
Widespread.
1, 2, 4, 5, 7–12, 15–17

T. naevosiforme Dahlst. [S]
Native
5 Ensay Burn, 1967, *U. Duncan* (BM).
In rocky clefts and on shores and by streamsides.
In addition to the above record known also from **2** Uisken, *Kenneth & A. Stirling*. **7** Lòn Biolaireach (BM); Kilbrenan, *Kenneth & A. Stirling*. **8** S Ardmeanach, *Richards*. **10** W of Carsaig (BM). **17** Lochdonhead (BM).

T. nordstedtii Dahlst. [S]
Native
7 Kilninian, 1968, *Jermy* (BM).
On mortared bridge.

T. praestans H. Linberg f. [S]
Native
2 Iona, the Dùn, 1970, *James* (BM).
On ledges in cliffs and gorges.
In addition to the above record, known also from **6** Ardow Burn waterfalls (BM). **7** Below Cnoc nan Uan (BM); Loch a' Ghael (BM).

T. spectabile Dahlst. [S]
Native
8 N of Coirc Bheinn, 1969, *Eddy & Moyes* (BM).
On rocks by streams and in runnels.
In addition to the above locality, known also from **5** Cruachan Odhar (BM) and **7** Cnoc an dà Chinn (BM). All stations are in W Mull.

T. unguilobum Dahlst. [S]
Native
3 Slugan Dubh, 1967, *Cannon & Bangerter* (BM).
On machair, roadsides and walls.
Widely distributed.
2, 3, 5–7, 10, 12, 13, 17

T. brachyglossum (Dahlst.) Dahlst. [E]
Native
3 Ardalanish Bay, 1971, *Kenneth & A. Stirling*.
On sand dunes.

T. fulviforme Dahlst. [E]
Native
2 Iona, Arduara, *Kenneth & A. Stirling*.
On sand dunes.

T. proximum (Dahlst.) Dahlst. [E]
Native
1 Cairn na Burgh Beg, 1968, *Jermy & T. Parker* (BM).
On cliff ledges.
In addition to the above locality, doubtfully from **2** Soa (BM).

T. pseudolacistophyllum van Soest [E]
Native
7 Kilninian, Allt a' Mhuilinn, 1968, *Jermy* (BM).
On roadside, by bridge.

LILIOPSIDA

ALISMATALES

ALISMATACEAE

Alisma L.

A. plantago-aquatica L.
Water-plantain. Native; W
Rare; known only from margins of freshwater loch above bridge at head of Bellart estuary at **6** Dervaig, 1958, *U. Duncan & Beattie* (BM).

HYDROCHARITACEAE

Elodea Michx.

E. canadensis Michx.
Canadian Waterweed
The only known record: **2** Iona, near Cobhain Cuildich, 1912, *Wilmott*; stated to be 'moribund in pool'. Not refound.

NAJADALES

JUNCAGINACEAE

Triglochin L.

T. palustris L.
Marsh Arrow-grass. Native; W
Keddie (1850).
In flushes, and marshy margins of streams and lochs, ascending to at least 800 m.
Occasional to frequent; widespread, especially at lower altitudes.
1–17

T. maritima L. [Map, Fig. 2.31]
Sea Arrow-grass. Native; OWE
Keddie (1850).
In upper saltmarsh levels, saline pools and creeks and in rock crevices in the splash zone.
Occasional; widespread around the coasts.
1–12, 14, 16, 17

ZOSTERACEAE

Zostera L.

Critical determinations by T. G. Tutin.

Z. marina L.
Eel-grass. Native; OWE
13 Dorling [Dòirlinn], 1877, *G. Ross* (BM).
In sandy beds of sea lochs and more sheltered areas of the open sea, at depths depending on the clarity of the water.
Rare; in scattered localities round the coast.
2 Iona, N end of sound, *Picken* (BM). **3** Fionnphort, *U. Duncan* (BM). **5** Sgeir na Cille (BM). **6** Loch Mingary, *G. Ross* (1879). **8** Inch Kenneth, *Picken* (BM); S Ardmeanach, *Flannigan.* **10** Carsaig, *Pope*; Loch Buie, *Picken* (BM). **13** As above.

Z. angustifolia (Hornem.) Reichb.
Narrow-leaved Eel-grass. Native; ON
17 NE side of Loch Don, 1966, *U. Duncan* (BM).
Mud flats of small sea-loch.
Rare; known only from Loch Don as above and Lochdonhead, *Kenneth & A. Stirling* (BM).

POTAMOGETONACEAE

Potamogeton L.

Critical determinations by J. E. Dandy.

P. natans L.
Broad-leaved Pondweed. Native; W
Keddie (1850).
In lochs and lochans, always in an appreciable depth of water, lacking in the more oligotrophic waters.
Occasional; widely scattered.
2, 3, 6, 7, 9, 11–13, 15–17
There is a doubtful record for **4** Ulva.

P. polygonifolius Pourr.
Bog Pondweed. Native; WSA
Syme, pre 1852, in Wats. *Cyb. Brit.* ms. slips.
In oligotrophic mires, small streams, ditches and shallow loch margins.
Frequent to abundant; widespread.
1–17

P. lucens L.
Shining Pondweed. Native; W
16 Loch Uisg, *Maclagan* (1855).
In areas of deeper muds in mesotrophic loch.
Rare; the only modern records are from **3** Loch Poit na h-I, *U. Duncan* (BM) and *BMMS* (BM).
The first record, from SE Mull, has not been confirmed.

P. gramineus L.
Various-leaved Pondweed. Native; CN
'Mull', 1897, *Macvicar* (BM).
Shallower water of lochs.
Rare; known only from two localities in the Ross, two in NE Mull and doubtfully from Ulva.
3 Loch Poit na h-I (BM); Loch near Bunessan, *U. Duncan* (BM). **4** Ulva in lochan on Beinn Chreagach, *Salzen* (needs confirmation). **6** Loch Peallach, *Kenneth* (BM). **13** Caol Lochan (BM).

P. × zizii Koch ex Roth
P. gramineus × lucens
Native; CN
Rare; known only from one loch **3** Loch Poit na h-I, 1966, *U. Duncan* (BM) where it occurs with both parents; only seen as cast up plants on the drift line.

P. × nitens Weber
P. gramineus × perfoliatus
Native; CN
6 Mishnish Lochs, 1893, *Macvicar.*
In 75–100 cm water at Aros; in a slow stream at Mishnish.
Rare; in NE Mull only, where both parents occur.
In addition to the above record, known also from Tobermory river, *Macvicar*, in the same area and from lake in **13** Aros Park (BM).
The Macvicar records were represented by specimens confirmed by Dandy, but subsequently destroyed during the war.

P. alpinus Balb.
Reddish Pondweed. Native; CN
6 Tobermory Reservoir, 1939, *BEC.*
In an oligotrophic lochan and in deeper water in a large loch.
Rare; known only from a few localities in N Mull.
In addition to the above locality known also from: **7** Lochan NW of Cnoc an dà Chinn (BM); **13** Lochnameal (BM); Lochan a' Ghurrabain (BM).

P. praelongus Wulf.
Long-stalked Pondweed. Native; CN
'Mull', 1897, *Macvicar* (CGE).
At Loch a' Ghleannain occurs in 1·25 m of water.
Rare; in addition to the old record above, known from the following modern records: **6** W side of Loch Peallach, *U. Duncan* (BM); **17** Loch a' Ghleannain (BM).

P. perfoliatus L.
Perfoliate Pondweed. Native
6 Loch an Tòrr, 1970, *Cannon, L. Ferguson & Jones* (BM).
In drift on loch shore.
Rare; in addition to the above record, recorded from **9** Cnoc Mòr and Loch Assopol, for both of which confirmation is desirable.

P. pusillus L.
Lesser Pondweed
An unconfirmed record from **15** Glen Forsa, *G. Ross* (1879) which is possible for Mull, but which may have been *P. berchtoldii.*

P. berchtoldii Fieb.
Small Pondweed. Native; W
2 Iona, 1955, *Ferreira.*
Often occurring in flowing water, sometimes under slightly brackish conditions.
Rare; in addition to the above record occurs in **3** Loch Poit na h-I, *U. Duncan* (BM); **6** Loch na Cuilce, *Kenneth* (BM); Dervaig, in river Bellart (BM).

P. filiformis Pers.
Slender-leaved Pondweed
Recorded as 'V.-c. 103' by Macvicar (in Ewing, 1899) and almost certainly refers to Coll or Tiree where the species is well known, but it should be looked for in Mull near the sea.

P. pectinatus L.
Fennel Pondweed. Native; W
Known only from **6** Loch na Cuilce, 1967, *Kenneth* (BM) where it occurs in shallow water, under slightly brackish conditions at the head of the loch.

RUPPIACEAE

Ruppia L.

Determinations by J. E. Dandy.

R. cirrhosa (Petagna) Grande
R. spiralis L. ex Dum.
Spiral Tasselweed
An immature specimen collected at **17** Lochdonhead may be this species but fruiting material is needed for confirmation.

R. maritima L.
Beaked Tasselweed. Native; W
2 Iona, 1955, *Ferreira.*
In salt marsh pools and brackish rock-pools.
Rare; in scattered localities.
1 Sgeir an Eirionnaich (BM); Fladda (BM); Bac Beag (BM). **2** As above; Soa (BM). **10** Rossal, *J. Robertson & Birse.* **14** near Garmony, *Kenneth.* **17** Lochdonhead (BM); Leth-fhon (BM).

NAJADACEAE

Najas L.

Determination by J. E. Dandy.

N. flexilis (Willd.) Rostk. & Schmidt
Slender Naiad. Native; ON
Known only from **3** Loch Poit na h-I, 1970, *U. Duncan* (BM).

ERIOCAULALES

ERIOCAULACEAE

Eriocaulon L.

E. aquaticum (Hill) Druce
E. septangulare With.
Pipewort
Recorded in the Comital Flora (Druce, 1932) for Iona from a source which we have been unable to trace. It is known from Coll and Tiree and has recently been discovered in several localities on the Ardnamurchan peninsula, so that its occurrence in our area cannot be discounted. Presumably the only possible locality in Iona would be Loch Staoineig, and it may be that the use of this small loch as a water supply for the village has caused changes in water level that have brought about the extinction of an *Eriocaulon* population.

LILIIFLORAE

LILIACEAE

Tofieldia Huds.

T. pusilla (Michx.) Pers.
Scottish Asphodel
Listed in Turner and Finlay (1967) on the basis of a record from **6** Mingary and needing confirmation.

Narthecium Huds.

N. ossifragum (L.) Huds.
Bog Asphodel. Native; ON
Keddie (1850).
In mires with *Scirpus cepitosus*, *Molinia* etc., ascending to at least 600 m.
Abundant; widespread.
1–17

Polygonatum Mill.

P. multiflorum (L.) All.
Solomon's-seal. Hortal
Known only from **5** Calgary Bay, 1969, *James* (BM) where it is established in a hedgerow near the graveyard.

Lilium L.

L. pyrenaicum Gouan
Pyrenean Lily. Hortal
Known only in **7** Torloisk House, 1969, *Cannon & Jermy* (BM) where it is regenerating naturally in the margins of the woods round the house.

Scilla L.

S. verna Huds. [Map, Fig. 2.37]
Spring Squill. Native; OWE
1, 2 Staffa and Iona, *Lightfoot* (1777).
In short well-drained turf on exposed cliff top, on rocks and cliff ledges.
Rare but where it occurs it may be locally frequent, as on Iona and the small islands.
1 Cairn na Burgh More (BM); Fladda (BM); Lunga, *Ferreira*; Staffa, *BMMS* and many early records. **2** Iona, Cnoc nam Bràdhan, *U. Duncan*; the Dùn (BM); S of Baile Mòr (BM); Cùl Bhuirg (BM). **3** Uisken, *Kenneth & A. Stirling*; Port nan Ròn. **8** Inch Kenneth.

Endymion Dumort.

E. non-scriptus (L.) Garcke
Bluebell. Native; OWE
G. Ross (1879).
In deciduous woodland, on roadside banks, wooded streamsides and cliffs.
Frequent to abundant; widespread.
1–17
White-flowered forms have been rarely noted by G. Ross and later observers.

Colchicum L.

C. autumnale L.
Meadow Saffron. Hortal
Naturalised at **17** Torosay Castle, 1967, *James* (BM) and near a house at **6** N end of Loch Frisa.

JUNCACEAE

Juncus L.

J. squarrosus L.
Heath Rush. Native; W
'Mull', 1848, *Syme* (BM).
A calcifuge species of grass-heath and open moorland.
Frequent; widespread.
1–3, 5–17

J. tenuis Willd.
Slender Rush. Colonist
13 Near Tobermory, 1939, *Chapple* (OXF).
On damp tracksides and roadsides.
Occasional; scattered but more frequent in the N of Mull.
2, 5–8, 10–17

J. compressus Jacq.
Round-fruited Rush
Described by the collector as 'in huge clumps' from **8** Gribun, 1929, *Gilmour* (BM) but not subsequently rediscovered. Its occurrence in Mull as a native is unlikely.

J. gerardii Lois. [Map, Fig. 2.29]
Saltmarsh Rush. Native; W
6 Bloody Bay, 1876, *G. Ross* (BM).
In saline or brackish inorganic soils at upper saltmarsh levels and round rock-pools.
Frequent; widespread round the coasts.
1–10, 12–14, 16, 17

J. bufonius L.
Toad Rush. Native; W
G. Ross (1877).
In muddy pools and by tracksides in upper levels of stony shores.
Occasional to frequent; widespread especially at lower altitudes.
1–17

J. inflexus L.
Hard Rush
Recorded from **6** Rareich and Loch Peallach, *G. Ross* (1879) but its occurrence in Mull is not likely.

J. effusus L.
Soft Rush. Native; W
G. Ross (1877).
In wet pastures and lowland flushes, characteristically associated with *Polytrichum commune* and *Sphagnum*, often colonising disturbed peaty areas and ascending to higher altitudes than *J. acutiflorus*.
Abundant; widespread.
1–17

J. conglomeratus L.
J. subuliflorus Drejer
Compact Rush. Native; W
Keddie (1850).
In similar situations to *J. effusus*, but avoiding the more extreme acid habitats and preferring some silt movement.
Frequent to abundant; widespread but certainly less frequent than *J. effusus*.
1–17

J. maritimus Lam.
Sea Rush. Native; W
6 Bloody Bay, *G. Ross* (1878).
At uppermost saltmarsh levels, or in silty or sandy soils within splash zone.
Rare; in scattered localities in the west.
2 Iona, *Ferreira*. **3** coast opposite Erraid; Fidden, *SSS*. **4** Ulva, near Ulva House (BM). **6** As above. **8** Tiroran, *Flannigan*. **10** An Lethfhonn (BM).

J. acutiflorus Ehrh. ex Hoffm.
Sharp-flowered Rush. Native; SSA
G. Ross (1877).
A characteristic dominant of the low-lying habitats, also in wet pastures and at streamsides.
Frequent; widespread.
1–17

J. articulatus L.
Jointed Rush. Native; CS
G. Ross (1877).
Lowland mires, but not dominating as the preceding species, also in wet pastures and in the margins of damp woodland.
Frequent; widespread.
1–17
Some of the numerous records may be errors for *J. acutiflorus* owing to the difficulty of distinguishing between these species early in the season.

J. bulbosus L.
Bulbous Rush. Native; NM
8 Tòn Dubh Sgairt, 1968, *Eddy* (BM).
Skeletal gravels at middle to high altitudes.
Rare.
8 Coire Buidhe (BM); as above; E summit of Bearaich (BM); Coirc Beinn; Creag Brimishgan. **10** Cnocan Buidhe (BM). **11** Beinn a' Ghràig. **12** Meall Damh Àrd.
The above treatment is based on specimens seen by and field studies made by A. Eddy who states: 'In general a much smaller plant than average *J. kochii* and its flowers and fruit mature later'.

J. kochii F. W. Schultz
Bulbous Rush. Native; W
11 Aird Kilfinichen, 1956, *Flannigan*.
In wet habitats from oligotrophic to eutrophic,

including totally submerged forms (*J. fluitans* auct.) in the shallower lochs.
Frequent to abundant; widespread.
1–17

J. triglumis L.
Three-flowered Rush. Native; AA
8 S Ardmeanach, 1956, *Flannigan.*
In wet gravelly flushes at streamsides at higher altitudes.
Rare; confined to the Ardmeanach peninsula.
8 Maol Mheadonach, *U. Duncan*; Beinn na h'Iolaire (BM); Bearraich (BM); Fionna Mhàm (BM); Creag Brimishgan; Coirc Bheinn.
There are also two records from **2** Iona, 1912, by Wilmott, but these are most unlikely.

Luzula DC.

L. pilosa (L.) Willd.
Hairy Woodrush. Native; W
G. Ross (1877).
In grass heath and Callunetum, shady banks and gullies and in deciduous woodland.
Occasional; widespread, more frequent in the north.
1, 3–9, 11–17

L. sylvatica (Huds.) Gaudin
Great Woodrush. Native; W
Keddie (1850).
Deciduous woodland, banks, ledges in gorges and on cliffs, and forming colonies in well-drained, acidic alpine grass heath.
Frequent; widespread.
1–17

L. spicata (L.) DC. [Map, Fig. 2.17]
Spiked Woodrush. Native; AA
11 Ben More, 1905, *Vachell.*
Scree margins, open stony summits and montane grassland.
Rare; at high altitudes on the mountains of the Ben More complex and Glen Forsa.
8 Coirc Beinn (BM). **11** Ben More, summit ridge (BM); Coire nam Fuaran, *Templeman* (BM); Beinn a' Ghraig. **15** Loch a' Mhaim, *Stace*; Mainnir nam Fiadh (BM); Beinn Chreagach Mhor; Sgùlan Beag; Sgùlan Mòr, *Birse & J. Robertson* (SSS); Maol a'Chearraidh; An Cruachan; Beinn Mheadon.

L. campestris (L.) DC.
Field Woodrush. Native; W
Keddie (1850).
In grass heath and open Callunetum, and on road and tracksides.
Occasional to frequent; widespread.
1–17

L. multiflora (Retz.) Lejeune
Heath Woodrush. Native; W
13 Erie [Erray], 1876, *G. Ross* (BM).
Grass heath and open deciduous woodland, gorge ledges and scree margins at higher altitudes.
Frequent to abundant; widespread.
1–17
Both compact (var. **congesta** (DC.) Lej.) and spreading forms are present.

AMARYLLIDACEAE

Allium L.

A. vineale L.
Wild Onion. Native or colonist; C
2 Iona, 1948, *Braid* (GL).
Only recorded habitat, 'cliff scrub'.
Rare; scattered localities in the north and west.
2 As above. **3** Erraid, *McAllister.* **5** Between Kilninian and Tostarie (BM). **6** Criadhach Mhòr. **13** Tobermory.

A. schoenoprasum L.
Chives. Denizen
Recorded only from **12** Glen Forsa House, 1958, *Gerrans* (BM).

A. ursinum L.
Ramsons. Native; W
Keddie (1850).
In damp shady habitats in deciduous woodland and on cliffs.
Occasional to locally frequent in the north-west.
1–13, 15–17

Narcissus L.

N. pseudonarcissus L.
Wild Daffodil. Hortal
Known only from **6** near Loch Peallach, 1967, *Melderis.*

IRIDACEAE

Sisyrinchium L.

S. californicum (Ker.-Gawl.) Ait. f.
Yellow-eyed-grass. Hortal
Known only from margins of roadside runnel between **6** Ardow and Druimghigha, 1972, *Gaster.*

Iris L.

I. versicolor L.
Purple Iris. Hortal
Recorded as an introduction only from **2** Iona, below Abbey, 1958, *Gerrans* (BM).

I. pseudacorus L.
Yellow Iris. Native
Stuart in *Lightfoot* (1777).
By lochsides and streamsides, in marshy pastures and damp areas at top of beaches.
Frequent to locally abundant in the west.
1–17

Crocosmia Planch.

C. × crocosmiflora (Lemoine) N. E. Br.
Montbretia. Hortal
2 Iona, 1955, *Harris.*
Along roadsides, streamsides and in derelict gardens.
Occasional; naturalised in scattered localities, but well established and spreading.
2–5, 7, 11–13, 17

Curtonus N.E.Br.

C. paniculatus (Klatt) N.E.Br.
Known only from **13** Tobermory, 1967, *Kenneth*, where it is described as 'persisting in native vegetation'. Could be confused with robust forms of Montbretia and possibly overlooked elsewhere.

ORCHIDALES

ORCHIDACEAE

Critical determinations and advice by P. F. Hunt and Jean K. Bowden.

Cephalanthera Rich.

C. longifolia (L.) Fritsch
Narrow-leaved Helleborine. Native; W
J. D. Hooker in Watson, *Cyb. Brit.* ms. slips, pre-1849.
In deciduous woodland and under scrub in gully.
Rare; in five scattered localities.
5 Treshnish House, 1974, *Clarke.* **7** Kilninian (BM); near Kilbrenan, *Kenneth & A. Stirling* (BM). **13** Portfield, S peninsula of Loch Spelve (BM). **17** E side of Loch Gourapin, Druimfin, 1877, *G. Ross* (BM).
Ross (1879) noted 'I have watched . . . for five years . . . here in Druimfin woods but never seen a mature fruit'.

Epipactis Sw.

E. helleborine (L.) Crantz
Broad-leaved Helleborine. Native; W
13 Bedarach Wood, 1876, *G. Ross* (BM).
In deciduous woodland.
Rare; in scattered localities in N Mull and Ulva.
4 Ulva, wood near church, *U. Duncan.* **5** Near Calgary, *Kenneth.* **13** Aros Park, *U. Duncan*; Tobermory, path to Rubha nan Gall (BM).

Listera R. Br.

L. ovata (L.) R. Br.
Common Twayblade. Native; W
13 Druimfin, 1876, *G. Ross* (BM).
In margins of deciduous woodland.
Rare; in scattered localities in N Mull, the Gribun and Ulva.
4 Ulva, *Salzen.* **5** Ensay Burn; Calgary House, *J. Duncan*; Port Langamull (BM). **6** Rubha nan Gall to Bloody Bay, *Stace.* **7** Kilninian (BM). **8** Port Uamh Beathaigh, *Flannigan.* **12** Estuary of Aros river, *Gerrans* (BM); Aros Castle. **13** Bedarach, *G. Ross* (1879); Tobermory.

L. cordata (L.) R. Br.
Lesser Twayblade. Native; NM
13 Aros Park, 1939, *BEC.*
Under dense cover of *Calluna*, often on slightly raised moss hummocks.
Occasional; in scattered localities. Probably under-recorded.
3, 5–8, 10–14, 16, 17

Neottia Ludw.

N. nidus-avis (L.) Rich.
Birdsnest Orchid. Native; W
13 Aros Park, 1939, *BEC.*
In deciduous woodland.
Rare; known only from **13** Aros Park woods, 1965, *U. Duncan*; woods SW shore of Tobermory Bay, 1968, *Kingston* and path from Tobermory to Aros woods, 1970; same locality, 1973, *Bamforth* (BM).

Hammarbya Kuntze

H. paludosa (L.) Kuntze
Bog Orchid. Native; CN
5 Treshnish, towards the Point, pre-1965, *Caulfield* in Turner and Findlay (1967).
Amongst *Sphagnum* and *Carex* in basic flush.
Rare; apart from the above record which we have been unable to confirm, known only from **16** SW of Dalnaha (BM).

Corallorhiza Chatel.

C. trifida Chatel.
Coralroot Orchid. Native; CN
Known only from **16** between Dalnaha and Lian Mòr, 1970, *Robson* (BM) in *Sphagnum* under open birch cover on wet slope.

Coeloglossum Hartm.

C. viride (L.) Hartm.
Frog Orchid. Native; CN
Keddie (1850).
On dry moorland and in fixed shell-sand pasture.
Rare; in scattered localities in N and W Mull.
2 Iona, *Ferreira.* **3** Near Fionnphort, *BEC.* **5** Haun; E of Calgary Bay, *Paish*; Rubha an Àrd (BM). **6** Glengorm Castle. **7** Cnoc nan Dubh Leitire (BM). **8** Caisteal Sloc nam Ban, *Ribbons.* **13** Runa-gal [Rubha nan Gall], 1877, *G. Ross* (BM).

Gymnadenia R. Br.

G. conopsea (L.) R. Br.
Fragrant Orchid. Native; W
13 Heatherfield, 1876, *G. Ross* (BM).
In open Callunetum, marshy pastures and by roadsides.
Occasional to frequent; widespread.
2–17

Pseudorchis Séguier

P. albida (L.) A. & D. Löve
Leucorchis albida (L.) E. Mey. ex Schur
Small-white Orchid. Native; NM
16 Loch Spelve, 1848, *Syme* (BM).
On well-drained banks and grass heath.
Rare; in scattered localities, but not recorded from Brolass, the Ross or Iona.
5 Ensay Burn, *U. Duncan.* **6** Dervaig, *Gerrans* (BM); Glengorm Castle. **7** Fanmore, *U. Duncan*; waterfalls near Tòrr an Damh (BM). **8** Port Uamh Beathaigh, *Flannigan*; Tiroran, *Bruce* (E). **11** Allt na Coire Mòire; Tenga

Brideig (BM). **13** Aros Park, *Clark*; Balliscate, *G. Ross* (1879). **15** Scallastle. **16** As above.

Platanthera Rich.

P. chlorantha (Custer) Reichb.
Greater Butterfly-orchid. Native; W
Erray, 1876, *G. Ross* (BM).
In damp pastures and in margins of *Salix aurita* carr and roadsides.
Occasional; in scattered localities at lower altitudes.
4, 5, 7–9, 11–13, 16, 17

P. bifolia (L.) Rich.
Lesser Butterfly-orchid. Native; W
Loch Mean [? Loch Meadhoin], 1876, *G. Ross* (BM).
In marshy pastures and on roadsides.
Occasional; widely scattered at lower altitudes.
3–11, 13, 15–17

Orchis L.

O. mascula L.
Early-purple Orchid. Native; W
G. Ross (1879).
In marshy pastures, shady cliff gullies and stream gorges, and along margins of *Salix* carr.
Occasional; widespread but apparently less common in SE Mull than elsewhere.
1–10, 12, 16, 17

Dactylorhiza Neck. ex Nevski

D. fuchsii (Druce) Soó subsp. **fuchsii**
Dactylorchis fuchsii (Druce) Vermeul.
Common Spotted-orchid. Native; W
8 Gribun, 1961, *Ribbons*.
In areas of richer base status, mainly coastally and especially at Carsaig and the Gribun, where calcareous rocks are exposed.
Occasional; widespread in scattered localities.
3, 5–13, 16

D. × **venusta** T. & T. A. Stephenson
D. fuchsii subsp. *fuchsii* × *purpurella*
Native; W
7 Fanmore, 1965, *U. Duncan* (BM).
In coastal pastures and stream bed on beach.
Rare; in addition to the above record known only from **5** Calgary Bay (BM).

D. maculata subsp. **ericetorum** (E. F. Linton) Hunt & Summerh.
Dactylorchis maculata subsp. *ericetorum* (E. F. Linton) Vermeul.
Heath Spotted-orchid. Native; ON
'Mull', 1834, *J. Shiels* (BM).
In wet and dry acid grassland, in Molinetum, open Callunetum and rough grazing on cliffs to 400 m.
Frequent; widespread.
1–17
Populations at higher altitudes are generally paler, i.e., nearly white, than those at lower altitudes.

D. × **transiens** Druce
D. maculata subsp. *ericetorum* × *fuchsii* subsp. *fuchsii*
Native; ON
7 Fanmore, 1965, *Duncan* (BM).
Pastures near the sea.
Rare; in addition to the above record known from **2** Iona, *U. Duncan*; **8** Creag a' Ghaill, *U. Duncan* and **17** Gorton (BM).

D. × **formosa** (T. & T. A. Stephenson) Soó
D. maculata subsp. *ericetorum* × *purpurella*
Native; ON
12 Salen to Gruline, 1958, *Gerrans* (BM).
In pastures near the sea and lowland mires.
Rare; in addition to the above locality known from **2** Iona, *U. Duncan*; Druim an Aoineadh (K). **6** Dervaig, *Gerrans*. **7** Fanmore, *U. Duncan*. **13** Near Lochnameal (BM).

D. incarnata (L.) Soó subsp. **incarnata**
Early Marsh-orchid. Native; W
2 Iona, 1951, *Ferreira*.
In boggy places and marshy pastures especially near the sea and sometimes in more eutrophic habitats such as the machair stream at Calgary.
Occasional; in scattered localities.
2–6, 8, 9, 11–13, 16, 17
Ranging in colour from deep purple-pink to pale flesh tints.

D. × **latirella** (P. M. Hall) Soó
D. incarnata × *purpurella*
Native; W
7 Fanmore, 1965, *U. Duncan* (BM).
In coastal pastures and open damp grass heath.
Rare; in addition to the above locality known from **9** Allt an Laoigh.

D. purpurella (T. & T. A. Stephenson) Soó
Northern Marsh-orchid. Native; NSA
'Mull', 1834, *J. Shiels* (BM).
In wet pastures and on roadsides, cliff ledges and in gullies.
Occasional; widespread, but apparently less common in SE Mull.
1–17

Anacamptis Rich.

A. pyramidalis (L.) Rich.
Pyramidal Orchid
Included in the list in Turner & Finlay (1967) on the basis of a 'pre-1965' record from **5** Sunipol by *Caulfield*. We have been unable to confirm this record during our own fieldwork. The Ewing (1899) record for v.c. 103 probably refers to Coll or Tiree, where the species is well known.

ARALES

ARACEAE

Arum L.

A. maculatum L.
Lords-and-ladies. ? Hortal
Known only in deciduous woodland and adjacent hedgerow from **7** Torloisk, 1968, *James*

(BM). Mull is beyond the natural range for *Arum* in Britain. Plants with spotted leaves are the usual introduced forms (C. T. Prime, pers. comm.); the Mull plants are without leaf spots.

LEMNACEAE

Lemna L.

L. minor L. [Map, Fig. 2.10]
Common Duckweed. Native; W
G. Ross (1877).
Along margins of streams and runnels.
Occasional; widespread in scattered localities in the west.
1, 3–10
Possibly associated with a locally higher base-status from spray and also sometimes with increased nitrogen levels round farms etc.

TYPHALES

SPARGANIACEAE

Sparganium L.

Critical determinations by C. D. K. Cook.

S. erectum L.
Branched Bur-reed. Native; W
G. Ross (1877).
By margins of lochs and under freshwater estuarine conditions.
Rare; four localities in N Mull.
6 Loch na Cuilce, *Kenneth* (BM); Loch an Tòrr (BM). **12** Killichronan, *Kenneth*. **13** Tobermory.
Our material is too young for subspecific determination, but Prof. Cook believes one specimen (from Loch an Tòrr) to be subsp. *microcarpum* (Neum.) Domin.

S. emersum Rehm.
Unbranched Bur-reed. Native; W
G. Ross (1877).
By margins of lochs and lochans.
Rare; in scattered localities.
6 Lochan a' Chuirn. **6** Loch na Cuilce, *Kenneth* (BM); Loch Peallach, *G. Ross* (1877). **12** Meall Dam Àrd. **16** Creach Beinn; Laggan Deer Forest.
The specimen from Loch na Cuilce has been determined as a possible hybrid with *S. angustifolium*.

S. angustifolium Michx.
Floating Bur-reed. Native; NSA
6 Loch Peallach, 1961, *A. Stirling*.
In shallow water of peaty lochs, more rarely in slow-flowing streams.
Occasional; widespread in scattered localities.
2, 6–12, 15–17
The most frequent *Sparganium* in the area.

S. minimum Wallr.
Least Bur-reed. Native; NC
G. Ross (1877).
By margins of lochs, in freshwater and brackish ditches near saltmarshes.
Rare to occasional; widely scattered; possibly over-recorded.
3, 6, 9–11, 14, 15, 17

CYPERALES

CYPERACEAE

Eriophorum L.

E. angustifolium Honck.
Common Cottongrass. Native; CN
Keddie (1850).
In a wide variety of wet habitats, particularly round bog pools and in mires. More tolerant to drought than *E. vaginatum* and long persistent in drained bogs.
Frequent to abundant; widespread.
1–17

E. vaginatum L.
Harestail Cottongrass. Native; CN
16 Kinlochspelve, 1848, *Syme* (BM).
Particularly in the drier areas of peat bogs, with *Scirpus cespitosus* and *Myrica*.
Frequent; widely distributed.
2–17

Scirpus L.

S. cespitosus L.
Trichophorum cespitosum (L.) Hartm.
Deergrass. Native; CN
G. Ross (1877).
In damp peaty areas, a common associate of *E. tetralix* on the Ross. Can withstand some grazing and burning.
Frequent to abundant; widespread.
1–17

S. maritimus L.
Sea Clubrush. Native; OWE
6 Bloody Bay, 1876, *G. Ross* (BM).
In upper levels of saltmarsh, pools in the splash zone and brackish ditches.
Occasional; widely distributed round the coast.
1–8, 10, 12, 16, 17

S. lacustris L.
Schoenoplectus lacustris (L.) Palla
Common Clubrush. Native; W
G. Ross (1877).
Freshwater lochs.
Rare; in scattered localities.
3 Loch Poit na h-I. **6** Loch Peallach, 1889, *Kidston & J. Stirling* (E). **9** Cnoc Mòr; Loch Assopol; Malcolm's Point; Creachan Mòr. **12** Lochan a' Ghurrabain (BM); Caol Lochan; Loch a' Ghualine Duibhe. **15** Loch Squabain (BM). **17** Loch a' Ghleannain (BM).

S. tabernaemontani C.C.Gmel.
Schoenoplectus tabernaemontani (C.C.Gmel.) Palla
Grey Clubrush. Native; OWE
Loch Cuan near Dervaig [? Loch a' Chumhainn], 1939, *BEC*.
In brackish estuarine habitats.
Rare; in a few scattered localities.
6 Dervaig, Loch na Cuilce, *Kenneth* (BM) and *BMMS* (BM); Loch a' Chumhainn (BM). **15**

Scallastle Bay (BM). **17** Craignure Golf Course, *CSSF*; Leth-fhon, Lochdonhead (BM).

S. setaceus L.
Isolepis setacea (L.) R. Br.
Bristle Clubrush. Native
3 Bunessan, 1877, *G. Ross* (BM).
In ditches, by wet tracksides, muddy streamsides; often near the sea.
Occasional; widespread.
2–13, 15–17

S. cernuus Vahl [Map, Fig. 2.7]
Isolepis cernua (Vahl) Roem. & Schult.
Slender Clubrush. Native; OS
2 Iona, 1873, *G. Ross* (BM).
Crevices in rocks, by shallow streams and in shady flushes, always coastal.
Rare; in scattered localities in the west.
2 Iona, 1873, *G. Ross* (BM). **3** Erraid (BM); coast opposite Erraid (BM). **4** Gometra (BM). **5** Haun (BM); Dùn Aisgain, *J. Robertson & Birse* (SSS); Port Burgh, *Birse* (BM); *Chapple* (OXF).

S. fluitans L.
Eleogiton fluitans (L.) Link
Floating Clubrush. Native; OWE
2 Iona, 1887, *Sommerville* (E, BM).
In runnels, streams, lochs. Usually in slow-flowing water, but can occur in fast flowing streams.
Occasional; widespread.
2–17
Can occur completely submerged in quite deep water, when it appears to be sterile.

Eleocharis R. Br.

E. quinqueflora (F. X. Hartm.) Schwarz
E. pauciflora (Lightf.) Link
Few-flowered Spikerush. Native; CN
'Mull', 1848, *Syme* (BM).
On streamsides, flushed areas, peaty rivulets and brackish pools, and in open communities especially on irrigated inorganic soils.
Occasional; widespread.
1–17

E. multicaulis (Sm.) Sm.
Many-stalked Spikerush. Native; CN
G. Ross (1877).
In closed communities on flushed peats, in Molinetum and damp heathland.
Occasional to frequent; widespread.
1–17

E. palustris (L.) Roem. & Schult.
Common Spikerush. Native; W
G. Ross (1877).
On peaty, or more rarely inorganic soils, in areas with a high water table. Can withstand slightly brackish conditions.
Occasional to frequent.
1–17

E. uniglumis (Link) Schult.
Slender Spikerush. Native; W
'Mull', 1848, *Syme* (BM).
In shallow water at lochsides, saltmarsh pools and runnels.
Occasional; widespread.
2, 3, 5–10, 12, 14, 16, 17

Blysmus Panz.

B. rufus (Huds.) Link
Saltmarsh Flatsedge. Native; ON
Lightfoot (1777).
Fringing brackish pools in the splash zone and uppermost saltmarsh pools.
Occasional; widely distributed around the coasts.
1–8, 10, 12, 14–17

Schoenus L.

S. nigricans L.
Black Bogrush. Native; WSA
Loch Buy [Buie], 1848, *Syme* (BM).
In flushes and rivulets, on coastal rocks and upper saltmarsh levels, ascending to c. 250 m.
Occasional to frequent; widespread, but more frequent coastally in the west.
1–13, 15–17

Rhynchospora Vahl

R. alba (L.) Vahl
White Beaksedge. Native; NSA
Tobermory, 1829, *G. Gordon* (1829).
In flushes and mires, especially *Molinia* associations.
Occasional; widespread.
3–5, 7–10, 11–17

Cladium Browne

C. mariscus (L.) Pohl
Great Fensedge. Native; WSA
1848 (presumed), *Syme* in Watson, *Cyb. Brit.* ms. slips.
Lochs of relatively high base-status and with deep, peat bottoms.
Rare; confined to the Ross.
3 Ardtun (BM); Loch an Sgalain; Loch an t'Suidhe, *BEC*; Loch Mòr Ardalanish; Meall nan Carn.
The *Atlas* dot (Perring & Walters, 1962) for square 17/42 is an error.

Carex L.

C. laevigata Sm.
Smooth-stalked Sedge. Native; OS
16 Guahalish [Gualacholish] Loch Spelve, 1848, *Syme* (BM).
On heavy soils and more limy soils in flushed situations, clay cliffs and coastal woodland.
Occasional; widely distributed in scattered localities.
3–13, 15–17

C. distans L.
Distant Sedge. Native; SSA
Keddie (1850).

On ledges within the spray zone, in upper saltmarsh communities and flushes and wet hollows near the sea.
Occasional to frequent; coastally in the west, but very rare on the Sound of Mull.
1–6, 8, 10, 12, 13, 16

C. hostiana DC.
Tawny Sedge. Native; WSA
'Mull', 1848, *Syme* (BM).
Species-rich flushes on hillsides and in cliff grassland.
Occasional to frequent; widespread and possibly under-recorded.
1–17

Carex binervis Sm.
Green-ribbed Sedge. Native; OWE
16 Kinlochspelve, 1848, *Syme* (BM).
In montane grassland and Callunetum, descending to similar coastal habitats and cliff ledges.
Abundant; widespread.
1–17
Very variable, from plants of up to 1·5 m on wet inland cliff ledges to small plants of grazed turf on mountain summits.

C. lepidocarpa Tausch
Long-stalked Yellow Sedge
Flannigan, 1956, recorded this species from **8** Tiroran. Other field records include **2** central Iona and **10** between Carsaig and the Nun's Pass, both of which localities could be calcareous enough for *C. lepidocarpa* but without seeing voucher specimens we must regard its occurrence as doubtful.

C. demissa Hornem.
Common Yellow Sedge. Native; C
Staffa, 1843, *Stables* (K-Wa).
In flushed areas at tracksides and streamsides and the stony margins of lochs.
Abundant; widespread.
1–17

C. demissa × distans
Native; SSA
Known only from **17** Grass Point, 1966, *Jermy* (BM); with parents.

C. serotina Mérat
Small-fruited Yellow Sedge. Native; CN
Cameron to Glen Byre, 1961, *Ribbons*.
In saltmarsh turf at high levels, usually at points where fresh water enters.
Rare to occasional; widespread in sheltered habitats round the coast.
3, 4, 6, 8–10, 12, 14, 15
A record for *C. flava* var. *oederi* [pre-1852], from **1** Staffa by Stables in the Watson *Cyb. Brit*. ms. slips is probably this species.

C. scandinavica E. W. Davies
Native; AS
Recorded only in a roadside depression with *Blysmus* at **8** Kilfinichen Bay, 1968, *David.*

C. extensa Gooden.
Long-bracted Sedge. Native; OS
Keddie (1850).
On upper levels of saltmarsh, cliff crevices within the splash zone.
Rare to occasional; widespread in scattered localities around the coasts, but not recorded from the S coast of Mull.
1, 2, 4–6, 10, 12, 15, 17

C. sylvatica Huds. [Map, Fig. 2.42]
Wood Sedge. Native; WSA
13 Bedarrach, 1876, *G. Ross* (BM).
Under woodland conditions, sometimes in shady gullies. Elsewhere as a relict from previous woodland.
Occasional; in scattered localities, mainly in N Mull.
3–8, 12–14

C. rostrata Stokes
Bottle Sedge. Native; W
Keddie (1850).
In sheltered parts of lochs, often filling peat pools and lochans; occasionally a flush plant.
Occasional; widespread.
2–17

C. vesicaria L.
Bladder Sedge. Native; CN
5 Reudle, 1968, *Eddy* (BM).
Streamside mire and in roadside ditch.
Rare; in addition to the above locality known only from **12** Salen, road to pier (BM).
The specimen collected by G. Ross and included by him in his 1879 list as *C. vesicaria × ampullacea* is *C. rostrata.*

C. pendula Huds.
Pendulous Sedge. Native; CS
5 Calgary, 1939, *Chapple* (OXF).
On clay soils on wooded cliffs.
Rare; in isolated localities.
5 Cliffs N of Calgary Bay (BM); near Frachadil, *Kenneth*. **10** Carsaig, *Kenneth & A. Stirling*.
At about the northern limit for this species on the W side of the country.

C. pallescens L.
Pale Sedge. Native; W
'Mull', 1848, *Syme* (BM).
Damp but well-drained soils in open woodland grassland and on cliff ledges.
Frequent; widely distributed.
1–17

C. panicea L.
Carnation Sedge. Native; W
G. Ross (1877).
In a wide range of habitats from mountain grassland and hypnoid moss associations to lowland streamsides and marshes.
Frequent to abundant; widespread.
1–17

C. limosa L.
Bog Sedge. Native; CN
2 Gull Glen, Dùn Cùl Bhuirg, 1912, *Wilmott.*
In basin mires with *Menyanthes* and *C. lasiocarpa.*
Rare; in addition to the above record **2**, also recorded from **3** Loch an t'Suidhe, *BEC*; Linne nan Ribheid; Tir Fhearagain (BM). **16** W of Loch Bhearnach.

C. paupercula Michx.
Tall Bog Sedge. Native; NM
Rare; known only from a *Sphagnum* bog on watershed of **16** Beinn a' Bhainne, 1961, *BSBI* where it was refound by *BMMS.*

C. flacca Schreb.
Glaucous Sedge. Native; WSA
G. Ross (1877).
In well-drained montane and lowland grassland in situations of higher base-status, also in some more acid coastal situations where sea-spray may raise the base level locally.
Frequent; widespread.
1–17

C. hirta L.
Hairy Sedge. Native; WSA
2 Iona, 1955, *Ferreira.*
On sandy shores with shell fragments and in moist ground at base of cliffs.
Rare; in addition to the above record also known from **3** Uisken Bay (BM) and **4** Gometra, W of Burial Ground (BM).

C. lasiocarpa Ehrrh.
Slender Sedge. Native; CN
6 Loch Peallach, 1877, *G. Ross* (BM).
In mire communities round lochs.
Rare; in scattered localities, not recently recorded from N Mull.
3 Loch an t'Suidhe, *BEC*; Tir Fhearagain (BM). **6** As above. **9** N of Loch Assopol (BM). **10** Coladoir river bog system. **15** Near Sgùrr Dearg, *Stace.*

C. pilulifera L.
Pill Sedge. Native; NSA
'Mull', 1848, *Syme* (BM).
In drier areas of montane grass heath and Callunetum and in similar lowland habitats, especially in the more open communities.
Frequent; widespread.
1–17

C. caryophyllea Latourr.
Spring Sedge. Native; W
'Mull', 1848, *Syme* (BM)
In well-drained base-rich grassland in upland and lowland situations.
Occasional; widespread; probably under-recorded.
1–13, 15–17

C. nigra (L.) Reichard
Common Sedge. Native; W
16 Kinlochspelve, 1848, *Syme* (BM).
In a wide range of bog and mire habitats, streamsides and boulder strewn shores, ascending to high altitude peaty pools and lochans.
Abundant; widespread.
1–17
The slender-leaved plants which occur as fringes to bog pools are var. **strictiformis** L. H. Bailey.

C. bigelowii Torr. ex Schwein. [Map, Fig. 2.19]
Stiff Sedge. Native; AA
11 Ben More, 1887, *Ewing* (BM, E).
In high altitude grass heath and on ledges in *Rhacomitrium* heath.
Occasional; generally distributed at higher elevations through the main central massif, the mountains of Glen Forsa, Ben Buie and Creach Beinn.
8, 11, 15, 16

C. paniculata L.
Greater Tussock Sedge. Native; C
Erie [Erray], 1876, *G. Ross* (BM).
In valley mires.
Rare; in scattered localities.
3 Loch Poit na h-I, *U. Duncan* (BM). **4** S of Ulva House (BM). **6** Dun Auladh, *McAllister.* **12** Meall Damh Àrd (BM); Feith Ban (BM); Lòn Biolaireach (BM); Killichronan, *Clark.* **13** Aros Park, *Clark*; and as above. **17** Lochan a' Doire Dharaich.

C. appropinquata Schumach.
Fibrous Tussock Sedge
The record in Clapham, Tutin and Warburg (1952) is an error.

C. otrubae Podp.
False Fox Sedge. Native; SSA
5 Calgary Bay, 1876, *G. Ross* (BM).
By damp roadsides, in rocky clefts and pools in the spray zone.
Occasional to frequent on the western coasts, rare in the east. Particularly common on the small islands.
1–10, 12, 16
Keddie's record for *C. vulpina* (1850) presumably should be referred to this species.

C. arenaria L. [Map, Fig. 2.32]
Sand Sedge. Native; OWE
5 Calgary Bay, 1876, *G. Ross* (BM).
In the loose sand of foredunes and blowouts and in fixed dunes.
Occasional in the west to locally frequent in the southwest; not recorded from the Sound of Mull.
2, 3, 5, 6, 8, 16

C. echinata Murr.
Star Sedge. Native; W
Keddie (1850).
In boggy situations on wet hillsides and by streamsides and wet tracksides.
Abundant; widespread.
1–17

C. remota L.
Remote Sedge. Native; WSA
13 Near Tobermory, 1876, *G. Ross* (BM).
In damp shady places, such as wet woodland and tree covered cliffs.
Rare to occasional; widespread in scattered localities.
3–10, 12, 13, 15, 16

C. curta Gooden.
White Sedge. Native; CN
16 Loch Buie area, 1956, *H. Milne-Redhead.*
In eutrophic mires, often near the sea.
Rare; scattered localities in N and E Mull.
6 Loch na Cuilce (BM). **12** Killbeg. **13** Lochnameal (BM). **15** Scallastle, *U. Duncan.* **16** As above. **17** Gorton, Loch Don (BM).

C. ovalis Gooden.
Oval Sedge. Native; CN
Keddie (1850).
On wet hill pastures, in Callunetum and by tracksides.
Frequent; widespread.
1–17

C. pauciflora Lightf. [Map, Fig. 2.42]
Few-flowered Sedge. Native; CN
16 Ben Buie, 1877, *G. Ross* (BM).
In upland bogs in association with *Sphagnum.*
Rare; in scattered localities in the east.
10 Beinn Chreagach (BM). **11** Ben More, *A. Stirling.* **12** Loch na Dairidh, *U. Duncan.* **13** Near Caol Lochan, *BEC*; near Tobermory, *Temperley* (BM). **16** As above; Beinn a' Bhainne (BM); Allt a' Bhaccain-Sheilich (BM); An Liathanach (BM).

C. pulicaris L.
Flea Sedge. Native; WSA
Kinlochspelve, 1848, *Syme* (BM).
In grassland on wet slopes, on ledges in steep valleys and ravines and in flushes.
Frequent to abundant; widespread.
1–17

C. dioica L.
Dioecious Sedge. Native; CN
6 Loch Peallach, 1876, *G. Ross* (BM).
On wet hillside flushes, grassland and wet pastures at lower altitudes.
Occasional to frequent; widespread.
2–17
Probably spreading much by vegetative means, all the plants in one patch being usually of the same sex.

GLUMIFLORAE

GRAMINEAE

Phragmites Adans.

P. australis (Cav.) Trin. ex Steud.
P. communis Trin.
Reed. Native; W
Keddie (1850).
In reedswamp, ditches, sometimes persisting as a relict of former swampy conditions.
Occasional to locally frequent; widespread.
1–17

Molinia Schrank

M. caerulea (L.) Moench
Purple Moor-grass. Native; W
G. Ross (1877).
The dominant grass of large areas particularly on the peat, usually in mires and avoiding areas with stagnant groundwater; favoured by burning. Forming a characteristic association with *Myrica* at altitudes up to 300 m.
Abundant; widespread.
1–17

Sieglingia Benh.

S. decumbens (L.) Bernh.
Heath-grass. Native; WSA
G. Ross (1877).
In species-rich grassland.
Frequent to abundant; widespread.
1–17

Glyceria R.Br.

G. fluitans (L.) R.Br.
Floating Sweet-grass. Native; W
Keddie (1850).
In ditches, small streams and in shallow standing water.
Occasional; widespread at low altitudes.
1–8, 10–12, 15–17

G. plicata Fr.
Plicate Sweet-grass. Native; SSA
Known only from a roadside ditch at **5** Calgary Bay, 1966, *Cannon & Groves* (BM). Probably reaching its northwestern limits in Mull.

G. declinata Bréb.
Small Sweet-grass. Native; WSA
6 Poll Athach, 1961, *Ribbons.*
In ditches and sides of small streams.
Rare; in a few scattered localities. In addition to the above known only from **10** Carsaig, *U. Duncan* (BM); **13** Lochnameal (BM); and **17** Lochdonhead, *U. Duncan.*

G. maxima (Hartm.) Holmberg
Reed Sweet-grass
Known only from *G. Ross* (1877, 1879) and stated to be 'common'. In view of this comment and the absence of any other record we regard this as an error.

Festuca L.

F. pratensis Huds.
Meadow Fescue. Native; W
G. Ross (1877).
By roadsides and tracks.
Rare; in scattered localities; probably under-recorded.
2 Iona, *Braid* (GL). **5** Calgary Bay, *G. Ross* (1879). **6** Poll Athach, *Ribbons*; Glengorm

Castle, *V. Gordon*; Loch Peallach. **9** Cnoc Mòr. **13** Tobermory. **14** Lower Glen Forsa; Cnoc an Teine.

F. arundinacea Schreb.
Tall Fescue. Native; W
G. Ross (1877).
By roadsides and in sheltered bushy habitats.
Rare; in scattered localities but not recorded from SE Mull.
2 Iona, 1955, *Ferreira* and *BMMS*. **3** Uisken Bay (BM). **6** Dervaig, *CSSF*. **8** Gribun, *McAllister*. **13** Tobermory.

F. gigantea (L.) Hill
Giant Fescue. Native; W
G. Ross (1877).
In shady deciduous woodland and gullies.
Rare; only recorded from N Mull.
5 Calgary, *U. Duncan & Beattie*. **7** Coille Chill' a' Mhoraire. **13** Tobermory; Rubha na Leip (BM); Rubha nan Gall (BM); Aros Park, *U. Duncan*.

F. rubra L.
Red Fescue. Native; W
G. Ross (1879).
In a very wide range of habitats, a constant species of most grassland types.
Abundant; widespread.
1–17

F. ovina L.
Sheep's Fescue. Native; W
Keddie (1850).
In a wide range of habitats but also occurring in drier habitats than *F. rubra* and possibly tending to replace that species at higher altitudes.
Abundant; widespread.
1–17

F. tenuifolia Sibth.
F. ovina subsp. *tenuifolia* (Sibth.) Peterm.
Fine-leaved Sheep's Fescue. Native
Aird Kilfinichen, 1956, *Flannigan*.
Possibly a characteristic species of older Callunetum and *Betula* woods.
Rare; in scattered localities, but certainly under-recorded.
1 Fladda. **5** Tostarie (BM). **8** Coire Buidhe; Fionna Mhàm; as above. **12** Salen Forest (BM); Cnoc an Teine. **14** Allt Mòr Coire nan Eunachair, *Stace*. **16** Allt a' Bhaccain-Sheilich.

F. vivipara (L.) Sm.
Viviparous Fescue. Native; AS
Tommigoullin [? Tom na Gualainne], 1848, *Syme*.
Characteristically at higher altitudes although occurring right down to sea level. Possibly absent from the more basic soils.
Occasional; widespread but absent from most of Brolass and W Mornish and Treshnish.
1–4, 6–17

Lolium L.

L. perenne L.
Perennial Rye-grass. Native and denizen; W
G. Ross (1877).
Probably not a true native grassland species.
In pastures and on roadsides.
Occasional to frequent; widespread at lower altitudes.
1–17

L. multiflorum Lam.
L. perenne subsp. *multiflorum* (Lam.) Husnot
Italian Rye-grass. Denizen
G. Ross (1877).
In pastures and on roadsides near farms, often in nitrogen-rich habitats.
Rare; in scattered localities.
3 An t'Àrd. **4** Near Ulva House (BM). **5** Haun (BM). **6** Poll Athach, *Ribbons*. **8** Inch Kenneth; S Gribun; S Ardmeanach, *Flannigan*. **13** Tobermory. **17** Duart.

Vulpia C. C. Gmel.

V. bromoides (L.) Gray
Squirrel-tail Fescue. Native; WSA
G. Ross (1877).
On dry disturbed roadsides and in rock crevices.
Rare to occasional; in scattered localities.
2–6, 9, 10, 12, 13, 16, 17

Puccinellia Parl.

P. maritima (Huds.) Parl. [Map, Fig. 2.30]
Common Saltmarsh-grass. Native; OWE
12 Salen, 1890, *Ewing* (BM).
In upper saltmarsh levels and on rocks in the splash zone.
Occasional to frequent around the coasts.
1–14, 16, 17
The record in Ross (1879) is probably an error as a specimen under this name at BM has been redetermined as *Agrostis stolonifera*.

P. distans (L.) Parl.
Reflexed Saltmarsh-grass. Native
2 Iona, Port Pollarain, 1912, *Wilmott*.
Known only from the above record and another from **8** Gribun, 'abundant', *Gilmour* (1932); not subsequently confirmed.

Catapodium Link

C. marinum (L.) C. E. Hubbard [Map, Fig. 2.33]
Sea Fern-grass. Native; M
2 Iona, 1948, *Braid* (GL).
On maritime rocks and shell sand at top of beaches.
Rare; scattered localities in the west, particularly in Iona.
1 Bac Bheag (BM). **2** Iona; Cnoc-nam-Bràdhan, *U. Duncan*; Cùl Bhuirg (BM); Arduara (BM); Eileann Annraidh (BM). **3** Cnoc an t'Suidhe (BM); Port nan Ròn. **5** Port Langamull (BM); Knoc, *Birse* (SSS).

Poa L.

P. annua L.
Annual Meadow-grass. Native; W
Keddie (1850).
On roadsides and tracksides, disturbed ground, river gravels and in short turf on cliff tops often associated with guano.
Frequent; widespread.
1–17
A perennial form occurs at higher altitudes in *Montia* flush associations.

P. alpina L.
Alpine Meadow-grass
Recorded by Ewing (1890) but not confirmed. Its presence on Mull is a possibility as the species has been recorded from adjacent Morvern.

P. nemoralis L.
Wood Meadow-grass. Native; W
13 Tobermory, 1939, *Chapple* (OXF).
In shady deciduous woodland, on cliffs and in gullies.
Occasional; widely distributed but apparently absent from Brolass, the Ross and Iona.
1, 4–8, 10–17

P. compressa L.
Flattened Meadow-grass. ? Native; C
13 Calve and Lovage islands, 1955, *Harris*.
Known only from the above record **13** and from Tobermory.

P. pratensis L.
Smooth Meadow-grass. Native and ? denizen; W
Keddie (1850).
On roadsides, round farms etc. Possibly not a natural constituent of Mull grasslands.
Occasional to locally frequent; widespread.
1–17

P. angustifolia L.
Narrow-leaved Meadow-grass. Native
Known only from dry walls at **13** Tobermory, 1970, *Eddy* (BM).

P. subcaerulea Sm.
Spreading Meadow-grass. Native; AS
2 Iona, S of Baile Mòr, 1967, *Kendrick & Moyes* (BM).
In machair and short turf on cliff tops.
Rare to occasional; scattered localities in the north and west.
1–3, 5–7, 10, 12, 13

P. trivialis L.
Rough Meadow-grass. Native; W
G. Ross (1877).
In ditches, streamsides, damp meadows and on waste ground.
Frequent; widespread.
1–17

P. flabellata Hook. f.
This species from the Falkland Islands was experimentally introduced by a Mr MacRae in 1961. (See *Shetland Times* 1 March 1963.)

Catabrosa Beauv.

C. aquatica (L.) Beauv. [Map, Fig. 2.6]
Whorl-grass. Native; W
8 Gribun, 1929, *Gilmour* (BM).
In shallow running water in coastal habitats.
Occasional; confined to SW Mull and Iona.
2, 3, 8, 9, 16

Dactylis L.

D. glomerata L.
Cocksfoot. Native; W
Keddie (1850).
On roadsides, disturbed ground, pastures and sea cliffs.
1–17
Cliff-top forms remain small and bluish after transplantation to garden conditions.

Cynosurus L.

C. cristatus L.
Crested Dogstail. Native; OWE
Keddie (1850).
Species-rich grassland on well-drained slopes.
Frequent; widespread.
1–17

Briza L.

B. media L.
Quaking-grass. Native; OWE
10 E of Carsaig, 1966, *James*.
In disturbed situations and species-rich basic grassland.
Rare; in widely scattered localities. In addition to the above known from **2** Iona, near Baile Mòr; **6** Dervaig, *CSSF*; and **15** Scallastle, *CSSF*.

Melica L.

M. uniflora Retz.
Wood Melick. Native; W
13 Bedarrach and Druimfin, *G. Ross* (1879).
In shady deciduous woodland.
Rare; in addition to the above old record recently recorded from another locality in the same area: Tobermory, path to lighthouse, *Clark* (BM).

M. nutans L.
Mountain Melick. Native; CN
5 Calgary, 1965, *Kenneth*.
In shady groups, sometimes on ledges within spray from waterfalls.
Rare; in addition to the above locality known from **7** Ardow Burn waterfalls (BM); **12** Allt Ardnacross (BM); and Lòn Biolaireach (BM).

Bromus L.

B. ramosus Huds.
Zerna ramosa (Huds.) Lindm.
Hairy-brome. Native; SSA
G. Ross (1877).

In deciduous woodland.
Rare; scattered localities in N Mull with an isolated record in the south.
5 Calgary, *Kenneth*. **6** Mingary, *CSSF*; Penmore, *G. Ross* (1878); Rubha nan Gall to Bloody Bay, *Stace*. **7** Torloisk, *CSSF*. **10** Carsaig to Nun's Pass, *Kenneth & A. Stirling*. **12** Salen. **13** Tobermory (BM); Rubha nan Gall (BM); Aros Park (BM).

B. sterilis L.
Anisantha sterilis (L.) Nevski
Barren Brome. ? Colonist
Keddie (1850).
On waste ground near farms and houses.
Rare; two localities in the north and two in the south-west.
2 Iona; Baile Mòr; Culbuirg, *Braid* (GL). **3** Uisken. **12** Ardnacross. **13** Tobermory.

B. mollis group (based on Smith, 1968).

B. hordeaceus L. subsp. **hordeaceus**
B. mollis L.
Soft-brome. ? Native; W
G. Ross (1877).
On waste ground and in arable fields.
Occasional; widespread in scattered localities, especially round farms and houses.
1–4, 7–10, 12, 13, 17
Some of the field records for this species may refer to the closely similar *B. × pseudothominii*.

B. × pseudothominii P. Smith
B. hordeaceus subsp. *hordeaceus × lepidus*
Native
5 Calgary Bay, 1966, *Cannon & Groves* (BM).
On roadsides, waste ground round farms and fixed sand above beach.
Rare; widespread in scattered localities.
5 As above. **6** Dervaig (BM). **8** Inch Kenneth (BM). **9** Tràigh nam Beach (BM). **12** Shingle spit at head of Loch na Keal (BM); Salen (BM).

B. hordeaceus subsp. **thominii** (Hardouin) Hylander
B. thominii Hardouin
Lesser Soft-brome. ? Native; OWE
8 Gribun, 1961, *Kenneth*.
Known from the above record (the basis of the 17/43 dot in the *Atlas*) and two records from **13** Tobermory and Erray House. All were made before the recognition of *B. × pseudothominii* and are probably referable to that hybrid.

B. lepidus Holmberg
Slender Brome. ? Native
12 Glen Forsa House, 1958, *Gerrans* (BM).
On waste ground, roadsides and cultivated ground.
5 Calgary, *U. Duncan & Beattie*. **9** Tràigh nam Beach (BM). **12** As above. **13** Tobermory (BM).

B. racemosus L.
Smooth Brome
Recorded by Keddie (1850) from **1** Staffa or **2** Iona, and also by G. Ross (1879) from **13** Erie [Erray] and Balliscate. There are no recent records and in view of the general distribution of the species, it is believed to be an error for Mull.

Brachypodium Beauv.

B. sylvaticum (Huds.) Beauv.
False-brome. Native; W
13 Tobermory, [1829], *G. Gordon* (K-Wa).
In deciduous woodland, on steep roadside banks and in mountain gullies at lower altitudes.
Frequent; widespread.
1–17

Agropyron Gaertn.

A. caninum (L.) Beauv.
Bearded Couch. Native; W
G. Ross (1877).
In deciduous woodland and sheltered gullies.
Rare; in scattered localities.
8 Balmeanach to Mackinnon's Cave, *Kenneth*; SW Ardmeanach. **10** Carsaig. **13** Druimfin, *G. Ross* (1877). **15** Scallastle, *CSSF*.

A. repens (L.) Beauv.
Common Couch. Native; W
G. Ross (1877).
On disturbed ground, waste places, gardens and on pebble beach.
Rare to occasional; in scattered localities, not recorded from SE Mull.
2–5, 8, 10–13

A. pycnanthum (Godr.) Gren. & Godr.
A. pungens auct.
Sea Couch
The specimen collected at **5** Calgary Bay by Ross in 1876 and presumably the basis for the record in Ross (1877), has been redetermined as *A. junceiforme × repens*.

A. junceiforme (A. & D. Löve) A. & D. Löve
Sand Couch. Native; OWE
I-Columb-Kill [Iona], *Lightfoot* (1777).
On loose dry sand on beaches above the strand line.
Rare; in scattered localities in the south and west.
2 Iona, N end (BM); near Calva (BM); Eileann Annraidh (BM); Arduara, *SSS*. **3** Uisken, *Kenneth & A. Stirling*. **5** Calgary Bay, *Kenneth*. **9** Cnoc Mòr; Cnoc nan Gabhar. **16** Laggan Sands, *Kenneth*.

A. × laxum (Fr.) Tutin
A. junceiforme × repens
Native; OWE
5 Calgary, 1876, *G. Ross* (BM).
Known only from the above and a recent record from the same area: 1966, *U. Duncan*.

Both parents are present in the area and the hybrid may have persisted there for at least 90 years.

Koeleria Pers.

K. cristata (L.) Pers. [Map, Fig. 2.13]
Crested Hair-grass. Native; OWE
Keddie (1850).
Mainly near the sea in species-rich, dry, grassland, especially in shallow stony skeletal soils.
Occasional to frequent in the west.
1–10, 12, 13, 16

Avena L.

A. strigosa Schreb.
Bristle Oat. Colonist
13 Balliscate, 1876, *G. Ross* (BM).
A cornfield weed.
Formerly recorded from two areas in NE Mull where it was apparently common.
12 Near Aros Castle, *BEC*; Salen, *Vachell* (NMW) – the latter presumably a voucher for the previous record. **13** As above.
Ross (1879) states 'among corn in all the crofts and farms'. The reduction in cereal growing, coupled with improved agricultural techniques, has no doubt caused the apparent disappearance of this species in common with other arable weeds.

Helictotrichon Bess.

H. pratense (L.) Pilg.
Meadow Oat-grass
Reported by Keddie (1850) but the absence of any subsequent records suggests that this may be an error for *H. pubescens*, which although common in the area, Keddie does not include in his list. However, the general distribution of the species does not rule out its occurrence in Mull.

H. pubescens (Huds.) Pilg.
Downy Oat-grass. Native; W
13 Errie [Erray], 1876, *G. Ross* (BM).
In rocky gullies, on stony sea shores, and well-drained relatively base-rich habitats.
Occasional; widely distributed in the west but rare in the east.
1–10, 12, 13, 15, 16

Arrhenatherum Beauv.

A. elatius (L.) Beauv. ex J. & C. Presl.
False Oat-grass. Native; W
G. Ross (1879).
By roadsides and wood margins, in waste places and disturbed areas.
Occasional to frequent; widely distributed.
1–17
Becoming dominant on some of the small islands, where it was probably introduced with sheep.

Holcus L.

H. lanatus L.
Yorkshire-fog. Native; W
Keddie (1850).
In grass and *Calluna* heaths, open deciduous woodland, meadows and sandy foreshores.
Abundant; widely distributed.
1–17

H. mollis L.
Creeping Soft-grass. Native; W
G. Ross (1877).
In deciduous woodland and shady gullies, occasionally in pastures.
Occasional; widespread, not nearly so frequent as *H. lanatus*, but possibly somewhat under-recorded.
1–17

Deschampsia Beauv.

D. cespitosa (L.) Beauv.
Tufted Hair-grass. Native; W
Keddie (1850).
In deciduous woodland, neutral mires, on alluvial terraces and in montane grassland, ascending to 900 m.
Frequent; widespread.
1–17
A specimen originating from **11** Beinn nan Gabhar (BM) may be var. *pseudalpina*, see note after next species.

D. alpina (L.) Roem & Schult.
Alpine Hair-grass. Native; AS
11 Ben More, 1887, *Ewing* (GL).
On cliffs above 300 m.
Rare; in addition to the above the following records are known: **11** Ben More, NE rocks, *Templeman* (BM); north corrie, *McAllister* (BM); Beinn nan Gabhar, *McAllister* (BM).
We are indebted to H. A. McAllister (pers. comm.) for the following note on this complex: 'Recent cytotaxonomic work suggests that there may be four entities within the *Deschampsia cespitosa/alpina* complex:

1. Seminiferous diploid $2n = 26$ is *D. cespitosa* (widespread).
2. Seminiferous tetraploid $2n = 52$ is *D. cespitosa* subsp. *littoralis?* Large glumes and often more than two florets per spikelet (mountains of N England and SW Scotland).
3. Southern viviparous $2n = 39$ (52). Probably is *D. cespitosa* var. *pseudalpina* (Syme) Druce. Like *D. cespitosa* but viviparous. Plantlet attachment to panicle not brittle. Plantlet small, leaf not keeled, apex acute, cucullate, leaf margin toothed to apex (SW Scottish Highlands).
4. Northern viviparous $2n = (39)$ 52 is *D. alpina* s.s. Small plant. Plantlet attachment to panicle very brittle. Plantlet large, glumes and lemmas appressed to plantlet. Leaves $\pm$ keeled with bluntly cucullate tip. Leaf margin near tip untoothed (NE Highlands in very open habitats).

Types 3 and 4 can be very difficult to separate in the wild, but are quite distinct in cultivation.

Both 3 and 4 occur on Beinn nan Gabhar, but I have only seen 4 from Ben More.'

D. flexuosa (L.) Trin.
Wavy Hair-grass. Native; W
13 Rareich, 1876, *G. Ross* (BM).
On well-drained slopes in soils containing some residual peat, from sea level to the highest altitudes.
Frequent; widely distributed.
1–17

Aira L.

A. praecox L.
Early Hair-grass. Native; WSA
13 Bedarrach, 1876, *G. Ross* (BM).
In shallow soil on rocks and walls, in machair and heaths.
Frequent; widespread.
1–17

A. caryophyllea L.
Silver Hair-grass. Native; W
G. Ross (1877).
On walls, boulders and in machair.
Occasional; widely distributed.
2–10, 12, 13, 15–17

Ammophila Host

A. arenaria (L.) Link
Marram. Native; OWE
2 Iona, *Keddie* (1850).
In loose wind-blown sand at top of beaches.
Rare to occasional; in scattered localities in the south and west.
2, 3, 5, 7, 9, 16

Calamagrostis Adans.

C. epigejos (L.) Roth
Wood Small-reed. Native; CN
13 MacLean of Coll's wood, Tobermory, 1829, *G. Gordon* (1839).
In damp open woodland and *Salix* carr.
Rare; in scattered localities in the north and west.
3 Uisken, *Kenneth & A. Stirling*. **4** SE of Ulva House (BM); near A' Chrannag (BM). **5** Sunipol; N shore of Calgary Bay, *Vachell*. **6** Croig, *G. Ross* (1879). **8** Balmeanach to Mackinnon's Cave, *Kenneth*. **13** Bedarrach, Druimfin and Heatherfield, *G. Ross* (1878); as above; Aros Park, *BEC*.

Agrostis L.

A. canina L.
Brown Bent. Native; W
G. Ross (1877).
In acid to medium base-rich grassland often in flushed or mire conditions, on streamsides and wet ledges.
Frequent; widespread.
1–17

A. tenuis Sibth.
Common Bent. Native; W
G. Ross (1877).
In more basic, drier situations than *A. canina*, on heaths and moorlands.
Frequent; widespread but less common than *A. canina*.
1–17

A. stolonifera L.
Creeping Bent. Native; W
6 Bloody Bay, 1876, *G. Ross* (BM).
In damp situations, especially near the sea; often spreading round pool margins and on waste ground.
Occasional in the east to locally frequent in the west.
1–17

Phleum L.

P. bertolonii DC.
P. nodosum auct.
Smaller Catstail. Native; OWE
12 Salen, 1966, *Cannon & Groves*.
On roadsides and disturbed ground.
Rare; scattered localities in N Mull, Iona and the Treshnish Isles.
1 Cairn na Burgh More. **2** Iona, central area. **5** Tostarie. **6** Dun Auladh, *McAllister*; Torloisk, *CSSF*. **11** Knock. **12** Salen. **13** Tobermory; cliffs N of Tobermory; Criadach Mhòr.

P. pratense L.
Timothy. ? Native and denizen; W
G. Ross (1879).
On roadsides, low-lying grassland and disturbed ground.
Rare to occasional; widespread in scattered localities.
2, 3, 5, 6, 9–14, 16, 17

P. arenarium L.
Sand Catstail. Native; OS
Only known from semi-stabilised shell sand at edge of machair on **2** Iona, N end of island, 1968, *Mullin & Gardiner*.

Alopecurus L.

A. pratensis L.
Meadow Foxtail. Native; W
Keddie (1850).
On roadsides, low-lying pastures and disturbed ground.
Rare to occasional; in scattered localities.
2, 3, 6–10, 12, 13, 16, 17

A. geniculatus L.
Marsh Foxtail. Native; W
Keddie (1850).
In wet meadows, muddy depressions at streamsides, especially in nitrogen rich habitats.
Occasional; in scattered localities, more frequent in the north and west.
1–17

Milium L.

M. effusum L.
Wood Millet. Native
13 Druimfin, *G. Ross* (1879).

In deciduous woodland.
Rare; in two widely separated areas.
10 Carsaig near Innamore Lodge, *Kenneth & A. Stirling*. **13** Aros Park (BM); *Henderson* (E).

Anthoxanthum L.

A. odoratum L.
Sweet Vernal-grass. Native; W
Keddie (1850).
In a great variety of habitats, but avoiding poorly drained situations.
Abundant; widespread.
1–17

Phalaris L.

P. arundinacea L.
Reed Canary-grass. Native
G. Ross (1877).
In species-rich mires and roadside ditches.
Occasional; widespread at lower altitudes.
1–13, 15, 17

P. canariensis L.
Canary-grass. Colonist
Known only from **2** Iona, near Baile Mòr, 1912, *Wilmott*. Presumably introduced with bird-seed.

Parapholis C. E. Hubbard

P. strigosa (Dumort.) C. E. Hubbard
Hard-grass. Native; OS
6 Dervaig, 1939, *BEC*.
Upper margins of saltmarsh.
Wilmott (1942) notes this record as 'needs confirmation', accepted by Clapham, Tutin & Warburg (1952) and (1962), but its occurrence has yet to be substantiated.

Nardus L.

N. stricta L.
Mat-grass. Native; W
G. Ross (1877).
In grass heath on peaty and alluvial terraces, montane grassland especially of N-facing slopes and in areas with late-lying snow.
Abundant; widespread.
1–17

References

ALLEN, D. E. (1969). *The Flowering Plants of the Isle of Man*. Douglas.

BORRER, W. (1810). *Scotch habitats*. MS at British Museum (Natural History).

CANNON, J. F. M. & BANGERTER, E. B. (1968). Plant Records from Mull and the adjacent small islands. *Proc. bot. Soc. Brit. Isl.*, **7**: 365–372.

———. (1970). Plant records from Mull and the adjacent small islands 2. *Watsonia*, **8**: 145–153.

———. (1972). Plant records from Mull and the adjacent small islands 3. *Watsonia*, **9**: 27–32.

CLAPHAM, A. R., TUTIN, T. G. & WARBURG, E. F. (1962). *Flora of the British Isles* ed. 2. Cambridge.

COUSENS, J. E. (1965). The status of the Pedunculate and Sessile Oaks in Britain. *Watsonia*, **6**: 161.

DANDY, J. E. (1958). *List of British Vascular Plants*. London.

———. (1969). Nomenclatural changes in the List of British Vascular Plants. *Watsonia*, **7**: 157–178.

DONY, J. G., ROB, C. M., & PERRING, F. H. (1974). *English names of wild flowers*. London.

DRUCE, G. E. (1920). The extinct and dubious plants of Britain. *Rep. botl. Soc. Exch. Club. Br. Isl.*, **5**: 731–799.

———. (1932). *The comital Flora of the British Isles*. Arbroath.

EWING, P. (1890). A contribution to the topographical botany of the west of Scotland. *Proc. Trans. nat. Hist. Soc. Glasgow*, n.s. **2**: 309–321.

———. (1892a). Second contribution to the topographical botany of the west of Scotland. *Proc. Trans. nat. Hist. Soc. Glasgow*, n.s. **3**: 159–160.

———. (1892b). Third contribution to the topographical botany of the west of Scotland. *Proc. Trans. nat. Hist. Soc. Glasgow*, n.s. **3**: 161–165.

———. (1892), (1899, ed. 2) *Glasgow catalogue of native and established plants*, being a contribution to the topographical botany of the western and central counties of Scotland. Glasgow.

———. (1896). Contribution to the topographical botany of the west of Scotland. *Proc. Trans. nat. Hist. Soc. Glasgow*, n.s. **4**: 199–214.

FINDLAY, C. (1971). *A study of the factors limiting the distribution of J. R. Matthews'*

phytogeographical elements in thirteen localities. Undergraduate dissertation, University of Aberdeen. Copy in Botany Department, British Museum (Natural History).
GARNETT, T. (1800). *Observations on a tour through the Highlands and a part of the Western Isles of Scotland*, etc. 2 vols. London.
GERRANS, M. B. (1960). Notes on the flora of the Isle of Mull. *Proc. bot. Soc. Br. Isl.*, **3**: 369–374.
GILMOUR, J. S. L. (1932). New county and other records (sub. *Glyceria*). *Rep. bot. Soc. Exch. Club. Br. Isl. 1931*, **9**: 678.
GORDON, G. (1829). *List of plants gathered chiefly in the Province of Moray*. MS at British Museum (Natural History).
———. (1839). *Collectanea for a Flora of Moray; or, a list of the phaenogamous plants and ferns hitherto found in the Province*. Elgin.
GORDON, S. (1961). Plant notes: *Koenigia islandica* L. *Proc. bot. Soc. Br. Isl.*, **4**: 160.
GROVES, E. W. (1971). *Hippophae rhamnoides* L. on the Isle of Mull, Argyllshire. *Watsonia*, **8**: 396.
HARLEY, R. M. (1956). *Rubus arcticus* L. in Britain. *Watsonia*, **3**: 237–238.
HESLOP HARRISON, J. W. & BOLTON, E. (1938). The rose flora of the Inner and Outer Hebrides and of other Scottish islands. *Trans. Proc. bot. Soc. Edinb.*, **32**: 424–431.
HOOKER, W. J. (1821). *Flora Scotica*. London.
———. (1830). *British Flora* ed. 1. London.
HUDSON, W. H. (1778). *Flora Anglica* ed. 2. London.
KEDDIE, W. (1850). *Staffa and Iona described and illustrated*. Glasgow.
KENNETH, A. G. & STIRLING, A. McG. (1970). Notes on the Hawkweeds (*Hieracium* sensu lato) of western Scotland. *Watsonia*, **8**: 97–120.
KENT, D. H. (1958). *British herbaria*. London.
LANJOUW, J. & STAFLEU, F. A. (1964). *Index herbariorum 1. The herbaria of the world* ed. 5. (Regnum Vegetabile vol. 31). Utrecht.
LIGHTFOOT, J. (1777). *Flora Scotica* ed. 1. London.
MACCULLOCH, J. (1819). *A description of the Western Isles of Scotland*. London.
MACLAGAN, D. P. (1855). Notice of plants in the neighbourhood of Oban and in part of the island of Mull. *Scottish Gardener*, **4**: 93–95.
MACLEAN, J. P. (1923), (1925). *History of the Island of Mull*, vols. 1 & 2. San Mateo, California.
MATTHEWS, J. R. (1923). Notes on Scottish Plants. *Trans. Proc. bot. Soc. Edinb.*, **28**: 170–173.
———. (1955). *Origin and distribution of the British flora*. London.
MILLAR, J. M. (1972). *Flowers of Iona*. Glasgow.
MOLONY, E. (ed.) (1951). *Portraits of islands*. [Chap. 5 (pp. 52–61) by Bruce Campbell]. London.
MURRAY, W. H. (1966). *The Hebrides*. Edinburgh.
PENNANT, T. (1774). *A tour in Scotland and voyage to the Hebrides 1772*. Chester & London.
PERRING, F. H. & SELL, P. D. (eds.) (1968). *Critical supplement to the Atlas of the British flora*. London & Edinburgh.
PERRING, F. H. & WALTERS, S. M. (ed.) (1962). *Atlas of the British flora*. London & Edinburgh.
POBEDIMOVA, E. G. (1968). Species novae generis *Cochlearia* L. *Nov. Sist. Visshikh. Rast.*, **5**: 131–139.
———. (1969). Revisio generis *Cochlearia* L. *Nov. Sist. Visshikh. Rast.*, **6**: 67–106.
POLUNIN, N. (1953). Arctic plants not yet found in Britain. *Watsonia* **3**: 34–35.
RATCLIFFE, D. (1960). *Koenigia islandica* in Mull. *Trans. Proc. bot. Soc. Edinb.*, **39**: 115–116.
———. (1963). BSBI field meeting in Mull 1961. *Proc. bot. Soc. Br. Isl.*, **5**: 94–95.
———. (1968). An ecological account of the Atlantic bryophytes in the British Isles. *New Phytol.*, **67**: 165–439.
RAVEN, J. E. (1952). *Koenigia islandica* in Scotland. *Watsonia*, **2**: 188–190.
RAVEN, P. H. (1963). *Circaea* in the British Isles. *Watsonia*, **5**: 262–272.
RAY, J. (1724). *Synopsis methodica stirpium Britannicarum* ed.3 by J. J. Dillenius. London.
RITCHIE, A. & RITCHIE, E. (1934, 1945). *Iona past and present* eds. 3 & 4. Edinburgh.
ROBERTS, R. H. (1968). The hybrids of *Mimulus cupreus*. *Watsonia*, **6**: 371–376.
ROSS, G. (1877). Isle of Mull, Mid Ebudes. *Rep. bot. Loc. Rec. Club for 1876:* 188–192.

———. (1879). On the flora of Mull. *Trans. bot. Soc. Edinb.*, **13**: 234–242.

SIBBALD, R. (1684). *Scotia illustrata . . . De plantis scotiae*, pars secunda **1**: 18. Edinburgh.

SLACK, A. A. (1969). Isle of Mull, 13–20 July. *Proc. bot. Soc. Br. Isl.*, **7**: 638.

SMITH, P. (1968). The *Bromus mollis* aggregate in Britain. *Watsonia*, **6**: 327–344.

SMITH, J. E. (1790–1814). *English Botany* ed. 1. London.

SYME, J. T. I. BOSWELL (1863–1886). *English Botany* ed. 1. London.

TEMPLEMAN, A. (1942). Report on excursions arranged in 1939: Iona and Lunga. *Rep. bot. Soc. Exch. Club. Br. Isl.*, **12**: 249–250.

TRAIL, J. W. H. (1909). Additional vice-county records from West Scotland. *Ann. Scot. nat. Hist.*, **18**: 250.

TURNER, N. & FINLAY, A. (1967). *The Isle of Mull.* Glasgow.

WALFORD, T. (1818). *The scientific tourist through England, Wales and Scotland*, vol. 2. London.

WALKER, J. (1808). *Essays on natural history and rural economy.* London & Edinburgh.

WALTERS, S. M. (1968). *Betula* in Britain. *Proc. bot. Soc. Br. Isl.*, **7**: 179–180.

WATSON, H. C. (1837). *The new botanist's guide to the localities of the rarer plants of Britain* vol. 2: *Scotland and adjacent isles.* London.

———. (1847–1859). *Cybele Britannica* 4 vols. London. The MS slips for Cybele Britannica are at the British Museum (Natural History).

———. (1873–1874). *Topographical botany* ed. 1. Thames Ditton.

WEBB, D. R. (1956). A new subspecies of *Pedicularis sylvaticus* L. *Watsonia*, **3**: 239–241.

WILMOTT, A. J. (1912). *The flora of the Island of Iona* (*Argyllshire*). MS at the British Museum (Natural History).

———. (1942). V.C. 103 The Isle of Mull. *Rep. bot. Soc. Exch. Club. Br. Isl.*, **12**: 236–249.

WOODELL, S. R. J. & KOOTIN-SANWU, M. (1971). Intraspecific variation in *Caltha palustris.* *New Phytologist*, **70**: 173–186.

12. Ferns and their allies

E. B. Bangerter, J. F. M. Cannon and A. C. Jermy

Introduction

The layout and notations used in this section are the same as in Chapter 11 as are the methods as to how records were collected. The nomenclature is based on *Flora Europaea* (Valentine, 1964) with some minor changes. The arrangement of genera is that in the herbarium at the British Museum (Natural History) based on Crabbe, Jermy & Mickel (1975).

We should like to thank H. V. Corley, J. D. Lovis and C. N. Page for checking identifications in *Dryopteris filix-mas* agg., *Asplenium trichomanes* agg. and *Equisetum*, respectively.

Index to genera

LYCOPODIOPHYTA

LYCOPODIOPSIDA

LYCOPODIALES

LYCOPODIACEAE

Lycopodium L.

L. annotinum L.
Interrupted Clubmoss. Native; NC
16 W side of Glen Liebetal [Libidil] 1848, *Syme* (BM).
In dwarf shrub communities on wet ravine edges.
Rare; recorded from one area only, on the Laggan Deer Forest.
16 Lag a' Mhuilinn-luaidh, *BSBI*; and the above record.

L. clavatum L.
Stagshorn Clubmoss. Native; NC
16 Hills south of Loch Spelve. 1848, *Syme* (BM).
On well-drained slopes on basalt soils, often in species-rich grassland.
Rare; in scattered localities; not known from the south-west.
6 Loch an Torr (BM); 'S Airde Beinn, *Kenneth* (BM); Reservoir [? Tobermory], *Bowerman.* **7** Loch a' Ghael (BM); Beinn na Drise (BM). **11** N corrie of Ben More, *U. Duncan.* **13** "Heatherfield", *G. Ross* (1879). **15** Beinn Chreagach Mhor and Bheag. **16** Allt a' Bhacain sheilich; Cnoc Shalachry, *Crabbe.*

Diphasiastrum Holub

D. alpinum (L.) Holub [Map, Fig. 2.25]
Lycopodium alpinum L.; *Diphasium alpinum* (L.) Rothm.
Alpine Clubmoss. Native; AA
11 S side of Ben More, 1966 *U. Duncan.*
In montane and usually species-poor grassland, normally above 600 m.
Rare to occasional; widely distributed in the central massif with outlying localities in central **9** Brolass and on **7** Cnoc an dà Chinn.
7–9, 11, 15, 16

Huperzia Bernh.

H. selago (L.) Bernh. ex Schrank & Mart.
Lycopodium selago L.
Fir Clubmoss. Native; NM
13 Tobermory, 1837, *Curtis* (BM).
On well-drained summit plateaux, drier ledges on open cliff faces and in more open heather communities on the granite. Eliminated by burning.
Occasional to frequent; widespread but not recorded from the Treshnish Isles or Staffa.
2–17

SELAGINELLALES

SELAGINELLACEAE

Selaginella Beauv.

S. selaginoides (L.) Link
Lesser Clubmoss. Native; AA
Without exact locality, 1848, *Syme* (BM).
On wet flushed hillsides and cliff ledges with impeded drainage in medium base-rich sites.
Frequent; widespread.
2–17

ISOETOPSIDA

ISOETALES

ISOETACEAE

Isoetes L.

I. lacustris L.
Quillwort. Native; NSA
13 Near Tobermory, *White* (1881).
On stony bottoms or occasionally in shallow, fine nekron mud of lakes at 50–205 cm; in more mesotrophic waters than the following species.
Rare; widely scattered.
6 W side of Loch Peallach, *U. Duncan* (BM). **7** Loch a' Ghael (BM). **8** Lochan between Fionna Mham and Beinn na Sreine (BM). **9** Above Carsaig Arches, *Templeman* (BM); Lochan Binnein Ghorrie (BM); Lochan on W side of Creachan Mòr (BM). **11** Loch Bà (BM). **16** Lochan an Daimh (BM); Loch Uisg, *BPS* (BM). **17** Loch Ghleannain (BM).
The population in Lochan an Daimh consists of very old, large and distinct-looking plants. They differ from *I. lacustris* both on Mull and elsewhere in leaf and velum structure and may prove to be a new taxon.

I. echinospora Durieu
I. setacea Lam.
Spring Quillwort. Native; WSA
13 Caol Lochan, Druimfin, 1876, *G. Ross* (BM).
In similar habitats to I. lacustris but in waters of lower cation content and with less competition from other macrophytes.
Rare; widely scattered.
3 Loch Poit na h-I (BM). **6** Loch Peallach (BM). **13** Small lochan, Upper Druimfin (BM); as above. **15** Loch Squabain (BM).

EQUISETOPHYTA

EQUISETOPSIDA

EQUISETALES

EQUISETACEAE

Equisetum L.

E. fluviatile L.
Water Horsetail. Native; NC
G. Ross (1877).
In herb-rich mire and *Juncus acutiflorus* swamp. An early coloniser of open water of mesotrophic lakes but not abundant in that rôle.
Occasional; widespread.
2, 3, 5–17

E. palustre L.
Marsh Horsetail. Native; WC
G. Ross (1877).
From sea level to 900 m in wet meadows and *Juncus* associations and in peaty high-level flushes.
Occasional; widespread but not recorded from SE Mull.
2–6, 8–10, 12–16

E. sylvaticum L.
Wood Horsetail. Native; WC
G. Ross (1879).
In shady peaty areas and more rarely on inorganic soils in *Vaccinium myrtillus* communties, *Salix* carr and damp woodland from sea level to 400 m.
Occasional to frequent; widespread.
2–17

E. arvense L.
Common Horsetail. Native; W
G. Ross (1877).
On glacial sand alluvial areas and river gravels, occasionally in wet flushes with considerable inorganic content.
Occasional to frequent; widespread except in SE Mull.
1–16

E. ×litorale Kühlew. ex Rupr.
E. arvense × fluviatile
Native; NC

2 Iona, Baile Mòr, 1960, Fereira (ABD).
In inorganic marshy soils where parents grow close together.
Rare, but probably un-recognised and under-recorded.
Also from **10** Carsaig (BM).

E. telmateia Ehrh.
Great Horsetail. Native; SSA
8 Near Mackinnon's Cave, 1961, *BSBI*.
In clay soil in water runnels flushed with water of high calcium/magnesium content.
Rare. Restricted by lack of base-rich sites.
8 As above; Creag a' Ghaill, *U. Duncan* (BM). **10** Between Carsaig and Nuns' Pass, *Kenneth & A. Stirling* (BM).

POLYPODIOPHYTA

OPHIOGLOSSOPSIDA

OPHIOGLOSSALES

OPHIOGLOSSACEAE

Ophioglossum L.

O. vulgatum L.
Adderstongue. Native; SC
3 Port nan Ròn, 1966, Eddy (BM).
In sandy loam under *Pteridium*.
Rare; confined to the Ross peninsula.
3 As above. **9** Aoineadh Beag; Tràigh Cadh an Easa, 1977, *Kelham*.

Botrychium Sw.

B. lunaria (L.) Sw.
Moonwort. Native; NC
17 Duart Castle, *Maclagan* (1855).
In well drained ledges and grassy slopes in open situations; often associated with local calcium-rich veins.
Occasional in scattered localities, but apparently absent from Brolass, the Ross and Iona.
4, 6, 8, 13, 16, 17

OSMUNDOPSIDA

OSMUNDALES

OSMUNDACEAE

Osmunda L.

O. regalis L.
Royal Fern. Native; WSA
Lightfoot (1777).
In higher rainfall areas, now confined to SW facing cliffs and ravines and only as remnants elsewhere. Possibly more widespread before peat cutting.
Occasional; mainly in the S and W but previously recorded from **6** Sorne, *G. Ross* (1879)
2–4, 8, 10, 11, 16, 17
Frequently collected for garden cultivation.

POLYPODIOPSIDA

PTERIDALES

ADIANTACEAE

Cryptogramma R. Br.

C. crispa (L.) R. Br. ex Hook. [Map, Fig. 2.18]
Parsley Fern. Native; AA
Without exact locality, 1848, *Syme* (BM).
A very strict calcifuge of mountain scree slopes, often in rock detritus with little soil; usually above 150 m.
Rare; locally distributed in the central massif and the mountains around Glen Forsa.
11 Ben More; Abhain Dhìseig and Coire nam Fuaran, *BEC*; N corrie, *U. Duncan*; Abhainn na h-Uamha; An Gearnal Braigh a' Chaoil, *Ribbons*. **15** Mainnir nam Fiadh (BM); Dun da Ghaoithe; Sgur Dearg, *Stace*; Beinn Talaidh, *Bibby*; Beinn Chreagach Mhor and Bheag; Beinn a' Chraig.

HYMENOPHYLLALES

HYMENOPHYLLACEAE

Hymenophyllum Sm.

H. tunbrigense (L.) Sm.
Tunbridge Filmy-fern. Native; SO
16 Ravine N of Loch Spelve, 1848, *Syme* (BM).
On coarse-grained rocks in deep shade.
Rare; confined to SE Mull with the exception of two localities on granite in the Ross.
3 Wood opposite Ardfenaig, Loch Caol (BM); Beinna Ghlinne Mhòir. **16** As above Creach Beinn, *Kenneth*; S side of Loch Buie, *BSBI*; S of Loch Uisg, *BPS*. **17** Rubha na Sròine (BM); Craignure, *BSBI*.

H. wilsonii Hook.
Wilson's Filmy-fern. Native; WO
13 Near Tobermory, 1829, *Christy* (E).
Amongst mosses, including *Sphagnum* spp., on rock ledges, boulders, tree boles etc.
Occasional to frequent; a widespread and locally abundant epiphyte but not recorded from Brolass.
3–8, 10–17

POLYPODIALES

POLYPODIACEAE

Polypodium L.

P. vulgare agg.
Polypodies
Keddie (1850).

P. vulgare L.
Common Polypody. Native; W
A common epiphyte on tree trunks and branches, especially where exposed to the SW; also on rocks and shallow soils on banks.
Frequent; widespread.
2–7, 9–17

P. interjectum Shivas
P. vulgare L. subsp. *prionodes* Rothm.
Intermediate Polypody. Native; WSA
17 Port nam Marbh, 1964, *Slack & A. Stirling*.
In similar situations to *P. vulgare*, but less epiphytic and in more clement situations.
Occasional; widespread.
1, 3–8, 10–13, 16, 17

P. × mantoniae Rothm.
P. interjectum × vulgare
Native
Known only from **12** between Salen and the Forsa river, 1958, *Gerrans* (BM) but may be more frequent; can usually be distinguished by the sterile sori and the vigorous and larger clones.

CYATHEALES

DENNSTAEDTIACEAE

Pteridium Scop.

P. aquilinum (L.) Kuhn
Bracken. Native; W
G. Ross (1877).
Most frequently on, and rapidly colonising, herb-rich and basic grassland; on the deeper soils up to 700 m.
Abundant; very widespread.
1–17
Spread phenomenally widely only this century partly due to the decrease in grazing by cattle. However, as elsewhere in Scotland, its sudden spread is not easily explained.

ASPLENIALES

THELYPTERIDACEAE

Phegopteris Fée

P. connectilis (Michx) Watt
Thelypteris phegopteris (L.) Slosson
Beech Fern. Native; NC
13 Tobermory, 1837, *Curtis* (BM).
In deep loam on ledges in ravines or on wet banks in woods.
Frequent; widespread.
3–17

Oreopteris J. Holub

O. limbosperma (All.) J. Holub
Thelypteris limbosperma (All.) H. P. Fuchs; *T. oreopteris* (Ehrh.) Slosson
Lemon-scented Fern. Native; WSA
G. Ross (1877).
In valleys, beside streams, and on steep slopes in the open; never in deep shade and requiring some underground water movement around the roots.
Frequent; widespread.
1–17

ASPLENIACEAE

Asplenium L.

A. scolopendrium L.
Phyllitis scolopendrium (L.) Newm.
Hartstongue. Native; WSA
Keddie (1850)
In crevices and underhangs in calcareous rocks, most frequently in mortared walls in damp shady situations.
Rare to occasional; widespread but scattered.
2, 5, 6, 8–11, 13, 16, 17

A. adiantum-nigrum L.
Black Spleenwort. Native; WSA
G. Ross (1877).
On drier rock faces, on stone walls more commonly on E and S facing aspects.
Frequent; widespread, especially around the coast.
11–17

A. marinum L. [Map, Fig. 2.36]
Sea Spleenwort. Native; WA
1, 2 Staffa and Iona, 1829, *Christy* (E).
In crevices and on rock ledges in exposed positions near the sea.
Locally frequent in the west and south and markedly absent on the S E and E seaboard.
1–11, 16

A. trichomanes L. agg.
Maidenhair Spleenwort. Native; W
Keddie (1850)
All specimens determined by Dr J. D. Lovis.

A. trichomanes subsp. **trichomanes**
Usually in crevices on vertical faces of fine-grained non-calcareous rocks in deep moist ravines.
Rare; A rare taxon in the British Isles as a whole with no obvious distribution pattern except that it is a calcifuge.
10 Between Beinn nan Gobhar and Cnoc na Bo Ruaidhe (BM); S of Cnoc a' Bhràghad (BM). **12** Glac an Lin (BM). **13** Aros Park, S end of lake (BM).

A. trichomanes subsp. **quadrivalens** D. E. Meyer emend. Lovis
On mortared walls facing SW; also in wetter ravines and on vertical faces of rocks containing calcium by streams.
Frequent; widespread, absence from **14** and **15** due to rain shadow and, lack of suitable wet habitats. The apparent coastal distribution of this species is due to its occurrence in walls along the roads.
2, 3, 5–13, 16, 17

A. viride Huds.
Green Spleenwort. Native; NM
11 Ben More, *Moore* (1859).
On mortared walls and in crevices of base-rich rocks.
Very rare; the only recent record is from area **8**.
2 Iona nunnery, 1906, *Davie* (E); near Baille Mòr, 1912, *Wilmott*. **8** Gribun, *MacAllister*.

A. ruta-muraria L.
Wall-rue. Native; W

Without exact locality, 1848, *Syme* (BM).
On mortared walls, less frequently on ledges and crevices of basic rocks in S or SW aspects.
Occasional; widespread.
2–13, 16, 17

A. ceterach L.
Ceterach officinarum DC.
Rustyback. Native; SSA
2 Iona, *Moore* (1855).
On S and SW aspects of mortared walls.
Very rare.
2 Iona, *DMS*. The origin of this record is uncertain and possibly a reiteration of Moore (1855); searching during the Survey failed to refind it.

ASPIDIACEAE

Athyrium Roth

A. filix-femina (L.) Roth
Lady Fern. Native; W
G. Ross (1877).
In wet lowland woods, wooded streamsides and streamsides in open moorland; occasionally in large block scree.
Frequent; widespread.
1–17

Gymnocarpium Newm.

G. dryopteris (L.) Newm.
Thelypteris dryopteris (L.) Slosson
Oak Fern. Native; NC
13 Tobermory, 1829, *Christy* (E.)
In similar habitats to *Phegopteris* but preferring wetter conditions.
Occasional; widespread.
2, 4, 9, 11–16

Cystopteris Bernh.

C. fragilis (L.) Bernh.
Brittle Bladder-fern. Native; W
16 W from Lochbuie House, *Maclagan* (1855).
In crevices in basalt usually near the sea or on sea-facing mountain sites; otherwise in damp mortared walls.
Rare to occasional; in scattered localities, but absent from Brolass and the Ross.
6–8, 11, 13, 16

Woodsia R.Br.

W. ilvensis (L.) R.Br.
Oblong Woodsia
Recorded for vice-county 103 by Druce (1932) with no substantiating specimen and most likely in error, although Mull is a more likely area than Coll or Tiree.

Polystichum Roth

P. aculeatum (L.) Roth
Hard Shield-fern. Native; WSA
G. Ross (1877).
In ravines or in dense vegetation on wet ledges; on calcareous rocks and mortared walls in more exposed situations.
Occasional; widely distributed.
2–13, 15, 16

The variety **cambricum** (S. F. Gray) Hyde & Wade, an ecotype of wet or very shady localities, has been recorded from the waterfall near **7** Kilbrennan and the adjacent cliffs.

P. × bicknellii (Christ) Hahne
P. aculeatum × setiferum
Has been recorded in the immature state from **13** Aros Park, and further confirmation is needed.

P. lonchitis (L.) Roth
Holly Fern. Native; NM
11 Ben More, 1859, *Moore.*
Also recorded from inland cliffs, **8** Ardmeanach Peninsula, *BSBI*.
Possible habitats exist but further confirmation is desirable.

P. setiferum (Forsk.) Woynar
Soft Shield-fern. Native; WSA
13 Bedarrach, *G. Ross* (1879).
In deep loam in steep shaded ravines.
Rare; in a few widely scattered localities.
3 Kintra, *Dyce*. **5** Ensay Burn, *J. Duncan*; Treshnish House, *Dyce*. **9** Allt Bun an Easa (BM). **13** As above. **15** Allt Mòr Coire nan Eunachair; Scallastle, *CSSF*. **16** Allt na Codha.

Dryopteris Adans.

D. filix-mas agg.
Male-ferns
G. Ross (1877).
Considerable hybridisation occurs between members of the *D. filix-mas* aggregate: *D. oreades* (2*x*), *D. filix-mas* sensu stricto (4*x*) and *D. pseudomas* (2*x*, 3*x*, apogamous). *D. tavelii* (5*x*, apogamous) is a morphologically distinct taxon, resulting from hybridisation between *D. filix-mas* and *D. pseudomas* but capable of reproducing itself from apogamous spores. Specimens of *D. pseudomas* which show signs of being tetraploid in spore and stomata size and which contain some abortive spores are possibly the result of hybridising of that cytotype with *D. oreades*. Such a hybrid would be difficult to distinguish morphologically from *D. pseudomas* but could well be frequent.

D. filix-mas (L.) Schott
Male-fern. Native; W
In well-drained situations at road sides, on more open streamsides and in ravines; occasionally in woods.
Occasional; apparently confined to the N and W of Mull apart from one record in **16** S of Loch Uisg.
2–6, 8, 12, 13, 16

D. tavelii Rothm.
von Tavel's Male-fern. Native; WSA
In shady wood on deep soil.
Recorded once only with putative parents in

13 Aros Park, 1970, *Jermy* but likely to occur elsewhere.

D. pseudomas (Woll.) Holub & Pouzar
D. borreri auct. angl.
Scaly Male-fern. Native; WSA
12 Between Aros bridge and the castle, 1958, *Gerrans* (BM).
In wooded valleys and amongst block scree in more open sites on well-drained soils and in exposed rocky gullies on W coast.
Frequent; widely distributed.
2–17

D. oreades Fomin
Mountain Male-fern. Native; NSA
11 Ben More, 1956, *Flannigan.*
In open stable scree and ravines, more rarely in woods and shady places. Absent in areas exposed to SW salt-laden wind.
Occasional; widespread.
3, 6–8, 10–13, 15–16
The plants on open hillsides in the east of the Island are mostly of the small ecotype, which maintains its form in cultivation.

D. × mantoniae Fraser-Jenkins & Corley
D. filix-mas × oreades
Native
Recorded only from steep rocky bank above **15** Lussa River on Garbh Leathad but possibly elsewhere.

D. × pseudoabbreviata Jermy
D. aemula × oreades
Native
16 Leachan Dubh, Cruach Ardura, 1967, *Eddy* (BM).
On a steep slope in an open, birch-oak wood, with *Vaccinium* and other fern species. Recorded only once with the putative parents. This is an unusual hybrid involving the diploid *D. aemula* with *D. oreades* (see Jermy, 1968; cytology since confirmed by M. Gibby, *pers. comm.*).

D. aemula (Ait.) Kuntze
Hay-scented Buckler-fern. Native; SO
13 Near Tobermory, *Tanner* in Newman (1851).
In moist mossy woodland in warmer parts of the Island, rarely reaching 600 m; on the granite often amongst dwarf heath.
Occasional; widely scattered but locally frequent in the Ross of Mull and round Loch Uisg. Possibly under-recorded.
2–4, 6, 8, 11–13, 16, 17

D. carthusiana (Vill.) H. P. Fuchs
D. lanceolatocristata (Hoffm.) Alston
Narrow Buckler-fern. Native; W
G. Ross (1877).
In *Molinia* mires around lochs and in seepage lines in lowland situations.
Rare; recorded recently from only two localities. One of the commonest ferns on the mainland of Argyllshire.
13 Druimfin, *G. Ross* (1879). **17** Near Torosay Castle (BM); Lochan Doire Dharaich (BM).

D. austriaca (Jacq.) Woynar
D. dilatata (Hoffm.) A. Gray
Broad Buckler-fern. Native; W
Keddie (1850).
Shady banks and in woods on deeper and richer soils, also in wet gullies at higher altitudes to 750 m.
Frequent; widely distributed and often in some quantity. The only *Dryopteris* recorded from the more exposed small islands of **1** Lunga, Fladda and **2** Soa.

D. × deweveri Jansen (Jansen & Wachter)
D. austriaca × carthusiana
Native
17 Near Torosay Castle, 1968, *Jermy* (BM).
In shady woods on wet peat, usually with parents.
Rare.
Known from two widely separated localities.
13 Aros Park (BM). **17** As above.
Possibly produces occasional fertile spores which may distribute the hybrid beyond the range of the parents.

D. expansa (C. Presl) Fraser-Jenkins & Jermy
D. assimilis S. Walker
Northern Buckler-fern. Native; NM
13 Aros Park, 1967, *Jermy* (BM).
In mature woods or amongst large boulders in deep soils. Often creeping in moss litter, occurring from sea-level to 150 m.
Rare; widely scattered in association with old woodlands.
3 Loch Caol (BM). **6** Dùn Auladh, *Dyce*. **13** As above. **16** Creach Beinn. (BM).

BLECHNACEAE

Blechnum L.

B. spicant (L.) Roth
Hard Fern. Native; WSA
Keddie (1850).
In woods, dwarf heathland and in ravines usually in well-drained situations.
Abundant; very widespread.
1–17
The var. *anomalum* Moore, with laminate fertile pinnae, is common on the Lagggan Peninsula but tends to revert in cultivation.

References

CRABBE, J. A., JERMY, A. C. & MICKEL, J. T. (1975). A new generic sequence for the pteridophyte herbarium. *Fern Gaz.*, **11**: 141–162.

Dandy, J. E. (1958). *List of British vascular plants.* London.
Druce, G. C. (1932). *The comital flora of the British Isles.* Arbroath.
Jermy, A. C. (1968). Two new hybrids involving *Dryopteris aemula. Br. Fern Gaz.*, **10**: 9–12.
Keddie, W. (1850). *Staffa and Iona described and illustrated . . .* Glasgow.
Lightfoot, J. (1777). *Flora Scotica* ed. 1. London.
Maclagan, D. P. (1855). Notice of plants in the neighbourhood of Oban and part of the Island of Mull. *Scottish Gardener*, **4**: 93–95.
Moore, T. (1855). *The ferns of Great Britain and Ireland.* London.
———. (1859–60). *Nature printed British ferns*, Vols 1 & 2. London.
Newman, E. (1851). *Lastrea recurva* in the Isle of Mull. *Phytologist*, **4**: 725.
———. (1854). *History of British ferns.* London.
Ross, G. (1877). Isle of Mull, Mid Ebudes. *Rep. Loc. Rec. Club*, **1876**: 188–192.
———. (1879). On the flora of Mull. *Trans. bot. Soc. Edinb.*, **13**: 234–242.
Valentine, D. H. (ed.) (1964). *Flora Europaea*, Vol. 1. pp. 1–25. Cambridge.
White, F. B. (1881). The cryptogamic flora of Mull. *Scottish Naturalist*, **6**: 155–162.

13. Liverworts and mosses

A. Eddy

Introduction

Collection of data

The method of sampling was similar to that used in the lichens and vascular plants, namely one of subjective selection of localities according to their habitat potential, and over the five years of the Survey a good coverage of the Island was obtained. Field record cards designed for the Bryophyte Mapping Scheme were used whenever possible. Records were stored in loose-leaf binders as with vascular plants; a total of 571 species and varieties has been recorded.

Distributions on Mull are given either by a list of recording area numbers or, in special cases such as uncommon taxa, by full details. These records have been accumulated from our own five-year Survey, augmented by other BM specimens which extend the range, and by literature records. The bryophytes of the Royal Botanic Gardens, Kew are now incorporated into the British Museum (Natural History) and contain the bulk of the J. Duncan, W. Young and H. N. Dixon collections. Specimens collected by G. Ross & T. King and by P. Ewing & D. Kennedy are partly at BM and mostly in Glasgow University (those seen are indicated by GL). Ross, King and others collected mosses during a conference on Mull of the Cryptogamic Society of Scotland in 1881; they are reported in Ross & King (1882) and include also 'a number observed by Mr Ross during his residence in the island'. Ewing, Kennedy, Macvicar and others are quoted in Macvicar (1910) but without dates of collecting.

Arrangement and nomenclature

To conform with other recently published Floras, the arrangement of genera and species in the present treatment follows with a few modifications that of Warburg (1963) for the mosses and Paton (1965b) for the hepatics. These works follow, in turn, the systems adopted by Richards & Wallace (1950) and Jones (1958). Where specific nomenclatural changes have been made in accordance with recently published works, the previously accepted name is given as a synonym; other synonyms are excluded.

The systematic position and taxonomic levels of divisions above that of genus have been subjected to many changes over recent years, and a settled system for the mosses is, to date, unavailable. Richards & Wallace (1950) base their system on the work of Fleischer (1915–1922), but maintain a wider definition of families (e.g., Ditrichaceae and Seligeriaceae are included within the Dicranaceae). At the same time, the Fissidentaceae, which are probably best included within the Dicranales, are placed in the separate order Fissidentales. Also of doubtful value are the orders Grimmiales, Funariales, Isobryales and Hookeriales, while the limits of the Hypnobryales are, at best, somewhat uncertain. Cavers (1964) recommends the sinking of most of the higher divisions of Fleischer into the single order Bryales (Eu-Bryales), recognizing 6 orders in all. With the exception of the Tetraphidales, Cavers' system does not necessitate changes in the sequence of presentation of families and genera, and is therefore adopted here. For simplicity subdivisions into sub-orders and series are ignored.

The Hepaticopsida have received rather more attention during recent years with a number of orders and families under revision. Grolle (1972) reviews the current status of family names, and those used below follow his nomenclature. In general this does not involve changes in the order of presentation of the genera as adopted by Paton (1965b) and the orders are the same as those in Jones (1958), with the exception of the Sphaerocarpales (now included in the Marchantiales). At generic level *Mylia* has been transferred from the Plagiochilaceae to the Jungermanniaceae and some minor rearrangement of genera has been carried out.

The following terms are used throughout the text to indicate the altitudinal zones in which the species is found. It is realised that the transition from the lowland to alpine zones is a continuum the subdivision of which is highly subjective. Further the zones themselves are subject to depression or elevation according to exposure.

Lowland:	below 200 m.
Lower subalpine:	200–500 m.
Higher subalpine:	500–900 m.
Alpine:	above 900 m.

Acknowledgements

I would like to thank Alan H. Norkett for identification of *Fissidens* and Bridget J. Ozanne for checking the manuscript; Dr U. K. Duncan has contributed a valuable set of data.

Index to genera

BRYOPHYTA

HEPATICOPSIDA

ANTHOCEROTALES

ANTHOCEROTACEAE

Anthoceros L.

A. husnotii Steph.
On damp, retentive inorganic soils in the lowlands; possibly transient.
Rare; fertile.
5 Ensay Burn, *Lobley & U. Duncan.* **6** Dervaig.

Phaeoceros Prosk.

P. laevis (L.) Prosk.
Anthoceros laevis L.
On damp inorganic soils in open situations; lowland.
Rare; fertile.
8 Kilfinichen, in field, *U. Duncan & James.* **17** Craignure.

MARCHANTIALES

TARGIONIACEAE

Targionia L.

T. hypophylla L.
Xerophilous and thermophilous; only seen on south-facing coastal rocks.
Rare; fertile but no mature spores present.
4 Ulva, foot of Glen Glass.
The Mull plant is near the northernmost limit of the British range of this species.

AYTONIACEAE

Reboulia Raddi

R. hemisphaerica (L.) Raddi
Basiphilous; on coastal rock-ledges and skeletal soils, e.g., calcareous sand; in similar situations to *Targionia* but of wider tolerance.
Occasional; fertile.
2 Iona, *M. Dalby & U. Duncan.* **3** Fidden; Erraid. **4** Ulva, *U. Duncan.* **6** Mingary Àrd. **7** Kilninian. **10** Cliffs east of Carsaig.

CONOCEPHALACEAE

Conocephalum Wiggers

C. conicum (L.) Underw.
Basiphilous and shade-loving; on damp rocks and mortar, occasionally on soil in lowland ravines and wooded valleys.
Frequent and widespread; fruit rare.
3–8, 10–13, 16, 17

LUNULARIACEAE

Lunularia Adans.

L. cruciata (L.) Dum.
Only seen near habitation on shaded rocks, walls and soil in the lowland.
Occasional; almost certainly introduced. Sterile or male only; gemmiferous.
4 Ulva, farmyard. **5** Kilninian, *U. Duncan*; Calgary. **7** Torloisk. **10** Carsaig, Innamore Lodge. **12** Salen, *U. Duncan.* **13** Tobermory. **17** Craignure.

MARCHANTIACEAE

Preissia Corda

P. quadrata (Scop.) Nees
Basiphilous; ledges of basic rocks and coastal exposures and on mortar of old walls, occasionally terrestrial; indifferent to altitude.
Frequent around the coast, local elsewhere; fertile.
3–17

Marchantia L. emend. Raddi

M. polymorpha L. var. **polymorpha**
Ruderal; near buildings on damp stonework and inorganic bare soil; adventive on burnt

ground; a weed in greenhouses and conservatories.
Occasional, probably introduced; fertile and gemmiferous.
5 Calgary Bay, *Lobley & U. Duncan*. **7** Torloisk. **11** Knock. **12** Salen. **13** Tobermory; Aros Park, *U. Duncan*. **14** Fishnish Bay. **15** Scallastle Bay. **17** Torosay Castle.

M. polymorpha var. **aquatica** Nees
Only found once in the higher subalpine: in low, mossy vegetation in a mesotrophic flush.
Rare; sterile; sparingly gemmiferous.
11 Ben More, north corrie.

RICCIACEAE

Riccia L.

R. sorocarpa Bisch.
On retentive, bare, neutral to mildly acid inorganic soils in the lowlands.
Rare; fertile.
3 Loch Poit na h-I, *Lobley & U. Duncan*; Uisken. **7** Laggan Bay, *Lobley & U. Duncan*. **14** Scallastle Bay.

R. glauca L.
Basiphilous: in small rosettes on fine-grained inorganic neutral soils in the lowlands; also on woodland tracks and paths.
Rare; fertile.
12 Salen. **17** Torosay Castle.
A form with very narrow bright green thalli was isolated from among the Torosay plants. It retained its habit in culture, eventually fruiting; the spores are identical to those of **R. glauca** var. **minor** Lindb., not listed in Paton (1965b).

R. beyrichiana Hampe
Basiphilous; only seen on compacted soil by path in wooded estate.
Rare; fertile.
17 Torosay Castle.

METZGERIALES

ANEURACEAE

Riccardia Gray

R. multifida (L.) Gray
Shade tolerant; saturated peaty soils and wet rocks in woodlands and ravines; also in subalpine springs and flushes. Indifferent to altitude.
Frequent; fertile.
1–17

R. sinuata (Dicks.) Trev.
Slightly basiphilous; in small patches on damp soil in shaded habitats. Lowland and mainly on the coast.
Occasional; fertile.
2 Iona, near the Nunnery. **3** Loch Poit na h-I, *Lobley & U. Duncan*; Fidden; Erraid. **5** Calgary Bay, *Corley*. **6** Loch a' Chumhainn. **8** Balmeanach, *Lobley & U. Duncan*. **10** Carsaig. **12** Salen, 1932, *J. Duncan & W. Young*. **16** Lochbuie, S. shore.

R. latifrons (Lindb.) Lindb.
Shade tolerant; on decorticated wet logs, or peat in woodlands and ravines in the lowland.
Occasional; fertile.
2 Iona. **3** Loch Poit na h-I, *Lobley & U. Duncan*. **6** Mishnish, *Macvicar* (1910). **12** Loch Frisa; Salen Forest. **13** Allt nan Torc. **15** Scallastle Bay. **16** Cruach Ardura.

R. palmata (Hedw.) Carruth.
In deep-green small tufts or creeping on stumps and decorticated rotting logs in dense shade. Lowland and lower subalpine.
Locally frequent; mainly sterile.
4–8, 11–16

R. pinguis (L.) Gray
In a variety of wet habitats by streams and in mires of various types; indifferent to base-status and to altitude.
Common and widespread; occasionally fruiting.
1–17

Cryptothallus Malmb.

C. mirabilis Malmb.
Saprophytic; colourless thallus immersed 3–7 cms in wet (not waterlogged) bryophyte carpets, including *Sphagnum* in acid birch woods and drained acid mires.
Rare; fertile.
3 Loch Poit na h-I, E side. **16** Loch Uisg; Loch Spelve, west of Dalnaha.

PELLIACEAE

Pellia Raddi

P. epiphylla (L.) Corda
On damp, usually shaded substrates in woodlands and ravines: densely crispate forms in flushes and mires and by streams and waterfalls. Indifferent to altitude.
Common; lowland, terricolous forms fruiting; aquatic forms sterile or male only.
1–17
P. borealis Lorbeer apud K. Müll. is a diploid race which cannot be distinguished in the field. Its presence on Mull is probable, but not yet established.

P. neesiana (Gottsche) Limpr.
In subalpine flushes and among irrigated rocks.
Rare or under-recorded; fertile.
8 Creag a' Ghaill. **10** Beinn Chàrsaig. **11** Ben More, north corrie.

P. endiviifolia Dicks.
Basiphilous; on irrigated coastal rocks and basic exposures inland: less commonly on peaty soil in eutrophic flushes and mires.
Frequent, especially by the sea; fruit uncommon.
1–17

METZGERIACEAE

Metzgeria Raddi

M. furcata (L.) Dum.
Mainly epilithic or epiphytic; in lax or compact yellowish patches among rocks on walls and on bark. Abundant in the lowlands, especially near the sea but ascends to over 300 m.
Common, locally abundant; fertile.
1–17

M. fruticulosa (Dicks.) Evans
In very small tufts on branches of shrubs, mainly *Sambucus*, in lowland woods and estates.
Rare; sterile; gemmiferous.
3 Bunessan, on *Salix*. **10** Pennyghael. **12** Kilichronan, *Lobley & U. Duncan*. **13** Aros Bridge, *Kenneth*; Aros Park.

M. conjugata Lindb.
Distinctly basiphilous; in lowland woods, ravines and rock-clefts, mainly along the sheltered east coast.
Locally frequent; occasionally fertile.
4, 5, 7, 8, 11–17

M. hamata Lindb.
Requiring constantly high atmospheric humidity and confined to densely shaded damp rocks in lowland wooded ravines and gullies; probably frost sensitive.
Rare; sterile.
11 Scarisdale Wood. **12** Allt na Searmoin, *Kennedy* (1910). **13** Near Tobermory, *Kennedy*. **16** Loch Spelve, west of Dalnaha. **17** Craignure Woods.

BLASIACEAE

Blasia L.

B. pusilla L.
On moist, neutral or basic inorganic soils and in low vegetation in the lowland zone.
Rare; fertile and gemmiferous.
3 Near Fidden; Bunessan, *Lobley & U. Duncan*. **5** Calgary Bay. **13** Aros Park, *U. Duncan*. **17** Duart Point.

CODONIACEAE

Fossombronia Raddi

F. foveolata Lindb.
On moist, retentive soils in open situations near the west coast.
Rare or transient; fertile.
3 Loch Poit na h-I, *Lobley*. **6** Calgary Bay.

F. wondraczekii (Corda) Dum.
In small or diffuse colonies on damp inorganic soils; semi-ruderal or amongst sparse low vegetation; pathside and ditches. Lowland.
Occasional; fertile.
6 Dervaig. **7** Achleck. **12** Kilichronan, *U. Duncan*; Salen. **13** Aros Park, *U. Duncan*. **17** Craignure.

F. caespitiformis De Not.
On damp, neutral mineral soil by road.
Rare; fertile.
17 Torosay Castle.

CALOBRYALES

HAPLOMITRIACEAE

Haplomitrium Nees

H. hookeri (Sm.) Nees
On flushed peaty gley soil with *Schoenus* and *Eleocharis*. A single record at 180 m.
7 Below Cruachan Druim na Croise, above Loch Frisa shore.

JUNGERMANNIALES

ANTHELIACEAE

Anthelia Dum.

A. julacea (L.) Dum.
Calcifuge; in small mounds or extensive mats on irrigated rocks and thin peat on basaltic slabs. Abundant in the alpine and subalpine zones and sometimes dominant in small, oligotrophic flushes descending almost to sea-level locally.
Abundant except in the lowlands; male plants common, fruit rare.
4–17

A. juratzkana (Limpr.) Trev.
A single, somewhat doubtful record, without locality, 1932, *J. Duncan & Young*, is based on sterile material.

HERBERTACEAE

Herberta Gray

H. adunca (Dicks.) Gray
Calcifuge; in orange tufts among bryophytes in dwarf-shrub communities on skeletal, peaty soils among rocks; and in ravines and sheltered hillsides. Subalpine down to 120 m.
Local but sometimes in quantity; sterile.
8 Maol Mheadonach; Creag a' Ghaill, *U. Duncan*. **10** Glen Byre. **11** Ben More, *Kennedy* (1910); Abhainn Bhagoair, *Corley*; Beinn nan Gabhar, *U. Duncan*. **12** Allt na Searmoin, *Kennedy* (1910). **13** Near Tobermory, *Kennedy* (1910). **15** Lussa River. **16** Cruach Ardura; ravines S of Lochbuie. **17** Craignure Woods, *M. Dalby & U. Duncan*.

LEPICOLEACEAE

Mastigophora Nees

M. woodsii (Hook.) Nees
Terricolous; in the 'hepaticosum' facies of dwarf-shrub communities, associated with, e.g., *Herberta adunca* in rocky gorges. Lower subalpine.
Rare; sterile.
11 Allt Coire nan Gabhar, *H. Milne-Redhead*; Abhainn Dhiseig, *Corley*. **16** Ravine N of Lochbuie.

PTILIDIACEAE

Ptilidium Nees

P. ciliare (L.) Hampe
Calcifuge; on thin peat or peaty mineral soil among bryophytes and siliceous rocks.

Rare; sterile.
8 Maol Mheadhonach, *U. Duncan.* **11** Abhainn Dhiseig. **16** N of Lochbuie, in ravine.

TRICHOCOLEACEAE

Trichocolea Dum. emend. Nees

T. tomentella (Ehrh.) Dum.
Among damp rocks in shaded woods and ravines; also amongst *Juncus* in mires. Mainly lowland.
Occasional; sterile.
3 Loch Poit na h-I, *Lobley & U. Duncan.* **6** Loch Frisa. **11** Scarisdale Wood. **13** Tobermory, 'in marsh', *Macvicar* (1910); Aros Park. **15** Lussa River valley. **17** Loch a' Ghleannain.

PSEUDOLEPICOLEACEAE

Blepharostoma (Dum.) Dum.

B. trichophyllum (L.) Dum.
Basiphilous; on a variety of substrates including damp, basic rocks, soil and bases of trees. Indifferent to altitude.
Occasional; sometimes fertile.
5 Loch a' Chumhainn. **7** Laggan Bay, *Lobley & U. Duncan.* **8** The Wilderness; Balmeanach, *Lobley & U. Duncan.* **10** Carsaig Bay. **11** Scarisdale River. **12** Allt na Searmoin, *Kennedy* (1910). **13** Tobermory, *Kennedy* (1910).

LEPIDOZIACEAE

Bazzania Gray emend. Carringt.

B. trilobata (L.) Gray
Calcifuge: peat-covered siliceous rocks, bases of trees and stumps in open woodlands (mainly *Quercus-Betula*) and block-screes. Lowland and lower subalpine.
Occasional; sterile.
8 Maol Mheadonach, E side; Balmeanach, *Lobley & U. Duncan.* **11** River Clachaig, N side. **13** Aros Park. **15** Lussa River valley. **16** Cruach Ardura, *Corley.*

B. tricrenata (Wahlenb.) Trev.
Calcifuge; on well drained, shallow peaty soils among rocks, particularly characteristic of the 'hepaticosum' facies of dwarf-shrub communities in the subalpine zone; on a variety of humus rich substrates in *Betula* woods, alpine and subalpine rock-ledges.
Frequent, locally common in the subalpine zones; sterile.
3–17
B. triangularis Pears. is included in *B. tricrenata* in Paton (1965b). On Mull this taxon has been found in one or two places in the alpine and higher subalpine zones, e.g., **11** North Corrie of Ben More and source of Scarisdale River. Forms intermediate between typical *B. triangularis* and *B. tricrenata* were not observed.

Lepidozia (Dum.) Dum.

L. reptans (L.) Dum.
On dry peat, thick-barked trees and among damp siliceous rocks, woodlands, ravines and under heather; also on subalpine block screes.
Common and locally abundant; widespread. Fertile.
1–17

L. pearsonii Spruce
Creeping among acidophilous bryophytes on peaty soils in open birch woods and oligotrophic dwarf-shrub associations. Lower subalpine and lowland.
Rare, perhaps under-recorded; fertile.
12 Allt na Searmoin, *Kennedy* (1910) and 1932, *Young*; Glen Aros. **16** Loch Spelve, S side, *Betula* wood east of Dalnaha.

L. pinnata (Hook.) Dum.
Calcifuge; in pale patches on humus-capped siliceous rocks in woods and ravines. Lowland and lower subalpine.
Rare; sterile.
8 Balmeanach, *Lobley & U. Duncan.* **11** Scarisdale River, *U. Duncan.* **16** S of Lochbuie, 1956, *Milne-Redhead.*

L. setacea (Weber) Mitt.
Calcifuge; in thin mats or mixed with *Sphagnum* and other mosses in bogs or occasionally on rotting wood in carr. Mainly lowland and lower subalpine.
Occasional but locally frequent; fertile.
3 Loch Poit na h-I, *Lobley & U. Duncan.* **6** Mishnish, *Macvicar* (1910); Loch na Criadhach Moire. **7** Torloisk. **9** Loch Assopol. **10** Lower Glen More, below Ulavalt. **11** Ben More, lower N slope. **12** Allt na Searmoin. **14** Fishnish Bay. **16** Loch Spelve, Dalnaha.

L. trichoclados K. Müll.
Calcifuge; ecologically close to *L. setacea* but preferring shallow peat among and overlying rocks; also on *Betula* stump. Mainly subalpine.
Rare.
3 Loch Poit na h-I, E shore. **8** Balmeanach, *Lobley & U. Duncan.* **11** Scarisdale River.

CALYPOGEIACEAE

Calypogeia Raddi

C. neesiana (Mass. & Carest.) K. Müll.
Shade tolerant; on peat soil or rotting wood under vegetation in scrub and in ravines. Mainly lowland.
Rare; sterile.
6 Mishnish, *Macvicar* (1910). **11** Abhainn Dhiseig. **13** Aros Park. **16** N side of Loch Uisg.

C. muellerana (Schiffn.) K. Müll.
On neutral to acid organic and inorganic substrates in sheltered situations in shade of trees or rocks. Lowland and lower subalpine.
Frequent, widespread; sterile; gemmiferous.
1–17
A growth form comparable to *C. submersa* (Arn.) Massal. sensu Macvicar (1926) occurs in considerable quantity in **6** Loch na Criadhach Moire.

C. trichomanis (L.) Corda emend. Buch
Strictly calcifuge; on wet peat among irrigated siliceous rocks, in bogs and acid mires. Alpine and subalpine.
Rare; sterile.
6 'S Airdhe Beinn. **11** Ben More, N side; Beinn Fhada. **16** Near Lochbuie, *M. Dalby & U. Duncan*; Laggan Deer Forest.

C. fissa (L.) Raddi
On soil, peat or damp rocks in dense shade or in open ground where consistently wet, e.g., among *Sphagnum*. Indifferent to altitude.
Common and widespread; rarely fertile; always gemmiferous.
1–17

C. sphagnicola (Arn. & Pers.) Warnst. & Loeske
Calcifuge; among *Sphagnum*.
Rare.
6 Beinn Chreagach, Lochan Dearg.

C. arguta Nees & Mont.
On damp but not wet inorganic soils in densely shaded places, usually under overhanging rocks or roots of trees. Lowland.
Frequent, in sheltered lowland areas only.
4–17

JUNGERMANNIACEAE

Lophozia (Dum.) Dum.

L. ventricosa (Dicks.) Dum.
Calcifuge; on a variety of substrates at all altitudes but particularly common on organic soils in bogs, heaths, woodlands and rock ledges.
Ubiquitous; occasionally fertile (male plants common); gemmiferous.
1–17

L. porphyroleuca (Nees) Schiffn.
Calcifuge; only recorded from a rotting *Betula* stump in a deep ravine.
Rare; fertile.
16 W of Dalnaha, S side of Loch Spelve.

L. alpestris (Schleich. ex Weber) Evans
Calcifuge, in reddish-brown diffuse patches on ledges and in crevices of mountain rocks. Alpine and subalpine.
Local; gemmiferous but rarely fertile.
11 Ben More, summit of north corrie; Beinn nan Gobhar, *U. Duncan*; Gleann Dubh, *U. Duncan*.
Reduced forms cf. var. **gelida** (Tayl.) K. Müll. were found creeping among *Gymnomitrion* on exposed rocks on **8** Creag a' Ghaill.

L. excisa (Dicks.) Dum.
Tendency towards calcifuge; peaty banks, walls and rock ledges. Lowland and subalpine.
Rare; fertile.
8 The Wilderness. **10** Carsaig Bay. **11** Ben More, *U. Duncan*.

L. incisa (Schrad.) Dum.
Calcifuge; on wet rotting wood and peat, rarely on inorganic soils, in a variety of habitats. Indifferent to altitude.
Common and widespread; gemmiferous; occasionally fertile.
1–17

L. bicrenata (Schmid.) Dum.
Calcifuge; in thin patches on inorganic or thin organic soils on terraces and among rocks in heath vegetation in rather open situations. Lowland and subalpine.
Rare or under-recorded; fertile.
7 Carn Mòr, E slope, with *Nardia scalaris*. **10** Glen Byre.

Leiocolea (K. Müll.) Buch

L. turbinata (Raddi) Buch
On calcareous, skeletal soils and damp basic rocks. Lowland and subalpine.
Occasional; fertile.
5 Calgary Bay sea cave, *M. Dalby & U. Duncan*. **6** Loch a' Chumhainn. **8** Balmeanach, *Lobley & U. Duncan*; The Wilderness. **10** Carsaig Bay. **13** Aros Park. **16** Loch Buie, S side.

L. badensis (Gottsche) Jorg.
Calcicolous; only seen on damp shell-sand in crevice of granite.
3 Erraid.

L. muelleri (Nees) Jorg.
On damp, basic rocks and calcareous soils. Indifferent to altitude.
Occasional; rarely fertile.
3 Loch Poit na h-I; Fidden. **8** The Wilderness; Mackinnon's Cave. **11** Abhainn Dhiseig. **13** Near Tobermory, *Kennedy* (1910). **16** Loch Spelve, ravine near Dalnaha.

L. heterocolpos (Thed.) Buch
Calcicolous; only recorded from a raised beach cliff.
Rare.
14 Between Craignure and Glenforsa, *Stirling & Kenneth*.

Barbilophozia Loeske

B. floerkii (Weber & Mohr) Loeske
Calcifuge; on shallow peat and humified soil among rocks. Indifferent to altitude.
Rare; sterile.
2 Iona, *M. Dalby & U. Duncan*. **8** Balmeanach, *E. Lobley*; Creag a' Ghaill. **11** Ben More, north corrie; Beinn Fhada. **15** Lussa River. **16** Laggan Deer Forest, ravine S of Lochbuie.

B. atlantica (Kaal.) K. Müll.
Calcifuge; only seen in birch scrub in crevice of basaltic exposure, 250 m.
16 Cruach Ardura.

B. attenuata (Mart.) Loeske
Calcifuge; on thin peat on rocks and on spongy

bark or rotting wood. Lowland and subalpine.
Rare; sterile; gemmiferous.
6 'S Airde Beinn, on block scree. **16** Cruach Ardura, at base of *Quercus*; Loch Spelve, ravine near Dalnaha, on *Betula* stump.

B. barbata (Schmid.) Loeske
Calcicolous; on skeletal shell-sand soils by the sea, creeping among other bryophytes in short turf.
Rare; sterile.
2 Iona, N coast. **3** Near Fidden. **5** Calgary Bay.

Tritomaria Schiffn.

T. quinquedentata (Huds.) Buch
In small, pure or mixed mats on calcareous detritus, basic strata and crevices of coastal rocks. Indifferent to altitude.
Frequent near the sea, local elsewhere.
2–8, 10, 11, 13–17

T. exsecta (Schmid.) Schiffn.
Calcifuge; on wet peat and rotting wood in shade in woodlands and ravines. Lowland and lower subalpine.
Rare or under-recorded; gemmiferous.
6 Near Glengorm Castle. **16** Loch Spelve, ravine near Dalnaha. **17** Craignure, *Crundwell & Wallace*.

T. exsectiformis (Breidl.) Schiffn.
Calcifuge; on wet peat and rotting wood in woodlands, ravines and under tall vegetation in oligotrophic mires and dwarf-shrub associations. Mainly lowland and subalpine but ascending to at least 650 m.
Frequent and widespread; always gemmiferous but not seen fertile.
2–4, 6–17

Sphenolobus (Lindb.) Steph.

S. minutus (Schreb.) Steph.
Terricolous: among mosses on skeletal, peaty soils, mainly in the dwarf-shrub associations on the sides of deep valleys and ravines. Mainly subalpine.
Rare; sterile.
11 Ben More, 1932, *J. Duncan & Young*; Scarisdale River gully, with *Bazzania* and *Herberta*; Beinn Fhada, S side. **16** Ravine S of Lochbuie.

Anastrepta (Lindb.) Schiffn.

A. orcadensis (Hook.) Schiffn.
Terricolous; in tufts or mixed with other hepatics, almost confined to the 'hepaticosum' facies of *Vaccinium-Calluna* dwarf-shrub associations.
Local; sterile.
11 Ben More, *Kennedy* (1910); head of Scarisdale River gully. **16** Cruach Ardura; ravine S of Lochbuie. **17** Loch a' Ghleannain, *Corley*.

Gymnocolea (Dum.) Dum.

G. inflata (Huds.) Dum.
Calcifuge; on wet peat or among *Sphagnum*, occasionally on wet, siliceous exposures by loch shores. Lowland and lower subalpine.
Rare; perianths present.
6 'S Airde Beinn, by the loch. **9** Creach Mòr. **16** Ben Buie.

Solenostoma Mitt.

S. triste (Nees) K. Müll.
Slightly basiphilous; in green mats on wet, basic rocks or thin, flushed soil among rock exposures in ravines, usually near streams. Indifferent to altitude.
Common and widespread; fertile.
3–17
Luxuriant forms in subaquatic habitats are frequent, apparently replacing *S. cordifolium*.

S. sphaerocarpoidea (De Not.) Paton & Warb.
Basiphilous; only seen on damp, basic stratum by small waterfalls.
11 Ben More, Abhainn Dhiseig (flagelliferous and fertile). **12** Allt na Searmoin.

S. pumilum (With.) K. Müll.
Basiphilous; in closely adherent patches or diffuse stems on damp basic rocks in ravines, characteristically in recesses or underhangs near streams and waterfalls. Lowland and subalpine.
Occasional becoming locally frequent; fertile.
4–17

S. cordifolium (Hook.) Steph.
Subaquatic, slightly basiphilous: in dark or nigrescent cushions by streams and springs. Alpine and subalpine.
Rare; sterile.
8 Beinn na Sreine, N side in a *Montia* spring.

S. crenulatum (Sm.) Mitt.
Calcifuge; on bare, retentive inorganic soils by roads, streams and on hillsides; ruderal in disturbed sites indifferent to altitude but most abundant in the lowland and lower subalpine zones.
Common; fertile.
This species occurs most abundantly as the attenuated form, called *Aplozia crenulata* var. *gracillima* Sm. in Macvicar (1926), which is probably an environmental modification which only occurs in the higher subalpine zone.
1–17

Plectocolea (Mitt.) Mitt.

P. obovata (Nees) Mitt.
In compact patches on neutral or slightly basic inorganic soil in crevices of rocks and in undercut stream-banks. Mainly subalpine.
Occasional; fertile.
8 Balmeanach. **11** Ben More, *Kennedy* (1910); Abhainn Dhiseig; Scarisdale River, *U. Duncan*. **13** Near Tobermory, *Kennedy* (1910); Aros Park, *U. Duncan*. **15** Lussa River valley. **16** Cruach Ardura.

P. hyalina (Lyell ex Hook.) Mitt.
On inorganic but humus-rich soils on stream-banks, ditches and consistently wet soils in woods and field-margins. Mainly lowland.
Frequent around the coast, elsewhere occasional; fertile.
3–14, 16, 17

P. paroica (Schiffn.) Evans
Only seen in a deep cleft of basic exposure at 300 m.
8 Above Mackinnon's Cave.

Nardia Gray

N. compressa (Hook.) Gray
Subaquatic and calcifuge; usually in rather large cushions in oligotrophic flushes and streams; occasionally submerged in peaty pools and lochans. Alpine and subalpine.
Frequent; sterile.
6–16

N. scalaris (Schrad.) Gray
On a variety of non-basic soil types at all altitudes; most abundant on inorganic drift soils on open hillsides in the subalpine zones; rarely on purely organic substrates.
Abundant; antheridia common, perianths rather infrequent.
1–17

Mylia Gray emend. Lindb.

M. taylori (Hook.) Gray
Calcifuge; in compact patches on well-drained peat in drier areas of bog or heath vegetation and rock ledges in open *Betula* woods and dwarf-shrub associations. Indifferent to altitude, mainly subalpine.
Locally frequent; sterile.
4, 6, 8, 10, 13, 15, 16

M. anomala (Hook.) Gray
Calcifuge; more or less confined to bog vegetation; creeping among *Sphagnum* or other bryophytes on wet peat; indifferent to altitude.
Common and widespread; sterile; gemmiferous.
1–17

M. cuneifolia (Hook.) Gray
Corticolous, rarely epilithic; on trunks and branches of trees (commonly *Betula*) in open *Betula* and *Betula-Quercus* woodlands, commonly mixed with epiphytic *Plagiochila* spp.
Local; frequent in some E coast areas; sterile.
11 Scarisdale Wood; Knock. **12** Allt na Searmoin. **13** Aros Park on *Rhododendron*, *Lobley & U. Duncan*. **15** Lussa River. **16** Cruach Ardura; Lochbuie, on tall *Calluna*.

GYMNOMITRIACEAE

Marsupella Dum.

M. ustulata (Hub.) Spruce.
Calcifuge; in thin fuscous or nigrescent patches on damp rocks in the alpine zone.
Rare; fertile.
11 Ben More, above 600 m, *H. Milne-Redhead*; several places in the north corrie; Beinn Fhada, 700 m.

M. funckii (Web. & Mohr) Dum.
Calcifuge and semi-ruderal; on bare, inorganic soils and fine gravel by streams, tracks, ditches and disturbed ground by roads. Lowland and lower subalpine.
Frequent; fertile.
4–17

M. alpina (Gottsche) Bernet
Calcifuge; in black patches on wet, siliceous rocks. Alpine.
Rare.
11 Ben More, north corrie; Gleann Dubh, *U. Duncan*. **17** Ben Buie, N summit.

M. emarginata (Ehrh.) Dum.
Calcifuge; in often extended patches on wet, siliceous rocks in a variety of habitats. In higher subalpine zone sometimes in extensive carpets on flushed peaty soils with *Andreaea alpina*.
Abundant; only occasionally fertile.
1–17

M. aquatica (Schrad.) Schiffn. var. **aquatica**
Calcifuge; in dull, brownish-green tufts on irrigated rocks in and by subalpine and alpine streams and ravines.
Occasional; sterile.
6 'S Àirde Beinn, *U. Duncan*. **11** Ben More, *Kennedy* (1910); 1932, *J. Duncan & Young*. **16** Ben Buie.

M. aquatica var. **pearsonii** (Schiffn.) E. W. Jones
In similar situations to the type variety.
Rare; sterile.
11 Ben More, *Kennedy* (1910); wet rock by a *Molinia-Sphagnum* mire at 480 m.
The variety is very doubtfully distinct from the type, which is itself held to be distinct from *M. emarginata* only on the basis of minor and variable structural details.

Gymnomitrion Corda

G. concinnatum (Lightf.) Corda
In yellowish patches on exposed basalt ledges and crags. Alpine and high subalpine, descending to 600 m.
Rare; sterile.
11 Ben More, 1932, *J. Duncan & Young*; north corrie, in several places; Beinn nan Gabhar, *U. Duncan.*

G. obtusum (Lindb.) Pears.
In similar habitats to the preceding species but descending to below 500 m.
Rare; fertile.
8 The Wilderness; Creag a' Ghaill. **11** Ben More, *Corley*. **15** Ben Buie, S side, 500 m. **16** Laggan Deer Forest, summit ridge.

G. crenulatum Gottsche
On exposed or shaded siliceous rocks. Alpine and subalpine, descending locally to below 300 m.
Frequent above 450 m becoming common in the alpine zone.
4–17

PLAGIOCHILACEAE

Plagiochila (Dum.) Dum.

P. carringtonii (Balf. ex Carringt.) Grolle
Intermixed with tall hepatics in the 'hepaticosum' facies of *Calluneto-Vaccinietum*. Subalpine.
Rare; sterile.
11 Scarisdale River, 225 m, *H. Milne-Redhead.* **16** Ravine S of Lochbuie.

P. asplenioides (L.) Dum. var. **asplenioides**
Slightly basiphilous; in crevices of coastal rocks, old walls and basic exposures inland; and on skeletal or basic inorganic, rarely humified, soils in ravines and woodlands; occasionally on logs or foots of trees. Indifferent to altitude.
Common in the lowlands, elsewhere occasional; fertile.
1–17

P. asplenioides var. **major** Nees
In colonies or mixed with mosses, e.g., *Rhytidiadelphus triquetrus* on basic to mildly acid damp soils in lowland *Quercus* woods.
Widespread below 200 m; locally frequent; fertile.
3, 5–7, 10–14, 16, 17

P. spinulosa (Dicks.) Dum.
Calcifuge; abundant on humified soils among rocks. Mainly lowland and lower subalpine, rare above 400 m.
Abundant; female plants frequent, male not recorded.
1–17
A variable plant occurring on a wide range of substrates: small forms on exposures of basalt in rather open situations; epiphytic forms, often minute, mixed with other corticolous mosses. Of the large number of varieties previously described of this protean species only var. *inermis* Carringt. is maintained in Paton (1965b). A proportion, at east, of the reduced epiphytic and epilithic Mull populations are referable to this variety, which grades into the type quite imperceptibly. It is without doubt an environmental modification.

P. punctata Tayl.
Corticolous; in small, compact pure or mixed tufts on trunks and branches of trees (principally *Betula*) in open woodlands. Lowland and lower subalpine.
Frequent in sheltered localities; sterile.
3, 6, 7, 10–17

P. tridenticulata (Hook.) Dum.
Corticolous; on trunks and main branches of trees (principally *Quercus* or *Betula*) in rather open woodlands. Lowland.
Occasional; but confined to sheltered situations, mainly on the E side of the Island. Sterile or male only.
12 Kellan Wood. **13** Tobermory, *Ewing* (GL); Aros Park. **15** Lussa River, An t-Sleaghach. **16** Cruach Ardura, An Coire; Loch Uisg, S side.

LOPHOCOLEACEAE

Lophocolea (Dum.) Dum.

L. bidentata (L.) Dum.
In patches or as scattered stems among grassy vegetation in a variety of habitats but absent from *Sphagum*-bogs or markedly basic soils. Indifferent to altitude.
Frequent; sterile or male only.
1–17

L. cuspidata (Nees) Limpr.
In extended, pale patches on rotting wood, raw humus and roots of trees in shade; also creeping under vegetation on inorganic soils and sometimes ruderal; rarely on damp rock faces in ravines, usually near water. Mainly lowland.
Common and widespread; fertile.
1–17
Infertile states of *L. bidentata* and *L. cuspidata* may be difficult to distinguish in the field. Possibly some records of *L. bidentata* are based on misidentifications, but the total distribution statement is unlikely to be affected.

L. heterophylla (Schrad.) Dum.
Only recorded from introduced trees and rotting wood in plantations or in townships; elsewhere in Britain sometimes epilithic.
Rare; fertile.
12 Salen, on *Acer pseudoplatanus*. **13** Aros Park, on *Rhododendron*, *Lobley & U. Duncan.* **17** Torosay Castle.

L. fragrans (Moris & De Not.) Moris & De Not.
Only seen as a small patch on damp, densely shaded rock in a deep ravine near the sea.
13 Aros Park, Druim Breac, opposite Calve Island; probably cryophobic.
A southern Atlantic species reaching its northern limit in this region; recorded previously in Tiree, 1897, *Macvicar* (BM).

Chiloscyphus (L.) Corda

C. polyanthos (L.) Corda var. **polyanthos**
In green or nigrescent loose tufts attached to rocks in and by streams, mainly in regions with basic rock-exposures. Lowland and lower subalpine.
Occasional becoming locally frequent; occasionally fertile.
4 Ulva House, *M. Dalby & U. Duncan.* **6**

Tobermory River; 'S Airde Beinn, *U. Duncan.* **7** Laggan Bay. **8** Tavool House. **10** Pennyghael, *U. Duncan.* **13** Tobermory, *Kennedy* (1910); Aros Park. **14** Fishnish Bay. **17** Torosay Castle.

C. polyanthos var. **rivularis** (Schrad.) Nees
Aquatic; attached to rocks in stream-beds and cascades. Lowland.
Rare; sterile.
6 Glen Gorm. **8** Balmeanach. **11** Loch Bà, R Clachaig tributary.

C. pallescens (Ehrh.) Dum.
Basiphilous; pale, tufted or creeping in eutrophic and mesotrophic flushes and mires and by small upland streams and springs. Indifferent to altitude but mainly lowland.
Occasional; fertile.
2 Iona, *M. Dalby & U. Duncan.* **5** Ensay Burn, *M. Dalby & U. Duncan.* **8** W of Creag a' Ghaill; Maol Mheadonach, *U. Duncan.* **12** Loch Frisa, S shore. **13** Near Tobermory, *Macvicar* (1910). **16** Fontinalis Lochs.

GEOCALYCACEAE

Harpanthus Nees

H. scutatus (Web. & Mohr) Spruce
On wet humus-rich soils or rotting wood in closed woodland areas and deep gullies. Lowland.
Rare.
3 Loch Poit na h-I. **10** Loch Buie, S shore, *H. Milne-Redhead.* **16** Loch Spelve, ravine near Dalnaha.

Saccogyna Dum.

S. viticulosa (L.) Dum.
On wide range of substrates including soil, in crevices of rocks and on rotting wood raw humus, preferring shade but tolerant to a fair degree of exposure. At all altitudes.
Abundant; nearly always sterile.
1–17

CEPHALOZIELLACEAE

Cephaloziella (Spruce) Schiffn.

C. pearsonii (Spruce) Douin
The only record is on perpendicular wet basaltic rock face in shade.
11 Beinn nan Gobhar, W side, 1966, *U. Duncan* (specimen det. J. Paton).

C. rubella (Nees) Warnst.
Calcifuge; in very thin reddish-brown patches on compacted peaty soils and rotting wood; occasionally creeping through spongy moss-hummocks.
Rare or under-recorded; fertile.
2 Iona, *U. Duncan.* **10** Carsaig. **14** Fishnish.

C. hampeana (Nees) Schiffn.
Calcifuge; in similar habitats to *C. rubella* and almost inseparable in the field.
Rare; fertile.
3 Loch Poit na h-I, *Lobley.* **7** Lagganulva. **12** Salen.

C. starkei (Funck) Schiffn.
Calcifuge; on bark of old *Quercus* trees and dry, rotting wood. Lowland.
Rare; fertile.
12 Salen. **14** Fishnish. **17** Torosay Castle.

C. stellulifera (Spruce) Schiffn.
A record of this species "Tobermory, on bank at the Battery" (Macvicar, 1910) is well outside the accepted range of the species. The fragment at BM is very poor and its identity has not been confirmed.

Cephalozia (Dum.) Dum.

C. bicuspidata (L.) Dum. var **bicuspidata**
More or less calcifuge; on damp or wet substrates of various types: on damp amorphous peat, clay and sandy soils in woods and ravines; pathsides, damp rocks, rotting wood and creeping among other bryophytes, including *Sphagnum* in a variety of oligotrophic vegetation types. Indifferent to altitude.
Common, locally abundant; fertile.
1–17

C. bicuspidata var. **lammersiana** (Hüb.) Breidl.
Hygrophytic; in sometimes rubescent patches on peat among irrigated rocks. Mainly alpine and subalpine.
Occasional; fertile.
3 Tòrr Mòr. **6** Loch Frisa. **8** Beinn na Sreine. **11** Ben More, *Kennedy* (1910); north corrie. **16** Loch a' Ghleannain, *Corley*; Laggan Deer Forest.

C. loitlesbergeri Schiffn.
Calcifuge; mixed with *Sphagnum papillosum* in bog.
Rare.
3 Between Loch Poit na h-I and Fidden.

C. connivens (Dicks.) Lindb.
Calcifuge; usually creeping among *Sphagnum* in bogs and oligotrophic mires; occasionally on raw humus or wet, rotting wood. Indifferent to altitude.
Frequent; fertile.
3–17

C. media Lindb.
Calcifuge, but somewhat base-tolerant; commonly epilithic occurring also on peat, rotting wood and silt by streams. Lowland and lower subalpine.
Frequent; fertile.
2–7, 10, 12–17

C. catenulata (Hüb.) Lindb.
Calcifuge; on raw humus and rotting wood.
Rare; fertile.
3 Loch Poit na h-I, *Lobley & U. Duncan.* **11** Coladoir River.

C. leucantha Spruce
Calcifuge; on well-drained peat overlying shaded rocks, rarely on rotting stumps.

3 Loch Poit na h-I. **6** Loch an Tòrr. **11** Ben More, N slope. **12** Allt na Searmoin, *Kennedy* (1910). **16** Loch Spelve, near Dalnaha.

Cladopodiella Buch

C. francisci (Hook.) Buch
Calcifuge; on damp, mainly inorganic soils in open situations.
Rare; fertile.
11 Lower Glen More, Ulavalt, *Lobley & U. Duncan*; SE slopes of Sgùlan Mòr.

C. fluitans (Nees) Buch
Calcifuge; scattered stems, rarely pure patches, among *Sphagnum* in saturated areas of bogs; occasionally as partly immersed mats in bog-pools. Only seen in lowland bogs.
Rare; sterile.
3 0.5 km E of Fidden. **6** Loch na Criadhach Mòire. **10** Lower Glen More, raised bog complex. **11** Ben More, base of N side. **13** Caol Lochan.

Nowellia Mitt.

N. curvifolia (Dicks.) Mitt.
Calcifuge; on decorticated rotting logs in shade or rather rarely on compacted peat in similar situations. Frequent in lowland woods, ascending to c. 250 m.
Frequent except in exposed areas; fertile.
3–17

Hygrobiella Spruce

H. laxifolia (Hook.) Spruce
In thin patches on wet rocks by mountain streams or seepages. Alpine and subalpine.
Rare; fertile.
11 Ben More, north corrie, *Kennedy* (1910).

ADELANTHACEAE

Odontoschisma (Dum.) Dum.

O. sphagni (Dicks.) Dum.
Calcifuge; in oligotrophic mire and bog vegetation as brown patches on wet peat or more frequently creeping among *Sphagnum* in bogs; a constant species in the *Myrica-Molinia* associations. Indifferent to altitude but most abundant at low elevations.
Common below 300 m; sterile, sometimes gemmiferous.
1–17

O. denudatum (Mart.) Dum.
Calcifuge; in small reddish-brown or greenish-yellow patches on rotting stumps in rather open situations; occasionally on drained peat hummocks in and around bogs or on raw humus in heath vegetation.
Occasional; sterile, gemmiferous.
3 Loch Poit na h-I, *Lobley & U. Duncan*. **5** Calgary. **6** Loch Frisa, *Macvicar*. **7** Bruach Mhòr. **10** Glen More. **11** Scarisdale Wood. **13** Aros Park. **15** Lussa River. **16** Cruach Ardura.

Adelanthus Mitt.

A. decipiens (Hook.) Mitt.
Dark green tufts or mixed with, e.g. *Plagiochila spinulosa*, on trunks and boles of trees, less frequently on siliceous rocks; in sheltered woodlands.
Local; mainly in SE. Sterile or male only.
15 Lussa River valley. **16** Cruach Ardura, An Coire, *Corley*; Loch Uisg, S shore; Loch Spelve, near Dalnaha.

SCAPANIACEAE

Douinia (C. Jens.) Buch

D. ovata (Dicks.) Buch
On shaded rock exposures or tree trunks near the ground, usually mixed with other low-growing bryophytes. Indifferent to altitude.
Rare or under-recorded; fertile.
11 Ben More, Gleann na Beinne Fada; Coire nan Ghabar. **17** Craignure, 1954, *Crundwell & Wallace*.

Diplophyllum Dum.

D. albicans (L.) Dum.
On base-poor substrates in a wide range of habitats; organic and inorganic soils, rocks and bases of trees in woods, ravines, corries, screes and under all types of acidophilous vegetation. Indifferent to altitude.
Abundant; fertile, frequently very gemmiferous.
1–17

D. taxifolium (Wahlenb.) Dum.
Only recorded from shaded basaltic slabs in the higher subalpine zone.
11 Ben More, base of north corrie.

Scapania (Dum.) Dum.

Scapania curta (Mart.) Dum.
Only found once on acid or slightly basic soils.
12 Salen, among rocks above road.
This specimen, although sterile, accords well with the species as described in Arnell (1956) but lacks perianths; some doubt remains as to the exact identity of the Mull plant. Similarly an early record "near Tobermory" (Macvicar, 1910) may not be *S. curta*.

S. scandica (Arn. & Buch) Macv.
On damp siliceous rocks or on wood in shaded positions. Indifferent to altitude.
Rare or under-recorded; sterile, gemmiferous.
13 Aros Park, associated with *Riccardia palmata*. **15** Sgurr Dearg, with *Campylopus* sp., *Corley*.
From the habitat and altitude in which it was found, there was some doubt as to the identity of the Aros plant. However, the oil bodies, leaf and stem anatomy agree well with var. **argutedentata** Buch *sensu* Arnell (1956). In Europe the variety is more widespread and more southern in its distribution than the type. *S. scandica* in Britain may consist of more than one species (Paton 1965b).

S. irrigua (Nees) Dum.
In tufts among *Sphagnum fallax* and other bryophytes in mesotrophic or oligotrophic mires, carr and woodlands. Lowland.
Occasional; sterile.
2 Iona. **3** Loch Poit na h-I, *Lobley & U. Duncan*; Bunessan; near Fidden. **6** Tobermory River. **8** Near Tavool House. **9** Loch Assopol. **12** Killichronan, *U. Duncan*. **13** Near Tobermory, *Macvicar* (1910). Aros Park, *U. Duncan*. **15** Lussa River. **16** Loch Spelve, S shore.

S. umbrosa (Schrad.) Dum.
Calcifuge; on rotting wood rarely peat in shaded habitats in sheltered woodlands and ravines. Lowland.
Rare; sterile, gemmiferous.
5 Calgary Bay, *M. Dalby & U. Duncan*. **8** Tiroran. **11** Scarisdale Wood. **13** Aros Park. **17** Torosay Castle.

S. aequiloba (Schwaegr.) Dum.
Only recorded from damp, calcareous detritus among basaltic rocks on the W coast.
3 Near Fidden.

S. aspera Bernet
Calcicolous; on coastal rocks or skeletal calcareous soils; confined to basic rocks elsewhere.
Rare; fertile and gemmiferous.
3 Fidden; Erraid. **5** Treshnish Point. **8** The Wilderness. **13** Aros Park. **17** Craignure Woods, *M. Dalby & U. Duncan*.

S. gracilis (Lindb.) Kaal.
Calcifuge; on a wide variety of substrates, preferably with some humus: heaths, woodlands, in ravines, on logs, trunks of trees, rocks and skeletal soils. Mainly lowland and lower subalpine but ascending into the alpine zone.
Common and widespread; fertile, gemmiferous.
1–17
Very variable in stature according to exposure.

S. nemorea (L.) Grolle
Calcifuge; on shaded, damp rocks, logs and bases of trees, occasionally on peaty soils among vegetation of drained mires and carr-woodlands. Lowland and lower subalpine.
Occasional to locally frequent; fertile, gemmiferous.
3–17

S. undulata (L.) Dum.
Perhaps calcifuge but some forms at least base-tolerant; in green patches under water in shallow streams in lowland gullies (var. *undulata*); in extended, dark purplish mats in alpine and subalpine springs and flushes (var. **dentata** (Dum.) Douin). Indifferent to altitude.
Common and widespread becoming locally abundant; frequently fertile.
1–17
A very variable species. Var. *dentata* is not listed in Paton (1965b). It is by far the more abundant form on Mull where, in general, there do not appear to be many intermediate states; var. *undulata* is more frequent in the lowland zone.

S. uliginosa (Swartz) Dum.
Calcifuge; in brown or ochraceous mats in oligotrophic flushes and springs. Alpine.
Rare; sterile.
11 Ben More, *Macvicar* (1910); north corrie, in *Sphagnum auriculatum* dominated flush.

S. subalpina (Nees) Dum.
On inorganic soils, mainly alluvial in origin, by streams and among periodically irrigated rocks. Mainly subalpine but descending to sea-level.
Rare; occasionally with perianths.
6 Dervaig, *Macvicar* (1910). **7** Loch Frisa, on NE shore. **8** Creag a' Ghaill, *U. Duncan*. **11** Ben More, S side. **13** Aros Park, *U. Duncan*. **15** Lussa River. **16** Laggan Deer Forest, ravine above southern cliffs.

S. compacta (Roth) Dum.
In small, pale patches on vertical faces of damp, shaded basalt in sheltered woodlands and ravines; rarely on skeletal soils. Lowland.
Occasional; fertile.
4 Near Ulva House. **6** Dervaig, *Macvicar* (1910). **7** Lagganulva. **12** Killichronan. **13** Aros Park. **16** Loch Buie; Loch Uisg.

S. ornithopodioides (With.) Pears.
On peaty soils among bryophytes in the 'hepaticosum' facies of *Vaccinieto-Callunetum* and on skeletal soils in subalpine ravines.
Rare; sterile.
11 Scarisdale River *H. Milne-Redhead*; Beinn nan Gabhar, *U. Duncan*. **16** Ravine S of Lochbuie.

RADULACEAE

Radula Dum.

R. complanata (L.) Dum.
Tufted (on twigs) or in closely adherent patches; on rocks, trees and shrubs in shaded habitats. Mainly lowland and lower subalpine, but ascends to at least 650 m.
Frequent; locally common woodlands. Fertile and gemmiferous.
2–17

R. lindbergiana Gottsche
In closely adherent patches on densely shaded rocks, apparently preferring more basic strata. Lowland and lower subalpine.
Rare; gemmiferous but not recorded with female organs.
11 Scarisdale Wood. **13** Tobermory, *Macvicar* (1910); Aros Park. **16** Loch Uisg, S side.

R. carringtonii Jack
Confined to densely shaded rocks near water in extremely sheltered lowland habitats; probably frost-sensitive.

Rare; sterile or male only.
16 Deep ravine S of Lochbuie, 1956, *H. Milne-Redhead*; Loch Spelve, ravine near Dalnaha.

R. aquilegia (Tayl.) Tayl. ex Gottsche & Lindenb.
Mainly corticolous, occasionally epilithic; in consistently humid areas in sheltered woodlands and deep ravines, usually in the vicinity of streams and waterfalls. Lowland.
Rare; sterile or female only.
6 Dervaig, Penmore Mill. **11** Loch Bà, Coille na Sroine. **12** Salen, *Young* (1932); Allt na Searmoin, *Kennedy* (1910). **13** Aros Park, opposite Calve Island; Allt nan Torc. **16** An Coire, *Corley*.

PLEUROZIACEAE

Pleurozia Dum.

P. purpurea Lindb.
Calcifuge; in nigrescent, purplish patches or scattered, creeping stems in *Sphagnum* bogs. Mainly subalpine.
Frequent; widespread. Sterile or with rudimentary perianths.
1, 4–17

PORELLACEAE

Porella L.

P. laevigata (Schrad.) Lindb.
On bark and sheltered rocks near the sea, confined to basic exposures inland. Lowland woods and ravines.
Occasional, locally frequent; sterile or female only.
4 Ulva House. **5** Calgary Bay. **7** Kilninian, *U. Duncan*. **10** Carsaig Bay. **12** Gruline; Knock Farm. **13** Tobermory, *Macvicar* (1910). **16** Loch Uisg; Loch Spelve, near Dalnaha.

P. thuja (Dicks.) C. Jens.
Basiphilous and xerophilous; on exposed rocks near the sea mainly on the west coast.
Rare.
1 Staffa, near Fingal's Cave. **2** Iona, NW coast. **3** Fidden. **5** Treshnish Point.

P. platyphylla (L.) Lindb.
Basiphilous; coastal rocks and mortar of walls. Lowland.
Rare; sterile.
5 Calgary Bay, *M. Dalby & U. Duncan*. **17** Craignure.

JUBULACEAE

Marchesinia Gray emend. Carringt.

M. mackaii (Hook.) Gray
Xerophilous; in dark, flat patches on vertical rock faces near the sea, mainly on cliffs with a southern aspect.
Locally frequent; fertile.
1, 4–10, 13, 16

Lejeunea Lib.

L. cavifolia (Ehrh.) Lindb.
Epilithic and epiphytic; lowland woods and ravines, in shade.
Rare; occasionally fertile.
6 Dervaig, *Macvicar* (1910). **12** Allt na Searmoin, *Kennedy* (1910). **13** Tobermory, 1881, *Ewing*. **16** Loch Uisg, S shore. **17** Torosay Castle.
The apparent frequency of this species implied by the early records is unsubstantiated by recent investigations. Possibly a proportion of the early records are referable to *L. patens*, although the latter is also recorded as being frequent by Macvicar (1910).

L. patens Lindb.
Mainly epilithic and epiphytic but occurring also on raw humus; preferring some shade but frequent among rather exposed rocks, abundant in woodlands, ravines and dwarf-shrub communities. Lowland and subalpine, ascending to at least 650 m.
Common; fertile.

L. lamacerina Gottsche ex Steph.
Rupicolous or terricolous (elsewhere sometimes epiphytic but not recorded from trees on Mull); confined to sheltered coastal habitats.
Rare; fertile.
5 Calgary Bay. **16** Lochbuie, *M. Dalby & U. Duncan* (identification confirmed by Paton).

L. ulicina (Tayl.) Tayl.
Corticolous, occasionally epilithic; in diffuse patches on bark of trees, in woodlands, frequently mixed with other corticolous bryophytes; also occasional on *Calluna* or *Myrica*. Lowland and lower subalpine.
Frequent and widespread; sterile.
2–17

Drepanolejeunea (Spruce) Schiffn.

D. hamatifolia (Hook.) Schiffn.
Epiphytic, less frequently epilithic; on damp, densely shaded basalt and bark of trees (mainly *Fraxinus*) in lowland woods and deep ravines.
Frequent and widespread; fertile.
4–17

Harpalejeunea (Spruce) Schiffn.

H. ovata (Hook.) Schiffn.
In small, closely adherent patches on smooth, wet shady rocks in lowland ravines, always in very sheltered habitats.
Rare; fertile.
8 Allt na Teangaidh, *U. Duncan*. **11** Scarisdale Wood. **12** Allt na Searmoin, *Kennedy* (1910). **13** Near Tobermory, *Macvicar* (1910). **16** Loch Spelve, ravine near Dalnaha.

Cololejeunea (Spruce) Schiffn.

C. calcarea (Lib.) Schiffn.
In diffuse patches or rather dense, minute tufts on basic rocks, commonly with other calcicolous bryophytes, usually on vertical surfaces or underhangs. Indifferent to altitude

but restricted edaphically.
Rare; sterile.
8 Below Creag a' Ghaill. **13** Aros Park, *U. Duncan*. **16** Lochbuie, 1963, *H. Milne-Redhead*. **17** Craignure.

C. minutissima (Sm.) Schiffn.
On bark of trees in very sheltered localities; once recorded on moss tuft on shaded rocks. Lowland.
Rare; fertile.
6 Dervaig, Penmore Mill. **13** Tobermory, *Macvicar* (1910). **15** Lussa River. **16** Loch Uisg; Loch Spelve, S side.

Aphanolejeunea Evans

A. microscopica (Tayl.) Evans
Saxicolous, shade tolerant; in minute, closely adherent pale patches on smooth, damp rocks in dense shade; confined to deep ravines in sheltered east-coast areas, nearly always near water.
Rare; mainly in the east. Fertile.
12 Allt na Searmoin, *Kennedy* (1910). **13** Near Tobermory, *Macvicar* (1910); Aros Park. **15** Lussa River valley. **16** Loch Spelve, ravine near Dalnaha.

Colura Dum.

C. calyptrifolia (Hook.) Dum.
Mainly on mosses on rocks and trees; occasionally epilithic and epiphytic in woodlands and ravines. Lowland.
Rare; mainly in the east. Sterile.
6 Loch an Tòrr. **11** Loch Bà, Coille na Sroine. **13** Upper Druimfin. **16** An Coire; Lochbuie; Loch Spelve, ravine near Dalnaha.

Jubula Dum.

J. hutchinsiae (Hook.) Dum.
Shade tolerant, hygrophytic; a frost-tender species only recorded on a permanently wet rock face, in a deep, narrow gully near the sea.
Rare.
8 Tiroran.

Frullania Raddi

F. tamarisci (L.) Dum. var. **tamarisci**
Facultatively xerophilous; on rather dry rocks, trees and shrubs in exposed or slightly shaded localities; rarely terrestrial on skeletal soils or among impoverished shrubs. Indifferent to altitude but most abundant around the coast.
Common; locally abundant and widespread. Fertile.
1–17

F. tamarisci var. **cornubica** Carringt.
Xerophilous; in thin, blackish patches on coastal rocks.
Occasional in SW but locally frequent; sterile.
1 Staffa; Lunga; Bac Mòr. **2** Iona. **3** Fidden; Erraid. **6** Mingary Àrd. **10** E of Carsaig Bay.

F. germana (Tayl.) Tayl. ex Gottsche, Lindenb. & Nees
In flat but rather easily detached mats on rocks and occasionally trunks of large trees. Lowland and mainly on or near the coast.
Frequent near the sea; rare elsewhere. Occasionally fertile.
1–17

F. microphylla (Gottsche) Pears.
Saxicolous, xerophilous; in thin or diffuse nigrescent patches on coastal rocks, mainly just above the *Grimmia maritima* zone.
Occasional; mainly in the west.
1 Bac Mòr. **2** Iona, *M. Dalby & U. Duncan*. **3** Fidden; Erraid. **5** Treshnish Point. **9** Cliffs W of Malcolm's Point. **10** Cliffs E of Carsaig. **13** Tobermory, *Macvicar* (1910). **16** Laggan Peninsula, near Lord Lovat's Cave.

F. fragilifolia (Tayl.) Tayl. ex Gottsche, Lindenb. & Nees
Saxicolous, rarely epiphytic and more or less xerophilous; resembling, and in similar habitats to *F. microphylla*, but less confined to proximity to the sea.
Frequent around the coast; elsewhere rare.
1–4, 6–8, 10, 13, 16, 17

F. dilatata (L.) Dum.
On trunks and branches of trees but commonly epilithic; avoiding extremes of exposure. Mainly lowland and lower subalpine.
Common and widespread; fertile.
2–17

BRYOPSIDA

SPHAGNALES

SPHAGNACEAE

Sphagnum L.

S. palustre L.
Calcifuge; in pure or mixed carpets in oligotrophic fen communities; frequently forming characteristic associations with *Juncus effusus* and *Polytrichum commune* in open situations and open *Betula-Salix* carr; disturbed peat, margins of bogs but not present on undisturbed bogs. Lowland and lower subalpine, occasional to rare above 650 m.
Common and widespread: occasionally fruiting.
1–17

S. magellanicum Brid.
Calcifuge, hygrophytic; almost confined to relatively undisturbed surfaces of mainly topogenous bogs in valleys and on low terraces; a hummock builder in the 'hummock-hollow' complex of bogs or in mats mixed with *S. nemoreum and S. papillosum*; apparently persistent for some time after draining, but sensitive to burning.

Local; but usually in quantity where it occurs. Sterile.
3 Near Bunessan, *Lobley & U. Duncan*; raised bog near Fidden. **6** Loch na Criadhach Moire. **7** Loch a' Gael. **10** Glen More raised bog complex. **15** Near Loch Bearnach, *Corley*. **16** Laggan Deer Forest, near Fontinalis Lochs.

S. papillosum Lindb.
Calcifuge; in extensive carpets or ochraceous patches in a wide variety of wet, oligotrophic peaty substrates; co-dominant over large areas of bogs and *Myrica-Molinia* associations. Indifferent to altitude.
Widespread and abundant; fruiting rather infrequently
1–17

S. imbricatum Hornsch. ex Russ.
An oceanic species found in the 'hummock complex' of undisturbed bogs. The species is probably extinct on Mull but occurs in peat as a recent sub-fossil in samples taken from 5–10 cms depth in the large bog-complex at the western extremity of the Ross of Mull, S of Torr Mòr.

S. compactum DC.
Calcifuge; forming low, compact, orange-brown cushions on thin peat overlying siliceous rocks; rarely found on deep, ombrogenous peat. Mainly lowland and subalpine but ascending to at least 600 m.
Abundant and widespread; especially in the lowland and lower subalpine zones. Commonly fruiting.
1–17

S. strictum Sull.
Calcifuge, possibly cryophobic; paralleling *S. compactum* in its edaphic requirements but geographically more restricted. Lowland and lower sub-alpine, not seen above 300 m.
Rare; in scattered localities. Fertile.
4 Ulva, north-east slope of Ben Chreagach. **6** S'Àirde Beinn. **8** East side of Maol Mheadhonach, *U. Duncan*. **11** Gleann Dubh, *U. Duncan*; Glen More, Ullavalt. **15** Loch Bearnach. **16** Loch Spelve, near Dalnaha.

S. teres (Schimp.) Ångstr.
Basiphilous; among grassy vegetation in base-rich mires associated with calcareous rocks or facing the sea.
Rare and local; sterile.
11 Ben More 1887, *Ewing* (GL). **15** Lussa River, An Coire.

S. squarrosum Pers. ex Crome
Similar to *S. palustre* in its ecology, but distinctly more base-demanding and much less frequent: streamsides, oligotropic mires, wet rocks in ravines. Mainly lowland, but ascending to 310 m on Ben More.
Frequent and widespread; occasionally fertile.
3–6, 9–17

S. fallax Klinggr.
S. recurvum sensu auct. brit.
Calcifuge, but slightly base-tolerant, hygrophytic; in wide, green or ochraceous mats in a variety of oligotrophic bog and mire communities, including *Betula* and mixed carr; characteristically associated with *Juncus effusus*, *J. acutiflorus* and *Polytrichum commune*; common on disturbed peat but not normally part of the flora of undisturbed *Sphagnum*-bog surfaces. Indifferent to altitude.
Widespread; common to locally abundant. Fertile.
1–17

S. tenellum Pers.
Calcifuge; in wet hollows in bogs and *Myrica-Molinia* associations and a variety of wet, oligotrophic habitats on hillsides, commonly with *S. compactum*. Indifferent to altitude.
Common and widespread; fertile.
1–17

S. cuspidatum Ehrh. ex Hoffm.
Calcifuge; forming lax mats in hollows in bogs and in stagnant bog pools; emergent around small lakes and by acid streams (replaced by *S. auriculatum* where there is water-movement). Indifferent to altitude.
Common and widespread; fertile. Male plants common; fruit infrequent.
2–17

S. contortum Schultz
Basiphilous: usually in lax patches or intermixed with sedges and other plants in mesotrophic bogs and mires. Mainly lowland and near the sea due to general lack of basic underlying strata.
Rare; in scattered localities. Sterile.
3 Uisken. **5** Kilninian; Ensay Burn, *E. Lobley & U. Duncan*. **8** Gribun, Balmeanach, *E. Lobley & U. Duncan*. **9** Allt Chaomhain. **17** Near Duart Point.

S. subsecundum Nees var. **subsecundum**
Calcifuge but slightly base-tolerant; oligotrophic flushes and wet, peaty slopes and hill-terraces.
Rare; scattered localities. Sterile.
4 Ben Chreagach. **5** North-east of Treshnish Point, *M. Dalby & U. Duncan*. **11** Gleann na Beinne Fada. **13** Aros, Upper Druimfin, *U. Duncan*. **16** Laggan Deer Forest.

S. subsecundum var. **inundatum** (Russ.) C. Jens.
Calcifuge but base-tolerant to some degree; in similar habitats to the type variety but much more frequent; mainly mixed with sedges and rushes in oligotrophic bogs and mires; on wet, grassy slopes and streamsides.
Uncommon in the lowlands; rare above 300 m.

Sterile or male plants only, not seen in fruit.
3 Loch Poit na h-I. **4** Relict mire near Ulva House. **5** Calgary. **8** Ardmeanach, Tavool House. **9** Loch Arm. **11** Northern slopes of Ben More, *U. Duncan & E. Lobley*. **15** North side of Beinn Talaidh, *M. Dalby & U. Duncan.* **16** Lochbuie. **17** Loch a' Gleannain.
European bryologists frequently accord this taxon specific rank but its distinguishing characters are purely quantitative. Holmen (1955) records a diploid chromosome number for this variety.

S. auriculatum Schimp.
S. subsecundum Nees var. *auriculatum* (Schimp.) Lindb.
Calcifuge; a variable, often rubescent species occurring in a wide variety of wet, oligotrophic habitats: bog-pools, ditches and mires; abundant in shallow, acid flushes at higher altitudes and a conspicuous component of the *Sphagnum-Carex* flushes. Indifferent to altitude.
Abundant and widespread; fruit very rare.
1–17
This taxon is not very closely related to *S. subsecundum* and is to be regarded as a distinct species.
Submerged plants with larger leaves are found in stagnant pools and small lakes. These merge with the typical plant and are habitat modifications not distinguished by varietal names in this treatment.

S. fimbriatum Wils.
Calcifuge; forming soft, pale tufts or dispersed in vegetation of open carr and wet areas of open *Betula* woods; frequent in grassy areas of oligotrophic mires and in *Myrica-Molinia* associations. Mainly lowland but ascending to 500 m.
Frequent and widespread; fertile.
3–6, 8–17

S. girgensohnii Russ.
Calcifuge but slightly base-tolerant; resembles *S. fimbriatum* in habitat and ecology but more robust, and shows a preference for more open ground in oligotrophic mires, streamsides and irrigated hill-slopes. Ascends to 700 m.
Occasional but locally frequent; widespread. Sterile.
4 Ulva, Ben Chreagach; **6,7** Loch Frisa, NE shore. **8** Gribun, above Mackinnon's Cave. **11** Loch Bà, s end of the lake. **12** Loch Frisa. **15** Beinn Talaidh, *M. Dalby & U. Duncan.* **16** Loch Buie, Maol a' Bhaird, *Ratcliffe.*

S. russowii Warnst.
Calcifuge; forming red or variegated cushions on wet mountain-sides, usually on the crests in **11**, peaty terraces. Mainly subalpine but extending its range into lowland and alpine zones in Scotland.
Rare and local; sterile.
11 Ben More, foot of north corrie. **16** Loch Buie, south side, Allt a' Bhaird.
The apparent scarcity of this species on Mull is puzzling in view of its relative frequency on the Scottish mainland.

S. fuscum (Schimp.) Klinggr.
Strictly calcifuge; forming brown hummocks in drier areas of ombrogenous bogs of the *Calluneto – Eriophoretum* type. Subalpine.
Rare; in scattered localities. Not seen in fruit but antheridia present in some plants.
7 2 km north of Kilninian. **10** Cruach Inagairt. **15** Beinn Creagach Mhor. **16** Laggan Deer Forest on terrace north of summit ridge.

S. warnstorfianum Russ.
Basiphilous; in deep crimson patches or mixed with other bryophytes in basic fens and flushes. Indifferent to altitude.
Rare and local; sterile or male plants only.
8 Ardmeanach, Maol Mheadonach. **11** Ben More, lower slopes, *Muirhead.*

S. nemoreum Scop. var. **nemoreum**
Strictly calcifuge. A variable species forming dense or lax cushions variously tinted crimson except in shade; on wet peat in bogs, grass-heaths and open, acid woodlands from sea-level to the summits of the higher mountains.
Abundant and widespread; fertile.
1–17

S. nemoreum var. **rubellum** (Wils.) Briz.
S. rubellum Wils.
In similar habitats to var. *nemoreum*, but absent from exposed sites and high altitudes; on Mull almost confined to lowland *Sphagnum*-bogs and oligotrophic mires.
Locally frequent; possibly under-recorded. Fertile.
1–6, 9, 11–17
S. rubellum is variously regarded as a synonym of *S. nemoreum* and as a good species. In this survey forms with closely imbricate, spirally arranged branch-leaves have been listed under *S. nemoreum* var. *nemoreum*; plants with pentastichous branch-leaves have been provisionally placed in *S. nemoreum* var. *rubellum.* Stem-leaf characters have proved to be inconsistent in both taxa.

S. subnitens Russ.
S. plumulosum Röll
Calcifuge, but with wider tolerance than *S. nemoreum*; in low, reddish or dull, pinkish-brown hummocks on pea in bogs, acid mires, by moorland streams and amongst irrigated mountain rocks. Indifferent to altitude.
Common and widespread; frequently fruiting.
1–17

S. molle Sull.
Calcifuge; in small patches on peat terraces on the lower slopes of hills.

Rare; in scattered localities. Rarely fruiting.
3 Tòrr Fada. **4** Ulva, east side of Beinn Chreagach. **6** 'S Àirde Beinn, *U. Duncan.* **7** Cnoc nan Uan, *U. Duncan.* **8** Ardmeanach, west of Mackinnon's Cave, *U. Duncan.* **10** Glen More. **15** Beinn Talaidh, north side, *M. Dalby* & *U. Duncan.* **16** Cruach Ardura.

ANDREAEALES

ANDREAEACEAE

Andreaea Hedw.

A. alpina Hedw.
Calcifuge; in small, blackish or dark red-brown tufts on rocks in the alpine zone; forming black carpets on damp fine-grained soils in association with *Anthelia julacea* and *Marsupella* spp.
Common and widespread except in the lowlands; locally abundant above 100 m. Fertile.
4, 6–8, 10–12, 15–17

A. rupestris Hedw.
In small, low brown tufts on exposed basaltic rocks. Mainly alpine and subalpine, descending occasionally to sea-level.
Frequent and widespread; fertile.
2–8, 10–12, 15–17
Some populations approach *A. rupestris* var. *alpestris* (Thed.) Sharp in structure, but these forms do not seem to be sufficiently well delimited from the typical form to warrant separate listing on Mull.

A. rothii Web. & Mohr. var. **rothii**
In dense or diffuse scattered black tufts on exposed rocks, less frequently as brownish, more luxuriant tufts on damp, north-facing alpine crags; in the subalpine zone shows a distinct preference for vertical rock faces opposed to the prevailing winds from the sea. Indifferent to altitude on Mull.
Common and widespread; fertile.
1–17

A. rothii var. **crassinervia** (Bruch) Mönk.
In similar situations to the type variety, but on somewhat wetter substrates.
Recorded only once, in the sterile state: **11** south side of Sròn Daraich, 1966, *U. Duncan.*

POLYTRICHALES

POLYTRICHACEAE

Atrichum P. Beauv.

A. undulatum (Hedw.) P. Beauv.
Gregarious or tufted green plants on neutral to acid soils; on woodland banks, streams sides and among rocks. Ascends to at least 650 m but is much more abundant in the lowlands.
Common and widespread; fertile.
1–17

Oligotrichum Lam. & DC.

O. hercynicum (Hedw.) Lam. & DC.
Calcifuge; in lax tufts or gregarious, on bare or skeletal inorganic soils; a colonist of mountain-ridge detritus and streamside alluvium occurring in a variety of disturbed habitats such as roadsides, cuttings and screes. Indifferent to altitude and exposure.
Common and widespread; frequently fruiting.
2–17

Polytrichum Hedw.

P. nanum Hedw.
Calcifuge; a colonist of skeletal soils in open situations; most frequent on level or slightly inclined sites, characteristically associated with *Oligotrichum* on fine summit-scree or ridge detritus. Alpine and subalpine but not rare at sea-level.
Common and widespread; locally abundant on mountain ridges. Established colonies always fertile.
4, 7, 8, 10–12, 14–17

P. aloides Hedw. var. **aloides**
Usually a calcifuge; a colonist of vertical or steeply inclined inorganic soils such as sides of ditches and undercut stream-banks; less commonly on level ground except at sites of recent disturbance. Most abundant in the lowlands but ascends to at least 100 m. (where, however, it is largely replaced by *P. urnigerum*).
Widespread; occasionally locally frequent. Established colonies abundantly fertile.
2–17

P. aloides var. **minimum** (Crome) Rich. & Wall.
In similar situations to the type variety.
A single record only. **7** Laggan Bay on a damp, shaded bank, 1969 *Lobley* & *U. Duncan.*
This variety is possibly a spontaneous hybrid between *P. aloides* and *P. nanum.*

P. urnigerum Hedw.
Distinctly basiphilous; in glaucous colonies on skeletal, inorganic soils by roads, streams and on mountain slopes. Most abundant in the alpine and subalpine zones but commonly descending to sea-level.
Common and widespread; frequently fertile.
1, 4–17
On unstable sites this species frequently remains in an immature (i.e., sterile, short-stemmed) condition resembling *P. juniperinum* or *P. aloides.*

P. alpinum Hedw.
Calcifuge?; in dull green colonies on acid soils in upland grass-heaths, screes, soil-capped ledges and disturbed or relict peaty soils on mountain slopes. Mainly alpine and subalpine, descending to 300 m. locally.
Widespread except in the lowlands; common at higher elevations. Fruit frequent.
7–13, 15, 16
This species replaces *P. urnigerum* on acid soils with high organic fraction, and is itself replaced by *P. commune* in wetter areas.

P. piliferum Hedw.
Xerophilous: on skeletal soils among rocks, open sites by roads, stream-gullies, walls, coastal rocks and dunes (where it can act as a soil-binding colonist). Indifferent to altitude and to soil-reaction.
Common and widespread; fertile.
1–17

P. juniperinum Hedw.
Often xerophilous; on dry soils in similar habitats to *P. piliferum* with which it often grows in associations; base tolerant to a degree but replaced by *P. piliferum* or *P. urnigerum* on strongly basic soils. Indifferent to altitude.
Common and widespread; fertile.
1–17

P. alpestre Hoppe
Calcifuge; in tufts on peat or supplanting compact cushions of *Sphagnum nemoreum.* Mainly alpine and subalpine, descending to the lowlands locally.
Rare; fertile.
6 Loch an Tòrr; Lochan Dearg. **7** Loch a' Ghael. **8** Maol Mheadonach. **10** Glen More. **11** Ben More. **15** Beinn Talaidh, north side, *M. Dalby & U. Duncan.* **16** Laggan Deer Forest.

P. aurantiacum Sw.
Calcifuge; on disturbed peat particularly in areas of recent ditching operations or afforestation; less frequent on peat or peaty soils in ravines and acid woodlands and not seen on undisturbed sites. Lowland.
Locally frequent; fertile.
3, 6–10, 12–15
This is possibly a relatively recent introduction in view of the lack of records from non-anthropogenic sites.

P. formosum Hedw.
Calcifuge; in dull green colonies or large, loose tufts on well drained inorganic or highly humified soils; common in the lowlands, mainly in somewhat sheltered sites such as woodland floors; replaced by *P. commune* on wetter, organic subtrates. Not reliably recorded above 300 m, above which it is replaced by *P. alpinum.*
Widespread; locally frequent in acid woodlands. Fertile.
1–17

P. commune Hedw.
Calcifuge; a variable species typically forming tall, loose hummocks on wet, peaty soils in bogs, grass-heaths and poor fen; persisting as dwarf forms in drained grasslands and peaty gravel slopes on hillsides; rare on undisturbed bog surfaces, otherwise abundant on acid, organic substrates, and a constant species in base-poor mire, e.g., the *Juncus effusus–Sphagnum recurvum* association.
Common and widespread; fertile.
1–17
This species is very variable in stature and forms approaching var. **humile** Sw. are common. No specimens which could be unequivocably assigned to the var. *perigoniale* (Michx) B.S.G. have been recorded.

BUXBAUMIALES

BUXBAUMIACEAE

Diphyscium Mohr

D. foliosum (Hedw.) Mohr
Calcifuge; on thin or well-drained base-poor soils on hillsides, wooded slopes and banks of streams in open or in shaded habitats. Most frequent in the subalpine zones, but descends to sea-level locally.
Frequent and widespread; fertile.
2–17
The sterile plant is inconspicuous but is more abundant and widespread than the fertile plant on Mull.

TETRAPHIDALES

TETRAPHIDACEAE

Tetraphis Hedw.

T. pellucida Hedw.
Corticolous, lignicolous and humicolous; a low-level epiphyte on trees (principally *Quercus*); on stumps, logs and thin, dry peat on rock-ledges; mainly in lowland woods.
Occasional; rarely fruiting; gemmiferous.
4 Near Ulva House. **7** Bruch Mhor, west of Lagganulva. **8** Tiroran. **12** Salen, Allt na Searmoin, *U. Duncan.* **16** Cruach Ardura. **17** Craignure, *U. Duncan*; Torosay, *M. Dalby & U. Duncan.*

T. browniana (Dicks.) Grev.
In diffuse, thin patches on damp rocks in dense shade: crevices and underhangs in deep ravines and lowland woods.
Uncommon; but readily overlooked when infertile. Fruiting.
4 Near Ulva House. **7** West of Kilninian. **8** Balmeanach, *Lobley & U. Duncan.* **10** Pennyghael. **11** Abhainn Dhiseig; Scarisdale River, *U. Duncan.* **13** Aros Park. **15** Lussa River valley. **16** Duibh Leitir. **17** Craignure Woods.

BRYALES

FISSIDENTACEAE

Fissidens Hedw.

F. viridulus Wahlenb.
Basiphilous; more or less confined to inorganic soils derived from calcareous rocks, or on the base rock where constantly damp; occasionally in crevices of siliceous rocks near the sea where calcareous matter accumulates.
Rare and local; fertile.
2 Iona, *Trotter.* **8** Gribun; in gully above Mackinnon's Cave. **10** Carsaig Bay; *M. Dalby & U. Duncan.* **16** Loch Buie, cliffs on southern shore.

F. minutulus Sull.
Basiphilous; in small colonies on damp basic rocks in ravines and woodlands.
Rare and local.
5 Calgary Bay, *Corley*. **8** Ardmeanach, steep bank by waterfall, *H. Milne-Redhead.*
All Mull material is var. **minutulus.**

F. bryoides Hedw.
Densely gregarious or scattered on vertical or steeply inclined inorganic soils which are consistently damp but not wet in woodlands, quarries and bases of walls, occasionally colonising soils in cultivated areas; indifferent to soil reactions and to altitude but most abundant on fine-grained, neutral soils. Lowland.
Common and widespread; usually fertile.
1–17
Although common, in many apparently suitable habitats this species is replaced by *T. taxifolius.*

F. curnowii Mitt.
Rupicolous or terricolous; in green patches on wet, shaded rocks near the sea and basic flushed soils by loch margins.
Rare; sterile.
3 Loch Poit na h-I, *E. Lobley*. **6** Loch Mingary, in cleft of sea-cliff. **16** Loch Buie, in entrance of sea cave.

F. exilis Hedw.
A colonist of bare, fine-grained mineral soils in ravines and lowland woods; also ruderal and often an ephemeral colonist of cultivated ground.
Rare; fertile.
7 Torloisk House. **10** Pennyghael. **12** Salen. **13** Aros Park. **14** Lower River Forsa.

F. celticus Paton
Terricolous; skiophilous: on damp retentive soils in shade, usually by streams in woodlands or ravines. Lowland.
Rather rare; not seen in fruit.
11 Scarisdale Wood. **13** Aros Park. **17** Torosay Castle, on stream bank, *Kenneth and Stirling.*

F. osmundoides Hedw.
Slightly but distinctly basiphilous; growing in rather dense tufts among irrigated rocks by streams in woodlands and ravines and in subalpine basic flushes.
Frequent; fertile.
2–17

F. taxifolius Hedw. subsp. **taxifolius**
Growing on a wide range of inorganic, rarely purely organic soils of varying base-status; in grasslands, woodlands, gullies and damp walls. Indifferent to altitude.
Common and widespread; frequently fruiting.
1–17

F. taxifolius subsp. **pallidicaulis** Mitt.
On flushed ground and irrigated rocks.
Sterile specimens of this taxon were found at: **12** Allt na Saille, 1968, *Norkett.* **17** Craignure, 1968, *Norkett.* It is the first record of this taxon for Britain.

F. cristatus Wils.
Basiphilous; in small tufts in clefts of basic rocks, or siliceous rocks by the sea; also on dry, skeletal soils and calcareous sand on the coast. Indifferent to altitude but edaphically restricted in the hills.
Common around the coast; rare inland. Fruit rare.
1–17

F. adianthoides Hedw.
Basiphilous; in tufts on flushed soils or in clefts of wet rocks in open situations or in shade. Indifferent to altitude but in the mountains largely replaced by luxuriant forms of *F. taxifolius.*
Frequent in the lowlands, elsewhere rather rare. Fertile.
1–17

ARCHIDIACEAE

Archidium Brid.

A. alternifolium (Hedw.) Mitt.
Ruderal and salt tolerant: in scattered colonies or loose tufts on damp soils by roads and paths, on bare damp soils near the sea and in the upper zones of halophytic vegetation on saltings.
Rare; but somewhat transient. Not found in fruit.
12 Allt na Searmoin, *U. Duncan*; Killichronan, *U. Duncan.* **13** Tobermory. **17** Lochdonhead.

DITRICHACEAE

Pleuridium Brid.

P. acuminatum Lindb.
Ruderal; in loose tufts or scattered colonies on bare sandy soils, usually in anthropogenic sites by paths, in fields and occasionally semi-natural open grasslands.
Rare; but somewhat ephemeral and probably under-recorded. Always fertile.
2 In Nunnery garden. **3** Bunessan, *E. Lobley & U. Duncan.* **5** Calgary Bay, *M. Dalby & U. Duncan.* **12** Salen. **13** Tobermory, *G. Ross et al.* (1881); above the town; on farm road. **17** Torosay Castle.

Ditrichum Hampe

D. cylindricum (Hedw.) Grout
Often ruderal; a transient or persistent colonist of bare sandy soils on disturbed or recently exposed ground.
Rare; possibly an introduction. Fertile.
2 Near the jetty. **5** Calgary Bay, *M. Dalby & U. Duncan.* **8** Gribun, Balmeanach, *E. Lobley & U. Duncan.* **12** Salen, by the river; Glenforsa House, *U. Duncan.*

D. heteromallum (Hedw.) Britt.
Calcifuge; in pure or mixed tufts or gregarious on inorganic or humified base-poor soils and open, gravelly soils on screes and mountain ridges, stream banks, roadsides and colonising disturbed ground by roadworks. Indifferent to altitude.
Frequent; and widespread; but seldom in quantity. Fertile.
4–17

D. zonatum (Brid.) Limpr.
Calcifuge; in low dense green tufts in clefts of basaltic rocks. Alpine and higher subalpine, descending to about 450 m.
Locally frequent in the higher mountains; sterile.
11 Ben More, 1932, *J. Duncan & W. Young*; ditto, in the north corrie; Beinn nan Gabhar, *U. Duncan*. **15** Beinn Bheag, *Corley*. **16** Ben Buie.
Among the Mull specimens are forms referable to the variety *scabrifolium* Dix. Such forms appear to grade into smooth-leaved plants, often within the same tuft, and are not listed separately here.

D. flexicaule (Schwaegr.) Hampe
Calcicolous and somewhat xerophilous; in diffuse to rather compact tufts, or as scattered stems on dry, basic soils or clefts of basic rocks; frequent in shell-sand turf and among coastal rocks where blown sand, containing calcareous grains, accumulates. Indifferent to altitude.
Common around the coastline; rare inland. Fruit rare.
1–6, 8–10, 12–17

Distichium B.S.G.

D. capillaceum (Hedw.) B.S.G.
Basiphilous; in rather dense tufts in clefts and on ledges of rocks, associated with basic veins of rock or enriched sites in areas of siliceous rocks; a subalpine species generally, and replacing *Ditrichum flexicaule* to some extent in the mountains, but also found at sea level on blown sand on coastal rocks.
Occasional; fertile.
1 Lunga; Staffa. **3** Fidden; Erraid. 5 Treshnish Point. **8** The Wilderness. **16** Laggan Deer Forest. **17** Torosay Castle Woods, *M. Dalby & U. Duncan*.
Very dwarf but fruiting specimens from exposed west-coast habitats sometimes strongly resemble *D. inclinatum*.

D. inclinatum (Hedw.) B.S.G.
Calcicolous; in short tufts on basic detritus among rocks. Alpine and subalpine.
A single fruiting record from **8** Creag a' Ghaill on exposed ledge of basic basalt at 350 m. 1967.
An unexpected species in view of the paucity of high-altitude basic exposures, but its identity is not in doubt.

Ceratodon Brid.

C. purpureus (Hedw.) Brid.
Calcifuge but somewhat base-tolerant; in tufts or somewhat diffuse colonies intermixed with other bryophytes on a variety of substrates, including sandy soils by paths, on heaths and dunes, a colonist of disturbed ground around habitation, and characteristic adventive sites of recent fires.
Common and widespread; fertile.
1–17
A very variable species probably represented by numerous genotypes on Mull. Of the many varieties described in the literature, the only one of recognised importance in Britain, var. *conicus* (Hampe) Husn., is not recorded.

SELIGERIACEAE

Brachydontium Fürnr.

B. trichodes (Web. f.) Fürnr.
Basiphilous; on basic but not highly calcareous rocks in damp, shaded ravines; on loose blocks and small stones by streams in sheltered places. Subalpine.
Rare; fertile.
8 The Wilderness. **13** Allt nan Torc.

Seligeria B.S.G.

S. doniana (Sm.) C. Müll.
Calcicolous; in diffuse colonies on vertical or underhanging calcareous rocks in shaded situations. Indifferent to altitude but edaphically restricted to low elevations.
Rare and local; fertile.
8 Below Creag a' Ghaill. **16** Between Loch Spelve and Port Donain on Càrn Bàn, *Stirling*.

S. recurvata (Hedw.) B.S.G.
Basiphilous; in similar situations to the preceding species but less confined to highly calcareous rocks.
Rare and local; fertile.
5 Treshnish Point. **8** Ardmeanach, *H. Milne-Redhead*. **12** Glen Aros, Allt na Saille. **16** Loch Spelve, near Dalnaha.

Blindia B.S.G.

B. acuta B.S.G. var. **acuta**
Oligotrophic to eutrophic but avoiding extremely acid or basic sites; on wet rocks by streams and open flushes. At all altitudes, abundant in the alpine and subalpine.
Common and widespread; frequently fertile.
1–17

B. acuta var. **arenacea** Mol.
In small dark tufts on irrigated rocks. Alpine and subalpine.
Occasional; widespread. Usually sterile.
8 Beinne na Sreine. **11** Gleann na Beinne Fada. **12** Allt na Searmoin. **16** Laggan Deer Forest, above the southern cliffs, *U. Duncan*.

DICRANACEAE

Pseudephemerum (Lindb.) Loeske

P. nitidum (Hedw.) Reim.
Ruderal and ephemeral; in small colonies or scattered on bare, muddy soils or fine silt by lowland lochs.
Very rare; but inconspicuous and transient, therefore possibly under-recorded. Fertile.
12 Near Salen, 1932, *J. Duncan & Young*. **6** Dervaig. Loch na Cuilce. **13** Aros Park.

Dicranella (C. Müll.) Schimp.

D. palustris (Dicks.) Crundw. ex E. F. Warb.
Calcifuge; growing in bright, pale green or yellowish green tufts in oligotrophic springs and flushes, and by streams and lochs. At all altitudes, but most abundant above 200 m.
Abundant and widespread; very rarely fertile (one specimen found with an old seta).
2–17

D. schreberiana (Hedw.) Dix.
In small tufts on silt by streams, and wet under-cut stream banks. Lowland and lower subalpine.
Rare; scattered. Occasionally fruiting but usually sterile.
7 Kilninian. **8** Balmeanach. **13** Allt nan Torc. **16** South of Loch Spelve, Dalnaha.

D. varia (Hedw.) Schimp
Basiphilous; on inorganic clay or sandy soils by streams and ditches and a colonist of disturbed ground, fields and irrigated sandy soils by the sea. Indifferent to altitude.
Occasional; scattered. Fertile.
2 Iona. **3** Bunessan. **5** Calgary Bay. **6** Mingary Àrd. **8** Gribun, *U. Duncan*. **11** Ben More, 600 m (var. **tenella** B.S.G. *sensu* Dixon (1924)). **12** Loch Bà. **13** Aros Park. **16** Loch Buie. **17** Torosay Castle.

D. rufescens (With.) Schimp.
In diffuse tufts or scattered stems on bare inorganic soils by streams and on clay banks; occasionally a colonist of recently exposed drift by roads and ditch-sides.
Occasional; widespread. Fertile.
4 Near Ulva House. **7** Achleck. **10** Carsaig. **13** Tobermory, *G. Ross et al.* (1881) Aros Park, *U. Duncan*. **14** Fishnish Bay. **15** Lussa River.

D. crispa (Hedw.) Schimp.
Only seen as scattered fertile stems among other bryophytes on compacted sand by Mingary Burn in **6** Glen Gorm.

D. subulata (Hedw.) Schimp.
Only seen on gravelly detritus by a small stream in the higher subalpine zone of **11** Ben More, near foot of the main corrie, 800 m where it was fertile.

D. cerviculata (Hedw.) Schimp.
Strictly calcifuge; in low green patches on wet peat, usually on the vertical sides of peat-cuttings and drainage ditches; rarely on inorganic, acid soils by streams. Indifferent to altitude.
Rare; scattered. Fertile.
6 Loch an Tòrr; Dervaig road 2 km west of Tobermory. **7** Above Torloisk. **9** Loch Assopol. **12** Salen Forest, in conifer plantation. **14** Fishnish Bay, by forestry road.

D. heteromalla (Hedw.) Schimp var. **heteromalla**
Calcifuge; in tufts or small patches on mildly to strongly acid soils and porous rocks in damp but not wet situations, mainly in shade. Mainly lowland but ascending to at least 350 m.
Common and widespread in the lowlands; fertile.
1, 3–7, 9–14, 16, 17

D. heteromalla var. **interrupta** (Hedw.) B.S.G.
A single sterile record from **15** Ben Buie in crevice of basaltic rocks at 700 m.
This variety is not included in Warburg (1963) but is very distinctive in appearance, resembling in the field tall forms of *Dicranum falcatum*. The Mull specimen is identical with specimens in Schimper's herbarium.

Rhabdoweisia B.S.G.

R. fugax (Hedw.) B.S.G.
Rupicolous and somewhat humicolous: on ledges and in crevices of basaltic rocks, usually with thin, organic detritus in shade. Mainly subalpine but descending to 50 m alt.
Rare and local; fertile.
4 Ulva, Ben Chreagach. **6** Quinish, Loch a' Chumhainn. **11** ? Ben nan Lus, 1932, *J. Duncan & Young*.

R. denticulata (Brid.) B.S.G.
Rupicolous; in situations ecologically similar to *R. fugax* but more robust and much more frequent; on rather dry humus in deep shade (i.e., constantly high atmospheric humidity) in woodlands, screes and ravines.
Occasional; widespread. Fertile.
3 Tòrr Mòr. **4** Ulva, wood near House. **8** Tiroran. **11** Abhainn Dhiseig, *Corley*; Scarisdale River, 600 m. **13** Aros Park. **16** Loch a' Ghleannain, *Corley*. **17** Cruach Ardura.

R. crenulata (Mitt.) Jameson
Saxicolous; calcifuge; somewhat hygrophytic: resembling the preceding species but more luxuriant and in somewhat wetter but otherwise similar habitats. Mainly subalpine.
Rare; usually fruiting but sometimes found sterile (unlike *R. denticulata*).
7 Achleck, Allt a' Mhuilinn. **8** Ardmeanach, Allt na Teangaidh. **11** Abhainn Dhiseig; Ben More, in the north corrie. **15** Lussa River valley, below Sgurr Dearg; Gortembuie, Glen Cannel, *Ratcliffe*, 1961.

Dichodontium Schimp.

D. pellucidum (Hedw.) Schimp. var. **pellucidum**
Basiphilous; in low rather dense tufts on irrigated rocks or thin alluvium overlying rocks mainly by streams. Indifferent to altitude.
Frequent and widespread in the lowlands; edaphically restricted in the mountains. Fruit rather rare.
2, 3, 5–8, 10–17

D. pellucidum var. **fagimontanum** (Brid.) Schimp.
Only seen, sterile, in irrigated turf in basic grassland near the sea at **5** Calgary Bay.

D. pellucidum var. **flavescens** (With.) Husn.
Basiphilous but less markedly so than the type variety; in larger tufts on streamside alluvium and on wet rock ledges, frequently in the spray from waterfalls.
Rather frequent; especially in the subalpine zones and more frequent than var. *pellucidum* above 500 m. Occasionally fruiting.
1–17

Dicranoweisia Lindb.

D. cirrata (Hedw.) Lindb.
In small green, rarely blackish cushions on trunks and branches of trees, stumps, logs, and sometimes on timber posts and planks, not rarely on siliceous rocks and walls where it may be darker in colour and resemble *D. crispula*; most abundant in woods or similarly sheltered areas, but also found in fairly exposed places. Mainly lowland and lower subalpine, ascending to about 500 m.
Common and widespread in the lowlands; fertile.
3–17

D. crispula (Hedw.) Lindb.
In small, neat, dark green cushions on basaltic rocks in rather exposed situations, usually on rock surfaces not far above vegetation or water, but apparently intolerant of continuous shade. Mainly subalpine.
Rare but locally frequent; fruit uncommon.
4 Beinn Chreagach. **6** 'S Airde Beinn. **8** Craig a' Ghaill. **10** Glen Byre. **11** Ben More, base of north corrie.

Arctoa B.S.G.

A. fulvella (Dicks.) B.S.G.
Alpine and thermophobic: in low, pure or mixed tufts on damp basalt crags. Above 800 m.
Rare and local; confined to Ben More massif. Fertile.
11 Ben More, 1932, *J. Duncan & Young*; main corrie, *Corley*; Beinn nan Gobhar, *U. Duncan.*

Dicranum Hedw.

D. falcatum Hedw.
In low, rather dense tufts among basaltic rocks in some shade or with a northerly aspect. Mainly alpine, descending locally to 600 m.
Rare and very local; but frequent near the summits of the highest mountains. Fertile.
11 Ben More, in the north corrie; Beinn Fhada, north facing crags. **16** Ben Buie.

D. starkei Web. & Mohr.
In similar situations to, and often associated with *D. falcatum*, on moist basaltic rocks and screes on north-facing slopes and crags; alpine.
Rare; and the only record not refound. Fertile.
11 Ben More, 1932, *J. Duncan & Young.*

D. fuscescens Sm.
Calcifuge; very variable in phenotype and in choice of habitat; on dry, peaty soils, screes, siliceous rocks, logs and epiphytic on fissured bark of some trees. Indifferent to altitude, and a constituent of high-altitude *Carex bigelowii* associations.
Frequent at lower altitudes; common in the alpine and higher subalpine zones. Low-altitude forms frequently fertile, high-altitude ones rather rarely fruiting.
1–17

D. majus Sm.
Calcifuge; humicolous, in tall bright green tufts on acid, humus rich soils among boulders, on the ground in woodland, on rotting wood, preferring shade; mainly lowland and lower subalpine, ascending to at least 450 m.
Abundant; fertile.
1–17

D. bonjeanii De Not.
Basiphilous; in pale green, soft tufts in basic grassland, marginal zones of eutrophic mires and occasional in open scrub: mainly lowland.
Occasional; but locally in some quantity. Sterile.
1 Bac Mòr. **2** Iona, *M. Dalby & U. Duncan.* **3** Loch Poit na h-I. **5** Calgary Bay. **6** Loch Mingary, *U. Duncan.* **6** Tobermory River, near the Dervaig road. **8** Balmeanach, *U. Duncan*; Kilfinichen, *U. Duncan.* **11** Loch Bà, Coille na Sroine. **14** Fishnish Bay.

D. scoparium Hedw.
Slightly calcifuge; a very variable species growing on a wide variety of substrates and in an equally wide range of habitats; in grasslands, heaths, drier areas in upland blanket-bogs, on humus in woodlands, a prominent member of the mixed bryophyte communities in screes, dwarf-shrub associations; substrates include acidic mineral soils, drained peat, rotting wood and tree-trunks. Indifferent to altitude.
Abundant and widespread; fertile.
1–17
The straight-leaved ochraceous form of upland dry blanket peat is referable to var. **spadiceum** sensu Dixon (1924). This form, abundant in similar habitats in N England, is somewhat local on Mull, but occurs in quantity on the hills above Kilninian.

Dicranodontium B.S.G.

D. uncinatum (Harv.) Jaeg.
Calcifuge; only seen on damp rocks on north-facing cliff of **8** The Wilderness just below 300 m. A single sterile record.

D. asperulum (Mitt.) Broth.
Calcifuge; a single sterile record on damp rock-ledge in the alpine zone of **11** Ben More, north-facing rock ledge in the north corrie.

D. denudatum (Brid.) Britt. var. **denudatum**
Calcifuge; in tufts on wet rocks, usually with deposits of acid humus, occasionally on stumps. Alpine and subalpine.
Rare; sterile.
7 Achleck, Allt Loch a' Ghael. **16** Cruach Ardura, *Corley*; Loch Spelve, near Dalnaha. Without exact locality, 1932, *J. Duncan & Young*.

D. denudatum var. **alpinum** (Schimp.) Hagen
On thin peaty soil in short turf in the higher subalpine zone of **11** Ben More, on exposed hillside, on west side, 100 m. Sterile.

Campylopus Brid.

C. subulatus Schimp.
Ruderal on sandy, gravelly detritus by roads and other waste ground areas. Lowland.
Rare; possibly introduced. Sterile.
12 Salen; Allt na Sermoin, by old road, *U. Duncan*.
The closely related taxon *C. schimperi* Milde is somewhat less continental in distribution than *C. subulatus* but is considered to be merely a variety of the latter by many European authors. *C. schimperi* has not been recorded, but would not be unexpected on Mull in, for example, mountain-ridge detritus.

C. schwarzii Schimp.
Calcifuge; in silky, golden tufts on rather dry, amorphous peat and compacted humus-rich siliceous soils in exposed situations. Alpine and higher subalpine, rarely below 500 m.
Rare; sterile.
7 Cnoc an dà Chinn. **8** Maol Mheadhonach. **11** Ben More, 1932, *J. Duncan*; western slopes. **15** Sgùrr Dearg, *Corley*. **16** Cruach Ardura.

C. fragilis (Brid.) B.S.G.
Calcifuge; in small, low bright-green tufts on dry peat, peaty sand or gravel, tree-stumps and on tree-roots. From sea-level to the mountain summits.
Common and widespread; fruit rare.
1–17

C. pyriformis (Schultz) Brid.
Calcifuge; in low, yellowish tufts on oxidised peat under *Calluna* and *Pteridium*; occasionally a colonist of burnt ground and recently disturbed peat in relatively dry situations. Lowland.
Occasional, locally in some quantity. Fertile.
2, 3, 5–7, 9, 12–14, 16–17

C. flexuosus (Hedw.) Brid.
Strictly calcifuge; shade-tolerant, almost confined to organic substrates, occasionally on sandy soils with high organic content; heaths and bogs, rotting stump and logs, and humus among rocks. Indifferent to altitude.
Common and widespread; frequently fertile.
1–17
C. flexuosus is notoriously variable. The varieties listed in Warburg (1963) are the more marked forms with a degree of stability of phenotype; even these intergrade to some degree and are not recorded separately here.

C. shawii Wils. ex Braithw.
Calcifuge; in rather loose tufts (with a *Dicranum*-like appearance) on peat in low altitude bogs.
Rare; sterile.
3 Kintra. **11** Glen More, 1 km east of Ulavalt. **13** Caol Lochan.

C. setifolius Wils.
Calcifuge; in large tufts on irrigated basaltic rocks.
Rare and local; sterile.
7 Loch Frisa, Coille na Dubh Leitre. **11** Ben More, 1932, *J. Duncan & Young*; Loch Bà, Doire Daraich. **16** Loch Spelve, ravine above Dalnaha.

C. atrovirens De Not.
Calcifuge; in dark green or black hummocks on wet siliceous rocks or in mats on shallow, wet, amorphous peat overlying basaltic slabs. Indifferent to altitude.
Abundant and widespread; always sterile.
1–17
C. atrovirens is stable in its characters over most of the island, but muticous forms are by no means rare, especially in exposed areas along the west coast (including the Treshnish Islands). The var. **falcatus** Braithw. was collected on **11** Ben More, in Gleann Dubh, 1970, *U. Duncan*.

C. introflexus (Hedw.) Brid.
Calcifuge; well-drained peat in acid sandy soils in open situations; invasive on peat thrown up in ditching operations. Lowland and lower subalpine.
Occasional to locally frequent; probably increasing. Fertile.
6 Mishnish. **7** Druim na Cille, above Kilninian. **8** E side of Maol Mheadonach. *U. Duncan*. **9** Ard Fada. **10** Ulavalt. **11** Gleann Dubh, *U. Duncan*. **12** Salen Forest.
Investigations by Gradstein et al. (pers. comm.) suggest that two, possibly subspecifically distinct taxa are represented in Britain under this name. That referred to in Dixon's Handbook behaves as a Mediterranean type in Europe, is more or less confined to the southern

part of Britain, and is usually sterile. The Mull form, which is the form rapidly extending its range in Britain and which is usually fertile, is thought to be an introduction from the southern hemisphere (probably New Zealand). If so, then the Mull plant must be regarded as a recent invader.

C. brevipilus B.S.G.
Calcifuge; in low, compact tufts on dry humus soils in exposed situations.
Rare; sterile.
1 Staffa; Bac Mòr. **4** Ulva, cliffs above the south-west coast. **8** The Wilderness, on summit of western cliffs. **11** Gleann Dubh, *U. Duncan.* **16** Laggan Deer Forest, gully above south coast.

LEUCOBRYACEAE

Leucobryum Hampe

L. glaucum (Hedw.) Schimp.
Calcifuge (but occurs in basic areas where there is local or surface leaching); in small compact cushions on *Calluna*- and grass-heaths in open and shaded situations and in larger hummocks in bogs; locally common in *Betula* woods. Ascends to at least 600 m.
Frequent; widely distributed. Always sterile.
1–17

ENCALYPTACEAE

Encalypta Hedw.

E. vulgaris Hedw.
Basiphilous; in small tufts on calcareous rocks, on mortar of walls and rock outcrops near the sea. Exclusively lowland.
Rare; fertile.
2 Iona, *M. Dalby & U. Duncan.* **3** Bunnesan. **7** Torloisk. **10** Carsaig Bay, *M. Dalby & U. Duncan.*

E. streptocarpa Hedw.
Calcicolous; on skeletal soils and in small, flat patches on limestone mortar of walls and calcareous detritus on coastal rocks; occasionally found on more basic veins in the basalt. Indifferent to altitude.
Occasional, but locally frequent on mortar and near the sea; rare inland. Sterile.
2, 3, 5–8, 11–17

POTTIACEAE

Tortula Hedw.

T. ruralis (Hedw.) Crome
Basiphilous; in low tufts or colonies on skeletal or sandy soils in short turf in open situations. Lowland.
Rare; in scattered localities near the sea. Fruit not seen.
3 Fidden; Uisken. **6** Dervaig. **12** Gruline. **13** Near Tobermory, *G. Ross et al.* (1881).

T. ruraliformis (Besch.) Rich. & Wall.
Usually basiphilous, photophilic; confined to the coast, a colonist and sand-binding on dunes and sandy rock-ledges by the sea.
Occasional and local; sterile.
2 Iona, 1932, *Trotter*. **3** Fidden; Erraid; Uisken. **5** Calgary Bay. **10** Carsaig. **16** Loch Buie.
An important pioneer species in the early sequence of sand-dune fixation.

T. intermedia (Brid.) Berk.
Usually basiphilous; photophilic; on basic rocks and mortar, becoming less restricted to basic rocks near the sea. Mainly lowland.
Occasional; fertile.
1 Staffa, basalt above Fingal's Cave. **3** Fidden. **6** Dervaig. **8** Balmeanach. **10** Glen Byre. **11** Knock, roof of outbuilding. **13** Tobermory, *G. Ross et al.* (1881); Aros Castle, 1932, *J. Duncan & Young.*

T. laevipila (Brid.) Schwaegr.
On trunks and lower branches of trees and large shrubs, mainly in long established plantations.
Rare; possibly introduced. Fertile.
8 Gribun, Balnahard, *Corley.* **12** Gruline. **15** Loch Buie, at edge of wooded estate. **17** Torosay Castle.

T. virescens (De Not.) De Not.
In low tufts on bark of trees. Only record on *Sambucus* at **10** Pennyghael, by House, *U. Duncan.*

T. papillosa Wils. ex Spruce
A single sterile record. On trunk of large *Acer pseudoplatanus* at **17** Grass Point, Auchnacraig, 1967, *Kenneth & Stirling.*
This and the two preceding species seem to be confined to 'artificial woodland' i.e., estates or similar plantations, being unrecorded from natural woodlands. Possibly, therefore, they are introductions, brought in with young trees from various parts of the mainland.

T. subulata Hedw.
Basiphilous; on mortar of old walls, rarely among coastal rocks. Lowland.
Occasional; except locally in the environs of towns and other buildings. Fertile.
6 Dervaig. **7** Lagganulva. **10** Carsaig Bay. **12** Salen. **13** Tobermory and Aros Park. **16** Loch Buie.

T. muralis Hedw.
Low tufts or more commonly gregarious on rocks and walls. Lowland.
Frequent around buildings; elsewhere rare, fertile.
2–5, 7, 10–14, 16, 17
This is a polymorphic species, of which at least two forms occur on Mull. That in natural habitats is a more robust type with long setae and larger leaves. The common form on walls has not been found in natural habitats on the Island.

Pottia Fürnr.

P. lanceolata (Hedw.) C. Müll.
Found once as an ephemeral colonist of bare soil in garden at **6** Dervaig.
This species is very local in W Scotland.

P. heimii (Hedw.) Fürnr.
In low tufts in open turf on sea-cliffs and machair; frequent but in small quantity in vegetation on the upper zones of saltmarshes.
Occasional; confined to the coast. Fertile.
1 Bac Mòr; Fladda; Staffa. **2** Iona, *U. Duncan.* **3** Uisken, *Lobley & U. Duncan.* **5** Calgary Bay, *Corley.* **6** Dervaig. **9** Ardchrishnish. **12** Salen. **16** Lord Lovat's Cave.

P. intermedia (Turn.) Fürnr.
Ruderal: in transient colonies on bare soil in fields and gardens; occasionally in the upper reaches of salt flats (but less halotolerant than *P. heimii*). Lowland.
Rare; only in the E, perhaps under-recorded. Fertile.
13 Tobermory, in bunker of golf course. **17** Saltmarsh near Torosay Castle.

P. truncata (Hedw.) Fürnr.
Ruderal; in small tufts or gregarious on bare soil in gardens, fields and disturbed soil by roads. Lowland.
Occasional; fertile.
2 Nunnery garden. **4** Ulva, north-east side, *U. Duncan.* **6** Dervaig. **11** Knock. **12** Salen. **13** Tobermory, *G. Ross et al.* (1881) **17** Torosay Castle.
With the exception of *P. heimii*, the species of *Pottia* listed above are mainly winter annuals which can be difficult to detect during summer months and are no doubt under-recorded. However, they tend to be somewhat basiphilous, preferring inorganic soils of neutral or basic reaction and are unlikely to be frequent on Mull except locallywhere soils of this type occur.

Phascum Hedw.

P. cuspidatum Hedw.
An ephemeral annual; in similar situations to *Pottia truncata* and commonly associated with that species.
Rare; perhaps under-recorded. Fertile.
4 Ulva, in farmyard. **10** Loch Buie, on edge of oat-field. **12** Salen. **13** Tobermory, in garden; Aros Park. *U. Duncan.*

Acaulon C. Müll.

A. muticum (Hedw.) C. Müll.
Ephemeral in minute colonies on bare, retentive inorganic soil. Lowland.
A single fertile record from **5** Calgary Bay, on mud by track near the sea, among sparse grass.

Cinclidotus P. Beauv.

C. fontinaloides (Hedw.) P. Beauv.
Attached to rocks in shallow, flowing water.
Very rare.
2 Iona, *M. Dalby & U. Duncan.* **13** Tobermory, near mouth of Tobermory river.

Barbula Hedw.

B. convoluta Hedw. var. **convoluta**
In small tufts or patches on inorganic soils, rock-ledges and old walls; ruderal on bare soils and paths, resistant to treading and a colonist of sandy soils by the sea. Mainly lowland.
Frequent; around habitations and not uncommon by the sea, elsewhere rare. Fertile.
2–17

B. convoluta var. **commutata** (Jur.) Husn.
In similar habitats to the type.
Rare.
2 Iona, *M. Dalby & U. Duncan.* **13** Tobermory, *U. Duncan*; Tobermory River. **17** Torosay Castle Woods, *M. Dalby & U. Duncan.*

B. unguiculata Hedw.
Basiphilous; in greenish or yellowish tufts on rock-ledges around the coast, on basic rocks inland, on mortar, in sandy and skeletal soils by paths and a colonist of sand-dunes. Lowland.
Frequent around the coast; common around habitation. Fertile.
1–17

B. revoluta Brid.
Calcicolous, somewhat xerophilous; in small, dense tufts on calcareous detritus on coastal cliffs and on mortar. Exclusively lowland.
Rare; almost confined to mortar. Sterile.
3 Fidden; Bunessan, *Lobley & U. Duncan.* **6** Dervaig, on bridge. **12** Killichronan, *Lobley & U. Duncan*; Salen. **13** Tobermory. **17** Torosay Castle, *M. Dalby & U. Duncan.*

B. hornschuchiana K. F. Schultz
Photophilic and somewhat xerophilous; gregarious on bare, dry sandy soils.
Rare and local; sterile.
2 Near the Abbey, *Lobley & U. Duncan.* **3** Uisken. **5** Calgary Bay, *Corley.* **8** Inch Kenneth.

B. fallax Hedw.
Basiphilous, terricolous or rupicolous on porous rocks; in brownish or yellowish tufts or colonies on basic, inorganic soils; ruderal on sandy soils around habitation and on coastal dunes and machair grasslands and alluvium in open situations; on mortar and dry, basic rocks and thin detritus among coastal exposures. Mainly lowland but altitude limit (c. 400 m) edaphically determined.
Frequent around the coast and environs of habitation; fertile.
1–6, 8–13, 16, 17

B. reflexa (Brid.) Brid.
Calcicolous; on skeletal soils and on basic rocks and calcareous detritus including shell-sand. Indifferent to altitude but mainly lowland.
Rare and local; sterile.

2 Iona, NE coast. **5** Calgary Bay, *M. Dalby & U. Duncan.* **8** The Wilderness.

B. spadicea (Mitt.) Braithw.
Basiphilous; in dull green tufts on damp, basic rocks or alluvium overlying rocks by streams in more or less basic areas.
Very rare; a sterile plant found only once at **8** Balmeanach, 1969, *Lobley & U. Duncan.*

B. rigidula (Hedw.) Milde
Calcicolous; and often xerophilous; in small brownish tufts on mortar, calcareous rocks and coastal exposures.
Occasional but locally frequent; fertile and usually gemmiferous.
2–6, 8, 10–14, 16, 17

B. tophacea (Brid.) Mitt.
Calcicolous; in brown or olivaceous tufts on wet rocks, commonly associated with re-deposited calcium carbonate; occasionally on irrigated coastal rocks and calcareous sand.
Locally frequent around the coast; local inland. Occasionally fruiting.
1–8, 12–17

B. cylindrica (Tayl.) Schimp.
Basiphilous; on soft, basic rocks or, more commonly, on slightly basic skeletal or mineral soils by streams or in woodlands; occasionally ruderal.
Rare; fruit scarce.
3 Kilfinichen, *U. Duncan.* **4** Ulva House. **5** Calgary Bay. **10** Carsaig Bay. **12** Salen, Allt na Searmoin. **13** Tobermory, Aros Park, *U. Duncan.* **17** Craignure Woods; Torosay Castle Woods, *M. Dalby & U. Duncan.*

B. vinealis Brid.
Basiphilous; on basic and coastal rock-exposures.
Rare; in two widely separate localities. Sterile.
3 Fidden. **6** Loch Mingary.

B. ferruginascens Stirt.
Calcicolous; gregarious on porous, basic rocks and calcareous mineral soils. Subalpine or alpine. Only on **8** The Wilderness, on fine talus on cliff-top, with *Saxifraga oppositifolia.* Sterile.

B. recurvirostra (Hedw.) Dix.
Basiphilous; on calcareous rocks, mortar of walls and (a single record) shell sand by the sea. Indifferent to altitude but restricted by acidity of the geology.
Frequent in the lowlands; elsewhere very rare. Fertile.
1–17

Gymnostomum Hedw.

G. aeruginosum Sm.
Basiphilous; in dense, dull green cushions on wet cliffs in ravines, wet sea cliffs and at the foot of some walls.
Occasional in scattered localities; sterile.
8 Balmeanach, *U. Duncan*; Mackinnon's Cave, on small tufa formations. **10** Carsaig Bay, *M. Dalby & U. Duncan.* **12** Near Salen, 1932, *J. Duncan & Young.* **16** Lochbuie.

G. recurvirostrum Hedw.
Basiphilous; deep-green tufts on wet rocks by streams and waterfalls and in irrigated gravels on mountains.
Common and widespread; usually sterile.
1, 4–8, 10–17

G. calcareum Nees & Hornsch.
Calcicolous; in small, dense, bright green tufts on dry calcareous rocks only found above **8** Mackinnon's Cave.

Eucladium B.S.G.

E. verticillatum (With.) B.S.G.
Calcicolous; on wet calcareous rocks, often associated with redeposition of carbonates; usually in deep clefts and underhangs. Mainly lowland as altitude limit edaphically determined.
Local; mainly sterile.
2 Iona, 1938, *Trotter.* **4** Ulva. **8** Balmeanach, *U. Duncan.* **10** Glen Byre; Carsaig, *Kenneth & Stirling*; Loch Buie, *M. Dalby & U. Duncan.* **11** Small cliff west of Scarisdale Wood. **13** Allt nan Torc.

Anoectangium Schwaegr.

A. aestivum (Hedw.) Mitt.
More or less basiphilous; growing in dense cushions on vertical, humid rocks in corries and ravines. Mainly alpine and subalpine in Britain, but descending almost to sea-level on Mull.
Occasional; widespread. Sterile.
4–8, 11–13, 15–17

Tortella (C. Müll.) Limpr.

T. tortuosa (Hedw.) Limpr.
Basiphilous (although less markedly so than in Southern Britain), more or less photophilic; clefts and ledges of rocks, usually associated with basic veins in the hills, less restricted around the coasts. Indifferent to altitude.
Frequent; common around coastal districts. Normally sterile, but fertile observed at Fidden.
1–17

T. flavovirens (Bruch.) Broth.
Anhalophobous; confined to cliffs by the sea not far above H.W.S.T.
Occasional; sterile.
1 Fladda; Bac Mòr; Staffa. **2** Iona, by the jetty, *U. Duncan.* **3** Erraid. **4** Ulva. **5** Treshnish Point. **8** Dishig. **10** Carsaig Bay, *Stirling & Kenneth.* **16** Lord Lovat's Cave.

Trichostomum Bruch

T. tenuirostre (Hook. & Tayl.) Lindb.
Calcifuge; in dark green loose tufts on wet rocks by streams and waterfalls, usually in shade in ravines and wooded hillsides. Indifferent to altitude.
Frequent; widespread. Sterile.
4–8, 10–17

T. hibernicum (Mitt.) Dix.
In field characteristics and ecology more or less identical to the preceding species, but less frequent. Mainly subalpine.
Occasional in scattered localities; sterile.
5 Ensay Burn. **8** Allt na Teangaidh. **11** Gortenbuie, *Ratcliffe*; Abhainn Dhiseig. **15** Sgùrr Dearg. **16** Ben Buie.

T. crispulum Bruch
Calcicolous; in tufts or as scattered short stems on basic rocks, ledges of coastal cliffs and on mortar; among sparse vegetation on skeletal or sandy soils by the sea.
Occasional but locally frequent; rare inland. Not seen in fruit.
1 Staffa. **2** Iona. **3** Fidden. **5** Treshnish Point. **7** Lagganulva. **10** Carsaig Bay, *U. Duncan.* **13** Tobermory, Aros Park, *U. Duncan.* **17** Craignure Woods.

T. brachydontium Bruch var. **brachydontium**
Somewhat basiphilous; in yellowish-green tufts on rock-ledges, walls and compacted inorganic soils; slightly shade-tolerant but preferring open situations, especially near the sea. Mainly lowland.
Frequent; common and locally abundant around the coast. Fertile.
1–17

T. brachydontium var. **littorale** (Mitt.) C. Jens.
In similar situations to var. *brachydontium* but more confined to sheltered positions by streams and near the sea, especially where there are accretions of sand or silt.
Occasional but locally frequent; fruit rather scarce.
1 Fladda. **4** Ulva, near House. **5** Treshnish Point, *M. Dalby & U. Duncan.* **6** Glen Gorm. **8** Mackinnon's Cave. **12** Kellan Wood. **13** Aros House. **16** Allt na Codha.

T. brachydontium var. **cophocarpum** (Schimp.) Rich. & Wall.
Basiphilous; somewhat hygrophytic; a luxuriant variety of the species growing on damp, shaded crevices and ledges in gullies near the sea.
Rare; fertile.
2 Iona, 1938, *Trotter.* **4** Little Colonsay. **5** Caliach Point, *Corley.* **8** Inch Kenneth.

Weissia Hedw.

W. controversa Hedw.
Somewhat calcifuge; gregarious or loosely tufted on steeply inclined inorganic, neutral to acidic fine-grained soils; distinctly skiophilous. Mainly lowland.
Locally frequent; fertile.
1–7, 9, 10, 12–14, 16, 17

W. rutilans (Hedw.) Lindb.
Recorded by G. Ross *et al.* (1881) (as *W. mucronata* B.S.G.) from near Tobermory but the specimen has not been traced.

W. microstoma (Hedw.) C. Müll.
Calcicolous; among other mosses or in patches on bare ground in exposed situations, on soft basic rocks.
Rare and local; fertile.
3 Uisken. **5** Treshnish Point, *M. Dalby & U. Duncan.* **8** Inch Kenneth, by the landing. **10** Carsaig Bay, *U. Duncan.*

Leptodontium Hampe

L. flexifolium (Sm.) Hampe
Calcifuge; in small tufts on porous acidic rocks, on thatch and rarely on peat.
Rare; in scattered localities. Sterile.
3 Fidden. **6** Dervaig. **10** Carsaig Bay, near Innamore Lodge.

GRIMMIACEAE

Grimmia Hedw.

G. maritima Turn.
Halophilous; in small, dense tufts on coastal rocks seldom more than a few metres from the upper tidal limit of the sea.
Common on the coast; virtually absent inland. Fertile.
1–17

G. apocarpa Hedw.
Basiphilous on exposed basic or coastal rocks and mortar. Mainly lowland.
Frequent near the sea and on mortar; occasional elsewhere. Fertile.
1–13, 15, 17

G. conferta Funck
Basiphilous; only recorded from exposed basaltic sea cliff at **5** Treshnish Point. Fertile.

G. stricta Turn.
Basiphilous; in similar situations to *G. apocarpa* but less xerophilous; on basic mountain rocks.
Rare and local; fertile.
8 The Wilderness. **11** Ben More; Beinn nan Gobhar, *U. Duncan.*

G. alpicola Hedw. var. **rivularis** (Brid.) Broth.
Sub-aquatic; in blackish tufts or mats on rocks in and by upland streams.
Rare; fertile.
8 Craig a' Ghaill. **11** Gleann Dubh.

G. laevigata (Brid.) Brid.
In yellowish tufts on exposed coastal rocks.
Rare; sterile.

8 Inch Kenneth, western cliffs. **10** Carsaig Bay, *U. Duncan.*

G. doniana Sm.
In small hoary tufts on exposed basalt above 650 m.
Rare; fertile.
11 Ben More, N corrie; cairn at summit, *Corley.* **15** Ben Buie, near summit.

G. pulvinata (Hedw.) Sm.
In small tufts mainly on horizontal rock surfaces and wall-tops. Mainly lowland.
Frequent around habitation; uncommon elsewhere. Fertile.
2–7, 10–14, 16, 17

G. funalis (Schwaegr.) B.S.G.
Saxicolous; on exposed siliceous rocks.
Rare; sterile.
6 'S Airde Beinn, *U. Duncan.* **8** Maol Mheadonach, *U. Duncan*; Balmeanach, *U. Duncan*; The Wilderness.

G. torquata Hornsch. ex Grev.
On exposed mountain rocks from 300 m upwards.
Rare; sterile
6 'S Airde Beinn, *U. Duncan.* **8** Maol Mheadonach, *U. Duncan.* **17** Craignure, Fraioch Mhoir, *Crundwell & Wallace.*

G. trichophylla Grev.
Somewhat basiphilous; in yellowish tufts on unshaded rocks, mainly near the sea.
Occasional; locally frequent. Fruit rare.
1 Staffa; Fladda. **2** Iona, *M. Dalby & U. Duncan.* **3** Fidden; Bunessan, *Lobley & U. Duncan.* **8** Gribun, *Corley*; Port na Croise, *U. Duncan*; Inch Kenneth. **10** Carsaig Bay, *M. Dalby & U. Duncan.* **12** Tobermory; Aros Glen, Allt na Saille.

G. hartmanii Schimp.
On basaltic rocks among hygrophytic vegetation or near water.
Rare; sterile but gemmiferous.
5 Calgary Bay, on shore, *M. Dalby & U. Duncan.* **6** Loch Frisa. **7** Lagganulva. **14** Glen Forsa.

G. decipiens (Schultz) Lindb. var. **robusta** (Ferg. ex Braithw.) Braithw.
In hoary tufts on exposed rocks which have a southern exposure. Subalpine and lowland.
Occasional; occasionally fruiting.
4 Ulva, south-facing cliffs. **8** The Wilderness; Maol Mheadhonach, *U. Duncan.* **9** Cliffs near Malcolm's Point. **10** Carsaig Bay, *U. Duncan.* **15** Loch Buie coast, *Lobley & U. Duncan.* **16** Laggan Peninsula, south coast.

G. patens (Hedw.) B.S.G.
Calcifuge; in dark tufts or patches (resembling *Rhacomitrium heterostichum* forms) on basalt exposures in the alpine and subalpine zones.
Occasional; widespread. Sterile.
4 Ulva, Beinn Chreagach. **7** Hillside above Kilninian. **8** Balmeanach, *Lobley & U. Duncan*; Maol Mheadonach. **10** Glen Byre. **11** Ben More, on N slopes. **13** Near Tobermory, *G. Ross et al.* (1881). **15** Glen Forsa. **16** Laggan Deer Forest.

Rhacomitrium Brid.

R. ellipticum (Turn.) B.S.G.
Calcifuge: in small tufts on basalt above 400 m.
Occasional but locally frequent; fertile.
6 'S Airde Beinn, *U. Duncan.* **8** Ardmeanach, summit ridge. **10** Beinn na Croise. **11** Ben More, near summit. **12** Beinn nan Lus, 1932, *J. Duncan & Young.* **15** Ben Buie; Laggan Deer Forest, summit ridge. **13** *G. Ross et al.* (1881).

R. aciculare (Hedw.) Brid.
In blackish-green mats on wet or irrigated rocks in and by small streams, by waterfalls and occasionally at margins of lochs. Indifferent to altitude.
Common and widespread; fertile.
3–1

R. aquaticum (Brid.) Brid.
Calcifuge; hygrophytic; shade tolerant; in yellowish tufts or mats on inclined, wet rocks. Indifferent to altitude.
Common, locally abundant; fertile.
1–17

R. fasciculare (Hedw.) Brid.
In yellowish mats on bare rock surfaces. Indifferent to altitude but mostly lower subalpine and lowland.
Frequent; fertile.
2–17

R. heterostichum (Hedw.) Brid. var. **heterostichum**
More or less calcifuge; in very variable, but usually somewhat hoary tufts on siliceous rocks. Indifferent to altitude.
Ubiquitous and common; fertile.
1–17

R. heterostichum var. **alopecurum** (Brid.) Hübn.
In similar habitats to the typical plant, but preferring some shade.
Rare; sterile.
1 Bac Mòr. **6** 'S Airde Beinn. **11** Abhainn Dhiseig, Ben More, 1932, *J. Duncan & Young.*

R. heterostichum var. **gracilescens** B.S.G.
In similar situations to the type, mainly above 200 m.
Common and widespread; occasionally fertile.
This variety is of doubtful taxonomic significance, and on Mull the multiplicity of intergrading forms defy categorization.
2–17

R. microcarpon (Hedw.) Brid.
Calcifuge; basaltic rocks about the summit of the higher mountains.
Rare; recorded only from **11** Ben More. Fertile.

R. canescens (Hedw.) Brid. var. **canescens**
On rocks in open situations. Mainly lowland and lower subalpine.
Frequent; fertile.
1–3, 6–8, 10–16

R. canescens var. **ericoides** (Hedw.) Hampe
Basiphilous; in pale yellowish tufts or extended, sometimes diffuse patches on skeletal, sandy or gravely soils: fine-grained scree on mountains and coastal sands.
Frequent; muticous forms rare. Occasionally fertile.
1–3, 5, 6, 9–12, 14, 16, 17

R. lanuginosum (Hedw.) Brid.
Calcifuge a constant species in the alpine and higher subalpine zones in association with *Alchemilla alpina* and co-dominant over large areas above 800 m; also on hummocks in peat-bogs and coastal rocks; somewhat intolerant of shade. Ubiquitous in montane areas, descending to sea-level.
Common, and abundant in the mountains. Occasionally fertile.
1–17
Over most of its range, *R. lanuginosum* behaves as a typical calcifuge species, but is found in calcareous habitats on occasion, particularly in the south and east of Britain.

FUNARIACEAE

Funaria Hedw.

F. hygrometrica Hedw.
Nitrophilous; in silt and mud, on rocks by streams; also ruderal and a colonist of bare soils and burnt ground; by paths and in fields and gardens.
Frequent by upland streams; very common in the lowlands where it is frequently invasive and transient. Always fertile.
1–17
The upland stream-side plants are probably native and possibly belong to a different race from the invasive form which may be an ancient introduction. The latter is more polymorphic but is consistently longer in the seta than the former.

F. attenuata (Dicks.) Lindb.
Gregarious on inorganic soils by streams and on banks in woodlands. Mainly lowland and lower subalpine.
Frequent; fertile.
2–17

F. fascicularis (Hedw.) Schimp.
Semi-ruderal and probably transient; on bare, inorganic soils. Lowland.
A single record from **2** Iona, by path in the Nunnery gardens. Fertile.

F. obtusa (Hedw.) Lindb.
Probably calcifuge; on peaty soils among heather and in grasslands, particularly on terraced slopes in the hills on rocks and alluvium by streams. Mainly upland and subalpine but descending to sea-level.
Common and widespread; always fertile.
7–17

Physcomitrium Brid.

P. pyriforme (Hedw.) Brid.
In scattered colonies on mud or silt among grass, usually near water. Lowland.
Rare, but probably transient and under-recorded. Fertile.
13 Tobermory, *G. Ross et al.* (1881). **17** Duart Bay, at upper level of salting.

EPHEMERACEAE

Ephemerum Hampe.

E. serratum (Hedw.) Hampe
Ephemeral; on mud between stones. Lowland.
Very rare; but transient and probably under-recorded. Fertile.
6 Dervaig, near road-bridge.

OEDIPODIACEAE

Oedipodium Schwaegr.

O. griffithianum (Dicks.) Schwaegr.
On damp ledges and inclined rocks with fine humus in the alpine zone in sheltered places in north-facing corries and crags; descends to 750 m.
Very local; gemmiferous and occasionally fertile.
11 Cruachan Beag, *Corley*; Ben More, at summit of north corrie, *U. Duncan.*

SPLACHNACEAE

Tetraplodon B.S.G.

T. mnioides (Hedw.) B.S.G.
Coprophilous, on weathered droppings of ungulates. Mainly subalpine and alpine.
Frequent; fertile.
3, 4, 6–11, 15, 16

Splachnum Hedw.

S. sphaericum Hedw.
Coprophilous, on dung of ungulates on wet moorland. Mainly at high altitudes but descending to 100 m.
Frequent; widespread. Fertile.
4, 7, 8, 10, 11, 15, 16

S. ampullaceum Hedw.
Coprophilous, on droppings of ungulates; indifferent to altitude, and the most frequent of the coprophilous mosses in the lowlands.
Common; widespread. Fertile.
1–17

BRYACEAE

Orthodontium Schwaegr.

O. lineare Schwaegr.
Only seen on base of rotting fence-post and by coniferous plantations.
Introduced; fertile.
12 Salen Hotel. **14** Fishnish. **17** Craignure, *Lobley & U. Duncan.*
Possibly introduced.

Pohlia Hedw.

P. acuminata Hornsch.
In rock crevices and fine scree in gully. Alpine. Very rare; fertile (see note under the next species).
11 Ben More, north corrie; Beinn nan Gabhar.

P. polymorpha Hornsch.
On skeletal soils and among rocks about mountain summits.
Very rare; fertile.
16 Ben Buie summit.
This taxon is very closely related to *P. acuminata*, which is considered by some authors to be merely an autoicous state. Collections which did not show both male and female organs have been omitted.

P. elongata Hedw.
Calcifuge; on humus-rich skeletal soils; occasionally on rotting wood; mainly on mountain slopes and stream-banks.
Frequent; fertile.
4–8, 11, 12, 15, 16

P. cruda (Hedw.) Lindb.
Basiphilous; in somewhat glaucous tufts in rock-crevices of basic strata inland and coastal rocks with basic accretions.
Frequent and widespread; frequently fertile.
1–3, 5–11, 13, 15, 16

P. nutans (Hedw.) Lindb.
Calcifuge; on organic substrates including peat, rotting wood, humus among rocks, leached sandy soils; in bogs, heaths and woodlands. At all altitudes.
1–17
Common and widespread; fertile.
The form most commonly met with on wet peat is var. **longiseta** (Brid.) B.S.G. and is of doubtful taxonomic status.

P. rothii (Correns) Broth.
Basiphilous; on open retentive gravelly or sandy soils by roads and other disturbed sites. Lowland.
Rare; in scattered localities. Sterile but gemmiferous.
12 Salen. **14** Fishnish.

P. camptotrachela (Ren. & Card.) Broth.
P. annotina (Hedw.) Loeske
Often ruderal; on moist, inorganic soils by roads, cuttings and by montane streams and detritus. Mainly lowland and lower subalpine.
Frequent to locally common; sterile, gemmiferous.
2, 4–17

P. proligera (Lindb.) Limpr. ex Arnell
Calcifuge; on skeletal gritty or sandy soils, in similar habitats to *P. camptotrachela* and sometimes mixed with it. Probably indifferent to altitude, but established records are lowland.
Rare; probably under-recorded. Sterile; gemmiferous.
3 Fionnphort, *U. Duncan.* **11** Gruline Bridge, *U. Duncan.*

P. wahlenbergii (Web. & Mohr) Andr.
Forming whitish colonies on wet earth or eutrophic peat by streams and in mires; occasionally ruderal on retentive wet soils in fields; perhaps nitrophilous. Indifferent to altitude.
Common and widespread; rarely fertile.
1–17

P. delicatula (Hedw.) Grout
In similar situations to the preceding species but preferring inorganic substrates. Lowland.
Scattered localities around the coast; sterile.
2 Iona, *Trotter.* **4** Ulva, near House. **6** Dervaig, in ditch. **7** Torloisk. **10** Carsaig Bay. **12** Salen. **14** Fishnish, in Forestry Commission plantation.

Plagiobryum Lindb.

P. zieri (Hedw.) Lindb.
Calcicolous; on damp, calcareous rocks, and basic veins in basalt in gullies. Descending to 200 m.
Rare and local; occasionally fertile.
8 The Wilderness; below Creag a' Ghaill; Balmeanach, *Lobley & U. Duncan.* **11** Gleann na Beinne Fada. **15** Sgùrr Dearg, *Corley.*

Anomobryum Schimp.

A. filiforme (Dicks.) Solms-Laub.
Basiphilous; on flushed rock ledges and crevices; basic rocks in fine spray from waterfalls. Mainly subalpine but not rare at sea-level.
Frequent; occasionally fertile.
4–8, 10–17

Bryum Hedw.

B. pendulum (Hornsch.) Schimp.
Basiphilous; in tufts on basic or coastal rocks and mortar; also on calcareous sandy soil by the sea. Lowland.
Occasional; locally frequent around townships. Fertile.
5 Calgary Bay. **6** Dervaig. **13** Tobermory. **12** Salen. **14** Fishnish Bay.

B. inclinatum (Brid.) Bland.
In similar situations to *B. pendulum* but more frequent, especially on skeletal sandy soils.
Occasional; locally frequent on the coast. Fertile.

2 Iona, in machair. **3** Fidden; Erraid. **5** Calgary Bay, *Corley*; *M. Dalby & U. Duncan.* **6** Quinish, on wall; 'S Airde Beinn, *U. Duncan.* **7** Cnoc nan Uan, ruderal, *Lobley & U. Duncan.* **8** Tavool House. **12** Salen. **13** Tobermory. **16** Loch Buie, eastern end.

B. pallens Sw.
Basiphilous; very variable, greenish to bright red small or luxuriant tufts on streamside alluvium, flushed coastal sands, ledges of wet rocks. Indifferent to altitude.
Frequent on basic substrates; rare elsewhere. Fruit uncommon.
3–8, 10–13, 15, 16

B. weigelii Spreng.
In pinkish, soft tufts in high-altitude spring of flush vegetation.
Rare; recorded once in **11** Ben More, north corrie, 1932, *J. Duncan & Young*. Sterile.

B. pseudotriquetrum (Hedw.) Schwaegr. var. **pseudotriquetrum**
In mires, springs, streamsides, margins of lochs and saltings. Indifferent to altitude.
Common and widespread; fertile but fruit sparse.
1–17

B. pseudotriquetrum var. **bimum** (Brid.) Lilj.
Basiphilous; in mesotrophic flushes in the subalpine zone.
Rare; possibly under-recorded. Fertile.
8 below Creag a' Ghaill.

B. intermedium (Brid.) Bland.
Basiphilous; on mortar and in turf on shell sand by the sea. Lowland.
Rare; fertile.
3 Uisken. **17** Craignure.

B. caespiticium Hedw.
More or less calcifuge; gritty soil by roads, occasionally on walls (not on mortar) and open heaths. Lowland, ascending to at least 350 m.
Occasional; widely distributed. Occasionally fertile.
5 Calgary, roadside verge. **6** Dervaig, churchyard. **9** Ard Fada. **10** Pennyghael, on the old boathouse. **12** Salen.

B. argenteum Hedw.
Nitrophilous; ruderal and adventive; on nitrogen-enriched soils on sea-cliffs, paths and walls around buildings; fields and saltmarshes on acid and basic soils.
Common in the lowlands and abundant near habitation; in the mountains seen only around triangulation points or similarly artificial habitats. Occasionally fertile.
1–17

B. bicolor Dicks.
Nitrophilous; almost identical in its ecological requirements to *B. argenteum* and commonly associated with it but less abundant and more restricted to anthropogenic habitats.
Frequent in the lowlands; rare elsewhere. Usually sterile; gemmiferous.
2, 3, 5–7, 11–14, 17
Non-fruiting plants of the **Bryum erythrocarpum** complex are extremely inconspicuous in the field. These listed here are certainly under-recorded. Furthermore, a specialist's eye would no doubt be able to detect species in addition to those recorded, e.g., *B. violaceum* Crundw. & Nyh., *B. klinggraeffii* Schimp.

B. ruderale Crundw. & Nyh.
More or less basiphilous, on neutral or basic soils; semi-ruderal, mainly around areas of habitation or other disturbance. Lowland.
Rare; probably under-recorded.
2 Iona, Nunnery garden, *Lobley & U. Duncan.*
6 Dervaig, edge of path to oatfield. **12** Salen.

B. radiculosum Brid.
In short turf on coastal sand.
Only recorded from single locality. Sterile; bulbiferous.
3 Fidden.

B. tenuisetum Limpr.
Calcifuge; adventive on disturbed peaty soil by road.
Rare; introduced?
14 Fishnish, Bailemeanach, site of recent tree-felling.

B. micro-erythrocarpum C. Müll & Kindb.
Calcifuge (fide Crundwell & Nyholm, 1964); terricolous and more or less ruderal.
Rare; probably under-recorded. Sterile; bulbiferous.
17 Torosay Castle Woods, by footpath.

B. bornholmense Winkelm & Ruthe
Calcifuge; on humus-rich soils; more or less adventive. Lowland and lower subalpine.
Rare; or under-recorded.
8 Maol Mheadhonach, E side, *U. Duncan.* **14** Fishnish Bay.

B. rubens Mitt.
Slightly basiphilous; more or less ruderal species of bare or recently disturbed inorganic soils. Lowland.
Probably frequent, but under-recorded. Occasionally fruiting.
2 Iona, Nunnery garden. **4** Farmyard near Ulva House. **6** Dervaig, by road. **8** Kilfinichen, *Lobley & U. Duncan.* **12** Salen. **13** Tobermory.
14 Fishnish, Doire Dorch, *Lobley & U. Duncan.*
17 Craignure.

B. muehlenbeckii B.S.G.
Among wet rocks at high altitudes.
Rare; fruit not seen.
11 Ben More, near summit, 1932, *J. Duncan & Young.*

B. alpinum With.
More or less calcifuge; in vinous-red, rarely green, tufts on wet stony detritus, silt and wet rocks by streams.
Alpine and subalpine but commonly descending to sea-level.
Common and widespread; locally abundant in the alpine and subalpine zones. Sterile.
1–17
Most green specimens collected on Mull were from uncharacteristically shaded sites, and do not belong to var. **viride** Husn., which has been recorded once only: **4** Ulva, 1966, *U. Duncan.*

B. capillare Hedw.
Basiphilous; on basic rocks inland, on coastal rocks, and on mortar of walls. Indifferent to altitude but much more frequent in the lowlands, especially on mortar.
Common in the lowlands; elsewhere local. Fertile.
Most or all of the Mull material of this species seems to belong to *B. capillare* s.s., but the aggregate is at present under revision.
1–17

B. riparium Hagen
Damp clefts and ledges of basaltic rocks. Subalpine.
Very rare; sterile.
11 Ben More, ridge east of summit. **15** Sgùrr Dhearg; Beinn Bheag, *Corley.*

Rhodobryum (Schimp.) Limpr.

R. roseum (Hedw.) Limpr.
Calcicolous; scattered on in colonies in short-turf on basic well-drained soils (machair).
Rare; sterile.
2 Iona, NE coast. **5** Calgary Bay.

MNIACEAE

Mnium Hedw.

M. hornum Hedw.
On acid to slightly basic (not calcareous) soils; on terraced slopes, rock ledges and roots of trees especially where fine humus accumulates; abundant on wooded slopes. Indifferent to altitude but much more abundant on the drier soils of the lowlands.
Common and widely distributed, locally abundant in lowland woods. Fertile.
1–17

M. marginatum (With.) P. Beauv. var. **marginatum**
Calcicolous; in small colonies on sheltered, basic exposures.
Rare; only found sterile.
8 Below Creag a' Ghaill.

M. stellare Hedw.
Calcicolous, saxicolous; in bluish tufts on ledges of damp basic rocks.
Rare; sterile.
8 Above Mackinnon's Cave; Balmeanach, *Lobley & U. Duncan.* **17** Craignure, on cliff, *Kenneth and A. Stirling.*

M. cuspidatum Hedw.
On walls, rocks and banks in woodlands and similarly sheltered habitats.
Rare; fertile.
5 Calgary Bay, *Lobley & U. Duncan.* **12** Killichronan, Allt na Saille. **13** Tobermory. **17** Craignure Woods.

M. longirostrum Brid.
Creeping on stony ground in woodlands and ravines; frequently on logs and coastal, sandy soils. Mainly lowland in basic-soil areas.
Frequent around the coast; elsewhere only occasional. Fruit rare.
2–8, 10, 12, 13, 15–17

M. affine Bland.
Terricolous or rupicolous; on damp, basic rocks or soil in woodlands and ravines, sometimes in drier areas of mires. Lowland.
Rare.
3 Bunessan. **8** Gribun, *Corley.* **14** Fishnish, Doire Dorch, *Corley.*

M. rugicum Laur.
Basiphilous; among vegetation and hygrophytic mosses in base-rich mires. Lowland.
Very rare; sterile.
16 Loch a' Ghleannain, by outlet stream, *Corley*

M. seligeri (Lindb.) Limpr.
Basiphilous; in patches, usually mixed with other bryophytes in deep, eutrophic springs flushes and mires on highly organic soils. Indifferent to altitude.
Rare; widespread. Not seen in fruit.
2 Iona, *M. Dalby & U. Duncan.* **9** Loch Assopol, at south end. **16** Fontinalis Lochs. **17** Loch a' Ghleannain.

M. undulatum Hedw.
Basiphilous; possibly nitrophilous; on base-rich, moist soils in shade, in woodlands, grasslands, drained mires and among rocks in ravines. Mainly lowland, but ascending into the subalpine and alpine zones where shelter is available.
Common and widely distributed; sexually mature plants commonly seen, but rarely found in fruit.
2–17

M. punctatum Hedw. var. **punctatum**
On firm moist inorganic soils in shade; on rotting wood and silt or clay-covered rocks by streams in gullies. Indifferent to altitude.
Common and widely distributed; fertile.
2–17
Extended, rust-colored patches of the persistent protonemata of this plant are frequently seen forming a felt-like coating on smooth, damp rocks by streams. Small plantlets may

develop on these, but mature plants seldom develop without an accumulation of soil, humus or silt.

M. punctatum var. **elatum** Schimp.
In tomentose tufts in bryophyte-flushes in the subalpine zone.
Rare; in two scattered localities. Fertile.
8 West of Craig a' Ghaill. **11** Ben More, North Corrie.
This species hitherto regarded as a variety of *M. punctatum* and not listed in the *Census Catalogue* strongly resembles *M. pseudopunctatum* in the field. It is perhaps rather more base-demanding than the latter.

M. pseudopunctatum B.S.G.
In wet flushes on irrigated peat among sedges in base-poor mires and occasionally in spongy vegetation around lochans. Mainly alpine and subalpine, descending almost to sea-level on Mull.
Occasional; fertile
2 Iona, near the south end. **4** Ulva, Beinn Chreagach. **8** N slope of Beinn na Sreine. **11** Ben More, at foot of the north corrie. **13** Caol Lochan. **15** Sgùrr Dearg, S slopes. **17** Craignure, below cliffs.

AULACOMNIACEAE

Aulacomnium Schwaegr.

A. palustre (Hedw.) Schwaegr.
Calcifuge; in tufts or scattered stems among *Sphagnum* or other mosses in bogs, oligotrophic flushes and fens. At all altitudes.
Common and widespread; occasionally fruiting.
1–17

A. androgynum (Hedw.) Schwaegr.
Lignicolous; in small tufts on rather dry rotting logs and stumps in lowland woods.
Rare; sterile, gemmiferous.
11 Scarisdale River gully, *U. Duncan.* **13** Tobermory, woodland below the golf course. **14** Fishnish; burnt log in Torosay plantation.

BARTRAMIACEAE

Bartramia Hedw.

B. hallerana Hedw.
Basiphilous; in soft tufts on more basic bands in basaltic exposures; in ravines and steep, wooded slopes. Mainly alpine and subalpine, descending into the lowlands locally.
Rare; fertile.
8 The Wilderness. **11** Allt na Beinne Fada; Ben nan Gabhar. **17** Craignure, cliffs above the pier; Fraioch Mhoir, *Crundwell & Wallace.*

B. pomiformis Hedw.
On damp but not wet clefts and underhangs of rocks in shaded situations. Indifferent to altitude.
Frequent; except in the granitic areas. Fertile.
1, 4–8, 10–17

B. ithyphylla Brid.
In identical situations to *B. pomiformis* with which it is often associated, although less common.
Frequent; fertile.
1, 4–8, 10–17

Conostomum Sw.

C. tetragonum (Hedw.) Lindb.
In small bright-green tufts on thin, peaty soil on basaltic rocks near the summits of the higher mountains.
Rare; fertile.
11 Ben More, summit of the north corrie.

Philonotis Brid.

P. fontana (Hedw.) Brid. var. **fontana**
In whitish tufts in a variety of wet habitats other than extremely calcareous mires or ombrogenous bogs; by streams, acid flushes, oligotrophic mires. Indifferent to altitude.
Common; locally abundant. Fertile.
1–17

P. fontana var. **adpressa** (Ferg.) Limpr.
In similar situations to var. *fontana* but mainly in alpine and subalpine flushes.
Frequent; fruit uncommon.
4, 6–8, 10–13, 15, 16

P. capillaris Lindb.
On damp soil in sheltered situations. Lowlands.
Rare; not found fertile.
4 Ulva, N of House, *U. Duncan.*
More than one taxon may be represented in Britain under this name, in which case some northern records may be refereable to *P. arnellii* Husn. The Ulva plant compares well with *P. capillaris* from SW Europe.

P. calcarea (B.S.G.) Schimp.
Basiphilous; in compact tufts in base-rich, flushed grasslands and eutrophic mires. Indifferent to altitude.
Occasional; local. Fertile.
2 Iona, *M. Dalby & U. Duncan.* **3** Loch Poit na h-I. **8** The Wilderness; Creag a' Ghaill. **10** Carsaig Bay, *U. Duncan.* **11** Gleann na Beinn Fada.

Breutelia Schimp.

B. chrysocoma (Hedw.) Lindb.
Basiphilous; in hummocks or straggling among vegetation on flushed terraces, by streams and among wet rocks. Indifferent to altitude.
Common; locally abundant. Rarely fruiting.
1–17

PTYCHOMITRIACEAE

Campylostelium B.S.G.

C. saxicola (Web. & Mohr) B.S.G.
Calcifuge; on shaded or overhung damp rocks in wooded valleys or ravines; in clumps or as scattered stems.
Rare; fertile.
12 Allt na Searmoin, near Salen. **13** Aros Park.

Ptychomitrium Fürnr.

P. polyphyllum (Sw.) Fürnr.
Saxicolous; mainly on basaltic rocks and particularly on old stone walls. Indifferent to altitude but uncommon above 500 m.
Common and widespread. Fertile.
1–17

Glyphomitrium Brid.

G. daviesii (With.) Brid.
Calcifuge; in small tufts on basalt, locally frequent on stone walls. Mainly lower subalpine.
Occasional; locally frequent. Fertile.
4 Ulva, Beinn Chreagach. **5** Ensay Burn. **6** 'S Àirde Beinn. **7** Lagganulva. **8** Gribun, *Corley*. **10** Carsaig Bay, *U. Duncan*. **11** Scarisdale Wood, *Corley*; Scarisdale River, *U. Duncan*. **12** Gruline. **13** Tobermory, *G. Ross et al.* (1881). **16** Lochbuie. **17** Craignure.

ORTHOTRICHACEAE

Amphidium Schimp.

A. mougeotii (B.S.G.) Schimp.
Calcifuge; yellowish tufts on wet cliffs and ravines. At all altitudes.
Frequent; especially in the hills. Fruit not seen.
1, 4–17

Zygodon Hook. & Tayl.

Z. viridissimus (Dicks.) R. Br. var. **viridissimus**
In small, dense tufts on bark, walls and rocks. Lowland.
Common; fertile.
1–8, 10–17

Z. viridissimus var. **stirtonii** (Schimp. ex Stirt.) Hagen
On coastal rocks and walls, rarely epiphytic.
Frequent; fertile.
1–7, 10–14, 16, 17

Z. conoideus (Dicks.) Hook. & Tayl.
On bark of trees, principally *Fraxinus* and *Sambucus*, in somewhat sheltered areas. Lowland.
Rare; fertile.
4 Ulva, near House. **8** Tiroran. **10** Pennyghael, *U. Duncan*. **12** Salen. **13** Aros House, 1898, *H. N. Dixon*. **14** Fishnish Bay, *Corley*. **16** Lochbuie. **17** Torosay Castle Woods.

Orthotrichum Hedw.

O. rupestre Schleich. ex Schwaegr.
Calcifuge?; on siliceous rocks, occasionally walls, in dry places. Lowland, ascending to about 300 m.
Common around the coasts; decreasing rapidly with altitude. Fertile.
1–5, 7–9, 11–17

O. anomalum Hedw.
Calcicolous; on calcareous rocks and mortar of walls; occasionally on siliceous rocks near the sea. Lowland.
Locally frequent; fertile.
2–13, 15–17

O. cupulatum Brid.
Basiphilous; in similar habitats to *O. anomalum* but much less frequent.
Fertile.
2 Iona, NE side, *M. Dalby & U. Duncan*. **3** Fidden. **5** Calgary. **8** Mackinnon's Cave. **10** Carsaig Bay, *M. Dalby & U. Duncan*. **12** Killichronan, *Lobley & U. Duncan*. **13** Tobermory; Aros Park, *U. Duncan*. **14** Fishnish Bay. **16** Lochbuie. **17** Craignure; Torosay Castle Woods, *M. Dalby & U. Duncan*.

O. affine Brid.
On trunks and branches of shrubs and trees mainly *Sambucus* and *Corylus*, in sheltered situations in woodlands and ravines. Lowland and lower subalpine.
Occasional; fertile.
6 Dervaig, Loch a' Chumhainn. **10** Pennyghael. **11** Coille na Sròine. **12** Killichronan, *Lobley & U. Duncan*. **13** Aros Park, *U. Duncan*. **16** Lochbuie. **17** Torosay Castle Woods, *M. Dalby & U. Duncan*.

O. striatum Hedw.
On branches of shrubs, mainly *Sambucus* and *Corylus* in semi-shaded sites. Lowland.
Occasional; fertile.
7 Woods east of Kilninian. **11** Coille na Sròine. **12** Allt na Searmoin. **16** Loch a' Ghleannain, *Corley*. **17** Torosay Castle, *M. Dalby & U. Duncan*.

O. lyellii Hook. & Tayl.
On trunks of trees in open but sheltered sites, mainly in the east and south-east. Lowland.
Rare; fertile and gemmiferous.
11 Gruline. **13** Aros Park, on *Fraxinus*. **16** Loch Uisg, on *Acer pseudoplatanus*, *U. Duncan*. **17** Torosay Castle.

O. stramineum Hornsch. ex Brid.
Corticolous; on trunks and main branches of large trees (*Fraxinus*, *Acer*, *Quercus*, *Sorbus*) in open woodland and wooded estates. Lowland.
Occasional; fertile.
4 Ulva, near House. **6** N of Dervaig. **8** Kilfinichen, *U. Duncan*. **10** Pennyghael. **11** Scarisdale Wood. **16** Loch Uisg, *U. Duncan*. **17** Auchnacraig, *Kenneth & A. Stirling*.

O. tenellum Bruch ex Brid.
In small tufts on trunks and branches of shrubs mainly *Sambucus* and *Corylus* in sheltered sites. Lowland.
Rare; fertile.
6 Dervaig. **13** Allt nan Torc; Upper Druimfin. **16** Lochbuie, *E. Lobley & U. Duncan*. **17** Craignure Woods, *M. Dalby & U. Duncan*.

O. pulchellum Brunton
In small tufts on stems of *Sambucus* in sheltered habitats. Lowland.
Rare; fertile.
10 Pennyghael, *U. Duncan*. **16** Lochbuie. **17** Torosay Castle.

O. diaphanum Brid.
In tufts or somewhat diffuse, mixed patches on trees, stumps and occasionally walls. Lowland.
Occasional; locally frequent. Fertile.
2–4, 7–14, 16, 17

Ulota Brid.

U. phyllantha Brid.
Anhalophobous; in yellowish tufts on trees and rocks, particularly abundant near the sea and only slightly less tolerant of salt spray than *Grimmia maritima*. Lowland and lower subalpine.
Common and widely distributed; sterile, gemmiferous.
1–17

U. vittata Mitt.
On branches of *Corylus* in very sheltered habitat. Lowland.
Very rare; fertile.
12 Kellan Mill, 1968, *U. Duncan.*

U. crispa (Hedw.) Brid.
Epiphytic on most trees and shrubs where they occur; particularly abundant and luxuriant in lowland ravines, overhanging streams.
Common and widespread; fertile.
2–17

U. crispula Brid.
In similar situations to *U. crispa.*
Rare or overlooked; fertile.
15 An Coire, on *Corylus.*

U. bruchii Hornsch. ex Brid.
In similar habitats to *U. crispa* and frequently mixed with that species. Subalpine and lowland.
Occasional; locally frequent. Fertile.
3 Bunessan, *Lobley & U. Duncan*; Loch Poit na h-I. **4** Ulva, west of House. **10** Pennyghael; Carsaig Bay. **11** Scarisdale Woods. **12** Killichronan, *Lobley & U. Duncan.* **13** Tobermory, *G. Ross et al.* (1881); Allt nan Torc. **15** Lussa River.
The records above are of plants which fit well into the concept of the species. Plants with sporophytes intermediate in structure between this species and *U. crispa* are frequent, and may be the result of hybridisation. Some authorities (cf. Nyholm, 1960) venture the opinion that *U. bruchii* may itself be of hybrid origin between e.g. *U. crispa* and *U. drummondii.*

U. drummondii (Hook. & Grev.) Brid.
Corticolous; on trunks and branches of trees, principally *Sorbus aucuparia* in ravines. Lower subalpine and lowland.
Rare; fertile.
3 Bunessan, *Lobley.* **10** Pennyghael, **11** Scarisdale Wood. **15** Beinn Talaidh, *M. Dalby & U. Duncan.* **16** Cruach Ardura.

U. hutchinsiae (Sm.) Hammar
On basaltic rocks in rather open situations but avoiding extremes of shade or exposure.
Rare; scattered. Fertile.
8 Inch Kenneth, east end. **10** W of Lochbuie, on shore, *M. Dalby & U. Duncan.* **14** Fishnish Bay. **15** Glen Forsa.

FONTINALACEAE

Fontinalis Hedw.

F. antipyretica Hedw. var. **antipyretica**
Basiphilous; in large, submerged skeins attached to rocks in streams, rivers and eutrophic lakes. Lowland and lower subalpine.
Frequent below 200 m; fruit not recorded.
2, 3, 6–17

F. antipyretica var. **gracilis** (Lindb.) Schimp.
Aquatic; attached to stones in high altitude streams with improved cationic content.
Very rare; sterile.
11 Allt Coire nan Gabhar, 500 m.

F. squamosa Hedw.
Attached to rocks in streams. Subalpine.
Very rare; sterile.
10 Glen Byre Burn tributary, 350 m.

CLIMACIACEAE

Climacium Web. & Mohr

C. dendroides (Hedw.) Web. & Mohr
Basiphilous; in damp basic grasslands and drier areas of eutrophic swamps and mires. Indifferent to altitude.
Locally frequent; especially around the coast. Sterile.
1–17

HEDWIGIACEAE

Hedwigia P. Beauv.

H. ciliata (Hedw.) P. Beauv.
In hoary patches on exposed, dry siliceous rocks. At all altitudes.
Common; widely distributed. Fertile.
1–17

H. integrifolia P. Beauv.
In similar habitats to the preceding species but more restricted to areas of higher average atmospheric humidity; mainly on basaltic slabs and boulders in wet heath.
Occasional; fertile.
4 Ulva, Beinn Chreagach. **7** Druim na Cille. **8** Below Creag a' Ghaill; W side of Maol Mheadonach. **11** Ben More, Sròn Daraich, *U. Duncan.* **15** Glen Forsa. **16** Lochbuie, *M. Dalby & U. Duncan*; Loch Spelve, S of the loch.

CRYPHAEACEAE

Cryphaea Mohr

C. heteromalla (Hedw.) Mohr
On bark of small trees (nearly always *Sambucus*) in sheltered areas. Lowland.

Occasional but local; fertile.
5 Kilninian. **6** Quinish House. **10** Pennyghael, *James & U. Duncan.* **12** Gruline. **17** Torosay Castle Woods, *M. Dalby & U. Duncan.*

LEUCODONTACEAE

Leucodon Schwaegr.

L. sciuroides (Hedw.) Schwaegr.
Only recorded once, on *Fraxinus*. Lowland. Sterile.
17 Torosay Castle woods.

Antitrichia Brid.

A. curtipendula (Hedw.) Brid.
On exposed rocks or dry skeletal soils at low to moderate elevations.
Very rare; sterile.
6 'S Àirde Beinn, 1968, *U. Duncan.* **13** Tobermory, *G. Ross et al.* (1881).

Pterogonium Sm.

P. gracile (Hedw.) Sm.
Rupicolous; on shaded, damp rocks in woods, ravines and on old walls small forms on somewhat exposed rocks near the sea. Mainly lowland and lower subalpine, ascending to at least 600 m.
Frequent in the lowlands; decreasing with altitude.
1–17

MYURIACEAE

Myurium Schimp.

M. hebridarum Schimp.
On skeletal soils among rocks just above the shore-line; at least tolerant of sea-spray, but possibly confined to the coast due to intolerance of freezing temperatures.
Rare; sterile.
4 Ulva, NE side, *U. Duncan.* **5** Coastal rocks between Calgary Bay and Caliach Point.

NECKERACEAE

Neckera Hedw.

N. crispa Hedw.
Basiphilous; on skeletal, basic soils; in shade or open positions on limestone, basic bands in basalt and on calcareous detritus. Indifferent to altitude.
Occasional; local. Occasionally fertile.
2 Iona, near the E coast, *M. Dalby & U. Duncan.* **5** Calgary. **8** In gully below Creag a' Ghaill. **10** Carsaig, sea-cliff. **12** Allt na Saille. **13** Aros Park; Tobermory, *G. Ross et al.* (1881); Allt nan Torc. **17** Craignure Woods.

N. pumila Hedw.
Corticolous, rarely saxicolous; on bark of trees in more or less sheltered situations. Lowland.
Rare; not seen in fruit.
8 Balnahard, Gribun. **10** Pennyghael, *James & U. Duncan.* **17** Torosay Castle woods, *M. Dalby & U. Duncan.*

N. complanata (Hedw.) Hüben.
On rocks, usually in somewhat skeletal localities; common on coastal rocks; occasionally around the lower parts of trees and stable terraces of woodland banks. Mainly lowland, ascending to about 300 m.
Common in the lowlands; especially near the sea. Fertile.
1–17

THAMNIACEAE

Thamnium B.S.G.

T. alopecurum (Hedw.) B.S.G.
Luxuriant forms on wet rocks in ravines near streams and waterfalls; also very small forms on rather dry basic and coastal rocks in exposed situations.
Frequent but local; sterile.
1–7, 10–13, 15–17

HOOKERIACEAE

Hookeria Sm.

H. lucens (Hedw.) Sm.
On wet earth in crevices and on ledges of wet rocks in ravines and basic woodlands. Lowland.
Widespread but seldom in quantity; fertile.
4–17

LESKEACEAE

Heterocladium B.S.G.

H. heteropterum (Bruch ex Schwaegr.) B.S.G. var. **heteropterum**
Rupicolous or terrestrial, seldom epiphytic: on rock ledges and crevices, earth and occasionally tree-roots; always in deep shade of woodlands and ravines. Mainly lowland and lower subalpine, but ascending to at least 800 m. alt.
Common and widespread; not seen in fruit.
1–17

H. heteropterum var. **flaccidum** B.S.G.
In similar habitats to var. *heteropterum.*
Rare; sterile.
12 Aros Park, on damp rock in wood, 1968, *James & U. Duncan.*

Anomodon Hook. & Tayl.

A. viticulosus (Hedw.) Hook & Tayl.
Basiphilous; on rocks or mortar of walls. Lowland.
Occasional and local; sterile.
2 Iona, *M. Dalby & U. Duncan.* **7** Laggan Bay, *Lobley & Duncan.* **10** Carsaig Bay, *M. Dalby & U. Duncan.* **14** Tobermory. **16** Loch Buie, on the S side of loch. **17** Torosay Castle.

THUIDIACEAE

Thuidium

T. tamariscinum (Hedw.) B.S.G.
Terricolous; creeping among vegetation in grasslands, woodlands and among rocks, avoiding only extremes of exposure, acidity

and excessively wet areas. Indifferent to altitude but rare in the alpine zone.
Common; locally abundant in the lowlands. Fruit very rare.
1–17

T. delicatulum (Hedw.) Mitt.
Basiphilous; in calcareous grasslands, eutrophic mires and on damp, basic rocks.
Occasional; fruit rare.
2 Iona, W coast. **3** Erraid. **5** Calgary Bay, *Corley*. **6** Dervaig, *U. Duncan*. **13** Aros House, 1898, *H. N. Dixon*. **16** An Coire, *Corley*. **17** Torosay (fruiting); Craignure, cliff above pier.

T. philibertii Limpr.
Calcicolous, rupicolous or terrestrial; on limestone rocks and in calcareous, coastal grassland.
Rare; fruit not recorded.
8 Below Creag a' Ghaill. **10** Carsaig Bay, *M. Dalby & U. Duncan*. **17** Craignure Woods.

AMBLYSTEGIACEAE

Cratoneuron (Sull.) Spruce

C. filicinum (Hedw.) Spruce
In deep carpets or mixed with other mosses in mesotrophic springs and mires: also on wet rocks and alluvium. Indifferent to altitude.
Common and widespread; fruit uncommon.
2–17
Very variable, with numerous varietal names, mostly of doubtful status, scattered in the literature. The var. **fallax** (Brid.) Roth is of some significance, and is recorded from **2** Iona, *M. Dalby & U. Duncan* and from **17** Torrosay.

C. commutatum (Hedw.) Roth var. **commutatum**
Basiphilous; on irrigated rocks, especially in the spray from waterfalls and turbulent streams, and along seepage lines on cliffs; frequently associated with redeposited carbonates; occasionally in flushes, but usually replaced there by the variety *falcatum*.
Frequent but local in basic areas; fertile.
3–8, 10–17

C. commutatum var. **falcatum** (Brid.) Mönk.
Calcicolous; in tufts or extended orange-brown mats in base-rich springs and flushes, rarely on wet rocks.
Occasional but local; occasionally fertile.
Possibly an environmental modification rather than a genetic variant.
2–8, 10–17

C. commutatum var. **virescens** (Schimp.) Rich. & Wall.
10 Carsaig Bay, on wet basic rocks, 1967, *M. Dalby & U. Duncan.*

C. commutatum var. **sulcatum** (Lindb.) Mönk.
11 Ben More, 1887, *Ewing* (GL).

Campylium (Sull.) Mitt.

C. stellatum (Hedw.) Lange & C. Jens.
In mesotrophic flushes and mires, by slow-running streams and lake margins. Indifferent to altitude but occasional to rare at high elevations.
Common; locally abundant. Fruit rare.
2–17

C. protensum (Brid.) Kindb.
Basiphilous; on irrigated basic rocks and occasionally in eutrophic mires, especially on the coast.
Occasional; sterile.
2 Iona, *M. Dalby & U. Duncan*. **3** Fidden. **5** Calgary Bay. **8** Inch Kenneth. **14** Fishnish Bay.
This taxon may be difficult to distinguish from weak forms of *C. stellatum* and its status seems to depend, to a degree, upon individual interpretation. Doubtful records are excluded.

C. chrysophyllum (Brid.) J. Lange
Calcicolous; on dry, calcareous sandy soils and rock ledges.
Rare; sterile.
3 Fidden.

C. polygamum (B.S.G.) J. Lange & C. Jens.
In mesotrophic mires and flushes and by lochs. Indifferent to altitude.
Rare; fertile.
3 Loch Poit na h-I by the S end. **6** Loch Frisa, at the NE end.

Leptodictyum (Schimp.) Warnst.

L. riparium (Hedw.) Warnst.
Nitrophilous: mainly lignicolous by lochs and streams. Lowland.
Rare; sterile.
6 Quinish, Poll Athach. **11** Gruline, shore of Loch Bà. **13** Aros Park, by the lake.

Hygroamblystegium Loeske

H. fluviatile (Hedw.) Loeske
Only recorded from wet basic rocks by a waterfall.
Rare; sterile.
8 Ardmeanach, Coireachan Gorma.

Amblystegium B.S.G.

A. serpens (Hedw.) B.S.G. var. **serpens**
On a variety of substrates; somewhat nitrophilous; in thin patches on damp rocks, tree-roots, rotting wood and earth banks in shaded sites; mainly near habitation. Lowland.
Occasional, but frequent locally. Fertile.
2–8, 10–13, 16, 17

A. serpens var. **salinum** Carr.
In small, dense tufts on damp sand in coastal turf and saltings.
Frequent; confined to the coast.

1–3, 5, 6, 8–17
Specimens intermediate between the variety and the typical plant are frequent in estuarine localities.

A. compactum (C. Müll.) Aust.
Terricolous; calcicolous?; only recorded from calcareous sand by the sea.
Rare; sterile.
2 Iona, northern extremity. **3** Uisken.
The taxonomic position of the British taxon is obscure and is further complicated by nomenclatural confusion (especially with *A. serpens* var. *salinum*). Furthermore, it is by no means certain that the European taxon is the same as the species originally described from N. America.

Drepanocladus (C. Müll.) Roth

D. aduncus (Hedw.) Warnst.
Basiphilous; seen only in coastal mires.
Rare; sterile.
3 Near Fidden. **17** Lochdonhead.

D. fluitans (Hedw.) Warnst.
Calcifuge; in bog-pools, lochs, montane streams, oligotrophic mires; sometimes submerged. Indifferent to altitude.
Common and widely distributed. Fruit rare.
1–17
Most of the sub-aquatic forms of this highly polymorphic species approach var. **falcatus** (B.S.G.) Warnst.

D. exannulatus (B.S.G.) Warnst.
Calcifuge; hygrophytic; in usually red-tinged cushions in oligotrophic springs and mires, frequently associated with *Sphagnum* species. Alpine and higher subalpine.
Rare; sterile.
4 Ulva, above Cragaig, *M. Dalby & U. Duncan.* **8** Maol Mheadonach, *M. Dalby & U. Duncan* (sub var. *rotae* (De Not.) Wynne). **11** Ben More, north corrie; Beinn Fhada. **15** Sgùrr Dearg, south side. **16** Laggan Deer Forest, Maol a' Bhaird; Ben Buie, north summit.

D. revolvens (Turn.) Warnst. var. **revolvens**
Hygrophytic or sub-aquatic; calcifuge but with a slight base-demand: in blackish purple masses in oligotrophic to somewhat mesotrophic flushes; frequently associated with *Scorpidium*, in the former, with e.g. *Acrocladium sarmentosum* and *Schoenus nigricans* in the latter. Indifferent to altitude.
Common and widespread; locally abundant. Occasionally fertile but fruit seldom abundant.
1–17

D. revolvens var. **intermedius** (Lindb.) Rich. & Wall.
Hygrophytic; distinctly basiphilous: in paler brownish hummocks in mesotrophic and eutrophic mires.
Rare; sterile.
8 Small spring below Creag a' Ghaill. **11** E side of Maol Mheadonach, *U. Duncan.*

D. uncinatus (Hedw.) Warnst.
Calcifuge; terricolous: on well drained inorganic damp soils, e.g., mountain detritus, streamside alluvium, gritty soils by roads and leached sandy soils by the sea.
Occasional; widespread. Fertile.
4 Ulva, by Culinish road. **8** Maol Mheadonach, *U. Duncan*; The wilderness, above cliff. **10** Glen Byre; Pennyghael. **11** Head of Scarisdale River gully. **12** Loch Frisa. **14** Fishnish Bay. **16** Laggan Deer Forest.

Hygrohypnum Lindb.

H. ochraceum (Turn. ex Wils.) Loeske
Basiphilous; attached to rocks in streams with a reasonably high base-content. Mainly subalpine.
Very rare; sterile.
8 The Wilderness, by small cascade on upper cliffs with *Eurhynchium riparioides.*

H. luridum (Hedw.) Jenn.
In brownish mats attached to rock surfaces intermittently submerged in shallow running water, mainly in ravines and gullies and lowland. Lower subalpine, ascending to at least 400 m.
Frequent; in scattered localities. Fertile when emergent.
4–17

H. eugyrium (Schimp.) Broth.
Elsewhere in Britain in similar habitats to the other species of the genus, but Mull habitat not recorded.
Rare; a single record not refound during the survey.
7/11 Loch na Keal, 1932, *W. Young.*

Scorpidium (Schimp.) Limpr.

S. scorpioides (Hedw.) Limpr.
Calcifuge; in half-submerged nigrescent mats or mixed with vegetation in bog-pools and oligotrophic mires. Indifferent to altitude.
Common and locally abundant; fruit very rare.
1–17

Acrocladium Mitt.

A. stramineum (Brid.) Rich. & Wall.
Calcifuge; mainly mixed with vegetation of oligotrophic mires and on wet, upland peat-covered rocks.
Indifferent to altitude.
Occasional; sterile.
6 'S Àirde Beinn. **8** Beinn na Sreine. **10** Beinn na Croise. **11** Ben More, north summit. **12** Near Salen, *Corley*; Allt na Searmoin, *U. Duncan.* **15** Sgùrr Dearg. **16** Fontinalis Lochs.

A. cordifolium (Hedw.) Rich. & Wall.
Basiphilous; straggling green stems among *Juncus acutiflorus* in mesotrophic mire.

Rare; sterile.
6 Loch Frisa, north-east shore.

A. giganteum (Schimp.) Rich & Wall.
Somewhat basiphilous; submerged or emergent in wetter parts of mesotrophic mires. Indifferent to altitude.
Frequent; widespread below 400 m. Fruit very rare.
2, 3, 6–17

A. sarmentosum (Wahlenb.) Rich. & Wall.
Terricolous; in vinous red, rarely green, patches in wet turf, mesotrophic gravel flushes and irrigated stony ground on hillsides; commonly associated with *Drepanocladus revolvens, Eleocharis* or *Schoenus*. Indifferent to altitude but mainly subalpine.
Frequent in scattered localities; fertile.
4, 5, 7, 8, 10–12, 15–17

A. cuspidatum (Hedw.) Lindb.
On damp soils in grasslands, fens, woodlands and among rocks on mountains and in ravines, avoiding only extremely base-poor habitats. At all altitudes.
Common and more or less ubiquitous; occasionally fertile.
1–17

BRACHYTHECIACEAE

Isothecium Brid.

I. myurum Brid.
At bases of larger trees and on rocks in woods and ravines. Lowland, ascending to 200 m.
Frequent; fertile.
3–8, 10–17

I. myosuroides Brid. var. **myosuroides**
In frequently extensive mats on rocks and bases of trees. Lowland and lower subalpine, ascending to at least 700 m.
Common; ubiquitous below 300 m. Fertile.
1–17

I. myosuroides var. **brachythecioides** (Dixon) C. Jens.
Irrigated rocks and flushed gravelly soils. Alpine and subalpine.
Rare; sterile.
8 Beinne na Sreine. **11** Ben More; Cruachan Dearg. **15** Beinn Bheag, *Corley.*

Brachythecium B.S.G.

B. albicans (Hedw.) B.S.G.
More or less xerophilous; on dry sandy soils in open turf and dunes, especially near the sea. Lowland.
Occasional; locally frequent. Sterile.
1 Lunga. **2** Iona, *M. Dalby & U. Duncan.* **3** Fidden; Uisken. **5** Calgary Bay. **8** Balmeanach. **12** Salen. **14** Fishnish Bay. **16** Loch Buie.

B. glareosum (Spruce) B.S.G.
Basiphilous; on basic rocks and soil in gullies, drained eutrophic mires and flushed calcareous grasslands.
Rare; sterile.
5 Calgary Bay. **8** Gleann Doire Dhubhaig. **11** Scarisdale River gully, *U. Duncan.*

B. rutabulum (Hedw.) B.S.G.
Nitrophilous; on disturbed ground in gardens, by paths and walls around dwellings; on rotting wood and tree-roots and on muddy soils around saltmarshes and in drained mires. Mainly lowland.
Common; abundant around farms and dwellings. Fertile.
1–17

B. rivulare B.S.G.
Basiphilous; usually as golden cushions or mixed with vegetation in flushes and occasionally on wet rocks by streams and waterfalls; a characteristic plant around alpine and subalpine springs in association with *Montia fontana* and *Chrysosplenium oppositifolium.*
Frequent and widely distributed; fruit rare in terrestrial forms less so in rupicolous ones.
3–17

B. velutinum (Hedw.) B.S.G.
On walls and tree-bases. Lowlands.
Rare; fertile.
4 Ulva, NE side, *U. Duncan.* **5** Torloisk. **12** Salen. **13** Tobermory. **17** Torosay Castle.

B. populeum (Hedw.) B.S.G.
Rupicolous on rocks and walls. Lowlands.
Rare. Fertile.
2 Iona, *M. Dalby & U. Duncan.* **4** Ulva House. **5** Calgary. **7** Torloisk. **11** Knock. **12** Allt na Saille. **5** An Coire, *Corley.* **16** Lochbuie.

B. plumosum (Hedw.) B.S.G.
On siliceous rocks and boulders by streams and waterfalls. Lowland and subalpine, rarely alpine.
Common and widely distributed; fertile.
1–17

Camptothecium B.S.G.

C. sericeum (Hedw.) Kindb.
Usually on siliceous rocks, but not strictly calcifuge; occasionally on tree trunks; frequent on walls. Mainly lowland, but rising to about 300 m.
Common; fertile.
1–17

C. lutescens (Hedw.) B.S.G.
Calcicolous; on dry calcareous soils and ledges of rocks with blown sand by the sea. Lowland.
Occasional; local. Sterile.
1 Staffa. **2** Iona, *M. Dalby & U. Duncan.* **3** Fidden; Erraid; Uisken. **5** Calgary Bay, *M. Dalby & U. Duncan.* **10** Carsaig Bay. **16** Loch Buie. **17** Torosay, on mortar rubble.

Cirriphyllum Grout

C. piliferum (Hedw.) Grout
Basiphilous; straggling among vegetation on base rich soils in woodlands, less commonly in more open situations. Lowland.
Locally frequent; sterile.
2–17

C. crassinervium (Tayl.) Loeske & Fleisch.
Rupicolous and distinctly basiphilous; in shaded situations in ravines and sheltered wood valleys. Only seen at low altitudes.
Rare; sterile.
6 Dervaig woods. **8** Mackinnon's Cave.

Eurhynchium B.S.G.

E. striatum (Hedw.) Schimp.
Basiphilous; on basic soils among vegetation in mixed woodlands (*Quercus*, *Corylus* and *Fraxinus*) and shaded areas in ravines, and at the foot of walls. Mainly lowland.
Locally frequent in basic areas; elsewhere rare or absent. Rarely fruiting.
1–17

E. praelongum (Hedw.) Hobk. var. **praelongum**
E. stokesii sensu auct. eur.; non Turner
On a variety of soils in shaded situations; usually creeping among grasses and other mosses in woodlands, on stream-banks, rotting wood and rocks; avoiding calcareous and extremely acid substrates. Indifferent to altitude but requiring protection in the alpine zone.
1–17
Frequent in the lowlands; elsewhere occasional. Occasionally fertile.

E. praelongum var. **stokesii** (Turn.) Hobk.
On humus-rich acid substrates in areas of constantly high atmospheric humidity; mainly by streams and in dwarf-shrub communities. Less frequent than the type variety.
3–17

E. swartzii (Turn.) Curn.
E. praelongum sensu auct. eur.; non (Hedw.) Hobk.
Basiphilous; on basic or neutral damp, inorganic soils in woodlands and sheltered rock-crevices; in basic grasslands.
Locally frequent; sterile.
1, 3–8, 10–17

E. speciosum (Brid.) Milde
Basiphilous; among vegetation of eutrophic mires near the sea and on flushed rocks.
Rare, fertile but fruit not developed.
5 Calgary Bay. **13** Aros Park.

E. riparioides (Hedw.) Rich.
Slightly base-demanding; in shallow streams, waterfalls and on various irrigated rock surfaces often in considerably extended carpets.
Common and widespread; fruiting when emergent.
3–16

E. murale (Hedw.) Milde
Calcicolous; growing in small patches or straggling on rocks.
Rare; sterile.
8 Balmeanach, Rebecca's Cave, 1969, *Lobley.*

E. confertum (Dicks.) Milde
Slightly basiphilous; on rocks and walls; rather rare in natural habitats and not seen far from habitation. Lowland.
Occasional; perhaps introduced. Fertile.
4 Ulva House. **6** Dervaig, Penmore Mill. **10** Carsaig, Innamore Lodge. **13** Tobermory. **13** Killichronan, *Lobley & U. Duncan.* **17** Torosay Castle, *M. Dalby & U. Duncan.*

Rhynchostegiella (B.S.G.) Limpr.

R. pumila (Wils.) E. F. Warb.
Basiphilous; on ledges under overhangs of limestone and basaltic rocks. Mainly lower subalpine and lowland, but ascending to at least 500 m.
Rare; in scattered localities. Not seen in fruit.
4 Ulva, Ben Chreagach. **8** Balmeanach, *Lobley & U. Duncan*; Mackinnon's Cave. **10** Carsaig Bay. **16** W of Lochbuie, cave-entrance, *M. Dalby & U. Duncan*; near Dalnaha.

R. tenella (Dicks.) Limpr.
Calcicolous; on basic rocks and mortar of walls, especially around the coasts. Lowland.
Locally frequent; occasionally fertile.
3–8, 10–14, 16, 17

ENTODONTACEAE

Orthothecium B.S.G.

O. rufescens (Brid.) B.S.G.
Calcicolous; in rubescent tufts on irrigated limestone cliff by small waterfalls.
Very rare; sterile.
8 Above Mackinnon's Cave.

O. intricatum (Hartm.) B.S.G.
Calcicolous, somewhat hygrophilous; in clefts of basic rocks. Subalpine.
Very rare; sterile.
8 The Wilderness. **17** Craignure, *Crundwell & Wallace* (GL).

Entodon C. Müll.

E. concinnus (De Not.) Paris
Calcicolous; on dry shell-sand among sparse grasses and other bryophytes by the sea.
Rare; sterile.
2 Iona, N coast. **3** Ardalanish Bay, *A. Stirling & Kenneth*; Fidden. **5** Calgary Bay, *Corley.*

Pseudoscleropodium (Limpr.) Fleisch.

P. purum (Hedw.) Fleisch.
Slightly basiphilous; creeping among vegetation in neutral and basic grasslands; in clearings in woodland, and turf and banks by roads and

ditches. Mainly lowland, but ascending into the alpine zone.
Common and widespread in the lowlands. Rarely fertile.
1–17

HYPNACEAE

Pleurozium Mitt.

P. schreberi (Brid.) Mitt.
Calcifuge; in tufts or mixed with other mosses and vegetation on well-drained but moist acid soils in a variety of habitats including woodlands, heaths, drier areas of bogs, and peaty detritus among boulders of screes; avoiding only waterlogged and alkaline soils. Indifferent to altitude.
Ubiquitous; sterile.
1–17

Isopterygium Mitt.

I. pulchellum (Hedw.) Jaeg. & Sauerb.
Basiphilous; on sheltered rocks, usually with some humus, in ravines, woodlands and alpine block-screes.
Occasional in scattered localities; fertile.
6 Quinish, Mingary. **8** Maol Mheadonach, *U. Duncan.* **10** Carsaig. **11** Ben More; Abhainn Disheig. **13** Aros Park. **16** Loch Spelve, above Dalnaha. **17** Torosay Castle, *M. Dalby & U. Duncan.*

I. elegans (Hook.) Lindb.
In flat, green patches on inorganic and organic soils in woodlands; roots of trees, rotting logs and damp shaded rocks in ravines, woodlands; also in crevices and shaded ledges of alpine and subalpine rocks.
Common and widely distributed; sterile.
1–17

I. muelleranum (Schimp.) Lindb.
More or less hygrophytic; in crevices and deep horizontal clefts in somewhat basic strata. Alpine and subalpine.
Very rare; sterile.
11 Abhainn Dhisheig. **17** Craignure, 1954, *Crundwell & Wallace.*

Plagiothecium B.S.G.

P. denticulatum (Hedw.) B.S.G. var. **denticulatum**
In green patches on humus-rich neutral to mildly acid soils in shade; woodlands, streamsides, block screes. Mainly lowland but ascending to at least 650 m.
Frequent; fertile.
3, 4, 6–8, 10–17

P. denticulatum var. **obtusifolium** (Turn.) Paris
On humus among rocks in block scree. Higher subalpine zone.
Rare; sterile.
11 Ben More, north corrie, 800 m.

P. platyphyllum Mönk.
On moist vegetation in high-altitude springs and flushes.
Rare; sterile.
8 Above The Wilderness. **11** Ben More, below north corrie.

P. roeseanum B.S.G.
On humus-rich soils on or among rocks in sheltered habitats; in woodlands, screes and ravines.
Rare; sterile.
8 Creag a' Ghaill. **10** Carsaig, *M. Dalby & U. Duncan.* **11** Cruachan Dearg, SW side. **16** Cruach Ardura. Unlocalized, *G. Ross et al.* (1881).

P. succulentum (Wils.) Lindb.
On inclined humus-rich neutral soils on woodland banks, among tree-roots and rocks in ravines. Mainly lowland and lower subalpine, becoming very local above 600 m.
Common; fertile.
1–17

P. sylvaticum (Brid.) B.S.G.
Basiphilous; on basic to mildly acid soils, rock ledges, woodland banks and ravines. Mainly lowland and lower subalpine, but recorded at 650 m.
Occasional but locally frequent; fertile.
4–8, 10, 12–17

P. undulatum (Hedw.) B.S.G.
Calcifuge; in flat, whitish mats or scattered stems in oligotrophic vegetation; drier areas of bogs, rock ledges, birch-woods and grass-heaths. Indifferent to altitude.
Ubiquitous; fertile.
1–17

Hypnum Hedw.

H. cupressiforme Hedw. var **cupressiforme**
In wide, green patches on rocks, walls, bases of trees. At all altitudes.
Common; locally abundant. Fertile.
1–17

H. cupressiforme var. **resupinatum** (Wils.) Schimp.
On bark in open situations, rarely on walls. Lowland and lower subalpine.
Frequent; sterile.
2, 4–17

H. cupressiforme var. **filiforme** Brid.
In extensive mats on vertical surfaces of tree-trunks, rarely rock slabs. Indifferent to altitude.
Abundant; fruit not rare.
2–17
This variety intergrades with the type, and may be a habitat form without genetic distinction.

H. cupressiforme var. **mamillatum** Brid.
In small green patches on trees, shrubs and logs in sheltered habitats.

Occasional; fertile.
4 Ulva House, *U. Duncan.* **6** Loch a' Chumhainn. **7** Achleck. **8** Tiroran. **10** Carsaig, *Kenneth & A. Stirling.* **11** Loch Bà, Coille na Sroine. **15** Lussa River. **16** Loch Uisg.

H. cupressiforme var. **ericetorum** B.S.G.
Calcifuge; in pale, extended mats among oligotrophic vegetation on peat, dwarf shrub heaths, grass-heaths, larch-woods and among alpine and subalpine rocks.
Ubiquitous; fruit very rare.
1–17

H. cupressiforme var. **tectorum** B.S.G.
Somewhat basiphilous; in mats on coastal rocks, limestone exposures and walls. Lowland.
Frequent; especially near the sea. Fruit not seen.
1–17

H. cupressiforme var. **lacunosum** Brid.
Calcicolous; in short turf or sparse vegetation on skeletal soils near the sea, mainly as scattered stems.
Locally frequent around the coast.
1–3, 5–8, 10, 13, 14, 16, 17

H. callichroum (Brid.) B.S.G.
Basiphilous; terricolous or epilithic; on skeletal basic soils and damp rocks, extending onto roots of trees.
Rare; sterile.
8 The Wilderness. **13** Near Tobermory, *G. Ross et al.* (1881, as *H. hamulosum*); Aros Park, *U. Duncan.*

H. hamulosum B.S.G.
Basiphilous; on skeletal calcareous soils and basic rocks. Subalpine.
Rare; sterile.
8 Creag a' Ghaill, 1965 *U. Duncan.*

H. lindbergii Mitt.
Somewhat hygrophilic; on wet inorganic soils and streamside alluvium; disturbed areas by roads and ditches.
Frequent; sterile.
3 Uisken. **5** Calgary Bay, *M. Dalby & U. Duncan.* **7** Torloisk, Allt a' Mhuilinn. **8** Tavool House. **14** Fishnish, Doire Dorch, *Lobley & U. Duncan.* **16** Loch Spelve, Dalnaha.

Ptilium (Sull.) De Not.

P. crista-castrensis (Hedw.) De Not.
In grassland among dwarf shrubs, in birchwoods and 'hepaticosum' facies of shrubheaths.
Occasional; fruit very rare.
7 Druim na Cille. **8** Ardmeanach, Coire nan Griogag. **11** Abhainn Dhiseig. **12** Allt na Searmoin, *U. Duncan.* **13** Tobermory, *G. Ross et al.* (1881); Aros Park. **16** Loch Spelve, near Dalnaha. **17** Craignure Woods, *Lobley & U. Duncan.*

Ctenidium (Schimp.) Mitt.

C. molluscum (Hedw.) Mitt. var **molluscum**
Calcicolous; on basic rocks and skeletal soils in ravines, woodlands and around the coast. Indifferent to altitude.
Frequent; fertile.
1–17

C. molluscum var. **fastigiatum** (Bosw. ex Hook.) Braithw.
In similar habitats to the type variety.
Rare; sterile.
2 Iona, *Trotter*, 1938.

C. molluscum var. **condensatum** (Schimp.) Britt.
Ecology similar to that of the type variety.
Occasional; sterile.
5 Treshnish, *M. Dalby & U. Duncan.* **8** Mackinnon's Cave. **17** Craignure, *Crundwell & Wallace*, 1954.

C. molluscum var. **robustum** Boul. ex Braithw.
In reddish, *Drepanocladus*-like patches or scattered on irrigated skeletal soils. Subalpine.
Locally frequent; sterile.
4, 7, 8, 9–16
The variety *robustum* is very distinctive, resembling a *Drepanocladus* in habit. Forms intermediate between this taxon and var. *molluscum* were not recorded on Mull, nor does there seem to be a significant overlap in habitat preferences.

Hyocomium B.S.G.

H. flagellare B.S.G.
Calcifuge? apparently avoiding extremely base-poor substrates, on shaded, irrigated rocks and soil by streams and seepages in woodlands and ravines. Lowland and lower subalpine.
Occasional; locally frequent. Not seen in fruit.
4–8, 10–16

SEMATOPHYLLACEAE

Sematophyllum Mitt.

S. novae-caesareae (Aust.) Britt.
In low patches in extremely sheltered sites in ravines and lowland woods near the sea.
Rare; sterile.
11 Scarisdale Wood, 1961, *Ratcliffe.* **16** Loch Uisg, S side.

HYLOCOMIACEAE

Rhytidium (Sull.) Kindb.

R. rugosum (Hedw.) Kindb.
Calcicolous; on dry calcareous soils in open situations; confined to shell-sand on dunes and among coastal rocks.
Very rare; sterile.
5 Treshnish Point. **10** Carsaig Bay, *U. Duncan*

Rhytidiadelphus (Lindb.) Warnst.

R. triquetrus (Hedw.) Warnst.
Basiphilous; among grasses in mixed woodland; base-rich grasslands and margins of fens. Mainly lowland and lower subalpine.
Frequent in the lowlands; rare elsewhere. Sterile.
1–17

R. squarrosus (Hedw.) Warnst.
In loose mats and creeping through grasses on a wide variety of inorganic and humus-rich soils; lowland grasslands and woodland clearings, including lawns, fields and plantations; in coastal turf; ascending into the alpine zone on rock ledges and margins of flushes: absent only on extremely acid soils or where exposed to desiccating agents.
Common and widespread; locally abundant at lower altitudes. Fruit rare.
1–17

R. loreus (Hedw.) Warnst.
Calcifuge; terricolous; in loose mats or creeping among other bryophytes on peat or peaty detritus; abundant in the ground flora of open birch-woods, *Calluna* heaths, rocky banks and screes; common in bogs of various kinds except where waterlogged. Indifferent to altitude and reaching the summits of the highest mountains.
Ubiquitous; fruit rare.
1–17

Hylocomium B.S.G.

H. brevirostre (Brid.) B.S.G.
Mainly rupicolous; on humus-capped rocks in shade. Woodlands and ravines. Lowland and lower subalpine.
Frequent; especially along the east coast. Fertile.
3–17

H. umbratum (Hedw.) B.S.G.
Basiphilous; among grasses or other vegetation in a base-rich area.
Rare; sterile.
13 Aros Park, near the Falls, 1898, *H. N. Dixon* but not refound during the Survey.

H. splendens (Hedw.) B.S.G.
In grass-heath vegetation, dwarf shrub communities and open woodlands; absent only from extremely acid, waterlogged areas. At all altitudes.
Ubiquitous; fruit not recorded.
1–17

References

ARNELL, S. (1956). *Illustrated moss flora of Fennoscandia. I. Hepaticae*. Lund.

CAVERS, F. (1910–11). The inter-relationships of the Bryophyta I–V, *New Phytol.*, **9**: 81–92, 93–112, 157–186, 193–234, 269–304 (1910); VI–XI, *New Phytol.*, **10**: 1–46, 84–85 (1911). Reprinted, with additional notes, by Dawsons of Pall Mall, London as *New Phytologist Reprint* No. 4 (1964).

COURTEJAIRE, J. (1967). Les hépatiques de quelques îles occidentales d'Écosse. *Revue bryol. lichen.*, n.s. **35**: 165–170.

CRUNDWELL, A. C. & NYHOLM, E. (1962). Notes on the genus *Tortella* I. *T. inclinata, T. densa, T. flavovirens* and *T. glareicola*. *Trans. Br. bryol. Soc.*, **4**: 187–193.

———. (1964). The European species of the *Bryum erythrocarpum* complex. *Trans. Br. bryol. Soc.*, **4**: 597–637.

DIXON, H. N. (1899). Bryological notes from the West Highlands. *J. Bot., Lond.*, **37**: 300–310.

———. (1924) (Reprint 1954). *The student's handbook of British mosses, with illustrations and keys to the genera and species by H. G. Jameson* ed. 3, revised and enlarged. Eastbourne.

FLEISCHER, M. (1922). Allgemeine Uebersicht des/eines natürlichen Systems der Laubmoose. In *Flore de Buitenzorg 5. Les muscinées (Die Musci der Flora von Buitenzorg) 4. Bryales*: xi–xxxi. Leiden.

GREENE, S. W. (1957). The British species of the *Plagiothecium denticulatum – P. silvaticum* group. *Trans. Br. bryol. Soc.*, **3**: 181–190.

GREIG-SMITH, P. (1954). Notes on Lejeuneaceae II. A quantitative assessment of criteria used in distinguishing some British species of *Lejeunea*. *Trans. Br. bryol. Soc.*, **2**: 458–469.

GROLLE, R. (1972). Die Namen der Familien und Unterfamilien der Lebermoose (Hepaticopsida). *J. Bryol.*, **7**: 201–236.

HOLMEN, K. (1955). Chromosome numbers of some species of *Sphagnum*. *Bot. Tidsskr.*, **52**: 37–42.

JONES, E. W. (1958). An annotated list of British hepatics. *Trans. Br. bryol. Soc.*, **3**: 353–374.

KENNEDY, D. (1910). Quoted in MACVICAR (1910) see p. 13.1.

LITTLE, E. R. B. (1968). The oil bodies of the genus *Riccardia* Gray. *Trans. Br. bryol. Soc.*, **5**: 536–540.

MACVICAR, S. M. (1910). The distribution of Hepaticae in Scotland. *Trans. Proc. bot. Soc. Edinb.*, **25**: 1–336.
———. (1926). (Reprint 1961). *The student's handbook of British hepatics, with illustrations by H. G. Jameson* ed. 2, revised and enlarged. Eastbourne.
NYHOLM, E. (1954–69). *Illustrated moss flora of Fennoscandia II. Musci.* Fascicle 1 (1954), 2 (1956), 3 (1958), 4 (1960), 5 (1965), 6 (1969). Lund (fascicles 1–5), Stockholm (fascicle 6).
PATON, J. A. (1962). The genus *Calypogeia* Raddi in Britain. *Trans. Br. bryol. Soc.*, **4**: 221–229.
———. (1965a). A new British moss, *Fissidens celticus* sp. nov. *Trans. Br. bryol. Soc.*, **4**: 780–784.
———. (1965b). *Census catalogue of British hepatics* ed. 4. Truro.
RATCLIFFE, D. A. (1968). An ecological account of the Atlantic bryophytes in the British Isles. *New Phytol.*, **67**: 365–439.
RICHARDS, P. W. & WALLACE, E. C. (1950). An annotated list of British mosses. *Trans. Br. bryol. Soc.*, **1**(4): Appendix i–xxxi.
ROSS, G. et al. (1881). Quoted in ROSS & KING (1882) see p. 13.1.
ROSS, G. & KING, T. (1882). Mosses. In F. B. White, The cryptogamic flora of Mull. *Scott. Nat.*, **6**: 210–212.
STIRTON, J. (1899). Lichens and mosses from Carsaig, Argyll. *Ann. Scot. nat. Hist*, no. **29**: 41–45.
WARBURG, E. F. (1963). *Census catalogue of British mosses* ed. 3. Oxford.
WHITEHOUSE, H. L. K. (1963). *Bryum riparium* Hagen in the British Isles. *Trans. Br. bryol. Soc.*, **4**: 389–403.
WILLIAMS, S. (1950). The occurrence of *Cryptothallus mirabilis* v. Malmb. in Scotland. *Trans. Br. bryol. Soc.*, **1**: 357–366.

Addenda

The following species have since been identified from material collected during the survey:

JUNGERMANNIACEAE (p. 13.7)

Eremonotus Kaal.

E. myriocarpus (Carringt.) Pears.
Hygrophilous. Only seen once, mixed with *Solenostoma triste* on flushed rocks at 820m.
11 Ben more, in the north corrie.

BARTRAMIACEAE (p. 13.34)

Plagiopus Brid.

P. oederi (Brid.) Limpr.
Very rare; a single record.
11 Ben More, crevice in the north corrie c. 850 m.

14. Lichens

P. W. James

Introduction

Unlike the bryophytes, the lichens of the Island of Mull had until 1965, received scant attention from collectors and specialists. This lack of interest is very strongly reflected in the paucity of records in both the literature and herbaria from which only 37 taxa could be recorded after a thorough survey of both sources. This number contrasts markedly with the total of 715 taxa, representing 50 per cent of the British lichen flora (James, 1965, 1966) which are now known to occur on Mull and are based on the records and collections of U. K. Duncan (1966, 1968, 1969), U. K. Duncan & James (1967), and the present Survey (1968 to 1972). Most of the early records refer to ubiquitous species, e.g., *Cladonia coccifera*, *Lichina pygmaea*, and give little indication of the richness of the flora and its ecological and phytogeographical features of particular interest.

Because of the large area and difficult terrain of the Island no attempt could be made in the time available to obtain an overall coverage of the lichens. Instead, a large series of widely divergent habitats was selected, based on different rock types, age of tree cover and its species composition, altitude, proximity of sea or freshwater. Sites modified by man, including buildings, were also examined in detail.

Arrangement and nomenclature

The systematic arrangement to family level is based on that outlined by Henssen & Jahns (1973) and Henssen (1976) with some minor modifications by the author. The nomenclature at generic and species level mainly follows that of James (1965, 1966), and Duncan (1970). Subsequent changes are indicated in the text, together with the relevant synonymy. In addition, recent critical taxonomic surveys of particular genera and other important source works are indicated where appropriate. General floras and keys consulted were Smith (1918, 1926), Poelt (1969), Duncan (1970), Dahl & Krog (1973) and Fletcher (1975a,b). Delimination of marine and maritime zones follows that of Fletcher (1973a,b). Representative collections of all species are lodged in the collections of the Museum (BM). As in the vascular plant account exact localities are given when ten or less records have been made for that species. Locality reference numbers (in bold) refer to the map at the end of the book.

Details of the lichen substances in all macrolichens and many samples of microlichens are appended. Much of this data is original and is particularly important in relation to the identification of different entities in such genera as *Ramalina* and *Usnea*. The standard basic chemical spot tests (K, KC, C, P, I) were supplemented by the application of thin-layer chromatography (TLC) according to the method outlined by Culberson & Kristinsson (1970) and Menlove (1974). The presence of accessory substances, not always present in the species, is indicated by the symbol (±) meaning more or less. Some difficult chemical complexes on TLC were resolved by the use of micro-crystal tests, following the method of Thomson (1968).

Acknowledgements

I am grateful to B. J. Coppins, Dr A. Fletcher, Dr D. L. Hawksworth, Prof. Dr A. Henssen, Dr H. Hertel and Prof. R. Santesson for assistance in the identification of some critical species; W. Birse, Dr H. M. Jahns, Dr M. R. D. Seaward, the late Miss N. Wallace and, particularly, Dr Ursula K. Duncan for their valuable help in the field; Janet E. Menlove (Mrs R. Brinklow) helped with the arduous task of checking the manuscript.

Index to genera

LICHENES

ASCOMYCETES

CALICIALES

CALICIACEAE

Calicium Pers.

C. glaucellum Ach.
In dry, sheltered aspects of decorticated boles, logs and stumps of broad-leaved and coniferous trees; mostly confined to mature woodland or parkland sites.
Occasional.
4, 6, 11–13, 16, 17

C. subtile Pers.
On dry, sheltered bark and decorticated wood of conifers in mature woodland and parkland sites.
Rare.
13 Tobermory, Aros Park, Drùim Breac. **17** Torosay Castle grounds, on base of *Picea abies*; Gleann Rainich, An Cairealach, on *Pinus*.

C. viride Pers.
In dry, sheltered aspects of mature wayside and woodland broad-leaved trees and conifers, particularly *Fraxinus*, *Quercus* and *Ulmus*.
Frequent; widespread in all lowland areas, especially in the north and west.
3, 5–7, 12–14, 16, 17

Chaenotheca (Th. Fr.) Th. Fr.

C. aeruginosa (Turn. ex Sm.) A. L. Sm.
On decaying bole of moribund *Sorbus aucuparia*.
Rare.
11 Loch Bà, Doire Daraich.

C. brunneola (Ach.) Müll. Arg.
In dry bark crevices and on decorticated wood of old trees in old woodland and parkland.
Rare.
12 Salen, decorticated stump of *Pinus*. **13** Tobermory, Aros Park, decorticated wood (?*Ilex*) in sheltered coastal woodlands. **17** Loch Spelve, Strathcoll, old fence post; grounds of Torosay Castle, on aged *Castanea*.

C. chrysocephala (Ach.) Th. Fr.
On mature, often isolated, conifers in open, exposed situations or at the edge of mixed plantations.
Rare.
7 Torloisk House grounds, on *Picea*. **12** Salen Forest, near Achadh nan Each, on rotting conifer stump. **17** Torosay Castle grounds, on *Pinus*; Gleann Rainich, An Cairealach, on *Pinus*; Loch a' Ghleannain, stump of *Quercus*.
Often poorly developed and sterile. Contains vulpinic acid.

C. ferruginea (Turn. ex Sm.) Mig.
On boles of mature broad-leaved trees and conifers, especially at the margins of old woodland in lowland areas.
Occasional.
10 Pennyghael, on *Pinus*. **12** Salen, on *Quercus*. **13** Tobermory, Aros Park, on *Quercus*. **17** Torosay Castle, on sheltered *Quercus*; Loch a' Ghleannain, on *Quercus*.
The brown staining of the thallus is due to an unidentified pigment.

Coniocybe Ach.

C. furfuracea (L.) Ach.
On dry bark, tree roots, lignum, pebbles, acidic rock and soil in sheltered recesses at the bases of ancient trees and under rock overhangs, especially in old mixed woodland; also in crevices at or near the base of the sheltered sides of old stone walls.
Frequent; in lowland sites and occasionally fertile.
Contains vulpinic acid.
2, 4, 5–10, 12, 13, 17
Fertile specimens were recorded from **5, 12, 13, 17.**

C. sulphurea (Retz.) Nyl.
In deep shade on sheltered bole of *Sambucus nigra* in derelict garden (with *Collema occultatum*).
Rare; but easily overlooked owing to the diminutive size of the ascocarps.
7 North side of Loch Tuath, half mile west of Ballygown, *U. Duncan & James.*

Sphinctrina Fr.

S. gelasinata (With.) Zahlbr.
A parasymbiont on the thalli of *Pertusaria hymenea* and *P. pertusa*, rarely on *P. albescens*, *P. amara* and *P. leioplaca*.
Occasional; throughout the range of distribution of the host species.
2, 4–6, 8–13, 15, 16

SPHAEROPHORACEAE

Sphaerophorus Pers.

S. fragilis (L.) Pers.
On mossy acidic boulders and rock outcrops in upland areas.
Frequent; widespread on all open moorland and scree slopes above 90 m; especially frequent on the Laggan Deer Forest.
3, 7–12, 15, 16
All collections are sterile. Contains sphaerophorin, fragilin and thamnolic acid.

S. globosus (Huds.) Vain.
On mossy acid boulders and rock outcrops in all areas; less frequently on boles of old trees, especially *Betula*, in lowland and coastal sites.
Frequent to abundant; in all areas.
1–17

Fertile specimens were recorded in **4, 7–9, 12, 13, 15.** Contains sphaerophorin, fragilin, thamnolic acid (±) and squamatic acid.

S. melanocarpus (Sw.) DC.
In mossy boulders, more rarely rock outcrops, in well-wooded lowland and coastal areas; rarely on old, often decaying boles of trees in sheltered, boggy situations.
Occasional; becoming locally frequent in the north-east.
4, 5, 7, 11–13, 15
Fertile specimens recorded in **5, 12, 13.** Contains sphaerophorin, fragilin, constictic acid; no stictic or norstictic acid.

MYCOCALICIACEAE

Stenocybe Nyl.

S. brophila Wats.
Associated with hepatics on acidic rocks; also on tree-boles in sheltered, lowland sites.
Rare to occasional; but probably overlooked.
12 By Allt na Searmoin, south of Salen, on *Scapania punctata* on bole of *Fraxinus*, *U. Duncan.* **15** An t-Sleaghach, by Lussa River. **16** Lochbuie, near Marian Cottage, on *Scapania* sp. on sheltered side of ancient *Fagus.* **17** Craignure Woods, among hepatics on the trunk of *Betula*, *U. Duncan.*

S. pullatula (Ach.) Stein
Specific to *Alnus glutinosa* and especially frequent on branches and twigs over-hanging lochs and streams and permanently wet habitats.
Frequent; limited by the distribution of the phorophyte **4, 7, 8, 12, 13, 16, 17**

S. septata (Leight.) Massal.
Mainly specific to *Ilex*, but also very rarely on *Corylus*, especially in old woodlands and well-wooded valleys near the coast, usually confined to boles and main branches of older trees.
Frequent; distribution limited by that of the phorophyte.
4, 7, 10, 12, 13, 15–17

LECANORALES

COLLEMATACEAE

Collema Web.

Bibliogr.: Degelius (1954).
C. auriculatum Hoffm.
On basic rocks, mortar of old walls of derelict buildings and bridges; also on granite and gneiss boulders partially covered in wind-blown shell-sand
Frequent; in all lowland areas.
2–6, 10, 12, 13, 16, 17

C. crispum (Huds.) Web.
Predominantly on mortar of old walls, bridges and buildings; more rarely on basic rocks and scree as well as gneiss or granite associated with wind-blown shell-sand.
Frequent; especially near villages and farmsteads.
2–5, 8, 9, 13, 14, 16, 17

C. cristatum (L.) Web.
On basic rocks near the coast and on acidic rocks associated with shell-sand.
Occasional; often locally frequent.
2 Iona Golf Course, on gneiss and a granite erratic partially buried in shell-sand. **8** Inch Kenneth, near the Chapel, on limestone; Gribun, below Creag a' Ghaill, on limestone. **10** Carsaig Bay, on calcareous sandstone cliffs and scree, *U. Duncan.*

C. fasciculare (L.) Web.
On *Corylus* in a sheltered valley and on *Fagus* in mature woodland.
Rare.
5 Kilninian, Allt Hostarie, on *Corylus.* **16** Lochbuie, near Moie Lodge, on *Fagus.*

C. flaccidum (Ach.) Ach.
Damp acidic or more or less basic rocks in sheltered, often shaded habitats in ravines often near freshwater; also on moist shaded boles of broad-leaved trees, especially *Fraxinus*, *Acer pseudoplatanus* and *Ulmus.*
Abundant to frequent; in all lowland areas.
2–5, 7–10, 12, 13, 16, 17

C. furfuraceum (Arnold) Du Rietz
Boles and branches of sheltered wayside, valley and woodland broad-leaved trees, especially *Fraxinus*, *Ulmus* and *Corylus*, often within the spray zone of waterfalls or near streams; also very rarely on gneiss outcrops which are partially submerged in shell-sand.
Abundant; in all lowland areas.
2–9, 11–17

C. multipartitum Sm.
On basic rocks near the coast, also gneiss rock partially submerged in wind-blown shell-sand.
Rare.
2 Golf Course, on gneiss in shell-sand. **10** Carsaig, Nuns' Pass, on calcareous sandstone cliffs and scree.

C. nigrescens (Huds.) DC.
On gneiss rocks partially buried in wind-blown shell-sand.
Rare.
2 Golf Course.

C. occultatum Bagl.
On sheltered damp boles of *Sambucus nigra* in a derelict cottage garden.
Rare.
7 North side of Loch Tuath, half mile west of Ballygown, *U. Duncan & James.*

C. polycarpon Hoffm.
On basic rocks near the coast.
Occasional, but very local.

8 Inch Kenneth, Port an Ròin, on limestone; Gribun, near Balmeanach. **10** Carsaig Bay, near Nuns' Pass, calcareous sandstone scree; Lochbuie, Rubha na h-Inbhire, *U. Duncan.*

C. subfurvum Degel.
On boles of mature broad-leaved trees, especially *Fraxinus*, *Acer pseudoplatanus* and *Ulmus*, in moist wayside, valley and mature woodland sites; also rarely on sheltered mossy rocks near streams.
Frequent; often abundant, in all lowland areas.
2–17 (saxicolous in **2**).

C. subnigrescens Degel.
On smooth bark of branches in crown of fallen *Fraxinus* in small valley.
Rare.
8 Ardmeanach, near Tiròran, *U. Duncan & James.*

C. tenax (Sw.) Ach. var. **tenax**
On basic soil and rocks near the coast; mortar of walls, bridges and farmsteads; acidic rocks associated with wind-blown shell-sand.
Frequent; especially in the west.
1–17

C. tenax var. **ceranoides** (Borr.) Degel.
Partially buried in drifting base-rich soil and wind-blown shell-sand.
Rare; chiefly confined to the west.
2 Golf Course; Arduara. **5** Head of Calgary Bay; Mornish, Sunipol.

Leptogium (Ach.) Gray

L. azureum (Ach.) Mont.
L. tremelloides (L.fil.) Gray
On moss-covered boles of wayside and woodland broad-leaved trees, especially *Fraxinus* and *Ulmus*, in sheltered, humid coastal and lowland sites.
Occasional; but often locally frequent.
7 Kilninian, on *Corylus*. **9** Ardmeanach, below Tiròran on coast, on *Fraxinus*, *U. Duncan & James*. **10** Carsaig, on *Fraxinus*; Pennyghael, on *Ulmus* and *Fraxinus*.
For discussion on the taxonomy and nomenclature of this species, see Jørgensen (1976).

L. brebissonii Mont.
On sheltered broad-leaved trees in old woodland as well as in thickets of old *Corylus.*
Rare.
11 Loch Bà, old *Fagus* near An Dubh Àrd. **13** Tobermory, Aros Park, on *Corylus*; near bridge over Allt nan Torc. **16** Lochbuie, on *Fagus*, near Moie Lodge.
The first records of this species for Scotland.

L. burgessii (L.) Mont.
On sheltered broad-leaved trees in well wooded sites, especially *Corylus* and *Fraxinus*, particularly in humid ravines on or near the coast; also rarely on isolated wayside trees and acidic rocks.
Frequent; often locally abundant.
4, 6–17

L. cyanescens (Rabenh.) Körb.
On sheltered, moss-covered tree boles and acidic rocks.
Occasional.
7 Kilninian, on *Fraxinus*. **8** Ardmeanach, Tavool House, Abhiann Beulath an Tairbh, base of isolated *Fraxinus*; below Tiròran on *Fraxinus*; on mossy rocks near Scobull Point. **10** Pennyghael on *Ulmus*; Carsaig, on isolated *Fraxinus* below Sròn nam Boc. **12** North side of Loch na Keal, Kellan Mill, mossy bole of *Fraxinus.*

L. hibernicum Mitch. ex P. M. Jørg.
On *Corylus* in a humid, sheltered gorge; also on *Fraxinus* near the coast.
Rare.
7 Kilninian, woods above Tràigh na Cille. **8** at edge of clearing between Scobull Point and Sgeir Mhòr, *U. Duncan & James.*
The first records for Scotland.

L. lichenoides (L.) Zahlbr.
On base-rich saxicolous substrates, including limestone, calcareous sandstones, mortar, cement and consolidated shell-sand; also on moss-covered boles of old woodland trees in sheltered lowland and coastal sites.
Frequent.
1–13, 15–17

L. massiliense Nyl.
On calcareous sandstones and mortar in exposed lowland and coastal sites.
Rare; but locally frequent in **10**.
10 Carsaig, between Aird Ghlas and Nuns' Pass. **12** Head of Loch na Keal, on farm buildings at Tòrr nam Fiann.

L. plicatile (Ach.) Leight.
On base-rich substrates including calcareous sandstones, also on granite rocks associated with consolidated wind-blown shell-sand.
Rare; but often locally frequent.
2 A'Mhachair, Cùl Bhuirg and Culbuirg, on granite rocks partially covered in shell-sand; opposite Eilean Mòr. **5** Mornish, Port Langamull, on basalt associated with shell-sand. **10** Carsaig, on sandstone between Aird Ghlas and Nuns' Pass.
This species is abundant on Lismore Island.

L. rivulare (Ach.) Mont.
On basalt rocks in swift-flowing stream; also on a nearby *Fraxinus.*
Rare.
9 Ardmeanach, near Tavool House, Abhainn Beulath an Tairbh, *U. Duncan & James.*

L. schraderi (Ach.) Nyl.
On base-rich substrates, especially crumbling mortar of derelict buildings.

Occasional, but very local.
2 Golf Course, on granite associated with wind-blown shell-sand. **3** Erraid, walls of ruined observatory. **7** Kilmore, ruins of Drimnacroish House. **13** Tobermory, walls of disused whisky distillery; walls of garden of Aros House, *U. Duncan & James.*

L. sinuatum (Huds.) Massal.
On basic substrates, especially calcareous sandstone, mortar and granite associated with wind-blown shell-sand.
Occasional to frequent.
2–5, 7, 10–13, 16, 17

L. teretiusculum (Wallr.) Arnold
On boles of wayside broad-leaved trees, especially *Ulmus* and *Fraxinus* in lowland and coastal sites; also rarely on basic rocks and mortar.
Occasional.
2–5, 13, 16, 17

L. tremelloides auct. angl. non (L. fil.) Gray
Loosely attached amongst mosses on rock outcrops and boulders as well as boles of aged broad-leaved trees on or near the coast.
Occasional; but locally abundant.
2–4, 6, 8, 10, 12–14, 16, 17
Fertile at **8** Scridain, loch shore at Scobull Point.
See Jørgensen (1977).

PARMELIACEAE

Alectoria Ach.

Bibliogr.: Hawksworth (1972a).
A. bicolor (Ehrh.) Nyl.
On tops of large moss-covered boulders of stable scree.
Rare.
11 Beinn nan Gabhar, locally abundant at 370 m.
Contains fumarprotocetraric acid.

A. chalybeiformis (L.) Gray
On exposed rock faces and mossy boulders near the coast.
Rare; confined to a few localities in the west.
2 Near Arduara, on granite erratic. **3** Fionnphort, east-facing side of large granite boulder near the jetty. **4** Gometra, mossy boulder at the head of Sailean Mòr.
Contains fumarprotocetraric acid (±).

A. fuscescens Gyeln.
(including var. **positiva** (Gyeln.) D. Hawksw.)
On broad-leaved and coniferous trees, including *Acer pseudoplatanus*, *Alnus*, *Betula*, *Larix*, *Pinus sylvestris* and *Quercus* in woodlands and plantations; also on tops of moss-covered boulders.
Frequent; widely distributed, especially in the east.
Ascending to 150 m in **11** S of Loch Bà, Coille na Sròine, on *Betula* but found at sea level on **2** Iona, at Cladh an Diseart.
2, 4, 6, 8, 11–12, 14, 17
Contains fumarprotocetraric acid.

A. pubescens (L.) Howe
On exposed tops of rock outcrops and boulders on or near mountain summits.
Rare; confined to the higher mountains.
11 Ben More at 950 m. **15** Sgùrr Dearg at 750 m.
No lichen substances detected.

A. subcana (Nyl. ex Stiz.) Gyeln.
On boles of old conifers and broad-leaved trees in sheltered, mature woodland and plantations.
Rare; only recorded from the east.
7 Salen Forest, near Lettermore, on *Larix* sheltered by evergreen conifers. **17** Craignure, Druim Mòr, on bole of *Quercus*.
Contains fumarprotocetraric acid. The record from near Lettermore corresponds to var. **subosteola** (Gyeln.) Motyka.

Cavernularia Degel.

Bibliogr.: James (1959); Swinscow (1960b).
C. hultenii Degel.
On boles of sheltered, mature *Betula* and conifers in moist valleys.
Rare.
13 Lochnameal, valley of the Allt nan Torc and in the vicinity of Lochnameal Farm, both at 100 m.
Contains physodic acid and atranorin.
This interesting species, first discovered in Wester Ross in 1958, is now known from the Ardnamurchan Peninsula (Ben Laga, on *Sorbus* and *Betula*, *James*) and from several sites in the Central Highlands of Scotland.

Cetraria Hoffm.

See also *Platismatia* Culb. & Culb.

C. chlorophylla (Willd.) Vain.
On twigs, branches and boles of broad-leaved trees and conifers, especially in sheltered or boggy woodland; particularly frequent on *Salix*; also rarely on rocks and boulders in upland scree.
Frequent; widespread in all lowland areas, becoming rare above 100 m.
2, 4–6, 8–17
Contains protolichesterinic and rangiformic (±) acids.

C. hepatizon (Ach.) Vain.
There is a specimen in (BM), ex herb. Dillwyn Llewelyn received 1889, named '*Parmelia tremelloides*' from 'Mull' which is this species. Although the underside is abnormally black and the pycnoconidia have tapered ends, the medulla is P+ red and contains stictic acid. (*C. commixta* (Nyl.) Th. Fr., not recorded for Mull, contains α-collatolic acid).

C. islandica (L.) Ach.
Confined to low *Calluna* and short turf on or near the summits of mountains, especiallly of the Ben More massif.
Occasional, but usually abundant and in extensive patches where it occurs.
7, 11, 15, 16
Contains fumarprotocetraric acid and an unknown substance (? protolichesterinic acid).

C. sepincola (Ehrh.) Ach.
On a single moribund *Betula.*
Rare.
11 Coladoir River, Teanga Brìdeig.
Contains protolichesterinic acid.
An interesting western extension to the range of a species with a predominantly northern distribution and otherwise confined to valleys in the Grampian mountains, The Cheviots and a single locality in South Wales.

Cetrelia Culb. & Culb.

Bibliogr.: Culberson and Culberson (1968).
C. olivetorum (Nyl.) Culb. & Culb.
C. cetrarioides (Del. ex Duby) Culb. & Culb.; *Parmelia cetrarioides* (Del. ex Duby) Nyl.
On mossy rocks as well as branches and boles of broad-leaved trees in sheltered woodland sites, especially in the east.
Occasional to frequent; especially in **12** and **13.**
3–5, 7, 8, 10, 12–17
All samples contain imbricaric acid and atranorin; neither olivetoric nor perlatolic acids were detected in any specimen.

Cornicularia Ach.

C. aculeata (Schreb.) Ach.
On dry, well-drained, peaty soils associated with *Calluna–Erica cinerea* and on humus rich soil in crevices of acid rocks, especially amongst boulders in stable scree on the Ben More massif; also occasionally on soil on old stone walls.
Occasional; but widespread in most open, upland sites. Ascending to 900 m on Ben More.
2–5, 10–13, 15–17
Contains protolichesterinic acid.
Thalli are seldom well-developed and often consist of a few, poorly developed, prostrate branches.

C. muricata (Ach.) Ach.
Occurring in similar habitats to *C. aculeata* but perhaps more frequent in pockets of soil on and among boulders on upland scree slopes.
Occasional; but widespread in most upland sites. Less frequent than *C. aculeata.*
3, 5, 6, 8, 11, 13, 16, 17
Contains protolichesterinic acid.

C. normoerica (Gunn) Du Rietz
On exposed acidic boulders and outcrops on or near the summits of mountains.
Rare; mainly confined to the Ben More massif.
11 Ben More, at c. 940 m; Beinn a' Mheadhoin, at 580 m; Cruachan Dearg, at 690 m. **15** Beinn Talaidh, at 740 m, *U. Duncan.*
No lichen substances detected. Thalli are mostly poorly developed and without ascocarps.

Evernia Ach.

E. prunastri (L.) Ach.
On wayside and woodland broad-leaved trees and conifers in all lowland areas, particularly on *Larix* in mature plantations on the eastern side of the Island.
Abundant, but seldom in quantity at any one site.
2–17
Fertile specimens were collected at **3** Loch Caol, Ardfenaig.
Contains usnic and evernic acids and atranorin(±).

Hypogymnia (Nyl.) W. Wats.

See also *Parmelia* Ach.

H. physodes (L.) Nyl.
On boles and branches of broad-leaved trees, especially *Quercus* and *Betula* in all wayside and woodland sites, also on conifers, especially *Larix*, at the margins of plantations; rarer on rock outcrops, boulders, conifer palings, and old stems of *Calluna* in both sheltered areas and cliff tops.
Frequent; widespread in lowland areas, becoming less so in upland sites.
1–17
Contains atranorin, physodalic and physodic acids and two unknown substances. Fertile specimens are not uncommon.

H. tubulosa (Schaer.) Hav.
Parmelia tubulosa (Schaer.) Bitt.
On boles and branches of broad-leaved trees and conifers in all wayside and woodland areas. More rarely on old stems of *Calluna*, old palings and rock outcrops.
Frequent and widespread; with a distribution and abundance similar to that of the preceding species. However, rarer than *H. physodes* on broad-leaved trees but more frequent than this species on conifers in some eastern areas of the Island.
2–17
Contains atranorin, physodic acid and two unknown substances. Fertile in **2, 3** and **16.**

Menegazzia Massal.

M. terebrata (Hoffm.) Massal.
On broad-leaved trees, rarely on conifers and boulders, in sheltered, damp lowland woodland; also on *Corylus* and *Salix* in carrs. Rare on sheltered mossy rocks near the coast.
Frequent, often locally abundant; often well-developed and forming extensive colonies in suitable habitats.
3, 4, 6–10, 12, 13, 16, 17
Contains stictic, constictic, norstictic (±) acids, atranorin and three unknown substances.

Parmelia Ach.

P. alpicola Th. Fr.
Confined to an exposed rock outcrop on a summit plateau.
Rare.
11 Ben More, at 940 m; three thalli in poor condition.
Contains alectorialic, protolichesterinic (±) acids.

P. britannica D. Hawksw. & P. James
Xeric supralittoral; acidic rocks, sometimes spreading to soil and low-growing plants.
Locally frequent in the west; seldom far from the coast.
2 Calva and Carraig an Daimh. **3** Erraid, near the observatory, on granite; Fionnphort, near jetty. **4** Ulva, Craigaig; Gometra, Slatham, near Burial Ground. **5** North side of Loch Tuath, Lòn Reudle.
Contains gyrophoric acid and atranorin.
This recently described species resembles *P. revoluta* Flörke but differs in the more elongated lobes with sinuate axils, the blue-black, granular, subapical soralia, and in the more or less exclusively saxicolous maritime habitat.

P. caperata (L.) Ach.
On boles and main branches of broad-leaved trees, more rarely conifers and mossy rocks in lowland wayside, valley and mature woodland sites; also rarely on exposed coastal rocks and associated soil in sunny situations.
Occasional to frequent; often locally abundant on individual trees.
2–10, 12–14, 16, 17
Rarely fertile; of the 154 collections examined, only 3 had ascocarps. Contains caperatic, usnic and protocetraric acids.

P. conspersa (Ehrh. ex Ach.) Ach.
On acid rocks in sunny, exposed situations; also on slate roofing-tiles and old walls.
Frequent in all areas from sea-level to the summit of Ben More.
1–4, 6–8, 10–17
Frequently fertile; contains usnic, stictic, constictic, norstictic (±) acids, three unknown substances and atranorin.

P. crinita Ach.
On mossy tree-boles, boulders and rock-faces, also sometimes entangled with *Calluna* and other low-growing woody vegetation. Abundant in coastal and lowland areas, becoming very rare in upland sites.
1–10, 12–14, 16, 17
Contains stictic, constictic acids and atranorin.

P. delisei (Duby) Nyl.
Xeric supralittoral; on sunny, exposed rocks, frequently associated with the *Ramalina siliquosa* aggr.
Occasional; especially on Iona, becoming rarer in the east.
1–3, 5, 7, 8, 13
Indistinguishable from *P. pulla* except in its chemistry.
Contains glomelliferic, perlatolic (±) acids.

P. discordans Nyl.
On exposed boulders in stable scree.
Rare; only recorded once but probably overlooked for *P. omphalodes.*
11 Ben More massif, Maol nan Damh, at 630 m, *U. Duncan.*
Contains lobaric, protocetraric acids and atranorin.

P. endochlora Leight.
On mossy boles of broad-leaved trees overhanging streams, or in humid or boggy sites.
Rare; but sometimes locally frequent in the sites recorded.
3 Beinna Ghlinne Mhòir, on *Corylus*; Bunessan, Loch Caol, Ardfenaig Woods. **12** Salen, on *Fagus.* **13** Tobermory, Cnoc Cappullach on *Betula.* **16** Loch Spelve, Rubha Riabhach, on *Betula.*
Easily overlooked for *P. laevigata*, a much more frequent species. The latter has larger lobes and is lighter grey in colour, especially when wet. In *P. endochlora* the medulla is pale primrose yellow whereas in *P. laevigata* it is pure white. *P. endochlora* contains atranorin, barbatic, obtusatic and norobtusatic acids.

P. exasperata DNot.
P. aspera Massal.
On smooth bark of branches and twigs of broad-leaved trees and shrubs, especially *Acer pseudoplatanus*, *Crataegus*, *Fagus*, *Fraxinus*, *Prunus spinosa*, *Salix* and *Sorbus aucuparia*, in open habitats.
Abundant; widespread in most coastal, lowland and upland sites.
2–6, 8–11, 13–17
No lichen substances detected.

P. exasperatula Nyl.
On *Tilia* near the coast.
Rare.
13 Tobermory, along the harbour front.
No lichen substances detected.

P. glabratula (Lamy) Nyl. subsp. **glabratula**
On boles and branches, rarely twigs, of broad-leaved trees, rarely on conifers, in wayside, valley and woodland sites; also rarely on sunny rocks and roofing slates.
Abundant; in all lowland and coastal areas, becoming less frequent in upland sites.
1–17
Contains lecanoric acid.

P. glabratula subsp. **fuliginosa** (Fr. ex Duby) Laund.
On acid rocks and boulders and especially old walls and roofing slates; rarely on boles of old wayside trees and decorticated wood (including driftwood); in exposed, sunny sites.

Frequent; often locally abundant, in all areas from sea-level to the summit of Ben More.
1–6, 8–17
Contains lecanoric acid.

P. incurva (Pers.) Fr.
Confined to boulders in consolidated scree on a north-facing slope.
Rare.
11 Ben More massif, Beinn Fhada, Coire nan Gabhar, at 370 m.
Contains usnic, alectoronic, fumarprotocetraric (±) acids and atranorin.

P. laciniatula (Flag. ex Oliv.) Zahlbr.
On *Tilia* near the coast.
Rare.
13 Tobermory, along the harbour front.
No lichen substances detected.

P. laevigata (Sm.) Ach.
On mossy boles and branches of old trees in sheltered valleys and mature woodland; particularly frequent in sheltered *Salix* carr and in mossy upland *Betula–Quercus* woodland; also more rarely on mossy boulders in sheltered lowland sites by streams and along loch shores.
Frequent; widespread.
2–17
Contains barbatic, obtusatic and norobtusatic acids and atranorin. Only three fertile thalli were recorded.

P. loxodes Nyl.
P. isidiotyla Nyl.
Xeric supralittoral; on acid rocks and boulders predominantly in sunny, exposed sites; less frequently near inland farmsteads and along shores of sea- and fresh-water lochs.
Frequent; more common than *P. verruculifera*, in all coastal areas.
1–13, 16, 17
Contains either glomelliferic, perlatolic and stenosporic (±) acids.

P. mougeotii Schaer.
On granite boulders and rocks on exposed hillsides.
Rare; confined to the Ross of Mull.
3 Erraid, above Erraid Sound; Bunessan, Ardtun, Tòrr Mòr; Fionnphort, Catchean; Kintra; Rubh' Ardalanish, Tòrr na Sealga.
Contains usnic, stictic acids and atranorin.

P. omphalodes (L.) Ach.
On acidic rocks, very rarely on wayside tree boles; particularly frequent on boulders in stabilised scree on the Ben More massif.
Frequent to abundant; especially in upland areas.
1–17
Contains salazinic, lobaric acids and atranorin; see also *P. discordans*.

P. pastillifera (Harm.) Schub. & Klem.
On sunny, exposed rocks and boles of isolated, broad-leaved trees in sunny lowland sites.
Rare; confined to the west.
3 Bunessan, on *Fraxinus*. **4** Ulva, on basalt rocks, Craigaig. **8** Inch Kenneth, rocks near chapel.
Contains lecanoric acid and atranorin. All specimens have dark, verruciform isidia; material with concolorous, elongate isidia, corresponding to *P. tiliacea* (Hoffm.) Ach., was not seen.

P. perlata (Huds.) Ach.
On mossy wayside and woodland trees and boulders; also on old walls and amongst *Calluna* and other low-growing vegetation, in sheltered sites on the coast.
Abundant.
Ascending to 340 m in **15** Craignure, above Coill an Fhraoich Mhòir.
1–17
Contains stictic, constictic, norstictic (±) acids, three unknown substances and atranorin. Fertile specimens were recorded from **3** and **7**.

P. pulla Ach.
P. prolixa (Ach.) Carroll
Predominantly xeric supralittoral. On sunny acid rocks in exposed sites, often in association with *Ramalina siliquosa* aggr.
Frequent; becoming abundant on the west coast.
1–6, 9, 10, 12, 13, 16
Contains stenosporic, gyrophoric (±) acids; one specimen with divaricatic and gyrophoric acids.

P. reddenda Stirt.
On broad-leaved trees and conifers in mature, mixed woodland in lowland sites.
Rare.
7 Torloisk House, luxuriant on boles of *Picea abies*. **13** Tobermory, Aros Park, on *Fraxinus*. **17** Torosay Castle grounds, on *Quercus*.
Contains several unknown fatty acids and atranorin.
These records are the most northern in Britain for this species.

P. reticulata Tayl.
On boles of old broad-leaved trees, especially *Fagus* in sheltered lowland valleys.
Rare, but often locally abundant where it occurs.
5 Kilninian, on *Fraxinus* beside Allt Hostarie. **7** Torloisk House on *Fagus*. **13** Tobermory, Aros Park, near Aros House, on *Ilex*. **16** North shore of Loch Uisg, on dead *Fagus*.
Contains salazinic acid and atranorin.

P. revoluta Flörke
On wayside and woodland trees in most sheltered lowland and, more occasionally, upland sites; most frequently recorded on *Fagus*, rarely on acid rocks in valley woodlands.

Frequent; widespread in lowland sites becoming rarer in upland areas.
3–8, 10–13, 15–17
Contains gyrophoric acid and atranorin.

P. saxatilis (L.) Ach.
On broad-leaved trees and conifers and acid rocks in all areas.
Abundant; especially in upland areas.
1–17
Occasionally fertile.
Contains salazinic, lobaric (±) acids and atranorin.

P. sinuosa (Sm.) Ach.
On broad-leaved trees, chiefly *Corylus*, *Betula*, *Salix* and *Sorbus aucuparia* in upland valley woodlands, especially near bogs; also on rocks by sheltered streams.
Occasional to frequent; especially in the east.
3–5, 8, 9, 11–14, 16
Contains salazinic, consalazinic, norstictic (±) and usnic acids.

P. subaurifera Nyl.
Predominantly on smooth bark of twigs and branches of broad-leaved trees and shrubs, especially *Fraxinus*, *Prunus spinosa*, *Salix* and *Tilia*.
Frequent in most lowland and upland areas.
2–6, 8–11, 13–17
Contains lecanoric acid, skyrin (rhodophyscin).

P. subrudecta Nyl.
On wayside and woodland trees, usually in well-illuminated sunny, lowland and coastal sites; recorded on *Fraxinus*, *Ilex*, *Prunus spinosa*, *P. cerasus* and *Quercus*.
Occasional.
5, 7, 9–11, 13, 15, 16
Contains lecanoric acid and atranorin.

P. sulcata Tayl.
On rocks, boulders, trees, roofing tiles in most coastal, lowland and upland sites.
Abundant; widespread.
1–17
Contains salazinic acid and atranorin.
Occasionally fertile.

P. taylorensis Mitch.
On mossy boles of old trees and on boulders in mature woodland and in sheltered, humid, well-wooded valleys. More rarely in upland *Quercus–Betula* woodland on south-facing slopes and on *Calluna* and other woody plant detritus on or near exposed coasts.
Frequent; widespread in coastal and lowland areas, becoming rarer in upland sites, and not seen above 100 m.
2, 4–7, 10–13, 15, 16
Contains lecanoric and evernic acids, and atranorin.

P. verruculifera Nyl.
P. glomellifera (Nyl.) Nyl.
Xeric supralittoral; on sunny exposed coastal rocks; rarely, in more inland sites, on walls, roofing slates, boulders in streams and along margins of lochs.
Frequent; in coastal areas, especially in the west.
1–5, 7, 9–12, 14, 17
Contains divaricatic, gyrophoric (±) acids.

Platismatia Culb. & Culb.

Bibliogr.: Culberson & Culberson (1968).
P. glauca (L.) Culb. & Culb.
Cetraria glauca (L.) Ach.
On branches and boles of wayside and woodland broad-leaved trees and conifers, being particularly frequent on older *Larix* plantations and on old *Betula* in upland valley sites; more rarely on stable boulder scree, rock outcrops, old stems of *Calluna*, vegetation detritus, soil, fence posts and rails.
Abundant; widespread in all areas.
1–17
Very polymorphic. Contains atranorin and caperatic acid.

Pseudevernia Zopf

Bibliogr.: Hawksworth & Chapman (1971).
P. furfuracea (L.) Zopf. var. **furfuracea**
On acid boulders in stable scree in a lowland site.
Rare.
10 Mornish, Glengorm Castle, near Standing Stones.

P. furfuracea var. **ceratea** (Ach.) D. Hawksw.
On boles of mature conifers in wayside sites and plantations, particularly in lowland areas away from the sea; also recorded on acid rocks and boulders in a few lowland and upland areas.
Occasional; chiefly confined to the east.
4 Ulva, below Dùn Bhioramuill, on *Betula*. **10** Mishnish, near ruins of Ardmore. **12** Salen Forest, on *Larix* in plantation below Guala Achach nan Each; Ardnacross, on branches of *Pinus*. **14** Salen Forest, near Garmony, on old boles of *Pinus* and conifer palings.
This taxon is represented by two chemical races in Britain: var. *furfuracea* with physodic acid in the medulla (med. C-, KC+ rose-red) and var. *ceratea* containing olivetoric acid in the medulla (med. C+ red, KC+ red). In agreement with the findings of Hawksworth & Chapman (1971), the latter is more frequent of the two races on Mull, var. *furfuracea* being mainly southern in distribution in Britain.

Usnea P. Browne ex Adans.

U. distincta Motyka
On boles and branches of *Larix* in a sheltered plantation amongst evergreen conifers.
Rare.
12 Salen Forest, near the eastern entrance.
Characterised by the numerous, stout, elongated branches arising from near the base of the

thallus and the comparative absence of lateral branchlets. The soralia are minute and fleck-like.
Rare in Scotland but also recorded from the Ardnamurchan Peninsula.
Contains stictic, constictic, norstictic, usnic acids and three unknown substances.

U. extensa Vain.
On boles and branches of broad-leaved and coniferous trees, usually in sheltered plantations or moist boggy situations.
Rare.
12 Salen Forest, near the eastern entrance, on *Larix*; Glenforsa Hotel, on *Salix aurita*. **13** Tobermory, Upper Druimfin, on boles of *Betula* in small *Pinus* stand.
In this species the thallus is pendulous to sub-pendulous, the few main branches terminating with straggling contorted apices. Lateral branches are usually densely developed near the blackened base. The soralia are widely spaced, rounded, excavate and very conspicuous even to the naked eye. Two chemical strains are present on Mull: i) Stictic, constictic, norstictic (in quantity), usnic acid and three unknowns; ii) Psoromic, conpsoromic and usnic acids; the latter strain is represented by the gathering from the Salen Forest.

U. filipendula Stirt.
On boles and branches of conifers and broad-leaved trees in sheltered, moist habitats, or more rarely on trees in windy, exposed situations.
Rare.
3 Beinne Ghlinne Mhoir, WSW of Bunessan, on wind-blown *Quercus*. **12** Salen Forest, near eastern entrance on *Larix*; Glen Forsa, near Tigh Ban, on *Salix aurita*. The main and secondary stems of the pendulous thallus are thickly coated in style-like isidia. The specimen from sheltered *Larix* in Salen Forest was fertile. Contains salazinic and usnic acids.

U. flammea Stirt.
U. rupestris Mot.
Scrambling over *Calluna*, more rarely attached to rocks and boulders, especially on exposed cliff tops in coastal areas.
Rare.
2 West of Baille Mòr; Cnoc Druiden; Cul Bhuirg. **3** Ardtun, Tòrr Mòr, on aged *Calluna* stems and associated rocks; Erraid, Cnoc. **16** South of Loch Uisg, Lochan nan Caorach, on *Calluna*.
Characterised by the elongate, epapillate, vermiform branches and the absence of lateral branches except towards the apices. The soralia are minute and fleck-like and the main branches have numerous and conspicuous annulations. Contains stictic, constictic, norstictic (±) usnic acids and 3 unknown substances.

U. fragilescens Stirt.
On boles, branches and twigs of broad-leaved trees and conifers, especially in sheltered valley and mature woodland sites. Abundant in all lowland areas, rarely seen above 90 m.
2–17
This species, the most frequently encountered of the genus on the Island, is characterised by the tufted habit, the more or less inflated main branches which are slightly narrowed where they join the main stems, the more or less lax medulla, and the lack of conspicuous blackening at the base. The isidia are in clusters and are similar to those found in *U. subfloridana*. Contains stictic, constictic, norstictic (±), usnic acids and three unknown substances.

U. fulvoreagens Räs.
On twigs, rarely branches, of broad-leaved trees and shrubs in very sheltered lowland sites. Rare; only recorded from a few localities in the west.
7 North side of Loch Tuath, Normann's Ruh, on *Prunus spinosa*; Torloisk House gardens, on *Malus*. **8** North side of Loch Scridain, Scobull Point on *Prunus spinosa*, *U. Duncan & James*. **9** Brolass, Killunaig, on *Crataegus* by road-side.
This species has a compact, small, richly-branched thallus with numerous twisted branches with contorted apices. The soredia are numerous, especially near the branch endings and there often become confluent giving the apices the appearance of skeletal fingers. Contains salazinic and usnic acids.

U. glabrescens Motyka
On branches and boles of wayside trees in sheltered lowland sites.
Rare, but possibly overlooked and confused with *U. subfloridana*.
10 South side of Loch Scridain, conifer woods one mile east of Pennyghael, on *Pinus*. **16** North side of Loch Uisg, Ath Leitir, on *Quercus*.
Apart from the presence of true soralia, this species seems identical and may be conspecific with *U. subfloridana*. Contains either salazinic and usnic acids and an unknown substance (? barbatic acid) or thamnolic, alectorialic (±) and usnic acids.

U. hirta (L.) Web. em. Motyka
On boles of conifers in open lowland sites.
Rare.
6 Mishnish, Glengorm Castle, near Ballimeanach and Sorne. **17** Torosay House, on *Pinus* near the pier.
This species has a totally flaccid appearance when wet. The main branches have a foveolate surface, a lax medulla and are sparsely to densely covered in short, spinulose isidia which sometimes enlarge into small branchlets. Contains protolichesterinic, usnic acids.

U. inflata Del. ex Duby
U. intexta Stirt; *U. subpectinata* Stirt.
On boles of deciduous trees especially at the margins of woods. Rarely on *Salix* in sheltered carr.
Frequent.
2, 5–8, 10–16
Two chemical strains are represented in the Mull populations of this species; the most frequent is that containing stictic, constictic, norstictic (±) and usnic acids and three unknown substances. A few collections contained salazinic, consalazinic and usnic acids corresponding to *U. intexta* var. *constrictula* (Stirt.) D. Hawksw.

U. rubiginea (Michaux) Massal.
On boles of broad-leaved and coniferous trees in sheltered wayside, valley and mature woodland sites in lowland areas. Particularly frequent on *Pinus*.
Occasional; widespread in all lowland areas, but never abundant in any one area.
3–6, 8, 10–13, 15–17
Very easily distinguished by the red-brown colouration of all or part of the thallus. Contains stictic, constictic, norstictic (±) and usnic acids and two or three unknown substances.

U. subfloridana Stirt.
On boles, branches and twigs of sheltered or exposed broad-leaved trees, shrubs and conifers in wayside, valley and mature woodland sites; more rarely on stems of old *Calluna* and on rocks and boulders. Not as generally abundant as *U. fragilescens*, but more widespread, occurring in more open upland habitats, especially in *Salix*, and *Quercus – Betula* woodland.
Frequent; widespread.
2–17
This species is characterised by the tufted, rarely subpendulous habit, the numerous branches and the blackened base of the thallus. The medulla is persistently compact and solid. The isidia occur in small clusters and are easily abraded and then simulate soralia. The thallus is very polymorphic; some specimens are well fertile and resemble *U. florida* which, however, is not known as far north as Mull. *U. subfloridana* occurs as two chemical strains on Mull: i) thamnolic, alectorialic (±) and usnic acids; ii) barbatic (±), alectorialic (±), salazinic (±) and usnic acids. The former strain is the most frequent, occurring in 75 per cent of the specimens.

LECANORACEAE

Haematomma Massal.

Bibliogr.: Laundon (1970); Hawksworth (1970).
H. elatinum (Ach.) Massal.
On boles of broad-leaved trees and conifers, especially *Fagus, Fraxinus, Ilex* and *Sorbus aucuparia* in damp, lowland, sheltered woodland and wayside sites; frequent on *Betula* and *Quercus* and rarely on mosses overgrowing acid boulders in upland woodlands.
Frequent; especially in sheltered, moist woodlands, becoming rare in more exposed upland sites.
3–5, 7, 9, 11–13, 15–17
Contains thamnolic acid.

H. ochroleucum (Neck.) Laund. var. **ochroleucum**
Often dominant under dry, acidic rock overhangs and along sheltered cliff-faces and on the underside of stabilised boulders in scree; rarely directly wetted by rain.
Abundant; widespread.
1–17
Fertile specimens were recorded in **2, 4, 6, 7, 13–15**

H. ochroleucum var. **porphyrium** (Pers.) Laund.
In similar situations to and often in association with the species but also on sheltered, dry sides of old wayside trees in lowland and coastal areas.
Frequent; widespread, but not as abundant as the species on rocks.
2, 4–11, 13, 14, 16, 17
The two varieties of *Haematomma ochroleucum* are conveniently separated in the field by the colour of their thalli: var. *ochroleucum* contains usnic acid and is yellow-green, whereas var. *porphyrium* lacks this acid and is grey-green. Both varieties contain porphyrilic acid and atranorin. Whilst the type variety is exclusively saxicolous and mainly confined to highland Britain, this variety is both saxicolous and corticolous and is also frequent in lowland Britain. Both taxa were occasionally found with ascocarps.

H. ventosum (L.) Massal.
On acid rocks and boulders, especially in stable scree on exposed upland areas; very rare below 90 m.
Occasional; but probably widespread.
7 Kilmore, Cruachan Ceann a' Ghairbh, on basalt. **11** Ben More, near summit 946m; Creag Mhic Fhionnlaidh, *c.* 300 m; Cruachan Dearg, *c.* 600 m. **12** Glen Aros, Cnoc an Tota, on glacial erratic at *c.* 150 m. **16** Lochbuie, above Loch Airdeglais, Creag na h-Iolaire at *c.* 300 m; Laggan Deer Forest, Allt Ohirnie, on basalt, 225 m.
Contains thamnolic, divaricatic and usnic (±) acids. Specimens from Ben More and Allt Ohirnie are without usnic acid.

Lecania Massal.

L. aipospila (Wahlenb. ex Ach.) Th. Fr.
Mesic supralittoral. Often near roosts of sea birds, in sheltered recesses in rock outcrops and cliffs.
Rare; very local.
2 West coast, near Eilean Maol Mhàrtuin. **8** Ardmeanach, Carraig Mhic Thòmais. **15**

Scallastle Bay, Alterich, on crumbling basalt at entrance to caves, *U. Duncan & James.*

L. cyrtella (Ach.) Th. Fr.
Chiefly on twigs and small branches of *Sambucus nigra* near farmsteads; more rarely on boles of *Acer pseudoplatanus*, *Fraxinus* and *Ulmus* in wayside inland and coastal sites.
Occasional to frequent; especially in the west.
2–6, 8, 10, 12, 13, 15, 17

L. erysibe (Ach.) Mudd
Mainly on mortar of old walls in coastal, lowland, and more rarely, upland sites; on coastal rocks, especially in base- or nutrient-rich or sheltered habitats.
Occasional to frequent; especially in coastal sites.
2, 3, 5, 6, 8, 10, 12, 13, 15, 17
Forma **sorediata** Laund. was recorded in **3** Kintra, on concrete bridge.

L. ralfsii (Salwey) A. L. Sm.
Xeric supralittoral; on rocks, mainly in the upper part of the *Verrucaria maura* zone.
Rare; possibly overlooked.
2 West coast, near Spouting Cave. **4** Gometra, Rubha na h-Àirde. **17** Craignure, Rubha na Sròine.

L. rupicola (Nyl.) P. James
Sheltered acidic rocks on the coast.
Rare.
2 West coast, near spouting cave.
Contains atranorin, chloratranorin and gangleoidin.

Lecanora Ach.

L. actophila Wedd.
Mesic supralittoral; strictly confined to a narrow zone on acidic rocks and boulders within or just above the *Verrucaria maura* zone on all exposed to moderately sheltered shores.
Abundant; on all sunny, exposed headlands, becoming rarer towards the heads of larger sea-lochs.
1–17
Often associated with *L. helicopis*. Contains unknown pigments.

L. andrewii B. de Lesd.
Xeric supralittoral; only known from a large granite erratic near the coast.
Rare.
2 North-east coast, Ardura.
Contains arthothelin and an unknown substance.

L. atra (Huds.) Ach.
On exposed or sheltered rocks on or near the coast and on old walls, roof tiles and rocky outcrops in inland areas; rarely on boles of old wayside trees in lowland and coastal sites.
Frequent.
1–5, 7–9, 11–17
Contains α-collatolic acid and atranorin.

Specimens seen were often without well-developed ascocarps.

L. badia (Hoffm.) Ach.
Chiefly restricted to granite rocks; elsewhere, on boulders on scree slopes on the Ben More massif.
Locally frequent in the south-west; occasional to rare elsewhere.
Contains lobaric acid and an unknown substance.
1–4, 7, 8, 11, 13, 16
In sites outside the granite exposures on the Ross of Mull the thalli are often very poorly developed and the ascocarps abortive.

L. caesiocinerea Nyl. ex Malbr.
On acidic rock outcrops and boulders, especially those enriched by organic nutrients, such as birds' perching stones and near farmsteads, in coastal, lowland and, more rarely, upland sites.
Frequent; especially near the coast.
1–14, 16, 17
No lichen substances detected.

L. calcarea (L.) Sommerf.
On mortar of old stone wall near the coast.
Rare.
12 Salen, Ard Mòr, near the pier.
This species is abundant on Lismore Island.
No lichen substances detected.

L. campestris (Schaer.) Hue
On exposed to sheltered boulders on the coast; also on old walls, roof tiles, rock outcrops, especially those in the vicinity of farmsteads and associated with nutrient- or base-rich substrata.
Frequent; widespread, but seldom abundant in any one site.
1–17
Contains atranorin.

L. carpinea (L.) Vain.
On smooth bark of twigs and young trees, especially *Acer*, *Fraxinus* and *Quercus*.
Frequent; especially in the east.
2, 4–7, 9, 11, 13–15, 17
Contains sordidone.

L. chlarona (Ach.) Nyl.
On boles of broad-leaved trees and conifers in wayside and woodland sites; especially frequent on conifer fence rails at the edge of woodlands and plantations; also on *Salix* in carr and bogs.
Frequent; widespread.
2–5, 7–13, 16, 17
Contains atranorin, chloratranorin and fumarprotectraric acid.

L. chlarotera Nyl.
On boles of broad-leaved trees, more rarely on conifers in all wayside and woodland sites.
Abundant to frequent; very widespread.
1–17

Very polymorphic. Contains atranorin, chloro-atranorin and gangleoidin (±).

L. confusa Almb.
On boles of broad-leaved trees and conifers in wayside and woodland sites in coastal, lowland and, more rarely, upland areas; also frequent on decorticated wood of fence rails and gates, especially near the coast.
Frequent to abundant.
1–17
Contains thiophaninic acid.

L. conizaeoides Nyl.
On bole of *Tilia*
14 Tobermory, harbour front.
Rare.
The presence of this lichen only within the largest conurbation in the Island is indicative of the very low levels of sulphur dioxide pollution (15 μgm/m^3) on Mull. The species is very rare on Lismore Island but more frequent in and around Oban on the mainland. Contains fumarprotocetraric and usnic acid. Records in areas with very low sulphur dioxide levels are nearly always on decorticated conifer fence-posts indicating that the species may have existed as a rare and unrecorded coloniser in such habitats prior to its rapid spread resulting from industrialisation and increased urbanisation of many lowland areas of Britain in the nineteenth century.

L. contorta (Hoffm.) Steiner
L. hoffmannii Ach., *nom. illegit.*
On limestone and other basic rocks on or near the coast.
Rare; limited by the scarcity of suitable habitats.
8 Inch Kenneth, Port an Ròin. **10** Carsaig, Sgeir Bhuidhe, around the old quarries.
This species is frequent on the limestone exposures of Lismore Island.
No lichen substances detected.

L. crenulata (Dicks.) Hook.
On calcareous sandstone and limestone outcrops on or near the coast; also on old cement, mortar and, more rarely, asbestos.
Occasional; widespread but seldom abundant in any site.
2–4, 6, 8, 10, 12, 13, 16, 17

L. dispersa (Pers.) Sommerf.
On rocks especially in mineral-, nutrient- or base-rich sites; also on mortar and cement, more rarely asbestos, associated with old buildings and farmsteads. Infrequent on boles of wayside and woodland trees.
Frequent; often locally abundant in suitable habitats.
1–17
Very polymorphic.

L. expallens Ach.
On boles and branches of broad-leaved trees and conifers in wayside and woodland sites; also frequent on decorticated wood of old posts, fence rails and gates, especially near the coast, often in associations with *L. confusa.*
Abundant; in all lowland and coastal sites, becoming rarer in upland areas.
1–17
Contains thiophaninic acid.

L. fugiens Nyl.
Xeric supralittoral. On rocks and boulders, usually on or near the coast.
Occasional.
2 Cùl Bhuirg; Eilean Didil; Rubha na Carraig-géire; Arduara. **3** Camas Tuath. **4** Ulva, Craigaig, *U. Duncan & James*; Gometra, Rubha Dùn Iasgain. **12** Salen. **15** Scallastle Bay, Alterich.
Contain arthothelin and an unknown substance.

L. gangaleoides Nyl.
On acidic rocks, especially in sheltered areas near the coast and in broad-leaved woodlands; more rarely on boulders on upland scree slopes and by the margins of fresh water lochs.
Frequent to occasional; widespread.
1–6, 8–10, 12–14, 16, 17
Contains gangaleoidin, atranorin and an unknown pigment.

L. gibbosa (Ach.) Nyl.
On acid rocks, especially in sunny situations on or near the coast; occasionally on roof tiles and exposed upland scree boulders.
Occasional; becoming more frequent near the coast.
1–4, 6, 8, 12, 13, 16

L. helicopis (Wahlenb. ex Ach.) Ach.
Mesic supralittoral, restricted to acid rocks and boulders within or just above the *Verrucaria maura* zone on all exposed to moderately sheltered shores, often in association with *L. actophila* but more tolerant of sheltered habitats.
Abundant; on all exposed headlands and shores, becoming rarer towards the heads of larger sea-lochs.
1–17

L. intricata (Ach.) Ach.
On exposed acid rocks and boulders in coastal, lowland and particularly, upland sites.
Frequent; widespread.
2–17
Contains usnic acid.

L. intumescens (Rebent.) Rabenh.
On smooth bark of twigs and young trees, especially *Acer pseudoplatanus*, *Fraxinus* and *Quercus* in mainly coastal and lowland sites.
Occasional; probably overlooked.

3, 4, 6, 7, 10–13, 17
Contains psoromic, conpsoromic acids, atranorin, chloratranorin and an unknown substance (±). Some specimens also contain lichexanthone.

L. jamesii Laund.
On smooth bark of young broad-leaved trees and especially *Salix* in or by carr and streams in sheltered, often well-wooded coastal and lowland sites. Rarely on decorticated wood in damp woodland.
Frequent; in most suitable lowland sites.
2–14, 16, 17
Approximately half the gatherings have ascocarps.
Contains atranorin, chloratranorin and gangaleoidin.

L. lacustris (With.) Nyl.
On intermittently inundated rocks and boulders in streams, rivers and lochs; also in seepage runnels on large boulders and rock outcrops. Most frequent in sunny sites but can also tolerate considerable shade.
Abundant.
2–17

L. laevata (Ach.) Nyl.
On wet, sometimes periodically inundated, acidic rocks, especially by streams and loch shores. Often with *L. lacustris* but generally preferring less frequently inundated habitats.
Frequent; widespread.
2–5, 7, 8, 10, 11, 13, 15–17

L. leprosescens Sandst.
Xeric supralittoral; on acidic rocks and boulders enriched by nutrients on or near the coast.
Locally frequent; especially in the southwest.
1, 5, 7, 12, 13, 16, 17
Fertile specimens found in **2**, **3**, **7** and **12**. Contains atranorin and chloratranorin.

L. muralis (Schreb.) Rabenh.
On smooth rocks and boulders in streams and along loch shores; on acidic rock outcrops and boulders manured by birds or near farmsteads, especially near the coast; more rarely on gneiss rocks partially covered by windblown shell-sand, as well as roof tiles and asbestos.
Frequent; especially in lowland and coastal areas.
1, 4, 6, 7, 9, 12, 13, 15–17
Specimens from upland sites are mainly confined to birds' perching rocks.
Contains usnic acid and zeorin.

L. pallida (Schreb.) Rabenh.
On smooth bark of twigs, branches and young boles, especially *Acer pseudoplatanus*, *Fraxinus*, *Salix* spp. and *Sorbus aucuparia* in most lowland and coastal wooded areas.
Frequent.
3, 5–9, 11–13, 15–17
Often confused with *L. carpinea*. Contains protocetraric acid, atranorin and chloratranorin.

L. piniperda Körb.
On bark, decorticated wood of moribund conifers and palings in old mixed woodland and plantations, particularly in lowland areas.
Rare; probably overlooked.
6 Glengorm Castle near Port Chill Bhraonain. **12** Salen on pine boles, *U. Duncan*. **13** Tobermory, Aros Park, on old pines and palings; north side of Loch Frisa, Salen Forest, boles of *Larix*. **16** Torosay Castle grounds on *Pinus sylvestris*.

L. poliophaea (Wahlenb. ex Ach.) Ach.
Mesic supralittoral; confined to nutrient-enriched acidic coastal rocks, especially in the vicinity of birds' perching stones and cliffs.
Occasional; widespread and locally abundant on smaller off-shore islands, e.g., **1** Treshnish Islands, Staffa.
1–4, 6, 12, 14, 15–17
No lichen substances detected.

L. polytropa (Hoffm.) Rabenh.
On acidic rock outcrops and boulders, especially in open moorland areas.
Frequent, but often poorly developed.
1–13, 15–17
Contains usnic, rangiformic (±) acids and zeorin.

L. recedens (Tayl.) Nyl.
In seepage track on large acidic boulder in scree.
Rare, the only record.
12 South of Salen village, Beinn Bhuidhe, 150 m.
No lichen substances detected.

L. rupicola (L.) Zahlbr.
Confined to acidic rocks, especially basalt, on or near the coast.
Occasional to frequent; particularly on Iona and Ulva.
1–4, 6, 10, 12–17
Contains atranorin, chloratranorin, thiophaninic, roccellic acids and one unknown substance.

L. sambuci (Pers.) Nyl.
On twigs and branches of *Sambucus nigra*, especially near farmsteads; very rarely on *Acer pseudoplatanus*. Confined to coastal and lowland sites.
Occasional, but widespread.
2 Baile Mòr, near Nunnery. **4** Ulva, near ferry. **7** side of Loch Tuath, Ballygown, on *Sambucus*; Fanmore. **13** Tobermory, near disused whisky distillery. **16** Croggan.

L. subcarnea (Liljebl.) Ach.
On sheltered, dry recesses of acid rocks on or near the coast.
Rare.

4 Ulva, small caves near Ardalum House. **16** Lochbuie, cliffs below Na h'Airichean. **17** Loch Don, on Druim Mòr Aird na Drochaide, *U. Duncan & James*.
Contains norstictic, connorstictic, fumarprotocetraric (±) acids, chloratranorin and atranorin.

L. superiuscula Nyl.
On wet, acidic rocks in an upland site.
Rare.
11 Sleibhte-coire, below Guibean Uluvailt, 280 m.
Contains miriquidic acid, also present in the holotype, appearing as a turquoise green spot on TLC plates treated with dilute sulphuric acid and heated.

L. sylvatica (Arnold) Sandst.
On smooth, acidic rocks, often in deep shade, in old woodland sites.
Occasional; frequent in the east.
5, 11–13, 16

L. tenera (Nyl.) Cromb.
Xeric supralittoral, confined to dry, sheltered recesses in acidic rocks on or near the coast.
Rare; but probably overlooked.
1 Fladda; Staffa. **2** Soa. **17** Craignure, south of Rubha na Sròine.
Contains stictic, norstictic (±) acids, atranorin, chloratranorin and one unknown substance.

Solenopsora Massal.

S. holophaea (Mont.) Samp.
In dry rock crevices or thin soil on rocks on sunny exposed cliffs on or near the coast.
Rare; local, mainly confined to the west.
2 Cùl Bhuirg. **3** Ardtun, coastal cliffs near Rubha Breac. **4** Gometra, Rubha Dùn Iasgain, on decomposing basalt cliffs. **8** Inch Kenneth, Bàgh an Iollaich. **12** Salen, on low cliffs opposite Sgaith Mhòr, *U. Duncan & James*. **14** Scallastle Bay, Alterich, *U. Duncan & James*.
Contains atranorin.

S. vulturiensis Massal.
On acid rocks and associated soil in sheltered crevices on or near the coast.
Occasional to frequent; especially in the west.
1–10, 12–14
Fertile specimens were found in **2, 4, 5, 7**.
Contains atranorin.

Squamarina Poelt

S. crassa (Huds.) Poelt
On basic sandstone near the coast.
Rare.
10 Carsaig Bay, below Sròn nam Boc, *U. Duncan & James*.
This species is frequent on Lismore Island.
Contains usnic acid and zeorin.

LECIDEACEAE

Arthrorhaphis Th.Fr.

(See also *Bacidia* De Not.em.Zahlbr.)

A. citrinella (Ach.) Poelt
Bacidia citrinella (Ach.) Deichm. Br. & Rostr.
In pockets of humus-rich acidic soil on exposed rock outcrops, stone walls and between boulders, especially on exposed moorland areas above 90 m.
Frequent; widespread.
2, 4, 6, 8, 11–16
Ascocarps were recorded on material from **4, 6, 8, 12–16**. Contains rhizocarpic acid and an unknown substance.

Bacidia De Not.emend. Zahlbr.

(See also *Micarea* Fr. and *Arthrorhaphis* Th.Fr.)
B. arceutina (Ach.) Arnold
On windswept *Sambucus nigra* near the coast.
Rare; only known from a single gathering.
16 Lochbuie, near the chapel, *U. Duncan & James*.
This species is easily recognised by the spiral arrangement of the spores in the asci and the almost colourless hypothecium and epithecium.

B. beckhausii Körb.
Amongst mosses on old broad-leaved trees and conifers, especially in mature woodland and at the edges of boggy clearings; also on rotting decorticated wood in boggy sites.
Rare, but possibly overlooked; mainly confined to upland areas.
3 Ardfenaig Woods. **7** Torloisk House woods, on *Quercus*. **10** Pennyghael, base of *Pinus*, *U. Duncan & James*; Carsaig, Innamore Lodge, on fallen *Pinus*. **13** Tobermory, Upper Druimfin, *Pinus* by the roadside, *U. Duncan & James*.
In this species the epithecium and thecium are K+ mauve.

B. endoleuca (Nyl.) Kickx
On wayside and woodland broad-leaved trees and conifers in sheltered boggy valleys and at the edge of woodland clearings and plantations.
Occasional to frequent.
3–8, 10, 13, 14, 16–17
An aggregate taxon in Britain in need of critical study.

B. friesiana Körb.
On boles and branches of broad-leaved trees in lowland sites, especially near farmsteads.
Rare.
5 Calgary Bay, on *Acer pseudoplatanus*. **7** Ballygown, on *Ulmus*, *U. Duncan*. **16** Lochbuie, on *Sambucus nigra*.
The ascocarps in this species may vary in colour from pale grey to black; see also James (1971).

B. herbarum (Stiz.) Arnold
On dried grass inflorescences and old rabbit

faeces amongst short grass on shell-sand in dune areas near the coast.
Rare.
2 Golf Course.
Differs from *B. muscorum* in the red-brown, not black, ascocarps.

B. inundata (Fr.) Körb.
On sheltered, often poorly illuminated, submerged rocks in streams and lochs. Rarely on periodically submerged tree roots at the margins of streams.
Occasional; but sometimes locally abundant as at **7** Kilbrenan and Ballygown.
3, 7–8, 11, 13, 16–17
The material is very polymorphic. In shaded sites the ascocarps become translucent and pale pink and the hypothecium pale straw yellow or nearly colourless. In more open habitats the ascocarps are black-brown and the hypothecium opaque and red-brown.

B. muscorum (Ach.) Mudd
On consolidated shell-sand and mosses in amongst short turf near the sea; also on old rabbit faeces in similar situations.
Rare; only in the west.
2 Cnoc-nam-bradhan, near Spouting Cave, *U. Duncan & James*. **5** Calgary Bay; Sunipol.

B. naegelii (Hepp) Müll. Arg.
On wayside trees, especially near houses and farmsteads, in lowland coastal areas; frequently on *Sambucus nigra* and there often forming associations with *Caloplaca cerina*, *C. cerinella* and *Lecania cyrtella*.
Frequent.
2–5, 7, 8, 10, 12, 13, 16, 17

B. nitschkeana (Lahm ex Rabenh.) Zahlbr.
On confined sheltered aspect of boles of conifers in old plantations.
Rare.
7 Near Lettermore, on *Larix*. **13** Upper Druimfin, on *Picea*. **17** Torosay House woods, *U. Duncan & James*.

B. phacodes Körb.
On broad-leaved trees, especially wayside *Acer pseudoplatanus*, *Fraxinus*, and *Ulmus*, often in the vicinity of farmsteads; rarely in small very shady ravines.
Occasional.
5 Calgary burial ground, on moribund *Acer pseudoplatanus*, *U. Duncan & James*. **6** West of Dervaig, Druimghigha, on *Ulmus*; Dervaig, on *Acer pseudoplatanus*. **10** An Leth-onn, on *Acer pseudoplatanus*. **17** Torosay Castle on *Ulmus*.
A related species *B. cuprea* (Massal.) Lett. is not recorded for Mull but has been found on shaded recesses in limestone on Lismore Island.

B. rubella (Hoffm.) Massal.
On mature wayside trees, in lowland and coastal areas, especially *Ulmus* spp. and *Acer pseudoplatanus*, near farmsteads and houses.
Occasional to frequent; usually abundant where it occurs.
2–6, 12, 13, 16, 17

B. sabuletorum (Schreb.) Lett.
On basic substrata or acid boulders where these are associated with mortar walls; also on shell-sand near the coast. Rarely on tree boles, especially *Ulmus*.
Frequent; widespread in all lowland areas, ascending to 80 m in **15** Glen Forsa, Tomsléibhe, *U. Duncan*.
2–5, 8, 10–13, 15–17

B. scopulicola (Nyl.) A. L. Sm.
Xeric supralittoral; in sheltered dry rock crevices and beneath overhangs on cliffs, particularly those obscured by summer vegetation.
Rare; only known from very sheltered habitats in the west.
2 Cùl Bhuirg, crevices in coastal gneiss. **3** West of Loch na Làthaick, on granite near Camas Tuath; Bunessan, Lochan Mòr.
The most northerly record for this species in Britain.
All the collections are without ascocarps.

B. sphaeroides (Dicks.) Zahlbr.
On moss-covered boles of broad-leaved trees and conifers in humid, often shady sites; also rarely on sheltered mosses and wet rocks near lochs and streams.
Rare; confined to lowland sites in the east.
12 On moss tuft on bridge over Allt Ardnacross; Salen Forest on *Larix* and palings. **13** Aros Park, on *Picea* and *Cupressus* near Spùt Dubh, *U. Duncan & James*. **16** Lussa River, An t-Sleaghach, on *Ilex*. **17** Craignure, Druim Mòr, encrusting mosses on boles of *Acer pseudoplatanus*.

B. umbrina (Ach.) Bausch
On acidic, rarely basic, rocks and boulders in open sites, especially upland scree; also occasionally on branches and twigs of broad-leaved trees especially *Fagus*, and on rotting, decorticated wood in bogs.
Frequent, possibly abundant; especially along the coast. Often well developed near farmsteads, and on upland scree boulders.
1–6, 8, 10, 12, 16, 17
Material with a blue-green epithecium seen only in **2, 4, 12** and **17**. Easily overlooked due to tiny ascocarps and the dull colour of the thallus.

Byssoloma Trevis.

B. subdiscordans (Nyl.) P. James
B. rotuliforme (Müll. Arg.) Santesson
Sparingly on damp rocks in sheltered, west-facing gorge.
Rare.
5 Calgary, woodland above the Burial Ground.

Catillaria (Ach.) Th. Fr.

C. atropurpurea (Schaer.) Th. Fr.
Mainly confined to moss-covered boles of mature broad-leaved trees, especially *Fraxinus*, *Quercus* and *Ulmus*, in sheltered, well-wooded, lowland valleys and ravines; rarely on wayside trees away from streamsides.
Locally frequent; widespread, especially in the north and east.
4, 6–8, 12, 13, 15–17

C. biformigera (Leight.) P. James
Under sheltered overhangs of basalt and plutonic rocks, including large boulders amongst scree.
Rare; recorded from both the coast and upland areas.
3 Loch Caol, granite boulders in woodland. **4** Ulva, Craigaig, on basalt cliffs. **11** An Gearna at 550 m, on cliffs and associated boulders on north-facing hillside.
Contains atranorin and an unknown substance.

C. chalybeia (Borr.) Massal.
On dry, partially inundated acidic or more or less basic boulders and pebbles in sheltered or exposed sites from the littoral zone on the seashore to the summit of Ben More; also on rocks and boulders in streams.
Abundant in all areas.
1–17
Polymorphic species with a wide ecological amplitude.

C. chloroscotina (Nyl.) Arnold
On boulders in streams and seepage tracks on rock outcrops on mountain sides and ravines.
Rare.
11 Ben More, Gleann Dubh, *U. Duncan*.
Possibly a shade morphotype of *C. chalybeia*.

C. chlorotiza (Nyl.) P. James
In sheltered recesses under boulders.
Rare.
4 Gometra, Tòrr Mòr, near burial ground.

C. dufourii (Nyl.) Vain.
On encrusting mosses associated in the basalt rocks in upland sites.
Rare.
8 Maol Mheadhonach, at 370 m, *U. Duncan & James*. **16** Laggan Deer Forest, Maol-a' Bhàird, on *Andraea* at 280 m.
The first recorded occurrence of this species in Britain, now also known from Cairngorm Mountain.

C. griffithii (Sm.) Malme
On wayside and broad-leaved trees and conifers; also on the stems of *Calluna* in sheltered sites.
Frequent to abundant.
2–17
Plants on broad-leaved trees tend to develop pycnidia only, whereas those on conifer trees are usually richly fertile.

C. lenticularis (Ach.) Th. Fr.
On basic rocks, sometimes on mortar of old walls, bridges and farmsteads; rarely on bones of sheep (Treshnish, Fladda).
Occasional, but widespread in suitable base-rich habitats.
1–3, 5, 8–10, 12, 13, 17

C. lightfootii (Nyl.) Zahlbr.
On a wide range of trees and shrubs, including *Fraxinus*, *Myrica*, *Salix*, and *Sorbus aucuparia*, in moist or boggy areas.
Abundant in all but mountain summit areas, especially so in shrub-dominated bogs and *Salix* carr.
2–17
A polymorphic species ranging from entirely sorediate, non-fertile thalli to those without soralia and with numerous ascocarps. Sorediate forms are generally characteristic of damper or more sheltered habitats.

C. littorella (Nyl.) Zahlbr.
In very sheltered, dry declivities and recesses in basalt on or near the coast; often partially obscured by summer vegetation.
Rare; only recorded from Gometra and Ulva.
4 Gometra, near the Burial Ground; Ulva, above Brionn Pholl; Craigaig, rock face partially obscured by *Hedera helix*.
The ascocarps vary in colour from dull red-brown to pale pink on the same thallus according to the degree of illumination.

C. micrococca (Körb.) Th. Fr.
On leaves and small twigs of *Picea*.
Rare.
13 Salen Forest, near Achad nan Each, *N. Wallace*.

C. pulverea (Sm.) Lett.
On mature tree boles in sheltered, moist valley woodland; also on *Larix* in sheltered plantations.
Occasional; in a few lowland sites, probably more frequent than the records suggest owing to the difficulty of identifying sterile material.
7 Torloisk House, on *Alnus* and *Fagus*. **11** Benmore Lodge, on *Sorbus aucuparia*. **12** Salen Forest, eastern entrance, on *Larix*. **16** Lochbuie, Moie Lodge on *Corylus* and *Fagus*.
Contains atranorin, chloratranorin, fumarprotocetraric acid (±) and zeorin; frequently sterile.

C. sphaeroides (Massal.) Schul.
On boles of mature wayside, valley and woodland trees, especially *Fraxinus* and *Ulmus*, in sheltered, often poorly illuminated habitats; usually by streams and lochs in lowland areas.
Frequent.
4–5, 7–13, 16, 17

C. cf. subviridis (Nyl.) Zahlbr.
On decorticated wood inside ancient *Fraxinus*, near the coast.

Rare.
17 Craignure.

Catinaria Vain.

C. grossa (Pers. ex Nyl.) Vain.
On wayside, rarely woodland, broad-leaved trees, often in sunny situations. Most frequent on *Fraxinus* and *Ulmus*.
Occasional; widespread but seldom frequent in any one locality.
3–5, 7, 11–13, 15, 17

Gomphillus Nyl.

G. calicioides (Del. ex Duby) Nyl.
On moss-covered boles of broad-leaved trees by streams or in sheltered humid woodland.
Occasional; especially in the west.
6 Dervaig, near Dùn Auladh, on *Fraxinus*. **7** North side of Loch Tuath, Torloisk House. **12** Allt Ardnacross, on *Fraxinus*. **16** Loch Buie, near Moie Lodge, on *Ulmus*.
The thalli of this species form extensive semi-translucent patches on mosses and are often considerably damaged by browsing snails (*Clausilia* spp.).

Fuscidea V. Wirth & Vězda

Bibliogr.: Wirth and Vězda (1972).

F. cyathoides (Ach.) V. Wirth & Vězda
Lecidea cyathoides (Ach.) Ach.
On acid boulders and outcrops in sunny situations.
Frequent in most areas; particularly on the Ben More massif. Often poorly developed on basalt rocks, particularly in the east.
1–4, 6, 8–11, 13, 15–17
Easily distinguished from all other species of the genus by the bean-shaped spores. Forma **corticola** (Fr.) Vain. was recorded from near Linndhu and Lochnameal farm, on *Sorbus aucuparia*.
Contains fumarprotocetraric acid.

F. kochiana (Hepp) V. Wirth & Vězda
L. kochiana Hepp.
On acid rocks in exposed, sunny situations.
Rare; locally abundant in two sites on **11** Ben More, An Gearna, at 550 m; A' Chioch, at *c.* 700 m.
Contains divaricatic acid.

F. periplaca (Nyl.) V. Wirth & Vězda
L. periplaca Nyl.
On smooth basalt rock face near the coast.
Rare; only recorded once.
5 Port na Caillich, near Caliach Point.
No lichen substances detected.

F. tenebrica (Nyl.) V. Wirth & Vězda
L. tenebrica Nyl.
On sunny, smooth acidic rock faces, especially on granite and basalt in exposed sites mainly in upland areas.
Frequent; easily overlooked.
3, 5–8, 11–13, 15, 17
No lichen substances detected.

Lecidea Ach. sect. **Lecidea**

See also *Fuscidea* V. Wirth & Vězda, *Lecidella* Körb. em. Hert. & Leuck. and *Trapelia* Choisy.

L. aglaea Sommerf.
On hard acidic rocks and boulders in upland sites.
Occasional on the Ben More massif; rare and very local elsewhere.
10 Loch Fuaron. **11** Ben More, exposed hillside near the summit cairn at 945 m, *U. Duncan*; Beinn nan Gabhar, at 460 m; Maol Mheadhonach at 470 m; Lag a' Bhàsdair, in short *Calluna-Erica tetralix* heath. **16** Port Ohirnie, near cliff edge at 230 m.
Contains atranorin and trace of usnic acid.

L. albocaerulescens (Wulf.) Ach. var. **albocaerulescens**
Huilia albocaerulescens (Wulf.) Hertel
On basalt and other acidic rock outcrops, especially in sheltered habitats near the coast and by lochs.
Frequent; in all areas.
1–17

L. albocaerulescens var. **flavocoerulescens** (Horn.) Schaer.
In similar habitats to the species but more characteristic of rocks rich in metals, especially iron and lead.
Occasional to frequent; mainly in the west.
1–8, 10, 11, 13, 14, 17
Probably a variant of the type species characteristic of iron-rich rocks.

L. atrata (Ach.) Wahlenb.
L. dicksonii (J. F. Gmel.) Ach., *L. pissodes* Stirt.
On acid rocks and boulders, especially those associated with stable scree. Rather less frequent on basalt than on other igneous rocks.
Frequent; in all areas, especially on the Ben More massif.
3–5, 7–11, 14–16

L. auriculata Th. Fr.
On pebbles, boulders and rock outcrops chiefly of quartzite.
Rare; confined to the Ben More massif.
11 A'Chioch at 700 m; Maol nan Damh at 650 m.

L. bauchiana (Körb.) Lett.
On pebbles, boulders and rock outcrops, particularly on sheltered banks by paths in woodland; also on stones and rarely old roots of fallen trees in well-wooded, sheltered lowland sites.
Occasional; more frequent in the east.
4, 6–10, 12, 13, 15–17

L. berengeriana (Massal.) Th. Fr.
Encrusting mosses on old broad-leaved trees in sheltered old woodland.
Rare.
7 Achleck, on *Fagus* near Post Office. **12** Aros Park, Spùt Dubh, on *Quercus*.

L. cinnabarina Sommerf.
On boles of mature or old broad-leaved trees and conifers, particularly in old woodland sites; also on palings, fence-posts and decorticated stumps.
Occasional to frequent; especially on old conifers in neglected plantations, becoming rarer but still widespread in upland areas.
3, 4, 6, 7, 9–13, 15, 17
Contains fumarprotocetraric acid. All gatherings are sterile in which state this taxon seems chemically and morphologically indistinguishable from *Mycoblastus fucatus.*

L. clavulifera Nyl.
On pebbles and in rock crevices protected from direct rain in sheltered, well wooded sites; also rarely on exposed roots of fallen trees.
Rare; confined to the east.
12 Salen, on pebbles in bank. **13** Aros Park, on roots of *Fagus.*

L. crustulata (Ach.) Spreng. aggr.
Huilia crustulata (Ach.) Hertel
On acidic boulders, pebbles and outcrops, especially in dry heathland.
Abundant in all areas.
1–17
A very polymorphic taxon which probably includes several entities.

L. endomeleana Leight.
In crevices of sheltered basalt by paths; also on schist boulders of sheltered old stone-walls.
Rare, but probably overlooked.
9 Glengorm, road cutting; near Mingary on old wall.

L. fuscoatra (L.) Ach.
On acidic rocks and boulders in sunny sites, especially in upland areas and on and near the coast.
Frequent; but mostly poorly developed and easily overlooked.
1–17
Distinguishable in the depauperate condition by the C+ rose-red reaction due to gyrophoric acid in the thallus cortex (not medulla). A very polymorphic species.

L. geophana Nyl.
On rotting sacking and decorticated wood.
Rare.
3 Port nan Ròn, on old sacking. **7** Head of Loch Mingary, rotting burnt wood (? conifer).

L. granulosa (Hoffm.) Ach.
On soil, rotting wood, tree stumps, old tree boles in woodlands and by streams and roads; also frequently forming extensive patches on *Calluna-Erica cinerea* moorland.
Abundant in all areas.
1–17
With *L. uliginosa*, an important primary coloniser and soil stabiliser after heathland burning. One of the commonest lichens on Mull and very polymorphic; very sorediate, mostly sterile, forms are chiefly confined to woodlands, whereas less sorediate specimens with numerous ascocarps are more frequent in open moorland areas. Contains gyrophoric acid.

L. hydrophila Fr.
Huilia hydrophilum (Fr.) Hertel
Semi-inundated rocks and boulders in or by lochs and streams in upland and a few lowland areas.
Rare, but perhaps overlooked as *L. albocaerulescens* and *L. contigua.*
6 Lochan's Airde Beinn. **7** Cnoc an dà Chinn, wet rock face below Loch Tràth. **16** Wet rocks in Abhainn a' Chaiginn Mhòir; E-facing slopes of Ben Buie at 230 m.
Easily distinguished from related species by the simple paraphyses and, above all, by the aeruginose-blue epithecium. This species is more frequent in western Ireland where it frequently replaces *L. albocaerulescens* (see James, 1958).
No lichen substances detected.

L. illita Nyl.
On weakly basic rock outcrops in sheltered, rather persistently damp, habitats.
Rare.
8 Creag a' Ghaill, *U. Duncan & James*; Coille na Dubh Leitire.
Contains gyrophoric acid.

L. insularis Nyl.
Parasitic on *Lecanora rupicola* in two coastal sites.
Rare; only recorded in the west.
2 Rock outcrops near Loch Staoineig. **4** Gometra, rocks below House.

L. lactea Flörke ex Schaer.
L. pantherina (Hoffm.) Th. Fr.
On acid rock outcrops and large boulders, especially in sunny lowland sites.
Occasional; most frequent in Mishnish and Mornish areas.
2–4, 6–9, 12, 13, 15
Contains norstictic acid.

L. lapicida (Ach.) Ach.
On acid rocks and boulders in more or less sunny, exposed sites; especially frequent on stable scree on the middle slopes of the Ben More massif.
1–13, 15–17
Often very poorly developed on weathered basalts.
Contains stictic, constictic acids and three unknown substances.

L. leucophaea (Flörke ex Rabenh.) Nyl.
L. sporotea Stirt.; *L. orphanaeilla* Stirt.

Acid rocks and boulders especially near fresh water.
Frequent and widespread; easily overlooked.
1–17
One of the most frequent lichens on Mull but frequently imperfectly developed especially on basalt outcrops on the wetter part of the Island. Prefers more sheltered, damper sites than *L. lapicida* and *L. lithophila.*

L. limborina (Nyl.) Lamy
Sunny, exposed rock outcrop.
Rare.
4 Ulva, by the House, *U. Duncan & James.*

L. lithophila (Ach.) Ach.
On acidic rocks, especially exposed outcrops and boulders on open moorland.
Abundant; very widespread in all but the most sheltered areas.
1–17
Although a very common species, specimens from more exposed western areas of Mull are often depauperate, the thalli often reduced to scattered fragments with only a few ascocarps.

L. lucida (Ach.) Ach.
On sheltered, more or less dry rock faces, particularly in mature or old broad-leaved woodland; rare on shaded walls or north or east-facing walls of houses.
Frequent; especially in the east, absent from Ross of Mull and Iona.
4, 5, 7, 9–13, 15–17
Contains rhizocarpic acid and atranorin (±).

L. macrocarpa (DC.) Steud.
L. contortula Stirt.; *Huilia macrocarpa* (DC.) Hertel
On acid rocks, boulders and pebbles particularly on open moorland.
Frequent; widespread in most uplands becoming rarer on or near the coast.
2–12, 14–17

L. mullensis Stirt.
On acidic rocks.
Rare; not refound.
11 Ben More, 1876, *Stirton.*
Contains norstictic acid.

L. nigrogrisea Nyl.
Temporary seepage tracks on basalt outcrops.
Rare.
8 South side Loch na Keal, Creag Mhòr, *U. Duncan & James.*

L. orosthea (Ach.) Ach.
On more or less sheltered, dry, vertical rock faces and underhangs of large boulders, rock outcrops and cliffs.
Frequent to abundant; widespread in all suitable sites.
1–17
Fertile specimens were recorded from **4** and **12.**
Contains usnic acid and zeorin.

L. paraclitica Nyl.
On rotting palings at margin of sheltered woodland.
Rare; only recorded from a single site.
13 Aros Park, Upper Druimfin, *U. Duncan & James.*

L. pelobotrya (Wahlenb.) Leight.
On damp rock outcrops and boulders, mainly in upland moorland sites; rarely in temporary seepage tracks and on boulders at the margins of lochs.
Frequent; especially in the parish of Kilmore.
2–5, 7–13, 15, 16.
Contains gyrophoric acid.

L. percontigua Nyl.
On acid rocks and boulders in exposed sites, especially in upland areas.
Occasional but widespread; frequent on the Ben More massif.
Probably overlooked for *L. macrocarpa.*
3, 5, 6, 11, 13, 14, 16
Contains norstictic and stictic acids.

L. phaeops Nyl.
On shaded, often permanently damp rock faces and boulders; also often in old woodland or by the margins of lochs in lowland areas.
Occasional; usually locally frequent.
4, 7, 12, 13, 17
Contains psoromic acid.

L. pycnocarpa (Körb.) Ohl.
On acid rock outcrops and boulders in walls in exposed situations.
Rare.
8 Derryguaig, on old wall. **11** Creag Mhic Fhionnlaidh, *c.* 330 m; Tòrr na h-Uamha, by side of small lochan, *c.* 385 m.

L. quernea (Dicks.) Ach.
Boles of old and mature broad-leaved trees and conifers in sunny wayside areas.
Rare.
4 Ulva, on *Fagus* near the House. **5** Grounds of Glengorm House, on *Larix.* **13** Aros Park, on *Quercus* by Lochan Churrabain.
This species, abundant in southern England and Wales, is very rare in Scotland.
Contains arthothelin and ?thiophanic acid.

L. semipallens Nyl.
On rock outcrops, boulders and pebbles in very sheltered, shaded sites, especially near streams and damp paths in old woodland, in lowland areas.
Occasional; widespread.
4, 7–9, 11–17

L. speirea (Nyl.) Nyl.
On hard, slightly base-rich rocks and boulders; also on acid rocks near calcareous seams or adjacent to mortar on old walls.
Occasional.

2, 4, 5, 7, 12, 13, 16
Usually associated with *Protoblastenia rupestris* and *Rhizocarpon petraeum.*

L. sulphurea (Hoffm.) Wahlenb.
On acid rocks of outcrops, boulders and stone walls in all and particularly, coastal areas.
Occasional to frequent; often poorly developed in inland sites.
1–4, 6, 7, 9, 10, 12, 13, 15, 17
Contains usnic, α-collatolic (±) acids, atranorin (±) and zeorin.

L. sylvicola Flot.
On rock outcrops, boulders and pebbles in sheltered, shaded sites, especially near streams and damp paths in woodland.
Occasional to frequent; eastern parts of the Island.
4, 6–10, 12, 13, 15, 16

L. symmicta (Ach.) Ach.
On twigs and young branches, rarely on boles of broad-leaved trees and shrubs and conifers; also on palings and decaying tree stumps and rotting logs.
Frequent; widespread.
1–17

L. taylorii (Salwey) Mudd
On shaded, frequently damp, rock faces and boulders particularly in old woodlands or by the margins of lochs in lowland areas. Often in associations with *L. phaeops.*
Occasional; sometimes locally frequent.
3, 4, 8–11, 13, 16, 17

L. tenebricosa (Ach.) Nyl.
On smooth bank of low-growing trees and shrubs, especially *Salix* and *Sorbus aucuparia* in boggy areas. Often in associations with *Lecanora jamesii.*
Frequent.
3–8, 10–17

L. tenebrosa Flot.
L. endocyanea Stirt.
On sunny rocks in open situations especially near the coast.
Frequent in the west becoming rare in the east.
1–7, 10–12, 14, 16, 17

L. tumida Massal.
On acid rock outcrops, boulders in open situations.
Abundant; a very common and widespread lichen in all upland and lowland saxicolous habitats.
1–17
Often poorly developed, especially on weathered basalt.

L. turgidula Fr.
Confined to boles and branches of mature conifers; also on conifer palings and stumps at the margins of old plantations.
Frequent.
4, 5, 7–9, 11–17

L. uliginosa (Schrad.) Ach.
On peaty soil, especially in exposed areas of moorland; also on rotting wood and boles of trees especially those in open, often wet situations.
Abundant.
1–17
Important, with *L. granulosa*, as a soil consolidator after heathland burning.

L. umbonella Nyl.
On basalt rocks and erratic boulders on Ben More massif.
Rare; very local.
11 Cruachan Dearg between Coir' Odhar and Màm a' Choir' Idhir, 370 m; Ben More, above Coir' Odhar, *c.* 610 m; An Cruachan, *c.* 500 m.

L. valentior Nyl.
On acidic boulders and rock outcrops in or near streams and lochs usually sheltered by overhanging trees.
Occasional; confined to a few lowland sites.
11 E. side of Loch Bà, Doire Daraich; **13** Loch Churrabain. **14** Lussa River, near Strathcoil.

L. wallrothii Flörke ex Spreng.
On soil in rock crevices, by paths or among scattered *Calluna* stands on or near the coast.
Occasional; confined to western coasts.
2 S of Baile Mòr, Druim Dhùgaill. **4** Ulva, S side of Tòrr Mòr. **9** Ardmeanach, Burgh.
Contains gyrophoric acid.

Lecidea sect. **Psora** (Hall.) Schaer.

L. cinereorufa Schaer.
On rock outcrops, boulders in scree, and, especially, wall tops in or near sheep pens in exposed situations.
Frequent and widespread; most upland areas but becoming much rarer in the west and absent from Iona and Ulva.
5, 7–11, 13–16

L. decipiens (Hedw.) Ach.
On consolidated shell-sand among acid coastal rocks.
Rare; only two thalli seen.
2 A' Mhachair.

L. demissa (Rutstr.) Ach.
On bare soil on or near mountain summits.
Occasional.
8–12, 14–16

L. friesii Ach.
On conifer palings.
Rare.
11 Coladoir River, Teanga Brideig.
Contains squamatic acid.

L. glaucolepidea Nyl.
On peaty soil in leached upland moorland; also on disused ant-hills.

Rare; but widespread.
5 Penmore House, near Dùn nan Gall, on peat mounds. **11** Glen Clachaig, Coille na Creige Duibhe, on old ant-hills. **13** Glac an Lin, Meall nan Gabhar and Meall Damh Àrd. **16** Laggan Deer Forest, near Lochan na Craoibhe-caoruinn.
An interesting extension of the range of this British endemic.
Contains squamatic acid.

L. lurida (With.) Ach.
On soil and shell-sand in crevices in sheltered, rather dry cliff faces on the coast.
Rare; confined to a few localities on the extreme west coast.
2 A' Mhachair, near Spouting Cave. **4** Gometra, Lochan a' Churraidh, *Thomas*. **5** Calgary Bay, near pier.

L. scalaris (Ach.) Ach.
Rotting logs, tree stumps and boles of living conifers; more rarely on old broad-leaved trees and palings.
Occasional; most frequent in lowland wayside areas but ascending to 150 m on **16** Sròn nam Boc (on dead stump of *Betula*).
3, 5, 7, 11–13, 16, 17
All specimens are sterile. Contains lecanoric acid.

Lecidella Körb. em. Hert. & Leuck.

L. elaeochroma (Ach.) Haszl. forma **elaeochroma**
Lecidea limitata auct.; *Lecidea parasema* Ach. var. *elaeochroma* Ach.
On boles, branches and twigs of broad-leaved trees and, more rarely, conifers in all areas.
Abundant; widespread in all suitable sites; especially common on smooth bark of wayside trees in lowland areas.
1–17

L. elaeochroma forma **soralifera** (Erichs.) D. Hawksw.
Lecidea elaeochroma Ach. var. *soralifera* Erichs.
Similar sites to, and often associated with, the species.
Occasional; but widespread, mainly confined to lowland areas.
3, 4, 6, 8, 9, 12, 17
The yellow delimited soralia which characterise this form are frequently sparingly produced and often confined to restricted areas of particular thalli which exhibit no supression in ascocarp production.

L. goniophila (Ach.) Schaer.
Lecidea goniophila Ach.
Confined to more or less vertical rock faces especially in sheltered old woodland in lowland areas.
Rather local but usually frequent in suitable sites, especially in the east.
4, 7, 12, 13, 15–17

L. scabra (Tayl.) Hert. & Leuck.
Lecidea scabra Tayl.
On sheltered or sunny acidic rocks and boulders in most lowland coastal sites and, more rarely, upland areas. Lignicolous in **3** Ardtun, Knockan, fallen fence-post in derelict garden.
Frequent to abundant.
1–15, 17
On basalt, frequently sterile and depauperate and thus easily overlooked.
Fertile specimens from **1–4, 6, 8, 12, 13, 17.**

L. stigmatea (Ach.) Hert. & Leuck.
Lecidea stigmatea Ach.
Confined to more or less base-rich rocks, acidic rocks and boulders on the coast or near limestone outcrops; also on mortar and asbestos especially of old derelict buildings and garden walls.
Abundant; widespread in suitable lowland and maritime sites, becoming less frequent in upland areas.
1–5, 7–13, 15–17

L. subincongrua (Nyl.) Hert. & Leuck.
Lecidea subincongrua Nyl.
Xeric supralittoral; confined to sunny acidic rocks and boulders.
Abundant; on all rocky coasts around the Island.
1–17

Lopadium Körb.

L. pezizoideum (Ach.) Körb.
Amongst mosses on rough boles of old broad-leaved trees in sheltered, damp, lowland woodland sites.
Rare.
7 Torloisk House, on *Quercus*. **13** Aros Park, on *Quercus*, *U. Duncan & James*.

Micarea Fr. em. Hedl.

Bibliogr.: Hedlund (1887); James (1971).
See also *Bacidia* De Not. and *Catillaria* (Ach.) Th. Fr.

M. asserculorum (Ach.) Hedl.
Lecidea asserculorum Ach.
Confined to decomposing bark and decorticated wood of conifers in old plantations.
Rare; but probably overlooked due to the minute size of ascocarps.
11 Loch Bà, near Benmore Lodge, on moribund *Pinus sylvestris*. **12** Loch Frisa, near Ledmore, decorticated conifer stumps. **15** Scallastle Bay, valley of Allt Achadh na Moine, near Doire Daraich.

M. chrysophthalma P. James
On ancient *Quercus* in a sheltered lowland site.
Rare; probably overlooked.
16 Torosay Castle grounds, Camas Mòr.
A recently described species also known from several localities in southern Britain and France (Bretagne).

M. cinerea (Schaer.) Hedl.
Encrusting mosses and hepatics on damp tree boles in sheltered valley sites.
Rare; probably overlooked.
4 Ulva, woods near the House, on *Acer pseudoplatanus*. **7** Achleck and Torloisk House, on *Alnus*, *Pseudotsuga* and *Sorbus aucuparia*.
Determinations confirmed by Vězda. This species has pale to dark grey semi-translucent ascocarps and seven septate spores with abruptly rounded apices.

M. gelatinosa (Flörke) Brodo
Lecidea gelatinosa Flörke
Often forming extensive colonies on undisturbed soil in old woodland clearings; also on recently exposed soil along streamsides and road cuttings.
Occasional; becoming frequent in the Mishnish and Quinish areas.
4, 5, 7–9, 11, 12

M. leprosula (Th. Fr.) Hedl.
Bacidia leprosula (Th. Fr.) Lett.
On humus-rich, acidic soil, especially in open moorland areas; associated with pockets of humus and mosses on boulders and rocky outcrops on steep, often partially wooded scree slopes.
Frequent; in all areas, ascending to 850 m on Ben More.
1, 3–6, 8–17

M. lignaria (Ach.) Hedl.
Bacidia lignaria (Ach.) Lett.
On humus-rich, base-poor soils, especially in open, rather dry moorland and on pockets of soil associated with boulders and rock outcrops; more rarely on decorticated and rotting wood, old stems of *Calluna* and other woody detritus, crumbling rock and edges of peat hags and blowouts.
Abundant; widespread. One of the commonest terricolous lichens on the Island, ascending to 940 m on Ben More.
1–17

M. melaena (Nyl.) Hedl.
Bacidia melaena (Nyl.) Zahlbr.
On humus-rich, acidic soils in wet moorland and open woodland sites and especially on rotting bark and wood of conifers in bogs; rarely on decorticated conifer stumps, fence-posts and rails in sheltered, boggy woodland.
Occasional to frequent; widespread.
3, 4, 6, 11, 12, 14, 16
This taxon may represent two distinct species: one widespread with a brown to fuscous-brown epithecium, and the other, much rarer, with an aeruginose blue epithecium and thecium corresponding to *Bacidia ilyophora* Stirt.

M. prasina Fr.
Catillaria prasina (Fr.) Th. Fr.
On decorticated and rotting boles of broad-leaved trees and conifers in sheltered lowland and coastal woodlands; also rarely on decorticated wayside trees.
Frequent; especially in the east.
4, 6, 11–14, 16, 17

M. violacea (Crouan ex Nyl.) Hedl.
Bacidia violacea (Crouan ex Nyl.) Arnold
Encrusting mosses on rock outcrops and boulders and old tree boles.
Rare; probably overlooked.
3 Beinna Ghlinne Mhòir. **12** Salen, woods near the village, *U. Duncan & James*. **14** Glen Forsa, on moribund pines. **16** Lochbuie, Laggan Lodge on crumbling basalt.

Mycoblastus Norm.

Bibliogr.: James (1971).
M. affinis (Schaer.) Schauer.
Encrusting mosses on old trees and boulders in boggy lowland and upland sites.
Rare.
12 Salen, on *Betula* below Bràigh a' Choire Mhòir. **13** Druimfin, on old *Betula* and mossy boulders, *U. Duncan & James*. **16** Lochan nan Caorach, on dead *Betula* and decorticated wood.

M. fucatus (Stirt.) Zahlbr.
On decorticated wood of old trees and fence railings in sheltered, damp woodland sites in lowland areas.
Rare; but probably overlooked when sterile.
7 Torloisk House woods, on fence railings. **16** Lochbuie, near Moie Lodge, on decorticated wood of large *Fagus*.
Both gatherings are fertile.
All British records of *M. fucatus* are from damp decorticated wood. This plant is indistinguishable from *Lecidea cinnabarina* when sterile; both contain fumarprotocetraric acid.

M. sanguinarius (L.) Norm.
On mossy and bare tree boles in upland valleys, often associated with *Parmelia laevigata* and *Menegazzia terebrata*; rare in lowland and coastal areas and there confined to acid bark trees in woodland.
Occasional to frequent; especially in more sheltered eastern areas.
3, 6, 7, 11–13, 15, 16

Rhizocarpon Ram. em. Th. Fr.

R. constrictum Malme
Xeric supralittoral; confined to rocks and boulders on or near the coast, often locally abundant in and just above the *Ramalina siliquosa* zone on the upper part of the shoreline.
Frequent.
1–17
Contains gyrophoric acid

R. geographicum (L.) DC.
On acid rocks and boulders, especially in upland areas but less common on basalt rock

where it is often poorly developed; and occasionally on roof tiles of old buildings.
Frequent, especially on the Ben More massif.
2–11, 14–17
Contains psoromic and rhizocarpic acids.

R. hochstetteri (Körb.) Vain.
On acid rock, especially in rather sheltered, shaded, moist sites, also on or near the margins of lochs.
Frequent; widespread but frequently overlooked.
3, 5–9, 11–13, 15–17
No lichen substances detected.

R. laevatum (Fr.) Hazl.
On boulder and rock outcrops in and by streams and loch sides; rarely on damp boulders in sheltered woodland near the coast.
Occasional to frequent.
3, 5, 7–9, 11–15, 17

R. obscuratum (Ach.) Massal.
On acidic rocks in sheltered or exposed sites in coastal, lowland and upland areas.
Frequent; often locally abundant.
1–17
A polymorphic species with a particularly wide ecological tolerance.
No lichen substances detected.

R. riparium Räs.
R. lindsayanum Räs.; *R. kittilense* Räs.
On exposed acid rocks.
Rare; in a few upland sites and on the Ross of Mull; probably overlooked.
3 Rubh' Ardalanish; Kintra and Rubha nan Cearc. **11** Ben More; A'Chioch. **15** Summit of Ben Talaidh, on stones of cairn.
Probably not distinct from *R. geographicum* from which it differs only in the bluish green epithecium and thecium.

Toninia Massal. em. Th. Fr.

T. aromatica (Sm.) Massal.
On calcareous rocks, mortar, and rarely asbestos; also on acid rocks on and near the coast.
Frequent; widespread.
1–8, 10–17

T. coeruleonigricans (Lightf.) Th. Fr.
On calcareous shell-sand and dunes near the sea; rarely on crumbling mortar of old walls.
Occasional; local.
2 Culbuirg; A'Mhachair; opposite Eilean Mòr; Calva; Ardura. **3** Erraid, Traigh Gheal; Fidden. **5** Calgary Bay, *U. Duncan*. **8** Kilfinichen.

T. leucophaeopsis (Nyl.) Th. Fr.
On metal-rich (lead) rocks in upland areas.
Rare.
6 Lochan's Airde Beinn. **16** By Allt Ohirne.
All collections are sterile.

T. lobulata (Sommerf.) Lynge
On pockets of calcareous shell-sand on acid boulders in lowland and coastal areas; also on crumbling mortar of old walls and buildings.
Occasional; local.
3–5, 10, 13, 17

T. mesoidea (Nyl.) Zahlbr.
Mesic supralittoral. In damp, sheltered basalt crevices on the upper shore.
Rare.
15 Scallastle Bay, near Alterich, *U. Duncan & James*. **16** Lochbuie, Rubh' a' Bharra Ghainmheachain.
Probably overlooked in error as *T. aromatica* and may only represent an ecotype of this taxon.

T. pulvinata (Tayl.) Oliv.
On encrusting mosses on boulders in a few upland sites.
Rare; chiefly in the east.
11 Beinn nan Gobhar, *c.* 300 m; above Allt Ghillecaluim, Coir' a' Mhàim. **12** above Salen Forest, Speinne Beag, 250 m. **15** Coire Bearnach, 320 m. **16** Laggan Deer Forest, Druim Fada.

CANDELARIACEAE

Candelaria Massal.

C. concolor (Dicks.) Stein
Confined to isolated broad-leaved trees, especially in villages and near farmsteads.
Occasional; only in a few lowland sites.
5 Calgary, Frachadil, on *Corylus*. **7** Lagganulva, on *Acer pseudoplatanus*. **8** Kilfinichen, on *Acer pseudoplatanus*, *U. Duncan & James*. **10** Pennyghael, near the boat house, on *Fraxinus*. **16** Lochbuie, *Acer pseudoplatanus* by telephone box.

Candelariella Müll. Arg.

Bibliogr.: James (1971); Laundon (1970).
C. aurella (Hoffm.) Zahlbr.
On cement and mortar of farm buildings, walls and asbestos roofing tiles; more rarely on acidic rocks, particularly those enriched by bird droppings or near sheep pens. Very rare on wayside broad-leaved trees.
Frequent; in lowland areas, becoming rare above 90 m.
1–6, 8–10, 12–15, 17

C. coralliza (Nyl.) Magnusson
Mainly confined to small crevices of sunny, acidic coastal rock outcrops and boulders, especially those enriched by bird droppings
Occasional; especially in the west
2 Clachanach, granite erratic. **3** S of Fionnphort, Knockvologan. **6** Dervaig, Ardow, on boulder. **8** Maol Mheadhonach, at 300 m.
Fertile in **2** and **6**.

C. reflexa (Nyl.) Lett.
On boles and branches of young and old wayside broad-leaved trees, especially *Sambucus nigra* near farmsteads, cattle pens and at the bases of wayside trees enriched by the urine of dogs.
Occasional; widely distributed.
2 Baile Mòr, on *Fraxinus* near the Nunnery. **4** On *Fraxinus* near Ulva House. **6** Penmore House, on *Sambucus nigra*. **7** Ballygown, on *Sambucus nigra*. **10** Pennyghael, on *Sambucus nigra*. **13** Tobermory, on *Tilia* by harbour front.
All specimens are sterile.

C. vitellina (Hoffm.) Müll. Arg.
On acidic rocks, old walls and roofing tiles, particularly along the coast and near lowland farmsteads; also on nutrient-enriched bird-rocks and standing stones in upland areas. Rare on boles of old wayside broad-leaved trees.
Frequent; often very abundant in coastal areas.
1–17

C. xanthostigma (Ach.) Lett.
On boles of old broad-leaved trees in wayside and woodland margin sites.
Rare; probably overlooked for *C. reflexa*, a much more frequent species.
4 Ulva, west of the House, on *Ulmus*. **6** Glengorm, near Sorne, on *Acer pseudoplatanus*.
Both specimens are sterile.

BAEOMYCETACEAE

Baeomyces Pers.

B. placophyllus Ach.
On drier areas of peat crust, especially at the margins of *Calluna-Erica tetralix* bog, more rarely at the edges of peat hags and blow-outs; also not uncommon on leached alluvial soils under conifers.
Occasional to frequent; widely distributed in upland areas above 100 m but seldom plentiful in any one locality.
3–5, 7, 8, 11, 12, 15, 16
Ascocarps were not observed. Contains stictic acid.

B. roseus Pers.
On peat or sandy soil associated with *Calluna – Erica cinerea* or drier parts of *Calluna – Erica tetralix* associations; also on intermittently damp, semi-stable soil at the sides of roads or earth banks.
Frequent; widespread, chiefly confined to upland sites.
3–9, 11, 12–17
Ascocarps rarely produced, only observed in **3–5, 11, 13, 14.**
Prefers damper sites to *B. rufus*. Contains baeomycesic acid and atranorin (±).

B. rufus (Huds.) Rebent.
On peat and sandy soils, often spreading to rocks and pebbles, in *Calluna–Erica tetralix* associations, stream sides, earth banks, and disturbed or burnt heathland and eroded streamsides.
Abundant; in all areas.
1–17
Ascocarps observed in **2–7, 9–15, 17**. Contains stictic acid (±) and atranorin.

Icmadophila Trevis.

I. ericetorum (L.) Zahlbr.
On rotting wood, damp peaty soils on moorland above 150 m; also at the margins of peat hags.
Occasional; widespread.
7, 9–12, 15, 16
Contains thamnolic acid; some populations also rich in perlatolic acid.

CLADONIACEAE

Cladonia Web.

C. arbuscula (Wallr.) Rabenh.
On the ground and amongst moss-covered boulders in open *Betula – Quercus* woodland; also amongst *Calluna*, *Erica tetralix* and *E. cinerea* on dry to wet heathland and rock ledges; frequent amongst mosses on scree boulders and on mountain summits.
Occasional; but widespread, less common than *C. impexa* and *C. tenuis*.
1, 2, 4–7, 10–13, 15–17
Contains fumarprotocetraric and usnic acids, occasionally ursolic acid (15%).

C. bacillaris (Ach.) Nyl.
On sandy soil and rotting logs and stumps, particularly at the margins of conifer plantations; rarely on dry peaty hummocks in heathland areas and on soil on old walls.
Occasional; locally abundant in conifer plantations in the east.
2–4, 6, 8–10, 15, 17
Contains barbatic acid and, rarely, traces of usnic acid (3%).

C. bellidiflora (Ach.) Schaer.
On mosses on scree boulders in upland areas and less frequently on peaty soil and at the margins of peat hags; rare amongst old *Calluna* in dwarf-shrub heath.
Frequent above 250 m; especially on the Ben More massif.
7, 9–11, 15, 16
Contains squamatic, usnic acids and bellidiflorin, and a yellow pigment (±).

C. caespiticia (Pers.) Flörke
On moss-covered boles of mature broad-leaved trees in sheltered, moist woodland, especially near streams and lochs; also rarely on damp, recently disturbed soil by roadsides and pathways.

Occasional; widely distributed in all lowland areas.
3, 5–7, 9, 10, 12, 13, 16, 17
Contains fumarprotocetraric acid.

C. capitata (Michx) Spreng.
On recently burnt peat on *Calluna–Erica tetralix* heathland.
Rare.
16 Laggan Deer Forest, Maol a' Bhàird.
Contains fumarprotocetraric acid.

C. cervicornis (Ach.) Flot.
Characteristic of relatively basic soils in sunny situations in coastal sites and on consolidated shell-sand. Also frequent in open dwarf-shrub communities and on moss-covered rocks and earth of banks at the margins of woods.
Frequent; widespread.
1–8, 10–17
This species is probably a morphotype of *C. verticillata*. Contains fumarprotocetraric acid, and sometimes traces of atranorin.

C. chlorophaea (Flörke ex Sommerf.) Spreng. aggr.
Amongst *Calluna*, especially on peat hummocks on recently burnt heathland; also on mosses and soil of old walls, roadside banks, scree boulders, tree stumps and boles; rarely on the ground in conifer plantations, tolerant of relatively low illumination.
Frequent; in all areas, possessing a very wide ecological amplitude.
1–17
This is an interesting aggregate of species, comprising several chemical strains that require further study. Populations, often intermixed, include individuals with fumarprotocetraric acid together with, rarely protocetraric acid or, more frequently, either cryptochlorophaeic or merochlorophaeic acids. Strains with grayanic acid are rarely encountered in Mull samples.

C. coccifera (L.) Willd.
On acid heathland and on pockets of soil in crevices of rock outcrops.
Abundant; widely distributed in most areas.
1–17
In spite of its apparent frequency, *C. coccifera* is often poorly developed and frequently only represented by basal squamules; fertile specimens were rarely encountered. Contains porphyrilic and usnic acids, and zeorin; also rarely barbatic and 4-0-demethylbarbatic acids and an unknown pigment (±).

C. coniocraea (Flörke) Spreng.
On boles of mature trees especially towards their bases; more rarely on acid heathland in association with *Calluna vulgaris* and on soil in small pockets in sheltered rock outcrops.
Frequent; especially in well-wooded areas.
1–17
This species appears to intergrade with *C. ochrochlora* and is probably conspecific. Contains fumarprotocetraric acid.

C. conista (Ach.) Robbins ex Allen
On recently disturbed soil on banks by pathways and roadsides; more rarely on recently consolidated soil on coastal cliffs.
Occasional to frequent; mainly confined to lowland areas.
1–4, 6, 7, 9, 12, 13, 16, 17
Contains fumarprotocetraric and bourgeanic acids and a large amount of atranorin.

C. crispata (Ach.) Flot. var. **cetrariiformis** (Del. ex Duby) Vain.
In open areas in dry to wet heathland; also at the margins of peat hags.
Occasional; locally frequent in a few upland localities.
5, 11, 15, 17
Contains squamatic acid.

C. cyathomorpha W. Wats.
On mosses on several large boulders on boggy *Erica tetralix – Scirpus* heath.
Rare.
15 Beinn Bheag, 320 m.
A British endemic characterised by the large basal squamules with more or less distinct red or pale brown veins on a white surface below. Contains fumarprotocetraric and an unknown substance (±).

C. digitata (L.) Hoffm.
On rotting tree-stumps and logs in sheltered woodland; rarely at the bases of mature trees, especially *Pinus* and *Betula* in open woodland.
Rare.
10 Carsaig, Innamore Lodge, on *Pinus* stump.
16 Cruach Ardura, rotting bole of *Betula*; Duibh Leitir, on *Pinus* and *Betula*.
Contains thamnolic and usnic acids, bellidiflorin and an unknown pigment.

C. fimbriata (L.) Fr.
On moss-covered tree boles, and on recently disturbed soil, especially near the coast and in lowland areas.
Occasional and widely distributed; never common in any one locality.
3, 7, 8, 12, 14, 17
Contains fumarprotocetraric acid and, rarely, an unknown substance (±).

C. floerkeana (Fr.) Flörke
On dry, rarely wet, heathlands dominated by *Calluna*; more rarely on decaying tree stumps and on humus in crevices on scree boulders in open situations.
Frequent in upland sites; becoming rarer near the coast.
3, 5, 7, 8, 10, 11, 13, 15–17
Contains barbatic acid, and rarely, didymic acid (±).

C. foliacea (Huds.) Willd.
On recently disturbed well-drained basic soils on or near the coast especially those of consolidated shell-sand.
Frequent in a few western localities; especially on Iona, Ulva and at Calgary Bay.
1–5, 13, 16, 17
Contains fumarprotocetraric and usnic acids.

C. furcata (Huds.) Schrad.
On soil and among mosses on old walls, banks and hedgerows and scree; on peat in rather dry open-heath areas; also frequent in either exposed or shaded woodland areas and there tolerating relatively low illumination.
Frequent; widespread.
1–17
Very polymorphic. Contains fumarprotocetraric acid; most specimens also contain atranorin (±).

C. gonecha (Ach.) Asah.
On soil between moss-covered boulders in scree and on relatively dry peat in low-growing stands of *Calluna* and *Erica* spp.; more rarely on the margins of peat hags.
Rare; confined to a few upland localities; most frequent on Ben More massif.
11 An Gearna, 460 m; Beinn Fada, 450 m; A'Chioch, 800 m, *U. Duncan*; Beinn a' Ghràig, 490 m. **15** Beinn Talaidh, 610 m.
All specimens contain squamatic acid (UV+), usnic acids and bellidiflorin.

C. gracilis (L.) Willd.
In *Calluna–Erica* associations on dry to moderately wet heathlands, especially in upland areas, also occasionally on soil on old walls.
Occasional but widespread, especially in upland sites, very rare in lowland areas and near the coast.
9, 11–15, 17
Often poorly developed and then sometimes difficult to distinguish from *C. furcata*. Contains fumarprotocetraric acid.

C. impexa Harm.
On the ground and amongst moss-covered boulders on scree slopes and in upland open *Betula–Querus* woodlands; also associated with *Calluna–Erica tetralix* and *E. cinerea* on heathland and on rock ledges and especially frequent in summit-heath communities of the higher mountains.
Frequent; widespread.
1–14, 16, 17
Contains usnic and perlatolic acids.

C. incrassata Flörke
On a sheltered, rotting tree stump in old woodland.
Rare.
12 Salen, near the village, *U. Duncan & James*.
Contains usnic, squamatic (±) acids and bellidiflorin; no didymic acid was observed in the Mull specimen.

C. luteoalba Wheld. & Wils.
Amongst mosses on basalt scree, walls and old trees; also on peat amongst *Calluna*.
Rare to occasional.
8 Near Loch Arish, 300 m. **9** Beinn an Lachain under *Calluna*, 230 m. **11** Ben More, eastern corrie, 680 m, *U. Duncan*. **13** Caol Lochan, 210 m. **16** Laggan Deer Forest, Druim Fada.
All specimens contain usnic and porphyrilic acids as well as zeorin.

C. macilenta Hoffm.
On bare or moss-covered tree boles especially *Betula* and *Alnus*, and on rotting logs and stumps; also on old walls and scree; more rarely on bare peat hummocks sheltered by *Calluna* on dry heath-land.
Abundant; widespread, especially in open, mature woodland.
1, 8, 10–17
Contains thamnolic acid and bellidiflorin (±); one sample also contained barbatic acid.
Grades into *C. polydactyla* from which it differs only in the absence of scyphi (see Rose & James, 1974).

C. ochrochlora Florke
Chiefly on rotting tree-stumps and logs; rarely at the bases of old trees and on moss-covered boulders of old walls in sheltered sites.
Frequent; in most lowland areas, becoming rare above 30 m.
3, 5–7, 9, 11–13, 15–17
Contains fumarprotocetraric acid.
Grades into *C. coniocraea* from which it differs only in the larger podetia with scyphi wider than podetial stalks (see Rose & James, 1974).

C. parasitica (Hoffm.) Hoffm.
On rotting tree-stumps and logs in old, sheltered woodland.
Rare.
7 Torloisk House. **13** Aros Park, *U. Duncan*. **16** Near Moie Lodge, *U. Duncan*.
Contains thamnolic acid.

C. pityrea (Flörke) Fr.
On more or less consolidated well-drained soils and amongst mosses and low vegetation on cliffs, out crops, old walls and banks, especially near or on the coast.
Occasional; especially in the north and west.
1, 4–6, 12, 13, 16, 17
Contains fumarprotocetraric acid.

C. pocillum (Ach.) O. J. Rich.
On soil and mosses associated with basic rocks, shell-sand and old mortar.
Occasional; becoming more frequent in the west.

1–5, 7, 12, 13, 17
Possibly an ecotype of *C. pyxidata*. Contains fumarprotocetraric, very rarely protocetraric acids.

C. polydactyla (Flörke) Spreng.
On rotting tree-stumps and logs, bases of old trees, especially *Betula*, and on moss-covered scree; also on peaty ground, associated with *Calluna–Erica cinerea.*
Abundant; especially in well-wooded lowland sites.
1–6, 8–13, 15–17
Contains thamnolic acid and bellidiflorin (±) and an unknown pigment.

C. pyxidata (L.) Hoffm.
On moss-covered boles and stumps of broad-leaved trees, soil and boulders in open situations.
Rare.
1 Bac Mòr, on soil. **5** Kilninian, Tostarie, on mossy boulders. **7** Torloisk House woods, on rotting tree stump. **8** Near Tiròran, on *Fraxinus*. **12** Near Ledmore, on mossy rocks.
Contains fumarprotocetraric acid.

C. rangiferina (L.) Web.
On rock-ledges and mosses amongst scree; rarely on peat in open areas amongst *Calluna* near mountain summits.
Rare; most frequent on Iona.
2 Cùl Bhuirg, amongst *Calluna*; Auchabhaich, on rock ledge; SE central Iona, *D. E. Hillcoat*; **4** Ulva, Ormaig, on basalt cliffs; **11** Maol Mheadhonach, on scree 380 m; A'Chioch, on mosses on scree 670 m.
Contains fumarprotocetraric acid and atranorin.

C. rangiformis Hoffm.
On basic soils associated with calcareous rocks, basalt, shell-sand and, more rarely, decomposing mortar.
Occasional; locally frequent but mainly confined to coastal regions and especially in the west.
1–6, 12, 17
See Laundon (1971, p. 175); contains fumarprotocetraric acid (±) and atranorin (±).

C. squamosa (Scop.) Hoffm. var. **squamosa**
Amongst mosses on old walls, bases of old trees, rotting stumps and logs in well wooded sites; also on peaty soil of *Calluna–Erica* heathland.
Abundant; widespread in all areas.
1–8, 10–17
Very polymorphic. Contains squamatic acid.

C. squamosa var. **allosquamosa** Hennipm.
In similar habitats to the species, but considerably rarer in all areas.
1, 4–6, 8–11, 15–17
Contains thamnolic acid.

C. strepsilis (Ach.) Vain.
On wet, frequently waterlogged, peaty gleys and gravels, especially in open areas in *Calluna–Erica tetralix* associations; also near or at the edges of seasonal, shallow pools in upland areas.
Frequent; widespread, but not frequently recorded below 300 m.
2–9, 11–17
Contains baeomycesic, squamatic acids and strepsilin (C + green).

C. subcervicornis (Vain.) Du Rietz
On dry peaty soils on *Calluna–Erica cinerea* heathland and pockets of humus in rock crevices and boulders, especially near the margins of woods.
Frequent; widespread.
1–4, 6–8, 10–17
Contains fumarprotocetraric acid and atranorin.

C. subulata (L.) Web.
On bare soil on open patches of dry *Calluna–Erica cinerea*-heathland; rarely on rotting tree-stumps and on soil on old walls and banks.
Occasional to frequent, especially in the vicinity of Glengorm.
3–6, 8–13, 17
Contains fumarprotocetraric acid.

C. tenuis (Flörke) Harm.
On the ground and amongst moss-covered boulders in open *Betula–Quercus* woodland and in heathland amongst *Calluna*, *Erica cinerea* and *E. tetralix*; also frequent amongst scattered boulders and scree near mountain summits.
Frequent to abundant.
1–6, 8–17
Contains fumarprotocetraric and usnic acids (±). Specimens without usnic acid, often cited as *C. leucophaea* des Abb. in the literature, are very rare on Mull.

C. uncialis (L.) Web.
On dry to permanently wet peat or peaty gleys, especially in *Calluna-Erica tetralix*-associations. More rarely on dry soils at the margins of lowland woods.
Frequent to widespread; rarely well-developed, often represented by a few scattered and often deformed fragments.
2–17
Contains usnic and squamatic acid.

C. verticillata (Hoffm.) Schaer.
On dry *Calluna–Erica cinerea* heathland and on roadside banks and walls, often at the margins of plantations.
Frequent.
4, 6, 8–13, 16
Contains fumarprotocetraric acid, rarely atranorin (±).

Pycnothelia (Ach.) Duf.

P. papillaria (Ehrh.) Duf.
On acid or peaty soils mainly in exposed upland situations and often an early coloniser of recently burnt areas of *Calluna* heath; more rarely on the margins of peat hags.
Frequent; widespread in most upland areas, but seldom well-developed.
3–13, 15–17

STEREOCAULACEAE

Pilophorus Th. Fr.

Bibliogr.: Jahns (1970).

P. strumaticus Nyl. ex Cromb.
P. distans (Hult.) Magnusson
On damp acid boulders and rock outcrops, often near water, particularly in upland areas.
Frequent above 90 m, but easily overlooked.
4, 6–13, 15, 16
Characteristically occurs near the bases of boulders and outcrops partially obscured by summer vegetation, especially in damp *Erica tetralix–Molinia* associations.
Rarely fertile. Contains atranorin and zeorin.

Stereocaulon Schreb.

S. dactylophyllum Flörke
Firmly attached to acid rocks in more or less exposed sites.
Occasional, widespread, especially in Laggan Deer Forest.
6, 8, 9, 11, 14–17
Often poorly developed and without ascocarps. Contains stictic, constictic, norstictic acids (±), atranorin, and three unknown substances.

S. delisei Bory ex Duby
Firmly attached to acid metalliferous rocks in a few upland sites.
Rare.
6 Mingarry Burn below Tòrr an Daimh. **8** Gleann Seilisdeir, acid rocks above Abhaim Bail' a' Mhuillinn. **16** Laggan Deer Forest, on rock outcrop near Beinn a' Bhairne.
Contains atranorin.

S. evolutum Graewe
On acid rock outcrops and boulders in most upland areas, especially in stream valleys and less exposed areas on open moorland.
Occasional to frequent.
3–5, 7, 8, 11–13, 15, 17
Contains lobaric acid and atranorin.

S. pileatum Ach.
On mineral-rich rocks in rather poorly drained, upland moorland.
Frequent; especially in the south and south-east; rare below 100 m.
3–5, 7, 9–12, 14, 16, 17
Contains lobaric acid and atranorin.

S. vesuvianum Pers.
On acidic, often mineral-rich, boulders and rock-outcrops in lowland and more sheltered upland sites.
Frequent; in all more open moorland areas, sometimes locally abundant, as on the Laggan Deer Forest.
2–17
Contains stictic acid, atranorin, and two unknown substances.

UMBILICARIACEAE

Umbilicaria Hoffm.

U. cylindrica (L.) Del. ex Duby
On isolated boulders and rock outcrops on or near mountain summits.
Rare; very local.
11 Cruachan Dearg, 700 m. **15** W side of Mainnir nam Fiadh, 730 m. **16** Creach Beinn, near bird roost 670 m.
Often poorly developed. No lichen substances found.

U. polyphylla (L.) Baumg.
On boulders and rock outcrops.
Rare; local.
15 Creag Dhubh, on erratic covered by bird droppings 580 m. **10** Cruach nan Con, 470 m.
Contains gyrophoric acid.

U. polyrrhiza (L.) Fr.
On boulders, especially those in exposed, isolated situations.
Rare.
3 Kintra, on granite; Ardalanish, Maol an t-Sithein on granite. **16** N side of Loch Spelve, above Fellonmore, at 300 m.

U. proboscidea (L.) Schrad.
On boulders and outcrops near or on mountain summits.
Rare; local.
11 Coire nam Fuaran, 625 m; Beinn nan Gabhar, 630 m; Corra-bheinn, boulders by Màm a' Choir' Idhir at 525 m; Màm Choireadail at 450 m. **15** Sgùrr Dearg, 673 m to 730 m.
Contains gyrophoric acid; no norstictic acid was detected.

U. pustulata (L.) Hoffm.
On tops of boulders, especially those enriched by bird droppings; also in nutrient-rich seepage tracks on exposed cliff faces.
Rare; but widespread, often abundant in the sites listed.
15 Ben Talaidh, NNW slope, c. 730 m. **16** Lochbuie, Cnoc-na-Braclaich. **17** Lochdonhead, Achnacraig; Abhainn Barr Chailleach.
Contains gyrophoric acid.

U. torrefacta (Lightf.) Schrad.
On large isolated boulder in heathland.
Rare.
11 Ben More, near the summit, 910 m, five thalli in poor condition.
Contains gyrophoric acid.

RAMALINACEAE

Ramalina Ach.

Bibliogr.: Krog & James (1977).

R. baltica Lett.
R. obtusata auct. angl., non (Arnold) Bitt.
On boles of broad-leaved trees in exposed, lowland, wayside situations, especially near the west coast, more rarely on boles of old conifers; very rare in dry, sheltered, rock-crevices near the coast.
Occasional; rather local but becoming more frequent in the vicinity of Lochbuie.
2–4, 6, 7, 9, 12, 13, 16, 17
British material of this taxon, including collections from Mull, may be conspecific with *R. canariensis* Steiner. Contains divaricatic acid.

R. calicaris (L.) Fr.
On twigs, branches, and, less frequently, boles of broad-leaved trees, especially in more exposed, lowland, wayside sites; very rarely on conifers.
Frequent; locally abundant but not recorded from the Treshnish Isles.
2–17
Both chemical strains of this species are present on Mull; 35% contain substances belonging to the sekikaic acid aggregate, the rest are without medullary substances.

R. duriaei (De Not.) Bagl.
R. evernioides Nyl.
On boles of broad-leaved trees especially in exposed, lowland, wayside sites; rarely in sheltered rock-crevices in coastal cliffs and rock outcrops.
Rare on Iona and Ulva; frequent elsewhere.
1, 4, 12, 15, 16
Bourgeanic acid is present in the medulla.

R. farinacea (L.) Ach.
On twigs, branches and boles of broad-leaved trees and, more rarely, conifers in low-land sites, also on shrubs, particularly *Prunus spinosa*; not infrequent on rocks and old *Calluna* stems, especially near the west coast.
Abundant; widespread.
1–17
Fertile specimens found in **3, 4, 15**
All four chemical strains of this taxon are present in Mull collections: *a* protocetraric and usnic acids (var. *farinacea*) *b* hypoprotocetraric and usnic acids (var. **hypoprotocetrarica** (W. Culb.) D. Hawksw.) *c* norstictic (±), salazinic and usnic acids (var. **reagens** (B. de Lesd.) D. Hawksw.) and *d* without medullary substances. The proportions (a:b:c) of the four strains were 70:2:10:5, a sample of 195 specimens was examined. See also *R. subfarinacea.*

R. fastigiata (Pers.) Ach.
On twigs, branches and boles of broad-leaved trees, and occasionally, conifers, particularly in windswept sites along or near the coast.
Abundant on or near the coast; becoming notably less frequent inland.
1–17
Contains undetermined substances of the evernic acid aggregate.

R. fenestrata Motyka
R. geniculata auct. non Nyl.
On stems of *Salix* in sheltered valley carr between mature broad-leaved woodland.
Rare.
7 Torloisk House.
Contains substances of the evernic acid aggregate. A doubtful species.

R. fraxinea (L.) Ach. var. **fraxinea**
On boles of *Acer pseudoplatanus* and *Quercus* in a wayside locality.
Rare.
7 Torloisk House grounds; above Achleck, by Allt Loch a' Ghael.

R. fraxinea var. **calicariformis** Nyl.
On branches, more rarely twigs and boles, of lowland wayside trees, especially frequent on *Crataegus*. Often intermixed with *R. farinacea* and *R. fastigiata.*
Very rare on conifers.
Occasional to frequent; but seldom abundant in any site.
3, 10, 12–17
No medullary substances detected.

R. pollinaria (Westr.) Ach.
Sheltered boulder near the coast.
Rare.
10 Carsaig Bay.
A small deformed specimen in a very sheltered habitat, which would have previously been referred to *R. intermedia* (Del. ex Nyl.) Nyl. is included under this species. *R. intermedia*, originally described from Newfoundland, contains acids of the sekikaic complex and has not been correctly recorded from Britain. *R. pollinaria* contains substances belonging to the evernic acid aggregate.

R. portuensis Samp.
On boles of *Fraxinus* in a wind-exposed wayside site.
Rare; but probably overlooked for *R. farinacea.*
3 Beinna Ghlinne Mhòir.
This species was first noted in Britain by R. Santesson on the Isles of Scilly, and subsequently collected by the author in Pembrokeshire. The Mull record, the first for Scotland, represents a considerable extension of the geographical range of the species.
Contains salazinic acid and an unidentified substance.

R. siliquosa (Huds.) A. L. Sm. aggr.
Xeric supralittoral; on exposed rocky shores, becoming less frequent and rare in inland areas where the species is confined to the tops of

rock outcrops, boulders or standing stones; rare on decorticated wood (e.g., fence-poles) and exceptionally on tree-boles (as at Calgary Bay).
Abundant; especially common in the west and south-west.
1–17
On the basis of chemical constituents and distributional patterns, *R. siliquosa sens. lato* has been divided into six species (Culberson, 1965). These are: *R. atlantica* W. Culb. – usnic acid only; *R. crassa* (Nyl.) Motyka – atranorin (±), salazinic, usnic acids; *R. curnowii* Cromb. ex Nyl. – atranorin, norstictic, stictic acids and two unknown substances; *R. druidarum* W. Culb. – atranorin, chloratranorin, hypoprotocetraric, usnic acids, and *R. stenoclada* W. Culb. – atranorin, norstictic and usnic acids. All chemical strains have been observed from Mull collections. However, recent studies on this morphologically and chemically diverse aggregate (Sheard & James, 1976; Sheard, pers. comm.) indicate only two taxa for which the names *R. cuspidata* (Ach.) Nyl. and *R. siliquosa* (Huds.) A. L. Sm. are suggested. *R. cuspidata* reaches maximum abundance lower on the shore than *R. siliquosa* and is characterised by either the acid deficient, or norstictic or stictic acid (with accessory norstictic acid) series of chemical strains; the thallus is glossy with non-foveolated branches which bear dark brown or black pycnidia towards their apices. Conversely, *R. siliquosa* encompasses the hypoprotocetraric or protocetraric or salazinic (with accessory protocetraric) acid series and is usually distinguished by its glaucous appearance and the foveolation of the flattened branches with colourless pycnidial warts.

R. subfarinacea (Nyl.) Nyl.
On acid boulders and rock outcrops especially on or near the coast, becoming much rarer inland; rarely on tree boles and stems of *Calluna* and *Ulex* on cliff tops overlooking the sea.
Frequent; becoming locally abundant in the W.
1–17
Contains hypoprotocetraric, or protocetraric, or norstictic, or salazinic acids as well as usnic acid. See *R. farinacea*.

ACAROSPORACEAE

Acarospora Massal

A. atrata Hue
Xeric supralittoral; on sunny but sheltered, basalt rocks.
Rare.
4 Gometra, Rubha Dùn Iasgain.
This collection is intermediate between *A. atrata* and *A. opaca* Magnusson and suggests that these two taxa may best be united under the former name.

A. fuscata (Nyl.) Arnold.
On acidic rocks and boulders enriched by bird droppings, or near sheep and cattle tracks in coastal or inland sites; also on walls surrounding grazing fields and sheep pens and in the vicinity of farmsteads.
Frequent in all areas; ascending to 950 m on Ben More.
2–5, 8, 10–12, 16, 17
Contains gyrophoric acid.

A. heppii Naeg. ex Körb.
On calcareous sandstone near the coast.
Rare; only recorded from a single locality.
10 Nuns' Pass, *U. Duncan & James.*

A. sinopica (Wahlenb. ex Ach.) Körb.
On acid rocks rich in heavy metals, especially iron; frequently associated with *Lecidea atrata* and *Rhizocarpon oederi.*
Occasional; widespread, ascending to 500 m on Ben More massif, Coirc Bheinn.
2, 4, 6, 8, 11, 16

A. smaragdula (Wahlenb. ex Ach.) Massal. var. **smaragdula.**
In crevices of sheltered acid rocks especially near the coast, often obscured by summer vegetation.
Frequent.
2–6, 10, 13–16
Most of the thalli examined are poorly developed and only sparingly fertile.
All specimens contain norstictic acid.

A. smaragdula var. **lesdainii** (Harm. ex A. L. Sm.) Magnusson
On sunny basalt rocks.
Rare.
4 Gometra, *U. Duncan.*

A. veronensis Massal.
On sunny gneiss and granite rocks in coastal areas.
Rare; confined to the south-west.
2 Cùl Bhuirg. **3** Erraid, near the Observatory, on *Calluna*; Uisken, on granite; Rubh 'Ardalanish; Àrd Dhughaill.

Biatorella DNot.

Bibliogr.: James (1971).
B. monasteriensis (Lahm ex Körb.) Lahm
On bole of old *Fraxinus* in a lowland site.
Rare.
14 Fishnish Bay, Doire Dorch.
Previously recorded only from Bala, N Wales and Yorkshire.

B. moriformis (Ach.) Th. Fr.
On rotting conifer fence-rails in *Calluna–Scirpus* bog.
Rare; probably overlooked.
6 Loch an Tòrr.

B. ochrophora (Nyl.) Arnold
On boles of mature broad-leaved trees in

wayside and valley sites; confined to sheltered, warm, lowland sites in the north-west.
Rare.
5 Kilninian, on *Fraxinus.* **7** Ballygown, on isolated *Ulmus, U. Duncan & James*; Achleck, Normann's Ruh, on mossy *Fraxinus, U. Duncan & James.*
The ascocarps of this species are K+purple (parietin).

Sarcogyne Flot.

S. privigna (Ach.) Anzi
On granite rocks near the coast.
Rare; confined to the Ross of Mull.
3 Kintra, near Gortain Ur; S of Loch Poit na h-I, Toba Bhreaca.

S. regularis Körb.
On basic rocks, near the coast and on mortar and cement of old, often derelict, buildings; very rarely on asbestos roofing.
Occasional; widespread, never frequent in any one locality.
2–4, 6, 7, 10–12, 16, 17

S. simplex (Dav.) Nyl.
On acidic rocks in open situations, especially in upland areas; frequent also on old walls and in crevices on granite rocks.
Occasional; widespread. Probably under-recorded due to the inapparent thallus.
1–4, 6, 11–13, 15–17

Thelocarpon Nyl.

T. epibolum Nyl.
On more or less moribund squamules of *Coriscium viride* (=*Omphalina hudsoniana* at the edge of peat hags and old cutting faces in upland areas.
Rare; probably overlooked.
8 Beinn na h-Iolaire, at 450 m; Maol Mheadhonach, at 450 m. **15** Beinn Bheag, at 460 m.

T. laureri (Flot.) Nyl.
On old leather boot above the shore line; also on bone in machair.
2 Iona, Golf Course. **3** Erraid, Traigh Gheal.

PANNARIACEAE

Pannaria Del.

Bibliogr.: James in Dahl & Krog (1973).
P. mediterranea Tav.
On mossy boles and branches of *Quercus* in sheltered, relatively humid, mature woodland.
Rare; probably overlooked.
13 Tobermory, Aros Park. **17** Loch Spelve, An Coire; Auchnacraig.

P. microphylla (Sw.) Massal.
On bases of old broad-leaved trees, especially *Fraxinus*, in wayside and woodland sites; also on soil in crevices between acidic rocks on or near the coast or more rarely on shaded rocks in very sheltered, valley woodlands.
Frequent; especially in lowland sites in the south-east.
2–8, 11–17
Frequently fertile: of the 98 specimens examined, 74 have ascocarps. The colour of the thalline squamules varies from pale grey in shaded habitats to deep purple-brown in exposed, sunny sites.

P. nebulosa (Hoffm.) Nyl.
On recently disturbed soil and on low vegetation on cliff edges and by coastal paths.
Rare; only recorded in the south and west.
8 SW Ardmeanach, near Burgh. **10** Carsaig, Nuns' Pass.
Both gatherings are sterile.

P. pezizoides (Web.) Trevis.
On mossy tree boles and rotting logs, especially *Betula*, in sheltered upland stream valleys; also on moist semi-stable soil, often near streams.
Rare.
8 Clachandhu, on soil and crumbling basalt rocks at entrance to gorge. **11** Gleann na Beinne Fada, locally frequent on *Betula.*

P. pityrea (DC.) Degel.
Predominantly on mossy, mature, broad-leaved tree boles in wayside and woodland sites, and on *Corylus* and *Salix* in damp, humid valley woodlands and carr, especially near the coast; rarely on conifers and mossy rocks and boulders.
Abundant; widespread in all coastal and lowland areas, becoming rarer in upland valleys, seldom occurring above 100 m.
1–17
All material examined is sterile.

P. rubiginosa (Thunb. ex Ach.) Del.
Similar to *P. pityrea* but more frequent on mossy boles of *Corylus* in sheltered, humid, valley woodlands near the coast.
Abundant; widespread in all lowland areas, becoming rarer in upland valleys and seldom occurring above 90 m.
2–17

P. sampaiana Tav.
On boles of mature broad-leaved trees in sheltered or exposed, mature woodlands and particularly on boles of exposed wayside trees, especially *Fraxinus* in lowland and coastal sites.
Occasional; becoming locally frequent in the east.
7 Ballygown on *Fraxinus.* **11** Coille na Sròine on *Fraxinus.* **12** Glenforsa House, on *Acer pseudoplatanus.* **13** Aros Park, on *Quercus* in very shaded site. **17** Torosay Castle grounds, on *Fraxinus* in sunny exposed hedgerow.

Parmeliella Müll. Arg.

P. atlantica Degel.
On boles and branches of wayside and woodland broad-leaved trees, especially *Fraxinus* and *Acer pseudoplatanus* in all lowland and

coastal areas; more rarely on mossy rock outcrops and old stems of *Calluna* on the coast.
Frequent; often locally abundant.
1–17

P. corallinoides (Hoffm.) Zahlbr.
On boles of wayside and woodland broad-leaved trees, especially *Fraxinus* and *Ulmus* in all lowland and coastal areas; very rarely on bare rocks in and by streams.
Frequent; ascending to 350 m in **15** above Scallastle Bay, on *Fraxinus* in steep-sided ravine.
2–4, 6–17

P. plumbea (Lightf.) Vain.
On boles and branches of wayside and woodland broad-leaved trees, especially *Fraxinus*, *Acer pseudoplatanus* and *Ulmus*, in all lowland and coastal areas; rarely on mossy rock outcrops on the coast.
Frequent; often locally abundant, ascending to 360 m on **11** Creag Dhubh.
1–17

Psoroma (Ach.) Ach. ex Michx

P. hypnorum (Vahl) Gray
Amongst mosses at the base of aged *Fagus* in old sheltered woodland.
Rare.
16 N side of Loch Uisg, Ath Leitir, near the Obelisk.

AGYRIACEAE

Xylographa (Fr.) Fr.

X. abietina (Pers.) Zahlbr.
On decorticated conifer stumps in sheltered, rather dry habitats, confined to mature, but mostly untended, conifer plantations and mixed woods.
Rare.
8 Glen Seilisdeir, on *Pinus* logs. **12** Salen, on *Pinus* logs. **13** Aros Park.

X. trunciseda (Th. Fr.) Reding.
On decorticated conifer logs and stumps in sheltered, humid, lowland sites.
Rare; but locally frequent in the localities listed.
13 Aros Park; Druimfin by the roadside. **17** Torosay Castle, on pine stumps in sawmill yard.

X. vitiligo (Ach.) Laund.
On decorticated conifer stumps, fence-rails and logs.
Occasional; often locally frequent in well-wooded areas especially in the east.
4, 7–11, 13–15, 17

LICHINACEAE

Ephebe Fr.

E. lanata (L.) Vain.
In seepage tracks on boulders and rock-faces and especially on wet rocks near swiftly flowing streams and waterfalls.
Frequent above 80 m; very rare in coastal and lowland sites.
3, 7, 9, 11, 12, 15, 16
Particularly abundant and often fertile above Lochbuie and on Laggan Deer Forest.

Euopsis Nyl.

E. pulvinata (Schaer.) Th. Fr.
Associated with mosses on old walls and acidic boulders.
Rare.
13 Near Lochan na Guailine Duibhe. **16** Port Ohirnie, 190 m.

Lempholemma Körb. ex Zahlbr.

L. chalazanellum (Nyl.) Zahlbr.
Amongst mosses on mortar of bridge.
Rare.
10 Pennyghael, bridge over Leidle River.

L. chalazanum (Ach.) B. de Lesd.
On old crumbling mortar of shaded wall.
Rare.
10 Pennyghael, on wall surrounding garden.

Lichina Ag.

L. confinis (O. F. Müll.) Ag.
Upper littoral fringe to lower mesic supra-littoral; confined to acidic rocks and boulders, typically with *Verrucaria maura* and *Caloplaca marina*; especially frequent on rock outcrops and boulders near the heads of sea-lochs and along sheltered sunny shores; rarely spreading up the shore in fresh-water seepage tracks.
Frequent; often locally abundant.
1–17

L. pygmaea (Lightf.) Ag.
Upper littoral fringe to lower littoral fringe; confined to acidic rocks, often occurring in dense swards in the upper part of the barnacle zone.
Occasional to frequent.
2, 4–6, 8, 10, 12, 14–17
Less frequent than *L. confinis* and preferring more sunny and wave-exposed shores and headlands.

Porocyphus Körb.

P. coccodes (Flot.) Körb.
In seepage tracks on acid rocks near the coast.
Rare; only recorded from the west.
2 Cùl Bhuirg, on gneiss. **3** Kintra, on granite. **5** Mornish, Port na Bà and Rubha an Àird, on ballast.

Psorotichia (Massal.) Forss.

P. schaereri (Massal.) Arnold
On exposed boulders and scree of calcareous sandstones near the coast.
Rare.
10 Several sites between Nuns' Pass and Aird Ghlas, *James & U. Duncan.*
Frequent on Lismore Island.

Thermutis Fr.

T. velutina (Ach.) Flot.
On boulders at the margin of an upland loch.
Rare.
13 Caol Lochan.

PLACYNTHIACEAE

Placynthium Gray

P. lismorense (Nyl. ex Cromb.) Vain.
Recorded from a single site associated with base-rich glauconitic sandstone scree near the coast.
Rare.
10 Carsaig, near Aird Ghlas.
Prior to this record this endemic British species was known only from the nearby Lismore Island where it is very abundant.

P. nigrum (Huds.) Gray
Restricted to base-rich saxicolous substrates, including limestones, calcareous sandstones, cement, mortar, asbestos and consolidated shell-sand covering granite outcrops and boulders.
Frequent; in all lowland and coastal sites, rarely above 90 m.
2–5, 7, 9–13, 15–17
Often found with ascocarps. *P. tremniacum* may be a form of this species with one-septate spores.

P. tantaleum (Körb.) Hue
In seepage tracks on exposed, moderately basic outcrops near the coast.
Rare.
8 Near Mackinnon's Cave. **10** Carsaig, between Uamh an Dùnain and Rubha Dubh.

P. tremniacum (Massal.) Jatta
On consolidated shell-sand associated with granite rocks near the coast.
Rare.
2 Iona, Golf Course, near Shian.
Probably a form of *P. nigrum* with one-septate spores.

PELTIGERACEAE

Massalongia Müll. Arg.

M. carnosa (Dicks.) Körb.
Overgrowing mosses on boulders in or by streams; also on rock outcrops and boulders along the margins of lochs.
Rare; probably overlooked.
8 Boulders in Abhainn Bail' a' Mhuilinn. **12** Near Ledmore. **15** Side of Loch Bearnach and in adjacent bog. **16** Lochbuie, in stream leading from Lochan na Leitreach.
Fertile specimens collected in **12** and **16**.

Nephroma Ach.

N. laevigatum Ach.
On broad-leaved trees in most lowland and coastal wayside and woodland areas; less frequently associated with mosses on boulders in *Corylus* woodland and on exposed rocky outcrops on the coast.
Frequent; locally abundant, ascending to 300 m in **11** Garbh Coire on *Sorbus aucuparia.*
1–17

N. parile (Ach.) Ach.
On broad-leaved trees and mossy rocks in lowland and coastal sites; rarely on sheltered rocky outcrops by the coast, but generally preferring more sheltered sites than *N. laevigatum.*
Frequent.
3, 4, 7–10, 12–14, 16, 17
Thalli with numerous aborted ascocarps were seen in **12** at Killichronan on a sheltered wayside *Acer pseudoplatanus.*

Peltigera Willd.

P. aphthosa (L.) Willd.
Amongst short grass on damp, often slightly basic, sheep-grazed hillsides; also on moss-covered rocks in sheltered, damp woodland, often near streams and more rarely on damp rocks along the margins of lochs and streams.
Occasional; mainly confined to lowland areas.
3, 5–7, 10, 12, 13, 17
The majority of the collections resemble var. *variolosa* (Massal.) Thoms. (*P. leucophlebia* (Nyl.) Gyeln.) but an accurate determination was not possible in the absence of ascocarps.

P. canina (L.) Willd.
On bases of boles of broad-leaved trees in sheltered, damp habitats, especially near water in lowland, well-wooded areas; also frequent on moss-covered boulders, old walls, and amongst mosses on soil in sheltered ravines; rarely recorded on garden lawns.
Occasional; widespread, but rare above 200 m; much less frequent than *P. praetextata.*
3–6, 10, 12, 13, 15–17
This species often appears to merge into *P. praetextata* and *P. rufescens*; further ecological investigations and transplant experiments may indicate that these three species are a range of ecotypes of a single taxon.

P. collina (Ach.) Schrad.
P. scutata (Dicks.) Duby
Wayside and woodland broad-leaved trees in lowland and coastal areas.
Frequent.
3–5, 7–13, 15–17
Fertile specimens found in **5, 12, 13, 16**

P. horizontalis (Huds.) Baumg.
On moss-covered boles of trees, fallen logs, boulders and rock outcrops often near water, in sheltered, damp woodland. More rarely on north-facing sides of walls.
Frequent; especially in the east.
4, 6–17

P. polydactyla (Neck.) Hoffm.
On moss-covered boles of trees, fallen logs, boulders and rock outcrops in sheltered, damp lowland and coastal woodland; also in open situations such as road and path verges, lawns and short turf in sheep-grazed pastures.
Frequent; often abundant in the east.
1–17

P. praetextata (Flörke ex Sommerf.) Zopf
On bases of boles, moss-covered boulders and rock outcrops, old walls, fallen logs or amongst mosses on soil in sheltered lowland and coastal woodland areas.
Abundant; especially in the east, becoming rare above 200 m.
2–17

P. rufescens (Weis) Humb.
On well-drained, often slightly basic soils, in exposed lowland areas; also on soil and mosses on the summits of old walls and on consolidated sand of dunes at or near the coast.
Frequent.
1–4, 6–8, 10, 13, 14, 17

P. spuria (Ach.) DC.
On soil by paths or roadsides; often on burnt soil in open woodlands and conifer plantations on consolidated shell-sand and near the coast.
Occasional.
2, 4, 6, 7, 9, 12, 13
Both sorediate and fertile stages are present on the Island.

Polychidium (Ach.) Gray

P. musicola (Sw.) Gray
Amongst mosses at the bases of old tree boles; also on boulders and rocks-faces, especially on N and E-facing aspects and near coast; most frequently observed near streams and lochs.
Occasional; in both lowland and upland sites.
5 Ensay Burn, *U. Duncan.* **6** East of Dervaig, an Tòrr, on old *Ulmus*; Lochan's Airde Beinn. **14** Scallastle Bay, Alterich, on mossy boulders. **16** Loch Uisg, on boulders by lake shore.

P. umhausense (Auersw.) Henss. See *Lobaria amplissima.*

STICTACEAE

Lobaria (Schreb.) Hue

L. amplissima (Scop.) Forss.
On boles and branches of broad-leaved trees in lowland and coastal wayside sites, especially at the margins of mature woodland; rarely amongst mosses on boulders on coast and along margins of lochs.
Frequent; widely distributed.
3–6, 8, 10–13, 15–17
All specimens have coralloid outgrowths of the blue-green morphotype (*Polychidium umhausense* (Auersw.) Henss.) Contains scrobiculin, two unknown substances (±) and, rarely atranorin; no lichen substances were detected in the blue-green excrescences.

L. laetevirens (Lightf.) Zahlbr.
On boles and branches of broad-leaved trees in wayside and woodland sites in lowland and coastal areas; also in *Salix* carr and *Corylus* scrub in sheltered valleys and ravines; rarely on mossy boulders on the coast or at the edge of lochs.
Frequent; often locally abundant.
1–17
Contains small quantities of atranorin (±).

L. pulmonaria (L.) Hoffm.
On boles and branches of broad-leaved trees in wayside and woodland sites in lowland and coastal areas; also in *Salix* carr and *Corylus* scrub in sheltered valleys and ravines; occasional on old *Calluna* stems and rock outcrops on the coast and along edges of lochs.
Frequent; often locally abundant.
1–17
Of the 347 specimens examined 169 have ascocarps.
Contains stictic, constictic, norstictic (±) acids and three unknown substances.

L. scrobiculata (Scop.) DC.
On boles and branches of broad-leaved trees and mossy boulders and rock outcrops in wayside and woodland sites in lowland and coastal areas; occasional on old *Calluna* stands near the coast.
Frequent; sometimes locally abundant in a few damp, sheltered sites.
3–17
Of the 275 specimens examined 12 have ascocarps.
Contains stictic, constictic, norstictic (±), usnic acids, three unknown substances and scrobiculin.

Pseudocyphellaria Vain.

P. crocata (L.) Vain.
On mossy boles and branches of broad-leaved trees, rarely conifers, in sheltered and mature woodland in lowland and coastal areas; also on *Corylus* in moist coastal valleys and near the ground on *Salix* in sheltered carr in lowland areas; rarely on moss-covered boulders and outcrops near and on the coast.
Occasional to frequent; especially in the east, locally abundant in a few areas, e.g., **13** Upper Druimfin. Rarely recorded above 30 m.
3–10, 12–14, 16, 17
Contains calycin, pulvinic dilactone, two unknown substances, stictic and norstictic acids and three triterpenoids. In *P. crocata*, the upper surface of the thallus is normally foveolate-ridged with numerous laminal and some marginal soralia. Occasionally, however, in some populations the upper surface remains smooth and the soralia entirely marginal. Such forms seem to approximate closely to *P. mougeotiana* (Del.) Vain.

P. intricata (Del.) Vain.
P. thouarsii (Del.) Degel.
On mossy boles and branches of broad-leaved trees, particularly *Corylus* and *Salix*, in sheltered and humid woodland in lowland and coastal areas; locally frequent on wayside trees and on mossy outcrops and boulders in coastal sites; rarely on the ground in old *Calluna* heath.
Frequent; sometimes locally abundant, especially in the east, more widespread than *P. crocata.*
2–13, 16, 17
Contains stitic, norstictic (±) acids, two unknown substances and three unknown triterpenoids.

P. lacerata Degel.
In crown of moss-covered *Quercus* in sheltered E-facing woodland near the coast.
Rare.
13 Aros Park, near Spùt Dubh.
Differs from *P. intricata* in the coarse marginal isidia and folioles. Previously recorded from SW Eire and N Wales, the present collection represents the first record for Scotland. Contains two unknown substances and 3 triterpenoids similar to those recorded for *P. intricata.*

Sticta Schreb.

S. canariensis Bory ex Del.
Sticta dufourii Del. (=blue-green morphotype, see James & Henssen, 1976).
Mossy rocks and boles of broad-leaved trees in sheltered, usually poorly lit, sites, often close to streams and freshwater lakes. Particularly frequent in sheltered gorges near the spray of waterfalls.
Frequent; especially in the east.
4, 6–8, 10–17
Examples of the blue-green phase (*S. dufourii*) with small marginal leaflets of *S. canariensis* attached, were collected in **7** Achleck, below Tòrr an Ogha and **15** Scallastle Bay, near Alterich, *U. Duncan & James.*
No lichen substances detected in either morphotype.

Sticta fuliginosa (Dicks.) Ach.
Mossy boles and branches of broad-leaved trees in most sheltered and humid, well-wooded lowland and coastal areas. Also on *Corylus* in sheltered valleys near the coast and on *Salix* in carr in more sheltered lowland sites. Locally frequent on wayside trees but rare on mossy outcrops in sheltered or exposed coastal sites.
Frequent; sometimes locally abundant, especially near the coast.
4–8, 10–17
No lichen substances detected.

S. limbata (Sm.) Ach.
Mossy boles of broad-leaved trees, especially *Fraxinus*, in wayside lowland and coastal areas. Rarer than *S. fuliginosa* and *S. sylvatica* in old woodlands. Frequent on *Salix* in open carr.
Frequent; sometimes locally abundant.
3–17
No lichen substances detected.

S. sylvatica (Huds.) Ach.
Mossy boles of broad-leaved trees in most sheltered and humid, often shaded well-wooded, lowland and coastal areas. Occasional on *Corylus* in sheltered valleys or near the coast. Also on *Salix* in carr, especially in sheltered sites. Much more frequent than *S. fuliginosa* on mossy rocks, especially in sheltered stream valleys.
Frequent; often locally abundant, in all lowland coastal woodland areas.
4–17
No lichen substances detected.

TELOSCHISTACEAE

Caloplaca Th. Fr.

Bibliogr.: Wade (1965); Nordin (1972): see also *Xanthoria* (Fr.) Th. Fr.

C. albolutescens (Nyl.) Oliv.
On crumbling basalt rocks on cliffs near the sea.
4 Gometra, below Tòrr Mòr.

C. arnoldii (Wedd.) Zahlbr. ex Ginzb.
Xeric supralittoral; dry, sheltered recesses and crevices in basic and acid rocks on or near the coast; often in associations with *C. saxicola.*
Rare, but locally frequent in the few sites recorded.
5 Creag a' Chaisteil, on basalt cliff. **8** Below Dishig and Derryguaig, under sheltered basalt overhangs. **10** Carsaig, abundant on sheltered cliffs of calcareous sandstone. **13** Salen. This species frequent on dry limestone cliffs near the sea on Lismore Island.

C. caesiorufa (Wibel) Flag.
Xeric supralittoral; on coastal acidic rock, more rarely on rocks enriched by bird-droppings or near farmsteads.
Frequent near the coast.
1–8, 10–14, 16–17
Distinguished from *C. ferruginea*, which prefers harder rock, by the duller orange-red ascocarps which become greenish brown when wet. In section, the phycobiont forms a continuous layer under a pseudoparenchymatous hypothecium.

C. cerina (Ehrh. ex Hedw.) Th. Fr. var. **cerina**
On boles and branches of wayside trees, especially *Fraxinus*, *Sambucus nigra* and *Ulmus*, near the coast or the vicinity of farmsteads.
Frequent; widespread in all lowland areas becoming rare above 60 m.
2–13, 15–17

C. cerina var. **chlorina** (Flot.) Müll. Arg.
In damp sheltered rock crevices in coastal areas and along loch shores.
Rare.
4 Ulva, Craigaig, on basalt, *U. Duncan.* **6** shore of Loch a' Chumhainn at Dervaig. **16** Loch Uisg, near Creach Bheinn Lodge.

C. cerinella (Nyl.) Flag.
On *Sambucus nigra* in wayside sites especially near the coast or lowland farmsteads; often in associations with *Bacidia naegelii*, *Caloplaca cerina*, *C. holocarpa* and *Lecania cyrtella.*
Occasional.
7 Lagganulva. **8** Inch Kenneth, near the Chapel. **10** Pennyghael. **12** Tobermory, Baydoun. **16** Lochbuie, near the Castle.

C. cirrochroa (Ach.) Th. Fr.
On dry, sheltered recesses of calcareous sandstone near the coast.
Rare. Probably limited by the absence of suitable natural substrates.
10 Carsaig, cliffs, *U. Duncan & James.*
This species is frequent on Lismore Island.

C. citrina (Hoffm.) Th. Fr.
Colonising a wide variety of substrates including wayside broad-leaved trees, basic, more rarely acidic, rocks and boulders, especially near the coast and near farmsteads, mortar and cement of bridges, walls and house-facings.
Abundant; in all lowland areas, becoming less common above 100 m.
1–17

C. decipiens (Arnold) Jatta
On cement of window lintel.
Rare.
13 Tobermory, Upper Church.
An unexpected record of a species with a predominantly eastern distribution in Britain; recently recorded from Kintyre in the west.

C. ferruginea (Huds.) Th. Fr.
On acid rocks in sheltered or exposed sites, especially on or near the coast or in proximity of lowland farmsteads, often associated with bird-droppings; also on smooth bark of broad-leaved trees and shrubs, especially *Corylus*, *Fraxinus*, *Salix* and *Sorbus aucuparia*, in mainly sheltered and boggy habitats.
Frequent.
1–17 (Corticolous in **2, 4–8, 11–13, 16, 17**).
Saxicolous forms are abundant in all areas of the Island; corticolous thalli less frequent and mainly confined to stream valleys. Polymorphic, especially in nutrient-enriched sites.

C. flavorubescens (Huds.) Laund.
C. aurantiaca auct.
Bole of *Fraxinus* in a sunny, exposed site.
Rare.
2 Baile Mòr, near St. Mary's Abbey.
This variety is more frequent on Lismore Island where it occurs on *Fraxinus* and *Acer pseudoplatanus.*

C. flavovirescens (Wulf.) D. T. & Sarnth.
C. aurantiaca var. *flavovirescens* (Wulf.) Th.Fr.
On basic substrates such as limestone, mortar of walls and bridges and gneiss and granite rocks partially covered by wind-blown shell-sand.
Occasional; but often abundant in the sites recorded.
2 Cùl Bhuirg, on gneiss rocks covered in shell sand; Camas Cuil an t-Saimh, on gneiss partially covered in shell sand. **3** Traigh nam Beach, old bridge on the Iona Road. **10** Carsaig Bay, coastal scree, *U. Duncan.* **17** Torosay Castle, on wall.

C. heppiana (Müll. Arg.) Zahlbr.
On mortar of wall by roadside.
Rare.
14 Chapel E of River Forsa outlet.
This species is frequent on Lismore Island.

C. holocarpa (Hoffm.) Wade
On a wide variety of substrates, including acid and basic rocks, especially on or near farmsteads, and on mortar, and stones of walls, bridges, monuments and houses; less frequently on bark of old wayside trees, especially at the margins of fields.
Abundant; widespread in lowland areas, becoming less frequent above 90 m.
1–17 (Corticolous in **2, 4–6, 8, 9–12, 17**).

C. irrubescens (Nyl.) Zahlbr.
On shaded basalt rock near the coast.
Rare.
12 Salen, low cliffs near village.

C. littorea Tav.
Mesic supralittoral; under dry, sheltered overhangs in gneiss and basalt rocks on or near the coast.
Rare; confined to a few extreme western localities.
2 Port an Duine Mhairbh, on gneiss; Eilean Maol Mhartuin; Port Chlacha Dubha, on gneiss. **4** Gometra, Rubha Maol na Mine, on basalt; Ulva, Craigaig. **5** Calliach Point, on basalt.
All the collections are sterile.

C. marina Wedd.
Mesic supralittoral; on exposed, sunny acidic and basic rocks.
Abundant; widespread on more exposed shores, becoming less common towards the heads of sea-lochs and there often replaced by *C. thallincola.*
1–17

C. microthallina Wedd.
Mesic supralittoral; confinde to the lower

part of the littoral zone, in crevices and minute declivities on sunny or sheltered, shady, acidic or basic rocks, often epiphytic on *Verrucaria maura* and other lichens. Often entirely replacing *Caloplaca marina* in shaded situations on the east coast.
Frequent; especially on the eastern coast of the Island.
1–5, 7, 9–12, 14–17
Specimens with ascocarps recorded in **2, 4, 7, 9–11, 15–17.**

C. obliterans (Nyl.) Jatta
On sheltered, rather moist basalt cliffs.
Rare.
12 Salen. **15** Ardura, Crùn Lochan.

C. sarcopisioides (Körb.) Zahlbr.
On broad-leaved trees in or near mature woodland sites.
Rare; but probably overlooked.
6 Penmore Mill, on *Quercus*. **14** Fishnish Bay, Coire Liath on *Fraxinus*.

C. saxicola (Hoffm.) Nordin
C. murorum (Ach.) Th. Fr.
In sheltered, more or less dry crevices and overhangs in basic and acidic rocks, especially near the coast; also on acidic stones in mortarstone facings of walls, bridges, houses and farmsteads.
Frequent.
2, 3, 5, 7, 8, 12–14, 16, 17
This species is very polymorphic; most of the material from Mull is of the variant often referred to as var. **pusillum** (Massal.) Malbr.

C. stillicidiorum (Vahl) Lynge
Encrusting mosses on calcareous sandstone boulders near the coast.
Rare.
10 Near Nuns' Pass, encrusting *Ctenidium molluscum.*
An unexpected record of a species normally confined to montane limestones in N England and Central Scotland.

C. thallincola (Wedd.) Du Rietz
Mesic supralittoral; confined to the littoral zone, on acidic and basic rocks favouring more sheltered sites than *C. marina.*
Abundant in all coastal areas, occasionally replacing *C. marina* in sheltered sites.
1–17

C. verruculifera (Vain.) Zahlbr.
C. granulosa auct. brit., non Müll. Arg.
Xeric supralittoral; on exposed or sheltered acidic rock shores in sites often liberally enriched with bird-droppings.
Rare.
1 Fladda. **3** Kintra, Rubha nan Cearc. **5** Port na Caillich; near Rubha an Àrd. **16** Lochbuie, Eilean Mòr.
According to Nordin (1972) *C. granulosa* (Müll. Arg.) Jatta is a smaller species with persistently narrower lobes, smaller isidia and an intense yellow-brown to red-brown colour. It is more characteristic of inland and more continental habitats and is a rare species in Britain mainly restricted to the Central Highlands of Scotland.

Protoblastenia (Zahlbr.) Steiner

P. calva (Dicks.) Zahlbr.
Confined to the hard mesozoic limestones and cornstones.
Rare.
10 Carsaig, above Sgeir Bhuidhe. **17** Craignure, Rubha na Sròine.
Often considered as a variety of the previous species, *P. calva* nevertheless remains distinct in assocations with *P. rupestris*. It is common on Lismore Island.

P. immersa (Hoffm.) Steiner
On sheltered, limestone outcrop near the coast.
Rare.
17 Craignure, Rubha na Sròine.
A common and widespread species on Lismore Island.

P. incrustans (DC.) Steiner
On calcareous sandstone near the coast.
Rare.
10 Above Sgeir Bhuidhe and Aoineadh nan Gamhna; above Carraig Mhor and Aoineadh a' Mhaide Ghil.
Recognised by the semi-immersed, minute ($>$ 0.3 mm) ascocarps with bright orange-coloured discs.
This species is frequent on Lismore Island.

P. monticola (Ach.) Steiner
On base-rich substrates, including limestone, calcareous sandstone, mortar and asbestos.
Frequent.
1–6, 8–10, 12, 13, 15–17

P. rupestris (Scop.) Steiner
On base-rich substrates, including limestone, calcareous sandstone, mortar, cement and asbestos.
Frequent; very widespread in suitable habitats.
2–15, 17
This common and widespread species is an important and easily identified indicator species of weak to moderately base-rich substrates.

Xanthoria (Fr.) Th. Fr.

X. aureola (Ach.) Erichs.
On acidic boulder near the head of sea-loch.
Rare.
6 Loch a' Chumhainn.

X. candelaria (L.) Th. Fr.
On trees and rocks, particularly old acidic-stone walls and rocks enriched bird droppings or near farmsteads; also on fence stumps and rails, more rarely on the boles of wayside trees, especially *Acer pseudoplatanus* and *Ulmus.*

Frequent; in all lowland and coastal areas, becoming rare above 100 m inland.
1–6, 8–12, 14–17
Very rarely fertile; seen only at **4** Ulva, Crag-aig, *U. Duncan & James*. **5** North side of Loch Tuath, Dùn Aisgain, Sloc an Neteogh.

X. elegans (Link) Th. Fr.
On concrete and asbestos near the west coast.
2 Baile Mòr, on asbestos roofing, *Crittenden*. **5** S side of Calgary Bay, on cottage wall.

X. fallax (Hepp) Arnold
On boles of *Ulmus* and spreading to acidic boulders near farmstead.
Rare.
7 Fanmore, *U. Duncan & James*.

X. parietina (L.) Th. Fr.
On trees and rocks, particularly those enriched with bird droppings and near farmsteads; also on both exposed and sheltered acidic coastal rocks.
Frequent; abundant on or near the coast, becoming rare in upland sites.
1–17
Polymorphic; specimens on the shore are generally sterile and correspond to var. **ectanea** (Ach.) Kickx.

X. polycarpa (Hoffm.) Oliv.
On twigs and branches of broad-leaved trees and shrubs, especially *Acer pseudoplatanus*, *Crataegus*, *Fraxinus* and *Prunus spinosa* in wayside and valley woodlands; very rarely on rock.
Frequent; widespread, especially in lowland and coastal areas.
2–14, 16–17

PHYSCIACEAE

Anaptychia Körb.

A. fusca (Huds.) Vain.
Pseudophysica fusca (Huds.) W. Wats.
Mainly xeric supralittoral; on sunny acid rocks, sometimes spreading to soil and plant detritus; rarely penetrating far inland and then mainly on the boles of mature wayside broad-leaved trees and tops of rock outcrops and standing stones in exposed sites.
Abundant; sometimes forming pure stands, in all coastal areas.
1–17
Frequently with well-developed ascocarps. No lichen substances detected.

A. mamillata (Tayl.) D. Hawksw.
Parmelia mamillata Tayl; *Anaptychia ciliaris* (L.) Körb. f. *melanosticta* (Ach.) Harm.
Submesic supralittoral; on sheltered, more or less permanently moist, acid rocks.
Rare.
4 Ulva, Port Bàta na Luinge. **13** Rubha nan Gall.
Both collections are sterile. No lichen substances detected.
Anaptychia mamillata differs from *A. ciliaris* (L.) Körb., common in eastern Britain, in the narrower, more branched and prostrate thallus lobes and their persistently darker colour. Furthermore, *A. mamillata* is confined to coastal rocks, a habitat not known for *A. ciliaris*.

Buellia De Not.

B. aethalea (Ach.) Th. Fr.
On sunny, acidic rocks and boulders, wall-tops, roofing tiles, in all areas but especially near the coast.
Frequent; in the west particularly on the granite, rather rare in the east.
1–6, 8, 12, 16, 17
Contains norstictic acid.

B. alboatra (Hoffm.) Deichm. Br. & Rostr. agg.
On basic rocks and boulders or acidic rock influenced by mortar or wind-blown shell-sand and often in sheltered, rather dry overhangs; rare on boles of broad-leaved trees, usually *Acer pseudoplatanus*, *Fraxinus* and *Ulmus*.
Occasional; in coastal and lowland areas, especially in the west.
2 On *Fraxinus* near The Nunnery; Culbuirg overhang of gneiss. **3** Knocknafenaig, wall of disused barn. **14** E of Glen Forsa, stone-mortar walls. **10** Carsaig, calcareous sandstone cliffs.

B. canescens (Dicks.) De Not.
On dry, leeward sides of wayside, broad-leaved trees, especially *Fraxinus* and *Ulmus*, in sheltered, but sunny, situations; also in dry rock-crevices of sunny, mortared farmstead walls and sheltered overhangs in cliff faces near the sea and sometimes on sea-bird roosts.
Occasional; widespread.
1–3, 5–10, 13, 17
Seldom well-developed; ascocarps were not seen on any of the material examined.
Contains atranorin and two unknown substances.

B. disciformis (Fr.) Mudd
On smooth, rarely rough, boles and branches of broad-leaved trees, particularly *Corylus*, in wayside, valley and woodland sites.
Abundant; widespread in lowland sites, less frequent above 40 m and there often confined to *Sorbus aucuparia*.
3–9, 12–17
A gathering from **7** Torloisk House on decorticated wood in sheltered site is characterised by large (20–26 × 8–13 μm), curved spores and numerous oil droplets in the thecium and hypothecium. Although related to *B. disciformis*, this taxon should be regarded as a new species.

B. erubescens Arnold
On branches of *Fagus* in a mature woodland site.

Rare.
13 Aros Park.
Contains norstictic acid and atranorin. See James (1971).

B. griseovirens (Turn. & Borr. ex Sm.) Almb.
On boles and branches of broad-leaved and coniferous trees, mostly in sheltered, humid sites. Most abundant in open *Betula* woods and on old *Salix* in damp carr.
Probably frequent but often sterile and therefore easily overlooked.
3–9, 12–17
Ascocarps were noted in **12** and **13**. See James (1971).

B. punctata (Hoffm.) Massal.
On wayside and woodland broad-leaved trees and conifers, especially near farmsteads; sheltered rocks, especially those enriched with bird droppings, on or near the coast.
Frequent; widespread in lowland areas, rarer in upland sites.
1–7, 9, 10, 12, 13, 16, 17

B. schaereri De Not.
On conifers and rotting logs in sheltered woodland.
Rare.
6 Glengorm Castle. **17** Torosay Castle on old *Pinus*.

B. stellulata (Tayl.) Mudd
On exposed and sunny rocks especially near or on the coast; on old walls and roofing tiles of houses and farm buildings.
Abundant; in all lowland sites, becoming rare in upland areas.
1–17
Often associated with *B. aethalea* on the coast but more frequent inland than that species.

B. subdisciformis (Leight.) Vain.
Xeric supralittoral; on acidic rocks, especially gneiss and granite.
Occasional; in Iona and the Ross of Mull.
2 Iona, Druim Dhùghaill; Eilean Maol Mhàrtuin, *U. Duncan & James*; Baile Mòr, Cladh an Diseart. **3** Erraid, Traigh Gheal; Port nan Ròn; N of Fionnphort, Bull Hole.
Contains norstictic acid.

B. verruculosa (Sm.) Mudd
On acidic, sunny rocks, often forming mosaics with *Buellia stellulata* and *B. aethalea*.
Occasional; but probably overlooked. Mainly confined to coastal areas.
1–4, 6, 8–12, 17

Heterodermia Trev.

H. obscurata (Nyl.) Trev.
Anaptychia obscurata (Nyl.) Vain.
On moss-covered branches of old broad-leaved trees, especially *Quercus*, in a few, well-wooded, lowland areas; also on boles of *Salix* spp. in wet carr near the coast.
Occasional; confined to the east.
13 Aros Park, by Lochan a' Ghurrabain; Upper Druimfin and Apper Mòr, on mossy logs and on *Salix* in carr. **16** Cruach Ardura, An Coire. **17** Craignure, Cnoc nan Cùbairean, on *Salix* in carr.
Contains zeorin and atranorin.

Physcia Ach.

P. adscendens (Th. Fr.) Oliv. em. Bitt.
On broad-leaved trees, especially *Acer*, *Fraxinus*, *Tilia* and *Ulmus*, near farmsteads; also on inland and coastal rocks and boulders, particularly near farmsteads, sheep pens and bird roosts.
Frequent; in all lowland areas, ascending to 690 m on **11** Cruachan Dearg.
1–13, 15–17

P. aipolea (Ehrh. ex Humb.) Hampe
On smooth bark of twigs, branches, more rarely boles, of broad-leaved trees, especially *Acer pseudoplatanus*, *Crataegus*, *Fraxinus*, and *Ulmus*.
Frequent; in all lowland areas, ascending to 500 m on **11** Gleann na Beinne Fada, on *Crataegus*.
1–6, 8–17

P. caesia (Hoffm.) Hampe
On rock outcrops and boulders, especially near farmsteads and on the coast.
Occasional and rather local; mainly confined to lowland and coastal areas.
1–4, 6, 8, 9, 12–14, 16, 17

P. orbicularis (Neck.) Poetsch
On broad-leaved trees, particularly *Fraxinus*, near farmsteads. Also on cement, asbestos, and nutrient enriched acidic rocks on or near the coast.
Frequent in all lowland areas; rare in upland areas and chiefly confined to lake shores and large boulders in streams.
1–17
This taxon is very polymorphic; material with broad lobes and predominately lip-shaped, marginal soralia, approximating to *P. labrata* Mereschk., is widespread in all lowland areas. Thalli with coarse, granular-coralloid isidia, resembling those in *P. sciastra* (Ach.) Du Rietz are rare and are mainly confined to the margins of lakes, e.g. **12** Ledmore, on wet rocks by lakeside, with *Dermatocarpon fluviatile*.

P. subobscura (Nyl.) Nyl.
Submesic supralittoral to xeric supralittoral; on more or less nutrient-enriched, acidic rocks, usually restricted to crevices.
Rare; probably overlooked.
4 S of Ulva House, Cùl a' Gheata; Craigaig, *U. Duncan & James*. **12** Salen, opposite Sgiath Mhòr.
This taxon may be a coastal variant of *P. tenella* with a darker grey, intricately branched,

closely appressed lobes forming regular rosettes, and an exclusively maritime distribution.

P. tenella (Scop.) DC.
On broad-leaved trees, especially *Fraxinus* and *Tilia*, particularly those enriched by dog urine or near farmsteads; also on inland and coastal rocks and boulders, especially those enriched by bird droppings.
Occasional; often associated with *P. adscendens* but much less frequent.
1, 2, 4–7, 9, 10–12, 13, 16, 17

P. teretiuscula (Ach.) Lynge
On roofs and walls of or near farmsteads.
Rare.
2 Walls of Iona Abbey and St. Mary's Abbey; Boineach, on asbestos roof. **3** S of Loch Caol, Tirghoil. **4** Ulva, Craigaig, on derelict cottage. **12** Salen, on derelict buildings.
The affinities between this species and *P. dubia* (Hoffm.) Lett. are in need of critical investigation.

P. tribacia (Ach.) Nyl.
On rock outcrops and farm walls, especially near the coast.
Rare.
2 Baile Mòr, St. Ronan's Bay, on rock outcrops. **3** S of Loch Caol, Tirghoil.

P. wainioi Räs.
On broad-leaved trees, especially *Acer pseudoplatanus*; also on outcrops, rocks and large boulders, mainly by loch shores and in streams, particularly those enriched by bird droppings.
Occasional; but widespread.
1–6, 9–15, 17
More frequent than *P. caesia* from which it differs in the development of apical and marginal, not laminal, soralia.

Physciopsis Choisy

Bibliogr.: Poelt (1965).
P. adglutinata (Flörke) Choisy
Confined to aged, wayside broad-leaved trees in well-illuminated lowland and coastal sites.
Rare; often poorly developed.
2 Baile Mòr, on *Acer pseudoplatanus* near the jetty, *U. Duncan & James*. **3** Ardtun, Tòrr Mòr, on old *Ligustrum* near farmstead; Bunessan, at dry base of ancient *Fraxinus* on road to Lower Ardtun. **10** Head of Loch Scridain, on *Acer pseudoplatanus* in grounds of the Highwayman Inn. **12** Salen, Kintallen, on *Fraxinus*.

Physconia Poelt

Bibliogr.: Poelt (1965, 1966).
P. enteroxantha (Nyl.) Poelt.
Physcia enteroxantha Nyl.
On isolated, aged *Fraxinus* near the coast.
Rare.
7 North side of Loch Tuath, Achleck. **10** Head of Loch Scridain, near the Highwayman Inn.

P. farrea (Ach.) Poelt
Physcia farrea (Ach.) Vain.
Recorded only from an aged *Ulmus* by roadside.
Rare.
7 Fanmore, *U. Duncan & James.*
The collection is well-developed, the thallus consisting of numerous, small imbricate squamules with numerous, dark, densely squarrose rhizinae on their pale undersides; larger marginal lobes are not developed in this species.

P. grisea (Lam.) Poelt
Physcia grisea (Lam.) Zahlbr.
Known from a single locality on aged *Acer pseudoplatanus* near the coast.
Rare.
10 Head of Loch Scridain, near the Highwayman Inn.
This species is unexpectedly scarce.

P. pulverulenta (Schreb.) Poelt
Physica pulverulenta (Schreb.) Hampe
On broad-leaved trees, especially *Fraxinus* and *Acer pseudoplatanus*, near farmsteads in lowland and coastal areas.
Occasional, but locally abundant.
2, 3, 5, 6, 8, 10–15, 17

Rinodina (Ach.) Gray

Bibliogr.: Sheard (1967).
R. atrocinerea (Dicks.) Körb.
Xeric supralittoral; mainly confined to exposed acidic rocks on or near the coast and most frequent within and just above the *Ramalina siliquosa* zone.
Frequent; widespread, especially on western coasts.
1–9, 12–14, 16, 17
Contains gyrophoric acid.

R. bischoffii (Trevis.) Massal.
On calcareous sandstones near the coast
Rare.
10 Carsaig, three sites between Rubh' a Chromam and Aird Ghlas.

R. exigua (Ach.) Gray
On bole of aged, wayside *Fraxinus* in sunny, sheltered site.
Rare.
7 Fanmore, *U. Duncan & James.*

R. luridescens (Anzi) Arnold
Xeric supralittoral; confined to sunny, acidic rocks on or near the coast; most frequent within and just above the *Ramalina siliquosa* zone.
Frequent; widespread, especially in the west.
1–8, 10, 12, 13, 15, 16
Contains gyrophoric acid and zeorin.

R. roboris (Duf. ex Nyl.) Arnold
On boles of aged, wayside *Fraxinus* in sheltered, sunny sites.
Rare; mainly confined to S Britain.
7 Achleck, Normann's Ruh. **17** near Duart Point; Camas Mòr.

R. sophodes (Ach.) Massal.
On twigs, branches, more rarely, boles of broad-leaved trees, especially *Fraxinus*, in lowland and coastal wayside and woodland areas.
Occasional to frequent; probably under-recorded due to the inaccessibility of the habitat.
2, 4–6, 8, 13, 16, 17

R. subexigua (Nyl.) Oliv.
On acid or slightly basic rocks and boulders, especially near the coast and associated with farmsteads and birds' perching sites; also on asbestos and acidic stones associated with mortar of old walls and bridges.
Frequent.
1, 7, 9–15, 17

R. subglaucescens (Nyl.) Sheard
In sheltered crevices in acidic rock on the coast.
Rare; confined to the south-west.
2 Iona, Carraig an Daimh; Eilean Maol Mhàrtuin. **3** Rubh' Ardalanish.
These are the most northern records of this species in Britain.
Contains atranorin.

R. teichophila (Nyl.) Arnold
On acidic, or sometimes slightly basic, more or less nutrient-enriched, rocks mainly near the coast.
Rare.
16 Lochbuie, Eilean Mòr, *U. Duncan & James*; An Learg and near Rocking Stone. **17** Walls of Duart Castle.
Contains a small quantity of atranorin.

PERTUSARIACEAE

Ochrolechia Massal.

O. androgyna (Hoffm.) Arnold
On boles of old, often moribund trees, and on mossy rocks and soil in old woodlands; also on the ground or encrusting rock or vegetable detritus amongst dry *Calluna-Erica cinerea* heathland and particularly frequent amongst mosses on basalt scree and boulders on the Ben More massif.
Abundant; widespread from sea-level to the summit of Ben More.
1–17
Contains gyrophoric acid and atranorin.

O. frigida (Sw.) Lynge
On the ground or on boulders, encrusting mosses especially *Rhacomitrium* spp., mainly in areas of late snow lie.
Rare and local; confined to the summits of the higher mountains.
7 Cnoc an dà Chinn; Cruachan Loch Tràth, 380 m. **11** Ben More, at 680–945 m, *U. Duncan*; Beinn nan Gabhar, *c.* 400 m; Cruach Choireadail, 610 m; Cruachen Dearg, at 600 m. **15** Beinn Bheag, 610 m; Sgùrr Dearg, *c.* 730 m. **16** Creach Beinn *c.* 675 m.
Contains gyrophoric acid.

O. inversa (Nyl.) Laund.
On isolated wayside trees frequently adjacent to bogs or fens in lowland areas.
Rarely in open glades of old woods.
Rare.
6 Glengorm Castle, on *Larix*. **7** Achleck, Normann's Ruh, on *Betula*. **16** An Coire, on dead *Quercus*. **17** Torosay Castle, on *Picea abies*.
Contains arthothelin (C+ orange), atranorin and an unknown substance.

O. pallescens (L.) Massal.
On smooth bark of trees, especially *Sorbus aucuparia*, in upland valleys; more rarely on the boles of isolated *Acer pseudoplatanus* near farmsteads in lowland areas.
Occasional; but widespread.
5, 7, 9, 12–14, 16, 17
Contains gyrophoric acid.

O. parella (L.) Massal.
On coastal rocks, often in association with *Ramalina siliquosa* aggr.; more rarely on boles of broad-leaved trees, gravestones, asbestos roofing, and mortar-stone walls of houses, bridges and monuments. Very rarely seen above 100 m.
Abundant on or near the coast becoming rarer inland; ascending to 100 m on **4** Ulva on Beinn Chreagach.
1–6, 8–14, 16, 17
Contains variolaric acid and two unknown substances (±); lecanoric acid in ascocarps.

O. tartarea (L.) Massal.
On old trees, especially *Betula*, rotting logs and mossy boulders in upland woodland; also on mossy scree and isolated boulders on heathland.
Frequent; in most upland areas, ascending to *c.* 900 m on **11** Ben More; becoming rarer in lowland areas.
3, 5–12, 15–17
Contains gyrophoric acid and atranorin.

O. turneri (Sm.) Hasselr.
On wayside and woodland trees, especially *Acer pseudoplatanas*, *Fraxinus* and *Ulmus*, in lowland valleys.
Abundant; in most lowland areas, especially in the south-east.
2–5, 7–17
Fertile specimens recorded in **3** and **15**.
No lichen substances detected.

O. yasudae Vain.
Mainly confined to wayside, broad-leaved trees, especially *Fraxinus* and *Ulmus* near farmsteads, in lowland areas.
Occasional; widespread, often very abundant on individual trees.
2–6, 12, 13, 16, 17
Fertile specimens recorded in **3, 12, 16** and **17.**
Contains gyrophoric acid.

Pertusaria DC.

Bibliogr.: Erichsen (1935).

P. albescens (Huds.) Choisy & Wern. var. **albescens**
On boles and branches of broad-leaved trees of all ages, also rarely on conifers at the margins of plantations. Occasionally encrusting mosses on acid or slightly basic rocks and boulders in exposed cliff sites.
Frequent; widespread. Saxicolous in **2, 4, 5, 10.**
2–17

P albescens var. **corallina** (Zahlbr.) Laund.
On similar sites and substrates to var. *albescens*, but not recorded as saxicolous.
Frequent; widespread, with similar distribution to that of the type variety.
2–8, 10–17

P. amara (Ach.) Nyl.
On boles and branches of broad-leaved trees, especially smooth bark of young *Fraxinus* and *Quercus*; rarely on mortar-stone walls and bridges. Also on consolidated shell-sand and associated gneiss rocks near the coast.
Frequent and often abundant.
1–17
Saxicolous in **10** and **16**; on shell-sand in **2.**
Contains picrolichenic acid (KC+ purple).

P. chiodectonoides Bagl. ex Massal.
On sunny, but sheltered, acidic rocks, especially basalt boulders and outcrops, near or on the coast.
Frequent in a few areas in the west; very rare in the east.
2–5, 7, 10, 17

P. coccodes (Ach.) Nyl.
On sunny, exposed side of ancient *Quercus* by roadside.
Rare.
16 Loch Don, Ceann na Lathaich.
A species with a predominantly eastern distribution in Britain. Contains norstictic and connorstictic acids.

P. corallina (L.) Arnold
On rock outcrops and scree boulders in exposed upland sites; rare on basalt below 90 m.
Occasional; rather local and most frequent on the Ben More massif.
2, 3, 8, 11–13, 15, 16
Contains stictic, constictic, norstictic (±) acids and two unknown substances.

P. dealbata Ach.
On acidic rock outcrops and smooth boulders in more or less exposed upland sites.
Rather local but common where it occurs.
1–3, 5–7, 9, 10, 12, 15, 16
Contains fumarprotocetraric acid.

P. flavicans Lamy
On sheltered but sunny rocks, especially basalt, on or near the coast. Occasionally on rocks on or near loch shores.
Occasional; most frequent in the west, not observed above 50 m.
2, 4, 5–7, 12, 13
Contains thiophaninic acid and small quantities of another substance (?arthothelin).

P. hemisphaerica (Flörke) Erichs.
On boles of broad-leaved trees, usually near the base, in open woodland and wayside sites.
Occasional; but widespread, most frequent in the east.
5 On *Fraxinus* by Allt Hostarie; Torloisk House, on *Fagus*. **15** Scallastle Bay, Alterich, on *Fraxinus*, *U. Duncan & James*. **16** An Coire, on *Quercus*. **17** Torosay Castle on *Castanea* and *Quercus*.
Contains lecanoric acid.

P. hymenea (Ach.) Schaer.
On boles and branches of broad-leaved wayside and woodland trees of all ages in most lowland and coastal areas; particularly frequent on *Quercus* and *Fagus*.
Abundant; in all lowland and coastal areas but decreasing with increasing altitude; rare above 90 m.
2–17
Contains thiophaninic acid and a xanthone.

P. lactea (L.) Arnold
On acid rock outcrops, especially basalt, in most lowland and upland areas; also frequently on acidic boulders of old walls and consolidated scree.
Frequent; in all areas, except the Treshnish Islands and Iona, but often poorly developed. Especially abundant on the SW-facing side of the Ben More massif.
3, 5, 7, 9, 11–13, 15, 16
Contains lecanoric acid.

P. leioplaca (Ach.) DC.
On smooth bark of broad-leaved trees, especially *Corylus* and *Fraxinus*, in well-wooded, lowland and coastal areas.
Abundant; in all lowland areas, becoming rare above 60 m but still persisting in small valleys to 150 m.
2–17
Contains stictic acid and a xanthone.

P. leucostoma (Bernh.) Massal
On smooth bark of *Corylus* and *Fraxinus* in a few sheltered valleys near the coast.
Rare; only in the west and north-west.

7 Valley of Port na Crìche, on *Corylus*; Tòrr an Daimh, near Kilbrenan, on young *Fraxinus*. **8** Near Tiròran, on *Fraxinus*, by streamside.

P. monogona Nyl.
On granite boulder in heathland at edge of loch.
Rare; known only from the Ross of Mull.
3 Near Loch Poit na h-I.
The first record of this species for Scotland.
Contains thamnolic acid.

P. multipuncta (Turn.) Nyl. agg.
On boles and, particularly, smooth branches of broad-leaved trees, rarely conifers, in well-wooded, lowland and coastal sites; also frequent on smooth bark of *Sorbus aucuparia*, and, more rarely, *Salix aurita*, in upland valleys.
Frequent; especially in the east.
2–17
The species is very polymorphic: forms approaching var. **ophthalmiza** Nyl. are largely confined to old conifers and *Betula* in upland woodlands and may represent a distinct taxon. *P. multipuncta* var. *multipuncta* contains variolaric acid, the var. *ophthalmiza*, two unknown substances.

P. pertusa (L.) Tuck.
On boles and branches of broad-leaved trees, especially *Fraxinus* and *Quercus* in all wayside and woodland areas.
Frequent; often locally abundant.
2–17
Contains stictic acid and a xanthone.

P. pseudocorallina (Liljebl.) Arnold
On acid rocks in all but heavily shaded or very damp habitats.
Abundant; often dominant in many upland areas. Frequently poorly developed, especially on basalt rocks in the west, due to weathering, damage by browsing snails, and competition with saxicolous algae.
1–17
Contains norstictic and connorstictic acids.

Phlyctis (Wallr.) Flot.

P. agelaea (Ach.) Flot.
On *Fraxinus* and *Salix* in damp, sheltered valleys near the coast.
Rare.
7 Kilninian, Tràigh na Cille, on *Salix*. **8** Tiróran, on *Fraxinus*, *U. Duncan & James*; near Tavool House, Abhainn Beulath an Tairbh, on *Fraxinus*.
Contains norstictic and connorstictic acids.

P. argena (Spreng.) Flot.
On broad-leaved trees, especially at the margins of woodland or by roadside; rarely saxicolous.
Frequent.
2–4, 6–7, 9–11, 13–16
Saxicolous, by the roadside in **12** near Salen.
Contains norstictic and connorstictic acids.

Placopsis Nyl.

P. gelida (L.) Lindsay
On large acidic boulders and rock outcrops in upland areas, often in temporary seepage tracks on boulders and rock outcrops in or by streams and lochs.
Occasional in most upland areas above 90 m; and more frequent and occasionally fertile on the Ben More massif; rare and mostly poorly developed in lowland and coastal sites.
4, 7–11, 13, 15, 16
Contains gyrophoric acid.

Trapelia Choisy

Bibliogr.: Hertel (1969, 1970).

T. coarctata (Sm.) Choisy
On sheltered, often moist, acidic rocks especially in or near seepage tracks or near streams and loch shores.
Frequent, often common; widespread.
2–7, 9–12, 14–17

T. obtegens (Th. Fr.) Hertel
On sheltered moist acidic rocks and pebbles, especially on paths and road cuttings.
Rare; probably overlooked.
12 Aros Park, on stony paths in clearings. **13** Salen, on pebbles by roadside.

T. ornata (Sommerf.) Hertel
On sheltered moist acidic rocks and especially damp bricks of derelict buildings.
Rare; probably overlooked.
4 Ulva, bricks of sheltered outbuilding near Ulva House. **11** Derryguaig, on sheltered basalt pebbles near the coast. **13** Killichronan, on rocks and boulders in field.

T. mooreana (Carroll) P. James
T. torellii (Anzi) Hert.; *Lecidea brujeriana* (Schaer. ex Dietr.) Arnold; *L. lopadioides* (Th. Fr.) Grumm.
On crumbling basalt, especially in damp frost hollows in upland sites.
Rare; local and frequent in a few sites.
8 Maol Mheadhonach, with *Koenigia islandica*; Maol na Coille Moire. **10** Ben Buie, 400 m. **15** Beinn Bheag, 350 m.

GYALECTALES

GYALECTACEAE

Dimerella Trevis.

D. diluta (Pers) Trevis.
On poorly illuminated boles of broad-leaved trees and conifers in mature woodland and valley sites, especially near the coast.
Rare to occasional; confined to a few lowland areas.
7 Near Lettermore, on *Larix*. **8** Scobull Point, on *Fraxinus*. **10** Carsaig, on *Fraxinus* near Innamore Lodge. **12** Calgary Bay, *U. Duncan.* **13** Aros Park on *Quercus*, *Picea* and *Cupressus macrocarpa*.

D. lutea (Dicks.) Trevis.
Chiefly on boles of mossy broad-leaved trees, conifers and logs especially in sheltered mature woodland and valley sites; rarely on soil and detritus on or near cliff edges overlooking the sea.
Frequent; widespread, especially in lowland woodland sites.
3, 5–7, 10, 12–14, 16, 17

Gloeolecta Lett. em. Vězda

G. bryophaga (Körb. ex Arnold) Vězda
On lead-rich (4.37 ppm; R. Nourish, pers. comm.) skeletal soils amongst *Calluna*.
Rare; recorded from a single locality.
11 By Allt Teanga Brìdeig, at 200 m.
This species, which is rare in Britain, favours substrates rich in lead or copper.

Gyalecta Körb.

See also *Gloeolecta* Lett. and *Gyalidea* Lett.

G. flotowii Körb.
On boles of mature wayside broad-leaved trees, especially *Fraxinus* and *Ulmus* on or near the coast.
Occasional.
4 Ulva, near the House, on *Ulmus*; close to Baligortan, on *Fraxinus*. **5** Ballygown, on *Fraxinus*. **8** Tiróran, opposite Sgeir Mhòr, on *Ulmus*; Torloisk House on *Fraxinus*. **14** Achadh Fada, on *Fraxinus*.

G. jenensis (Batsch) Zahlbr.
In damp sheltered crevices in basic rocks near the coast and gneiss rocks partially covered with wind blown shell-sand; also on damp mortar of walls, bridges, and old buildings, usually near the ground.
Occasional.
2 Iona, Golf Course, on shell-sand. **12** Salen village, on mortar, *U. Duncan & James*; Glenforsa House, on damp mortar. **16** Kinlochspelve churchyard. **17** Craignure, Rubha na Sròine, on damp limestone.

G. truncigena (Ach.) Hepp. var. **truncigena**
On boles of wayside trees, especially *Fraxinus*, *Sambucus nigra* and *Ulmus*, frequently near farmsteads.
Occasional; confined to lowland sites, particularly in the west.
7 Ballygown, on *Sambucus nigra*. **8** W of Tavool House, on *Sambucus nigra*. **10** Carsaig Bay, on *Fraxinus* and *Ulmus* near the pier.

G. truncigena var. **derivata** (Nyl.) Boist.
On wayside and woodland trees, especially near farmsteads.
Rare; probably overlooked.
5 Kilninian, on *Ulmus*. **7** Lagganulva, on *Sambucus nigra*. **8** Tavool House, on *Sambucus nigra*, *U. Duncan.*

Gyalidea Lett.

Bibliogr.: Vězda (1966).
G. fritzei (Stein) Vězda
On decomposing basalt pebbles in small frost hollows in an exposed upland area with *Koenigia islandica* and *Trapelia mooreana.*
Rare.
8 Maol Mheadhonach and Coire nan Each at *c.* 230 m.
Previously noted on mica-schist rocks near the summit of Ben Lawers and on basalt pebbles on the Ardnamurchan Peninsula.

Two other species of this genus have been recorded from the Ardnamurchan Peninsula: *G. roseola* (Arnold) Lett. ex Vězda from Strontian, Belsgrove, and *G. hyalinescens* (Nyl.) Vězda (*Lecidea hyalinescens* (Nyl.) Zahlbr.) on the E facing slope of Ben Hiant; both could also occur on Mull.

Gyalideopsis Vězda

Bibliogr.: Vězda (1972); James (1975).
G. anastomosans Vězda & P. James
On boles of *Salix aurita* in a damp, upland carr and *Fraxinus* in moist woodland near the coast; also on decorticated tree stumps in moist woodland.
Rare; possibly overlooked.
3 Beinna Ghlinne Mhòir, on *Salix*. **12** Near Knock, on *Fraxinus*. **16** Lochbuie, near Moie Lodge, on decorticated stumps.
Young, sterile thalli often having minute, evenly dispersed flattened isidia.

G. muscicola Vězda & P. James.
Overgrowing bryophytes (*Isothecium myosuroides*) on rocks in a sheltered, damp woodland site.
Rare.
3 Loch Caol, Ardfenaig.
This endemic British species has previously been recorded only from Caernarvonshire.

Pachyphiale Lönnr.

P. cornea (With.) Poetsch
On wayside and woodland broad-leaved trees and shrubs in sheltered lowland sites; more rarely on *Salix* and *Sorbus aucuparia* in wet habitats. Mostly confined to *Corylus* in upland sites.
Abundant; widespread in lowland and coastal areas, becoming scarcer in upland sites. Ascending to 240 m in **15** on south-west edge of Coile na Sròine.
3–13, 15–17

Petractis Fr.

P. clausa (Hoffm.) Kremp.
On damp limestone rocks sheltered by low herbaceous vegetation near the coast.
Rare.
17 Craignure, Rubha na Sròine.
This species is frequent on Lismore Island.

OSTROPALES

THELOTREMATACEAE

Diploschistes Norm.

D. caesioplumbeus (Nyl.) Vain.
Xeric supralittoral; on sunny acidic rocks in sheltered sites.
Occasional to frequent in coastal areas on Iona, Ross of Mull and Ulva, very rare elsewhere.
1–5, 16
The most northern records for this species in Britain.

D. scruposus (Schreb.) Norm.
Encrusting mosses and other lichens, e.g., *Cladonia pocillum*, or directly on rock, soil or shells and more rarely on old mortar and stone walls.
Occasional; but widespread in all coastal and lowland areas.
2–5, 10 12, 15–17
The Mull records include var. *bryophilus* (Ach.) Müll. Arg.

Ocellularia Meyer

O. subtilis (Tuck.) Riddle
On smooth bark of boles and branches of broad-leaved trees, especially *Corylus*, *Fagus* and *Sorbus aucuparia*, in lowland and coastal wayside and woodland areas; particularly abundant on stands of *Corylus* in sheltered, moist lowland valleys.
Abundant; widespread in all lowland areas.
3–17

Thelotrema Ach.

T. lepadinum Ach.
On smooth or rough boles of broad-leaved trees in well-wooded lowland and coastal areas, also rarely on conifers and then only at the margins of plantations.
Rarely on decomposing basalt rock on the coast.
Frequent; widespread.
3–17
Var. *scutelliforme* Ach., which is here considered to be an ecotype characteristic of damp, mossy boles in sheltered sites, is frequent, especially in the north-east.

LITHOGRAPHACEAE

Lithographa Nyl.

L. dendrographa Nyl.
On bole of wayside *Ulmus* near the coast.
Rare; the only record.
7 Kilninian.
First record for Scotland, subsequently recorded from Kintyre in 1972.

L. tesserata (DC.) Nyl.
On acidic or slightly basic rocks in upland and coastal areas.
Rare.
10 Carsaig Bay, above Nuns' Pass. **11** Ben More, on rocks towards summit, *c.* 800 m, *U. Duncan*. **15** Scallastle Bay, Alterich, on basalt.
Contains norstictic and connorstictic acids.

MELASPILEACEAE

Melaspilea Nyl.

M. lentiginosa (Lyell ex Leight.) Müll. Arg.
On boles of old *Corylus* in sheltered, shaded ravines.
Rare.
7 Kilninian, Allt Hostarie, *U. Duncan & James*.
17 Lochdonhead, Hazelbank, abundant in small thicket of *Corylus*.
Previously recorded on *Corylus* from Drimnin, Morvern. This and the Mull records are the most northerly for the species in Britain.

M. ochrothalamia Nyl.
In crevices in boles of old, sheltered broad-leaved trees, mainly *Quercus*, in mature woodland; rarely on *Salix* in boggy flushes.
Rare.
13 Aros Park, on *Quercus*. **16** Northern side Loch Uisg, near monument, on *Salix*. **17** Torosay Castle on *Quercus*.

GRAPHIDACEAE

Graphina Müll. Arg.

G. anguina (Mont.) Müll. Arg.
On smooth bark of sheltered woodland, more rarely wayside, broad-leaved trees and particularly on boles of *Corylus* in lowland valleys in the west.
Frequent to abundant; in all coastal and lowland areas, ascending to 220 m on the **16** Laggan Deer Forest, on *Fraxinus* by Allt na Cuinneige.
3–10, 12–14, 16, 17

Graphis Adans. em. Müll. Arg.

G. elegans (Borr. ex Sm.) Ach.
On smooth to moderately rough bark of boles and branches of wayside and woodland broad-leaved trees and conifers; especially frequent on *Betula*, *Fagus* and *Ilex* in lowland and coastal areas in the west and also recorded on old stems of *Calluna* in several lowland heathland sites.
Occasional in the west; becoming more frequent in the north-east.
3, 5–7, 9–14, 16, 17
Contains norstictic acid (±).

G. scripta (L.) Ach.
On smooth boles and branches of a wide range of broad-leaved wayside and woodland trees but rare on boles of conifers; also recorded on decorticate wood of *Fagus*, and, on one occasion, on rock.
Generally abundant.
1–17
Very polymorphic; a collection was made from rock at **14** Fishnish Bay, near Corrynachenchy, gorge of the Allt Mòr Coire nan Eunachair.
Contains norstictic acid (±).

Phaeographis Müll. Arg.

P. dendritica (Ach.) Müll. Arg.
including *P. inusta* (Ach.) Müll. Arg.
On smooth bark of broad-leaved trees especially in small sheltered valleys in all lowland areas.
Frequent; becoming abundant in well-wooded areas in the east.
1–17
Contains norstictic acid (±).

P. ramificans (Nyl.) Lett.
On smooth bark of broad-leaved trees especially *Ilex* in a few lowland sites.
Occasional; easily confused with *Graphis scripta* and therefore probably under-recorded.
4 Ulva, near the House, on *Ilex*. **7** Near Lettermore, on *Larix*. **10** Pennyghael, on *Ilex*; Carsaig, Pennycross House, on *Ilex*. **16** Lochbuie, Duibh Leitir, on *Crataegus*; north side of Loch Uisg, near Obelisk, on *Sorbus*.
Contains norstictic acid.

SPHAERIALES

PORINACEAE

Gongylia Körb.

Bibliogr.: Swinscow (1961).
G. incarnata (Th. Fr. & Graewe) Zahlbr.
Encrusting patches of moss protonemas at the edge of a recently eroded streamside.
Rare; recorded from a single locality.
16 Laggan Deer Forest, Allt na Cuinneige, at 250 m.
Previous records for this species in the British Isles are from Ben Lawers, and Teesdale. Vězda (1959) transfers this species to *Belonia* Körb.

G. sabuletorum (Fr.) Stein
On damp, leached soil amongst *Erica tetralix* on exposed hillsides.
Rare.
5 Brolass, above Gleann Airigh na Searsain (Beach River). **16** Allt Ohirnie.
This species is now considered to be a parasymbiont on the thallus of *Baeomyces rufus* (see Coppins, 1972, p. 327).

Porina Müll. Arg.

Bibliogr.: Swinscow (1962).
P. ahlesiana (Körb.) Zahlbr.
P. septemseptata (Hepp ex Zwackh.) Swinsc.
On moist, sheltered acidic rocks in poorly illuminated habitats by streams in a few lowland woodland and valley sites; also on damp rocks under over-hanging trees in the splash-zone at the edge of lochs.
Rare.
12 Salen, small ravine north of the village. **13** Aros Park; Upper Druimfin, near Linndhu, near the sea. **15** Ardura, Crùn Lochan, on damp boulders.
The most northern records for this species in Britain.

P. chlorotica (Ach.) Müll. Arg. var. **chlorotica**
On moist, more or less shaded, crevices in acidic or slightly basic rocks or on boulders by streams in sheltered woodland and valley sites; rare on damp rocks in coastal areas. A species with a very wide habitat tolerance.
Abundant; widespread.
1–17

P. chlorotica var. **carpinea** (Pers.) Keissl.
On mainly smooth bark of boles and branches of young or mature trees, especially *Fraxinus*, *Corylus* and *Acer pseudoplatanus*, in more or less sheltered lowland mature woodland and valley sites; especially frequent on sheltered old *Corylus* and young *Fraxinus* in small coastal valleys and ravines.
Frequent.
4–6, 8, 10–13, 15–17

P. chlorotica var. **persicina** (Körb.) Zahlbr.
On sheltered, rather damp, basic rock outcrop near the coast.
Rare.
17 Craignure, Rubha na Sròine.
This species is abundant on Lismore Island.

P. coralloidea P. James
On sheltered bole of aged *Quercus* in dense mixed woodland.
Rare.
13 Aros Park, near Lochan a' Churrabain.
First record for Scotland; see James, 1971.

P. guentheri (Flot.) Zahlbr. var. **guentheri**
On sheltered or sunny, periodically submerged, acidic rocks, especially near loch-margins and streams in upland areas.
Occasional; easily overlooked and probably more frequent than the records suggest.
6 Mingary Burn. **8** boulder by Abhainn Bail a' Mhuillinn. **11** E of Gruline House. **15** Lussa River below Doir a' Chuilinn. **16** Several sites along the N and S shores of Loch Uisg.

P. guentheri var. **grandis** (Körb.) Swinsc.
On inundated rocks in small stream in sheltered woodland.
Rare.
13 Aros Park, on basalt of the Allt nan Torc.

P. interjungens (Nyl.) Zahlbr.
On sheltered periodically submerged, basalt rocks in streams or along the margins of lochs in lowland areas.
Rare.
5 Langamull. **11** Loch Bà, near Benmore Lodge.

P. lectissima (Fr.) Zahlbr.
On sheltered, often vertical, damp rock faces in wooded ravines and occasionally in deep shade on the undersides of pearched scree boulders, often in association with *P. chlorotica*

aggr. Also sheltered sides of old walls and on boulders near streams and fresh-water lakes.
Occasional to frequent; in lowland and upland sites, rarely encountered in wet places along the coast.
3–7, 10–13, 16

P. leptalea (Dur. & Mont.) A. L. Sm.
On smooth bark of broad-leaved trees, especially young *Corylus*, *Salix* and *Fraxinus* in sheltered, humid valleys in lowland areas; rarely saxicolous.
Occasional; probably more frequent than the records indicate.
3–5, 10–14, 16.
Saxicolous on sheltered, basalt boulders in *Corylus-Fraxinus* woodland at **10** Carsaig.

P. olivacea (Pers.) A. L. Sm.
On mossy boles and branches of a wide variety of broad-leaved trees, especially *Corylus* and *Quercus* in sheltered, lowland, mature woodland and valley sites.
Occasional; but probably overlooked.
4, 7–11, 13, 16, 17
Specimens with macroconidia are more frequent than those with perithecia.

P. taylorii (Carroll ex Nyl.) Swinsc.
On *Fraxinus* at the edge of sheltered woodland site.
Rare.
7 Ballygown.
The determination of this collection must be considered tentative as only pycnidia are present. However, the structure of these and other features of the thallus conform with fertile material of this species collected on Lismore Island, on moribund *Acer pseudoplatanus* at Kilcheran, and on *Fraxinus* on the west side of Loch Baile a' Ghobhainn. Previously this species was known only from SW England and SW Eire.

PYRENULACEAE

Leptorhaphis Körb.

L. epidermidis (Ach.) Th. Fr.
Specific to smooth bark of mature *Betula*.
Frequent to abundant; in all areas where the host tree is present but absent from the Treshnish Isles and Iona.
3–17

Pyrenula Ach. em. Massal.

P. laevigata (Pers.) Arnold aggr.
Mainly confined to *Corylus*, rarely on *Salix* and *Fraxinus*, in sheltered, lowland valleys.
Occasional, but locally abundant in a few valleys in the west.
4–6, 10, 13

P. nitida (Weig.) Ach.
On smooth bark of broad-leaved trees, particularly in sheltered, shaded valleys, in all coastal and lowland areas; more rarely on wayside trees.
Frequent; widespread.
2–17

P. nitidella (Flörke) Müll. Arg.
On smooth bark of broad-leaved trees, particularly in sheltered, shaded valleys; more frequent on wayside trees than *P. nitida.*
Occasional to frequent; widespread in all coastal and lowland areas.
2, 4–10, 12–15, 17
Often in association with *P. nitida.*

Pyrenula sp.
On smooth bark of broad-leaved trees, especially *Corylus* in very sheltered lowland and coastal ravines.
Occasional; especially in the north-west.
4, 6–8, 12, 13
This is possibly a new taxon which differs from *P. nitida* in the absence of minute white spots (as seen with a ×10 lens) and in the larger spores.

MICROGLAENACEAE

Microglaena Körb.

Bibliogr.: Morgan-Jones & Swinscow (1965).
M. corrosa (Körb.) Arnold
On boulders and rock outcrops in a river.
Rare; probably overlooked.
15 Lussa River, near An t-Sleaghach.

M. larbalestieri A. L. Sm.
Moist, crumbling, basalt rocks at side of small stream in old, dense, broad-leaved woodland.
Rare; the only record.
13 Aros Park, Spùt Dubh.
This species, endemic to the British Isles is otherwise only known from Connemara, Galway and Loch Moidart, Inverness-shire.

M. muscorum (Fr.) Th. Fr.
Encrusting mosses and hepatics on tree boles, especially *Fraxinus*, in wayside lowland and coastal areas; also on weathered basalt detritus in upland areas.
Occasional; probably overlooked.
7 Achnacraig, on rock outcrop. **8** Scobull, on *Fraxinus*; Tavool House, on *Fraxinus*; Creag Mhòr, on mosses on basalt. **10** Carsaig, near Aoineadh a' Mhaide Ghil, on *Acer pseudoplatanus*, *U. Duncan & James.*
Also recorded from Lismore Island.

Tomasellia Massal.

T. gelatinosa (Chev.) Zahlbr.
On smooth bark of *Corylus*, *Crataegus*, *Quercus* and *Sorbus aucuparia* and *Fagus*, in most lowland and coastal sites; and on *Sorbus aucuparia*, in a few upland valleys.
Frequent and widespread; ascending to 300 m in **11** Garbh Coire.
2, 6, 8–13, 15–17

T. ischnobela (Nyl.) Keissl.
Confined to a few lowland and coastal sheltered

ravines on the smooth bark of *Corylus* and, rarely, *Betula*.
Occasional and local in the west; probably overlooked.
4 Grounds of Ulva House. **5** Kilninian, gorge of Allt Hostarie. **7** Torloisk; Normann's Ruh; Ballygown; Achronich, *U. Duncan & James*.

VERRUCARIALES

VERRUCARIACEAE

Dermatocarpon Eschw.

D. cinereum (Pers.) Th. Fr.
On consolidated soil on the top of an old stone-mortar wall near the coast.
Rare.
8 Kilfinichen.

D. fluviatile (Web.) Th. Fr.
Chiefly confined to almost submerged acidic rocks and boulders in and by swiftly flowing streams and loch sides; more rarely in semi-
Occasional; widespread but less frequent near the coast.
perment seepage tracks on cliffs faces.
1 Fladda, in seepage track. **3** Loch Poit na h-I, on rocks. **5** Calgary Bay, Ensay Burn. **7** Loch Frisa, Eilean Dubh. **8** Abhainn Bail a' Mhuil-linn. **12** Salen, Aros River, near Aros Cottage. **13** Caol Lochan. **16** W Loch Uisg, near pier.

D. hepaticum (Ach.) Th. Fr.
On soil in crevices associated with basic rocks; also on mortar and cement on old garden walls and buildings. Also recorded on accumulated shell-sand on gneiss rocks.
Locally frequent and widespread.
2–4, 6, 8, 10, 12–14, 16, 17

D. miniatum (L.) Mann var. **miniatum**
On basic or acidic rocks especially in coastal and lowland sites, often restricted to small crevices or seepage tracks.
Occasional; widespread, especially on coastal rocks.
1–4, 6, 10, 12–17

D. miniatum var. **complicatum** (Lightf.) Hellb.
In similar habitats to var. *miniatum* with which it often grows but preferring more sheltered and permanently moist substrates.
Rare to occasional.
6 Quinish, E side of Loch a' Chumhainn. **12** Glenaros House, on basalt cliffs. **13** Aros Park on cliff faces above Lochan a' Ghurrabain, *U. Duncan*.

Dermatocarpon sp. [aff. **arnoldianum** Degel.]
Semi-inundated boulders and rocks in streams and at the margins of lochs in upland sites.
Occasional.
11 above Loch Bà, Glen Clachaig. **12** Loch Frisa, near Ledmore; Aros River, near Achad nan Each.

Several other species of peltate *Dermatocarpon* were collected from boulders in streams and from loch shores. More detailed studies of the genus are needed before this material can be accurately identified.

Placidiopsis Beltr.

Bibliogr.: James (1959); Swinscow (1960).
P. custnanii (Massal.) Körb.
Known only from a single locality where it overgrows mosses and hepatics on a limestone outcrop near the coast.
Rare.
8 Near Mackinnon's Cave.
All thalli observed were badly damaged by browsing molluscs (*Clausilia* sp.)

Polyblastia Massal.

Bibliogr.: Swinscow (1971).
P. cruenta (Körb.) P. James & Swinsc.
P. henscheliana (Körb.) Lönnr.
More or less submerged acidic rocks in streams and rivers in lowland and upland areas.
Frequent; often dominant on boulders in upland streams. Ascending to 600 m on **11** Ben More.
4, 6–11, 13–17

P. cupularis Massal.
On basic rocks in damp sheltered coastal sites.
Rare.
10 Carsaig Bay, on calcareous sandstone scree. **17** Craignure, Druim Mòr, on limestone.

P. deminuta Arnold
On sheltered limestone outcrop near the coast.
Rare.
17 Craignure, Rubha na Sròine.
This species is frequent on limestone outcrops and boulders on Lismore Island.

P. dermatodes Massal.
Recorded only once from calcareous sandstones near the coast.
Rare; probably limited by the scarcity of suitable habitats.
10 Below Nuns' Pass, *U. Duncan & James*.
This species is frequent on Lismore Island.

P. gelatinosa (Ach.) Th. Fr.
Encrusting mosses on calcareous sandstone and old crumbling mortar wall.
Rare.
10 Near Nuns' Pass. **12** Aros Castle, on mosses on mortar.

P. scotinospora (Nyl.) Hellb.
On moist, acid or slightly basic, rocks in sheltered valleys, on rarely submerged boulders and rock at the margins of freshwater lochs; also in seepage tracks on rock outcrops.
Occasional; probably overlooked. Present in both lowland and upland areas.
4, 8, 9, 12, 16, 17
Resembles *P. theleodes* but with persistently smaller, often paler spores, 25–40 × 15–25 μm.

P. theleodes (Sommerf.) Th. Fr.
On shaded, often decomposing rocks, especially basalt, mainly in sheltered sites; most frequent on or near the shores of freshwater lochs.
Frequent; widespread, ascending to 400 m on **16** Creach Beinn.
3, 5, 8, 10, 12, 13, 15–17
Spores 60–95 × 35–60 μm, dark brown.

P. tristicula (Nyl.) Arnold
In crevices containing soil and mosses on a variety of mainly base-rich substrates, including boulders and rock outcrops, cement and mortar on old walls.
Rare to occasional; often sterile and then easily overlooked and probably more frequent than the records indicate.
3 Gribun, near Mackinnon's Cave, in limestone. **10** Carsaig, on calcareous sandstone. **17** Craignure, on mortar. **13** Tobermory, on crumbling wall of disused distillery.

P. wheldonii Travis
Sheltered ledges of calcareous sandstone near the coast.
Rare.
10 Carsaig, near Nuns' Pass.
The specimen is well-fertile, the perithecia containing asci with 8 brown, strongly muriform spores, 80–120 × 40–62 μm in size. *P. gelatinosa* has colourless spores.

Staurothele Norm.

Bibliogr.: Swinscow (1963a).
S. fissa (Tayl.) Zwackh
On frequently inundated hard, acidic rocks in streams and lochs.
Occasional; probably widespread.
5, 9, 12, 13, 16, 17

S. hymenogonia (Nyl.) Th. Fr.
On dry, basic rocks of calcareous sandstone on and near the coast. Also once recorded from old mortar of wall.
Rare; probably limited by the scarcity of suitable habitats.
8 Inch Kenneth, Port an Ròin and Bàgh an Iollaich. **10** Near Nuns' Pass; Pennyghael, on mortar, *U. Duncan & James.*

Thelidium Massal.

T. cataractarum Lönnr.
On sheltered acidic rocks and boulders in or by streams in mature woodland.
Rare.
13 Aros Park; lower part of Allt nan Torc; Tobermory River, near Ledaig. **16** S side of Loch Uisg, Sròn nam Boc.

T. decipiens (Hepp ex Arnold) Kremp.
On base-rich rocks in coastal areas. Rarely on mortar of old buildings and garden walls.
Occasional; restricted by scarcity of suitable habitats.
2–3, 6–7, 9–10, 12, 13, 16, 17

T. fumidum (Nyl.) Hazsl.
On smooth acidic rocks in swiftly flowing streams in ravines near the coast.
Rare.
7 Kilninian, lower ravine of Allt Hostarie. **12** Gorge of the lower Allt Ardnacross.

T. impressum (Stiz.) Zsch.
On calcareous sandstone rocks near the coast.
Rare.
10 Near Nuns' Pass.

T. papulare (Fr.) Arnold
On acidic and slightly basic rocks in sheltered stream valleys and ravines in lowland and coastal areas. Also on rocks and boulders at the edges of lochs in upland sites.
Occasional to frequent.
3, 5–7, 11–13, 15, 16

T. pyrenophorum (Ach.) Mudd
On slightly basic rocks by small stream in an upland locality.
Rare.
8 Creag a' Ghaill; head of waterfall at Caisteal Sloc nam Ban.

Thelopsis Nyl.

T. rubella Nyl.
On boles of broad-leaved trees e.g. *Hippocastanus*, *Fraxinus*, *Acer pseudoplatanus* and *Quercus*, in wayside and woodland lowland and coastal areas.
Occasional.
7, 10, 12, 13, 15–17

Thrombium Wallr.

T. epigaeum (Pers.) Wallr.
On consolidated soil by track-sides and rock crevices near the coast.
Rare, but probably overlooked.
6 Track to Glengorm, near Ballimeanach. **17** Java Point.

Verrucaria Schrad.

Bibliogr.: Swinscow (1968); Fletcher (1975a).
V. aethiobola Wahlenb. ex Ach.
On acidic rocks and boulders in or beside streams and lochs and in temporary seepage tracks on rocks in sheltered woodland; tolerant of shade and tending to occupy a zone above that of other predominantly aquatic species of the genus. Present in coastal sites where freshwater runs over rocks into the sea.
Frequent, often abundant; sometimes dominant in mountain streams and the most abundant and widely distributed of the freshwater species of *Verrucaria* on Mull.
3–8, 10–17

V. amphibia R. Clem.
Confined to a zone on the shore below that of

V. maura and often associated with *Pelvetia canaliculata.*
Rare to occasional; but probably overlooked, appears to be most frequent on more exposed western coasts.
1 Lunga. **2** Iona, Eilean Didil. **6** Quinish Point. **8** Burgh, Port na Croise.
Thallus often rather thick, black, rimose; surface with a fine, anastomosing pattern of minute ridges.

V. aquatilis Mudd
Bachmannia maurula (Müll. Arg.) Zsch.
On permanently submerged pebbles, more rarely boulders, in swift-flowing, often sheltered streams. Rarely on submerged boulders in lochs.
Frequent; probably more widespread than the records indicate, appears to be confined to lowland sites.
3, 6, 8–11, 13, 17
Spores small, 6–8 × 4–7 μm.

V. coerulea DC.
On sheltered basic rocks near the coast.
Rare; limited by the distribution of natural basic rock.
8 Inch Kenneth, on limestone, *U. Duncan.* **10** Carsaig Bay, on calcareous sandstone. **17** Craignure, Rubha na Sròine.
This species is frequent to abundant on Lismore Island.

V. ditmarsica Erichs.
Forming inconspicuous colonies in the lower littoral fringe.
Occasional; very inconspicuous and easily overlooked.
1–3, 5
Characterized by the very thin, maculated, brown sub-gelatinous thallus and small perithecia (<200 μm).

V. elaeomelaena (Massal.) Arnold
On partially or totally submerged rocks and boulders in streams and lochs.
Occasional; especially in the east.
5, 9–11, 13–15, 17
Superficially, the freshwater counterpart to *V. mucosa.*

V. fusconigrescens Nyl.
Xeric supralittoral; on sunny acidic rocks and boulders, especially those enriched by bird-droppings, on or near the coast.
Rare; possibly overlooked or confused with *V. nigrescens.*
4 Ulva, Cragaig. **7** Ballygown Bay. **8** Above Mackinnon's Cave, *U. Duncan & James*; Inch Kenneth, Bàgh an Iollaich, on limestone. **12** Salen, on rocks in saltmarsh.
Differs from *V. nigrescens* in the more discrete areolae and smaller spores (18–23 × 7–9 μm).

V. glaucina Ach.
On basic rock and scree including mortar, cement and asbestos and then associated with buildings around farmsteads.
2, 4–6, 8, 12, 13, 15, 17
Material with a pale fawn coloured thallus was observed in several localities. In British floras this variant has been named *V. fuscella* (Turn.) Winch but field observations suggest that it is only a colour variant of *V. glaucina.*

V. hochstetteri Fr.
On basic rock and associated scree and mortar and cement especially of derelict buildings near the coast.
Frequent; widespread.
2, 4–8, 10, 12–14, 16, 17

V. hydrela Ach.
On partially to totally submerged often weakly basic rocks and boulders in streams and freshwater lochs; also in seepage tracks in very sheltered places, often beside streams in ravines. More tolerant of shade than other aquatic species of the genus.
Frequent; but mainly confined to well-wooded sites in lowland areas.
6–8, 10–13, 15, 16
Spores 18–24 × 7–9 μm.

V. internigrescens (Nyl.) Erichs.
Xeric supralittoral; on sheltered rocks in seepage tracks.
Rare; probably overlooked.
4 Ulva, Cragaig.
Characterised by the pale brown thallus which is finely rimose-cracked, and the brackish habitat.

V. kernstockii Zsch.
Confined to more or less totally submerged rocks and pebbles in lower stretches of fast flowing streams.
Rare; but probably overlooked.
10 Carsaig, near Carraig Mhor. **16** Lochbuie.
Specimens have a minutely punctate surface corresponding to forma **minutipuncta** (Erichs) Serv. The spores are small, 8–11 × 6–8 μm.

V. margacea Wahlenb.
On more or less permanently submerged rocks and boulders in streams and lochs; rarely extending into the spray-zone by waterfalls.
Occasional to frequent; probably in all sheltered streams, especially in lowland areas.
3–5, 7, 8, 10–15
Spores 25–40 × 10–18 μm.

V. maura Wahlenb. ex Ach.
In the upper fringe of the littoral zone, often extending up the shore in small sheltered crevices and along local freshwater seepage tracks on the shore.
Abundant; often dominant along rocky shores in all coastal areas.
1–17

V. microspora Nyl.
In the upper littoral fringe below and extending into the lower part of the *V. maura* zone; often associates with *V. mucosa.*
Frequent; occurring in small patches seldom exceeding 3 cm diameter.
1–9, 11, 14, 15

V. mucosa Wahlenb. ex Ach.
In the upper littoral fringe below and extending into the lower part of the *V. maura* zone.
Frequent; often abundant on exposed shores, less common on sheltered coasts and towards the heads of sea-lochs.
1–14, 16, 17
The smooth, oily thallus with immersed perithecia are distinctive features; the hypothallus is always pale.

V. muralis Ach.
On basic rocks and associated scree and on mortar, cement and asbestos.
Frequent, often abundant; widespread.
2–17

V. nigrescens Pers.
On basic rocks and associated scree and a mortar, cement and asbestos. Also but less frequently, on acidic rock, especially on the coast and near farmsteads.
Abundant; widespread, becoming rare in upland areas.
1–17

V. prominula Nyl.
Mesic supralittoral; in shaded acid rock crevices of very sheltered, often friable rocks above the *V. maura* zone.
5 Calgary Bay near the mouth of Ensay Burn, *U. Duncan & James.*
The spores in this species have characteristic bluntly truncate apices.

V. sandstedei B. de Lesd.
Littoral, forming small patches between *V. striatula* in exposed sunny sites.
Rare; probably overlooked.
4 Ulva, Eilean Reilean, *U. Duncan & James.*
Characterised by the narrow (3–4 μm), often bent, spores and the minute (100–200 μm) perithecia.

V. sphinctrina Ach.
On hard basic rocks and associated scree.
Rare; limited by the availability of suitable natural substrata.
10 Carsaig, on calcareous sandstone. **14** Chapel near Glenforsa House on limestone gravestone.

V. striatula Wahlenb. ex Ach.
In littoral fringe below the *V. maura* zone and often extending down to the midlittoral zone; often in association with *V. amphibia* and *V. mucosa.*
Abundant; on all exposed shores, becoming rarer on sheltered coasts.
1–17
The thallus is markedly ornamented by a darker mosaic of ridges and papillae. Collected on **1** Staffa by *Groschke*, 1787 (herb. Smith in Linnean Society of London).

V. viridula (Schrad.) Ach.
On basic rock, scree, and pebbles and on concrete, mortar and asbestos; most frequent on mortar of old buildings.
Frequent; often abundant in all lowland sites, becoming scarcer in upland areas, probably due to the absence of suitable substrates.
1–17

ARTHONIALES

ARTHONIACEAE

Arthonia Ach.

Bibliogr.: Redinger (1936).
A. aspersella Leight.
On smooth bark of *Corylus*, rarely young *Quercus*, in sheltered mature woodland and valley ravines in lowland and coastal sites.
Locally frequent.
3–5, 8, 10, 13, 16, 17
Apart from the smaller size of the ardellae, *A. aspersella* is closely allied to *A. lurida* and may prove to be, on further study, conspecific.

A. didyma Körb.
On dry sheltered aspects of boles of old *Quercus* and *Castanea* in mature mixed plantations and parkland in lowland sites.
Rare; mainly confined to the north-east.
6 Quinish, near Home Farm, on dead *Quercus.* **13** Aros Park, on *Quercus.* **16** Lochbuie, near Moie Lodge, on *Castanea*; Lussa River, An t-Sleaghach, on *Quercus.* **17** Torosay Castle on *Quercus.*
This species, widespread in S England, is very rare in Scotland.

A. dispersa (Schrad.) Nyl. var. **dispersa**
On smooth bark of low-growing trees and shrubs on sheltered hillsides; usually near the coast.
Rare.
7 1 mile east of Achronich, on *Sambucus*, *U. Duncan.* **8** Near Scobull Point, on *Prunus spinosa.* **12** Near Pennygown, on *Crataegus*; Scallastle, near Cnoc Bhacain, on *Crataegus.*

A. dispersa var. **excipienda** (Nyl.) Oliv.
Habitats similar to var. *dispersa.*
Rare.
8 Inch Kenneth, on *Hedera helix* on Chapel.
12 Kellan Wood, on *Quercus.*

A. lapidicola (Tayl.) Deichm. Br. & Rostr.
On basic rocks and boulder in scree near the coast.
Rare.

10 Near Nuns' Pass, on calcareous sandstone. **17** Rubha na Sròine, on limestone.

A. leucopellaea (Ach.) Almqu.
Confined to mature *Picea abies* in old mixed plantations in two lowland sites.
Rare.
7 Torloisk House. **13** Aros Park, near Spùt Dubh.
The thalli of these two specimens are Pd+ orange and contain atranorin, thamnolic acid (±) and an unknown substance.

A. lobata (Flot.) Massal.
Under sheltered dry overhangs and entrances to caves on or near the coast; strongly hydrophobous and photophobous.
Locally frequent especially in the west.
2–6, 8, 10, 13, 16, 17
All gatherings are sterile and C+ rose-red (lecanoric acid), but the collections are very polymorphic and two distinct taxa may be represented.

A. lurida Ach.
On smooth bark, chiefly shaded boles of broad-leaved trees, including *Corylus*, *Fraxinus*, *Quercus*, *Ilex* and *Sorbus aucuparia*, especially in small stream ravines and dense mature woodland in lowland and coastal sites.
Frequent; especially in stands of *Corylus* in western areas.
1–17

A. phaeobaea Ach.
Mesic supralittoral; in crevices in acid rocks (basalt) at high-watermark on moderately exposed shores.
Recorded from a single locality.
1 Fladda.

A. punctiformis Ach.
On smooth bark of boles, branches and twigs of broad-leaved trees; rarely on smooth bark of young conifers especially at the southern and western edges of plantations. Ascending to 300 m in **8** gorge above Clachandhu, on *Sorbus aucuparia*.
Frequent to abundant; still probably under recorded.
2–8, 10, 12–17

A. radiata (Pers.) Ach.
On smooth bark of boles, branches and twigs of a wide range of broad-leaved trees, especially *Fraxinus*, in ravines and wayside and mature woodland sites; rarely on boles of conifers at the margins of plantations.
Abundant; widespread.
2–17
Most material is referable to var. *schwartziana* (Ach.) Almqu. which is here considered to represent a habitat modification of the species. The most frequent species of the genus on the Island.

A. spadicea Leight.
On very smooth bark, especially *Corylus*, in sheltered and shady ravines in lowland areas; less frequent on mature *Quercus* in dense old woodland.
Frequent; becoming more so in the east.
4–6, 8, 13, 14, 16, 17
A notably photophobous species.

A. stellaris Kremp.
On sheltered boles of old broad-leaved trees, especially *Ilex* and *Quercus*, and stream-sides.
Rare; confined to lowland sites, especially in the east.
7 Torloisk House on *Corylus*. **13** Aros Park, on *Quercus* and *Ilex*. **15** Near Strathcoil, on *Ilex*.
Known also from the Ardnamurchan Peninsula; that area and Mull represent the most northerly stations for this species in Britain.

A. tumidula (Ach.) Ach.
A. cinnabarina (DC.) Wallr.
On shaded, smooth bark of broad-leaved trees in moist, sheltered, wayside, valley and woodland sites; rarely on young conifers at the margins of plantations.
Abundant; especially in small coastal valleys in the west.
3–17
This species is very polymorphic.

Arthothelium Massal.

A. ilicinum (Tayl.) P. James
On smooth, rarely rough, bark of a wide range of broad-leaved trees in sheltered wayside, valley and woodland sites; recorded on *Betula*, *Corylus*, *Crataegus*, *Fagus*, *Fraxinus*, *Hedera*, *Quercus*, *Salix* spp., and *Sorbus aucuparia*. Rare on the young boles of *Picea*, and then at the margins of plantations.
Frequent, often locally abundant; ascending to 300 m on **11** Garbh Choire, on *Sorbus aucuparia*.
2–17

A. spectabile Flot. ex Massal.
On smooth bark of an old *Corylus* in a sheltered well-wooded ravine.
Rare; only recorded once, but probably overlooked for the previous species which it resembles externally.
7 Torloisk House, *U. Duncan & James*.
Also recorded for the Ardnamurchan Peninsula, the most northerly record for this species.

LECANACTIDACEAE

Lecanactis Eschw.

L. abietina (Ach.) Körb.
On dry sheltered boles of old conifers and broad-leaved trees in mature woodland in lowland and coastal areas; rarely on old *Betula* above 90 m.

Occasional; locally frequent in a few well-wooded sites.
3, 4, 7–13, 16, 17
Recorded with ascocarps in **4, 7, 8, 12, 13, 16, 17.**
Contains lecanoric and schizopeltic acids.

L. homalotropa (Nyl.) Gill
Conotrema homalotropum (Nyl.) A. L. Sm.
On boles of broad-leaved trees, especially *Acer pseudoplatanus* and *Fraxinus*, more rarely conifers in well-wooded, sheltered, lowland and coastal sites.
Frequent; locally abundant especially in the east.
3–7, 9, 10, 12, 13, 16, 17
No lichen substances detected.

Sclerophyton Eschw.

S. circumscriptum (Tayl.) Zahlbr.
Under shaded, dry overhangs of acid rocks on or near the coast.
Rare to occasional; confined to a few localities in the west.
2 Cùl Bhuirg; near Spouting Cave; An t-Ard; Calva. **3** Kintra. **4** Gometra, Tòrr Mòr, near the burial ground; Ulva, near Ardalum House. **10** Aoineath a' Mhaide Ghil.
These are the most northern records for this species in Britain. Contains psoromic and conpsoromic acids.

OPEGRAPHACEAE

Enterographa Fée

E. crassa (DC.) Fée
In shady, dry recesses in boles of broad-leaved trees in very sheltered valleys in lowland and coastal sites.
Rare and very local.
12 Salen, near the village, on *Fagus*. **13** Aros Park, on old *Fagus*. **17** Entrance to Java Lodge, on *Quercus*; Lochdonhead on young *Fraxinus*.
Contains confluentic acid. This species, frequent in southern and western England and Wales, is very rare in Scotland.

E. hutchinsiae (Leight.) Massal.
Confined to damp, often crumbling, basalt rocks, by or near streams in very shaded valley and woodland sites; very rare on sheltered, damp rock on cliff above the sea.
Occasional; only on or near the coast.
4 Ulva, Port Bàta na Luinge. **5** Calgary Bay, Ensay Burn. **12** Salen, near the village; Kellan Mill.
Contains confluentic acid.

Opegrapha Ach.

Bibliogr.: Redinger (1938).

O. atra Pers.
On boles, branches and twigs of smooth, rarely rough-barked, trees in wayside valley and woodland sites.
Abundant; frequently dominant, in lowland areas in all parts of the Island. Ascending to 300 m on **11** Garbh Choire, on *Betula* and *Sorbus aucuparia*.
1–17
Very polymorphic.

O. atricolor Stirt.
On bole of old partly decorticated *Ilex* in deep shade.
Rare.
15 Strathcoil, An t-Sleaghach at 15 m.

O. cesareensis Nyl.
Under dry overhangs in basalt cliffs on the coast.
Rare.
3 Achnahaird; Lochan Mòr; Uisken, near Ardchiavaig.
The most northerly records for this species in Britain.

O. chevallieri Leight.
On crumbling mortar of walls of derelict buildings and farmsteads; rarely on gneiss rocks partially covered by shell-sand.
Rare; confined to a few localities in the west.
2 Iona, Golf Course, on gneiss partially covered by shell-sand; Baile Mòr, walls of the Nunnery. **4** Ulva, Cragaig. **5** Calgary, walls around the burial ground. **8** Walls around Kilfinichen House and church.

O. confluens (Ach.) Stiz.
On more or less exposed, acidic boulders and rocks of walls on or near the coast; usually in sheltered site and under overhangs.
Rare.
3 Ardtun, Eilean an Duilisg; **4** Gometra, Rubha Dùn Iasgain. **8** Burgh; Tavool House, on basalt. **10** Carsaig, walls near the pier.

O. gyrocarpa Flot.
Under overhangs and in damp sheltered recesses in cliffs and acid rock outcrops mainly in shady coastal sites; more rarely on north and east-facing sides of old walls.
Frequent in most lowland areas; becoming rarer in upland sites.
2–6, 8, 10–17
Fertile specimens recorded in **3, 5, 6, 12** and **13.**

O. herbarum Mont.
On rough or smooth bark of broad-leaved trees in sheltered woodland and valley sites.
Occasional in lowland areas.
5, 7, 10–13, 17

O. lithyrga Ach.
On acid rock crevices and under overhangs, often partially obscured by summer vegetation or *Hedera helix*.
Rare; mainly confined to lowland and coastal areas.
4 Ulva, Cragaig. **12** Salen, near the village. **17** Torosay Castle.

These are close to the most northerly record for this species in Britain, at Loch Moidart.

O. niveoatra (Borr.) Laund.
On mature wayside and woodland trees especially *Fagus*, in lowland areas; rarely on decorticated wood of knot holes and standing dead trees.
Occasional; probably overlooked.
6 Quinish, near the old pier on *Fagus*; Druimghigha, on *Fagus*. **7** Torloisk House, on *Fagus*. **12** Salen, near village, on *Alnus*. **17** Torosay Castle, on *Quercus*.

O. ochrocheila Nyl.
On bark and decorticated wood of old broad-leaved trees in sheltered old woodland and rarely on dry acid or basic pebbles in sheltered rock recesses and at the base of old trees; recorded only from lowland sites.
Occasional.
4 Ulva, on *Fraxinus*, at entrance to Ardalum House. **7** Lagganulva, Oskamull, on *Quercus*. **8** Near Mackinnon's Cave, on basic pebble. **12** Carsaig House, on decorticated *Fraxinus*. **13** Aros Park, on base of *Quercus*.
The gold to ochre pruina on the ascocarps reacts K+ wine-red.

O. saxatilis DC.
In sheltered crevices in base-rich rocks near the coast.
Rare; probably overlooked.
8 Mackinnon's Cave. **10** Carsaig, on crevices in calcareous sandstone cliffs.

O. saxicola Ach.
In sheltered crevices in acidic or basic rocks in valley woodlands and on old walls.
Occasional; mainly confined to lowland habitats but ascending to 250 m at **6** Mishnish, Meall an Inbhire.
3, 4, 9, 11–13, 16

O. saxigena Tayl.
In sheltered, very shaded rock crevices in moist often wooded valleys and sheltered coastal sites, associated with *O. zonata* and *O. gyrocarpa*.
Rare.
6 Glengorm Castle, Port Chill Bhraonain. **7** Lagganulva, Bruach Mhòr, *U. Duncan & James*. **12** Salen, near the village, *U. Duncan & James*. **16** Ben Buie.

O. sorediifera P. James
On smooth, rarely rough, bark of broad-leaved trees, especially *Acer pseudoplatanus* and *Salix*, in sheltered, moist, valley woodlands.
Occasional; probably overlooked.
3 Beinna Ghlinne Mhòir, on *Salix*; Ardtun, Tòrr Mòr, on *Salix* and *Corylus*. **5** Calgary, on young *Acer pseudoplatanus* near the Burial Ground, *U. Duncan & James*. **13** Aros Park, on *Escallonia* sp. and *Acer pseudoplatanus*. **16** Lochbuie, Moie Lodge, on *Fagus*.

O. varia agg.
On dry sides of often very sheltered boles of old wayside and valley woodland trees, especially *Fraxinus* and *Ulmus*.
Occasional; but widespread in coastal areas and sheltered lowland stream gullies.
2, 4–7, 12–14, 16, 17
This aggregate includes *O. actophila* Nyl., **O. diaphora** (Ach.) Ach., **O. lichenoides** Pers., **O. pulicaris** auct., non (Hoffm.) Schrad. and *O. rimalis* Pers. best considered as microspecies pending further study. *O. pulicaris* is rare and *O. actophila* and *O. rimalis* are not recorded.

O. viridis (Ach.) Nyl.
On smooth, rarely rough, bark of broad-leaved trees, especially in sheltered, but lowland, open valley woodlands; recorded on *Acer pseudoplatanus*, *Corylus*, *Fagus*, *Fraxinus*, *Ilex* and *Quercus*.
Occasional; but widespread.
3, 5–9, 12, 15, 17

O. vulgata (Ach.) Ach.
On smooth or rough bark of broad-leaved and coniferous trees in wayside, valley and mature woodland sites; very rarely on acidic rocks.
Abundant; widespread in all lowland and coastal areas, becoming scarcer in upland sites, ascending to 300 m on **12** Cruach Tòrr an Lochain.
1–17
Saxicolous in **6** Dervaig.

O. zonata Körb.
Under sheltered, rather moist recesses and overhangs, especially near streams; often associated with *O. gyrocarpa* and *O. saxigena*.
Frequent and widespread; ascending to 400 m on **16** Creag an h-Iolaire.
2, 4–7, 12, 13, 15–17
Fertile specimens recorded in **4, 6, 13** and **17.**

PLEOSPORALES

ARTHOPYRENIACEAE

Acrocordia Massal.
Bibliogr.: Swinscow (1970).

A. alba (Schrad.) Massal.
Arthopyrenia alba (Schrad.) Zahlbr., *Acrocordia gemmata* (Ach.) Massal.
On rough, rarely smooth, bark of broad-leaved trees, especially *Fraxinus* and *Ulmus*, in wayside and woodland, lowland and coastal sites; very rarely on smooth bark of conifers.
Frequent to abundant in all areas below 100 m; ascending to 300 m in **15**.
2–10, 12–17
A specimen with pale pink perithecia (f. **carnea** B. de Lesd.) was recorded from **7** Torloisk House, on bole of mature *Fraxinus* in deep shade.

A. biformis (Borr.) Arnold
Arthopyrenia biformis (Borr.) Massal.
On rough or smooth bark of broad-leaved trees, especially *Corylus*, *Fraxinus*, *Quercus* and *Ulmus* in wayside and woodland lowland and coastal sites.
Abundant; especially on wayside trees in all lowland areas below 100 m, ascending to 230 m in **11** Coille na Creige Duibhe, on *Betula.*
2–8, 10–17

A. conoidea (Fr.) Körb.
Arthopyrenia conoidea (Fr.) Zahlbr., *Acrocordia epipolea* (Borr.) A. L. Sm.
On sheltered, base-rich sandstone face near the coast.
Rare.
10 Nuns' Pass.
Frequent on limestone on Lismore Island.

A. salweyi (Leight. ex Nyl.) A. L. Sm.
Arthopyrenia salweyi (Leight. ex Nyl.) Zahlbr.
On acidic or basic rock outcrops and scree on or near the coast, and on mortar of derelict buildings.
Locally frequent especially in the west.
3 Erraid, on mortar of disused Observatory. **4** Ulva, Cragaig. **8** Inch Kenneth, on walls of the ruined chapel. **10** Nuns' Pass, dominant on calcareous sandstone. **13** Aros Park near Spùt Dubh. **17** Rubha na Sròine.
This species is very rare on Lismore Island.

Arthopyrenia Massal.

See also *Acrocordia* Massal.

A. antecellans (Nyl.) Arnold
A. stigmatella (Gray) A. L. Sm.
On smooth bark of broad-leaved trees in sheltered, well-wooded sites, especially in mature woodland and pure stands of old *Corylus* in ravines near or on the coast.
Frequent, often locally abundant; widespread.
3–7, 10, 12, 13, 16, 17
Recognised by the unusually (for the genus) well-developed thallus and the numerous perithecia of different sizes on the same plant.

A. cerasi (Schrad.) Massal.
On smooth bark of wayside trees of *Betula* and *Prunus cerasus.*
Rare.
7 Torloisk House, on *Prunus.* **10** Pennyghael, on *Prunus.* **16** Near Strathcoil, on *Betula*; Coille Sron nam Boc, on *Betula.*

A. cinereopruinosa (Schaer.) Körb.
On broad-leaved, mature woodland trees, especially *Quercus.*
Occasional; locally abundant, especially in the Tobermory and Craignure areas.
7 Torloisk House on *Quercus.* **13** Tobermory, on *Sambucus nigra* near harbour, *U. Duncan*; Aros Park, on *Quercus.* **16** Cruach Ardura, on *Quercus.* **17** Torosay Castle, on *Quercus, U. Duncan.*

A. fallax (Nyl.) Arnold
On the smooth bark of branches and twigs of broad-leaved trees; also occasionally on boles of conifers, especially at the margins of plantations.
Abundant; widespread in all wayside and woodland areas.
1–17

A. halodytes (Nyl.) Arnold
Confined to littoral fringe; common on barnacles and limpets, more rarely (or overlooked) on coastal acidic and basic rocks.
Abundant; on all coasts except at the heads of sea lochs.
1–10, 13, 14, 16, 17
The interpretation of this species follows that of Swinscow (1965). Although a large number of specimens from different substrata (including 28 species of littoral mollusca, basic and acidic rock) were examined there was no evidence that the wide degree of polymorphism encountered could be interpreted in any other way than by habitat variation due to the hardness and texture of the substrate.

A. laburni Arnold
Confined to *Laburnum* in lowland areas.
Rare; probably limited by the distribution of the host.
8 Kilfinichen, *U. Duncan & James.* **10** Pennyghael, *U. Duncan & James.* **13** Aros Park.
This species is also recorded for Lismore Island.

A. punctiformis (Pers.) Massal.
On smooth bark of boles, branches and twigs, rarely on rough bark, of broad-leaved trees in wayside, valley and woodland sites; rarely on conifers, and then at the margins of plantations.
Abundant in all sites.
1–17

A. pyrenastrella (Nyl.) Norm.
A. cembrina (Anzi) Grumm.
Confined to young bark of boles and branches of *Ulmus* in sunny lowland situations.
Rare; only recorded from the north-west.
7 Half a mile W of Ballygown, *U. Duncan & James.* **8** Above Slochd Bay, *U. Duncan & James.*
The characteristic clustering of the perithecia is well marked in these gatherings.

A. saxicola Massal.
On damp, shaded limestone outcrop.
Rare.
17 Rubha na Sròine.

A. strontianensis Swinsc.
On periodically inundated rocks in small upland stream.
Rare.
12 Torranlochain, stream running into Féith Bhàn, at 90 m.

Arthopyrenia sp. nov.
On granite rocks in the littoral zone.
Rare; confined to the Ross of Mull, but probably overlooked.
3 Port nan Ròn; Fionnphort; Rubha nan Cearc.
This species superficially resembles *Verrucaria striatula* Wahlenb. ex Ach.

Microthelia Körb.

M. micula Körb.
On smooth bark of boles and branches of broad-leaved trees and shrubs, especially *Corylus* in lowland and coastal areas; rare in a few wooded upland valley sites.
Frequent; widespread on or near the coast becoming rare above 30 m but ascending to 300 m in **17** S edge of Coill an Fhraoich Mhòir on *Corylus*. Especially abundant in sheltered ravines in the west.
3–8, 10–17

DOTHIDEALES

MYCOPORACEAE

Dermatina Almqu.

D. quercus (Massal.) Zahlbr.
On smooth bark of twigs and small branches of young and mature *Quercus* or, very rarely, *Corylus*, in sunny wayside and woodland margin habitats.
Occasional; probably frequent on twigs in the upper canopy of all lowland *Quercus* trees in woodlands.
4–8, 10–13, 16, 17

D. swinscowii Riedl
On smooth bark of old *Corylus* in a small sheltered ravine near the coast.
Rare.
12 Lower ravine of Allt Ardnacross.

BASIDIOMYCETES

AGARICALES

TRICHOLOMATACEAE

Coriscium Vain.

C. viride (Ach.) Vain.
On damp peat, especially at the edges of old peat-hags, blow-outs or stream-cuttings in most upland sites above 100 m.
Locally frequent and widespread.
3, 4, 6–11, 13, 14, 17
The name refers to the lichenised basal squamules of the basidiomycete fungus *Omphalina hudsoniana* (Jennings) Bigelow (p. 15.12). Moribund squamules are occasionally host to *Thelocarpon epibolum* Nyl. Other species of *Omphalina* also form associations with algae, e.g. *O. ericetorum* (Pers. ex Fr.) M. Lange and *O. luteovitellina* (Pilát & Nannf.) M. Lange (see p. 15.12). In the sterile state these are often regarded as lichens of the genus *Botrydina* Bréb.

Normandina Nyl. em Vain.

N. pulchella (Borr.) Nyl.
On overgrowing mosses, hepatics and other lichens on tree boles and rocks in all lowland and coastal wayside and woodland areas.
Abundant; ascending to 130 m on **15** Coill an Fhraoich Mhòir, on *Betula.*
1–17
In line with current concepts, this taxon and *Coriscium viride* are here considered as basidiolichens.
It is worth nothing that in the case of *Normandina pulchella* the perithecia described in Swinscow (p. 167, 1963b) and elsewhere, and thought to be fruits of this taxon, are of a parasymbiont fungus (*Sphaerulina chlorococca* (Leight.) R. Sant.). When attacked by the parasymbiont the squamules of the lichen are often reduced to a sorediate rim around the developing perithecium.

LICHENES IMPERFECTI

Cystocoleus Thwaites

C. ebeneus (Dillw.) Thwaites
C. niger (Huds.) Hariot
On permanently dry, acidic rock faces, particularly in sheltered gorges in mature mixed woodland.
Rare; overlooked.
12 Gruline, Guala Buidhe. **13** Aros Park.
Apparently much rarer than *Racodium rupestre*, which it superficially resembles, and which occurs in similar habitats.

Lepraria Ach.

L. candelaris (L.) Th. Fr.
On dry, sheltered sides of boles of broad-leaved trees and conifers in well-wooded and wayside lowland, rarely upland, sites.
Occasional to frequent; especially in the east.
4, 6, 7, 10, 12, 13, 16, 17

L. crassissima (Hue) Lett.
In dry, often sheltered recesses below overhangs in more or less base-rich cliffs in lowland and coastal areas; rare on sheltered aspects of old mortar-stone walls especially those surrounding derelict gardens.
Occasional; widespread.
4, 5, 10, 12, 17

L. incana (L.) Ach.
On sheltered, usually dry, sides of boles of old broad-leaved trees and conifers; also in sheltered recesses of acid or slightly basic rocks and boulders.
Abundant; widespread.
1–17

L. membranacea (Dicks.) Vain.
In dry sheltered recesses, often beneath overhangs of acid or slightly base-rich cliffs in

a few coastal and lowland areas. Rare on the sheltered aspects of old trees especially *Fraxinus* in well-wooded lowland areas.
Occasional; widespread, especially in the east.
4, 9, 10, 12–14, 16, 17

L. neglecta auct., non (Nyl.) Lett.
On acid, often peaty, soils and associated mosses, especially in rock crevices, in most upland and a few lowland areas.
Frequent above 60 m; rare in a few exposed coastal and lowland sites especially in **7** and **9**.
3, 6–11, 15–17

Leprocaulon Nyl.

L. microscopicum (Vill.) Gams ex D. Hawksw.
On pockets of soil in dry, sheltered crevices of acid rocks and boulders in lowland and coastal areas, prefering sunny sites.
Occasional; becoming frequent in the west.
1–5, 7, 8, 10, 12, 13, 16, 17
Contains usnic acid, zeorin (±) and destrictin.

Leproplaca (Nyl.) Hue

Bibliogr.: Laundon, 1974
L. chrysodeta (Vain. ex Räs.) Laund.
On dry, very sheltered mortar at the base of an east-facing wall.
Rare; the only record.
8 Kilfinichen.
This species is frequent on the dry sheltered limestone cliffs of Lismore Island.

Racodium Pers.

R. rupestre Pers.
Cystocoleus rupestris (Pers.) Rabenh.
On sheltered, dry, often vertical rock faces, especially in well-wooded low-land areas.
Occasional to frequent; more frequently recorded than *Cystocoleus eburneus*.
4, 6, 7, 12, 13, 15, 16

Thamnolia Ach. ex Schaer.

T. vermicularis (Sw.) Ach. ex Schaer. var. **subuliformis** (Ehrh.) Schaer.
On bare earth or low vegetation on or near the exposed summits of higher mountains.
Occasional to frequent on the Ben More massif; rare elsewhere.
7 Beinn na Drise, 430 m; Beinn Buidhe, 380 m; **11** Maol nan Damh, 760 m; Cruachan Dearg, summit scree, *c.* 700 m; Coirc Bheinn, 690 m. **15** Sgùrr Dearg, exposed ridge near summit, 725 m. **16** Ben Buie, *c.* 700 m; Creag na h-Iolaire, *c.* 420 m.
Contains squamatic (UV+) and baeomycesic acids.

References

COPPINS, B. J. (1972). Field meeting at Richmond, Yorkshire. *Lichenologist*, **5**: 326–336.

CRUNDWELL, A. C. (1956). A lichen new to Scotland (*Sticta sinuosa*). *Glasg. Nat.*, **17**: 279.

CULBERSON, C. F. (1965). Some constituents of the lichen *Ramalina siliquosa*. *Phytochemistry*, **4**: 951–961.

———. (1969). *Chemical and botanical guide to lichen products*. Chapel Hill, North Carolina.

———. (1970). Supplement to 'Chemical and botanical guide to lichen products'. *Bryologist*, **75**: 177–377.

———. (1972). Improved conditions and new data for the identification of lichen products by a standardised thin-layer chromatographic method. *J. Chromatog.*, **72**: 113–125.

———. & KRISTINSSON, H. (1970). A standardised method for the identification of lichen products. *J. Chromatog.*, **46**: 85–93.

CULBERSON, W. L. & CULBERSON, C. F. (1968). The lichen genera *Cetrelia* and *Platismatia* (Parmeliaceae). *Contr. U.S. natn. Herb.*, **34**: 449–558.

DAHL, E. & KROG, H. (1973). *Macrolichens of Denmark, Finland, Norway and Sweden*. Oslo.

DEGELIUS, G. (1954). The lichen genus *Collema* in Europe. *Symb. Bot. Upsaliens.*, **13(2)**: 1–499.

DUNCAN, U. K. (1970). *Introduction to British lichens*. Arbroath.

ERICHSEN, C. F. E. (1935). *Pertusaria. Rabenh. Krypt.-Fl.*, **9**, 5(1): 333–680.

FLETCHER, A. (1973a). The ecology of marine (littoral) lichens on some rocky shores of Anglesey. *Lichenologist*, **5**: 368–400.

———. (1973b). The ecology of maritime (supralittoral) lichens on some rocky shores of Anglesey. *Lichenologist*, **5**: 401–422.

———. (1975a). Key for the identification of marine and maritime lichens I. Siliceous rocky shore species. *Lichenologist*, **7**: 1–52.

———. (1975b). Key for the identification of marine and maritime lichens II. Calcareous and terricolous species. *Lichenologist*, **7**: 73–115.

HAWKSWORTH, D. L. (1970). The chemical constituents of *Haematomma ventosum* (L.) Massal. in the British Isles. *Lichenologist*, **4**: 248–255.
———. (1972*a*). Regional studies in *Alectoria* (Lichenes) II. The British species. *Lichenologist*, **5**: 181–261.
———. (1972*b*). The natural history of Slapton Ley Nature Reserve IV. Lichens. *Field Studies*, **3**: 535–578.
———. & CHAPMAN D. S. (1971). *Pseudevernia furfuracea* (L.) Zopf and its chemical races in the British Isles. *Lichenologist*, **5**: 51–58.
HEDLUND, T. (1892). *Lecanora* (Ach.), *Lecidea* (Ach.) and *Micarea* (Fr.). *K. svenska Vetensk Akad. Handl.* **18**: 1–104.
HENSSEN, A. (1976). Studies in the developmental morphology of lichenised ascomycetes. Pp. 107–138 in D. H. Brown, D. L. Hawksworth & R. H. Bailey (eds.), *Lichenology: progress and problems*. London & New York.
———. & JAHNS, H. M. (1973). *Lichenes: Eine Einführung in die Flechtenkunde*. Stuttgart.
HERTEL, H. (1969). Die Flechtengattung *Trapelia* Choisy. *Herzogia*, **1**: 111–130.
———. (1970). Trapeliaceae – eine neue Flechtenfamilie. *Vort. bot. Ges.* (*Dtsch. bot. Ges.*), n.f. **4**: 171–185.
JAHNS, H. M. (1970). Remarks on the taxonomy of the European and North American species of *Pilophorus* Th. Fr., *Lichenologist*, **4**: 199–213.
JAMES, P. W. (1959). New British records. *Lichenologist*, **1**: 113.
———. (1965). A new check-list of British lichens. *Lichenologist*, **3**: 95–153.
———. (1966). A new check-list of British lichens: additions and corrections 1. *Lichenologist*, **3**: 242–247.
———. (1970). The lichen flora of shaded acid rock crevices and overhangs in Britain. *Lichenologist*, **4**: 309–322.
———. (1971). New or interesting British lichens 1. *Lichenologist*, **5**: 114–148.
———. (1975). The genus *Gyalideopsis* Vězda in Britain. *Lichenologist*, **7**: 155–161.
———. & HENSSEN, A. (1976). The morphological and taxonomic significance of cephalodia. Pp. 27–77 in D. H. Brown, D. L. Hawksworth & R. H. Bailey (eds.) *Lichenology: progress and problems*. London & New York.
JØRGENSEN, P. M. (1977). The foliose and fruticose lichens of Tristan da Cunha. Skr. norske Vidensk.-Acad. Mat.-nat. Kl., n.s., **36**: 1–40.
KROG, H. & JAMES, P. W. (1977). The genus *Ramalina* in Fennoscandia and the British Isles. *Norw. J. Bot.*, **24**: 15–43.
LAUNDON, J. R. (1970). Lichens new to the British flora 4. *Lichenologist*, **4**: 297–308.
———. (1971). Fumarprotocetraric acid in *Cladonia rangiformis*. *Lichenologist*, **5**: 175–176.
———. (1974). *Leproplaca* in the British Isles. *Lichenologist*, **6**: 102–105.
MENLOVE, J. E. (1974). Thin-layer chromatography for the identification of lichen substances. *Bull. Brit. Lich. Soc.*, **34**: 3–5.
MORGAN-JONES, G. & SWINSCOW, T. D. V. (1965). On the genus *Microglaena* Körb. *Lichenologist*, **3**: 42–54.
NORDIN, I. (1972). Caloplaca *sect.* Gasparrinia *i Nordeuropa Taxonomiska och Ekologiska Studier*. Uppsala.
POELT, J. (1965). Zur Systematik der Flechtenfamilie Physciaceae. *Nova Hedwigia*, **9**: 21–32.
———. (1966). Zur Kenntnis der Flechtengattung *Physconia* in der Alten Welt und ihrer Beziehungen zur Gattung *Anaptychia*. *Nova Hedwigia*, **12**: 102–135.
———. (1969). *Bestimmungsschlüssel europäischer Flechten*. Lehre.
REDINGER, K. (1936). Arthoniaceae. *Rabenh. Krypt.-Fl.* 9, **2**(**1**): 1–180.
———. (1938). *Opegrapha*. *Rabenh. Krypt.-Fl.* 9, **2**(**1**): 246–404.
ROSE, F. & JAMES, P. W. (1974). Regional studies on the British lichen flora I. The corticolous and lignicolous species of the New Forest, Hampshire. *Lichenologist*, **6**: 1–72.
SALISBURY, G. (1965). A monograph of the lichen genus *Thelocarpon* Nyl. *Lichenologist*, **3**: 175–196.
SHEARD, J. W. (1967). A revision of the lichen genus *Rinodina* (Ach.) Gray in the British Isles. *Lichenologist*, **3**: 328–367.

——— & JAMES, P. W. (1976). Typification of the taxa belonging to the *Ramalina siliquosa* species aggregate. *Lichenologist*, **8**: 35–46.

SMITH, A. L. (1918, 1926). *A monograph of the British lichens* ed. 2. Vols. 1 & 2. London.

STIRTON, M. D. (1877). Descriptions of new lichens. *Scot. Nat.*, **4**: 164–168.

SWINSCOW, T. D. V. (1960a). Pyrenocarpous lichens 1. *Lichenologist*, **1**: 169–178.

———. (1960b). *Cavernularia hultenii* Degelius in Scotland. *Lichenologist*, **1**: 179–183.

———. (1961). Pyrenocarpous lichens 2. *Gongylia* Körb. in the British Isles with first British record of *G. incarnata*. *Lichenologist*, **1**: 242–246.

———. (1962). Pyrenocarpous lichens 3. The genus *Porina* in the British Isles. *Lichenologist*, **2**: 6–56.

———. (1963a). Pyrenocarpous lichens 4. Guide to the British species of *Staurothele*. *Lichenologist*, **2**: 152–166.

———. (1963b). Pyrenocarpous lichens 5. *Lichenologist*, **2**: 167–171.

———. (1965). Pyrenocarpous lichens 8. The marine species of *Arthopyrenia* in the British Isles. *Lichenologist*, **3**: 55–64.

———. (1968). Pyrenocarpous lichens 13. Fresh-water species of *Verrucaria* in the British Isles. *Lichenologist*, **4**: 34–54.

———. (1970). Pyrenocarpous lichens 14. *Arthopyrenia* Massal. sect. *Acrocordia* (Massal.) Müll. Arg. in the British Isles. *Lichenologist*, **4**: 278–233.

———. (1971). Pyrenocarpous lichens 15. Key to *Polyblastia* Massal. in the British Isles. *Lichenologist*, **5**: 92–113.

THOMSON, J. W. (1968). *The lichen genus* Cladonia *in North America*. Toronto.

VĚZDA, A. (1959). K taxonomii rozsireni a ekologii lisenjniku *Belonia* Kbr. ve stredni Europe. *Prirodov. Cas. slezsky*, **20**: 241–253.

———. (1966). Flechtensystematische Studien IV. Die Gattung *Gyalidea* Lett. *Folia geobot. phytotax.*, **1**: 311–340.

———. (1969). Neue taxa und Kombinationen in der Familie Gyalectaceae (Lichenisierte fungi). *Folia geobot. phytotax*, **4**: 443–446.

———. (1972). Flechtensystematische studien VII. *Gyalideopsis*, eine neue Flechtengattung. *Folia geobot. phytotax.*, **7**: 203–215.

WADE, A. E. (1965). The genus *Caloplaca* Th. Fr. in the British Isles. *Lichenologist*, **3**: 1–28.

WIRTH, V. & VĚZDA, A. (1972). Zur Systematik der *Lecidea cyathoides* - gruppe. *Beitr. naturk. Forsch. SüdwDtl.*, **31**: 91–92.

Addenda

The following species have since been identified from material collected on the Survey.

LECANORACEAE

Haematomma species nova
On smooth bark of wayside and woodland broad-leaved trees, especially *Ilex*, *Salix* spp. and *Sorbus aucuparia*, in most lowland sites.
Frequent; widespread, especially in the east.
2, 4–8, 12–14, 16–17
Distinguished by the thin blue-grey thallus and scattered or more or less confluent bluish soralia. Contains ?divaricatic acid.

Lecanora subfuscata Magnusson
On broad-leaved trees.
Rare.
13 Glen Aros, on immature *Quercus*. **16** Craignure, Druim Mòr, on *Acer pseudoplatanus*.

LECIDEACEAE

Bombyliospora pachycarpa (Del. ex Duby) Massal.
On bole of wayside *Fraxinus*. The gathering is sterile.
Rare.
14 Fishnish Bay, Doire Dorch.
Contains usnic acid, zeorin, and an unknown substance. Old specimens appear glaucous due to the deposition of numerous pale blue, needle-shaped crystals.

Catillaria delutula (Nyl.) Zahlbr.
On very shaded damp pebbles in a path by the coast.
Rare.
13 Aros Park, Spùt Dubh.

Lecidea diducens Nyl.
On hard, acidic rocks in sunny situations on or near the coast.
Rare; probably overlooked.
5 Calliach Point. **9** Near Scoor and on walls of nearby chapel.
A species related to *L. auriculata* easily distinguished by the C+ red-purple reaction of the exciple due to 2′-0-methylanziaic acid. Atranorin is also present in the thallus of the Mull specimens.

L. glaucophaea Körb.
On both shaded and well-lit vertical acidic rocks.
Rare.
12 Bràigh a' Choire Moire. **13** Aros Park.

Micarea species nova
Encrusting mosses and hepatics on sheltered boles of broad-leaved trees in well-wooded situations.
Rare.
13 below Gualann Dhubh, on *Betula*. **15** Cruach Ardura, on *Quercus*. **17** Torosay, near Camas Mòr, on *Betula*.
Distinguished by the numerous, stalked, pink pycnidia. Most collections are without ascocarps. No lichen substances detected.

Rhizocarpon lecanorinum (Flörke ex Körb.) Anders
On hard acidic rock outcrops in sunny situations.
Rare; probably overlooked.
11 Coladoir River, Teanga Brìdeig, on erratic. **17** Loch Spelve, Croggan.

R. oederi (Web.) Körb.
On rock outcrops and especially stabilised acidic boulders in scree. Particularly characteristic of rocks rich in iron oxides.
Frequent; widespread.
2–8, 10–13, 15–17
Differs from *Lecidea atrata* in the sub-gyrose and matt, not concave and shining, ascocarps.

R. petraeum (Wulf.) Massal.
On sub-basic rock outcrops or acidic rocks associated with stone-mortar walls.
Frequent; widespread in most lowland and a few upland areas.
2–4, 6–8, 10–15, 17

R. polycarpon (Grogn.) Th.Fr.
On acidic rocks in a few upland sites.
Rare; local, mainly confined to the Ben More massif.
11 above Glen Cannel; Scarisdale River; Ben More, near summit cairn; Na Binneinean and Sròn nam Boc. **15** Coire Leth Dhearc-ola. **16** Beinn Fhada; Ben Buie.

CLADONIACEAE

Cladonia fragilissima Østh. & P. James
On peaty soil in open situations and at the margins of peat hags.
Rare.
9 Beinn Bhùgan. **10** Beinn Chreagach. **14** An Carnais.
Distinguished by the very brittle basal squamules, the sub-corymbose apices of the podetia and the presence of grayaninic acid. The material from **9** and **14** is without podetia.

PANNARIACEAE

Parmeliella species nova 1
On boles of broad-leaved trees, especially *Alnus*, *Corylus* and *Quercus*, in well-wooded sites; also on *Salix* in carr.
Frequent; widespread, especially in the east.
3–7, 10–13, 15–17
Resembles *Pannaria mediterranea* but the squamules lack the olivaceous tinge of that species. All the Mull gatherings are sterile.

P. species nova 2
On boles of broad-leaved trees in woodland sites.
Rare; mainly in the east.
13 Aros Woods, on *Corylus* and *Quercus*, **12** Allt Ardnacross, on *Corylus*. **17** Craignure, on *Fraxinus* near Inn.

LICHINACEAE

Porocyphus kenmorense (Holl ex Nyl.) Henss.
On wet, slightly basic rocks at the sheltered edge of a loch.
Rare.
17 Loch a' Ghleannain.
Distinguished by the effigurate margin of the thallus.

ARTHOPYRENIACEAE

Arthopyrenia willeyana R. Harris
On broad-leaved trees and conifers in sheltered, well-wooded sites.
Rare.
7 Torloisk House, on bole of *Cupressus macrocarpa*. **10** Carsaig, on *Sambucus nigra*.
Resembles *Acrocordia biformis* but the ascocarps are smaller and without an involucrellum; the spores are consistently longer, 14–20 μm.

15. Fungi

D. M. Henderson* and R. Watling*

Introduction

Collection of data

Fieldwork for the present Survey was carried out by the authors during the years 1967–1970. The seasonal coverage over these four years is shown in Table 15.1 and the areas covered are shown in Figs. 15.1–15.4 (see p. 15.71). Foam and soil samples (1 gm oven-dry), were taken as shown in Fig. 15.4 and were used to inoculate the following media: Ohio Agricultural Station agar, Cellulose agar, and Soil Extract agar. Culture was carried out at room temperature, 22°C, 38°C and 45°C. We are indebted to Dr A. E. Apinis, University of Nottingham, who single-handed took on the task of culturing and the subsequent analysis of these samples, and the results are shown in Tables 15.2 and 15.3.

A total of 61 species was recorded from incubated plant-debris at various temperatures in a moist chamber. The results from ten samples are given in Table 15.4.

Table 15.1 Seasonal coverage of fieldwork during the period 1967–1970

Month	Period(s)
March	12–20th
April	1–13th
May	1–17th; 22–30th
June	7–28th
July	1–12th; 14–31st
August	1–31st
September	1–30th
October	1–10th; 22–29th

* Royal Botanic Garden, Edinburgh.

Table 15.2 Number of fungi from soil samples cultured on Ohio Agricultural Station agar at various temperatures.

Sample number	Temperature 22 °C	38 °C	45 °C
1(A)b	1,600,000	< 16,000	< 160
2(B)b	1,300,000	< 17,000	500
3(C)c	7,600,000	< 20,000	< 200
4(D)b	12,500,000	< 12,500	< 125
4(D)c	6,750,000	< 67,500	< 675
5(E)b	4,000,000	< 40,000	< 400
6(F)b	4,900,000	12,400	< 120

Table 15.3 Number (in thousands) of bacteria, Actinomycetes and fungi at 22°C on Soil Extract agar (figures in light face) and on Cellulose agar (figures in **bold***)*

Sample number	Bacteria	Actinomycetes	Fungi
1(A)b	83,200	16,000	1,500
	216	**81**	**54**
2(B)b	53,300	10,700	1,300
	215	**108**	**162**
3(C)c	39,900	17,100	12,500
	160	**60**	**20**
4(D)b	74,000	18,000	12,000
	1,900	**32**	**400**
4(D)c	24,300	17,550	10,000
	13.5	**67.5**	**148.5**
5(E)b	84,000	236,000	4,400
	23.4	**< 23**	**23.5**
6(F)b	9,920	1,240	4,960
	< 12	**< 12**	**12.4**

Those species recorded at 10–12°C represent psychrophiles and the transitional psychrophile-mesophile group, while those recorded at 22 °C represent the common mesophiles. Species recorded at 38 °C represent fungi belonging to the thermophiles in the broad sense and the transitional mesophiles able to grow at the body temperature of mammals (Apinis, 1972).

Table 15.4 Number of species recorded from incubated plant-debris

Sample number	**Temperatures of moist chamber** 10–12 °C	22 °C	38 °C
1	3	7	4
2	10	2	7
3	4	3	6
4	2	5	4
5	4	5	6
6	5	9	5
7	0	2	4
8	0	3	3
9	0	0	4
10	3	4	5
Total	23	29	17

Voucher and herbarium material

Collections made by the authors and by P. D. Orton have been deposited in the Herbarium of the Royal Botanic Garden, Edinburgh (E); those made by M. C. Clark may be located at E or at the Royal Botanic Gardens, Kew (K) or in Clark's personal herbarium (Clark). Specimens collected by Dr R. W. G. Dennis are at K with duplicates at E; his hyphomycete material in the main has been deposited in the Commonwealth Mycological Institute, Kew (IMI). No herbarium material

Table 15.5 Total species of fungi collected. A = Total taxa (species, varieties and forms); B = Unconfirmed old records; C = Percent recorded as single collections

	A	B	C
EUMYCOTA			
Basidiomycotina			
Hymenomycetes			
Agaricales	613	3	32
Aphyllophorales, Tremellales	136	–	42
Gasteromycetes	15	–	43
Hemibasidiomycetes	92	1	37
Ascomycotina	534[1]	4	47
Deuteromycotina	274[2]	6	51
Zygomycotina	28	–	42
Mastigomycotina	19	–	76
MYXOMYCOTA	75	–	30
Insertae sedis	1	–	–
TOTAL	1,787[3]	14	44

Notes: 1. Includes 25 recorded as conidial state only.
2. Includes 15 recorded as sterile mycelia only.
3. Includes 21 varieties, 5 forms and 66 taxa of dubious rank.

quoted by Buchanan White (1881) has been traced; it is not housed in the Perth Museum where many of his other plant collections are to be found.

The total number of species recorded is 1,787 and a break-down of this given in Table 15.5. Fifty-three percent of the species recorded for Mull are represented in the above herbaria and the herbarium abbreviation is given in brackets after the record in the following text. Specimens formerly housed in the British Museum (Natural History) have been transferred to the Royal Botanic Gardens, Kew, under a re-organisation of curatorial and research responsibilities. Specimens collected by the Museum Survey parties on Mull are deposited in the Royal Botanic Garden, Edinburgh.

Arrangement and nomenclature

The systematic arrangement follows that of G. C. Ainsworth (1966) except that the Agaricales are separated from the other Hymenomycetes. Within families, genera and species are alphabetical. Because of the large number of species recorded, geographical localities are, for the most part, omitted being considered irrelevant due to the spasmodic occurrence of most fungi.

The reference for nomenclatural and taxonomic concept varies with the group and is indicated in the text under the appropriate hierarchical heading.

Acknowledgements

We are indebted to many mycologists for their help in allowing us to consult the field notes made by them whilst collecting on Mull. We are especially grateful to Dr R. W. G. Dennis, Royal Botanic Gardens, Kew for his extensive lists of the fungi he has collected and without which the final compilation would have certainly lacked or been rather sparse in certain areas. We would like to thank P. D. Orton and M. C. Clark, who have both found time to assist the Mull project, and also the Museum staff who have collected specimens whilst involved in surveying other plant-groups. Our particular thanks are extended to P. W. James who not only collected a great amount of material but who also assisted us in many other ways especially making us feel at home whenever we have been on the Island. It is also important to recall the work contributed by the late J. B. Evans who determined and carefully annotated specimens collected during the early part of the project. Dr Dennis has kindly commented on the completed list and indicated several points of interest. Dr A. E. Apinis undertook the analysis of the soil samples and is solely responsible for the identifications and the tables showing the number of micro-fungi isolated from these soils; we are deeply indebted to him.

Dr M. B. Ellis, W. D. Graddon, J. Hafellner, B. Ing, Professor C. T. Ingold, Dr D. Reid and Dr E. Zogg have all played a part in the determination of certain groups and we are grateful for their assistance. We finally wish to acknowledge the part played by Janis Sweeney in the large amount of work involved in preparing the file-cards and compiling the manuscript; and to Janet Brinklow (née Menlove) and B. J. Coppins, and the editors for finally checking the manuscript.

Index to Genera

EUMYCOTA: BASIDIOMYCOTINA

HYMENOMYCETES

AGARICALES

Bibliogr.: Dennis, Orton & Hora (1960); Lange (1955, 1957), Watling (1970, 1972).

CANTHARELLACEAE*

Cantharellus Adans. ex Fr.

C. cibarius Fr.
Common and widespread (K); recorded by White

C. cinereus Pers. ex Fr.

C. infundibuliformis Scop. ex Fr.
Widespread (E)

C. tubaeformis Bull. ex Fr.
Recorded by White, possibly the same as *C. infundibuliformis*

Craterellus Pers.

C. cornucopioides (L. ex Fr.) Pers.
Two collections from Aros Park; recorded by White

C. sinuosus (Fr.) Fr.
(E)

* Sometimes placed in Aphyllophorales.

BOLETACEAE

Boletus Dill. ex Fr.

B. badius Fr.
Widespread

B. calopus Fr.
Widespread (K)

B. chrysenteron Bull. ex St. Amans
Widespread (K); recorded by White

B. edulis Bull. ex Fr.
Common and widespread (E); several colour forms found on the main island

B. erythropus (Fr. ex Fr.) Secr.
Common (E, K)

B. luridus Schaeff. ex Fr.
Two recent collections (E); recorded by White

B. piperatus Bull. ex Fr.
Common

B. porosporus Watling
(E); not in the British List but widespread in the British Isles

B. spadiceus Quél.
(E); in the British List as *B. subtomentosus*

B. subtomentosus L. ex Fr.
Widespread (E, K)

Gyroporus Quél.

G. cyanescens (Bull. ex Fr.) Quél.
(K)

Leccinum Gray

L. aurantiacum (Bull. ex Fr.) Gray
Not in the British List; collections from Dervaig and Ulva (E) constitute the best documented finds of this species, sensu stricto, in the British Isles.

L. carpini (Schulz.) Pearson
Widespread (E, K)

L. holopus (Rostk.) Watling
(E)

L. roseotinctum Watling
(E)

L. scabrum (Bull. ex Fr.) Gray
Common (E, K); recorded by White as *Boletus scaber*.

L. variicolor Watling
Common in *Betula* woods (E)

Porphyrellus E. J. Gilbert

P. pseudoscaber (Secr.) Sing.

Suillus Micheli ex Gray

S. aeruginascens (Secr.) Snell

S. bovinus (L. ex Fr.) O. Kuntze
Widespread (E); recorded by White

S. grevillei (Klotsch) Sing.
Boletus elegans Schum. ex Fr.
Common and widespread, accompanied in some localities by var. **badius** (Sing.) Sing. (K); recorded by White as *Boletus flavus*.

S. luteus (L. ex Fr.) Gray
Common (E, K); recorded by White.

S. variegatus (Sow. ex Fr.) O. Kuntze
(E, K)

Tylopilus Karst.

T. felleus (Bull. ex Fr.) Karst.

PAXILLACEAE

Paxillus Fr.

P. involutus (Batsch ex Fr.) Fr.
Very common and widespread in woods and copses (K)

GOMPHIDIACEAE

Gomphidius Fr.

G. glutinosus (Schaeff. ex Fr.) Fr.
Common and widespread in plantations (E); as well as the normal, a pink-coloured form resembling *G. roseus* was collected under *Picea*; recorded by White .

G. maculatus Fr.
Frequent and widespread under *Larix* (E, K)

G. roseus (Fr.) Karst.

HYGROPHORACEAE

Hygrophorus Fr.

H. agathosmus (Fr. ex Secr.) Fr.
(E)

H. chrysaspis Métrod
Common

H. chrysodon (Batsch ex Fr.) Fr.

H. cossus (Sow. ex Berk.) Fr.

H. hypothejus (Fr. ex Fr.) Fr.

H. olivaceoalbus (Fr. ex Fr.) Fr.

Hygrocybe (Fr.) Wünsche

H. aurantiosplendens Haller

H. calyptraeformis (Berk. & Br.) Fayod

H. cantharellus (Schw.) Schw.
Common and widespread (E, K)

H. ceracea (Wulf. ex Fr.) Karst.
(E)

H. chlorophana (Fr.) Karst.
Frequent (E)

H. citrinovirens (Lange) J. Schaeff.

H. clivalis (Fr.) Orton & Watling
(E)

H. coccinea (Schaeff. ex Fr.) Kummer
Common in grassland (K); a closely related although distinct taxon was found associated with *Sphagnum* at Loch Bà.

H. collemannianus (Blox.) Orton & Watling

H. conica (Scop. ex Fr.) Kummer
Common and widespread

H. conicoides (Orton) Orton & Watling
(E)

H. flavescens (Kauffm.) A. H. Smith & Hesler
Frequent

H. fornicata (Fr.) Sing.
(E)

H. aff. **fuscescens** (Bres.) Orton & Watling
Dervaig and Loch na Keal (E). Not in the British List; the best documented British record of this distinctive agaric, as yet unnamed.

H. insipidus (Lange ex Lundell) Moser

H. intermedia (Pass.) Fayod

H. lacma (Schum. ex Fr.) Orton & Watling
Common and widespread (E)

H. laeta (Pers. ex Fr.) Karst.
Very common and widespread (E)

H. langei Kühner
Common. A 4-spored agaric, closely related to *H. langei*, is frequent in sand-dunes: spores elongate ellipsoid, 7–9.5 × 4–5 μm.

H. lilacina (Laest.) Moser
Summit of Ben More on peaty banks

H. marchii (Bres.) Sing.
Common

H. metapodia (Fr. ex Fr.) Moser

H. miniatus (Fr.) Kummer
(K)

H. aff. **miniatus** (cf. *H. coccineocrenata* (Orton) Moser)
In wet Molinietum (E); although not uncommon an as yet undescribed species growing on peaty soil.

H. nigrescens (Quél.) Kühner
(K)

H. nitrata (Pers. ex Pers.) Karst.

H. nivea (Scop. ex Fr.) Orton & Watling
Common in grassland (E)

H. obrussea (Fr.) Wünsche

H. ovina (Bull. ex Fr.) Kühner
(E)

H. pratensis (Pers. ex Fr.) Donk
Very common (E); recorded by White

H. psittacina (Schaeff. ex Fr.) Karst.
Very common and widespread (K)

H. punicea (Fr.) Kummer
Very common and widespread in grasslands (E)

H. quieta (Kühner) Sing.
(E)

H. reai Maire
Common (E)

H. russocoriacea (Berk. & Miller) Orton & Watling
Common in grassland (E)

H. splendidissima (Orton) Moser
Common in grassland (E)

H. strangulata (Orton) Kriessel

H. subradiata (Schum. ex Secr.) Orton & Watling

H. unguinosa (Fr.) Karst.

H. vitellina (Fr.) Karst.
Common

H. virginea (Wulf. ex Fr.) Orton & Watling
Common in grassland (E)

PLEUROTACEAE

Pleurotus (Fr.) Kummer

P. ostreatus (Jacquin ex Fr.) Kummer
Common and widespread on hardwood (K)

P. pulmonarius (Fr.) Quél.
(K)

TRICHOLOMATACEAE

Armillaria (Fr.) Staude

A. mellea (Vahl ex Fr.) Kummer
Widespread (K); recorded also by White

Calocybe Kühner ex Donk

C. carneum (Bull. ex Fr.) Donk
Tricholoma carneum (Bull. ex Fr.) Kummer

C. gambosum (Fr.) Sing.
Iona (Clark)

Cellypha Donk

C. griseopallida (Weinm.) W. B. Cooke
(K)

Chamaemyces Batt. ex Earle

C. fracidus (Fr.) Donk
A single collection from Iona (E); often called *Drosella fracida* (Fr.) Sing. (nom. illeg.).

Clitocybe (Fr.) Staude

C. brumalis (Fr. ex Fr.) Quél.

C. clavipes (Pers. ex Fr.) Kummer

C. dicolor (Pers.) Lange

C. dealbata (Sow. ex Fr.) Kummer

C. ditopus (Fr. ex Fr.) Gill.

C. flaccida (Sow. ex Fr.) Kummer
Widespread (E)

C. fragans (Sow. ex Fr.) Kummer
Widespread (E)

C. langei Sing. ex Hora
(E); a mixture of at least two taxa

C. metachroa (Fr.) Kummer

C. nebularis (Batsch ex Fr.) Kummer
Widespread (E)

C. obsoleta (Batsch ex Fr.) Quél.
(E)

C. phyllophila (Fr.) Kummer

C. quercina Pearson
(E)

C. rivulosa (Pers. ex Fr.) Kummer

C. suaveolens (Schum. ex Fr.) Kummer
(E)

C. umbilicata (Schaeff. ex Fr.) Kummer
(E)

C. vibecina (Fr.) Quél.
Frequent but often confused with *C. langei* (E)

Clitopilus (Fr. ex Rabenh.) Kummer

C. hobsonii (Berk. & Br.) Orton
(E)

C. prunulus (Scop. ex Fr.) Kummer
Widespread (E)

Collybia (Fr.) Staude
See also *Tephrocybe*

C. butyracea (Bull. ex Fr.) Kummer
Frequent (K)

C. cirrhata (Schum. ex Fr.) Kummer
Frequent (E)

C. confluens (Pers. ex Fr.) Kummer
Widespread (E, K); recorded by White

C. dryophila (Bull. ex Fr.) Kummer
Common and widespread (E, K)

C. maculata (Alb. & Schw. ex Fr.) Kummer
Common and widespread; recorded by White

C. peronata (Bolt. ex Fr.) Kummer

C. striaepilea (Fr.) Orton

C. tuberosa (Bull. ex Fr.) Kummer
(E)

Cyphellopsis Donk

C. anomala (Pers. ex Fr.) Donk
On *Fagus* (K)

C. monacha (Speg.) Reid
On *Ulex* (K)

Cystoderma Fayod

C. amianthinum (Scop. ex Fr.) Fayod
Very common and widespread (E, K)

C. carcharias (Pers. ex Secr.) Fayod
(E)

C. granulosum (Batsch ex Fr.) Fayod
Recorded only by White; possibly nothing more than *C. amianthinum*

Delicatula Fayod

D. cephalotricha (Joss.) Cejp
Mycena cephalotricha Joss.
(E)

Dermoloma Lange ex Herink

D. atrocinereum (Pers. ex Pers.) Orton
(E)

Flammulina Karst.

F. velutipes (Curt. ex Fr.) Karst.
Rare

Hygrophoropsis (Schroet.) Maire

H. aurantiaca (Wulf. ex Fr.) Maire var. **aurantiaca**
Very common and widespread (E)

H. aurantiaca var. **pallida** (Cooke) Kühner & Romagn.
Common and widespread (E) a very distinct taxon

Laccaria Berk. & Br.

L. amethystea (Bull. ex Mérat) Murrill
Very common and widespread in woods (E)

L. bicolor (Maire) Orton
Common (E)

L. laccata (Scop. ex Fr.) Cooke
Very common and ubiquitous (K); recorded by White. A very variable fungus

L. proxima (Boud.) Pat.
Common in damp areas (E)

L. striatula (Peck) Peck
(E)

L. tortilis (Bolt. ex S. F. Gray) Cooke
(E)

Lachnella Fr.

L. alboviolascens (Alb. & Schw. ex Pers.) Fr.
On *Ulex* (K)

L. villosa (Pers. ex Fr.) Gill.
On *Cirsium palustre* (K)

Lepista (Fr.) W. G. Smith

L. irina (Fr.) Bigelow

L. nuda (Bull. ex Fr.) Cooke

L. sordida (Fr.) Sing.

Leptoglossum (Cooke) Sacc.

L. retiruge (Fr.) Kühner & Romagn.
On moss, Aros Park Forest (E)

L. rickenii Sing. ex Hora
On decaying *Peltigera*, an unusual habitat (E)

Lyophyllum Karst.

L. decastes (Fr. ex Fr.) Sing.

L. loricatum (Fr.) Kühner

Marasmius Fr.

M. androsaceus (L. ex Fr.) Fr.
Common and widespread (K); recorded by White

M. bulliardii Quél.
(Clark)

M. graminum (Libert) Berk.

M. oreades (Bolt. ex Fr.) Fr.
Common in grassland

M. ramealis (Bull. ex Fr.) Fr.
(K)

M. rotula (Scop. ex Fr.) Fr.
Common in mixed woodland (K)

Melanoleuca Pat.

M. strictipes (Karst.) Schaeff.
(E)

Mycena Pers. ex Gray

M. acicula (Schaeff. ex Fr.) Kummer
(K)

M. aetites (Fr.) Quél.
(E)

M. alcalina (Fr. ex Fr.) Kummer
Common on conifer wood

M. amicta (Fr.) Quél.
Common in conifer plantations

M. bulbosa (Cejp) Kühner
On *Juncus* (E)

M. capillaripes Peck
Common in conifer plantations (E)

M. chlorantha (Fr. ex Fr.) Kummer

M. citrinomarginata Gill.
(E)

M. coccinea Sow. ex Quél.

M. corticola (Pers. ex Fr.) Gray

M. delectabilis (Peck) Sacc.

M. epipterygia (Scop. ex Fr.) Gray
Common and widespread

M. fibula (Bull. ex Fr.) Kühner
Frequent and ubiquitous (E)

M. filopes (Bull. ex Fr.) Kummer
Common in mixed woodland (E)

M. flavoalba (Fr.) Quél.
Common in grassland

M. galericulata (Scop. ex Fr.) Gray
Frequent and widespread on old wood

M. galopus (Pers. ex Fr.) Kummer
Common and ubiquitous

M. haematopus (Pers. ex Fr.) Kummer
Common and widespread on hardwood stumps (E)

M. hiemalis (Osbeck ex Fr.) Quél.

M. inclinata (Fr.) Quél.
Common and widespread on stumps of *Quercus* (E)

M. lactea (Pers. ex Fr.) Kummer

M. leptocephala (Pers. ex Fr.) Gill.
Common and ubiquitous (K)

M. leucogala (Cooke) Sacc.
(K)

M. lineata (Bull. ex Fr.) Kummer

M. maculata Karst.

M. metata (Fr. ex Fr.) Kummer
Common and ubiquitous (K)

M. olivaceomarginata (Massee) Massee
(K)

M. pearsoniana Dennis

M. pelianthina (Fr.) Quél.

M. polygramma (Bull. ex Fr.) Gray
Frequent; attached to twigs (K)

M. pterigena (Fr. ex Fr.) Kummer
Abundant on fern litter, Aros Park Forest (E)

M. pura (Pers. ex Fr.) Kummer var. **pura**
Very common (E, K); recorded by White

M. pura var. **alba** Gill.
Ulva

M. purpureofusca Peck
(E)

M. rorida (Scop. ex Fr.) Quél.

M. rubromarginata (Fr. ex Fr.) Kummer
Occasional; recorded by White

M. sanguinolenta (Alb. & Schw. ex Fr.) Kummer
Very common and ubiquitous (K)

M. speirea (Fr. ex Fr.) Gill.
(E)

M. stylobates (Pers. ex Fr.) Kummer
(E)

M. swartzii (Fr. ex Fr.) A. H. Smith

M. tenerrima (Berk.) Sacc.
(E)

M. uracea Pearson

M. vitilis (Fr.) Quél.
Frequent in woodlands

M. vulgaris (Pers. ex Fr.) Kummer

Nyctalis Fr.

N. asterophora Fr.
(E)

N. parasitica (Bull. ex Fr.) Fr.
(E)

Omphalina Quél.

See *Botrydina* & *Coriscium* (Lichens, p. 14.58) and *Leptoglossum*

O. epichysium (Pers. ex Fr.) Quél.
(K)

O. ericetorum (Pers. ex Fr.) M. Lange
Very common and widespread (E, K). There is little doubt that this is a complex of micro-species.

O. hepatica (Fr. ex Fr.) Orton
(K)

O. hudsoniana (Jennings) Bigelow
O. luteolilacina (Favre) Henderson
(E)

O. luteovitellina (Pilát & Nannf.) M. Lange
Frequent on algal scum on wet exposed peat (K)

O. mutila (Fr.) Orton
(K)

O. oniscus (Fr. ex Fr.) Quél.
(E)

O. philonotis (Lasch) Quél.
(K)

O. pyxidata (Bull. ex Fr.) Quél.

O. sphagnicola (Berk.) Moser
In *Sphagnum*, Grass Point (E)

O. velutina (Quél.) Quél.

Oudemansiella Speg.

O. mucida (Schrad. ex Fr.) Höhn.
Common, particularly on *Fagus*

O. radicata (Relhan ex Fr.) Sing.

Panellus Karst.

P. stipticus (Bull. ex Fr.) Karst.
Frequent on hardwood (K)

Phyllotus Karst.

P. porrigens (Pers. ex Fr.) Karst.
Pleurotellus porrigens (Pers. ex Fr.) Kühner & Romagn.
Common and widespread on conifer wood (E, K).

Pleurotellus Fayod

P. dictyorrhizus (DC. ex Fr.) Kühner
(K)

Resupinatus Nees ex Gray

R. applicatus (Batsch ex Fr.) Gray

Squamanita Imbach

S. paradoxa (A. H. Smith & Sing.) Bas
(E)
Gruline House (E). The first British record.

Strobilurus Sing.

S. tenacellus (Pers. ex Fr.) Sing.
Pseudohiatula tenacella (Pers. ex Fr.) Métrod.
(E)

Tephrocybe Donk
(*Collybia* p.p.)

T. atrata (Fr. ex Fr.) Donk

T. carbonaria (Velen.) Donk
(E)

T. palustris (Peck) Donk
Common and widespread (E, K)

Tricholoma (Fr.) Staude

T. acerbum (Bull. ex Fr.) Quél.
Frequent and widespread, particularly under *Quercus* (E)

T. albobrunneum (Pers. ex Fr.) Kummer

T. argyraceum (Bull. ex St. Amans) Gill.
Common in mixed woodland (K)

T. aurantium (Schaeff. ex Fr.) Ricken

T. columbetta (Fr.) Kummer
(E, K)

T. fulvum (DC. ex Fr.) Sacc.
Common and widespread under *Betula* (E)

T. imbricatum (Fr. ex Fr.) Kummer

T. lascivum (Fr.) Gill.

T. pessundatum (Fr.) Quél.

T. portentosum (Fr.) Quél.
Frequent in mixed woodland

T. psammopus (Kalchbr.) Quél.
(E)

T. resplendens (Fr.) Karst.

T. saponaceum (Fr.) Kummer var. **saponaceum**
Common and widespread in mixed woodland (E)

T. saponaceum var. **squamosum** (Cooke) Rea
(E)

T. sejunctum (Sow. ex Fr.) Quél.
(K)

T. sulphureum (Bull. ex Fr.) Kummer

T. terreum (Schaeff. ex Fr.) Kummer

T. ustale (Fr. ex Fr.) Kummer

T. vaccinum (Pers. ex Fr.) Kummer
(K)

T. aff. **viridolutescens** Moser
Salen (E). Further collections of this agaric are required to ascertain whether it is the same as Moser's species.

Tricholomopsis Sing.

T. decora (Fr.) Sing.
Penmore Mill (E)

T. rutilans (Schaeff. ex Fr.) Sing.
Common on conifer wood; recorded by White

ENTOLOMATACEAE

Entoloma (Fr.) Kummer

E. ameides (Berk. & Br.) Sacc.
Widespread (E)

E. fuscomarginatum Orton
(E)

E. cf. **griseoluridum** Kühner & Romagn.
This fungus needs further examination in the field

E. helodes (Fr.) Kummer
(E)

E. jubatum (Fr.) Karst.

E. nidorosum (Fr.) Quél.

E. nitidum Quél.
Common

E. porphyrophaeum (Fr.) Karst.
Widespread (K)

E. sericatum (Britz.) Sacc.
Common in wet areas

E. turbidum (Fr.) Quél.
Common (K)

Leptonia (Fr.) Kummer

L. anatina (Lasch) Kummer
(E)

L. atromarginata (Romagn. & Favre) Konrad & Maubl.
(E)

L. caerulea Orton
(E)

L. caesiocincta (Kühner) Orton
(E)

L. chalybaea (Pers. ex Fr.) Kummer
Two collections only (E); recorded by White

L. corvina (Kühner) Orton
(E)

L. cyanoviridescens Orton
(E)

L. euchroa (Pers. ex Fr.) Kummer
Two collections from Aros (E)

L. fulva Orton
Frequent and widespread (E)

L. griseocyanea (Fr. ex Fr.) Orton
(E)

L. lampropus (Fr. ex Fr.) Quél.
Common although may be a mixture of taxa (E)

L. lazulina (Fr.) Quél.

L. lividocyanula (Kühner) Orton
(E)

L. polita (Pers. ex Fr.) Konrad & Maubl.

L. aff. **pyrospila** Romagn. ex Orton

L. querquedula (Romagn.)
Rhodophyllus querquedulus Romagn. Penmore Mill. Apparently unrecorded from the British Isles; not yet transferred to *Leptonia.*

L. rhombispora (Kühner & Bois., sub *Rhodophyllus)*
Ulva (E)

L. aff. rosea Longyear
A distinctive fungus combining features of both *L. sericella* and *L. rosea*; found elsewhere in Scotland

L. sericella (Fr. ex Fr.) Barbier (incl. var. *laevipes* Maire)
Common and widespread (E)

L. turci Bres.
(E)

L. xanthochroa Orton
Salen woods (E). Additional to the British List

Leptonia species nova
Ardura (E); a distinctive fungus considered by Orton as undescribed

Nolanea (Fr.) Kummer

N. cetrata (Fr. ex Fr.) Kummer
Very common and widespread (E)

N. cucullata (Favre) Orton
(K)

N. cuneata Bres.
Common and widespread

N. cuspidifer (Kühner & Romagn.) Orton
(E)

N. farinolens Orton

N. icterina (Fr.) Kummer
(E)

N. infula (Fr.) Gill.

N. inutilis (Britz.) Sacc. & Trav.

N. radiata (Lange) Orton

N. sericea (Bull. ex Mérat) Orton
Frequent in grassy areas

N. solstitialis (Fr.) Orton
Frequent and widespread (E)

N. staurospora Bres.
Very common and ubiquitous (K)

N. vernus Lundell
Cnoc an Teine (E). Not in the British List although apparently widespread.

CORTINARIACEAE

(Includes Crepidotaceae sensu Singer)

Cortinarius Pers. ex Gray

C. anomalus (Fr.) Fr.

C. bibulus Quél.
Under *Alnus*, Aros Park Forest (E); a new record for the British Isles.

C. bolaris (Pers. ex Fr.) Fr.
Common.

C. caninus (Fr.) Fr.

C. cinnabarinus Fr.

C. cinnamomeobadius R. Henry
(E)

C. cinnamomeus (L. ex Fr.) Fr.

C. croceofolius Peck
(E)

C. crocolitus Quél.

C. cyanites Fr.
Recorded by White as *C. cyanipes* Fr.

C. cyanopus (Secr.) Fr.

C. decoloratus (Fr.) Fr.
Salen woods and Loch Bà (E); the only recent authentic British records

C. evernius (Fr. ex Fr.) Fr.

C. fasciatus Scop. ex Fr.

C. flexipes (Pers. ex Fr.) Fr.

C. gentilis (Fr.) Fr.
(E)

C. glandicolor (Fr.) Fr.

C. helvelloides (Fr.) Fr.
(K)

C. hemitrichus (Pers. ex Fr.) Fr.

C. herpeticus Fr.
A very interesting find from Salen wood (E)

C. lepidopus Cooke
Common and widespread

C. largus Fr.

C. limonius (Fr. ex Fr.) Fr.
Two interesting records from Ardura and Salen of a poorly known agaric (E)

C. microspermus Lange

C. mucosus (Bull. ex Fr.) Kickx

C. obtusus (Fr.) Fr.
Common and widespread; probably a mixture of several taxa

C. pinicola Orton
(E); a similar taxon was found near Ardnadro-shet in *Quercus* woods.

C. porphyropus Fr.

C. pseudosalor Lange
Common and widespread (K)

C. puniceus Orton

C. purpurascens Fr.
Recorded only by White

C. rigidus (Scop. ex Fr.) Fr.

C. rubicundulus (Rea) Pearson

C. saniosus (Fr.) Fr.

C. scaurus (Fr. ex Fr.) Fr.

C. semisanguineus (Fr.) Gill.

C. simulatus Orton

C. speciosissimus Kühner & Romagn.

C. subtriumphans R. Henry ex Orton

C. tabularis (Bull. ex Fr.) Fr.

C. torvus (Fr. ex Fr.) Fr.
(E)

C. turbinatus (Bull. ex Fr.) Fr.

C. uliginosus Berk.

C. varius (Schaeff. ex Fr.) Fr.

C. vibratilis (Fr.) Fr.

C. violaceus (L. ex Fr.) Fr.
(E)

Cortinarius species novae
Two specimens (E) are considered to be new species by P. D. Orton

Crepidotus (Fr.) Staude

C. applanatus (Pers. ex Pers.) Kummer

C. autochthonus Lange

C. calolepis (Fr.) Karst.
(K)

C. cesatii (Rabenh.) Sacc.
(K)

C. luteolus (Lambotte) Sacc.

C. mollis (Schaeff. ex Fr.) Kummer
Common and widespread (K)

C. subsphaerosporus (Lange) Kühner and Romagn.

C. subtilis Orton
(E)

Flammulaster Earle

F. ferruginea (Maire) Watling
A single collection (K)

F. subincarnata (Joss. & Kühner) Watling

Galerina Earle

G. calyptrata Orton
(E)

G. clavata (Vel.) Kühner

G. graminea (Vel.) Kühner
(K)

G. hypnorum (Schrank ex Fr.) Kühner
Very common; recorded by White

G. mniophila (Lasch) Kühner
Common (E)

G. mutabilis (Schaeff. ex Fr.) Orton
Rare (E)

G. mycenopsis (Fr. ex Fr.) Kühner
Frequent

G. paludosa (Fr.) Kühner
Frequent and widespread (E, K)

G. sphagnorum (Pers. ex Fr.) Kühner
(E)

G. tibiicystis (Atk.) Kühner
(K)

G. vittaeformis (Fr.) Moser
Common and widespread (K); a form with larger spores illustrated in Lange (1939) was also collected.

Gymnopilus Karst.

G. fulgens (Favre & Maire) Sing.
(E)

G. junonius (Fr.) Orton
(K)

G. penetrans (Fr. ex Fr.) Murr.
Common and widespread

Hebeloma (Fr.) Kummer

H. crustuliniforme (Bull. ex St. Amans) Quél.
(E); probably a mixture of taxa

H. leucosarx Orton

H. aff. leucosarx Orton
Differs in spore-size and morphology

H. mesophaeum (Pers.) Quél.
(E)

H. sacchariolens Quél.
Common (E)

H. sinapizans (Paulet ex Fr.) Gill.

H. sinuosum (Fr.) Quél.

Inocybe (Fr.) Fr.

I. abjecta (Karst.) Sacc.
(E)

I. asterospora Quél.
(K)

I. bongardii (Weinm.) Quél.

I. calamistrata (Fr.) Gill.
Common and widespread (E)

I. calospora Quél. apud Bres.
(E)

I. casimiri Vel.
(K)

I. eutheles (Berk. & Br.) Quél.
(E)

I. fastigiata (Schaeff. ex Fr.) Quél. var. **fastigiata**
(K)

I. fastigiata var. **umbrinella** (Bres.) Heim.
(K)

I. fibrosoides Kühner
(K)

I. flocculosa (Berk.) Sacc.
(K)

I. geophylla (Sow. ex Fr.) Kummer var. **geophylla**
Common and widespread (E, K)

I. geophylla var. **lilacina** Gill.

I. godeyi Gill.

I. grammata Quél.
(K)

I. hystrix (Fr.) Karst.
(Clark, K)

I. lucifuga (Fr. ex Fr.) Kummer

I. maculata Boud.

I. mixtilis (Britz.) Sacc.
(E)

I. napipes Lange
(K)

I. obscuroides Orton
(E)

I. petiginosa (Fr. ex Fr.) Gill.

I. posterula (Britz.) Sacc.
(E)

I. praetervisa Quél.
(E)

I. pudica Kühner
In Aros Park (E)

I. pusio Karst.
(E)

I. salicis Kühner

I. umbrina Bres.

Naucoria (Fr.) Kummer

N. bohemica Vel.
(E)

N. celluloderma Orton
(E)

N. escharoides (Fr. ex Fr.) Kummer
Common under *Alnus* (E)

N. scolecina (Fr.) Quél.
Frequent under *Alnus* (E)

N. striatula Orton
(K)

N. subconspersa Kühner ex Orton
(K)

N. zetlandica Orton

Pellidiscus Donk

P. pallidus (Berk. & Br.) Donk
On *Ulex* (K)

Phaeomarasmius Scherffel

P. erinaceus (Fr.) Kühner
(E)

Pholiota (Fr.) Kummer

P. alnicola (Fr.) Sing.

P. aff. **dissimulans** (Berk. & Br.) Sacc.
(E)

P. flammans (Fr.) Kummer
Frequent on conifer wood (E, K)

P. highlandensis (Peck) A. H. Smith
P. carbonaria (Fr. ex Fr.) Sing.
Frequent at sites of old bonfires (K)

P. lubrica (Pers. ex Fr.) Sing.

P. myosotis (Fr. ex Fr.) Sing.
Frequent in boggy areas (E)

P. scamba (Fr. ex Fr.) Moser
(E)

P. squarrosa (O. Müll. ex Fr.) Kummer
(E)

BOLBITIACEAE

Agrocybe Fayod

A. arvalis (Fr.) Sing.
Widespread

A. dura (Bolt. ex Fr.) Sing.

A. erebia (Fr.) Kühner
Widespread (E)

Bolbitius Fr.

B. vitellinus (Pers. ex Fr.) Fr.
Widespread (K)

Conocybe Fayod

C. coprophila (Kühner) Kühner

C. dunensis Wallace

C. filaris (Fr.) Kühner
(K)

C. ochracea (Kühner) Sing. forma **ochracea**

C. ochracea forma **macrospora** Kühner

C. pseudopilosella (Kühner) Kühner & Romagn.

C. pubescens (Gill.) Kühner
Common and widespread (E)

C. rickeniana Sing. ex Orton
(K)

C. tenera (Schaeff. ex Fr.) Kühner

C. vexans Orton

Conocybe species nova
A very distinct species apparently undescribed: characterised by a striking contrast between gill and pileus colours (E)

Conocybe species nova
A widespread apparently undescribed species growing on dung. It is characterised by distinctly angled spores.

STROPHARIACEAE

Deconica (W. G. Smith) Karst.

D. coprophila (Bull. ex Fr.) Karst.
Common and widespread (K)

D. subcoprophila (Britz.) Sacc.
On dung (E)

Hypholoma (Fr.) Kummer

H. capnoides (Fr. ex Fr.) Kummer
Rare (E)

H. elongatum (Pers. ex Fr.) Ricken
Common and widespread

H. ericaeum (Pers. ex Fr.) Kühner
(E)

H. ericaeoides Orton

H. fasciculare (Huds. ex Fr.) Kummer
Very common and widespread (K); recorded by White

H. marginatum (Pers. ex Fr.) Schroet.

H. polytrichi (Fr. ex Fr.) Ricken
(K)

H. radicosum Lange
(E)

H. udum (Pers. ex Fr.) Kühner

Psilocybe (Fr.) Kummer

P. semilanceata (Fr. ex Secr.) Kummer
Frequent in grassy areas; recorded by White

Stropharia (Fr.) Quél.

S. aeruginosa (Curt. ex Fr.) Quél.
Two recent collections; recorded by White

S. coronilla (Bull. ex Fr.) Quél.

S. fimetaria Orton
Glen Aros, Penmore Mill and Ulva (E); recently described species possibly better placed in *Psilocybe*; found previously in central Scotland

S. merdaria (Fr.) Quél.
(E, K); the source of material for a study in polymorphism (see Watling, 1971)

S. semiglobata (Batsch ex Fr.) Quél.
Abundant and widespread on dung (E, K); recorded by White

S. squamosa (Pers. ex Fr.) Quél.
(K)

COPRINACEAE

Coprinus (Pers. ex Fr.) Gray

C. atramentarius (Bull. ex Fr.) Fr.

C. bisporus Lange
(K)

C. cineratus Quél.
cf. *C. semitalis* Orton
Salen and Knockvologan

C. cinereus (Schaeff. ex Fr.) Gray

C. congregatus (Bull. ex St. Amans) Fr.
Widespread and frequent (E)

C. ephemeroides (Bull. ex Fr.) Fr.
(E)

C. exstinctorius (Bull. ex St. Amans) Fr.

C. flocculosus DC. ex Fr.
(E)

C. friesii Quél.

C. hemerobius Fr.

C. heptemerus M. Lange & A. H. Smith

C. lagopus (Fr.) Fr.
(E)

C. leiocephalus Orton
(E); in the British List as *C. plicatilis* var. *microsporus* Kühner

C. lipophilus Heim & Romagn.
Dishig (E). Not in the British List; a member of the *C. patouillardii* complex.

C. macrocephalus (Berk.) Berk.

C. micaceus (Bull. ex Fr.) Fr.
Frequent and widespread

C. miser (Karst.) Karst.
Common (E, K)

C. narcoticus (Batsch ex Fr.) Fr.

C. niveus (Pers. ex Fr.) Fr.

C. nudiceps Orton
(E)

C. patouillardii Quél.
Widespread (E, K); this is a complex of taxa

C. pellucidus Karst.

C. plicatilis (Curt. ex Fr.) Fr.
See also *C. leiocephalus* above

C. radiatus (Bolt. ex Fr.) S. F. Gray
(K)

C. sclerotiger Watling
A member of the *C. tuberosus* Quél. complex Glen Forsa (E)

C. stellatus Bull.

C. stercoreus (Bull.) Fr.
This is included under *C. stercorarius* Fr. in the British List. A form with almost smooth sphaerocytes was also collected.

C. sterquilinus (Fr.) Fr.

C. velox Godey

C. vermiculifer Joss. ex Dennis

Panaeolina Maire

P. foenisecii (Pers. ex Fr.) Maire
Common and widespread in grassy areas (E, K)

Panaeolina species nova?
Resembling *P. foenisecii* but for the almost smooth spores; found on the Treshnish Isles (K)

Panaeolus (Fr.) Quél.

P. acuminatus (Schaeff. ex Secr.) Quél.

P. ater (J. Lange) Kühner & Romagn.

P. campanulatus (Bull. ex Fr.) Quél.

P. olivaceus Möller

P. papilionaceus (Bull. ex Fr.) Quél.
One recent record, Salen (K); recorded by White

P. rickenii Hora
Frequent in grassy areas (E)

P. semiovatus (Sow. ex Fr.) Lundell
Very common and widespread on dung (E, K)

P. speciosus Orton
A recently described species from Central Scotland; a closely related taxon with a strongly purple-flushed but smaller pileus was collected on dung at Pennyghael (E)

P. sphinctrinus (Fr.) Quél.
Very common and widespread on dung; a closely related taxon but with smaller basidiospores (11–12 × 7–8 × 6–5–7.0 μm) was collected on rabbit pellets at Calgary (K)

P. subbalteatus (Berk. & Br.) Sacc.
(E)

P. aff. subbalteatus
Apparently undescribed although a widespread fungus; see Watling & Richardson (1971)

Psathyrella (Fr.) Quél.

P. albidula (Romagn.) Moser

P. atomata (Fr.) Quél.

P. candolleana (Fr.) Maire
Frequent on rotten stumps (K)

P. conopilea (Fr.) Pearson & Dennis
(K)

P. coprobia (Lange) A. H. Smith
(E)

P. cotonea (Quél.) Konrad & Maubl.

P. flexispora Orton

P. gracilis (Fr.) Quél.

P. hydrophila (Bull. ex Mérat) Maire
Frequent and widespread (K)

P. obtusata (Fr.) A. H. Smith
(K)

P. pennata (Fr.) Pearson & Dennis

P. pilulaeformis (Romagn.) Orton
Aros Park (E); not in the British List

P. spadiceogrisea (Fr.) Maire
(E)

P. aff. stercoraria Kühner & Joss.

P. subnuda (Karst.) A. H. Smith
(E)

P. xanthocystis Orton

AGARICACEAE

Agaricus L. ex Fr.

A. altipes (Möller) Möller
(E); apparently a new British record

A. arvensis Schaeff. ex Secr.

A. augustus Fr.
(E)

A. bisporus (Lange) Pilát

A. campestris L. ex Fr.
Widespread (K); includes var. *equestris* Möller

A. fissuratus (Möller) Möller
Frachadie. Not in the British List but widespread in Scotland in maritime pastures

A. haemorrhoidarius Schulzer
(E)

A. langei (Möller) Möller
Widespread (E)

A. macrosporus (Möller ex Schaeff.) Pilát

A. porphyrocephalus Möller
Widespread

A. silvaticus Schaeff. ex Secr.

A. silvicola (Vitt.) Peck

Agaricus sp.
An undetermined species from the upper shore grasslands of Loch Spelve and Loch na Keal (E)

LEPIOTACEAE

Lepiota Pers. ex Gray

L. cristata (Fr.) Kummer
(E)

L. felina (Pers. ex Fr.) Karst.

L. cf. **rhacodes** (Vitt.) Quél.
(E); differs in structure of cap-scales and larger spores (13–15.5 × 9–10 μm)

AMANITACEAE

Amanita Pers. ex Hook.

A. citrina Schaeff. ex Gray

A. excelsa (Fr.) Kummer

A. fulva Schaeff. ex Secr.
Widespread (E, K); fairly constant in colour but a yellow ochraceous form similar in some respects to *A. crocea* (Quél.) Kühn. & Romagn. was found.

A. muscaria (L. ex Fr.) Hook.
Widespread; recorded by White

A. rubescens (Fr.) Gray
Common (K); recorded by White

A. umbrinolutea Secr.
Loch Buie. Not in the British List but apparently widespread in the British Isles

A. vaginata (Bull. ex Fr.) Vitt.
(K)

A. virosa Secr.
Widespread (K)

PLUTEACEAE

Pluteus Fr.

P. cervinus (Schaeff. ex. Fr.) Kummer
Very common and widespread on hardwood stumps (K)

P. phlebophorus (Ditmar ex Fr.) Kummer

P. salicinus (Pers. ex Fr.) Kummer
Frequent on hardwood (K)

RUSSULACEAE

Lactarius Pers. ex Gray

L. acris (Bolt. ex Fr.) Gray

L. aspideus (Fr. ex Fr.) Fr.
(E)

L. blennius (Fr. ex Fr.) Fr.
Common and widespread: recorded by White

L. camphoratus (Bull. ex Fr.) Fr.
Frequent in woodland

L. chrysorrheus Fr.
Common in *Quercus* woods

L. cimicarius (Batsch ex Secr.) Gill.
(E)

L. circellatus Fr.

L. cyathula (Fr.) Fr.
(E)

L. deliciosus (L. ex Fr.) Gray
Common, widespread (K); recorded by White. Some confusion exists in respect to the differences between this species and *L. deterrimus* Gröger (see Gröger, 1968).

L. flavidus Boud.

L. flexuosus (Pers. ex Fr.) Gray
(E)

L. fluens Boud.
Dervaig. Not previously recorded for Britain but sufficiently different from *L. blennius* to be recognised as distinct.

L. fuliginosus (Fr.) Fr.

L. glaucescens Crossland
Probably as common as *L. piperatus* with which it is often confused

L. glyciosmus (Fr. ex Fr.) Fr.
Common in *Betula* woods (E)

L. helvus (Fr.) Fr.

L. hysginus (Fr. ex Fr.) Fr.
(E)

L. lacunarum Romagn. ex Hora
(E)

L. mitissimus (Fr.) Fr.
Common and widespread (E); recorded by White

L. obscuratus (Lasch.) Fr.
Frequent under *Alnus* (E, K)

L. pallidus (Pers. ex Fr.) Fr.

L. piperatus (Scop. ex Fr.) Gray
(K)

L. pterosporus Romagn.
Although not in the British List a fairly widespread fungus. Aros Park (E)

L. pyrogalus (Bull. ex Fr.) Fr.
Common in woodland

L. quietus (Fr.) Fr.
Very common in *Quercus* woods

L. representanaeus Britz.
(E)

L. resimus (Fr.) Fr.

L. rufus (Scop. ex Fr.) Fr.
Common in woodland on podsols (E, K)

L. serifluus (DC. ex Fr.) Fr.

L. subdulcis (Pers. ex Fr.) Gray
Common and widespread (K); recorded by White

L. subsalmoneus Pouzar
(E)

L. tabidus Fr.
Very common and widespread (K)

L. torminosus (Schaeff. ex Fr.) Gray
Common in *Betula* woods; recorded by White

L. turpis (Weinm.) Fr.

L. uvidus (Fr. ex Fr.) Fr.
Common and widespread

L. vellereus (Fr.) Fr.
Frequent and widespread; recorded by White

L. vietus (Fr.) Fr.
Common in wet *Betula* woods.

L. volemus (Fr.) Fr.
Frequent and widespread (E, K)

Russula (Pers, ex Fr.) Gray

R. adusta (Pers. ex Fr.) Fr.
(K)

R. aeruginea Lindblad ex Fr.
(K)

R. albonigra (Krombh.) Fr.
(K); this is *R. anthracina* sensu Romagnesi (1967)

R. aquosa Leclair

R. atropurpurea (Krombh.) Britz.
Frequent particularly in *Quercus* woods

R. aurantiaca (J. Schaeff.) Moser
A rare fungus growing with *Betula* (E). Not in the British List.

R. aurata With. ex Fr.
(K)

R. azurea Bres.

R. betularum Hora
Common and widespread under *Betula* (K)

R. caerulea Fr.
Common and widespread under conifers (E, K)

R. claroflava Grove
Common and widespread in wet *Betula* woods (K)

R. cyanoxantha (Schaeff ex Secr.) Fr. forma **cyanoxantha**
Very common and widespread in woodland (K)

R. cyanoxantha forma **pallida** Sing.
(K)

R. delica Fr.
Very common and widespread in woodland (K)

R. densifolia (Secr.) Gill.
R. acrifolia Romagn.

R. emetica (Schaeff. ex Fr.) Gray
Recorded by White. A much confused species. Only records based on carefully examined material have been included.

R. emeticella (Sing.) Hora
Frequent and widespread in mixed woodland

R. erythropus Peltereau

R. fellea (Fr.) Fr.
Very common and widespread under *Fagus* (K)

R. firmula J. Schaeff.

R. foetens (Pers. ex Fr.) Fr.
Frequent and widespread (E, K); recorded by White

R. fragilis Pers. ex Fr. forma **fragilis**
Common and widespread in mixed woodland (K)

R. fragilis forma **nivea** (Pers.) Cooke

R. gracillima J. Schaeff.

R. grisea (Pers. ex Secr.) Fr.

R. heterophylla (Fr.) Fr.
Frequent and widespread; recorded by White

R. integra L. ex Fr.

R. knauthii (Sing.) Hora

R. laurocerasi Melzer

R. lepida Fr.
(K)

R. lilacea Quél.
(E)

R. lutea (Huds. ex Fr.) Gray
Common and widespread in woodland (K)

R. mairei Sing. forma **mairei**
Very common and widespread under *Fagus*

R. mairei forma **fageticola** (Melzer) Romagn.
Differs in only minor details from the type form (K)

R. nigricans (Bull. ex Mérat) Fr.
Common and widespread; recorded by White (K)

R. nitida (Pers. ex Fr.) Fr.
Frequent and widespread in wet woodland (K)

R. ochroleuca (Pers. ex Secr.) Fr.
Very common and widespread in woodland (K)

R. paludosa Britz.

R. parazurea J. Schaeff.

R. pectinata (Bull. ex St. Amans) Fr.
(K)

R. persicina Kromb.
Penmore Mill. Not previously recorded from Britain but apparently widespread

R. puellaris Fr.
Common and widespread under conifers

R. queletii Fr.
Frequent under conifers (K); recorded by White

R. rosea Quél.
(K)

R. rutila Romagn.
Kilninian (E). Not in the British List

R. sanguinea (Bull. ex St. Amans) Fr.
Frequent under conifers (K)

R. sardonia Fr. var. **sardonia**
Common and widespread under conifers (K, E)

R. sardonia var. **mellina** Melzer
A common variety lacking purple colours in pileus

R. sororia (Fr.) Romell
R. amoenolens Romagn. (see Romagnesi, 1967) (K)

R. velenovskyi Melzer & Zvara

R. versicolor J. Schaeff.

R. vesca Fr.
Very common and widespread in mixed woodland (K)

R. veternosa Fr.

R. xerampelina (Schaeff. ex Secr.) Fr.
(K)

APHYLLOPHORALES

Bibliogr.: Christiansen (1960); Donk (1964).

LENTINELLACEAE

Lentinellus Karst.

L. cochleatus (Pers. ex Fr.) Karst.
Frequent in *Fagus* woods (E)

AURISCALPIACEAE

Auriscalpium Gray

A. vulgare S. F. Gray

BANKERACEAE

Phellodon Karst.

P. niger (Fr. ex Fr.) Karst.
Croggan (E)

P. tomentosus (L. ex Fr.) Banker
(E)

CLAVARIACEAE

Bibliogr.: Corner (1950, 1970).

Clavaria Vaill. ex Fr.

C. acuta Fr.
(E)

C. fumosa Pers. ex Fr.
(E)

C. vermicularis Fr.
Common throughout the Island in grassland areas

C. zollingeri Lév.

Clavariadelphus Donk

C. fistulosus (Fr.) Corner var. **contortus** Corner

Clavulinopsis van Overeem

C. corniculata (Fr.) Corner
Very common and widespread in grassland areas (E)

C. fusiformis (Fr.) Corner
Frequent throughout the Island in grassland areas (E)

C. aff. **griseola** (Rea) Corner

C. helvola (Fr.) Corner
Common throughout the Island in grassland areas (E)

C. luteoalba (Rea) Corner

C. umbrinella (Sacc.) Corner
(E)

Pistillaria Fr.

P. micans (Pers.) Fr.

P. setipes Grev.
On moss

Ramariopsis (Donk) Corner

R. kunzei (Fr.) Donk
(E)

Typhula (Pers.) Fr.

Typhula species—see Deuteromycotina: Sclerotium

CLAVULINACEAE

Clavulina Schroet.

C. cinerea (Bull. ex Fr.) Schroet.
Very common and widespread in woodland (E, K)

C. cristata (Holmsk. ex Fr.) Schroet. var. **cristata**
Common and widespread in woodland (E); recorded by White

C. cristata var. **coralloides** Corner
(E)

C. rugosa (Bull. ex Fr.) Schroet var. **rugosa.**
Common and widespread in woodlands (E); recorded by White

C. rugosa var. **alcynoria** Corner
Frequent (E)

CONIOPHORACEAE

Serpula Pers. ex Gray

S. lacrymans (Wulf. ex Fr.) Schroet.
Suspected to be present in several house timbers; typical cubical rotting observed

CORTICIACEAE

Bibliogr.: Donk (1959); Julich (1972).

Athelia Pers.

A. epiphylla Pers.
(K)

Athelopsis Parm.

A. lembospora (Bourd.) Parm.
On old wood (K)

Botryobasidium Rick

B. conspersum John Erikss.
(K) Associated with *Acladium:* Deuteromycot.

B. laeve (John Erikss.) Parm.
(K)

Corticium Pers. ex Gray

C. evolvens (Fr.) Fr.
On *Corylus* (K)

C. fuciforme Wakef.
On grass (E)

Cristella Pat.

C. farinacea (Pers. ex Fr.) Donk
On *Fagus* (K)

Gloeocystidiellum Donk

G. porosum (Berk. & Curt.) Donk
(E)

Grandinia Fr.

G. helvetica (Pers.) Fr.
On *Quercus* (K)

G. mutabilis (Pers.) Bourd. & Galz.
On *Alnus* (K)

Hyphoderma Fr.

H. pallidum (Bres.) Donk
On *Larix* (K)

H. roseocremeum (Bres.) Donk
On *Quercus* and *Salix* (K)

H. setigerum (Fr.) Donk
On old wood (E, K)

H. tenue (Pat.) Donk
On *Quercus* (K)

Hyphodontia John Erikss.

H. alutaria (Burt) John Erikss.
On log (K)

H. hastata (Litsch.) John Erikss.

H. pallidula (Bres.) John Erikss.
On *Alnus* (K)

H. papillosa (Fr.) John Erikss.
On logs (K)

H. sambuci (Pers. ex Pers.) John Erikss.
On *Sambucus* (K)

H. subalutacea (Karst.) Erikss.
On *Quercus* (K). A similar taxon apparently intermediate with *H. hastata* also found (E)

Hypochnicium John Erikss.

H. punctulatum (Cooke) John Erikss.
(E)

Meruliopsis Bond. apud Parm.

M. taxicola (Pers.) Bond.
(E)

Merulius Fr.

M. corium Fr.
(K)

Mycoacia Donk

M. fuscoatrata (Fr. ex Fr.) Donk
On *Pinus sylvestris* logs (K)

Odontia Pers. ex Gray

O. bicolor (Alb. & Schw. ex Fr.) Quél.
(K); considered by Parmasto to belong to the genus *Metulodontia*

O. corrugata (Fr.) Bres.
On *Ulex* (K)

Peniophora Cooke

P. cinerea (Fr.) Cooke
On *Betula* (K)

P. incarnata (Pers. ex Fr.) Cooke
On woody debris (E, K)

P. limitata (Chaill. ex Fr.) Cooke
On canker outgrowth of *Fraxinus* (K)

P. lycii (Pers.) Höhn. & Litsch.
Frequent and widespread on old branches (E, K)

P. quercina (Pers. ex Fr.) Cooke
On *Quercus* (E, K)

Phlebia Fr.

P. hydnoides (Cooke & Massee) Christiansen
Frequent and widespread on old branches (K)

P. livida (Pers. ex Fr.) Bres.
On *Fagus* (K)

P. radiata Fr.
(K); records include *P. merismoides* Fr.

P. roumeguerii (Bres.) Donk
On *Quercus* (K)

Radulomyces Christiansen

R. confluens (Fr.) Christiansen
On *Betula* (K)

Sistotrema Fr.

S. brinkmannii (Bres.) Bourd. & Galz.
On *Quercus* (K)

S. commune John Erikss.
On plant debris (K)

Sistotremastrum John Erikss.

S. niveocremeum (Höhn. & Litsch.) John Erikss.
On conifer wood (K)

Vuilleminia Maire

V. comedens (Nees ex Fr.) Maire
On old wood (K)

Xenasma Donk

X. filicinum (Bourd.) Christiansen
On ferns (K)

GANODERMATACEAE

Ganoderma Karst.

G. adpressum (S. Schultz) Donk
G. europaeum Stey.
(K)

G. applanatum (Pers. ex Wallr.) Pat.
(K)

HYDNACEAE

Bibliogr.: Maas Geesteranus (1975).

Hydnum L. ex Fr.

H. repandum L. ex Fr. var. **repandum**
Very common and widespread in woodland (K); recorded by White

H. repandum var. **rufescens** Pers.
Common in woodland (E, K); recorded by White

HYMENOCHAETACEAE

Coltricia Gray

C. perennis (L. ex Fr.) Murr.
Only one recent record (K); recorded by White

Hymenochaete Lév.

H. corrugata (Fr. ex Fr.) Lév.
Common particularly on *Corylus* (E, K)

H. rubiginosa (Dicks.) Lév.

Inonotus Karst.

I. radiatus (Sow. ex Fr.) Karst.
Common and widespread on several hardwoods (K)

Phaeolus Pat.

P. schweinitzii (Fr.) Pat.

Phellinus Quél.

P. ferruginosus (Schrad. ex Fr.) Bourd. & Galz.

P. igniarius (L. ex Fr.) Quél.

P. laevigatus (Fr.) Bourd. & Galz.
On *Betula* (K)

P. nigricans (Fr.) Bourd. var. **subresupinatus** (Lund) Jahn
(K)

P. pomaceus (Pers.) Maire
On *Prunus spinosa* (K)

POLYPORACEAE
Bibliogr.: Pegler (1966).

Bjerkandera Karst.

B. adusta (Willd. ex Fr.) Karst.
On wood (K)

Ceriporia Donk

C. reticulata (Pers. ex Fr.) Domanski
On *Quercus* (K)

C. viridans (Berk. & Br.) Donk
On *Quercus* (K)

Coriolellus Murrill

C. campestris (Quél.) Bond.
On *Corylus* (K)

Coriolus Quél.

C. versicolor (L. ex Fr.) Quél.
Very common on stumps, trunks of hardwoods (K); recorded by White

Daedalea Pers. ex Fr.

D. quercina L. ex Fr.
(K)

Datronia Donk

D. mollis (Sommerf.) Donk
Common on *Fagus* trunks (K)

Fomes (Fr.) Fr.

F. fomentarius (L. ex Fr.) Kickx
Frequent and widespread on *Betula* (K). A second record on *Fagus*, near Gruline.

Gloeophyllum Karst.

G. sepiarium (Wulf. ex Fr.) Karst.

Heterobasidion Bref.

H. annosum (Fr.) Bref.
Very common and widespread on timber of all kinds (K)

Hirschioporus Donk

H. abietinus (Dicks. ex Fr.) Donk
Frequent on conifer branches (K); recorded by White

H. fuscoviolaceus (Ehrenb. ex Fr.) Donk

Oxyporus (Bourd. & Galz.) Donk

O. populinus (Schum. ex Fr.) Donk
Inside hollow stump of *Quercus* (K)

Piptoporus Karst.

P. betulinus (Bull. ex Fr.) Karst.
Common and widespread on *Betula* (K); recorded by White

Podoporia Karst.

P. confluens Karst.
(E)

Polyporus Fr. ex Fr.

P. brumalis Pers. ex Fr.
Frequent and widespread particularly on *Betula* (K); recorded by White

P. lentus Berk.
On *Ulex* (K)

P. melanopus (Swartz ex Fr.) Fr.

P. nummularius (Bull.) Pers.
(K)

P. picipes Fr.
On *Quercus* (K)

P. squamosus Huds. ex Fr.
Common on standing hardwood trees (K)

P. varius Pers. ex Fr.
Common on fallen branches of *Salix* spp. (K)

Pseudotrametes Bond. & Sing. ex Sing.

P. gibbosa (Pers.) Bond. & Sing.
On log of *Fagus* (E, K)

Schizopora Velen.

S. paradoxa (Schrad. ex Fr.) Donk
Xylodon versisporus auct.
Very common and widespread on decaying hardwood (K)

Tyromyces Karst.

T. caesius (Schrad. ex Fr.) Murrill
Common and widespread on hardwood timber (E, K)

T. semipileatus (Peck) Murrill
Common and widespread on hardwood (E, K)

For a conidial state of this genus see *Ptychogaster albus*: Deuteromycotina

SCHIZOPHYLLACEAE

Henningsomyces Sacc.

H. candidus (Pers.) O. Kuntze
On plant debris (K)

SPARASSIDACEAE

Sparassis Fr.

S. crispa (Wulf.) Fr.

STEREACEAE

Stereum Pers. ex S. F. Gray

S. gausapatum Fr.
On *Quercus* (K)

S. hirsutum (Willd. ex Fr.) Fr.
Very common and widespread on hardwood timber (K); recorded by White

S. purpureum (Pers. ex Fr.) Fr.

S. rugosum (Pers. ex Fr.) Fr.
Common and widespread particularly on *Corylus* (K)

S. sanguinolentum (Fr.) Fr.
Frequent and widespread on conifer timber (K)

THELEPHORACEAE

Hydnellum Karst.

H. concrescens (Pers. ex Schw.) Banker
(K)

Thelephora Ehrh. ex Fr.

T. terrestris Ehrh. ex Fr.
(E)

Tomentella Pat.

T. hydrophila Bourd. & Glaz.
On *Salix* (K)

T. mucidula (Karst.) Höhn. & Litsch.
On *Larix* (K)

TULASNELLACEAE

Ceratobasidium D. P. Rogers

C. cornigerum (Bourd.) D. P. Rogers
(K)

Tulasnella Schroet.

T. violea (Quél.) Bourd. & Galz.
On road (K)

TREMELLALES

Bibliogr.: Donk (1966).

AURICULARIACEAE

Basidiodendron Rick

B. eyrei (Wakefield) Luck.
On *Fagus* (K)

Hirneola Velen.

H. auricula-judae (Bull. ex St. Amans) Berk.
Frequent particularly on *Sambucus* (K)

Platygloea Schroet.

P. vestita Bourd. & Galz.
On *Corylus* (K)

TREMELLACEAE

Exidia Fr.

E. albida (Huds. ex Hook.) Bref.
On *Hedera* (K)

E. glandulosa (Bull. ex St. Amans) Fr.
On *Quercus* (K)

Myxarium Wallr.

M. hyalinum (Pers.) Donk

Pseudohydnum Karst.

P. gelatinosum (Scop. ex Fr.) Karst.
Common and widespread on conifer wood (E, K)

Sebacina Tul.

S. effusa (Bref. ex Sacc.) Pat.
(E)

S. fugacissima Bourd. & Galz.
(K)

Stypella Möll.

S. papillata Möll.
On conifers (K)

Tremella Dill. ex Fr.

T. exigua Desm.
On *Ulex* (K)

T. foliacea (Pers. ex S. F. Gray) Pers.
Frequent (K); recorded by White

T. frondosa Fr.
On *Betula* (K)

T. mesenterica Retz. ex Hook.
On woody debris (E, K)

DACRYMYCETALES

Bibliogr.: Reid (1974).

DACRYMYCETACEAE

Calocera (Fr.) Fr.

C. cornea (Batsch ex Fr.) Fr.
Very common particularly on *Fagus* trunks (E, K)

C. viscosa (Pers. ex Fr.) Fr.
Frequent on conifer stumps.

Dacrymyces Nees ex Fr.

D. stillatus Nees ex Fr. (incl. records of *D. deliquescens* (Bull. ex Mérat) Duby)
(E, K)

GASTEROMYCETES

LYCOPERDALES

LYCOPERDACEAE

Bovista Pers.

B. nigrescens Pers.
(E)

B. plumbea Pers.
(K)

Calvatia Pers.

C. excipuliformis (Pers.) Perdeck

Lycoperdon Pers.

L. foetidum Bon.
L. perlatum var. *nigrescens* Pers.
Abundant

L. molle Pers.

L. perlatum Pers. var. **perlatum**
Abundant (E) recorded by White as *L. gemmatum*

L. pyriforme Pers.
Frequent on woody debris (K)

L. spadiceum Pers.
(K)

Vascellum Smarda

V. depressum (Bon.) Smarda
(E)

SCLERODERMATALES

SCLERODERMATACEAE

Scleroderma Pers.

S. citrinum Pers.
S. aurantium L. ex Pers.
Occasional

S. verrucosum Pers.
(E)

NIDULARIALES

NIDULARIACEAE

Cyathus Haller ex Pers.

C. striatus (Huds. ex Pers.) Pers.
Only a single recent collection (K); recorded by White

Nidularia Fr.

N. farcta (Roth ex Pers.) Fr.
(K)

SPHAEROBOLACEAE

Sphaerobolus Tode ex Pers.

S. stellatus Tode ex Pers.
(E)

PHALLALES

PHALLACEAE

Phallus Pers.

P. impudicus L. ex Pers.

HEMIBASIDIOMYCETES

UREDINALES

Bibliogr.: Jørstad (1951, 1958); Wilson (1934); Wilson & Henderson (1966).

COLEOSPORIACEAE

Coleosporium Link

C. tussilaginis (Pers.) Lév.
Common and widespread on various hosts (E, K); recorded by White

MELAMPSORACEAE

Chrysomyxa Unger

C. empetri Cummins
(Clark)

Kuehneola Magn.

K. uredinis (Link) Arth.
(E)

Melampsora Cast.

M. capraearum Thüm.
Frequent; recorded by White as *M. salicina* Lév.

M. epitea Thüm. var. **epitea**
(E, K)

M. hypericorum Wint.
Common on *Hypericum androsaemum* (Clark, E, K)

M. larici-pentandrae Kleb.
(E)

M. lini var. **lini** (Ehrenb.) Lév.
(Clark, E, K)

Melampsorella Schroet.

M. caryophyllacearum Schroet.
Galls seen on *Abies*.

Melampsoridium Kleb.

M. betulinum (Fr.) Kleb.
Frequent on *Betula*, one record on *Alnus* from Ulva (E); recorded by White as *Melampsora*

Milesina Magn.

M. blechni Syd.
(E)

M. dieteliana (Syd.) Magn.
Only one record during the survey (E); recorded by White as *M. polypodii* and noted in Wilson (1934)

Pucciniastrum Otth

P. circaeae (Wint.) De Toni
Common and widespread on *Circaea* spp. (Clark, E, K)

P. epilobii Otth
(E)

P. guttatum (Schroet.) Hyl., Jørst. & Nannf.

P. vaccinii (Wint.) Jørst.
(Clark)

PUCCINIACEAE

Phragmidium Link.

P. bulbosum (Str.) Schlecht.

P. fragariae (DC.) Rabenh.
Common on *Potentilla sterilis* (Clark, K)

P. mucronatum (Pers.) Schlecht.
Frequent; recorded by White

P. rubi-idaei (DC.) Karst.
Frequent (Clark, E, K); recorded by White as *P. gracile*

P. violaceum (C. F. Schultz) Wint.
Widespread and common on *Rubus* spp.

Puccinia Pers.

P. acetosae Körn.
Common and widespread on *Rumex acetosa* (Clark); recorded by White under *Uredo bifrons* DC.

P. aegopodii (Str.) Röhl.
(Clark, E)

P. annularis (Str.) Röhl.
Frequent and widespread on *Teucrium scorodonia* (Clark, E); recorded by White as *P. scorodoniae* Link

P. brachypodii Otth
(E)

P. buxi DC.
On *Buxus* (Clark, K)

P. calcitrapae DC.
Abundant and widespread on various hosts (Clark); recorded by White as *P. compositarum* Schroet.

P. calthae Link
(E)

P. calthicola Schroet.
(Clark, E)

P. caricina DC. var. **caricina**
Common and widespread on various *Carex* spp. (Clark, E, K)

P. caricina var. **uliginosa** (Juel) Jørst.

P. caricina var. **urticae-acutae** (Kleb.) Henderson

P. chaerophylli Purton

P. chrysosplenii Grev.
(E)

P. circaeae Pers.
Frequent on *Circaea lutetiana* and *C. intermedia*; recorded by White on *C. alpina*

P. clintonii Peck
(E)

P. cnici Mart.
(Clark, E)

P. cnici-oleracei Pers. ex Desm.
Frequent on *Cirsium arvense* (E)

P. commutata P. & H. Syd.

P. coronata Corda
(K)

P. deschampsiae Arth.
Only one recent record; recorded by White

P. difformis Kunze
(E)

P. dioicae Magn.
(Clark, E)

P. epilobii DC.

P. fergussonii Berk. & Br.
Frequent on *Viola palustris* (E)

P. festucae Plowr.
(Clark, E)

P. hieracii Mart. var. **hieracii**
Very common and widespread on various composite hosts (Clark, E, K)

P. hieracii var. **hypochoeridis** (Oud.) Jørst.

P. hydrocotyles Cooke
(E)

P. lagenophorae Cooke

P. lapsanae Fuckel
(Clark)

P. luzulae Lib.
Only one recent record; recorded by White as *Trichobasis oblongata* Berk.

P. magnusiana Körn.
(E)

P. menthae Pers.
(Clark, E).

P. microsora Körn. ex Fuckel
(E)

P. obscura Schroet.
(Clark, E)

P. poae-nemoralis Otth
Occasional on various grass hosts (E)

P. poarum Niels.

P. polygoni-amphibii Pers.

P. primulae Duby
Frequent on *Primula vulgaris* (Clark, E, K)

P. pulverulenta Grev.
(E, K)

P. punctata Link
Occasional on *Galium* sp. (E); recorded by White as *P. galiorum* Link

P. punctiformis (Str.) Röhl.

P. pygmaea J. Erikss.

P. recondita Rob. & Desm.
(E)

P. saniculae Grev.
Two collections during the present survey (K); recorded by White

P. saxifragae Schlecht.

P. sessilis Schroet.
(Clark, E)

P. tumida Grev.
(E, K)

P. variabilis Grev.
(K)

P. violae DC.
Frequent on *Viola riviniana* (Clark, E, K)

Trachyspora Fuckel

T. intrusa (Grev.) Arth.
Abundant on various *Alchemilla* spp. (Clark, E, K)

Tranzschelia Arth.

T. anemones (Pers.) Nannf.
Frequent on *Anemone nemorosa*

Triphragmium Link

T. ulmariae (DC.) Link
Abundant and widespread on *Filipendula ulmaria* (Clark, E, K)

Uromyces (Link) Unger

U. acetosae Schroet.
(Clark)

U. airae-flexuosae Ferd. & Winge

U. anthyllidis Schroet.
(E)

U. armeriae Kickx

U. behenis (DC.) Unger
(E)

U. dactylidis Otth
Abundant and widespread on *Ranunculus ficaria* and various grasses
(Clark, E, K)

U. muscari (Duby) Lév.
(Clark, K)

U. nerviphilus (Grognot) Hotson

U. rumicis (Schum.) Wint.
Common on *Rumex* spp. (Clark, K); recorded by White

U. trifolii (DC.) Lév.

U. valerianae (DC.) Lév.
Frequent on *Valeriana officinalis* (Clark, E); recorded by White as *Lecythea valerianae* Berk.

U. viciae-fabae (Pers.) Schroet. var. **viciae-fabae**
(E)

U. viciae-fabae var. **orobi** (Schum.) Jørst.
(K)

TILLETIALES

TILLETIACEAE

Entorrhiza Weber

E. aschersoniana (Magn.) Lagerh.
(K)

Tilletia Tul.

T. decipiens (Pers.) Körn.

Urocystis Rabenh.

U. anemones (Pers.) Wint.
Only one record (E); recorded by White as *U. pampholygodes* Schlech.

U. filipendulae (Tul.) Schroet.
(E)

USTILAGINALES

USTILIGINACEAE

Cintractia Cornu

C. caricis (Pers.) Magn.
Only one recent record (K); recorded by White as *Ustilago urceolorum* Tul.

Sphacelotheca De Bary

S. hydropiperis (Schum.) De Bary
(K)

Ustilago (Pers.) Rousell

U. violacea (Pers.) Fuckel
(E)

EUMYCOTA: ASCOMYCOTINA

HEMIASCOMYCETES

ENDOMYCETALES

ENDOMYCETACEAE

Hansenula Syd.

Hansenula sp.
Isolated from brown earth and wet peat.

TAPHRINALES

TAPHRINACEAE

Taphrina Fr.

T. amentorum (Sadeb.) Rostrup
On female catkins of *Alnus glutinosa* (K); causes thickening of the scale leaves

T. betulae (Fuckel) Johan.
Frequent and widespread on *Betula*

T. caerulescens (Mont. & Desm.) Tul.
Parasitic on leaves of *Quercus*; frequently causing spotting

T. populina Fr.
Parasitic on leaves of *Populus tremula*

T. potentillae (Farl.) Johan.
Parasitic on stems and leaves and causing distortion of *Potentilla erecta* (Clark, E)

T. sadebeckii Johan.
Parasitic on *Alnus*; causes yellow spots on under-surface of leaves

T. tosquinetii (West.) Magn.
Frequent and widespread on *Alnus glutinosa* (Clark, E)

PROTOMYCETACEAE

Protomyces Unger

P. macrosporus Unger
Parasitic on *Aegopodium podagraria*

EUASCOMYCETES

PLECTASCALES

GYMNOASCACEAE

Arachniotus Schroet.

A. albicans Apinis
Isolated from soil samples from hill pasture and under *Myrica*

Arthroderma Currey

A. curreyi Berk.
Isolated from two soil samples

Byssochlamys Westling

B. fulva Olliver & G. Smith
Isolated from woodland soil

Gymnoascus Baran.

G. umbrinus Boud.
Isolated from three soil samples

Myxotrichum Kunze

M. deflexum Berk.
Isolated from wet leaves

EUROTIACEAE

Allescheria Sacc. & Syd.

A. boydii Shear
Isolated from brown earth sample and from submerged leaves

A. terrestris Apinis
Isolated from several soil samples

Emericella Berk.

E. nidulans (Eidam) Vuill.
Isolated from soil sample

Eurotium Link ex Fr.

E. chevalieri Mangin
Isolated from brown earth sample

E. repens De Bary
On wet organic material

Pseudoeurotium van Beyma

P. zonatum van Beyma
Isolated from several soil samples

Thielavia Zopf

T. leptodermus Booth
On submerged leaves of *Myrica*

T. terricola (Gilman & Abbot) Emmons
Isolated from several soil samples

ELAPHOMYCETACEAE

Elaphomyces Nees ex Fr.

E. granulatus Fr.
Hypogeous in mixed woodland (Clark)

E. muricatus Fr.
Hypogeous in mixed woodland (E)

ONYGENACEAE

Onygena Pers. ex Fr.

O. corvina Alb. & Schw. ex Fr.
Frequent and widespread on feathers

O. equina (Will.) Pers. ex Fr.
Frequent and widespread on sheep-horn

ERYSIPHALES

ERYSIPHACEAE

Erysiphe Hedw. f. ex Fr.

E. cichoracearum DC. ex Mérat
Conidial state only; on various composite hosts

E. galii Blumer
On *Galium aparine*

E. graminis DC. ex Mérat
Conidial state only; on various *Gramineae*

E. polygoni DC. ex Mérat
Conidial state only; retained in the strict sense of Junell (1967), on *Scabiosa succisa*

E. ranunculi Grev.
On *Ranunculus flammula*

E. sordida Junell
On *Plantago maritima*

E. ulmariae Desm.
On *Filipendula* (K)

Microsphaera Lév.

M. alphitoides Griff. & Maubl.
Conidial state only

M. penicillata (Fr.) Lév.
It is uncertain to what this early record by White refers

Podosphaera Kunze ex Lév.

P. clandestina (Fr.) Lév.
P. oxycanthae (DC.) De Bary
Conidial state only

Sphaerotheca Lév.

S. alchemillae (Grev.) Junell
Parasitic on various rosaceous plants

S. epilobii (Link) Sacc.
Parasitic on *Epilobium* spp.

S. plantaginis (Cast.) Junell
Parasitic on *Plantago lanceolata*

Uncinula Lév.

U. bicornis (Fr.) Lév.
On living leaves of *Acer pseudoplatanus*

HYPOCREALES

HYPOCREACEAE

Hypocrea Fr.

H. lactea (Fr.) Fr.
(Clark)

H. pulvinata Fuckel
Frequent and widespread on *Piptoporus betulinus*: Basidiomycotina

H. rufa (Pers. ex Fr.) Fr.
See *Trichoderma*: Deuteromycotina

H. sp. aff. **rufa**
(K); described by Webster (1964) as "aff. *rufa* species 1"

Hypocreopsis Karst.

H. rhododendri Thaxter
Several collections on twigs and branches of *Corylus*, Croggan and Knockan (Clark, K); the first British records

Protocrea Petch

P. farinosa (Berk. & Br.) Petch
On old log of *Fagus*

Sphaeronaemella Karst.

S. fimicola March.
On horse dung (K)

MELANOSPORACEAE

Chaetomium Kunze ex Fr.

C. anguipilum Ames
Isolated from soil under *Myrica*

C. globosum Kunze ex Fr.
Dervaig (Clark)

C. seminudum Ames
Isolated from woodland soil

C. thermophile La Touche
Isolated from brown earth sample

Melanospora Mudd

M. caprina (Fr.) Sacc.
(Clark)

M. fimicola Hansen
Isolated from several soil samples

NECTRIACEAE

Byssonectria Karst.

B. lateritia (Fr.) Petch
On *Lactarius deliciosus*: Basidiomycotina

Gibberella Sacc.

G. zeae (Schw.) Petch
See *Fusarium graminearum*: Deuteromycotina

Hypomyces (Fr.) Tul.

H. aurantius (Pers. ex Fr.) Tul.
On decaying Polyporaceae: Basidiomycotina (Clark, K)

H. rosellus (Alb. & Schw. ex Fr.) Tul.
On decaying *Stereum*: Basidiomycotina (K)

Nectria Fr.

N. cinnabarina (Tode ex Fr.) Fr.
Frequent on wood; particularly as *Tubercularia* stage (see Deuteromycotina) (E, K); recorded by White

N. coccinea (Pers. ex Fr.) Fr.
On boughs of *Fagus* (K)

N. ditissima Tul.
Isolated from wet leaves

N. episphaeria (Tode ex Fr.) Fr.
On old stromata of Pyrenomycetes: Ascomycotina (E, K)

N. fuckeliana Booth
On conifer wood (K)

N. lugdunensis Webster
See *Heliscus lugdunensis*: Deuteromycotina

N. galligena Bres.
Forming canker on *Fraxinus*

N. gliocladioides Smalley & Hansen
See *Gliocladium*: Deuteromycotina

N. haematococca Berk. & Br.
See *Fusarium solani*: Deuteromycotina

N. radicicola Gerlach & Nilss.
See *Cylindrocarpon destructans*: Deuteromycotina

CLAVICIPITALES

CLAVICIPITACEAE

Claviceps Tul.

C. nigricans Tul.
Frequent on *Eleocharis* (E); only sclerotia present

C. purpurea (Fr.) Tul.
Frequent and widespread on various grasses (E, K); see host list for individual formae speciales; recorded by White

Cordyceps (Fr.) Link

C. capitata (Holm. ex Fr.) Link
On *Elaphomyces* (Clark, K)

C. militaris (L. ex St. Amans) Link
On insect pupae (E); two recent collections and recorded by White as *Torrubia militaris* Fr.

C. ophioglossoides (Ehrenb. ex Fr.) Link
On *Elaphomyces* (K)

Epichloe (Fr.) Tul.

E. typhina (Pers. ex Fr.) Tul.
Frequent and widespread on various grasses (Clark)

SPHAERIALES

AMPHISPHAERIACEAE

Apiorhynchostoma Petrak

A. curreyi (Rab.) Müller
On decorticated conifer wood (K)

Paradidymella Petrak

P. perforans Munk
On *Ammophila arenaria*

P. tosta (Berk. & Br.) Petrak
On dead stems of *Epilobium angustifolium* (E)

Vialaea Sacc.

V. insculpta (Fr.) Sacc.
On twigs of *Ilex aquifolium*

DIAPORTHACEAE

Apioporthe Höhn.

A. vepris (de Lacr.) Wehmeyer.
On dead stems of *Rubus* spp. (Clark)

Calosporella Schroet.

C. innesii (Curr.) Schroet.
On *Acer pseudoplatanus* (K)

Caudospora Starb.

C. taleola (Fr.) Starb.
On *Quercus* (K)

Diaporthe Nits.

D. arctii (Lasch) Nits.
On dead woody herbaceous stems

D. eres Nits.
On dead twigs (K)

D. leiphaemia (Fr.) Sacc.
On *Quercus* (K)

Diaporthopsis Fabre

D. angelicae (Berk.) Wehmeyer
On dead umbelliferous stems

D. pantherina (Berk.) Wehmeyer
On dead stems of *Pteridium*

Ditopella De Not.

D. ditopa (Fr.) Schroet.
On twigs of *Alnus* (K)

Gaeumannomyces von Arx & Olivier

G. graminis (Sacc.) von Arx & Olivier var. **avenae** (Turn.) Dennis
On grass stems

Gnomonia Ces & De Not.

G. cerastis (Reiss) Ces. & De Not.
On *Acer pseudoplatanus* (K)

G. inclinata (Desm.) Auersw.
(Clark)

G. padicola Kleb.
See *Actinonema padi*: Deuteromycotina

G. rubi (Rehm) Wint.
On *Rubus fruticosus* (K)

G. setacea (Pers. ex Fr.) Ces. & De Not.
On leaves of *Betula* and *Corylus* (K)

G. vulgaris Ces. & De Not.
On fading leaves of *Corylus* (Clark, K)

Gnomoniella Sacc.

G. tubiformis (Fr.) Sacc.
See *Leptothyrium alneum*: Deuteromycotina

Linospora Fuckel

L. capreae (DC.) Fuckel
On dead leaves of *Salix*

L. vulgaris Fuckel
On *Salix* (K)

Mamiania Ces. & De Not.

M. coryli (Batsch ex Fr.) Ces. & De Not.
Recorded by White as *Sphaeria coryli*

Melanconis Tul.

M. alni Tul.
See *Melanconium apiocarpum*: Deuteromycotina

M. stilbostoma (Fr.) Tul.
On twigs and branches of *Betula* spp. (E, K)

Phomatospora Sacc.

P. arenaria Sacc., Bomm. & Rouss.
Iona (K). Not previously recorded for the British Isles although it has been found in several coastal localities on the European continent

P. berkeleyi Sacc.
On dead culms of *Phalaris arundinacea*

Plagiostoma Fuckel

P. devexa (Desm.) Fuckel
On dead stems of *Rumex* (K)

P. pustula (Pers. ex Fr.) von Arx
On dead leaves of *Quercus* (Clark, K)

Sillia Karst.

S. ferruginea (Pers. ex Fr.) Karst.
On small stems at base of *Corylus*

Sydowiella Petrak

S. fenestrans (Duby) Petrak
On dead stems of *Epilobium angustifolium*

DIATRYPACEAE

Cryptosphaeria Grev.

C. eunomia (Fr.) Fuckel
On twigs and branches of *Fraxinus* (K)

Diatrype Fr.

D. disciformis (Hoffm. ex Fr.) Fr.
Frequent on twigs and branches of *Fagus*

D. stigma (Hoffm. ex Fr.) Fr.
On branches and large twigs (E, K)

Diatrypella (Ces. & De Not.) Sacc.

D. favacea (Fr.) Sacc.
On dead branches of *Betula* (K)

D. quercina (Pers. ex Fr.) Cooke
On dead branches of *Quercus*

Endoxyla Fuckel

E. rostrata (Tode ex Fr.) Munk
On *Fagus* (K)

Eutypa Tul.

E. acharii Tul.
On branches of *Acer pseudoplatanus*

E. flavovirens (Fr.) Tul.
On old wood

E. lata (Pers. ex Fr.) Tul.
Frequent and widespread on hardwoods (K)

E. scabrosa (Bull. ex Fr.) Fuckel
On *Fagus* (K)

Eutypella (Nits.) Sacc.

E. prunastri (Pers. ex Fr.) Sacc.

Valsa Fr.

V. ambiens (Pers. ex Fr.) Fr.
On branches of *Fagus* (K); see also *Cytospora ambiens*: Deuteromycotina

V. curreyi Sacc.
See *Cytospora curreyi*: Deuteromycotina

V. fuckelii Nits.
On branches of *Corylus*; see also *Cytospora fuckelii*: Deuteromycotina

LASIOSPHAERIACEAE

Cercophora Fuckel

C. coprophila (Fr.) Chen.
Common on dung (E, K)

C. mirabilis Fuckel
On dung (K); see Lundquist (1972)

Coniochaeta (Sacc.) Massee

C. leucoplaca (Berk. & Rav.) Cain
On sheep droppings

C. ligniaria (Grev.) Massee
Isolated from bare soils under *Ulex* and from hill pasture soils (E); it has been recorded as *C. discospora* (Auersw.) Cain on dung

C. scatigena (Berk. & Br.) Cain
On dung.

Gelasinospora Dowding

G. cerealis Dowding
Isolated from several soil samples

Helminthosphaeria Fuckel

H. corticiorum Höhn.
Associated with old resupinate fungus: Basidiomycotina

Lasiophaeria Ces. & De Not.

L. glabrata (Fr.) Munk
Killiechronan (Clark, E)

L. ovina (Fr.) Ces. & De Not.

L. spermoides (Hoffm. ex Fr.) Ces. & De Not.
Frequent and widespread on wood (K)

Podospora Ces.

P. appendiculata (Auersw.) Niessl
On cow dung

P. curvula (De Bary) Niessl
Frequent on dung (K)

P. decipiens Wint. ex Fuckel
Isolated from rotting debris in swamp.

P. intestinacea Lundq.
On dung (K); a newly described species whose paratypes include two collections from Mull (see Lundquist, 1972)

P. minuta (Fuckel) Niessl
Frequent and widespread on dung in soils (E, K)

P. setosa Wint.
Isolated from rotting debris in swamp; a similar taxon was found on dung (E)

Sordaria Ces. & De Not.

S. fimicola (Rob.) Ces. & De Not.
Isolated from soils and frequent on dung (K)

S. hirta Hansen
Isolated from soil

S. humana (Fuckel) Wint.
Isolated from soil

S. papyricola Wint.
Isolated from soil

S. superba De Not.
Isolated from soil (K); see comments by Lundquist (1972)

Zygospermella Cain

Z. insignis (Mont.) Cain
On dung (E)

Z. setosa (Cain) Cain
On dung

CERATOSTOMELLACEAE

Ceratostomella Sacc. emend. Höhn.

C. ampullasca (Cooke) Sacc.
On rotten wood; Salen and Croggan

C. rostrata Fuckel
Isolated twice from wet organic debris

Ceratostomella sp.
Isolated from wet organic debris.

POLYSTIGMATACEAE

Endodothella Theiss. & Syd.

E. junci (Fr.) Theiss. & Syd.
Common on *Juncus* (Clark, K); recorded by White

Phyllachora Nits. apud Fuckel

P. graminis (Pers. ex Fr.) Fuckel
On dead stems of *Dactylis glomerata* (Clark)

P. sylvatica Sacc. & Speg.
On dead stems of *Festuca rubra*

Physalospora Niessl

P. empetri Rostrup
On *Empetrum nigrum* (Clark)

Polystigma DC. ex Chev.

P. fulvum DC. ex St. Amans
One recent collection; recorded by White

SPHAERIACEAE

Chaetosphaeria Tul.

C. callimorpha (Mont.) Sacc.
On *Rubus fruticosus*
(K)

C. myriocarpa (Fr.) Booth
On wood (E, K)

C. pulviscula (Cutt.) Booth
On *Fagus* (K)

Niesslia Auersw.

N. pusilla (Fr.) Schroet.
On wet leafy debris (K)

Zignoella Sacc.

Z. ovoidea (Fr.) Sacc.
On woody debris of *Rubus idaeus* (K)

XYLARIACEAE

Anthostoma Nits.

A. microsporum (Karst.) Karst.
On *Corylus* (K)

Anthostomella Sacc.

A. phaeosticta (Berk.) Sacc.
Entosordaria ammophilae (Phil. & Plowr.) Höhn.
On *Ammophila arenaria* (Clark)

A. tumulosa (Rob. & Desm.) Sacc.
On *Luzula sylvatica* (K)

Bolinia (Nits.) Sacc.

B. lutea (Alb. & Schw. ex Fr.) Miller
On *Quercus* (K); commonly placed in the genus *Nummularia*

Daldinia Ces. & De Not.

D. concentrica (Bolt. ex Fr.) Ces. & De Not.
On *Fraxinus* (K)

Hypocopra Fr.

H. equorum (Fuckel) Wint.
Isolated from soil of hill-pasture

Hypoxylon Bull. ex Fr.

H. cohaerens (Pers. ex Fr.) Fr.
On old wood of *Fagus*

H. fragiforme (Pers. ex Fr.) Kickx
Frequent and widespread on *Fagus* (E, K); recorded by White under the synonym *H. coccineum* Bull. ex. Fr.

H. fuscum (Pers. ex Fr.) Fr.
Frequent and widespread on *Corylus* (Clark)

H. multiforme (Fr.) Fr.
On very old wood of deciduous trees (Clark)

H. rubiginosum (Pers. ex Fr.) Fr.
Frequent and widespread on dead wood of various trees especially *Fraxinus* (K)

H. semi-immersum Nits.
H. confluens auctt.
On rotting wood of deciduous trees (E)

H. serpens (Pers. ex Fr.) Kickx
On rotting wood of deciduous trees (E)

Lopadostoma (Nits.) Trav.

L. turgidum (Pers. ex Fr.) Trav.
On branches of *Fagus*

Rosellinia De Not.

R. aquila (Fr.) De Not.
On dead branches of *Ulex* (K)

R. mammiformis (Pers. ex Fr.) Ces. & De Not.
On wet branches of *Salix*

Ustulina Tul.

U. deusta (Fr.) Petrak
On old trunk of *Acer pseudoplatanus* (Clark)

Xylosphaera Dumort.

X. carpophila Pers. ex Dumort.
Tavool

X. hypoxylon (L. ex Fr.) Dumort.
Common and widespread on rotten hardwood (K)

X. polymorpha (Pers. ex Mérat) Dumort.
On old wood

CORONOPHORALES

CORONOPHORACEAE

Bertia De Not.

B. moriformis (Tode ex Fr.) De Not.
On conifer wood (E, K)

VERRUCARIALES

VERRUCARIACEAE*

Muellerella Hepp. ex Müll. Arg.

M. polyspora Hepp
Isolated from wet leaves

PHACIDIALES

PHACIDIACEAE

Didymascella Maire & Sacc.

D. thujina (Durand) Maire
On *Thuja plicata*

Micraspis Darker

M. strobilina Dennis
A new species based on material from Loch Bà, growing on cones of *Pinus sylvestris* (K)

Phacidiostroma Höhn.

P. multivalve (DC. ex St. Amans) Höhn.
(K); see also *Ceuthospora phacidiodes*: Deuteromycotina

Schizothyrioma Höhn.

S. ptarmicae (Desm.) Höhn.
Parasitic on leaves of *Achillea ptarmica*

HYPODERMATACEAE

Coccomyces De Not.

C. coronatus (Schum. ex Fr.) Karst.
See *Leptothyrium medium*: Deuteromycotina

C. dentatus (Kunze & Schmidt) Sacc.
On dead leaves of *Quercus* with or without *Leptothyrium quercinum*: Deuteromycotina

Colpoma Wallr.

C. crispum (Pers. ex Fr.) Sacc.
(Clark)

C. quercinum (Pers.) Wall.
Frequent on *Quercus* twigs (K)

Hypoderma DC. emend. De Not.

H. commune (Fr.) Duby
On dead herbaceous stems

* See also lichenised forms, pp. 14.50–53.

H. pinicola Brunch.
See *Hendersonia acicola*: Deuteromycotina

H. rubi De Not.
On *Rubus fruticosus* (K); see *Leptothyrium rubi*: Deuteromycotina

H. virgultorum DC. ex St. Amans
On *Rubus*

Lophodermium Chev.

L. apiculatum Duby

L. arundinaceum (Schrad. ex Fr.) Chev.
Frequent and widespread on various grasses (Clark, E)

L. gramineum (Fr.) Chev.
Isolated from litter in soils

L. juniperinum (Fr.) De Not.
On needles of *Juniperus* (Clark)

L. piceae (Fuckel) Höhn.
On needles of *Picea* (K)

L. pinastri (Schrad. ex Fr.) Chev.
(Clark); see also *Leptostroma pinastri*: Deuteromycotina

L. rhododendri Ces. ex Sacc.
On *Rhododendron ponticum* (Clark)

L. cf. **vagulum** Wilson & Robertson
On *Rhododendron ponticum*, Aros Park

Rhytisma Fr.

R. acerinum (Pers. ex St. Amans) Fr.
Frequent and widespread on *Acer pseudoplatanus* (Clark, K); recorded by White

R. salicinum Fr.
Two recent records (Clark, E); recorded by White

HELOTIALES

ORBILIACEAE

Orbilia Fr.

O. auricolor (Blox. ex Berk.) Sacc.
On bark of *Fraxinus*

O. epipora Karst.
On bark of *Betula*

O. leucostigma (Fr.) Fr.
On dead wood.

O. luteorubella (Nyl.) Karst.
On old wood.

O. sarraziniana Boud.
On damp woody debris (K)

O. xanthostigma (Fr.) Fr.
Common and widespread on rotten wood (Clark, K)

DERMATACEAE

Actinoscypha Karst.

A. graminis Karst.
(Clark)

A. punctum (Rehm) Müller
Only rarely recorded previously for the British Isles; on *Nardus*, Ben More (K)

Belonium Sacc.

B. hystrix (De Not.) Höhn.
On *Molinia*; a much confused discomycete although apparently an autonomous and easily recognized species. (Dennis, 1971)

Belonopsis (Sacc.) Rehm

B. excelsior (Karst.) Rehm
On *Phragmites*. This species and the following may be considered to be in *Niptera* Fr.

B. filispora (Cooke) Nannf.
On *Brachypodium* (Clark)

Calloria Fr.

C. carneo-flavida Rehm
On *Urtica*

C. fusarioides (Berk.) Fr.
On *Urtica* (Clark); see *Cylindrocolla urticae*: Deuteromycotina

Diplocarpon Wolf

D. agrostemmatis Nannf.
See *Marssonina delastrei*: Deuteromycotina

D. earliana (Ellis & Everh.) Wolf
See *Marssonina potentillae*: Deuteromycotina

D. rosae Wolf
See *Actinonema rosae*: Deuteromycotina

Fabraea Sacc.

F. ranunculi (Fr.) Karst.
On living leaves of *Ranunculus* spp. (Clark)

Mollisia (Fr.) Karst.

M. benesuada (Tul.) Phill.
(Clark)

M. cinerea (Batsch ex Mérat) Karst.
Frequent and widespread on various dead wood and branches (Clark, K)

M. culmina Sacc.

M. fallax (Desm.) Gill.
(K)

M. cf. **fallens** Karst.
An interesting record which needs further examination

M. ligni (Desm.) Karst.
On decorticated wood of deciduous trees (K)

M. melaleuca (Fr.) Sacc.
On rotting wood of deciduous trees (K)

M. palustris (Rob.) Karst.
On rotting stems of marsh plants (E)

M. aff. **palustris** (Rob.) Karst.

M. ramealis (Karst.) Karst.
On dead twigs of *Betula* (Clark)

M. revincta (Karst.) Rehm
On stems of *Filipendula ulmaria*

M. urticicola Phill.
On stems of *Urtica dioica*

M. ventosa (Karst.) Karst.
On very old wood

Mollisia sp.
Isolated from wet leaves; it is interesting to note that the aquatic hyphomycete *Anguillospora crassa* has a *Mollisia* perfect state; see Deuteromycotina

Niptera Fr.

See also *Belonopsis* (Sacc.) Rehm

N. melatephra (Lasch.) Rehm
On *Carex flacca*, Glen Aros

N. phaea Rehm
On *Scirpus caespitosus*, Ross of Mull (K)

N. aff. **phaea** Rehm
On a number of Cyperaceae

N. pilosa (Crossl.) Boud.
On *Carex* (K)

N. pulla (Phill. & Keith) Boud.
On dead leaves of *Phragmites australis* (K)

N. ramincola Rehm
On wood of *Swida* (K)

N. umbonata Fuckel
(Clark, as *Mollisia amenticola* Pers.)
There is some disagreement as to the correct generic placing for this fungus on *Alnus* catkins; it has been placed in *Mollisiella* Boud.

Pezicula Tul.

P. livida (Berk. & Br.) Rehm
On rotting wood of conifers (K)

P. rubi (Lib.) Niessl
On dead stems of *Rubus fruticosus* (K)

Ploettnera P. Henn.

P. exigua (Niessl) Höhn.
On dead stems of *Rubus fruticosus* (K)

Propolomyces Sherwood

P. farinosus (Pers.) Sherwood
Propolis versicolor auct., non (Fr.) Fr.
Frequent on various dead woods, branches and on woody stems (Clark, K)

Pseudopeziza Fuckel

P. calthae (Phill.) Massee
Fabraea roussearia Sacc. & Bamm.
On *Caltha palustris*.

P. ribis Kleb.
See *Gloeosporium ribis*: Deuteromycotina

P. trifolii (Biv.-Bern.) Fuckel
On *Trifolium repens* (Clark)

Pyrenopeziza Fuckel

P. arenivaga (Phill.) Boud.
On old culms of *Ammophila arenaria*

P. digitalina (Phill.) Sacc.
On leaves of *Digitalis purpurea*, Killiechronan (K).

P. fuckelii Nannf.
On leaves of *Salix* (Clark, K)

P. petiolaris (Alb. & Schw. ex Fr.) Nannf.
On old petioles of fallen leaves of *Acer pseudoplatanus* (Clark, K)

P. urticicola (Phill.) Boud.
Glen Aros and near Croggan (Clark, E)

Tapesia (Pers. ex Fr.) Fuckel

T. cinerella Rehm.
Killiechronan (Clark)

T. fusca (Pers. ex Mérat) Fuckel
On various types of woody debris (K)

T. melaleucoides Rehm
On dead twigs of *Calluna* (Clark, E)

T. retincola (Rab.) Karst.
On dead culms of *Phragmites* (Clark, E)

T. rosae (Pers. ex Mérat) Fuckel
On dead branches of *Rosa*

Trichobelonium (Sacc.) Rehm

T. obscurum (Rehm) Rehm
On old stems of *Calluna* (Clark)

Trochila Fr.

T. craterium Fr.
On old leaves of *Hedera*.

T. ilicina (Nees ex Fr.) Greenhalgh & Morgan-Jones
On dead leaves of *Ilex* (K)

T. laurocerasi (Desm.) Fr.
On dead leaves of *Prunus laurocerasus* (K); see *Ceuthospora laurocerasi*: Deuteromycotina

HYALOSCYPHACEAE

Arachnopeziza Fuckel

A. aurata Fuckel
On dead wood (Clark)

A. aurelia (Pers.) Fuckel
Near Salen (E)

Cistella Quél.

An unidentified species of this genus was found on *Populus tremula* wood at Dishig.

Clavidisculum Kirschst.

C. kriegerianum Kirschst.
On *Picea* needles (K)

Dasyscyphus Gray

D. acutipilus (Karst.) Sacc.
Widespread on *Phragmites* (Clark)

D. apalus (Berk. & Br.) Dennis
Widespread on *Juncus* (Clark, E)

D. bicolor (Bull. ex Mérat) Fuckel var. **rubi** (Bres.) Dennis
On *Rubus idaeus* (Clark)

D. brevipilus le Gal
On old wood of broad-leaf trees (K)

D. capitatus (Peck.) le Gal
On dead leaves of *Quercus* (Clark)

D. carneolus (Sacc.) Sacc. var. **longisporus** Dennis
On dead grass culms (Clark)

D. ciliaris (Schrad. ex Fr.) Sacc.
On fallen leaves of deciduous trees

D. clandestinus (Bull. ex Mérat) Fuckel
On old herbaceous stems (Clark)

D. controversus (Cooke) Rehm
On dead culms of *Phragmites* (Clark)

D. corticalis (Pers. ex Fr.) Massee
On old wood of deciduous trees (K)

D. diminutus (Rob.) Sacc.
On dead culms of *Juncus* (Clark, E)

D. dumorum (Rob.) Massee
On lower surfaces of leaves of *Rubus fruticosus* (Clark)

D. eriophori (Quél.) Sacc.
On dead leaves of *Eriophorum*, near Dervaig

D. fugiens (Buck.) Massee
On dead culms of *Juncus*

D. fuscescens (Pers.) Gray
On fallen leaves of *Quercus* (Clark)

D. grevillei (Berk.) Massee
On dead herbaceous stems (Clark)

D. nidulus (Schmidt & Kunze) Massee
On dead herbaceous stems (Clark)

D. niveus (Hedw. ex Fr.) Sacc.
Common and widespread on decorticated wood, especially of *Quercus* (Clark, K)

D. nudipes (Fuckel) Sacc.
Frequent on *Filipendula ulmaria* (Clark)

D. papyraceus (Karst.) Sacc.
On decorticated wood of *Salix* and *Betula* (K)

D. pteridis (Alb. & Schw. ex Pers.) Massee
On dead stems of *Pteridium* (Clark)

D. pulveraceus (Alb. & Schw. ex Fr.) Höhn.
On dead branches of deciduous trees and shrubs (Clark, E)

D. pulverulentus (Lib.) Sacc.
On fallen needles of *Pinus* (K)

D. pygmaeus (Fr.) Sacc.
On damp woody debris (K)

D. salicariae Rehm
Base of dead stems of *Lythrum salicaria* (Clark)

D. sulphurellus (Peck) Sacc.
On dead twigs of *Myrica* (Clark, E)

D. sulphureus (Pers.) Massee
On old herbaceous stems (Clark)

D. virgineus Gray
Abundant and widespread on various plant substrates (Clark, E, K)

Eriopeziza (Sacc.) Rehm

E. caesia (Pers.) Rehm
On old rotten wood of *Quercus*

Hyaloscypha Boud.

H. flaveola (Cooke) Nannf.
Amongst the leaves of *Pteridium* (Clark)

H. hyalina (Pers.) Boud.
On decorticated branches of *Quercus* (K)

H. leuconica (Cooke) Nannf.
On old conifer wood (Clark)

H. stevensonii (Berk. & Br.) Nannf.
On wood of *Pinus sylvestris*

H. velenovskyi Graddon
Glen Aros (E)

Lachnellula Karst.
Trichoscyphella Nannf.

L. hahniana (Seav.) Dennis
On dead twigs and branches of *Larix* (Clark, K)

L. subtilissima (Cooke) Dennis
Three recent collections on *Pinus* (Clark, K); recorded by White as *Peziza calycina* Schum.

L. willkommii (Hartig) Dennis
Forming cankers on *Larix*

Microscypha Syd.

M. grisella (Rehm) Syd.
Frequent and widespread on *Pteridium* (Clark)

Mollisiopsis Rehm

M. dennisii Graddon
On *Ulex europaeus* stems (K); see Graddon (1972)

Phialina Höhn.

P. ulmariae (Lasch) Dennis
On dead stems of *Filipendula*

Torrendiella Boud. & Torrend

T. eucalypti (Berk.) Boud
On *Eucalyptus* leaves, Calgary (Clark)

Unguicularia Höhn.

U. cirrhata (Crouan) le Gal
On dead wood of *Sambucus*

U. incarnatina (Quél.) Nannf.
Rarely collected in the British Isles; Lettermore, Salen Forest (Clark) (Dennis, 1971)

U. scrupulosa (Karst.) Höhn.
Possibly identical with *U. cirrhata*

SCLEROTINIACEAE

Botryotinia Whetzel

B. calthae Henn. & M. B. Ellis
On *Caltha*, Penmore Mill (E); the second British record, see Henderson (1970)

Ciboria Fuckel

C. amentacea (Balbis ex Fr.) Fuckel
On catkins of *Salix* (K)

C. caucus (Rebent. ex Fr.) Fuckel

C. viridifusca (Fuckel) Höhn.
On catkins of *Alnus* (Clark)

Rutstroemia Karst. incl. *Poculum* Velen.

R. firma (Pers. ex Gray) Karst.
On old wood of *Quercus* (K)

R. plana Henderson
On old stems of *Eleocharis* (E); first described from Penmore Mill, see Henderson (1970)

R. sydowiana (Rehm) White
On fallen leaves of *Quercus* (Clark)

Sclerotinia Fuckel

S. curreyana (Berk.) Karst.
On old culms of *Juncus* (K)

S. eleocharidis Henderson
(E); first described from Penmore Mill, see Henderson (1970)

S. fuckeliana (De Bary) Fuckel
See *Botrytis cinerea*: Deuteromycotina

S. globosa (Buchwald) Webster
See *Botrytis globosa*: Deuteromycotina

S. gregoriana Palmer
On *Scirpus cespitosus* (E)

GEOGLOSSACEAE

Geoglossum Pers. ex Fr.

G. cookeianum Nannf.
Two records from Calgary, in grassland (Clark, E)

G. fallax Durand
In grassland (E)

G. glutinosum Pers. ex Fr.
Near Loch Bà

G. nigritum Cooke
Frequent in grasslands on sandy soil (Clark, E)

G. starbaeckii Nannf.
Three records from grassy places on Ulva (E)

Leotia Pers. ex Fr.

L. lubrica Pers.
Common and widespread in woodland (Clark, K); recorded by White

Microglossum Sacc.

M. olivaceum (Pers. ex Fr.) Gill.
In grassland near Glen Forsa Hotel (E)

M. viride (Pers. ex Fr.) Gill.
Frequent in woodland (E, K)

Mitrula Pers. ex Fr.

M. paludosa Fr.
On wet leaves (Clark)

Spathularia Pers. ex Wallr.

S. flavida Pers. ex Fr.
One recent collection (K); recorded by White

Thuemenidium Kuntze

T. atropurpureum (Batsch ex Fr.) Kuntze
Frequent in grasslands (E)

Trichoglossum Boud.

T. hirsutum (Pers. ex Fr.) Boud.
In pastures (E); several records

T. walteri (Berk.) Durand.
Amongst short grass (K)

HELOTIACEAE

Ascocoryne Groves & Wilson

A. cylichnium (Tul.) Korf
On old wood (Clark)

A. sarcoides (Jacq. ex Gray) Korf
Frequent and widespread on wood (Clark, K); recorded by White

Ascotremella Seav.

A. faginea (Peck) Seav.
On *Sambucus*; previously confused with *Coryne foliacea* Bres. (Dennis, 1971)

Bisporella Sacc.

B. citrina (Batsch ex Fr.) Korf & Carp.
On rotten wood (Clark, K); recorded by White under *Helotium*

B. fuscocincta (Graddon, sub *Calycella*)
On wood (K)

B. sulfurina (Quél.) Carp.
On wood (Clark, E, K)

Bulgaria Fr.

B. inquinans Fr.
Frequent and widespread on fallen trunks and branches (Clark, E, K)

Catinella Boud.

C. olivacea (Batsch ex Pers.) Boud.
On wet wood (E)

Cenangium Fr.

C. cf. **glaberrimum** Rehm
(K)

Chlorociboria
Seav. ex Ram., Korf & Bat. emend. Dixon

C. aeruginascens (Nyl.) Kan. ex Ram., Korf & Bat.
Chlorosplenium aeruginascens (Nyl.) Karst.
Frequent on hardwood (Clark, K); recorded by White under *Helotium*

Claussenomyces Kirscht.

C. atrovirens (Pers.) Korf & Abawi
On rotten wood of *Quercus* (Clark)

C. prasinula (Karst.) Korf & Abawi
On decaying wood

Cudoniella Sacc.

C. acicularis (Bull. ex Fr.) Schroet.
On wood (Clark)

C. clavus (Alb. & Schw. ex Fr.) Dennis var. **clavus**
On wet debris (Clark, E)

C. clavus var. **grandis** (Boud.) Dennis
Gruline, near Dervaig and Glen Forsa (Clark)

Cyathicula De Not.
(incl. *Phialea* (Fr.) Gill.)

C. cyathoidea (Bull. ex Mérat) Thüm.
Common and widespread on dead stems of various plants (E)

C. pteridicola (Crouan) Dennis
On dead stems of *Pteridium* (K)

C. turbinata (Syd.) Dennis
On dead stems and leaves of *Ranunculus repens* (Clark)

Durella Tul.

D. connivens (Fr.) Rehm.
On *Salix* (K)

D. macrospora Fuckel
On *Betula* (K)

Encoelia (Fr.) Karst.

E. furfuracea (Roth ex Pers.) Karst.
On twigs of *Corylus* (E, K)

E. glauca Dennis
Near Croggan (Clark, K)

Godronia Moug. & Lév.

G. callunigera (Karst.) Karst.
On twigs of *Calluna*

G. cassandrae Peck. var. **callunae** Groves
On *Calluna*, in five localities (Clark)

Heterosphaeria Grev.

H. patella (Tode ex Fr.) Grev.
Common and widespread on herbaceous stems especially on *Angelica* (Clark, E); recorded by White

Hymenoscyphus Gray
(incl. *Helotium* Fr. p.p.)

H. calyculus (Sow. ex Fr.) Phill.
On rotten wood (Clark)

H. caudatus (Karst.) Dennis
On decaying leaves (Clark)

H. cf. **caudatus** (Karst.) Dennis
Differs from the typical form in having a pale amethyst apothecium (K)

H. fagineus (Pers. ex Fr.) Dennis
On fallen cupulus of *Fagus* (K)

H. fructigenus (Bull. ex Mérat) Gray
On fallen fruits of *Corylus* (Clark, K)

H. imberbis (Bull. ex St. Amans) Dennis
On twigs of *Alnus*, Laggan Ulva (K)

H. phyllogenus (Rehm) O. Kuntze
On fallen, rotting leaves of broad-leaf trees (Clark)

H. robustior (Karst.) Dennis
On dead herbaceous stems (Clark)

H. scutula (Pers. ex Fr.) Phill.
Widespread on herbaceous stems (Clark, K)

H. vernus (Boud.) Dennis
On wet twigs (K)

Hymenoscyphus species indet.
Four distinct species from scattered localities, on wood (Clark, K)

Neobulgaria Petrak

N. lilacina (Wulf ex Fr.) Dennis
On conifer bark, Lettermore, Salen Forest (E)

Ombrophila Quél.

O. violacea Fr.
(Clark)

Pezizella Fuckel

P. alniella (Nyl.) Dennis
On fallen female catkins of *Alnus* (K)

P. amenti (Batsch ex Fr.) Dennis
On fallen catkins of *Salix* (K)

P. chrysostigma (Fr.) Sacc.
On dead stems of ferns (Clark)

P. eburnea (Rob.) Dennis
On dead culms of grasses (Clark)

Pezizella sp.
On *Iris pseudacorus*, Loch an Tòrr

Pezoloma Clem.
Sphagnicola Vel.

P. ciliifera (Karst.) Korf
Amongst wet marsh plants and mosses (Clark)

Phaeangellina Dennis

P. empetri (Phill.) Dennis
On dead leaves of *Empetrum nigrum* (Clark)

Pseudohelotium Fuckel

P. alaunae Graddon
Glen Aros (Clark)

Tympanis Tode ex Fr.

T. conspersa Fr.
On dead twigs of *Sorbus aucuparia*

Uncinia Vel. non Pers.

U. laricionis Vel.
Glen Aros (Clark)

Velutarina Korf

V. rufo-olivacea (Alb. & Schw. ex Pers.) Karst.
On dead twigs and branches of *Ulex* (K)

LECANORALES*

LECIDEACEAE

Agyrium Fr.

A. rufum (Pers. ex Pers.) Fr.
On various substrates (E, K)

Karschia Körb.

K. lignyota (Fr.) Sacc.
Frequent and widespread on hardwood (E, K)

K. nigerrima Sacc.
On old wood (K)

---

* See also lichenised forms, pp. 14.16–25.

K. stygia (Berk. & Curt.) Müller
On decorticated *Fraxinus*, Glen Forsa (E)

Mniaecia Boud.

M. jungermanniae (Fr.) Boud.
On moss in wet ravines (E)

OSTROPALES

OSTROPACEAE

Acrospermum Tode ex Fr.

A. compressum Tode ex Fr.

Apostemidium Karst.

A. fiscellum (Karst.) Karst.
On wet *Salix* (Clark)

A. guernisaci (Crouan) Boud.
(Clark)

Stictis Pers. ex Fr.

S. friabilis (Phill. & Plowr.) Sacc. & Trav.
S. sulfurea Rehm.
On fruit body of *Peniophora*: Basidiomycotina and on *Rubus idaeus* (K) (see Dennis, 1971)

S. luzulae Lib. var. **junci** Karst.
An unusual record on *Juncus conglomeratus* (K) (Dennis, 1971)

S. radiata Pers. ex Gray
On dead wood of various deciduous trees (K)

S. stellata Wall.
On dead stems of *Scrophularia*

Vibrissea Fr.

V. truncorum Fr.
On wet wood of deciduous trees (Clark)

PEZIZALES

THELEBOLACEAE

Thelebolus Tode ex Fr.

T. nanus Heim
On cow dung

T. stercoreus Tode ex Fr.
On old dung

ASCOBOLACEAE

Ascobolus Pers. ex Fr.

A. albidus Crouan
On dung (K)

A. furfuraceus Pers. ex Fr.
Frequent and widespread on dung (Clark, K)

Coprotus Korf & Kimbr.

C. aurora (Crouan) Kimbr., Luck-Allen & Cain
On cow dung (K)

C. granuliformis Kimbr.
Ascophanus argenteus (Curr.) Boud.
On cow dung (K)

C. lacteus (Cooke & Phill.) Kimbr., Luck-Allen & Cain
On dung

Iodophanus Korf

I. carneus (Pers. ex Pers.) Korf
On old straw, dung etc. (E)

Lasiobolus Sacc.

L. ciliatus (Schmidt ex Fr.) Boud.
Frequent and widespread on dung

Pyronema Carus

P. domesticum (Sow. ex Fr.) Sacc.
On burnt patch in sand dunes (E)

Saccobolus Boud.

S. depauperatus (Berk. & Br.) Hansen
Recorded only by White, for the first time in Scotland

HUMARIACEAE

Aleuria Fuckel

A. aurantia (Fr.) Fuckel
Common on bare soil in woods and open spaces

Anthracobia Boud.

A. melaloma (Alb. & Schw. ex Fr.) Boud.
On burnt ground (Clark)

Cheilymenia Boud.

C. coprinaria (Cooke) Boud.
Common on dung (Clark, E, K)

C. vitellina (Pers. ex Fr.) Dennis
Croggan, Killiechronan (Clark)

Coprobia Boud.

C. granulata (Bull. ex Fr.) Boud.
Common and widespread on dung (K); recorded by White under *Peziza*

Melastiza Boud.

M. chateri (W. G. Smith) Boud.
On sandy soil (Clark)

Octospora Hedw.

O. alpestris (Sommerf.) Dennis & Itzerott
O. carneola (Saut.) Dennis

Pachyella Boud.

P. babingtonii (Berk.) Boud.
On very wet branches and twigs (Clark, E)

Scutellinia (Cooke) Lamb.

S. hirta (Schum.) Lamb.
(Clark)

S. pseudotrechispora (Schroet.) le Gal.
Glen Aros (K)

S. scutellata (L. ex St. Amans) Lamb.
Frequent and widespread on rotten wood usually in standing water (Clark, E, K,)

S. umbrarum (Fr.) Lamb.
On damp earth and debris (Clark)

Sphaerosporella (Svrček) Svrček & Kubička

S. brunnea (Alb. & Schw. ex Fr.) Svrček & Kubička
Aros Park.

Trichophaea Boud.

T. gregaria (Rehm) Boud.
On soil in conifer wood, Loch Buie (K).

PEZIZACEAE

Otidea Fuckel

O. alutacea (Pers. ex Gray) Massee
On the ground in woodland

O. bufonia (Pers.) Boud.
On soil in woodland (E)

O. cochleata (L. ex St. Amans) Fuckel
On the ground

O. umbrina (Pers. ex Gray) Bres.
Amongst moss (K)

Peziza Dill. ex Fr.

P. badia Pers. ex Mérat
Several recent records (Clark, E, K); recorded by White under the aggregate name which could be one of several closely related species. A member of this group from Loch Bà and with distinctly reticulate spores 16–18 × 2.5–8.5 μm should be compared with *Galactinia phlebospora* le Gal.

P. fimeti (Fuckel) Seav.
On dung

P. subviolacea Svrček
On burnt soil

P. succosa Berk.
On soil in woodland (E)

P. vesiculosa Bull. ex St. Amans
Frequent and widespread on dung and straw

HELVELLACEAE

Cyathipodia Boud.

C. macropus (Pers. ex Fr.) Dennis
Common in woodland (K); recorded by White

Helvella L. ex St. Amans

H. crispa Fr.
On bare ground in woodland

H. lacunosa Afz. ex Fr.
On bare ground in woodland (K)

Leptopodia Boud.

L. elastica (Bull. ex St. Amans) Boud.
Laggan Ulva

Rhizina Fr.

R. undulata Fr.
On the ground under conifers (E)

LOCULOASCOMYCETES

PLEOSPORALES

PLEOSPORACEAE

Chaetosphaerella Müller & Booth

C. phaeostroma (Dur. & Mont.) Müller & Booth
On dead wood (K)

Didymella Sacc.

D. scotica Dennis
A species based on a collection (K) on *Ammophila arenaria* from Iona (Dennis, 1971)

Eudarluca Speg.

E. filum (Cast.) Speg.
See *Darluca filum*: Deuteromycotina

Herpotrichia Fuckel

H. macrotricha (Berk. & Br.) Sacc.
On rotten plant debris

Leptosphaeria Ces. & De Not.

L. acuta (Fr.) Karst.
On dead stems of *Urtica* (Clark, K)

L. ammophilae (Lasch) Ces. & De Not.

L. arundinacea (Sow. ex Fr.) Sacc.
On dead stems of *Phragmites* (E)

L. avenaria Weber
See *Septoria avenae*: Deuteromycotina

L. dolioloides (Auersw.) Karst.
On dead stems of Compositae

L. doliolum (Fr.) De Not.
On herbaceous stems (Clark)

L. fuckelii Niessl

L. graminis (Fuckel) Sacc.
On dead leaves of *Phragmites*

L. jaceae Holm
On *Centaurea nigra*

L. juncina (Auersw.) Sacc.
On dead culms of *Juncus*

L. nardi (Fr.) Ces. & De Not.
On dead *Nardus*

L. nigrans (Rob.) Ces. & De Not.
On dead grass stems

L. ogilviensis (Berk. & Br.) Ces. & De Not.
On dead stem of composite

L. silenes-acaulis De Not.
Gribun (E)

Leptosphaeria sp.
On *Nowellia curvifolia*, Torloisk House

Melanomma Nits. ex Fuckel

M. pulvis-pyrius (Pers. ex Fr.) Fuckel
Widespread and frequent on various substrates (E, K); recorded by White.

Metasphaeria Sacc.

M. clypeosphaerioides Bomm., Rouss. & Sacc.
On *Rubus fruticosus* (K)

Ophiobolus Riess

O. acuminatus (Sow. ex Fr.) Duby
On dead stems of *Cirsium palustre* (K)

Pleospora Rabenh.

P. herbarum (Pers. ex Fr.) Rabenh.
Frequent on various herbaceous stems (E, K)

P. vagans Niessl
Frequent and widespread on various herbaceous stems (E, K)

Pteridiospora Penzig & Sacc.

P. scoriadea (Fr.) Dennis
On twigs of *Betula*

Sporormia De Not.

S. intermedia Auersw.
Frequent on dung (E)

S. megalospora Auersw.
On dung (E)

S. minima Auersw.
On dung and isolated from soil

Teichospora Fuckel

T. obducens (Fr.) Fuckel
On wood of *Fraxinus* (K)

Trematosphaeria Fuckel

T. pertusa (Pers. ex Fr.) Fuckel
On old wood (K)

Trichodelitschia Munk

T. bisporula (Crouan) Munk
On dung of all kinds (E)

VENTURIACEAE

Coleroa (Fr.) Rabenh.

C. alchemillae (Grev.) Wint.
On *Alchemilla xanthochlora*

Stigmatea Fr.

S. robertiani (Fr.) Fr.
One recent collection on *Geranium robertianum*; recorded by White

S. silvatica Sacc.
On needles of *Juniperus*

Venturia Sacc.

V. maculaeformis (Desm.) Wint.
On leaves of *Epilobium montanum*

V. rumicis (Desm.) Wint.
On leaves of *Rumex* sp.

V. inaequalis (Cooke) Wint.
See *Fusicladium dendriticum*: Deuteromycotina

HERPOTRICHIELLACEAE

Berlesiella Sacc.

B. nigerrima (Blox. ex Curr.) Sacc.
On stromata of *Eutypa*: Ascomycotina

BOTRYOSPHAERIACEAE

Botryosphaeria Ces. & De Not.

B. festucae (Lib.) von Arx & Müller

B. quercuum (Schw.) Sacc.
On *Fagus*

Guignardia Viala & Ravaz

G. punctoidea (Cooke) Schroet.
On *Quercus* (K)

LOPHIOSTOMATACEAE

Lophiosphaera Trev.

L. ulicis (Pat.) E. Müller
A new record for the British Isles (Dennis, 1971)

Lophiostoma (Fr.) Ces. & De Not.

L. angustilabrum (Pers. ex Fr.) Chest. & Bell var. **crenatum** (Pers. ex Fr.) Chest. & Bell
On *Hedera* (K)

L. caulinum (Fr.) Ces. & De Not.
On dead herbaceous stems

L. vagabundum (Sacc.) Chest. & Bell
On dead herbaceous stems (K); often placed in *Lophiotrema*

Lophiotrema Sacc.

An unidentified species isolated from soil

Platystomum Trev.

P. compressum (Pers. ex Fr.) Trev.
Lophidium compressum (Pers. ex Fr.) Sacc.

HYSTERIALES

HYSTERIACEAE

Farlowiella Sacc.

F. carmichaeliania (Berk.) Sacc.
(E, K); see also *Montospora megalospora* var. *megalospora*: Deuteromycotina

F. cf. **carmichaeliania** (Berk.) Sacc.
Also see *Monotospora megalospora* var. *fusispora*. On *Fraxinus*, Glen Forsa (E); differs in having boat-shaped spores with more pointed ends.

Gloniopsis De Not.

G. praelonga (Schw.) Zogg
Common and widespread on various substrates (E, K)

Glonium Mühlenb. ex Fr.

G. lineare (Fr.) De Not.
On decorticated wood (K)

Hysterium Tode ex Fr.

H. angustatum Alb. & Schw. ex Mérat
On bark of deciduous trees (K)

Lophium Fr.

L. mytilinum (Pers. ex Fr.) Fr.
On dead wood and bark of conifers (Clark)

Mytilidion Duby

M. acicola Wint.
On needles of *Juniperus communis*, Loch Bà; not in Bisby & Mason (1940): see Dennis (1974)

M. decipiens (Karst.) Sacc.
On needles of *Juniperus communis*

M. gemmigenum Fuckel
On bark of conifer trees

DOTHIDEALES

DOTHIDEACEAE

Plowrightia Sacc.

P. ribesia (Pers. ex Fr.) Sacc.
On dead stems of *Ribes nigrum*

Rhopographus Nits.

R. filicinus (Fr.) Nits.
Common and widespread on *Pteridium* (K) accompanying *Leptostroma filicinum*: Deuteromycotina

Scirrhia Nits.

S. agrostidis (Fuckel) Wint.
On leaves of *Agrostis tenuis*

S. rimosa (Alb. & Schw. ex Fr.) Fuckel
On leaves of *Phragmites communis*

MYCOSPHAERELLACEAE

Cymadothea Wolf

C. trifolii (Killian) Wolf
See *Polythrincium trifolii*: Deuteromycotina; often placed in *Mycosphaerella*

Mycosphaerella Johans.

M. aegopodii Pot.
See *Phleospora aegopodii*: Deuteromycotina

M. iridis (Desm.) Schroet.
On old stems of *Iris pseudacorus* (K)

M. isariphora Johans.
See *Septoria stellariae*: Deuteromycotina

M. juncaginacearum Schroet.
See *Asteroma juncaginacearum*: Deuteromycotina

M. lineolata (Rob. & Desm.) Schroet.
On *Molinia caerulea*

M. cf. **longissima** (Fuckel) Lind.
On *Phalaris arundinacea*

M. maculiformis (Pers. ex Fr.) Schroet.
On dead leaves of broad-leaf trees (K)

M. montellica Sacc.
On *Molinia caerulea*

M. punctiformis (Pers. ex Fr.) Starbäck
Croig (Clark)

M. tassiana (De Not.) Johans.
See *Cladosporium*: Deuteromycotina

Mycosphaerella indet. sp. 1
On *Filipendula ulmaria*, near Salen

Mycosphaerella indet. sp. 2
On *Lythrum salicaria* at Killichronan

CAPNODIALES

CAPNODIACEAE

Strigopodia Bat.

S. resinae (Sacc. & Bres.) Hughes
On resinous exudates of conifers (K); also see *Helminthosporium resinae*: Deuteromycotina

MICROTHYRIALES

MICROTHYRIACEAE

Asterina Lév.

A. veronicae (Lib.) Cooke
One recent record on *Veronica officinalis* (Clark); recorded by White

Aulographum Lib.

A. vagum Desm.
On fallen leaves

Morenoina Theiss

An unidentified species on *Rubus fruticosus*, Aros Park (Clark, IMI)

Stomiopeltis Theiss.

An unidentified species on needle of *Juniperus*

EUMYCOTA: DEUTEROMYCOTINA

Unidentified sterile mycelia isolated in culture appear in the substrate lists.
The classification of this group is under revision and is therefore not further subdivided here.

Acladium Link ex Pers.

A. conspersum Link
On woody debris of deciduous trees (IMI, K)

Acremonium Link ex Fr.

A. album Preuss
On sporangia of Myxomycotina (IMI)

Actinonema Fr.

A. padi Fr.
Reported in Europe to be the conidial state of *Gnomonia padicola*: Ascomycotina; recorded by White

A. rosea Fr.
Reported to be the conidial state of *Diplocarpon rosae*: Ascomycotina; recorded by White

Actinothyrium Kunze

A. graminis Kunze
On culms of *Molinia* (K)

Aegerita Pers. ex Fr.

A. candida Pers. ex Fr.
On woody debris of *Quercus* (IMI)

Alatospora Ingold

A. acuminata Ingold
In foam samples from streams

Anguillospora Ingold

A. crassa Ingold
In foam samples; also see note under *Mollisia*: Ascomycotina

A. longissima (Sacc. & Syd.) Ingold
In foam samples from streams

Arthrinium Kunze ex Fr.

A. curvatum (Link ex Corda) Kunze var. **minus** M. B. Ellis
On *Carex* debris (K)

A. morthieri Fuckel
(Clark) (Dennis, 1974)

A. pucciniodes (DC. ex Mérat) Kunze
First British record

Arthrobotrys Corda

A. oligospora Fres.
On rotting leaves

Articulospora Ingold

A. tetracladia Ingold
In foam samples from streams

Articulospora sp.
Isolated from soil under *Pinus*

Ascochyta Lib.

A. graminicola Sacc.
On fading leaves of grasses

A. typhoidearum (Desm.) Cunnell
On culms of *Sparganum* (K)

A. viciae Lib.
On leaves of *Vicia sepium*

Aspergillus Mich. ex Fr.

A. flavus Link ex Fr.
Isolated from soil and vegetable debris

A. fumigatus Fres.
Common and widespread in soils

A. niger van Tiegh.
Isolated from wet peat

A. terreus Thom
Isolated from soil

Asteroma DC. ex Fr.

A. juncaginacearum Rab
On *Triglochin maritimum*; conidial state of *Mycosphaerella juncaginacearum*: Ascomycotina

Bactridiopsis Frag. & Cif.

An unidentified species isolated from soil

Botrytis Pers. ex Fr.

B. cinerea Pers. ex Fr.
Common and widespread on various substrates (E, IMI, K); some isolates supposed conidial state of *Sclerotinia fuckeliana*: Ascomycotina

B. globosa Raabe
On leaves of *Allium ursinum*; conidial state of *Sclerotinia globosa*: Ascomycotina

Brachysporium Sacc.

B. britannicum Hughes
On woody debris of deciduous trees (IMI)

B. graminis Boy. & Jacz.
Isolated from several soil samples

B. obovatum (Berk.) Sacc.
Isolated from soil

B. sativum Pam, King & Back.
On rotting leaves

Brachysporium sp.
Isolated from brown earth

Camarographium Bubák

C. stephensii Bubák
Aros Park (IMI)

Candida Berkh.

An unidentified species was isolated from soil samples

Cephalosporiopsis Peyron.

C. alpina Peyron.
Isolated from soil and vegetable debris

Cephalosporium Corda

C. acremonium Corda
Frequent; isolated from soil and vegetable debris

Cercospora Fres.

C. mercurialis Pass.
On *Mercurialis perennis*

C. murina Ell. & Kell.
On *Viola palustris*

Ceuthospora Grev., non Fr.

C. laurocerasi Grove
Supposed imperfect stage of *Trochila laurocerasi*: Ascomycotina recorded by White as *C. lauri* (Grev.) Grev.

C. phacidiodes Grev.
On various substrates (K); supposed imperfect stage of *Phacidostroma multivalve:* Ascomycotina

Chalara (Corda) Rabenh.

C. fusidioides Corda
On *Senecio aquaticus* (E)

Chloridium Link

C. chlamydosporus (van Beyma) Hughes
Isolated from several soil samples

Chrysosporium Corda

C. asperatum Carmich.
Frequent and widespread in soils

C. pannorum (Link) Hughes
Frequent and widespread; includes *Sporotrichum carnis* Brookes & Hansford

Cladosporium Link ex Fr.

C. cladosporioides (Fres.) de Vries
Isolated from wet peat

C. herbarum Link
Common and widespread; some strains supposed imperfect stage of *Mycosphaerella tassiana*: Ascomycotina; recorded by White

Clathrosphaerina van Beverw.

C. zalewskei van Beverw.
Isolated from old leaves

Clavariopsis de Wild

C. aquatica de Wild
In foam sample from stream

Clavatospora S. Nillss.

C. stellata Ingold & Cox
In foam sample from stream

Colletotrichum Corda
Vermicularia Tode ex Fr.

C. atramentarium (Berk. & Br.) Taubenh.
Isolated from wet peat.

C. dematium (Pers. ex Fr.) Grove
On herbaceous stems

C. graminicolum Wilson
On remains of grass

C. lineola Corda
On *Ammophila arenaria*

Coniothyrium Corda

C. psammae Oud.
On dead leaves of *Ammophila*

Cordana Preuss emend. Sacc.

C. pauciseptata Preuss
Isolated from soil

Coryneopsis Grove

C. rubi (Westd.) Grove
Seimatosporium lichenicola (Corda) Shoem. & Müller
On dead stems of *Rubus fruticosus* (K)

Cryptocline Petrak

C. paradoxa (De Not.) von Arx
On *Hedera* (K)

Cylindrium Bon.

C. griseum Bon.
On wood of *Quercus*

Cylindrocarpon Wollenw.

C. destructans (Zins.) Scholt.
C. radicicola Wollenw.
Isolated from soil samples; conidial state of *Nectria radicicola*: Ascomycotina

Cylindrocolla Bon.

C. urtica (Pers. ex Fr.) Bon.
On old stems of *Urtica* (K); conidial state of *Calloria fusarioides*: Ascomycotina

Cytospora Ehrenb. ex Fr.

C. ambiens Sacc.
On twigs of *Quercus* (K); conidial state of *Valsa ambiens*: Ascomycotina

C. curreyi Sacc.
On branch of *Larix* (K); conidial state of *Valsa curreyi*: Ascomycotina

C. dubyi Sacc.
On *Juniperus*

C. fuckelii Sacc.
On twigs of *Corylus* (K); conidial state of *Valsa fuckelii*: Ascomycotina

C. massariana Sacc.
On *Sorbus aucuparia* (K)

Dactylella Grove

An unidentified species isolated from wet peat

Dactylium Nees ex Fr.

D. dendroides Fr.
On old fruiting body of *Polyporus squamosus*: Basidiomycotina (IMI)

D. varium (Nees) Fr.
Cladobotryum varium Nees
On moss on bank (IMI)

Dendryphion Wallr.

D. comosum Wallr.
Glen Aros (Clark, E)

Diheterospora Kamysk.

D. catenulata Kamysk.
Isolated from soil

Dinemasporium Lév.

D. graminum Lév. var. **graminum**
Isolated from soil

D. graminum var. **strigosulum** Karst.
On culms of *Juncus conglomeratus*

Discosia Lib.

D. artocreas (Tode ex Fr.) Fr.
On debris of *Carex* sp.

Ectostroma Fr.

E. iridis Fr.
On stems of *Iris pseudacorus*

Endophragmia Duvemoy & Maire

E. atra (Berk. & Br.) M. B. Ellis
(Clark)

Flagellospora Ingold

F. curvula Ingold
In foam samples from streams

Fumago Pers.

F. vagans Pers. ex Sacc.
On needles of *Pinus nigra* (E)

Fusarium Link ex Fr.

F. culmorum (W. G. Smith) Sacc.
Isolated from soil and on grass remains in rabbit pellets

F. graminearum Schwabe
On rotting leaves; conidial state of *Gibberella zeae*: Ascomycotina

F. poae (Peck) Wollenw.
Isolated from soil

F. solani (Mart.) Sacc.
Isolated from several soils; conidial state of *Nectria haematococca*: Ascomycotina

F. sporotrichoides Sherb.
On *Iris pseudacorus* and forming red patches at base of stems of *Hygrophorus*: Basidiomycotina

Fusicladium Bon.

F. depressum (Berk. & Br.) Sacc.
Cercosporidum depressum Berk. & Br.
On *Angelica sylvestris*

F. dendriticum (Wallr.) Fuckel
On cultivated apples; conidial state of *Venturia inaequalis*: Ascomycotina

Fusidium Link ex Fr.

F. griseum Link ex Wallr.
Isolated from soil samples

Gibellula Cav.

G. araneanum (Schw.) Syd.
On spider (Clark)

Gilmaniella Barron

G. humicola Barron
Isolated from soil samples

Gliocladium Corda

G. deliquescens Sopp
Isolated from soil samples

G. roseum (Link ex Fr.) Banier
Frequent in soils; some strains are imperfect states of *Nectria gliocladiodes*: Ascomycotina

Gliomastix Gueguen

G. convoluta (Harz) Mason
Isolated from soil and wet peat

Gloeosporium Desm. & Mont.

G. rhododendri Briosi & Cav.
On leaves of *Rhododendron ponticum*

G. ribis Mont. & Desm.
On *Ribes nigrum*; conidial state of *Pseudopeziza ribis*: Ascomycotina

G. veronicarum Ces.
Discogloeum veronicae (Lib.) Petrak
On fading leaves of *Veronica beccabunga*

Graphium Corda

G. calicioides (Fr.) Cooke & Massee
On conifer bark, Lettermore Forest (E)

Helicosporium Nees ex Fr.

H. vegetum Nees ex Fr.
On leaves of *Fagus*, in stream

Heliscus Sacc. & Therry

H. lugdunensis Sacc. & Therry
In foam samples in stream; conidial state of *Nectria lugdunensis*: Ascomycotina

Helminthosporium Link ex Fr.
(See also *Pseudospiropes*)

H. cylindricum Corda
On *Betula* (K)

H. folliculatum Corda
Sporidesmium folliculatum (Corda) Link ex Fr.
On *Sorbus* (K)

H. hirudo Sacc. var. **anglicum** Grove
Sporidesmium anglicum (Grove) M. B. Ellis
On decorticated wood (K)

H. resinae Bres.
On exudate of *Larix* (IMI), conidial state of *Strigopodia resinae*: Ascomycotina

H. velutinum Link ex Fr.
On *Corylus* (IMI)

Helminthosporium sp.
Isolated from soil samples

Hendersonia Sacc.

H. acicola Münch. & Tub.
(E); connected with needle blight, probably conidial state of *Hypoderma pinicola*: Ascomycotina

H. crastophila Sacc.
On culms of *Phalaris*

Heterocephalum Thaxt.

An unidentified species isolated from soil samples

Heterosporium Klotz. ex Cooke

H. ossifragi (Rostr.) Lind.
On leaves and stems of *Narthecium*

Hobsonia Berk.

H. mirabilis (Peck) Lind.
On wood, first European record (Dennis, 1971)

Hughesiella Bat. & Vital

An unidentified species from wet peat under *Myrica*; possibly synonymous with *Thielaviopsis*

Humicola Traaen

H. brevis (Gilm. & Abbott) Gilm.
Isolated from several soil samples

H. fuscoatra Traaen
Common and widespread in soils

H. grisea Traaen
Common and widespread in soils

H. nigrescens Omvik
Isolated from soil

Humicola sp.
Isolated from several soil samples

Illosporium Mart. ex Fr.

I. cf. **coccineum** Fr.
On thalli of *Parmelia exasperata*: Lichenes (K)

Isaria Pers. ex Fr.

I. farinosa Dicks. ex Fr.
Isolated from soil

Isaria sp.
Isolated from soil under *Myrica*; see also *Paecilomyces*

Isariopsis Fres.

I. alborosella (Desm.) Sacc.
On debris of *Cerastium*

I. carnea Oud.
On leaves and stems of *Lathyrus pratensis*

Leptostroma Fr. ex Fr.

L. donacinum Sacc. var. **majus** Trail
On culms of *Ammophila arenaria*

L. filicinum Fr. ex Fr.
On dead petioles of *Pteridium* (K); conidial state of *Rhopographus filicinus*: Ascomycotina

L. pinastri Desm.
On pine needles (K); conidial state of *Lophodermium pinastri*: Ascomycotina

Leptothyrium Kunze ex Wallr.

L. acerinum Corda
On *Acer pseudoplatanus* (K)

L. alneum (Lév.) Sacc.
On *Alnus glutinosa*; supposed conidial state of *Gnomiella tubiformis*: Ascomycotina recorded by White under *Melasmia* Sacc.

L. medium Cooke
On leaves of *Quercus* (K); possibly connected to *Coccomyces coronatus*: Ascomycotina

L. pinastri Karst.
On *Pinus sylvestris*

L. quercinum Sacc.
On old leaves of *Quercus* (K); conidial state of *Coccomyces dendatus*: Ascomycotina

L. rubi Sacc.
On stems of *Rubus* (K); possibly conidial state of *Hypoderma virgultorum*: Ascomycotina

Linodochium Höhn.

L. hyalinum (Lib.) Höhn.
On needles of *Pinus* (K); an interesting British record (Dennis, 1971)

Macrophoma (Sacc.) Berl. & Vogl.

M. graminella (Sacc.) Berl. & Vogl.
On culms of *Phalaris*

Mammaria Ces.

M. echinobotryoides Ces.
Isolated from a soil sample

Marssonina Magn.

M. delastrei (Delacr.) Magn.
On leaves of *Silene dioica*; conidial state of *Diplocarpon agrostemmatis*: Ascomycotina

M. potentillae (Desm.) Magn.
On *Potentilla anserina*

Melanconium Link ex Fr.

M. apiocarpum Link
On twigs of *Alnus*; conidial state of *Melanconis alni*: Ascomycotina

M. bicolor Nees
On twigs (E)

Menispora Pers. ex Chev.

M. caesia Preuss
On wood

Monilia Bon.

An unidentified species isolated from peat under *Myrica*

Monodictys Hughes

M. paradoxa (Corda) Hughes
Isolated from two soil samples

Monodictys sp.
Isolated from two soil samples

Monotospora Sacc.

M. megalospora Berk. & Br. var. **megalospora**
On wood; conidial state of *Farlowiella carmichaeliana*: Ascomycotina

M. megalospora var. **fusispora** Sacc.
On wood (E); conidial state of *Farlowiella cf. carmichaeliana*

Mycogone Link ex Chev.

M. rosea Link ex Chev.
Isolated from wet leaves

Myrioconium Syd.

M. tennellum (Sacc) Höhn.
On culms of *Juncus* (K)

Myrothecium Tode ex Fr.

M. roridium Tode ex Fr.
Isolated from a soil sample

M. verrucaria (Alb. & Schw.) Dit. ex Fr.
On bark and isolated from wet vegetable debris

Myxosporium Link ex Corda

M. lanceola Sacc. & Roum.
Fusicoccum quercinum Sacc.
Conidial state of *Diaporthe leiphaemia*: Ascomycotina.

Oidiodendron Robak

O. flavum Szilvinyi
On *Fagus* (IMI)

O. fuscum Robak
Isolated from soil

O. griseum Robak
Isolated from soil and on wood (K)

Oidiodendron sp.
Isolated from soil under *Myrica*

Ovularia Sacc.

O. destructiva (Phill. & Plowr.) Massee
On leaves of *Myrica*

O. obliqua (Cooke) Oud.
On leaves of *Rumex* spp.

O. primulana Karst.
On leaves of *Primula*

O. sphaeroidea (Sacc.) Sacc.
On leaves of *Lotus uliginosus*

Paecilomyces Bain.

P. elegans (Corda) Mason & Hughes
Isolated from soil

P. varioti Bain.
Isolated from soil

Papulospora Preuss

An unidentified species on submerged leaves under *Myrica*

Penicillium Link ex Fr.

P. capsulatum Raper & Fennell
Isolated from soil

P. claviforme Bain.
On leaves of *Quercus*

P. fellutanum Biourge
Isolated from soil samples

P. frequentans Westling
Isolated from soil

P. implicatum Biourge
Isolated from wet peat

P. nigricans Bain. ex Thom
Isolated from two soils

P. piceum Raper & Fennell
Isolated from rotting vegetation and soil

P. purpurogenum Stoll
Isolated from soil

P. rubrum Stoll
Isolated from soil

P. simplicissimum (Oud.) Thom
Isolated from soil

P. spinulosum Thom
Isolated from soil

P. steckii Zaleski
Isolated on three occasions from peat and wet leaf debris

P. stoloniferum Thom
On submerged leaves of *Myrica*

P. thomii Maire
Isolated on several occasions from soil and vegetable debris

P. verruculosum Peyronel
Isolated from soil

P. waksmannii Zaleski
Isolated from soil

Penicillium spp.
Five as yet undetermined isolates from soil samples belong to this genus.

Periconia Tode ex Schw.

P. atra Corda
Aros Park (Clark)

P. byssoides Pers. ex Corda
Isolated from soil

P. hispidula (Pers.) Mason & Ellis
On culms of *Ammophila arenaria*

Phaeoseptoria Speg.

P. aireae (Grove) Sprague
On leaves of *Deschampsia caespitosa* (K)

Phialophora Medlar

P. cf. **richardsiae** (Nannf.) Const.
On straw debris in midden (E)

Phialophora sp. 1
On *Fagus* (IMI)

Phialophora sp. 2
Isolated from soil under *Fagus*

Phleospora Wallr.

P. aegopodii (Preuss) Grove
On *Aegopodium podagraria*; probably a state in the life cycle of *Mycosphaerella aegopodi*: Ascomycotina

P. pseudoplatani (Rob. ex Desm) Bub.
On leaves of *Acer campestre*

Phoma Sacc.

P. arundinacea Sacc.
On various grasses (K)

P. complanata Desm.
On dead leaves of *Ligusticum* (E)

P. herbarum Westd.
On herbaceous stems (K); records include *P. oleracea* var. *scrophulariae* Sacc.

P. lingam (Tode ex Fr.) Desm.
On stems of Kale in field (K)

P. longissima Berk.
On dead herbaceous stems of *Oenanthe*

Phoma sp.
Isolated from soil under *Fagus*

Phomopsis (Sacc.) Sacc.

P. achilleae Höhn.
On dead stems of *Senecio aquaticus*

P. controversa (Sacc.) Trav.
On *Fraxinus*; pycnidial stage of *Diaporthe controversa*: Ascomycotina

P. crustosa (Bomm., Rouss. & Sacc.) Trav.
On *Ilex*; conidial stage of *Diaporthe crustosa*: Ascomycotina

P. ligulata Grove
On *Ulex* debris (K); conidial stage of *Diaporthe nucleata*: Ascomycotina

P. occulta (Sacc.) Trav.
On cones of *Pinus* (K); conidial stage of *Diaporthe conorum*: Ascomycotina

P. revellens (Sacc.) Höhn.
On *Corylus* (K); conidial stage of *Diaporthe revellens*: Ascomycotina

P. stictica Trav.
On *Buxus sempervirens* (K); conidial stage of *Diaporthe retecta*: Ascomycotina

Phomopsis sp. 1
An unidentified species on old stems of *Hieracium* sp. (K)

Phomopsis sp. 2
An unidentified species on submerged *Fagus* leaves

Phyllosticta Pers. ex Desm.

P. hedericola Dur. & Mont.
On leaves of *Hedera helix*

P. roboris Oud.
On leaves of *Quercus* (K)

P. violae Desm.

Pleurophragmium Cost.

P. simplex (Berk. & Br.) Hughes
On debris of *Urtica dioica* (K)

Pollaccia Bald. & Cif.

P. radiosa (Lib.) Bald., Cif. & Rudh.
On wood of *Populus tremula*

Polymorphum Chev.
Dichaena Fr., *Psilospora* Rabenh.

P. rugosum (Fr.) D. Hawksw. & Punithalingam
Dichaena faginea (Fr.) Fr.
On bark of *Fagus* (K) and *Quercus* (K)
White recorded *Dichaena strobilina* (Fr.) Fr. but it is impossible to know, in the absence of voucher material, what species was meant.

Polyscytalum Reiss

P. foecundissimum Reiss
Isolated from soil samples

Polythrincium Kunze & Schum. ex Fr.

P. trifolii Kunze & Schum. ex Fr.
On *Trifolium repens*; conidial state of *Cyamadothea trifolii*: Ascomycotina

Pseudospiropes M. B. Ellis

P. longipilus (Corda) Hol.-Jech.
On *Betula*

P. nodosus (Wallr.) M. B. Ellis
On woody hosts

Ptychogaster Corda

P. albus Corda
On mossy wood; see *Tyromyces*: Hymenomycetes: Basidiomycotina

Pullularia Berkh.

P. pullulans (De Bary) Berkh.
Isolated from soil samples

Pycnostysanus Lindau

P. azaleae (Peck) Mason
Sporocybe azaleae (Peck) Sacc.
On buds and leaf spots of *Rhododendron* (K)

Pycnothyrium Died.

P. litigiosum (Desm.) Died.
On debris of *Pteridium*

Ramularia Sacc.

R. ajugae (Niessl) Sacc.
On leaves of *Ajuga reptans* (K)

R. anagallidis Lindroth
On leaves of *Veronica anagallis-aquatica*

R. arvensis Sacc.
On leaves of *Potentilla anserina*

R. heraclei (Oud.) Sacc.
On leaves of *Heracleum sphondylium*

R. lactea (Desm.) Sacc.
On leaves of *Viola riviniana*

R. obducens Thüm.
On leaves of *Pedicularis* sp.

R. pruinosa Speg.
On leaves of *Senecio jacobaea*

R. ranunculi Peck
On leaves of *Ranunculus repens*

R. valerianae (Speg.) Sacc.
On leaves of *Valeriana officinalis*

R. variabilis Fuckel
On leaves of *Digitalis*

Rhabdospora (Sacc.) Sacc.

R. ramealis (Rob. & Desm.) Sacc.
On living stems of *Rubus fruticosus* (K)

Rhabdospora sp.
On *Senecio jacobaea* (K)

Rhinotrichum Corda

An unidentified species isolated from soil under *Pinus*

Rhizoctonia DC. ex Fr.

An unidentified species isolated from soil

Rhizosphaera Mang. & Har.

R. kalkhoffi Bub.
On needles of *Picea* (K)

Sclerotium Tode ex Fr.

S. complanatum Tode
On herbaceous debris (IMI)

S. hyacinthi Guepin
On *Endymion non-scriptus* capsules

S. muscorum Gray
Amongst *Leucobryum*

S. papyricola auct.
A soil isolate

S. rolfsii Sacc.
Isolated from a soil sample

Sclerotium sp. 1
On *Eleocharis*; sclerotia of *Claviceps nigricans*: Ascomycotina

Sclerotium sp. 2
On *Scirpus cespitosus*; sclerotia of *Sclerotinia gregoriana*: Ascomycotina

Sclerotium sp. 3
On *Mercurialis perennis*; sclerotia of *Typhula*: Basidiomycotina

Sclerotium sp. 4
On Gramineae; sclerotia of *Claviceps purpurea*: Ascomycotina

Scopulariopsis Bain.

S. brevicaulis (Sacc.) Bain.
Isolated from a soil sample

Sepedonium Link ex Fr.

S. chrysospermum Fr.
One recent collection; recorded by White on a disintegrating fruit-body of *Boletus* sp.; conidial state of *Apiocrea chrysosperma:* Ascomycotina

Septoria Sacc.

S. acetosa Oud.
On leaves of *Rumex acetosa* (K)

S. anemones Desm.
On leaves of *Anemone nemorosa* (K)

S. avenae Frank
On leaves of *Anthoxanthum odoratum*; conidial state of *Leptosphaeria avenaria*: Ascomycotina

S. gei Rob. ex Desm.
On leaves of *Geum* sp.

S. hydrocotyles Desm.
On leaves of *Hydrocotyle*, three recent collections; recorded by White

S. lysimachiae West.

S. oenanthes Ell. & Ev.
On *Oenanthe crocata*

S. cf. **oudemansii** Sacc.
On *Poa nemoralis*

S. polygonorum Desm. var. **persicaria** Trail
On leaves of *Polygonum persicaria* (E)

S. rosae Desm.
On *Rosa canina* agg.

S. scabiosicola Desm.
Two recent collections (K); recorded by White

S. senecionis-scleraticus Syd.
Resembling a *Rhabdospora*

S. stachydis Rob. & Desm.
One recent collection; recorded by White

S. stellariae Rob. & Desm.
On *Stellaria media*; supposed conidial state of *Mycosphaerella isariphora*: Ascomycotina

S. tormentillae Rob. & Desm.
On *Potentilla erecta*

S. virgaureae Desm.

Septoria sp.
On *Scirpus cespitosus*

Sesquicillium W. Gams

S. candelabrum (Bon.) W. Gams
Isolated from wet peat

Sphaceloma De Bary

S. necator (Ell. & Ev.) Jenk. & Shear
On leaves of *Rubus fruticosus*

Sphaerellopsis Cooke
Darluca Cast., non Raf.

S. filum (Biv.-Bern. ex Fr.) Sutton
In uredosori of rust fungi: Basidiomycotina. Reported to be conidial state of *Eudarluca filum*: Ascomycotina.

Spirosphaera van Beverw.

S. floriforme van Beverw.
On wet leaves

Sporotrichum Link ex Fr.
See also *Chrysosporium*

S. thermophile Apinis
Isolated from a soil sample

Stachylidium Link ex Fr.

S. cyclosporum Grove ex Wallr.
Verticillium cyclosporum (Grove) Nees ex Wallr.
On *Betula*

Stagonospora (Sacc.) Sacc.

S. compta (Sacc.) Died.
On living leaves of *Trifolium repens*

S. subseriata (Desm.) Sacc.
On leaves of grasses (K)

S. vexata Sacc.
On leaves of *Phalaris arundinacea*

Stysanus Corda

S. microsporus Sacc.
Doratomyces m. (Sacc.) Morton & Smith
Isolated from soil and wet peat

S. stemonitis (Pers. ex Fr.) Corda
Doratomyces s. (Pers. ex. Fr.) Morton & Smith
Isolated from soil

Sympodiella Kendr.

S. acicola Kendr.
Isolated from soil

Tetracladium De Wild

T. setigerum (Grove) Ingold
In foam from stream

Tetraploa Berk. & Br.

T. aristata Berk. & Br.
On wet straw

Thermoidium Miehe

T. sulphureum Miehe
Isolated from wet peat

Thermomyces Tsiklinsk.

T. lanuginosus Tsiklinsk.
Isolated from soil

Thielaviopsis Went

T. basicola (Berk. & Br.) Ferraris
Isolated from blackened root of composite (E)

Thyriostroma Died.

T. spiraea (Fr.) Died.
On stems of *Filipendula ulmaria* (K)

Tiarospora Sacc. & March.

T. perforans (Rob.) Höhn.

Tilachlidium Preuss

T. tomentosum (Schrad. ex Fr.) Lindau
On sporangia of *Trichia varia*: Myxomycota (K)

Torula Pers ex Fr.

See also *Xylohypha*

T. herbarum (Pers.) Link ex Gray
Glen Aros (Clark)

T. lucifuga Oud.
Isolated from soil

Trichocladium Harz

Trichocladium sp. 1
On decorticated *Betula* (E)

Trichocladium sp. 2
Isolated from soil samples

Trichoderma Pers ex Fr.

T. album Preuss
Isolated from soil

T. koningii Oud.
Frequent and widespread in soils

T. viride Pers. ex Fr.
Includes records of *T. lignorum* (Tode) Harz; widespread in soils and on vegetable debris (IMI). Some isolates are conidial stage of *Hypocrea rufa*: Ascomycotina

Trichophyton Malmst.

T. ajelloi (van Breus.) Ajello
Isolated from soil and vegetable debris

T. aff. **simii** (Pinoy) Stockdale
Isolated from soil

T. terrestre Durie & Frey
Isolated from soil

T. vanbreuseghemii Rioux
On submerged leaves of *Quercus*

Trichosporon Behrend

An unidentified species isolated from soil samples

Tricladium Ingold

T. gracile Ingold
In foam samples from streams

T. splendens Ingold
In foam samples from streams

Trimmatostroma Corda

T. betulinum (Corda) Hughes
Coniothecium betulina Corda
From soil under *Ulex*

Trimmatostroma sp.
On *Quercus*

Triposporium Corda

T. elegans Corda
On *Populus tremula*

Triscelophorus Ingold

T. monosporus Ingold
In foam samples from streams

Tubercularia Tode ex Fr.

T. vulgaris Tode ex Fr.
On branches and twigs (E, K); conidial state of *Nectria cinnabarina*: Ascomycotina

Ulocladium Preuss

U. atrum Preuss
On damp ceiling paper

Varicosporium Kegel

V. elodeae Kegel
In foam sample from a stream

Varicosporium sp.
On submerged leaves

Verticillium Nees ex Wallr.

V. terrestre (Pers. ex Fr.) Sacc.
Isolated from soils

Volutella Tode ex Fr.

V. arundinis Desm. ex Fr.
On *Luzula* leaves from several localities (K)

V. buxi (DC. ex Fr.) Berk.
On *Buxus* (K)

V. ciliata (Alb. & Schw.) Fr.
Isolated from soils

Xylohypha (Fr.) Mason

X. nigrescens (Pers. ex Fr.) Mason
On woody debris of deciduous trees (IMI, K); recorded by White as *Torula ovalispora* Berk.

EUMYCOTA: ZYGOMYCOTINA

ZYGOMYCETES

MUCORALES

MUCORACEAE

Absidia van Tiegh.

A. corymbifera (Cohn) Sacc. & Trotter

A. cylindrospora Hagem
Isolated on four occasions

A. glauca Hagem
Isolated from seven soil samples

A. orchidis (Vuill.) Hagem

A. ramosa (Lindt) Lendner
Isolated from seven soil samples

Mucor Mich. ex Fr.

M. fragilis Bain.

M. genevensis Lender

M. hiemalis Wehmer
Isolated from four soil samples

M. pusillus (Lindt) Hagem

M. racemosus Fres.

M. ramannianus Møller
Often placed in *Mortierella*

M. silvaticus Hagem
Isolated from five soil samples

Zygorhynchus Vuill.

Z. moelleri Vuill.

Z. psychrophilus Schip. & Hintikka

MORTIERELLACEAE

Mortierella Coemans

M. alpina Peyronel
Isolated on four occasions

M. bainieri Costan.
Isolated from five soil samples

M. humilis Linnem.

M. hygrophila Linnem.

M. macrocystis W. Gams

PIPTOCEPHALIDACEAE

Piptocephalis De Bary

P. cylindrospora Bain.

Syncephalis van Tiegh. & Le Monn.
An unidentifiable species only

PILOBOLACEAE

Pilaira van Tiegh.

P. anomala Schroet.

Pilobolus Tode ex Fr.

P. crystallinus (Tode ex Hook.) van Tiegh.
On dung (K)

P. kleinii van Tiegh.
(K)

ENTOMOPHTHORALES

ENTOMOPHTHORACEAE

Entomophthora Fres.

E. aphidis Fres.

CHYTRIDIOMYCETES

MONOBLEPHARIDALES

MONOBLEPHARIDACEAE

Monoblepharella Sparrow

M. taylorii Sparrow
On submerged leaves of *Myrica*

CHYTRIDIALES

SYNCHYTRIACEAE

Synchytrium De Bary & Woron.

S. aff. **aureum** Schroet.
(K)

S. erieum Karling
(K)

EUMYCOTA: MASTIGOMYCOTINA

OOMYCETES

SAPROLEGNIALES

SAPROLEGNIACEAE

Achlya Nees

A. racemosa Hildebr.

Saprolegnia Nees

S. asterophora De Bary

S. ferax (Gruith.) Thuret

PERONOSPORALES

PYTHIACEAE

Phytophthora De Bary
An unidentifiable species from soils of hill pasture

Pythiogeton Minden

P. utriforme Minden
On submerged leaves of *Myrica* and on rotting debris in a swamp

PERONOSPORACEAE

Bremia Regel

B. centaureae Syd.

B. lactucae Reg.
(K)

B. sonchi Saw.

Peronospora Corda

P. alta Fuckel
(K)

P. lotorum H. Syd.
(K)

P. obovata Bon.
(K)

P. ranunculi Gaum.
(E, K)

P. tomentosa Fuckel
(K)

P. violacea (Berk.) Cooke
(K)

Plasmopara Schroet.

P. densa Schroet.
(K)

P. nivea (Unger) Schroet.
Frequent on *Angelica sylvestris* and *Oenanthe crocata* (K)

P. pusilla Schroet.
(K)

ALBUGINACEAE

Albugo (Pers.) Rouss. ex Gray

A. caryophyllacearum (Wallr.) Cif. & Biga
Cystopus lepigoni De Bary
(K)

A. tragopogonis (Pers.) Schroet. var. **cirsii** Cif. & Biga
Cystopus cubicus De Bary
(K)

MYXOMYCOTA

Bibliogr.: Ing (1968)

CERATIOMYXOMYCETES

CERATIOMYXALES

CERATIOMYXACEAE

Ceratiomyxa Schroet.

C. fruticulosa (O. Müll.) Macbr.
Frequent and widespread (Clark, E, K)

MYXOMYCETES

LICEALES

LICEACEAE

Licea Schard. emend. Rostaf.

L. kleistobolus Martin
Developed in moist chamber culture, Croig (Clark); third Scottish record

L. minima Fr.
On conifer debris (Clark)

L. parasitica (Zukal) Martin
Developed in moist chamber cultures

L. pedicellata (H. C. Gilb.) H. C. Gilb.
Developed in moist chamber cultures, Croig (Clark); third Scottish record

L. variabilis Schrad.
On pine sticks (Ing)

RETICULARIACEAE

Lycogala Mich. ex Adans.

L. epidendrum (L.) Fr.
Common and widespread on rotting logs (BM, Clark, E, Ing, K)

TRICHIALES

TRICHIACEAE

Arcyria Hill ex Wiggers

A. cinerea (Bull.) Pers.
Common and widespread on rotten wood (K)

A. denudata (L.) Wettst.
Frequent and widespread on decaying wood

A. incarnata (Pers) Pers.
On *Quercus* (K)

A. nutans (Bull.) Grev.
On woody debris (Clark, K)

A. oerstedtii Rostaf.
At base of stump Gruline (Clark)

Perichaena Fr.

P. chrysosperma (Currey) Lister
Developed in moist bark-culture, Kintra (Clark)

Reticularia Bull. emend. Rostaf.

R. lycoperdon Bull.
Frequent on woody debris (K); recorded by White

Tubifera Gmel.

T. ferruginosa (Batsch) Gmel.
On decaying conifer wood (Clark, Ing)

CRIBRARIACEAE

Cribraria Pers. emend. Rostaf.

C. argillacea (Pers.) Pers.
Frequent and widespread on rotten conifer wood (BM, K)

C. aurantiaca Schrad.
Frequent and widespread on dead conifer wood (Clark, Ing)

C. cancellata (Batsch.) Nann.-Bremek.
On conifer wood

C. persoonii Nann.-Bremek.
On conifer wood, Carsaig Bay

P. corticalis (Batsch) Rostaf. var. **corticalis**
On conifer wood; Glen Aros (K)

P. minor (G. List.) Hagelst.
Developed in moist chamber culture, Loch Buie (Clark)

Trichia Haller emend. Rostaf.

T. affinis De Bary
(Clark, Ing, K)

T. botrytis (Gmel.) Pers.
Frequent on old twigs especially of conifers (Clark, K)

T. contorta (Ditm.) Rost. var. **contorta**
On *Fagus* (K)

T. contorta var. **inconspicua** (Rostaf.) Lister
On dead *Polygonum* stems, Aros Park (Clark)

T. decipiens (Pers.) Macbr.
Frequent and widespread on dead wood (Clark, K)

T. floriformis (Schw.) G. Lister
Common and widespread on wet rotten wood (Clark, K)

T. persimilis Karst.
Common and widespread on dead wood (Clark, K)

T. varia (Pers.) Pers.
Frequent and widespread on wet rotten wood

ECHINOSTELIALES

ECHINOSTELIACEAE

Echinostelium De Bary

E. minutum De Bary
Developed in moist chamber culture, Calgary House (Clark)

STEMONITALES

STEMONITACEAE

Amaurochaete Rostaf.

A. fuliginosa (Sow.) Macbr.
On newly fallen conifer logs, Craignure

Collaria Nann.-Bremek.

C. lurida (List.) Nann.-Bremek.
On litter Aros Park (Clark); third Scottish record

C. insessum (G. Lister) B. Ing
Developed in moist chamber culture, Croggan (Clark)

Calloderma G. Lister

C. oculatum (Lipp.) G. Lister
Developed in moist chamber culture, Aros Park (Clark)

Comatricha Preuss emend. Rostaf.

C. elegans (Racib.) Lister
On conifer sticks, Fishnish (Clark)

C. laxa Rostaf.
On sticks, Loch na Keal (Clark)

C. nigra (Pers.) Schroet.
Common and widespread on woody debris (K)

C. pulchella (Bab.) Rostaf.
On conifer litter, Lettermore (Clark)

C. typhoides (Bull.) Rostaf.
Frequent and widespread on damp logs (Clark, K)

Enerthenema Bowman

E. papillatum (Pers.) Rostaf.
On sticks and bark of conifer and *Quercus* (Clark)

Lamproderma Rostaf.

L. arcyrioides (Sommerf.) Rostaf.
On litter in *Iris* marsh, Gruline (Clark)

L. columbinum (Pers.) Rostaf.
Frequent and widespread on mossy wood and rocks

Macbrideola H. C. Gilb.

M. cornea (G. Lister & Cran.) Alexop.
Developed in moist chamber culture, Aros Park (Clark)

Stemonitis Gled. ex Wiggers

S. axifera (Bull.) Macbr.
Frequent and widespread on old logs (Clark, K)

S. fusca Roth
Common and widespread on logs and stumps (Clark)

Symphytocarpus Ing & Nann.-Bremek.

S. flaccidus (Lister) Ing & Nann.-Bremek.
On *Pinus*, Loch Bà (K)

PHYSARALES

PHYSARACEAE

Badhamia Berk.

B. globosa (A. & G. Lister) B. Ing
On wet mossy rocks Aros Park; third British record

B. foliicola List.
On bark, Croig (Clark)

B. lilacina (Fr.) Rostaf.
At edge of *Sphagnum* patches

B. panicea (Fr.) Rostaf.
On bark of logs, Aros Woods (Clark)

Craterium Trentep.

C. minutum (Leers) Fr.
On litter of all kinds; frequent and widespread

Fuligo Haller emend Pers.

F. muscorum Alb. & Schw.
On terrestrial mosses, Aros Park (Clark)

F. septica (L.) Web. var **septica**
Recorded by White without data

F. septica var. **flava** (Pers.) Morg.
Common and widespread on logs and stumps

Leocarpus Link

L. fragilis (Dicks.) Rostaf.
Common and widespread on plant litter

Physarum Pers. emend. Rostaf.

P. bivalve Pers.
On grass litter, Aros Park

P. citrinum Schum.
On terrestrial mosses, Loch Spelve

P. compressum Alb. & Schw.
On litter, Kintra

P. leucophaeum Fr.
Frequent and widespread on dead wood

P. nutans Pers. var. **nutans**
Frequent and widespread on dead wood; recorded by White

P. pusillum (Berk. & Curt.) G. Lister
On marsh vegetation, Carsaig; fifth Scottish record

P. viride (Bull.) Pers.
Occasional on conifer litter and branches

Physarum species nova
An undescribed species superficially resembling *Didymium laxifilum* G. Lister & Ross.
On mossy stump of *Acer pseudoplatanus* (Clark)

DIDMYIACEAE

Diderma Pers.

D. effusum (Schw.) Morg.
In leaf litter, Glen Aros (Clark)

D. globosum Pers.
On leaves and stems of herbaceous plants, Croig (Clark); fourth British record

D. lucidum Berk. & Br.
Aros Park (Clark); known only from a few localities in Britain (see Ing, 1974)

D. montanum (Meyl.) Meyl.
On mosses, Aros Park (Clark); first British record

D. ochraceum Hoffm.
On mosses on wet rocks, Aros Park (Clark)

D. sauteri (Rostaf.) Macbr.
On wet mossy rocks, Aros Park (Clark); the second British record. The only other record is from Appin, Argyll, in the early part of the century.

Didymium Schrad.

D. clavus (Alb. & Schw.) Rabenh.
Occasionally formed on bark in culture (Clark)

D. difforme (Pers.) Gray
On plant litter, Craignure Bay (Ing)

D. melanospermum (Pers.) Macbr.
On litter especially in conifer woods, Craignure Bay (Clark)

D. nigripes (Link) Fr.
On litter especially *Ilex* leaves, Croig (Clark)

D. squamulosum (Alb. & Schw.) Fr.
Occasional on plant litter (Clark)

Lepidoderma De Bary

L. tigrinum (Schrad.) Rostaf.
Occasional; on decaying wood and wet moss often accompanied by *Nostoc* spp.

INCERTAE SEDIS

Frankia alni (Wor.) Brunch. Widespread on roots of *Alnus*

Fungal associates arranged in alphabetical order of hosts

PTERIDOPHYTA

Dryopteris carthusiana *Saprophyte:* Xenasma filicinum.

D. filix-mas agg. *Saprophyte:* Pezizella chrysostigma.

Pteridium aquilinum *Parasite:* Tulasnella anceps. *Saprophytes:* Craterium minutum, Cyathicula pteridicola, Dasycyphus pteridis, Diaporthopsis pantherina, Didymium squamulosum, Hyaloscypha flaveola, Leocarpus fragilis, Leptostroma filicinum, Microscypha grisella, Pycnothyrium litigosum, Rhopographus filicinus, Sistotrema commune, Tubulicrinis accedens, Xenasma filicinum.

CONIFEROPHYTA

Abies alba *Parasites:* Melampsorella caryophyllacearum (aecidia), Heterobasidion annosum.

Cupressus macrocarpa *Saprophyte:* Tricholomopsis rutilans.

Juniperus communis *Parasites:* Stigmatea silvatica, Lophodermium juniperinum. *Saprophytes:* Cytospora dubyi, Mytilidion acicola, M. decipiens, Stomiopeltis sp.

Larix decidua *Parasites:* Cytospora curreyi, Heterobasidion annosum, Lachnellula willkommii. *Saprophytes:* Athelia epiphylla, Botryobasidium sp., Colpoma crispum, Helminthosporium resinae, Hyphoderma pallidum, H. setigerum, Hyaloscypha leuconica, H. stevensonii, Lachnellula hahniana, Lophium mytilinum, Mycena purpureofusca, Odontia bicolor, Pholiota flammans, Propolomyces farinosus, Stereum sanguinolentum, Strigopodia resinae, Tomentella mucidula, Tremella foliacea, Tubulicrinis subulatus. *Suspected mycorrhizals:* Gomphidius maculatus, Suillus grevillei, S. luteus.

Picea abies *Parasites:* Heterobasidion annosum, Lophodermium piceae, Rhizosphaera kalkhoffi, Stereum rugosum. *Saprophytes:* Clavidisculum kriegerianum, Hypholoma radicosum, Lachnellula subtilissima, Nectria fuckeliana, Pezicula livida, Stereum sanguinolentum. *Suspected mycorrhizals:* Boletus edulis *forma*, Gyroporus cyanescens.

Pinus spp. *Parasites:* Heterobasidion annosum, Lophodermium pinastri. *Saprophytes:* Auriscalpium vulgare, Calocera viscosa, Ceratobasidium cornigerum, Dasycyphus pulverulentus, Gymnopilus penetrans, Hyphoderma tenue, Leptostroma pinastri, Linodochium hyalinum, Micraspis strobilina, Pholiota flammans, Phomopsis occulta, Pleurotellus porrigens, Strobilurus tenacellus, Stereum sanguinolentum, Tremella foliacea. *Suspected mycorrhizals:* Boletus badius, Gomphideus roseus, Suillus bovinus, S. luteus, S. variegatus

Thuja plicata *Parasite:* Didymascella thujina

Undetermined conifer wood Apiorhynchostoma curreyi, Bertia moriformis, Calocera viscosa, Dacrymyces stillatus, Gloeophyllum sepiarium, Graphium calicioides, Hypholoma capnoides, Lachnellula subtilissima, Mollisia ventosa, Mycena rubromarginata, Odontia bicolor, Podoporia confluens, Psathyrella cotonea, Pseudohydnum gelatinosum, Sistotremastrum niveocremeum, Stypella papillata, Tricholomopsis decora, T. rutilans, Tubulicrinis accedens, T. subulatus. (See also Myxomycota.)

MAGNOLIOPHYTA

Acer campestre *Parasite:* Phleospora pseudoplatani.

A. pseudoplatanus *Parasites:* Rhytisma acerinum, Uncinula bicornis. *Saprophytes:* Clitopilus hobsonii, Diaporthe eres, Eutypa acharii, Gnomonia cerastis, Licea parasitica, Mycosphaerella maculiformis, Pleurotus ostreatus, Polyporus squamosus, Pyrenopeziza petiolaris.

Achillea ptarmica *Parasite:* Schizothyrioma ptarmicae.

Aegopodium podagraria *Parasites:* Phleospora aegopodii, Protomyces macrosporus, Puccinia aegopodii (teleutospores).

Agropyron repens *Parasite:* Erysiphe graminis.

Agropyron sp. *Saprophytes:* Ascochyta graminicola, Colletotrichum graminicolum.

Agrostis canina *Parasite:* Epichloe typhina.

A. tenuis *Parasites:* Epichloe typhina, Tilletia decipiens. *Saprophytes:* Ascochyta graminicola, Scirrhia agrostidis.

Agrostis spp. *Saprophyte:* Pleospora vagans.

Ajuga reptans *Parasite:* Ramularia ajugae.

Alchemilla glabra *Parasite:* Sphaerotheca alchemillae.

A. xanthochlora *Parasite:* Coleroa alchemillae.

Alchemilla sp. *Parasite:* Trachyspora intrusa (telutosori).

Allium ursinum *Parasites:* Botrytis globosa, Puccinia sessilis (aecidia).

Alnus glutinosa *Parasites:* Inonotus radiatus, Leptothyrium alneum, Melampsoridium betulinum (uredo and teleutospores), Microsphaera penicillata, Taphrina amentorum, T. sadbeckii, T. tosquinetii. *Saprophytes:* Ciboria viridifusca, Clavariadelphus fistulosus var. contortus, Ditopella ditopa, Grandinia mutabilis, Hyphodontia pallidula, Hymenoscyphus imberbis, H. vernus, Melanconium apiocarpum, Melanconis alni, Mollisia benesauda, Niptera umbonata, Pezizella alniella, Sclerotium complanatum. *Suspected mycorrhizals:* Frankia alni, Lactarius obscuratus. *Possible mycorrhizals:* Cortinarius helvelloides, Naucoria celluloderma, N. escharoides, N. scolecina, N. striatula, N. subconspersa. A special relationship obviously exists between *Alnus* and certain of the larger *Salix* spp. and members of the genus *Naucoria.* The nature of this relationship in unknown. *Alnus* is often taken as an endotrophic mycorrhizal.

Ammophila arenaria *Parasite:* Puccinia pygmaea (uredospores). *Saprophytes:* Colletotrichum lineola, Coniothyrium psammae, Didymella scotica, Entosordaria ammophilae, Leptosphaeria ammophilae, Leptostroma donacium var. majus, Lophodermium arundinaceum, Paradidymella perforans, Periconia hispidula, Phomatospora arenaria, Pleospora herbarum, Pyrenopeziza arenivaga, Tiarospora perforans.

Anemone nemorosa *Parasites:* Septoria anemones, Tranzschelia anemones.

Angelica sylvestris *Parasite:* Plasmopara nivea. *Saprophytes:* Botrytis cinerea, Diaporthe arctii, Diaporthopsis angelicae, Fusicladium depressum, Heterosphaeria patella, Leptosphaeria doliolum.

Anthoxanthum odoratum *Parasites:* Claviceps purpurea, Puccinia poae-nemoralis (uredospores), P. recondita (uredospores), Septoria avenae.

Anthyllis vulneraria *Parasite:* Uromyces anthyllidis (uredo and teleutosori).

Armeria maritima *Parasite:* Uromyces armeriae (uredo and teleutospores).

Arrhenatherum elatius *Saprophyte*: Leptosphaeria nigrans.

Arum maculatum *Parasite:* Puccinia sessilis (aecidia and spermagonia).

Aster tripolium *Parasite:* Puccinia cnici-oleracei (teleutospores).

Avena sativa (cultivated oats) *Parasite:* Gaeumannomyces graminis var. avenae.

Bellis perennis *Parasite:* Puccinia obscura (aecidia).

Betula spp. (*B. pendula* and *B. pubescens* have not been distinguished). *Parasites:* Inonotus radiatus, Melanconis stilbostoma, Melampsoridium betulinum (uredo and teleutospores), Piptoporus betulinus, Taphrina betulae. *Saprophytes:* Arachnopeziza aurata, Botryobasidium laeve, Ceuthospora phacidiodes, Coprinus micaceus, Coriolus versicolor, Dasyscyphus papyraceus, D. pulveraceus, Diatrypella favacea, Durella macrospora, Fomes fomentarius, Gnomonia setacea, Gymnopilus junonius, Helminthosporium cylindricum, Hyphoderma tenue, Hypoxylon multiforme, Melanconium bicolor, Monodictys paradoxa, Orbilia epipora, Peniophora cinerea, Phellinus igniarius, P. laevigatus, Polyporus brumalis, Psathyrella xanthocystis, Pseudospiropes longipilus, Petridospora scoriadea, Radulomyces confluens, Schizopora paradoxa, Stachylidium cyclosporum, Stereum hirsutum, Tremella foliacea, Trichocladium sp., Tyromyces caesius. *Suspected mycorrhizals:* Cortinarius crocolitus, C. cyanopus, C. decoloratus, C. evernius, C. largus, C. lepidopus, C. porphyrochrous, C. subtriumphans, C. tabularis, Lactarius fuliginosus, L. glyciosmus, L. helvus, L. mitissimus, L. tabidus, L. torminosus, L. turpis, L. vietus, Leccinum holopus, L. roseo-tinctum, L. scabrum, L. variicolor, Russula betularum, R. claroflava, R. foetens, R. persicina, R. versicolor, Tricholoma fulvum, T. resplendens. *Non mycorrhizal:* Inocybe casimiri.

Brachypodium sylvaticum *Parasite:* Puccinia brachypodii. *Saprophyte:* Belonopsis filispora.

Brassica sp. 'Kale' *Parasites:* Botrytis cinerea, Phoma lingam.

Buxus sempervirens *Parasite:* Puccinia buxi (teleutosori). *Saprophytes:* Phomopsis stictica, Volutella buxi.

Calluna vulgaris *Saprophytes:* Godronia callunigera, G. cassandrae var. callunae, Phlebia hydnoides, Tapesia fusca, T. melaleucoides, Trichobelonium obscurum.

Caltha palustris *Parasites:* Botryotinia calthae, Pseudopeziza calthae, Puccinia calthae, P. calthicola.

Campanula rotundifolia *Parasite:* Coleosporium tussilaginis (uredo and teleutospores).

Carex binervis *Parasite:* Puccinia caricina (uredospores).

C. demissa *Parasite:* Puccinia dioicae (uredo and teleutospores).

C. flacca *Saprophyte:* Arthrinium morthieri, Hysteropezizella sp., Niptera melatephra.

C. nigra *Parasite:* Puccinia caricina.

C. pallescens *Parasites:* Cintractia caricis, Puccinia caricina (uredospores).

C. vesicaria *Parasite:* Puccinia microsora (uredo, amphi and teleutospores).

Carex spp. *Parasites:* Discosia astrocreas. *Saprophytes:* Arthrinium acervatum var. minus, A. morthieri, Niptera pilosa.

Cerastium glomeratum *Parasite:* Peronospora tomentosa.

Cerastium sp. *Saprophyte:* Isariopsis alborosella.

Chrysosplenium oppositifolium *Parasite:* Puccinia chrysosplenii (teleutosori).

Circaea × intermedia *Parasites:* Puccinia circaeae (teleutosori), Pucciniastrum circaeae (uredospori).

C. lutetiana *Parasites:* Puccinia circaeae (teleutosori), Pucciniastrum circaeae (uredosori).

Cirsium arvense *Parasite:* Puccinia punctiformis (uredo and teleutospores).

C. heterophyllum *Parasites:* Puccinia calcitrapae (uredo and teleutospores), P. cnici-oleracei (teleutospores), P. hieracii var. hieracii (uredo and teleutospores). *Saprophyte:* Leptosphaeria jaceae.

C. palustre *Parasites:* Albugo tragopogonis var. cirsii, Puccinia calcitrapae (uredo and teleutospores), P. dioicae (aecidia). *Saprophytes:* Diaporthe arctii, Lachnella villosa, Leptosphaeria dolioloides, Ophiobolus acuminatus, Unguicularia incarnatina.

C. vulgare *Parasite:* Puccinia cnici (uredo and teleutospores).

Conopodium majus *Parasite:* Puccinia tumida (teleutosori).

Corylus avellana *Parasites:* Cytospora fuckelii, Encoelia furfuracea, Stereum rugosum. *Saprophytes:* Anthostoma microsporum, Cenangium cf. glaberrimum, Coriollelus campestris, Corticium evolvens, Diatrypella favacea, Gnomonia setacea, G. vulgaris, Helminthosporium velutinum, Hymenochaete corrugata, Hymenoscyphus caudatus, H. fructigenus, Hyphoderma setigerum, Hypocrea sp. 1, Hypoxylon fuscum, Inonotus radiatus, Mamiana coryli, Phomopsis revellens, Platygloea vestita, Sillia ferruginea, Stereum hirsutum, Stictis radiata, Tremella frondosa, Valsa fuckelii. *Suspected mycorrhizal:* Leccinum carpini. *Non-mycorrhizal:* Inocybe fibrosoides.

Crataegus monogyna *Parasite:* Podosphaera oxyacanthae. *Saprophytes:* Diatrype stigma, Hysterium angustatum.

Dactylis glomerata *Parasite:* Uromyces dactylidis (uredo and teleutospores). *Saprophytes:* Phyllachora graminis, Pleospora vagans.

Deschampsia caespitosa *Parasites:* Phaeoseptoria aireae, Puccinia deschampsiae.

D. flexuosa *Parasite:* Uromyces airae-flexuosae (uredo and teleutospores).

Digitalis purpurea *Parasite:* Ramularia variabilis. *Saprophytes:* Phoma herbarum, Pyrenopeziza digitalina.

Eleocharis palustris *Parasites:* Claviceps nigrescens, Sclerotinia eleocharidis. *Saprophytes:* Rustroemia plana, Sclerotium sp.

Empetrum nigrum (records include **E. hermaphroditum**) *Parasite:* Chrysomyxa empetri (uredosori). *Saprophytes:* Aulographum vagum, Ceuthospora phacidiodes, Phaeangellinia empetri, Physalospora empetri.

Endymion non-scriptus *Parasite:* Uromyces muscari (teleutospores). *Saprophyte:* Sclerotium hyacinthi.

Epilobium angustifolium *Parasite:* Pucciniastrum epilobii (uredosori). *Saprophytes:* Paradidymella tosta, Sydowiella fenestrans.

E. montanum *Parasites:* Puccinia epilobii (teleutosori), Venturia maculaeformis.

E. palustre *Parasites:* Puccinia epilobii (teleutosori), Sphaerotheca epilobii.

Epilobium sp. *Parasites:* Puccinia pulverulenta, Sphaerotheca epilobii.

Eriophorum sp. *Saprophyte:* Dasyscyphus eriophori.

Fagus sylvatica *Parasite:* Fomes fomentarius. *Saprophytes:* Antrodia mollis, Athelopsis lembosporum, Botryosphaeria quercuum, Basidiodendron eyrei, Calocera cornea, Chaetosphaeria pulviscula, Crepidotus calolepis, Cristella farinacea, Cyphellopsis anomala, Dacrymyces stillatus, Dasyscyphus brevipilus, Diatrype disciformis, D. stigma, Endoxyla rostrata, Eutypa flavovirens, E. scabrosa, Fomes fomentarius, Ganoderma adpressum, Ganoderma applanatum, Glonium lineare, Heterobasidion annosum, Henningsomyces candidus, Hymenoscyphus fagineus, Hypochnicium punctulatum, Hypoxylon cohaerens, H. fragiforme, H. rubiginosa, H. serpens, Lasiosphaeria spermoides, Lentinellus cochleatus, Lopadostoma turgidum, Mollisia fallax, Nectria coccinea, Oidodendron flavum, Phialophora sp. 1, Phlebia hydnoides, P. livida, Podoporia confluens, Polymorphum rugosum, Protocrea farinosa, Pseudotrametes gibbosa, Sebacina fugasissima, Stereum hirsutum, Tubercularia vulgaris, Tyromyces semipileatus, Valsa ambiens, Vuilleminia comedens, Xylohypha nigrescens, Xylosphaera longipes. *Suspected mycorrhizals:* Cortinarius torvus, Lactarius blennius, Russula aurata, R. erythropus, R. fellea, R. mairei. *Non-mycorrhizals:* Clavulinopsis cf. griseola, Craterellus sinuosus, Nolanea cucullata.

Festuca gigantea *Parasite:* Puccinia coronata (uredospores).

F. rubra *Parasite:* Corticium fuciforme. *Saprophytes:* Colleotrichum graminicolum, Phyllachora sylvatica, Pleospora vagans.

Filipendula ulmaria *Parasites:* Erysiphe ulmariae, Sphaerotheca alchemillae, Triphragmium ulmariae (aecidia, uredo and teleutosori), Urocystis filipendulae. *Saprophytes:* Dasycyphus nudipes, Mollisia revincta, Mycosphaerella sp. *1*, Phialina ulmariae, Thyristroma spireae.

Fraxinus excelsior *Parasites:* Armillaria mellea, Nectria galligena. *Saprophytes:* Coriolus versicolor, Cryptosphaeria eunomia, Daldinia concentrica, Farlowiella aff. carmichaeliana, Hypoxylon rubiginosum, Hysterium angustatum, Karschia stygia, Monotospora megalospora, M. megalospora var. fusispora, Orbilia auricolor, Peniophora limitata, Phlebia radiata, Phomopsis controversa, Polyporus squamosus, Teichospora obducens, Tyromyces caesius, T. semipileatus, Xylohypha nigrescens.

Galium aparine *Parasites:* Erysiphe cichoracearum, E. galli, Puccinia difformis (teleutospores).

G. palustre *Parasite:* Puccinia punctata (uredo and teleutospores).

G. saxatile *Parasites:* Puccinia punctata (uredo and teleutospores), Pucciniastrum guttatum (uredo and teleutospores).

G. verum *Parasite:* Puccinia punctata (uredo and teleutospores).

Geranium pratense *Parasite:* Plasmopara pusilla.

G. robertianum *Parasite:* Stigmatea robertiani.

Geum sp. *Parasite:* Septoria gei.

Glyceria fluitans *Parasite:* Claviceps purpurea.

Hedera helix *Parasites:* Cryptocline paradoxa, Phyllosticta hedericola, Trochila craterium. *Saprophytes:* Exidia albida, Lophiostoma angustilabrum var. crenatum, Peniophora lycii.

Heracleum sphondylium *Parasites:* Ramularia patella, Leptosphaeria doliolum.

Heracleum sp. *Saprophytes:* Hypoderma commune, Phomopsis sp.

Hieracium sp. *Parasite:* Thielaviopsis basicola.

Holcus lanatus *Parasite:* Puccinia coronata (uredospores).

H. mollis *Parasites:* Puccinia coronata (uredospores), P. poae-nemoralis (uredospores), P. recondita (uredo and teleutospores).

Hydrocotyle vulgaris *Parasites:* Puccinia hydrocotyles (uredosori), Septoria hydrocotyles.

Hypericum androsaemum *Parasite:* Melampsora hypericorum (aecidia and teleutosori).

H. pulchrum *Parasite:* Melampsora hypericorum (teleutosori).

Hypochoeris radicata *Parasite:* Puccinia hieracii var. hypochoeridis (uredo and teleutospores).

Ilex aquifolium *Saprophytes:* Exidia albida, Phacidiostroma multivalve, Phomopsis crustosa, Pleurotus pulmonarius, Trochila ilicina, Vialaea insculpta.

Iris pseudacorus *Parasites:* Fusarium sporotrichoides, Mycosphaerella iridis. *Saprophytes:* Ectostroma iridis, Pezizella sp., Pezoloma ciliifera.

Juncus acutiflorus *Saprophytes:* Dasyscyphus diminiatus, Leptosphaeria juncina.

J. bufonius *Parasite:* Entorrhiza aschersoniana.

J. conglomeratus *Saprophytes:* Dinemasporium gramineum var. strigosulum, Stictis luzulae var. junci.

J. effusus *Parasites:* Myrioconium tenellum, Sclerotinia curreyana. *Saprophytes:* Dasyscyphus apalus, Delicatula cephalotricha, D. fugiens, Endodothella junci, Mycena bulbosa.

Lapsana communis *Parasites:* Erysiphe cichoracearum, Puccinia lapsanae (uredo and teleutospores).

Lathyrus montanus *Parasite:* Uromyces viciae-fabae.

L. pratensis *Saprophyte:* Isariopsis carnea.

Leontodon autumnalis *Parasite:* Puccinia hieracii var. hieracii (uredo and teleutospores).

Ligusticum scoticum *Saprophyte:* Phoma complanata.

Linum catharticum *Parasite:* Melampsora lini (uredo and teleutosori).

Lonicera periclymenum *Parasite:* Puccinia festucae (aecidia). *Saprophytes:* Peniophora lycii, Platystomum compressum.

Lotus corniculatus *Parasite:* Peronospora lotorum.

L. uliginosus *Parasite:* Ovularia sphaeroidea.

Luzula campestris *Parasite:* Puccinia obscura (uredospores).

L. sylvatica *Parasites:* Puccinia luzulae (uredospores), Puccinia obscura (uredospores). *Saprophytes:* Anthostomella tumulosa, Niesslia pusilla.

Lysimachia nemorum *Parasite:* Septoria lysimachiae.

Lythrum salicaria *Saprophytes:* Dasyscyphus nidulus, D. salicariae, Lophiostoma caulium, Mycosphaerella sp. 2, Sistotrema commune.

Malus sp. (cultivated apple) *Parasite:* Fusicladium dendriticum.

Mentha aquatica *Parasite:* Puccinia menthae (uredo and teleutospores).

Mercurialis perennis *Parasite:* Cercospora mercurialis. *Saprophyte:* Typhula sp.

Molinia caerulea *Saprophytes:* Actinothyrium graminis, Belonium hystrix, Didymella sp., Lophodermium apiculatum, Mycosphaerella lineolata, M. montellica, Phoma arundinacea, Pleospora vagans, Stagonospora subseriata.

Myrica gale *Parasite:* Ovularia destructiva. *Saprophytes:* Crepidotus subtilis, Dasyscyphus sulphurellus.

Myrrhis odorata *Parasite:* Puccinia chaerophylli (uredosori).

Nardus stricta *Parasite:* Claviceps purpurea. *Saprophytes:* Actinoscypha punctum, Leptosphaeria nardi.

Narthecium ossifragum *Parasite:* Heterosporium ossifragi.

Odontites verna *Parasite:* Plasmopara densa.

Oenanthe crocata *Parasites:* Plasmopara nivea, Septoria oenanthes.

Oenanthe sp. *Saprophytes:* Dasyscyphus grevillei, Pistillaria micans, Phoma longissima, Sistotrema commune.

Pedicularis sylvatica *Parasite:* Puccinia clintonii (teleutospores).

Pedicularis sp. *Parasite:* Ramularia obducens.

Petasites hybridus *Parasite:* Coleosporium tussilaginis (uredo and teleutospores).

Phalaris arundinacea *Parasite:* Puccinia sessilis (uredo and teleutospores). *Saprophytes:* Hendersonia crastophila, Leptosphaeria fuckelii, L. graminis, L. nigrans, Lophodermium arundinaceum, Macrophoma graminella, Mollisia aff. palustris, Mycosphaerella cf. longissima, Phoma arundinacea, Phomatospora berkeleyi, Pleospora vagans, Stagonospora vexata.

Phragmites australis *Parasite:* Puccinia magnusiana (uredo and teleutospores). *Saprophytes:* Belonopsis excelsior, Dasyscyphus acutipilus, D. controversus, Leptosphaeria arundinacea, Niptera pulla, Scirrhia rimosa, Tapesia retincola.

Plantago lanceolata *Parasites:* Sphaerotheca plantaginis, Synchytrium aff. aureum, S. erieum.

P. major *Parasite:* Peronospora alta.

P. maritima *Parasite:* Erysiphe sordida.

Poa nemoralis *Parasites:* Puccinia poae-nemoralis (uredospores), P. poarum (uredospores), Septoria cf. oudemansii, Uromyces dactylidis.

P. trivialis *Parasite:* Puccinia poarum (uredospores).

Polygonatum sp. *Parasite:* Botrytis cinerea.

Polygonum amphibium *Parasite:* Puccinia polygoni-amphibii (uredo and teleutospores).

P. cuspidatum *Saprophyte:* Hymenoscyphus robustior.

P. hydropiper *Parasite:* Sphacelotheca hydropiperis.

P. persicaria *Parasite:* Septoria polygonorum var. persicaria.

Polygonum sp. *Saprophyte:* Plagiostoma devexa.

Populus tremula *Parasites:* Pollaccia radiosa. *Saprophytes:* Lasiosphaeria ovina, Lophidium compressum, Triposporium elegans. *Suspected mycorrhizal:* Leccinum aurantiacum.

Populus sp. *Parasite:* Taphrina populina.

Potentilla anserina *Parasites:* Marssonina potentillae, Ramularia arvensis.

P. erecta *Parasites:* Septoria tomentillae, Sphaerotheca alchemillae, Taphrina potentillae.

P. sterilis *Parasite:* Phragmidium fragariae (uredo and teleutosori).

Primula vulgaris *Parasite:* Puccinia primulae (teleutospores).

Primula spp. *Parasite:* Ovularia primulana.

Prunus laurocerasus *Parasite:* Trochila laurocerasi. *Saprophytes:* Phellinus ferruginosus, Stictis radiata.

P. padus *Parasites:* Actinonema padi, Polystigma fulvum.

P. spinosa *Parasites:* Phellinus pomaceus, Uncinula prunastri.

Prunus sp. *Saprophyte:* Hymenochaete corrugata.

Quercus spp. (*Q. petraea* and *Q. robur* have not been distinguished) *Parasites:* Colpoma quercinum, Microsphaera alphitoides, Stereum gausapatum, Taphrina caerulescens. *Saprophytes:* Aegerita candida, Athelopsis lembosporum, Bolinia lutea, Brachysporium britannicum, Calocera cornea, Caudospora taleola, Ceriporia reticulata, C. viridans, Ceuthospora phacidioides, Chlorociboria aeruginascens, Crepidotus mollis, Coccomyces dentatus, Coniothecium sp., Coriolus versicolor, Daedalea quercina, Dasyscyphus capitatus, D. ciliaris, D. fuscescens, D. niveus, Diaporthe leiphaemia, Diatrype stigma, Diatrypella quercina, Eriopeziza caesia, Exidia glandulosa, Grandinia helvetica, Guignardia punctoidea, Hypoxylon rubiginosum, H. semi-immersum, H. serpens, Hyphoderma roseocremeum, H. tenue, Hyphodontia subalutacea, Hysterium angustatum, Karschia lignyota, Lasiosphaeria spermoides, Leptothyrium medium, L. quercinum, Mollisia ligni, Myxosporium lanceola, Oxyporus populinus, Peniophora quercina, Phellinus ferruginosus, Phlebia hydnoides, P. roumeguerii, Phyllosticta roboris, Plagiostoma pustula, Pleurotus ostreatus, Polymorphum rugosum, Polyporus picipes, Rutstroemia firma, R. sydowiana, Schizopora paradoxa, Sistotrema brinkmanii, Stereum hirsutum, Tremella mesenterica, Trimmatostroma sp., Tyromyces semipileatus, Vuilleminia comedens, Xylohypha nigrescens, Zignoella sp. *Suspected mycorrhizals:* Amanita excelsa, A. virosa, Boletus luridus, B. porosporus, Cortinarius bolaris, C. herpeticus, C. limonius, C. rubicundulus, C. scaurus, Lactarius quietus, Leccinum carpini, Russula albonigra, R. cyanoxantha, R. mairei var. fageticola, R. rosea, Tricholoma acerbum, T. aff. viridolutescens. *Non-mycorrhizal:* Craterellus sinosus.

Ranunculus ficaria *Parasite:* Uromyces dactylidis (aecidia).

R. flammula *Parasites:* Erysiphe ranunculi, Peronospora ranunculi.

R. repens *Parasites:* Fabraea ranunculi, Peronospora ranunculi, Ramularia ranunculi, Urocystis anemones. *Saprophyte:* Cyathicula turbinata.

Rhinanthus minor *Parasite:* Coleosporium tussilaginis (uredo and teleutospores).

Rhododendron ponticum *Parasites:* Gloeosporium rhododendri, Pycnostysanus azaleae. *Saprophytes:* Lophodermium rhododendri, Lophodermium cf. vagulum.

Ribes nigrum *Parasites:* Gloeosporium ribis, Puccinia caricina var. caricina (aecidia). *Saprophyte:* Plowrightia ribesia.

R. uva-crispa *Parasite:* Puccinia caricina (aecidia).

Rosa canina *Parasite:* Phragmidium mucronatum (uredo and teleutosori).

Rosa sp. *Saprophyte:* Tapesia rosae.

Rubus fruticosus agg. *Parasites:* Kuhneola uredinis, Phragmidium violaceum (aecidia, uredo and teleutosori), Rhabdospora ramealis, Sphaceloma necator. *Saprophytes:* Apioporthe vepris, Chaetosphaeria callimorpha, Coryneopsis rubi, Dasyscyphus melatheja, Gloniopsis praelonga, Gnomonia rubi, Hypoderma virgultorum, Leptothyrium rubi, Lophiostoma vagabundum, Metasphaeria clypeosphaerioides, Orbilia sarraziniana, Pezicula rubi, Ploettnera exigua.

R. idaeus *Parasite:* Phragmidium rubi-idaei (uredo and telutosori). *Saprophytes:* Dasyscyphus bicolor var. rubi, Herpotrichia macrotricha, Zignoella ovoidea.

Rumex acetosa *Parasites:* Puccinia acetosae (uredospores), Septoria acetosae, Uromyces acetosae.

R. obtusifolius *Parasites:* Ovularia obliqua, Uromyces rumicis (uredo and teleutospores).

Rumex sp. *Parasites:* Ovularia obliqua, Uromyces rumicis, Venturia rumicis. *Saprophyte:* Plagiostoma devexa.

Salix aurita *Parasites:* Melampsora epitea (uredospores), Venturia microspora. *Saprophyte:* Pyrenopeziza fuckelii.

S. cinerea *Parasite:* Melampsora capraearum (uredospores).

S. cinerea × aurita *Parasite:* Melampsora capraearum (uredospores).

S. herbacea *Parasites:* Melampsora epitea (uredospores), Rhytisma salicinum.

S. pentandra *Parasite:* Melampsora larici-pentandrae (uredo and teleutospores).

S. repens *Parasite:* Rhytisma salicinum. *Saprophyte:* Linospora capreae.

Salix spp. (various tree species of uncertain identity) *Saprophytes:* Bjerkandera adusta, Ciboria amentacea, Dasyscyphus papyraceus, Durella connivens, Flammulaster subincarnata, Hyphoderma roseocremeum, Linospora vulgaris, Nectria episphaeria, Peniophora cinerea, P. lycii, Pezizella amenti, Phaeomarasmius erinaceus, Polyporus varius, Pyrenopeziza fuckelii, Tomentella hydrophila, Vuillemenia comedens.

Sambucus nigra *Saprophytes:* Ascotremella faginea, Dacrymyces stillatus, Hirneola auricula-judae, Hyphodontia sambuci, Unguicularia cirrhata.

Sanicula europaea *Parasite:* Puccinia saniculae (aecidia and uredosori).

Saxifraga stellaris *Parasite:* Puccinia saxifragae (teleutosori).

Scirpus cespitosus *Parasites:* Sclerotinia gregoriana, Septoria sp. *Saprophytes:* Niptera phaea, N. aff. phaea, Sclerotium sp.

Scrophularia nodosa *Saprophytes:* Phoma oleracea var. scrophulariae, Stictis stellata.

Sedum anglicum *Saprophyte:* Pleospora herbarum.

Senecio aquaticus *Saprophytes:* Botrytis cinerea, Chalara fusidioides, Phomopsis achilleae.

S. jacobaea *Parasite:* Ramularia pruinosa. *Saprophyte:* Rhabdospora sp.

S. viscosus *Parasite:* Coleosporium tussilaginis (uredo and teleutospores).

S. vulgaris *Parasite:* Puccinia lagenophorae (teleutospores).

Silene dioica *Parasite:* Marssonina delastrei.

S. maritima *Parasites:* Uromyces behenis (teleutosori), Ustilago violacea.

Solidago virgaurea *Parasite:* Septoria virgaureae.

Sorbus aucuparia *Parasite:* Cytospora massariana. *Saprophytes:* Crepidotus mollis, Tympanis conspersa.

Sorbus sp. *Saprophytes:* Dasyscyphus corticalis, Diaporthe eres, Helminthosporium folliculatum.

Sparganium sp. *Parasite:* Ascochyta typhoidearum.

Spergula arvensis *Parasites:* Peronospora obovata, Ustilago violacea.

Spergularia sp. *Parasite:* Albugo caryophyllacearum.

Stachys sylvatica *Parasite:* Septoria stachydis.

Stellaria graminea *Parasite:* Ustilago violacea.

S. media *Saprophyte:* Septoria stellariae.

Succisa pratensis *Parasites:* Erysiphe polygoni, Peronospora violacea, Septoria scabiosicola. *Saprophyte:* Leptosphaeria ogilviensis.

Swida sanguinea *Saprophyte:* Niptera ramincola.

Taraxacum spp. *Parasites:* Puccinia hieracii var. hieracii (uredo and teleutospores), P. variabilis (uredo and teleutospores).

Teucrium scorodonia *Parasite:* Puccinia annularis (teleutospores).

Trifolium pratense *Parasite:* Uromyces trifolii (uredospores).

T. repens *Parasites:* Pseudopeziza trifolii, Stagonospora compta, Uromyces nerviphilus (teleutospores). *Saprophyte:* Polythrincium trifolii.

Triglochin maritimum *Parasite:* Asteroma juncaginacearum.

Tussilago farfara *Parasites:* Coleosporium tussilaginis (uredo and teleutospores), Puccinia poarum (aecidia).

Ulex europaeus *Saprophytes:* Cyphellopsis monacha, Dacromyces lacrymalis, Dasyscyphus pulverulentus, D. pygmaeus, Hymenochaete corrugata, Lachnella alboviolascens, Lophiosphaera ulicis, Merulius corium, Mollisiopsis dennisii, Pellidiscus pallidus, Peniophora incarnata, Phomopsis lingulata, Pleospora herbarum, Polyporus lentus, Rosellinia aquila, Tapesia fusca, Velutarina rufo-olivacea.

Ulmus glabra *Saprophytes:* Crepidotus mollis, Polyporus squamosus, Tyromyces semipileatus.

Ulmus sp. *Saprophytes:* Dasyscyphus corticalis, Stictis radiata.

Urtica dioica *Parasite:* Puccinia caricina var urticae-acutae (aecidia). *Saprophytes:* Acrospermum compressum, Botrytis cinerea, Calloria carnea, C. fusarioides, Cylindrocolla urticae, Dasyscyphus sulphureus, Didymium difforme, Leptosphaeria acuta, L. doliolum, Mollisia urticiola, Phoma herbarum, Pleurophragmium simplex.

Vaccinium myrtillus *Parasite:* Pucciniastrum vaccinii (ureodospores).

Valeriana officinalis *Parasites:* Puccinia commutata (teleutospores), Ramularia valerianae.

Veronica anagallis-aquatica *Parasite:* Ramularia anagallidis.

V. officinalis Parasites: Asterina veronicae, Gloeosporium veronicarum.

Viola palustris *Parasites:* Cercospora murina, Puccinia fergussonii (teleutospores).

V. riviniana *Parasites:* Puccinia violae (uredo and teleutospores), Ramularia lactea.

Unspecified hardwoods (mainly Fagaceae, Betulaceae and Corylaceae). *Saprophytes:* Acladium conspersum, Apostemidium guernisaci, Armillaria mellea, Bisporella citrina, B. fuscocincta, B. sulphurina, Botryobasidium conspersum, B. laeve, Bulgaria inquinans, Calocera cornea, Catinella olivacea, Ceratostomella ampullacea, Chaetosphaeria myriocarpa, C. pulviscula, Chlorociboria aeruginascens, Coprinus atramentarius, C. exstinctorius, C. prasinula, Crepidotus applanatus, C. autochthonus, C. calolepis, C. subsphaerosporus, Cudoniella acicularis, C. clavus, Dasyscyphus brevipilus, D. niveus, D. pulveraceus, D. pygmaeus, Eutypa lata, Farlowiella carmichaeliana, Flammulina velutipes, Galerina mutabilis, Gloeocystidiellum porosum, Hobsonia mirabilis, Hyaloscypha hyalina, Hymenochaete rubiginosa, Hyphoderma tenue, Hyphodontia alutaria, H. papillosa, Hypholoma fasciculare, Hypoxylon multiforme, Lasiosphaeria spermoides (often associated with Armillaria mellea rhizomorphs), Lycoperdon pyriforme, Marasmius ramealis, M. rotula, Melanomma pulvis-pyrius, Menispora caesia, Mollisia benesuada, M. cinerea, M. fallax, M. melaleuca, Mycena acicula, M. alcalina, M. galericulata, M. haematopus, M. inclinata, M. polygramma, M. speirea, M. tenerrima, Myriothecium roridum, Nectria cinnabarina,

Oidiodendron griseum, Orbilia leucostigma, O. luteorubella, O. xanthostigma, Oudemansiella mucida, O. radicata, Pachyella babingtonii, Panellus stipticus, Pholiota alnicola, P. squarrosa, Pleurotellus dictyorhizus, Pleurotus ostreatus, Pluteus cervinus, P. salicinus, Psathyrella candolleana, P. hydrophila, P. pilulaeformis, P. spadiceogrisea, Rosellinia mammiformis, Scutellinia scutellata, Sebacina effusa, Sporodesmium anglicum, Stereum purpureum, Thaxteria phaeostroma, Torula ovalispora, Trematosphaeria pertusa, Tulasnella violea, Tyromyces caesius, Vibrissea truncorum, Xylosphaera hypoxylon, X. polymorpha (see also Myxomycota).

Species found in some of the more common habitats

Species on decaying herbaceous stems and debris and on monocotyledonous stems, straw, stubble, etc.

Cladosporium herbarum, Crepidotus cesatii, Cudoniella clavus var. grandis, Cyathicula cyathoidea, Dasyscyphus nidulus, Heterosphaeria patella, Hymenoscyphus scutula, Leocarpus fragilis, Mollisia aff. palustris, Mycena stylobates, Pezizella elburnea, Physarum bivalve, Phialophora aff. richardsiae, Tetraploa aristata, Vermicularia dematium.

Species associated with bryophytes

On individual moss species Fuligo muscorum (on *Dicranum*), Leptosphaeria sp. (on *Nowellia curvifolia*), Sclerotium muscorum (on *Leucobryum glaucum*), Pistillaria setipes (on *Rhytidiadelphus squarrosus*).
The following on *Sphagnum:* Badhamia lilacina, Galerina paludosa, G. sphagnorum, G. tibiicystis, Hygrocybe cantharella, H. aff. coccinea, Hypholoma elongatum, Omphalina epichysium, O. ericetorum, O. oniscus, O. philonotis, Tephrocybe palustris.

In moss-cushions and moss-beds (bryophytes not determined): Galerina calyptrata, G. hypnorum, Hypholoma polytrichii, Melastiza chateri, Mniaecia jungermanniae.

On various mosses on bark: Colloderma oculatum, Dactylium varium, Lamproderma columbinum, Macbrideola cornea, Mycena corticola, M. hiemalis, Trichia affinis.

On moss-covered rocks: Badhamia globosa, Diderma lucidum, D. sauteri, Physarum citrinum. On moss-covered rocks usually associated with *Nostoc* spp.: Lepidoderma tigrinum.

On bonfire-sites: Tephrocybe atrata, T. carbonaria, Conocybe ochracea f. macrospora, Pholiota highlandensis (amongst *Funaria*), Psathyrella pennata, Pyronema domesticum.

Species found on algal scum

Omphalina ericetorum, O. hepatica (on log in conifer woodland), O. luteovitellina (in mountainous regions).

Species on fungi and lichenised fungi (substrate in brackets)

Hypomyces aurantius (Bjerkandera adusta), Sepedonium chrysospermum (Boletus sp.), Henningsomyces candidus (Coriolus versicolor), Hypomyces aurantius (C. versicolor), Acremonium album (Comatricha sp.), Nectria episphaeria (Diatrype sp.), Cordyceps capitata (Elaphomyces sp.), C. ophioglossoides (E. granulatus), Berlesiella nigerrima (Eutypa sp.), Mucor genevensis (Heterobasidion annosum), Dactylium dendroides (Hypomyces rosellus, Inonotus radiatus), Trichoderma viride (I. radiatus), Byssonectria lateritia (Lactarius deliciosus), Illosporium? coccineum (Parmelia exasperata), Leptoglossum rickenii (Peltigera canina), Helminthosphaeria corticiorum (Peniophora sp.), Stictis friabilis (Peniophora sp.), Acremonium album (Perichaena corticalis), Trichoderma viride (Phellinus ferruginosus), Phlebia hydnoides (P. laevigatus), Hypocrea pulvinata (Piptoporus betulinus), Sphaerellopsis filum (Puccinia dioicae; P. poae-nemoralis), Nyctalis asterophora (Russula sp.), N. parasitica (R. nigricans), Myxarium hyalinum (Stereum sanguinolenta), Tilachlidium tomentosum (Trichia varia), Nectria episphaeria (Unidentified pyrenomycetous stroma).

Species associated with animal debris

On sheep-horn: Onygena equina.

On dung (dung/soil or dung/straw mixtures): Agaricus bisporus, A. macrosporus (sawdust/straw-mixture), Ascobolus albidus, A. furfuraceus, Bolbitius vitellinus, Cercophora coprophila, C. mirabilis, Cheilymenia coprinaria, C. stercorea, C. theleboloides (straw-mixture), Collybia cirrhata (very old dung), Coniochaeta leucoplaca (sheep dung), C. scatigena, Conocybe coprophila, C. pubescens (straw mixture), Coprinus cinereus, C. congregatus (straw-mixture), C. ephemeroides (cow dung), C. lipophilus, C. flocculosus (straw mixture), C. macrocephalus, C. miser, C. niveus, C. nudiceps, C. patouillardii, C. pellucidus, C. radiatus, C. sclerotiger, C. stellatus, C. stercoreus, C. sterquilinus, C. triplex, C. velox, C. vermiculifer, Coprobia granulata (cow dung), Deconica coprophila, Fusarium culmorum (straw in dung), Iodophanus carneus, Laccaria amethystea (undoubtedly an accidental fruiting), Lasiobolus ciliatus (cow dung), Octospora alpestris, Onygena corvina (feathers), Panaeolus semiovatus, P. speciosus, P. sphinctrinus, P. subbalteatus (straw-mixture), Peziza fimeti (cow dung), P. vesciculosa (straw-mixture), Pilaira anomala, Pilobolus crystallinus (horse dung), P. kleinii, Podospora appendiculata, P. curvula, P. intestinacea, P. minuta (cow dung), Psathyrella conopilea, P. coprophila, P. coprobi (cow dung), P. aff. stercoraria (cow dung), Sordaria fimicola, S. humana, S. superba, Sphaerobolus stellatus, Sphaeronomella fimicola, Sporormia intermedia, S. megalospora, S. minima, Stropharia fimetaria (cow dung), S. merdaria (dung/straw mixture), S. semiglobata, Thelebolus nanus, T. stercoreus, Trichodelitschia bisporula, Zygospermella insignis, Z. setosa.

Entomophagous species: On Lepidoptera pupa: Cordyceps militaris; on aphids (on *Prunus avium*): Entomophthora aphidis; on moribund spider: Gibellula araneanum.

Species found in woodland

In conifer plantations * = suspected mycorrhizal species
Boletus badius*, B. edulis*, Cantharellus cibarius, Clavulina cinerea, C. cristata, C. rugosa, Clitocybe bicolor, C. flaccida, C. fragrans, C. langei, C. metachroa, C. obsoleta, C. quercina, C. vibecina, Collybia butyracea, C. maculata, Conocybe pseudopilosella, C. pubescens, Cortinarius cinnamomeobadius*, C. flexipes*, C. pinicola*, C. speciossissimus*, C. uliginosus*, Cystoderma amianthinum, Entoloma nitidum, Galerina vittaeformis, Gomphidius glutinosus* (grey-lilac and pink forms), Hebeloma crustuliniforme*, H. longicaudum, H. sacchariolens, H. sinapizans*, Hygrophoropsis aurantiaca, Hygrophorus agathosmus*, Inocybe asterospora, I. flocculosa, I. grammata, I. lucifuga, I. mixtilis, I. napipes, Lactarius camphoratus*, L. deliciosus*, L. rufus*, Marasmius androsaceus, Mycena capillaripes, M. lineata, M. vulgaris, Nolanea cetrata, Psathyrella obtusata, Russula caerulea*, R. emetica*, R. queletii*, R. sanguinea*, R. sardonia*, R. sardonia var. mellina*, Stropharia squamosa, Suillus bovinus*, S. luteus*, S. variegatus*, Tricholoma pessundatum*, T. vaccinum*, Trichophaea gregaria.

In mixed oak woods at Leachan Dubh, Cruach Ardura.
Amanita virosa, Cantharellus cibarius, C. infundibuliformis, Cortinarius pseudosalor, C. limonius, Craterellus sinuosus, Entoloma turbidum, Inocybe petiginosa, Lactarius tabidus, L. torminosus, L. uvidus, L. volemus, Leptonia aff. rosea, L. sericella, Mycena haematopus, M. sanguinolenta, Peniophora quercina, Russula emeticella, Stereum gausapatum.

In *Alnus* and *Salix* associations around N side of Loch na Keal.
Laccaria tortilis (in ditch), Naucoria celluloderma, N. escharoides, N. scolecina, N. striatula, N. subconspersa.

In mixed woodland/plantation with scattered coniferous trees * = suspected mycorrhizal species.
Agaricus altipes, A. haemorrhoidarius, A. langei, A. silvaticus, A. silvicola, Agrocybe erebia, Amanita citrina*, A. excelsa*, A. fulva*, A. muscaria* (usually with *Betula*), A. rubescens*, A. vaginata*, A. virosa* (usually with *Betula*), Boletus calopus*, B. chrysenteron*, B. edulis*, B. erythropus*, Boletus luridus*, B. piperatus*, B. spadiceus*, B. subtomentosus*, Calvatia excipuliformis, Cantharellus cibarius, C. cinereus, C. infundibuliformis, Clavulina cinerea, C. cristata, C. cristata var. coralloides, Clitocybe brumalis, C. ditopus, C. nebularis, C. suaveolens, C. umbilicata, Clitopilus prunulus, Collybia confluens, C. dryophila, C. maculata, C. peronata, Coltricia perennis, Conocybe filaris, C. pubescens, Coprinus lagopus, C. leiocephalus, C. narcoticus, Cortinarius bolaris*, C. cinnabarinus, C. croceofolius, C. cyanopus, C. decoloratus, Cortinarius fasciatus, C. gentilis, C. glandicolor, C. hemitrichus, C. lepidopus, C. limonius (especially under *Quercus*), C. mucosus*, C. obtusus, C. pseudosalor*, C. rigidus, C. saniosus, C. semisanguineus, C. tabularis, C. turbinatus, C. varius, C. vibratilis, C. violaceus*, Craterellus cornucopioides,

Cyathipodia macropus, Cystoderma amianthinum, Elaphomyces granulatus, E. muricatus, Entoloma griseoluridum, E. jubatum, E. nidorosum, E. sericatum (usually in wet areas with *Betula*), E. turbidum, Galerina mniophila, G. mycenopsis (mossy areas), G. vittaeformis, G. vittaeformis var. major, Hebeloma crustuliniforme*, H. leucosarx (in wet areas), H. mesophaeum, H. sacchariolens, H. sinuosum, Helvella crispa, H. lacunosa, Hydnellum concrescens, Hydnum repandum, H. rufescens, Hygrophoropsis aurantiaca, Hygrophorus agathosmus, H. chrysaspis (under *Fagus sylvatica*), H. hypothejus (under *Pinus sylvestris*), H. olivaceoalbus, Inocybe abjecta, I. bongardii, I. calamistrata, I. calospora, I. eutheles, I. fastigiata, I. geophylla, I. geophylla var. lilacina, I. godeyi, I. hystrix, I. maculata, I. obscuroides, I. petiginosa, I. posterula, I. praetervisa, I. umbrina, Laccaria amethystea, L. bicolor, L. laccata, L. striatula, Lactarius aspideus*, L. chrysorrhaeus* (under *Quercus*), L. cimicarius*, L. circellatus*, L. flavidus*, L. flexuosus*, L. fluens*, L. glaucescens*, L. pallidus*, L. piperatus (under *Quercus*), L. pterosporus* (under *Quercus* and *Fraxinus*), L. pyrogalus* (under *Corylus*), L. representanaeus* (under conifers), L. resimus*, L. serifluus*, L. subdulcis*, L. uvidus* (under *Betula*), L. vellereus*, L. vietus, L. volemus*, Leotia lubrica, Lepiota felina, L. aff. rhacodes, Leptonia euchroa, L. sericella, L. serrulata, Lycoperdon perlatum, L. perlatum var. nigrescens, Lyophyllum decastes, Microglossum viride, Mycena aetites, M. amicta, M. chlorantha, M. citrinomarginata, M. epipterygia, M. fibula, M. filopes, M. galopus, M. lactea, M. leptocephala (under *Pinus*), M. leucogala, M. maculata, M. metata, M. pelianthina, M. pura, M. pura var. alba, M. sanguinolenta, M. swartzii (also in rough pasture), M. vitilis, Naucoria bohemica, Nolanea cetrata, N. cuneata, N. icterina, Omphalina ericetorum, O. pyxidata, Otidea alutacea, O. umbrina, Paxillus involutus, Peziza badia, P. succosa, Phellodon niger, P. tomentosus, Pholiota dissimulans, P. lubrica, P. scamba, Porphyrellus pseudoscaber*, Psathyrella gracilis, Rhizina undulata (under conifers), Russula acrifolia*, R. adusta, R. aeruginea* (under *Betula*), R. aquosa* (wet places under *Betula*), R. atropurpurea*, R. aurantiaca*, R. azurea*, R. cyanoxantha*, R. delica*, R. densifolia*, R. emeticella*, R. firmula*, R. foetens*, R. fragilis*, R. gracillima*, R. grisea*, R. heterophylla*, R. integra*, R. knauthii*, R. laurocerasi*, R. lepida*, R. lilacea*, R. lutea*, R. mairei* (under *Fagus sylvatica*), R. nigricans*, R. nitida* (under *Betula*), R. ochroleuca*, R. parazurea*, R. pectinata*, R. puellaris*, R. rutila*, R. sororia*, R. velenovskyi*, R. vesca*, R. veternosa*, R. xerampelina*, Scleroderma aurantium, S. verrucosum, Spathularia flavida, Thelephora terrestris*, Tricholoma acerbum*, T. albobrunneum*, T. argyraceum, T. aurantium, T. columbetta, T. imbricatum, T. lascivum, T. portentosum, T. psammopus, T. saponaceum, T. sejunctum, T. sulphureum, T. terreum, T. ustale, Tylopilus felleus*, Tubaria conspersa, T. furfuracea.

Species in grassland

Grassland in maritime situations and associations on shallow soils on maritime rocks (e.g. cliff-tops, land-slip areas at bases of cliffs); the majority of these species are also found on rough grassland on less basic soils.
Agaricus campestris, A. porphyrocephalus, Clavaria fumosa, Clavulinopsis corniculata, C. helvola, Cystoderma amianthinum, Entoloma ameides, Geoglossum cookeianum, G. fallax, G. nigritum, G. starbaeckii, Hygrocybe aurantiosplendens, H. ceracea, H. chlorophana, H. coccinea, H. conica, H. fornicata, H. flavescens, H. intermedia, H. lacma, H. laeta, H. langei, H. marchii, H. nitrata, H. nivea, H. nigrescens, H. ovina, H. pratensis, H. psittacina, H. punicea, H. reai, H. russocoriacea, H. splendidissima, H. strangulata, H. subradiata, H. unguinosa, H. virginea, Leptonia chalybaea, L. fulva, L. lampropus, L. lividocyanula, Lycoperdon foetidum, L. spadiceus, Melanoleuca strictipes, Mycena flavoalba, M. olivaceomarginata, Nolanea cuspidifier, N. farinolens, N. radiata, N. sericea, N. solstitialis, N. staurospora, Panaeolus olivaceus, Trichoglossum hirsutum.

Base-rich grassland: *Agrostis-Festuca*-herb-rich grasslands often in upland well drained sites; also lowland pasture of *Agrostis-Holcus-Arrhenatherum* association.
Agaricus fissuratus, Agrocybe arvalis, Bovista nigrescens, B. plumbea, Clavulinopsis fusiformis, C. umbrinella, Clitocybe dealbata, C. rivulosa, Conocybe rickeniana, C. tenera, Cystoderma carcharias, Entoloma fuscomarginatum, E. helodes, E. porphyrophaeum, Galerina clavata, G. mniophila, G. vittaeformis, Geoglossum starbaeckii, Hygrocybe calyptraeformis, H. fuscesens, H. laeta, H. miniata, H. psittacina, H. vitellina, Hypholoma ericaeum, H. ericaeoides, H. udum, Leptonia atromarginata, L. caerulea, L. caesiocincta, L. corvina, L. cyanoviridescens, L. griseocyanea, L. aff pyrospila, L. querquedula, L. rhombispora, L. aff. rosea, L. sericella, L. serrulata, L. turci, L. xanthochroa, Lycoperdon perlatum, Microglossum olivaceum, Mycena flavoalba, M. olivaceomarginata, Nolanea infula, N. inutilis, Omphalina ericetorum, Panaeolina foenisecii, Panaeolus campanulatus, Psilocybe semilanceata, Stropharia aeruginosa, S. coronilla, Thuemenidium atropurpureum Trichoglossum walteri.

Species on submerged or rotting debris

***Fagus* leaves submerged in a small stream** E end of Ulva, collected on 17 April, 1969; analysed in culture.
Ascomycotina: Ceratostomella rostrata, Mollisia sp., Muellerella polyspora, Myxotrichum deflexum, Nectria ditissima, Pseudopeziza arenivaga.
Deuteromycotina: Aspergillus fumigatus, Cephalosporium acremonium, Chloridium chlamydosporis, Clathrosphaerina zalewskei, Helicosporium vegetum, Humicola fuscoatra, Mycogone rosea, Penicillium steckii, P. stoloniferum, P. thomii, Penicillium spp., Phomopsis sp., Sclerotium papyricola, Sporotrichum thermophile, Trichoderma album. Also mycelia sterilia: hyaline and septate: brown and septate.
Zygomycotina: Absidia glauca, A. orchidis, A. ramosa,
Mastigomycotina: Saprolegnia ferax.

***Quercus* leaves submerged in a small stream** Coille na Sròine, Loch Bà, collected 16 April 1969; analysed in culture.
Ascomycotina: Ceratostomella rostrata, Ceratostomella sp. Eurotium repens, Sordaria fimicola, S. hirta, Sporormia intermedia.
Deuteromycotina: Aspergillus flavus, A. fumigatus, Chrysosporium pannorum, Fusarium solani, Heterocephalum sp., Humicola fuscoatra, H. grisea, Penicillium claviforme; also mycelia sterilia: hyaline and septate; hyaline and thermophilic.
Zygomycotina: Absidia ramosa, Mortierella alpina, M. baineri, Mucor pusillus.
Mastiogomycotina: Saprolegnia ferax, Sclerotium cf. papyricola, Spirosphaera floriforme, Trichocladium sp., Trichopyton ajelloi, T. vanbreuseghenii, Varicosporium sp.

Submerged leaves of *Myrica* in ditch, from Salen, collected on April 17, 1969.
Ascomycotina: Allescheria boydii, Gelasinospora cerealis, Nectria coccinea, Thielavia leptodermus.
Deuteromycotina: Aspergillus fumigatus, Chrysosporium asperatum, Clathrosphaerina zalewskei, Humicola fuscoatra, H. grisea, Humicola sp., Isaria farinosa, Mycogone rosea, Papulospora sp., Penicillum spp., Spirosphaera floriforme, Trichoderma koningii. Also mycelium sterile: hyaline and septate.
Zygomycotina: Absidia glauca, Syncephalis sp.
Mastigiomycotina: Monoblepharella taylori, Pythiogeton utriforme.

Rotting debris of *Potamogeton* and *Molinia*, between Glen Forsa and Loch Bà, 16 April 1969.
Ascomycotina: Allescheria terrestris, Coniochaeta ligniaria, Hypomyces aurantius, Podospora curvula, P. decipiens, P. setosa, Pseudopeziza arenevaga, Sporormia intermedia.
Deuteromycotina: Arthrobotrys oligospora, Aspergillus fumigatus, Brachysporium sativum, Cephalosporiopsis alpina, Chloridium chlamydosporis, Cladosporium cladosporioides, C. herbarum, Clathrosphaerina zalewskei, Coniothryium sp., Fusarium graminearum, F. solani, Gliocladium roseum, Humicola fuscoatra, H. grisea, Myrothecium verrucaria, Penicillium piceum, P. steckii, P. thomii, Polyscytalum foecundissimum, Sclerotium rolfsii, Volutella ciliata.
Zygomycotina: Absidia ramosa, Mucor silvaticus.
Mastigomycotina: Pythiogeton utriforme, Saprolegnia asterophora.

Species cultured from soil samples

Shallow peaty gley from under superficially burned hill-pasture with *Molinia*, collected between Glen Forsa and Loch Bà, c. 105 m, 16 April 1969.
Ascomycotina: Allescheria terrestris, Arachniotus albicans, Arthroderma curreyi, Coniochaeta ligniaria, Gelasinospora cerealis, Gymnoascus umbrinus, Hypocopra equorum, Melanospora fimicola, Pseudoeurotium zonatum, Sordaria fimicola, S. humana, S. papyricicola, Sporormia minima, Thielavia terricola.
Deuteromycotina: Aspergillus fumigatus, A. terreus, Bactridiopsis sp., Candidia sp., Chloridium chlamydosporis, Chrysosporium asperatum, C. pannorum, C. pannorum var. carneum, Cladosporium herbarum, Coniothyrium sp., Fusarium culmorum, F. poae, Gliocladium deliquescens, Gliomastix convoluta, Humicola brevis, H. fuscoatra, H. grisea, Humicola sp., Monodictys sp., Paecilomyces varioti, Penicillium fellutanum, P. nigricans, P. thomii, Penicillium spp., Rhizoctonia sp., Sclerotium papyricola, Trichocladium sp., Trichoderma album, T. koningi, T. viride, Verticillium terrestre, Volutella ciliata. Also mycelia sterilia, hyaline and sepate; hyaline, sepate, and ropy; hyaline and clamp connected; brown and septate.
Zygomycotina: Absidia cylindrospora, A. glauca, A. orchidis, Mortierella hydrophila, Mucor ramannianus, M. silvaticus, Piptocephalis cylindrospora.
Mastigomycotina: Phytophthora sp.

Wet peat from under *Myrica* collected at Salen, 17 April 1969.
Ascomycotina: Allescheria terrestris, Arachniotus albicans, Chaetomium anguipilum, Emericella nidulans, Gelasinospora cerealis, Hansenula sp., Lophodermium gramineum, Podospora minuta, Pseudoeurotium zonatum, Sordaria fimicola.
Deuteromycotina: Aspergillus fumigatus, A. niger, Brachysporium graminis, Cephalosporium acremonium, Chloridium chlamydosporis, Chrysosporium pannorum, Cladosporium cladosporioides, C. herbarum, Colletotrichum atramentarium, Dactylella sp., Gliocladium roseum, Gliomastix convoluta, Heterocephalum sp., Hughsiella sp., Humicola brevis, H. fuscoatra, H. grisea, Isaria farinosa, Monilia sp., Myrothecium verrucaria, Oidiodenderon sp., Penicillium fellutanum, P. implicatum, P. nigricans, P. steckii, P. thomii, Penicilium spp., Polyscytalum foecundissimum, Pullularia pullulans, Scopulariopsis brevicaulis, Sesquicillium candelabrum, Stysanus microsporus, Thermoidium sulphureum, Trichocladium sp 2., Trichoderma koningii, T. viride, Trichsporon sp. Also mycelia sterilia: hyaline and septate, brown and septate.
Zygomycotina: Absidia cylindrospora, A. glauca, A. ramosa, Mortierella bainieri, M. hydrophila, M. macrocystis, Mucor hiemalis, M. ramannianus, M. silvaticus, Syncephalis sp.

Bare soil under dense thicket of *Ulex europaeus*, collected from Salen, 17 April 1969.
Ascomycotina: Allescheria terrestris, Coniochaeta ligniaria, Gelasinospora cerealis, Lophiotrema sp., Pseudoeurotium zonatum, Sordaria fimicola.
Deuteromycotina: Aspergillus fumigatus, A. flavus, Brachysporium graminis, Candida sp., Chrysosporium asperatum, C. pannorum, Cylindrocarpon destructans, Fusarium solani, Gilmaniella humicola, Gliocladium roseum, Gliomastix convoluta, Humicola brevis, H. fuscoatra, H. grisea, H. nigrescens, Myrothecium roridium, Oidiodendron fuscum, O. griseum, Paecilomyces elegans, Penicillium capsulatum, P. fellutanum, P. frequentans, P. purpurogenum, P. spinulosum, P. verruculosum, P. waksmannii, Penicillium spp., Periconia byssoides, Polyscytalum foedundissimum, Pullularia pullulans, Rhizoctonia sp., Sclerotium papyricola, Scopulariopsis brevicaulis, Stysanus microsporus, S. stemonitis, Thermomyces lanuginosus, Trichoderma koningii, T. viride, Trichophyton? simii, Trichosporon sp., Trimmatostroma betulinum, Volutella ciliata. Also mycelia sterilia: hyaline and sepate; hyaline septate and ropy; hyaline and clamp-connected; brown and septate; yellow and septate.
Zygomycotina: Absidia cylindrospora, A. glauca, A. ramosa, Mucor racemosus, M. silvaticus.

Brown earth with turf of *Agrostis*, collected in Glen Aros, 18 April 1969.
Ascomycotina: Allescheria boydii, A. terrestris, Chaetomium thermophile, Eurotium chevalieri, Gelasinospora cerealis, Hansenula sp., Melanopora fimicola, Pseudoeurotium zonatum, Thielavia terricola.
Deuteromycotina: Aspergillus flavus, A. fumigatus, Brachysporium graminis, Brachysporium sp., Candida sp., Chrysosporium asperatum, C. pannorum, Cladosporium herbarum, Dinemasporium graminum, Gilmaniella humicola, Gliocladium roseum, Gliomastix convoluta, Helminthosporium sp., Heterocephalum sp., Humicola brevis, H. fuscoatra, H. grisea, H. nigrescens, Monodictys sp., Penicillium simplicissimum, Sclerotium papyricola, Trichoderma album, T. viride, Trichophyton simii, T. terrestre, Verticillium terrestre. Also mycelia sterilia: hyaline and septate; hyaline, septate and ropy, and clamp-connected; brown and septate.
Zygomycotina: Absidia glauca, Mortierella alpina, M. bainieri, M. humilis, Mucor fragilis, M. hiemalis, Zygorhynchus moelleri.

Under stand of old *Fagus*, near Ulva House, collected 17 April 1969.
Ascomycotina: Allescheria terrestris, Arthroderma curreyi, Byssochlamys fulva, Chaetomium semi-nudum, Gelasinospora cerealis, Gymnoascus umbrinus, Melanospora fimicola, Pseudoeurotium zonatum.
Deuteromycotina: Aspergillus fumigatus, Brachysporium obovatum, Candida sp., Cephalosporiopsis alpina, Chloridium chlamydosporis, Chrysosporium asperatum, C. pannorum, Cladosporium herbarum, Cylindrocarpon destructans, Fusarium graminearum, F. solani, Fusidium griseum, Gliomastix convoluta, Humicola brevis, H. grisea, Humicola sp., Mammaria echinobotryoides, Penicillium rubrum, Penicillium spp., Phialophora sp., Phoma sp., Pullularia pullulans, Sporotrichum thermophile, Stysanus stemonitis, Trichoderma viride, Trichophyton ajelloi, T. terrestre, Trichosporon sp., Varicosporium sp. Also mycelia sterilia: hyaline and septate; hyaline and ropy; hyaline and clamp-connected; brown and septate.
Zygomycotina: Absidia corymbifera, A. cylindrospora, A. ramosa, Mortierella alpina, M. bainieri, M. humilus, M. ramannianus, Mucor hiemalis, M. racemosus, M. silvaticus.

Under *Pinus* with thin leaf cover, collected in Glen Aros, 18 April 1969.
Basidiomycotina: Coprinus friesii.

Ascomycotina: Gymnoascus umbrinus, Pseudoeurotium zonatum, Pyrenopeziza arenivaga, Thielavia terricola.
Deuteromycotina: Articulospora sp., Aspergillus flavus, A. fumigatus, Cephalosporium acremonium, Chrysosporium asperatum, C. pannorum, Cordana pauciseptata, Diheterospora catenulata, Dinemasporium graminum, Gilmaniella humicola, Gliocladium roseum, Gliomastix convoluta, Helminthosporum sp., Mammaria echinobotryoides, Oidiodendron fuscum, Paecilomyces elegans, Penicillium piceum, P. purpurogenum, P. wakesmannii, Penicillium spp., Rhinotrichum sp., Rhizoctonia sp., Sympodiella acicola, Torula lucifuga, Trichoderma album, T. koningii, T. viride. Also mycelia sterilia: hyaline and septate; hyaline and clamp-connected, brown and septate; yellow and septate.
Zygomycotina: Absidia glauca, A. ramosa, Mortierella alpina, M. bainieri, Mucor hiemalis, M. pusillus, Zygorhynchus psychrophilus.

Species determined from samples of foam from streams

In deciduous (mainly *Fagus*) wood near Ulva House, collected 17 April 1969.
Alatospora acuminata, Anguillospora crassa, A. longissima, Articulospora tetracladia, Clavariopsis aquatica, Clavatospora stellata, Dactylella aquatica, Flagellospora curvula, Heliscus lugdunensis, Tetracladium setigerum, Tricladium gracile, T. splendens, Varicosporium elodeae.

In oak-woodland between Salen and Aros, collected on 14 April, 1969.
Alatospora acuminata, Anguillospora crassa, A. longissima, Articulospora tetracladia, Heliscus lugdunensis, Tricladium gracile.

In the lower stretch of Allt Mòr, near Kellan Mill, collected on 15 April 1969.
Alatospora acuminata, Anguillospora crassa, A. longissima, Articulospora tetracladia, Flagellospora curvula, Triscelophorus monosporus.

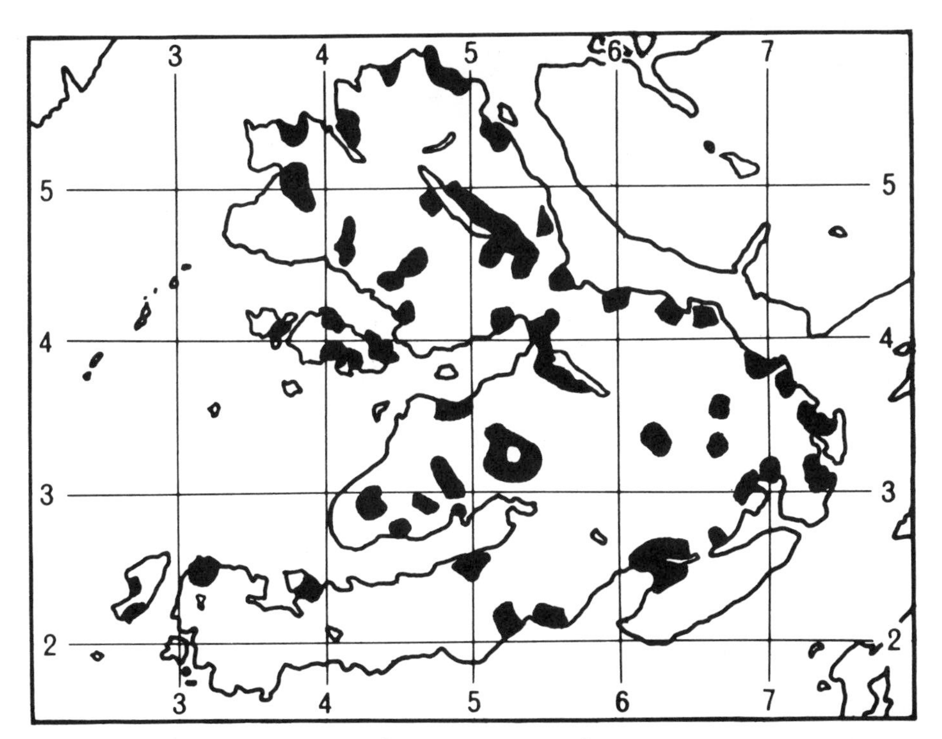

Fig. 15.1 Areas collected by mycologists on autumn forays.

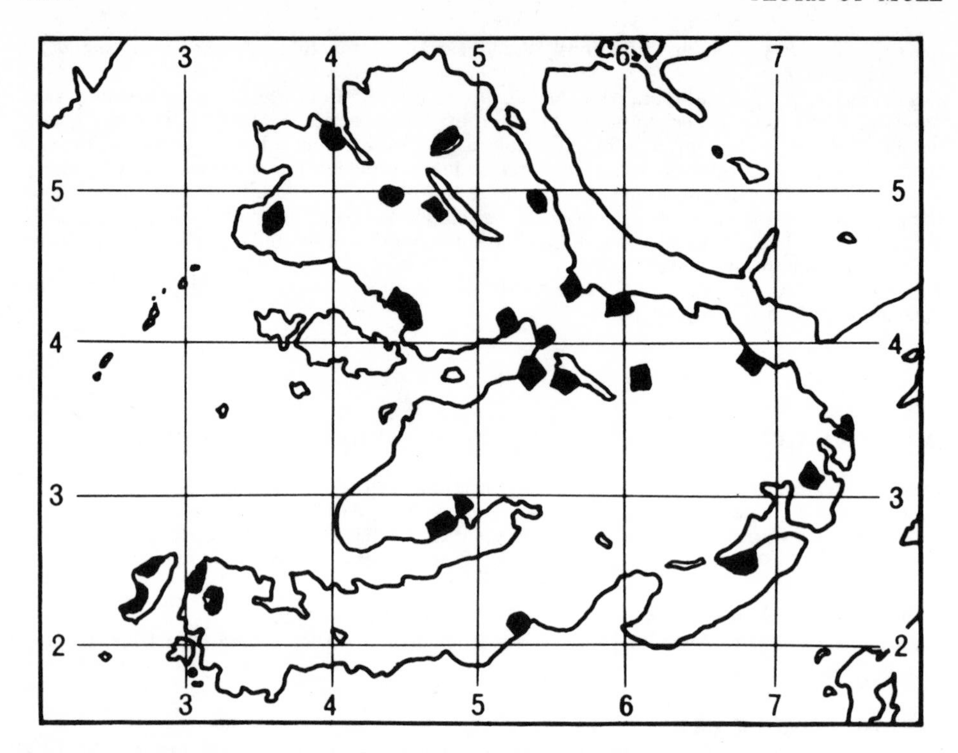

Fig. 15.2 Areas collected by mycologists on spring and early summer forays.

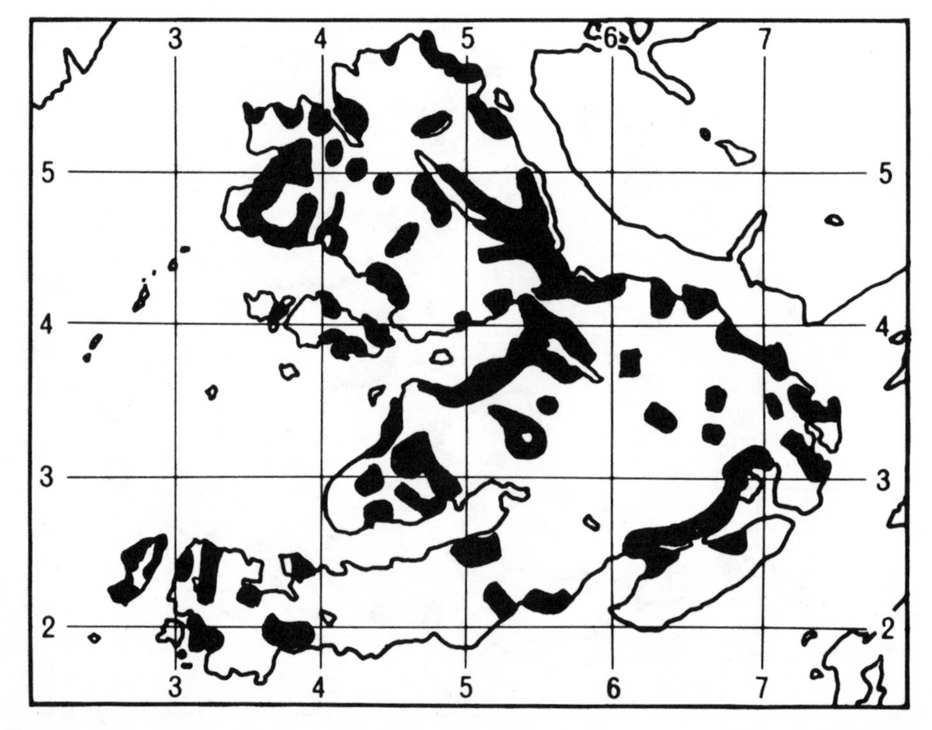

Fig. 15.3 Total area collected by mycologists, 1966–1970.

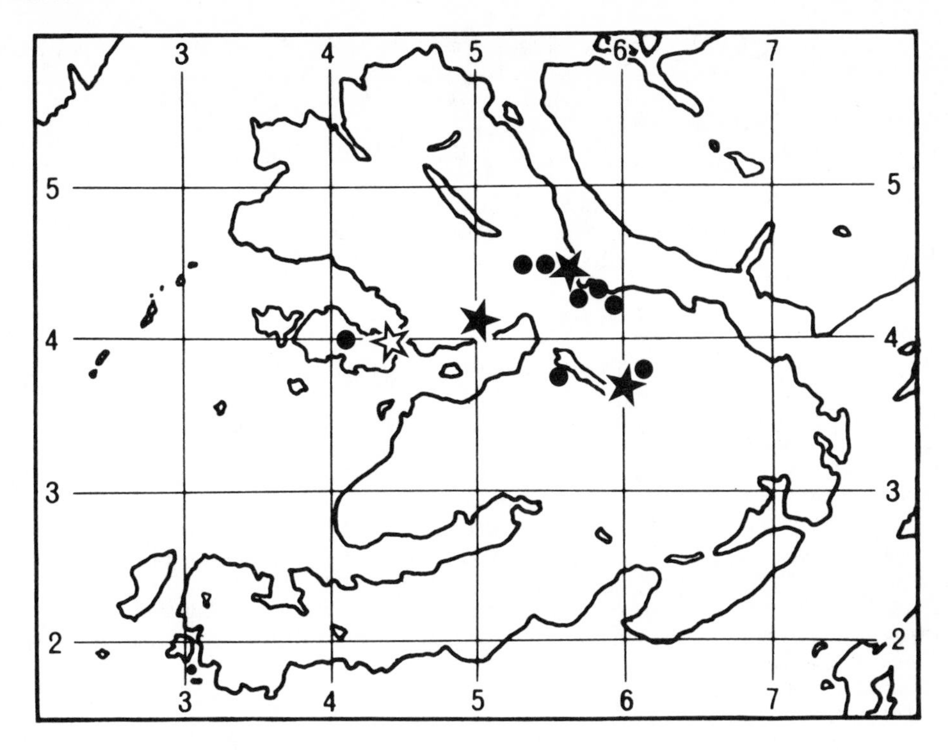

Fig. 15.4 Sites of collection for culture of microspecies: ● *soil;* ★ *soil and foam;* ☆ *foam.*

References

AINSWORTH, G. C. (1966). A general purpose classification for fungi. *Bibl. syst. Mycol.*, No **1**: 1–4.

APINIS, A. E. (1972). Thermophilous fungi in certain grasslands. *Mycopath. Mycol. appl.*, **48**: 63–74.

BISBY, G. R. & MASON, E. W. (1940). List of Pyrenomycetes recorded for Britain. *Trans. Br. Mycol. Soc.* **24**: 127–243.

CHRISTIANSEN, M. P. (1960). Danish resupinate fungi. *Dansk bot. Arkiv.*, **19**: 1–388.

CORNER, E. J. H. (1950). A monograph of *Clavaria* and allied genera. *Ann. Bot. Mem.*, **1**, 1–740.

———. (1970). Supplement to "A monograph of *Clavaria* and allied genera". *Nova Hedwigia Beihefte*, **33**: 1–299.

DENNIS, R. W. G.. (1971). New or interesting British microfungi. *Kew Bull.*, **25**: 335–374.

———. (1972). Some forgotten names among British Helotiales. *Kew Bull.*, **26**: 469–476.

———. (1974). New or interesting British microfungi II. *Kew Bull.*, **29**: 157–179.

———, ORTON, P. D. & HORA, F. B. (1960). New check list of British Agarics and Boleti. *Trans. Br. mycol. Soc. Suppl.*, **1960**: 1–225.

DIXON, J. R. (1974). *Chlorosplenium* and its segregates I. Introduction and the genus *Chlorosplenium*. *Mycotaxon*, **1**: 65–104.

DONK, M. A. (1959). Notes on Cyphellaceae – I. *Persoonia*, **1**: 23–110.

———. (1964). A conspectus of the Families of Aphyllophorales. *Persoonia*, **3**: 199.

———. (1966). Check list of European Hymenomycetous Heterobasidiae. *Persoonia*, **4**: 145–335.

GRADDON, W. G. (1972). Some new Discomycete species – 2. *Trans. Br. mycol. Soc.*, **58**: 147–159.

GRÖGER, F. (1968). Zur Kenntnis von *Lactarius semisanguifluus* Heim & Le Clair. *West. Pilzbriefe*, **7**: 3–11.

HENDERSON, D. M. (1970). Notes on three British Ascomycetes. *Notes Roy. bot. Gdn. Edinb.*, **30**: 203–206.

ING, B. (1968). Myxomycetes from Scottish Islands. *Trans. bot. Soc. Edinb.*, **40**: 395–415.

———. (1974). Rediscovery of a rare Welsh fungus. *Nature in Wales*, **14**: 4–6.

JØRSTAD, I. (1951). The Uredinales of Iceland. *Skr. norske Vidensk. – Akad. mat.-nat. Kl.*, 1951 (**2**): 1–87.

———. (1958). Uredinales of the Canary Islands. *Skr. norske Vidensk. – Akad. mat.-nat. Kl.* 1958 (**2**): 1–182.

JULICH, W. (1972). Monographie der Athelieae (Corticiaceae: Basidiomycetes). *Willdenowia*, **7**: 1–283.

JUNELL, L. (1967). Erysiphaceae of Sweden. *Symb. Bot. Upps.*, **19**: 1–117.

LANGE, J. (1939). *Flora Agaricina Danica*, vol. 4. Copenhagen.

LANGE, M. (1955). Den Botaniske Ekspedition til vestgrönland 1946, Macromycetes II, Greenland Agaricales (pars). *Medd. Grønland*, **147 (11)**: 1–69.

———. (1957). Ibid. Macromycetes, III: (i) Greenland Agaricales (pars), Macromycetes caeteri; (ii) Ecological and Plant Geographical studies. *Medd. Grønland*, **148 (2)**: 1–125.

LUNDQUIST, N. (1972). Nordic Sordariaceae s. lat. *Symb. Bot. Upsal.*, **20**: 1–374.

MASS GEESTERANUS, R. A. (1975). *The terrestrial hydnums of Europe*. Amsterdam.

PEGLER, D. N. (1966). 'Polyporaceae' – Part I with a Key to British Genera. *Bulletin, Br. mycol. Soc.* No. 26.

REID, D. (1974). A monograph of the British Dacrymycetales. *Trans. Br. mycol. Soc.* **62**: 433–494.

ROMAGNESI, H. (1967). *Les Russules d'Europe et d'Afrique du Nord*. Paris.

WATLING, R. (1969). Records of Boleti and notes on their taxonomic position II. *Notes Roy. bot. Gdn Edinb.*, **29**: 265–272.

———. (1970). *British fungus flora*, part 1: Boletaceae, Gomphidiaceae, Paxillaceae. Edinburgh.

———. (1971). Polymorphism in *Psilocybe merdaria*. *New Phytol.*, **70**: 307–326.

———. (1972). Notes on some British Agarics III. *Notes Roy. bot. Gdn. Edinb.*, **32**: 127–134.

———. & RICHARDSON, M. J. (1971). The Agarics of St Kilda. *Trans. bot. Soc. Edinb.*, **41**: 165–187.

WEBSTER, J. (1964). Culture studies on *Hypocrea* & *Trichoderma*, I. Comparison of perfect and imperfect states of *H. gelatinosa*, *H. rufa* & *Hypocrea* sp. 1. *Trans. Br. mycol. Soc.*, **47**: 75–96.

WHITE, F. B. (1881). The cryptogamic flora of Mull. *Scott. Nat.*, **6**: 155–162.

WILSON, M. (1934). Uredineae in Scotland. *Trans. bot. Soc. Edinb.*, **31**: 345–449.

———. & HENDERSON, D. M. (1966). *British rust fungi*. Cambridge.

16. Freshwater algae
(excluding diatoms)

D. J. Hibberd*

Introduction

The aim of the freshwater team of the Mull Survey was to investigate the freshwater algal flora of the Island over as wide an ecological range as possible. The diatoms, which form a prominent feature of the flora, were studied from the outset of the Survey by Patricia A. Sims and are presented separately and in greater detail in Chapter 17. The enumeration of other species of freshwater algae presented in this chapter is the result of work carried out by the author between 1969 and 1971. Particular attention was given to desmids, an important and characteristic element in the flora of acid and peaty pools. Some groups are under-represented because of difficulties in identification. For example, members of the Zygnemataceae and Oedogoniaceae occurred commonly but were never found in a fertile condition and were therefore impossible to identify with certainty. Flagellate algae were mostly not well preserved and are in any case difficult to identify when dead, especially when only a few specimens are present, though scales of species of *Synura* and *Mallomonas* were identified using electron microscopy of shadowcast whole mounts. Blue-green algae, species of which occurred in every possible habitat, are also under-represented since only species occurring in relatively large quantities and those which could be clearly ascribed to a named taxon have been included. In addition to these limitations, seasonal variations may account for the apparent absence of many species, as might also the relatively small numbers of samples it was possible to study in the time available. A total of 390 species and varieties are recorded.

Sampling and collection methods

Euplankton (open water plankton) was collected using a 200 mesh net and was preserved as soon as possible after collection. The remainder of the material was collected simply either by scraping from rocks, submerged wood and plant stems, squeezing tychoplankton (unattached species caught amongst other plants) from submerged mosses and filamentous algae or by sampling directly from pool bottoms with a collecting tube. Material was preserved in glutaraldehyde buffered to pH 7 with 0.1 M phosphate buffer, though Transeau's solution (6:3:1::water:95% alcohol:formaldehyde + 5 ml glycerine/100 ml) was used for filamentous forms.

* Now at the Culture Centre of Algae and Protozoa (NERC), 36 Storey's Way, Cambridge.

Samples of this preserved material were subsequently examined without further treatment.

Arrangement and nomenclature

The systematic arrangement of the text is based, with some minor amendments and additions, on that of Christensen (1962). Determinations have been made from the standard monographs wherever possible and the work used is listed in the text under the appropriate class, family or generic heading. Except for those species occurring in the plankton of the deepwater lochs or those which are obviously rare or show a marked preference for a particular habitat or locality, exact sampling localities are not given. However, the number of *samples* in which the species has been recorded is placed at the end of the entry; this is not to be confused with a locality number.

Acknowledgements

I am grateful to Patricia A. Sims for collecting a number of samples in the early part of the Survey and to Pamela Cambridge for technical help. Jenny A. Moore gave considerable help in checking the manuscript and bibliography.

Index to genera

CYANOPHYTA

Bibliogr.: Geitler (1932).

CYANOPHYCEAE

CHROOCOCCALES

CHROOCOCCACEAE

Aphanocapsa Näg.

A. montana Cramer
In a maritime pool at **10** Carsaig.
(1)

A. muscicola (Menegh.) Wille
Amongst *Sphagnum* and on rock in waterfall spray.
(4)

Aphanothece Näg.

A. clathrata West & West
In a bog-pool.
(1)

A. microscopica Näg.
On wet rocks.
(1)

A. nidulans P. Richt.
Amongst *Sphagnum.*
(1)

Chroococcus Näg.

C. limneticus Lemm.
In euplankton of **3** Loch Poit na h-I; **17** Loch a' Ghleannain; and tychoplankton of a lochan margin.
(4)

C. minor (Kütz.) Näg.
Amongst *Sphagnum.*
(3)

C. minutus (Kütz.) Näg.
Amongst other blue-green algae on maritime cliffs at **10** Carsaig.
(1)

C. turgidus (Kütz.) Näg.
Amongst *Sphagnum* and in bog-pools; one of the most characteristic species in these habitats.
(32)

C. westii (W. West) Boye-Petersen
On marine cliffs at **10** Carsaig.
(1)

Dactylococcopsis Hansgirg

D. smithii R. & F. Chodat
Amongst *Sphagnum*; in a bog-pool; at a loch edge.
(3)

Gloeocapsa Kütz.

G. alpina Näg. ex Cramer
On maritime cliffs at **10** Carsaig.
(1)

G. compacta Kütz.
On maritime cliffs at **10** Carsaig in spray of waterfall.
(1)

G. decorticans (A. Br.) P. Richt.
On maritime cliffs at **10** Carsaig and in spray of waterfall.
(2)

G. itzigsohnii Bornet in Zopf
Amongst scrapings from wet rocks.
(2)

G. magma (Bréb.) Kütz.
Amongst scrapings from wet rocks.
(2)

G. polydermatica Kütz.
Amongst rock scrapings in spray of waterfalls and maritime cliffs at **10** Carsaig.
(3)

G. quaternaria Kütz. ex Geitler
Amongst scrapings from wet rock.
(1)

G. ralfsiana (Harv.) Kütz.
Amongst scrapings from wet rocks.
(1)

G. rupestris Kütz.
On maritime cliffs at **10** Carsaig.
(1)

Gloeothece Näg.

G. membranacea (Rabenh.) Bornet
On maritime cliffs at **10** Carsaig.
(1)

G. rupestris (Lyngb.) Bornet
On maritime cliffs at **10** Carsaig.
(1)

Merismopedia Meyen

M. elegans A. Br.
Amongst tychoplankton at margin of a lochan.
(1)

M. glauca (Ehrenb.) Näg.
Amongst *Sphagnum* and in a bog-pool.
(2)

M. punctata Meyen
In bog-pools and amongst *Sphagnum.*
(6)

M. tenuissima Lemm.
In a bog-pool.
(1)

Microcystis Kütz.

M. aeruginosa Kütz.
In euplankton of **3** Loch Poit na h-I; **6** Loch Peallach and Loch Carnain an Amais; **7** Loch Frisa; **11** Loch Bà; **13** Lochan a' Ghurrabain; **16** Loch Uisg; **17** Loch a' Ghleannain.
(8)

M. incerta Lemm.
In euplankton of **3** Loch Poit na h-I.
(1)

M. pulverea (Wood) Forti
In a maritime pool.
(1)

Rhabdoderma Schmidle & Lauterborn

R. gorskii Wolosz.
In euplankton of **7** Loch Frisa.
(1)

R. lineare Schmidle & Lauterborn
Amongst *Sphagnum* and in bog-pools.
(3)

Synechococcus Näg.

S. aeruginosus Näg.
Amongst *Sphagnum*, in bog-pools and in hollows in wet rock but probably overlooked in many samples.
(8)

CHAMAESIPHONALES

CHAMAESIPHONACEAE

Chamaesiphon A. Br. & Grun.

C. curvatus Nordst.
On floating debris in **7** Loch Frisa.
(1)

HORMOGONALES

NOSTOCACEAE

Anabaena Bory

A. cylindrica Lemm.
In maritime pool at **10** Carsaig.
(1)

A. flos-aquae (Lyngb.) Bréb.
In euplankton of **9** Loch Assopol.
(1)

A. lapponica Borge
In maritime pool at **10** Carsaig.
(1)

A. oscillarioides Bory
In euplankton of **16** Loch Uisg.
(1)

Nostoc Vauch.

N. commune Vauch.
Amongst wet moss on a rock face.
(1)

N. humifusum Carm. sec. Harvey in Hook.
On boulders in stream.
(1)

N. microscopicum Carm. sec. Harvey in Hook.
On wet or dripping rocks including maritime cliffs at Carsaig.
(2)

OSCILLATORIACEAE

Lyngbya Ag.

L. limnetica Lemm.
On rocks in waterfall and in a bog-pool.
(2)

L. perelegans Lemm.
On wet rocks near a stream and waterfall.
(2)

Oscillatoria Vauch.

O. amoena (Kütz.) Gom.
On rocks near streams and waterfalls, and dripping rock faces.
(6)

O. amphibia Ag.
In a slow running flush.
(1)

O. irrigua Kütz.
In euplankton of **7** Loch Frisa and in a maritime pool at **10** Carsaig.
(3)

O. sancta (Kütz.) Gom.
In a maritime pool at **10** Carsaig.
(1)

O. tenuis Ag.
On wet rocks, a bog-pool and amongst green filamentous algae.
(4)

Phormidium Kütz.

P. tenue (Menegh.) Gom.
In a maritime pool at **10** Calgary Bay.
(1)

Schizothrix Kütz.

S. lacustris A. Br.
In mats on wet rock.
(1)

SCYTONEMATACEAE

Petalonema Berk.

P. alatum Berk.
On maritime cliffs at **10** Carsaig.
(3)

P. velutinum (Rabenh.) Migula
On boulders in waterfall spray.
(2)

Scytonema Ag.

S. mirabile (Dillw.) Born.
On dripping rock and boulders splashed by streams.
(9)

Tolypothrix Kütz.

T. letestui Frémy
On boulders in a river, a stream and around a loch margin.
(3)

T. limbata Thuret
On a boulder in a lochan.
(1)

T. tenuis Kütz.
In maritime pools at **10** Carsaig and **14** Salen.
(2)

STIGONEMATACEAE

Hapalosiphon Näg. in Kütz.

H. fontinalis (Ag.) Born.
Amongst *Sphagnum.*
(1)

H. hibernicus West & West
Amongst *Sphagnum.*
(1)

Stigonema Ag.

Species of **Stigonema** are characteristic of intermittently wet rock surfaces, that is receiving some moisture e.g. spray from streams and waterfalls or windblown rain, but never subject to continuous wetting or permanent drying out. The following were recorded: **S. hormoides** (Kütz.) Born. & Flah. (2); **S. informe** Kütz. (2); **S. lavardei** Frémy (4); **S. mamillosum** (Lyngb.) Ag. (10); **S. minutum** (Ag.) Hass. (7); **S. ocellatum** Thuret (7); **S. panniforme** Born. & Flah. (1); **S. tomentosum** (Kütz.) Hieron. (1).

RIVULARIACEAE

Calothrix Ag.

C. braunii Born. & Flah.
On rocks intermixed in *Phormidium tenue* in a maritime pool at **5** Calgary Bay.
(1)

Dichothrix Zanardini

D. gypsophila (Kütz.) Born. & Flah.
On dripping rock face mixed with *Stigonema* and *Gloeocapsa* spp.
(1)

D. meneghiniana (Kütz.) Forti
On boulders at edge of stream.
(1)

D. orsiniana Born. & Flah.
In scrapings from boulders in streams and loch margins.
(7)

Rivularia Ag.

R. baccariana (De Not.) Born. & Flah.
On boulders in streams.
(4)

RHODOPHYTA

FLORIDEOPHYCEAE

Growths of **Batrachospermum** species (Batrachospermaceae) occurred on the Island in relatively slow-flowing streams, often in those cutting through bare peat. A **Lemanea** species (Lemaneaceae) was found on slabs in the rapidly flowing streams in the NW of the Island.

CRYPTOPHYTA

CRYPTOPHYCEAE

Cryptomonads occurred relatively rarely, usually as isolated cells. Identification of this taxonomically difficult group was therefore mostly not attempted.

Cryptomonas ovata Ehrenb. (Crytomonadaceae) was found in euplankton of **9** Loch Assopol; and in a pool in root pit of a fallen tree.

DINOPHYTA

DINOPHYCEAE

Dinoflagellates usually occurred only in small numbers in the samples taken; since they constitute a taxonomically difficult group, only the species which occurred in any quantity (mainly in the euplankton of the deepwater lochs) are listed here.

PERIDINIALES

PERIDINIACEAE

Peridinium Ehrenb.

P. inconspicuum Lemm.
In bog-pools and amongst *Sphagnum.*
(7)

P. willei Huitf. – Kaas
In euplankton of deepwater lochs.
(13)

CERATIACEAE

Ceratium Schrank

C. carolinianum (Bailey) Jörgensen
In euplankton of **17** Loch a' Ghleannain.
(2)

C. hirundinella (Müll.) Schrank
In the euplankton of deepwater lochs.
(10)

DINOCOCCALES

DINOCOCCACEAE

Cystodinium Klebs

C. cornifax (Schilling) Klebs
In bog-pools.
(4)

RHAPHIDOPHYTA

RHAPHIDOPHYCEAE

RHAPHIDOMONADALES

RHAPHIDOMONADACEAE

Gonyostomum Diesing

G. semen (Ehrenb.) Diesing
In a bog-pool.
(1)

Vacuolaria Cienk.

V. viriscens Cienk.
In bog-pools.
(2)

CHRYSOPHYTA

CHRYSOPHYCEAE

RHIZOCHRYSIDALES

STYLOCOCCACEAE

Lagynion Pascher

L. ampullaceum (Stokes) Pascher
In a marine supralittoral pool at **10** Carsaig, on *Oedogonium.*
(1)

CHROMULINALES

CHRYSOCOCCACEAE

Chrysopyxis Stein

C. bipes Stein
In a pool formed by the root pit of a fallen tree, on *Mougeotia.*
(2)

C. stenostoma Lauterb.
In a maritime pool at **10** Carsaig, on *Oedogonium.*
(1)

CHRYSAMOEBACEAE

Chrysamoeba Klebs

C. radians Klebs
In a maritime pool at **10** Carsaig.
(1)

Chrysostephanosphaera Scherff.

C. globulifera Scherff.
In a bog-pool.
(1)

OCHROMONADALES

DINOBRYACEAE

Dinobryon Ehrenb.

Bibliogr.: Huber-Pestalozzi (1941); Ahlstrom (1937).

D. bavaricum Imhof
In euplankton of **3** Loch an Sgalain; **7** Loch Frisa; **10** Loch Fuaron; **11** Loch Bà; **17** Loch a' Ghleannain.
(7)

D. cylindricum Imhof
In euplankton and tychoplankton of lochs and lochans; rarely in bog-pools.
(13)

D. divergens Imhof
In euplankton of **3** Loch Poit na h-I; **9** Loch Assopol; and **10** Loch Fuaron; rarely in bog-pools.
(7)

D. pediforme (Lemm.) Steinecke
In a bog-pool.
(1)

D. sertularia Ehrenb.
Amongst submerged *Sphagnum* and in bog-pools.
(4)

D. utriculus Stein var. **acutum** Schiller
In a bog-pool on *Mougeotia.*
(1)

SYNURACEAE

Mallomonas Perty

Bibliogr.: Harris & Bradley (1960).

M. cratis Harris & Bradley
In a brackish saltmarsh pool.
(1)

Mallomonas spp. occurred relatively commonly in this habitat.

Occasional cells of unidentified species occurred in the euplankton of some deepwater lochs.

Synura Ehrenb.

S. spinosa Korsh. forma **spinosa** sensu Petersen & Hansen
In euplankton of **3** Loch Poit na h-I; **7** Loch Frisa; **9** Loch Assopol; **17** Loch a Ghleannain

ISOCHRYSIDALES

DEREPYXIDACEAE

Rhipidodendron Stein

R. huxleyi Kent
Amongst submerged *Sphagnum* and in bog-pools and edges of lochans; in supralittoral pools.
(6)

XANTHOPHYTA

XANTHOPHYCEAE

MISCHOCOCCALES

MISCHOCOCCACEAE

Mischococcus Näg.

M. confervicola Näg.
In a bog-pool.
(1)

CHLOROPEDIACEAE

Ophiocytium Näg.

Unidentified species of *Ophiocytium* occurred in small numbers embedded in detritus on the bottom of bog-pools and in maritime pools.
(6)

CHARACIOPSIDIACEAE

Peroniella Gobi

P. hyalothecae Gobi
Epiphytic on *Hyalotheca dissiliens*.
(1)

TRIBONEMATALES

TRIBONEMATACEAE

Tribonema Derbes & Sol.

Tribonema species occurred commonly in a variety of habitats but mostly could not be identified from preserved material. Each of the two following were recorded once: **T. aequale** Pasch.; **T. viride** Pasch.

VAUCHERIALES

VAUCHERIACEAE

Vaucheria DC.

Bibliogr.: Blum (1972); Christensen (1969).
Vaucheria species occurred commonly in flushes and on other wet substrates. Though the majority of collections were found to be sterile, the following species have been identified from sexual material: **V. aversa** Hass.; **V. bursata** (O. F. Müller) Ag.; **V. canalicularis** (L.) T. Chr.; **V. frigida** (Roth) Ag.; **V. racemosa** (Vauch.) DC.

EUSTIGMATOPHYTA

EUSTIGMATOPHYCEAE

Bibliogr.: Hibberd & Leedale (1971); Hibberd (1974).

Chlorobotrys regularis (West) Bohlin (Chlorobotrydaceae) is one of the commonest and most characteristic species found amongst *Sphagnum* and in bog-pools; also in maritime pools at **10** Carsaig. (21)

EUGLENOPHYTA

EUGLENOPHYCEAE

EUGLENALES

EUGLENACEAE

Euglena Ehrenb.

E. gracilis Klebs
In a bog-pool.
(1)

E. granulata (Klebs) Lemm.
In a bog-pool.
(1)

E. lucens Gunther
In a bog-pool.
(1)

E. proxima Dang.
In a vehicle track on farmland; also in a bog-pool and amongst *Sphagnum*.
(3)

E. spirogyra Ehrenb.
In a vehicle track on farmland.
(1)

Lepocinclis Perty

L. ovum (Ehrenb.) Lemm.
In a vehicle track on farmland.
(1)

Phacus Dujard.

P. curvicauda Swirenko
In a vehicle track on farmland.
(1)

P. platyaulax Pochm.
In a bog-pool.
(1)

P. pleuronectes (Müll.) Dujard.
In a bog-pool.
(1)

Trachelomonas Ehrenb.

The following species, except *T. hispida*, occurred in bog-pools:

T. abrupta (Swirenko) Deflandre
(6)

T. cylindrica (Ehrenb.) Playfair
(1)

T. granulosa Playfair
(3)

T. hispida (Perty) Stein emend. Deflandre
Along the margin of a lochan.
(1)

T. volvocina Ehrenb.
(2)

CHLOROPHYTA

CHLOROPHYCEAE

VOLVOCALES

VOLVOCACEAE

Bibliogr.: Huber-Pestalozzi (1961).

Eudorina Ehrenb.

E. elegans Ehrenb.
In euplankton of deepwater lochs.
(12)

Gonium Müller

G. formosum Pascher forma **formosum**
In euplankton of **3** Loch Assopol.
(1)

G. formosum forma **suecicum** Huber-Pestalozzi
In a marine supralittoral pool at **10** Carsaig.
(1)

TETRASPORALES

TETRASPORACEAE

Schizochlamys A. Br. in Kütz.

S. compacta Prescott
In a marine supralittoral pool at **10** Carsaig.
(1)

CHLOROCOCCALES

CHLOROCOCCACEAE

Bibliogr.: Korshikov (1953).

Characium A. Br.

C. falcatum Schroeder
On filamentous algae on stones in stream splash.
(1)

Desmatractum West & West

D. bipyramidatum (Chodat) Pascher
In bog-pools.
(2)

Tetraedron Kütz.

T. minimum (A. Br.) Hansg.
In a marine supralittoral pool at **10** Carsaig and an inland lochan margin.
(2)

PALMELLACEAE

Bibliogr.: Korshikov (1953).

Sphaerocystis Chodat emend. Korsh.

S. schroeteri Chodat
In euplankton of **6** Loch an Tor, Loch Peallach, Loch Carnain an Amais; **7** Loch Frisa; and **17** Loch a' Ghleannain.
(5)

HORMOTILACEAE

Bibliogr.: Korshikov (1953).

Palmodictyon Kütz.

P. varium (Näg.) Lemm.
Amongst *Sphagnum* and in bog-pools.
(3)

OOCYSTACEAE

Bibliogr.: Korshikov (1953).

Ankistrodesmus Corda

A. falcatus (Corda) Ralfs
In bog-pools and margins of lochans; in a supralittoral pool at **10** Carsaig.
(12)

Eremosphaera De Bary

E. oocystoides Prescott
Amongst *Sphagnum.*
(1)

E. viridis De Bary
On mud film, amongst *Sphagnum* and in a bog-pool.
(4)

Oocystis Näg.

Species of *Oocystis* occurred commonly in samples from bog-pools, maritime pools and amongst *Sphagnum* but could not be identified satisfactorily from preserved material.

MICRACTINIACEAE

Bibliogr.: Korshikov (1953).

Goleokinia Chodat

G. radiata Chodat
In a marine supralittoral pool at **10** Carsaig.
(1)

DICTYOSPHAERIACEAE

Bibliogr.: Korshikov (1953).

Botryococcus Kütz.

B. braunii Kütz.
Amongst *Sphagnum.*
(3)

B. protruberans West & West
In euplankton of **3** Loch Poit na h-I; **6** Loch Peallach; **7** Loch Frisa; **10** Loch Fuaron; and **11** Loch Bà.
(5)

Botryosphaera Chodat emend. Korsh.

B. sudetica (Lemm.) Chodat
In euplankton of **6** Loch Peallach, **7** Loch Frisa, **10** Loch Fuaron and **11** Loch Bà,
(4)

Dictyosphaerium Näg.

D. pulchellum Wood
In euplankton of **3** Loch Poit na h-I and **8** Loch Bà; in open bog-pool; in a marine supra-littoral pool at **10** Carsaig.
(5)

SCENEDESMACEAE

Bibliogr.: Korshikov (1953).

Coelastrum Näg.

C. sphaericum Näg.
In a marine supralittoral pool at **10** Carsaig.
(1)

Crucigenia Morren

C. irregularis Wille
In marine supralittoral pools at **10** Carsaig.
(2)

C. quadrata Morren
Amongst *Sphagnum.*
(1)

Scenedesmus Meyen

Bibliogr.: Uherkovich (1966).

S. acuminatus (Lagerh.) Chodat
In euplankton of **3** Loch Poit na h-I; in a marine supralittoral pool.
(2)

S. denticulatus Lagerh. var. **linearis** Hansg.
In marine supralittoral pools at **5** Calgary Bay and **10** Carsaig.
(2)

S. ovalternus Chodat var. **graevenitzii** (Bernard) Chodat
In a bog-pool.
(1)

S. quadricauda (Turpin) Bréb. sensu lato
In bog-pools and margins of lochans; in a supralittoral pool at **10** Carsaig.
(6)

S. serratus (Corda) Bohlin forma **minor** Chodat
In euplankton of **3** Loch Poit na h-I; also along margins of lochans and in bog-pools.
(4)

S. tibiscensis Uherkov.
In a marine supralittoral pool at **10** Carsaig.
(1)

HYDRODICTYACEAE

Bibliogr.: Korshikov (1953).

Pediastrum Meyen

Bibliogr.: Sulek (1969).

P. boryanum (Turpin) Menegh. var. **boryanum**
In euplankton of **9** Loch Assopol; amongst filamentous algea on dripping rocks.
(2)

P. boryanum var. **cornutum** (Racib.) Sulek
In euplankton of **3** Loch Poit na h-I; in a marine supralittoral pool at **10** Carsaig.
(2)

P. boryanum var. **longicorne** Reinsch
In a marine supralittoral pool with var. *cornutum.*
(2)

P. tetras (Ehrenb.) Ralfs
By margin of lochan; in a bog-pool and maritime pools.
(3)

ULOTHRICALES

ULOTHRICACEAE

Bibliogr.: Printz (1964).

Binuclearia Wittr.

B. tectorum (Kütz.) Berger in Wichm.
Amongst *Sphagnum* and in bog-pools; in supralittoral pools.
(26)

Geminella Turpin

G. interrupta (Turpin) Lagerh.
On rocks in waterfall.
(1)

Ulothrix Kütz.

U. subtilissima Rabenh.
On boulders in streams; in a bog-pool.
(6)

U. variabilis Kütz.
In a bog-pool.
(1)

U. zonata (Web. & Mohr) Kütz.
In waterfalls and streams and on rocks within splash zones.
(13)

MICROSPORACEAE

Bibliogr.: Printz (1964).

Microspora Thuret

M. aequabilis Wichm.
On rocks in stream.
(1)

M. amoena (Kütz.) Rabenh. var. **amoena**
Amongst moss in roadside ditch; on drying mud by roadside.
(2)

M. amoena var. **gracilis** (Wille) De Toni
In a wet depression in rock.
(1)

M. irregularis (West & West) Wichm.
On rocks in stream.
(2)

M. palustris Wichm.
In a bog-pool.
(1)

CYLINDROCAPSACEAE

Bibliogr.: Printz (1964).

Cylindrocapsa Reinsch

C. conferta West
On rocks in fast flowing stream.
(1)

CHAETOPHORALES

CHAETOPHORACEAE

Draparnaldia Bory

D. glomerata (Vauch.) Ag.
In a bog-pool.
(1)

Stigeoclonium Kütz.

S. attenuatum (Hazen) Collins
On stones at loch margin.
(1)

OEDOGONIALES

OEDOGONIACEAE

Bulbochaete species commonly occurred in bog-pools and were also found in streams, at the rocky edges of lochs and in supralittoral pools. **Oedogonium** species were found more rarely, nearly always in bog-pools. All material from both genera was found to be sterile and therefore could not be identified from preserved material.

SIPHONOCLADALES

CLADOPHORACEAE

Bibliogr.: Heering (1921).

Rhizoclonium Kütz. emend. Brand

R. hieroglyphicum (Ag.) Kütz. emend. Stockmayer
In waterfalls and streams.
(6)

ZYGNEMATALES

ZYGNEMATACEAE

Species of **Mougeotia, Spirogyra** and **Zygnema** occurred ubiquitously in the tychoplankton of pools, lochs and lochans, in streams, on dripping rocks and in supralittoral pools. Since these were invariably sterile in our collections, however, they could not be identified from preserved material.

Zygogonium Kütz.

Z. ericetorum Kütz.
In shallow, slow-flowing trickles in bogs and mires.
(6)

MESOTAENIACEAE

Cylindrocystis Menegh.

Bibliogr.: Krieger (1937).

C. brebissonii Menegh.
Amongst *Sphagnum* and in bog-pools.
(44)

Gonatozygon De Bary

Bibliogr.: West & West (1904).

G. brebissonii De Bary
Amongst *Sphagnum* and in bog-pools; in maritime pools at **10** Carsaig.
(5)

G. monotaenium De Bary
At a Lochan margin.
(1)

Netrium (Näg.) Itzigs. & Rothe

Bibliogr.: Krieger (1937).

N. digitus (Ehrenb.) Itzigs. & Rothe var. **digitus**
Amongst *Sphagnum* and on mud surface in bog-pools; in maritime pools at **10** Carsaig.
(21)

N. digitus var. **latum** Hust.
Amongst *Sphagnum.*
(3)

N. digitus var. **naegelii** (Bréb.) Krieger
Amongst *Sphagnum* and in bog-pools.
(4)

Spirotaenia Bréb.

Bibliogr.: West & West (1904).

S. condensata Bréb.
Amongst *Sphagnum.*
(1)

DESMIDIACEAE

Arthrodesmus Ehrenb.

Bibliogr.: West & West (1912).

A. bifidus Bréb. var. **truncatus** West
In a bog-pool.
(1)

A. octocornis Ehrenb.
Amongst *Sphagnum* and in bog-pools.
(5)

A. trispinatus West & West
In bog-pools; at a lochan margin.
(3)

Bambusina Kütz.

Bibliogr.: West, West & Carter (1923).

B. brebissonii Kütz.
Gymnozyga moniliformis Ehrenb.
In bog-pools and amongst *Sphagnum*.
(11)

Closterium Nitzsch

Bibliogr.: Krieger (1937).

C. abruptum West
In bog-pools.
(2)

C. acerosum (Schrank) Ehrenb. var. **acerosum**
In a bog-pool.
(1)

C. acerosum var. **angolense** West & West
In euplankton of **7** Loch Frisa.
(1)

C. acutum Bréb. var. **acutum**
In bog-pools and amongst *Sphagnum*; in maritime pools at **10** Carsaig.
(6)

C. acutum var. **linea** (Perty) West & West.
In bog-pools and amongst *Sphagnum*.
(3)

C. baillyanum Bréb.
Amongst *Sphagnum*.
(2)

C. costatum Corda
Amongst *Sphagnum* and in bog-pools.
(2)

C. dianae Ehrenb. var. **dianae**
In bog-pools and amongst *Sphagnum*; in maritime pools at **10** Carsaig.
(19)

C. dianae var. **pseudodianae** (Roy) Krieger
In bog-pools and along loch margin.
(3)

C. gracile Bréb.
In bog-pools and amongst *Sphagnum*; in maritime pools at **10** Carsaig.
(14)

C. intermedium Ralfs var. **intermedium**
Amongst *Sphagnum* and in bog-pools.
(11)

C. intermedium var. **hibernicum** West & West
Amongst *Sphagnum*.
(1)

C. kuetzingii Bréb.
In euplankton of deepwater lochs and around margins of two lochans.
(8)

C. libellula Focke
In a bog-pool.
(1)

C. lunula (Müll.) Nitzsch
Amongst *Sphagnum* and in bog-pools.
(4)

C. striolatum Ehrenb.
Amongst *Sphagnum* and in bog-pools.
(5)

C. subscoticum Gutw.
In a bog-pool and around a loch margin.
(2)

Cosmarium Corda

Bibliogr.: West & West (1905, 1908, 1912); Krieger & Gerloff (1962, 1965, 1969) marked *.

C. abbreviatum Racib. var. **minus** Roy & Bisset
In a bog-pool.
(1)

C. abbreviatum var. **planctonicum** Reverdin*
In euplankton of **7** Loch Frisa and **9** Loch Assopol.
(2)

C. amoenum Bréb.
Amongst *Sphagnum* and in bog-pools.
(6)

C. angulosum Bréb.
In bog-pools and in maritime pools at **10** Carsaig.
(3)

C. bioculatum Bréb. var. **depressum** (Schaarschm.) Schmidle*
At margins of lochans.
(2)

C. blyttii Wille var. **novae-sylvae** West & West
In bog-pools and in maritime pools at **10** Carsaig.
(6)

---

* Determined after Krieger & Gerloff.

C. botrytis Menegh.
In euplankton of **7** Loch Frisa and **9** Loch Assopol, in bog-pools and on dripping rocks.
(7)

C. brebissonii Menegh.
Amongst *Sphagnum* and in bog-pools.
(6)

C. caelatum Ralfs
In a bog-pool and amongst *Sphagnum.*
(2)

C. capitulum Roy & Bisset var. **groenlandicum** Börgesen*
In euplankton of **3** Loch Poit na h-I and **9** Loch Assopol.
(2)

C. contractum Kirchn. var. **ellipsoideum** (Elfv.) West & West
In a bog-pool.
(1)

C. contractum var. **retusum** (West & West) Krieger & Gerloff*
In a bog-pool.
(1)

C. crenatum Ralfs

In bog-pools.
(2)

C. cucumis Corda ex Ralfs var. **cucumis***
Amongst *Sphagnum.*
(1)

C. cucumis var. **magnum** Racib.*
Amongst *Sphagnum.*
(1)

C. cucurbita Bréb. ex Ralfs var. **cucurbita***
Amongst *Sphagnum*; in open pools.
(26)

C. cucurbita var. **attenuatum** G. S. West
On a wet rock face.
(1)

C. cucurbita var. **latius** (West & West) Krieger & Gerloff*
Amongst *Sphagnum.*
(1)

C. decedens (Reinsch) Racib.*
In a bog-pool.
(1)

C. depressum (Näg.) Lundell var. **depressum***
In euplankton of deepwater lochs.
(5)

C. depressum var. **planctonicum** Reverdin*
In euplankton of **6** Loch Peallach; **7** Loch Frisa; and **9** Loch Assopol.
(3)

* Determined after Krieger & Gerloff.

C. difficile Lütkemuller var. **difficile***
In bog-pools, on a wet rock face and around the margin of **11** Loch Bà.
(4)

C. difficile var. **subimpressulum** Messik.*
Amongst *Sphagnum* and detritus in a bog-pool; in maritime pools at **10** Carsaig.
(1)

C. difficile var. **sublaeve** Lütkemuller*
In bog-pools.
(2)

C. formosulum Hoffm.
(including var. *nathorstii* (Boldt) West & West)
In bog-pools and on dripping rocks; in maritime pools at **10** Carsaig.
(6)

C. granatum Bréb.*
Around the margins of lochans.
(3)

C. humile (Gay) Nordst. var. **humile**
In bog-pools and in maritime pools at **10** Carsaig.
(3)

C. humile var. **danicum** (Börg.) Schmidle
In bog-pools.
(4)

C. impressulum Elfv.*
On stones in streams and in a waterfall; in maritime pools at **10** Carsaig.
(2)

C. margaritiferum Menegh.
In bog-pools; amongst *Sphagnum* and along margin of **11** Loch Bà.
(10)

C. melanosporum Arch.*
In bog-pools and amongst *Sphagnum.*
(4)

C. meneghinii Bréb.*
In a bog-pool and at the margin of two Lochans.
(3)

C. notabile De Bary*
In a marine supralittoral pool at **10** Carsaig.
(1)

C. novae-semliae Wille var. **sibiricum** Boldt
In bog-pools and amongst *Sphagnum.*
(4)

C. nymannianum Grun.*
Amongst *Sphagnum.*
(4)

C. obliquum Nordst. var. **tatricum** (Gutw.) Krieger & Gerloff*
Amongst *Sphagnum.*
(3)

C. oblongum Bennett
Amongst *Sphagnum.*
(1)

C. ornatum Ralfs
Amongst *Sphagnum.*
(2)

C. parvulum Bréb.*
Amongst *Sphagnum* and on wet rocks.
(6)

C. plicatum Reinsch*
Amongst *Sphagnum* and in a bog-pool.
(2)

C. portianum Arch.
In bog-pools.
(3)

C. pseudoexiguum Racib.
Amongst *Sphagnum.*
(2)

C. pseudopyramidatum Lundell
In a bog-pool.
(1)

C. punctulatum Bréb. var. **punctulatum**
Amongst *Sphagnum* and in bog-pools.
(4)

C. punctulatum var. **subpunctulatum** (Nordst.) Börg.
In similar habitats to the type variety.
(6)

C. pygmaeum Arch.*
Amongst *Sphagnum* and in bog-pools.
(6)

C. pyramidatum Bréb. *
Amongst *Sphagnum* and in bog-pools.
(10)

C. quadratum Ralfs*
In bog-pools and amongst *Sphagnum.*
(3)

C. quadrifarium Lundell*
Amongst *Sphagnum.*
(1)

C. ralfsii Bréb. var. **ralfsii***
In bog-pools.
(2)

C. ralfsii var. **montanum** Racib.*
Amongst *Sphagnum* and in open bog-pools.
(6)

C. regnellii Wille var. **pseudoregnellii** (Messik.) Krieger & Gerloff*
In a bog-pool.
(1)

C. reniforme (Ralfs) Arch. var. **reniforme**
Around lochan margins and in a marine supralittoral pool at **10** Carsaig.
(3)

C. reniforme var. **compressum** Nordst.
In a bog-pool.
(1)

C. repandum Nordst. var. **minus** (West & West) Krieger & Gerloff*
In a bog-pool.
(1)

C. sexangulare Lundell var. **minus** Roy & Bisset*
In bog-pools and in marine supralittoral pools at **10** Carsaig.
(5)

C. sphagnicolum West & West*
In bog-pools and amongst *Sphagnum*; in maritime pools at **10** Carsaig.
(3)

C. speciosum Lundell
On dripping rocks.
(1)

C. sportella Bréb.
In marine supralittoral pool at **10** Carsaig.
(1)

C. subarctoum (Lagerh.) Racib.*
In euplankton of **7** Loch Frisa and **9** Loch Assopol.
(2)

C. subcrenatum Hantzsch
In bog-pools and amongst *Sphagnum.*
(7)

C. subdanicum West
In a bog-pool.
(1)

C. subtumidum Nordst.*
In bog-pools and in maritime pools at **10** Carsaig.
(18)

C. tinctum Ralfs var. **tinctum***
In bog-pools and in maritime pools at **10** Carsaig.
(9)

C. tinctum var. **subretusum** Messik.*
At the edge of a lochan.
(1)

---

* Determined after Krieger & Gerloff.

C. truncatellum (Perty) Rabenh.*
In bog-pools and amongst *Sphagnum.*
(4)

C. tumidum Lundell*
Around the margin of a lochan.
(1)

C. turpinii Bréb.
On stones in a stream.
(1)

C. venustum (Bréb.) Arch. var. **venustum**
Amongst *Sphagnum* and in bog-pools.
(6)

C. venustum var. **minus** (Wille) Krieger & Gerloff*
Amongst *Sphagnum.*
(1)

Desmidium Ag.

Bibliogr.: West, West & Carter (1923).

D. coarctatum Nordst. var. **cambricum** West
Amongst *Sphagnum* and in a bog-pool.
(2)

D. swartzii Ag.
Amongst *Sphagnum* and in a bog-pool; in euplankton of **16** Loch Uisg.
(3)

Docidium Bréb.

Bibliogr.: Krieger (1937).

D. undulatum Bailey var. **dilatatum** (Cleve) West & West
Amongst *Sphagnum.*
(2)

Euastrum Ehrenb.

Bibliogr.: Krieger (1937).

E. ampullaceum Ralfs
Amongst *Sphagnum* and in bog-pools; at margins of lochans.
(10)

E. ansatum Ehrenb. var. **ansatum**
Amongst *Sphagnum* and in maritime pools at **10** Carsaig.
(1)

E. ansatum var. **dideltiforme** Ducell.
In bog-pools.
(9)

E. bidentatum Näg.
(including var. *speciosum* (Boldt) Schmidle)
In bog-pools.
(8)

E. binale (Turpin) Ehrenb. var. **binale**
In bog-pools; amongst *Sphagnum* and around margins of lochs.
(8)

* Determined after Krieger & Gerloff.

E. binale var. **gutwinskii** Schmidle
Amongst *Sphagnum* and in a bog-pool.
(6)

E. binale var. **hians** West
In bog-pools and amongst *Sphagnum*; in maritime pools at **10** Carsaig.
(10)

E. boldtii Schmidle
In bog-pools and amongst *Sphagnum.*
(6)

E. crassum (Bréb.) Kütz.
(including var. *scrobiculatum* Lundell)
Amongst *Sphagnum* and in bog-pools.
(15)

E. cuneatum Jenner
Amongst *Sphagnum.*
(5)

E. denticulatum (Kirschn.) Gay
In bog-pools.
(2)

E. didelta Ralfs var. **didelta**
In bog-pools and amongst *Sphagnum.*
(7)

E. didelta var. **truncatum** Krieger
In bog-pools.
(2)

E. dubium Näg.
Amongst *Sphagnum* and in bog-pools.
(9)

E. elegans (Bréb.) Kütz. var. **elegans**
Amongst *Sphagnum* and in bog-pools; in maritime pools at **10** Carsaig.
(21)

E. elegans var. **ornatum** West
Amongst *Sphagnum.*
(2)

E. gayanum De Toni
In bog-pools.
(2)

E. inerme (Ralfs) Lundell
In bog-pools; around a lochan margin.
(3)

E. insigne Hass.
Amongst *Sphagnum.*
(1)

E. oblongum (Grev.) Ralfs
Amongst *Sphagnum* and in bog-pools; and in maritime pools at **10** Carsaig.
(5)

E. pectinatum Bréb.
In similar habitats to the previous species but also along the margins of lochans.
(11)

E. pingue Elfv.
Amongst *Sphagnum.*
(1)

E. pinnatum Ralfs
In a bog-pool.
(1)

E. subalpinum Messik.
Amongst *Sphagnum.*
(1)

E. sublobatum Bréb. var. **dissimile** (Nordst.) West & West
On wet rock.
(1)

E. subrostratum West & West
In a bog-pool.
(1)

E. ventricosum Lundell
In bog-pools and amongst *Sphagnum.*
(4)

E. verrucosum Ehrenb. var. **alatum** Wolle
In euplankton of **6** Loch an Tòrr, **9** Loch Assopol, and **11** Loch Bà; in a bog-pool.
(4)

Hyalotheca Ehrenb.

Bibliogr.: West, West & Carter (1923).

H. dissiliens (W. Smith) Bréb.
Amongst *Sphagnum* and in bog-pools; in maritime pools at **10** Carsaig.
(34)

H. undulata Nordst.
In tychoplankton at the margin of **11** Loch Bà.
(1)

Micrasterias Ag.

Bibliogr.: Krieger (1939).

M. americana (Ehrenb.) Ralfs
In a bog-pool.
(1)

M. denticulata Bréb.
(including var. *intermedia* Norst.)
In euplankton of **6** Loch Peallach; amongst *Sphagnum* and in a bog-pool.
(4)

M. jenneri Ralfs
In bog-pools.
(4)

M. oscitans Ralfs var. **mucronata** (Dixon) Wille
Amongst *Sphagnum.*
(1)

M. papillifera Bréb. var. **glabra** Nordst.
In euplankton of **16** Loch Uisg.
(1)

M. rotata (Grev.) Ralfs
In a bog-pool.
(1)

M. sol (Ehrenb.) Kütz.
In euplankton of **16** Loch Uisg.
(1)

M. thomasiana Archer
Amongst *Sphagnum* and in a bog-pool.
(2)

M. truncata (Corda) Bréb.
In bog-pools and amongst *Sphagnum*; in euplankton of **16** Loch Uisg.
(8)

Penium Bréb.

Bibliogr.: Krieger (1937).

P. cylindrus (Ehrenb.) Bréb.
In bog-pools.
(4)

P. exiguum West
In bog-pools and amongst *Sphagnum*; in maritime pools at **10** Carsaig.
(7)

P. polymorphum Perty
Amongst *Sphagnum.*
(7)

P. spirostriolatum Barker
Amongst *Sphagnum* and in bog-pools.

Pleurotaenium Näg.

Bibliogr.: Krieger (1937).

P. ehrenbergii (Bréb.) De Bary
In bog-pools and loch margins; in maritime pools at **10** Carsaig.
(6)

P. minutum (Ralfs) Delp.
Amongst *Sphagnum* and in bog-pools.
(16)

P. tridentatum (Wolle) West
Amongst *Sphagnum.*
(3)

Sphaerozosma Corda emend. Bourr.

Bibliogr.: Bourrelly (1964).

S. filiformis (Ehrenb.) Bourr.
Onychonema filiforme (Ehrenb.) Roy & Bisset
At margin of **11** Loch Bà; in euplankton of **16** Loch Uisg.
(2)

Spondylosium Bréb.
Bibliogr.: West, West & Carter (1923).

S. planum (Wolle) West & West
In euplankton of **6** Loch an Torr.
(1)

S. pulchellum Arch.
Amongst *Sphagnum*.
(1)

Staurastrum Meyen.
Bibliogr.: West & West (1912); West, West & Carter (1923).

S. alternans Bréb.
In bog-pools.
(2)

S. anatinum Cooke & Wills (sensu Brook, 1959)
In euplankton of deepwater lochs. No attempt has been made to identify the material with the various forms distinguished by Brook.
(6)

S. arctiscon (Ehrenb.) Lundell
In euplankton of **7** Loch Frisa.
(1)

S. arnellii Boldt
In bog-pools and amongst *Sphagnum*.
(2)

S. avicula Bréb.
Around a lochan margin.
(1)

S. bieneanum Rabenh. var. **bieneanum**
At the edge of **12** Loch Ghurrabain, and in a bog-pool.
(2)

S. bieneanum var. **ellipticum** Wille
In a bog-pool.
(1)

S. brachiatum Ralfs
In bog-pools and margins of lochans.
(3)

S. brasiliense Nordst. var. **lundellii** West & West
In euplankton of **11** Loch Bà.
(1)

S. chaetoceros (Schröder) G. M. Smith (sensu Brook, 1959)
In euplankton of **9** Loch Assopol and **11** Loch Bà.
(2)

S. chavesii Bohlin
Amongst *Sphagnum*.
(1)

S. controversum Bréb.
In bog-pools.
(2)

S. cumbricum West var. **cambricum** West
In euplankton of **3** Loch Poit na h-I and **7** Loch Frisa; in a ditch amongst *Zygnema* and *Ulothrix*.
(4)

S. denticulatum (Näg.) Arch.
In euplankton of **3** Loch Poit na h-I and **6** Loch Peallach; also in a bog-pool and supra-littoral pools.
(3)

S. diplacanthum De Not var. **anglicum** Turner
In a bog-pool.
(1)

S. forficulatum Lundell
In bog-pools.
(1)

S. furcatum (Ehrenb.) Bréb.
Along the margin of **11** Loch Bà.
(1)

S. gatniense West & West
In euplankton of **16** Loch Uisg.
(1)

S. gladiosum Turner
Around the margin of a lochan.
(1)

S. gracile Ralfs (sensu Brook, 1959)
In bog-pools; in euplankton of **7** Loch Frisa.
(3)

S. hexacerum (Ehrenb.) Wittr.
In a bog-pool and amongst *Sphagnum*; at the margin of a lochan.
(3)

S. hirsutum (Ehrenb.) Bréb.
In bog-pools.
(2)

S. inconspicuum Nordst.
Around the margin of a lochan.
(1)

S. irregulare West
In a bog-pool.
(1)

S. lunatum Ralfs var. **lunatum**
In bog-pools and in tychoplankton around loch margin; in maritime pools at **10** Carsaig.
(3)

S. lunatum var. **planctonicum** West & West
In euplankton of lochs and lochans.
(11)

S. margaritaceum (Ehrenb.) Menegh.
Amongst *Sphagnum* and in bog-pools.
(13)

S. minutissimum Reinsch var. **convexum** West & West
In a bog-pool.
(1)

S. orbiculare Ralfs var. **depressum** Roy & Bisset
Amongst *Sphagnum* and in bog-pools.
(6)

S. orbiculare var. **ralfsii** West & West.
In bog-pools and at the margin of a lochan.
(4)

S. pelagicum West & West
In euplankton of **3** Loch Poit na h-I and **7** Loch Frisa.
(2)

S. pilosum (Nag.) Arch.
In a bog-pool.
(1)

S. pingue Teil. (sensu Brook, 1959)
In euplankton of **3** Loch poit na h-I, **7** Loch Frisa and **17** Loch a' Ghleannain.
(3)

S. polymorphum Bréb.
In bog-pools; amongst *Sphagnum* and around margins of lochans.
(7)

S. proboscidium (Bréb.) Arch.
Amongst *Sphagnum* and in a bog-pool.
(2)

S. pseudopelagicum West & West
In euplankton of **17** Loch a' Ghleannain and **7** Loch Frisa.
(2)

S. punctulatum Bréb. var. **punctulatum**
In bog-pools.
(4)

S. punctulatum var. **kjellmanii** Wille
In bog-pools and in maritime pools at **10** Carsaig.
(4)

S. scabrum Bréb.
In bog-pools.
(1)

S. sebaldii Reinsch
In a bog-pool.
(1)

S. sexangulare (Bulnh.) Lundell
In euplankton of **11** Loch Bà and **16** Loch Uisg.
(2)

S. simonyi Heimerl.
In a bog-pool and amongst *Sphagnum.*
(6)

S. striolatum (Näg.) Arch.
In a bog-pool.
(1)

S. subavicula West & West
In a bog-pool.
(1)

S. subcruciatum Cooke & Wills
In euplankton of **17** Loch a' Ghleannain; in a bog-pool.
(3)

S. teliferum Ralfs
In bog-pools and around margins of lochans.
(3)

S. tetracerum Ralfs
In bog-pools; and in tychoplankton around margins of lochs and lochans.
(11)

Staurodesmus Teil.

Bibliogr.: Teiling (1967).

S. connatus (Lundell) Thom.
In euplankton of **9** Loch Assopol.
(1)

S. convergens (Ehrenb.) Teil. var. **laportei** Teil.
In a bog-pool.
(1)

S. crassus (West) Florin
In euplankton of **3** Loch Poit na h-I and **9** Loch Assopol; around margin of **11** Loch Bà.
(4)

S. cuspidatus (Bréb.) Teil.
In euplankton of **6** Loch Carnain an Amais and **9** Loch Assopol.
(2)

S. dejectus (Bréb.) Teil.
In euplankton of **3** Loch Poit na h-I; in bog-pools; around margin of a lochan.
(3)

S. extensus (Borge) Teil.
In bog-pools and amongst *Sphagnum.*
(11)

S. glaber (Ehrenb.) Teil. var. **limnophilus** Teil.
In euplankton of **16** Loch Uisg; in bog-pools; around lochan margins.
(3)

S. jaculiferus (West) Teil.
In euplankton of **3** Loch Poit na h-I.
(1)

S. megacanthus (Lundell) Thunn. var. **megacanthus**
In euplankton of **3** Loch Poit na h-I and **13** Lochan a' Ghurrabain.
(2)

S. megacanthus var. **scoticus** (West) Lillier
In euplankton of **16** Loch Uisg and **17** Loch a' Ghleannain.
(2)

S. omearii (Archer) Teil.
Amongst *Sphagnum.*
(2)

S. patens (Nordst.) Croasd.
In euplankton of **3** Loch Poit na h-I and in a bog-pool.
(2)

S. quadratus (Schmidle) Teil.
In a bog-pool.
(1)

S. sellatus Teil.
In euplankton of **6** Loch Carnain an Amais and Loch an Torr; **7** Loch Frisa; **10** Loch Fuaron.
(5)

S. spencerianus (Mask.) Teil.
In euplankton of **3** Loch Poit na h-I; **6** Loch an Torr and Loch Carnain an Amais; **7** Loch Frisa; **17** Loch a' Ghleannain.
(5)

S. subtriangularis (Borge) Teil. var. **subtriangularis**
In euplankton of **7** Loch Frisa.
(1)

S. subtriangularis var. **inflatus** (West) Teil.
In euplankton of **6** Loch Carnain an Amais and Loch Peallach; **7** Loch Frisa; **11** Loch Bà; **16** Loch Uisg; **17** Loch a' Ghleannain.
(7)

S. subtriangularis var. **limneticus** Teil.
In euplankton of **7** Loch Frisa and **11** Loch Bà.
(2)

Teilingia Bourr.

Bibliogr.: Bourrelly (1964); West, West & Carter (1923).

T. granulata (Roy & Bisset) Bourr.
Sphaerozosma granulatum Roy & Bisset
In bog-pools and around margins of lochans.
(4)

Tetmemorus Ralfs

Bibliogr.: Krieger (1937).

T. brebissonii (Menegh.) Ralfs var. **brebissonii**
In bog-pools amongst *Sphagnum.*
(9)

T. brebissonii var. **minor** De Bary
In bog-pools and amongst *Sphagnum.*
(8)

T. granulatus (Bréb.) Ralfs
In bog-pools and amongst *Sphagnum*; in maritime pools at **10** Carsaig.
(43)

T. laevis (Kütz.) Ralfs
In bog-pools and amongst *Sphagnum.*
(13)

Xanthidium Ehrenb.

Bibliogr.: West & West (1912).

X. antilopaeum (Bréb.) Kütz. var. **antilopaeum**
In bog-pools and detritus around **13** Lochan a' Ghurrabain.
(3)

X. antilopaeum var. **depauperatum** West & West
In euplankton of deepwater lochs.
(18)

X. antilopaeum var. **hebridarum** West & West
In euplankton of **6** Loch an Torr and around a lochan margin.
(2)

X. antilopaeum var. **polymazum** Nordst.
In euplankton of **16** Loch Uisg.
(1)

X. armatum (Bréb.) Rabenh.
Amongst *Sphagnum.*
(5)

X. smithii Arch.
Amongst *Sphagnum* and in bog-pools.
(6)

CHAROPHYTA

CHAROPHYCEAE

CHARALES

CHARACEAE

Bibliogr.: Wood & Imahori (1964, 1965); synonymy given is that of Allen (1950) and Groves & Bullock-Webster (1920, 1924).

Chara L. emend. Ag.

C. globularis Thuill. var. **virgata** (Kütz.) R. D. Wood
C. delicatula auctt.
In summit pools and nutrient-poor lochans and

in shallow slow-flowing streams through the deeper peat areas. Occasional and widespread. (16)

This is a distinct variant which has been treated at specific level by Agardh (1824), Kützing (1834), Migula (1897), Groves & Bullock-Webster (1924) and Allen (1950).

C. vulgaris L. forma **contraria** (A. Br. ex Kütz. emend. R. D. Wood) R. D. Wood

C. contraria A. Br. ex Kütz.
In the shallower water of **3** Loch Poit na h-I; the only record.

Nitella (Ag. emend. A. Br.) Leonhardi

N. flexilis (L.) Ag.
N. opaca (Ag. ex Bruz.) Ag. pro parte
In shallower peaty lochs; occasional.
(6)

References

AGARDH, C. A. (1824). *Systema algarum*. Lund.

AHLSTROM, E. H. (1937). Studies on variability in the genus *Dinobryon* (Mastigophora). *Trans. Am. microsc. Soc.*, **56**: 139–159.

ALLEN, G. O. (1950). *British stoneworts (Charophyta)*. Arbroath.

BLUM, J. L. (1972). Vaucheriaceae. In *North American Flora* (ser. 2), **8**: 1–64.

BOURRELLY, P. (1964). Une nouvelle coupure générique dans la famille des Desmidiées: le genre *Teilingia*. *Revue algol.* n.s., **7**: 187–191.

BROOK, A. J. (1959). *Staurastrum paradoxum* Meyen and *S. gracile* Ralfs in the British freshwater plankton, and a revision of the *S. anatinum* - group of radiate desmids. *Trans. R. Soc. Edinb.*, **63**: 589–628.

CARTER, N. (1923). *See* West & West (1904–12).

CHRISTENSEN, T. (1962). Systematisk botanik: alger. *Botanik* **2(2)**: 1–178.

———. (1969). *Vaucheria* collections from Vaucher's region. *Biol. Skr.*, **16(4)**: 1–36.

GEITLER, L. (1930–32). Cyanophyceae (Blaualgen) Deutschlands, Österreichs und der Schweiz mit Berucksichtigung der übrigen Länder Europas sowie der angrenzenden Meeresgebiete, 1 (1930); 2, 3 (1931); 4, 5, 6 (1932). In R. Kolkowitz, Die Algen, in G. L. Rabenhorst, *Kryptogamen-Flora von Deutschland, Österreich und Schweiz*, vol. 14, 2nd. ed., Leipzig.

GROVES, J. & BULLOCK-WEBSTER, G. R. (1920–24). *The British Charophyta*, vol. 1: *Nitelleae* (1920); vol. 2: *Chareae* (1924); London. Johnson reprint 1971, London & New York.

HARRIS, K. & BRADLEY, D. E. (1960). A taxonomic study of *Mallomonas*. *J. gen. Microbiol.*, **22**: 750–777.

HEERING, W. (1921). Chlorophyceae IV. Siphonocladiales, Siphonales. In A. Pascher, *Die Süsswasserflora Deutschlands, Österreichs und der Schweiz*, Heft 7. Jena.

HIBBERD, D. J. (1974). Observations on the cytology and ultrastructure of *Chlorobotrys regularis* (West) Bohlin with special reference to its position in the Eustigmatophyceae. *Br. phycol. J.*, **9**: 37–46.

——— & LEEDALE, G. F. (1971). A new algal class – the Eustigmatophyceae. *Taxon*, **20**: 523–525.

HUBER-PESTALOZZI, G. (1941). Das Phytoplankton des Süsswassers, Systematik und Biologie, Teil 2(1) Chrysophyceen: Farblose Flagellaten Heterokonten. In A. Thienemann, *Die Binnengewässer Einzeldarstellungen aus der Limnologie und ihren Nachbargebieten*, Teil 16. Stuttgart.

———. (1961). *Ibid.* Teil 5 Chlorophyceae (Grünalgen) Ordnung Volvocales.

KORSHIKOV, O. A. (1953). *Viznachnik prisnovodnikh vodorostei Ukrayins'koyi RSR* [Freshwater algae of SSR Ukraina] V. *Pidklas Protokokovi (Protococcineae) Vakuol'ni (Vacuolales) ta Protokokovi (Protococcales)*. Kiev.

KRIEGER, W. (1933–39). Die Desmidiaceen Europas mit Berucksichtigung der aussereuropäischen Arten. In G. L. Rabenhorst, *Kryptogamen-Flora von Deutschland, Österreich und der Schweiz*, vol. 13 (1/1[1, 1933: 2, 1935; 3, 4, 1937], 1/2[1, 1939]), 2nd ed. Leipzig.

——— & GERLOFF, J. (1962–69). *Die Gattung Cosmarium* 1 (1962), 2 (1965), 3 & 4 (1969). Lehre.

KÜTZING, F. T. (1834). Beschreibung einiger neuen Arten der Gattung *Chara*. *Flora*, **17**: 705–707.

MIGULA, W. (1890–97). Die Characeen Deutschlands, Oesterreichs und der Schweiz, unter Berücksichtigung aller Arten Europas. In G. L. Rabenhorst, *Kryptogamen-Flora von Deutschland, Oesterreich und der Schweiz*, Teil 5/1, 2, 3, 4 (1890); 5, 6 (1891); 7 (1892); 8 (1893); 9 (1894); 10 (1895); 11 (1896); 12 (1897); 2nd ed. Leipzig.

PRINTZ, H. (1964). Die Chaetophoralen der Binnengewässer, eine systematische Übersicht. *Hydrobiologia*, **24**: 1–376.

SULEK, J. (1969). Taxonomische Übersicht der Gattung *Pediastrum* Meyen. Pp 197–261 in B. Fott (Ed.), *Studies in phycology*. Stuttgart.

TEILING, E. (1967). The desmid genus *Staurodesmus*, a taxonomic study. *Ark. Bot.*, **6**: 467–630.

UHERKOVICH, G. (1966). *Die Scenedesmus – Arten Ungarns*. Budapest.

WEST, W. & WEST, G. S. (1904–12). *A monograph of the British Desmidiaceae*, vols. 1 (1904), 2 (1905), 3 (1908), 4 (1912), [5 (1923) by Nellie Carter], London. Johnson reprint 1971, London & New York.

———. (1905). A further contribution to the freshwater plankton of the Scottish lochs. *Trans. R. Soc. Edinb.*, **41**: 477–518.

——— & CARTER, N. (1923). *See* West & West (1904–12).

WOOD, R. D. & IMAHORI, K. (1964–65). *A revision of the Characeae 1. Monograph of the Characeae* (1965), *2. Iconograph of the Characeae* (1964). Weinheim.

17. Freshwater diatoms

Patricia A. Sims

Introduction

Diatoms (Bacillariophyta) form a prominent feature of the algal flora and can be found as a rich brown film coating wet rocks, stones and aquatic vegetation in most localities. This chapter presents an account of these diatoms which, because they were studied in greater detail than the other freshwater algae, have been treated more fully and out of systematic sequence.

Sampling and collecting methods

Collections were made in May–June 1967, June–July 1968, April 1969 and October 1970. Sites were selected subjectively to give as wide a coverage of the Island as possible; many sites were revisited over the four years.

Euplankton (open water plankton) was collected using a 200 mesh net and was preserved as soon as possible after collection. The remainder of the material was collected by either scraping from rocks and other substrata, squeezing tychoplankton (unattached species caught amongst other plants) from submerged *Sphagnum* and other aquatics, including mosses and filamentous algae, or by sampling directly from pool bottoms with a collecting tube.

A total of over 900 samples of freshwater algae was collected and examined briefly on the day of collection or the following day; 352 samples selected for the determination of the diatoms were preserved in Pocock's solution (0.5 g iodine, 1.0 g potassium iodide, 4 ml acetic acid, 24 ml formaldehyde, 400 ml distilled water) and have been retained in the National Collections. From these samples c. 460 species and varieties of diatoms have been identified.

Preparation and identification of specimens

A portion of each gathering was cleaned by boiling in concentrated sulphuric acid followed by the cautious addition of potassium chlorate crystals in a fume cupboard. After thorough washing, a few drops of the cleaned material were dried down on a coverslip and mounted in GBI 163 (a synthetic thermoplastic resin produced by G.B.I. (Laboratories) Ltd, Heaton Mills, Denton, Manchester).

The species have been identified using the descriptions and figures of Hustedt (1930, 1927–66), or where other works have been used mention of these is given in square brackets after the species name.

Arrangement and nomenclature

The arrangement of the families and genera follow that of Hendey (1964) with some minor amendments. Species are arranged alphabetically within genera. The arrangement of other data is similar to that of the terrestrial groups: localities have been listed if under eight sites contained the species. Where the species occurred more often a general ecological comment is given where possible, followed by the relative density of the species in each gathering (in five categories: isolated specimens, sparse, in small numbers, numerous and in large numbers). This is followed by a statement on its frequency (occasional, frequent, common, and abundant) and its general distribution throughout the Island. The habitats in which the species have been found are summarised in the following terms:

Acidophilous – species confined to acid oligotrophic habitats such as peaty pools and bogs.

Basiphilous – species which require a measurable concentration of cations (i.e. calcium, potassium, magnesium) and which are therefore absent from base-poor habitats.

pH indifferent – tolerant species which are commonly found in both acidic and basic habitats.

Anhalophobous – species which, although not requiring the presence of seawater, nevertheless tolerate it in low concentrations and may be more abundant in coastal situations than elsewhere because of the increase in nutrients from sea spray.

Oxyphilous – species which probably require a high oxygen level and are mainly confined to sub-aerial or turbulent aquatic habitats, e.g., fast-flowing streams, amongst irrigated mosses, in seepages and water films.

Acknowledgements

I am grateful to Dr Penelope Dawson for contributing the accounts of *Gomphonema* and *Gomphoneis*, Dr R. Crawford for help with *Melosira*, and Theresa Parker (Mrs M. Brendall) for cleaning and mounting the material. J. R. Carter and R. Ross assisted with difficult identifications and nomenclatural problems respectively.

Index to genera

BACILLARIOPHYTA

BACILLARIOPHYCEAE

BACILLARIALES

MELOSIRACEAE

Melosira Ag.

M. ambigua (Grunow) O. Müller var.
In small numbers in **3** Loch Assopol, on submerged rocks at margin and **13** Caol Lochan, amongst decaying moss and submerged debris at margin.

M. arenaria Moore ex Ralfs
Sparse from **8** near Mackinnon's Cave waterfall, on moss covered boulder in fast flowing stream; **7** Allt Loch a' Ghael, c. 180 m, in

moss flush; and **11** Loch Bà, on small stones in shallow marginal water.

M. dendrophila (Ehrenb.) Ross & Sims ined.
[Van Heurck, 1882: t. 89, figs. 19–20]
M. roeseana Rabenh. var. *epidendron* (Ehrenb.) Grunow f. *porocyclia* Grunow
Often in large numbers on exposed wet rocks and irrigated mosses, sparse elsewhere.
7 River Bellart near source, on *Myriophyllum*. **8** Creag a' Ghaill, c. 240 m, on wet vertical side of gully; in irrigated moss on rock slab; on flushed rocks. **11** Ben More, north corrie c. 870 m, in gelatinous matrix at base of small gully; on underhang of wet rock.

M. distans (Ehrenb.) Kütz. var. **distans**
Acidophilous. Small numbers present in high altitude lochans.
Occasional; scattered inland.

M. distans var. **africana** O. Müller
Acidophilous. In large numbers in high altitude bog pools and lochans, sparse on wet rocks and in streams.
Occasional; scattered inland.

M. distans (Ehrenb.) Kütz. **var. nov.**
Acidophilous – neutral. In mountain streams, on irrigated rocks, in pools, flushes and in the plankton of **3** Loch Assopol, **6** Loch an Druim an Aoinirth and **7** Loch Frisa; rarely in large numbers.
Occasional; scattered but more commonly found at the higher altitudes.

M. granulata (Ehrenb.) Ralfs
Isolated specimens on **8** Bearraich, c. 400 m., on mud surface at edge of small stony pool with *Koenigia islandica*.

M. italica (Ehrenb.) Kütz. subsp. **italica**
Weakly basiphilous. Usually in large numbers in shallow mesotrophic streams and flushes.
Occasional; scattered.

M. italica subsp. **subarctica** O. Müller
Abundant in the euplankton of **3** Loch Poit na h-I, less numerous in the plankton of **3** Loch Assopol. Present also in littoral samples from these lochs and in some mesotrophic flushes.

M. juergensii Ag.
In small numbers on coastal cliffs receiving windblown sea spray.
Occasional; mainly confined to the **1** Treshnish Isles and the S and W coastal areas.

M. varians Ag.
Sparse, on surface of mud in the **6** estuary of the River Bellart and in a pool on top of the **8** Creag a' Ghaill cliffs.

THALASSIOSIRACEAE

Cyclotella Kütz.

C. antiqua W. Smith
An isolated specimen found in a plankton tow from **7** Loch Frisa.

C. comensis Grunow
Often in large numbers in plankton and littoral samples from **3** Loch Poit na h-I; **6** Lochan's Airde Beinn and Loch Meadhorn; **7** Loch Frisa and Loch a' Ghael; **11** Loch Bà; **13** Lochan a' Churrabain; **16** Loch an t'sithein; and **17** Loch a' Ghleannain.

C. comta (Ehrenb.) Kütz.
pH indifferent. In plankton and littoral samples from the larger lochs, often in large numbers. Occurring in small numbers in mesotrophic flushes and streams and on wet coastal rocks.
Frequent; widespread.

C. glomerata Bachmann
Numerous in the littoral of **3** Loch Poit na h-I and Loch Assopol. Less frequent in **7** Loch Frisa, **11** Loch Bà and **17** Loch a' Ghleannain.

C. kuetzingiana Thwaites
In acid to neutral waters. Occasionally in large numbers in plankton and littoral samples from the oligotrophic and some mesotrophic lochs; more rarely from streams.
Occasional; widespread inland.

C. meneghiniana Kütz.
Basiphilous and anhalophobous. Small numbers in the littoral of **3** Loch Poit na h-I and **13** Caol Lochan and in large numbers in a small lochan on **9** Creachan Mòr. Present sparingly in the Rivers Forsa and Bellart, in coastal streams and seepages and in the splash zone of some waterfalls.
Frequent but rarely found in quantity; widespread.

C. socialis Schütt
Small numbers present in littoral samples from **6** Loch Meadhorn; **7** Loch Frisa; **11** Loch Bà; and **17** Loch a' Ghleannain.

Stephanodiscus Ehrenb.

S. hantzschii Grunow
Small numbers present in three gatherings from **3** Loch Poit na h-I taken in April 1969. Probably seasonal as not noted from the same locality in June 1967, July 1968 and October 1970.

S. rotula (Kütz.) Hendey var. **rotula**
S. astrea (Ehrenb. ex Kütz.) Grunow
Sparse from **8** Bearraich, c. 400 m., on mud surface at edge of small stony pool with *Koenigia islandica*; Tiroran, squeezings from leafy liverwort at side of waterfall.

S. rotula var. **minutula** (Kütz.) Ross & Sims ined.
S. astrea var. *minutula* (Kütz.) Grunow
Basiphilous. Planktonic and dispersed in

littoral vegetation and submerged debris in the following lochs but never in large numbers. **2** Loch Staoneig. **3** Loch Poit na h-I; Loch Assopol. **7** Loch Frisa. **11** Loch Bà.

RHIZOSOLENIACEAE

Rhizosolenia Ehrenb.

R. eriensis H. L. Smith
Numerous in plankton samples from **7** Loch Frisa.
Specimens of *Rhizosolenia* were also noted when fresh material was examined from other plankton tows but these were lost in the cleaning process.

DIATOMACEAE

Diatoma Bory

D. hiemale (Roth) Heiberg var. **hiemale**
Weakly basiphilous. In large numbers, often the dominant species, amongst mosses covering boulders in streams and in the spray zone of waterfalls on the **8** Creag a' Ghaill cliffs and Mackinnon's Cave areas; in small numbers in streams, flushes and seepages elsewhere.
Occasional; widespread.

D. hiemale var. **mesodon** (Ehrenb.) Grunow
In acid waters but occurring in greater numbers in the base-rich habitats. In large numbers, occasionally in almost pure gatherings, from the dripping sides of waterfalls and shallow muddy mesotrophic streams; less numerous in other streams, flushes, peaty pools and on flushed rock surfaces; sparse in lochs and rivers.
Abundant; widespread.
One of the most frequently encountered diatoms in the flora.

D. tenue Ag. var. **tenue**
D. elongatum (Lyngb.) Agr. var. *tenue* (Ag.) Van Heurck
Basiphilous and anhalophobous. Amongst submerged aquatics in mesotrophic streams, on wet rocks and in the littoral of lochs, occasionally in large numbers.
Occasional; scattered.
A distorted form as figured by Fricke (in Schmidt, *Atlas*: t. 268, figs 66 & 66a) occurred commonly in **11** Loch Bà.

D. tenue var. **elongatum** Lyngb.
D. elongatum (Lyngb.) Ag.
Basiphilous and anhalophobous. Often in large numbers in the deeper inland lochs, in coastal seepages and in the spray zone of waterfalls.
Frequent in coastal areas but rare inland.

Meridion Ag.

M. circulare (Grev.) Ag. var. **circulare**
Weakly basiphilous. Often the dominant species in samples from mesotrophic flushes and streams, notably those with a muddy substrate; in small numbers in oligotrophic streams; sparse in rivers and lochs. Able to withstand sea spray in moderate amounts.
Frequent; widespread.

M. circulare var. **constrictum** (Ralfs) Van Heurck
Ecological requirements similar to those of var. *circulare* although mixed populations not recorded; apparently confined to moving water.
Not as common as var. *circulare*; scattered.

Tabellaria Ehrenb.

Bibliogr.: Knudson (1952).

T. fenestrata (Lyngb.) Kütz.
Weakly acidophilous to neutral. Favouring the littoral of the larger lochs where it may occur in large numbers; occasionally present in the plankton of these lochs. Small numbers also present in streams and mires.
Occasional; scattered inland.

T. flocculosa (Roth) Kütz.
Most numerous in inland oligotrophic habitats but wide ranging ecologically. In streams, rivers, mires, flushes, amongst irrigated mosses including *Sphagnum*, on wet rocks, in pools and lochs; often in very large numbers.
The most commonly encountered diatom in the flora; ubiquitous.

T. quadriseptata Knudson
Acidophilous. Often in large numbers in high altitude bog-pools and lochans and amongst *Sphagnum*.
Occasional; scattered inland.

Asterionella Hassal

A. formosa Hassal var. **formosa**
Basiphilous. Planktonic or dispersed amongst submerged vegetation in mesotrophic lochs. Dominant species in plankton samples from Loch a' Ghleannain.
3 Loch Poit na h-I; Loch Assopol. **7** Loch Frisa. **17** Loch a' Ghleannain.

A. formosa var. **gracillima** (Hantzsch) Grunow
A. gracillima (Hantzsch) Heiberg
Basiphilous. In large numbers in the plankton and tychoplankton of **3** Loch Assopol and **7** Loch Frisa and in a pool beside **2** Loch Staoneig. Also numerous on flushed rock surfaces at **5** Calgary Bay and on the **8** Creag a' Ghaill cliffs. In small numbers in the plankton and tychoplankton of **13** Lochan a' Churrabain and **17** Loch a' Ghleannain.

Opephora Petit

O. martyi Hérib.
Basiphilous and anhalophobous. Sparse in shallow mesotrophic streams on the **8** Creag a' Ghaill ridge and in littoral samples from **3** Loch Poit na h-I and Loch Assopol; more numerous

in lightly brackish habitats e.g., **1** Treshnish Isles.
Occasional; coastal and the Ross of Mull.

Fragilaria Lyngb.

F. bicapitata Mayer
Basiphilous and oxyphilous. In shallow mesotrophic streams, moss flushes, seepages and the littoral of lochs; rarely in large numbers.
Occasional; scattered.

F. brevistriata Grunow
Basiphilous, oxyphilous and anhalophobous. Sparse in the shallow littoral zone of the larger lochs; numerous on flushed rock surfaces of coastal cliffs.
3 Loch Poit na h-I; Loch Assopol. **5** Calgary Bay. **7** Loch Frisa. **8** Burgh. **10** Carsaig.

F. capucina Desmaz.
Basiphilous?; its occurrence in acid waters being probably due to the flushing effects of rapid water-flow. Often in large numbers amongst mosses and aquatic vegetation in fast flowing rivers and streams; less numerous in the littoral of the larger lochs.
Frequent; widespread.

F. constricta Ehrenb. (including f. *stricta* (A. Cleve) Hust.)
Sparse in submerged littoral vegetation in **6** Loch an Tòrr, **13** Caol Lochan and **16** Loch Uisg.

F. construens (Ehrenb.) Grunow var. **construens.**
Basiphilous and anhalophobous. Most numerous in the littoral of mesotrophic lochs and pools; present also in coastal seepages and on wet rocks at side of waterfalls; sparse in streams and rivers.
Occasional; scattered.
Specimens referable to var. *venter* (Ehrenb.) Grunow also occur infrequently but have not been separately recorded.

F. construens var. **binodis** (Ehrenb.) Grunow
Sparse in littoral samples from **3** Loch Poit na h-I, Loch Assopol and **13** Caol Lochan; also present in seepages and shallow streams on **8** Creag a' Ghaill.

F. crotonensis Kitton
Basiphilous. Often in large numbers in the plankton and amongst submerged vegetation in the larger lochs, in the River Forsa and in some shallow mesotrophic streams and flushes.
Occasional; scattered.

F. lapponica Grunow
Sparse in one sample from **13** Lochan na Guailne Duibhe, on submerged grass in shallow marginal water.

F. leptostauron (Ehrenb.) Hust. var. **leptostauron**

F. harrisonii (W. Smith) Grunow
A few specimens recorded from two shallow-water, bottom mud samples from **13** Caol Lochan and from a seepage on coastal cliffs at **10** Carsaig.

F. leptostauron var. **dubia** (Grunow) Hust.
A solitary specimen recorded from the mud surface of a shallow puddle at **5** Calgary Bay.

F. pinnata Ehrenb.
Basiphilous and anhalophobous. In shallow water of streams, moss flushes, coastal seepages and mesotrophic lochs.
Occasional but rarely in large numbers; scattered.

F. vaucheriae (Kütz) B. Peters. var. **vaucheriae**
F. intermedia Grunow
Synedra vaucheriae (Kütz.) Kütz.
Possibly basiphilous, anhalophobous. Often in large numbers amongst mosses and submerged vegetation in fast flowing rivers and streams; also in moss flushes, on the wet rocks beside waterfalls, in coastal seepages and in the littoral of the larger mesotrophic lochs.
Common; widespread particularly in coastal areas.
Very variable in size and shape, notably in the lochs.

F. vaucheriae var. **capitellata** (Grunow) Patr.
Found with var. *vaucheriae* in the base-enriched habitats, e.g., coastal seepages.

F. virescens Ralfs
pH indifferent but favouring base-rich areas, anhalophobous.
Often in large numbers, in streams, rivers, pools, on wet rocks and in the littoral of the mesotrophic lochs.
Common; widespread.
The size and shape of the specimens is very variable and includes the var. **capitata** Østrup, var. **elliptica** Hust., and var. **subsalina** Grunow, with the small forms predominating.
Mull specimens are 5–56 × 4–8 μm with 14–19 transapical striae in 10 μm.

Synedra Ehrenb.

S. amphicephala Kütz.
Sparse in base-poor areas, more numerous in base-rich habitats.
Occasionally in large numbers in mesotrophic streams, moss flushes, on flushed rocks and mosses; sparse in acid pools.
Frequent; widespread in coastal areas but scattered inland.

S. delicatissima W. Smith [Patrick & Reimer, 1966: 136]
Sparse in: **3** Loch Poit na h-I, stone at margin; **6** SW Ardmore Bay, boulder in fast-flowing stream; **7** Loch Frisa, in plankton tow; **13**

Lochan na Guailne Duibhe, on submerged grass at margin.

S. famelica Kütz.
Weakly basiphilous. Occasionally in large numbers coating stones and vegetation in streams and rivers; also in the splash zone of some waterfalls.
Occasional: scattered.

S. fasciculata (Ag.) Kütz.
S. tabulata (Ag.) Kütz.; *S. affinis* Kütz.
Sparse in the following: **5** Calgary Bay, on wet rock face; **7** Loch Frisa, in plankton tow; **10** Rubha Dubh, in thick yellow gelatinous matrix on coastal cliffs.

S. minuscula Grunow
Basiphilous. In mesotrophic pools and on flushed rock surfaces, often numerous.
Occasional; scattered.
Mull specimens are 18–32 × 2–3 μm, with 16–18 transapical striae in 10 μm.

S. nana Meister
In small numbers from **9** Cnoc Reamhar, c. 250 m, in slowly flowing stream; c. 280 m, in soaked *Sphagnum*; **15** Allt na Dubh Choire, c. 460 m, in bog-pool; and **16** Loch Uisg, amongst *Sphagnum* at margin.

S. parasitica (W. Smith) Hust. var. **parasitica**
Basiphilous. In small numbers in mesotrophic streams, flushes and pools, on the wet rocks beside waterfalls and in the littoral of the following large lochs: **3** Loch Poit na h-I, Loch Assopol and **7** Loch Frisa.
Occasional; widespread.

S. parasitica var. **subconstricta** (Grunow) Hust.
Occurring in similar still-water habitats as var. *parasitica* but less numerous.

S. pulchella Ralfs ex Kütz.
Basiphilous and anhalophobous. In large numbers in seepages on coastal rocks and mud banks, in mesotrophic lochs and on submerged rocks beneath waterfalls. Small numbers present in mesotrophic streams and rivers and on wet mud.
Rare inland; widespread and frequent in coastal areas.

S. radians Kütz.
S. acus Kütz. var. *radians* (Kütz.) Hust.
pH indifferent. Most commonly among submerged bryophytes and aquatics but also on rocks in the splash zone of waterfalls, on boulders in streams, in pools and flushes and in the littoral of lochs; often in large numbers.
Frequent; widespread.

S. rumpens Kütz.
Weakly basiphilous, anhalophobous. Occasionally in large numbers in the littoral of some of the mesotrophic lochs and on flushed rock surfaces, especially those in coastal areas.
Occasional; scattered.

S. tenera W. Smith
In weakly acid to neutral waters. Most commonly found amongst moss covering boulders in streams and on rock slabs; also in mires, rivers and the littoral of some lochs; occasionally in large numbers.
Occasional; widespread inland.

S. ulna (Nitzsch) Ehrenb. var. **ulna**
pH indifferent but more numerous in base-rich areas, anhalophobous.
Frequently in large numbers in streams, rivers, lochs, on flushed rocks and soaked mosses, in pools and on muddy substrates.
One of the most commonly encountered diatoms in this flora; widespread.

S. ulna var. **danica** (Kütz.) Grunow
Numerous in plankton and littoral samples from **3** Loch Poit na h-I, Loch Assopol and **7** Loch Frisa.

Hannaea Patr.

Ceratoneis Ehrenb.

H. arcus (Ehrenb.) Patr.
Ceratoneis arcus (Ehrenb.) Kütz.
In acid to neutral waters, halophobous. Occurs in large numbers as an epiphyte on mosses covering rocks in streams and on the sides of waterfalls; in small numbers in rivers; sparse in lochs.
Frequent; widespread.

EUNOTIACEAE

Peronia Bréb. & Arnott ex Kitton

P. fibula (Bréb. ex Kütz.) R. Ross
P. erinacea Bréb. & Arnott ex Kitton; *P. heribaudii* Brun & M. Perag. ex Hérib.
Acidophilous and halophobous. In oligotrophic streams, mires, bog pools and *Sphagnum*; rarely in large numbers.
Frequent; widespread inland.

Eunotia Ehrenb.

E. arcus Ehrenb. var. **arcus**
Indifferent or weakly basiphilous. Occasionally in large numbers in mesotrophic streams and flushes and in coastal seepages; sparse in the larger lochs. More base-tolerant than most other species of the genus.
Occasional; scattered.

E. arcus var. **bidens** Grunow
Sparse in mesotrophic flushes and coastal seepages.
Occasional; scattered.

E. arcus var. **uncinata** (Ehrenb.) Grunow
Sparse in the larger mesotrophic lochs and in mesotrophic flushes.
Occasional; scattered.

E. bactriana Ehrenb.
Isolated specimens only in **8** lochan on Beinn na Srèine, c. 480 m, on stems of *Juncus*.

E. bidentula W. Smith
Acidophilous. Often in large numbers in oligotrophic mires and peaty pools, commonly amongst *Sphagnum*.
Occasional; scattered inland.
Very variable in size, the smallest specimens being 14 μm long and the largest 52 μm in length.

E. curvata (Kütz.) Lagerst.
E. lunaris (Ehrenb.) Grunow non Bréb. ex Rabenh.
Acidophilous; ubiquitous and numerous in oligotrophic habitats. In bogs, streams, rivers, and lochs at all altitudes.
Abundant; widespread.

E. denticulata (Bréb. ex Kütz.) Rabenh.
Acidophilous. In moorland streams, peaty pools and *Sphagnum* bogs, rarely in large numbers.
Occasional; scattered inland.
Mull specimens are 22–48 × 3.5–8 μm with 16–21 transapical striae in 10 μm.

E. diodon Ehrenb.
Acidophilous. Occurring very sparingly in slow moving waters, in oligotrophic mires and on irrigated siliceous rocks, mainly above 300 m.
Occasional; scattered.

E. elegans Østrup
Acidophilous but slightly base-tolerant. In small numbers in oligotrophic and mesotrophic lochs, bog-pools and mires, often at high altitudes.
Occasional; scattered.

E. exgracilis A. Cleve [Cleve-Euler 1953a: 106]
Acidophilous. In small numbers, occasionally in quantity, in bog-pools, on peat banks of streams through mires and amongst *Sphagnum* and other mosses at loch margins.
Occasional; scattered.
Very close to *E. paludosa* Grunow which is barely arched and has parallel margins whereas the Mull specimens are strongly arched, with the dorsal margin being the more strongly reflexed. The capitate apices of *E. paludosa* are more abrupt than those of the Mull specimens which are gently rounded. Striae number 18–19 in 10 μm at the centre and are indistinct at the apices.

E. exigua (Bréb. ex Kütz.) Rabenh.
Acidophilous. Often in very large numbers in mires, less numerous in oligotrophic pools and lochans, commonly in association with *Sphagnum* species.
Frequent; widespread.

E. fallax A. Cleve
Base tolerant but probably calcifuge and oxyphilous. Occurring in small numbers mainly in shallow swift flowing streams, also in flushes and at the sides of small waterfalls.
Occasional; scattered.

E. flexuosa Bréb. ex Kütz. var. **flexuosa**
Acidophilous. Sparse in base-poor lochs and slow flowing streams.
Occasional; scattered.

E. flexuosa var. **eurycephala** Grunow [Patrick & Reimer, 1966: 188]
Acidophilous. Often in large numbers in *Sphagnum*; less numerous amongst vegetation in base-poor lochs.
Occasional; scattered.

E. formica Ehrenb.
Sparse in acid or slightly base-enriched waters; in large numbers among wet mosses in a few mesotrophic habitats.
Occasional; scattered.

E. glacialis Meister
E. gracilis (Ehrenb.) Rabenh.
Acidophilous. Sparse in oligotrophic mires, lochs and streams often at high altitudes; not found in dystrophic habitats.
Occasional; widespread inland.

E. iatriaensis Foged [Foged, 1970: 180]
Isolated specimens in **16** Loch an t'sithein, with filamentous algae covering loch floor, c. 30 cms deep. Specimens reaching up to 14 μm in length.

E. incisa W. Smith ex W. Gregory [Gregory, 1854: 96, t. 4, fig. 4]
E. veneris sensu Hust. (1930; 1932).
Ubiquitous in acid habitats throughout the island; usually in small numbers but in large numbers in *Sphagnum* at loch and pool margins and in moss flushes.
Frequent; widespread.
See also Patrick, 1958: 3–4.

E. kocheliensis O. Müller
Isolated specimens from **6** Loch Carnain an Amais, vegetation beside loch and **7** River Bellart, near source, on mud bank with *Betrachospermum* sp.

E. major (W. Smith) Rabenh.
Acidophilous. Favouring irrigated mosses and rocks, often at high altitudes; also in oligotrophic mires and bog-pools; occasionally in large numbers.
Frequent; widespread inland.

E. naegelii Migula
E. alpina (Naegeli ex Kütz.) Hust.
Mostly in small numbers in streams through

mires, in rivers and in high altitude bog-pools and lochans.
Occasional; scattered

E. nymanniana Grunow [Cleve-Euler 1953a: 108]
Acidophilous. In small numbers in *Sphagnum* and bog-pools.
Occasional; scattered.

E. parellela Ehrenb. var. **parallela**
E. angusta A. Cleve
Isolated specimens in a variety of habitats including streams, rivers, lochs and sedge flushes.
Occasional; scattered.

E. parellela var. **media** (A. Cleve) A. Berg [Cleve-Euler, 1953a: 83]
A few specimens only from **8** Creag a' Ghaill in seepage at foot of cliffs and **9** near Loch a' Charraigein, c. 240 m, in slow-moving stream.

E. pectinalis (O. F. Müller) Rabenh. var. **pectinalis**
Acidophilous but slightly base-tolerant. Sparse, occasionally in large numbers, in oligotrophic to mesotrophic lochs, in flushes and slow moving streams; also amongst submerged bryophytes including *Sphagnum.*
Occasional; scattered inland.

E. pectinalis var. **minor** (Kütz.) Rabenh. (incl. forma *impressa* (Ehrenb.) Hust.)
pH indifferent. Found in a wide variety of habitats but favouring flushed rocks and aquatic vegetation in shallow water habitats; never in large numbers.
Abundant; widespread.

E. pectinalis var. **undulata** (Ralfs) Rabenh.
Acidophilous. Often in large numbers in oligotrophic lochs, mires and rivers; sparse in bog-pools.
Present in most samples from **14**, the Glen Forsa area, but otherwise rare and scattered.

E. pectinalis var. **ventricosa** (Ehrenb.) Grunow
E. pectinalis var. *ventralis* (Ehrenb.) Hust.
Acidophilous. Usually in small numbers in oligotrophic lochs and streams; in quantity in littoral samples from **16** Loch an t'sithein.
Occasional; scattered inland.

E. perpusilla Grunow. var. **perminuta** (Grunow) R. Ross
E. tridentula Ehrenb. var. *perminuta* Grunow; *E. polydentula* Brun var. *perpusilla* sensu Hust. (1930).
Isolated specimens at **7** Allt Loch a' Ghael, 180 m, in sedge flush by stream and **8** Mackinnon's Cave, on rock slab at side of waterfall.

E. praerupta Ehrenb. var. **praerupta**
Indifferent to base status. In small numbers in a wide variety of habitats including flushed rocks, bog-pools, mires, lochs and rivers.
Occasional; scattered.

E. praerupta var. **bidens** (Ehrenb.) Grunow
Indifferent to base status. Sparse in bog-pools, mires, streams, on flushed rocks and in the larger lochs.
Occasional; scattered.

E. praerupta var. **muscicola** B. Peters.
In small numbers in three localities: **6** Loch Carnain an Amais, in mire vegetation; **7** River Bellart, near source, on mud bank with *Batrachospermum* sp. and **8** Creag a' Ghaill, amongst detritus in small pool at top of cliffs.

E. scandinavica A. Cleve ex Fontell [Cleve-Euler, 1953a: 129]
Acidophilous. Sparse and sometimes occuring as isolated specimens in the shallow water of oligotrophic bog-pools and lochs, often at high altitudes.
Occasional; scattered inland.

E. septentrionalis Østrup
In acid or slightly base-enriched waters. In small numbers in flushes, mires and streams and on irrigated rocks; sparse in the larger lochs.
Occasional; scattered.

E. serra Ehrenb. var. **diadema** (Ehrenb.) Patr.
E. robusta Ralfs var. *tetraodon* (Ehrenb.) Ralfs
Acidophilous. In small numbers in bog-pools, oligotrophic streams and lochans; frequently amongst *Sphagnum.*
Occasional; scattered inland.

E. suecica A. Cleve
Numerous in two samples from **11** Ben More, c. 900 m on flushed vertical rocks and sparse on **8** Creag a' Ghaill, 150 m, wet slab in gully.

E. tenella (Grunow) A. Cleve
Acidophilous. Sparse in oligotrophic streams, lochs and flushes.
Frequent; widespread inland.

E. triodon Ehrenb.
Acidophilous. Sparse, occasionally in large numbers, in bog-pools and lochans at high altitudes.
Occasional; scattered.

E. valida Hust.
Acidophilous. Mainly dispersed amongst vegetation in oligotrophic mires, bog-pools and streams but rarely in large numbers.
Occasional; scattered.

E. vanheurckii Patr.
E. faba (Ehrenb.) Grunow
Acidophilous. Occasionally in large numbers amongst submerged vegetation in bogs and

in the shallow marginal waters of oligotrophic lochs and pools; mainly at high altitudes.
Occasional; scattered inland.
The shape of the specimens is often distorted.

ACHNANTHACEAE

Achnanthes Bory

A. affinis Grunow
Basiphilous, anhalophobous and oxyphilous. Often in large numbers on flushed rock surfaces, especially those at the sides of waterfalls and in coastal areas; sparse in the littoral of mesotrophic lochs and streams.
Common; widespread.
Mull specimens are 8–26 × 3–4 μm, with 27–30 transapical striae in 10 μm.

A. austriaca Hust.
Recorded as a littoral species from the inland larger lochs and from some high altitude lochans and pools. Present in small numbers only.
Occasional; in scattered localities.

A. biasolettiana (Kütz.) Grunow
In a wide variety of shallow water habitats including mud, streams, mires, irrigated rocks and mosses and the littoral of lochs, rarely in large numbers. Not markedly halophilous, present in coastal areas and well inland.
Common; widespread.

A. clevei Grunow
Basiphilous. In small numbers in the littoral of the mesotrophic lochs; sparse in mesotrophic seepages and streams.
3 Loch Poit na h-I; Loch Assopol. **7** Loch Frisa. **8** Creag a' Ghaill cliffs. **11** Loch Bà. **13** Caol Lochan.

A. coarctata (Bréb. ex W. Smith) Grunow
Oxyphilous. On flushed rock surfaces and irrigated mosses; also in shallow pools and lochs. Generally sparse but the dominant species in one sample from an exposed rock face.
Occasional; scattered, most frequently observed on the **8** Creag a' Ghaill cliffs.

A. conspicua A. Mayer
Basiphilous. A generally sparse species apparently confined to the littoral of the mesotrophic lochs.
3 Loch Poit na h-I; Loch Assopol. **9** Lochan on Creachan Mòr. **11** Loch Bà.

A. depressa (Cleve) Hust.
Recorded as isolated specimens. **6** Lochan's Airde Beinn, on mud in wavesplash zone on shore. **7** Loch Frisa, on *Juncus*, *Myriophyllum* and muddy shingle. **8** Creag a' Ghaill, on vertical slab at side of gully. **13** Caol Lochan, on *Fontinalis*; on mud from loch floor; on decaying moss and submerged debris.

A. didyma Hust.
Basiphilous. Small numbers in the littoral of mesotrophic lochs, on coastal rocks and in streams.
3 Loch Poit na h-I; Loch Assopol. **7** Allt Loch a' Ghael, c. 180 m, on moss-covered boulder in stream. **10** Carsaig cliffs, on flushed rock. **13** Caol Lochan.

A. exigua Grunow var. **heterovalvata** Krasske
Recorded in small numbers in littoral samples from **3** Loch Poit na h-I, on *Juncus* and *Fontinalis*; on *Nitella*, c. 35 cms deep; with *Pellia* sp., at margin.

A. exilis Kütz.
Sparse in samples from **8** Mackinnon's Cave, on slab at side of waterfall; **11** Loch Bà, on small stones in shallow water; **13** Lochan na Guailne Duibhe, on submerged grass at edge and Caol Lochan, amongst decaying moss and submerged debris.

A. flexella (Kütz.) Brun var. **flexella**
Most frequently found at higher altitudes in shallow water habitats; of wide ecological tolerance but seldom in large numbers.
Common; widespread.

A. flexella var. **alpestris** Brun
Mainly lacustrine but also occurring in high altitude pools, seepages and on flushed rocks and mosses; rarely in large numbers.
Common; widespread.

A. hauckiana Grunow
Halophilous. Sparse in littoral samples from **2** Loch Staoneig and **3** Loch Poit na h-I.

A. holstii Cleve
Sparse from **6** Lochan's Airde Beinn, on mud on shore and **7** Loch Frisa, in plankton tow with marginal detritus.

A. kryophila B. Peters.
Small numbers in the littoral of the larger lochs and in subalpine pools.
3 Loch Poit na h-I. **7** Loch Frisa. **8** Creag a' Ghaill, c. 240 m, in pools on *Fontinalis*. **13** Caol Lochan. **15** Coire Mòr, 530 m, in *Sphagnum* pool. **16** Loch an t'sithein.

A. laevis Østrup [Østrup, 1910: 130; Carter, 1963a: 237].
Sparse in four samples from the litttoral of **3** Loch Poit na h-I.

A. lanceolata (Bréb.) Grunow
Anhalophobous and possibly oxyphilous. Most numerous in the base-rich habitats and not present in strongly acid conditions. In a wide variety of shallow water habitats but in greater numbers in moving water; often the dominant species in some coastal gatherings.
Common; widespread.

Gatherings include forma **capitata** O. Müller, f. **ventricosa** Hust., var. **elliptica** Cleve, and var. **rostrata** (Østrup) Hust., often equally common with the type and often in mixed populations.

A. lapponica (Hust.) Hust.
pH indifferent. Favours irrigated rocks and mosses but also present in the littoral of the larger lochs and in rivers; never in large numbers.
Common; widespread.

A. laterostrata Hust.
Recorded as isolated specimens in littoral samples from **3** Loch Poit na h-I; Loch Assopol; **13** Caol Lochan, Lochan a' Churrabain; **17** Loch a' Ghleannain.

A. lemmermannii Hust.
Recorded as isolated specimens from **10** Loch Fuaron, on immersed peat forming boggy patch beside loch and **15** Glen Forsa, floating strands in drainage channel.

A. levanderi Hust.
Possibly basiphilous; said to be subalpine but descends to sea level on Mull. Sparse, coating stones and vegetation in the littoral of lochs and on flushed coastal rocks.
3 Loch Poit na h-I; Loch Assopol. **5** Calgary Bay, on flushed rocks. **7** Loch Frisa. **8** Beinn na Srèine, small lochan; Creag a' Ghaill, on dripping rock. **11** Loch Bà. **13** Caol Lochan.

A. linearis (W. Smith) Grunow
pH indifferent. Recorded from a wide range of shallow water habitats but never in large numbers.
Ubiquitous, occurs in 90% of all the Mull samples.

A. marginulata Grunow
Weakly basiphilous. Favouring shallow water habitats; in pools, on flushed rocks and mosses and in the littoral of some lochs; rarely in large numbers.
Frequent; widespread.

A. microcephala (Kütz.) Grunow
In the spray zone of waterfalls and as an epiphyte on *Myriophyllum*, *Juncus* and *Fontinalis* etc. in lochs and rivers, often in large numbers. Appears to be more base-demanding than *A. minutissima* and consequently neither as common nor so numerous.
Common; widespread.

A. minutissima Kütz.
pH indifferent. Most numerous on flushed rock surfaces and irrigated mosses covering boulders in fast flowing streams.
Ubiquitous, occurs in 90% of all the Mull samples and often the dominant or co-dominant species. The entity referred to as var. *cryptocephala* Grunow is as common and often mixed with var. *minutissima* and is not separated here.

A. oestrupii (A. Cleve) Hust.
Basiphilous? Sparse in littoral samples from the following lochs.
3 Loch Poit na h-I. **7** Loch Frisa. **13** Caol Lochan.
Mull specimens are 9–21 μm in length (see also Carter 1961: 326).

A. peragalloi Brun & Hérib.
Isolated specimens from **10** Cnocan Buidhe, on submerged moss at edge of lochan and **13** Lochan na Guailne Duibhe, on submerged grass at edge.

A. petersenii Hust. [Hustedt, 1937: 179]
Sparse in exposed shallow water habitats often at high altitudes; on flushed rocks and mosses and in very shallow streams.
Rare; scattered, mainly observed on the Creag a' Ghaill cliffs.

A. pseudoswazi Carter [Carter, 1963b: 201]
Weakly basiphilous. On bottom mud and detritus in the shallow marginal waters of pools and lochs; also present on flushed rocks and mosses, occasionally on aquatics; rarely in large numbers.
Frequent and widespread at all altitudes.
Ends of raphe curve in the same direction at both poles.

A. rossii Hust. [Hustedt 1954: 467]
In small numbers in littoral samples from **3** Loch Poit na h-I; Loch Assopol. **13** Lochan na Guailne Duibhe.
Mull specimens are: 10–20 × 5–7.5 μm with c. 33 transapical striæ in 10 μm.

A. rostellata A. Cleve [Cleve-Euler 1953b: 35]
Isolated specimens recorded in four marginal, shallow water samples from **3** Loch Poit na h-I.

A. rupestris Krasske
Basiphilous? and subalpine. Small numbers in seepages and on flushed rock surfaces.
Recorded in seven samples from the **8** Creag a' Ghaill cliffs, 250–260 m and in two samples from **11** Ben More at 870 m.
Mull specimens are 12–19 × 5–6 μm with 15–17 transapical striae in 10 μm.

A. saxonica Krasske
Not present in strongly acid habitats. In large numbers, occasionally the dominant or co-dominant species in coastal streams and flushes; less numerous in inland streams, rivers and in the littoral of lochs.
Common in coastal areas; widespread.
Mull specimens are 10–16 × 5–7.5 -m with 10–14 transapical striae in 10 μm on the rapheless valve. [See also Carter, 1961: 17].

A. sublaevis Hust. [Carter 1961: 37]
Basiphilous? In small numbers in the littoral of some mesotrophic lochs.
3 Loch Poit na h-I; Loch Assopol. **7** Loch Frisa. **13** Caol Lochan; Lochan na Guailne Duibhe.

A. subsaloides Hust.
Halophilous. Large numbers in areas receiving sea spray on the W coast and on the **1** Treshnish Isles; small numbers present in a sample from **8** Creag a' Ghaill cliffs at 150 m.

A. suchlandtii Hust.
Sparse. **3** Loch Poit na h-I, on *Juncus* and *Fontinalis* in shallow water; in bottom sample with *Pellia* sp.; Loch Assopol, in plankton sample with marginal detritus. **7** Allt Loch a' Ghael, c. 180 m, on moss covered boulder in stream. **13** Allt nan Torc, on floor of shallow stream.

A. trinodis (W. Smith) Grunow
Occasional in two samples from S facing maritime cliffs. **8** Burgh, in black gelatinous matrix of blue-green algae on wet rock face at top of shore. **10** Rubha Dubh, in thick yellowish gelatinous matrix on cliffs.

Achnanthes sp.
Small numbers coating vegetation and substrates in the shallow marginal waters of some lochs, high altitude lochans, peaty pools, streams and mires.
Occasional; scattered inland.
Mull specimens are: 12–17 × 5–6 μm with c. 32 transapical striae in 10 μm on the raphe valve.
These specimens resemble those figured by Cleve-Euler, 1953b: fig. 592, which she names as *A.*? *altaica* (Poretzky) A. Cleve.

Cocconeis Ehrenb.

C. diminuta Pant.
Basiphilous. Occasionally in large numbers in the littoral of some mesotrophic lochs.
3 Loch Poit na h-I; Loch Assopol. **11** Loch Bà. **13** Caol Lochan.

C. placentula Ehrenb. var. **placentula**
Weakly basiphilous and anhalophobous. Occasionally in large numbers as an epiphyte on aquatic vegetation in mesotrophic streams and rivers, and in the larger lochs; present also on flushed coastal rocks.
Occasional; widespread but predominately coastal.

C. placentula var. **euglypta** (Ehrenb.) Grunow
In similar habitats to var. *placentula* but more numerous and widespread; also in large numbers in the splash zone of waterfalls.
Frequent; widespread but predominately coastal.

C. placentula var. **lineata** (Ehrenb.) Van Heurck
This variety is more numerous than the above and is present in habitats similar to and occasionally with var. *placentula.*

C. scutellum Ehrenb. var. **parva** (Grunow) Cleve
Isolated specimens only from **3** Loch Assopol in littoral samples; **8** Creag a' Ghaill, in shallow stony stream and Burgh, on wet rocks on shore; and **16** Loch Uisg, in littoral samples.

C. thumensis A. Mayer
Sparse in littoral samples from **1** Sgeir a' Chaisteil, in brackish pool; **3** Loch Poit na h-I, on *Juncus, Fontinalis, Myriophyllum, Pellia* and filamentous algae, and similarly in Loch Assopol; and **7** Loch Frisa.

Rhoicosphenia Grunow

R. curvata (Kütz.) Grunow ex Rabenh.
Basiphilous and anhalophobous (halophilous?). Small numbers in mesotrophic lochs and in a woodland pool on decaying vegetation. In large numbers on wet cliff faces and in pools receiving sea spray on the Treshnish Isles.
1 Bac Mòr; Bac Beag; Sgeir a' Chaisteill. **3** Port nan Ròn, in woodland pool. **7** Loch Frisa. **13** Caol Lochan. **17** Loch a' Ghleannain.

NAVICULACEAE

Diatomella Grev.

D. balfouriana Grev.
Acidophilous and calcifuge. Sparse in oligotrophic streams and moss flushes; mainly at high altitudes.
Occasional; scattered inland.

Navicula Bory

N. acceptata Hust.
Small numbers present as a bottom living form in the shallow water of mesotrophic lochs and streams.
3 Loch Poit na h-I. **8** Creag a' Ghaill. **11** Loch Bà.

N. angusta Grunow
N. cari Ehrenb. var *angusta* (Grunow) A. Cleve
N. cincta (Ehrenb.) Ralfs var. *angusta* (Grunow) Cleve
Calcifuge and halophobous. On various substrates in mainly oligotrophic rivers, streams, pools and loch margins.
Absent from dystrophic bogs.
Frequent inland but never in large numbers; widespread.

N. avenacea Bréb. ex Grunow
N. viridula Kütz. var. *avenacea* (Bréb.) Van Heurck
Basiphilous and anhalophobous. In small numbers in mesotrophic flushes, seepages and shallow streams; present also in the littoral of the larger lochs and in brackish habitats.

Frequent in coastal areas but rare inland; widespread.

N. bacillum Ehrenb.
Occurring as isolated specimens only. **3** Loch Poit na h-I, on *Pellia* and filamentous alga at loch margin. **7** Loch Frisa, on *Juncus*, *Myriophyllum* and muddy shingle in c. 18 cms water. **11** Loch Bà, on small stones at margin. **13** Lochan na Guailne Duibhe, on submerged grass at edge.

N. bryophila B. Peters.
Oligotrophic to mesotrophic habitats, halophobous. On wet rocks and soaked mosses at the sides of waterfalls and gorges, often in large numbers; also occurring sparingly on moss covered rocks in streams and in the littoral of lochs.
Occasional; scattered inland.

N. calida Hendey [Hendey, 1964: 198]
Basiphilous and halophilous. Sometimes very numerous on the surface mud of shallow coastal pools and streams, and on coastal cliffs. In small numbers in littoral samples from **3** Loch Poit na h-I and **6** Loch Peallach.
Frequent in coastal areas.

N. capitata Ehrenb. var. **capitata**
N. hungarica Grunow var. *capitata* (Ehrenb.) Cleve
Basiphilous and anhalophobous. In littoral samples from base-rich lochs and from coastal seepages; mostly in small numbers.
3 Loch Poit na h-I, in seven littoral samples. **7** Loch Frisa. **10** Calgary Bay, in seepages on coastal rocks.

N. capitata var. **hungarica** (Grunow) R. Ross
N. hungarica Grunow var. *hungarica*
Sparse in **3** Loch Poit na h-I, on stone at margin and at **7** Kilbrenan, in brown gelatinous coating on underhang near base of waterfall.

N. cari Ehrenb.
Sparse in **7** Allt Loch a' Ghael, 180 m, in sedge flush at side of stream and on moss-covered boulder in stream, **13** Caol Lochan, on bottom mud in shallow water and Lochan na Guailne Duibhe, on submerged grass at margin.

N. cincta (Ehrenb.) Ralfs
Basiphilous and anhalophobous (halophilous ?). Small numbers in mesotrophic streams, in flushes and on wet coastal rocks; also in the littoral of the larger lochs.
Frequent on the **1** Treshnish Isles and in coastal areas, rare in base-rich inland areas.

N. clementioides Hust. [Hustedt, 1943a: 285]
Sparse in **3** Loch Poit na h-I, on stone in very shallow water at margin.
Forming part of a taxonomic complex with *N. clementis* (see Carter, 1962).

N. clementis Grunow var. **clementis** [Schmidt, *Atlas*: t. 398, figs. 8–12]
Basiphilous and anhalophobous. Sparse in the littoral of mesotrophic lochs, in brackish pools and on estuarine mud.
1 Sgeir a' Chaisteil, pool. **3** Loch Poit na h-I, at margin on *Pellia* and filamentous algae; on peat of bank; on *Juncus* and *Fontinalis*. **6** Near Loch na Cuilce, mud bank. **7** Loch Frisa, on *Juncus* and *Myriophyllum*. **10** Loch Fuaron, in muddy patch beside loch. **13** Caol Lochan, on bottom mud in shallow water. **17** Loch Don, on mud bank of stream.

N. clementis var. **linearis** Brander [Schmidt, *Atlas*: t. 403, fig. 43; Cleve-Euler 1953b: 148]
Sparse in the littoral of some mesotrophic lochs, also on wet coastal rocks.
3 Loch Poit na h-I, on *Juncus* and *Fontinalis*. **7** Loch Frisa, on *Juncus* and *Myriophyllum*. **10** Carsaig, coastal cliffs. **13** Caol Lochan, on decaying moss and submerged debris at margin; on *Fontinalis*.
Mull specimens have 0–4 isolated puncta present to one side of the central area (the number may differ on the two valves of the same frustule); at the poles the ends of the raphe turn in opposite directions.

N. cocconeiformis W. Gregory ex Grev.
Absent from base-poor habitats and possibly halophobous. In small numbers in streams, flushes, on wet rocks, in pools and in the littoral of lochs.
Frequent; widespread inland.
Some of the lake forms have an orbicular central area with the median ends of the raphe widely spaced.

N. contenta Grunow
Acidophilous to neutral. Mostly on wet rocks and soaked exposed mosses but also in rivers, streams and lochans; usually in small numbers. In large numbers in an exposed soaked moss sample where *Melosira dendrophila* was the dominant species.
Occasional; widespread.

N. cryptocephala Kütz.
Weakly basiphilous. On dripping rocks and soaked mosses, on moss covered boulders in streams and rivers; also in the littoral of lochs.
Frequent but seldom in large numbers; widespread notably inland.

N. cuspidata (Kütz.) Kütz.
Basiphilous and anhalophobous. Isolated specimens in the littoral of some mesotrophic lochs; also in the upper estuaries of the Rivers Bellart and Forsa.
2 Loch Staoneig. **3** Loch Assopol. **6** River Bellart. **7** Loch Frisa. **12** River Forsa.

N. detenta Hust.
Sparse in **13** Caol Lochan, on decaying moss and submerged debris at margin.

N. digitoradiata (W. Gregory) Ralfs
Recorded as isolated specimens in two littoral samples from **3** Loch Poit na h-I and in one from Loch Assopol.

N. digna Hust.
Sparse from **4** near Ulva House, in moss covering concrete step in stream beneath dam; Cragaig, on rocks in fast-flowing stream; and **5** Calgary Bay, on surface of mud and dead leaves in (permanent?) puddle.
Mull specimens are 11–15 × 4–5.5 μm with c. 35 transapical striae in 10 μm.

N. disjuncta Hust.
In small numbers in **3** Loch Poit na h-I, on *Pellia* and filamentous alga at margin and **13** Lochan na Guailne Duibhe, covering submerged grass at margin.

N. elegans W. Smith
In small numbers on muddy substrates in the weakly brackish water of the upper estuaries of the Rivers Coladoir, Bellart and Abhainn Lirien, and on coastal cliffs. One inland record from a small lochan on **9** Creachan Mòr, c. 270 m, on *Myriophyllum.*

N. elginensis (W. Gregory) Ralfs var. **elginensis**
N. dicephala sensu Hust. (1930), non (Ehrenb.) W. Smith
Basiphilous and anhalophobous. Small num bers in coastal areas, sparse elsewhere.
3 Port nan Ròn, detritus in woodland pool; Loch Assopol, on rocks at margin. **5** Calgary Bay, on surface of mud and dead leaves in (permanent?) puddle. **7** Allt Loch a' Ghael, 180 m, in moss flush. **13** Lochan na Guailne Duibhe, on submerged grass at margin.
Valves very variable in shape: margins parallel or gently rounded with the apices ranging from capitate to rostrate (see Cholnoky, 1959: t. 5, figs. 202–204).

N. elginensis var. **neglecta** (Krasske) Patr.
N. dicephala var. *neglecta* (Krasske) Hust.
Basiphilous and anhalophobous. Small numbers in mesotrophic streams, on flushed rocks and in the littoral of some lochs.
3 Loch Poit na h-I, on *Juncus* and *Fontinalis*. **4** Ulva, Ormaig, on roots in fast-flowing shallow stream; Glen Glass, c. 150 m, on wet rock slab. **5** Calgary Bay, on the surface of mud and dead leaves in (permanent?) puddle. **13** Lochan na Guailne Duibhe, on submerged grass at margin.

N. explanata Hust.
Sparse in littoral samples from **3** Loch Assopol, **6** Loch Peallach and **13** Lochan na Guailne Duibhe and Caol Lochan.

N. festiva Krasske
N. vitrea (Østrup) Hust.
Acidophilous and halophobous. Often numerous in *Sphagnum*, in bog-pools and in slow-moving peaty streams and peat cuttings. Sparse in rivers and lochs.
Frequent; widespread inland.

N. flotowii Grunow [Carter, 1960b: 287]
Sparse in **9** Loch Arm, on aquatics and **17** Gleann Rainich, in squeezings from moss in stream.

N. fossalis Krasske
Sparse in **3** Loch Poit na h-I, on *Phragmites* stem, and on *Pellia* and filamentous alga at margin; Loch Assopol, in scrapings from rocks at margin and **8** Creag a' Ghaill, covering small stones in muddy substrate of shallow stream.

N. fragilarioides Krasske
Oxyphilous. Small numbers in seepages and in the splash zone of waterfalls.
8 Creag a' Ghaill, in seepage at foot of cliffs. **10** Rubha Dubh, on rock face beneath small waterfall on coastal cliffs. **11** Ben More, 370 m, in gelatinous matrix at base of small gully. **12** Allt Ardnacross, on dripping slabs at side of waterfall.

N. gastrum (Ehrenb.) Kütz.
Isolated specimens recorded from the littoral of **2** Loch Staoneig, on *Myriophyllum* and **3** Loch Poit na h-I, on *Nitella*, and on *Pellia* and filamentous alga at loch margin.

N. gottlandica Grunow [Patrick & Reimer, 1966: t. 48, fig. 14]
Weakly basiphilous. In the littoral of some mesotrophic lochs on a variety of substrates but commonly on vegetation; rarely in large numbers.
3 Loch Poit na h-I. **6** Loch an Tòrr. **7** Loch Frisa; Loch a' Ghael. **11** Loch Bà. **13** Caol Lochan.

N. graciloides A. Mayer
Basiphilous and anhalophobous (halophilous?). In small numbers on a variety of shallowly submerged substrates in mesotrophic lochs; more commonly found on flushed coastal rocks in the S and W of the Island and on the **1** Treshnish Isles.
1 Bac Beag. **2** Loch Staoneig. **3** Loch Poit na h-I. **5** Calgary Bay. **10** Carsaig. **11** Loch Bà.

N. gregaria Donkin
Basiphilous and anhalophobous (halophilous?). Often in large numbers on mud in the shallow mesotrophic streams of the coastal fringe; in small numbers in the littoral of **2** Loch Staoneig, **3** Loch Poit na h-I, Loch Assopol, **7** Loch Frisa and **17** Loch a' Ghleannain. Abundant in a polluted stream near **2** Iona Abbey.
Occasional; mainly coastal fringe and the larger lochs.

N. grimmei Krasske
Dominant species in large numbers in one sample only: **5** Calgary Bay, on the surface of mud and dead leaves in (permanent?) puddle.

N. hassaica Krasske
Sparse in the following: **3** Loch Poit na h-I, on peat bank; Loch Assopol, in scrapings from rocks at margin; **4** Ulva, Ormaig, on roots in fast-flowing shallow stream and **7** Allt Loch a' Ghael, 180 m, in moss flush.

N. heufleri Grunow
N. cincta var. *heufleri* (Grunow) Van Heurck
Sparse in sample from **4** Ulva, Ormaig, on roots in fast-flowing shallow stream.

N. hoefleri Cholnoky
Acidophilous. Often in large numbers in high altitude bog-pools and lochans.
6 Lochan's Airde Beinn, c. 285 m, on mud in wave-splash zone at margin. **9** Allt Bun an Easa, c. 250 m, in bog-pool. **10** Beinn Chreagach, c. 360 m, on dead stems submerged in lochan. **15** Maol nam Fiadh, c. 375 m, coating floor of shallow bog-pool; Allt an Dubh Choire, c. 450 m, in bog-pool.

N. hustedtii Krasske
In small numbers in littoral samples from **3** Loch Poit na h-I and **13** Caol Lochan.

N. imbricata Bock
Isolated specimens from **7** Loch Frisa, Lettermore, on stones beneath 15 cms water in sheltered inlet and **8** Creag a' Ghaill, in wet trickle at foot of cliffs.
Mull specimens are 38–44 × 10–12 μm with 13–15 transapical striae (distinctly punctate) in 10 μm.

N. ingrata Krasske
In small numbers in one sample only from **13** Caol Lochan, amongst decaying moss and submerged debris at margin.

N. insociabilis Krasske
Sparse on flushed rocks and mosses, in shallow streams and pools. **3** Port nan Ròn, shallow pool in woodland. **4** Ulva, Ormaig, in fast-flowing shallow stream. **5** Calgary Bay, on wet rocks at N side of bay. **6** Loch Carnain an Amais, in boggy patch beside loch. **7** Allt Loch a' Ghael, 180 m, in moss flush. **8** Creag a' Ghaill, in fast-flowing shallow stream.

N. integra (W. Smith) Ralfs
Isolated specimens from **13** Lochan na Guailne Duibhe, coating submerged grass at edge of loch.

N. jaagii Meister
In small numbers in **13** Caol Lochan, on mud floor under c. 8 cms water; coating *Fontinalis* and on decaying moss and submerged debris at margin.

N. jaernefeltii Hust.
Occasionally in large numbers in the littoral of mesotrophic lochs.
2 Loch Staoneig. **3** Loch Poit na h-I; Loch Assopol. **7** Loch Frisa. **13** Caol Lochan; Lochan na Guailne Duibhe.

N. krasskei Hust.
Sparse in **6** Lochan's Airde Beinn, in wave-splash zone at margin; **15** Glen Forsa, on roots and debris in small stream through mire and in Gleann Rainich, in squeezings from moss covering rocks in stream.

N. lacustris W. Gregory
A solitary valve observed in scrapings from submerged peat forming bank of **3** Loch Poit na h-I.

N. laevissima Kütz.
Occurring mainly as isolated specimens.
3 Loch Poit na h-I, on *Juncus* and *Fontinalis*; Loch an Sgalain, *Sphagnum* at edge of loch. **7** Loch Frisa, on *Myriophyllum*; Allt Loch a' Ghael, 180 m, in moss flush. **13** On stones in stream SW of Loch na Meal; Lochan na Guailne Duibhe, submerged grass at edge. **15** Glen Forsa, on roots in stream.

N. lanceolata (Ag.) Kütz.
Basiphilous. In small numbers in mesotrophic lochs and pools.
2 Iona, in flocculent material from shallow pool near Loch Staoneig. **3** Loch Assopol, on shallowly submerged rocks at margin; Port nan Ròn, shallow pool in woodland. **7** Loch Frisa, on *Juncus* and *Myriophyllum* in shallow water at margin.

N. levanderi Hust.
Isolated specimens from **13** Lochan na Guailne Duibhe, on submerged grass at margin.

N. mediocris Krasske
Acidophilous. Mostly in small numbers in bog-pools, *Sphagnum*, oligotrophic flushes and high altitude lochans.
Occasional but frequent in the **11** Allt Teanga Brideig area.

N. menisculus Schumann
Small numbers in coastal seepages and pools and in lochans with a muddy substrate. **1** Bac Beag, mud ledge on coastal cliff; Sgeir a' Chaisteil, pool. **5** Calgary Bay, on mud and dead leaves in (permanent?) puddle. **8** Creag a' Ghaill, seepage at foot of cliffs. **13** Caol Lochan; Lochan na Guailne Duibhe.

N. minima Grunow
Basiphilous. In small numbers in the shallow marginal waters of the larger lochs, in shallow streams, moss flushes and on wet rocks.
Occasional; scattered.

N. minuscula Grunow
In the shallow water of pools, soaked mosses and streams; occasionally in large numbers.
4 Ulva, Ormaig. **8** Creag a' Ghaill.

N. mutica Kütz.
Basiphilous and anhalophobous. In small numbers on wet rocks of the **8** Creag a' Ghaill cliffs also sparingly on wet coastal rocks and in coastal pools on **1** Bac Beag, **3** Port nan Ròn, **5** Calgary Bay and **10** Carsaig cliffs.

N. naumannii Hust.
In acid to neutral waters. In small numbers in streams, mires, on wet rocks and in the littoral of lochs. Often associated with *Fontinalis* spp.
Occasional; scattered inland.

N. nivalis Ehrenb.
Isolated specimens recorded from **11** Loch Bà, on small stones in shallow water at margin.

N. oblonga (Kütz.) Kütz.
Numerous in two samples from **13** Caol Lochan, on bottom mud in shallow water and on decaying moss and submerged debris at margin.

N. occulta Krasske
Sparse at **5** Calgary Bay, on the surface of mud and leaves in (permanent?) puddle and **8** Creag a' Ghaill, on vertical slab under film of water.

N. opportuna Hust. [Hustedt, 1950: 436]
In small numbers in the littoral of **3** Loch Poit na h-I, on stones, on *Juncus*, *Fontinalis*, *Myriophyllum* and *Pellia*; also in **7** Loch Frisa, on *Juncus* and *Myriophyllum*.
Mull specimens are 10–17 × 6–8.5 μm with 16–20 transapical striae (faintly punctate) in 10 μm.

N. pelliculosa (Bréb. ex Kütz.) Hilse
Sparse, in one sample only on **2** NE Iona, on mud floor of shallow stream through coastal pasture.

N. peregrina (Ehrenb.) Kütz.
Small numbers in littoral samples from **3** Loch Poit na h-I and **13** Lochan na Guailne Duibhe. Present in large numbers in weakly brackish mud cliff samples from **1** the Treshnish Isles; less numerous on westerly facing coastal cliffs on the mainland.

N. perpusilla (Kütz.) Grunow
Acidophilous to neutral, oxyphilous. Often in very large numbers on exposed wet rocks and mosses at high altitudes; less numerous in streams, loch margins and on dripping mosses and rocks at the sides of waterfalls.
Common; widespread.

N. petersenii Hust. [Schmidt, *Atlas:* t. 402]
Acidophilous. Sparse in slow-flowing streams, on wet rocks and in bog-pools at high altitudes; also present in mires.
Occasional; scattered inland.

N. placenta Ehrenb.
Sparse in samples.
3 Loch an Sgalain, on *Sphagnum* at margin; Port nan Ròn, in shallow pool in woodland. **4** Ulva, N coastal area, *Sphagnum* squeezings. **6** Loch an Tòrr, on *Sparganium* stem. **17** Meall Reamhar, squeezings from moss.

N. placentula (Ehrenb.) Kütz. forma **rostrata** (A. Mayer) Hust.
Recorded as isolated specimens from **2** NW Iona, bottom mud of shallow stream through coastal pasture.

N. protracta (Grunow) Cleve forma **elliptica** Gallik [Werff & Huls, 1957: 109]
Isolated specimens in two littoral samples from **3** Loch Poit na h-I.

N. pseudoscutiformis Hust.
Often numerous. In **3** Loch Poit na h-I, on stones, on *Pellia* and filamentous alga at margin and from a shallowly submerged peat bank; also in Loch Assopol, on shallowly submerged rocks at margin.

N. pseudotuscula Hust.
Basiphilous. Mostly in small numbers in the littoral of mesotrophic lochs especially those with bottom mud; often in association with *Navicula tuscula*.

N. pseudoventralis Hust.
N. modica Hust.
In small numbers in the littoral of **3** Loch Poit na h-I, on *Myriophyllum*, **13** Caol Lochan, on decaying moss and submerged debris at margin and in Lochan na Guailne Duibhe, on submerged grass at margin.

N. pupula Kütz.
In neutral to basic habitats. In small numbers in mesotrophic lochs, pools, flushes and streams.
Occasional; mainly coastal and the larger lochs.
Forma *capitata* Skvortzow and forma *rectangularis* (W. Gregory) Grunow are present with the type and are not separated here.

N. pusilla W. Smith
Basiphilous, anhalophobous (halophilous?) and oxyphilous. Occuring sparingly on flushed rocks and in a woodland pool; most numerous on muddy substrates in shallow brackish habitats on the Treshnish Isles.
1 Bac Mòr, coastal gully; Bac Beag, on mud in marshy ground near shore: Sgeir a' Chaisteil, pool. **3** Port nan Ròn, on bottom detritus in woodland pool. **8** Creag a' Ghaill, on wet vertical slab forming side of gully.

N. radiosa Kütz. var. **radiosa**
pH indifferent and anhalophobous. In small numbers, occasionally in quantity, in lochs, pools, rivers, streams, mires and on flushed rocks and mosses.
One of the most frequently encountered diatoms on the island; ubiquitous.

N. radiosa var. **tenella** (Bréb. ex Kütz.) Grunow
Basiphilous and anhalophobous. Small numbers in fast flowing mesotrophic streams, in coastal seepages and in the littoral of the larger mesotrophic lochs.
Occasional; coastal fringe and the larger lochs.

N. rhynchocephala Kütz.
Basiphilous and anhalophobous. Often in large numbers in the littoral zone of the larger mesotrophic lochs; also in the lower reaches of the River Forsa and in some fast flowing streams in coastal areas.
Occasional; coastal fringe, and larger lochs.

N. rotula A. Cleve [Cleve-Euler, 1953b: 112]
Small numbers recorded in two littoral samples from **13** Caol Lochan, on *Fontinalis* in water c. 15 cms deep, and on decaying moss and submerged debris.

N. salinarum Grunow var. **intermedia** (Grunow) Cleve
N. cryptocephala var. *intermedia* Grunow
Basiphilous. Numerous in the littoral of mesotrophic lochs.
3 Loch Poit na h-I; Loch Assopol. **6** Loch an Tòrr. **11** Loch Bà. **13** Caol Lochan.
Mull specimens are 25–38 × 6–8 μm, with 14–16 transapical striae in 10 μm.

N. seminulum Grunow
In small numbers, occasionally in quantity, in shallow water habitats; on flushed rocks and soaked mosses, coating stones in shallow streams and at the margins of lochs, also in the splash zone of waterfalls.
Occasional; scattered.

N. soehrensis Krasske
Sparse, only recorded from **6** Lochan's Airde Beinn, on exposed mud at loch edge.

N. sponsa Carter [Carter, 1964: 226]
Sparse in littoral samples from **3** Loch Poit na h-I, **7** Loch Frisa and **13** Caol Lochan.

N. stroemii Hust.
Sparse from **7** Loch Frisa, on *Juncus* and *Myriophyllum* in c. 18 cms water and **8** Creag a' Ghaill, in seepage at foot of cliffs.

N. subatomoides Hust.
Sparse. **3** Loch Poit na h-I, on *Juncus* and *Fontinalis*. **4** Ulva, Ormaig, on roots in fast-flowing shallow stream. **8** Creag a' Ghaill, on dead grass in peaty pool. **12** River Forsa, near Pennygown on *Juncus kochii*. **13** Caol Lochan, on decaying moss and submerged debris at margin.

N. subbacillum Hust.
Numerous in the Loch Frisa sample otherwise sparse.
2 S Iona, on soaked moss covering boulder in stream. **7** Loch Frisa, on *Juncus* and *Myriophyllum* c. 18 cms. water; Kilbrenan, wet rocks at side of waterfall. **8** Creag a' Ghaill, seepage at foot of cliffs.

N. subtilissima Cleve
Acidophilous. Occasionally in large numbers in bog-pools, mires and soaked mosses, especially *Sphagnum* spp.
Frequent; widespread inland.

N. submuralis Hust.
Sparse in **3** Loch Poit na h-I, on *Pellia* and filamentous alga at margin and **4** Ulva, Cragaig, with filamentous alga sample in shallow stream.

N. tantula Hust.
pH indifferent but favouring base-rich areas. Often in large numbers on the surface mud of shallow pools and streams, in moss flushes and in the littoral of the following lochs: **3** Loch Poit na h-I; **9** lochan on Creachan Mòr; and **13** Caol Lochan.
See also Schoeman (1973) for comments on size and variation.

N. tridentula Krasske
Sparse. **3** Loch Poit na h-I, on *Pellia* and filamentous alga at loch margin. **5** Calgary Bay, on surface of mud and dead leaves in (permanent?) puddle; wet rock slabs at side of path. **7** Loch Frisa, on *Juncus* and *Myriophyllum* in c. 18 cms water. **13** Lochan na Guailne Dubhe, on submerged grass at loch margin.
These Mull specimens are up to 23 μm long.

N. tripunctata (O. F. Müller) Bory
N. gracilis Ehrenb.
Isolated specimens from **7** Allt Loch a' Ghael, 180 m, moss flush and **8** Mackinnon's Cave, on wet rock surface.

N. tuscula (Ehrenb.) Van Heurck
Often numerous in littoral samples from mesotrophic lochs, especially those with bottom mud or shingle.
3 Loch Poit na h-I; Loch Assopol. **7** Loch Frisa. **10** Lochan on Cnoc an Buidhe. **11** Loch Bà. **13** Caol Lochan.

N. utermoehlii Hust.
Basiphilous. Mostly in small numbers in the littoral of mesotrophic lochs; also in a mesotrophic stream.

2 Loch Staoneig. **3** Loch Poit na h-I. **4** Ulva Ormaig, on roots in fast flowing stream. **13** Caol Lochan.

N. variostriata Krasske
pH indifferent and anhalophobous. In small numbers, occasionally in quantity, in shallow water habitats; on soaked mosses, flushed rocks, on mud of river banks, in streams and mires and in the littoral of lochs. Found in large numbers in a mud sample from a weakly brackish habitat.
Frequent; widespread.

N. vanheurckii Patr.
N. rotaeana sensu Hust. (1930) non (Rabenh.) Grunow
In small numbers on flushed rocks and mosses and in the shallow marginal waters of lochs.
Occasional; scattered.

N. ventralis Krasske
Only recorded from **13** Caol Lochan, in small numbers on decaying moss and submerged debris at loch margin.

N. viridula (Kütz.) Kütz. emend. Van Heurck
Sparse: **2** SW Iona, on moss covered boulder in stream; **7** Loch Frisa, on *Juncus* and *Myriophyllum.*

N. vulpina Kütz.
Basiphilous. Mostly in small numbers on bottom mud and small stones, rarely on vegetation, in the larger mesotrophic lochs; also in mesotrophic coastal streams with a muddy substrate and in flushes.
Occasional; coastal and larger lochs.

Navicula species novae
A Isolated specimens.
3 Loch Poit na h-I, on *Myriophyllum;* Loch Assopol, on stones at margin. **6** SW Ardmore Bay, in gelatinous matrix of algae on wet mud. **8** Creag a' Ghaill, in wet trickle at foot of cliffs. **13** Caol Lochan, on *Fontinalis.*
B Sparse in samples: **2** N Iona, on mud floor of stream through coastal pasture and **16** Loch Uisg, amongst submerged *Sphagnum* at margin.

Neidium Pfitzer

N. affine (Ehrenb.) Pfitz. var. **affine**
Isolated specimens in **13** Caol Lochan, amongst decaying moss and submerged debris at margin.

N. affine var. **longiceps** (W. Gregory) Cleve
Sparse on aquatics in pool beside **10** Loch Fuaron and in **13** Caol Lochan, amongst decaying moss and submerged debris at margin.

N. affine var. **amphirhynchus** (Ehrenb.) Cleve
Basiphilous. In small numbers in streams with a muddy substrate, in mires and in the littoral of some mesotrophic lochs
Occasional; scattered.

N. bisulcatum (Lagerst.) Cleve var. **bisulcatum**
Acidophilous to neutral. Small numbers in streams, mires, flushed mosses and in the littoral of some lochs; often at high altitudes.
Frequent; widespread inland.
Mull specimens are 34–74 × 8–12 μm with 20–30 transapical striae (usually 22–26) in 10 μm. The number of striae in 10 μm does not appear to be related to the length-breadth ratio nor to convex or straight margins.

N. bisulcatum var. **subundulatum** (Grunow) Reimer
N. bisulcatum f. *undulata* (O. Müller) Hust.
Isolated specimens recorded from **2** Iona, on moss covered rock in slow flowing stream and amongst flocculent material in shallow peaty pool close to Loch Staoneig.

N. dilatatum (Ehrenb.) Cleve
Sparse in bottom samples and on littoral vegetation in **3** Loch Poit na h-I and in **7** Loch Frisa.
Mull specimens are 50–56 × 21–23 μm with 18–20 transapical striae in 10 μm. These are distinctly punctate with 18–20 puncta in 10 μm. Apices often weakly apiculate.

N. dubium (Ehrenb.) Cleve
Basiphilous. Small numbers on bottom mud and vegetation in the littoral of some mesotrophic lochs.
3 Loch Poit na h-I; Loch Assopol. **11** Loch Bà. **13** Lochan na Guailne Duibhe; Caol Lochan.

N. iridis (Ehrenb.) Cleve var. **iridis**
Sparse in **3** Loch Assopol, in plankton tow with slight marginal contamination; **4** Ulva, Ormaig, on roots in fast-flowing shallow stream; **10** pool beside Loch Fuaron, amongst bottom vegetation.

N. iridis var. **amphigomphus** (Ehrenb.) Mayer
Sparse in littoral samples from **3** Loch an Sgalain, amongst *Sphagnum* and **13** Caol Lochan, on bottom mud.

N. iridis var. **ampliatum** (Ehrenb.) Cleve
Sparse in **3** Loch an Sgalain, amongst *Sphagnum* at edge; **7** Loch Frisa, in plankton tow; and in **13** Caol Lochan, on bottom mud and amongst decaying moss and submerged debris at margin.

Anomoeoneis Pfitzer

A. follis (Ehrenb.) Cleve
Isolated specimens from a pool beside **10** Loch Fuaron, on submerged aquatics.

A. serians (Bréb. ex Kütz.) Cleve var. **serians**
Acidophilous, halophobous. In high altitude bog pools, mires and among *Sphagnum*, often in large numbers; also present in the littoral of oligotrophic lochs, but sparsely.

Widespread; much less common than var. *brachysira*.

A. serians var. **brachysira** (Bréb. ex Rabenh.) Hust.
Acidophilous, halophobous. In similar habitats to var. serians but of wider tolerance and more numerous.
One of the most frequently encountered diatoms in this flora; widespread.

A. styriaca (Grunow) Hust.
Small numbers. **6** Lochan's Airde Beinn, on mud in shallow water. **7** Loch Frisa, on *Juncus* at 15 cms depth; Kilbrenan waterfall. **8** Creag a' Ghaill, on irrigated rocks; in *Sphagnum* pool. **9** Cnoc Mòr, in stream and on *Sphagnum*. **10** Carsaig, in seepage on cliffs. **11** Loch Bà, on stones in shallow water.

A. vitrea (Grunow) R. Ross
A. exilis Cleve pro parte, non *Navicula exilis* Kütz. nec Hust. et auct. al.
Indifferent to base status. In streams, rivers, on wet rocks, in pools, flushes, bogs and the littoral of lochs. Ubiquitous inland; one of the most common freshwater diatoms on Mull. Very variable in size and shape notably in lochs where it reaches 32 μm in length. The var. *lanceolata* Mayer occurs with the type and is not separated here.

A. zellensis (Grunow) Cleve
Base tolerant and slightly halo-tolerant. Mainly on irrigated substrates, e.g. waterfalls and flushed rocks, occasionally on flushed peat; often in large numbers.
Occasional; scattered.

Stauroneis Ehrenb.

S. agrestis B. Peters.
Small numbers on roots in shallow flowing water in two localities: **4** Ulva, Ormaig; **15** Beinn Talaidh, Coire Ghaibhre, c. 210 m.

S. anceps Ehrenb.
In oligotrophic to mesotrophic waters. Sparse in streams, rivers, on flushed rocks, in peaty pools and the littoral of lochs.
Occasional; widespread.
The forma *linearis* (Ehrenb.) Rabenh. and varieties *hyalina* Brun & Perag. and *siberica* Grunow are included here with the type as intermediate forms occur.

S. ignorata Hust.
S. parvula Grunow var. *prominula* Grunow; *S. prominula* (Grunow) Hust.
Basiphilous and anhalophobous. Recorded as isolated specimens on mud in the littoral of the larger mesotrophic lochs, in shallow streams and in coastal seepages.
1 Bac Mòr, coastal seepage. **3** Loch Poit na h-I. **4** Ulva, Ormaig, in stream. **7** Loch Frisa. **17** Loch Don, in stream.

S. kriegeri Patr.
S. pygmaea Krieger
Isolated specimens from shallow water habitats at **4** Ulva, Ormaig, on roots in fast-flowing shallow stream and **5** Calgary Bay, on flushed rocks and in (permanent?) puddle.

S. legumen (Ehrenb.) Kütz.
Basiphilous. Sparsely distributed in mesotrophic streams, coastal seepages and pools and in the littoral of some lochs.
Occasional; scattered.

S. obtusa Lagerst.
Sparse. **1** Bac Beag, mud on sea cliffs. **4** Ulva, Ormaig, on rocks in fast-flowing stream. **6** Loch na Cuilce, on surface mud of marsh. **7** Ballygown, on flushed peat. **8** Creag a' Ghaill, on wet rocks in gully.

S. phoenicenteron (Nitzsch) Ehrenb.
pH indifferent but more numerous in acid habitats. In bog-pools, amongst *Sphagnum*, on wet mud, in streams and in the littoral of some lochs but rarely in large numbers.
Frequent; widespread.

S. producta Grunow
Basiphilous and anhalophobous. Sparsely distributed in samples from the littoral of mesotrophic lochs and pools, and in shallow streams and flushes.
Occasional; scattered.

S. smithii Grunow
Basiphilous and anhalophobous. In the littoral of mesotrophic lochs and streams also in brackish pools but rarely in large numbers.
Occasional; mainly coastal.

Diploneis Ehrenb.

D. elliptica (Kütz.) Cleve
Basiphilous but can tolerate base-poor conditions, anhalophobous. In large numbers in coastal seepages and in the littoral of the mesotrophic lochs; small numbers in bog pools, mires, moss-covered boulders in streams, on wet rocks and mud banks.
Frequent; widespread.

D. finnica (Ehrenb.) Cleve
Sparse; found only in **3** Loch Poit na h-I, on *Pellia* mixed with filamentous alga at loch margin.

D. marginestriata Hust.
Small numbers present on littoral peat from **3** Loch Poit na h-I and in two littoral samples from **13** Caol Lochan, on mud floor and amongst decaying moss and submerged debris.

D. minuta B. Peters.
Sparse; in two localities **3** Loch Poit na h-I, on submerged peat forming bank and **8** Creag a' Ghaill, in wet trickle at foot of cliffs. Mull

specimens are 13–20 × 3.5–4.5 μm, with 35–40 transapical striae in 10 μm.

D. oblongella (Naegeli ex Kütz.) R. Ross
D. ovalis (Hilse) Cleve var. *oblongella* (Naegeli ex Kütz.) Cleve
pH indifferent, anhalophobous. In mires, flushes and pools, on wet rocks and mosses, moss-covered boulders and aquatics in streams; rarely in lochs.
Frequent, but rarely in large numbers; widespread;
Small forms often collected.

D. oculata (Bréb.) Cleve
Sparse in samples from **3** Loch Poit na h-I, on *Chara* and *Phragmites* stems and on *Pellia* mixed with filamentous alga at loch edge.

D. puella (Schumann) Cleve
Basiphilous and anhalophobous. In small numbers in mesotrophic lochs, pools and streams also in seepages on coastal rocks and mud.
Occasional; coastal, and mainly western but also in **7** Loch Frisa and **13** Caol Lochan.

D. petersenii Hust.
pH indifferent. In small numbers on flushed rocks and mosses, in streams and rivers also in the littoral of some lochs.
Occasional; widespread inland.
Most specimens are 14–20 × 5–7 μm, with 22–24 transapical striae in 10 μm, but specimens from **3** Loch Poit na h-I reach 25–32 × 7 μm with 22 transapical striae in 10 μm.

Caloneis Cleve

C. alpestris (Grunow) Cleve
Basiphilous and anhalophobous. In mesotrophic pools and seepages, on wet coastal rocks and in the littoral of **3** Loch Poit na h-I and **13** Caol Lochan; rarely in large numbers.
Occasional; widespread in S and W coastal areas.

C. amphisbaena (Bory) Cleve
Isolated specimens from **1** Bac Beag, on mud of sea cliffs facing SE and **5** Treshnish Point, in moss covering vertical slabs at side of coastal waterfall.

C. bacillum (Grunow) Mereschowsky
Weakly basiphilous and anhalophobous. Sparse, occasionally in large numbers, in mesotrophic seepages, on irrigated coastal rocks and in the littoral of some mesotrophic pools and lochs.
Frequent and widespread in coastal areas, rare inland.

C. clevei (Lagerst.) Cleve
Recorded with certainty from only one locality where it occurred as isolated specimens: **13** Caol Lochan, on mud from loch floor.

C. fasciata (Lagerst.) Cleve [Lund, 1946: 58]
Isolated specimens from **6** SW Ardmore Bay, on moss covered boulder in fast-flowing stream; on wet surface of mud; **8** Bearraich, c. 400 m, on surface of wet mud at margin of small stony pool with *Koenigia islandica.*

C. hebes (Ralfs) Patr.
C. obtusa Cleve
Acidophilous, mainly subalpine. In small numbers in slowly flowing streams through moorland areas, in peaty pools and in the littoral of the high altitude lochans.
Occasional; scattered inland.

C. latiuscula (Kütz.) Cleve
In small numbers in the larger lochs. **3** Loch Poit na h-I on stem of *Phragmites* c. 2 m from shore. **7** Loch Frisa, on *Juncus* and *Fontinalis* on muddy shingle. **11** Loch Bà, covering small stones in shallow water.

C. limosa (Kütz.) Patr.
C. schumanniana (Grunow) Cleve
Basiphilous. Isolated specimens in mesotrophic flushes, shallow streams and pools. Also present in the littoral of **3** Loch Poit na h-I, Loch an Sgalain and **13** Caol Lochan.
Occasional; scattered.
Dimensions: 54–98 × 13–17 μm, 16–20 transapical striae in 10 μm. Some specimens have poorly developed crescent-shaped markings.

C. pulchra Messikommer
Isolated specimens from **3** Loch Poit na h-I, on peat bank shallowly submerged; **8** Creag a' Ghaill, in flush and seepage at foot of cliffs; **10** Rubha Dubh, on rock face beneath small freshwater waterfall and **16** Loch an t'sithein, amongst green filamentous alga on loch floor.

C. ventricosa (Ehrenb.) Meister var. **ventricosa**
C. silicula (Ehrenb.) Cleve
Usually occurring as isolated specimens. **3** Loch Poit na h-I, three littoral samples. **4** Ulva, Glen Glass, on wet rocks. **5** Port Haunn, moss on flushed rock. **7** Allt Loch a' Ghael, c. 180 m, moss covered boulder in stream. **10** Rubha Dubh, on flushed rock.

C. ventricosa var. **subundulata** (Grunow) Patr. [Patrick & Reimer, 1966: 584]
Sparse, only on **2** Iona, Druim Dhughaill on moss covered rock in slowly flowing stream; pool near Loch Staoneig, in bottom sediment.

C. ventricosa var. **truncatula** (Grunow) Meister
C. silicula var. *truncatula* (Grunow) Cleve
pH indifferent. In small numbers in peaty pools, mesotrophic streams, amongst *Sphagnum* and in the littoral of some lochs.
Occasional; scattered.

Pinnularia Ehrenb.

P. abaujensis (Pant.) R. Ross [Ross, 1947: 199–200]

P. gibba sensu Hustedt (1930) et auct., non (Ehrenb.) Ehrenb.
Acidophilous. In bog-pools, on flushed peat, amongst *Sphagnum*, in oligotrophic mires, inland streams and in the littoral of some lochs and high altitude lochans; rarely in large numbers.
Common; widespread inland.

P. acrosphaeria W. Smith
Weakly acidophilous to neutral. In small numbers, occasionally in quantity, on flushed rocks and mosses, on the bottom mud of pools and more rarely in mesotrophic streams and rivers.
Occasional; scattered.

P. acuminata W. Smith
P. hemiptera sensu Hust. (1930) et auct., non (Kütz.) Cleve
Isolated specimens in: **3** Loch Poit na h-I, three bottom samples in shallow water; **4** Ulva, Ormaig, in shallow *Sphagnum* pool near coast; **6** Lochan's Airde Beinn, on mud in wave-splash zone; **7** Allt Loch a' Ghael c. 180 m, in *Sphagnum* flush.

P. alpina W. Smith
Sparse in samples: **1** Bac Mòr, on bottom of peaty pool; **9** Allt Bun an Easa c. 270 m, in *Sphagnum* bog; **10** near Loch Fuaron, in pool.

P. appendiculata (Ag.) Cleve
Acidophilous to neutral, favouring acid shallow water habitats. Large numbers in *Sphagnum* and in bog-pools; less numerous in the littoral of lochs, in flushes and amongst soaked mosses.
Frequent; widespread inland.

P. balfouriana Grunow ex Cleve
pH indifferent and oxyphilous. Mostly in small numbers in shallow water habitats; on wet rocks, flushed mosses and in spray-zone vegetation.
Occasional, most frequent in sub-alpine localities; widespread.

P. biceps W. Gregory
P. interrupta W. Smith
Acidophilous. Often in large numbers in *Sphagnum* and bog-pools; also in slow-flowing streams, mires and lochs.
Frequent; widespread.
Forma **petersenii** R. Ross [Ross, 1947: 201] has been found in small numbers at **4** Ulva, Ormaig, on roots in fast-flowing shallow stream, **13** Caol Lochan on decaying moss and submerged debris and at Lochan na Guailne Duibhe, on submerged grass at margin.

P. borealis Ehrenb. var. **borealis**
Acidophilous. In small numbers on rocks and mosses in streams, on mud banks and floors of rivers, in mires and bog-pools.
Occasional; widespread.

P. borealis var. **brevicostata** Hust.
Sparse. Only from **10** Rubha Dubh, coastal cliffs beneath small waterfall.

P. borealis var. **rectangulata** Hust. [Schmidt, *Atlas*: t. 385, fig. 28]
Sparse in **3** Port nan Ròn, in small pool in woodland; **4** Ulva, Glen Glass, c. 150 m, on rock face with seepage; **8** Creag a' Ghaill, on vertical slab with seepage.

P. brandelii Cleve
Sparse in the littoral of mesotrophic lochs, mostly on rocks and mud but also on vegetation.
3 Loch Poit na h-I. **7** Loch Frisa. **13** Caol Lochan; Lochan na Guailne Duibhe.

P. braunii (Grunow) Cleve var. **amphicephala** (Mayer) Hust.
Isolated specimens recorded from a pool near **10** Loch Fuaron.

P. brebissonii (Kütz.) Rabenh. var. **brebissonii**
P. microstauron var. *brebissonii* (Kütz.) Hust.
Sparse in sample from **11** Loch Bà, on small stones at margin.

P. brebissonii var. **diminuta** (Grunow) Cleve
P. microstauron var. *brebissonii* f. *diminuta* (Grunow) Hust.
Isolated specimens from **5** Calgary Bay, on surface of mud and dead leaves in (permanent?) puddle.

P. cleveana R. Ross
P. undulata W. Gregory
Acidophilous. In small numbers in shallow water habitats: peaty pools, streams, mires, *Sphagnum* and on the bottom mud in the littoral of some lochs.
Occasional; scattered inland, mainly above 300 m.

P. dactylus Ehrenb.
Isolated specimens recorded from a *Sphagnum* pool at c. 300 m on **8** Creag a' Ghaill.

P. divergens W. Smith var. **divergens**
Acidophilous. In small numbers, occasionally in quantity, in *Sphagnum* pools, mires, moss flushes, streams and in the littoral of some lochs.
Frequent; widespread.

P. divergens var. **elliptica** (Grunow) Cleve
Sparse from **9** Allt Bun an Easa, c. 250 m, in slow-moving shallow stream and **15** Allt an Dubh Choire, c. 525 m, beside waterfall.

P. divergens var. **undulata** (M. Perag. & Hérib.) Hust.
Acidophilous. In small numbers amongst *Sphagnum*, in mires and on roots and aquatic vegetation in rivers and in the littoral of lochs.
Rare; scattered.

P. divergentissima (Grunow) Cleve
Acidophilous. Small numbers in shallow peaty pools, slow moving streams, mires and amongst *Sphagnum*.
Occasional; widespread inland.

P. episcopalis Cleve
Isolated specimens in **7** Loch Frisa, on *Juncus* and *Myriophyllum*, 25–30 cms deep.

P. esox Ehrenb.
Very sparse in one sample from **16** Loch an t'sithein, amongst green filamentous algae on bottom, 30 cms deep.

P. gentilis (Donkin) Cleve
In acid to neutral waters. Sometimes in large numbers in bottom littoral samples from lochs; also in mires and flushes.
Occasional; scattered.

P. gracillima W. Gregory
Weakly basiphilous or pH indifferent. In small numbers in streams, mires, on flushed rocks and mosses, amongst *Sphagnum* and in the littoral of lochs; large numbers in a gelatinous matrix covering wet rock slabs at 870 m on **11** Ben More.
Frequent; widespread.

P. hilseana Janisch
P. subcapitata var. *hilseana* (Janisch) O. Müller
In large numbers in a *Sphagnum* gathering from the edge of **3** Loch an Sgalain. Sparse in **13** Caol Lochan, amongst decaying moss and submerged debris and in a *Sphagnum* gathering from margin; also in small numbers in **16** Lochan on Maol na Dubh-leitreach, in *Sphagnum* from edge.

P. intermedia (Lagerst.) Cleve [Patrick & Reimer, 1966: 617]
Acidophilous. In small numbers in shallow water habitats: on flushed rocks and mosses in shallow streams and the margins of lochs.
Rare; scattered.

P. lata (Bréb.) W. Smith
Acidophilous. In small numbers in *Sphagnum* squeezings and on peat in shallow pools and lochans.
Occasional; scattered.

P. laticeps A. Cleve [Schmidt, *Atlas*: t. 44, fig. 33; Cleve-Euler, 1955: 43]
Sparse in the littoral of **13** Lochan na Guailne Duibhe.

P. leptosoma (Grunow) Cleve
Sparse at **6** Loch Carnain an Amais, in mire beside loch.

P. lundii Hust. [Hustedt 1954b: 474]
P. globiceps W. Gregory var. *crassior* Grunow
Basiphilous. In small numbers in the littoral of **2** Loch Staoneig, **3** Loch Poit na h-I and **13** Caol Lochan and Lochan na Guailne Duibhe. Sparse on mud and dead leaves in a (permanent?) puddle at **3** Calgary Bay.

P. major (Kütz.) W. Smith
Acidophilous. Generally in small numbers amongst *Sphagnum*, in mires, streams, pools and the littoral of lochs.
Occasional; scattered inland.

P. mesogongyla Ehrenb.
P. gibba var. *mesogongyla* (Ehrenb.) Hust.
Isolated specimens occurring in a littoral sample from **13** Lochan na Guailne Duibhe and in a stream through a mire near source of **15** River Forsa.

P. mesolepta (Ehrenb.) W. Smith
4 Ulva, Ormaig, sparse on roots in fast-flowing shallow stream. **13** Caol Lochan, numerous in three bottom mud littoral samples; sparse in Lochan na Guailne Duibhe on submerged grass at margin.

P. microstauron (Ehrenb.) Cleve
Acidophilous to neutral. Occasionally in large numbers in the littoral of some of the larger lochs, in streams, pools and mires.
Frequent; widespread inland.

P. nobilis (Ehrenb.) Ehrenb.
Sparse in **15** Glen Forsa, in stream through mire and **16** Loch Uisg, amongst submerged *Sphagnum* in c. 90 cms water near margin.

P. nodosa (Ehrenb.) W. Smith
Small numbers in **3** Loch Poit na h-I, on stem of *Phragmites*; Port nan Ròn, in shallow pool in woodland. **6** Dervaig, on mud from upper estuary of the River Bellart. **7** Loch Frisa, Lettermore, on stones at margin; **17** Gleann Rainich, moss in stream.

P. obscura Krasske [Krasske 1932: 117]
Sparse in acidophilous to neutral habitats.
4 Ulva, Ormaig, on roots in fast-flowing shallow stream. **6** Loch Carnain an Amais, boggy patch beside loch; SW Ardmore bay, on boulder submerged in fast-flowing shallow stream, on surface of mud and in seepage; River Bellart, on mud bank. **11** Loch Bà, on small stones in shallow water at margin.

P. parvula (Ralfs) A. Cleve-Euler [Patrick & Reimer, 1966: 625]
Acidophilous. In bog-pools, amongst *Sphagnum* and on flushed peat: occasionally in large numbers.
Occasional; scattered inland.

P. platycephala (Ehrenb.) Cleve
Sparse in **3** Loch Poit na h-I, on stem of *Phragmites*, in Loch Assopol, on rocks at edge and in **13** Caol Lochan, on bottom mud.

P. rupestris Hantzsch [Patrick & Reimer, 1966: 630]
Acidophilous. Often in large numbers in *Sphagnum*; present in small numbers in bog-pools and peat cuttings.
Occasional; scattered inland.

P. semicruciata (A. Schmidt) A. Cleve [Schmidt, *Atlas*: t. 44, fig. 43]
Sparse in **3** Loch Poit na h-I, on *Myriophyllum* and on eroded peat bank and in **13** Caol Lochan, on *Fontinalis*.

P. stomatophora (Grunow) Cleve
Acidophilous. Occasionally in large numbers in *Sphagnum*, on submerged mosses, in streams, mires, peaty pools and the littoral of lochs.
Common; widespread.

P. stomatophoroides Mayer [Mayer, 1940: 157]
Isolated specimens in **13** Caol Lochan, on *Fontinalis*.

P. streptoraphe Cleve
Isolated specimens recorded in a single sample from **5** Calgary Bay, on wet rock slab.

P. subcapitata W. Gregory
Acidophilous. Often in large numbers in *Sphagnum*, less numerous in inland oligotrophic streams, mires, moss-flushes and bog-pools.
Frequent; widespread inland.

P. subsolaris (Grunow) Cleve
Acidophilous. Mostly in small numbers in *Sphagnum* and soaked mosses, in streams and peaty pools.
Occasional; scattered inland.

P. sudetica Hilse [Patrick & Reimer, 1966: 611]
Acidophilous. Often the dominant species amongst *Sphagnum* from the margins of pools and lochs; small numbers in peaty streams and bog-pools; often in association with *P. rupestris*.
Occasional; scattered inland.

P. viridis (Nitzsch) Ehrenb.
Sparse in a wide range of ecological conditions but most numerous in slightly acid habitats; present in *Sphagnum*, bog-pools, streams, rivers, flushes and in the littoral of lochs; rarely in large numbers.
Common; widespread.

Mastogloia Thwaites ex W. Smith

M. elliptica (Ag.) Cleve var. **dansei** (Thwaites) Cleve
Basiphilous and anhalophobous. On bottom mud in the littoral of lochs and in coastal seepages. Often in association with *M. smithii* var. *amphicephala* but usually less numerous.
3 Loch Poit na h-I; Loch Assopol. **10** Carsaig cliffs. **13** Caol Lochan.

M. grevillei W. Smith
Possibly basiphilous. Occurring as isolated specimens in moss samples from flushed habitats at **4** Ulva, Glen Glass and **7** near Loch a' Ghael.

M. smithii Thwaites ex W. Smith var. **smithii**
Basiphilous and anhalophobous. In small numbers on exposed mud and on wet cliff faces of **1** Bac Beag and on bottom mud in the littoral of **3** Loch Poit na h-I.

M. smithii var. **amphicephala** Grunow
Basiphilous and anhalophobous. In large numbers in the bottom littoral of Loch Poit na h-I, less numerous in the other two lochs also in small numbers in some mesotrophic flushes.
3 Loch Poit na h-I; Loch Assopol. **4** Ulva, Glen Glass. **5** Calgary Bay. **13** Caol Lochan.

M. smithii var. **lacustris** Grunow
Basiphilous and anhalophobous. In large numbers as a bottom living form in **13** Caol Lochan, less numerous in mesotrophic seepages and on estuarine mud.
Occasional; widespread, mainly in coastal areas.

Frustulia Rabenh.

F. rhomboides (Ehrenb.) De Toni var. **rhomboides**
Acidophilous. In large numbers in *Sphagnum*, bog pools and on flushed peat; less numerous in the shallow marginal waters of some oligotrophic lochans.
Occasional; widespread.

F. rhomboides var. **amphipleuroides** (Grunow) De Toni
Acidophilous to neutral. Large numbers in the river and mires of the Glen Forsa area. Less numerous in rivers, streams, flushes and amongst irrigated mosses elsewhere.
Occasional; scattered inland.

F. rhomboides var. **saxonica** (Rabenh.) De Toni
Acidophilous, halophobous. Often in very large numbers amongst the flocculent detritus present in bog pools and ditches, also on flushed peat, in mires, streams and in the littoral of some lochs. Often the dominant diatom in *Sphagnum* squeezings.
Ubiquitous in inland areas, one of the most commonly encountered diatoms in this flora.
Forma *undulata* Hust. and f. *capitata* (A. Mayer) Hust. occur with the variety *saxonica* and are not separable.

F. spicula Amossé var.? **alpina** Amossé [Amossé, 1972: 306–308]
In small numbers in Loch Poit na h-I, sparse elsewhere.
3 Loch Poit na h-I, three littoral samples on mud. **8** Creag a' Ghaill, wet trickle at foot of

cliffs. **10** Rubha Dubh, on wet cliffs. **13** Caol Lochan, three bottom mud littoral samples; Allt nan Torc, mud floor of stream. **15** Allt an Dubh Choire, 458 m, amongst bottom detritus in stream.

F. vulgaris (Thwaites) De Toni
Most numerous in base-rich habitats but present also in weakly acid conditions, anhalophobous. Often in large numbers in seepages over mud, less numerous in streams, irrigated mosses, moss and sedge flushes and on flushed rocks; sparse in lochs.
Frequent; widespread.

Amphipleura Kütz.

A. kriegeriana (Krasske) Hust. [Krasske, 1943: 84]
Frustulia kriegeriana Krasske
Acidophilous to neutral. Sparse in samples from subalpine localities; in the shallow water of pools, flushes, seepages and streams.
Occasional; scattered, most frequently observed on the **8** Creag a' Ghaill cliffs.
Many of the specimens have capitate apices.

A. pellucida (Kütz.) Kütz.
Weakly basiphilous, halophobous. In the shallow water of pools, lochs and streams, in soaked mosses, on exposed mud and in water films on rocks but rarely in large numbers.
Frequent; widespread inland.

A. rutilans (Trentep.) Cleve
Halophilous. Often in large numbers in brackish habitats on the **1** Treshnish Isles; sparse on S facing sea cliffs at **8** Burgh.
Occasional; scattered in coastal areas.

Gyrosigma Hassal

G. acuminatum (Kütz.) Rabenh.
Small numbers in bottom littoral samples from **3** Loch Poit na h-I, Loch Assopol and **11** Loch Bà. Abundant on mud and dead leaves in a (permanent?) puddle at **5** Calgary Bay.

G. attenuatum (Kütz.) Rabenh.
Sparse in **3** Loch Poit na h-I, on *Phragmites*.

G. spenceri (Quek.) Griff. & Henfr.
Sparse in **3** Loch Poit na h-I, on stems of *Phragmites* and **17** Loch a' Ghleannain, on *Myriophyllum.*

GOMPHONEMACEAE

Determined by Penelope A. Dawson

Gomphonema Agardh

G. acuminatum Ehrenb. var. **acuminatum** [Mayer, 1928: 94]
G. acuminatum Ehrenb. var. *genuinum* Mayer
Weakly basiphilous, epiphytic and epilithic. Occasionally in large numbers in fast flowing mesotrophic streams, generally sparse in flushes and the littoral of some lochs. Usually present with *G. acuminatum* var. *coronatum.*
Occasional; scattered.

G. acuminatum var. **brebissonii** (Kütz.) Grunow [Mayer, 1928: 94]
Epiphytic and epilithic in littoral samples. Sparse in three littoral samples from **3** Loch Poit na h-I and in large numbers in two from **13** Caol Lochan.

G. acuminatum var. **coronatum** (Ehrenb.) W. Smith [Mayer, 1928: 96]
pH indifferent, epiphytic and epilithic. Usually sparse but occasionally in large numbers in streams, rivers, lochs, on dripping rock surfaces and from bottom mud and peat.
Widespread and ubiquitous, one of the most frequently encountered diatoms in the flora.

G. acuminatum var. **elongatum** (W. Smith) Rabenh. [Mayer, 1928: 96]
Sparse in **3** Loch Poit na h-I on submerged stones; on *Pellia* and filamentous alga at margin; **6** Loch Carnain an Amais, boggy patch beside loch; and **8** The Wilderness, on *Chara* in pool.

G. acuminatum var. **trigonocephalum** (Ehrenb.) Grunow [Schmidt, *Atlas:* t. 239, figs. 16–18; Mayer, 1928: 95]
Sparse on **2** Iona, near St Mary's Abbey, epiphytic on *Ulothrix* in stream.

G. angustatum (Kütz.) Rabenh. [Mayer, 1928: 106]
Basiphilous, anhalophobous, epiphytic and epilithic. On mud, stones and irrigated mosses in streams and on coastal cliffs; also in the littoral of some mesotrophic lochs.
Occasional but rarely in large numbers; scattered in southern and western localities.

G. auritum A. Br. ex Kütz.
G. longiceps Ehrenb. var. *subclavatum* Grunow; *G. montanum* Schum. var. *subclavatum* (Grun.) Grun.; *G. subclavatum* (Grun.) Grun.
Basiphilous, anhalophobous, epiphytic, epipelic and epilithic, favouring shallow water habitats. Sparse, occasionally numerous on flushed habitats, in mesotrophic streams, on mud, in coastal seepages and in lochs.
Common; widespread in coastal areas, scattered inland.

G. bohemicum Reichelt & Fricke [Schmidt, *Atlas*: t. 235, figs 18–25]
Acidophilous to neutral, epipelic and epiphytic. Sparse in shallow water of streams and rivers, on flushed rocks and in the littoral of some lochs.
Occasional; scattered.

G. constrictum Ehrenb. var. **constrictum** [Mayer, 1928: 91]
G. constrictum Ehrenb. var. *genuinum* Mayer
Basiphilous, anhalophobous, epiphytic and epilithic. In large numbers in a coastal stream and a coastal puddle, less numerous in mesotrophic

streams, sparse in lochs and on flushed rock surfaces.
Occasional; scattered.

G. constrictum var. **capitatum** (Ehrenb.) Grunow [Mayer, 1928: 91]
5 Calgary Bay, small numbers on surface mud and dead leaves in (permanent?) puddle. **6** Loch Peallach, sparse on surface of submerged peat at margin. **13** Allt nan Torc, in large numbers on mud floor of stream. **15** In stream joining Lussa River near mouth, sparse in scrapings from stone.

G. gracile Ehrenb. [Carter, 1960a: 260]
pH indifferent but more commonly found in weakly oligotrophic habitats, epiphytic and epilithic. In the shallow water of streams, rivers, pools, mires, on flushed rocks and often as an epiphyte on aquatics in the littoral of lochs; rarely in large numbers.
Common; widespread.

G. intricatum Kütz. var. **intricatum** [Schmidt, *Atlas*: t. 234, figs. 47–50]
Weakly acidophilous to weakly basiphilous, epiphytic and epilithic in shallow water. In small, rarely in large numbers, in mires, on irrigated rocks and mosses, in mesotrophic streams and rivers; sparse in the littoral of mesotrophic lochs.
Common; widespread.

G. intricatum var. **dichotomum** (Kütz.) Grunow [Schmidt, *Atlas*: t. 234, figs. 52–54]
Basiphilous and anhalophobous, epilithic and epipelic in shallow water. Occurs only in the October 1970 gatherings.
5 Calgary Bay, sparse on surface of mud and dead leaves in (permanent?) puddle. **8** Creag a' Ghaill, sparse in brownish gelatinous matrix on dripping rock face; Burgh, large numbers in blackish gelatinous matrix on coastal rock. **10** Rubha Dubh, small numbers on rock face beneath small waterfall, sparse on wet coastal rock face. **11** Loch Bà, large numbers on small stones at margin. **16** Cnoc Carraich, sparse on stones in stream.

G. intricatum var. **pumilum** Grunow [Schmidt, *Atlas*: t. 234, fig. 57]
Basiphilous, anhalophobous, epiphytic and epilithic. Often in large numbers on dripping rock faces of inland and coastal waterfalls, on coastal flushed rocks and in the littoral of the larger mesotrophic lochs; small numbers in mesotrophic streams.
Common; widespread in coastal localities, scattered inland.

G. intricatum var. **vibrio** (Ehrenb.) Cleve [Schmidt, *Atlas*: t. 235, figs. 4–14]
Weakly basiphilous, epilithic or epiphytic. Favouring fast flowing mesotrophic streams, running surfaces of waterfalls and seepages on coastal cliffs, rarely in lochs. Generally sparse but found in large numbers in a mesotrophic moss flush.
Frequent and widespread in the N of the Island; scattered elsewhere.

G. lanceolatum Ag. [Schmidt, *Atlas*: t. 237, figs. 1–8; Mayer, 1928: 114]
pH indifferent but more widely distributed in oligotrophic regions, epiphytic and epilithic. Usually sparse on flushed rocks and mosses, in streams and rivers, peaty pools, mires and in the littoral of lochs.
Common; widespread inland, scattered in coastal localities.

G. parvulum (Kütz.) Kütz. var. **parvulum** [Mayer, 1928: 103]
G. parvulum (Kütz.) Kütz. var. *genuinum* Mayer
Basiphilous, epiphytic, epilithic and epipelic in shallow water. In mesotrophic streams, coastal seepages, coastal pools and in the littoral of mesotrophic lochs; occasionally in large numbers.
Frequent; widespread in coastal areas, rare inland.

G. parvulum var. **aequalis** Mayer [Mayer, 1928: 104]
Basiphilous, anhalophobous, epipelic and epiphytic. Sparse, occasionally in large numbers, on rocks and amongst mosses and aquatics in coastal streams and coating wet mud.
Occasional; common in **4** S Ulva and **6** Ardmore Bay, scattered in other coastal areas, rare inland.

G. parvulum var. **exilis** (Grunow) Mayer [Mayer, 1928: 105]
More numerous in the base-rich areas but present in neutral and weakly acid habitats. Occasionally in large numbers in mires, streams, rivers, on dripping rock faces, seepages on basic rocks and in the littoral of lochs.
Common; widespread.

G. parvulum var. **exilissimum** Grunow [Mayer, 1928: 104]
Weakly basiphilous but more numerous in the base-rich habitats, anhalophobous, epilithic and occasionally epiphytic. Occasionally in very large numbers in the slackwater of streams and rivers, on the dripping rock faces of waterfalls, in seepages, pools, on wet mud and in the littoral of lochs.
Common; widespread particularly in coastal areas.

G. parvulum var. **micropus** (Kütz.) Cleve [Mayer, 1928: 105]
Acidophilous – neutral. Small numbers in pools and lochans. Sparse in streams and rivers.
Rare; scattered.

G. sagitta Schumann [Mayer, 1928: 102]
Sparse in **12** River Forsa, near Pennygown, in scrapings from rocks in shallow river bed.

G. subtile Ehrenb. [Mayer, 1928: 101]
Sparse in **7** Loch a' Ghael, on *Myriophyllum* and **13** Caol Lochan, on mud floor.

G. tergestinum (Grunow) Fricke [Carter, 1960a: 263]
Basiphilous and anhalophobous. In large numbers on coastal cliffs and on basic outcrops. Found only in the October 1970 gatherings. **8** Creag a' Ghaill, c. 220 m, in brownish gelatinous matrix. **10** Rubha Dubh, on rock face beneath small waterfall; in yellowish-brown patches on coastal cliff; on wet rock ledge. **11** Loch Bà, sparse, covering small stones in shallow water at margin.

Didymosphenia M. Schmidt

D. geminata (Lyngb.) M. Schmidt [Schmidt, *Atlas*: t. 214, fig. 10]
Basiphilous and possibly anhalophobous and oxyphilous. In large numbers as an attached form to rocks in fast-flowing coastal streams and on waterfalls but sparse on attached mosses; most numerous on the wet rock surfaces of **7** Kilbrenan and **8** Mackinnon's Cave waterfalls.
Frequent; scattered localities but more frequent in coastal areas.

Gomphoneis Cleve

Bibliogr.: Dawson (1974).

G. olivaceum (Hornemann) P. Dawson ex Ross & Sims ined. [Schmidt, Atlas: t. 233 figs. 9–16]
Gomphonema olivaceum (Hornemann) Bréb.
Sparse in a mesotrophic coastal stream at **6** SW Ardmore Bay, in squeezings from *Fontinalis* and *Scapania undulata*; also on exposed mud at side of stream.

G. quadripunctatum (Østrup) P. Dawson ex Ross & Sims ined. [Carter, 1960a: 263]
Gomphonema quadripunctatum (Østrup) Wislouch; *Gomphonema olivaceoides* Hust.
Basiphilous. In large numbers in the splash zone of waterfalls, sparse in mesotrophic lochs, streams and rivers and on wet rocks at **8** Creag a' Ghaill.
Occasional; scattered.
Specimens matching the form from Kentmere (Schmidt, Atlas: t. 233, fig. 35) are present in a gathering of *Myriophyllum* from a stream entering **6** Loch an Tòrr.

CYMBELLACEAE

Cymbella Ag.

C. aequalis W. Smith ex Greville [Hustedt, 1955: 52]
Acidophilous. In large numbers in high altitude bog-pools and lochans, sparse elsewhere. **2** Pool near Loch Staoneig, in bottom detritus. **8** Beinn na Srèine, 480 m, on *Juncus* stems in lochan. **10** Near Loch Fuaron, 240 m, on aquatics from pool. **15** Coire Mòr, 510 m, on dead *Sphagnum subsecundum* in bog-pool; Allt an Dubh Choire, 460 m, in bottom mass in bog-pool; central Glen Forsa, on rock face forming side of drainage channel.

C. affinis Kütz.
Abundant and sometimes the dominant species on irrigated rocks and vegetation principally by waterfalls; occurs in fewer numbers in streams and in mesotrophic lochs.
Common; widespread.

C. amphicephala Naegeli ex Kütz. [Schmidt, *Atlas*: t. 377, figs 31–34]
Basiphilous. Numerous in the littoral of **7** Loch Frisa and **13** Caol Lochan, sparse in littoral samples from **2** Loch Staoneig, **3** Loch Poit na h-I and Loch Assopol.

C. angustata (W. Smith) Cleve
Basiphilous, oxyphilous and anhalophobous. Often in large numbers on dripping rock surfaces in base-rich areas e.g. **8** Creag a' Ghaill ridge and on coastal cliffs; sparse on inland waterfalls.
Occasional; scattered.

C. aspera (Ehrenb.) Cleve
pH indifferent and anhalophobous. Mainly occurring sparingly in slow moving streams, mires and seepages.
Occasional; widespread.

C. cesatii Grunow ex A. Schmidt var. **cesatii**
Acidophilous to neutral. On mud and rocks in base-poor pools and lochs, on flushed rock surfaces and streams at all altitudes, always rather sparingly.
Frequent; widespread inland.

C. cesatii var. **capitata** Krieger
A pH-indifferent variety occurring, often in large numbers, in a wide range of habitats including lochs, mires and on flushed rocks and mosses.
Common; widespread.
A larger form of this species (48–62 × 6–7 μm with 17–20 transapical striae in 10 μm) is present in quantity in **3** Loch Poit na h-I and Loch an Sgalain, **7** Loch Frisa, Loch a' Ghael and **13** Caol Lochan. It has a wider axial area than var. *cesatii*, the central area varies from small circular to lanceolate whilst the raphe is oblique for a part of its length at least.

C. cistula (Hemprich & Ehrenb.) Kirchner
Weakly basiphilous and anhalophobous. In a variety of habitats but most commonly found on dripping rock faces especially those at the sides of waterfalls; also present in mesotrophic lochs and streams and in streams and rivers

where rapid water-flow compensates for low base status; rarely in large numbers.
Frequent; widespread but favouring coastal areas.

C. cornutum (Ehrenb.) R. Ross
C. lanceolata (Ehrenb.) Kirchner
In large numbers in the littoral of **3** Loch Assopol; sparse in **3** Loch Poit na h-I, **7** Loch Frisa and **13** Lochan na Guailne Duibhe.

C. cuspidata Kütz.
Sparse in littoral samples from **3** Loch Poit na h-I, **7** Loch Frisa and **13** Caol Lochan.

C. cymbiformis Ag.
Weakly basiphilous and anhalophobous. On irrigated rocks and mosses principally by waterfalls, in coastal seepages, mesotrophic streams and lochs, rarely in large numbers.
Occasional; widespread but favouring coastal areas.

C. delicatula Kütz.
Basiphilous. In shallow running water mainly on flushed rock surfaces and among mosses in similar habitats, sometimes very numerous locally, e.g., **8** Mackinnon's Cave and Creag a' Ghaill; infrequent in the larger lochs.
Frequent in coastal fringe, rare inland.

C. ehrenbergii Kütz.
Sparse in **10** near Loch Fuaron, 240 m, on aquatics in pool; **13** Caol Lochan, on decaying moss, submerged vegetation and bottom mud and Lochan na Guailne Duibhe on submerged grass at margin.

C. gaeumannii Meister [Meister, 1935: 88]
Sparse in littoral samples from lochs; present also in irrigated moss samples from stream.
3 Loch an Sgalain; Loch Assopol. **7** Loch Frisa; Allt Loch a' Ghael, 180 m. **13** Caol Lochan; Lochan na Guailne Duibhe. **16** Loch Uisg.

C. hebridica (Grunow) Cleve
Acidophilous and sub-alpine. Most frequently observed in small numbers in high altitude lochans and bog pools; occasionally in *Sphagnum*.
Occasional; scattered.

C. helvetica Kütz.
Weakly basiphilous and anhalophobous; in a wide variety of habitats but most abundant in eutrophic-mesotrophic waters or where rapid waterflow compensates for low base status. Present in large numbers in the littoral zone of mesotrophic lochs, on coastal cliffs and in waterfalls.
Common; widespread.

C. heteropleura (Ehrenb.) Kütz. [Schmidt, *Atlas*: t. 374, figs. 11, 12; Cleve-Euler, 1955: 149]
Sparse in **3** Loch Poit na h-I, on *Juncus* and *Fontinalis*; **10** near Loch Fuaron, 240 m, on aquatics in pool and **13** Caol Lochan, on *Fontinalis.*
Mull specimens are 80–100 × 22–26 μm, with 9–11 transapical striae in 10 μm and with rostrate to almost apiculate apices in some specimens.

C. hustedtii Krasske
Basiphilous and anhalophobous. Present on wet rock faces on S-facing coastal cliffs. Large numbers at Port Haunn, occasional elsewhere. **5** Port Haunn, coastal seepage. **8** Burgh, on shore rocks, in gelatinous matrix of blue-green algae. **10** Rubha Dubh, in seepage from *Schoenus* tussocks above cliff face and on rock beneath small freshwater waterfall.

C. hybrida Grunow ex Cleve
Sparse in the littoral of **3** Loch Poit na h-I, **7** Loch Frisa and **13** Caol Lochan.

C. laevis Naegeli
Basiphilous. Sparse in littoral samples from **3** Loch Poit na h-I, **7** Loch Frisa, **11** Loch Bà and **13** Caol Lochan; small numbers on wet rock surfaces at side of waterfall, **8** Mackinnon's Cave and in seepage, Creag a' Ghaill.

C. leptoceras (Ehrenb.) Grunow
Basiphilous and possibly anhalophobous. Often large numbers on base-rich flushed rock surfaces beside waterfalls and in coastal areas; occurring sparingly in the littoral of mesotrophic lochs.
Occasional; scattered.

C. microcephala Grunow
Basiphilous and probably anhalophobous and oxyphilous. Preferring moving water in shallow streams and on flushed rock surfaces where it often occurs in large numbers; present also in shallow water at loch margins.
Common; widspread.

C. naviculiformis Auersw. ex Heiberg
pH indifferent, of wide tolerance. Small numbers on submerged aquatics in lochs, on irrigated mosses in streams, on flushed rocks and in mires and amongst *Sphagnum.*
Frequent but never in large numbers; widespread.

C. norvegica Grunow ex A. Schmidt
pH indifferent. A mainly subalpine species occurring principally on irrigated rocks in very small numbers.
Occasional; scattered inland.

C. parva auct. [Hustedt, 1930: fig. 675]
sensu Cleve (1894: 172) pro parte; non *C. parva* (W. Smith sub *Cocconema*) Wolle
Cocconema pachycephala Rabenh.
Weakly acidophilous to basiphilous, anhalophobous. Most numerous in the base-rich

habitats. Often in very large numbers on wet coastal rocks, sometimes mixed in populations of blue-green algae forming blackish or yellowish gelatinous matrices; numerous in the splash zone of waterfalls and on inland flushed rocks; sparse in coastal streams, flushes, wet mud, pools and lochs.
Common; widespread in coastal areas, scattered inland.

C. perpusilla A. Cleve
pH indifferent. In a wide variety of habitats including lochs, irrigated rocks and oligotrophic and mesotrophic mires; seldom in quantity.
Common; widespread.

C. prostrata (Berk.) Grunow
Basiphilous. Most commonly found in the littoral of the larger lochs but also present on flushed rocks especially those in the splash zone of waterfalls; rarely in large numbers.
Occasional; widespread.

C. rabenhorstii R. Ross var. **rabenhorstii**
C. gracilis (Rabenh.) Cleve
Acidophilous. Occurring in a wide variety of habitats but most numerous in slow-moving oligotrophic streams, bog-pools and mires.
Abundant; widespread. One of the most frequently encountered diatoms in this flora.
The var. **girodii** (Hérib.) A. Cleve [Cleve-Euler, 1955: 129] is present in small numbers, often with var. *rabenhorstii*, in oligotrophic lochans and pools; mainly subalpine.
Mull specimens are 50–70 × 8–10 μm with 11–12 transapical striae in 10 μm.

C. scotica W. Smith
C. incerta (Grunow) Cleve
Acidophilous. Dispersed amongst submerged vegetation and on various substrates in oligotrophic mires, pools and slow-moving streams, occasionally in large numbers.
Occasional; scattered inland, more commonly at high altitudes.

C. sinuata W. Gregory
pH indifferent, possibly anhalophobus and oxyphilous. Occurring sparingly on rocks and vegetation in shallow streams, on flushed rocks and in the littoral of lochs.
Frequent; widespread.

C. subaequalis Grunow
C. aequalis W. Smith
pH indifferent. In small numbers in streams and lochs, on flushed rocks and in mires.
Common; widespread.
This is a variable species and the range of form is well illustrated by Mayer (1928: t. 4, figs. 1–10).

C. tumidula Grunow ex A. Schmidt
Basiphilous. Usually sparse, occasionally in large numbers, on irrigated rocks and mosses; also on vegetation in the larger lochs.
Occasional; scattered.

C. turgida (W. Gregory) Cleve
Recorded as isolated specimens from the littoral of mesotrophic lochs and streams, also on irrigated rocks.
Occasional; scattered.

C. ventricosa Ag.
In large numbers in shallow, base-enriched habitats, often coastal; less numerous in mesotrophic streams, rivers and pools.
Common; widespread.

Amphora Ehrenb.

A. coffeiformis (Ag.) Kütz. var. **borealis** (Kütz.) Cleve
In two localities only: **1** Bac Beag, in large numbers on mud ledge of coastal cliffs. **2** Loch Staoneig, sparse on *Myriophyllum.*

A. holsatica Hust.
Numerous in one sample from **1** Sgeir a' Chaisteil, in a brackish pool near centre of island.

A. normanii Rabenh.
Sparse in two gatherings from shallow somewhat base-enriched streams. **2** W Iona, through coastal pasture. **8** plateau above Creag a' Ghaill, 360 m, covering small stones.

A. ovalis (Kütz.) Kütz. var. **ovalis**
Sparse, in the bottom littoral of **2** Loch Staoneig, **3** Loch Poit na h-I and **11** Loch Bà.

A. ovalis var. **libyca** (Ehrenb.) Cleve
Basiphilous and anhalophobous. In the bottom littoral of mesotrophic lochs, in shallow streams, coastal pools, moss flushes and rock surfaces receiving sea spray; never numerous.
Frequent; widespread, especially on the coastal fringe.

A. ovalis var. **pediculus** (Kütz.) Van Heurck
Basiphilous and anhalophobous. Most commonly found as a bottom living form in the wave-splash zone at the margin of the larger mesotrophic lochs; also present in shallow base-enriched streams and in coastal seepages, occasionally very numerous.
Occasional; mainly coastal and the larger lochs.

A. veneta Kütz. var. **veneta**
Basiphilous and anhalophobous. In small numbers in bottom littoral samples from mesotrophic lochs and in seepages on coastal cliffs.
Occasional; scattered.

A. veneta var. **capitata** Haworth [Haworth, 1974: 48]
Basiphilous and anhalophobous. Occasionally in large numbers in bottom littoral samples from mesotrophic lochs; present also in seepages on W- and S-facing cliffs on the **1** Treshnish Isles.
Occasional; scattered.
More frequent than the type variety although the two occasionally occur together.

EPITHEMIACEAE

Denticula Kütz.

D. elegans Kütz.
Basiphilous and anhalophobous. In small numbers in streams, the littoral of mesotrophic lochs and on irrigated rocks; often in large numbers on coastal flushed rocks and seepages.
Occasional; scattered but more frequent in coastal areas.

D. tenuis Kütz.
Weakly acidophilous to basiphilous, anhalophobous. Notably more numerous in the higher-base-status samples. Often in large numbers on dripping rock surfaces, in shallow mesotrophic streams, in rivers and in the littoral of the larger lochs.
Common; widespread in coastal areas, scattered inland.

Epithemia Bréb.

E. adnata (Kütz.) Kütz.
E. zebra (Ehrenb.) Kütz.
Basiphilous and anhalophobous. Mostly in small numbers on vegetation in mesotrophic streams and rivers, in flushes, coastal seepages and on wet rock surfaces; also present in deeper water as an epiphyte on aquatic vegetation in **3** Loch Poit na h-I, Loch Assopol and **13** Caol Lochan.
Frequent; mainly coastal but scattered inland.
Forms referable to var. *porcellus* (Kütz.) Grunow and var. *saxonia* (Kütz.) Grunow are present with the type, intermediates occurring.

E. argus (Ehrenb.) Kütz.
Basiphilous and anhalophobous. In large numbers on coastal rock seepages especially those which have a yellowish or blackish gelatinous matrix of blue-green algae. Present in smaller numbers in **3** Loch Poit na h-I, Loch Assopol, **7** Loch Frisa and **13** Caol Lochan; sparse in inland flushed habitats.
Occasional; mainly coastal.

E. hyndmannii W. Smith
Basiphilous and anhalophobous. Locally in large numbers in coastal areas, sparse elsewhere. **3** Loch Assopol, on rocks in c. 25 cms water. **5** Haunn, sea cliff gully, in water film on wet rock face. **8** Burgh, in black gelatinous matrix of blue-green algae on wet rock face at top of shore. **12** Allt Ardnacross, dripping rock slabs at side of coastal waterfall.

E. reticulata Naegeli ex Kütz.
E. muelleri Fricke
Basiphilous and anhalophobous. Dominant species at Burgh, mostly in small numbers elsewhere.
3 Loch Poit na h-I, on *Pellia* and filamentous algae at margin. **8** Burgh, in black gelatinous matrix of blue-green algae on wet rock face at top of shore. **10** Rubha Dubh, rock beneath small freshwater waterfall. **13** Caol Lochan, N shore, on *Fontinalis*.

E. smithii Carruthers [W. Smith, 1853: 13]
E. proboscidea W. Smith *non* Kütz.
In large numbers in a littoral sample from **13** Caol Lochan. Sparse in gatherings from wet coastal rocks at **8** Burgh and **10** Carsaig.

E. sorex Kütz.
Basiphilous and anhalophobous. Often in large numbers in base-rich habitats i.e. in coastal seepages and some mesotrophic lochs; small numbers on inland flushed rocks and mosses also in rivers and pools.
Frequent in coastal areas; rare inland.

E. turgida (Ehrenb.) Kütz.
Basiphilous and anhalophobous. Usually in small numbers on wet coastal rocks, in mesotrophic streams, flushes and flushed rock surfaces; also in **3** Loch Poit na h-I and Loch Assopol.
Occasional; mainly coastal.

Rhopalodia O. Müller

R. gibba (Ehrenb.) O. Müller
Basiphilous and anhalophobous. Often in large numbers on dripping rock faces beside waterfalls and on coastal cliffs; generally sparse in streams and in the littoral of lochs.
Common; widespread in coastal areas, scattered inland.

R. gibberula (Ehrenb.) O. Müller
Basiphilous and anhalophobous. Often in large numbers on dripping rock surfaces and amongst irrigated mosses; less numerous in rivers, streams, flushes and lochs.
Common; widespread in coastal areas, scattered inland.

NITZSCHIACEAE

Nitzschia Hass.

N. acicularis (Kütz.) W. Smith
Small numbers in plankton and littoral samples taken from **3** Loch Poit na h-I.

N. amphibia Grunow forma **amphibia**
Basiphilous. Sparse in mesotrophic streams and on flushed rocks and mosses; also as a

bottom littoral form in **3** Loch Poit na h-I and **13** Caol Lochan.
Occasional; scattered.

N. amphibia Grunow forma **abbreviata** Manguin
In small numbers in **3** Loch Poit na h-I, in the bottom littoral and on *Juncus* and *Fontinalis*; **4** Ulva, Ormaig, on rocks in fast flowing streams and **7** Allt Loch a' Ghael, c. 180 m, in moss flush.
No intermediate forms between this and f. *amphibia* were seen.

N. angustata (W. Smith) Grunow v. **angustata**
Numerous in two samples only: **3** Loch Poit na h-I, on *Juncus* and *Fontinalis* and **6** Ardmore Bay, amongst filamentous green algae covering stones in swiftly-flowing shallow stream.

N. angustata var. **acuta** Grunow
Weakly basiphilous. In small numbers in most of the larger lochs, also in mesotrophic streams, in moss flushes and on flushed rocks.
Occasional; scattered.

N. communis Rabenh.
Recorded from one sample only where it was the dominant species **2** W Iona, on bottom mud of shallow stream through coastal pasture.

N. denticula Grunow
Basiphilous and possibly anhalophobous. In small numbers in bottom littoral samples from the listed lochs; in large numbers at Burgh and sparse from the mud floor of a shallow mesotrophic stream.
3 Loch Poit na h-I; Loch Assopol. **6** Ardmore Bay, stream. **8** Burgh, in seepage on coastal cliffs. **11** Loch Bà. **13** Caol Lochan; Lochan na Guailne Duibhe.

N. dissipata (Kütz.) Grunow var. **dissipata**
Weakly basiphilous. In small numbers in fast-flowing coastal streams, mesotrophic pools and in the littoral of the larger mesotrophic lochs.
Occasional; coastal and larger lochs.

N. dissipata var. **acula** (Hantzsch) Grunow
N. acula Hantzsch ["acuta" in Hustedt, 1930: 412]
Sparse in **13** Caol Lochan, on decaying moss and submerged debris and in **14** River Forsa, on *Myriophyllum.*

N. epithemoides Grunow
As isolated specimens from **4** Ulva, Ormaig, on roots in fast-flowing shallow stream; **6** SW Ardmore Bay, on wet mud beside stream; **10** Rubha Dubh, on rock beneath small freshwater waterfall and **13** Lochan na Guailne Duibhe, on submerged grass at margin.

N. filiformis (W. Smith) Schütt
Sparse in samples from **4** Ulva, on roots in fast-flowing shallow stream; **6** SW Ardmore Bay, on *Fontinalis* in swiftly flowing stream; and **17** Gleann Rainich, moss in stream.

N. fonticola Grunow
Basiphilous. Small numbers in the littoral and plankton of the larger lochs; also in mesotrophic streams and on flushed coastal rocks.
Occasional; scattered.

N. frustulum (Kütz.) Grunow
Sparse in a moss flush beside **7** Allt Loch a' Ghael at 180 m; numerous in a coastal seepage at **8** Burgh.

N. gracilis Hantzsch
Weakly basiphilous. In small numbers in the littoral of the larger mesotrophic lochs, sparse in pools, flushes and mesotrophic streams.
Occasional; scattered.

N. hantzschiana Rabenh.
In quantity in base-rich habitats, sparse in base-poor conditions. In large numbers on muddy substrates in streams, margins of lochs, flushes and mesotrophic pools; sparse in stony streams, on wet rock faces and the stony margins of lochs and pools.
Frequent; widespread.

N. ignorata Krasske
Weakly acidophilous to weakly basiphilous. Occasionally in large numbers on flushed rocks and mosses, in streams and pools, rarely in lochs.
Occasional; scattered.
The forma **longissima** Manguin is present in small numbers in *Sphagnum* from the margin of **3** Loch an Sgalain and in a bog-pool on **4** S Ulva; also sparingly from a peat bank of **3** Loch Poit na h-I and from a pool beside **10** Loch Fuaron.

N. kuetzingiana Hilse
Sparse in five littoral samples from **3** Loch Poit na h-I and in small numbers from roots in fast-flowing shallow stream at **4** Ormaig.

N. linearis W. Smith
Basiphilous. In large numbers on the muddy substrates of shallow mesotrophic streams and in the littoral of the mesotrophic lochs. Sparse in mesotrophic flushes and on wet rock surfaces at Mackinnon's Cave waterfall.
Frequent; widespread.

N. palea (Kütz.) W. Smith
Tolerant of a wide pH but absent from extremely acid habitats. In streams, pools, flushes, seepages and in the littoral of lochs. Generally sparse but occurring in large numbers in one habitat, a stream near **2** Iona Abbey, which was probably polluted.
Frequent; widespread.

N. paleacea Grunow
Sparse in two localities: **4** Ormaig, from roots in fast-flowing shallow stream and **5** Calgary Bay, in scrapings from rock slab.

N. palustris Hust. [Hustedt, 1934: 396]
Occurring as isolated specimens. **3** Erraid, Port nan Ròn, woodland pool; Loch Poit na h-I, from stone at margin. **4** Ulva, moss covering concrete step by reservoir. **8** Creag a' Ghaill, vertical slab of gully; Tiroran, on liverwort beneath shady waterfall.

N. recta Hantzsch
Basiphilous. Numerous in the littoral of mesotrophic lochs; in small numbers in mesotrophic streams.
2 Iona, Loch Staoneig; irrigated moss in stream. **3** Loch Poit na h-I. **4** Ulva, Ormaig, roots in fast-flowing shallow stream; Cragaig, in similar habitat. **7** Loch Frisa. **13** Caol Lochan.

N. sigmoidea (Nitzsch) W. Smith
Sparse, found only on a wet rock slab in gully at **5** Calgary Bay.

N. sinuata (W. Smith) Grunow
Anhalophobous and oxyphilous, favouring base-rich habitats. In large numbers in coastal seepages especially those which have a gelatinous matrix of blue-green algae. Sparse in soaked mosses, on flushed rocks and in shallow streams.
Occasional; mainly coastal.

N. stagnorum Rabenh.
Sparse in the following samples. **2** N Iona, moss-covered rock in slowly flowing stream. **4** Ulva, Cragaig, on rocks in stream. **6** SW Ardmore Bay, in algal matrix in slowly flowing water; on wet mud. **7** Kilbrenan, on wet rocks at top of waterfall. **13** Allt nan Torc, stones in stream near Lochnameal.

N. sublinearis Hust.
Weakly basiphilous. Numerous on shallowly submerged muddy substrates of streams and loch margins; in small numbers on flushed rocks and in soaked mosses.
Occasional; scattered.

N. subtilis (Kütz.) Grunow
In small numbers in one sample from a shallow peaty pool at **2** Druim Dhùghaill, in brown scum on surface.

N. suchlandtii Hust. [Hustedt, 1943b: 233]
Sparse in three littoral samples from **3** Loch Poit na h-I and in one from **3** Loch Assopol.

N. terrestris (B. Peters.) Hust. [Petersen, 1928: 418]
N. vermicularis var. *terrestris* B. Peters.
In small numbers.
3 Port nan Ròn, in shallow woodland pool; Loch Poit na h-I on deeply shelving bank of peat; Loch Assopol, on rocks at margin. **4** Ulva, Ormaig, on roots in fast-flowing shallow stream. **7** Allt Bun an Easa, 150 m, on irrigated moss on rock in steep wooded gorge. **13** Caol Lochan, on decaying moss and submerged debris at loch margin.

N. thermalis (Ehrenb.) Auersw. var. **minor** Hilse ex Rabenh.
Isolated specimens from **3** Loch Poit na h-I, on *Juncus* and *Fontinalis* and **15** Glen Forsa, on *Sphagnum* in drainage channel through mire.

N. tryblionella Hantzsch var. **debilis** Mayer
In large numbers in brackish habitats but present in smaller numbers elsewhere.
4 Ulva, Ormaig, on rock in fast-flowing shallow stream; Glen Glass, 150 m, on flushed rock. **6** SW Ardmore Bay, on surface of wet mud. **10** Rubha Dubh, on cliffs, with yellow gelatinous matrix of blue-green algae in freshwater seepage.

N. vermicularis (Kütz.) Hantzsch
Sparse in a single sample from **3** Loch Assopol, in plankton tow with slight marginal contamination.

Hantzschia Grunow

H. amphioxys (Ehrenb.) Grunow
Sparse in bottom samples from **3** Erraid, Port nan Ròn, on detritus in woodland pool; **7** Loch Frisa, Lettermore, on stones in sheltered inlet c. 15 cms water and **11** Loch Bà, on small stones in shallow water at edge.

SURIRELLACEAE

Cymatopleura W. Smith

C. elliptica (Bréb. ex Kütz.) W. Smith
Sparse in two bottom littoral samples from **3** Loch Assopol, in one from **3** Loch Poit na h-I and in another from the mud floor of **13** Caol Lochan.

C. librile (Ehrenb.) Pant.
C. solea (Bréb.) W. Smith
Recorded as isolated specimens in five littoral samples from **3** Loch Poit na h-I.

Stenopterobia Bréb. ex Van Heurck

S. intermedia (Lewis) Van Heurck
Calcifuge, halophobous. Sparsely distributed in oligotrophic peaty streams and bog pools; small numbers in *Sphagnum* squeezings.
Frequent; widespread inland.

Surirella Turpin

S. angusta Kütz.
Sparse in two samples only: **5** Calgary Bay, on surface of mud and dead leaves in (permanent?) puddle; and **11** Loch Bà, on small stones in shallow water.

S. biseriata Bréb.
Acidophilous. In high altitude bog-pools and lochans, rarely in large numbers.
8 Creag a' Ghaill, c. 300 m, in flush; Beinn na Srèine, c. 480 m, on stems on *Juncus* in small lochan. **15** Allt an Dubh Choire, c. 456 m, in bogpool; Coire Mòr, c. 600 m, amongst dead *Sphagnum subsecundum* in bog-pool.

S. delicatissima Lewis
Acidophilous to neutral waters. Small numbers in streams, rivers, bog-pools, mires and in the littoral of some high altitude lochans.
Frequent; scattered inland.

S. elegans Ehrenb.
Generally sparse but large numbers present in a plankton sample from Loch Assopol (possibly contaminated by marginal disturbances). **3** Loch Poit na h-I, on *Typha* stem; Loch Assopol, plankton tow; submerged rocks c. 25 cms. **7** Loch Frisa, in plankton tow. **11** Loch Bà, on stones at margin.

S. islandica Østrup [Carter, 1962: 157]
Sparse in a littoral sample from **7** Loch Frisa.

S. linearis W. Smith
pH indifferent but more often encountered in acid habitats. Usually sparse, occasionally in large numbers, in mires, streams, flushes, pools and in the littoral of some lochs.
Common; widespread inland.
The var. **constricta** (Ehrenb.) Grunow occurs with the type.

S. moelleriana Grunow
pH indifferent. Most commonly found in streams and in the littoral of lochs but also present on flushed rocks and mosses; rarely in large numbers.
Frequent; widespread.

S. ovata Kütz. var. **ovata**
Basiphilous and anhalophobous. Small numbers in coastal streams and on wet coastal rocks; also present in the littoral of some mesotrophic lochs.
Occasional; scattered.

S. ovata var. **pinnata** (W. Smith) Hust.
Sparse on wet rocks on the **8** Creag a' Ghaill cliffs and numerous in a (permanent?) puddle at **5** Calgary Bay.

S. robusta Ehrenb. var. **robusta**
Small numbers present in **3** Loch Poit na h-I, on *Typha* stem, in Loch Assopol, submerged rocks at margin and **4** Ulva, Ormaig, on roots in fast-flowing shallow stream.

S. robusta var. **splendida** (Ehrenb.) Van Heurck
Small numbers present in **2** W Iona, stream through coastal pasture and **3** Loch Assopol, plankton tow.

S. spiralis Kütz.
Basiphilous and anhalophobous. Small numbers present in coastal areas; a solitary frustule recorded from Loch Poit na h-I.
1 Bac Mòr, on mud ledge at top of gully. **2** W Iona, stream through coastal pasture. **3** Loch Poit na h-I, in bottom sample. **5** Port Haunn, on rock slab of cliffs; in irrigated moss beside small coastal waterfall; Calgary Bay, on surface of mud and dead leaves in (permanent?) puddle.

S. tenera Gregory
Sparse in littoral samples from **3** Loch Poit na h-I.

Campylodiscus Ehrenb.

C. noricus Ehrenb. var. **hibernicus** (Ehrenb.) Grunow
Sparse in littoral samples from **3** Loch Poit na h-I and Loch Assopol.

References

Amossé, A. (1972). Note sur le genre *Frustulia*. *Revue algol.*, (n.s.) **10**: 306–308.
Carter, J. R. (1960a). Diatom notes: British freshwater forms of the genus *Gomphonema*. *Microscope*, **12**: 255–264.
———. (1960b). Diatom notes: some minute forms of the genus *Navicula*. *Microscope*, **12**: 283–289.
———. (1961). Diatom notes: the genus *Achnanthes* as it occurs in British fresh waters [1–3]. *Microscope*, **12**: 320–326, **13**: 15–22, **13**: 37–45.
———. (1962). Diatom notes: some unusual British forms. *Microscope*, **13**: 156–162.
———. (1963a). Diatom notes: more unusual British forms. *Microscope*, **13**: 231–237.
———. (1963b). Some new diatoms from British waters 1. *J. Quekett microsc. Club*, **29**: 199–203.
———. (1964). Some new diatoms from British waters 2. *J. Quekett microsc. Club*, **29**: 225–228.

CHOLNOKY, B. J. (1959). Neue und seltene Diatomeen aus Afrika IV. Diatomeen aus der Kaap-Provinz. *Öst. bot. Z.*, **106**: 1–69.

CLEVE-EULER, A. (1951–53, 55). Die Diatomeen von Schweden und Finnland. 1 (1951), 2 (1953a), 3 (1953b), 4 (1955), 5 (1952). *K. svenska Vetensk Akad. Handl.* (4 ser.), 2 (1), 4 (1), 4 (5), 5 (4), 3 (3).

DAWSON, P. A. (1974). Observations on diatom species transferred from *Gomphonema* C. A. Agardh to *Gomphoneis* Cleve. *Br. phycol. J.*, **9**: 75–82.

FOGED, N. (1970). The diatomaceous flora in a postglacial kieselguhr deposit in south western Norway. In J. Gerloff and B. J. Cholnoky, Diatomaceae II. *Nova Hedwigia*, **31**: 169–202.

GREGORY, W. (1854). Notice of the new forms and varieties of known forms occurring in the diatomaceous earth of Mull, with remarks on the classification of the Diatomaceae. *Q. Jl microsc. Sci.* **2**: 90–100.

HAWORTH, E. Y. (1974). Some problems of diatom taxonomy in Scottish lake sediments. *Br. phycol. J.*, **9**: 47–55.

HENDEY, N. I. (1964). An introductory account of the smaller algae of British coastal waters V. Bacillariophyceae (Diatoms). *Fishery Invest., Lond.* (4 ser.)

———. (1974). A revised check-list of British marine diatoms. *J. mar. biol. Ass. U.K.*, **54**: 277–300.

HUSTEDT, F. (1927–59, 61–66). Die Kieselalgen Deutschlands, Österreichs und der Schweiz mit Berucksichtigung der übrigen Länden Europas sowie der angrenzenden Meeresgebiete 1. (1927–30), 2. (1931–59), 3. (1961–66). In G. L. Rabenhorst, *Kryptogamen-Flora von Deutschland, Österreich und der Schweiz* 7 (1), 7 (2), 7 (3). Leipzig.

———. (1930). Bacillariophyta (Diatomeae). In A. Pascher, *Die Süsswasserflora Mitteleuropas*, Heft 10. Jena [Second edition. of H. v. Schönfeldt, Bacillariales (Diatomeae), in Die Süsswasserflora Deutschlands, Österreich und der Schweiz, Heft 10 (1913)].

———. (1934). Die Diatomeenflora von Poggenpohls Moor bei Dötlingen in Oldenburg. *Abh. Vortr. Bremer wiss. Ges.*, Jahrg. 8/9: 362–403.

———. (1937). Süsswasserdiatomeen von Island, Spitzbergen und den Färöer-Inseln. *Bot. Arch., Berlin* **38**: 152–207.

———. (1943a). Neue und wenig bekannte Diatomeen. *Ber. dt. bot. Ges.*, **61**: 271–290.

———. (1943b). Die Diatomeenflora einiger Hochgebirgsseen der Landschaft Davos in den Schweizer Alpen. *Int. Revue ges. Hydrobiol. Hydrogr.*, **43**: 225–280.

———. (1950). Die Diatomeenflora norddeutscher Seen mit besonderer Berücksichtigung des holsteinischen Seengebiets. *Arch. Hydrobiol.* **43**: 329–458.

———. (1954). Die Diatomeenflora der Eifelmaare. *Arch. Hydrobiol.*, **48**: 451–496.

———. (1955). Neue und wenig bekannte Diatomeen. 8. *Abh. naturw. Ver. Bremen*, **34**: 47–68.

KNUDSON, B. M. (1952). The diatom genus Tabellaria 1. Taxonomy and morphology. *Ann. Bot.* n.s. **16**: 421–440.

KRASSKE, G. (1932). Beiträge zur Kenntnis der Diatomeenflora der Alpen. *Hedwigia*, **72**: 92–134.

———. (1943). Zur Diatomeenflora Lapplands. *Ber. dt. bot. Ges.*, **61**: 81–88.

LUND, J. W. G. (1945–46). Observations on soil algae 1. The ecology, size and taxonomy: of British soil diatoms, part 1 (1945), part 2 (1946). *New Phytol.*, **44**: 196–219, **45**: 56–110.

MAYER, A. (1928). Die bayerischen Gomphonemen. *Denkschr. K. bayer. bot. Ges. Regensb.* **17**: 83–128.

———. (1940). Die Diatomeenflora von Erlangen. *Denkschr. regensb. bot. Ges.*, **21**: 113–224.

———. (1947). Die bayerischen *Encyonema* – und *Cymbella* – Arten mit ihren Formen. *Ber. bayer. bot. Ges.*, **27**: 226–281.

MEISTER, F. (1935). Seltene und neue Kieselalgen. *Ber. schweiz. bot. Ges.*, **44**: 87–108.

ØSTRUP, E. (1910). *Danske diatomeer.* Copenhagen.

PATRICK, R. (1958). Some nomenclatural problems and a new species and a new variety in the genus *Eunotia* (Bacillariophyceae). *Notul. Nat.*, no. **312**: 1–15.

PATRICK, R. & REIMER, C. W. (1966). The diatoms of the United States, Vo . 1. *Monogr. Acad. nat. Sci. Philad.*, **13**.

PETERSEN, J. B. (1928). The aërial algae of Iceland. In L. Kolderup Rosenvinge & E. Warming, *The botany of Iceland*, vol. 2, part 2, no. 8: 325–447. Copenhagen & London.

ROSS, R. (1947). Freshwater Diatomeae (Bacillariophyta) in N. Polunin, Botany of the Canadian Eastern Arctic, Part II Thallophyta and Bryophyta. *Bull. natn. Mus. Can.*, 97 (Biol. Series 26): 178–233.

SCHMIDT, A. et al. (1874–1959). *Atlas der Diatomaceen-Kunde*. Leipzig. Index by G. D. Hanna in *Bibliotheca Phycologica*, 10 (1969).

SCHOEMAN, F. R. (1973). *A systematical and ecological study of the diatom flora of Lesotho with special reference to the water quality*. Pretoria.

SMITH, W. (1853, 1856). *A synopsis of the British Diatomaceae*, vols. I & II. London.

VAN HEURCK, H. (1880–85). *Synopsis des diatomées de Belgique*, Atlas tt. 1–30 (1880), 31–77, 53-bis (1881), 78–103 (1882), 104–132, 22-bis, 82-bis, 83-bis, 83-ter, 95-bis (1883), Table alphabétique des noms (1884), Texte (1885). Anvers.

WERFF A. VAN DEN & HULS, H. (1957). *Diatomeeënflora van Nederland* I. The Hague.

18. Marine diatoms

T. B. B. Paddock

Introduction

Diatoms were collected from a range of habitats within the marine littoral from sandy and rocky foreshores under direct tidal governance, to mildly saline drainage channels, pans and pools which are only occasionally inundated by the tide, or which derive some salinity from heavy spray or from the salt-water table. Some of these may be described most conveniently as brackish, i.e., a fresh-water community influenced by salt-water or the converse. Plankton net-samples were not studied; neither were benthic and periphytic (tychoplankton) samples collected from the infra-littoral.

Sampling and collecting methods

Gatherings were made in April 1969 (94 samples) and October 1970 (44 samples). The method of sampling was not quantitive. Scrapings were taken from various substrates; wood, rock, mud, plant stems etc., which were coloured with a film of diatoms or where diatoms were thought likely to occur in significant numbers. The enumeration and recording of diatoms which occur only rarely is much neglected because it is time consuming, and because the rarer forms are masked by the common. Single cells are recorded here: they may be fortuitous specimens of "drift" and from a great distance; or they may represent the extreme fringe of a larger population centered elsewhere, e.g., in the sublittoral; or be the sole off-peak survivors of a species more abundant at other times of the year. While there is no reason to suppose that the gatherings made are typical of the diatom flora throughout the year, the relatively large number of samples examined does suggest that, subject to the limitations of the analysis, a record of the principal components of the flora has been obtained for the two periods in which the gatherings were made. The preponderence of pennate diatoms over centric diatoms is only in part due to the sampling methods used, and it is worth noting that dominant, or even noticeable growths of centric diatoms were lacking. A total of 212 species and varieties have been identified.

Preparation and identification of specimens

Portions of the sample were washed with distilled water, digested with concentrated hydrochloric acid and then re-washed in distilled water. This removes soluble salts and calcium carbonate. The material was then boiled with concentrated sulphuric acid and crystals of potassium chlorate carefully added until the mixture cleared, i.e., until all organic matter destroyed. (This reaction is dangerous and should be carried out in a fume cupboard and great care must be taken to add only small amounts of potassium chlorate at any one time.) The sample was finally centrifuged and washed several times with distilled water. Drops of the cleaned material were air dried onto cover-glasses and permanent slides prepared using the thermoplastic resin GBI 163 (G.B.I. (Labs) Ltd., Heaton Mills, Denton, Manchester), which combines a refractive index high enough for general use with ease of preparation and good long-term stability.

The species concepts of Hustedt (1927–66) are the main source of comparison in the following account. Other sources are Hendey (1964) indicated by an asterisk (*) after the species name, and further authors as stated in parentheses.

Arrangement and nomenclature

The arrangement of families follows that of Hendey (1974a); the genera and species are alphabetical. The frequency given at the beginning of each entry is that of the average relative density of the species in the prepared slide, estimated by eye. A high proportion of the species found were recorded only rarely or as single cells, and most were present in low numbers (i.e., occasional). Because of the subjective sampling and relatively small number of sites sampled, the listing of geographical localities would be misleading. The number of samples in which the species has been recorded is given at the end of each entry to indicate the frequency that the species was encountered in the Survey.

Index to genera

BACILLARIOPHYTA

BACILLARIOPHYCEAE

BACILLARIALES

MELOSIRACEAE

Melosira Ag.

M. juergensii Ag.
Abundant in two spring gatherings; from pools in the upper littoral.
(3)

M. moniliformis (O. F. Müller) Ag. *
Frequent; in spring gathering from an upper tidal pool dominated by *Okedenia.*
(1)

M. nummuloides Ag.
Occasional; in saltmarsh and upper littoral pools as epiphytes, and on estuarine mud.
(17)

M. sulcata (Ehrenb.) Kütz.
Occasional; from intertidal seaweed washing.
(2)

M. varians Ag.
A single specimen in a brackish saltmarsh pool.
(1)

THALASSIOSIRACEAE

Cyclotella Kütz.

C. meneghiniana Kütz.
Rare; from supralittoral stones.
(1)

COSCINODISCACEAE

Coscinodiscus Ehrenb.

C. nitidus W. Gregory *
A planktonic species found as single cells in two shore gatherings.
(3)

ACTINODISCACEAE

Actinoptychus Ehrenb.

A. senarius (Ehrenb.) Ehrenb.
Occasional; on intertidal mud.
(1)

EUPODISCACEAE

Actinocyclus Ehrenb.

A. octonarius Ehrenb. *
Occasional; as epiphytes on littoral algae.
(3)

Auliscus Ehrenb.

A. sculptus (W. Smith) Ralfs ex Pritch.
Single specimen from bottom of a littoral pool.

BIDDULPHIACEAE

Biddulphia Gray

B. alternans (J. W. Bailey) Van Heurck *
Rare; very small specimens from littoral algal washings.
(1)

B. antediluviana (Ehrenb.) Van Heurck *
Isolated specimens in littoral mud and gravel
(1)

B. aurita (Lyngb.) Bréb.
Rare; in littoral mud, and amongst epiphytes.
(3)

B. pulchella S. F. Gray
Occasional; in littoral mud and gravel.
(1)

Cerataulus Ehrenb.

C. turgidus (Ehrenb.) Ehrenb.
A single specimen, in coarse intertidal sand.

DIATOMACEAE

Fragilaria Lyngb.

F. pinnata Ehrenb. *
Frequent; on stones in saltmarsh and estuaries.
(3)

F. striatula Lyngb. *
Abundant; on intertidal stones, and algae.
(4)

Licmophora Ag.

L. ehrenbergii (Kütz.) Grunow *
Scraped from a metal plate on the foreshore.
(1)

L. flabellata (Grev.) Ag. *
Frequent; in washings from intertidal algae.
(2)

L. gracilis (Ehrenb.) Grunow *
Occasional; epiphytic on intertidal algae.
(4)

L. hyalina (Kütz) Grunow
Occasional; in washings of intertidal algae.
(3)

L. juergensii Ag. *
Occasional; in a washing from intertidal algae.
(1)

L. lyngbyei (Kütz.) Grunow ex Van Heurck *
Occasional; in washings of intertidal algae.
(2)

L. nubecula (Kütz.) Grunow ex Schneider
Occasional; in washings of intertidal algae.
(1)

L. paradoxa (Lyngb.) Ag.
Occasional; from algae and stones in intertidal zone.
(5)

Striatella Ag.

S. unipunctata (Lyngb.) Ag. *
A single cell in washings from intertidal algae.
(1)

Dimeregramma Ralfs ex Pritch.

D. minor (W. Gregory) Ralfs var. **nana** (W. Gregory) Van Heurck [Van Heurck 1896]
Occasional; in coarse littoral sand.
(1)

Plagiogramma Grev.

P. staurophorum (W. Gregory) Heiberg
Rare; on coarse intertidal sand.
(1)

Grammatophora Ehrenb.

G. angulosa Ehrenb.
Occasional; from washings of intertidal algae, and scraped from jetty.
(3)

G. marina (Lyngb.) Kütz.
Always rare; among epiphytes of intertidal algae and on intertidal sand and rock.
(11)

G. oceanica (Ehrenb.) Grunow
Rare; epilithic and epiphytic in intertidal zone; on stones in saltmarsh and upper littoral.
(9)

G. serpentina (Ralfs) Ehrenb.
Occasional; among epiphytes of littoral algae.
(7)

Rhabdonema Kütz.

R. arcuatum (Lyngb.) Kütz. *
Occasional; among saltmarsh pool epiphytes and washings from intertidal algae.
(4)

R. minutum Kütz. *
Rare; from stones in saltmarsh and from surface of intertidal mud.
(2)

ACHNANTHACEAE

Achnanthes Bory

A. brevipes Ag. *
Occasional to frequent; on stones, mud, and as epiphytes in littoral.
Widespread around the island in low salinity conditions.
(18)

A. delicatula (Kütz.) Grunow
Occasional; on saltmarsh and estuarine stones and mud and in habitats subject to freshwater seepage.
(8)

A. flexella (Kütz.) Brun
Occasional; this is a freshwater species from estuarine and saltmarsh habitats.
(5)

A. hauckiana Grunow
Frequent; among epiphytes in upper littoral and saltmarsh pools.
(13)

A. lanceolata (Bréb.) Grunow
Occasional; from littoral, where freshwater flushing present, supralittoral stones and mud and on saltmarshes.
(9)

A. longipes Ag. *
Occasional; and mainly in the October gatherings. That this diatom was never encountered in quantity probably reflects the less than fully saline conditions which were sampled.
(8)

A. microcephala (Kütz.) Grunow
Occasional; from mud and stones in saltmarsh pools and from the surface of intertidal mud.
(5)

A. pseudogroenlandica Hendey *
Occasional; in washings from seaweeds (1). Only hitherto published record for Britain is the type locality in Pembrokeshire.

A. saxonica Krasske
Common; from the top of the littoral zone in saltmarshes, on stones, mud, in pools and among epiphytes.
(21)

A. subsalsoides Hust.
Occasional; from shallow pools and near freshwater outlets.
(5)

A. taeniata Grunow
Occasional; from stones in estuary.
(1)

Cocconeis Ehrenb.

C. diminuta Pantocsek
Rare; among epiphytes of littoral and saltmarsh pools and in coarse littoral sand.
(3)

C. dirupta W. Gregory
Common in one gathering but otherwise sporadic and in low numbers; intertidal on stones, mud and algae and among saltmarsh pool epiphytes.
(8)

C. japonica (Hust.) A. Cleve [Cleve-Euler 1953]
Rare; on estuarine stones
(1)

C. molesta Kütz. var. **crucifera** Grunow ex Van Heurck
Occasional; on stones on tidal shore.
(1)

C. pediculus Ehrenb.
Among epiphytes in high level pool.
(1)

C. peltoides Hust.
Occasional; on littoral stones, sand, mud and algae, and also in saltmarsh pools on mud and as epiphytes.
(11)

C. pinnata W. Gregory [Peragallo 1897–1908]
A single valve among epiphytes on *Fucus* at Dervaig. Does not appear to have been refound in Britain since Gregory's original report, from Lamlash Bay, Arran.

C. placentula Ehrenb.
Occasional; generally distributed through the intertidal zone and saltmarshes.
(14)

C. pseudomarginata W. Gregory
Occasional; on estuarine and intertidal sediments and in washings from intertidal algae.
(6)

C. scutellum Ehrenb.
Common and sometimes abundant; from a wide range of coastal habitats.
(64)

C. speciosa W. Gregory *
On littoral stones and mud, and in algal washings.
(8)

Rhoicosphenia Grunow

R. curvata (Kütz.) Grunow ex Rabenh.
Common; epiphytic and epilithic in saltmarsh and estuary.
(16)

R. marina (W. Smith) A. Schmidt
Common; on intertidal and estuarine mud and

stones; in washings of intertidal algae and in saltmarsh and upper littoral pools.
(28)

NAVICULACEAE

Navicula Bory

N. aboensis (Cleve) Hust.
Found only in one gathering from a saltmarsh pool.

N. abrupta (W. Gregory) Donkin
Rare; on intertidal and estuarine mud and sand, among algal epiphytes.
(9)

N. arenaria Donkin *
Occasional; from coarse tidal sand and among epiphytes of littoral algae.
(3)

N. atlantica A. Schmidt *
Occasional; on tidal mud and among algal epiphytes.
(2)

N. avenacea Bréb. ex Grunow [Brockmann 1950]
Occasional; in upper littoral and saltmarsh pools, among sediments and epiphytes and on stones.
(16)

N. calida Hendey *
Common in April; on stones in brackish saltmarsh pool.
(1)

N. cari Ehrenb.
Occasional; in supralittoral pools.
(2)

N. carinifera Grunow ex A. Schmidt *
Rare; in a gathering of upper littoral epiphytes. First recorded for Britain from Anglesey by Hendey (1964) who says that it is usually considered a warm water species.

N. cincta (Ehrenb.) Ralfs
Frequent; on estuarine and saltmarsh mud, in association with freshwater runoff.
(6)

N. clementis Grunow *
As solitary specimens; among littoral epiphytes and on sediments, in rock and saltmarsh pools. A species of wide ecological tolerance also found well inland in Mull.
(21)

N. cluthensis W. Gregory forma **rostrata** (Simonsen) Hust.
Occasional; in a gathering from high level marsh pool.
(1)

N. comoides (Ag.) Perag.
Occasional; in gatherings of algal washings and from estuarine stones.
(2)

N. creuzburgensis Krasske
Occasional; in saltmarsh pools, and channels on estuarine mud.
(4)

N. crucicula (W. Smith) Donkin
Occasional to frequent; in saltmarshes, on stones and in pools and channels.
(7)

N. crucigera (W. Smith) Cleve
Rare; among littoral epiphytes.
(2)

N. crucigeroides Hust.
A single specimen on stone in a saltmarsh pool.

N. cryptolyra Brockm.
A single specimen among epiphytes on *Fucus*.

N. crystallina Hust.
Rare; on coarse tidal sand from a single gathering from Loch Beg [17/531293].
A new British record.

N. digitoradiata (W. Gregory) Ralfs *
Frequent, sometimes common; variable in form and tolerant of a wide range of salinity.
(28)

N. directa (W. Smith) Ralfs *
An isolated specimen among epiphytes of littoral algae.

N. distans (W. Smith) A. Schmidt. *
A single cell, among course littoral sand.

N. dithmarsica König ex Hust.
Occasional; among epiphytes of *Fucus*, in low salinity saltmarsh pool, on intertidal sand and on stones in estuaries (6). A new British record [Erraid 17/200308; Loch Don 17/329727; Dervaig 17/518429].

N. elegans W. Smith *
Occasional and sometimes frequent; in salt-marsh pools, on stones and mud, and among epiphytes of intertidal algae. (11) Large and characteristic brackish water species.

N. finmarchica (Cleve & Grunow) Cleve *
Occasional; in coarse littoral sand.
(1)

N. flanatica Grunow [Brockmann 1950]
Occasional; among intertidal and saltmarsh pool epiphytes.
(2)

N. florinae Moller
Rare; in a gathering of coarse littoral sand.
(1)

N. forcipata Grev. *
Common; among intertidal and saltmarsh epiphytes and on sediments and estuarine stones.
(23)

N. gregaria Donkin
Often frequent and widespread in various littoral habitats.

N. halophila (Grunow) Cleve
Common; among a gathering of epiphytes of littoral algae.
(1)

N. hennedyi W. Smith *
Occasional; among epiphytes of intertidal algae.
(3)

N. humerosa Bréb.
Occasional; in saltmarsh and upper littoral pools, among epiphytes and on intertidal sand.
(7)

N. hyalina Donkin *
Occasional; in a gathering of saltmarsh epiphytes.
(1)

N. hyalinula De Toni [Hendey 1974b]
Occasional; in coarse littoral sand and amongst epiphytes in saltmarsh pool.
(3)

N. integra (W. Smith) Ralfs
Frequent; on surface of intertidal sand.
(1)

N. jamalinensis Cleve [Schmidt, t. 394, figs 15–18]
Frequent in one gathering from intertidal sand. Cleve-Euler (1953) records this as an Arctic marine form.

N. jarnefeltii Hust.
Occasional; on stones in upper saltmarsh pools and in estuarine conditions.
(6)

N. lapidosa Krasske
Rare; amongst epiphytes in saltmarsh pool.
(1)

N. latissima W. Gregory *
Occasional, sometimes common; on intertidal mud and sand, on soil and stones in saltmarsh and among littoral epiphytes.
(7)

N. lundstroemii Cleve *
Rare; on stones in a saltwater channel.
(1)

N. marina Ralfs *
Occasional; in saltmarsh on soil and stones and in pools and on stones in estuary.
(6)

N. marnierii Manguin
Rare; on estuarine mud and stones. The first British record [Loch Don 17/329727] of this species originally described from the Antarctic.
(2)

N. meniscus Schum. *
Occasional; in saltmarshes on mud and stones and in runnels and high-level pools.
(8)

N. monilifera Cleve
Occasional; on estuarine stones and in a high-level pool, among epiphytes.
(2)

N. mutica Kütz.
Often frequent; intertidal among epiphytes in saltmarsh pools and in estuaries on stones and mud. Some range of variation seen but most records can be attributed to forma **mutica** or forma **cohnii** (Hilse) Grunow.
(15)

N. palpebralis Breb. ex W. Smith *
Occasional to frequent; on stones and sediments, in saltmarsh pools and channels and on intertidal mud and sand.
(14)

N. peregrina (Ehrenb.) Kütz. [Patrick & Reimer 1966]
Frequent; in saltmarsh pools and channels, on mud, stones and intertidal sand and in higher rock pools.
(10)

N. plicata Donkin
Single valves in two gatherings; amongst saltmarsh epiphytes and on stones.

N. pseudoscutiformis Hust.
Occasional; in upper saltmarsh on stones and in runnels; on intertidal surfaces flushed with freshwater and amongst epiphytes of littoral algae.
(6)

N. pusilla W. Smith
Rare; in saltmarsh and upper littoral pools, on stones and among epiphytes.
(12)

N. pygmaea Kütz.
Occasional; on estuarine and saltmarsh mud and stones and on sea-spray-blown rock faces.
(8)

N. ramosissima (Ag.) Cleve *
Occasional; epiphytic and epilithic in littoral pools. In autumn collections.
(4)

N. rhombica W. Gregory
Occasional; in saltmarsh pools among epiphytes and on stones, among epiphytes of intertidal algae and on intertidal mud and sand.
(11)

N. rhombicula Hust.
Isolated specimens in one gathering from coarse intertidal sand.

N. rhyncocephala Kütz.
In saltmarsh and upper littoral pools, on estuarine and intertidal sand.
(4)

N. salarinarum Grunow
Common; on stones in saltmarsh pool and among washings of intertidal algae.
(2)

N. schonkenii Hust.
Isolated specimen from intertidal sand at Erraid (NGR 17/200308). Not recorded in Hendey (1974b), though Hustedt (1959) observes that O'Meara (1867) has recorded this species from Ireland.

N. scutelloides W. Smith
Rare; in a slightly saline saltmarsh pool.
(1)

N. subforcipata Hust.
Occasional; on stones in estuary, and on intertidal sand and mud. (4) A new British record [Loch Don 17/329727; Loch Beg 17/53/293; Laggan Bay 17/453402] ; previously known only from the coast of Norway.

N. subinflatoides Hust.
Isolated specimen; in tidal sand.

N. tuscula (Ehrenb.) Van Heurck *
Occasional; among epiphytes of *Fucus*.
(1)

N. variostriata Krasske
Sometimes frequent; in less saline saltmarsh pools and channels and on seacliff seepages.
(10)

N. viridula (Kütz.) Kütz. emend. Van Heurck var. **viridula** [Patrick & Reimer 1966]
Sometimes frequent; in saltmarsh pools and channels and on estuarine mud.
(9)

Anomoeoneis Pfitzer

A. vitrea (Grunow) R. Ross
Occasional; in the upper littoral in fresh-water seepages.
(4)

Stauroneis Ehrenb.

S. africana Cleve *
Occasional; among epiphytes of intertidal algae and on intertidal sand.
(2)

S. amphioxys W. Gregory var. **amphioxys** *
Occasional; on saltmarsh mud.
(1)

S. amphioxys var. **obtusa** Hendey *
Occasional; from muddy bottoms and among epiphytes of saltmarsh pools.
(9)

S. producta Grunow
Occasional; in water of low salinity from saltmarsh pools and runnels and on littoral surfaces affected by freshwater outflow.
(7)

S. rossii Hendey *
Rare; amongst epiphytes from saltmarsh pool.
(1)

S. salina W. Smith *
Occasional; amongst epiphytes of littoral algae and on intertidal mud.
(2)

S. spicula Hickie
Occasional; on mud in a saltmarsh pool and amongst epiphytes of intertidal algae.

Diploneis Ehrenb.

D. bombus (Ehrenb.) Cleve *
Occasional; on littoral mud, in rock and saltmarsh pools, and in spray-blown moss.
(8)

D. chersonensis (Grunow) Cleve *
Occasional; among algal epiphytes.
(1)

D. crabro Ehrenb. var. **crabro** *
Occasional; amongst epiphytes of intertidal and saltmarsh pool algae, and on intertidal sand.
(4)

D. crabro var. **perpusilla** Cleve [Peragallo 1897–1908]
Occasional among washings from intertidal algae.
(1)

D. didyma (Ehrenb.) Cleve *
Occasional; in the intertidal and in saltmarshes on stones, mud, sand and algae.
(18)

D. fusca (W. Gregory) Cleve *
Occasional; among epiphytes of intertidal algae and in a saltmarsh pool.
(3)

D. interrupta (Kütz.) Cleve
Occasional; on saltmarsh, on mud and epiphytic.
(2)

D. notablis (Grev.) Cleve [Peragallo 1897–1908]
Occasional; on saltmarsh mud and amongst epiphytes.
(4)

D. littoralis (Donkin) Cleve
Rare; on spray-blown rockface and among coarse littoral sand.
(2)

D. ovalis (Hilse) **Cleve**
Occasional; on stones in saltmarsh pools, on intertidal stones and sand.
(5)

D. smithii (Breb.) Cleve
Frequent; on various intertidal substrates.
(20)

D. stroemii Hust. [Brockmann 1950]
Occasional; on intertidal sand and mud, and in saltmarshes as an epiphyte and on sediments.
(5)

D. suborbicularis (W. Gregory) Cleve
Frequent; on intertidal sand and in a saltmarsh pool.
(2)

D. vacillans (A. Schmidt) Cleve
Rare; on estuarine mud and stone.
(3)

Caloneis Cleve

C. amphisboena (Bory) Cleve
In association with freshwater outflow onto beach.
(2)

C. brevis (W. Gregory) Cleve *
Rare; among water surface scum and among epiphytes in a saltmarsh pool.
(2)

C. perlepida (Grunow) A. Berg [Berg 1951]
Rare; from littoral algal washings and estuarine mud.
(3)

C. subsalina (Donkin) Hendey *
Occasional, sometimes frequent; in saltmarsh and upper littoral pools and amongst epiphytes in estuaries and in similar habitats of lower salinity.
(14)

C. westii (W. Smith) Hendey *
Solitary specimens occasional; On saltmarsh epiphytes and sediments and estuarine mud.
(12)

Scoliopleura Grunow

S. tumida (Breb. ex Kütz.) Rabenh. *
Occasional in saltmarsh mud.
(1)

Scoliotropis Cleve

S. latestriata (Breb. ex Kütz.) Cleve *
Occasional; on intertidal sand and mud.
(3)

Trachyneis Cleve

T. aspera (Ehrenb.) Cleve *
Solitary; on littoral and estuarine stones, mud, and among epiphytes of intertidal algae.
(3)

Mastogloia Thwaites ex W. Smith

M. binotata (Grunow) Cleve
Represented by a solitary specimen from littoral mud.

M. elliptica (Ag.) Cleve
Sometimes frequent; on mud and stones in estuaries and on plants, stones and in saltmarshes.
(24)

M. exigua Lewis
Frequent; in a sample of epiphytes from a littoral pool; also from tidal mud.
(2)

M. pumila (Grunow) Cleve
Occasional; among epiphytes from saltmarsh pools.
(3)

M. smithii Thwaites ex W. Smith
Sometimes frequent; amongst epiphytes and epiliths of the littoral saltmarsh and estuaries.
(18)

Frustulia Rabenh.

F. rhomboides (Ehrenb.) De Toni *
Frequent; in saltmarsh and upper littoral and estuarine locations.
(21)

Amphipleura Kütz.

A. rutilans (Trentep.) Cleve *
Common; throughout the littoral and estuarine habitats.
(24)

Brebissonia Grunow

B. boeckii (Ehrenb.) Grunow *
Often very abundant; in saltmarsh, estuarine and upper littoral habitats.
(29)

Pleurosigma W. Smith

P. aestuarii (Breb. ex Kütz.) W. Smith *
Occasional; on intertidal and estuarine mud and stones and among epiphytes of intertidal algae.
(8)

P. angulatum (Quekett) W. Smith *
Occasional; on saltmarsh, estuarine and intertidal mud.
(5)

P. elongatum W. Smith [Smith 1853]
Occasional; from mud and stones in saltmarsh and in foam and scum.
(2)

P. formosum W. Smith [Smith 1853]
Occasional; from surface of intertidal mud.
(1)

P. intermedium W. Smith [Smith 1853]
Rare; amongst epiphytes from a rockpool.
(1)

P. strigosum W. Smith *
Occasional; among littoral and saltmarsh pool epiphytes, on intertidal sand and on spray-beaten rockface.
(4)

Gyrosigma Hass.

G. acuminatum (Kütz.) Rabenh.
Occasional; on stones in estuary.
(1)

G. balticum (Ehrenb.) Cleve *
Very occasional; on tidal mud.
(2)

G. fasciola (Ehrenb.) Cleve var. **fasciola** *
Occasional; on tidal mud.
(3)

G. fasciola var. **sulcata** (Grunow) Cleve *
Occasional; on tidal mud.
(1)

G. hippocampus (Ehrenb.) Hass. *
Rare; on saltmarsh mud.
(1)

G. littorale (W. Smith) Cleve
Rare; on estuarine stones.
(1)

G. spenceri (Quek.) Griff. & Henfr.
Occasional; on tidal mud.
(1)

G. wansbeckii (Donkin) Cleve *
Occasional; on mud and among epiphytes in littoral zone.
(2)

Tropidoneis Cleve

T. lepidoptera W. Gregory var. **minor** Cleve [Peragallo 1897–1908]
Occasional; on intertidal sand and among epiphytes of intertidal algae.
(3)

T. pusilla (W. Gregory) Cleve *
Occasional; on coarse littoral sand.
(2)

T. vitrea (W. Smith) Cleve *
Occasional; on coarse littoral sand and on estuarine mud.
(2)

Amphiprora Ehrenb.

A. alata (Ehrenb.) Kütz. *
Occasional; from estuarine mud and saltmarsh.
(2)

A. decussata Grunow [Van Heurck 1880–85]
Occasional; in estuarine mud.
(1)

AURICULACEAE

Auricula Castrac.

A. complexa (W. Gregory) Cleve *
Occasional; in littoral mud and gravel and amongst algal epiphytes.
(3)

GOMPHONEMACEAE

Gomphonema Ag.

G. kamtschaticum Grunow [Grunow 1878]
Rare; in washings from intertidal algae.
(1)

CYMBELLACEAE

Amphora Ehrenb.

A. arenicola (Grunow) Cleve *
Occasional; from littoral epiphytes and among coarse sand.
(2)

A. decussata Grunow *
Occasional; among seaweed washings.
(1)

A. exigua W. Gregory [Brockmann 1950]
Occasional; on coarse littoral sand, estuarine stones and among epiphytes in a saltmarsh pool.
(3)

A. ostrearia Bréb. var. **lineata** Cleve *
Occasional; in the littoral on various sediments and among algal washings.
(4)

A. pseudohyalina Simonsen
A single specimen found among epiphytes from high-level pool. Rather broader in valve view than those which Simonsen (1960) figures.
(1)

A. tenerrima Aleem & Hust.
Rare; from littoral algal washings.
(1)

Okedenia Eulens. ex De Toni

O. inflexa (Breb. ex Kütz.) De Toni *
Abundant; on *Corallina* in a rock pool.
(1)

NITZSCHIACEAE

Nitzschia Hass.

N. acuminata (W. Smith) Grunow *
Sometimes frequent; on intertidal and estuarine mud, sand and stones.
(5)

N. angularis W. Smith *
Rare; from a rock face within reach of sea-spray.
(1)

N. apiculata (W. Gregory) Grunow *
Common; in saltmarsh and intertidal habitats on sand, mud, stones and as epiphytes.
(20)

N. armoricana Perag.
Frequent; from saltmarshes on mud and stones and in channels.
(4)

N. bilobata W. Smith *
Occasional; in saltmarsh and estuarine mud and among intertidal and rock pool epiphytes.
(5)

N. clausii Hantzsch
Common; from stones in brackish pool.
(1)

N. debilis (Arnott) Grunow [Van Heurck 1880–1885]
Occasional to frequent; in estuarine mud, with epiphytes on soil and stones in saltmarsh pools.
(10)

N. frustulum (Kütz.) Grunow
Common; among epiphytes and on intertidal sand.
(3)

N. hungarica Grunow *
Occasional; on mud in stream outflow.
(1)

N. levidensis (W. Smith) Van Huerck *
Occasional; on estuarine mud, saltmarsh stones and among epiphytes of littoral algae.
(3)

N. littoralis Grunow [Cleve-Euler 1952]
Rare; from tidal mud surface.
(2 adjacent gatherings)

N. navicularis (Breb. ex Kütz.) Grunow *
Occasional; from a rockpool and among epiphytes of intertidal algae.
(2)

N. palea (Kütz.) W. Smith
Common; near a sewage outfall onto a sandy beach and in a saltmarsh pool.
(2)

N. punctata (W. Smith) Grunow *
Rare; in saltmarshes on stones and among epiphytes in pools, in estuarine conditions on stones and mud and on intertidal sand and stones.
(19)

N. sigma (Kütz.) W. Smith *
Occasional, sometimes frequent; among epiphytes of intertidal algae and on estuarine mud and stones.
(9)

Bacillaria Gmelin

B. paxillifer (O. F. Müller) Hendey *
Occasional; from littoral sand and stones and among algal epiphytes.
(6)

Hantzschia Grunow

H. virgata (Roper) Grunow *
Occasional; from a sandy beach
(1)

SURIRELLACEAE

Surirella Turpin

S. armoricana Perag. *
Occasional; on surface of tidal mud.
(1)

S. comis A. Schmidt *
Occasional; among epiphytes of intertidal algae and on intertidal mud and gravel.
(2)

S. fastuosa (Ehrenb.) Kütz. *
Rare; on littoral mud and among epiphytes of intertidal algae.
(8)

S. gemma (Ehrenb.) Kütz. *
Frequent; among epiphytes of intertidal algae and on surface of intertidal mud.
(8)

S. hispida Ross & Abdin [Ross & Abdin 1949]
Common; from intertidal mud and sand and among algal washings.
(5)

S. lata W. Smith [Smith 1853]
Rare; among algal washings and in shore pool. The name is given without prejudice to the relationship of *S. lata* with *S. hybrida* Grun. and similar entitites.
(3)

S. ovalis Bréb. *
Common, sometimes abundant; on saltmarsh, estuarine and intertidal mud and stones and among algal washings.
(25)

S. ovata Kütz. *
Common; on littoral, superlittoral and salt-

marsh surfaces. Very closely related to, and often confused with, the previous species.
(36)

S. smithii Ralfs *
Frequent; from low salinity habitats on saltmarsh and estuarine muds and stones.
(23)

Campylodiscus Ehrenb.

C. echeneis Ehrenb.
Single specimen from adjacent spring samples of estuarine mud.
(2)

C. fastuosus Ehrenb.
Isolated specimens; among washings from littoral algae, amongst saltmarsh epiphytes and from littoral mud and gravel; in Spring gatherings.
(4)

C. ralfsii W. Smith
Occasional; from washings from littoral algae and in an intertidal rock-pool.
(2)

References

BERG, A. (1952). Eine Diatomeengemeinschaft an der schwedischen Ostküste: eine ökologische Studie. *Ark. Bot.*, (2 ser.) **2**: 1–39.

BROCKMANN, C. (1950). Die Watt-Diatomeen der schleswig-holsteinischen Westküste. *Abh. senckenberg. naturf. Ges.* **478**: 1–26.

CLEVE-EULER, A. (1951–53, 55). Die Diatomeen von Schweden und Finnland 1 (1951), 2 (1953), 3 (1953), 4 (1955), 5 (1952). *K. svenska VetenskAkad. Handl.* (4 ser.), 2 (1), 4 (1), 4 (5), 5 (4), 3 (3).

GRUNOW, A. (1878). Algen und Diatomaceen aus dem Kaspischen Meere. In O. Schneider, *Naturwiss. Beitr. z. Kennt. d. Kaukasusländer*. Dresden.

HENDEY, N. I. (1964). An introductory account of the smaller algae of British coastal waters. V. Bacillariophyceae (Diatoms). *Fishery Invest., Lond.* (4 ser.).

———. (1974a). A revised check-list of British marine diatoms. *J. mar. biol. Ass. U.K.*, **54**: 277–300.

———. (1974b). Some benthic diatoms from the coast of Cornwall in the neighbourhood of Porthleven. *Nova Hedwigia, Beih.*, **45**: 291–327.

HUSTEDT, F. (1927–59, 1961–66). Die Kieselalgen Deutschlands, Österreichs und der Schweiz mit Berucksichtigung der übrigen Länder Europas sowie der angrenzenden Meeresgebiete 1 (1927–30), 2 (1931–59), 3 (1961–66). In G. L. Rabenhorst, *Kryptogamen-Flora von Deutschland, Österreich und der Schweiz* 7 (1), 7 (2), 7 (3). Leipzig.

O'MEARA, E. (1867). On new forms of Diatomaceae from dredgings off the Arran Islands, Co. Galway. *Q. Jl microsc. Sci.*, n.s. **7**: 245–247.

PATRICK, R. & REIMER, C. W. (1966). The diatoms of the United States, vol. 1. *Monogr. Acad. nat. Sci. Philad.*, **13**.

PERAGALLO, H. & M. (1897–1908). *Les diatomées marines de France et des districts maritimes voisins*. Grez-sur-Loing.

ROSS, R. & ABDIN, G. (1949). Notes on some diatoms from Norfolk. *Jl R. microsc. Soc.*, **69**: 225–230.

SCHMIDT, A. et al. (1874–1959). *Atlas der Diatomaceen-Kunde*. Leipzig. Index by G. D. Hanna in *Bibliotheca Phycologica*, 10 (1969).

SMITH, W. (1853, 1856). *A synopsis of the British Diatomaceae*, vols. I & II. London.

VAN HEURCK, H. (1880–85). *Synopsis des diatomées de Belgique*, *Atlas* tt. 1–30 (1880), 31–77, 53–bis (1881), 78–103 (1882), 104–132, 22–bis, 82–bis, 83–bis, 83–ter, 95–bis (1883), Table alphabétique des noms (1884), Texte (1885). Anvers.

———. (1896). (English translation by W. E. Baxter, *A treatise on the Diatomaceae*. London); (1899) (*Traité des diatomées*. Anvers; facsimile reprint 1963 Brussels).

19. Marine algae
(excluding diatoms)

J. H. Price and I. Tittley

Introduction

Collection of data

Marine field work was carried out on seven visits during the period of the Survey. Major expeditions usually included at least five personnel, a number which enabled several tasks to be undertaken simultaneously. During the 5 years' field work, shores were visited during the period April to October. Unfortunately, no winter field work was undertaken, thus the winter-ephemeral elements of the marine flora, and variation in plant-community structure involving the die-back of summer annual species, were missed. A total of 247 species has been recorded during the Survey.

Small outboard motor-boats enabled investigation of less accessible shores, and were invaluable in undertaking diving and dredging operations. Helicopter facilities used by other members of the Survey were also used by us to visit otherwise inaccessible offshore islands. Boat-based expeditions, using the facilities provided by the Scottish Marine Biological Association, were also undertaken to visit inaccessible parts of Mull, Gometra and the Treshnish Isles. In all, 101 sites were investigated in detail and these were spread as evenly as accessibility would allow around the island group (see Fig. 19.1). As wide a range of ecological sites as possible was selected for investigation. The floras of areas with differing exposures, geology and topography were compared and contrasted.

Shore collecting techniques largely involved the bottling of material and preserving in 4% formalin solution for subsequent critical determination; crustose specimens were air dried attached to their substrate. Detailed field notes, particularly ecological and habitat descriptions as well as the usual species lists, were made and are filed at the Museum. The extent of ecological information would have been reflected more fully in this species catalogue but for restrictions on space.

Infralittoral techniques, during the earlier part of the Survey, involved dredging from a boat; subsequently, SCUBA diving techniques were employed, material being collected in nylon mesh bags. During the 1968 field-trip a technique was used, originally employed by Glasgow University divers for work elsewhere in Scotland, whereby collecting was from vertical sections measured from the water-line at the time. Collections were kept separate for recording purposes. Thus in the ecological introduction and species list, depth records for that year generally appear in terms of collecting sections 0–3 m, 3–6 m, 6–9 m etc. The greatest depth to which dives were made was 30 m. In other years, chiefly 1970, diving was carried out on a spot-dive basis

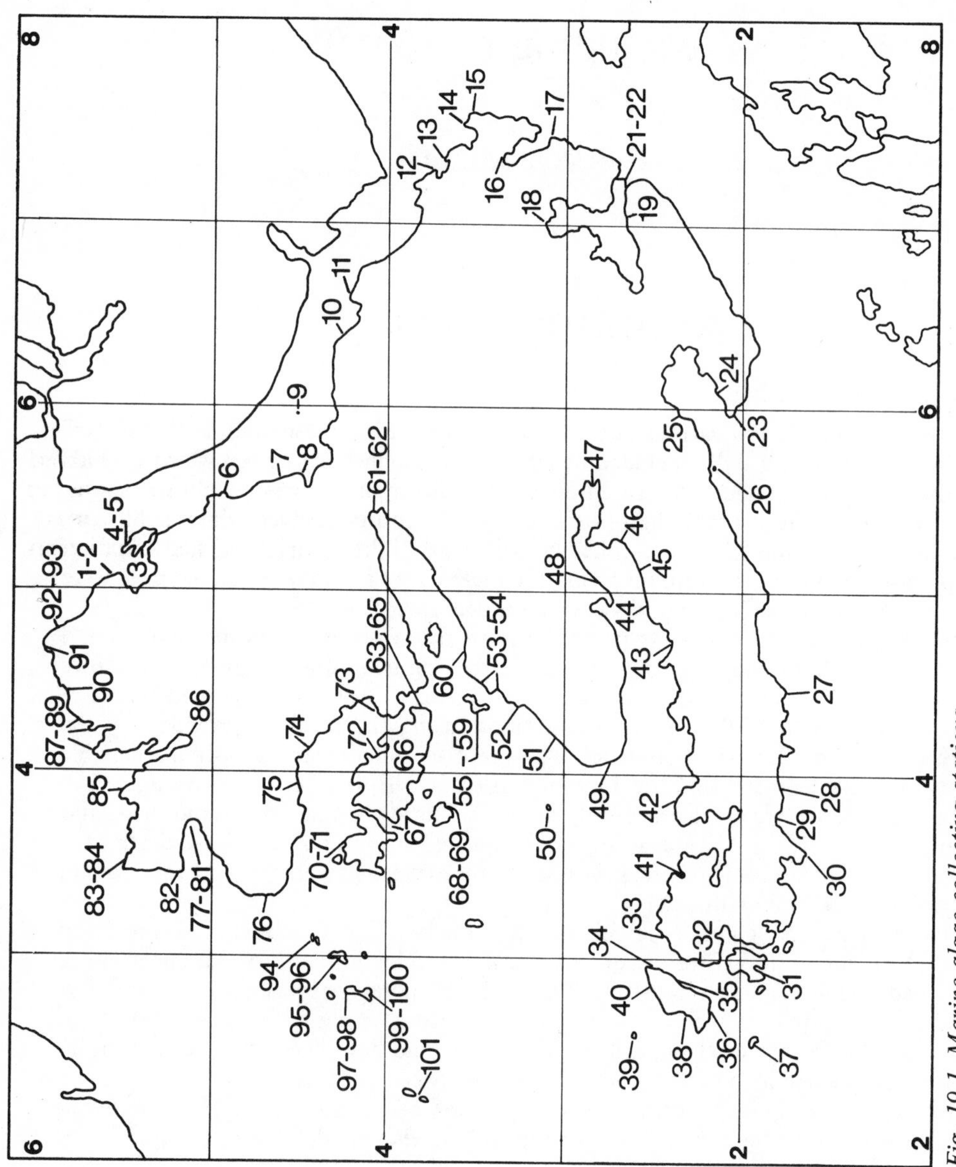

Fig. 19.1 Marine algae collecting stations.

since sufficient ecological descriptive work had already been done. Certain of these later dives were to greater depths than previously.

In addition to those acknowledged elsewhere, vacation students at the Museum undertook projects based on collections made from Mull. A comparison was made between the floras of exposed and less exposed sites at stations 6 and 52. Particular attention was paid to epiphytes on *Laminaria* stipes (G. Dwyer). An investigation was also made on epiphyte-host relationships on the basis of material collected at station 54 where *Cystoseira* was the major host (Elizabeth Bevan).

Arrangement and nomenclature

The systematic arrangement and nomenclature generally follow Parke & Dixon (1976). Ecological terminology largely follows Lewis (1964), except that we omit "eu" from eu-littoral. We can be no more objective about abundance scales since we undertook no quantitative distribution work. A list of station numbers is given when a species has been recorded for less than 20 sites. Those species recorded for 20 and more stations have been noted as "generally distributed". For further data on collecting stations see Table 19.1. We have not consistently maintained citation of fruiting records; the cases cited merely represent instances in which the stated fruiting condition was an outstanding facet of the material. Fruiting records have been represented by the following symbols:

♂ Male gametangia
♀ Female gametangia
⚲ Carposporophytes
⊕ Tetrasporangia
● Monosporangia
⊞ Plurilocular sporangia
○ Unilocular sporangia

The following abbreviations are used for herbaria consulted for marine algal records from Mull:—

BM: British Museum (Natural History)
E: Royal Botanic Garden, Edinburgh
GL: Botany Department, University of Glasgow
K in BM: Herb. Kew (now housed in BM)
STA-G: St Andrews University, Gatty Herbarium.

In addition to the distribution in Mull we have given a brief statement on the distribution of each species (except in Cyanophyta) in Scotland as a whole and, in context, in the north-east Atlantic. We felt this necessary because of the lack of any comprehensive work giving these data for marine algae. For the Scottish records, Batters (1902) has been quoted as a major reference and many recent records have been taken from the following papers which have not been quoted individually in the text: Burrows, 1960 (Kirkcudbrightshire), 1963 (Fair Isle); Gilbert, Holligan & Holligan, 1973 (North Rona); Irvine, Guiry, Tittley & Russell, 1975 (Shetland); Krinos, 1970 (Kintyre); McAllister, Norton & Conway, 1967 (Argyll, Ayr); Norton, 1972 (Lewis and Harris); Norton, McAllister, Conway & Irvine, 1969 (Colonsay); Norton & Milburn, 1972 (Argyll); Watling, Irvine & Norton, 1970 (St Kilda); Wilkinson, 1975 (Orkney). For the north-east Atlantic the following works were consulted: Ardré, 1970 (Portugal); Børgesen, 1903 (Faeroes); Børgesen & Jónsson, 1905 (N Atlantic); Caram & Jónsson, 1972 (Iceland); Jaasund, 1965 (N Norway); Jónsson, 1901, 1912 (Iceland); Jorde, 1966 (Norway); Jorde & Klavestad, 1963 (Norway); Kylin, 1907 (Norway), 1944, 1947 (Sweden); Levring, 1937 (Norway), 1940 (Sweden); Lund, 1959

(E Greenland); Printz, 1952 (Norway); Rosenvinge, 1909, 1924 (Denmark); Sundene, 1953 (Norway); Svendsen, 1959 (Spitzbergen).

Certain difficult groups were examined by specialists (see below), nevertheless there are still some problems outstanding. *Vaucheria*, for example, was not cultured to a fertile state, thus collections of sterile material were not determined to species. Representative material of all groups is held in BM. Routine material was pressed for herbarium specimens; critical material was mounted as slide preparations. Cyanophyta were investigated only to a limited degree. Material of this group has been lodged with, and awaits determination by, the SMBA, Oban.

Acknowledgements

We are indebted for help in critical identifications to: W. Adey and Yvonne M. Butler (crustose Corallinaceae); Linda M. Irvine (Rhodophyta); G. Russell (Phaeophyta). Elsie Conway, J. Healey, J. Milburn and T. Norton (Glasgow University) joined the field team in 1968 and we are grateful for their considerable assistance. The cooperation of the Scottish Marine Biological Association staff under the direction of H. T. Powell is mentioned elsewhere but we should like to record our sincere gratitude to M. Browne and, in particular, to M. J. Picken who spent many hours diving on our behalf. S. I. Honey has given considerable help in checking the manuscript.

Table 19.1 Collecting stations for marine algae

Tobermory	1	Rubha na Leip, diving to 20 m off north headland. June 1968.
	2	Port na Coite, steep rocky-boulder shore N of MacBrayne's Pier. June 1968.
Calve Island	3	Several very sheltered sites between Calve Island and Eilean na Beithe, shore largely of shingle and mud; strong freshwater influence. May 1969.
	4	Rubh an Righ, steep, N-facing, moderately exposed rocky shore. May 1969.
	5	Bogha nan Gèodh – Sgeir Calve – Bogha na Sruthlaig, slightly exposed E-facing boulder-shingle shores. May 1969.
Sound of Mull	6	Ardnacross, Rubh' a' Ghlaisich, intertidal, a gently sloping rocky headland; offshore diving to 16 m. September 1970.
	7	Rocky shore half-way between Rubha Àrd Ealasaid and Kintallen. June 1968.
	8	Aros River estuary, saltmarsh and shingle beach either side of river. May 1969.
	9	Eileanan Glasa, exposed skerry, 2 km offshore. September 1970.
	10	Doire Dorch near Fishnish Bay, steeply sloping, sheltered rocky shore. August 1966.
	11	Fishnish Point and Fishnish Bay, estuary saltmarsh, and boulder-strewn lower intertidal of sand and mud. September 1970.
	12	Craignure Bay, Java Point, slightly exposed rocky headland. September 1970.
	13	Craignure, steeply sloping slightly exposed shore at Rubha na Sròine; diving off headland to 20 m. June 1968.

Table 19.1 (continued)

Sound of Mull (*cont.*)	14	Duart Point, moderately exposed rocky headland with almost perpendicular slopes. September 1970.
	15	Dun Fuaraidh near Duart Point, diving to 20 m. September 1970.
Loch Don	16	Near Auchnacraig by outfall from Leth Fhonn, extremely sheltered and shallow areas of boulder-shingle substrate. May 1967.
	17	Near Grass Point, inner sheltered and outer more exposed sites at Port an t-Sasunnaich. August 1966.
Loch Spelve	18	NW arm, near Sàilean nan Each, extensive intertidal of consolidated shingle. August 1966.
	19	Croggan and sites seawards showing an increase in exposure. August 1966.
	20	Dredging to 5 m in the narrows between Rubha na Faoilinn and Port na Saille, and between Leckruah and Rubha nan Sailthean. May 1967.
	21	Diving to 20 m in the narrows near Grob a' Chuthaich headland W of Rubha na Faoilinn. June 1968.
	22	Rubha nan Sailthean, Bogh' an Taillir and Port nan Crullach, steep, incised, moderately exposed sites at the mouth of the Loch. August 1966; May 1967.
Loch Buie	23	Very exposed, SW-facing rocky headland at Rubha na Faoilinn. May 1968.
	24	E shore, below Aoineadh Fada S of Rubha Liath, a moderately exposed intertidal of rock slabbing and boulders. May 1968.
	25	Below Coill' a Chaiginn, diving to 6.5 m just offshore on the sheltered NW side. May 1969.
Carsaig Bay	26	Gamhnach Mhòr, diving at sites on the E and N sides; intertidal observations at sheltered and exposed sites at Gamhnach Mhòr and below Aoineadh a' Mhaide Ghil. May, June 1968.
Ross of Mull	27	Rubha nam Bràithrean, diving to 18 m. September 1970.
	28	Port Uisken, rocky outcroppings in an otherwise sandy area. August 1966.
	29	Ardalanish Bay, rocky outcroppings in an otherwise sandy area. August 1966.
	30	Rubh' Ardalanish, a rocky semi-exposed littoral; offshore diving to 18.5 m. September 1970.
	31	Traigh Gheal, moderately exposed granitic rocky shore. September 1967.
	32	Small bay by Iona ferry jetty, Fionnphort. May 1968; May 1969.
	33	Rocky shores of varying exposure, Kintra. September 1970.
Iona	34	NE coast, diving in *Zostera* bed at 6.5 m off S end of Traigh Bahn, 400 m NE of Cladh an Diseart, May 1969.
	35	St Ronan's Bay, diving just N of Baile Mòr pier. June 1968; May 1969.
	36	S coast, exposed shore and pools between Port na Curaich and Port Goirtean Iar. May 1969.

Table 19.1 (continued)

Iona (*continued*)	37	Soa Island, sheltered gully on N side of island. July 1969.
	38	Camas Cuil an t-Saimh, and A' Mhachair, rocky promontories with varying degrees of exposure. May 1967; June 1968.
	39	Rèidh Eilean, a very exposed island shore. July 1967.
	40	NW coast near Calva and Port Chlacha Dubha, incut rocky shores of varying exposure. May 1968; May 1969.
Loch na Làthaich	41	Camas Tuath, diving to 23 m. August 1970.
	42	E of Port Uamh na Gaibhre, a dissected, very exposed rocky headland. July 1969.
Loch Scridain	43	Near An Carraigean, diving to 9 m; and E of Eilean nam Caorach, intertidal. June 1968.
	44	Port na Clòidheig, diving to 20 m. June 1968.
	45	Near Pennyghael Boat House. September 1968.
	46	Sgeir Mhòr and Rubha Buidhe, sheltered rocky shore with pools at all levels. May 1967.
	47	An Leth-onn, saltmarsh at estuary of Coladoir River. August 1966; May 1967.
	48	Dredging between Aird Kilfinichen and Dererach. May 1967.
Ardmeanach	49	Carrachan Mòr. September 1967.
	50	Erisgeir, very exposed island 3 km offshore; diving to 22 m. August 1970.
	51	Sites at Coireachan Gorma, very exposed NW facnig rocky shore and shaded, sheltered cave; diving to 23 m. August 1970.
Gribun	52	Near Balmeanach Farm and Port Uamh Beathaig. June 1965; September 1970.
	53	Rubha Baile na h-Airde, an extensive sheltered intertidal area of slabbing with rock pools and vertical rock faces. August 1966; May 1967.
	54	Dredging and diving in shallow lagoon area between Rubh' a' Ghearrain, Rubha Baile na h-Airde and Samalan Island. June 1968; October 1968; May 1969; September 1970.
Inch Kenneth	55	An Fhearsard (extreme E tip), an extensive area of sheltered, gently sloping intertidal traversed by ridges. May 1969.
	56	SE end of island, diving at 3-4 m in *Zostera* bed. May 1969.
	57	Sheltered intertidal and offshore diving site (to 8 m) at the SW end. May 1969.
	58	N coast, open rocky shore at Bàgh an Iollaich. May 1969.
	59	N coast, shallow diving site in *Laminaria digitata* beds. May 1969.
Loch na Keal	60	S shore near Tràig Doire Dhubhaig, steeply shelving shore (45°) below cliffs; diving to 35 m. June 1968.
	61	Shingle-boulder shore at head of loch. July 1968.
	62	Shingle-boulder shore by River Bà estuary. July 1968.
Ulva	63	Rocky headlands and inlets on S side of bay NE of Ulva House. May 1967.

Table 19.1 (continued)

Ulva (*continued*)	64	S coast, slightly exposed at Sgeir nan Leac. May 1967.
	65	Ard Dearg, 400 m SW of Dùn Bhioramuill, S of headland, diving to 27 m. June 1968.
	66	Between Cragaig and Tòrr Mòr, sheltered, shallow, sandy bay. May 1968.
	67	Baile A'Chalaidh harbour, diving to 15 m. August 1968.
Staffa	68	S end of island, moderately exposed intertidal of rocky pools, ridges and channels near the landing stage. July 1969.
	69	S end of island, very exposed shore by Fingal's Cave. July 1969.
Gometra	70	Acairseid Mhòr, sheltered littoral; diving in infralittoral *Zostera* beds at 4–5 m. June 1969.
	71	N coast near Rubha na Sròine; diving to 30 m. August 1970.
Loch Tuath	72	1 km NW of Sgeir Ruadh; diving to 15 m. June 1968.
	73	Dredging in 4 m near Eilean Garbh. May 1967.
	74	Tòrr an Àrd, moderately exposed littoral and infralittoral shallow lagoon system. July 1969.
	75	Sgeir na Cille, Allt Hostarie stream, shingle beach and reef-lagoon system. May 1969.
Calgary Bay	76	Port Haunn, moderately exposed area of intertidal slabbing with pools and gullies. September 1970.
	77	Creag a' Chaisteil, moderately exposed, steeply sloping rocky shore. May 1967.
	78	Lainne Sgeir, rocky promontory with pools at various levels. May 1967; May 1968; October 1968.
	79	Lagoons and rock pools with an outer N-facing reef area; also dredging in infralittoral at 1-2 m. May 1968.
	80	Rocky shore sites of increasing exposure seawards from the pier. August 1966.
	81	Dredging W of the pier. May 1967.
	82	Below Croc Udmail, exposed rocky promontory. August 1966.
Mornish	83	Caliach Point. August 1966.
	84	Port na Caillich, slightly exposed littoral with deep pools; diving offshore to 27 m. June 1968.
	85	NW of loch by Croig, irregular coastline of pools, lagoon systems and sandy bays. July 1969.
Loch a' Chumhainn	86	Brackish water and saltmarsh by outfall of the River Bellart. June 1969.
Mishnish	87	Loch Mingary approaches, W side, rocky intertidal; diving to 6 m. June 1969.
	88	Loch Mingary approaches, E side. June 1969.
	89	Laorin Bay, sites on rocky shores of varying exposure from inner sheltered head-end to exposed outer W shore. August 1966.
	90	Port Chill Bhraonain, moderately exposed rocky promontories with gullies and boulder beaches. May 1967.
	91	Ardmore Point, slightly exposed rocky terracing and slabbing with shallow pools. May 1969.

Table 19.1 (continued)

Mishnish (*continued*)	92	Near Rudh na Sealhaigh, Bloody Bay, intertidal of rocky terracing with shallow pools. May 1969.
	93	Bloody Bay, diving to 5 m in *Laminaria hyperborea* beds. May 1969.
Treshnish Islands	94	Cairn na Burgh Beg, rocky boulder shore. July 1968.
	95	Fladda, NE coast, offshore diving to 20 m. August 1970.
	96	Fladda, SE coast. July 1968.
	97	Sgeir an Eirionnaich', SE-facing shore of rocky outcrops and pools. July 1968.
	98	Sgeir a' Chaisteil. June 1968.
	99	Lunga, SE coast, offshore diving to 20 m. August 1970.
	100	Lunga, SW coast, very exposed, offshore diving to 20 m. August 1970.
	101	Bac Beag and Bac Mòr, causeway of rocky pools and lagoons connecting the two islands. June 1969.

Index to genera

CYANOPHYTA

CYANOPHYCEAE

CHROOCOCALES

ENTOPHYSALIDACEAE

Entophysalis Kütz.

E. conferta (Kütz.) Drouet & Dailey
Littoral and infralittoral; epiphytic on *Enteromorpha*, *Sphacelaria*, *Ceramium*, *Laurencia*, *Polysiphonia* and *Pterosiphonia*.
4, 14, 26, 31, 33, 36, 40, 51, 55, 57, 68, 76, 78, 92, 94, 98

NOSTOCALES

OSCILLATORIACEAE

Microcoleus Gom.

M. lyngbyaceus (Kütz.) Crouan
Saltmarsh.
86

M. vaginatus (Vauch.) Gom.
Infralittoral.
51

Oscillatoria Gom.

O. lutea Ag.
Infralittoral.
13

Porphyrosiphon Kütz.

P. notarisii (Menegh.) Kütz.
Littoral. Epiphytic on *Asperococcus*, *Dictyota* and *Sphacelaria*.
6, 14, 33, 42, 78, 97, 101

Schizothrix Gom.

S. calcicola (Ag.) Gom.
Saltmarsh.
78, 84, 86

RIVULARIACEAE

Calothrix Born. & Flah.

C. crustacea Born. & Flah.
Littoral (high level pools).
43

Rivularia Born. & Flah.

R. atra Born. & Flah.
Littoral. Epiphytic on *Cladophora* and *Corallina*.
43, 78

R. nitida Born. & Flah.
Saltmarsh.
86

RHODOPHYTA

FLORIDEOPHYCEAE

NEMALIALES

ACROCHAETIACEAE

Audouinella Bory

Current opinion is that all species hitherto included in *Acrochaetium* should be recombined in this genus. We have accepted the recombinations in Woelkerling (1972, 1973a, 1973b,) and Parke & Dixon (1976); determinations are based on the key in volume 1 of the *British Marine Flora* (Dixon & Irvine, 1977).

A. daviesii (Dillw.) Woelkerl.
Littoral; infralittoral to 9 m. Epiphytic on *Laurencia pinnatifida*, *Palmaria palmata*, *Polysiphonia nigrescens*, and *Rhodomela confervoides*.
♀ June 1968; ● June 1968 and May 1969.
Frequent. 26, 31, 33, 40, 43, 44, 49, 59, 68, 92, 95
Scotland: frequent but less so in the west.
NE Atlantic: widely distributed except Greenland, Iceland and Spitzbergen.

A. floridula (Dillw.) Woelkerl.
Littoral; infralittoral to 18 m. Epilithic (sand-binding).
Abundant. 4, 6, 13, 22, 32, 38, 40, 44, 48, 53, 54, 63, 66, 69, 72, 79, 81
Scotland: very frequent.
NE Atlantic: not recorded north of Shetland.

A. membranacea (Magn.) Papenf.
Infralittoral, 15–19 m. Endozoic, in *Dynamena ? pumila*. Immature ⊕ Aug. 1970.
Rare; probably more frequent than detected. 50
Scotland: Shetland, Orkney, Lewis and Argyll, but probably more widespread. Always endozoic in colonial hydroids or similar marine invertebrates, and noted in all the more recent diving studies in the area. Sometimes also recorded from the lower littoral.
NE Atlantic: north to Bear Island, W Greenland, and White Sea.

A. microscopica (Näg.) Woelkerl.
Littoral, on wave beaten surfaces. Epiphytic on *Ceramium shuttleworthianum*.
Rare. 24
The determination of this species is tentative only; the name and concept may need revising when the type of *Kylinia scapae* Lyle has been

examined. We accept this name on the basis of studies made by Woelkerling (v.s.) who did not see the type of *K. scapae*.

A. purpurea (Lightf.) Woelkerl.
Supralittoral; littoral; infralittoral, to 8 m. Epilithic, particularly on damp and shaded cave walls or similar surfaces within reach of spray or wind-carried spume; epiphytic on *Laminaria hyperborea* stipes and holdfasts; epizoic on *Patella*. In estuarine areas on and among mud around the bases of phanerogams. Frequent. 8, 34, 43, 47, 51, 54, 60, 65, 69, 74, 84, 85, 100. Previous records: Iona, ". . . upon the base of Abbot Mackinnon's tomb in the ruined Abbey . . .", *Lightfoot* (1777), as *Byssus purpurea*, *Trevelyan*, 1823 (E) as *Trentepohlia purpurea*; "ruins of the Cathedral", June/July 1826, Herb. Greville (E) and Herb. Hooker (GL, K in BM), as *T. purpurea; Hooker* (GL); rocks in sea cave, *Hedge*, 2 May 1958 (E) as *Rhodochorton purpureum;* Fingal's Cave, Staffa, *Harvey*, in Hooker (1833), as *T. purpurea*.

These records were variously repeated in many subsequent works as *Byssus purpurea*, *Conferva purpurea*, *Trentepohlia purpurea*, *Callithamnion purpureum*, *C. rothii*, *C. rothii β purpureum*, *Rhodochorton purpureum* and *R. rothii*.
Scotland: widespread.
NE Atlantic: widespread, north to Spitzbergen and Greenland.

A. secundata (Lyngb.) Dixon
Littoral; infralittoral to 21 m. Epiphytic on *Porphyra umbilicalis*, *Callithamnion hookeri*, *Spermothamnion* and *Zostera*. ● June 1968 and May 1969.
Occasional. 13, 34, 54, 92, 97. Previous record: Cairn na Burgh Mòr, Treshnish Islands, June 1953, *Tindal* (STA-G).
Scotland: frequent in the west (Shetland, Fair Isle, Orkney, North Rona, Lewis, Argyll, Bute and Ayr).
NE Atlantic: widespread and frequent, but not in E Greenland or Spitzbergen.

A. seirolana (Gibs.) Dixon
Infralittoral, to 6 m. Epiphytic on *Callithamnion hookeri*, itself on *Chondrus crispus*. Sporangia (possibly ● or immature ⊕) June 1968.
Rare. 26
Scotland: in the east.
NE Atlantic: only in Faeroes.

A. thuretii (Bornet) Woelkerl.
Littoral, in pools. Epiphytic on *Cladophora* sp.
Rare. 2
Scotland: not previously reported.
NE Atlantic: all except Iceland, Greenland, Spitzbergen, N Norway, Faeroes and S Sweden.

GELIDIACEAE

Gelidium Lamour.

Although the morphological variation is probably clinal, it is usually possible to recognise a wide form, *G. latifolium*, and a narrow form, *G. pusillum*, representing the opposite extremes. Leaving aside the more southerly upright forms hitherto known as *Gelidium sesquipedale* (Clem.) Born. & Thur. and *G. cartilagineum* (L.) Gaill. it is probable that our treatment includes all those growth forms present in various niches throughout the whole of western Scottish island and mainland coasts.

G. latifolium (Grev.) Born. & Thur. agg.
Littoral, shallow pools and adjacent ridges, in lower littoral around *Fucus serratus* level; occasionally in shaded conditions in higher *Pelvetia/Fucus spiralis* level pools or in damp circumstances under *Ascophyllum*. Infralittoral: shallow infralittoral, 0–3 m. under *Laminaria digitata* and *L. hyperborea*, whence it may spread upwards under dense *Fucus serratus*; occasionally forming luxuriant and abundant sward, overlapping the whole infralittoral fringe area as underflora. Epilithic.
Frequent. Probably even more widespread than recorded. 22, 33, 36, 40, 53, 60, 63, 75, 78, 79, 92. Previous record: Gribun rocks, Aug 1943, *Newton* (1949).
Scotland: Relatively widely recorded for western mainland and islands, but the difficulty of disentangling the nomenclature employed renders some records uncertain. There is no firm evidence for Shetland or Orkney but there are identifiable records from Lewis and Harris, Galloway, Argyll, Ayr and Bute.
NE Atlantic: only otherwise recorded from south of Bergen, Norway.

G. pusillum (Stackh.) Le Jol. agg.
Littoral, in pools at *Ascophyllum* and *Pelvetia/Fucus spiralis* boundary levels, occasionally emerging in damp niches and similar conditions on adjacent rock surfaces, especially in shaded lower littoral; when particularly luxuriant, it may form dense continuous turf under *Laurencia/Colpomenia/Leathesia* communities at all levels up to midlittoral, where it continues beneath dense *Ascophyllum*. Infralittoral fringe; may spread down into the immediate fringe areas from the dense lower littoral turf mentioned above. Epilithic; occasionally detected spreading epiphytically over densely associated *Corallina officinalis*.
Abundant. Generally distributed.
Scotland: in mainland and island areas under *G. crinale*, *G. pulchellum*, *G. pusillum*, and *Gelidium* spp.
NE Atlantic: In western Norway Børgesen and Jónsson report a form under the name *G. crinale*, but more recent works record only the occurrence of *G.* cf. *latifolium*. Scottish records of this aggregate seem otherwise to lie at the northern periphery of *G. pusillum* distribution in the E Atlantic.

HELMINTHOCLADIACEAE

Nemalion Duby

N. helminthoides (Velley) Batt.
Littoral, on shores exposed to wave action. Epilithic; epizoic on *Balanus* and *Patella* in open and exposed conditions. ♀ Aug. and Sept. 1970.
Occasional; probably more widespread. 16, 51, 52, 76, 81
Scotland: Bute, Ayr, Argyll, Harris, Orkney, Moray Firth.
NE Atlantic: Denmark and S and W Norway.

BONNEMAISONIACEAE

Asparagopsis Montagne

A. armata Harv.
Falkenbergia rufolanosa (Harv.) Schm. (= tetrasporic phase)
Infralittoral, to 15 m; commonly associated with pebbles, sand and *Laminaria* bases; often enmeshed in basal detritus over *Pecten*, or other attached macroalgae than *Laminaria*.
Occasional. Possibly overlooked since detected elsewhere whenever particularly sought. Spreading in northern Europe as both the *Asparagopsis* (gametophyte) and *Falkenbergia* (tetrasporophyte) stages.

The spread of this organism northward is of considerable interest. It has analogies with the spread of *Colpomenia* earlier in the century. Walker, Burrows & Lodge (1954: 315) reported *Falkenbergia* as in quantity in the harbour at Port St Mary, Isle of Man; they indicated interest in whether it would be followed, as elsewhere, by appearance of the *Asparagopsis* phase. Conway (1960) first reported the *Falkenbergia* phase from western Scotland (The Strand, Colonsay - Oronsay, and Castle Bay, SE Colonsay, on *Cladophora rupestris*, Sept. 1959, sterile). Norton *et al.* (1969) reported *Asparagopsis armata* (presumably the *Falkenbergia* phase) from six stations, lower littoral to 6 m deep, sometimes on *Cladophora rupestris* and *Jania rubens*, on Colonsay. McAllister *et al.* (1967) record the infertile *Falkenbergia* phase on *Heterosiphonia plumosa* from Easdale Quarry, Argyll, in 6–24 m depth. Similarly, recent field work has established two localities on Orkney (Wilkinson, 1975) and two on Shetland (Irvine *et al.*, 1975) where *Falkenbergia* has been detected, in all cases in association with the *Trailliella* phase of *Bonnemaisonia hamifera*. The Orkney material was from lower littoral pools, and in this respect differs from all our own material, from the Easdale Quarry records, and from the Shetland material, all of which were substantially sublittoral in origin. Lower littoral material of *Falkenbergia* was also detected from Colonsay.

All the northern material seems to be sterile and to have been the result of vegetative spread. This, however, may simply be the result of the season when collections were made (mostly summer) since McLachlan (1967) found British *Falkenbergia* to be winter-fertile. There are no records of either phase from further north, and the Shetlands record is the limit known. The current period of worsening climate in northern Europe may halt the spread northwards.

As commented by Conway (1960), southern records of the alga indicated the appearance of *Falkenbergia* and *Asparagopsis* phases almost simultaneously. Recent evidence from the northern spread indicates strongly that *Falkenbergia*, at this northern end of the range of *Asparagopsis*, is much the more rapid and actively colonising phase. Thomas (1955) has already suggested that there are differences in the ecological requirements of the two forms.

Bonnemaisonia Ag.

B. asparagoides (Woodw.) Ag.
Infralittoral, 6–23 m. Epilithic; epiphytic on *Phycodrys rubens* and *Phyllophora crispa*. June to Sept.
Frequent. 1, 13, 27, 44, 51, 65, 72, 99, 101
Scotland: for Orkney, Argyll, Bute and Ayr (Batters). The Orkney meeting of the British Phycological Society detected only drift material. Burrows added Kirkudbrightshire, again only as drift, and Norton *et al.* recorded the species from Colonsay and from the Argyll mainland.
NE Atlantic: S and W Norway to as far north as Trondheim fjord, from W Sweden and doubtfully from Iceland.

B. hamifera Hariot
Trailliella intricata Batt. (= tetrasporic phase)
Littoral, pools at lower levels; infralittoral, to 23 m. Epilithic; epiphytic on *Dictyota dichotoma*, *Corallina officinalis*, *Gelidium*, *Phyllophora crispa*, *P. pseudoceranoides;* epizoic on *Pecten*. Local habitat requirements seem to be very similar to those outlined by Floc'h (1969) and Haugen (1970); both indicated areas of comparative shelter in generally wave-washed habitats, usually protected by rocky points or low skerries, with associated patchy or continuous detritus (often sand) in shallow depths. These are exactly the circumstances in the coastal area around Calva, where protection from the direct effect of the western long fetch is provided by Eilean Chalba and numerous offshore skerries and rocks, many of which clear at low water springs to reveal associated areas of fine sand.
Frequent. Generally distributed.
Scotland: until very recently, only the fertile or infertile sporophyte (*Trailliella*) phase had been recorded, appearing relatively widespread (Galloway; Colonsay; Argyll mainland around and to the south of Oban; Lewis and Harris; Orkney; Shetland; Fair Isle). In 1970, Krinos recorded the gametophytic *Bonnemaisonia hamifera* stage as a fairly common member of

the Westport to Tangy flora (Kintyre), growing sterile in both summer 1967 and spring 1968, on rocky ledges just below the water surface. Surprisingly, she did not record the *Trailliella* phase, despite correctly noting its being reported by McAllister *et al.* (1967). The most common form of this alga recorded by us was also *Trailliella*, exclusively nonfertile. However, the *Bonnemaisonia* phase was also detected (station 40) on exposed rocks at Calva, on each of the two visits made there. The plants were sporadic and were growing in the shallow sublittoral and in deep lower littoral pools. Our material was sterile and epilithic in 1968, but attached to *Ceramium rubrum* in 1969.
NE Atlantic: Trailliella occurs widely in Scandinavia. *Bonnemaisonia* has been recorded from Brittany to Norway and, during the period 1962–1970, has spread northwards (see McLachlan *et al.*, 1969; Haugen, 1970).

GIGARTINALES

FURCELLARIACEAE

Furcellaria Lamour.

F. lumbricalis (Huds.) Lamour.
Littoral, in shallow pools, channels, or lagoons with sandy or detrital bottoms in the lower littoral; occasionally in deep shaded pools at higher (*Pelvetia*) levels or middle (*Ascophyllum*) levels; infralittoral, in shallow depths to 14 m mostly on pebbles buried in sand on shallow slopes. Epilithic on unstable substrata; often associated with bases of *Cystoseira nodicaulis* and *Halidrys*. Tolerates the presence of freshwater flowing over the intertidal. ♀ May to Oct.; ♂ May 1967; ⊕ Oct. 1968.
Abundant. Generally distributed. Previous record: Gribun Rocks, Aug. 1943, *Newton* (1949).
Scotland: widely distributed.
NE Atlantic: frequent, but not from Greenland, Iceland, Jan Mayen. Infrequent in N Norway.

Halarachnion Kütz.

H. ligulatum (Woodw.) Kütz.
Infralittoral, 18–21 m. Epilithic.
Rare. 1
Scotland: only in Orkney and Shetland. Batters, however, gives a long list of previous Scottish records, including Aberdeen (Peterhead), Banff (Macduff), Kintyre (Southend), Arran, Bute (Cumbrae) and Ayr (Portincross), as well as the well-founded Orkney records. There is material in BM from Kirkwall Bay to support the latter, but none in support of the other cited localities. However, the Batters statements concerning western Scottish localities are largely acceptable in view of our own find and of the known spasmodic appearance of this species.
NE Atlantic: there are a few doubtful records from W Norway; recorded as rare from W Sweden. Specimens are also known from the lower sublittoral of Heligoland, and from Denmark.

POLYIDEACEAE

Polyides Ag.

P. rotundus (Huds.) Grev.
Littoral; infralittoral to 15 m. Epilithic. ♀ (nemathecia) June 1968.
Frequent. 5, 21, 22, 38, 44, 45, 47, 53, 54, 75, 92
Scotland: abundant.
NE Atlantic: widespread.

RHABDONIACEAE

Catenella Grev.

C. caespitosa (With.) L. Irvine in Parke & Dixon
Littoral, on vertical faces in shade along side of pool at *Pelvetia*/*Fucus spiralis* boundary level, under overhanging *Pelvetia*. Epilithic.
Rare, but probably more widespread. 78
Scotland: widespread on the mainland and western islands.
NE Atlantic: W Norway (south of Bergen) and Iceland (Vestmannaeyjar).

RHODOPHYLLIDACEAE

Calliblepharis Kütz.

C. ciliata (Huds.) Kütz.
Infralittoral, 1–21 m. Epilithic; epiphytic on *Cystoseira nodicaulis.*
Rare; probably more widespread than our records indicate. 21, 54, 72. Previous record: Iona, "not common" *Lightfoot* (1777) as *Fucus ciliatus.*
Scotland: Orkney (Elwick Harbour, Shapinsay), Argyll (Loch Etive), Bute (Cumbrae) and Arran (Batters). Also observed in Aug. 1973 by W. A. A. Robertson (*pers. comm.*) in Dunstaffnage Straits, Argyll, at about 12–15 m depth, and by Norton *et al.* on Colonsay (Scalasaig).
NE Atlantic: not recorded.
Our material represents one of the northern limits for the species in Great Britain and in Europe.

C. jubata (Good. & Woodw.) Kütz.
Littoral, mid-level, under dense *Ascophyllum–Fucus serratus* cover, and therefore in damp conditions; very shallow infralittoral, in sheltered areas on exposed rocks. Epilithic.
Rare. 40, 66, 92
Scotland: a dubious record for Orkney: Shapinsay, Kirkwall (Batters). With that reservation, the present records are again the furthest north in Europe, being well north of the Easdale Quarry, Argyll, location from which (in 6–12 m) the specimens reported by McAllister *et al.* were collected. Also recorded recently from Balacary Point, Kirkudbrightshire; Batters previously reported it from Kincardine, Dumbarton, Bute and Ayr.

Cystoclonium Kütz.

C. purpureum (Huds.) Batt.
Littoral, lower level (*F. serratus–F. vesiculosus* boundary); infralittoral, all depths 0–18 m. Epilithic, often on small fragments in unstable substrata such as sand (in *Zostera* beds) and general detritus; epiphytic on *Laminaria hyperborea* stipes; epizoic on spider crabs. ♀ June to Sept.
Abundant. Generally distributed. Previous record: Gribun Rocks, Aug. 1943, *Newton* (1949).
Scotland: consistently recorded in the west.
NE Atlantic: widely distributed, but absent from Spitzbergen and E Greenland.

PLOCAMIACEAE

Plocamium Lamour.

P. cartilagineum (L.) Dixon
Littoral, in shaded deep pools on lower shore; infralittoral to 23 m. Epilithic; epiphytic on *Cystoseira nodicaulis, C. tamariscifolia, Desmarestia aculeata, Halidrys, Laminaria digitata* (stipes), *L. hyperborea* (stipes), *Gigartina stellata*, and *Phyllophora crispa;* epizoic on *Pecten*. May 1969 and Aug. 1970.
Abundant. Generally distributed. Previous records: Iona, *Keddie* (1846), (GL); as *P. coccineum, Ritchie & Ritchie* (1947); Cairn na Burgh Mòr, Treshnish, drift, *Tindal*, June 1953 (STA-G).
Scotland: Abundant and widespread.
NE Atlantic: widely reported but not known from Spitzbergen or Greenland.

SPHAEROCOCCACEAE

Sphaerococcus Stackh.

S. coronopifolius Stackh.
Usually epilithic.
Rare, detected only once as drift.
Scotland: recorded only scantily from Ayr, Colonsay and Orkney (Clouston, 1862).
NE Atlantic: not recorded further north and, if present attached to substrate on Mull, probably nearing its northern limit.

GRACILARIACEAE

Gracilaria Grev.

G. verrucosa (Huds.) Papenf.
Lower littoral, in sand and detritus at bases of rocks in standing water and under overhanging major flora; infralittoral, 0–22 m. Epilithic, on pebbles in sand and detritus, and on or in compacted detritus; epizoic on *Pecten*. ♀ June to Aug.; ♂ Oct. 1968.
Occasional. 22, 38, 41, 44, 52, 54, 67
Scotland: recorded for the western area (including Orkney and Shetland) in all studies that have employed diving techniques, and mentioned as drift in some other cases. Probably much more widespread than the records indicate, even ubiquitous in suitable conditions.
NE Atlantic: from N Norway (near Trondheim) to W Sweden.

PHYLLOPHORACEAE

Ahnfeltia Fr.

A. plicata (Huds.) Fr.
Littoral, epilithic in damp conditions and pools; in the shallow infralittoral (0–6 m.) on pebbles associated with *Cystoseira nodicaulis* bases; occasionally epiphytic on *Halidrys siliquosa* in *Zostera* beds.
Locally abundant. Generally distributed.
Scotland: widely recorded.
NE Atlantic: from E Greenland south to Portugal and S Spain.

Phyllophora Grev.

P. crispa (Huds.) Dixon
Littoral, in deep shaded pools at *Fucus serratus* level; infralittoral, to 21 m. Epilithic.
Abundant. Generally distributed. Previous record: doubtfully that from Iona, *Lightfoot* (1777), as *Fucus rubens*.
Scotland: consistently recorded.
NE Atlantic: Denmark, W Norway and W Sweden.

P. pseudoceranoides (S. G. Gmel.) Newr. & A. R. A. Taylor
Littoral, in pools and deep shade; infralittoral 18 m. Epilithic; occasionally epiphytic on *Laminaria hyperborea*. ♀ June to Sept.
Abundant. Generally distributed.
Scotland: A consistent facet of the lower littoral pool flora.
NE Atlantic: Denmark, W. Norway, W Sweden, the Faeroes and Iceland.

P. traillii Holmes ex Batt.
Littoral, lower levels, in pools and deep shade. Epilithic.
Rare. 78
Scotland: Batters lists the species for East Lothian, Mid-Lothian, Fife, Orkney and Bute. Recently collected in Lewis and Harris, and in Orkney and Shetland.
NE Atlantic: not otherwise recorded, Shetland being the northern limit of this species in Europe.

P. truncata (Pallas) Zinova
Littoral; infralittoral to 15 m. Epilithic.
Rare. 21, 46, 53
Scotland: recorded frequently for the west and recently for Shetland.
NE Atlantic: widely distributed and recorded recently for Iceland, Greenland and Spitzbergen; extending to the Murman and White Seas.

GIGARTINACEAE

Chondrus Stackh.

C. crispus Stackh.
Littoral, widespread and often abundant in all damp situations in the lower littoral, occasionally extending up-shore to include large amounts in deeper pools at *Pelvetia/Fucus spiralis* level; tolerates the outflow of fresh

water. On rare occasions, in wave-beaten areas as an underflora of *F. serratus*; infralittoral, to 12 m. Epilithic on rocks and pebbles; epiphytic on stipes of *Laminaria digitata* and *L. hyperborea*, occasionally associated with and perhaps epiphytic on *L. saccharina* bases. ♀ May to Oct.; ⊕ June 1968.
Abundant. Generally distributed. Previous records: Iona, *Keddie* (1850); Rubh an Righ, Calve Island, *Kitching* (1935); Cairn na Burgh Mòr, Treshnish Islands, June 1953, *Tindal* (STA-G).
Scotland: Consistently recorded for all areas of mainland, western islands and Orkney and Shetland.
NE Atlantic: widely distributed, but absent from Spitzbergen, Greenland and similar arctic areas.

Gigartina Stackh.

G. stellata (Stackh.) Batt.
Littoral, common and luxuriant on rocky shores exposed to considerable wave-action, often band-forming with *Himanthalia* and *Laurencia pinnatifida* in lower littoral; sometimes in pools on shores of lesser wave-action, middle shore and below, or in similar damp circumstances as underflora in *Fucus serratus*; occasionally stunted and bleached, in high pools in *F. spiralis* zone; similarly on pebbles in beds of fresh water streams flowing over shore; shallow infralittoral, to 6 m. Epilithic; occasionally epizoic on *Mytilus*. ♀ May to Oct.
Abundant. Generally distributed. Previous records: Calve Island, Aug. 1934, *Kitching* (1935); Gribun Rocks, Aug. 1943, *Newton* (1949); Cairn na Burgh Mòr, Treshnish Islands, June 1953, *Tindal* (STA-G).
Scotland: widespread in the west.
NE Atlantic: equally widespread but absent from Greenland, Jan Mayen, and Spitzbergen.

CRYPTONEMIALES

CORALLINACEAE

Corallina L.

C. officinalis L.
Littoral, common in middle and lower littoral in pools, standing water, or damp areas under major flora, occasional in deep shaded higher level pools, turf-forming with *Laurencia hybrida*; infralittoral, shallow depths to 12 m. Epilithic; epiphytic on *Cystoseira nodicaulis*, *C. tamariscifolia*, *Halidrys siliquosa*, and *Laminaria hyperborea* stipes and holdfasts; epizoic on *Ensis* and on living spider crabs.
Abundant. Generally distributed. Previous record: Iona, *Keddie* (1850), Cairn na Burgh Mòr, Treshnish Islands, June 1953, *Tindal* (STA-G).
Scotland: all areas of the western mainland and islands.
NE Atlantic: widely distributed, except for Greenland, Jan Mayen and Spitzbergen.

Dermatolithon Fosl.

D. hapalidioides (Crouan & Crouan) Fosl.
Littoral; infralittoral, 2–3 m. Epilithic; epiphytic on *Laminaria hyperborea*.
Occasional. 22, 38, 47, 53, 79
Scotland: widespread in the north.
NE Atlantic: reaching SW Norway.

The following species have been identified by Y. M. Butler; because of taxonomic uncertainty further distribution is not given.

D. corallinae (Crouan & Crouan) Fosl.
Littoral. Epiphytic on *Corallina officinalis*. 79

D. littorale (Suneson) Lemoine
Littoral. Epiphytic on *Furcellaria lumbricalis*. 85

D. pustulatum (Lamour.) Fosl.
Littoral. Epiphytic on *Chondrus crispus*, *Fucus serratus*, *Furcellaria lumbricalis*, *Gigartina stellata* and *Palmaria palmata*. 4, 13, 40, 92

Fosliella Howe

All specimens, other than *F. tenuis*, were determined by Y. M. Butler. Further distribution is not given because of taxonomic uncertainty.

F. farinosa (Lamour.) Howe
Littoral; shallow infralittoral. Epiphytic on *Cladostephus spongiosus*, *Cystoseira tamariscifolia*, *Furcellaria lumbricalis* and *Polysiphonia elongata*.
Rare. 57, 78, 79

F. limitata (Fosl.) Ganesan
Littoral. Epiphytic on *Chondrus crispus*, *Cystoseira nodicaulis*, *Furcellaria lumbricalis* and *Polysiphonia elongata*.
Occasional. 40, 54, 78, 79, 85

F. minutula (Fosl.) Ganesan
Littoral. Epiphytic on *Furcellaria lumbricalis*.
Rare. 79

F. tenuis Adey & Adey
Littoral; infralittoral to 22 m. Epiphytic.
Rare, but probably more widespread than recorded here. 53, 100
Specimens determined by W. H. Adey. For data on distribution and other aspects, see Adey & Adey (1973). They comment that the species is often overlooked in fresh collections, although it is frequent in moderately deep water. The species is said not to compete well with other crustose corallines, which may explain the low mean relative abundances noted for Britain.

Jania Lamour.

J. rubens (L.) Lamour.
Littoral; infralittoral to 3 m. Epiphytic on *Cladostephus spongiosus*, *Cystoseira tamariscifolia*, *C. nodicaulis*, *Dictyota dichotoma*, *Fucus*

serratus, *Halidrys siliquosa* and *Furcellaria lumbricalis*.
Common. 16, 20, 22, 44, 53, 75, 78, 79, 81, 84
Scotland: Cumbrae, Bute, Argyll (Lismore), Orkney, Aberdeen and Fife (Batters); Colonsay; Arran (BM); Orkney (L. M. Irvine, *pers. comm.*)
NE Atlantic: Denmark, W Norway and W Sweden. The records for Orkney and southern Scandinavia (on similar latitudes) represent the northern limit of the species in Europe.

Lithophyllum Phil.

L. incrustans Phil.
Littoral. Epilithic.
Common. 2, 26, 30, 38, 51, 70, 75, 87, 90, 97, 101
Scotland: Harris, Lewis, North Rona and Sula Sgeir, Orkney, Fair Isle and Shetland.
NE Atlantic: Norway and Faeroes.

L. orbiculatum (Fosl.) Fosl.
Infralittoral to 24 m. Epilithic.
Occasional. 26, 31, 100, 101
Scotland: widespread (Adey & Adey, 1973).
NE Atlantic: widespread.

Lithothamnion Phil.

L. glaciale Kjellm.
Littoral, in pools and lagoon systems down to the infralittoral fringe; also infralittoral. Epilithic.
Rare. 22, 79, 97
Scotland: widespread, Adey & Adey (1973) found 36 per cent relative abundance in the Lismore/Mull area of Loch Linnhe.
NE Atlantic: widespread and probably the most abundant mid-to deeper water coralline in these northern and subarctic waters.

L. sonderi Hauck
Infralittoral to 23 m. Epilithic.
Rare. 51, 101
Scotland and *NE Atlantic:* widespread (Adey & Adey, 1973).

Melobesia Lamour.

All specimens identified by Y. M. Butler.

M. membranacea (Esper) Lamour.
Littoral. Epiphytic on *Chondrus crispus*, *Furcellaria lumbricalis*, *Gigartina stellata*, *Laminaria* sp. and *Palmaria palmata*.
Occasional. 13, 21, 40, 44, 85, 92
Further distribution is not given because of taxonomic uncertainty.
Species belonging to this genus have been reported previously for Scotland, Denmark, N Norway and S Sweden.

Mesophyllum Lemoine

M. lichenoides (L.) Lemoine
Littoral. Epiphytic on *Corallina officinalis*.
Occasional. 22, 31, 53, 79, 92
Scotland: sporadically recorded: Kintyre, Rockall (Lemoine, 1923), Lewis and Harris and Orkney (L. M. Irvine *pers. comm*).
NE Atlantic: sporadic but Adey & Adey (1973) indicate its northern distribution extends to Iceland and N Norway.

Phymatolithon Fosl. emend. Adey

P. calcareum (Pall.) Adey & McKibbin
Littoral; infralittoral to 18 m. Epilithic.
Locally frequent. 31, 40, 51, 54. Previous record: Gribun Rocks, Aug. 1943, *Newton* (1949), as *Lithothamnion calcareum.* Found as extensive maerl deposits at station 54.
Scotland: similarly found as maerl at other localities, including the Outer Hebrides and Shetland.
NE Atlantic: Norway and Denmark.

P. lenormandii (Aresch.) Adey
Littoral. Epilithic.
Frequent. 2, 7, 22, 46, 55, 75, 79, 97. Previous records: Gribun Rocks, Aug. 1943, *Newton* (1949), as *Lithothamnion lenormandii f. typica.*
Scotland: frequent.
NE Atlantic: widely distributed to Iceland and N Norway, its range extending south to subtropical waters according to Adey & Adey (1973).

P. polymorphum (L.) Fosl.
Littoral; infralittoral to 23 m. Epilithic.
Occasional. 22, 38, 47, 97, 100
Scotland: frequent.
NE Atlantic: widely distributed to Iceland and N Norway. Adey & Adey (1973) record the species as a dominant crustose coralline in shallow to middle depths throughout Europe.

P. rugulosum Adey
Infralittoral to 24 m. Epilithic.
Occasional. 26, 100, 101
Scotland and *NE Atlantic:* widespread. Adey & Adey (1973) suggest that this species occurs in abundance only at exposed or island sites. Our records for Lunga, Bac Mòr (Treshnish Isles) and Carsaig would tend to confirm this observation.

DUMONTIACEAE

Dilsea Stackh.

D. carnosa (Schmidel) O. Kuntze
Littoral, in deeper shaded pools at most levels and occasionally in caves; most luxuriant in upper infralittoral as underflora and in shade, in 0–3 m; found down to 16 m on stabilized stones. Sometimes massive plants occur where the *Laminaria hyperborea* forest thins out. Epilithic.
Abundant. Generally distributed. Previous records: Iona, *Keddie* (1850), as *Iridaea edulis*; Staffa, *Dendy* (1859, 1860), as *Iridaea*; Gribun Rocks, Aug. 1943, *Newton* (1949).
Scotland: in the west and outer islands, Orkney and Shetland.
NE Atlantic: widely distributed, including Iceland, Spitzbergen, W and S Norway and W Sweden.

Dumontia Lamour.

D. incrassata (O. F. Müll.) Lamour.
Littoral, in damp conditions and shallow pools or channels at all levels on shores of up to moderately strong wave action; infralittoral, on rocks and pebbles from 0–15 m. Epilithic commonly; occasionally epiphytic on larger algae, such as laminae of *Fucus serratus*.
⊕ May to July. Upper littoral plants in Mull, and elsewhere in western Scotland, seem generally to die out in the July/August period, with increasing temperature and insolation. Sporelings, presumably left to over-winter, appear in the following Spring. Degenerating specimens were noted in July 1969 and no material was detected in October 1968 in Calgary Bay where there had been prolific growths in May of that same year. Similar sequences have been noted elsewhere during the Survey. Apart from being slightly later in the year, this closely resembles circumstances we have noted in Kent, SE England.
Abundant. Generally distributed. Previous records: Iona, *Keddie* (1850); "Mull", Aug. 1899, *T. Wise* (GL); both as *D. filiformis*.
Scotland: widely distributed in the west.
NE Atlantic: widespread.

KALLYMENIACEAE

Callophyllis Kütz.

C. laciniata (Huds.) Kütz.
Littoral, occasional in damp circumstances on *Fucus serratus*; infralittoral fringe and infralittoral, 0–22 m. Epilithic on rock or consolidated mixed stones at all depths; epiphytic on *Cystoseira nodicaulis*, *Laminaria digitata* stipes, *L. hyperborea* stipes and holdfasts, *Plocamium cartilagineum*; epizoic on *Pecten*. Juvenile forms June 1968; ♀ June to Oct.
Abundant. Generally distributed. Previous records: Iona coasts, *Hooker* (1821), as *Sphaerococcus laciniatus; Greville* (1830), as *Rhodomenia laciniata; Keddie* (1850), as *Rhodymenia laciniata.*
Scotland: on western islands and mainland, Orkney and Shetland.
NE Atlantic: W and S Norway, W Sweden and Faeroes.

Kallymenia J.Ag.

K. reniformis (Turn.) J.Ag.
Infralittoral to 18 m. Epilithic.
Rare. 84, 100
Scotland: Batters and others have recorded the species on a number of occasions from Orkney (Skaill and Papa Westray), and from Argyll (Machrihanish Bay). Subsequently the species was recorded again for Argyll (Carsaig Island, Sound of Jura) and for Caithness (Ackergill).
NE Atlantic: Morocco to Britain. More recently the most northern finds in Europe have been discovered in the infralittoral on Shetland and as far north as Out Stack beyond Muckle Flugga.

CHOREOCOLACACEAE

Choreocolax Reinsch

C. polysiphoniae Reinsch
Littoral; mid-littoral. Epiphytic on *Polysiphonia lanosa*, itself on *Ascophyllum nodosum.*
Rare but probably more widespread. 5, 14, 92
Scotland: rarely recorded; Lewis and Harris and Shetland but probably simply ignored everywhere.
NE Atlantic: N and W Norway and possibly synonymous with *C. odonthaliae* on Skåne (Levring, 1935).

HILDENBRANDIACEAE

Hildenbrandia Nardo

H. rubra (Sommerf.) Menegh.
Littoral, at all levels, even in high littoral caves where shaded damp conditions are present; rarely recorded from the infralittoral and then only in shallow depths. Epilithic, on pepples in sand or over rock, or on rocks; frequently peripheral to freshwater influence, as in tricklets of streams. ⊕ Sept. 1970.
Frequent; becoming locally abundant and luxuriant. 1, 3, 5, 11, 12, 14, 33, 35, 40, 51, 55, 57, 68, 69, 76, 79, 92
Scotland: common in the west.
NE Atlantic: E Greenland, W Sweden and N Norway. These records are often from fairly deep (35 m) infralittoral conditions. The apparent restrictions on Mull and Scotland may therefore need revision, and infralittoral records from elsewhere may represent a different taxonomic concept.

PALMARIALES

PALMARIACEAE

Palmaria Stackh.

P. palmata (L.) O. Kuntze
Littoral; infralittoral to 13 m. Epilithic; epiphytic on *Laminaria hyperborea* and *L. digitata.*
Abundant. Generally distributed. Previous record: Iona, *Keddie* (1850), as *Rhodymenia palmata.*
Scotland: abundant around the coast.
NE Atlantic: widely distributed as far north as Spitzbergen.

RHODYMENIALES

CHAMPIACEAE

Chylocladia Grev. ex Hook.

C. verticillata (Lightf.) Bliding
Littoral, in pools and damp niches in steep rock surfaces; infralittoral, most commonly down to 22 m. Epilithic, in sand or small stones buried in sand; epiphytic on *Codium fragile* subsp. *atlanticum*, *Cystoseira nodicaulis*, *C. tamariscifolia*, *Halidrys siliquosa*, *Laminaria hyperborea* and *Furcellaria lumbricalis;* epizoic on *Pecten*. ⊕ Sept. 1970; ♂ ♀ June 1968.
Abundant. Generally distributed. Previous

records: "Mull", Aug. 1899, *T. Wise*, (GL); Gribun Rocks, Aug. 1943, *Newton* (1949); both as *C. kaliformis*.
Scotland: widely distributed in the west, Orkney and Shetland, but not recorded for Fair Isle or St Kilda.
NE Atlantic: restricted to W and S Norway and W Sweden.

Gastroclonium Kütz.

G. ovatum (Huds.) Papenf.
Littoral, shaded shallow pools in lower littoral at *Fucus serratus* level, and in shaded damp circumstances and towards the infralittoral fringe, occasionally extending into pools at the *Laminaria digitata* level. Epilithic, occasionally epiphytic on *Cystoseira tamariscifolia*, *F. serratus*, *L. digitata* and in higher (*F. spiralis*) level pools on fairly wave exposed shores.
Frequent. 22, 36, 38, 40, 52, 53, 54, 68, 76, 78, 90, 101. Previous record: Iona, May 1948, *Drew*, cystocarps (BM).
Scotland: reported from the west as far north as Orkney.
NE Atlantic: not recorded futher north or east of Scotland.

Lomentaria Lyngb.

L. articulata (Huds.) Lyngb.
Littoral; infralittoral to 6 m. Epilithic; epiphytic on *Ascophyllum*, *Cystoseira tamariscifolia*, *Fucus serratus*, *Laminaria digitata* (hapteron overgrown with *Membranipora*), *L. hyperborea*, *Corallina officinalis* and *Cystoclonium purpureum*.
Abundant. Generally distributed. Previous records: Iona, 1826, Hb. *Greville* (E); *Keddie* (1850), as *Chylocladia articulata*; Gribun Rocks, Aug. 1943, *Newton* (1949).
Scotland and *NE Atlantic:* widespread.

L. clavellosa (Turn.) Gaill.
Littoral; infralittoral to 22 m. Epilithic and epiphytic on *Ascophyllum*, *Halidrys*, *Himanthalia*, *Laminaria digitata*, *L. hyperborea*, *L. saccharina*, *Corallina*, *Phyllophora crispa*. ♀⊕ May and June.
Frequent. Generally distributed.
Scotland: widespread.
NE Atlantic: Denmark, W Sweden, S and W Norway and Faeroes, but not found in the colder water areas of N Norway, Spitzbergen, Greenland or Iceland.

L. orcadensis (Harv.) Coll. ex Taylor
Infralittoral to 19 m. Epilithic. ♀ June 1968.
Rare. 21, 30
Scotland: Orkney and Shetland, St Kilda, Lewis and Harris, and Argyll.
NE Atlantic: Denmark, W Sweden, and W Norway, near Bergen.

CERAMIALES

CERAMIACEAE

Antithamnion Näg.

A. plumula (Ellis) Thur.
Littoral, pools in lower reaches; infralittoral, most frequent habitat 0–22 m. Epilithic; epiphytic on *Chorda filum*, *Cladostephus spongiosus*, *Laminaria hyperborea*, *Desmarestia aculeata*, *Ceramium rubrum*, *Heterosiphonia plumosa*, *Phyllophora crispa*, *Polysiphonia nigrescens*. Epizoic on *Chlamys*, *Ensis* and *Pecten*. ⊕ ♀ June 1968, Aug. 1970; ♂ June 1968.
Frequent; generally distributed. 6, 13, 21, 39, 41, 43, 51, 54, 65, 67, 71, 79, 99, 113
Scotland: spasmodically reported for the west, but probably more widespread than indicated.
NE Atlantic: consistently recorded, although literature records may be misleading because of variations in taxonomic concept and misdeterminations.

A. spirographidis Schiffner
Infralittoral, in 14–22 m. Epiphytic on *Desmarestia aculeata*, *Ceramium rubrum*, *Heterosiphonia plumosa* and *Phyllophora crispa*.
Occasional; possibly more widespread. 34, 41, 67, 99
Scotland: very rarely recorded; noted by McAllister *et al.* from Argyll at Dunollie and Carsaig West but probably more widespread at least further south.
NE Atlantic: not recorded north of Scotland.

Callithamnion Lyngb.

C. arbuscula (Dillw.) Lyngb.
Littoral, upper half of the littoral most frequently; occasionally lower on shore, associated with *Gigartina*. Epilithic; epizoic on *Balanus*, *Mytilus* and *Patella*. Occurs exclusively on rocky shores subject to considerable wave-action, most luxuriantly in circumstances which permit the retention of slight residual moisture. ⊕ May to Aug.; ♂ June to Aug.; ♀ June to Sept.
Abundant. Generally distributed. Previous records: Staffa, July 1826, *Greville* (BM); 1832, *Harvey* (GL); Treshnish Islands, Cairn na Burgh Mòr, June 1953, *Tindal* (STA-G).
Scotland: recorded widely on all open western shores.
NE Atlantic: the most widespread and consistently present of the species of *Callithamnion*; widely distributed, but absent from the inner Skagerrak and Kattegat, from Denmark and from cold northern areas. The northward and eastward limits appear to be at or very near North Cape.

C. byssoides Arnott ex Harv.
Littoral, midlittoral ridges, as epiphyte on underflora to dense *Ascophyllum nodosum*, in retained moisture in shade; infralittoral, 3–18 m. Epilithic; epiphytic on *Cladophora*

rupestris and on *Laminaria hyperborea* stipes. ⊕ May, June.
Occasional; possibly more widespread. 1, 55, 66
Scotland: rarely recorded, but possibly because of difficulty in determination. Recently established in Argyll in areas just to the south of Oban and Colonsay; Batters recorded it from Orkney (Kirkwall), Arran and Cumbrae.
NE Atlantic: S and W Sweden, Denmark and S and W Norway.

C. corymbosum (Sm.) Lyngb.
Littoral, pools in the lower levels; infralittoral, 0–14 m. Epiphytic on *Fucus* spp., *Laminaria digitata*, *Brongniartella byssoides*, *Ceramium rubrum*, *Corallina officinalis* and *Gracilaria verrucosa* ⊕ May to Oct.; ♂ May 1969, ♀ May to Aug.; **♀** May to Sept.
Frequent. 26, 32, 34, 52, 54, 67
Scotland: not widely reported recently, probably by omission. Batters reported it from various localities on both east and west coasts, including Argyll (Loch Etive) and the most northerly, Orkney (Kirkwall); recently reported in the littoral of Colonsay, the Westport-Tangy area of Kintyre, and Shetland.
NE Atlantic: Denmark to N Norway and Faeroes.

C. decompositum J. Ag.
Infralittoral to 9 m. Epiphytic on *Acrosorium uncinatum*, *Callophyllis laciniata* and *Laminaria hyperborea*. May; parasporangia June and August.
Rare; but possibly more widespread. 51, 54, 65
There are no previous authentic records of this growth-form in U.K. In Scandinavian seas the same growth-form is often referred to *C. bipinnatum* Crouan frat.; distinguishing between these two taxa is difficult. Both are very close to *C. tripinnatum auct.*, as described, although all material examined (including Harvey's) of the latter has been shown to be *C. hookeri*. The northern material here named *C. decompositum* is often parasporangial, and the few British records of parasporangia in *C. hookeri* (q.v.) may have involved material of the former. Until the relationships between *C. decompositum*, *C. bipinnatum* and *C. tripinnatum* are better understood (Price, in prep.), the former name is retained for the Mull material.

C. granulatum (Ducluz.) Ag.
Littoral, shallow pools in lower levels, commonly in areas of strong swell and/or wave-action: sometimes mixed with *C. arbuscula* in more open conditions on exposed cliff faces. Epilithic; often epiphytic on *Corallina*. ⊕**♀** Sept. 1970; ♀ May 1969.
Rare; but probably more widespread. 52, 58, 76
Scotland: not widely reported, probably being mistaken for *C. arbuscula*. Reported by Batters from Ayr, Bute, Orkney, Forfar, Fife and East Lothian. Recently confirmed for Orkney and Lewis.
NE Atlantic: widespread in the southern half of Europe as far north as Faeroes.

C. hookeri (Dillw.) Gray
Middle and lower littoral, rarely higher, in damp situations but tolerates some drying; infralittoral, 0–27 m; when in positions exposed to high degrees of wave action usually in pools with some protection. Epilithic, only rarely; mostly epiphytic, on *Ascophyllum nodosum*, *Cladostephus spongiosus*, *Fucus serratus*, *Laminaria hyperborea* (stipes), *Acrosorium uncinatum*, *Callophyllis laciniata*, *Corallina officinalis*, *Ceramium flabelligerum*, *Lomentaria articulata*, *Phycodrys rubens*, *Plumaria elegans*, *Polysiphonia lanosa;* epizoic on *Pecten*. ⊕**♀** May to Sept.; ♂ May and July 1969; ♀ June 1968, May 1969, and Sept. 1970.
Abundant. Generally distributed.
Scotland: widely recorded on western islands and mainland coasts; probably overlooked.
NE Atlantic: N Norway (Nordland) to W Sweden and Denmark; further records of *C. polyspermum* and *C. scopulorum* frequently relate to forms of *C. hookeri* and such records exist for Faeroes, Finmark, and Iceland. *C. brodiaei* auct., also involved in the *C. hookeri* complex, is known from a similarly wide Scottish distribution and is reported for Denmark, W Sweden and S Norway.

C. roseum (Roth) Lyngb.
Littoral, lower midlittoral downwards; infralittoral (0 –) 12–15 m. Epiphytic on *Fucus* and *Polysiphonia lanosa;* epizoic on *Scrupocellaria*. ⊕ **♀** May to June; ♀ May to Sept.; ♂ Sept. 1970.
Occasional. 2, 26, 32, 33
The variable application of this name makes statements about the distribution of the species difficult.
Scotland: Batters included many localities in both east and west Scotland from as far north as Orkney, south to Ayr. Recent Shetlands work included collections which may be this species. Apart from McAllister *et al.*, who cite, under the name *C. tenuissimum*, records from just south of Oban which may be applicable to this species or to *C. byssoides*, there are no other recent records from the area.
NE Atlantic: reported from N and W Norway and Denmark and probably widespread within its range.

C. tetragonum (With.) Gray
Infralittoral. Epiphytic, on *Laminaria hyperborea* stipes and laminae. **♀** ♂ ♀ July 1968, all on same plant.
Occasional; probably more widespread in some years than detected. 38, 101. Previous record: Staffa, July 1826, *Carmichael* (K in BM)
Scotland: in the lower littoral and the infra-

littoral on various hosts: Colonsay; Argyll mainland; Lewis; St Kilda (drift only); North Rona; Orkney and Shetland. Earlier, Batters listed Aberdeen (Peterhead), Orkney (Skaill etc.), Bute and Arran, and Ayr (Portencross) for the species, and Orkney (Skaill), Argyll (Appin, Machrihanish Bay), Bute (Cumbrae), and Ayr (Saltcoats) for the variety *brachiatum* J. Ag., which has no taxonomic significance. Batters' Orkney material is in the BM and since the species is easily recognisable, there is no reason to doubt other locations. BM holds additional material from Kintyre (Carradale), S Ayr (Ballantrae) and from Argyll (Loch Fyne).
NE Atlantic: S Norway, W Sweden, Denmark and recently reported from the Vestmannaeyjar, Iceland.

Ceramium Roth

C. ciliatum (Ellis) Ducluz.
Littoral, in deep, laterally shaded pools in high to midlittoral. Epilithic, amongst *Corallina;* epiphytic on *Corallina.* ⊕ July.
Occasional; possibly more widespread. 22, 68, 78, 101. Previous record: Iona, 1826, Hb. *Greville* (E).
Scotland: Galloway, Ayr (Saltcoats), Arran, Colonsay, Isle of Lewis, St Kilda and Orkney; records from Shetland have not recently been confirmed.
NE Atlantic: from the Faeroes to the Canaries.

C. diaphanum (Lightf.) Roth agg. (incl. *C. strictum* Harv.)
Littoral, in pools in mid- to lower littoral, also epiphytic in lower littoral pools; infralittoral, to 15 m. Epilithic; epiphytic on *Fucus serratus*, *Halidrys siliquosa*, *Laminaria digitata*, *L. hyperborea;* epizoic on *Ensis* and *Venus.* ♀ May to July; ⊕ May.
Abundant. Generally distributed. Previous records: Iona, *Keddie* (1850); Gribun Rocks, Aug. 1943, *Newton* (1949), as *C. strictum.*
Scotland: generally recorded from island and mainland areas in the west.
NE Atlantic: only recorded for W Norway (Nordland) and S Sweden.

C. flabelligerum J. Ag.
Littoral, in pools, sometimes in exposed damp places. Epilithic; occasionally epiphytic on various algae. ⊕ June 1968.
Occasional, but probably more widespread than recorded. 7, 16, 22
Scotland: Lewis; Ayrshire; Colonsay; Galloway. Batters records it for Midlothian (Joppa), Bute (Cumbrae), Ayr (Largs) and Ailsa Craig.
NE Atlantic: a record unconfirmed recently from W Norway only and, in the absence of confirmation, the Lewis locality may be the furthest north.

C. rubrum (Huds.) Ag. agg.
Littoral, lower levels, in pools and damp places, occasionally in pools higher on shore; infralittoral, on rocks to at least 21 m. in favourable circumstances. Epilithic; epiphytic on many other green, brown and red algae; occasionally epizoic. ⊕ May to Oct.; ♀ May to Sept.; ♀ May 1969; ♂ Aug. 1970.
Abundant. Generally distributed. Previous record: Iona, *Keddie* (1850).
Scotland: throughout the west.
NE Atlantic: widely distributed as far as N Norway, Spitzbergen, and W (but not E) Greenland through to the Barents Sea (Murman) coast.

C. shuttleworthianum (Kütz.) Rabenh.
Littoral, lower levels, usually below mid-tide level; commonly on wave-beaten shores, but also present elsewhere in lesser amounts. Epilithic; epiphytic occasionally, on *Ascophyllum nodosum;* epizoic on *Mytilus* and *Patella.* ⊕ May to Sept.; ♀ May 1969.
Abundant. Generally distributed. Previous record: Iona, 1826, Hb. *Greville* (E), as *C. ciliatum.*
Scotland: widespread in the west, common in the north.
NE Atlantic: Iceland; W Norway (Nordland), Faeroes.

C. strictum Harv.
See *C. diaphanum*

Compsothamnion (Näg.) Schmitz

C. thuyoides (Sm.) Schmitz
Infralittoral, in 6–21 m. Epilithic, although other authors report the species as epiphytic. ⊕ Sept. 1970.
Rare; but probably more widespread. 15, 50, 95
Dixon (1960) considers that the apparent rarity is probably due to confusion of the taxon with young *Callithamnion;* no such confusion is possible in the presence of reproductive organs (tetrasporangia; carposporangia) which are terminal on short laterals in *Compsothamnion. C. gracillimum* De Toni, considered by many authors a variety of *C. thuyoides*, is sometimes difficult to separate.
Scotland: Batters reports it from Orkney, Fife (Kincraig, Earlsferry) and Bute (Arran; Cumbrae). Specimens also known from Orkney (Risa Island, Scapa Flow, Kirkwall Bay), from Aryshire and from Easdale and Carsaig, Argyll.
NE Atlantic: infrequent, being reported from W Norway in 1905 but not confirmed in either of the most recent studies of that area; also recorded from Kristinaberg (W Sweden) and from a few localities in Oslofjord (S Norway), but very rarely.

Griffithsia Ag.

G. corallinoides (L.) Batt.
Infralittoral, 3–18 m. Epilithic; epizoic on *Pecten.* ⊕ June 1968.

Occasional. 13, 48, 65, 72
Scotland: infrequently recorded for the western mainland and islands: Argyll; Cumbrae; Orkneys; Cromarty; Shetlands (where abundant in 1973), but probably overlooked in the past since it grows sublittorally and quite possibly more frequent and widespread.
NE Atlantic: S and W Norway as far north as Bergen and the western (Bohuslän) coast of Sweden.

G. flosculosa (Ellis) Batt.
Littoral in pools at *Fucus serratus – Laminaria digitata* level; infralittoral in 12–23 m. Epilithic on pebbles and rock surfaces; epiphytic on *Laminaria hyperborea.* ⊕ ♂ June 1968.
Frequent. 21, 30, 33, 38, 53, 74, 79, 84, 85, 87, 99, 100, 101
Not common as an intertidal plant on Mull, the very large majority of collections being infralittoral.
Scotland: frequent on almost every part of the coast, in western Scotland, including the Shetlands (where it is generally less abundant than *G. corallinoides*) and the Orkneys (Batters).
NE Atlantic: other than the Børgesen & Jónsson (1905) statement of W Norway which is unsubstantiated in recent works for the area, Faeroes is the northern limit.

Pleonosporium (Näg.) Näg. ex Hauck.

P. borreri (Sm.) Näg. ex Hauck.
Littoral, shaded pools in standing water at mean low water springs level; infralittoral, more frequently in 3–22 m. Epilithic; epiphytic on *Laminaria digitata*, and *L. hyperborea.* Epizoic on *Pecten.* ⊕ June to Sept.
Occasional; possibly more widespread. 6, 31, 40, 51, 52, 65, 76
Scotland: not widely recorded, Orkney and Argyll (Falls of Lora) (Batters); recent records from that county have been established widely for the mainland. Detected in the Shetlands and elsewhere, probably indicating that the species is widespread and common but needs to be looked for particularly. Apart from our own littoral records, Norton (1972) records *P. borreri* epiphytic in the littoral of both Lewis and Harris. The familiar concept of this species as exclusively infralittoral is therefore not correct, although it is certainly more frequent there and, in our own experience, needs shaded standing water conditions to appear in the littoral.
NE Atlantic: tenuously recorded only for W Norway. The recently established Shetland record may well represent the northern limit for the species.

Plumaria Schmitz

P. elegans (Bonnem.) Schmitz
Littoral (particularly in shaded locations); infralittoral to 6 m. Epilithic; epiphytic on *Ascophyllum*, *Cystoseira tamariscifolia*, *Laminaria digitata* and *Corallina officinalis.*
Abundant. Generally distributed. Previous record: Gribun, *Newton* (1949).
Scotland: frequent and widespread.
NE Atlantic: widely distributed to N Norway and Iceland.

Ptilota Ag.

P. plumosa (Huds.) Ag.
Littoral in pools; infralittoral to 27 m. Epilithic; epiphytic on *Laminaria digitata*, *L. hyperborea*, *L. saccharina*, *Cystoseira nodicaulis*, *Fucus serratus* and *Palmaria palmata.* ♀ May 1967, June 1968 and Aug. 1966.
Abundant. Generally distributed. Previous records: Iona, *Lightfoot* (1777), as *Fucus plumosus*; Staffa, Sept. 1846, *W. Keddie* (GL); *Greville* (1830); Gribun Rocks, Aug. 1943, *Newton* (1949).
Scotland: a consistent element of the lower- and infralittoral.
NE Atlantic: from Spitzbergen to England (Yorkshire) and S Ireland.

Ptilothamnion Thur. in Le Jol.

P. pluma (Dillw.) Thur.
Infralittoral, 6–19 m. Epiphytic on *Laminaria hyperborea.* ⊕ ♂ Aug. 1970.
Rare; but probably sometimes overlooked. 100
Scotland: Shetland (Dixon, 1963; and 1973) and the Isle of Harris; Batters reported it also from Appin, Argyll, but we cannot trace material to support this record. The Shetland records represent the northern limit of the species in the NE Atlantic.

Seirospora Harv.

S. seirosperma (Harv.) Dixon
Infralittoral to 20 m. Epilithic; epiphytic on *Desmarestia aculeata*; epizoic on *Pecten.* Seirospores June to Aug.
Occasional; probably more widespread than records indicate. 13, 41, 54, 67, 71, 72
Scotland: Batters reports it from Aberdeen (Peterhead), Cromarty, Orkney (Kirkwall Bay) and Bute (Arran); BM material substantiates the Orkney records only. Recently detected in Shetland, and there are records for the Argyll mainland adjacent to Mull.
NE Atlantic: S and W Norway, W Sweden and Denmark.

Spermothamnion Aresch.

S. repens (Dillw.) Rosenv.
Infralittoral in 12–21 m. Epiphytic on *Phyllophora crispa*, *Ceramium rubrum* and other algae; epizoic on *Pecten.* ♀ Aug. 1970.
Rare. 13, 52, 67, 99
Scotland: Argyll, Colonsay, Lewis and Harris, and Shetland; probably overlooked elsewhere.
NE Atlantic: Finmark, Nordland, and S and W Norway, and along the whole coast of Sweden; also common and widespread in Denmark.

DELESSERIACEAE

Acrosorium Zanard. ex Kütz.

A. reptans (Crouan & Crouan) Kylin
See *Cryptopleura ramosa*

A. uncinatum (Turn.) Kylin
Infralittoral, 3–23 m. Epilithic; epiphytic on *Laminaria hyperborea* and *Phyllophora crispa*. ⊕ Aug. 1970.
Occasional; possibly more widespread. 15, 26, 27, 51, 99, 101

The difficulties of distinction between *Acrosorium uncinatum*, *A. reptans* and *Cryptopleura ramosa*, the latter two of which are sometimes alternative forms of the same plant, have been emphasised by Hoek & Donze (1966) and commented on by Ardré (1970). Our records relate to the *A. uncinatum* form, except as otherwise detailed under *Cryptopleura ramosa*.
Scotland: There are few previous records of *A. unciniatum*: Batters records it (as *Nitophyllum uncinatum*) from Skaill, Orkney, and it has been recently recorded from Lewis and from West Carsaig, Argyll.
NE Atlantic: not recorded further north than Orkney.

Cryptopleura Kütz.

C. ramosa (Huds.) Kylin ex Newton
Littoral, in deep pools high on shore and in shallow lower littoral pools, also on damp rocks under *Ascophyllum nodosum* and *Fucus serratus;* infralittoral in 0–23 m. Epilithic; epiphytic, in lower littoral, on *Corallina officinalis*, *Furcellaria lumbricalis*, *Lomentaria articulata*, *Cystoseira nodicaulis*, and *C. tamariscifolia*; epiphytic, in infralittoral, on *Laminaria digitata*, *L. hyperborea* (stipes and holdfasts) and *L. saccharina*; epizoic on *Pecten* in lower end of infralittoral range. ⊕ ♀ May to Oct.; ♂ June 1968.
Abundant. Generally distributed. Previous records: Iona, *Lightfoot* (1777), as *Fucus endiviaefolius*; July 1851, Hb. *Traill* (E) as *Nitophyllum laceratum.*
Scotland: consistently recorded from the west, Orkney and Shetland.
NE Atlantic: recorded only for the Faeroes.

In Mull, deeper infralittoral finds of the species often involved, sometimes exclusively, the *Acrosorium reptans* phase which was commonly fertile and not infrequently epiphytic on *Phyllophora crispa.* Batters has previously recorded *Nitophyllum* (= *Acrosorium*) *reptans* from Arran and Cumbrae. Material of "normal" *Cryptopleura*, collected in August 1970, bore tetrasporangia and carposporangia on the same plants.

Delesseria Lamour.

D. sanguinea (Huds.) Lamour.
Littoral, occasional in deep pools in lower littoral (*Himanthalia* level) of exposed rocky headlands or in deep and shaded pools at *Fucus serratus* level in more sheltered areas; infralittoral, common between 0–24 m, in the upper 3 m of which it is most common in protected niches, although also occurring epiphytically. Epilithic; epiphytic on *Laminaria digitata* and *L. hyperborea* stipes; epizoic on *Pecten* in depths from 9–21 m.
Abundant. Generally distributed. Previous records: Iona, *Lightfoot* (1777), as *Fucus sanguineus*; Staffa, *Dendy* (1859, 1860); Gribun Rocks, Aug. 1943, *Newton* (1949).
Scotland: virtually everywhere in the west but St Kilda.
NE Atlantic: widely recorded but absent from Greenland and Spitzbergen.

Hypoglossum Kütz.

H. woodwardii Kütz.
Littoral, in pools or damp places at most levels, but only in deep pools when high on shore; infralittoral, 0–21 m. Epilithic; epiphytic on stipes of *Laminaria saccharina*; epizoic on *Buccinum* and *Pecten.* ⊕ June.
Frequent, but never abundant in growth. 1, 13, 20, 21, 35, 40, 43, 44, 53, 65, 66, 79, 91. Previous record: Gribun Rocks, Aug. 1943, *Newton* (1949).
Scotland: frequent in the west.
NE Atlantic: absent further to the north-east, the Shetlands representing the northward limit in the E Atlantic.

Membranoptera Stackh.

M. alata (Huds.) Stackh.
Littoral; infralittoral to 27 m. Epilithic; epiphytic on *Ascophyllum*, *Desmarestia aculeata*, *Fucus serratus*, *F. vesiculosus*, *Laminaria digitata*, *L. hyperborea*, *L. saccharina*, *Chondrus*, and *Gigartina stellata.* ⊕ June 1968; ♀ June 1968, May 1969.
Common. Generally distributed. Previous records: Iona, *Keddie* (1850), as *Delesseria alata*; Gribun Rocks, *Newton* (1949).
Scotland: widespread.
NE Atlantic: widely distributed as far north as Norway (Nordland) and Iceland.

Myriogramme Kylin

M. bonnemaisonii (Ag.) Kylin
Infralittoral, 6–9 m. Epilithic.
Rare. 20, 99
Scotland: widespread.
NE Atlantic: Shetland is the most northerly record.

Nitophyllum Grev.

N. punctatum (Stackh.) Grev.
Littoral, lower levels; infralittoral to 24 m. Epilithic; epiphytic on *Cystoseira nodicaulis*, *Laminaria hyperborea*, *L. digitata* stipes and *Plocamium.* ⊕ June to Sept.; ♀ Aug. 1970.
Common. 1, 13, 21, 22, 26, 30, 43, 44, 54, 60, 65, 67, 72, 79, 84, 100, 101. Previous record: Iona, *Keddie* (1850).

Scotland: frequent.
NE Atlantic: sporadically in W Norway only.

Phycodrys Kütz.

P. rubens (L.) Batt.
Littoral; infralittoral to 27 m. Epilithic; epiphytic on *Jania rubens*, *Nitophyllum punctatum*, *Cystoseira nodicaulis*, *Laminaria digitata*, *L. hyperborea* and *L. saccharina*. Epizoic on *Pecten* and Spider Crab. ⊕ May and June; ♀ June 1968.
Abundant. Generally distributed. Previous record: Iona, *Hooker* (1821), as *Delesseria sinuosa*.
Scotland: A common cold water species, widely distributed and abundant.
NE Atlantic: widely distributed and abundant.

Polyneura Kylin

P. gmelinii (Lamour.) Kylin
Infralittoral to 9 m. Epiphytic on *Laminaria hyperborea* stipes. ♂ June 1968.
Rare. 26. Previous record: Gribun Rocks, Aug. 1943, *Newton* (1949), as drift.
Scotland: Argyll and Orkney (Batters; Newton, 1931, 1949).
NE Atlantic: the Batters record for Orkney (confirmed by material in BM) represents the northern limit of the species in Europe.

DASYACEAE

Heterosiphonia Mont.

H. plumosa (Ellis) Batt.
Littoral in pools; infralittoral, to 22 m. Commonly epilithic at all depths, on rocks or pebbles in sand; epiphytic on *Laminaria hyperborea* and *Phyllophora crispa*; occasionally epizoic on *Pecten*. ⊕ June to Oct.; ♀ June and Aug.
Abundant. 1, 15, 21, 22, 26, 27, 30, 33, 40, 41, 44, 45, 50, 51, 54, 67, 72, 79
Scotland: widely reported for the west, Orkney and Shetland.
NE Atlantic: around Bergen, W Norway and the Bohuslän coast, W Sweden. Possibly part of the northward spreading of the southerly floristic element (cf. Printz, 1952).

RHODOMELACEAE

Brongniartella Bory

B. byssoides (Gooden. & Woodw.) Schmitz
Littoral, occasionally in lower level standing water; infralittoral, 3–22 m. Epilithic; epiphytic on *Desmarestia aculeata*, *Laminaria hyperborea* and *Delesseria sanguinea*; epizoic on *Pecten* in depths. ♂ ♀ ♀ June to Aug.
Abundant. Generally distributed.
Scotland: widespread on both east and west coast, including Bute, Ayr, Galloway, Argyll, Lewis, Orkney and Shetland.
NE Atlantic :Denmark to Nordland, N Norway.

Chondria Ag.

C. dasyphylla (Woodw.) Ag.
Littoral, in pools or standing water; infralittoral, 0–3 m. Epilithic (infralittoral); epiphytic (littoral) on *Cystoseira tamariscifolia* and *Corallina officinalis*. Oct. 1968.
Occasional; but possibly more widespread. 40, 44, 54, 78, 79
Scotland: only recorded in recent years from Orkneys and Colonsay; Batters listed Aberdeen (Peterhead), Elgin (Lossiemouth), Orkneys, Bute (Arran, Cumbrae) and Ayr (Ardrossan) as known localities. The record for Cumbrae may be in error, specimens in BM being of *Chylocladia*.
NE Atlantic: Denmark; possibly the Orkneys represent the northern limit for this species.

Laurencia Lamour.

L. hybrida (DC.) Lenorm. ex Duby
Littoral; infralittoral to 8 m. Epilithic; epiphytic on *Fucus* spp. and *L. hybrida*. ⊕ May to July.
Abundant. Generally distributed. Previous records: Gribun Rocks, Aug. 1943, *Newton* (1949), as *L. caespitosa*. Cairn na Burgh Mòr, Treshnish Islands, June 1953, *Tindal* (STA-G).
Scotland: recorded frequently as far north as Shetland which appears to be the northern limit in Europe.

L. obtusa (Huds.) Lamour.
Infralittoral, 0–3 m. Epilithic. ⊕ June 1968; ♀ July 1970.
Rare. 13, 52
Scotland: infrequent. Batters lists Firth of Forth, Orkney, Argyll (Loch Etive, Falls of Lora, Kintyre), Bute, Ayr and Solway Firth; recently confirmed for Colonsay and Ayr. The northern limit of its distribution in Europe is the Orkneys.

L. pinnatifida (Huds.) Lamour.
Littoral. Epilithic; occasionally epiphytic on *Laurencia hybrida*. ⊕ May to July; ♂ May 1968.
Abundant. Generally distributed. Previous records: Iona, *Keddie* (1850); Rubh an Righ, Calve Island, *Kitching* (1935); Gribun Rocks, Aug. 1943, *Newton* (1949); Cairn na Burgh Mòr, Treshnish Islands, June 1953, *Tindal*, (STA-G).
Scotland: recorded widely.
NE Atlantic: Denmark, W and S Norway and the Faeroes.

Odonthalia Lyngb.

O. dentata (L.) Lyngb.
Littoral; infralittoral, to 18 m. Epilithic; epiphytic on *Laminaria hyperborea*; epizoic on *Pecten*.
Abundant. Generally distributed. Previous record: Iona, *Lightfoot* (1777), as *Fucus dentatus* L.
Scotland: frequent.
NE Atlantic: frequent, Børgesen & Jónsson indicate its range extends into the Siberian, Murman and White Seas. The southern limit

of distribution in the British Isles is Isle of Man, Yorkshire and Down.

Polysiphonia Grev.

P. brodiaei (Dillw.) Grev. ex Harv. in Hook.
Littoral, particularly in pools on exposed shores. Epilithic; epiphytic (rarely) on *Corallina*. ⊕ May to Sept.; ♂ May to Sept.; ♀ July to Sept.
Abundant. Generally distributed. Previous record: Staffa, July 1826, *Carmichael* (E), as *P. brodiaei β*.
Scotland: widespread.
NE Atlantic: becoming less frequent in the colder northern waters. Tromsö, in N Norway, represents its northern limit of distribution.

P. elongata (Huds.) Spreng.
Littoral, in shaded standing water or damp circumstances; infralittoral, to 18 m. Epilithic; epiphytic on stipes of *Laminaria hyperborea* and *Zostera*; epizoic on *Pecten*. ♀ June 1968.
Frequent. 13, 20, 40, 44, 54, 64, 67, 72, 78, 84, 86, 92, 99
Scotland: frequent throughout.
NE Atlantic: widespread with the exception of the colder waters of N Norway, Spitzbergen, Iceland and Greenland.

P. fibrata (Dillw.) Harv.
Littoral, in pools; infralittoral to 9 m. Epilithic; epiphytic on stipes of *Laminaria hyperborea*. ♀ June 1968.
Occasional. 49, 53, 79, 97
Scotland: recorded only scantily: Bute, Argyll, Orkney and Shetland.
NE Atlantic: not recorded further and Shetland is thus the northern limit in Europe.

P. fibrillosa (Dillw.) Spreng.
Infralittoral, 6–9 m. Epilithic. ⊕ June 1968.
Rare. 13
Scotland: recorded frequently: East Lothian, Fife, Aberdeen, Orkney, Argyll, Bute and Ayr (Batters); recently reported for Shetland (drift, Dixon, 1963), Argyll, Colonsay, and Kintyre.
NE Atlantic: infrequent records which may be in part due to taxonomic confusion. *P. fibrillosa* has been considered to be *P. violacea* by Børgesen & Jónsson (1905), and also *P. violacea* f. *fibrillosa* (Dillw.) Aresh. (Sundene, 1953).

P. fruticulosa (Wulf.) Spreng.
Littoral; infralittoral to 7 m. Epilithic; epiphytic on *Cystoseira nodicaulis*, *C. tamariscifolia*, *Halidrys*, and *Laminaria saccharina*.
Frequent. 7, 33, 34, 54, 56, 64, 68, 78, 79, 92
Scotland: reported from Ayr and Bute (Batters), Shetland (Dixon, 1963), Lewis and Harris, Colonsay, and Kintyre.
NE Atlantic: Shetland forms the northern limit of distribution.

P. lanosa (L.) Tandy
Littoral. Epiphytic on *Ascophyllum nodosum* and less frequently on *Fucus serratus* and *F. vesiculosus*. ♀ Sept. 1970.
Abundant. Generally distributed. Previous records: Iona, *Keddie* (1850); Gribun Rocks, Aug. 1943, *Newton* (1949), both as *P. fastigiata*; Cairn na Burgh Mòr, Treshnish Islands, June 1953, *Tindal* (STA-G).
Scotland: throughout.
NE Atlantic: widespread excepting in the Arctic waters around Spitzbergen and Greenland.

P. nigra (Huds.) Batt.
Infralittoral to 21 m. Epilithic; epizoic on *Pecten*. ⊕ Aug. and Sept. 1970.
Occasional. 13?, 48, 52, 67, 84
Scotland: Colonsay, Lewis and Harris.
NE Atlantic: Denmark, N Norway (not recently recorded) and the Faeroes.

P. nigrescens (Huds.) Grev.
Littoral; infralittoral to 21 m. Epilithic; epiphytic on *Laminaria hyperborea*. Epizoic on *Pecten*. ♀ June to Sept.; ♂ ⊕ May and June.
Abundant. Generally distributed.
Scotland: consistently found.
NE Atlantic: widespread.

P. urceolata (Lightf. ex Dillw.) Grev.
Littoral; infralittoral to 21 m. Epilithic; epiphytic on *Desmarestia aculeata*, *Laminaria digitata*, *L. hyperborea* and *L. saccharina*; epizoic on *Pecten*. ⊕ May to Aug.; ♀ May to Aug.; ♂ May 1969.
Abundant. Generally distributed. Previous records: Iona, *Keddie* (1850); Staffa, July 1826, *Carmichael* (E).
Scotland: frequent.
NE Altantic: widely distributed to N Norway and the White and Murman Seas.

P. violacea (Roth) Spreng.
Infralittoral to 3 m. Epiphytic on *Laminaria hyperborea* stipes.
Rare. 44
Scotland: occasional (Batters; Lyle, 1929; McAllister *et al.*).
NE Atlantic: widely distributed, but absent from N Norway, Spitzbergen, Iceland and Greenland.
See note under *P. fibrillosa*.

Pterosiphonia Falkenb.

P. parasitica (Huds.) Falkenb.
Littoral; infralittoral to 21 m. Epilithic; epiphytic on *Corallina officinalis*, *Phyllophora crispa* and *Laminaria hyperborea*. ♀ ⊕ June 1968.
Abundant. 1, 15, 21, 22, 26, 27, 41, 43, 50, 54, 59, 60, 65, 66, 67, 71, 83, 84, 99
Scotland: abundant.
NE Atlantic: widely distributed, north to S Norway and Iceland.

P. thuyoides (Harv.) Schmitz
Infralittoral to 12 m. Epilithic.
Rare. 22, 31, 72
Scotland: infrequently recorded from East Lothian, Bute and Ayr (Batters).
NE Atlantic: no other records and Mull records therefore represent the northern limit of this species in Europe.

Rhodomela Ag.

R. confervoides (Huds.) Silva
Littoral; infralittoral to 18 m. Epilithic; epiphytic on *Laminaria digitata* and *L. hyperborea*. ⊕ May and June; ♀ June 1968.
Frequent. 13, 20, 21, 22, 25, 26, 27, 40, 41, 43, 44, 52, 53, 56, 57, 65, 72, 88, 100. Previous record: Iona, *Keddie* (1850), as *R. subfusca.*
Scotland: a common constituent of most shores.
NE Atlantic: widely distributed, although not recorded for the colder waters around northern Norway, Spitzbergen, and Greenland.

R. lycopodioides (L.) Ag.
Littoral; infralittoral. Epiphytic on *Laminaria hyperborea* stipes.
Rare. 33. Previous records: Staffa, 1826, Hb. *Greville* (E); *Greville* (1830)
The taxonomic status of this species has yet to be established; many authors have described plants of intermediate morphologies between *R. lycopodiodes* and *R. confervoides*, and have consequently reduced the former species to varietal status of the latter.

BANGIOPHYCEAE

PORPHYRIDIALES

GONIOTRICHACEAE

Goniotrichum Kütz.

G. alsidii (Zanard.) Howe
Infralittoral, 3–18 m. Epilithic. Epiphytic on *Cladophora*, *Sphacelaria plumula*, *Heterosiphonia plumosa* and *Phycodrys rubens.*
Rare; but probably more frequent and widespread than records indicate. 12, 44, 72
Scotland: few previous records: Loch Etive, Argyll (Batters) and Cumbrae, Bute, July 1891 (BM).
NE Atlantic: probably overlooked in many studies and recorded rarely from SW Norway and sparsely along the whole Swedish west coast. Børgesen & Jónsson include *G. alsidii* in their warm-boreal element, with northern limit in Scotland and the southern limit in N Africa or further south.

BANGIALES

ERYTHROPELTIDACEAE

Erythrotrichia Aresch.

E. carnea (Dillw.) J. Ag.
Littoral, in damp circumstances or high level pools; infralittoral, to 22 m. depth. Epiphytic on *Ceramium rubrum*, *Polysiphonia lanosa*, and *Sphacelaria fusca*, in the littoral; on *Cladophora*, *Membranoptera alata*, *Polysiphonia nigrescens*, and *Rhodomela confervoides* in the infralittoral.
Frequent; but probably often overlooked. 2, 13, 25, 32, 33, 40, 41, 44, 51, 52, 84, 85
Scotland: Shetland, Orkney, St Kilda, North Rona, Lewis, Argyll, Ayr and Bute.
NE Atlantic: sparse, being found in N and W Norway, Faeroes and W Sweden.

E. investiens (Zanard.) Born.
Littoral, in pools. Epiphytic on *Cladophora* sp.
Rare, but probably more frequent. 2
Scotland: not often recorded; Batters cites Bute (Arran), Ayr (Saltcoats and Fairlie), Edinburgh (Joppa) and Fife (Kilrenny).
NE Atlantic: no other record; this may mean little or nothing, in view of the taxonomic and observational problems attached to accurate determination.

BANGIACEAE

Bangia Lyngb.

B. atropurpurea (Roth) Ag.
Littoral, mid and lower shores predominantly; occasionally infralittoral, in very shallow depths. Epilithic; epiphytic on *Plumaria elegans*, *Polysiphonia fruticulosa* (itself on *Cystoseira nodicaulis*) and *Polysiphonia nigra.*
Frequent. 3, 5, 22, 23, 24, 38, 51, 67, 68, 90. Previous record: Cairn na Burgh Mòr, Treshnish Islands, on *Laurencia pinnatifida*, June 1953, *Tindal* (STA-G).
Scotland: General throughout the west.
NE Atlantic: N Norway to W Greenland.

Porphyra Ag.

P. leucosticta Thur.
Littoral; infralittoral to 9 m. Epiphytic on *Ascophyllum*, *Cystoseira nodicaulis*, *Fucus serratus*, *F. vesiculosus*, *Himanthalia*, *Laminaria hyperborea*, *Chondrus*, *Gelidium corneum*, *Gigartina stellata*, *Laurencia pinnatifida. Polysiphonia lanosa*, *P. nigrescens*, *Palmaria palmata* and *Ulva lactuca.*
Abundant. Generally distributed. Previous record: Cairn na Burgh Mòr, Treshnish Islands, June 1953, *Tindal* (STA-G).
Scotland: frequent.
NE Atlantic: widely distributed in the south; Denmark, W Norway, W Sweden and Faeroes.

P. miniata (Ag.) Ag.
Littoral, lower levels; infralittoral to 15 m. Epilithic; epiphytic on *Desmarestia aculeata*, *Fucus serratus*, *F. vesiculosus*, *Laminaria digitata*, *L. hyperborea*, *L. saccharina*, *Chondrus crispus*, *Gigartina stellata* and *Zostera marina.*
Frequent. 1, 4, 6, 13, 21, 26, 34, 40, 56, 58, 65, 101
Scotland: frequent.

NE Atlantic: from W Sweden to Spitzbergen, Iceland and E Greenland.

P. purpurea (Roth) Ag.
Littoral. Epilithic; epizoic on *Mytilus.*
Rare. 4, 11, 23, 24, 30. Previous record: Iona, *Lightfoot* (1777), as *Ulva laciniata.*
Scotland: frequent (often as *P. umbilicalis f. lacinata*).
NE Atlantic: widespread from N Norway to Iceland.

P. umbilicalis (L.) J.Ag.
Littoral. Epilithic; epizoic on *Patella.*
Abundant. Generally distributed. Previous records: Iona, *Keddie* (1850), as *P. vulgaris*; Staffa, *Dendy* (1859, 1860), as *P.* [?*umbilicalis*]; Gribun Rocks, Aug. 1943, *Newton* (1949).
Scotland: abundant.
NE Atlantic: widespread as far north as N Norway and Iceland.

XANTHOPHYTA

XANTHOPHYCEAE

VAUCHERIALES

VAUCHERIACEAE

Vaucheria DC.

Vaucheria spp.
Littoral.
Found in saltmarsh around bases of phanerogams; on one occasion found among *Sphacelaria*; epiphytic on *Cystoseira tamariscifolia.*
Occasional; but probably more widespread. 3, 33, 38, 78, 86
Vaucheria spp. are scantily reported for Scotland and the NE Atlantic.

PHAEOPHYTA

PHAEOPHYCEAE

ECTOCARPALES

ECTOCARPACEAE

Acinetospora Born.

A. crinita (Carm. ex Harv.) Kornm.
Littoral. Epilithic. ⊞ 1970.
Rare. 39
Scotland: frequent in the west; Batters records the species for Ayr, Renfrew, Bute, Argyll, Shetland, Aberdeen, Edinburgh and East Lothian.
NE Atlantic: there are no records north of Scotland and the Mull material is close to the northern limits of the species.

Ectocarpus Lyngb.

E. fasciculatus Harv.
Littoral; infralittoral to 15 m. Epilithic; epiphytic on *Desmarestia aculeata, Fucus serratus, Himanthalia lorea, Laminaria digitata, L. hyperborea* and *Scytosiphon lomentaria.* ⊞ May to Sept.; ○ June 1968.
Frequent. Generally distributed.
Scotland: frequent.
NE Atlantic: widely distributed.

E. siliculosus (Dillw.) Lyngb.
Littoral; infralittoral to 21 m. Epilithic; epiphytic on *Codium tomentosum, Chorda filum, Halidrys siliquosa, Laminaria hyperborea, Ceramium rubrum, Chondrus crispus,* and *Zostera.* ⊞ May to Sept.; ○ June to Sept.
Frequent. Generally distributed. Previous record: Iona, *Keddie* (1850).
Scotland: frequent.
NE Atlantic: widely distributed.

Feldmannia Hamel

F. globifera (Kütz.) Hamel
Infralittoral, 3–6 m. Epilithic.
Rare. 60
Scotland: Ayr, Bute, East Lothian (Dunbar) (BM) and Arran (Lamlash Bay) (BM).
NE Atlantic: W Norway and probably more widespread than the records indicate.

F. irregularis (Kütz.) Hamel
Littoral. Epilithic.
Rare. 66
Scotland: scantily recorded: Colonsay, Angus (Auchimithie) (BM) and Ayrshire (Sea Mill) (BM).
NE Atlantic: Norway.

F. simplex (Crouan & Crouan) Hamel
Littoral. Epiphytic on *Codium tomentosum.* ⊞ July 1968 and May 1969; ○ July 1968.
Rare. 22, 62, 75, 94
Scotland: once recorded: Arran (Batters).
NE Atlantic: not reported for areas further north, the Mull record being an extension to the distribution. Cardinal (1964) reports the species on *Codium* and once on *Taonia*, for the N coast of France. It is also known from Cornwall, Devon, Scilly Islands and Mayo (Clare Island).

Giffordia Batt.

G. granulosa (Sm.) Hamel
Infralittoral to 12 m. Epiphytic on *Chorda filum, Heterosiphonia plumosa* and *Polysiphonia nigrescens.* ⊞ June 1968.
Rare. 13, 44, 54
Scotland: Shetlands, Lewis, Colonsay and Argyll (Dunstaffnage).
NE Atlantic: Faeroes and Iceland.

G. hincksiae (Harv.) Hamel
Littoral; infralittoral to 9 m. Epilithic; epiphytic on *Palmaria palmata* and *Laminaria hyperborea.* ⊞ June 1968.

Rare. 20, 22, 43
Scotland: frequent.
NE Atlantic: widely distributed.

G. sandriana (Zanard.) Hamel
Infralittoral, 24 m to 27 m. Epilithic. ⊞ June 1968.
Rare. 65
Scotland: recorded infrequently for Argyll and Shetland.
NE Atlantic: widely distributed.

G. secunda (Kütz.) Batt.
Infralittoral to 18 m. Epiphytic on *Laminaria saccharina* blades. ⊞ May and June 1968.
Rare. 1, 40, 44
Scotland: not recorded previously but recently discovered on Shetland (1973, D. E. G. Irvine, *pers. comm.*).
NE Atlantic: Iceland.

Pilayella Bory

P. littoralis (L.) Kjellm.
Littoral; infralittoral. Epilithic in depths greater than 9 m; epiphytic on *Ascophyllum, Cystoseira tamariscifolia, Fucus ceranoides, F. serratus, F. vesiculosus, Halidrys, Laminaria hyperborea, L. saccharina* and *Polysiphonia lanosa.*
Abundant. Generally distributed. Previous record: Gribun Rocks, Aug. 1943, *Newton* (1949).
Scotland and *NE Atlantic:* widespread.

Spongonema Kütz.

S. tomentosum (Huds.) Kütz.
Littoral; infralittoral 6 m. Epiphytic on *Fucus spiralis, F. vesiculosus, Himanthalia, Laminaria digitata,* and *Laminaria hyperborea.* ⊞ May to Sept.; ○ July 1968.
Abundant. Generally distributed. Previous record: Staffa, June 1826, Hb. *Greville* (E), as *Ectocarpus tomentosus.*
Scotland: very frequent.
NE Atlantic: widely distributed.

RALFSIACEAE

Pseudolithoderma Svedel.

P. extensum (Crouan & Crouan) S. Lund
Littoral; infralittoral to 15 m. Epilithic.
Rare. 11, 17, 25, 33
Scotland: abundant in 1973 in Shetland and Orkney; Newton records it from Cumbrae, probably on the basis of *Batters*, 1891 (BM).
NE Atlantic: widely distributed.

Ralfsia Berk.

R. verrucosa (Aresch.) J. Ag.
Littoral. Epilithic. ○ May 1969.
Frequent. 6, 12, 24, 36, 51, 52, 55, 58, 75, 79, 87, 92
Scotland: common.
NE Atlantic: Norway and Sweden.

MYRIONEMATACEAE

Hecatonema Sauv.

H. maculans (Coll.) Sauv.
Littoral. Epiphytic on *Cladophora, Enteromorpha* and *Ceramium.* ⊞ May 1969.
Rare. 40
Scotland: occasional.
NE Atlantic: widely distributed.

Microspongium Reinke

M. globosum Reinke
Littoral. Epiphytic on *Enteromorpha.*
Rare. 49
Parke & Dixon (1968) note that *M. globosum* is probably the sporophytic phase of *Scytosiphon lomentaria.*
Scotland: first record and only recently noted from Orkney and Shetland.
NE Atlantic: widely distributed.

Myrionema Grev.

M. strangulans Grev.
Littoral; infralittoral to 9 m. Epiphytic on *Enteromorpha, Ulva lactuca, Chylocladia verticillata* and *Palmaria palmata.*
Frequent. 40, 48, 51, 54, 57, 63, 94, 97, 101
Scotland: frequent.
NE Atlantic: widely distributed.

Protectocarpus Kuck.

P. speciosus (Børg.) Kuck.
Littoral. Epiphytic on *Laurencia pinnatifida.*
Rare. 58
Scotland: previously recorded only for Shetlands (Dixon, 1963).
NE Atlantic: Faeroes and throughout Norway; otherwise known from Mayo, Cornwall, Dorset and Brittany.

Ulonema Fosl.

U. rhizophorum Fosl.
Littoral, in pools. Epiphytic on *Dumontia.* ○ June 1968 and May 1969.
Occasional. 13, 26, 40, 75, 84, 92
Scotland: East Lothian (Cove), Cumbrae and Shetland.
NE Atlantic: frequently recorded for N and W Norway and Sweden. Probably a species often overlooked and hence under-recorded.

ELACHISTACEAE

Elachista Duby

E. flaccida (Dillw.) Aresch.
Infralittoral to 2 m. Epiphytic on *Cystoseira nodicaulis.* ○ Sept. 1970.
Rare. 40, 54
Scotland: occasional.
NE Atlantic: not recorded elsewhere, its present northern limit being Shetland.

E. fucicola (Vell.) Aresch.
Littoral. Epiphytic on *Fucus serratus, F. spiralis* and *F. vesiculosus.* ⊞ May 1968; ○ Oct. 1968 and July 1969.

Abundant. Generally distributed. Previous record: Gribun Rocks, Aug. 1943, *Newton* (1949).
Scotland: frequent.
NE Atlantic: widely distributed.

E. scutulata (Sm.) Aresch.
Littoral, lower levels. Epiphytic on *Himanthalia elongata.* ○ Sept., 1970.
Occasional. 16, 68, 79, 90, 96
Scotland: frequent.
NE Atlantic: Norway and Faeroes.

CORYNOPHLAEACEAE

Corynophlaea Kütz.

C. crispa (Harv.) Kuck.
Littoral, higher levels. Epiphytic on *Chondrus crispus.*
Rare. 40
Scotland: Cumbrae, but not recorded for the mainland.
NE Atlantic: Isle of Man, Lundy, Dorset and the Channel Islands. Recently recorded for Clare (Pybus, 1975).
The Mull record represents the northern limit of distribution of this species in Europe.

Cylindrocarpus Crouan & Crouan

C. berkeleyi (Grev.) Crouan & Crouan
Littoral, lower levels. Epiphytic on *Ralfsia.*
Rare. 52
Scotland and *NE Atlantic:* occasional, as far north as Orkney.

Leathesia Gray

L. difformis (L.) Aresch.
Littoral; infralittoral to 3 m. Epilithic; epiphytic on *Cladophora*, *Cystoseira tamariscifolia*, *Halidrys siliquosa*, *Ceramium rubrum*, *Furcellaria lumbricalis*, and *Gigartina stellata.*
Abundant. Generally distributed. Previous records: Calve Island, Rubh an Righ and Bogha nan Gèodh, *Kitching* (1935); Gribun Rocks, Aug. 1943, *Newton* (1949).
Scotland and *NE Atlantic:* widespread.

CHORDARIACEAE

Chordaria Ag.

C. flagelliformis (O. F. Müll.) Ag.
Littoral, at all levels; shallow infralittoral. Epilithic. ○ Sept., 1968.
Frequent. 12, 14, 20, 22, 40, 45, 51, 52, 53, 68, 74, 76, 79, 81, 85, 101. Previous record: Gribun Rocks, Aug. 1943, *Newton* (1949).
Scotland: frequent.
NE Atlantic: widely distributed.

Cladosiphon Kütz.

C. zosterae (J.Ag.) Kylin
Infralittoral, 3–4 m. Epiphytic on *Zostera* blades.
Rare. 56
Norton *et al.* (1969) record ○ in July and August in Argyll; our material was collected in May.

Scotland: only scantily recorded from Bute, Ayr and Argyll.
NE Atlantic: Norway.

Eudesme J.Ag.

E. virescens (Carm. ex Harv.) J.Ag.
Littoral; infralittoral to 12 m. Epilithic; epiphytic on *Chorda filum*, *Halidrys*, *Furcellaria*, *Gigartina* and *Laurencia pinnatifida.* ○ May to Sept.
Frequent. 1, 6, 22, 36, 43, 44, 52, 54, 66, 68, 72, 75, 79, 85
Scotland: frequent.
NE Atlantic: widely distributed.

Mesogloia Ag.

M. vermiculata (Sm.) Gray
Littoral; infralittoral at 15 m. Epilithic. ○ July 1968.
Rare. 67, 74. Previous record: Iona, Hb. *Greville* (E).
Scotland: recorded widely.
NE Atlantic: Spitzbergen(?), Norway and W Sweden.

Sphaerotrichia Kylin

S. divaricata (Ag.) Kylin
Littoral. Epilithic. ○ Sept. 1970.
Rare. 33
Scotland: recorded scantily from Ayr, Bute (Fairlie) (Batters); and Fair Isle.
NE Atlantic: Norway and Sweden.

ACROTRICHACEAE

Acrothrix Kylin

A. gracilis Kylin
Infralittoral to 12 m.
Rare. 1
Scotland: Argyll (Easdale Quarry) and Fair Isle.
NE Atlantic: W and N Norway and W coast of Sweden. Probably more widespread.

SPERMATOCHNACEAE

Spermatochnus Kütz.

S. paradoxus (Roth) Kütz.
Infralittoral, 9–12 m. Epiphytic on *Chylocladia verticillata.*
Rare. 54
Scotland: although scantily recorded in the literature (Colonsay; and various localities by Batters) the species is represented in BM by material from other localities e.g. Kintyre, Arran, Cumbrae, Orkney and Shetland.
NE Atlantic: W coast of Norway (Hardangerfjord, Oslofjord, and areas just south of Bergen) and Sweden.

STRIARIACEAE

Isthmoplea Kjellm.

I. sphaerophora (Carm. ex Harv.) Kjellm.
Littoral. Epiphytic on (occasionally entwined among) *Cladophora rupestris*, *Callithamnion*, *Palmaria palmata*, *Plumaria*, *Polysiphonia lanosa;* occasionally part of complex underflora

to fucoids such as *Ascophyllum*. ○ May to July.
Frequent. 2, 5, 7, 26, 28, 37, 51, 52, 53, 55, 68, 97
Scotland: occasional.
NE Atlantic: widely distributed.

Stictyosiphon Kütz.

S. griffithsianus (Le Jol.) Holm. & Batt.
Littoral; infralittoral, 24–27 m. Epiphytic on *Palmaria palmata*. ⊞ July 1968; ○ June 1968.
Rare. 44, 65, 97
Scotland: infrequently recorded: Edinburgh, Fife, Aberdeen, Orkney, Bute, Ayr, Renfrew (Batters); Fair Isle; Lewis and Harris; Orkney (BM).
NE Atlantic: Børgesen & Johnson (1905) place the northern limit of distribution as Scotland.

S. tortilis (Rupr.) Reinke
Littoral; infralittoral. Epilithic. ⊞ May 1968; ○ June 1969.
Occasional. 11, 13, 25, 44, 55, 56, 61, 72
Scotland: Batters records this species for a number of localities; also recently reported for Dunstaffnage and is represented in BM from Shetland, Orkney, and other mainland localities.
NE Atlantic: widely distributed.

Striaria Grev.

S. attenuata (Grev.) Grev.
Infralittoral to 12 m. Epilithic. ○ May and June 1968.
Occasional but luxuriant. 13, 54, 66
Scotland: Orkney (Damsay Bay and Kirkwall) (BM); Cumbrae; Caithness (Ackergill); Ayr (Ardrossan); Argyll (Easdale Quarry) and also for Arran (Batters).
NE Atlantic: W Norway and Sweden.

MYRIOTRICHIACEAE

Myriotrichia Harv.

M. clavaeformis Harv.
Littoral, lower level pools. Epiphytic on *Asperococcus*, *Cystoseira tamariscifolia* and *Scytosiphon*.
Occasional. 25, 33, 40, 107, 135
Scotland: occasional.
NE Atlantic: Norway

M. filiformis Harv.
Littoral; shallow infralittoral. Epiphytic on *Scytosiphon* and *Zostera*. ⊞ May, 1969.
Rare. 6, 56
Scotland: occasional.
NE Atlantic: Norway and Sweden.

M. repens Hauck
Littoral; shallow infralittoral. Epiphytic on *Punctaria*, *Stictyosiphon* and *Zostera*.
Rare. 44, 47, 56, 61
Scotland: only once previously recorded: Lossiemouth (BM), and recently found on Shetland.
NE Atlantic: Norway and Sweden.

PUNCTARIACEAE

Asperococcus Lamour.

A. compressus Griff. ex Hook.
Littoral; infralittoral to 12 m. Epiphytic on *Halidrys* and *Polysiphonia nigrescens*.
Occasional. 22, 38, 44, 54, 64, 72
Scotland: scantily recorded, including Orkney and Shetland.
NE Atlantic: Shetland represents the northern limit, otherwise scantily recorded.

A. fistulosus (Huds.) Hook.
Littoral, at all levels; infralittoral to 18 m. Epiphytic on *Ulva lactuca*, *Fucus vesiculosus*, *Halidrys*, *Corallina officinalis*, *Laurencia* and *Polysiphonia nigrescens*; epizoic on *Ensis* and *Pecten*. ○ May to July. Observations made during the period May to October indicate that this species occurs abundantly during the summer and dies back later in the season.
Abundant. Generally distributed. Previous records: Iona, *Keddie* (1850); *Ritchie & Ritchie* (1947), both as *Asperococcus echinatus*.
Scotland: frequent.
NE Atlantic: widely distributed.

A. turneri (Sm.) Hook.
Littoral; infralittoral to 12 m. Epilithic. ○ May to July; ⊞ June 1968.
Occasional. 54, 63, 66, 67, 73
Scotland: Bute, Argyll, Orkney and Shetland.
NE Atlantic: Norway.

Desmotrichum Kütz.

D. undulatum (J. Ag.) Reinke
Infralittoral (shallow). Epiphytic on *Zostera*.
Rare. 56
Scotland: Fair Isle.
NE Atlantic: Norway, Sweden and Faeroes.
The taxonomic status of this entity is at present in doubt; there is evidence to indicate it may be a young stage of *Punctaria*.

Litosiphon Harv.

L. filiformis (Reinke) Batt.
Infralittoral, 3–10 m. Epiphytic on *Laminaria* and *Laurencia*. ○ June 1968.
Occasional. 26, 40, 60, 66, 72
Scotland: Colonsay, Lewis and Harris, Fair Isle, Renfrew.
NE Atlantic: widely distributed.

L. laminariae (Lyngb.) Harv.
Infralittoral to 3 m. Epiphytic on *Alaria esculenta* and *Zostera*.
Occasional. 26, 55, 56, 67, 74, 88
Scotland: frequent.
NE Atlantic: widely distributed, except Iceland.

L. pusillus (Carm. ex Hook.) Harv.
Littoral. Epiphytic on *Scytosiphon lomentaria*.
Rare. 32, 40. Previous record: Iona, *Keddie* (1850).

Scotland: Argyll, Fair Isle, Orkney, Shetland, Forfar, Fife, Edinburgh, East Lothian.
NE Atlantic: Norway and W Sweden.

Punctaria Grev.

P. latifolia Grev.
Infralittoral to 9 m. Epilithic. ○ June, 1968.
Rare. 54, 72
Scotland: occasional (Batters and others).
NE Atlantic: from N Norway and Iceland and areas further south, although not frequent.

P. plantaginea (Roth) Grev.
Infralittoral to 18 m. Epilithic; epiphytic on *Zostera.* ⊞ May 1968, 1969.
Occasional. 35, 40, 44, 54, 56, 78
Scotland: frequent.
NE Atlantic: widely distributed.

P. tenuissima (Ag.) Grev.
Littoral; infralittoral. Epiphytic on *Codium, Laminaria digitata* and *Zostera.*
Occasional. 40, 56, 57, 75
Scotland: Shetland (1973, D. E. G. Irvine, *pers. comm.*), Orkney, Fife, Renfrew, Argyll, Bute, Ayr and elsewhere.
NE Atlantic: not recorded.

DICTYOSIPHONACEAE

Dictyosiphon Grev.

D. foeniculaceus (Huds.) Grev.
Littoral, at all levels; infralittoral to 6 m. Epilithic; epiphytic on *Codium, Cystoseira tamariscifolia, Fucus serratus, Halidrys, Laurencia pinnatifida.* ○ June 1968.
Frequent. 13, 25, 26, 33, 40, 44, 54, 60, 70, 75, 79, 84, 86, 101
Scotland: frequent.
NE Atlantic: widely distributed.

TILOPTERIDACEAE

Tilopteris Kütz.

T. mertensii (Turn.) Kütz.
Infralittoral, 18–21 m. Epilithic. ○ June, 1968.
Rare. 65
Scotland: Ayr, Renfrew, Bute, Argyll, Shetland, Aberdeen, Edinburgh (Batters); Orkney (Clouston, 1862).
NE Atlantic: Norway and Sweden (Bohuslän, rare). The present records may be near the northern limits for the species.

SCYTOSIPHONACEAE

Colpomenia Derbès & Sol.

C. peregrina Sauv.
Littoral, mid and lower levels; infralittoral. Epiphytic on *Cladophora rupestris, Cladostephus spongiosus, Cystoseira nodicaulis, Ceramium diaphanum, Chondrus crispus, Chylocladia verticillata, Corallina officinalis, Cystoclonium purpureum, Gigartina stellata.* ⊞ Oct. 1968.
Frequent; in considerable abundance in May (1968–70). Generally distributed.
Scotland: scattered, largely on islands, including Shetland.
NE Atlantic: Norway (Hardangerfjord) is the northern limit of this species.

Petalonia Derbès & Sol.

P. fascia (O. F. Müll.) O. Kuntze
Littoral; shallow infralittoral. Epilithic; epiphytic on *Codium* and *Laminaria hyperborea.* ⊞ June 1969.
Frequent. 3, 22, 23, 24, 38, 40, 58, 69, 78, 79, 90, 91 Previous record: Staffa, June, 1826, *Greville* (E), as *Laminaria saccharina* var. *phyllitis*; *Dendy* (1859), as *Phyllitis.*
Scotland: frequent.
NE Atlantic: widely distributed.

Scytosiphon Ag.

S. lomentaria (Lyngb.) Link
Littoral; infralittoral to 12 m. Epilithic; epiphytic on *Codium, Ascophyllum, Fucus vesiculosus, Halidrys, Furcellaria* and *Zostera;* occasionally epizoic on *Patella vulgaris.*
Abundant. Generally distributed. Previous records: Iona, *Keddie* (1850); *Ritchie & Ritchie* (1947) both as *Chorda lomentaria*; Gribun Rocks, Aug. 1942, *Newton* (1949).
Scotland: frequent.
NE Atlantic: scattered records.

CUTLERIALES

CUTLERIACEAE

Cutleria Grev.

Cutleria multifida (Sm.) Grev.
Infralittoral to 12 m. Epilithic; epiphytic on *Laminaria hyperborea* holdfast.
Occasional. 15, 21, 26, 43, 44, 65
Found in Mull only as the *Aglaozonia* (sporophytic) phase; however, *Cutleria* gametophytes have been found in Britain as far north as Shetland.
Scotland: frequent.
NE Atlantic: W Norway and W Sweden.

DESMARESTIALES

ARTHROCLADIACEAE

Arthrocladia Duby

A. villosa (Huds.) Duby
Infralittoral, 8 m. Epiphytic on *Laminaria hyperborea* stipe.
Rare. 43
It is most unlikely that such an isolated record represents a really discontinuous distribution. The strong possibility is that *A. villosa* would be found on prolonged search in all suitable adjacent localities.
Scotland: from Orkney (Kirkwall) and Bute (Lamlash Bay, Arran) (BM); Batters records it for East Lothian and Argyll.
NE Atlantic: primarily southern in distribution being absent from Iceland, Faeroes and Norway.

DESMARESTIACEAE

Desmarestia Lamour.

D. aculeata (L.) Lamour.
Littoral; infralittoral to 21 m. Epilithic; epiphytic on *Laminaria saccharina* bases and *L. hyperborea* stipes. Sometimes codominant in the infralittoral with *L. saccharina*.
Frequent. Generally distributed.
Scotland: frequent.
NE Atlantic: widely distributed.

D. ligulata (Lightf.) Lamour.
Infralittoral to 12 m. Epilithic.
Rare. 21, 52, 79, 101
Scotland: frequent.
NE Atlantic: widely distributed.

D. viridis (O. F. Müll.) Lamour.
Infralittoral to 18 m. Epilithic; epiphytic on *Laminaria digitata*, *L. hyperborea* stipes and less frequently on *Zostera marina;* epizoic on *Pecten*.
Frequent. 1, 13, 20, 21, 25, 35, 43, 44, 48, 54, 57, 59, 60, 65, 72, 75. Previous record: Gribun Rocks, as drift, Aug. 1943, *Newton* (1949).
Scotland: frequent.
NE Atlantic: widely distributed.

LAMINARIALES

CHORDACEAE

Chorda Stackh.

C. filum (L.) Stackh.
Littoral (lower levels) and infralittoral to 12 m. Epilithic; often codominant with *Zostera* and *Laminaria saccharina* on boulders and pebbles on sandy bottom in sheltered, shallow, infralittoral.
Abundant. Generally distributed. Previous records: Iona, *Keddie* (1850), *Ritchie & Ritchie* (1947); Gribun Rocks, Aug. 1943, *Newton* (1949).
Scotland: frequent.
NE Atlantic: widely distributed.

LAMINARIACEAE

Laminaria Lamour.

L. digitata (Huds.) Lamour.
Littoral; infralittoral to 9 m. Epilithic; occasionally epiphytic on *Fucus serratus* and *Laminaria hyperborea*.
Abundant. Generally distributed. Previous records: Iona, *Keddie* (1850); *Ritchie & Ritchie* (1947); Rubh an Righ, Bogha nan Gèodh, Calve Island, Aug. 1934, *Kitching* (1935); Gribun Rocks, Aug. 1943, *Newton* (1949).
Scotland: consistently recorded.
NE Atlantic: widely distributed.

L. hyperborea (Gunn.) Fosl.
Infralittoral to 26 m. Epilithic; occasionally young plants epiphytic on older *L. hyperborea* plants.
Abundant. Generally distributed.
The typical western Scottish *L. hyperborea* beds are present in all suitable circumstances around Mull and offshore islands. Sporangia were found in May 1969.
Scotland: recorded consistently.
NE Atlantic: widely distributed.

L. saccharina (L.) Lamour.
Littoral; infralittoral to 21 m. Epilithic; occasionally young plants epiphytic on *L. digitata*, *L. hyperborea*, older plants of *L. saccharina* and *Phycodrys rubens*.
Abundant. Generally distributed. Previous records: Iona, *Keddie* (1850); *Ritchie & Ritchie* (1947); Gribun Rocks, Aug. 1943, *Newton* (1949).
Scotland: frequent.
NE Atlantic: widely distributed.

Saccorhiza Pyl.

S. polyschides (Lightf.) Batt.
Infralittoral to 22 m. Epilithic.
Occasional. 16, 18, 54, 60, 65, 71, 79, 81. Previous record: Iona, *Lightfoot* (1777), as *Fucus polyschides*.
Scotland: frequent.
NE Atlantic: Norway. The world and British distributions have been analysed by Norton & Burrows (1969).

ALARIACEAE

Alaria Grev.

A. esculenta (L.) Grev.
Littoral, lower level pools; infralittoral to 3 m. Epilithic; occasionally younger plants epiphytic on larger *Alaria* plants, *Fucus serratus*, *Laminaria digitata*, *L. hyperborea*, *Ceramium*, *Chondrus crispus* and *Phycodrys rubens*.
Abundant. Generally distributed. Previous records: Iona, *Keddie* (1850); Staffa, *Dendy* (1859); Gribun Rocks, Aug. 1943, *Newton* (1949).
Scotland: frequent.
NE Atlantic: widely distributed.

SPHACELARIALES

SPHACELARIACEAE

Sphacelaria Lyngb.

S. bipinnata (Kütz.) Sauv.
Littoral. Epiphytic on *Cystoseira nodicaulis*, *C. tamariscifolia* and *Halidrys*.
Frequent. 38, 40, 44, 46, 52, 54, 65, 73, 79
Scotland: generally distributed.
NE Atlantic: W and N Norway and Sweden.

S. britannica Sauv.
Littoral, upper level. Epilithic.
Rare. 51
Scotland: only from Fife (St Andrews), Orkney and Shetland but according to Irvine (1956) probably around all the British Isles. Possibly not a good species (see Parke & Dixon, 1976).
NE Atlantic: widely distributed.

S. cirrosa (Roth) Ag.
Littoral; infralittoral to 12 m. Epiphytic on *Cladophora*, *Cladostephus*, *Cystoseira tamar-*

iscifolia, *Laminaria hyperborea*, *L. saccharina*, *Corallina*, *Chylocladia*, *Furcellaria*, *Gracilaria*, *Polyides* and *Polysiphonia elongata*.
Propagules present June to October.
Abundant. Generally distributed. Previous record: Gribun Rocks, Aug. 1943, *Newton* (1949).
Scotland: frequent.
NE Atlantic: W and N Norway and Sweden.

S. fusca (Huds.) Gray
Littoral; infralittoral 0–3 m. Epiphytic on *Ascophyllum*, *Cystoseira*, *Laminaria hyperborea*, *Corallina*, and *Furcellaria*. ○ June 1968; propagules present, May to October.
Frequent. 7, 12, 32, 40, 43, 52, 54, 60, 65, 68, 72, 79
Scotland: Argyll, Orkney, Shetland.
NE Atlantic: A taxonomic complex probably more widespread than the few records would indicate.

S. plumula Zanard.
Infralittoral, 12–18 m. Epilithic; epizoic on *Pecten*.
Rare. 13, 90
Scotland: Argyll (Easdale Quarry) and Colonsay; also from Arran (BM). The Mull record therefore represents an extension in the range for Britain.
NE Atlantic: Norway.

S. radicans (Dillw.) Ag.
Littoral; infralittoral to 6 m. Epilithic; ? epiphytic. ○ June, 1968.
Occasional. 13, 21, 40, 44, 60, 100
Scotland: frequent.
NE Atlantic: Norway, Sweden, Iceland and Greenland.

STYPOCAULACEAE

Halopteris Kütz.

H. scoparia (L.) Sauv.
Infralittoral to 9 m. Epilithic.
Rare. 54, 72
Scotland: Ayrshire (Portencross).
NE Atlantic: Norway, W Sweden; Iceland (Levring, 1937; not substantiated by Caram & Jónsson, 1972).
The Mull records represent an extension in the range of distribution for Britain.

CLADOSTEPHACEAE

Cladostephus Ag.

C. spongiosus (Huds.) Ag. forma **spongiosus**
Littoral. Epilithic.
Abundant. Generally distributed. Previous record: Gribun Rocks, Aug. 1943, *Newton* (1949).
Scotland: recorded frequently.
NE Atlantic: widely distributed, except Iceland.

C. spongiosus f. verticillatus (Lightf.) P. v. R.
Littoral, mid and lower levels; infralittoral to 6 m. Epilithic; epiphytic on *Laminaria digitata* holdfast.
Frequent. 13, 22, 44, 54, 63, 67, 74, 79, 84, 85, 90
Scotland: Frequently reported.
NE Atlantic: W Norway and S Sweden.

DICTYOTALES

DICTYOTACEAE

Dictyota Lamour.

D. dichotoma (Huds.) Lamour.
Littoral; infralittoral to 23 m. Epilithic; epiphytic on *Cystoseira nodicaulis*, *C. tamariscifolia*, *Halidrys siliquosa*, *Laminaria hyperborea*, *Corallina officinalis*, *Furcellaria lumbricalis*, *Heterosiphonia plumosa*, *Laurencia pinnatifida*, and *Phyllophora crispa*. ⊕ August and September; ♀ ♂ June, 1968.
Abundant: Generally distributed. Previous records: Iona, *Keddie* (1850); *Ritchie & Ritchie* (1947).
Scotland: recorded frequently.
NE Atlantic: S Norway and Sweden.

FUCALES

FUCACEAE

Ascophyllum Stackh.

A. nodosum (L.) Le Jol.
Littoral. Epilithic.
Abundant. Generally distributed. Previous records: Iona, *Keddie* (1850); *Ritchie & Ritchie* (1947), both as *Fucus nodosus*; Rubh an Righ, Bogha nan Gèodh (Calve Island) and Eilean Mòr (Loch Buie), Aug. 1934, *Kitching* (1935); Salen Bay, Loch Spelve, Loch na Làthaich, An Leth Onn and Ulva Sound, *Gibb* (1957), as ecad *mackaii*.

Two morphological forms of this species have been found on Mull: ecad *mackaii* and ecad *scorpioides*. The ecological significance of these is discussed in Chapter 8 and by Gibb (1957).
Scotland: abundant.
NE Atlantic: widely distributed.

Fucus L.

F. ceranoides L.
Littoral. Epilithic in localities where there exists an adequate freshwater flow (rocky shore and saltmarsh).
Locally frequent. 3, 4, 7, 11, 12, 13, 16, 43, 47, 62, 70, 75, 86. Previous record: near the castle of Aros, *Turner & Hooker* (Greville, 1830).
Scotland: frequent.
NE Atlantic: widely distributed.

F. serratus L.
Littoral; infralittoral. Epilithic; epiphytic on *Laminaria digitata*.
Abundant. Generally distributed. Previous records: without locality, *Turner* (1809); Iona, *Keddie* (1850); *Ritchie & Ritchie* (1947); Rubh an Righ, Bogha nan Gèodh (Calve Island), Camas Cuil an t-Saimh (Iona), Ardmore Point (Mishnish), Aug. 1934, *Kitching* (1935); Gribun Rocks, Aug. 1943, *Newton* (1949).

Scotland: consistently recorded.
NE Atlantic: widely distributed.

F. spiralis L.
Littoral, higher levels. Epilithic.
Abundant. Generally distributed. Previous records: Rubh an Righ, Bogha nan Gèodh (Calve Island), Fionnphort, Camas Cuil an t-Saimh (Iona), Kellan (Loch na Keal), Ardmore Point (Mishnish), 1934, *Kitching* (1935); Gribun Rocks, Aug. 1943, *Newton* (1949).
Scotland: consistently recorded.
NE Atlantic: widely distributed.

F. vesiculosus L.
Littoral. Epilithic.
Abundant. Generally distributed. Previous records: Iona, *Keddie* (1850); *Ritchie & Ritchie* (1947); Rubh an Righ, Bogha nan Gèodh (Calve Island), Eilean Mòr (Loch Buie), Camas Cuil an t-Saimh and St Ronan's Bay (Iona), Staffa, Ardmore Point (Mishnish), Aug. 1934, *Kitching* (1935); Gribun Rocks, Aug. 1943, Salen Bay, Jan. 1943, *Newton* (1949).
Scotland: consistently recorded.
NE Atlantic: widely distributed.

Pelvetia Dcne & Thur.

P. canaliculata (L.) Dcne & Thur.
Littoral, higher levels. Fertile conceptacles were noted in May, 1969.
Abundant. Generally distributed. Previous records: Iona, *Keddie* (1850); *Ritchie & Ritchie* (1947); Rubh an Righ and Bogha nan Gèodh (Calve Island), Camas an t-Seilisdeir (Loch Spelve), Eilean Mòr (Loch Buie), Fionnphort, St Ronan's Bay, Camas Cuil an t-Saimh (Iona), Rubh' a' Ghearrain, Kellan (Loch na Keal), Ardmore Point (Mishnish), Aug. 1934, *Kitching* (1953); Gribun Rocks, Aug. 1943, *Newton* (1949).
Scotland: consistently recorded.
NE Atlantic: widely distributed.

HIMANTHALIACEAE

Himanthalia Lyngb.

H. elongata (L.) Gray
Littoral, lower levels; shallow infralittoral. Epilithic; on one occasion the "buttons" were found epiphytic on *Fucus*.
Abundant. Generally distributed.
Scotland: consistently recorded.
NE Atlantic: widely distributed.

CYSTOSEIRACEAE

Cystoseira Ag.

C. nodicaulis (With.) Roberts
Littoral, lower levels; infralittoral to 2 m. Epilithic in areas of very gently shelving volcanic sand substrate with buried or partially buried pebbles and boulders. Plants, up to 45 cm long, were attached to pebbles and boulders in depths of 20 cm to 1.5 m below MLWS. The discontinuous bed thus formed stretches over an area of about 100 m × 50 m in the bay between north side of Rubha Baile na Aird and landward facing shore of Inch Kenneth. Patches are also present as far north as the area between Rubh' a' Ghearrain and Samalan Island. This is the only population detected, and is clearly well established and dominant in the shallow infralittoral (0–2 m). It is a perennial plant; remains of bases and axes with restricted apical new growth were found.
Rare; but locally abundant. 26, 27
The discovery (in 1968) of *C. nodicaulis* represents a northward extension of range in Britain and in Europe.
Scotland: Colonsay.
NE Atlantic: generally distributed S of Scotland, at least on W-facing coastlines.

C. tamariscifolia (Huds.) Papenf.
Littoral, mid to upper littoral pools. Epilithic in shallow to moderately deep mid-littoral pools as extensive populations. Plants were growing on firm rock at bases of the shallower areas (to 45 cm) of the pools, or in detritus over rock in the lagoon at stn 78. The latter population was at about mid littoral level, but in other areas (stns 79 and 101), populations were in mid to lower *Pelvetia*-level pools. Plants over 45 cm in length were located. There was some evidence of die-back to older branches and axes in the October collections and in material from the shallower parts of pools. Proximal lower laterals remained vegetatively vigorous and healthy looking wherever plants were growing. In those parts of pools greater in depth than 66 cm, *Cystoseira* plants had an identical appearance in winter as in summer. Iridescence is less obvious in October than in May.
Rare. 78, 79, 101
Scotland: nearing the northern limit on Mull; previously recorded from South Uist, Barra, Canna, Sound of Sleat and Mallaig, Tiree, Colonsay, Gigha, Kintyre, Cumbrae and Ayrheads.
NE Atlantic: W Norway.

Halidrys Lyngb.

H. siliquosa (L.) Lyngb.
Littoral; infralittoral to 30 m but mostly in the upper 3 m.
Abundant. Generally distributed. Previous records: Iona, *Keddie* (1850); *Ritchie & Ritchie* (1947); Ardmore Point, Aug. 1934, *Kitching* (1935); Gribun Rocks, Aug. 1943, *Newton* (1949).
Scotland: consistently recorded.
NE Atlantic: W and N Norway, Sweden and Faeroes.

CHLOROPHYTA

CHLOROPHYCEAE

PRASIOLALES

PRASIOLACEAE

Prasiola (Ag.) Menegh.

P. stipitata Suhr
Supralittoral. Epilithic, abundant in areas of strong nitrogenous influence, e.g. bird-roosts.
Occasional; but probably overlooked. 2, 12, 26, 51, 55, 88
Scotland: Galloway, Lewis and Harris, St. Kilda, Orkney, Fair Isle and Shetland.
NE Atlantic: widespread.

ULOTHRICALES

ULOTHRICACEAE

Ulothrix Kütz.

U. flacca (Dillw.) Thur.
Littoral. Epiphytic on *Fucus serratus* and *Laminaria hyperborea* stipes.
Occasional. 4, 6, 22, 38, 40, 52
Scotland: Galloway, Colonsay, Lewis and Harris, Fair Isle and Shetland.
NE Atlantic: widespread.

Urospora Aresch.

U. penicilliformis (Roth) Aresch.
Littoral; infralittoral to 12 m. Epiphytic on *Fucus serratus* and *Palmaria palmata*; epizoic on *Pecten.*
Occasional, but probably overlooked. 22, 67, 78, 92
Scotland: rare.
NE Atlantic: widespread.

MONOSTROMATACEAE

Monostroma Thur.

M. grevillei (Thur.) Wittr.
Littoral, in pools often with *Ulva.* Epiphytic on *Corallina, Gelidium* and *Laurencia pinnatifida.*
Occasional (?). A late winter to early spring ephemeral species, absent during the summer (thus probably under-recorded during the Survey). 26, 36, 38, 48, 58, 91, 92
Scotland: rare.
NE Atlantic: widespread, especially in colder waters.

M. oxyspermum (Kütz.) Doty
Littoral. Epilithic, occasionally associated with *Corallina.*
Occasional. 24, 38, 40
Scotland: only from Fifeshire (as *Monostroma wittrockii* Born.)
NE Atlantic: Norway and Sweden.

ACROSIPHONIACEAE

Spongomorpha Kütz.

(Includes life history phases formerly treated as *Chlorochytrium inclusum* Kjellm.)

S. aeruginosa (L.) Hoek
Littoral, in pools at all levels. Epilithic; epiphytic on *Ulva, Halidrys, Chondrus* and *Gigartina.*
Frequent. 4, 5, 13, 26, 40, 53, 54, 68, 79, 101
Previous records: Gribun Rocks, 1943, *Newton* (1949), as *Cladophora lanosa;* Fingals Cave, Staffa, (?) *J. E. Smith* in Savage (1963) as *Conferva aeruginosa.*
Scotland: occasional.
NE Atlantic: widespread.

S. arcta (Dillw.) Kütz.
Littoral; infralittoral to 3 m. Epilithic; epiphytic on *Ulva, Ascophyllum, Cystoseira nodicaulis, Fucus serratus, Laminaria hyperborea, Gigartina stellata, Laurencia pinnatifida* and *Polysiphonia lanosa;* epizoic on *Patella.*
Frequent. Generally distributed, all years.
Scotland: frequent.
NE Atlantic: widespread.

ULVALES

ULVACEAE

Blidingia Kylin

B. minima (Näg. ex Kütz.) Kylin
Supralittoral; littoral, higher levels. Epilithic; epizoic on *Balanus.*
Frequent (probably more widespread than our records indicate). 46, 51, 52, 68, 70
Scotland: frequent.
NE Atlantic: widespread.

Enteromorpha Link in Nees

E. clathrata (Roth) Grev.
Littoral, lower levels, in sheltered conditions. Epilithic.
Rare. 66
Scotland: frequent.
NE Atlantic: widespread.

E. intestinalis (L.) Link [includes *E. compressa* (L.) Grev.]
Littoral; occasionally infralittoral to 21 m. Epilithic; epiphytic on *Fucus, Laminaria digitata,* and *Himanthalia*; also epizoic on *Patella.* Particularly abundant in areas of strong freshwater outfall, e.g. in river estuaries, saltmarshes, and high-level brackish pools.
Abundant. Generally distributed. Previous records: Iona, *Keddie* (1850); Calve Island, 1934, *Kitching* (1935); Gribun Rocks, 1943, *Newton* (1949).
Scotland: recorded widely.
NE Atlantic: widespread.

E. linza (L.) J.Ag.
Littoral: damp circumstances. Epilithic; epiphytic occasionally on *Laminaria hyperborea.*
Generally distributed.
Scotland: frequent.
NE Atlantic: widespread.

E. prolifera (O. F. Müll.) J.Ag.
Littoral; infralittoral to 21 m. Epilithic; and epiphytic on *Ascophyllum* and *Fucus serratus*. Often present in areas of freshwater influence.
Frequent. 2, 5, 7, 12, 13, 16, 36, 40, 52, 54, 69, 76, 79, 86, 90, 101
Scotland: occasional.
NE Atlantic: widespread.

Percursaria Bory

P. percursa (Ag.) Rosenv.
Littoral, on saltmarsh, mat-forming among *Rhizoclonium* and *Vaucheria.*
Rare, but probably overlooked. 86
Scotland: occasional.
NE Atlantic: widespread.

Ulva L.

U. lactuca L.
Littoral; infralittoral to 21 m. Epilithic; epiphytic on *Ascophyllum*, *Chorda filum*, *Fucus serratus*, *Cystoseira tamariscifolia*, *Laminaria digitata*, *L. hyperborea*, *Corallina*, *Gigartina stellata*, and *Laurencia pinnatifida;* epizoic on *Ensis* and *Pecten.*
Often occurs as underflora, but less often as major flora. Huge plants were occasionally found in sheltered localities, e.g. drift material at 5 was 1 m in diameter.
Abundant. Generally distributed. Previous records: Iona, *Keddie* (1850), also as *U. latissima*; Iona, *Dendy* (1859, 1860).
Scotland: abundant.
NE Atlantic: widespread.

CHAETOPHORALES

CHAETOPHORACEAE

Pringsheimiella Höhn.

P. scutata (Reinke) Marchew.
Littoral. Epiphytic on *Polysiphonia nigrescens.*
Occasional. 78
Scotland: rare, in Argyll and Shetland (D. E. G. Irvine, *pers. comm.*).
NE Atlantic: widespread.

CLADOPHORALES

CLADOPHORACEAE

Chaetomorpha Kütz.

C. capillaris (Kütz.) Børg.
Littoral, high and low level pools. Epilithic, or entwined as skeins with *Codium tomentosum*, *Rhizoclonium riparium*, *Halidrys siliquosa*, *Punctaria plantaginea*, *Ceramium rubrum* and *Corallina officinalis.*
Frequent. Maximum appearance as skeins during summer months, declining in abundance later in the season. Generally distributed, all years.
Scotland: frequent.
NE Atlantic: widespread.

C. linum (O. F. Müll.) Kütz.
Littoral, in upper midlittoral and higher level pools as coarse entangled masses. Epilithic.
Occasional. 14, 23, 57, 66, 76, 78, 91
Scotland: frequent.
NE Atlantic: widespread.

C. melagonium (Web. & Mohr) Kütz.
Littoral, mid- and lower level pools; infralittoral, 3–6 m. Epilithic; on one occasion epiphytic on *Laminaria hyperborea* stipes.
Frequent. 14, 22, 25, 31, 33, 38, 40, 51, 52, 53, 57, 76, 84, 92
Scotland: frequent.
NE Atlantic: widespread.

Cladophora Kütz.

C. pellucida (Huds.) Kütz.
Littoral, in pools. Epilithic.
Rare (possibly more widespread). 52, 58. Previous record: Iona, 1948, *Drew* (BM).
These two records represent the northern limit of this species in Europe, previously cited (Hoek, 1963) as being the Isle of Man. Not otherwise recorded for Scotland.

C. rupestris (L.) Kütz.
Littoral; infralittoral, to 15 m. Epilithic; epiphytic on *Cystoseira tamariscifolia*, *Laminaria hyperborea*, *Gigartina stellata*, *Furcellaria lumbricalis*, *Laurencia hybrida*, and *L. pinnatifida.*
Abundant. Generally distributed. Previous record: Iona, *Keddie* (1850), as *Conferva rupestris.*
Scotland: frequent.
NE Atlantic: widespread.

C. sericea (Huds.) Kütz.
Littoral, in consistently damp conditions; infralittoral, to 9 m. Epilithic, epiphytic on *Cladostephus spongiosus*, and occasionally epizoic on *Patella.*
Abundant. Generally distributed.
Scotland: occasional.
NE Atlantic: widespread.

Rhizoclonium Kütz.

R. riparium (Roth) Harv.
Littoral, in estuaries, saltmarsh and rock pools. Epilithic.
Frequent. 3, 8, 11, 52, 67, 68, 78, 84, 86, 91, 92, 102. Previous record: Iona, *Keddie* (1850), as *Conferva implexa.*
Scotland: frequent.
NE Atlantic: widespread.

CODIALES

BRYOPSIDACEAE

Bryopsis Lamour.

B. plumosa (Huds.) Ag.
Littoral; infralittoral, to 21 m. Epilithic.
Occasional, but a consistent facet of the flora and may be more widespread than our collections indicate. Observations throughout the summer period indicate a decline in relative abundance late in the season. 5, 14, 22, 51, 65, 79, 85
Scotland: sporadically recorded.
NE Atlantic: widespread.

CODIACEAE

Codium Stackh.

C. **fragile** (Sur.) Hariot subsp. **atlanticum** (Cotton) Silva
Littoral, on damp open surfaces and in pools at all levels. Epilithic, and occasionally epiphytic on *Cystoseira tamariscifolia* and *Laurencia*. Frequent. Generally distributed. Previous record: Iona, June 1826, Hb. *Greville* (E).
Scotland: frequent.
NE Atlantic: widespread.

C. **tomentosum** Stackh.
Littoral, mid to low level, in shaded pools and damp crevices, often associated with closely clumped underflora. Epilithic, often on pebbles. Occasional 16, 18, 22, 53, 78, 79, 81. Previous records: Iona, *Greville* (1830); Iona, *Berkeley* in Hooker (1833); Staffa, *Harvey* in Hooker (1833).
C. tomentosum has not been reported widely for Scotland, and there are no Scandinavian records. Silva (1955), in a consideration of the distribution of *C. tomentosum* in northern Europe, cites only "Orkney Islands, *J. H. Pollexfen* (BM)" as indicating the northern limit of distribution in Europe. Since Newton (1949) and Norton & Milburn (1972) record *C. tomentosum* for Skye and Colonsay respectively this species may well be detected further north on other western Scottish islands.

References

Adey, W. H. & Adey, P. J. (1973). Studies on the biosystematics and ecology of the epilithic crustose Corallinaceae of the British Isles. *Br. phycol. J.*, **8**: 343–407.

Ardré, F. (1970). Contribution a l'étude des algues marines de Portugal 1–la flore. *Port. Acta biol.*, Sér. B, **10**: 1–415.

Batters, E. A. L. (1902). A catalogue of the British marine algae being a list of all the species of seaweeds known to occur on the shores of the British Islands, with the localities where they are found. *J. Bot., Lond.*, **40**, Suppl.: 1–107.

Børgesen, F. (1903) (preprint 1902). The marine algae of the Faeroes. Pp. 339–681 in Warming, E., *Botany of the Faeröes based upon Danish investigations*. Copenhagen, Christiania & London.

——— & Jónsson, H. (1905). The distribution of the marine algae of the Arctic Sea and of the northernmost part of the Atlantic. Appendix, pp. I–XXVIII in Warming, E., *Botany of the Faeröes based upon Danish investigations*. Copenhagen, Christiania & London.

Burrows, E. M. (1960). A preliminary list of the marine algae of the Galloway coast. *Br. phycol. Bull.*, **2**: 23–25.

———. (1963). A list of the marine algae of Fair Isle. *Br. phycol. Bull.*, **2**: 245–246.

Caram, B. & Jónsson, S. (1972). Nouvel inventaire des algues marines de l'Islande. *Acta bot. isl.*, **1**: 5–31.

Cardinal, A. (1964). Étude sur les Ectocarpacées de la manche. *Nova Hedwigia, Beih.* **15**: 1–86.

Clouston, C. (1862). *Guide to the Orkney Islands*. Edinburgh, Kirkwall & Stromness.

Conway, E. (1960). Occurrence of *Falkenbergia rufolanosa* on the west coast of Scotland. *Nature, Lond.* **186**: 566–567.

Dendy, W. C. (1859). *The wild Hebrides*. London.

———. (1860). *The beautiful islets of Britaine*. London.

Dixon, P. S. (1960). Taxonomic and nomenclatural notes on the Florideae II. *Bot. Notiser* **113**: 295–319.

———. (1963). Marine algae of Shetland collected during the meeting of the British Phycological Society, August 1962. *Br. phycol. Bull.*, **2**: 236–243.

——— & Irvine, L. M. (1977). *Seaweeds of the British Isles*. Vol. 1, Rhodophyta Part 1. London.

Floc'h, J.-Y. (1969). On the ecology of *Bonnemaisonia hamifera* in its preferred habitats on the western coast of Brittany (France). *Br. phycol. J.*, **4**: 91–95.

Gibb, D. C. (1957). The free-living forms of *Ascophyllum nodosum* (L.) Le Jol. *J. Ecol.*, **45**: 49–83.

Gilbert, O. L., Holligan, P. M. & Holligan, M. S. (1973). The flora of North Rona 1972. *Trans. Proc. bot. Soc. Edinb.*, **42**: 43–68.

Greville, R. K. (1830). *Algae britannicae*. Edinburgh & London.

GUIRY, M. D. (1974). A preliminary consideration of the taxonomic position of *Palmaria palmata* (Linnaeus) Stackhouse = *Rhodymenia palmata* (Linnaeus) Greville. *J. mar. biol. Ass. U.K.*, **54**: 509–528.

HALOS, M. T. (1964). *Étude morphologique et systématique de quelques Ceramiaceés de la Manche*. Thèse, 3e cycle, Paris.

HAUGEN, I. N. (1970). The male gametophyte of *Bonnemaisonia hamifera* Hariot in Norway. *Br. phycol. J.*, **5**: 239–241.

HOEK, C. VAN DEN. (1963). *Revision of the European species of* Cladophora. Leiden.

——— & DONZE, M. (1966). The algal vegetation of the rocky Côte Basque (SW France). *Bull. Cent. Étud. Rech. scient. Biarritz*, **6**: 289–319.

HOOKER, W. J. (1821). *Flora scotica*. Part II. Edinburgh & London.

———. (1833). Cryptogamia Algae. Div. I. Inarticulatae. Pp. 264–322 in Hooker, W. J., *The English Flora of Sir James Edward Smith. Class XXIV. Cryptogamia*. Vol. V. (or Vol. II of Dr Hooker's British Flora.) Part I. London.

IRVINE, D. E. G. (1956). Notes on British species of the genus *Sphacelaria* Lyngb. *Trans. Proc. bot. Soc. Edinb.*, **37**: 24–45.

———, GUIRY, M. D., TITTLEY, I. & RUSSELL, G. (1975). New and interesting marine algae from the Shetland Islands. *Br. phycol. J.*, **10**: 57–71.

———, SMITH, R. M., TITTLEY, I., FLETCHER, R. L. & FARNHAM, W. F. (1972). A survey of the marine algae of Lundy. *Br. phycol. J.*, **7**: 119–136.

JAASUND, E. (1965). Aspects of the marine algal vegetation of North Norway. *Botanica gothoburg.*, **4**: 1–174.

JÓNSSON, H. (1901). The marine algae of Iceland (I. Rhodophyceae). *Bot. Tidsskr.*, **24**: 127–155.

———. (1912). The marine algal vegetation of Iceland. Pp. 1–186 in Kolderup-Rosenvinge, J. L. A. & Warming, J. E. B., *The botany of Iceland*, Vol. 1, part 1. Copenhagen & London.

JORDE, I. (1966). Algal associations of a coastal area south of Bergen, Norway. *Sarsia*, **23**: 1–52.

——— & KLAVESTAD, N. (1963). The natural history of the Hardangerfjord 4 – The benthonic algal vegetation. *Sarsia*, **9**: 1–99.

KEDDIE, W. (1850). *Staffa and Iona described and illustrated.* Glasgow, Edinburgh & London.

KITCHING, J. A. (1935). An introduction to ecology of intertidal rock surfaces on the coast of Argyll. *Trans. R. Soc. Edinb.*, **58**: 351–374.

KRINOS, M. H. G. (1970). Some marine algae from the Mull of Kintyre region of Argyll. *Trans. Proc. bot. Soc. Edinb.*, **40**: 545–556.

KYLIN, H. (1907). *Studien über die Algenflora der norwegischen Westküste*. Uppsala.

———. (1944). Die Rhodophyceen der schwedischen Westküste. *Acta Univ. Lund.*, n.f. 2, **40**(2): 1–104.

———. (1947). Die Phaeophyceen der schwedischen Westküste. *Acta Univ. Lund.*, n.f. 2, **43**(4): 1–99.

LEMOINE, P. (1923). Mélobésiées recuillies à Rockall par la croisière Charcot en 1921. *Bull. Mus. natn. Hist. nat., Paris*, **29**: 405–406.

LEVRING, T. (1935). Undersökningar över Öresund . . . XIX. Zur Kenntnis der Algenflora von Kullen an der schwedischen Westküste. *Acta Univ. Lund.*, n.f. 2, **31**(4): 1–64.

———. (1937). Zur Kenntnis der Algenflora der norwegischen Westküste. *Acta Univ. Lund.*, n.f. 2, **33**(8): 1–148.

———. (1940). *Studien über die Algenvegetation von Blekinge, Südschweden*. Uppsala.

LEWIS, J. R. (1964). *The ecology of rocky shores*. London.

LIGHTFOOT, J. (1777). *Flora scotica*. Vol. 2. London.

LUND, S. (1959). The marine algae of east Greenland. *Meddr Grønland*, **156** (1): 1–247.

LYLE, L. (1929). Marine algae of some German warships in Scapa Flow and of the neighbouring shores. *J. Linn. Soc. Bot.*, **48**: 231–257.

MCALLISTER, H. A., NORTON, T. A. & CONWAY, E. (1967). A preliminary list of sublittoral marine algae from the west of Scotland. *Br. phycol. Bull.*, **3**: 175–184.

MCLAUCHLAN, J. (1967). Tetrasporangia in *Asparagopsis armata*. *Br. phycol. Bull.*, **3**: 251–252.